REINFORCED CONCRETE STRUCTURES

REINFORCED CONCRETE STRUCTURES

(For the Students of B.E./B.Tech. of Civil Engineering for Different Technological Universities of India)

Dr. I.C. SYAL

Chartered Engineer
Ph.D. (Structures)
Retd. Professor & Head,
Department of Civil Engineering,
Punjab Engineering College,
Chandigarh

Dr. A.K. GOEL

Professor of Civil Engineering,
Faculty of Engineering & Technology,
Manav Rachna International University,
Sector-43, Aravali Hills,
Faridabad (Haryana)

S. CHAND
PUBLISHING

S Chand And Company Limited

(ISO 9001 Certified Company)

S Chand And Company Limited

(ISO 9001 Certified Company)

Head Office: Block B-1, House No. D-1, Ground Floor, Mohan Co-operative Industrial Estate, New Delhi – 110 044 | Phone: 011-66672000

Registered Office: A-27, 2nd Floor, Mohan Co-operative Industrial Estate, New Delhi – 110 044 Phone: 011-49731800

www.**schandpublishing.com**; e-mail: **info@schandpublishing.com**

Branches

Ahmedabad	:	Ph: 27542369, 27541965; ahmedabad@schandpublishing.com
Bengaluru	:	Ph: 22354008, 22268048; bangalore@schandpublishing.com
Bhopal	:	Ph: 4274723, 4209587; bhopal@schandpublishing.com
Bhubaneshwar	:	Ph: 2951580; bhubaneshwar@schandpublishing.com
Chennai	:	Ph: 23632120; chennai@schandpublishing.com
Guwahati	:	Ph: 2738811, 2735640; guwahati@schandpublishing.com
Hyderabad	:	Ph: 40186018; hyderabad@schandpublishing.com
Jaipur	:	Ph: 2291317, 2291318; jaipur@schandpublishing.com
Jalandhar	:	Ph: 4645630; jalandhar@schandpublishing.com
Kochi	:	Ph: 2576207, 2576208; cochin@schandpublishing.com
Kolkata	:	Ph: 23357458, 23353914; kolkata@schandpublishing.com
Lucknow	:	Ph: 4003633; lucknow@schandpublishing.com
Mumbai	:	Ph: 25000297; mumbai@schandpublishing.com
Nagpur	:	Ph: 2250230; nagpur@schandpublishing.com
Patna	:	Ph: 2260011; patna@schandpublishing.com
Ranchi	:	Ph: 2361178; ranchi@schandpublishing.com
Sahibabad	:	Ph: 2771238; info@schandpublishing.com

Third Edition 1992
Subsequent Editions and Reprints, 2001, 2004, 2007, 2009, 2012
Revised Edition, 2013

ISBN : 978-81-219-2353-8

PRINTED IN INDIA

By Vikas Publishing House Private Limited, Plot 20/4, Site-IV, Industrial Area Sahibabad, Ghaziabad – 201 010 and Published by S Chand And Company Limited, A-27, 2nd Floor, Mohan Co-operative Industrial Estate, New Delhi – 110 044.

To Our Parents

PREFACE TO THE FIFTH EDITION

The purpose of bringing 5th edition is to update the book with advancement in technology namely use of cements from 33 to 43 and 53 grades, reinforcement steel from Fe 250 to Fe 415 and now to Fe 500, use of flyash as pozzolona and use of high performance concretes with self compacting properties. The book now includes all the new concepts based on IS : 456-2000 Code of Practice for plain and Reinforced Concrete, IS : 1893-2002 (Part I) Criteria for Earthquake Resistant Design of Structures and latest addition in IS : 3370 Code for liquid retaining structures. All example problems are updated to match the latest developments. Four new chapters namely Concrete Bridges, Yield Line Theory, Masonry Structures, Design Examples on limit state Method are added.

The book now caters the needs of students of Degree, Diploma, AMIE, competitive examinations and practicing engineers.

Chandigarh **Dr. I.C. SYAL**
Dr. A.K. GOEL

PREFACE TO THE THIRD EDITION

It has been gratifying to find the earlier editions of the book read and used in so many parts of the country. The new edition owes much to the useful comments and suggestions of the teachers, students and the practising engineers to whom the authors express their grateful thanks.

A new chapter on "Prestressed Concrete" has been added to the new edition. In particular, the chapter discusses various aspects of prestressing, like types of prestressing, various methods of prestressing, materials used, losses in prestress, layout of cable profiles, analysis and methods of design of various elements and the detailed analysis and design of End Block. Numerous solved examples have been presented for better understanding of the prestressing fundamentals.

It is hoped, that apart from its use, to those taking competitive engineering examinations, the new edition will be a useful guide to the engineering community.

Chandigarh **Dr. I.C. SYAL**
1 November, 1991 **Dr. A.K. GOEL**

PREFACE TO THE FIRST EDITION

Reinforced Concrete Structures are now acquiring greater importance because of their easy adaptability to suit varying requirements. Advanced countries have started using material for nuclear structures and structures subjected to dynamic loads. Advances in knowledge have been manifold during the last few years, as a result of which analysis and design criteria are fast changing.

The purpose of this text is to present fundamental design concepts for the elementary structural components based on IS : 456-1978 Code of Practice for Plain and Reinforced Concrete which has been revised very recently to keep abreast with the latest developments in the field of concrete technology. The book consists of basically three parts. First part deals with introduction to concrete as material, service loads, properties and types of different constituent materials of concrete, grades and properties of concrete, different basis of design, concreting under special conditions, special concretes, forms, inspection and testing and stability of structures. Second part deals with resistance of sections for bending, shear, bond and torsion—basic concepts, design of simple R.C.C. component members such as singly and doubly reinforced beams, flanged beams, rectangular one way and two way slabs, non-rectangular slabs and flat slabs, axially and eccentrically loaded columns, foundations—isolated, strap and combined footings, stairs, retaining walls—cantilever type and counterfort type, spherical domes, beams curved in plan, reinforced brick work and design of formwork. Third part deals with Limit State Concept Design which is gaining international acceptance for being more rational. It includes concept of partial safety factors and values of design loads, Limit State of collapse: flexure for rectangular, singly and doubly reinforced sections, flanged sections, shear, torsion, compression in axially and eccentrically loaded columns, serviceability conditions such as deflection and cracking and development and anchorage lengths.

Appendices presenting live loads on different floors in accordance with IS : 875, Basic Wind pressure, SI unit equivalents, properties of round bars and unit weights of common types of materials, have been added.

This book presents design of simple elements with sufficient number of example problems. As High Strength Ribbed or Steel has practically replaced the Mild Steel bars, some of the example problems are worked out in Tor Steel so as to enable the readers to become conversant with the use of Tor Steel as reinforcement. The reinforcement details are represented in drawings in the manner as is done in actual field working drawings. SI units are mainly used in this text since IS : 456-1978 has already adopted this system of units. The book is mainly written for Civil Engineering students of Degree classes and practicing engineers. However, students of Diploma and AMIE can also equally benefit from it.

Chandigarh **Dr. I.C. Syal**

March, 1984 **Dr. A.K. Goel**

CONTENTS

PART III : LIMIT STATE METHOD OF DESIGN

PART IV : WATER STORAGE TANKS

PART V : PRESTRESSED CONCRETE

PART - I

MATERIALS

Concrete, a homogeneous mixture of cement, sand, gravel and water, is very strong in carrying compressive forces and hence is gaining increasing importance throughout the world. It is extremely weak in resisting tensile stresses. To covercome this weakness, concrete is reinforced with Mild steel bars, high strength steel deformed bars or steel fibres in the tension zone. Concrete of different strengths can be obtained by varying the proportions of its constituent materials. Still higher strengths for same proportions can be obtained by mixing concrete with mechanical mixers, use of vibrators and control of temperature and moisture.

Though concrete is not perfectly homogeneous material, yet by making some reasonable assumptions, it is treated as homogeneous and all elastic theories for analysis and design have been made applicable to it.

This Page has been intentionally left blank

Introduction

CHAPTER **1**

1.1 GENERAL INTRODUCTION

Framed multistorey structures, water reservoirs bridge decks, floors and roofs of different types of buildings, offishore structures, shells, folded plates, airfield pavements, flyovers, etc., are some of the very common structures made in concrete (Fig. 1.1) A structure is an assembly of a number of members such as slabs, beams, columns, walls, domes, etc., proportioned to resist the loads and forces; changes in climatic conditions such as temperature, frost, chemical attack, etc. Design of a structure includes assessing the loads and providing members of sufficient proportions to resist the assessed loads with a sufficient margin of safety. A structure is said to be efficiently desinged all the component members are so arranged that they transmit their self weight and other imposed loads to foundation or supporting structure by cheapest means so as to satisfy the requirements of architecture, structural stability and the nature of the site with sufficient safety. In addition to engineering calculations, experience and good judgment may also do much towards safety and economy of the structure. Highly complex mathematics though required in sohisticated structures, should not normally be allowed because such an analysis may not be possible by all engineers in their day to day design work. However, with the advent of electronic digital computers, it has now become possible to do fairly involved calculations in the design office also.

Fig. 1.1a. Shel

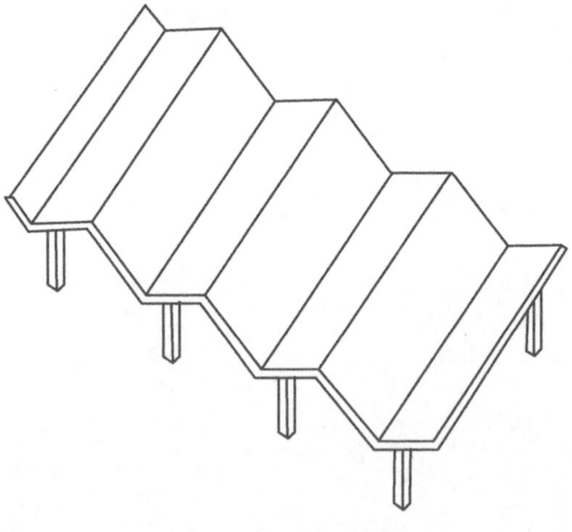

Fig. 1.1b. Folded plate

Structural design is controlled by certain regulations or codes of practices for the specific regions, still the designer has always the freedom to exercise his own judgment. In India, the design of reinforced concrete structures is mostly based upon Indian Standard Codes of Practice in force from time to time.

1.2 LOADS AND FORCES

An efficient designer would like to estimate the loads and forces depending upon the location of the structure, climatic conditions, purpose for which the structure is intended to be designed and the life of the structure. Most frequently the loads are assessed by referring to IS : 875–1987 (Parts 1 to 5) 'Code of Practice for Design Loads for Buildings and Structures'. However, the type of loading to which the structure is expected to be subjected during its life time, if somehow known, then actual loads must be considered. Code of practice gives loads which may be taken as a good guide but the decision to take actual loading is purely the discretion of the designer. Correct assessment of the loading is very important because, otherwise, either the structure would be unsafe if the load assessment is on the underside or uneconomical if the loads are assessed on the higher side of the loads to which the structure may be subjected during its service period.

A structure may encounter any one or a combination of the following loads for computation of the maximum stresses in the structure.

1.2.1 Dead Load

Dead load in a structure comprises the self weight of the walls, partitions if any, floors and roofs and includes the weights of all permanent fixtures (partitions, finihes, super-structures) in the building. Where positioning of partitions in a building is not known precisely, a uniformly distributed load of one-third of weight of partition subject to a minimum of 1000 N/m^2 (100 kg/m^2) in the case of office floors may be assumed. It is interesting to note that the dead loads are the predominant loads that are to be resisted be common engineering structures.

1.2.2 Live Load or Imposed Load

Live loads on floors and roofs shall comprise all loads which are placed temporily (such as people, shelves, tables, chairs, snow, rain, movable and non-movable loads, impact and vibrations) on it. Minimum live loads on different floors for different uses as suggested by the Indian are given in Appendix-A. However, in the case of multistoreyed buildings, following reductions in assumed total live loads may be made in designing columns, walls piers, their supports and foundations.

Number of floors carried by Member under consideration	*Percentage reduction of total Live Load on all floors above the member under consideration*
1	0
2	10
3	20
4	30
5 to 10	40
Over 10	50

No reduction shall be made in the case of warehouses, garages and other buildings used for storage purposes and factories and workshops designed for 3000 N/m^2 (500 kg/m^2).

The live loads as stipulated in the Indian Codes (as well as other codes) are only approximate and are very much on the higher side. Hence more research is being conducted all over the world to find exact live loads based on probability theory.

1.2.3 Wind Load

Very strong winds are encountered sometimes during the life time of the structure. The liability of a structures to high wind pressure depends not only on the geographical location and proximity of obstructions to wind flow but also on the characteristics of the structure (like size, shape, etc.) itself. The basic wind pressure on a structure is decided by referring to IS : 875-1987 (Part 3).

1.2.4 Seismic Load

Earthquacke shocks cause a movement of ground on which the structure is situated. This movement causes the structure to vibrate. The vibrations may be resolved in the three directions simultaneously and should be considered for the safety of the structures unless otherwise specifically stated. The earthquake forces shall be calculated in accordance with IS : 1893-2002 (Part 1).

1.2.5 Soil and Hydrostatic Pressure

In the design of structures or part of structures below ground level, such as basement walls, water tank walls, the pressure exerted by the soil or water or both shall be duly accounted for on the basis of established theories.

1.2.6 Erection Loads

All loads required to be carried by the structure or any part of it due to placing or storage of construction materials and erection equipment including all loads due to operation of such equipment shall be considered as 'erection loads'.

1.2.7 Miscellaneous Loads

In addition to the above mentioned loads, account should be taken of loads such as shrinkage, creep, temperature, foundation movements, elastic axial shortening, vibrations, impact, fatigue, stress concentration and the like.

The effects of wind, earthquake and erection loads have not been considered in the example problems included in this book.

1.3 CONCRETE

In the past, most of the structures were made either in masonry, steel or timber depending upon the availability of the materials and the nature of the structure. Very recently concrete, as building material, has come into being. In a very short period, concrete has gained so much importance that today more than 65 per cent of the structures coming up in the world, are constructed with concrete. Concrete is composed of inorganic materials called aggregates such as gravel, sand, crushed stone, slag, etc., cemented together with some cementing materials (most commonly used Portland Cement) and water. When the above materials are mixed together so as to form a workable mixture, it can be given any shape such as beams, slabs or columns. Cement and water react chemically to bind sand (known as fine aggregate) and gravel or broken stone (known as coarse aggregate). When the constituent materials are mixed together, they get hardened after a few hours and a hard stone-like product is obtained. This is called 'Concrete'. Concrete is very strong compression but extremely weak in tension and is a brittle substance. The properties of concrete are dependent upon several factors such as amount of cement, fine and coarse aggregate, water/cement ratio (wc/ratio), temperature at the time of mixing, humidity during moulding and temperature and humidity maintained subsequently. The process of keeping the concrete moist at desired temperature is called 'Curing of Concrete'.

1.4 REINFORCED CEMENT CONCRETE

Concrete being extremely weak in tension, has the tendency to crack when tensile stresses are induced in any part of the concrete. To arrest these cracks, reinforcement is provided in the tension zone. It may, however, be noted that reinforcement does not avoid cracks but it simply arrests further widening of cracks. So by providing tensile reinforcement at a proper place, the microcracks do develop but generally they are so small in width that these are not visible to the naked eye. Such a concrete which has reinforcement in tension zone, is known as "Reinforced Cement Concrete". Concrete develops in fact, a very good bond with steel.

The ratio between the amounts of fine and coarse aggregate necessarily depends on the grading and other characteristics of the materials in order that the volume of sand is sufficient to fill the voids in the coarse aggregate to produce a dense concrete. Water used for mixing and curing should be clean and free from injurious amounts of oils, acids, alkalies, salts, sugar, organic materials or other substances that may be deleterious to concrete or steel. Potable water is generally considered satisfactory for mixing concrete. Mixing or curing of concrete with sea water is not recommended because of the presence of the harmful salts in sea water. The presence of tannic acid or iron compounds in water is objectionable.

1.5 ADVANTAGES AND DISADVANTAGES OR REINFORCED CEMENT CONCRETE CONSTRUCTION

The reinforced cement concrete construction has the following advantages as compared to steel or timber constructions.

1. Concrete components can be cast in any desired size and shape.
2. Constituent materials are easily available and are quite cheap.
3. Grade of concrete suitable for a particular structure can be selected to provide an appropriate degree of durability as well as strength.
4. Concrete with richer cement content and with proper proportioning can be made water proof.
5. Its maintenance cost is very small as compared to that of timber or steel.
6. It is fire resistance and hence concrete construction is superior to steel and timber construction.
7. Monolithic construction in concrete is possible which provides greater flexibility in planning and design.
8. Strength of concrete increases with age.

The only disadvantage in this type of construction is that member once cast cannot be dismantled as in the case of steel and timber construction. It also requires strict quality control as opposed to steel or timber.

1.6 CONCRETE MATERIALS

1.6.1 Types of Cements

There is a large variety of cements available in the market. A few of them are mentioned below :

1. *Ordinary Portland Cement* (*OPC*) (IS : 269). This type of cement is most commonly used in concrete construction. Portland cement consists of the following constituents :

(*a*) Lima	60% — 67%
(*b*) Silica	17% — 25%
(*c*) Alumina	3% — 8%
(*d*) Iron oxide	0.5% — 6%
(*e*) Magnesia	0.1% — 4%
(*f*) Soda and Potash	0.2 % — 1%
(*g*) Sulphur trioxide	1% — 3%
(*h*) Free lime	0% — 1%

2. *Rapid Hardening Portland Cement* (IS : 8041). This type of cement is used where the strength of concrete is desired at an early date, *e.g.*, Hospitals, public buildings and road work. Shuttering can be removed considerably earlier if this cement is used, thus saving considerable time and expenses. This cement is particularly useful under conditions of frost.

3. *Low Heat Cement* (IS : 269). Low heat cement is particularly used in massive constructions like abutments, retaining walls, dams, etc., where less heat of hydration will be evolved. The heat generated if ordinary Portland cement is used at the end of 3 days, is generally of the order of 80 calories per gram of cement used whereas it is less than 65 calories per gram at the end of 7 days and less than 25 calories per gram at the end of 28 days in case of low heat cement. The temperature rise is an important factor because the shrinkage that may take place during cooling process introduces tensile stresses in concrete. This cement offers a better resistance to chemical deterioration than Ordinary Portland Cement (OPC).

4. *Portland Blast Furnace Slag Cement* (IS : 455). This cement is obtained by intergrinding Portland Cement clinker with blast furnace slag. The blast furnace slag content should not exceed 65%. It has low heat of evolution as compared to OPC and can be used for all massive concrete structures like dams, retaining walls, abutments, etc. It is sulphate resistant and generally used in constructions in sea water.

5. *Portland Pozzolana Cement* (IS : 1489). This cement is made either by intergrinding Portland Cement clinker with pozzolana (finely powdered form of burnt bricks, flyash, burnt clay, etc.). The pozzolana content generally varies from 10% to 25% by weight of cement. Pozzolanas themselves do not possess any cementing property but have the property of combining with lime to obtain a stable lime-pozzolana compound which possess the cement porperty. The pozzolana concretes have a greater resistance to chemical agencies. They can resist attack of sea water better than OPC.

6. *Other Cements*:

(*a*) White cement is mainly used for decorative works like facing slabs, floorings, ornamental concrete products, swimming pools, etc.

(*b*) Coloured concretes are obtained by adding 5% to 10% of ground pigments to white or ordinary cements. These are used for designed coloured floors.

(*c*) Sulphate resisting cement is used in culverts, syphons, canal lining, etc., where ground water deposits a good amount of sulphates in the soil.

(*d*) High strength ordinary Portland cement is used where very high strength is desired as compared to OPC.

1.6.2 Coarse Aggregate

Coarse aggregate are inert materials used for increasing the volume and strength of concrete, *e.g.*, gravel, crushed stone, etc. The nominal size of the coarse aggregate for reinforced concrete work is 20 mm which is generally considered satisfactory. However, for mass concreting work higher size upto 40 mm can be used and still higher sizes upto 75 mm may be used for lean concrete mixes to be used as levelling courses below the structural foundations.

1.6.3 Fine Aggregate

The particles of sand stone which fall between 0.15 mm to 4.8 mm are called fine aggregates. Coarser sands give better strength to concrete. Silt content in sands is injurious to concrete strength.

1.6.4 Water

Water used for mixing and curing should be clean and free from injurious amounts of oils, acids, alkalies, salts, sugar, organic materials or other substances that may be deleterious to concrete or steel. Potable water is generally considered satisfactory.

1.7 REINFORCING MATERIALS

Concrete being extremely weak in tension, reinforcement is provided to take up the tensile stresses in R.C.C. (Reinforced Cement Concrete) structure. Reinforcement can be mild steel bars, ribbed tor steel, steel fibres, glass fibres or bamboo reinforcement. Steel is generally used because of its easy availability, high strength, good bond with concrete and its coefficient of expansion being nearly same as for concrete. Different types of steel bars are available such as Mild Steel, Hot Rolled Deformed bars, Cold Twisted bars, Hard drawn steel wire fabric and Rolled Steel. All reinforcement should be free from loose rust and coats of paints, oils, mud or other coatings which may destroy or reduce bond. The HYSD steel bars, generally used as reinforcing bars are of different diameters varying from 5 mm to 50 mm. However, minimum size of mild steel bar available in the market is 6 mm and of tor steel is 8 mm. The Young's modulus of elasticity of steel should be taken as 200 kN/mm^2, *i.e.*, 2×10^6 kg/cm^2 for Mild as well as another steels.

1.8 GRADES OF CONCRETE

Because the concrete properties vary so much with composition and methods of mixing, many different types of concrete can be obtained, each suited for a specific purpose.

Concrete mixes have been classified into various grades by the Indian Standards Institution as given in Table 1.1. The characteristic strength is defined as the strength of material below which not more than 5% of the test results are expected to fall.

In the designation of a concrete mix, letter *M* refers to the mix, and the number to the specified characteristic compressive strength of cube at 28 days, expressed in N/mm^2. M5 and M7.5 grades of concrete may be used for lean concrete bases and simple foundations for masonry walls. These mixes need not be designed. Grades of concrete lower than M15 should not be used in reinforced concrete. Most commonly used concrete mix for beams, slabs, columns, framed structures, walls and footings is M20. Higher mixes than M20 are used for water tanks, shells, folded plates, grid structures and heavily loaded structures. Today concrete of compressive strength as high as 80 N/mm^2 have been obtained without much difficulty.

Table 1.1. Grades of Concrete

Group	*Grade Designation*	*Specified characteristic compressive strength at 28 days (N/mm^2)*
Ordinary	M10	10
Concrete	M15	15
	M20	20
Standard	M25	25
	M30	30
	M35	35
	M40	40
	M45	45
	M50	50
	M55	55
High	M60	60
Strength	M65	65
Concrete	M70	70
	M75	75
	M80	80

MINIMUM CEMENT CONTENT IN MIX

The minimum cement content that must be present in a concrete mix as recommended in IS : 456-2000, is given in Table 1.3.

1.9 PROPERTIES OF CONCRETE

1.9.1 Weight

The unit weights of plain concrete and reinforced concrete made with sand, gravel of crushed natural stone aggregate may be taken as 24000 N/m^3 (2400 kg/m^3) and 25000 N/m^3 (2500 kg/m^3) respectively.

1.9.2 Compressive Strength

With given properties of aggregate the compressive strength of concrete depends primarily on age, cement content and the water cement ratio. The characteristic compressive strengths of a range of concrete are given in Table 1.1. Characteristic strengths are based on the strength at 28 days. The strength at 7 days is about two-thirds of that at 28 days with Ordinary Portland Cement and generally is a good indication of strength likely to be attained.

Table 1.3. Minimum Cement Content, Maximum Water-Cement Ratio and Minimum Grade of Concrete for Different Exposure with Normal Weight Aggregates of 20 mm Nominal Maximum Size

S.No.	*Exposures*	*Pain Concrete*			*Reinforced Concrete*		
		Min. Cement Content kg/m^3	*Maximum Free Water Cement Ratio*	*Minimum Grade of Concrete*	*Minimum Cement Content kg/m^3*	*Maximum Free Water Cement Ratio*	*Minimum Grade of Concrete*
(1)	(2)	(3)	(4)	(5)	(6)	(7)	(8)
(*i*)	Mild	220	0.60		300	0.55	M20
(*ii*)	Moderate	240	0.60	M15	300	0.50	M25
(*iii*)	Severe	250	0.50	M20	320	0.45	M30
(*iv*)	Very Severe	260	0.45	M20	340	0.45	M35
(*v*)	Extreme	280	0.40	M25	360	0.40	M40

Notes: 1. Cement content prescribed in this table is irrespective of the grades of cement and it is inclusive of additions mentioned in 5.2 of IS : 456-2000. The additions such as fly ash or ground granuled blast furnace slag may be taken into account in the concrete composition with respect to the cement content and water-cement ratio if the suitability is established and as long as the maximum amounts taken into account do not exceed the limit of pozzolona and slag specified in IS : 1489 (Part I) and IS : 455 respectively.

2. Minimum grade for plain concrete under mild exposure condition is not specified.

1.9.3 Increase in Strength with Age

There is normally gain of strength beyond 28 days. The quantum of increase depends upon the grade and type of cement, curing and environmental conditions etc. The design should be based on 28 days characteristic strength of concrete unless there is a evidence to justify a higher strength for a particular structure due to age.

For concrete of grade M30 and above, the rate of increase of compressive strength with age shall be based on actual investigations. However, for guidance, following increase in strength with age can be taken.

After Months	*Age Factor*
3 Months	1.10
6 Months	1.15
1 Year	1.20

1.9.4 Tensile Strength of Concrete

The flexural and split tensile strengths of various concrete, are given in IS : 516-1959 and IS : 5816-1970 respectively. When the designer wishes to use an estimate of the tensile strength from compressive strength, the following formula may be used.

$$\text{Flexural strength } f_{cr} = 0.7\sqrt{f_{ck}} \ \text{ N/mm}^2 \qquad \text{... (1.1)}$$

where f_{ck} is the characteristic compressive strength of concrete.

The ratio of direct tensile strength f_{ct} to the compressive strength f_{ck} varies from 0.05 to 0.10.

1.9.5 Elastic Deformation

The modulus of elasticity is primarily influenced by the elastic properties of the aggregate and to lesser extent on the conditions of curing and age of the concrete, the mix proportions and the type of cement. The modulus of elasticity is normally related to the compressive characteristic strength of concrete.

$$E_c = 5000\sqrt{f_{ck}} \qquad \text{... (1.2)}$$

where E_c = the short-term static modulus of elasticity in N/mm², and

f_{ck} = characteristic cube strength of concrete in N/mm²

1.9.6 Shrinkage, Creep and Thermal Expansion

1.9.6.1 Shrinkage

The total shrinkage of concrete depends upon the constituents of concrete, size of member and environmental conditions. For a given environment, the total shrinkage in concrete is mostly influenced by the total amount of water present in concrete at the time of mixing, and to a lesser extent, by the cement content. In the absence of test data, the approximate value of the total shrinkage strain for design may be taken as 0.0003.

1.9.6.2 Creep

Creep of concrete depends, in addition to the factors that effect shrinkage on the stress in concrete, on age at loading and the duration of loading. As long as the stress in concrete does not exceed one-third of its compressive characteristic strength, creep may be assumed to be proportional to stress. In the absence of test data, and detailed information on the effect of variables, the ultimate creep strain may be estimated from the following values of creep coefficient (i.e., ultimate creep strain/elastic strain at the age of loading) as given in Table 1.4.

Table 1.4. Creep Coefficients

Age of loading	*Creep coefficient*
7 days	2.2
28 days	1.6
1 year	1.1

1.9.6.3 Thermal Expansion

The coefficient of thermal expansion depends upon the nature of the cement, the aggregate, the cement content, the relative humidity and the size of the sections. The value of the coefficient of thermal expansion for concrete with different aggregates be taken as given in Table 1.5.

Table 1.5. Coefficient of Thermal Expansion

Type of aggregate	*Coefficient of thermal expansion for concrete/°C*
Quartzite	1.2 to 1.3×10^{-5}
Sand stone	0.9 to 1.2×10^{-5}
Granite	0.7 to 0.95×10^{-5}
Basalt	0.8 to 0.95×10^{-5}
Lime stone	0.6 to 0.9×10^{-5}

1.9.7 Poisson's Ratio

Poissons ratio enables to determine tile secondary stresses that occur is a result of primary stresses, and has a value of 0.15 to 0.3 for concrete. In many cases, this effect is ignored.

1.9.8 Fire Resistance

The fire resistance of reinforced concrete construction depends upon the type of aggregates, the thickness of various parts comprising the member and cover of concrete over the reinforcement. The degree of resistance required depends upon the size of building and on the use of which it is to be put.

1.9.9 Modular Ratio

It is defined as the ratio of modulus of elasticity of steel (E_s) to that of concrete (E_c).

$$\text{Modular ratio } m = \frac{E_s}{E_c} \qquad \text{... (1.3)}$$

The values of E_s for all types of steel, has been adopted as 200 KN/mm^2 and for concrete M15 the value of E_c is 22.076 KN/mm^2 as obtained from Eq. 1.2. The modular ratio works out to be 9.06. However, IS Code recommends the value of 'm' as 280/(3 σ_{cbc}), from which the modular ratio 'm' works out to be 18.67 for M15 concrete. This value is adopted in 'working stress design method' because it compensates in part for the errors involved in considering reinforced concrete as theoretically elastic substance and for the neglect of any tensile resistance of the concrete in bending. The modular ratio is used only in the range of service stresses plays no part in the ultimate limit state design.

Table 1.6. Optional Tests Requirements of Concrete for Modulus of Rupture

Grade of Concrete	*Compressive strength of 15 cm cubes Min at 7 days*	*Modulus of Rupture by Beam test, Min at 72 ± 2 hours*	*and 7 days*
(1)	*(2)* N/mm^2	*(3)* N/mm^2	*(4)* N/mm^2
M10	7.0	1.2	1.7
M15	10.0	1.5	2.1
M20	13.5	1.7	2.4
M25	17.9	1.9	2.7
M30	20.0	2.1	3.0
M35	23.5	2.3	3.2
M40	27.0	2.5	3.4

1.9.10 Modulus of Rupture

The tensile resistance of concrete in bending is termed as 'Modulus of Rupture'. This resistance is normally neglected in the design of ordinary structural members but is taken into account in the design of liquid retaining structures. The minimum modulus of structural concrete at 3 days and 7 days are given in Table 1.6.

1.10 BASIS FOR DESIGN

Structures and structural members should be designed to have, at all sections, strengths at least equal to the structural effects of design loads and forces that occur during construction and use as determined by the relevant method of design. Structures and structural elements should also meet all other requirements of the relevant code.

The reinforcement alongwith the constituents (cement, sand, gravel and water) of concrete, makes the composite concrete highly anisotropic and non-homogeneous. In the earlier stages of its development; though structures were built using concrete it was difficult to formulate theories that simulated the exact behaviour of concrete structures. As such, they were designed using approximate methods based on experimental values or by using empirical or semi-empirical formulae. Due to the exhaustive research work carried out all over the world, it is now possible to predict the behaviour of concrete structures in a more rational manner.

The design should always be made in accordance with the principles of mechanics, recognised methods of design and sound engineering practice. In particular, adequate consideration should be given to the effects of continuity on the distribution of bending moments and shears due to monolithic construction. Reasonable factors of safety must be taken in designing the members. The ratio of strength of the section of a member at which it becomes unserviceable (may be due to failure or because of excessive deflection or cracking or some other similar reason) to actual force (may be bending moment, shear, axial force of torsional moment) to which the section is subjected, is termed as 'Factor of Safety'.

Various methods have been adopted in various codes of practices in different countries for the design. Some of the known methods of design are briefly explained subsequently.

1.10.1 Working Stress Method

This method is also known as 'Modular Ratio Method' or 'Elastic Stress Method'. In this method, the moment and forces acting on a structure are computed from the actual values of service loads but the stresses in concrete and the reinforcing steel are restricted to only a fraction of their true strengths in order to provide an adequate safety factor. A factor of safety of 3 in concrete and 1.78 to 1.80 in steel is generally adopted in this method. A greater factor of safety in concrete is adopted because in the event of failure, crushing of concrete takes place unexpectedly and explosively, since it is a brittle material. Whereas steel yields gradually and hence gives advance warning.

In the elastic approach, the design is based upon set theories available in large number of books because of the basic assumption that the reinforced concrete is homogeneous. All the forces such as bending moments, shear forces, torsional moments and axial loads can be easily computed assuming the material to behave perfectly elastic and correspondingly the structures are proportioned to develop stresses upto a fraction of the ultimate stress of concrete and yield stress of steel. A large number of structures such as water retaining structures, high rise buildings and other complex structures are still designed by the working stress method of design whereas in the other methods either design procedures are not developed or are not possible at present.

Since reinforced concrete is not a perfectly elastic material, the methods of analysis do not give real behaviour of the structure, highly stressed parts of the structure start deforming unproportionately to the loading and distribution of moments is also not same as for a perfectly elastic material. Concrete members never have uniform moment of inertia due to variable cracking along their length and varying areas of reinforcement. Because of these reasons, working stress design method is unable to give any real guide to the safety of a structure. There is absolutely no guarantee that a structure designed by elastic method, keeping the maximum stress in steel at working load to one half of the yield stress and maximum stress in concrete at working load to one-third the ultimate stress, will be able to carry an ultimate load of twice the working load. Mostly this procedure resulted in safe structures only because large stress factors of safety were used and many structures had also extra inherent reserves of strength.

1.10.2 Ultimate Strength Method

This method is also known as 'Load Factor Method'. It involves the analysis of sections at failure, the actual strength of a section is related to the actual load causing failure, with latter being determined by 'factorising' the design load. Service loads are multiplied with load factors to get the ultimate design loads and accordingly bending moment, shear and torsion are evaluated and the section is proportioned to carry stresses upto ultimate strength of materials.

The designer must satisfy himself that under full working load, no part of the structure is excessively cracked, none of the members have suffered excessive deflection and the structure will not vibrate or oscillate excessively under moving or fluctuating loads.

In this method of design, the true margin of safety is accurately known. The ultimate strength procedure is based upon the results obtained from the experimental investigations depicting the true behaviour of the structures. It permits the use of lower load factor for loads definitely known, such as dead load, and a higher factor for less certain loading. This gives a total margin of safety consistent with type of loading. A better evaluation of critical moment thrust ratio for members subjected to combined bending and axial load is obtained by this method.

In this design procedure, moments and shears are still established by elastic analysis mostly rather than by the formation of plastic hinge theory. Strict quality control is required in this method.

1.10.3 Limit State Method

In this method of design based on limit state concept, the structure shall be designed to withstand safely all loads liable to act on it throughout its life; it shall also satisfy the serviceability requirements such as prevention of excessive deflection, excessive cracking and excessive vibrations. The acceptable limit for the safety and serviceability requirements before failure ocurs is called 'Limit State'. The aim of the design is to achieve accpetable probabilities that the structure will not become unfit for the use for which it is intended, that is, that it will not reach a limit state. For ensuring the above objective, various partial factors of safety are employed in the limit state design. The characteristic values should be based on statistical data if available; where such data is not available they should be based upon experience. The design values are derived from characteristic values through the use of partial factors of safety. Design loads are obtained by multiplying a partial factor of safety for loads with characteristic loads and in a similar manner design strength of materials are obtained by dividing the characteristic strength with respective partial factors of safety for materials. Members are thus proportioned to carry limit design loads and materials are stressed to limiting design strengths.

This approach is more logical and rational and has the merit of simplicity and this method is based on propabilistic approach. This method includes the merits of both the design methods discussed above. By using this method both strength and serviceability requirements of structures are satisfied.

1.10.4 Methods Based on Experimental Investigations

Design based on experimental investigations on models or full size structures or elements are acceptable subject to experimental details and analysis connected therewith. Where the design is based on experimental investigations, load tests should be carried out to ensure that requirements of deflection and cracking are satisfied when the structure is subject to a load for 24 hours equal to the characteristic load multiplied by 1.33 times the corresponding partial factor of safety for that type of load for the limit state of serviceability.

1.10.5 Load and Resistance Factor Design Method

This method is rapidly emerging in many countries as the latest method of designing reinforced cement concrete structures and is said to be more rational and economical. This method of design is still in the development stage, and hence has not found its place in most of design books.

1.11 CONCRETING UNDER SPECIAL CONDITIONS

1.11.1 Concreting in Hot Weather

Normally, the concrete should be mixed at a normal temperature of 16 to 32°C. A few problems arise if the concreting is done in hot weather (temp. more than 32 degree C) because of the increased rate of evaporation from the fresh mix. This accelerates setting, lowers the workability, and hence lower strength of hardened concrete is obtained. Rapid evaporation also causes plastic shrinkage and subsequent cooling of hardened concrete would introduce initial tensile stresses. In any case, thin sections such as shell roofs, folded plates should not normally be cast under hot and dry conditions.

1.11.2 Concreting in Cold Weather

If concreting is done in cold weather the water freezes before the concrete is set. No water is then available for chemical reactions. This delays the setting and hardening of concrete. If low temperature remains for a long period, the process of setting will remain suspended. Because of the expansion of the mixing water on freezing, a lack of revibration would allow the concrete to set with a large volume of pores present and consequently the strength of concrete would be very low. Re-vibration on thawing would produce a satisfactory concrete.

If freezing takes place after the concrete has set but before it develops appreciable strength, expansion associated with formation of ice causes disruption and an irreparable loss of strength. If, however, the concrete has acquired sufficient strength, it can resist the freezing temperature without damage, not only by virtue of the higher resistance to the pressure of the ice but also because a large part of the mixing water will have become combined with cement or located in the gel pores and would thus not be able to freeze.

1.11.3 Concreting Under Water

When it is necessary to deposit concrete under water, it should be of at least M15 grade in case of plain concrete and M20 grade in case of reinforced concrete. The concrete should be densest possible; slag broken brick, soft lime stone, soft sand stone or other porous or weak aggregates must not be used. The concrete should contain at least 10% more cement than that required for same mix placed under dry conditions. The various methods employed for concreting under water are listed below :

(*a*) *Tremie*. In this process the concrete is poured through a hopper placed at the top of a tremie pipe whose lower end is placed in position at the site of concreting.

(*b*) *Drop Bottom Bucket*. In this process a bucket completely filled with concrete having a canvas flap cover at top and openable doors at the bottom, is lowered to the site in water and the concrete is discharged.

(*c*) *Bags*. Bags of at least 0.028 cubic metre capacity of jute or other coarse cloth shall be filled about two-thirds full of concrete, the spare end turned under so that the bag is square ended and securely tied. They shall be placed carefully in header and stretcher coarses so that the whole mass is interlocked.

(*d*) *Grouting*. A series of round cages made from 50 mm mesh of 6 mm steel and extending over the full height to be concreted shall be placed, over the area to be concreted at a centre to centre distance not exceeding 1 m and also to the faces of concrete. Stone aggregate of not less than 50 mm nor more than 200 mm size shall be deposited outside the cages. A stable 1 : 2 cement grout (w/c $\neq$ 0.6 to 0.8) is forced under pressure (about 0.2 N/mm^2) through 38-50 mm diameter pipe terminating into steel cages.

1.12 SPECIAL CONCRETES

1.12.1 Ready Mixed Concrete (Pre-Mixed Concrete)

If instead of being batched or mixed at site, concrete is delivered ready for placing from a central plant, it i referred to as ready mixed or pre-mixed concrete. Ready mixed concrete offers numerous advantages. It is used on congested sites, road construction where little space for mixing is available and where concreting is to be done at intervals. It is made under better conditions of control and requires less cement content, site organisation and supervisory staff. Ready-mixed concrete is a little costlier than site-mixed concrete. Since this concrete is to be pumped at site, its slump is more and has invariably larger amounts of fines (namely sand and fly ash of mixed). About 1% of water mixed is evaporated in 24 hours, some shrinkage cracks develop next day which may be filled with shrinkage resisting chemicals or by grouting.

1.12.2 Light Weight Concrete

Appreciable savings in weight and accompanying savings in steel costs and other costs are often obtainable through the use of light in weight concretes. These concretes are lighter in weight (1.6 kN/m^3 to 22 kN/m^3) as compared to normal concrete (24 kN/m^3). Often such concretes are obtained by putting in light aggregates instead of normal heavy weight aggregates.

Lighter concretes, in general, have higher absorptions, lower strengths, lower elastic modulii, lower thermal conductivity and higher shrinkage than heavier concretes. Light weight concretes can be produced as expanded concretes, air-entrained concretes or foamed concretes. One of the methods to obtain expanded concrete is to mix aluminium powder. The expansion may be as much as 100% of the unexpanded volume. Foamed concrete can be made by mixing cement and fine filler such as pulverised stone, silica flour, flyash, saw dust, cinder, etc., with a stabilised foam. By proper proportioning of the foam and solids, concretes weighing as little as 1.6 kN/m^3 have been produced. These materials have very low strength but are well adopted for production of insulated and fire-resistanc roof slabs and other members where strength is of little importance.

1.12.3 Fibre Reinforced Concrete

This is a composite material consisting of concrete and fibres of asbestos, glass or steel. In building structures mostly steel fibres are used. Randon fibre reinforcement (wire) presents an effective crack arrest mechanism in concrete, *i.e.*, it improves the tensile resistance of the concrete to a great extent in addition to some improvement in compressive strength. The concrete has low tensile sterngth due to the internal flaws and micro cracks present in it. To avoid the growth of these micro cracks, steel fibres are now mixed thoroughly in concrete. Such a concrete with high tensile strength is extremely useful in structures subjected to fatigue and impact loading, *e.g.*, air field construction, nuclear reactors, furnace supports subjected to thermal shocks and water tanks, etc.

Ordinary concrete has a tensile strength of 7% to 11% of its compressive strength. By mixing a suitable proportion of fibres, it has been observed that there is an increase of 2 to 3.5 times in the tensile strength and 1.5 to 1.7 times in the compressive strength. Fibre reinforced concrete is more ductile as compared to ordinary concrete.

1.12.4 Polymer Concrete

This is ordinary concrete made with portland cement, wet, cured and subsequently impregnated with a liquid or gaseous monomer under pressure and polymerized by gamma-radiation by chemical means. This type of concrete has much higher compressive, tensile and impact strengths and higher resistance to freezing and thawing, abrasion and to chemical attack. All these are due to the very low void content of the polymer concrete. The cost involved in polymerisation is quite high and hence its use is restricted.

1.12.5 Pumped Concrete

If the site of concreting is not easily accessible such as congested sites, tunnel lining, sites at high levels, etc., pumping delivers the concrete direct from the mixer to the form and so avoids double handling. Such a concrete is called "pumped concrete".

While pumping sharp bends should be avoided. Pumping can be done either by direct-acting concrete pumps where concrete gets into the pipe through a hopper by gravity and is sucked at site or by squeeze pumps. Where the concrete is pushed partly by rotating blades and partly by vacuum pump sucking, the pipe should always be full during the pumping operation. Difficulty of blockade is generally encountered.

1.13 FORM-WORK AND ITS REMOVAL

For concrete construction, forms or shuttering in the form of wooden planks or steel plates are used for keeping the green concrete in position till it hardens. This helps in giving the desired shape to the different components of the structure. The contractor is concerned with a fast turnover of form reuse for economy with safety whereas the owner wants quality, continued curing for maximum inplace strength and an adequate strength and modulus of elasticity to minimise initial deflection and cracking. Both want a satisfactory surface. To have the maximum strength and satisfactory surface, formwork should be removed after a fortnight. However, IS : 456-2000 says that the forms shall not be struck until the concrete has reached a strength at least twice the stress to which the concrete may be subjected at the time of removal of formwork. The strength referred to shall be that of concrete using the same cement and aggregates, with the same proportion and cured under conditions of temperature and moisture similar to those existing on the work. Where possible, the formwork shall be left longer as it would assist curing.

In normal circumstances and where ordinary portland cement is used, forms may generally be removed after the expiry of the following periods :

(*a*) Walls, columns and vertical faces of all structure members	16 to 24 hours as may be decided by the Engineer-In-Charge
(*b*) Slabs (props left under)	3 days
(*c*) Beam soffits (props left under)	7 days
(*d*) Removal of props under slabs	
(*i*) Spanning upto 4.5 m	7 days
(*ii*) Spanning over 4.5 m	14 days

(*e*) Removal of props under beams and arches

(*i*) Spanning upto 6 m — 14 days

(*ii*) Spanning over 6 m — 21 days

For other cements and lower temperature, the stripping time recommended for ordinary portland cement may be suitably modified.

The number of props left under, their sizes and disposition shall be such as to be able to safely carry the full dead load of the slab, beam or arch as the case may be together with any live load likely to occur during curing or further construction.

Where the shape of the element is such that the formwork has reentrant angles, the formwork, shall be removed as soon as possible after the concrete has set, to avoid shrinkage, cracking occurring due to the restraint imposed.

1.14 INSPECTION AND TESTING OF STRUCTURES

1.14.1 Inspection

Immediately after stripping the formwork, all concrete should be carefully inspected and any defective work or small defects either removed or made good before concrete has thoroughly hardened. In case of any doubts regarding the grade of concrete either due to poor workmanship or based on results of compressive strength tests of concrete cubes, core test or load tests may be carried out.

1.14.2 Testing

Load tests should be carried out as soon as possible after the expiry of 28 days from the time of placing of concrete. The structures or a part thereof, may be subjected to a load equal to full dead load of the structures plus 1.25 times the imposed loads for a period of 24 hours and then the imposed loads should be removed. The deflection due to imposed load only should be observed. If within 24 hours of removal of the imposed loads, the structure does not recover at least 75 per cent of the deflection under super-imposed loads, the test may be repeated after a lapse of 72 hours. If the recovery is less than 80%, the structure shall be deemed to be unacceptable. If the maximum deflection in mm, observed during 24 hours under load is less than $40l^2/D$ where l is effective span in metres and D the overall depth of the section in mm, it is not necessary to measure the recovery of the deflection.

Concrete tests can be carried out, if found necessary by the engineer-in-charge, as per the provision of IS : 516-1959.

Other nondestructive tests can also be adopted, *e.g.*, Rebound Hammer Tests or Ultra-Sonic Pulse Tests. Rebound hammer tests are carried out by a special type of hammer having a piston tied to the cylinder by a spring. The movement of the piston is recorded on a scale, the reading of which indicates the strength of concrete at that time. The ultrasonic test is based on the basic principle of velocity of ultrasonic pulse in concrete. The wave velocity is not determined direct but is calculated from the time taken by a pulse to travel a measured distance (*i.e.*, the thickness of the member).

1.15 STABILITY OF STRUCTURES

In designing any structure, care should be taken to external stability of the structure or its component members individually in addition to its internal resistance. For stability, generally two conditions are to be satisfied, *i.e.*, the structure should be safe against overturning and also against sliding.

1.15.1 Overturning

The stability of a structure as a whole against overtuning should be ensured so that the restoring moment shall be not less than the sum of 1.2 times the maximum overturning moment due to characteristic dead load and 1.4 times the maximum overturning moment due to the imposed loads. In case where dead load provides the restoring moment, only 0.9 times the characteristic dead load should be considered. Restoring moment due to imposed loads should be ignored.

The anchorages or counter weights provided for overhanging members (during construction and service) should be such that static equilibrium should remain, even when overturning moment is doubled.

1.15.2 Sliding

The structure should have a factor of safety against sliding of not less than 1.4 under the most adverse combination of the applied characteristic forces. In this case only 0.9 times the characteristic dead load should be taken into account.

To ensure stability at all times, account should be taken of probable variation in dead load during construction, repair or any other temporary measures. Wind or seismic loading should be treated as imposed loadings.

1.16 SI UNITS

Metric system has practically replaced the FPS (Foot Found Second) system. Lates IS : 456-1978 Code for plain and reinforced concrete has adopted SI system of units. SI is an abbreviation of the French "Systeme International d'system." In SI system, the units of force are Newtons in place of Kilograms as in Metric system. A Newton is the force which when applied to a body having a mass of one kilogram gives it an acceleration of one metre per second.

Thus, $$1\ \text{N} = \frac{1\ \text{kg}}{\text{g}} \quad \text{where g} = 9.806\ \text{m/sec}^2$$

$$1\ \text{kg} = 9.806\ \text{N} = 10\ \text{N (approximately)}$$

The units of stress shall be N/mm^2.

Some of the important conversions, from metric to SI units, are presented in Appendix-C.

1.17 STRENGTHENING OF STRUCTURES

Sometimes cracks develop on the concrete surface in a structure because of weakness designing, incorrect placement of bars of reinforcement, temperature variations, change in use of building or due to any other unforeseen reason, epoxy resins are available which can be put on or injected in the cracks. These resins have very high compressive and tensile. The engineer is to take decision if use of epoxy in the distressed section can serve the purpose.

EXERCISE 1

1. Write in brief the type of loads and forces generally, encountered in RCC structures.
2. (*a*) List advantages and disadvantages of RCC construction as compared to other types of construction.

   (*b*) Name various reinforcing materials.
3. (*a*) What are different grades of concrete?

   (*b*) What do you understand by Design Mix and Nominal Mix of concrete?
4. Write short notes on :

   (*a*) Curing (*b*) Batching (*c*) Mixing

5. Discuss briefly the important properties of concrete.
6. Discuss briefly the various methods of design used for RCC construction.
7. What are different types of cements available for use in RCC construction? Explain briefly each type of cement.
8. Explain the following :
   (*i*) Concreting in hot weather
   (*ii*) Concreting in cold weather
   (*iii*) Concreting under water
   (*iv*) Ready mixed concrete
   (*v*) Light weight concrete
   (*vi*) Fibre reinforced concrete
   (*vii*) Polymer concrete
9. What do you understand by Formwork and at what timings it may be removed for different types of RCC members?
10. Explain briefly the testing of structures.

This Page has been intentionally left blank

PART – II

WORKING STRESS METHOD OF DESIGN

Working stress method is based on the behaviour of a section under the loads expected to be encountered by it during its serive period. The strength of concrete in the tension zone of the member is neglected although the concrete does have some strength for direct tension and flexual tension. The materials, both concrete and steel, are assumed to behave perfectly elastically, that is stress is proportional to strain. The distribution of strain across a section is assumed to be linear. The sections that are plane before bending remain plane after bending. Thus the strain, hence the stress at any point is proportional to the distance of the point from the neutral axis. With this, a triangular stress distribution in concrete is obtained, ranging from zero at the neutral axis to a maximum at the compressive face of the section. It is further assumed in this method that there is perfect bond between the steel and the surrounding concrete. Since no slipping is to take place at any point between steel and the surrounding concrete, the strains in both materials at that point are same and hence the ratio of stresses in steel and concrete will be the same as the ratio of elastic moduli of steel and concrete. This ratio being known as modular ratio, the method is also called 'Modular Ratio Method'.

In this method, external forces and moments are assumed to be resisted by the internal compressive forces developed in concrete and tensile resistive forces in steel and the internal resistive couple due to the above two forces, in concrete acting through the centroid of triangular distribution of the compressive stresses and in steel acting at the centroid of the tensile reinforcement. The distance between the lines of action of resultant resistive forces is known as 'lever arm'.

Moments and forces acting on the structure are computed from the service loads. The section of the component member is proportioned to resist these moments and forces such that the maximum stresses developed in materials are resitricted to a fraction of their true strengths. The factors of safety used in getting maximum permissible stresses are as follows:

Material	*Factor of safety*
For concrete	3
For Steel	1.78

This Page has been intentionally left blank

Reinforced Concrete Beam Analysis

CHAPTER **2**

2.1 INTRODUCTION

Beams are members that are subjected to bending. Bending causes comprehensive as well as tensile stresses in the same cross section depending upon the position of particles and the type of end supports. In a simply supported beam, top half of cross section is under compression and the bottom half under tension. In a cantilever beam, top half of the section will be under tension and the bottom half under compression under the vertically downwards acting-loads. The variation of stresses in rectangular concrete section for a simply supported beam under vertically downward loads, is shown in Fig. 2.1. Stresses shall be indicated negative if compressive and positive if tensile.

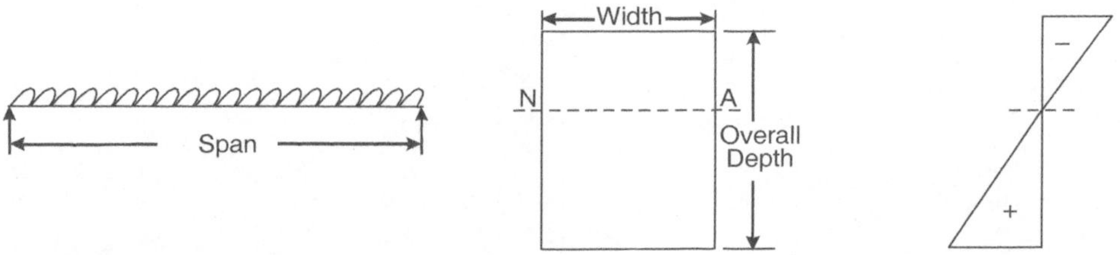

(a) Simply Supported Beam (b) Rectangular section (c) Variation of stresses

Fig. 2.1

In order to make the concrete beams capable of carrying tensile stresses (because concrete is extremely weak in tension) they have to be reinforced with a material which can take the tenion alone. Steel being very strong in tension, is mostly used for reinforcement purposes. Steel may be Mild Steel in the form plain round or square bars or high strength steel in the form of deformed bars. High strength deformed bars are more popular as steel requirement is less as compared to mild steel bars. Cost difference in two types of bars is insignificant.

The stresses vary in a homogeneous material in accordance with the simple theory of bending. The stress at any point in a material can be calculated from the following formula.

$$f = \frac{M}{I} \cdot y \qquad ...(2.1)$$

where f = stress at distance y from the neutral axis,

M = bending moment at the section,

I = moment of inertia of cross section about the neutral axis.

If may be noted that M and I both refer to the section where stresses are required to be found out.

Concrete alone is practically a homogenous material but when steel is placed in it, it becomes non-homogenous. As such it is difficult to apply the theory of simple bending to reinforced concrete structures. Hence a new theory should be evolved for such structures which is out of scope of this text. However, the theory of simple bending can be applied to reinforced concrete structures with certain modifications and assumptions.

2.2 END SUPPORTS

The ends of the structural members have to be supported so that the member is under equilibrium. The various types of end supports and the placement of reinforcing steel in tension zone of the member depending on the behaviour of the member are shown in Figs. 2.2 to 2.5.

(*i*) *Beams Simply Supported at Ends.* The beam rests on either walls or columns in a simple manner as shown in Fig. 2.2. The ends of the beam can have rotation in either direction, clockwise or anticlockwise, in the vertical plane. It permits movement in one direction only, i.e., along A-A as shown. The reactions act always at right angles to the longitudinal axis of the beam depending upon the type of loading.

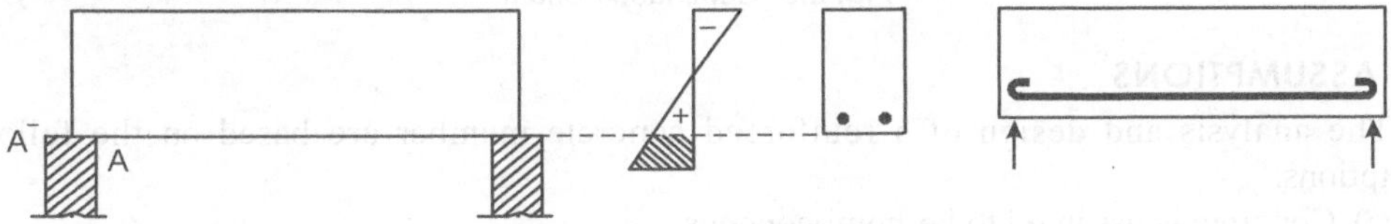

Fig. 2.2. Simply Supported Ends

(*ii*) *Beam with Hinged Supports.* This type of support permits rotation in either direction, clockwise or anticlockwise. It does not allow any movement over supports. The reactions can act at any angle depending upon the loading conditions. Beam with hinged supports at ends is shown in Fig. 2.3.

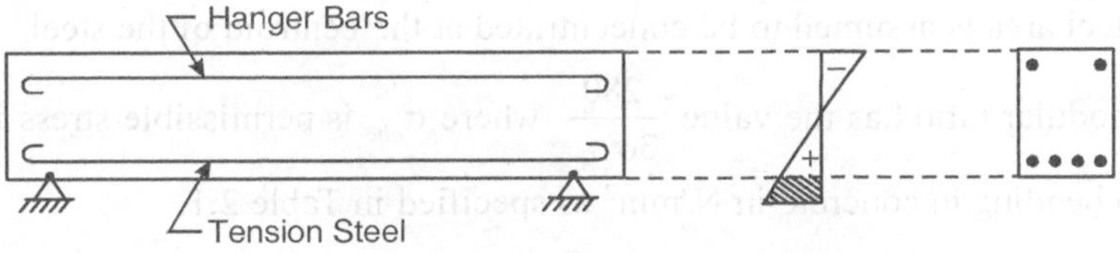

Fig. 2.3. Hinged Ends

(*iii*) Fixed Support. This type of support does not permit any translation or rotation in any direction. Loading products reactions and moments over support of this type. The placement of reinforcement in a beam built-in or fixed at ends is shown in Fig. 2.4.

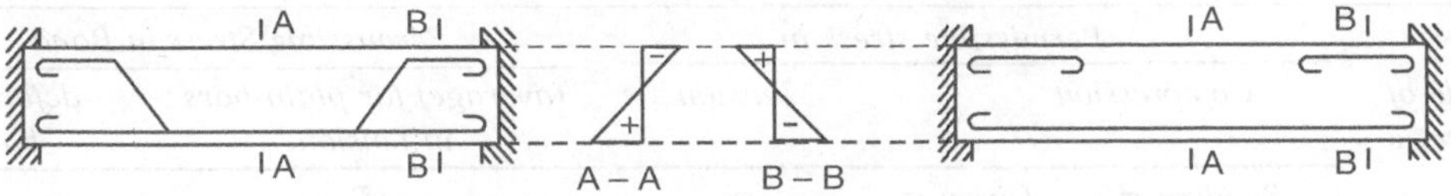

Fig. 2.4. Fixed Supports

Members in actual structures have combination of the supports as described-above. An overhanging beam has always one end fixed and the other completely free as shown in Fig. 2.5*a*. A beam which is continuously supported at many supports, is hanged at one place and simply rests (or placed on rollers) over other supports. Two span continuous beam (with vertical downward loading showing placement of steel in tension zones, is shown in Fig. 2.5*b*.

(a)

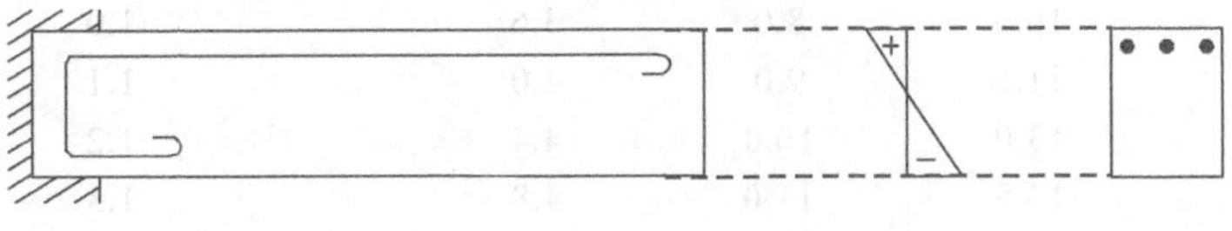

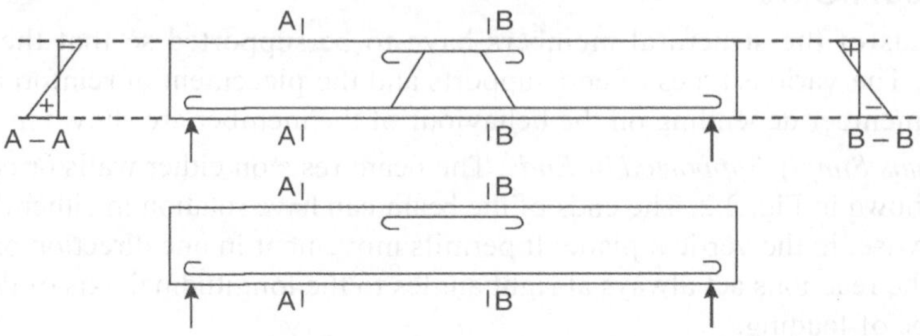

Fig. 2.5. Continuous Beam

2.3 ASSUMPTIONS

The analysis and design of a reinforced concrete member are based on the following assumptions.

(*i*) Concrete is assumed to be homogeneous.

(*ii*) At any cross-section, plane sections before bending remain plane after bending.

(*iii*) The stress-strain relationship for concrete is a straight line, under working loads.

(*iv*) The stress-strain relationship for steel is a straight line, under working loads.

(*v*) Concrete area on tension side is assumed to be ineffective.

(*vi*) All tensile stresses are taken up by reinforcement and none by concrete except when specialy permitted.

(*vii*) The steel area is assumed to be concentrated at the centroid of the steel.

(*viii*) The modular ratio has the value $\frac{280}{3\sigma_{cbc}}$ where σ_{cbc} is permissible stress is compression due to bending in concrete in N/mm^2 as specified in Table 2.1.

2.4 PERMISSIBLE STRESSES IN CONCRETE

The stress-strain curve for concrete is shown in Fig. 2.6. The permissible stresses for various grades of concrete shall be taken as those specified in Table 2.1 and Table 2.2.

Table 2.1. Permissible Stresses in Concrete

Grade of Concrete	*Permissible stress in* Compression: *Bending* σ_{cbc} *N/mm*^{2}	*Compression: Direct* σ_{cc} *N/mm*^{2}	*Tension* σ_{ct} *N/mm*^{2}	*Permissible Stress in Bond (average) for plain bars in tension* T_{bd} *N/mm*^{2}	*deformed bars*
(1)	(2)	(3)	(4)	(5)	(6)
M10	3.0	2.5	1.2	—	—
M15	5.0	4.0	2.0	0.6	0.96
M20	7.0	5.0	2.8	0.8	1.28
M25	8.5	6.0	3.2	0.9	1.44
M30	10.0	8.0	3.6	1.0	1.60
M35	11.5	9.0	4.0	1.1	1.76
M40	13.0	10.0	4.4	1.2	1.92
M45	14.5	11.0	4.8	1.3	2.08
M50	16.0	12.0	5.2	1.4	2.24

Note: 1. The bond stress given in col. 5 shall be increased by 25% for bars in compression.

2. The bond stress given in col. 5 shall be increased by 60% for deformed bars (col. 6)

Direct Tension in Concrete: For members in direct tension, when full tension is taken by the reinforcement alone, tensile stress shall be not greater than the values given in Table 2.1.

The tensile stresses shall be calculated as $\frac{F_t}{A_c + m.A_{st}}$.

where F_t = total tension in the member minus pre-tension in steel if any before concreting,

A_c = cross sectional area of concrete excluding any finishing material and reinforcing steel,

m = modular ratio,

A_{st} = cross sectional area of reinforcing steel in tension.

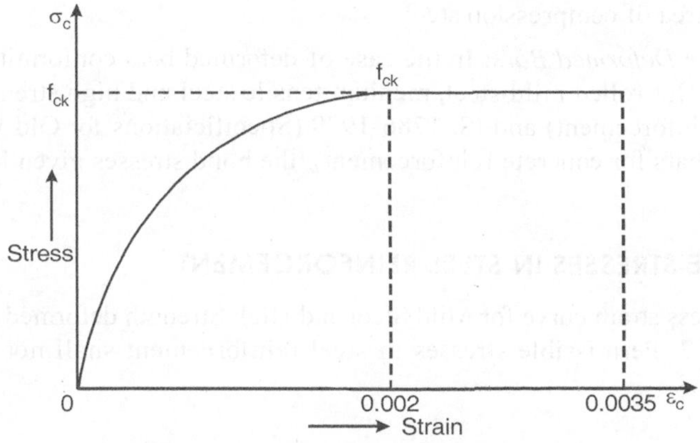

Fig. 2.6. Stress Strain Curve for Concrete

Table 2.2. Permissible Shear Stress in concrete

100 A_s / b.d (1)	*Permissible Shear Stress in concrete Tc, N/mm² Grade of Concrete* *M15* (2)	*M20* (3)	*M25* (4)	*M30* (5)	*M35* (6)	*M40 and above* (7)
≤ 0.15	0.18	0.18	0.19	0.20	0.20	0.20
0.25	0.22	0.22	0.23	0.23	0.23	0.23
0.50	0.29	0.30	0.31	0.31	0.31	0.32
0.75	0.34	0.35	0.36	0.37	0.37	0.38
1.00	0.37	0.39	0.40	0.41	0.42	0.42
1.25	0.40	0.42	0.44	0.45	0.45	0.46
1.50	0.42	0.45	0.46	0.48	0.49	0.49
1.75	0.44	0.47	0.49	0.50	0.52	0.52
2.00	0.44	0.49	0.51	0.53	0.54	0.55
2.25	0.44	0.51	0.53	0.55	0.56	0.57
2.50	0.44	0.51	0.55	0.57	0.58	0.60
2.75	0.44	0.51	0.56	0.58	0.60	0.62
3.00 and above	0.44	0.51	0.57	0.60	0.62	0.63

Note: A_s is that area of longitudinal tension reinforcement which continues at least one effective depth beyond the section being considered except at supports where the full area of tension reinforcement may be used.

For other values of $\frac{A_s \times 100}{b.d}$, the permissible stresses shall be obtained by interpolation.

Direct compression in Concrete: For members in direct compression, permissible stress in direct compression shall be taken from Table 2.1. The direct compression stress shall be calculated from the relation :

$$\frac{P}{A_c + m.\ A_{sc}}$$

where P = Direct compressive force,

A_{sc} = Area of compression steel.

Bond Stress for Deformed Bars: In the case of deformed bars conforming to IS:1139-1966 (Specificiations for Hot-rolled mild steel, medium tensile steel and high strength steel deformed bars for concrete reinforcement) and IS: 1786-1979 (Specificiations for Old Worked steel, High Strength deformed bars for concrete reinforcement), the bond stresses given in Table 2.1 may be increased by 60%.

2.5 PERMISSIBLE STRESSES IN STEEL REINFORCEMENT

The typical stress strain curve for Mild Steel and High Strength deformed bars, *i.e.*, Tor steel, is shown in Fig. 2.7. Permissible stresses in steel reinforcement shall not exceed the values specified in Table 2.3

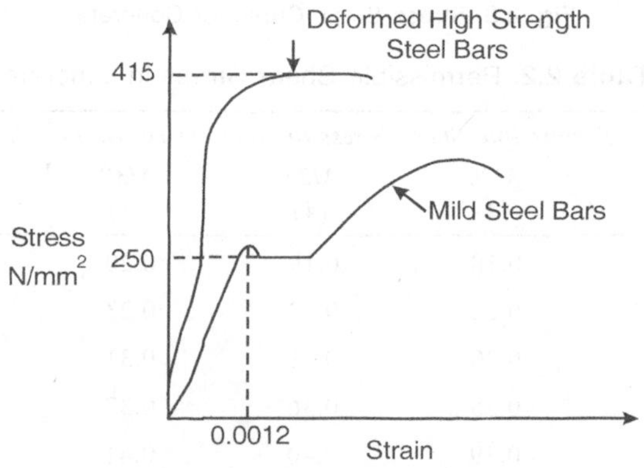

Fig. 2.7. Stress-Strain Curve for Steel

In flexural members, the value of σ_{st} given in Table 2.3 is applicable at the centroid of the tensile reinforcement subject to the condition that when more than one layer of the tensile reinforcement is provided, the stress at the centroid of the outermost layer shall not exceed by more than 10% the value given in Table 2.3.

Table 2.3. Permissible Stresses in Steel Reinforcement

Sl. No.	*Type of stress in Steel reinforcement*	*Permissible stress in N/mm²*			
		Mild Steel Bars Conforming to Grade I of IS: 432 (part I)-1996 or Deformed Mild Steel bars Conforming to IS: 1139-1966 (Fe 250)	*Medium Tensile Steel conforming to IS:432 (part I) -1966 or Deformed Medium Tensile Steel bars Conforming to IS: 1139-1966 (Fe 250)*	*High Yield Strength Deformed bars conforming to IS: 1139-1966 or IS:1786-1979 (Grade Fe 415)*	*(Grade Fe 500)*
(1)	Tension (σ_{st} or σ_{sv}):				
	(a) upto and including 20 mm	140	Half the guaranteed yield stress subject to a maximum of	230	275
	(b) over 20 mm	130	190	230	275
(2)	Compression in Column bars (σ_{sc})	130	130	190	190
(3)	Compression in bars in a beam of slab when the compressive resistance is taken into account.	The calculated compressive stress in the surrounding concrete multiplied by 1.5 times the modular ratio or σ_{sc} whichever is lower			
(4)	Compression in bars in a beam or slab where the compressive resistance of the concrete is not taken into account:				
	(a) upto and including 20 mm	140	Half the guaranteed yield stress subject	190	190
	(b) over 20 mm	130	to a maximum of 190	190	190

Note: 1. For high yield strength deformed bars of Grade Fe 500 the permissible stress in direcrt tension and flexural tension shall be 0.55fy. The permissible stresses for shear and compression reinforcement shall be as for Grade Fe 415.

2. For welded wire fabric conforming to IS: 1566-1967, the permissible value in tension σst is taken as 230 N/mm².

3. For the purpose of this code, the yield stress of steels for which there is no clearly defined yield point should be taken to be 0.2 per cent proof stress.

4. When Mild Steel conforming to Grade II of IS 432-1966 (Part I) is used the permissible stresses shall be 90% of the permissible stresses in Col. 3, or if the design details have already been worked out on the basis of Mild steel conforming to Grade I of IS:4321966 (Part I); the area of reinforcement shall be increased by 10% of that required for Grade I steel.

2.6 INCREASE IN PERMISSIBLE STRESSES

Where stresses due to wind (or earthquake), temperature and shrinkage effects are combined with those due to dead, live and impact load, the usual permissible stresses in concrete and steel may be exceeded upto a limit of $33\frac{1}{3}\%$. Wind and seismic forces need not be considered as acting simultaneously.

2.7 POSITION OF NEUTRAL AXIS

Neutral axis is the axis at which the stresses are zero in the section. The areas above and below it are subjected to compressive and tensile stresses respectively in the case of simply supported beams with vertical loads acting downwards while reverse is the case when the beam is a cantilever. Locating the position of neutral axis is of great importance to the Structural Engineer. This helps in knowing the amount of strains both in concrete and steel. The position of neutral axis in a section can be found be either of the following two ways depending upon the situation.

2.7.1 Position of Neutral Axis when Stresses Developed in the Section are known

Consider a beam section with overall dimensions as *b*, the width and *D*, the overall depth of the section. Let the effective depth be *d* as shown in Fig. 2.8, '*x*' be the distance of neutral axis (N.A.) from farthest compression fibre.

Let ε_c = Maximum strain in concrete,

ε_s = Maximum strain at the centroid of the steel,

σ_{cbc} = maximum compressive stress in concrete, in bending

σ_{st} = Stress in steel

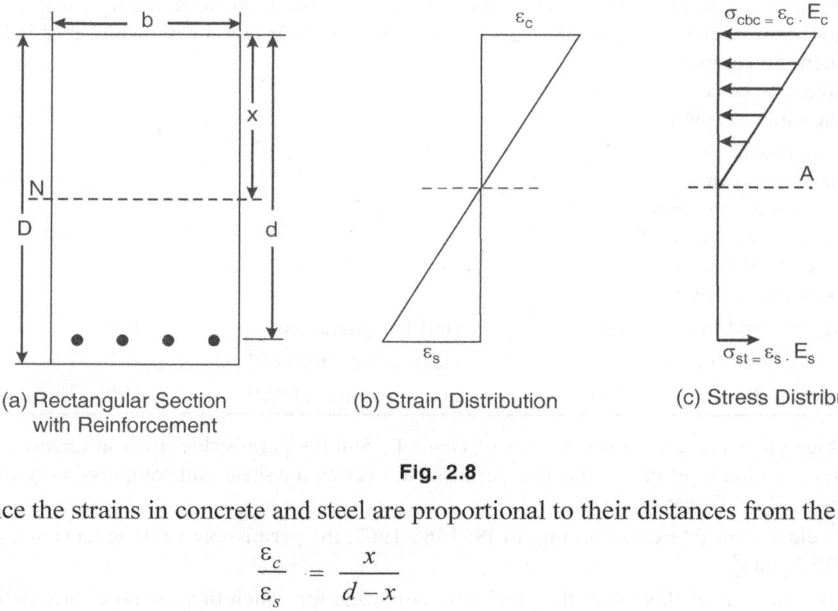

(a) Rectangular Section with Reinforcement (b) Strain Distribution (c) Stress Distribution

Fig. 2.8

Since the strains in concrete and steel are proportional to their distances from the neutral axis,

$$\frac{\varepsilon_c}{\varepsilon_s} = \frac{x}{d-x}$$

$$\frac{d-x}{x} = \frac{\varepsilon_s}{\varepsilon_c}$$

$$\frac{d}{x} - 1 = \frac{\sigma_{st}}{E_s} \cdot \frac{E_c}{\sigma_{cbc}}$$

$$= \frac{\sigma_{st}}{\sigma_{cbc}} \cdot \frac{1}{m} \qquad \text{where } \frac{E_s}{E_c} = m \text{ (modular ratio)}$$

$$\frac{d}{x} = 1 + \frac{\sigma_{st}}{m \cdot \sigma_{cbc}}$$

$$x = \frac{1}{1+\sigma_{st}/m\cdot\sigma_{cbc}}\cdot d$$

$$= k \cdot d \qquad \text{... (2.2)}$$

where k = neutral axis constant

$$= \frac{1}{1+\sigma_{st}/m\cdot\sigma_{cbc}}$$

2.7.2 Position of Neutral Axis when Section is known

The maximum moments developed in a section are resisted by the compression in concrete and tension in steel. The amount of tension taken by concrete is neglected. Neutral axis will be centre of gravity of composite section consisting of concrete area in compression and area of tensile steel.

Let b = width of section be

D = overall depth of section

d = effective depth of section

A_s = area of tensile steel

The compressive force in concrete = Total tensile force in steel.

Average stress above N.A × Area of concrete = Stress in steel × Area of steel.

$$\frac{\sigma_{cbc}}{2}\cdot b\cdot x = \sigma_{st}\cdot A_s$$

$$= m\cdot\sigma_{cbc}\cdot\frac{d-x}{x}\cdot A_s$$

$$b\cdot x^2 = 2\cdot m\cdot(d-x)\cdot A_s$$

or
$$b\cdot x^2 + 2.\, m\cdot A_s\cdot x - 2\cdot m\cdot A_s\cdot d = 0$$

or
$$x = \frac{-2\cdot m\cdot A_s \pm\sqrt{((2\cdot m\cdot A_s)^2 + 8\cdot m\cdot A_s\cdot b\cdot d)}}{2\cdot b}$$

$$x = \frac{-m\cdot A_s \pm\sqrt{((m\cdot A_s)^2 + 2\cdot m\cdot A_s\cdot b\cdot d)}}{b} \qquad \text{...(2.3)}$$

Note: 'x' cannot be negative.

$m\cdot A_s$ is called 'equivalent concrete area'.

Equivalent Area of Section: The area of concrete plus modular ratio times area of steel'. Because $b\cdot x\cdot x/2$ = moment of concrete area and $m\cdot A_s\cdot(d-x)$ is the moment of m times steel area about neutral axis, therefore the neutral axis is the centre of gravity of concrete area in compression and m times steel area concentrated at the centroid of steel.

2.8 MOMENT OF RESISTANCE

(*a*) *For Balanced Section*: When the maximum stresses in steel and concrete simultaneously reach their allowable values, the section is said to be a 'Balanced Section'. The moment of resistance shall be provided by the couple developed by compressive force acting at the centroid of stress diagram on the area of concrete in compression and tensile force acting at the centroid of reinforcement multiplied by the distance between these forces. This distance is termed as 'lever arm'.

$$\text{Total compressive force} = \frac{b \cdot x \cdot \sigma_{cbc}}{2}$$

$$\text{Total tensile force} = \sigma_{st} \cdot A_s$$

$$Z = \text{lever arm} = d - \frac{x}{3} \qquad \text{(Refer Fig. 2.9)}$$

$$= d - \frac{k \cdot d}{3}$$

$$= d\left(1 - \frac{k}{3}\right)$$

$$= j \cdot d \qquad ...(2.4)$$

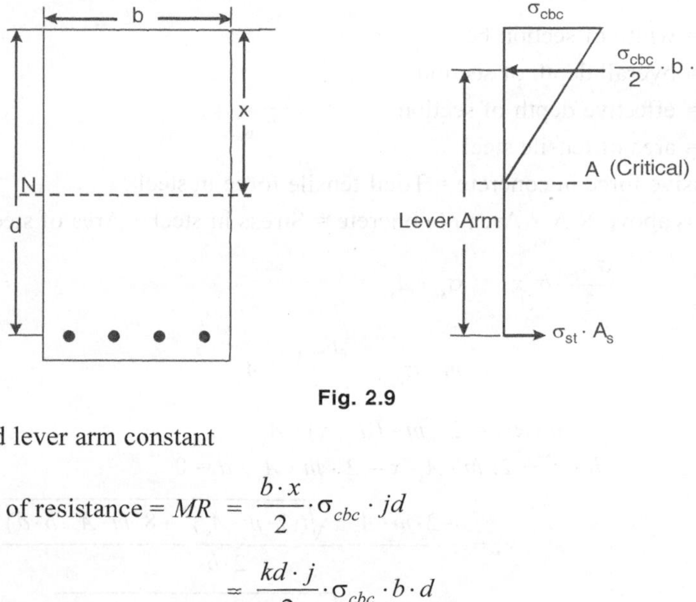

Fig. 2.9

where j is called lever arm constant

$$\text{Moment of resistance} = MR = \frac{b \cdot x}{2} \cdot \sigma_{cbc} \cdot jd$$

$$= \frac{kd \cdot j}{2} \cdot \sigma_{cbc} \cdot b \cdot d$$

$$= \frac{1}{2} \cdot k \cdot j \cdot \sigma_{cbc} \cdot b \cdot d^2$$

$$= Q \,.\, b \,.\, d^2 \qquad ...(2.5)$$

where Q is called moment of resistance constant, and is equal to $\frac{1}{2} \cdot k \cdot j \cdot \sigma_{cbc}$. The quantities k, j and Q are constant for a particular grade of concrete. For M20 Concrete and HYSD bars.

$$m = \frac{280}{3 \cdot \sigma_{cbc}} \qquad \sigma_{cbc} = 7\text{N/mm}^2 \qquad \sigma_{st} = 230\text{N/mm}^2$$

$$= \frac{280}{3 \times 7} = 13.33 \text{ (sometimes taken as 13)}$$

$$k = \frac{1}{1 + \sigma_{st} / m \cdot \sigma_{cbc}} = 0.29$$

Lever arm constant $j = 1 - \frac{k}{3}$

$= 0.9$

Moment of resistance constant $Q = \frac{1}{2} \cdot k \cdot j \cdot \sigma_{cbc}$

$$= \frac{1}{2} \times 0.29 \times 0.90 \times 7$$

$$= 0.91$$

$MR = Q \cdot b \cdot d^2 = 0.91 = b \cdot d^2$ for M20 concrete and HYSD bars

Alternatively, Moment of resistance = Total tension × Lever arm moment of resistance $= \sigma_{st} \cdot A_s \cdot j \cdot d$

$$\sigma_{st} \cdot A_s \cdot jd = Q \cdot b \cdot d^2$$

$$\frac{A_s}{b \cdot d} = \frac{Q}{\sigma_{st} \cdot j} = \frac{\frac{1}{2} \cdot \sigma_{cbc} \cdot k \cdot j}{\sigma_{st} \cdot j} = \frac{1}{2} \cdot \frac{\sigma_{cbc}}{\sigma_{st}} \cdot k \qquad \text{...(2.6)}$$

If P is the percentage of steel area $= \frac{A_s \times 100}{b \cdot d}$

A balanced section has definite percentage of steel termed as "Economic Percentage".

For M20 concrete and HYSD bars

$$P = \frac{1}{2} \cdot \frac{\sigma_{cbc}}{\sigma_{st}} \cdot k \cdot 100 = \frac{\frac{1}{2} \times 7 \times 0.29 \times 100}{230}$$

$$= 0.4413$$

The constants m, k, j, Q and p for different grades of concrete and different types of steels are presented in Table 2.4.

Table 2.4. Constants for Critical or Balanced Section

Grade of Concrete	*Permissible Stress in Concrete in compression* σ_{cbc} *(N/mm²)*		*Fe415 Steel with* σ_{st} *= 230 N/mm²*				*Fe500 Steel with* σ_{st} *= 275 N/mm²*			
		m	*k*	*j*	*Q*	*p*	*k*	*j*	*Q*	*p*
(1)	(2)	(3)	(4)	(5)	(6)	(7)	(8)	(9)	(10)	(11)
M15	5.0	18	0.29	0.9	0.65	0.315	0.2534	0.9155	0.580	0.2304
M20	7.0	13	0.29	0.9	0.91	0.440	0.2534	0.9155	0.812	0.3225
M25	8.5	11	0.29	0.9	1.11	0.535	0.2534	0.9155	0.986	0.3916
M30	10.0	9.0	0.29	0.9	1.30	0.630	0.2534	0.9155	1.160	0.4607
M35	11.5	8.0	0.29	0.9	1.50	0.725	0.2534	0.9155	1.334	0.530
M40	13.0	7.0	0.29	0.9	1.70	0.820	0.2534	0.9155	1.508	0.599
M45	14.5	6.5	0.29	0.9	1.89	0.914	0.2534	0.9155	1.682	0.668
M50	16.0	5.8	0.29	0.9	2.10	1.009	0.2534	0.9155	1.856	0.737

(*b*) *Under reinforced section*: When the percentage of steel in a section is less than that required for a balanced section, the section is called 'Under-reinforced section'. In this case concrete stress does not reach its maximum allowable value while the stress in steel reaches its maximum permissible vlaue. The position of the neutral axis will shift upwards, *i.e.*, the neutral axis depth will be smaller than that in the balanced section as shown in Fig. 2.10. The moment of resistance of such a section will be governed by allowble tensile stress in steel.

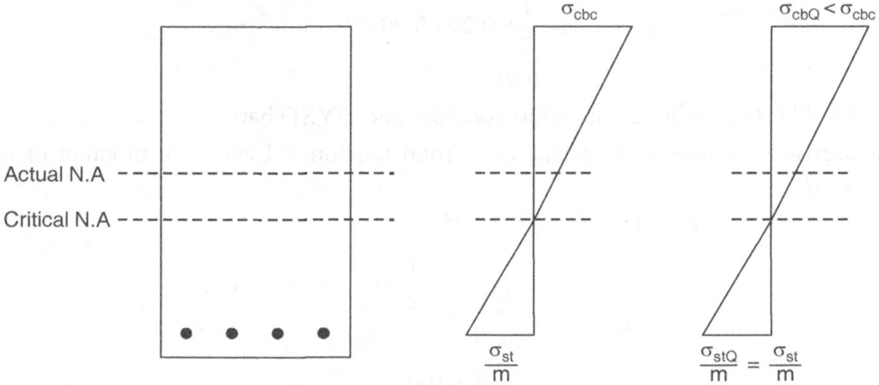

Fig. 2.10. Under-reinforced section

$$\text{Moment of resistance} = \sigma_{st} \cdot A_s \cdot \left(d - \frac{x}{3}\right)$$

$$= \sigma_{st} \cdot A_s \cdot j' \cdot d \quad \text{where } j' = 1 - \frac{k'}{3}$$

$$\text{Since } p = \frac{4_s \times 100}{b \cdot d}$$

$$\text{Moment of resistance} = \sigma_{st} \cdot p \cdot \frac{b \cdot d}{100} \cdot j' \times d$$

$$= \frac{\sigma_{st} \cdot p \cdot j'}{100} \cdot b \cdot d^2$$

$$= Q' \cdot b \cdot d^2 \quad \text{where } Q' = \frac{\sigma_{st} \cdot p \cdot j}{100} \qquad ...(2.7)$$

(*c*) *Over-reinforced section*: When the percentage of steel in a section is more than that required for a balanced section, the section is called 'Over-reinforced section'. In this case the stress in concrete reaches its maximum allowable value earlier than that in steel. As the percentage steel is more, the position of the neutral axis will shift towards steel from the critical of balanced neutral axis position. Thus the neutral axis depth will be greater than that in case of balanced section as shown in Fig. 2.11.

Moment of resistance of such a section will be governed by compressive stress in concrete

$$\text{Moment of resistance} = b \cdot x \cdot \sigma_{cbc} \cdot \frac{1}{2} \cdot \left(d - \frac{x}{3}\right)$$

$$= \frac{\sigma_{cbc}}{2} \cdot b \cdot x \cdot d \cdot \left(1 - \frac{k'}{3}\right)$$

$$= \frac{\sigma_{cbc}}{2} \cdot b \cdot x \cdot d \cdot j'$$

$$= \frac{1}{2} \cdot \sigma_{cbc} \cdot k' \cdot j' \cdot b \cdot d^2$$

$$= Q' \cdot b \cdot d^2$$

$$\text{where } Q' = \frac{\sigma_{cbc}}{2} \cdot k' \cdot j' = \text{constant} \qquad ...(2.8)$$

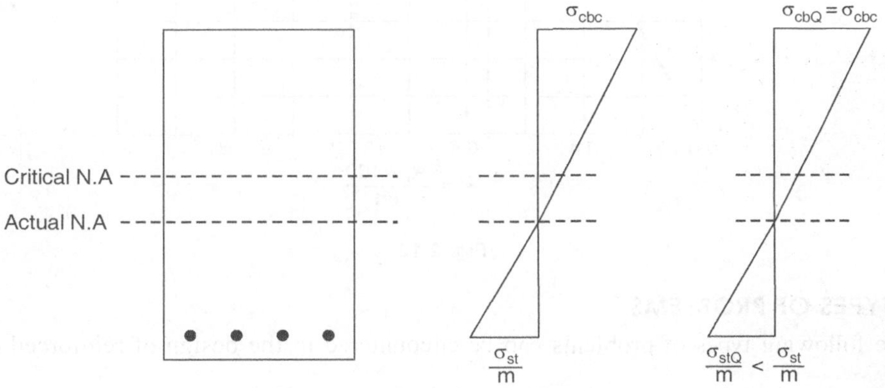

Fig. 2.11 Over-reinforced Section

In the under-reinforced and over-reinforced sections, there is wastage of one of the two materials. In under-reinforced section, stress in a steel reaches its permissible value earlier, thus some concrete remains un-utilised. Whereas in the over-reinforced section, stress remains in concrete reaches its permissible value earlier, thus some steel remains un-utilised. In case of over reinforced sections failure in concrete occurs earlier. Concrete being brittle, this happens suddenly and explosively without any warning. But in case of under-reinforces sections, stress in steel reaches its ultimate stress earlier. Steel being ductile, it takes some time to break. This gives sufficient warning before the final collapse of the structure. Hence it is always desirable to design under-reinforced sections.

2.9 EFFECT OF VARYING PERCENTAGE OF STEEL ON MOMENT OF RESISTANCE

The moment of resistance is proportional to the percentage of steel present in the section. The effect of varying percentage of steel on moment of resistance is shown in Fig. 2.12 where $Q' = \frac{M}{b \cdot d^2}$ and $p = \frac{A_s \times 100}{b \cdot d \cdot}$.

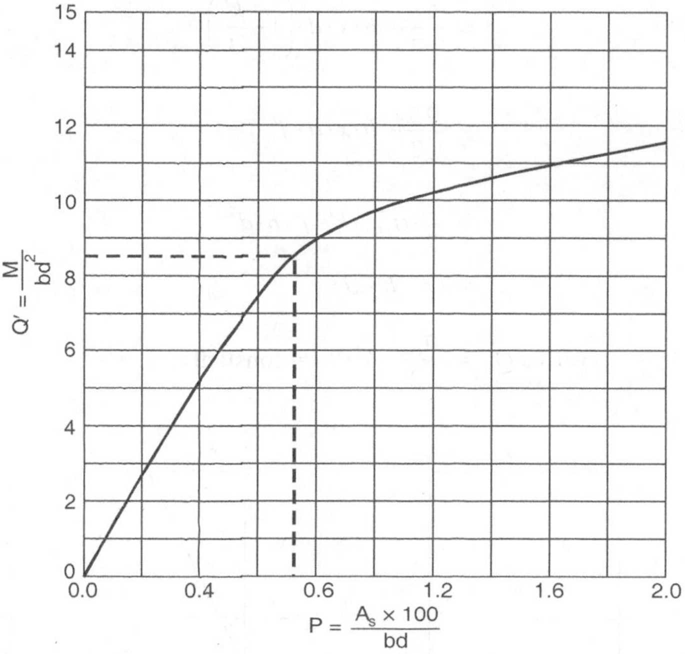

Fig. 2.12

2.10 TYPES OF PROBLEMS

The following types of problems can be encountered in the design of reinforced concrete beams.

2.10.1 Determination of Area of Tensile Reinforcement

The section, bending moment to be resisted and the maximum stresses in steel and concrete are given.

Steps to be followed

(*i*) Determine k, j, Q (or Q') for the given stresses.

(*ii*) Find the critical moment of resistance, $M = Q \cdot b \cdot d^2$ from the dimensions of the beam.

(*iii*) Compare the bending moment to be resisted with M, the critical moment of resistance.

(*a*) If *B.M.* is less than M, design the section as under reinforced.

$$M = \sigma_{st} \cdot A_s \cdot \left(d - \frac{x}{3}\right)$$

To find A_s in terms of x, take moments of areas about *N.A.*

$$b \cdot x \cdot \frac{x}{2} = m \cdot A_s \cdot (d - x)$$

$$A_s = \frac{b \cdot x^2}{2(m)(d - x)}$$

$$\therefore \quad M = \frac{\sigma_{st} \cdot b \cdot x^2}{2 \cdot m \cdot (d-x)} \left(d - \frac{x}{3}\right) = B.M. \text{ to be resisted}$$

Solve for 'x', and then A_s can be calculated.

(*b*) If *B.M.* is more than *M*, design the section as over-reinforced.

$$M = \frac{\sigma_{cbc}}{2} \cdot b \cdot x \left(d - \frac{x}{3}\right) = \text{B.M. to be resisted}$$

Determine 'x'. Then A_s can be obtained by taking moments of areas (compressive and tensile) about *N.A.* using the following expression

$$A_s = \frac{b \cdot x^2}{2 \cdot m \cdot (d-x)}$$

2.10.2 Design of Section for a Given Loading

Design the section as balanced section for the given loading.

Steps to be followed

(*i*) Find the maximum bending moment (*B.M.*) due to given loading.

(*ii*) Compute the constants k, j, Q for the balanced section for known stresses.

(*iii*) Fix the depth to breadth ratio of the beam section as 2 to 4.

(*iv*) From $M = Q \cdot b \cdot d^2$, find 'd' and then 'b' from depth to breadth ratio.

(*v*) Obtain overall depth 'D' by adding concrete cover to 'd' the effective depth.

(*vi*) Calculate A_s from the relation.

$$A_s = \frac{B \cdot M}{\sigma_{st} \cdot j \cdot d}$$

2.10.3 To Determine the Load Carrying Capacity of a Given Beam

The dimensions of the beam section, the material stresses and area of reinforcing steel are given.

Steps to be followed

(*i*) Find the position of the neutral axis from section and reinforcement given.

(*ii*) Find the position of the ciritical *N.A.* from known permissible stresses of concrete and steel.

$$x = \frac{1}{1 + \dfrac{\sigma_{st}}{m \cdot \sigma_{cbc}}} \cdot d$$

(*iii*) Check if

(*i*) > (*ii*) – the section is over-reinforced

(*i*) < (*ii*) – the section is under-reinforced

(*iv*) Calculate *M* from the relation

$$M = b \cdot x \cdot \frac{1}{2} \sigma_{cbc} \cdot \left(d - \frac{x}{3}\right) \text{ for over-reinforced section}$$

and $$M = \sigma_{st} \cdot A_s \cdot \left(d - \frac{x}{3}\right)$$ for under-reinforced section

(*v*) If the effective span and the support conditions of the beam are known, the load carrying capacity can be computed.

2.10.4 To Check the Stresses Developed in Concrete and Steel

The section, reinforcement and bending moment are given.

Steps to be followed

(*i*) Find the position of the *N.A* using the following relation.

$$b \cdot \frac{x^2}{2} = m \cdot A_s (d - x)$$

(*ii*) Determine lever arm, $z = d - \frac{x}{3}$

(*iii*) $BM = \sigma_{sa} \cdot A_s \cdot z$ is used to find out the actual stress in steel, σ_{sa}.

(*iv*) To compute the actual stress in concrete σ_{cba}, use the following relation.

$$BM = \frac{\sigma_{cba}}{2} b \cdot x \cdot z$$

2.11 SOLVED EXAMPLES

Example 2.1. Design a reinforced rectangular beam to resist a *B.M* of 200 KN.m (20 *t.m*). The width of the beam is to be kept at 450 mm. Use M20 concrete (σ_{cbc} = 7N/mm²) and (*i*) Mild Steel (s_{st} = 140N/mm²) and (*ii*) High strength deformed bars (σ_{st} = 230N/mm²).

Solution. Compute $m = \frac{280}{3\sigma_{cbc}}$

$$= \frac{280}{3 \times 7} = 13.33$$

(*i*) *Mild Steel*

$$k = \frac{1}{1 + \frac{\sigma_{st}}{m \cdot \sigma_{cbc}}}$$

$$= \frac{1}{1 + \frac{140}{7 \times 13.33}} = 0.4$$

$$j = 1 - \frac{k}{3} = 0.87$$

$$Q = \frac{1}{2} \cdot \sigma_{cbc} \cdot k \cdot j$$

$$= \frac{1}{2} \times 7 \times 0.4 \times 0.87$$

$$= 1.21$$

$$BM = Q \cdot b \cdot d^2$$
$$200 \times 1000 \times 1000 = 1.21 \times 450.d^2$$
$$d = 606 \text{ mm}$$

Assuming 20 mm diameter bars to be used

$$D = d + \frac{1}{2} \cdot \text{dia. of bar} + \text{cover}$$
$$= 606 + 10 + 25 = 641 \text{ mm} \qquad \text{say } 750 \text{ mm}$$
$$\therefore \quad d = 750 - 10 - 25 = 715 \text{ mm}$$

Area of steel required $\quad A_s = \dfrac{BM}{\sigma_{st} \cdot j \cdot d}$

$$= \frac{200 \times 10^2 \times 10^2}{140 \times 0.87 \times 715}$$
$$= 2297.39 \text{ mm}^2$$

Provide 8 bars of 20 mm dia., thus providing $A_s = 2514 \text{ mm}^2$ (Fig. 2.13).

(*ii*) *High Strength Deformed Bars*

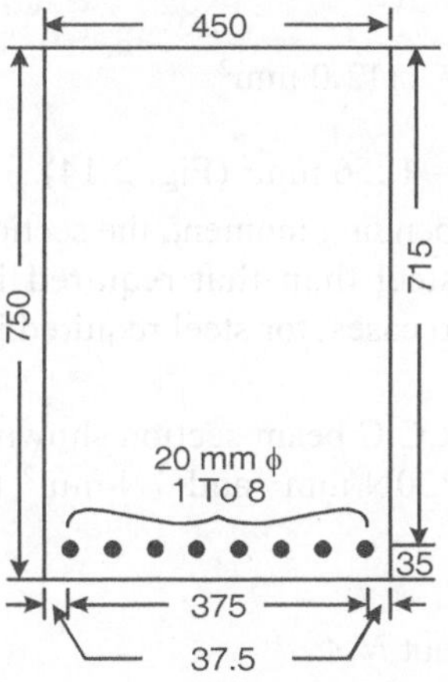

Fig. 2.13

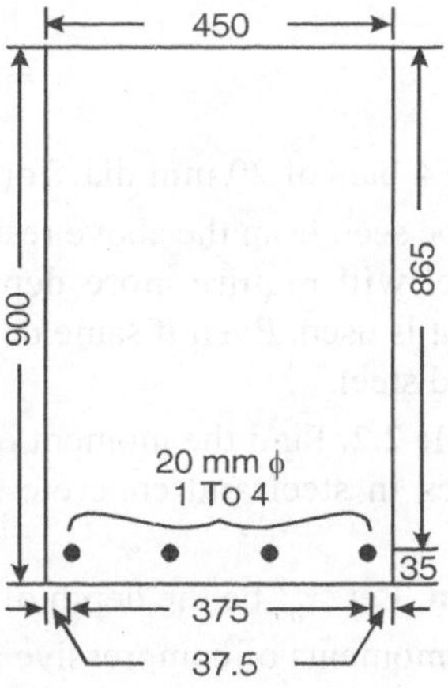

Fig. 2.14

$$x = k \cdot d = \frac{1}{1 + \dfrac{\sigma t}{m \cdot \sigma_{cbc}}} \cdot d$$

$$k = \frac{1}{1 + \dfrac{230}{7 \times 13.33}}$$

$$= 0.289$$

$$j = 1 - \frac{k}{3}$$

$$= 1 - \frac{0.289}{3}$$

$$= 0.904$$

$$Q = \frac{1}{2} \cdot \sigma_{cbc} \cdot k \cdot j$$

$$= \frac{1}{2} \times 7 \times 0.289 \times 0.904 = 0.914$$

$$BM = Q \cdot b \cdot d^2$$

$$200 \times 10^3 \times 10^3 = 0.914 \times 450 \times d^2$$

$$d = 697 \text{ mm}$$

Provide $D = 900$ mm $\therefore d = D - \text{cover} - \frac{1}{2}$ dia. of bar

$$= 900 - 25 - 10$$

$$= 865$$

Area of steel required $A_s = \dfrac{BM}{\sigma_{st} \cdot j \cdot d}$

$$= \frac{200 \times 10^6}{230 \times 0.904 \times 865} = 1112.0 \text{ mm}^2$$

Provide 4 bars of 20 mm dia. Tor, thus providing $A_s = 1256$ mm² (Fig. 2.14).

It may be seen from the above results that for a given bending moment, the section reinforced with tor steel will require more depth and much less steel than that required if mild steel reinforcement is used. Even if same depth is kept in the two cases, tor steel required is about 35% less than mild steel.

Example 2.2. Find the moment of resistance of the R.C.C beam section shown in Fig. 2.15 if the stresses in steel and concrete are not to exceed 230N/mm² and 7N/mm² respectively. $m = 13.33$.

Solution. Let 'x_a' be the depth of actual *N.A.*

Taking moments of compressive and tensile areas about *N.A.*,

$$\frac{b \cdot x^2 a}{2} = m.A_s(d - x_a)$$

$$450 \cdot \frac{x^2 a}{2} = 13.33 \times 2514(715 - x_a)$$

$$x^2 a + 148.94 x_a - 106492.5 = 0$$

Solving the above equation for 'x_a' and neglecting the negative value of 'x_a'.

$$x_a = 260.25 \text{ mm}$$

Depth of critical *N.A.* for the given stresses = 0.29 × 715 = 207.35 mm.

Depth of actual *N.A.* is more than that of the critical *N.A.*

The section is over-reinforced.

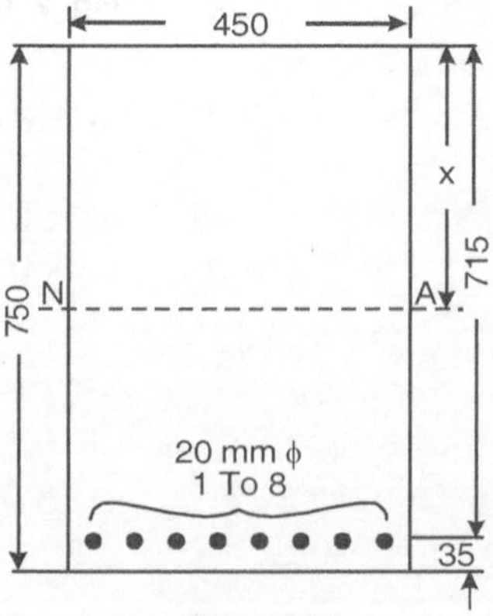

Fig. 2.15

$$\text{Moments of resistance} = b \cdot x_a \cdot \frac{\sigma_{cbc}}{2}\left(d - \frac{x_a}{3}\right)$$

$$= 450 \times 260.25 \times \frac{7}{2}\left(715 - \frac{260.25}{3}\right)$$

$$= 2.575 \times 10^8 \text{ N.mm}$$

$$= 257.5 \text{ KN.m}$$

Example 2.3. Determine the maximum superimposed load the beam section in Ex. 2.2 (Fig. 2.15) can carry if the effective span of the beam is 5 m and is simply supported at its ends, as shown in Fig. 2.16.

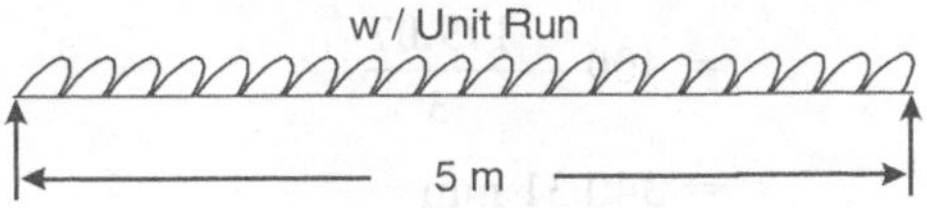

Fig. 2.16

Solution. Maximum $B.M. = \frac{w.L^2}{8}$

$$= \frac{w \times 5^2}{8} \text{ N.m if } w \text{ is in N/m}$$

BM = Moment of resistance

$$\frac{w \times 5^2}{8} = 257.5 \times 10^3$$

$$w = 82400.0 \text{ N/m}$$

Unit weight of reinforced cement concrete = 25000 N/m^3

$$\text{Self weight of beam per metre length} = \frac{450}{1000} \times \frac{750}{1000} \times 1 \times 25000 = 8437.5 \text{ N/m}$$

Total superimposed load which the beam can carry excluding self weight = 82400.0 – 8437.5 = 73962.5 N/m

Example 2.4. A rectangular beam 225 mm × 450 mm is simply supported over a span of 5 m. It is provided with 4 Nos. 20 mm Dia. HYSD bars as reinforcement. Calculate the maximum stresses developed in steel and concrete if the beam (Fig. 2.17) carries a *u. d. l.* of 9000 N/m (900 K g/m) including the self weight of beam, $m = 13$.

Solution.

$$\text{Bending Moment at mid span of beam} = \frac{w.L^2}{8}$$

$$= \frac{9 \times 5 \times 5}{8} = 28.125 \text{ K.N.m}$$

Let 'x' be the depth of *N.A.* from the top concrete fibre. Taking moment of compressive and tensile areas about *N.A.*

$$b \cdot \frac{x^2}{2} = m. A_s (d - x)$$

$$225 \cdot \frac{x^2}{2} = 13 \times 1256(400 - x)$$

$$x^2 + 145.14x - 58055 = 0$$

$$x = 179.07 \text{ mm}$$

$$\text{Lever arm} = d - \frac{x}{3}$$

$$= 400 - \frac{179.07}{3}$$

$$= 340.31 \text{ mm}$$

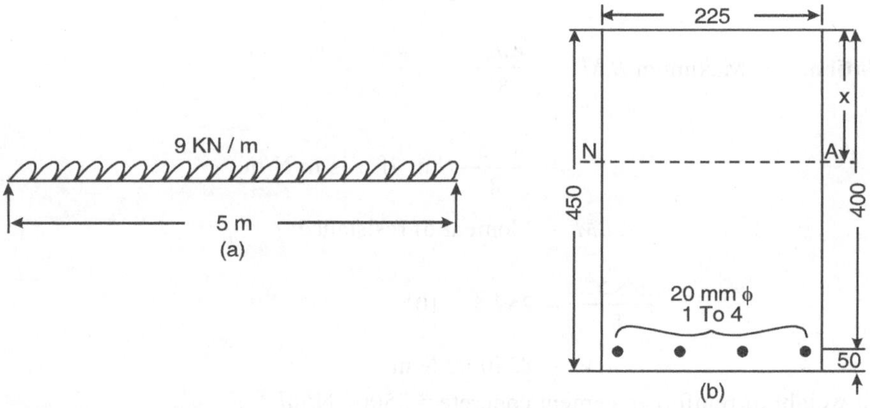

Fig. 2.17

Maximum Stress in Concrete

$$B.M. = \text{Moment of resistance}$$

$$= b \cdot x \cdot \frac{\sigma_{cba}}{2} \cdot \text{lever arm}$$

$$28125000 = 225 \times 179.07 \times \frac{\sigma_{cba}}{2} \times 340.31$$

$$\sigma_{cba} = 4.1025 \text{ N/mm}^2$$

Maximum Stress in Steel

$$B.M. = \text{Moment of resistance}$$

$$= A_s \cdot \sigma_{sta} \cdot \text{lever arm}$$

$$28125000 = 1256 \cdot \sigma_{sta} \cdot 340.31$$

$$\sigma_{sta} = 65.8 \text{ N/mm}^2$$

Example 2.5. Determine the area of tensile reinforcement required in a R. C. C. beam 225 mm × 450 mm subjected to a *B.M.* of 28125 N.m. Use M20 concrete and (*i*) Mild Steel (*ii*) High strength deformed bars (Tor steel).

Solution. Permissible stresses in concrete and steel as per IS : 456 – 2000 (Refer Table 2.1 and Table 2.3)

$$\sigma_{cbc} = 7\text{N/mm}^2 \text{ for M20 concrete}$$
$$\sigma_{st} = 140 \text{ N/mm}^2 \text{ for milld steel}$$
$$\sigma_{st} = 230 \text{ N/mm}^2 \text{ for tor steel,} \qquad m = 13.33$$

Assuming an effective cover of 50 mm,

$$d = 450 - 50 = 400 \text{ mm}$$

(*i*) Mild Steel

$$Q = 0.87$$
$$M = Q \cdot n \cdot d^2$$
$$= \frac{1.21 \times 225 \times 400 \times 400}{10^6} \text{KN.m}$$
$$= 43.56 \text{ KN.m} > 28.125 \text{ KN.m}$$

Beam is to be designed as under-reinforced, *i.e.*, the stress in steel reaches its maximum permissible value earlier than that in concrete.

$$M = A_s \cdot \sigma_{st} \cdot \left(d - \frac{x}{3}\right)$$

Taking moments of compressive and tensile areas N.A.

$$b \cdot \frac{x^2}{2} = m.A_s\,(d - x)$$

$$A_s = 225 \cdot \frac{x^2}{2} \cdot \frac{1}{13.33\,(400 - x)}$$

Substituting this value of A_s in M (moment of resistance)

$$M = 225 \cdot \frac{x^2}{2} \cdot \frac{1}{13.33\,(400 - x)} \cdot 140\left(400 - \frac{x}{3}\right)$$

where $\quad M = B.M. = 28125000$ N.mm

$\therefore \quad x^3 - 1200 \cdot z^2 - 71410 \cdot x + 2.856(10^7) = 0$

Solving by trial and error

$$x = 133.5 \text{ mm}$$

Substituting this vlaue of 'x' in 'A_s',

$$A_s = 564.4 \text{ mm}^2$$

(*ii*) HYSD Steel

$$Q = 0.91$$
$$M = Q \cdot b \cdot d^2$$
$$= 0.91 \times 225 \times 400 \times 400$$
$$= 32760000 \text{ N.mm}$$
$$= 32760 \text{ N.m} > 28125 \text{ N.m}$$

Beam is to be designed as under reinforced, *i.e.*, the stress in steel will reach its maximum permissible value earlier than that of concrete.

$$M = \sigma_{st} \cdot A_s \left(d - \frac{x}{3} \right)$$

$$b \cdot x \cdot \frac{x}{2} = m \cdot A_s(d - x)$$

$$\therefore \quad A_s = \frac{b \cdot x^2}{2m\,(d - x)}$$

$$\therefore \quad M = \frac{\sigma_{st} \cdot bx^2}{2m\,(d - x)} \left(d - \frac{x}{3} \right)$$

Substitue the values and simply, we get

$$x_3 - 1200x^2 - 42390\,x + 16956000 = 0$$

Solving for 'x'

$$x = 106.6 \text{ mm}$$

Taking moments of compressive and tensile areas about *N.A.*

$$225 \times \frac{106.6 \times 106.6}{2} = 13\,A_s\,(400 - 106.6)$$

$$A_s = 335.17 \text{ mm}^2$$

Example 2.6. A reinforced to concrete symmetrically tapered beam having cross section 300 mm wide, 600 mm depth at mid span and 300 mm wide, 400 mm depth at ends, has a span of 5 m simply supported. Determine the necessary tension reinforcement at mid span and at quarter sections if the beam is required to carry a superimposed load of 5000 N/m (500 Kg/m) in addition

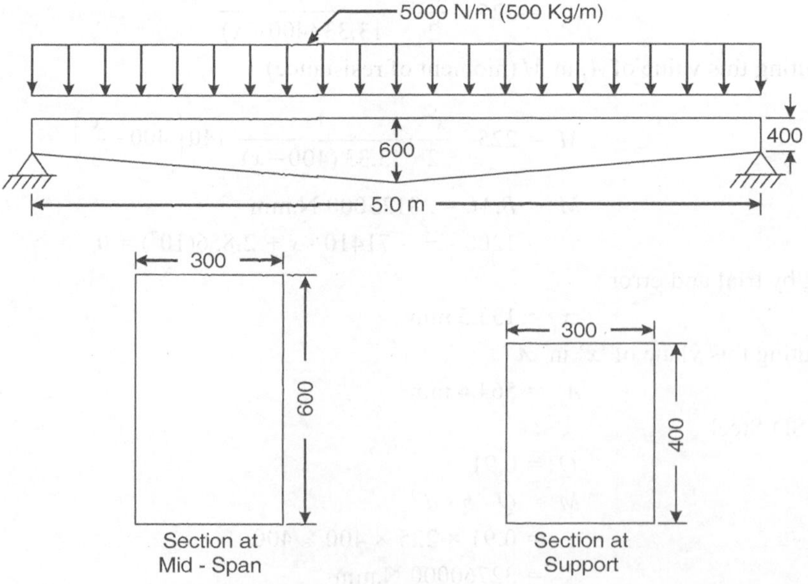

Fig. 2.18

to its self weight. Given M 20 concrete and HYSD bars of grade Fe 415 (σ_{st} = 230 N/mm²). (Fig. 2.18).

Solution.

Self weight of beam of uniform portion of section

$$= 0.3 \times 0.4 \times 25000 = 3000 \text{ N/m}$$

Self weight of beam of triangular portion of the section

$$= \frac{1}{2} \times 0.3 \times 0.2 \times 5 \times 25000$$

$$= 3750 \text{ N}$$

Total *u. d. l.* $w = 5000 + 3000 = 8000$ N/m

(*i*) *Mid Span*

$$\text{Maximum bending a midspan} = \frac{w \cdot L^2}{8} + \frac{w \cdot L}{6}$$

$$= \frac{8000 \times 5^2}{8} + \frac{3750 \times 5}{6}$$

$$= 28125 \text{ N.m}$$

Moment of resistance of the section at midspan

$$= \frac{0.91 \times 300 \times 560^2}{1000} = 85612.8 \text{ N.m greater than } 28125 \text{ N.m.}$$

Design as under reinforced section

$$M = A_s \, \sigma_{st} \left(d - \frac{x}{3} \right)$$

Taking moments of compressive and tensile areas about N.A.

$$b\frac{x^2}{2} = m \cdot A_s \, (d - x)$$

$$A_s = \frac{bx^2}{2m\,(d-x)} = \frac{300x^2}{2 \times 13.33\,(560 - x)}$$

$$\therefore \quad M = \frac{300x^2}{2 \times 13.33\,(560 - x)} \times 230 \left(560 - \frac{x}{3} \right)$$

$$M = 28125 \text{ N.m.}$$

Solving, $x = 98$ mm

$$\therefore \quad A_s = \frac{300}{2 \times 13.33} \times \frac{(98)^2}{560 - 98} \approx 234.45 \text{ mm}^2$$

Provide 12 mm dia. *M.S.* bars 2 Nos., thus providing A_s = 226 mm²

(ii) Quarter Span

The bending moment at quarter span is calculated as per the Fig. 2.19.

$$w = u.d.l = 8000 \text{ N/m},$$

Intensity of load at midspan due to self weight of triangular portion of the section

$$= \frac{3750 \times 2}{5} = 1500 \text{ N/m}$$

Maximum *B.M.* at 1.25 m from support

$$= 21875 \times 1.25 - \frac{8000 \times 1.25^2}{2} - 750 \times \frac{1.25}{2} \times \frac{1.25}{3}$$

$$= 20898.43 \text{ N.m} = 20.898 \text{ KN.m}$$

Moment of resistance at quarter span

$$= 0.91 \times 300 \times \frac{(500 - 40)^2}{10^6}$$

$$= 57.817 \text{ KN.m. greater than } 20.898 \text{ K.N.m}$$

$$\text{Area of steel required} = \frac{20989 \times 1000}{230 \times 0.90 \times 460}$$

$$= 218.00 \text{ mm}^2$$

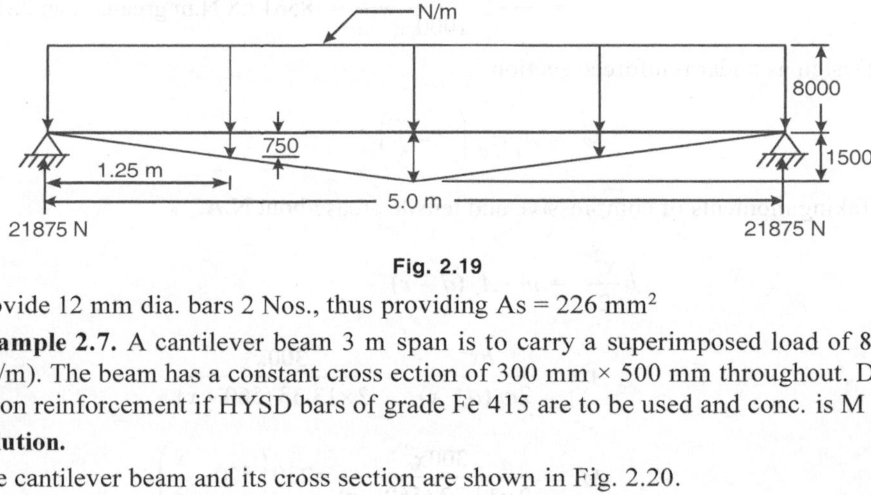

Fig. 2.19

Provide 12 mm dia. bars 2 Nos., thus providing As = 226 mm^2

Example 2.7. A cantilever beam 3 m span is to carry a superimposed load of 8000 N/m (800 Kg/m). The beam has a constant cross ection of 300 mm × 500 mm throughout. Determine the tension reinforcement if HYSD bars of grade Fe 415 are to be used and conc. is M 20.

Solution.

The cantilever beam and its cross section are shown in Fig. 2.20.

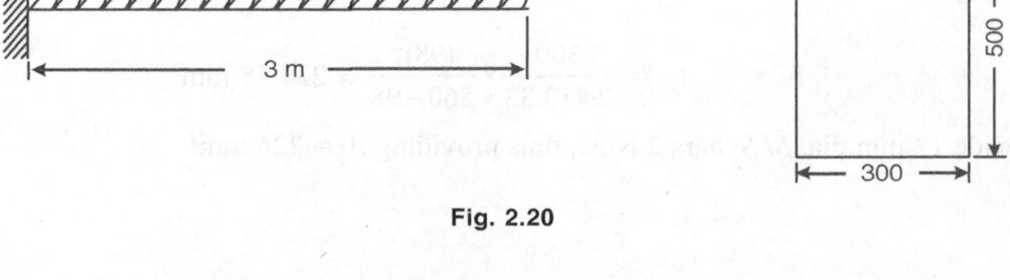

Fig. 2.20

$$\text{Self weight of the beam} = 0.3 \times 0.5 \times 25000 = 3750 \text{ N/m}$$

$$\text{Superimposed load} = 8000 \text{ N/m}$$

$$\text{Total } u.d.l. \text{ over beam} = 8000 + 3750 = 11750 \text{ N/m}$$

$$\text{Maximum } B.M. \text{ at fixed end} = 11750 \times \frac{3^2}{2} = 52875 \text{ N.m}$$

For a balanced section (taking effective cover to reinforcement = 50 mm)

M.O.R. (moment of resistance) = $0.91 \times 300 \times 450^2 = 55282500$ N.mm

$= 55282.5$ N.m

B.M. is approximately equal to moment of resistance of the section.

$$\text{Area of steel required at fixed end} = \frac{52875 \times 1000}{230 \times 0.90 \times 450} = 561.4 \text{ mm}^2$$

Provide 12 mm dia. HYSD 5 No. bars at top face of beam.

$$\text{Maximum } B.M. \text{ at 1.5 m from fixed end} = \frac{11750 \times 1.5a}{2} = 13218.75 \text{ N.m.}$$

Area of steel required at 1.5 m from fixed end

$$= \frac{13218.75 \times 1000}{230 \times 0.90 \times 450}$$

$$= 140.35 \text{ mm}^2$$

Provide 12 mm dia 2 No. bars, *i.e.*, curtail rest three bars at 1.5 m from the fixed end.

EXERCISE 2

1. Explain breifly 'Working Stress Method' of designing R. C. C. structures.
2. (*a*) Describe the different types of end supports.
   (*b*) What do you understand by (*i*) Balanced section (*ii*) Under-reinforced section (*iii*) Over-reinforced section?
3. (*a*) State the assumptions made in design of R. C. C. structures.
   (*b*) What is the effect of varying percentages of steel on the moment of resistance of R. C. C. sections?
4. Derive expressions for the position of neutral axis and moment of resistance for a balanced rectangular section and compute the same for M 20 concrete and HYSD bars of grade Fe 500.
5. Calculate the moment of resistance of a balanced section 300 mm × 600 mm taking effective cover as 50 mm if concrete used is M20. also calculate the amount of Fe 415 reinforcement in tension required to resist a bending moment of 75 KN.m; for the same section.
   (Ans: *MR* = 78.95 KN.m; A_s = 1120 mm²)
6. Calculate the moment of resistance of a R. C. section 300 mm × 450 mm reinforced with 4 bars of 20 mm dia. (HYSD, Fe 415). Concrete used is M 20. Take clear cover = 25 mm.
   (Ans: MR = 118.15 KN.m)
7. Compute the maximum stresses developed in concrete and steel in a rectangular cross-section 200 mm × 400 mm with effective depth as 365 mm. It is reinforced with 4 Nos. 16 mm dia. Fe500 bars on tension side. The beam is simply supported over a span of 6 m and carries a load of 8000 N/m inclusive of self weight. Concrete used is M 20, take m = 13.
   (Ans: σ_{cba} = 7.62 N/mm², σ_{ca} = 142 N/mm²)
8. Determine the maximum superimposed distributed load which the beam section 220 mm × 440 mm (Effective cover = 40 mm) reinforced with 4 Nos. 20 mm dia. Fe415 bars, can carry, if the effective span is 5 m. Concrete used M 20.
   (Superimposed load = 9365.6 N/m)

Beam Reinforced In Tension and Compression

CHAPTER
3

3.1 INTRODUCTION

Very frequently it becomes essential for a section to carry bending moment more than it can resist as a balanced section. Such a situation is encountered when the dimensions of the cross section are limited because of structural, head room or architectural reasons. Although a balanced section is the most economical section but because of limitations of size, section has to be sometimes over-reinforced by providing extra reinforcement on tension face than that required for a balanced section and also some reinforcement on compression face. Such sections reinforced both in tension and compression are also known as "Doubly Reinforced Sections". In some loading cases reversal of stresses in the section take place (this happens when wind blows in opposite directions at different timings), the reinforcement is required on both faces. Unless or otherwise it is a absolutely necessary, doubly reinforced sections should not be recommended.

3.2 MOMENT OF RESISTANCE OF DOUBLY REINFORCED SECTIONS

Consider a rectangular section reinforced on tension as well as compression faces as shown in Fig. 3.1.

Let b = width of section,

d = effective depth of section,

D = overall depth of section,

d' = cover to centre of compressive steel,

M = Bending momnt or total moment of resistance,

M_{bal} = Moment of resistance of a balanced section with tension reinforcement,

A_{st} = Total area of tensile steel,

A_{st1} = Area of tensile steel required to develop M_{bal},

A_{st2} = Area of tensile steel required to develop M_{ub},

A_{sc} = Area of compressive steel,

σ_{st} = Stress in tensile steel, and

σ_{sc} = Stress in compressive steel

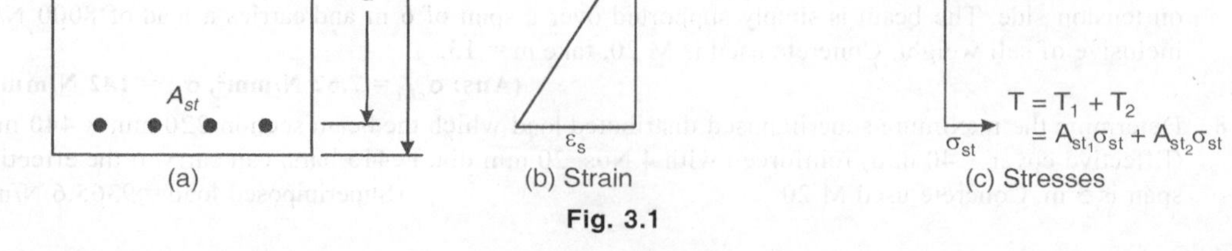

Fig. 3.1

Since strains are proportional to the distance from *N.A.*,

$$\frac{\text{Strains in top fibre of concrete}}{\text{Strain in compression steel}} = \frac{x}{x-d'}$$

$$\frac{\sigma_{cbc}/E_c}{\sigma_{sc}/E_s} = \frac{x}{x-d'}$$

$$\frac{\sigma_{cbc}}{\sigma_{sc}} \cdot \frac{E_s}{E_c} = \frac{x}{x-d'}$$

$$\sigma_{sc} = \sigma_{cbc} \cdot \frac{x-d'}{x} \cdot m \quad \text{...(3.1)}$$

Since $\sigma_{bc} \cdot \dfrac{x-d'}{x}$ is the stress in concrete at the level of compression steel, it can be denoted as σ'_{cbc}

$$\sigma_{sc} = m \times \sigma'_{cbc} \quad \text{...(3.2A)}$$

The IS : 456-2000 Code in Table 22 gives that the permissible compressive stress in bars, in a beam or slab when compressive resistance of the concrete is taken into account, can be taken as 1.5 m times the compressive stress in surrounding concrete (1.5m σ'_{cbc}) or permissible stress in steel in compression (σ_{sc}) whichever is less. (Reproduced in Table 2.3)

$$\sigma_{sc} = 1.5\text{m}\, \sigma'_{cbc} \quad \text{...(3.2B)}$$

The equivalent concrete area resisting compression

$$= (x \cdot b - A_{sc}) + 1.5\text{m}\, A_{sc}$$

$$= x \cdot b + (1.5\text{m} - 1)A_{sc} \quad \text{...(3.3)}$$

Taking moments about centre of tensile steel

$$\text{Moment of resistance } M = C_1 \cdot \left(d - \frac{x}{3}\right) + C_2\,(d - d')$$

where C_1 = total compressive force in concrete,

C_2 = total compressive force in compression steel,

$$M = b \cdot x \cdot \frac{\sigma_{cbc}}{2} \cdot \left(d - \frac{x}{3}\right) + (1.5\text{ m} - 1)A_{sc} \cdot \sigma_{cbc} \cdot \frac{x-d'}{x} \cdot (d-d')$$

$$= Q \cdot b \cdot d^2 + (1.5\text{m} - 1)A_{sc} \cdot \sigma_{cbc} \cdot \frac{x-d'}{x}\,(d - d') \quad \text{...(3.4)}$$

$$= M_1 + M_2$$

where M_1 = Moment of resistance of the balanced section = M_{bal}

M_2 = Moment of resistance of the compression steel for unbalanced moment

$$\text{Area of tension steel} = A_{st1} = \frac{M_1}{\sigma_{st} \cdot j \cdot d} \quad \text{...(3.5A)}$$

$$\text{Area of tension steel equivalent to compression steel} = A_{st2} = \frac{M_2}{\sigma_{st}(d-d')} \quad \text{...(3.5B)}$$

Thus the total tensile steel A_{st} shall be:

$$\therefore \quad A_{st} = A_{st1} + A_{st2} \qquad \text{...(3.5C)}$$

The area of steel required on compression face can be obtained from Eq. (3.4). Alternately, this steel 'A_{sc}' can also be obtained by taking moments of extra equivalent concrete area given by compression steel and equivalent area of concrete given by extra tensile steel about N.A.

$$(1.5\text{m} - 1)A_{sc}\,(x - d') = \text{m}A_{st2} \cdot (d - x)$$

$$A_{sc} = A_{st2} \cdot \frac{\text{m}}{(1.5\text{m} - 1)} \cdot \frac{d - x}{x - d'} \qquad \text{...(3.6)}$$

3.3 TYPES OF PROBLEMS

The following types of problems can be encountered in the design of doubly reinforced concrete beams.

3.3.1 Determination of Areas of Tension and Compression Reinforcement

The maximum bending moment, allowable stress in steel and concrete, modular ratio and dimensions of beam section are given.

Steps to be followed

(*i*) For given allowable stresses in materials, depth of critical neutral axis, lever arm constant, coefficient of moment of resistance can be obtained.

(*ii*) Assume some concrete cover to compression steel.

(*iii*) Moment of resistance M_1 of the balanced section canbe computed as

$$M_1 = Q \cdot b \cdot d^2$$

Balance of moment M_2 can be calculated as :

$$M_2 = M - M_1$$
$$= M - Q \cdot b \cdot d^2$$

(*iv*) Tension steel can be obtained from :

$$A_{st1} = \frac{M_1}{\sigma_{st} \cdot j \cdot d}$$

$$A_{st2} = \frac{M - M_1}{\sigma_{st} \cdot (d - d')}$$

$$\therefore \quad A_{st} = A_{st1} + A_{st2}$$

(*v*) Compression steel is determined from :

$$A_{sc} = (M - M_1)\,\frac{1}{1.5\text{m} - 1} \cdot \frac{x}{(x - d')} \cdot \frac{1}{\sigma_{cbc}(d - d')}$$

3.3.2 To Determine Moment of Resistance of Doubly Reinforced Beam Section

The dimensions of section, area of both tension and compression steels, grade of concrete and the type of steel are given.

Steps to be followed

(*i*) Determine actual position of the neutral axis 'x_a' by taking moments of concrete and equivalent steel areas about neutral axis.

(*ii*) For the given grade of concrete and the type of steel, permissible stresses can be known. Then obtain the position of the criticial neutral axis 'x_c'.

(*iii*) By comparing depth of actual neutral axis with the critical *N.A.*, determine whether concrete can be found out by taking moments about tensile steel.

If x_a is greater than x_c, the concrete reaches maximum allowable stress σ_{cbc} first and the moment of resistance can be found out by taking moments about tensile steel.

$$M = b \cdot x \cdot \frac{\sigma_{cbc}}{2} \cdot \left(d - \frac{x}{3}\right) + (1.5\text{m} - 1)A_{sc} \cdot \sigma_{cbc} \cdot \frac{x - d}{x} \cdot (d - d')$$

If x_a is less than x_c steel reaches maximum stress σ_{st} earlier. Moment of resistance can be obtained by taking moment about tensile steel.

$$M = \frac{b \cdot x}{2} \cdot \frac{\sigma_{st}}{\text{m}} \cdot \frac{x}{d - x}\left(d - \frac{x}{3}\right) + (1.5\text{m} - 1) \cdot A_{sc} \cdot \frac{\sigma_{st}}{\text{m}} \cdot \frac{x}{d - x} \cdot \frac{x - d'}{x} , (d - d')$$

3.3.3 To Check for Stresses

The dimensions of section, areas of compression and tension steel and the bending moment on section are given.

Steps to be followed

(*i*) Determine position of neutral axis by equating areas of concrete and equivalent areas of steel about *N.A.*

(*ii*) Taking moments of compressive forces about tensile steel and equating to external bending moment B.M.

$$M = (1.5\text{m} - 1)A_{sc} \cdot \frac{x - d'}{x} \cdot \sigma'_{cbc} \cdot (d - d') - b \cdot x \cdot \frac{\sigma'_{cbc}}{2} \cdot \left(d - \frac{x}{3}\right)$$

From this equation, value of σ'_{cbc}, stress in top fibre of concrete can be found.

(*iii*) Stress in tensile steel shall be :

$$\sigma'_{st} = m \cdot \sigma'_{cbc} \cdot \frac{d - x}{x}$$

(*iv*) Stress in compression steel can be computed from :

$$\sigma'_{sc} = 1.5\text{m}\sigma'_{cbc} \cdot \frac{x - d'}{x}$$

3.4 SOLVED EXAMPLES

Example 3.1. A reinforced concrete beam is simply supported over a span of 5 m and it carries a uniformly distributed load of 25000 N/m including its own weight. If the size of the beam is restricted to 300 × 500 mm, determine the areas of tension and compression steel if required. Given M20 concrete and HYSD rebars of Fe415.

Sol. Assume effective cover to steel = 40 mm

$$\text{Max. } B.M. = \frac{25000 \times 5^2}{8}$$
$$= 78125 \text{ N.m}$$
$$M_1 = Q \cdot b \cdot d^2$$
$$= 0.91 \times 300 \times 460^2$$
$$= 57766800 \text{ N.mm} = 57766.8 \text{ N.m.}$$
$$M_2 = M - M_1$$
$$= 78125 - 57766.8 = 20358 \text{ N.m}$$

Depth of critical neutral axis $x_c = 0.29 \times 460 = 133.4$

$$M_2 = (1.5\text{m} - 1) \cdot A_{sc} \cdot \sigma_{cbc} \cdot \frac{x - d'}{x}(d - d')$$

$$20358 \times 1000 = (1.5 \times 13 - 1)A_{sc} \cdot 7.0 \cdot \frac{133.4 - 40}{133.4} (460\text{–}40)$$

$$A_{sc} = 534 \text{ mm}^2$$

Provide 16 mm dia. 3 Nos., thus providing $A_{sc} = 603 \text{ mm}^2$ (Fig. 3.2)

$$A_{st1} = \frac{M_1}{\sigma_{st}\ j \cdot d} = \frac{57766.8 \times 1000}{230 \times 0.91 \times 460} = 600 \text{ mm}^2$$

$$A_{st2} = \frac{M_2}{\sigma_{st}(d - d')}$$

$$= \frac{20358 \times 1000}{230 \times 420}$$

$$= 210 \text{ mm}^2$$

$$\text{Total } A_{st} = A_{st1} + A_{st2} = 600 + 210 = 810 \text{ mm}^2$$

Provide 16 mm dia. 5 Nos., thus providing $A_{st} = 1000 \text{ mm}^2$ (Fig. 3.2)

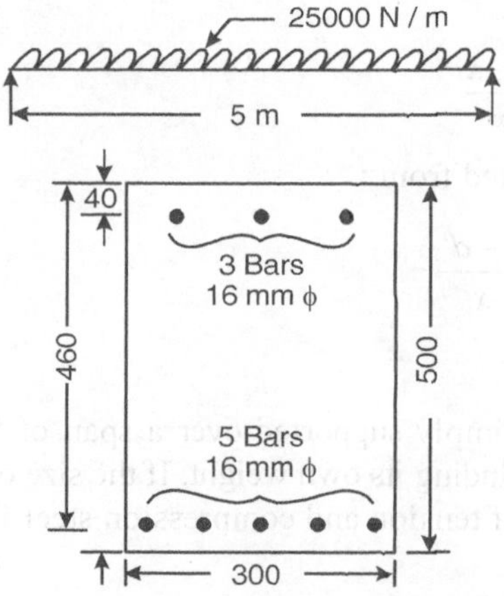

Fig. 3.2

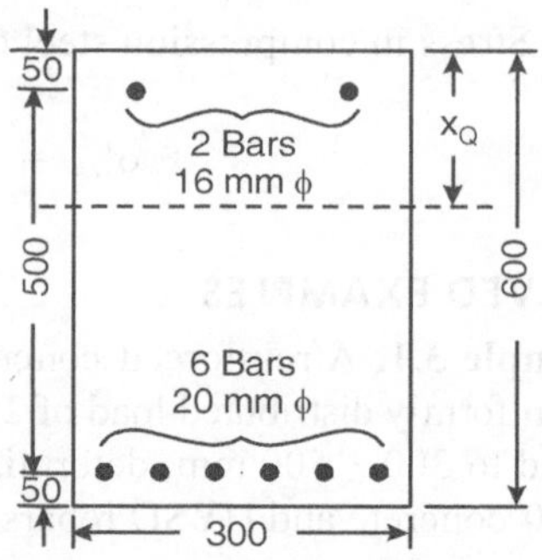

Fig. 3.3

Example 3.2. Determine the moment of resistance of the beam whose section is 300 × 600 mm and reinforced with 2 bars of 16 mm dia. at top and 6 bars of 20 mm dia. HYSD of Fe415 bars at bottom. Take effective cover = 50 mm for tension and compression steels. Use M 20 concrete (Fig. 3.3).

What superimposed load this beam can carry if its effective span is 8 m and it is simply supported at its ends?

Solution. Let x_a be the depth of actual neutral axis.

Taking moments of compressive and tensile areas about the neutral axis,

$$b \cdot x_a^2 + (1.5\text{m} - 1)A_{sc} \cdot (x_a - d) = \text{m} \cdot A_{st} \cdot (d - x_a)$$

$$\frac{300}{2} . x^2{}_a + (1.5 \times 13 - 1) \times 402\,(x_a - 50) = 13 \times 6 \times 314 \times (550 - x_a)$$

$$x_a = 215 \text{ mm}$$

$$\text{Depth of critical neutral axis} = 0.29 \times 550 = 160 \text{ mm less than } x_a$$

The section is over reinforced.

$$M = \frac{b \cdot x_a \cdot \sigma_{cbc}}{2} \cdot \left(d - \frac{x_a}{3} \right) + (1.5\text{m} - 1)A_{sc} \cdot \sigma_{cbc} \cdot \frac{x_a - d'}{x_a} (d - d')$$

$$= \frac{300 \times 215.7 \times 7.0}{2} \left(550 - \frac{215.7}{3} \right) + (1.5 \times 13 - 1) \times 402 \times 7.0$$

$$\times \left\{ \frac{215.7 - 50}{215.7} \right\} (550 - 50)$$

$$= 128278239 \text{ N.mm} = 128278.239 \text{ N.m}$$

$$\text{Max. } B.M. \text{ in a simply supported beam} = \frac{w \cdot L^2}{8}$$

$$\frac{w \cdot 8^2}{8} = 128278.239$$

$$w = 160348 \text{ N/m}$$

$$\text{Self weight of beam} = 0.3 \times 0.6 \times 1 \times 25000 = 4500 \text{ N/m}$$

$$\text{Superimposed load} = 160348 - 4500 = 155848 \text{ N/m}$$

Example 3.3. A reinforced concrete beam 300 mm wide and 550 mm deep from compression top edge to centre of tensile reinforcement. The compression steel on top face is 3 Nos. 16 mm. HYSD bars and tensile steel on bottom face is 6 Nos. 20 mm dia. of Fe415 bars. The center of compression steel is at a distance of 50 mm from top edge (Fig. 3.4). If concrete used is M 20 and the bending moment the section is to carry is 1.018 × 10^5 N.m (1.018 × 10^4 kg.m), determine the stresses developed in concrete and steel.

Solution. Let x_a be the depth of the *N.A.*

Taking moments of compressive areas and tension area about *N.A.*

$$\frac{b \cdot x^2}{2} + (1.5\text{m} - 1)A_{sc} \cdot (x_a - d') = m \cdot A_{st}(d - x_a)$$

$$\frac{300 \cdot x_a^2}{2} + (1.5 \times 13 - 1) \times 201 \times 3(x_a - 50) = 13 \times 6 \times 314\,(550 - x_a)$$

$$x_a = 209 \text{ mm}$$

Let σ'_{cbc} be the compressive stress in top fibre of concrete

Stress in concrete at level of compression steel

$$= \sigma'_{cbc} \cdot \frac{209 - 50}{209} = 0.76\sigma'_{cbc}$$

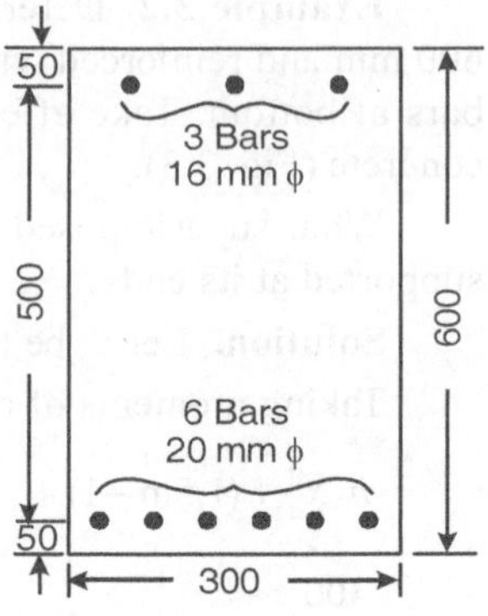

Fig. 3.4

$$M = \frac{300 \times 209}{2}\sigma'_{cbc}\left(550 - \frac{209}{3}\right) + (1.5 \times 13 - 1) \times 3 \times 201 \times 0.76\sigma'_{cbc}\,(550 - 50)$$

$$1.018 \times 10^8 = 0.193 \times 10^8\,\sigma'_{cbc}$$

$$\sigma'_{cbc} = 5.27 \text{ N/mm}^2$$

$$\text{Stress in compression steel} = 0.76 \times 5.27 \times 1.5 \times 13 = 78.1 \text{ N/mm}^2$$

$$\text{Stress in tension steel} = 13 \times 5.27\left(\frac{550 - 209}{209}\right)$$

$$= 111.8 \text{ N/mm}^2$$

3.5 STEEL BEAM THEORY

If the amount of compression reinforcement required equals or exceeds the amount of tension reinforcement obtained by using the theory as discussed in the preceding articles, the beam section may be designed by the 'STEEL BEAM THEORY'. In this theory, the compressive resistance provided by the concrete is neglected. The areas of steel in compression A_{sc} and that in tensile zone A_{st} are equal in magnitude. These areas of steel act analogous to the tension and compression flanges of an I-beam section, which resist bending.

The moment of resistance offered by the compressive and tensile steels placed at a distance of $(d - d')$ apart is given by the following expression:

$$M = A_{st} \cdot \sigma_{st} \cdot (d - d') \qquad \text{...(3.7)}$$

From the above Eq. (3.7)

$$A_{st} = A_{sc} = \frac{M}{\sigma_{st} \cdot (d - d')} \qquad \text{...(3.8)}$$

When this method of design is used, the spacing of stirrups/rings should not exceed eight times the diameter of bars forming the compression reinforcement. The permissible tensile stress in steel σ_{st} should be taken equal to the permissible compressive stress in steel σ_{sc} for the type of steel used. The indiscriminate application of 'Steel Beam Theory' is not advisable.

In this theory, the concrete serves only as a web and compression and tension steels as flanges of an imaginary steel joist. In fact, there is no proper justification of this theory. By taking $\sigma_{st} = \sigma_{sc}$, the stress in concrete fibres in compression is much more than otherwise permissible. The beams designed on this theory have very little margin for accidental overloading, poor workmanship, etc. Thus this theory is not usually recommended.

Example 3.4. A beam 250 mm × 600 mm is to resist a bending moment of 80 KN.m. Determine the areas of compression and tensile steel using Steel Beam Theory. (Fig. 3.5) (*i*) If HYSD steel with σ_{st} = 275 N/mm^2 and σ_{sc} = 190 N/mm^2 is used.

(*ii*) If Tor steel with σ_{st} = 230 N/mm^2 and σ_{sc} = 190 N/mm^2 is used.

Solution. Assuming 40 mm cover from extreme concrete fibres to c. g. of steels

(*i*) *HYSD Steel*

$$A_{st} = A_c = \frac{M}{\sigma_{st} \cdot (d - d')} \qquad \sigma_{st} = \sigma_{sc} = 190 \text{ N/mm}^2$$

$$= \frac{80 \times 10^6}{190 \times (600 - 80)}$$

$$= 809.7 \text{ mm}^2$$

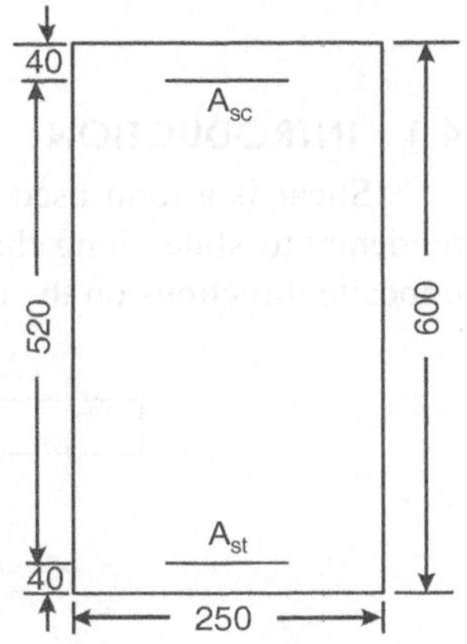

Fig. 3.5

(*ii*) *Tor Steel*

Take $\sigma_{st} = \sigma_{sc}$ = 190 N/mm^2

$$A_{st} = A_{sc} = \frac{M}{\sigma_{st} \cdot (d - d')}$$

$$= \frac{80 \times 10^6}{190 \cdot (600 - 80)}$$

$$= 809.7 \text{ mm}^2$$

EXERCISE 3

1. Derive an expression for moment of resistance of a rectangular section reinforced both in compression and tension.
2. A reinforced concrete beam 250 mm × 500 mm is simply supported over a span of 4 m and carries *u.d.l.* of 30 KN/m. Concrete used is M 20 and steel used is HYSD bars of grade Fe500. Determine the areas of tensile and compressive steel required. Take effective cover = 40 mm.

   (Ans: A_{st} = 518 mm2 A_{sc} = 498 mm^2)
3. Find the moment of resistance of a *R. C.* rectangular section 225 mm × 450 mm, reinforced with 4 bars of 20 mm dia. at bottom and 3 bars of 16 mm dia. at top. Take effective covers at top and bottom = 40 mm and m = 13 and HYSD bars of Fe 415. **(Ans: x_a = 139 mm, *MR* = 60.28 KN.m)**
4. A doubly reinforced concrete beam 250 mm × 550 mm size, is reinforced with HYSD Fe 500 5 bars of 16 mm ϕ at top and 4 bars 20 mm ϕ at bottom. Determine the stresses in steel and concrete if the beam is simply supported ove a span of 5 m and carries a load 22.4 KN/m inclusive of self weight. Concrete used M 20.

   (Ans: x_a = 164 mm, σ_{cba} = 3.875 N/mm^2, σ_{sca} = 57.12 N/mm^2, σ_{sta} = 106.8 N/mm^2)
5. Explain Steel Beam Theory. Why this method of design is not recommended by I.S. Code?

Shear, Bond and Torsion

CHAPTER
4

4.1 INTRODUCTION

Shear is a term used when the two sections on either side of a plane in a member have a tendency to slide along that plane. This tendency is caused by the shear forces that act in the opposite directions on the two sides of the shear plane as shown in Fig. 4.1.

Fig. 4.1

This shear is caused by the variation of bending moment along the span. The relationship between shear '*V*' and bending moment '*M*' is given below:

$$V = \frac{dM}{dx} \qquad ...(4.1)$$

Once the shear *V* is known, then shear stress 'τ_v' can be obtained by dividing the shear force with the area of cross section on which the shear is acting. This would be possible only if the variation along that cross section is uniform but it is not actually so. The shear stress variation for homogeneous and non-homogeneous materials is discussed in the subsequent article.

4.2 SHEAR STRESSES IN BEAMS

Beams may be made of one material only (such as concrete) known as 'homogeneous material' or they may be of different materials (such as concrete and steel reinforcement) known as 'non-homogeneous materials'. The shear stress variation for these two cases is discussed below.

4.2.1 Shear Stress Variation in Homogeneous Rectangular Beams

The shear stresses are caused by the variation of the bending moments as bending moment is zero at supports and maximum at the centre in case of simply supported beams loaded with uniformly distributed load. This variation from supports to any point on the beam will depend upon the type of loading and end supports of the beam.

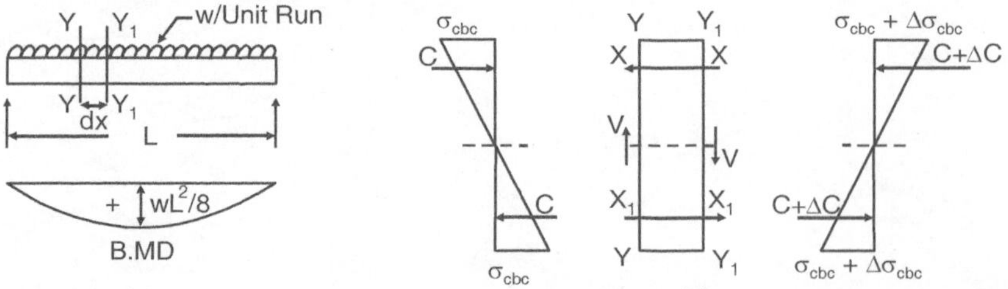

Fig. 4.2

As can be seen from Fig. 4.2, the bending moments are different at different sections. Consider two sections YY and Y_1Y_1 separated by distance dx. Bending moment at YY is less than that at Y_1Y_1. Therefore, the stresses developed in the material at YY shall be less than those developed at Y_1Y_1 which is shown in Fig. 4.2b. Due to this variation on both sides of the plane XX, there is a tendency of the fibres to slide over the adjacent fibres. The tendency of sliding is prevented by the horizontal shear stresses as shown with arrows on XX and X_1X_1. The horizontal shear stresses give rise to vertical shear stresses being complementary stresses.

Shear stress 'τ_v' at any section is given by:

$$\tau_v = \frac{V.A.\bar{y}}{b.I} \quad \text{...(4.2)}$$

where τ_v = shear stress,

V = shear force at the section,

A = area above the point under consideration (shown shaded in Fig. 4.3),

$\bar{y}$ = distance of centre of gravity of area A from neutral axis,

b = width of section,

I = moment of inertia,

$A = b\cdot(D/2 - y)$

$$\bar{y} = y + \frac{1}{2}(D/2 - y) = \frac{1}{2}(D/2 + y)$$

$$\therefore \quad \tau_v = \frac{v.b.(d/2 - y)\cdot\frac{1}{2}(D/2 + y)}{b.I}$$

$$= \frac{V}{2I}\left(\frac{D^2}{4} - y^2\right) \quad \text{...(4.3)}$$

Eq. (4.3) shows that the variation of shear stress is parabolic.

At $y = D/2,\ \tau_v = 0$

At $y = 0,\ \tau_{v\,max} = \dfrac{V.D^2}{8I}$

$$= \frac{V.D^2}{8\cdot\left(\frac{b.D^3}{12}\right)}$$

$$= \frac{3}{2}\cdot\frac{V}{b.D}$$

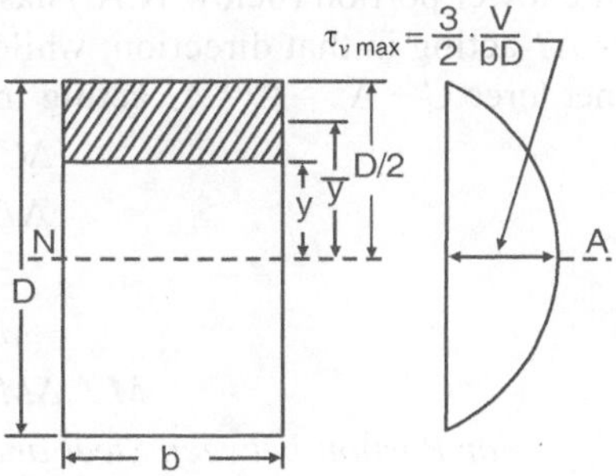

Fig. 4.3

$$\therefore \quad \tau_{v\,max} = \frac{3}{2}\cdot\frac{V}{b.D}$$

$$= \frac{3}{2}\cdot\tau_{v\,\text{average}} \quad \text{... (4.4)}$$

4.2.2 Shear Stress Variation in Rectangular Beams with Tension Reinforcement

Consider two sections YY and Y_1Y_1 separated by distance dx as shown in Fig. 4.4 with the following assumptions.

1. Concrete does not take any tension.
2. Shear stress is constant in width dx.
3. The bending moment at Y_1Y_1 is greater than the bending moment at YY.
4. Perfect bond exists between steel and the surrounding concrete.

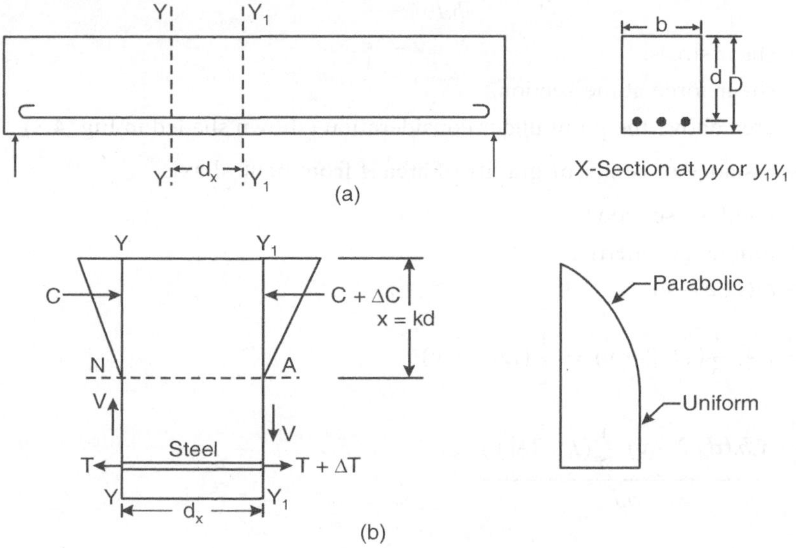

Fig. 4.4

Forces acting on the element are also shown in the Fig. 4.4b. The part of the section above neutral axis is in compression and below neutral axis is in tension. As $(T + \Delta T)$ is greater than T, the lower portion (below N.A.) has a tendency to slide towards right due to net force $T + \Delta T - T = \Delta T$ acting in that direction, while the upper portion has a tendency to slide towards left due to net force $C + \Delta_C - C = \Delta_C$ acting in that direction.

ΔC = small change in compression force in length dx,

ΔT = small change in tension force in length dx,

z = lever arm = $d - x/3$

M = moment at section YY,

$M + \Delta M$ = moment at section Y_1Y_1,

For Portion between Steel and Neutral Axis

At YY, T = tension in reinforcement

$$= \frac{M}{z}$$

At Y_1Y_1, $T + \Delta T$ = tension in reinforcement

$$= \frac{M + \Delta M}{z}$$

$\therefore$ Net force in reinforcement $= \Delta^T$

$$\Delta^T = \frac{M + dM}{z} - \frac{M}{z}$$

$$= \frac{dM}{z}$$

$\therefore$ $\Delta^T = \dfrac{dM}{z}$ tends the plane $(b \,.\, dx)$ to slid in the direction of dT. Δ^T induces shear stress in concrete since a perfect bond between steel and the surrounding concrete is assumed.

$\therefore$ Horizontal Shear stress, $\tau_H = \dfrac{\Delta T}{b \cdot dx} = \dfrac{dM}{z} \cdot \dfrac{1}{b \cdot dx}$

$$= \frac{1}{z \cdot b \cdot} \cdot \frac{dM}{dx}$$

$$\tau_H = \frac{V}{z \cdot b}$$

A complementary of this horizontal shear stress will be vertical shear stress τ_v of the same magnitude.

$\therefore$ Vertical shear stress $\tau_v = \dfrac{V}{z \cdot b}$ ...(4.5)

in the portion below neutral axis upto steel reinforcement.

For the Portion above Neutral Axis

This portion consists of one material, *i.e.*, concrete only which is assumed to be homogenous. Therefore, the shear stress variation can be assumed to follow the parabolic law given by the Eq. (4.3).

4.2.3 Shear Stress Variation in Rectangular Beams with both Tension and Compression Reinforcement

The shear stress variation in rectangular beams with both tension and compression reinforcement can be divided into three parts as given below:

(*i*) Shear stress variation in homogenous beams which is given by Eq. (4.1).

(*ii*) Shear stress variation in rectangular beams with tension reinforcement shown C_2 in Fig. 4.5.

(*iii*) Shear stress variation in rectangular beams with compression reinforcement in addition to tension reinforcement. Then the factor A. $\bar{y}$ suddenly below the level of compression steel and thus the curve will bulge out at the level of compression steel by the increased amount of $A.\,\bar{y}$ shown as C_1 in Fig. 4.5.

From the curve C_1, it is clear that the maximum shear stress occurs between the neutral axis and tension reinforcement. It will be given by

$$\tau_v = \frac{V}{z_1 \cdot b} \qquad \text{...(4.6)}$$

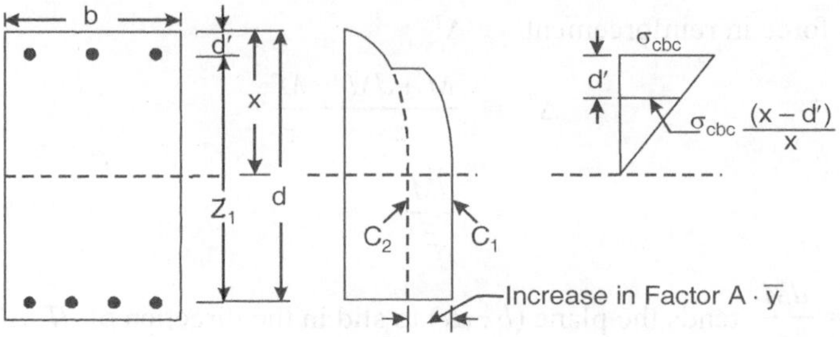

Fig. 4.5

This formula is more or less same as that for singly reinforced beam with a difference that 'z' is replaced by 'z_1' where z_1 is the distance between the c.g. of total compressive force and c.g. of tensile steel. The value of 'z_1' can be found as follows:

Total compressive force = Compressive force due to concrete above *N.A.* + compressive force in steel

$$\text{Total compressive force} = \frac{1}{2}\cdot b\cdot x\cdot\sigma_{cbc} + (1.5\text{m}-1)A_{sc}\cdot\sigma_{cbc},\frac{x-d'}{x}$$

The c.g. of the above total compressive force from top of the beam can be found by equation its moments to the moments of the compressive force due to concrete and compressive force in steel about top of the beam.

Let α = distance of c.g. of total compressive force from top of beam, then

$$\left[b\cdot x\cdot\frac{1}{2}\cdot\sigma_{cbc} + (1.5\text{m}-1)A_{sc}\cdot\sigma_{cbc}\cdot\frac{x-d'}{x}\right]a = b\cdot x\cdot\frac{1}{2}\cdot\sigma_{cbc}\cdot x/3 + (1.5\text{m}-1)A_{sc}\cdot\sigma_{cbc}\cdot\frac{x-d'}{x}\cdot d'$$

$$\therefore \quad a = \frac{b\cdot x\cdot\frac{1}{2}\cdot\sigma_{cbc}\cdot x/3 + (1.5\text{m}-1)\cdot A_{sc}\cdot\sigma_{cbc}\cdot\frac{x-d'}{x}\cdot d'}{b\cdot x\cdot\frac{1}{2}\cdot\sigma_{cbc} + (1.5\text{m}-1)A_{sc}\cdot\sigma_{cbc}\cdot\frac{x-d'}{x}}$$

$$\therefore \quad \text{Lever arm} \quad z_1 = d - a \qquad \text{...(4.7)}$$

ALTERNATIVELY

The lever arm z_1 can be found by another way as follows :

$$M = \sigma'_{st}\cdot z_1\cdot A_{st}$$

$$\therefore \quad z_1 = \frac{M}{\sigma'_{st}\cdot A_{st}} \qquad \text{...(4.8)}$$

where M = B.M. to be resisted by tension steel,

A_{st} = area of steel in tension

σ'_{st} = Stress in tension steel.

4.2.4 Shear in T-Beams

T-beams are the beams constructed monolithically with the slabs. Shear stress distribution in *T*-sections with tension reinforcement is shown in Fig. 4.6, which is same as the case of beams with tension and compression reinforcement. The only difference here is that in place of compression steel, large compression area is provided by the flange which gives the same effect as the compressive steel in doubly reinforced beam.

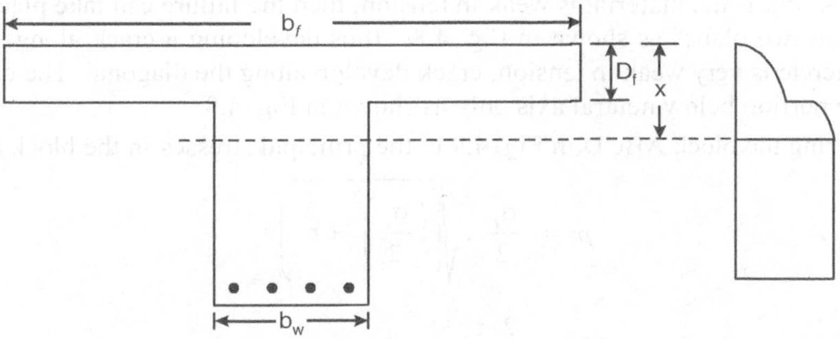

Fig. 4.6

$$\text{Shear stress} = \frac{V}{z \cdot b_{\omega}} \qquad ...(4.9)$$

where z = actual lever arm $= d - \frac{Df}{3}\left[\frac{3 \cdot x - 2Df}{2 \cdot x - Df}\right]$

Note: For calculation of 'z', refer to Eq. (2.4) in Chapter 2.

4.2.5 Shear in L-Beams

The calculation of shear stress in a *L*-beam section is similar to that of *T*-beam section without any difference (Fig. 4.7).

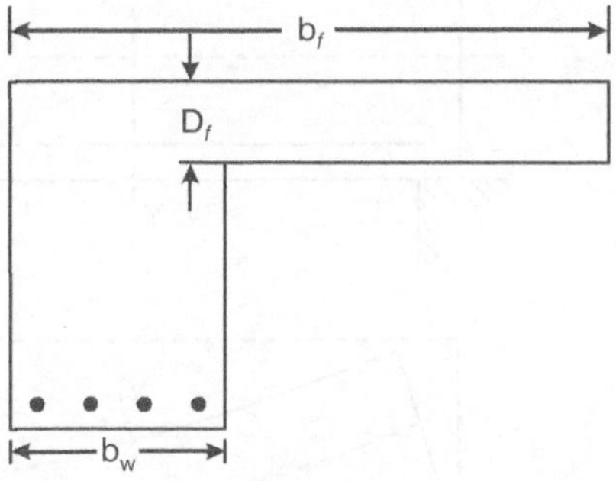

Fig. 4.7

$$\text{Shear stress} = \frac{V}{z \cdot b_{\omega}}$$

where z = lever arm

4.3 DIAGONAL TENSION

Diagonal tension is the tension that develops across the diagonal of a block taken from that part of a beam which is subjected to tension. This part is generally below the neutral axis (Fig. 4.8).

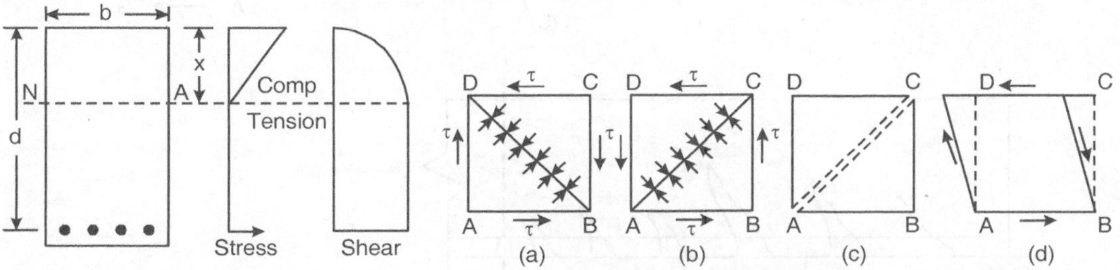

Fig. 4.8

Consider a block ABCD taken from this part subjected to shear stresses as shown in Fig. 4.8. Diagonal BD is subjected to compressive stresses (Fig. 4.8*a*). If the material is weak in compression, then the failure can take place by crushing. Diagonal AC is subjected to tensile stresses (Fig. 4.8*b*). If the material is weak in tension, then the failure can take place by splitting diagonal AC in two planes as shown in Fig. 4.8*c*, thus developing a crack along this diagonal. Since the concrete is very weak in tension, crack develop along the diagonal. The cracks shall be limited to the portion below neutral axis only as shown in Fig. 4.9.

Considering the block ABCD in Fig. 4.9 b, the principal stresses in the block are:

$$p = \frac{\sigma_t}{2} \pm \sqrt{\left(\left(\frac{\sigma_t}{2}\right)^2 + \tau^2\right)} \qquad ...(4.10)$$

and

$$\tan 2\theta = \frac{2 \cdot \tau}{\sigma_\tau} \qquad ...(4.11)$$

where p = principal stresses

σ_t = tensile stress on block

τ = shear stress on block

θ = inclination of principal planes with vertical (inclination of crack with the vertical).

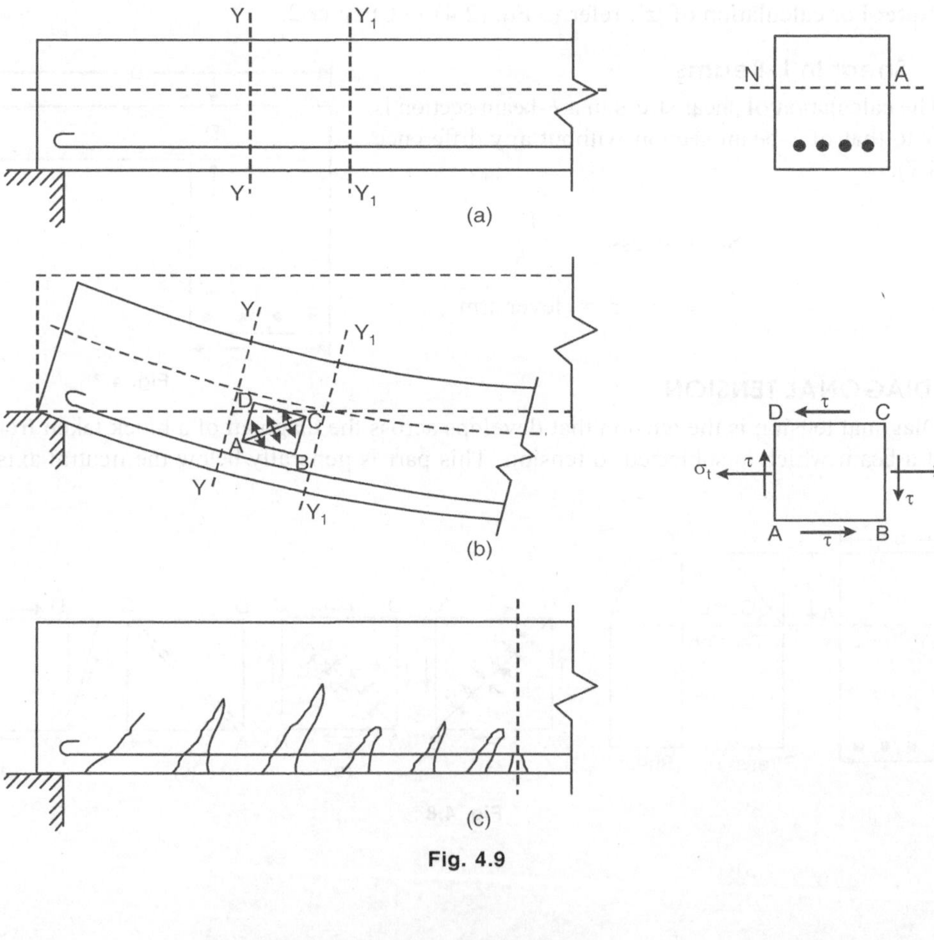

Fig. 4.9

To know the Inclination of Crack

For the principal stress developed, we have

$$\tan 2\theta = -\frac{2\tau}{\sigma_t}$$

(*i*) *At a Section where B.M.* = 0

$$\text{Tensile stress} = \sigma_t = 0$$

Maximum principal stress (tensile) = τ

$$\tan 2\theta = -\frac{2\cdot\tau}{0} = -\infty$$

$$\therefore \quad \theta = 45° \text{ or } 135°$$

Therefore, at or near the supports of a simply supported beam where bending moment is zero, the inclination of cracks shall be 45°.

(*ii*) *At a section where B.M. is maximum*

$$\text{Shear stress } \tau = 0$$

Maximum principal stress (tensile = σ_t

$$\tan 2\theta = 0$$

$$\therefore \quad \theta = 0$$

Therefore, at or near the midspan of a simply supported beam where the bending moment is maximum, the inclination of cracks with the vertical is zero, i.e., they are vertical as shown in Fig. 4.9 *c*.

The cracks can be prevented by providing reinforcement for diagonal tension. To avoid such cracks, the reinforcement called 'shear reinforcement' can be provided in either or both of the following ways.

(*i*) As vertical stirrups passing around the main reinforcement.

(*ii*) As bent up bars inclined at an angle at right angles to the plane of diagonal tension.

4.4 SHEAR REINFORCEMENT (VERTICAL STIRRUPS)

(*A*) *Assumptions made in the Design of Reinforcement for Diagonal Tension:*

1. Diagonal tension is equal to shear at the section.
2. Diagonal tension crack is inclined at 45° to the longitudinal axis of the member.
3. The crakcs extend over a maximum horizontal distance equal to the effective depth '*d*' of the beam.
4. Normal tension is zero.

(*B*) *Shear Reinforcement*: It is the reinforcement provided to take up the diagonal tension. It is also known as 'web reiforcement'. Whenever the value of actual shear stresses exceeds the permissible shear stress of the concrete used, (Table 2.2) the shear reinforcement is to be provided. The shear reinforcement shall be provided to carry a shear equal to the difference of the actual shear and the permissible shear strength of concrete section.

$$V_s = V - \tau_c . b.d \qquad ...(4.12)$$

where V_s = Shear force for shear reinforcement,

V = Actual shear force due to design loads,

τ_c = Permissible shear stress of concrete (Table 2.2),

b = breadth of section of member,

d = effective depth of section of member,

Let A_{sv} denote area of one stirrup (all legs) in a distance s_v.

s_v = spacing of stirrups.

σ_{sv} = permissible tensile stress in shear reinforcement which shall not exceed 230 N/mm²

Number of stirrups in the horizontal distance equal to effective depth of section.

$$= \frac{d}{s_v} \quad ...(4.13)$$

Total load on stirrups = No. of stirrups × Area of one stirrup × permissible stress in shear reinforcement

$$= \frac{d}{s_v} \cdot A_{sv} \cdot \sigma_{s\tau}$$

This should be equal to the shear force for shear reinforcement, *i.e.*, V_s.

$$\frac{d}{s_v} \cdot A_{sv} \cdot \sigma_{sv} = V_s$$

$$\therefore \quad s_v = \frac{A_{sv} \cdot \sigma_{sv} \cdot d}{V_s} \quad ...(4.14)$$

This gives the spacing of shear stirrups at section subjected to maximum shear force. Since the shear force decreases at other sections, the spacing of stirrups can be increased according to the requirement, subject to a maximum of 0.75*d* where *d* is the effective depth of the section.

$s_v \not> 0.75d$ or 450 mm whichever is less.

The vertical shear stirrups may be one, two, three or four legged depending upon the requirement with adjustment of spacing (Fig. 4.10).

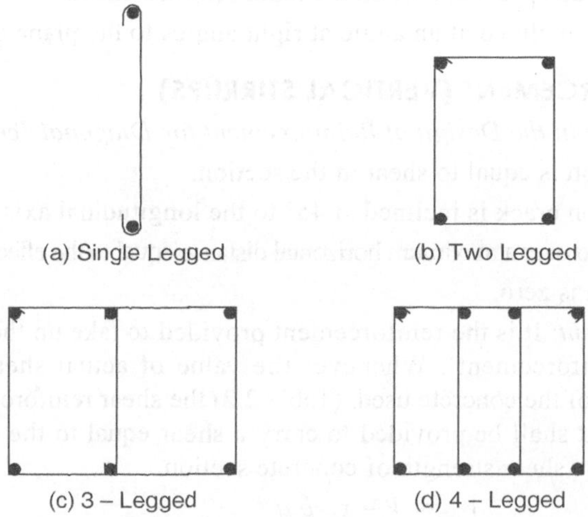

Fig. 4.10

These stirrups may be provided in a distance in which the actual shear stress exceeds the permissible value for the concrete. As for example, a simply supported beam with uniformly distributed load will have maximum shear force at supports and zero at the centre of beam as shown in Fig. 4.11. The distance *X* in which concrete is able to resist shear can be calculated as:

$$\frac{V_{max}}{\frac{1}{2}\cdot L} = \frac{\tau_c.b.d}{X}$$

$$\therefore \quad X = \frac{\tau_c.b.d}{V_{max}}\cdot\frac{L}{2}$$

$\therefore$ The distance in which shear reinforcement is required $= \frac{1}{2}L - X$ ...(4.15)

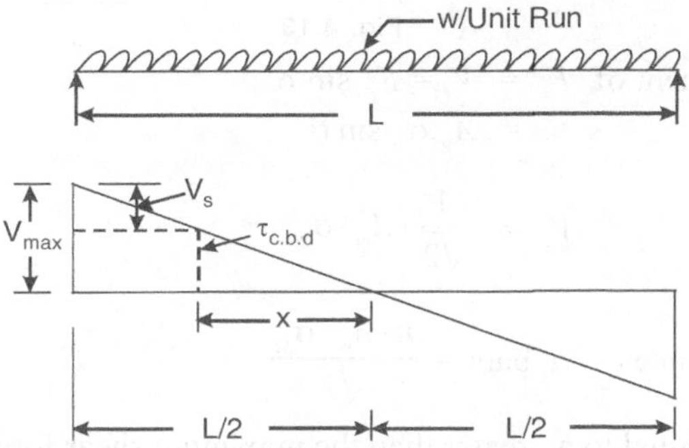

Fig. 4.11

4.5 SHEAR REINFORCEMENT (INCLINED OR DIAGONAL BARS)

Shear reinforcement can also be provided as inclined bars. These bars may be main tension reinforcement bent up from a place where they are no longer required for bending moment. The longitudinal bars can be bent up at a distance from support which varies from $L/5$ to $L/7$ where L is the effective span of beam. The reinforcement may be bent at any angle to the longitudinal axis of the beam, but in practices they are bent at 45° to 60°. These bars help in taking stresses in two ways.

(*i*) They take care of diagonal tension.

(*ii*) They act as support to hanger bars (Fig. 4.12) which are provided to keep vertical shear stirrups. Hanger bars are provided to keep vertical stirrups in position otherwise they may get displaced while concreting.

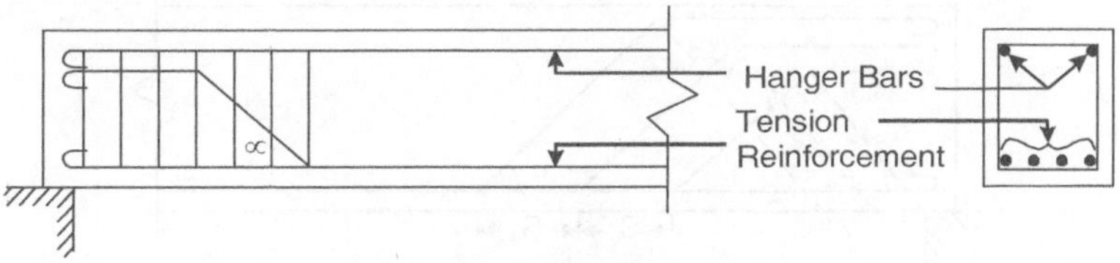

Fig. 4.12

4.5.1 Bars Bent at the Same Cross Section

Let $\propto$ = inclination of the bent up bar,

n = number of bars bent at the same cross section,

A_w = area of bent up bar,

σ_{st} = permissible, tensile stress in the reinforcement

= 140 N/mm^2 for Mild Steel,

= 230 N/mm^2 for High Strength Deformed bars (Fe 415)

Considering a bar bent up a angle $\propto$ (Fig. 4.13).

Tensile force which the bar can resist $F_t = A_w . \sigma_{st}$

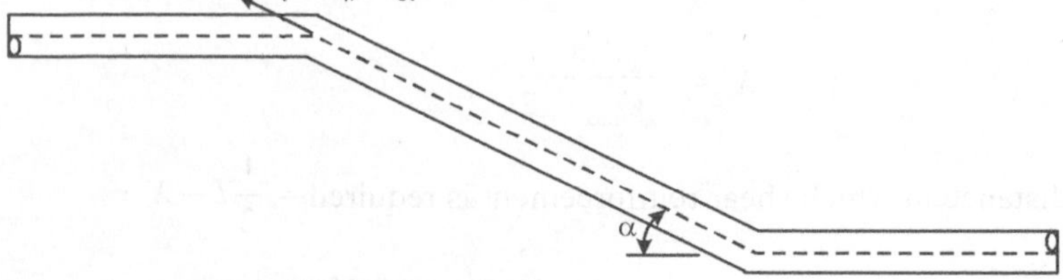

Fig. 4.13

The vertical component of $F_t = V_s = F_t . \sin\alpha$

$$= A_w . \sigma_{st} . \sin\alpha \quad ...(4.16)$$

For $\alpha = 45°$

$$V_s = \frac{1}{\sqrt{2}} \cdot A_w \cdot \sigma_{st}$$

Total vertical resistance of 'n' bars $= \dfrac{n \cdot A_w \cdot \sigma_{st}}{\sqrt{2}}$ ...(4.17)

If this resistance is equal to a greater than the maximum shear force V_s in the beam, then the bent up bars are sufficient to resist the shear safely. If it is less than the maximum shear force V_s, then the vertical stirrups have also to be provided for a shear force $= \left(V - \tau_c . b . d - \dfrac{n \cdot A_w \cdot \sigma_{st}}{\sqrt{2}}\right)$ in addition to the bent up bars in a distance $(L/2 - X)$.

4.5.2 Bent up Bars at Different Sections

When the bars are bent up at different cross sections as shown in Fig. 4.14, all at an angle a, then the strength of shear reinforcement shall be calculated from:

$$V_s = \frac{\sigma_{sv} \cdot A_{sv} \cdot d}{s_v} (\sin\alpha + \cos\alpha) \quad ...(4.18)$$

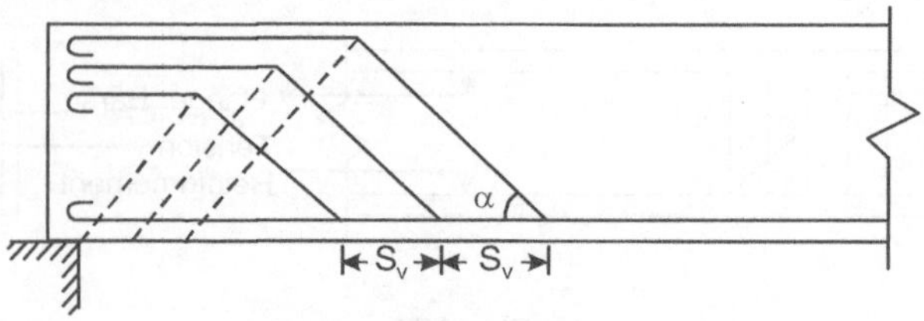

Fig. 4.14

where σ_{sv} = permissible tensile stress in shear reinforcement,

A_{sv} = total cross sectional area of bent up bars within a distance s_v,

s_v = spacing of bent up bars along the length of the member,

α = angle of bent up bars with the axis of the member.

If the shear force V is less than V_s, then no extra reinforcement is required. If V is more than V_s, then vertical stirrups have to be provided for a shear force $= (V - V_s - \tau_c . b . d)$. The vertical shear reinforcement is to be provided in a distance $\left(\frac{1}{2}L - X\right)$ from the supports where X is the distance from the point of zero shear to the point of maximum shear strength of the particular grade of concrete as explained earlier.

The dotted lines in Fig. 4.14 show the direction of cracks likely to develop due to diagonal tension. This gives a lattice girder effect in which the beam is considered equivalent to a truss consisting of compression members of concrete (along the crack line) and tension members of steel (at an angle α) with its height equal to the effective depth 'd' as shown in Fig. 4.15.

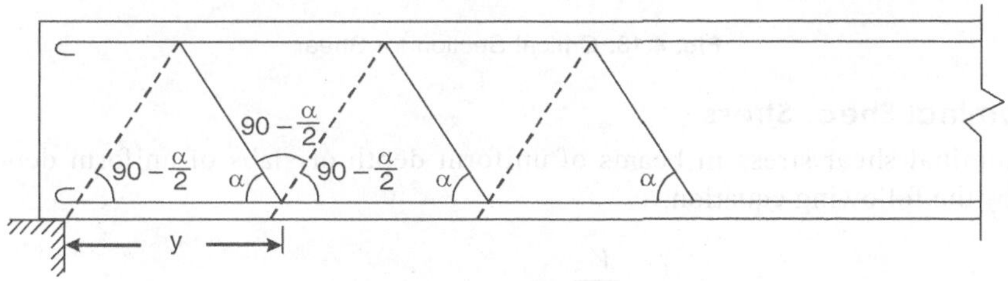

Fig. 4.15. Single System

If α = 45°, then $\left(90° - \frac{1}{2}\alpha\right) = 67\frac{1}{2}°$. This is done to make the shear reinforcement economical. With this, the first bar is to be bent at a distance $y = AB = BC = \frac{d}{\sin 45°} = d.\sqrt{2}$ from the supports as shown in Fig. 4.16.

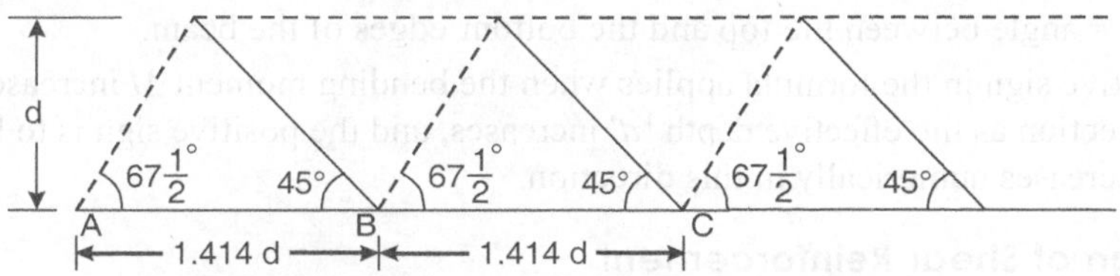

Fig. 4.16. Single System Equivalent Lattice Girder

If the bars are bent at half the above distance, *i.e.*, 0.707 d, then two bars shall provide resistance to diagonal tension and thus known as 'double system' as shown in Fig. 4.17.

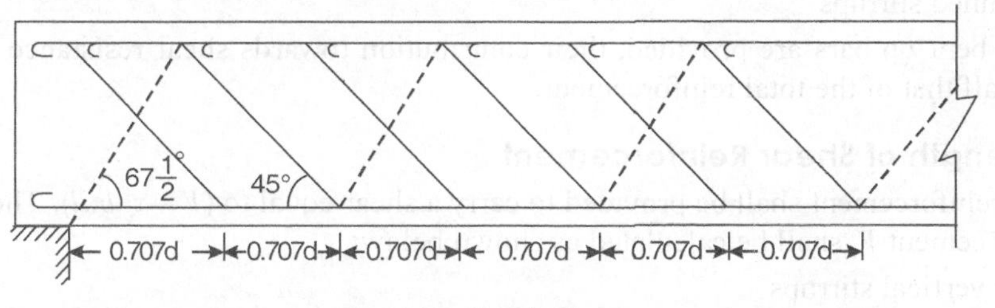

Fig. 4.17. Double System

4.6 DESIGN REQUIREMENTS FOR SHEAR REINFORCEMENT AS PER IS: 456-2000

4.6.1 Critical Section for Shear

The shears computed at the face of the support shall be used in the design of the member at that section except when the section in the direction of the applied shear introduces compression into the end region of the member, sections located at a distance less than '*d*' from the face of the support may be designed for the same shear as that computed at distance '*d*' (Fig. 4.18).

Fig. 4.18. Critical Section for Shear

4.6.2 Nominal Shear Stress

The nominal shear stress in beams of uniform depth or slabs of uniform depth shall be calculated by the following equation,

$$\tau_v = \frac{V}{b.d} \qquad \text{...(4.19)}$$

Nomainal shear stress in beams of varying depth shall be calculated as:

$$\tau_v = \frac{V \pm M/d.\text{Tan}\,\beta}{b.d} \qquad \text{...(4.20)}$$

where τ_v = nominal shear stress in beams or slabs,

M = bending moment at the section,

β = angle between the top and the bottom edges of the beam.

The negative sign in the formula applies when the bending moment M increases numerically in the same direction as the effective depth '*d*' increases, and the positive sign is to be taken when the moment decreases numerically in this direction.

4.6.3 Design of Shear Reinforcement

When τ_v exceeds τ_c, shear reinforcement shall be provided in any of the following forms:

(*a*) Vertical stirrups

(*b*) Bent up bars along with stirrups

(*c*) Inclined stirrups

Where bent up bars are provided, their contribution towards shear resistance shall not be more than half that of the total reinforcement.

4.6.4 Strength of Shear Reinforcement

Shear reinforcement shall be provided to carry a shear equal to $(V - \tau_c.b.d)$. The strength of shear reinforcement V_s shall be calculated as shown below:

(*a*) For vertical stirrups

$$V = \frac{\sigma_{ST}.A_{SV}.d}{S_v} \qquad \text{...(4.21)}$$

(*b*) For inclined stirrups or a series of bars bent up at different cross sections.

$$V_s = \frac{\sigma_{ST}.A_{SV}.d}{S_v}(\sin\alpha + \cos\alpha) \qquad \text{...(4.22)}$$

(*c*) For single bar or single group of parallel bars, all bent up at the same cross section.

$$V_s = \sigma_{sv} . A_{sv} . \sin \alpha \qquad ...(4.23)$$

where σ_{sv} = permissible tensile stress in shear reinforcement.

A_{sv} = total cross sectional area of stirrups legs or bent-up bars within a distance s_v,

s_v = spacing of stirrups or bent-up bars along the length of the member,

V = shear force due to design loads.

4.6.5 Minimum Shear Reinforcement

When τ_v is less than τ_c, minimum shear reinforcement in the form of stirrups shall be provided such that

$$\frac{A_{sv}}{b.s_v} \geq \frac{0.4}{0.87 f_y} \qquad ...(4.24)$$

where A_{sv} = total cross sectional area of stirrups legs effective in shear,

s_v = stirrup spacing along the length of the member,

b = breadth of beam or breadth of web of flanged beam,

f_y = characteristic strength of the stirrup reinforcement in N/mm^2 which shall not be taken greater than 415 N/mm^2

However, in members of minor structural importance such as lintels, or where the maximum shear stress calculated is less than half the permissible value, the provision of minimum shear reinforcement need not be complied with.

4.6.6 Maximum Spacing of Shear Reinforcement

The maximum spacing of shear reinforcement measured along the axis of the member shall not exceed 0.75 *d* for vertical stirrups and *d* for the inclined stirrups at 45°, where d is the effective depth of the section under consideration. In no case shall the spacing exceed 450 mm.

Where more than one type of shear reinforcement is used to reinforcement the same portion of the beam, the toal shear resistance shall be calculated as the sum of the resistance for various types separately. The area of the stirrups shall not be less than the minimum specified in 4.6.5.

4.6.7 Anchoring Shear Reinforcement

(*a*) *Inclined Bars*: The development length L_d of the bars shall be:

$$L_d = \frac{\phi.\sigma_s}{4.\tau_{bd}} \qquad ...(4.25)$$

where ϕ = nominal diameter of the bar,

σ_s = stress in bar at the section considered at design load,

τ_{bd} = design bond stress.

The development length includes anchorage values of hooks in tension reinforcement. This length shall be measured as under:

(*i*) In tension zone, from the end of the sloping or inclined portion of the bar, and

(*ii*) In the compression zone, from the mid depth of the beam.

(*b*) *Stirrups*: Notwithstanding any of the provisions of this standard, in the case of secondary reinforcement, such as stirrups and transverse ties, complete development lengths and anchorage shall be deemed to have been provided when the bar is bent through an angle at least 90° round a

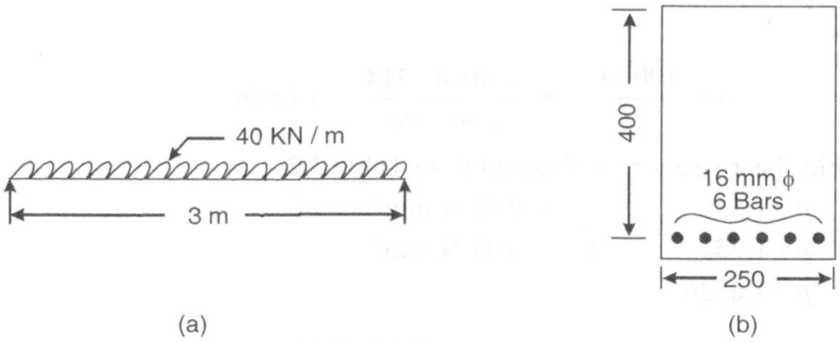

Fig. 4.20

$$= 44000 \text{ N}$$

$$p = \frac{100\,A_s}{b.d} = \frac{100 \times 6 \times 201}{250 \times 400} = 1.206$$

For $p = 1.206$ and M 20 concrete, the permissible shear stress τ_c is calculated from Table 2.2. as 0.414 N/mm^2.

Shear strength of concrete section = $\tau_c.b.d$

$$= 0.414 \times 250 \times 400$$

$$= 41400 \text{ N less than } 44{,}000 \text{ N}$$

∴ Shear reinforcement is to be provided.

Check for the Size of the Beam

$$\text{Nominal shear stress} = \frac{44{,}000}{250 \times 400} = 0.44 \text{ N/mm}^2$$

τ_c maximum for M20 concrete = 1.8 N/mm^2 from Table 4.1 (Page 90)

∴ Size of the beam is O.K.

To find the distance in which shear reinforcement is requried (Fig. 4.21)

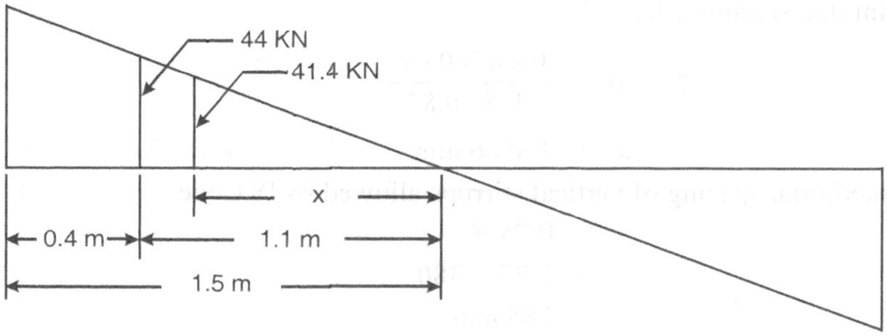

Fig. 4.21

Distance X from the point of zero shear where $S.F.$ is 41.4 KN = $\frac{1.1}{44} \times 41.4 = 1.035$ m

∴ Shear reinforcement is requried in a distance

$$\text{from support} = 1.5 - 1.035 = 0.465 \text{ m}$$

Shear force for shear reinforcement = $V - \tau_c.\,b.d$

$$= 44 - 41.4 = 2.6 \text{ KN}$$

The strength of vertical stirrups as recommended by IS Code is :

$$V_s = \frac{A_{sv}.\sigma_{sv}.d}{s_v}$$

If 8 mm ϕ (HYSD bars) 2 legged stirrups are used.

$$\therefore \text{ Spacing} \qquad s_v = \frac{A_{sv}.\sigma_{sv}.d}{V_s}$$

$$= \frac{2\times 50\times 230\times 400}{2.6\times 1000}$$

$$= 3538 \text{ mm}$$

Satisfying the requirement of minimum shear reinforcement, the spacing of 8 mm ϕ 2 legged stirrups is :

$$s_v = \frac{A_{sv}.f_y \times 0.87}{0.4\,b}$$

$$= \frac{2\times 50\times 415\times 0.87}{0.4\times 250}$$

$$= 361 \text{ mm less than 3538 mm}$$

Max spacing = $0.75\,d = 0.75 \times 400 = 300$ mm

∴ Provide 8 mm ϕ 2 legged vertical shear stirrups @ 300 mm c/c throughout.

Example 4.3. A reinforced concrete rectangular beam 400 mm wide and 700 mm overall depth is simply supported over a beam of 6 m (Fig. 4.22). It carries a concentrated load of 150 KN (15t) at 0.5 m from one end in addition to a superimposed load of 18 KN/m (1.8 t/m). The section is reinforced with 6 Nos. bars 20 mm ϕ. Using M20 concrete, calculate the shear reinforcement if (*a*) only vertical stirrups are used,(*b*) two bars are bent up at 45° at the same cross section, and (*c*) two groups of bars are bent-up at 45°, each group consisting of two bars, each group separated by 600 mm.

Solution.

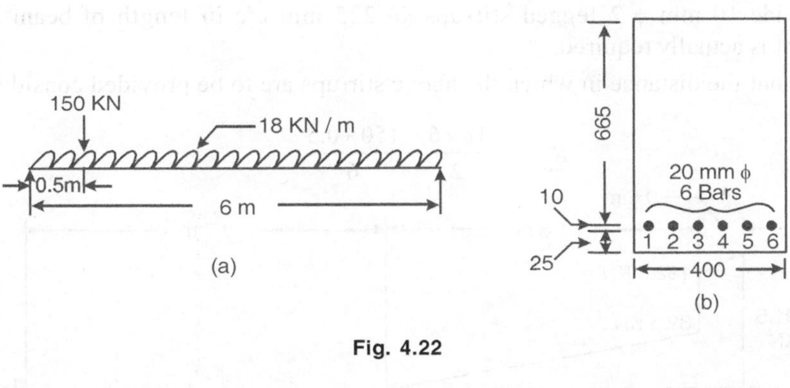

Fig. 4.22

$$\text{Effective depth } d = 700 - \text{clear cover } - \frac{1}{2}\phi$$

$$= 700 - 25 - 10 = 665 \text{ mm}$$

$$\text{Maximum shear force} = \frac{18\times6}{2}+150\times\frac{(16-0.5)}{6}$$
$$= 54 + 136.5 = 191.5 \text{ KN}$$
$$= 191{,}500 \text{ N}$$

$$p = \frac{100\times A_s}{b.d} = \frac{100\times6\times314}{400\times665} = 0.708$$

Permissible shear stress τ_c = 0.3416 N/mm^2

∴ Shear strength of concrete = $\tau_c, b.d$

$= 0.3416 \times 400 \times 665$

$= 90865.6$ N less than 191500 N

Shear reinforcement is required for shear force $V_s = 191500 - 90865.6 \text{ N} = 100634.4 \text{ N}$

(*a*) *When only Vertical Stirrups are Used*

$$V_s = 100634.4 \text{ N}$$

Using 10 mm ϕ 2 legged stirrups at

$$\text{Spacing } s_v = \frac{A_{sv}.\sigma_{sv}.d}{V_s}$$

$$= \frac{2\times78.5\times230\times665}{100634.4} = 238.6 \text{ mm}$$

Maximum spacing as per minimum shear reinforcement requirement is :

$$s_v = \frac{A_{sv}.f_y\times0.87}{0.4\,b}$$

$$= \frac{2\times78.5\times415\times0.87}{0.4\times400}$$

$$= 354.2 \text{ mm}$$

Maximum spacing is also given by $0.75\,d = 0.75 \times 665 = 498.75$ or 450 mm

∴ Provide 10 mm ϕ 2 legged stirrups @ 225 mm c/c in length of beam where shear reinforcement is actually required.

To find out the distance in which the above stirrups are to be provided consider Fig. 4.23.

$$R_B = \frac{18\times6}{2}+\frac{150\times0.5}{6}$$

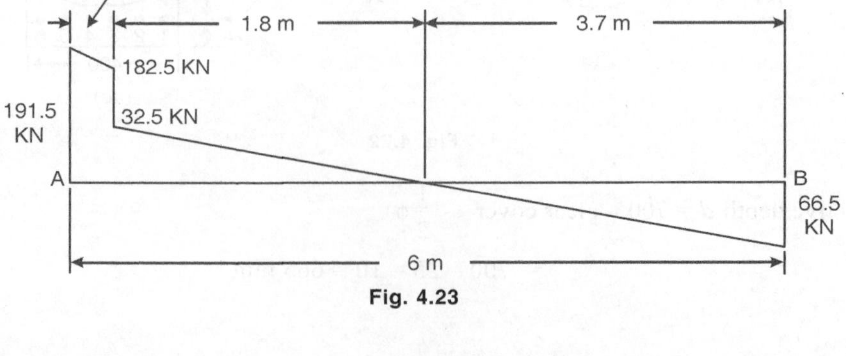

Fig. 4.23

$= 54 + 12.5 = 66.5$ KN

Point of zero shear from B is at $66.5 - 18X = 0$ $\quad X = 3.7$ m

∴ Design shear stirrups are actually required in 0.5 m from end A as shown in Fig. 4.23.

Hence provide 10 mm ϕ 2 legged stirrups @ 225 mm *c/c* from end A (support) and at 400 mm *c/c* in the remaining portion of the beam.

(b) When Two Bars Bent up at the Same Cross-Section

Considering Fig. 4.24, in which bars 2 and 5 are bent up at 45°. Shear strength of bent up bars

$$V_s = A_{sv}.s_v. \sin a$$
$$= 2 \times 314 \times 230 \times 0.707$$
$$= 102119.1 \text{ N}$$

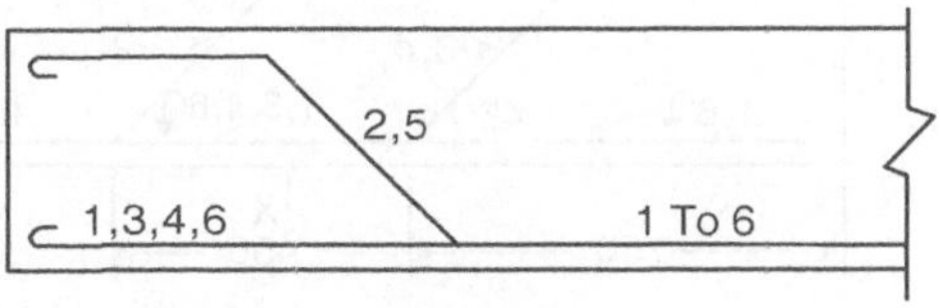

Fig. 4.24

(c) When two Bars bent up at the same cross-section (Fig. 424)

$$p = \frac{4 \times 314 \times 100}{400 \times 665} = 0.47, \tau_c = 0.290 \text{ N/mm}^2$$

Shear resistance of concrete section $= 0.29 \times 400 \times 665$

$= 77140$ N

Max. shear $= 191500$ N

Shear to be resisted by total shear reinforcement

$= 191500 - 77140 = 114360$ N

Actual shear strength of 2 Nos. bent up bars $= A_{sv}\ s_{sv}\ \sin \alpha$

$= 2 \times 314 \times 230 \times 0.707$

$= 102119.1$ N

Max, contribution of shear possible by bent up bars

$= 1/2 \times 114360$ N

$= 57180 \text{ N} < 62168.4$ N

∴ Net shear to be resisted by vertical stirrups

$'V'_s = 114360 - 57180 = 57180$ N.

Spacing of 8 mm dia. 2 legged vertical stirrups

$$S_v = \frac{A_{sv}.\sigma_{sv}.d}{V_s} = \frac{2 \times 50 \times 230 \times 665}{57180}$$
$$= 267.5\ mm$$

$$\text{Max, spacing} = S_{vmax} = \frac{(A_{sv} \times f_y \times 0.87)}{0.4b} = \frac{2 \times 50 \times 415 \times 0.87}{0.4 \times 400}$$
$$= 225.65\ mm$$

Provide 8 *mm* ϕ 2 legged vertical stirrups @ 225 mm *c/c* throughout.

(d) When two groups of (2 bars 20 mm ϕ) bars are bent at 45° and each group is separated by 450 mm (Fig. 4.25).

$$P = \frac{2 \times 314 \times 100}{400 \times 665} = 0.235, \tau_v = 0.22 \text{ N/mm}^2$$

Shear resistance of concrete section = 0.22 × 400 × 665

= 58520 N

Max, shear force = 191500 *N*

Shear to be resisted by total shear reinforcement = 191500 – 58250 = 132980 *N*

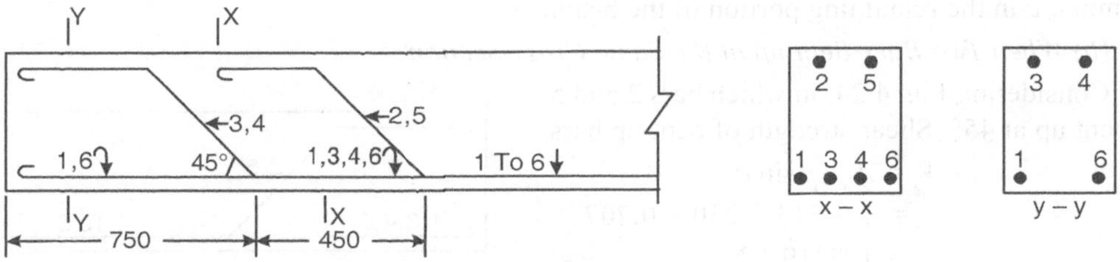

Fig. 4.25

Actual shear resistance of bent up bars

$$= \frac{\sigma_{sv} A_{sv}.d}{s_v}(\sin\alpha + \cos\alpha)$$

$$= \frac{230 \times 2 \times 314 \times 665}{450}(2 \times 0.707)$$

$$= 302285.7\ N$$

Max. contribution of shear possible by bent up bars

$$= 1/2 \times 132980 = 66490\ N < 302285.7\ N$$

∴ Net shear to be resisted by vertical stirrups.

$$V_s = 66490 \text{ N}$$

Spacing of 8 mm ϕ 2 legged vertical stirrups.

$$s_v = \frac{A_{sv}.\sigma_{sv}.d}{V_s} = \frac{2 \times 50 \times 230 \times 665}{66490} = 230 \text{ mm c/c}$$

$$< S_{v\,max} \text{ in } (b)_{max}\ i.e.\ 259.37 \text{ mm}$$

∴ Provide 8 mm ϕ 2 legged vertical stirrups @ 230 mm c/c throughout.

Note: From the example it is evident that bending of too many bars prove uneconomical.

4.8 BOND

The main assumption in the theory of reinforced concrete design is that there is a perfect bond of reinforcement with concrete, *i.e.*, there is no slipping of reinforcement bars whenever some forces are acting on it. this bond develops due to shrinkage of concrete on drying. Without this bond, the reinforcement will not serve any purpose to arrest the widening of cracks in the tension zone. The intensity of the adhesive force at the junction of steel and concrete is called the 'bond stress'. The bond is porivided by anchoring the bars properly and extending the bars beyond the point of maximum shear to a length which would be able to resist the shear at that section in the case of all flexural members. This length is called as 'development length' of bars.

4.9 CALCULATION OF BOND STRESS

Consider two sections *YY* and Y_1Y_1 at a distance *dx* apart in a simply supported beam as shown in Fig. 4.26. The difference in tensile force in the reinforcement at the sections *YY* and Y_1Y_1 is to be resisted by the bond stress which develops in the length *dx* of the beam.

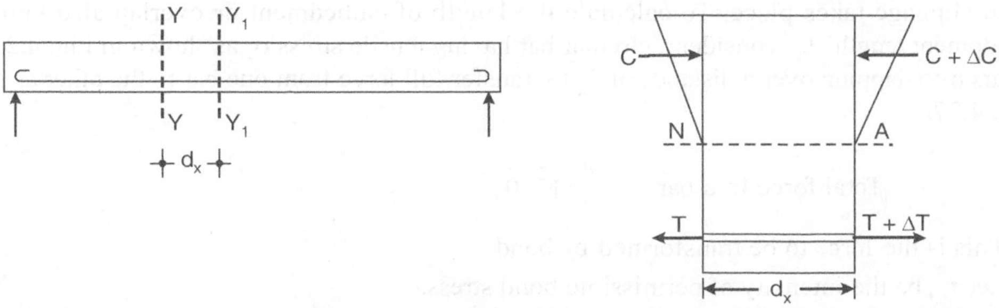

Fig. 4.26

Let ΔT be difference in the tensile forces in the reinforcement at the two sections. τ_{bd} be the intensity of bond stress. $\Sigma 0$ be the sum of perimeters of all bars resisting tension.

$$\Delta T = \text{(Intensity of bond stress)} \times \text{(Surface area providing the bond)}$$

$$= \tau_{bd}.dx.\Sigma_0$$

$$\Delta T = \frac{dM}{z}$$

$$= \frac{dM}{dx}\cdot\frac{dx}{z}$$

$$= V\cdot\frac{dx}{z}$$

Equating the two expressions for Δ^T

$$\tau_{bd}.dx.\Sigma 0 = V\cdot\frac{dx}{z} \qquad ...(4.26)$$

$$\therefore \qquad \tau_{bd} = \frac{V}{z.\Sigma 0}$$

τ_{bd} calculated on the basis of maximum shear force at any section is known as 'local bond stress' and τ_{bd} calculated on the basis of average shear force in the beam is known as 'average bond stress'.

4.10 DEVELOPMENT LENGTH OF BARS

Whenever some reinforcement bar is to be anchored or two bars have to be given an overlap, it is essential that they must get sufficient length of embedment or overlap as the case may be so

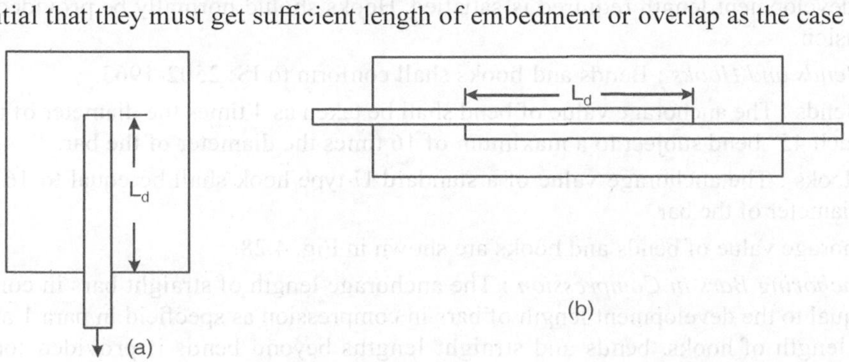

Fig. 4.27

that no slippage takes place. To calculate the length of embedment or overlap also known as 'development length' L_d, consider a circular bar having tensile stress σ_s as shown in Fig. 4.27 a, or two bars over-lapping over a distance of L_d to transfer full force from one bar to the other as shown in Fig. 4.27*b*.

$$\text{Total force in a bar} = \frac{\pi}{4} \cdot \phi^2 \cdot \sigma_s$$

This is the force to be transformed by bond.

Let τ_{bd} be the intensity of permissible bond stress.

$$\text{Bond force} = \tau_{bd} \cdot \text{p}\, \phi .\, L_d$$

Equating the total force in the bar to the bond force for no slippage to occur

$$\frac{\pi}{4} \cdot \phi^2 \cdot \sigma_s = \tau_{bd} \cdot \pi\phi .\, L_d$$

$$\therefore \quad L_d = \frac{\phi . \sigma_s}{4 . \tau_{bd}} \qquad \text{...(4.27)}$$

4.11 DESIGN CONSIDERATIONS FOR BOND AND ANCHORAGE

1. The calculated tension or compression in any bar at any section shall be developed on each side of the section by an appropriate development length or by a combination thereof

$$\text{Development length } L_d = \frac{\phi . \sigma_s}{4 . \tau_{bd}}$$

where ϕ = nominal diameter of the bar,

σ_s = Stress in bar at the section considered at design load,

τ_{bd} = design bond stress as given in Table 2.1.

The development length includes anchorage values of hooks in tension reinforcement.

For bars of sectons other than circular the development length should be sufficient to develop the stress in the bar by bond.

2. *Bars Bundled in Contact* : The development length of each bar of bundled bars shall be that for the individual bar, increased by 10% for two bars in contact, 20% for three bars in contact and 33% for four bars in contact.

3. *Anchoring Bars in Tension* : (*a*) Deformed bars may be used without end anchorage provided development length required is satisfied. Hooks should normally be provided for plain bars in tension.

(*b*) *Bends and Hooks* : Bends and hooks shall conform to IS: 2502-1963.

(*i*) Bends : The anchorage value of bend shall be taken as 4 times the diameter of the bar for each 45° bend subject to a maximum of 16 times the diameter of the bar.

(*ii*) Hooks : The anchorage value of a standard U-type hook shall be equal to 16 times the diameter of the bar.

Anchorage value of bends and hooks are shown in Fig. 4.28.

4. *Anchoring Bars in Compression* : The anchorage length of straight bars in compression shall be equal to the development length of bars in compression as specfieid in para 1 above. The projected length of hooks, bends and straight lengths beyond bends if provided for a bar in compression shall be considered for development length.

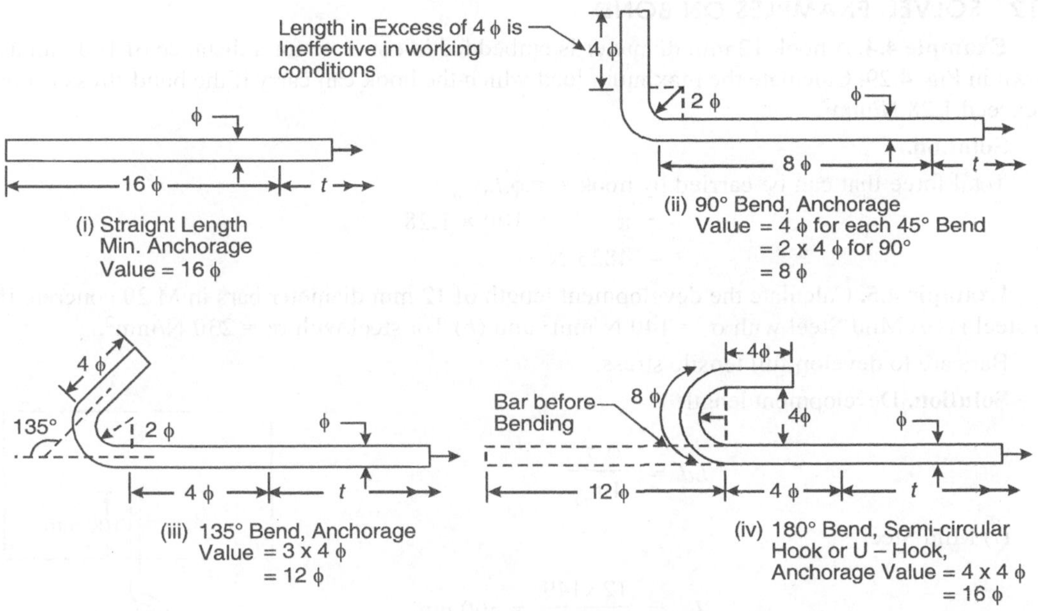

Fig. 4.28

5. *Lap Splices* : (*a*) Lap slices shall not be used for bars larger than 36 mm; for larger diameters, bars may be welded; in cases where welding is not practicable, lapping of bars larger than 36 mm may be permitted in which case additional spirals should be provided around the lapped bars.

(*b*) Lap splices shall be considered as staggered if the centre to centre distance of the splices is not less than 1.3 times the lap length calculated as described in (*c*) below.

(*c*) Lap length including anchorage value of hooks in flexural tension shall be L_d or 30 ϕ whichever is greater and for direct tension to 2 L_d or 30 ϕ whichever is greater. The straight length of the lap shall not be less than 15 ϕ or 200 mm, where L_d is the development length.

Note: Splices in tension members shall be enclosed in spirals made of bars not less than 6 mm diameter with pitch not more than 100 mm. Hooks shall be provided at the ends of bars in tension members.

(*d*) The lap length in compression shall be equal to the development length in compression calculated as described in para 1, but not less than 24 ϕ.

(*e*) When bars of two different diameters are to be spliced, the lap length shall be calculated on the basis of diameter of the smaller bar.

(*f*) When splicing of welded wire fabric is to be carried out, lap splices of wires shall be made so that overlap measured between the extreme cross wires shall be not less than the spacing of cross wires plus 100 mm.

(*g*) In case of bundled bars, lapped splices of bundled bars shall be made by splicing one bar at a time; such individual splices within a bundle shall be staggered.

Where splices are provided in reinforcing bars, they shall as far as possible be away from the sections of maximum stress and be staggered. It is recommended that splices in flexual members should not be at sections where bending moment is more than 50% of the moment of resistance, and not more than half the bars shall be spliced at a section.

Where more than one-half of the bars are spliced at a section or where splices are made at points of maximum stress, special precautions shall be taken, such as increasing the length of lap and/or using spirals or closely spaced stirrups around the length of the splice.

4.12 SOLVED EXAMPLES ON BOND

Example 4.4. A hook 12 mm diameter is embedded in concrete for a distance of 100 mm as shown in Fig. 4.29. Calculate the maximum load which the hook can carry if the bond stress is not to exceed 1.28 N/mm².

Solution.

Total force that can be carried by hook $= \pi.\phi.L.\tau_{bd}$

$$= \pi \times 12 \times 100 \times 1.28$$

$$= 4823 \text{ N}$$

Example 4.5. Calculate the development length of 12 mm diameter bars in M 20 concrete if the steel is (*a*) Mild Steel with $\sigma_s = 140$ N/mm² and (*b*) Tor steel with $\sigma_s = 230$ N/mm².

Bars are to develop full tensile stress.

Solution. Development length

$$Ld = \frac{\phi.\sigma_s}{4.\tau_{bd}}$$

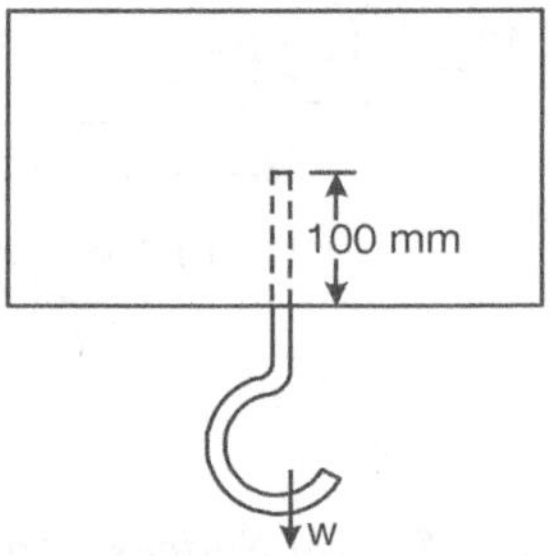

Fig. 4.29

(*i*) Mild Steel

$$L_d = \frac{12 \times 140}{4 \times 0.6} = 500 \text{ mm}$$

(*ii*) Tor Steel, τ_{bd} for tor steel = 1.6×0.8 N/mm²

$$L_d = \frac{12 \times 230}{4 \times 0.8 \times 1.6} = 539 \text{ mm}$$

4.13 TORSION

Torsion is the effect of loads and reactions acting on an axis which lies away from the axis of the flexural member. In such cases, torsional bending moments or twisting moments also come into picture in addition to the flexural bending moment and shear force. The effect of torsional moment is quite significant in structures like *L*-beams, beams curved in plan, *e.g.*, ring beam to support circular water tanks, curved balconies or other similar structures having curved boundaries. Additional steel is required to be provided to take care of the torsional moments. This additional steel is provided because torsional moments are assumed in the form of equivalent shear and equivalent bending.

4.14. DESIGN CONSIDERATIONS FOR TORSION AS PER IS:456-2000

1. Torsional reinforcement is not calculated separately from that required for bending and shear. Instead the total longitudinal reinforcement is determined for a fictitious bending moment which is a function is actual bending moment and torison; similarly web reinforcement is determined for a fictious shear which is a function of actual shear and torsion.

2. *Critical Section* : Sections located less than a distance '*d*' from the face of the support may be designed for the same torsion as computed at a distance '*d*', where *d* is the effective depth.

3. *Equivalent Shear* : Equivalent shear, V_e, shall be calculated from the formula :

$$V_e = V + 1.6\ T/b \qquad ...(4.28)$$

where V_e = equivalent shear,

V = shear,

T = torsional moment, and

b = breadth of beam, in case of flanged beams $b = b_w$.

The equivalent nominal shear stress τ_{ve}, in this case shall be calculated by :

(*i*) $\tau_{ve} = \frac{V_e}{b.d}$ for beams of uniform depth

(*ii*) $\tau_{ve} = \frac{V_e \pm M/d.\tan\beta}{b.d}$ for beams of varying depths,

where d = effective depth,

M = bending moment at the section,

β = angle between the top and the bottom edges of the beam.

The negative sign in the formula applies when the moment M increases numerically in the same direction as the effective depth 'd' increases and the positive sign when the moment M decreases numericaly in this direction.

The value of τ_{ve} shall not exceed the values of $\tau_{c,max}$ as given in Table 4.1.

Table 4.11. Maximum Shear Stress, $\tau_{c,max}$ N/mm²

Concrete Grade	M15	M20	M25	M30	M35	M40 and above
$\tau_{c,max}$ N/mm²	1.6	1.8	1.9	2.2	2.3	2.5

4. For flanged beams, *i.e.*, *T* and *L* beams, substitute b_w for b in the above expressions.

5. If the equivalent nominal shear stress τ_{ve} does not exceed τ_c given in Table 2.2, minimum shear reinforcement shall be provided as given in article 4.6.5.

6. If r_{ve} exceed τ_c given in Table 2.2, both longitudinal and transverse reinforcement shall be provided in accordance with para 7.

7. *Reinforcement in Members Subjected to Torison*

(*i*) Reinforcement for torsion, when required, shall consist of longitudinal and tranverse reinforcement.

(*ii*) *Logitudinal Reinforcement* : The longitudinal reinforcement shall be designed to resist an equivalent bending moment M_{e1}, given by

$$M_{e1} = M + M_1 \quad \text{...(4.29)}$$

where M = bending moment at the cross section, and

$M_t = T\,\frac{(1+D/b)}{1.7}$, where T is the torsional moment, D is the overall depth of the beam and b is the breadth of the beam.

(*iii*) If the numerical value of M_t as defined in the preceding para (*ii*), exceeds the numerical value of the moment M, longitudinal reinforcement shall be provided on the flexural compression face, such that the beam can also withstand an equivalent moment M_{e2} given by $M_{e2} = M_t - M$, the moment M_{e2} being taken as acting in the opposite sense of the moment M.

(*iv*) *Transverse Reinforcement* : Two legged closed hoops enclosing the corner longitudinal bars shall have an area of cross section A_{sv} given by

$$A_{sv} = \frac{T.sv}{b_1 d_1\, \sigma_{sv}} + \frac{v.s_v}{2.5 d_1\, \sigma_{sv}} \quad \text{...(4.30)}$$

but the total reinforcement shall not be less than

$$\frac{(\tau_{sv} - \tau_c) b.s_v}{\sigma_{sv}}$$

where T = torsional moment,

V = shear force,

s_v = spacing of the stirrup reinforcement,

b_1 = centre to centre distance between corner bars in the direction of the width,

d_1 = centre to centre distance between corner bars in the direction of the depth,

b = breadth of the member,

σ_{sv} = permissible tensile stress in shear reinforcement,

τ_{ve} = equivalent shear stress as given in para 3.

τ_c = shear strength of the concrete as specified in Table 2.2.

8. *Distribution of Torsion Reinforcement* : When a member is designed for torsion, torsion reinforcement shall be provided as below :

(*a*) *Transverse Reinforcement* for torsion shall be rectangular closed stirrups placed perpendicular to the axis of the member. The spacing of the stirrups shall not exceed the least of x_1, $\frac{x_1 + y_1}{4}$ and 300 mm, where x_1 and y_1 are respectively the short and long dimensions of the stirrup.

(*b*) *Longitudinal Reinforcement* shall be placed as close as practicable to the corners of the cross section and in all cases, there shall be at least one longitudinal bar in each corner of the ties. When the cross sectional dimension of the member exceeds 450 mm, additional longitudinal bars shall be provided to satisfy the requirements of the minimum reinforcement and spacing of side face reinforcement. The total area of such reinforcement shall not be less than 0.1 per cent of the web area and shall be distributed equally on two faces at a spacing not exceeding 300 mm or web thickness whichever is less.

4.15 SOLVED EXAMPLES ON TORSION

Example 4.1. A rectangular R. C. beam in M20 concrete is 300 mm wide and 350 mm deep and reinforced with 4 bars of 16 mm dia. on tension side. Calculate the steel for torsion if the torsional moment acting on the beam is 21 KN.m. (2100 Kg.m). The bending moment at the section is 30 KN.m and Shear Force = 25 KN

Solution. (Fig. 4.30)

$$\text{Equivalent shear } V_e = V + 1.6\frac{T}{b}$$

$$= 25000 + 1.6 \times \frac{21000 \times 1000}{300}$$

$$= 137000 \text{ N}$$

$$\text{Nominal shear stress} = \tau_{ve} = \frac{V_e}{b.d}$$

$$= \frac{137000}{300 \times 315}$$

$$= 1.45 \text{ N/mm}^2$$

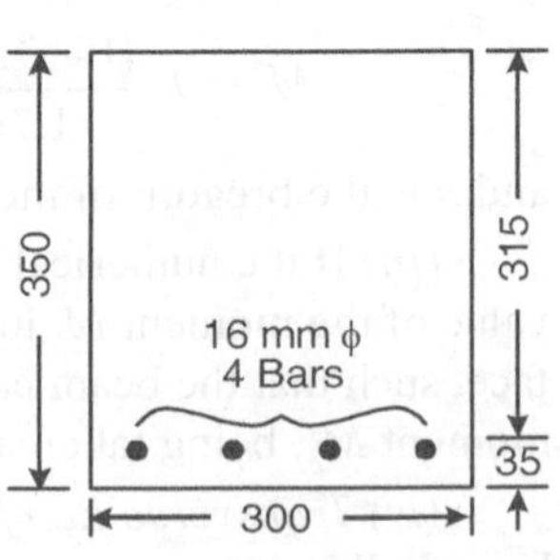

Fig. 4.30

Check

$\tau_{ve} < \tau_{c,max}$, *i.e.*, 1.8 N/mm² for M20

$$p = \frac{100A_s}{b.d}$$

$$= \frac{100 \times 4 \times 201}{300 \times 315}$$

$$= 0.85$$

From Table 2.2, for $\pi = 0.85$, τ_c shall be calculated as follows:

For $p = 0.75$, $\tau_c = 0.35$

For $p = 1.00$, $\tau_c = 0.39$

For $p = 0.85$, $\tau_c = 0.35 + \frac{0.39 - 0.35}{1.00 - .75} \times (0.85 - 0.75)$

$$= 0.366 \text{ N/mm}^2$$

$\therefore \quad \tau_{ve} > \tau_c$

Since τ_{ve} is more than τ_c, longitudinal as well as transverse reinforcement for torsion is needed.

Longitudinal Steel

$$M_1 = \frac{T(1 + D/b)}{1.7}$$

$$= \frac{21000}{1.7}\left(1 + \frac{350}{300}\right) = 26764.7 \text{ N.m}$$

M_t is less than the bending moment M

$$\therefore \quad M_{e1} = M + M_t$$

$$= 30,000 + 26764.7 = 56764.7 \text{ N.m}$$

Area of steel requried for equivalent moment

$$A_{st} = \frac{M_{e1}}{\sigma_{st} \cdot j \cdot d}$$

$$= \frac{56764.7 \times 1000}{230 \times 0.90 \times 315}$$

$$= 870.00 \text{ mm}^2$$

Area required for torsion only

$$= 870.00 - 4 \times 201 = 66.00 \text{ mm}^2$$

Provide 2 Nos. 12 mm dia. HYSD, Fe 415 bars as torsion steel as shown in Fig. 4.31.

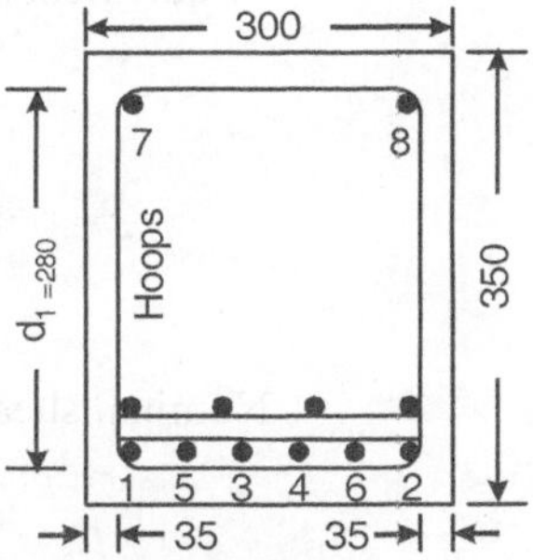

Fig. 4.31

Transverse Reinforcement

Using 2 legged 12 mm ϕ closed hoops enclosing the corner longitudinal bars,

$$A_{sv} = \frac{T \cdot s_v}{b_1 \cdot d_1 \cdot \sigma_{sv}} + \frac{V \cdot s_v}{2.5 d_1 \cdot \sigma_{sv}}$$

$$b_1 = 300 - 35 - 35 = 230 \text{ mm}$$

$$d_1 = 350 - 35 - 35 = 280 \text{ mm}$$

$$\therefore \quad s_v = \frac{A_{sv}}{\left[\dfrac{T}{b_1 \cdot d_1 \cdot \sigma_{sv}} + \dfrac{V}{2.5\, d_1 \cdot \sigma_{sv}}\right]}$$

$$= \frac{2 \times 113}{\left[\dfrac{21000 \times 1000}{230 \times 280 \times 220} + \dfrac{25000}{2.5 \times 280 \times 230}\right]}$$

$$= 143.62 \text{ mm}$$

$\therefore$ Provide 12 mm ϕ 2 legged closed hoops @ 140 mm *c/c* throughout.

Check for Minimum Transverse Reinforcement

(*i*) $$A_{st,\, min} = \frac{(\tau_{ve} - \tau_c).b.S_v}{\sigma_{sv}}$$

$$= \frac{(1.45 - 0.66) \times 300 \times 140}{230}$$

$$= 198 \text{ mm}^2 \text{ less than } 2 \times 113 \text{ mm}^2$$

$\therefore$ O.K.

(*ii*) Spacing $\ngtr$ the least of the following

(*a*) $x_1 = 230 + 2 \times 16/2 + 2 \times 12/2 = 258$ mm

(*b*) $y_1 = 280 + 16/2 + 2 \times 12/2 + 12/2 = 306$ mm $\qquad (x_1 + y_1)/4 = 141$ mm

(*c*) 300 mm

$\therefore$ Spacing of 140 mm c/c is O.K.

Example 4.2. A rectangular R. C. beam in M 20 concrete is 300 mm wide and 450 mm deep (overall) and is reinforced with 3 bars 16 mm ϕ on tension side. The beam cross section is subjected to a maximum bending moment of 30 KN.m. (3 t.m) and a maximum shear force of 30 KN (3 *t*). In addition the beam cross section is subjected to a torsional moment of 36 KN.m (3.6 t.m). Calculate the reinforcement for torsion. Fe 415.

Solution. (Fig. 4.32)

$$\text{Equivalent shear } V_e = V + 1.6\frac{T}{b}$$

$$= 30000 + \frac{1.6 \times 36000 \times 1000}{300}$$

$$= 150{,}000 \text{ N}$$

$$\text{Nominal shear stress} = \tau_{ve} = \frac{V_e}{b\,.\,d}$$

$$= \frac{150{,}000}{300 \times 410}$$

$$= 1.22 \text{ N/mm}^2$$

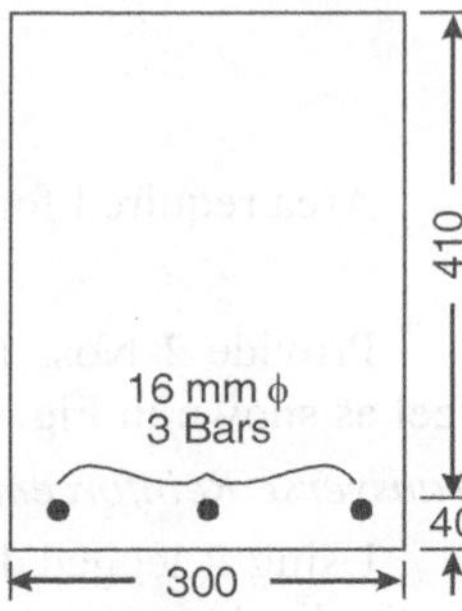

Fig. 4.32

Check

(*i*) $\tau_{ve} < \tau_{c,max}$ *i.e.*, 1.8 N/mm² for M20 concrete

(*ii*) τ_{ve} against t_c

$$p = \frac{100\,A_s}{b.d}$$

$$= \frac{100 \times 3 \times 201}{300 \times 410} = 0.49$$

$$\therefore \quad \tau_c = 0.287 \text{ N/mm}^2 \text{ from Table 2.2}$$

Since τ_{ve} is more than τ_c, longitudinal as well as transverse reinforcement for torsion is required.

Longitudinal Reinforcement

$$M_1 = \frac{T \cdot (1 + D/b)}{1.7}$$

$$= \frac{36000\,(1 + 450/300)}{1.7}$$

$$= 52941.2 \text{ N.m.} > M = 30{,}000 \text{ N.m}$$

Since M_t exceeds the numerical value of M, longitudinal steel should be provided on flexural compression face for $M_{e2} = M_t - M$, in addition of the steel required for $M_{e1} = M + M_t$.

$$M_{e1} = 30{,}000 + 52941.2 = 82941.2 \text{ N.m}$$

$$\text{Area of steel for } M_{e1} = \frac{M_{e1}}{\sigma_{st} \cdot j \cdot d}$$

$$A_{st} = \frac{82941.2 \times 1000}{230 \times 0.90 \times 410}$$

$$= 977.3$$

Area of steel for torsion only on tension face = 977.3 – 3 × 201 = 374.3 mm² Provide additional 7 Nos. 10 mm ϕ bars on tension face for torsion.

$$\text{Area of steel required for } M_{e2}, A_{sc} = \frac{M_{e2}}{\sigma_{st} \cdot j \cdot d} \quad \text{where } M_{e2} = 52941.2 - 30000 = 22941.2 \text{ N.m}$$

$$A_{sc} = \frac{22941.2 \times 1000}{230 \times 0.90 \times 410}$$

$$= 270.3 \text{ mm}^2$$

Provide 2 Nos. 12 mm ϕ plus 1 No. 10 mm ϕ on compression face for torsion, as shown in Fig. 4.33.

Transverse Reinforcement

Using 12 mm ϕ 2 legged closed hoops,

$$s_v = \frac{A_{sv}}{\left[\dfrac{T}{b_1 \cdot d_1 \cdot \sigma_{sv}} + \dfrac{V}{2.5\,d_1 \cdot \sigma_{sv}}\right]}$$

$$= \frac{2 \times 113}{\left[\dfrac{36000 \times 1000}{220 \times 370 \times 230} + \dfrac{30000}{2.5 \times 370 \times 230}\right]}$$

$$= 109.5 \text{ mm}$$

Provide 12 mm ϕ 2 legged closed hoops @ 100 mm c/c as shown in Fig. 4.33.

Check for Minimum transverse reinforcement

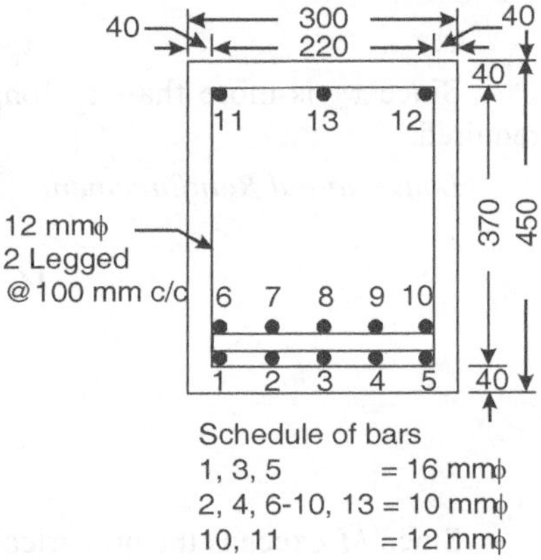

(*i*) $$A_{sv,\,min} = \frac{(\tau_{ve} - \tau_c) \cdot b \cdot s_v}{\sigma_{sv}}$$

$$= \frac{(1.22 \times 0.287) \times 300 \times 100}{230}$$

$$= 121.70 \text{ mm}^2 \text{ less than } 226 \text{ mm}^2$$

∴ O.K.

(*ii*) Spacing ≯ the least of the following

(*a*) $x_1 = 220 + 2 \times 16/2 + 2 \times 12/2$

$= 248$ mm

(*b*) $y_1 = 370 + 2 \times 10/2 + 2 \times 12/2$

$= 392$ mm

$$\frac{x_1 + y_1}{4} = 160.0 \text{ mm}$$

(*c*) 300 mm

∴ Spacing of 100 mm c/c is O.K.

EXERCISE 4

1. Explain briefly the shear stress distribution in a homogenous rectangular beam section and also in rectangular beams of concrete with tension reinforcement.
2. What is diagonal tension? Show that in a R. C. beam cracks appear at angle of 45° near supports and vertical cracks appear near mid span.
3. How vertical stirrups help in carrying shear? Discuss briefly different types of shear stirrups.
4. (*a*) What do you understand by 'Local Bond Stress' and 'Average Bond Stress'? Calculate the development lenght.

   (*b*) Sketch standard hook with dimensions in terms of diameter of bar.
5. What is torsion? How do you account for torsion in the design of a R. C. beam as per IS-specifications?

Design of Rectangular Beams

CHAPTER **5**

5.1 INTRODUCTION

In proportioning a beam it is necessary to ensure that the permissible stresses in concrete and in reinforcement are not exceeded but it is also desirable generally for maximum stresses developed in the materials to be equal to the permissible values. Thus, a section designed in the balanced condition, is considered to be the best. The percentage of steel, thus put in, is called 'economic percentage', but this may be misleading since the relative amounts of steel and concrete in the most economical design of beam depend not only on the permissible stresses but also on the cost of the materials.

The dimensions of the beams are primarily determined from considerations of moments and shear forces to be resisted. However, before the bending moment or shear force in a beam is computed, some dimensions of the member are assumed for assessing the self weight. The beam is then analysed to compute the maximum bending moment and maximum shear force under superimposed loads combined with self weight of the member. The breadth of the rectangular beam is, generally, kept one-third of total depth. In buildings such a residential and office type, a reasonable breadth is one half to two-thirds of total depth whereas in industrial buildings beams having proportions of breadth to depth of one half to one-third are often convenient. This may be taken as a guiding factor only. Once the maximum bending moment in a beam is known, width of beam can be decided by certain architectural reasons and support conditions. Regarding depth, it may be any value to satisfy the structural requirement, bending moment in particular or it may be limited depending upon the architectural or head room restrictions.

5.2 GENERAL NOTES ON THE DESIGN OF BEAMS (IS:456-2000)

1. *Effective Span* : Unless otherwise specified, the effective span of a member shall be as follows:

(*a*) *Simply Supported Beam* : The effective span of a beam that is not built integrally with its supports shall be taken as clear span plus the effective depth of beam or centre to centre distance of supports, whichever is less.

(*b*) *Continuous Beams* : If the width of support is less than 1/12 of the clear span, the effective span shall be same as given in para (*a*) above. If the width of support is greater than 1/12 of the clear span or 600 mm whichever is less, the effective span shall be taken as under:

(*i*) For the end span with one end free fixed and the other continuous or for intermediate spans, the effective span shall be the clear span between the supports.

(*ii*) For end span with one end free and the other continuous, the effective span shall be the clear span plus half the effective depth of the beam or the clear span plus half the width of the discontinuous support, whichever is less.

(*iii*) *Cantilever* - The effective length of a Canitlever shall be taken as its length to the face of the support plus half the effective depth except where it forms the end of a continuous beam where the length to centre of support shall be taken.

(*iv*) *Frames* - In the analysis of a continuous frame, centre to centre distance shall be used.

2. *Effective Depth* : Effective depth of a beam is the distance between the centroid of the area of tension reinforcement and the extreme compression fibre, excluding the thickness of

finishing material not placed monolithically with the member and the thickness of any concrete provided to allow for wear. This will not apply to the deep beams.

3. *Stiffness* : The relative stiffness of a beam may be based upon the moment of inertia of the section determined on the basis of any one of the following definitions:

(*a*) *Gross Section* : the gross section of the member ignoring reinforcement.

(*b*) *Transformed Section* : The concrete cross section plus the area of reinforcement transformed on the basis of modular ratio.

(*c*) *Cracked Section* : The are of concrete in compression plus the area of reinforcement transformed o n the basis of modular ration.

The assumptions made shall be consistent for all the members of the structures throughout the analysis.

4. *Control of Deflection* : The vertical deflection limits of beams may generally be assumed to be satisfied provided that the span to effective depth ratios are not greater than the values obtained as below:

(*a*) For spans upto 10 m

Cantilever	7
Simply supported	20
Continuous	26

(*b*) For spans above 10 m, the values in para (a) may be multiplied by 10/span in meters, except for cantilevers in which case deflection calculations should be made.

(*c*) Depending on the area and the type of steel for tension reinforcement, the values in (a) or (b) shall be modified as per Fig. 5.1

(*d*) Depending on the area of compression reinforcement, the value of span to depth ratio be further modified as per Fig. 5.2.

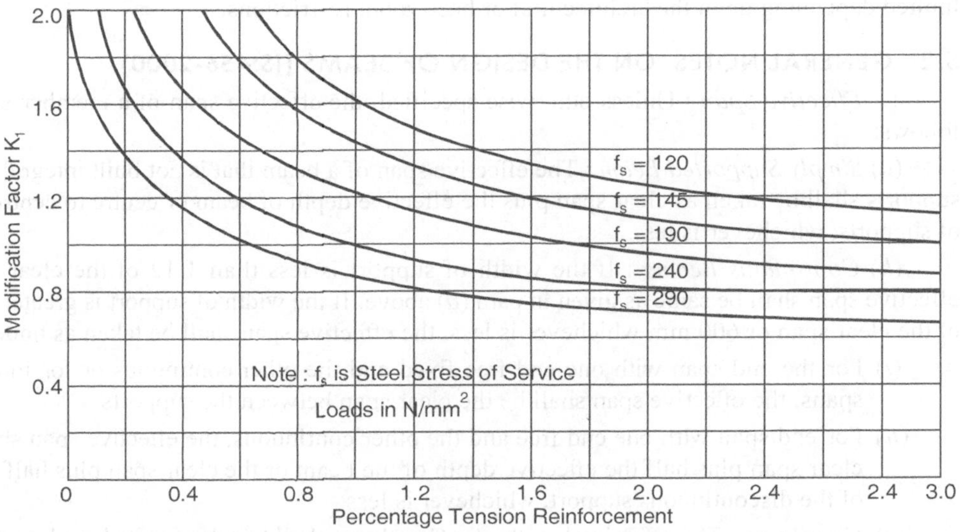

$$f_s = 0.58 f_y \times \frac{\text{Area of cross-section of steel required}}{\text{Area of cross-section of steel provided}}$$

Fig. 5.1. Modification Factor for Tension Reinforcement

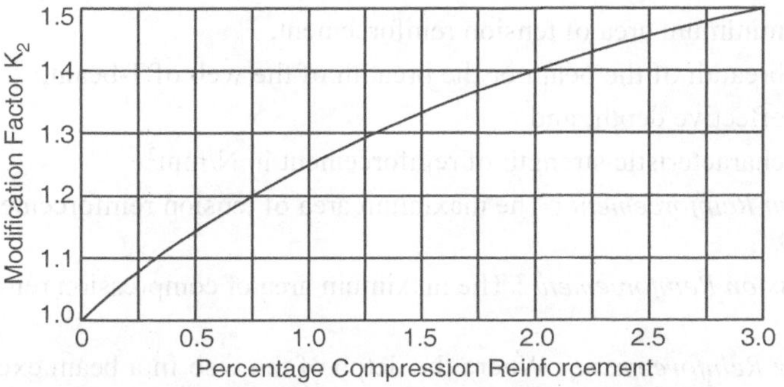

Fig. 5.2. Modification Factor for Compression Reinforcement

(*e*) For flanged beams, the values of (*a*) or (*b*) be modified as per Fig. 5.3 and the reinforcement percentage for use in Figs. 5.1 and 5.2 should be based on the area of section equal to $b_f \cdot d$.

5. *Slenderness Limits for Beams to Ensure Lateral Stability* : A simply supported or continuous beam shall be so proportioned that the clear distance between the lateral restraints does not exceed $60.b$ or $250.b^2/d$ whichever is less, where d is the effective depth of the beam and b the breadth of the compression face midway between the lateral restraints.

For a cantilever, the clear distance from the free end of the cantilever to the lateral restraint shall not exceed $25.b$ or $100.b^2/d$ whichever is less.

6. *Tension Reinforcement*

(*a*) *Minimum Reinforcement* : The minimum area of tension reinforcement shall not be less than that given by the following:

$$\frac{A_s}{b.d} = \frac{0.85}{f_y} \quad ...(5.1)$$

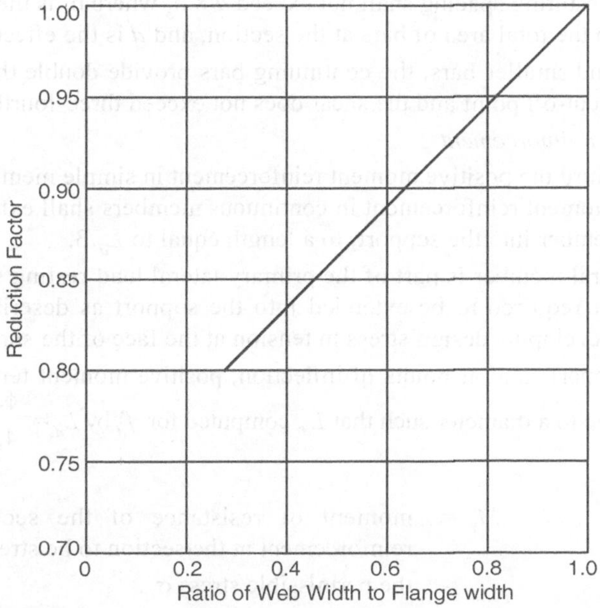

Fig. 5.3. Reduction Factors for Ratios of Span to Effective Depth for Flanged Beams

where As = minimum area of tension reinforcement,

b = breadth of the beam or the breadth of the web of T-beam,

d = effective depth, and

f_y = characteristic strength of reinforcement in N/mm^2

(*b*) *Maximum Reinforcement* : The maximum area of tension reinforcement shall into exceed 0.04 $b.D.$

7. *Compression Reinforcement* : The maximum area of compression reinforcement shall not exceed 0.04 $b.D.$

8. *Side face Reinforcement* : Where the detph of the web in a beam exceeds 750 mm, side face reinforcement shall be provided along the two faces. The total area of such reinforcement shall not be less than 0.1 per cent of the web area and shall be distributed equally on two faces at a spacing not exceeding 300 mm or web thickness whichever is less.

9. *Curtailment of Tension Reinforcement in Flexural Members* : For curtailment, reinforcement shall extend beyond the point in which it is no longer required to resist flexure or a distance equal to the effective depth of the member or 12 times the bar diameter, whichever is greater except at simple support or end of cantilever.

A point at which reinforcement is no longer required to resist flexure is where the resistance moment of the section, considering only the continuing bars, is equal to the design moment.

Flexural reinforcement shall not be terminated in a tension zone unless any one of the following conditions is satisfied:

(*a*) The shear at the cut-off point does not exceed two-thirds that permitted, including the shear strength of web reinforcement provided.

(*b*) Stirrup area in excess of that required for shear and torsion is provided along each terminated bar over a distance from the cut-off point equal to three-fourths the effective depth of the member. The excess stirrup area shall not be less than $0.4.s/f_y$, where b is the breadth of beam, s is the spacing, and f_y is the characteristic strength of reinforcement in N/mm^2. The resulting spacing shall not exceed $d/8.b_b$ where b_b is the ratio of the area of bars cut-off to the total area of bars at the section, and d is the effective depth.

(*c*) For 36 mm and smaller bars, the continuing bars provide double the area required for flexure at the cut-off point and the shear does not exceed three-fourths that permitted.

Positive Moment Reinforcement

(*a*) At least one-third the positive moment reinforcement in simple members and one-fourth the positive moment reinforcement in continuous members shall extend along the same face of the member into the support, to a length equal to $L_d/3$.

(*b*) When a flexural member is part of the primary lateral load resting system, the positive reinforcement required to be extended into the support as described in (a) shall be anchored to develop its design stress in tension at the face of the support.

(*c*) At simple supports and at points of inflection, positive moment tension reinforcement shall be limited to a diameter such that L_d, computed for f_d by $L_d = \frac{\phi.\sigma_s}{4.\tau_{bd}}$ does not exceed $\frac{M_1}{V} + L_0$

where M_1 = moment of resistance of the section assuming all reinforcement at the section to be stressed to fd,

f_d = the permissible stress σ_{st}.

V = shear force at the section due to design loads,

L_0 = the sum of the anchorage beyond the centre of the support and the equivalent anchorage value of any hook or mechanical anchorage at simple support; and at point of inflection L_0 is limited to the effective depth of the members or 12ϕ, whichever is greater.

ϕ = diameter of bar

The value of M_1/V in the above expression may be increased by 30% when the ends of the reinforcement are confined by a compressive reaction.

Negative Moment Reinforcement

Negative moment reinforcement in continuous, restrained, or cantilever members should be developed in or through the supporting member.

Negative moment reinforcement should have sufficient distance between the face of the support and the end of each bar to develop its full working strength.

At least one-third of the required negative moment reinforcement at the face of the support shall extend beyond the point of inflection for a distance not less than the effective depth of a member or 12ϕ or one-sixteenth of the clear span whichever is greater.

10. *Cover to Reinforcement* : Reinforcement shall have concrete cover and the thickness of such cover (exclusive of plaster or other decorative finish) shall be as follows:

(*a*) At each end of reinforcing bar not less than 25 mm, nor less than twice the diameter of such bar;

(*b*) For a longitudinal reinforcing bar in a beam, not less than 25 mm, nor less than the diameter of such bar;

(*c*) For any other reinforcement; not less than 15 mm, nor less than the diameter of such bar.

Increased cover thickness may be provided when surfaces of concrete members are exposed to the action of harmful chemicals (as in the case of concrete in contact with earth faces contaminated with such chemicals), acid, vapour, saline atmosphere, sulphurous smoke (as in the case of steam-operated railways), etc., and such increase of cover may be between 15 mm and 50 mm beyond the figures given above as may be specified by the engineer-in-charge.

For reinforcement concrete members totally immuered in sea water, the cover shall be 40 mm more than that specified above.

For reinforcement concrete members, periodically immersed in or subjected to sea spray, the cover of concrete shall be 50 mm more than that specified above.

For concrete of grade M25 and above, the additional thickness of cover specified in para 2 to 4 may be reduced to half.

In all such cases the cover should not exceed 65 mm.

11. *Spacing of Reinforcement*

(*A*) *Minimum Distance between Individual Bars*

The following shall apply for spacing of bars :

(*a*) The horizontal distance between two parallel main reinforcement bars shall usually be not less than the greatest of the following :

(*i*) The diameter of the bar if the diameters are equal.

(*ii*) The diameter of the larger bar if the diameters are unequal, and

(*iii*) 5 mm more than the nominal maximum size of coarse aggregate.

(*b*) When needle vibrators are used the horizontal distance between bars of a group may be reduced to two-thirds the nominal maximum size of the coarse aggregate, provided that sufficient space is left between groups of bars to enable the vibrator to be immersed.

(*c*) Where there are two or more rows of bars, the bars shall be vertically in line and the minimum vertical distance between the bars shall be 15 mm, two-thirds of nominal maximum size of aggregate or the maximum size of bar, whichever is greatest.

(*B*) *Maximum Distance between Bars in Tension*

The horizontal distance between parallel reinforcement bars, or groups near tension face of a beam shall not be greater than the value given the Table 5.1, depending on the amount of redistribution carried out in the analysis and the characteristic strength of the reinforcement.

Table 5.1. Clear Distance Between Bars

f_y	*Percentage Redistribution To or From Section Considered*				
	–30	–15	0	+15	+30
	CLEAR DISTANCE BETWEEN BARS				
N/mm^2	mm	mm	mm	mm	mm
250	215	260	300	300	300
415	125	155	180	210	235
500	105	130	150	175	195

12. *Moments and Shear Coefficients for Continuous Beams*

Unless more exact estimates are made, for beams of uniform cross-section which support substantially uniformly distributed loads over three or more spans which do not differ by more than 15% of the longes, the bending moments and shear forces used in design may be obtained using the coefficients given in Table 5.2 and Table 5.3 respectively.

For moments at supports where two unequal spans meet or in case where the spans are not equally loaded, the average of the two values for the negative moment at the support may be taken for design.

Table 5.2. Bending Moment Coefficients

Type of load	*Span Moments*		*Support Moments*	
	Near middle of end support	*At middle of interior span*	*At Support Next to the End Support*	*At other Interior Supports*
Dead Load and Imposed load (fixed)	$+\frac{1}{12}$	$+\frac{1}{16}$	$-\frac{1}{10}$	$-\frac{1}{12}$
Imposed Load	$+\frac{1}{10}$	$+\frac{1}{12}$	$-\frac{1}{9}$	$-\frac{1}{9}$

Note: For obtaining the bending moment, the coefficient shall be multiplied by the total design load and effective span.

Table 5.3. Shear Force Coefficients

Type of load	*At End Support*	*At Support Next to the End Support*		*At other Interior*
		Outer Side	*Inner Side*	*Supports*
Dead Load and Imposed Load (fixed)	0.4	0.6	0.55	0.5
Imposed Load (not fixed)	0.45	0.6	0.6	0.6

Note: For obtaining the shear force, the coefficient shall be multiplied by the total design load.

Beam and Slabs over Free End Supports

Where a member is build into a masonry wall which develops only partial restraint, the member shall be designed to resist a negative moment at the face of the support of $W.L/24$ where W is the total design load and L is the effective span, or such other restraining moment as may be shown to be applicable. For such a condition shear coefficient given in Table 5.3 at the end support may be increased by 0.05.

Critical Section for Moment

For monolitic construction, the moments computed at the face of the supports shall be used in the design of the members at those sections. For non-monolithic construction, the design of the member shall be done keeping in view the effective span as given in(1) above.

13. *Increase in Permissible Stresses*: Where stresses due to wind (or earthquake), temperature and shrinkage effects are combined with those due to dead, live and impact load, the usual permissible stresses in concrete as given in Tables 2.1, 2.2, 4.1 and in steel as given in Table 2.3, may be exceed upto a limit of $33\frac{1}{2}$ per cent. Wind and seismic forces need not be considered as acting simultaneously.

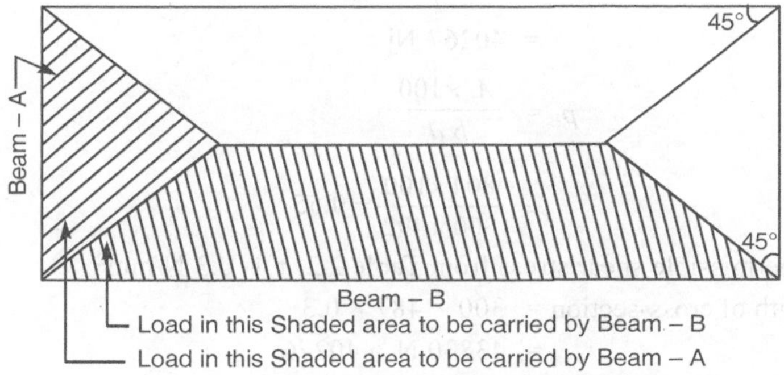

Fig. 5.4. Load Carried by Supporting Beams

14. *Checking for Shear and Anchorage*: For checking of beam sections for shear and end anchorage (or development length), procedure explained in chapter 4 may be followed.

15. *Loads Transmitted to Supporting Beams from Slabs*: The loads on beams supporting solid slabs spanning in two directions at right angles and supporting uniformly distributed loads, may be assumed to be in accordance with Fig. 5.4.

5.3 SOLVED EXAMPLES

Example 5.1. Design a R.C. rectangular beam for a simply supported span of 5 m and carrying a superimposed load of 20000 N/m (2000 Kg/m) inclusive of self weight of the beam. Use M 20 concrete and HYSD of Fe 415 bars. Take width of the beam = 300 mm.

Sol. Total load on the beam = 20000 N/m

$$\text{Max. B.M} = \frac{wl^2}{8}$$

$$= \frac{20000 \times 5 \times 5}{8}$$

$$= 62500 \text{ N.m}$$

Assuming the section to be balanced

$$MR = Qbd^2$$

$$\therefore \quad d = \sqrt{\left(\frac{M}{b\,Q}\right)} = \sqrt{\left(\frac{62500\times1000}{300\times0.91}\right)} = 478 \text{ mm}$$

Provide overall depth $D = 525$ mm

Use 16 mn ϕ bars

$$\therefore \quad d_{\text{provided}} = 525 - 30 - \frac{16}{2} = 487 \text{ mm}$$

$$\text{Area of steel required,} \quad A_{st} = \frac{M}{\sigma_{st}\, jd} = \frac{62500\times1000}{230\times0.90\times487} = 620 \text{ mm}^2$$

Provide 4 bars of 16 mm ϕ, thus providing A_{st} = 804 mm^2

Check for shear at critical section i.e., at distance 0.487 m from support.

$$\text{Max. } S.F. = \frac{wl}{2} - 0.487\times20000$$

$$= \frac{20000\times5}{2} - 9740$$

$$= 40267 \text{ Nj}$$

$$p = \frac{A_s\times100}{b\,d}$$

$$= \frac{804\times100}{300\times492} = 0.55$$

Maximum permissible shear stress from Table 2.2, τ_c = 0.3 N/mm^2

Shear strength of cross-section = 300 × 487 × 0.3

= 43830 N > 40276

∴ Provide minimum shear reinforcement of 8 mm ϕ, 2 legged stirrups at spacing,

$$s_v = \frac{A_{sv}\; 0.87 f_y}{0.4\, b} = \frac{25\times50\times415\times0.87}{0.4\times300} = 300.875 \text{ mm}$$

Max. permissible spacing of stirrups = 0. 75 d

= 0.75 × 492 = 369 mm

∴ Provide 8 mm ϕ, 2 legged stirrups @ 200 mm *c/c* throughout.

Check for End Anchorage

Development length, $L_d \not> \frac{M_1}{V} + L_0$

Assuming 2 bars to be bent up and 2 bars taken straight,

M_1 = 2 × 201 × 230 × 0.90 × 487 = 40525218 N.mm

V = 40267 N

L_0 = effective depth or 12 ϕ whichever is greater

= 487 mm or 12 × 16 = 192 mm

Adopt L_0 = 487 mm

$$\therefore \quad \frac{M_1}{V} + L_0 = \frac{40525218}{40267} + 487 = 1496 \text{ mm}$$

$$L_d = \frac{\phi . \sigma_s}{4\tau_{bd}} = \frac{16 \times 230}{4 \times 1.28}$$ (τ_{bd} = 1.28 N/mm² for M20 concrete from Table 2.1 for feformed bars)

$$= 718.75 \text{ mm}$$

The condition $L_d \ngtr \frac{M_1}{V} + L_0$ is satisfied. Therefore, end anchorage is *O.K.*

Check for Deflection

$$p = 0.55, \text{ From Fig. 5.1 modification factor } K_1 = 1.2$$

∴ For simply supported beam, and for deflection to be within limits permissible span to effective depth ration $= 20\ K_1$

$$= 20 \times 1.2 = 24$$

Actual span to depth ration $= \frac{5 \times 1000}{487} = 10.26 < 24 \therefore O.K.$

The reinforcement details are shown in Fig. 5.5

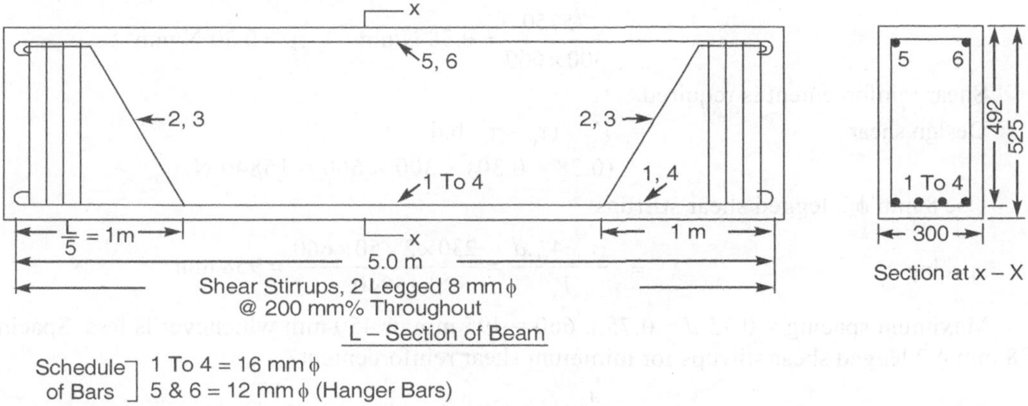

Fig. 5.5

Example 5.2. Design a cantilever beam over a span of 3 m, carrying a superimposed load of 20,000 N/m (2000 kg/m). Use M20 concrete and HYSD bars of Fe415 as reinforcement. The width of the cross section of beam is given as 300 mm.

Sol. Assume the size of the beam as 300 mm × 700 mm

∴ Self weight of beam = 0.3 × 0.7 × 1 × 25000 = 5250 N/m

Superimposed load = 20000 N/m

Total *u.d.l.* on beam = 25250 N/m

Maximum B.M. at fixed end $= \frac{1}{2} . w . L^2 = \frac{25250 \times 3^2}{2}$

$$= 113{,}625 \text{ N.m}$$

$$M = Q.b.d^2$$

For a balanced section $d = \sqrt{\left(\frac{113625 \times 1000}{0.91 \times 300}\right)} = 645 \text{ mm}$

$$D_{\text{overall}} = 700 \text{ mm } O.K.$$

$$d = 700 \text{ — ffective cover as 40 mm} = 660 \text{ mm}$$

Area of steel required $= \dfrac{M}{\sigma_{st} \cdot j.d}$

$$= \frac{113625 \times 1000}{230 \times 0.90 \times 660}$$

$$= 823 \text{ mm}^2$$

Provide 16 mm ϕ 5 Nos., thus providing A_{st} = 1005 mm²

Check for Shear

$$p = \frac{100\,A_{st}}{b.d} = \frac{100 \times 1005}{300 \times 660} = 0.507$$

From Table 2.2, Permissible shear stress = τ_c = 0.30 N/mm²

Max. S.F. at fixed end = 25250 × 3 = 75750 N

Nominal shear stress $t_v = \dfrac{V}{b.d}$

$$= \frac{75750}{300 \times 660} = 0.38 \text{ N/mm}^2 > \tau_c = 0.30 \text{ N/mm}^2$$

∴ Shear reinforcement is required.

Design shear $= V_s = (\tau_v - \tau_c).b.d$

$(0.38 - 0.30) \times 300 \times 660 = 15840$ N

Use 8 mm ϕ 2 legged shear stirrups

spacing $= \dfrac{\sigma_{sv}.A_{sv}.d}{V_s} = \dfrac{230 \times 2 \times 50 \times 660}{15840} = 958$ mm

Maximum spacing = 0.75 d = 0.75 × 660 = 495 mm or 450 mm whichever is less. Spacing of 8 mm ϕ 2 legged shear stirrups for minimum shear reinforcement

$$= \frac{A_{sv} \cdot f_y}{0.4\,b}$$

$$= \frac{2 \times 50 \times 415}{0.4 \times 300} = 345 \text{ mm}$$

Adopt spacing less than or equal to the least of the above. Provide shear stirrups 8 mm ϕ 2 legged @ 200 mm c/c in 2 m distance from fixed support and @ 300 mm c/c in a distance of 1 m from free end as in this 1 m the actual shear stress is less than half that permissible.

Check for End Anchorage

M_1 = 5 × 201 × 230 × 0.90 × 660 = 138, 828, 690 N.mm, V = 75750 N

L_0 = 660 mm or 12 × 20 = 240 mm whichever is more.

$$L_d = \frac{\phi.\sigma_s}{4.\tau_{bd}} = \frac{16 \times 230}{4 \times 1.28} = 718.75 \text{ mm}$$

$$\frac{M_1}{V} + L_0 = \frac{138828690}{75750} + 660 = 2493 \text{ mm}$$

∴ The condition $L_d \ngtr \dfrac{M_1}{V} + L_0$ is satisfied

Provide anchorage length from face of support to a distance of 1166.7 mm.

Check for Deflection

Actual span to effective depth ration = 3000/660 = 4.55

Percentage Steel p = 0.507 From Fig. 5.1 modification factor = K_1 = 1.18

$\therefore$ Permissible span to effective depth ratio = 5 × 1.18 = 8.26 > 4.55

The reinforcement details are shown in Fig. 5.6

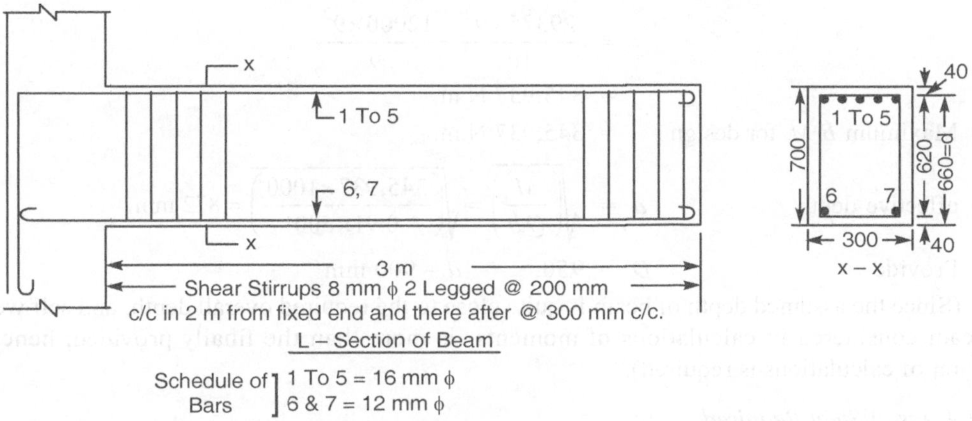

Fig. 5.6.

Example 5.3. Design a R.C. rectangular three span continuous beam with each span = 9 m effective, carrying a slab 125 mm thick over it (slab not cast monolithically with beam). The live load over the slab may be taken as 4000 N/m² (400 Kg/m²) and finishes as 2500 N/m² (250 Kg/m²). The width of slab which transfers load to beam may be taken as 3 m. Use M20 concrete and HYSD Fe 415 bars as reinforcement. Assume maximum size of coarse aggregate as 20 mm.

Sol. Assume the size of the cross section of the beam as 500 mm × 1000 mm (overall).

$\therefore$ Self weight of beam = 0.5 × 1.0 × 25000 = 12500 N/m

Dead Loads

Self weight of slab = 0.125 × 1 × 1 × 25000 = 3125 N/m²

Finishes = 2500 N/m²

Total = 5625 N/m²

Dead load on beam including self weight = 5625 × 3 + 12500 = 29375 N/m

$$w_D = 29375 \text{ N/m}$$

Live load on beam w_L = 4000 × 3 = 12000 N/m

Bending Moments

(*i*) Maximum *B.M.* near middle of end span $= \dfrac{w_D.L^2}{12} + \dfrac{w_L.L^2}{10}$

$$= \frac{29375 \times 9^2}{12} + \frac{12000 \times 9^2}{10}$$

$$= 295{,}481 \text{ N.m.}$$

(*ii*) Maximum *B.M.* at middle of interior span $= \dfrac{w_D.L^2}{16} + \dfrac{w_L.L^2}{12}$

$$= \frac{29375 \times 9^2}{16} + \frac{12000 \times 9^2}{12}$$

$= 148710 + 81000$

$= 229{,}710$ N.m

(*iii*) Maximum *B.M.* at support next to the end support

$$= \frac{w_D.L^2}{10} - \frac{w_L.L^2}{9}$$

$$= \frac{29375 \times 9^2}{10} - \frac{12000 \times 9^2}{9}$$

$= 345{,}937$ N.m

∴ Maximum *B.M.* for design $= 345{,}937$ N.m.

effective depth $d = \sqrt{\left(\frac{M}{Q.b}\right)} = \sqrt{\left(\frac{345{,}937 \times 1000}{0.91 \times 500}\right)} = 872$ mm

Provide $D = 950$ $d = 900$ mm

(Since the assumed depth of beam is quite close to the required overall depth, and self weight of beam considered in calculations of moments is more than the finally provided, hence no revision of calculations is required).

Areas of Steel Required

A_{st} required near middle of end spans $= \frac{295481 \times 1000}{230 \times 0.90 \times 900} = 1586 \text{ mm}^2$ at bottom

Provide 16 mm ϕ 9 Nos. (Area provided = 1809 mm^2)

A_{st} required at middle of interior span $= \frac{229710 \times 1000}{230 \times 0.90 \times 900} = 1233 \text{ mm}^2$ at bottom

Provide 16 mm ϕ 6 Nos. (Area provided = 1206 mm^2)

A_{st} at support next to end support $= \frac{345937.0 \times 1000}{230 \times 0.9 \times 900} = 1856 \text{ mm}^2$ at top

Provide 16 mm ϕ 11 Nos. (Area provided = 2211 mm^2)

Shear Force Computations

(*i*) Shear force at end support $= 0.4\ w_D.L + 0.45\ w_L.L$

$= 0.4 \times 29375 \times 9 + 0.45 \times 12000 \times 9$

$= 154{,}350$ N

(*ii*) Shear force at support next to end supports $= 0.6\ w_D.L + 0.6\ w_L.L$

$= 0.6 \times 29375 \times 9 + 0.6 \times 12000 \times 9$

$= 223{,}425$ N

Max. Shear Force $V = 223{,}425$ N

Nominal shear stress $= \tau_v = \frac{V}{b.d}$

$$= \frac{223{,}425}{500 \times 900}$$

$= 0.4965$ N/mm^2 less than $\tau_{c,max}$ for M20 = 1.8 N/mm^2

$$p = \frac{A_s \times 100}{b.d} = \frac{2200 \times 100}{500 \times 900} = 0.49$$

From Table 2.2, $\tau_c = 0.3 \text{ N/mm}^2$

∴ Shear reinforcement is required.

$$V_s = (0.49 - 0.3) \times 500 \times 900 = 85500 \text{ N}$$

Use vertical stirrups 10 mm ϕ 2 legged (neglect the effect of bending of bars if any).

$$s_v = \frac{\sigma_{sv}.A_{sv}.d}{V_s}$$

$$= \frac{230 \times 2 \times 78.5 \times 900}{85500}$$

$$= 389 \text{ mm}$$

Spacing required of 10 mm ϕ 2 legged stirrups as per minimum shear reinforcement

$$s_v = \frac{A_{sv}.0.87 f_y}{0.4\, b} = \frac{2 \times 78.5 \times (0.87 \times 415)}{0.4 \times 500} = 282.75$$

Provide 10 mm ϕ 2 legged shear stirrups @250 mm *c/c* throughout.

Side Face Reinforcement : Since depth of the beam is greater than 750 mm, side face reinforcement is required at the rat of 0.1% of the cross-sectional area.

$$\text{Area of Side reinforcement} = \frac{0.1 \times 500 \times 950}{100} = 475 \text{ mm}^2$$

Provide 4 bars of 10 mm ϕ + 2 bars of 12 mm ϕ

Area provided = 540 mm^2

Check for Maximum Reinforcement

Maximum tension reinforcement provided = 2200 mm^2

Allowable maximum tension reinforcement = $0.04 \times b \times D$

$= 0.04 \times 500 \times 950 = 19000 \text{ mm}^2 > 2200 \text{ mm}^2$

∴ *O.K*

Check of Minimum Distance between Main Bars: Horizontal distance between two main parallel bars is not to be less than greatest of the following:

(*i*) Diameter of main bar = 16 mm

(*ii*) 5 mm more than nominal maximum size of coarse aggregate = 5 + 20 = 25 mm

∴ Minimum clear distance required between two main bars = 25 mm

Clear distance provided = 34.5 mm as shown in Fig. 5.7 (*b*)

Check for Deflection

$$\text{Actual span/depth ratio} = \frac{0.7 \times 9000}{900} = 7.0$$

$$p = \frac{1800 \times 100}{500 \times 900} = 0.40 \text{ near centre of end span.}$$

Modification factor, K_1 = 1.35 from Fig. 5.1 where $f_s = 0.58 \times 415 \times \frac{1856}{2211} = 203 \text{ N/mm}^2$

∴ Allowable span/depth ratio = $1.35 \times 26 = 35.1 > 7.0$

∴ *O.K*

Check for Lateral Stability : For continuous beams, clear distance between lateral supports should not exceed

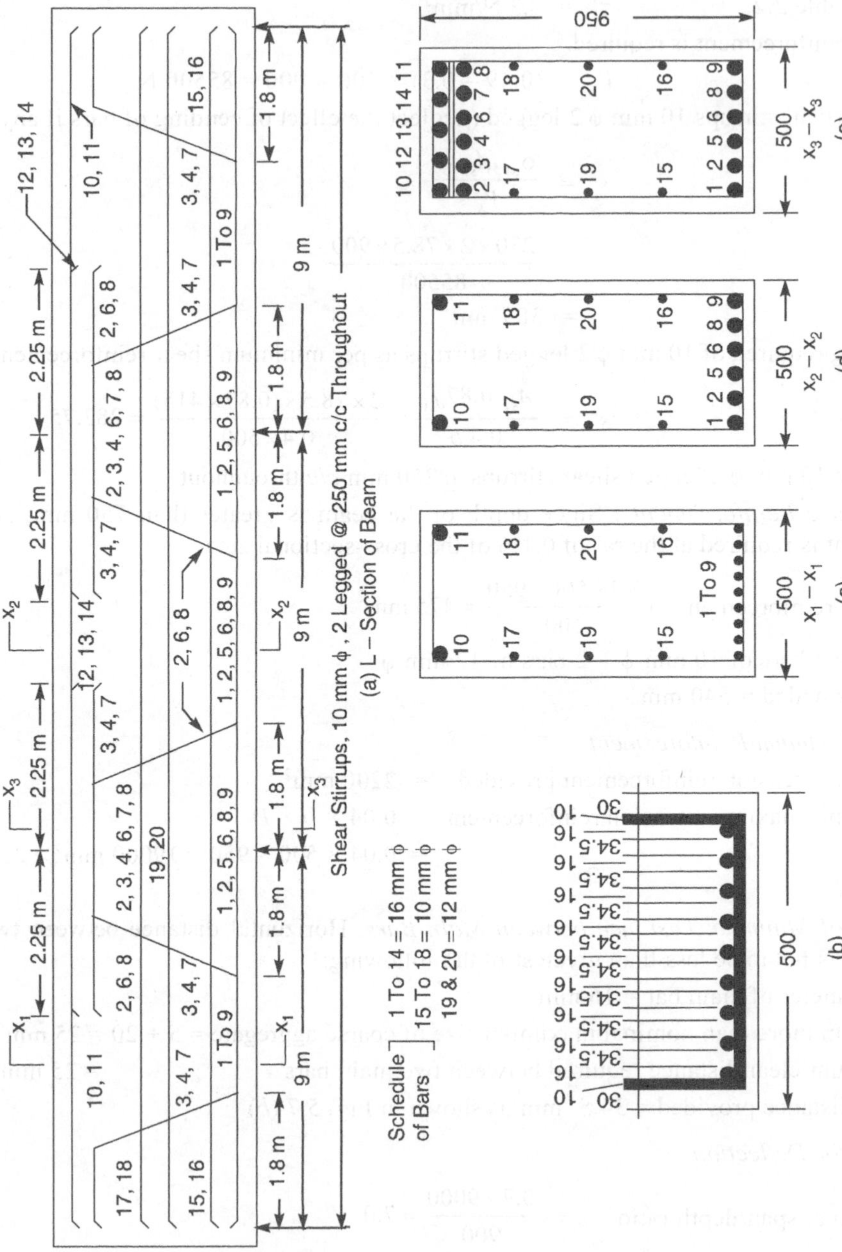

Fig. 5.7.

(*i*) $60\ b = 60 \times 0.5 = 30$ m

(*ii*) $\dfrac{250\,b^2}{d} = \dfrac{250 \times 0.5^2}{0.9} = 69.4$ whichever is less, i.e., 30 m

Since the total length of the beam for the 3–spans is 27m, < 30 m, lateral restraint should be provided at two extreme ends of the beam.

Check for Anchorage and Development Length: In continuous beams, the points of inflection are assumed to be located at 0.15 times the effective span from the centre line of each interior support, which in the present case shall be located at $0.15 \times 9 = 1.35$ m from the centre line of each interior support.

In continuous beams, one-fourth the positive moment reinforcement shall extend along the same face, *i.e.*, 9/4 = 2.25, say 3 nos. Should extend into the support. At points of inflection, positive moment tension reinforcement shall be limited to a diameter such that:

$$L_d \not> \frac{M_1}{V} + L_0$$

$$L_d = \frac{\phi.\sigma_s}{4\tau_{bd}} = \frac{16 \times 230}{4 \times 1.28} = 718 \text{ mm}$$

$$M_1 = 3 \times 201 \times 230 \times 0.90 \times 900 = 112338900 \text{ N.mm}$$

$$V = 223425 \text{ N}$$

$$L_0 = 12 \times 16 \text{ or } 900 \text{ mm whichever is greater}$$

$$= 900 \text{ mm}$$

$$\therefore \quad \frac{112338900}{223425} + 900 = 1402 \text{ mm} > 718 \text{ mm} \quad \therefore \textit{O.K.}$$

At least one-third of the total reinforcement provided for negative moment at support, i.e., 11/3 = 3.67, say 4 nos; shall extend beyond the point of inflection for a distance not less than (*i*) $d = 900$ mm (*ii*) $12\,\phi = 12 \times 16 = 192$ mm or (*iii*) clear span/16 = 9000/16 = 562.5 mm whichever is greater, i.e., 900 mm. Therefore provide 4 bars (2 hanger bars running throughout 2 extra pieces of 16 mm ϕ) upto a distance of $1.35 + 0.9 = 2.25$ m on either side of centre line of interior supports. Reinforcement details are shown in Fig. 5.7.

Example 5.4. Design a rectangular R.C. beam simply supporting over a span of 6 m (effective) and clear span of 5.25m. The superimposed load is 40,000 N/m (4000 Kg/m). The size of the beam is restricted to 300 mm wide × 800 mm deep (overall). Use M 20 concrete and HYSD (Fe 415) bars.

Sol.

Self weight of beam = 0.3 × 0.8 × 1 × 25000	=	6000 N/m
Superimposed load	=	40,000 N/m
Total *u.d.l.*	=	46,000 N/m

Maximum *B.M.* '*M*' $= \frac{46000 \times 6^2}{8} = 207000$ N.m., i.e., 2.07×10^8 N.mm

Assume effective depth as 750 mm

The moment of resistance of balanced section if reinforced on tension side 'M_1' $= 0.91 \times 300 \times (750)^2 = 1.5356 \times 10^8$ N.mm $< 2.07 \times 10^8$ N mm

$\therefore$ Unbalanced moment $M_2 = M - M_1$

Areas of steel required $= 2.07 \times 10^8 - 1.5356 \times 10^8 = 0.5344 \times 10^8$ N.mm

$$A_{st1} = \frac{M_1}{\sigma_{st} \cdot j.d}$$

$$= \frac{1.5356 \times 10^8}{230 \times 0.90 \times 750} = 990 \text{ mm}^2$$

$$x_c = 0.4 \times 750 = 300 \text{ mm}$$

$$A_{st_2} = \frac{M_2}{\sigma_{st}(d-d')}$$

$$= \frac{0.5344\times10^8}{230\times(750-50)} = 332 \text{ mm}^2$$

Total A_{ct}, i.e., steel on tension face $= A_{st1} + A_{st2}$

$$= 990 + 332 = 1322 \text{ mm}^2$$

Provide 16 mm φ 8 Nos.

$$M_2 = (1.5\,m - 1)\ A_{sc}.\frac{x_c - d'}{x_c}(d-d')$$

$$A_{sc} = \frac{0.5344\times10^8\times300}{(1.5\times13-1)\times7\times(300-50)\times(750-50)}$$

$$= 707 \text{ mm}^2$$

Provide 16 mm φ 4 Nos. on compression face.

Check for Shear

$$\text{Maximum Shear force} = \frac{46000\times5.25}{2}$$

$$= 120750 \text{ N}$$

$$\text{Nominal shear stress } \tau_v = \frac{120750}{300\times750} = 0.5367 \text{ N/mm}^2$$

$$\frac{A_s\times100}{b.d} = \frac{201\times8\times100}{300\times750} = 0.715$$

From Table 2.2, $\tau_c = 0.34 \text{ N/mm}^2$

$$V_s = (0.5367 - 0.34)\times 300\times 750 = 44257 \text{ N}$$

Providing 8 mm φ 2 legged shear stirrups @ $\frac{230\times2\times50\times750}{44257} = 390$ mm less than 0.75 × 750 = 562.5 mm or 450 mm. Spacing of 8 mm φ 2 legged bars required for minimum shear reinforcement

$$= \frac{2\times50\times0.87\times415}{0.4\times300} = 300 \text{ mm}$$

∴ Provide 8 mm φ 2 legged Shear stirrups @ 200 mm *c/c* in 1.5 m from each support and thereafter @ 400 mm *c/c*.

Check for Side Face Reinforcement : Since the depth of the beam is greater than 750 mm, side face reinforcement @ 0.1 per cent of cross-sectional area of beam, is required.

$$A_s = \frac{0.1\times300\times800}{100} = 240 \text{ mm}^2$$

Provide 10 mm φ 4 Nos.

Check for Deflection

$$\text{Span/Effective depth ratio} = \frac{6000}{750} = 8$$

p = 0.715 (tension reinforcement) Modification factor from Fig. 5.1 K_1 = 1.1. For compression reinforcement

$p = \frac{4\times200\times100}{300\times750} = 0.355$ Modification factor from Fig. 5.2, K_2 = 1.1.

Permissible Span/Effective depth ration = 1.1 × 1.1 × 20 = 24.2 > 8

Hence the deflection shall be within permissible limits.

Check for End Anchorage and Development length

$$L_d \ngtr \frac{M_1}{V} + L_0$$

$$L_d = \frac{\phi.\sigma_s}{4.\tau_{bd}} = \frac{16 \times 230}{4.0 \times 1.28} = 718 \text{ mm}$$

Assuming 4 bars are bent up.

$$M_1 = 4 \times 201 \times 230 \times 0.90 \times 750 = 124821000$$

$$V = 120750 \text{ N}, L_0 = 12\ \phi \text{ or } d \text{ whichever is greater, i.e., } 750 \text{ mm}$$

$$\frac{M_1}{V} + L_0 = 1783 \text{ mm} \therefore \text{ The condition for anchorage is satisfied.}$$

Hence extend 4 Nos. 16 mm ϕ bars beyond centre line of support by L_d = 718 mm.

The reinforcement details are shown in Fig. 5.8.

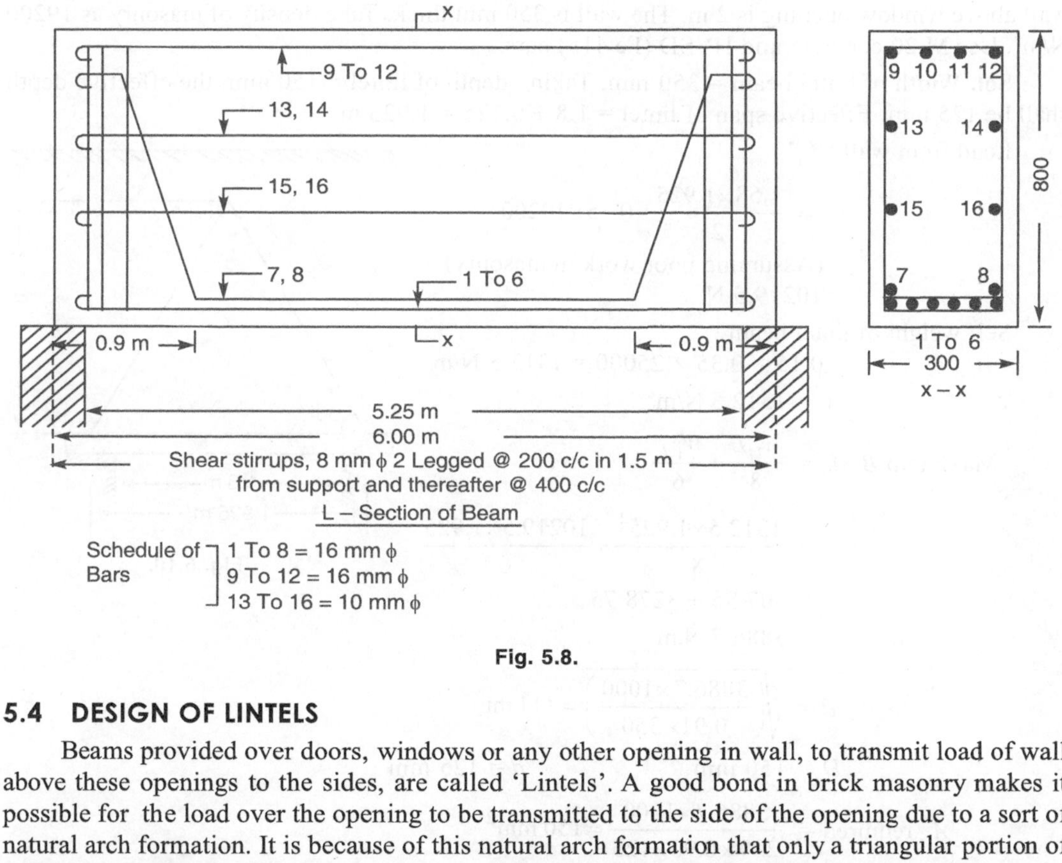

Fig. 5.8.

5.4 DESIGN OF LINTELS

Beams provided over doors, windows or any other opening in wall, to transmit load of wall above these openings to the sides, are called 'Lintels'. A good bond in brick masonry makes it possible for the load over the opening to be transmitted to the side of the opening due to a sort of natural arch formation. It is because of this natural arch formation that only a triangular portion of the masonry loads the lintel beam. The size of the triangle depends upon the nature of the bonding and type of the masonry. Normally, the base angles of this triangle are taken as 45° for good masonry work but in poor work, the same may be taken as 60° as shown in Fig. 5.9. If the height of the wall above lintel is not sufficient to form triangle as explained above, full rectangular load over the opening may be considered. In case a slab comes within the rectangular area, the rectangular load of masonry above opening, load from slab plus triangular portion of masonry above slab over opening span is to be taken into account.

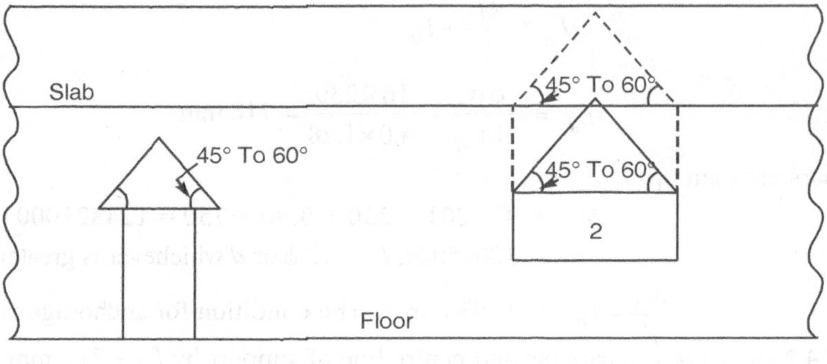

Fig. 5.9.

Example 5.5. A lintel beam is to be provided over a window 1.8 m wide. The height of the wall above window opening is 2 m. The wall is 350 mm thick. Take density of masonry as 19200 N/m^3. Use M 20 concrete and HYSD (Fe 415) bars.

Sol. Width of lintel beam = 350 mm. Taking depth of lintel = 150 mm, the effective depth shall be 125 mm. Effective span of lintel = 1.8 + 0.125 = 1.925 m

Load from wall 'W_1'

$$= \frac{1.58 \times 1.925}{2} \times 0.35 \times 19200$$

(Assuming poor work in masonry)

$$= 10219.5 \text{ N}$$

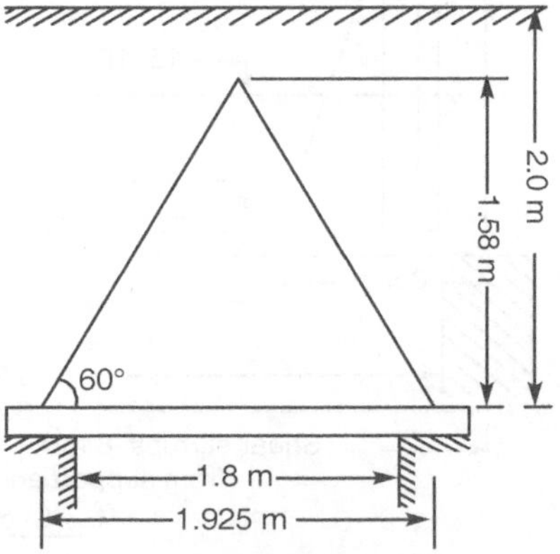

Fig. 5.10.

Self weight of lintel beam

$$= 0.15 \times 0.35 \times 25000 = 1312.5 \text{ N/m}$$

$$\therefore \quad w = 1312.5 \text{ N/m}$$

$$\text{Maximum } B.M. = \frac{w_1.L}{8} + \frac{W_{1.L}}{6}$$

$$= \frac{1312.5 \times 1.925^2}{8} + \frac{10219.5 \times 1.925}{6}$$

$$= 607.95 + 3278.75$$

$$= 3886.7 \text{ N.m}$$

$$d = \sqrt{\left(\frac{3886.7 \times 1000}{0.91 \times 350}\right)} = 111 \text{ mm}$$

$D = 150$ mm $\qquad d = 125$ mm

$$A_{st} \text{ required} = \frac{3886.7 \times 1000}{230 \times 0.90 \times 125} = 150 \text{ mm}^2$$

Provide 4 Nos. 8 mm ϕ at bottom and 2 Nos. 8 mm ϕ at top as hanger bars.

Check for Shear

$$\text{Maximum } S.F. = \frac{10219.5}{2} + \frac{1312.5 \times 1.8}{2}$$
$$= 6291 \text{ N}$$

$$\text{Nominal shear stress} = \tau_v = \frac{6291}{350 \times 125} = 0.1.43 \text{ N/mm}^2$$

$$\frac{A_s \times 100}{b.d} = \frac{50 \times 4 \times 100}{350 \times 125} = 0.44 \quad \therefore \quad \tau_c = 0.28 \text{ N/mm}^2$$

Since the nominal shear stress is half the permissible value, no shear reinforcement in the form of stirrups is required. However, from practical point of view 8 mm ϕ 2 legged stirrups @ 250 mm *c/c* may be provided; also bend 2 Nos. bottom bars out of 4 Nos. 8 mm ϕ at a distance span/5 from end supports for taking care of negative moment due to partial fixity. The arrangement of reinforcement is shown in Fig. 5.11

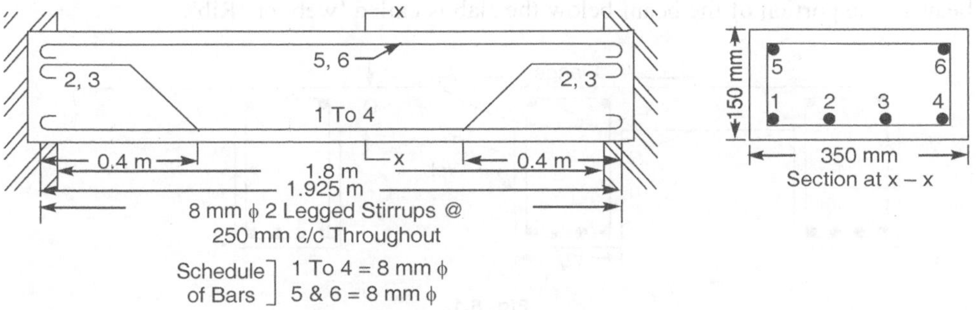

Fig. 5.11.

EXERCISE 5

1. A lintel beam is to be provide over a window opening of 1.5 m. The height of opening is 2.06 m and the bottom of the roof slab is 3.2 m. Design the lintel beam using M20 concrete and Tor steel. Take wall thickness as 225 mm. (Fe 415)
2. Design a lintel beam over a door 1.5 m wide at a height of 2.06 m. The slab height is 3.2 m. Use M20 concrete and HYSD (Fe 415). Take wall thickness as 225 mm.
3. Design a R.C. rectangular beam having a simply supported span of 6 m and carrying a superimposed load of 30 KN/m (3000 Kg/m) including self weight of the beam. Take width of beam as 300 mm. Use M20 concrete and HYSD (Fe500) bars.
4. Design a R.C. rectangular beam 250 mm × 500 mm over an effective span of 5 m simply supported. The beam carries a superimposed load of 20 KN/m (2000 Kg/m) excluding self weight of the beam. Use M25 concrete and HYSD(Fe415) bars.
5. Design a 3-span continuous beam to carry a superimposed load of 25 KN/m excluding self weight of the beam. Each span is 6 m. The cross-section of the beam may be taken as 250 mm × 750 mm. Use M25 concrete and HYSD (Fe500) bars.
6. Design a cantilever beam of 2.5 m span and carrying a load of 20 KN/m including self weight of the beam. The size of the beam is 300 mm × 600 mm. Use M20 concrete and Tor steel. (Fe415).

CHAPTER

6

Flanged Beams

6.1 INTRODUCTION

Flanged beam sections include. T-beams and L-beams. In practice, the slabs and beams are cast monolithically making no demarkation between the beams and slabs. Consequently the beams and slabs are so closely tied that when the beam deflects under applied loads it drags along with it a portion of the slab also as shown in Fig. 6.1 by dotted lines. This portion of the slab assists in resisting the effects of the loads and is called the 'Flange' of the T-beams. For design of such beams, the profile is similar to a T-section for intermediate beams and an inverted L-section for end beams. The portion of the beam below the slab is called 'web' or 'Rib'.

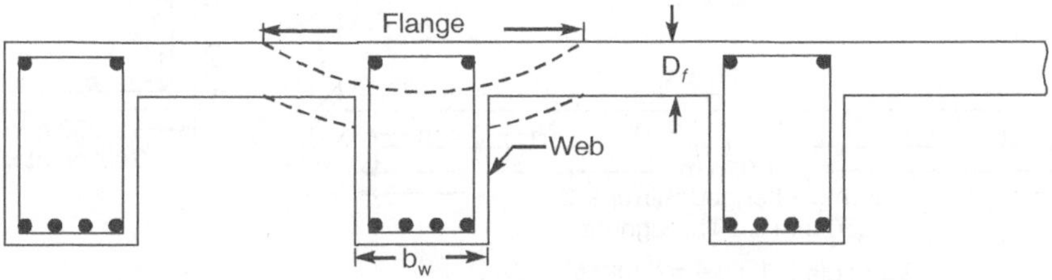

Fig. 6.1.

A slab which is assumed to act as flange of a T-beam or L-beam shall satisfy the following conditions:

(a) the slab shall be cast intergrally with the web or the web and the slab shall be effectively bonded together in any other manner; and

(b) if the main reinforcement of the slab is parallel to the beam, transverse reinforcement shall be provided as shown in Fig. 6.2, such reinforcement shall not be less than 60% of the main reinforcement at mid-span of the slab.

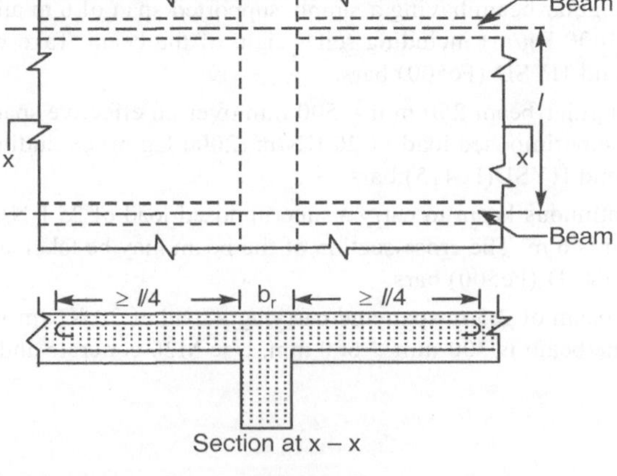

Fig. 6.2.

6.2 EFFECTIVE WIDTH OF FLANGE

In the absence of more accurate determination, the effective width of the flange may be taken as the following but in no case greater than the breadth of web plus half the sum of clear distances to the adjacent beams on either side.

(a) For T-beams, $b_f = \frac{l_0}{6} + b_w + 6.D_f$

(b) For L-beams, $b_f = \frac{l_0}{12} + b_w + 3.D_f$

(c) For Isolated beams, the effective flange width shall be obtained as below but in no case greater than the actual width.

T-beam, $b_f = \frac{l_0}{l_0 / b + 4} + b_w$

L-beam, $b_f = \frac{0.5\, l_0}{l_0 / b + 4} + b_w$

where b_f = effective width of flange

l_0 = distance between points of zero moment in beam

bw = breadth of web

D_f = thickness of flange

and b = actual width of flange

For continuous beams and frames, l_0 may be assumed as 0.7 times the effective span.

Fig. 6.3 explains the effective flange widths in L-beams and T-beams when the beams are cast monolithically with the slab.

To illustrate the procedure to compute effective flange width for an isolated beam section, take a specific example of a simply supported beam having span l_0 = 6m as shown in Fig. 6.4 (a) The cross section of this beam is either of the following:

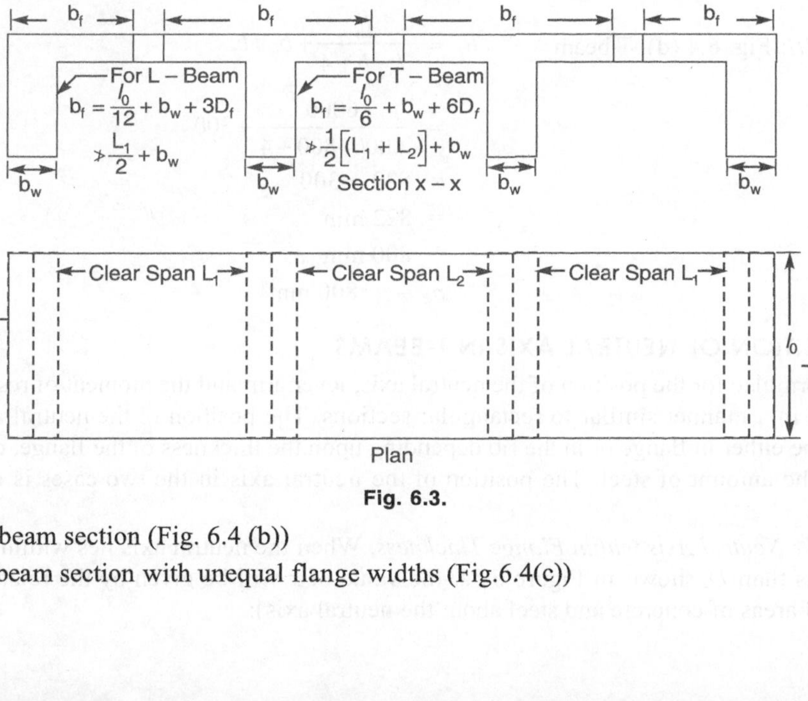

Fig. 6.3.

(*i*) L-beam section (Fig. 6.4 (b))

(*ii*) T-beam section with unequal flange widths (Fig.6.4(c)) (*iii*) T-beam section with equal flange widths (Fig. 6.4(d)).

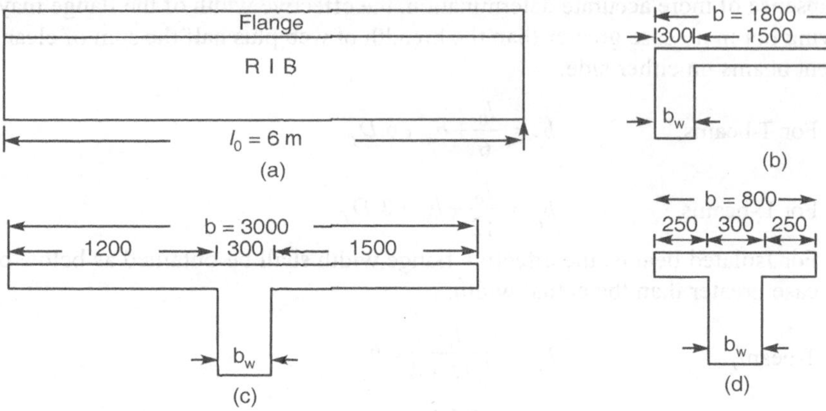

Fig. 6.4.

The effective flange width in the beam shown in Fig. 6.4(a) shall be:

Case I: Fig. 6.4 (b), L-beam, $b_f = \dfrac{0.5\, l_0}{l_0/b+4} + b_w \ngtr b$

$$= \frac{0.5 \times 6000}{6000/1800+4} + 300$$

$$= 710 \text{ mm} \ngtr 1800 \text{ mm}$$

Case II: Fig. 6.4 (c), T-beam, $b_f = \dfrac{l_0}{l_0/b+4} + b_w \ngtr b$

$$= \frac{6000}{6000/3000+4} + 300$$

$$= 1300 \text{ mm} \ngtr 3000 \text{ mm}$$

Case III: Fig. 6.4 (d), T-beam $b_f = \dfrac{l_0}{l_0/b+4} + b_w \ngtr b$

$$= \frac{6000}{6000/800+4} + 300$$

$$= 522 + 300$$

$$= 822 \text{ mm}$$

$$> 800 \text{ mm}$$

$\therefore$ $b_f = 800$ mm

6.3 POSITION OF NEUTRAL AXIS IN T-BEAMS

The formulae for the position of the neutral axis, lever arm and the moment of resistance can be obtained in a manner similar to rectangular sections. The position of the neutral axis of a T-beam may be either in flange or in the rib depending upon the thickness of the flange, depth of the beam and the amount of steel. The position of the neutral axis in the two cases is obtained as follows :

Case I : *Neutral Axis within Flange Thickness*: When the neutral axis lies within the flange, i.e., x is less than D_f shown in Figure 6.5a, the value of x will be given by the relation (taking moments of areas of concrete and steel about the neutral axis):

$$\frac{b.x^2}{2} = m.A_s(d-x) \qquad \text{... (6.1)}$$

Case II: Neutral Axis in Rib. (Fig. 6.5 (b))

(a) *When compression in Rib is taken into account* :

Taking the moments of effective areas of concrete and steel about neutral axis, the position of the neutral axis is given by the following relation :

$$b_f.D_f\left(x-\frac{D_f}{2}\right)+b_w.\frac{(x-D_f)^2}{2}=m.\ As.\ (d-x) \qquad \text{... (6.2)}$$

(b) *When compression in Rib is neglected* :

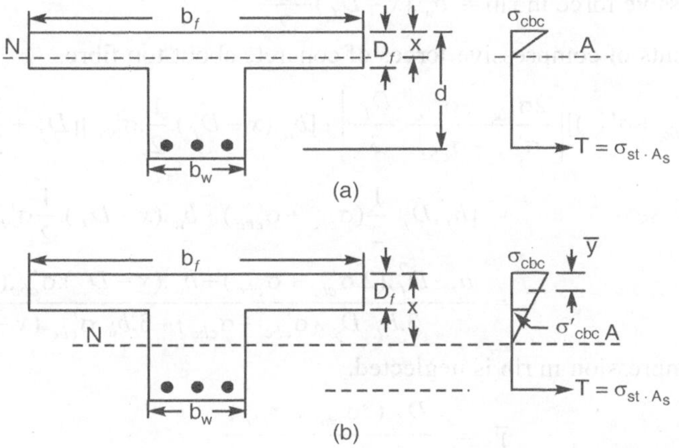

Fig. 6.5.

Taking the moments of effective areas of concrete and steel about neutral axis, the position of *N.A* is given by the following relation :

$$b_f.D_f\left(x-\frac{D_f}{2}\right) = m.\ As.\ (d-x) \qquad \text{... (6.3)}$$

6.4 MOMENT OF RESISTANCE OF T-BEAM AS BALANCED SECTION

Case I : *N.A. Within Flange Thickness*. Fig. 6.5 (*a*).

The lever arm '*z*' and moment of resistance are obtained in a similar manner as in the case of a rectangular beam an dare given below :

$$\text{lever arm} = z = d-\frac{x}{3} \qquad \text{... (6.4)}$$

$$M = b_f\ x.\frac{\sigma_{cbc}}{2}\left(d-\frac{x}{3}\right) \qquad \text{... (6.5)}$$

Since it is a balanced section, moment of resistance can also be obtained by :

$$M = A_{st}.\sigma_{st}\left(d-\frac{x}{3}\right) \qquad \text{... (6.6)}$$

Case II : *N. A. Within Rib, Fig. 6.5 (b)*: The lever arm for such a case can be obtained as follows:

Let σ_{abc} be the permissible compressive stress in bending in concrete at the top of flange, and σ'_{cbc} be the compressive stress at bottom of flange

$$\sigma'_{cbc} = \sigma_{cbc}\left(\frac{x-D_f}{x}\right) \qquad \text{... (6.7)}$$

Let $\bar{y}$ = the distance of *c. g.* of compressive force from the top of flange,

Determination of Lever Arm :

Total compressive force in flange = $b_f.\left(\frac{\sigma_{cbc}+\sigma'_{cbc}}{2}\right)$

Total compressive force in rib = $b_w.(x-D_f)\frac{\sigma'_{cbc}}{2}$

Taking moments of compressive forces of concrete about top fibre.

$$[b_f.D_f.\frac{1}{2}(\sigma_{cbc}+\sigma'_{cbc})]\left[\frac{2\sigma'_{cbc}+\sigma_{cbc}}{\sigma'_{cbc}+\sigma_{cbc}}.\frac{D_f}{3}\right]+[b_w.(x-D_f).\frac{1}{2}.\sigma'_{cbc}][D_f+\frac{1}{3}(x-D_f)]$$

$$= [b_f.D_f.\frac{1}{2}(\sigma_{cbc}+\sigma'_{cbc})+b_w.(x-D_f).\frac{1}{2}\sigma'_{cbc}].\bar{y}$$

$$\bar{y} = \frac{b_f\cdot D_f^2.(2.\sigma'_{cbc}+\sigma_{cbc})+b_w.(x-D_f).\sigma'_{cbc}(x-2.D_f)}{3.b_f.D_f.(\sigma'_{cbc}+\sigma_{cbc})+3.b_w.\sigma'_{cbc}.(x-D_f)} \qquad \text{... (6.8)}$$

When the compression in rib is neglected.

$$\bar{y} = \frac{D_f.(2\sigma'_{cbc}+\sigma_{cbc})}{3(\sigma'_{cbc}+\sigma_{cbc})} \qquad \text{... (6.9)}$$

Substituting the value of σ'_{cbc} in terms of σ_{cbc} from Eq. (6.7) in Eq. (6.9)

$$\bar{y} = \frac{D_f}{3}\left[\frac{3.x-2\,D_f}{2.x-D_f}\right] \qquad \text{... (6.10)}$$

$$\text{Lever arm 'z'} = d-\bar{y} \qquad \text{... (6.11)}$$

Substituting $\bar{y}$ from Eq. (6.10) in Eq. (6.11)

$$z = d-\frac{D_f}{3}\left[\frac{3x-2.D_f}{2.x-D_f}\right] \qquad \text{...(6.12)}$$

Moment of resistance can be found by multiplying total compressive force or total tensile force by the lever arm. The moment of resistance neglecting the contribution of rib is given be the following relation.

$$M = b_f.D_f.\frac{\sigma_{cbc}+\sigma'_{cbc}}{2}\left[d-\frac{D_f}{3}\left(\frac{3.x-2.D_f}{2.x-D_f}\right)\right]$$

$$= b_f.D_f.\sigma_{cbc}\left(\frac{2.x-D_f}{2.x}\right)\left[d-\frac{D_f}{3}\left(\frac{3.x-2.D_f}{2.x-D_f}\right)\right] \qquad \text{... (6.13)}$$

6.5 APPROXIMATE ANALYSIS

In case an approximate analysis of moment of resistance of a T-beam is to be made, the stress in flange thickness may be assumed to be uniform, i.e... $\sigma'_{cbc} = \sigma_{cbc}$ and also compressive stress in the rib is neglected.

$$\text{Total compressive force} = b_f.D_f.\sigma_{cbc}$$

$$\text{Lever arm, } z = d - \frac{D_f}{2}$$

$$M = b_f.D_f.\sigma_{cbc}\left(d - \frac{D_f}{2}\right) \quad ...(6.14)$$

6.6 SHEAR, END ACHORAGE AND CONTROL OF DEFLECTION IN T-BEAMS

The shear reinforcement in the case of T-beams is to be provided in exactly the same manner as in the case of rectangular beams taking the area of the rib as $b_w.\ d$.

The proper anchorage length may be provided as discussed in the section 4.10.

The deflection shall be controlled by checking span to effective depth ratio as mentioned in section 5.2 (4).

6.7 TYPES OF PROBLEMS

Type I : Determine Stresses in Concrete and Steel.

Given – the dimensions of the section, the applied bending moment *M* and the amount of tension steel.

Steps.

(*i*) Determine the position of *N. A.*

(*ii*) Equate moment of resistance of compressive forces with actual bending moment, and obtain maximum compressive stress in concrete (σ''_{cbc})

(*iii*) Obtain actual stress (σ'_{st}) in steel either by

$$\sigma'_{st} = m.\sigma''_{cbc}\left(\frac{d-x}{x}\right)$$

$$\text{or } \sigma'_{st} = \frac{M}{A_{st}.z}$$

Type II : Moment of Resistance of a Particular Section.

Given – the dimensions of the section, amount of tension steel and maximum permissible stresses.

Steps.

(*i*) Obtain position of actual *N. A.* "x_a"

(*ii*) Determine the position of critical or balanced *N.A.* from

$$x_c = K.d = \frac{1}{1+\dfrac{\sigma_{st}}{\sigma_{cbc}.m}}.d$$

(*iii*) Obtain moment of resistance by the following formula if 'x_a' is greater than 'x_c' (the section is over-reinforced)

(a) if *N. A.* lies within the flange thickness

$$M = b_f.x_a.\frac{1}{2}.\sigma_{cbc}\left(d - \frac{x_a}{3}\right)$$

(b) if *N. A.* lies in rib

$$M = b_f.D_f.\sigma_{cbc}.\left(\frac{2.x_a - D_f}{2.x_a}\right)\left[d - \frac{D_f}{3}\frac{3.x_a - 2.D_f}{2.x_a - D_f}\right]$$

(*iv*) Obtain moment of resistance by the following formula if x_a is less than x_c (the section is under reinforced)

$$M = A_{st}.\sigma_{st}.\left(d - \frac{x_a}{3}\right)$$

Type III : Design of T-section

Given – Mix, thickness of slab, *B.M.* and *S.F.*

Steps.

(*i*) Determine width of flange b_f.

(*ii*) Fix-up a suitable depth arbitrarily.

(*iii*) Calculate moment of resistance of the section assuming the failure of concrete to occur. If the moment of resistance is greater than *B.M.*, assumed depth is O.K., otherwise either increase the depth of the beam or reinforce section in compression side also in addition to tension side.

(*iv*) Calculate the amount of tension steel for actual *B.M.* treating it as a balanced section.

(*v*) Check for shear and provide shear reinforcement accordingly.

(*vi*) Check for deflection.

(*vii*) Check for end anchorage and development length.

6.8 SOLVED EXAMPLES

Example 6.1. A reinforced concrete beam Section shown in Fig. 6.6 is subjected to a bending moment of 3 × 10[5] N.m (3 × 10[4] Kg.m). Calculate the maximum stresses developed in concrete and steel if the concrete is of M20 grade.

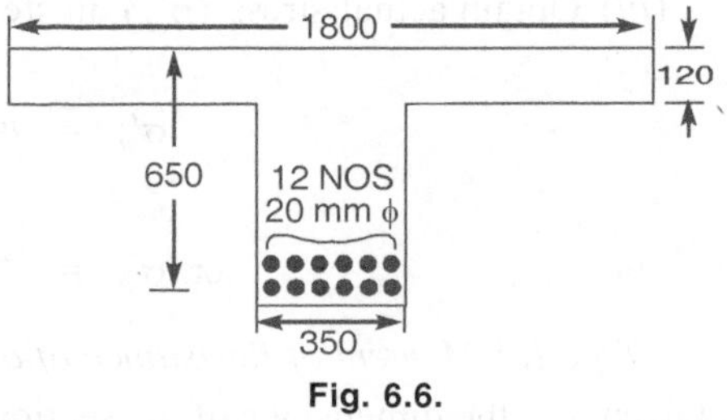

Fig. 6.6.

Sol. For M20 concrete, $\sigma_{cbc} = 7\text{ N/mm}^2$

$$m = \frac{280}{3.\sigma_{cbc}} = \frac{280}{3 \times 7} = 13.33$$

Let x be the depth of Neutral axis, assumed to lie in rib, from top concrete fibre. Take moments of compressive and tensile areas, neglecting the compression in web, about *N. A.*

$$1800 \times 120 \times \left(x - \frac{120}{2}\right) = \frac{280}{3 \times 6} 12 \times 314 \times (650 - x)$$

$$x = 171.32 \text{ mm}$$

$$> 120.00 \text{ mm}$$

Neutral axis lies in rib.

Let σ_{cbc} be actual compressive stress in the concrete at top fibre.

Then stress at bottom of flange

$$= \sigma'_{cbc} = \sigma_{cbc}\cdot\left(\frac{x - D_f}{x}\right)$$

$$= \sigma_{cbc}\left(\frac{x-120}{x}\right)$$

Using Eqn. 6.13

$$M = b.D_f.\sigma_{cbc}\cdot\left(\frac{2x - D_f}{2x}\right).\left\{d - \frac{D_f}{3}\left(\frac{3x-2D_f}{2x-D_f}\right)\right\}$$

$$3 \times 10^5 \times 1000 = 1800\times120\times\sigma_{cbc}\left(\frac{2\times171.32-120}{2\times171.32}\right)$$

$$\left\{650 - \frac{120}{3}\left(\frac{3\times171.32-2\times120}{2\times171.32-120}\right)\right\}$$

$$= 1388.89 = 390.57\ \sigma_{cbc}\ \text{N/mm}^2$$

$$\therefore \quad \sigma_{cbc} = 3.5566\ \text{N/mm}^2$$

Stress in steel can be found by using the equation of neutral axis.

$$x = \frac{1}{1+\dfrac{\sigma_{st}}{m.\sigma_{cbc}}}.d$$

$$\therefore \quad \sigma_{st} = m.\sigma_{cbc}\left(\frac{d-x}{x}\right)$$

$$= 13.33\times3.5566\left(\frac{650-171.32}{171.32}\right)$$

$$= 132.495\ \text{N/mm}^2$$

Example 6.2 Determine the moment of resistance of a T-beam section reinforced as shown in Fig. 6.7. Use M20 Concrete. Also determine the same if the section (300 × 400) is to be treated as rectangular.

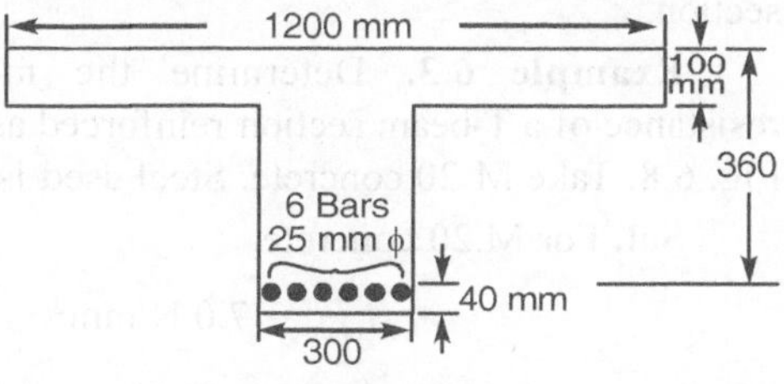

Fig. 6.7.

Sol. For M20 concrete

$$\sigma_{cbc} = 7\ \text{N/mm}^2$$

$$m = \frac{280}{3\times7} = 13.33$$

For HYSD steel, $\sigma_{st} = 230$ N/mm²

(*i*) *Moment of Resistance as T-section*

Let x = the position of neutral axis assumed to be in rib.

To determine the position of the neutral axis, take moments of compressive and tensile areas about *N.A.* (assuming no contribution from rib)

$$1200\times100\left(x - \frac{100}{2}\right) = 13.33 \times 6 \times 491 \times (360 - x)$$

$$x = 126.427\ \text{mm} > \text{Flange thickness } 100\ \text{mm}$$

Depth of critical neutral axis $x_c = 0.29\ d$

$= 0.29 \times 360$

$= 104.4$ mm

x is greater than x_c

The section is over-reinforced. Concrete reaches its maximum permissible stress earlier.

Moment of resistance $$M = \frac{b_f.D_f.\sigma_{cbc}}{2}\left(1+\frac{x-D_f}{x}\right)\left\{\frac{D_f}{3}\left(\frac{3.x-2\ D_f}{2.x-D_f}\right)\right\}$$

$$= \frac{1200\times100\times7}{2}\left(1+\frac{147.44-100}{126.427}\right)\left\{360-\frac{100}{3}\left(\frac{3\times147.44-2\times100}{2\times126.427-100}\right)\right\}$$

$= 162955680$ N.mm $= 162{,}955.680$ N.m

(*ii*) *Moment of Resistance as Rectangular Section*

$$\frac{b.x_a^{\ 2}}{2} = m.\ A_s\ (d - x_a)$$

$$300\times\frac{1}{2}.x_a^{\ 2} = 13.33 \times 6 \times 491\ (360 - x_a), \quad \therefore x_a = 202.84 \text{ mm}$$

$$x_c = 0.29\ d = 0.29 \times 360 = 104.4 \text{ mm}$$

$$x_a > x_c$$

The section is over-reinforced.

Moment of resistance $$= \frac{b.x_a.\sigma_{cbc}}{2}\left(d-\frac{x_a}{3}\right)$$

$$= \frac{300\times202.84\times7}{2}\left(360-\frac{202.84}{3}\right)$$

$= 62273095$ N. mm

$= 62273.095$

$\therefore$ Moment of resistance of T-section is 2.617 times that of corresponding rectangular section.

Example 6.3. Determine the moment of resistance of a T-beam section reinforced as shown in Fig. 6.8. Take M 20 concrete. Steel used is tor steel.

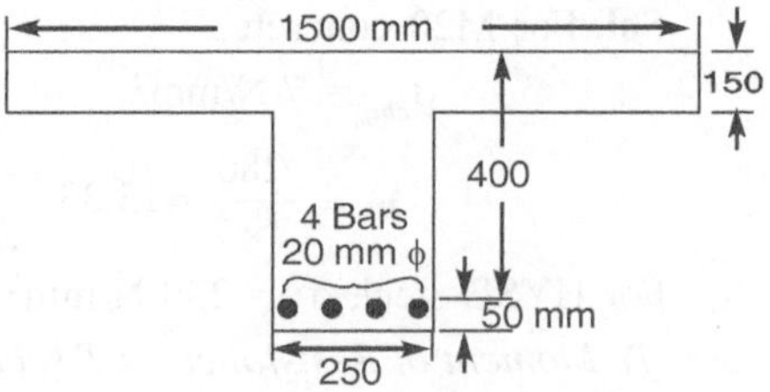

Fig. 6.8.

Sol. For M.20 concrete,

$$\sigma_{cbc} = 7.0 \text{ N/mm}^2$$

$$m = \frac{280}{3.\sigma_{cbc}} = 13.33$$

For deformed bars,

$$\sigma_{st} = 230 \text{ N/mm}^2$$

Let x be the position of *N.A.* assumed to lie in the rib.

To determine its position, take moments about *N.A.*

$$1500\times150\left(x-\frac{150}{2}\right) = 13.33 \times (4 \times 314)\ (400 - x)$$

$$x = 97.51 \text{ mm} \qquad \text{less than 150 mm}$$

∴ Neutral axis lies within the flange thickness.

$$\frac{1500\,x^2}{2} = 13.33 \times 4 \times 314\,(400 - x)$$

$$x = 83.99 \text{ mm}$$

Depth of critical $N.A.$ $= 0.289\,d$

$$x_c = 0.289 \times 400 = 115.6 \text{ mm}$$

$$x \text{ is less than } x_c$$

∴ The section is under-reinforced, steel reaches the maximum permissible stress earlier.

Lever arm $= z = d - \dfrac{x}{3}$

$$= 400 - \frac{83.99}{3} = 372 \text{ mm}$$

$$M = A_{st} \cdot \sigma_{st} \cdot z$$

$$= 4 \times 314 \times 230 \times 372 = 1.0746 \times 10^8 \text{ N.mm, } i.e., 1.0746 \times 10^5 \text{ N.m}$$

Example 6.4. Design a T-beam section in a hall 5 m × 15 m with the position of the beam as shown in Fig. 6.9. The thickness of the slab may be assumed as 125 mm. The concrete may be taken as M20 and reinforcement as HYSD of Fe415 bars. The superimposed live load on slab may be taken as 2000 N/m^2 (200 Kg/m^2) and the weight of finishes as 2500 N/m^2 (250 Kg/m^2) which includes screeding and flooring.

Sol. Assume the size of the beam as 225 mm wide and 325 mm hang from slab. The load on the intermediate T-beam shall be from an area as shown shaded in Fig. 6.9.

Self weight of the beam $= 0.325 \times 0.225 \times 1 \times 2500 = 1828.1$ N/m

Slab Load/m^2 $= 0.125 \times 1 \times 1 \times 25000 + 2000 + 2500 = 7625$ N/m^2

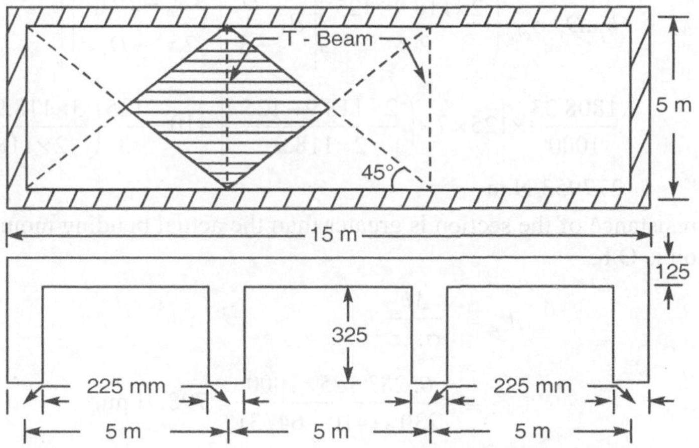

Fig. 6.9.

Total load from slab on beam $= 2 \times \dfrac{5 \times 2.5}{2} \times 7625 = 95312.5$ N

Equivalent slab load as $u.d.l.$ on beam $= \dfrac{95312.5}{5} = 19062.5$ N/m

Total load from slab and self weight of beam $= 19062.5 + 1828.1 = 20890.6$ N/m

$$\text{Max. } B.M. \text{ at mid-span} = \frac{20890.6 \times 5 \times 5}{8} = 65283.125 \text{ N.m}$$

Moment of resistance as T-beam (Fig. 6. 10)

Effective width of flange shall be the least of the following:

(*i*) $$b_f = \frac{l_0}{6} + b_w + 6\,D_f$$

$$= \frac{5000}{6} + 225 + 6 \times 125 = 1808.33 \text{ mm}$$

(*ii*) $$b_f = b_w + \frac{1}{2}(\text{clear span on either side})$$

$$= 225 + \frac{1}{2}(5000 - 225) \times 2 = 5000 \text{ mm}$$

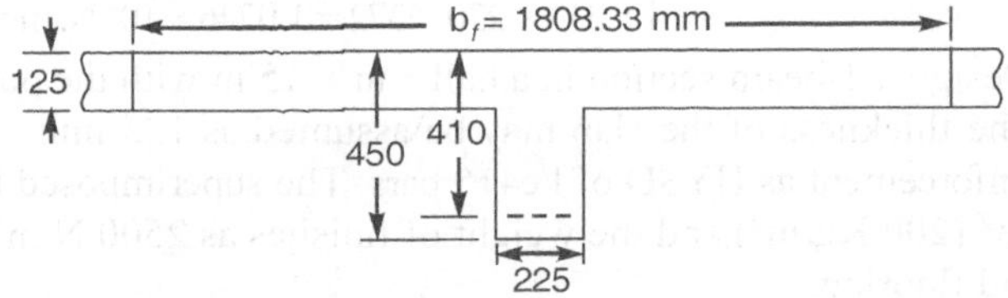

Fig. 6.10.

∴ Width of flange = 1808.33 mm

Assuming section to be balanced,

$$x_c = 0.29\,d = 0.29 \times 410 = 118.9 \text{ mm}$$

$$M = b_f . D_f . \sigma_{cbc}\left(\frac{2 \times 118.9 - 125}{2.x_c}\right)\left[d - \frac{D_f}{3}\left(\frac{3.x_c - 2.D_f}{2.x_c - D_f}\right)\right]$$

$$= \frac{1808.33}{1000} \times 125 \times 7 \times \left(\frac{2 \times 118.9 - 125}{2 \times 118.9}\right)\left[410 - \frac{125}{3}\left(\frac{3 \times 118.9 - 2 \times 125}{2 \times 118.9 - 125}\right)\right]$$

$$= 277953 \text{ N.m}$$

Moment of resistance of the section is greater than the actual bending moment. Therefore as *T*-beam, the section is O.K.

∴ $$A_{st} = \frac{M}{\sigma_{st} . z}$$

$$= \frac{65283.125 \times 1000}{230 \times (410 - 64/3)} = 798.91 \text{ mm}$$

Provide 3 Nos. 16 mm ϕ + 2 Nos. 12 mm ϕ as main steel in tension zone and use 12 mm ϕ 2 Nos. as hanger bars.

$$A_{st} \text{ provided} = 3 \times 201 + 2 \times 113 = 829 \text{ mm}^2$$

Check of Shear

$$\text{Max. } S.F. = V = \frac{20890.6 \times 5}{2} = 52226.5 \text{ N}$$

$$\frac{A_s \times 100}{b.d} = \frac{820 \times 100}{225 \times 410} = 1.021$$

$$\tau_c = 0.3924 \text{ N/mm}^2$$

$$\text{Shear strength} = 0.3924 \times 225 \times 410 = 36198.9$$

$$V_s = 52226.5 - 36198.9 = 16027.6 \text{ N}$$

Provide 8 mm ϕ 2 legged stirrups @

$$s_v = \frac{\sigma_{sv}.A_{sv}.d}{V_s} = \frac{230 \times 2 \times 50 \times 410}{16027.6} = 588.4$$

Max. spacing as per minimum shear reinforcement requirement

$$s_v = \frac{A_s.f_y \times 0.87}{0.4\,b} = \frac{2 \times 60 \times 415 \times 0.87}{0.4 \times 225} = 401.2$$

$$\text{Max. spacing} = 0.75d = 0.75 \times 410 = 307.5 \text{ mm}$$

Provide 8 mm ϕ 2 legged vertical stirrups @300 mm c/c as shown in Fig. 6.11.

Check for Deflection

$$\text{Actual} \quad \frac{span}{d} = \frac{5}{0.41} = 12.2$$

$$\frac{A_s \times 100}{b_f.d} = \frac{829 \times 100}{1808.33 \times 410} = 0.1120$$

$$\frac{b_w}{b_f} = \frac{225}{1808.33} = 0.124$$

Allowable span/d ratio for limiting deflection = 20. $K_1.K_2$

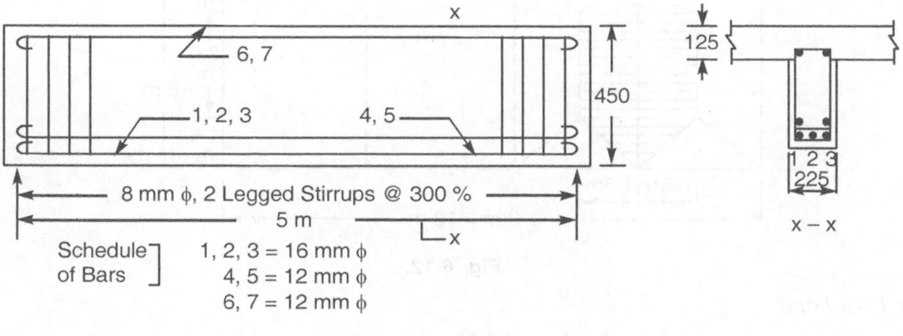

Fig. 6.11.

where K_1 = modification factor depending upon percentage tension steel and type of steel (Fig. 5.1)

and K_2 = reduction factor depending upon ratio of web width to flange width (Fig. 5.3) = 0.8

Allowable span/d ratio is 20 × 2.0 × 0.8 = 32 O.K.

Check for End Anchorage

$$L_d = \frac{M_1}{V} + L_0$$

where

$$L_d = \frac{\sigma_s.\phi}{4.\tau_{bd}} = \frac{20 \times 230}{4 \times 1.28} = 898.4 \text{ mm}$$

$$M_1 = 827 \times 230 \times 0.90 \times 410 = 79947540 \text{ N.mm}$$
$$V = 52226.5 \text{ N}$$

L_0 may be taken as $12\ \phi = 12 \times 20 = 240$ mm or 410, whichever is greater, i.e., 410 mm

$$\frac{M_1}{V} + L_0 = \frac{79947540}{52226.5} + 410 = 1940.8 \text{ greater than } 898.4 \text{ mm} \therefore \text{ O.K.}$$

End anchorage of $L_d = 898.4$ mm may be provided beyond centre of support.

Example 6.5. Design a *T*-beam section for a room 6 m × 12 m having three beams spanning 6 m at spacing of 4 m *c/c*. The width of rib is 225 mm and the hang of the beam is 200 mm. The slab is 100 mm thick cast monolithically with beam. Use M25 concrete and High Strength Steel deformed bars Fe415. Take the superimposed live load on slab as 4000 N/m² (400 Kg/m²) and finishes as 2500 N/m² (250 Kg/m²).

Sol. Area contributing load to an intermediate *T*-beam is shown shaded in Fig. 6.12 and is equal to $2 \times \frac{1}{2}(3+6) \times 1.5 = 13.5 \text{ m}^2$.

Self weight of the beam = 0.225 × 0.2 × 1 × 25000 = 1125 N/m

Loading

(*i*) *Dead Loads*

Self weight of slab = 0.1 × 1 × 1 × 25000 = 2500 N/m²

Finishes = 2500 N/m²

Total dead load on beam = (2500 + 2500) × 13.5 = 67500 N

Dead load/m $= \frac{67500}{6} = 11250$ N/m

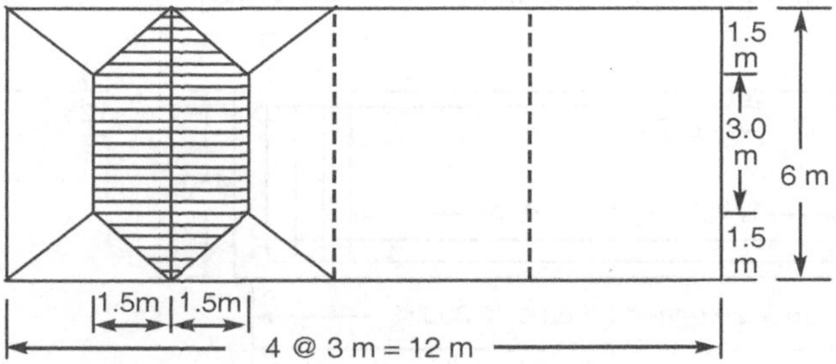

Fig. 6.12.

(*ii*) *Live Load*

Live load = 400 N/m²

Total Live load = 4000 × 13.5 = 54000 N

Live load per meter = 54000/6 = 9000 N/m

Total load/m of beam = 1125 + 11250 + 9000 = 21375 N/m

$$\text{Max. } B.M. = \frac{w.L^2}{8} = \frac{21375 \times 6^2}{8} = 96187.5 \text{ N.m}$$

Moment of Resistance of T-Beam as Balanced Section.

Effective width of flange shall be least of the following:

(*i*) $$b_f = \frac{l_0}{6} + b_w + 6\,D_f$$

$$= \frac{6000}{6} + 225 + 6 \times 100 = 1825 \text{ mm}$$

(*ii*) $$b_f = b_w + \frac{1}{2}(\text{clear span on either side})$$

$$= 225 + \frac{1}{2}(3000 - 225) \times 2 = 3000 \text{ mm}$$

$\therefore$ $$b_f = 1825 \text{ mm}$$

$$x_c = k.d = \frac{1}{1 + \frac{\sigma_{st}}{m.\sigma_{cbc}}}.d \quad \text{where } m = \frac{280}{3.\sigma_{cbc}} = \frac{280}{3 \times 8.5} = 11$$

$$= \frac{1}{1 + \frac{230}{11 \times 8.5}} d = 0.289\, d$$

$$x_c = 0.289 \times 250 = 72.25 \text{ mm}$$

$$M = b_f.x\frac{1}{2}\sigma_{cbc}.\left(\frac{d - x_c}{3}\right)$$

$$= 1825 \times 72.25 \times \frac{8.5}{2}\left(250 - \frac{72.25}{3}\right)$$

$$= 1.266 \times 10^8 \text{ N.mm}$$

$$> 96187.5 \times 10 \text{ N. mm}$$

$\therefore$ T-beam section is O.K.

$$A_{st} = \frac{M}{\sigma_{st}.z} = \frac{96187.5 \times 1000}{230\left(250 - \frac{72.25}{3}\right)} = 1851 \text{ mm}^2$$

Provide 10 Nos. 16 mm ϕ (tor), A_{st} provided = 2011 mm² and 2 Nos. 10 mm ϕ (tor) as hanger bars.

Check for shear

$$\text{Max. S.F.} = V = \frac{21375 \times 6}{2} = 64125 \text{ N}$$

$$\text{Nominal shear stress } \tau_v = \frac{V}{b_w \cdot d} = \frac{64125}{225 \times 250} = 1.14 \text{ N/mm}^2$$

$$p = \frac{100\, A_s}{b_w \cdot d} = \frac{100 \times 2011}{225 \times 250} = 3.575 \qquad \tau_c = 0.57 \text{ N/mm}^2$$

$\tau_c = \tau_v$ $\quad\therefore$ Shear reinforcement is necessary

$$V_c = V - \tau_c.\text{b.d} = 64125 - 0.57 \times 225 \times 250 = 32062.5 \text{ N}$$

Provide 8 mm.ϕ 2 legged stirrups at spacing least of the following.

(*i*) $$s_v = \frac{\sigma_{sv}.A_{sv}.d}{V_s} = \frac{230 \times 2 \times 50 \times 250}{32062.5} = 180 \text{ mm}$$

(*ii*) Max. spacing as per min. shear reinforcement requirement

$$s_v = \frac{A_{sv} \cdot 0.87 f_y}{0.4\, b_w} = \frac{2 \times 50 \times 415 \times 0.87}{0.4 \times 225} = 418 \text{ mm}$$

(*iii*) $0.75\,d = 0.75 \times 250 = 187.5$ mm

∴ Provide 8 mm ϕ (tor), 2 legged vertical stirrups @ 180 mm *c/c* throughout.

Check for Deflection

Actual span/*d* ration $= \dfrac{6}{0.25} = 24$

$$P = \frac{100 \times A_s}{b_f.d} = \frac{100 \times 2011}{1825 \times 250} = 0.44 \quad K_1 = 1.45 \text{ from Fig. 5.1}$$

$$\text{where } fs = 0.58 \times 415 \times \frac{1851}{2011} = 221.55 \text{ N/mm}^2$$

$$\frac{b_w}{b_f} = \frac{225}{1825} = 0.123, \qquad K_2 = 0.8 \text{ from Fig. 5.3}$$

Permissible span/*d* ratio = 20 × 1.45 × 0.8 = 23.2 less than 24

Hence unsafe in deflection.

To bring deflection within permissible limits, increase the hang from 200 mm to 300 mm.

∴ Self weight of the beam = 0.225 × 0.3 × 25000 = 1687.5 N/m

Total load/*m* of beam = 1687.5 + 11250 + 9000 = 21937.5 N/m

$$\text{Max. } B.M. = \frac{21937.5 \times 6 \times 6}{8} = 98718.75 \text{ N.m}$$

$$x_c = 0.289 \times 250 = 72 \text{ mm}$$

$$\therefore \quad M = \frac{1825 \times 100 \times 8.5}{1000}\left(\frac{2 \times 72 - 100}{2 \times 72}\right)\left\{250 - \frac{100}{3} \times \frac{3 \times 72 - 2 \times 100}{2 \times 72 - 100}\right\}$$

$$= 110276 > 98718.5 \text{ N.m}$$

$$A_{st} = \frac{98718.5 \times 1000}{230 \times (250 - 72/3)} = 1900 \text{ mm}^2$$

Provide 7 Nos. 20 mm ϕ (tor), Ast provided = 2198 mm^2

Check for Shear

$$\text{Max. } S.F. = V = \frac{21937.5 \times 6}{2} = 65812.5 \text{ N}$$

$$\tau_v = \frac{65812.5}{225 \times 250} = 1.17 \text{ N/mm}^2$$

$$p = \frac{2198 \times 100}{225 \times 250} = 12.9, \ \tau_c = 0.57 \text{ N/mm}^2$$

$$V_s = 65812.5 - 0.57 \times 225 \times 250 = 33750 \text{ N}$$

Provide 8 mm ϕ (tor), 2 legged stirrups at a spacing least of the following:

(*i*) $\dfrac{230 \times 2 \times 50 \times 250}{33750} = 170$ mm

(*ii*) 480 mm

(*iii*) $0.75 \times 250 = 187.5$ mm

∴ Provide 8 mm ϕ (tor), 2 legged vertical stirrups at the rate of 150 mm *c/c* throughout.

Check for Deflection

$$\text{Actual span}/d \text{ ratio} = \frac{6}{0.35} = 17.15$$

$$p = \frac{100 \times 2198}{1825 \times 250} = 0.48, K_1 = 1.3 \text{ from Fig. 5.1}$$

$$\text{where } fs = 0.58 \times 415 \times \frac{1900}{2198} = 208 \text{ N/mm}^2$$

$$\frac{b_w}{b_f} = 0.123, K_2 = 0.8 \text{ from Fig. 5.3}$$

$$\therefore \text{ Permissible span}/d \text{ ratio} = 20 \times 1.3 \times 0.8 = 20.18 > 17.15 \quad \text{O.K.}$$

Check for End Anchorage

$$L_d \not> \frac{M_1}{V} + L_0$$

Now $$L_d = \frac{230 \times 16}{0.4 \times 0.9} = 1022.22 \text{ mm}$$

$$M_1 = 2198 \times 230 \times 0.9 \times 250 = 113746500 \text{ N.mm}$$

$$L_0 = 12 \times 16 = 192 \text{ or } d = 250 \text{ mm whichever is greater}$$

$$\frac{M_1}{V} + L_0 = \frac{113746500}{65812.5} + 350 = 2078 \text{ mm O.K.}$$

The details of reinforcement are shown in Fig. 6.13

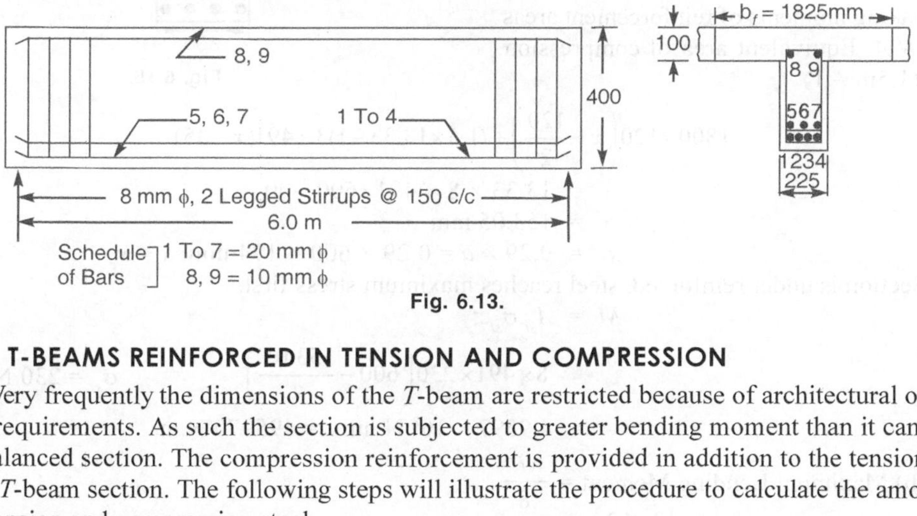

Fig. 6.13.

6.9 T-BEAMS REINFORCED IN TENSION AND COMPRESSION

Very frequently the dimensions of the *T*-beam are restricted because of architectural or head room requirements. As such the section is subjected to greater bending moment than it can resist as a balanced section. The compression reinforcement is provided in addition to the tension steel in the *T*-beam section. The following steps will illustrate the procedure to calculate the amount of total tension and compression steel.

1. Moment of resistance of *T*-beam as balanced section is given by Eq. 6.13

$$M_1 = b_f . D_f . \sigma_{cbc} \left(\frac{2.x_c - D_f}{2.x_c} \right) \left\{ d - \frac{D_f}{3} \left(\frac{3.x_c - 2D_f}{2.x_c - D_f} \right) \right\}$$

2. Assuming M to be the maximum Bending Moment at a section, the unbalanced moment

$$M_2 = M - M_1$$

3. Determine the area of total tension steel, A_{st}.

$$A_{st} = A_{st1} + A_{st2}, \qquad ...(6.15)$$

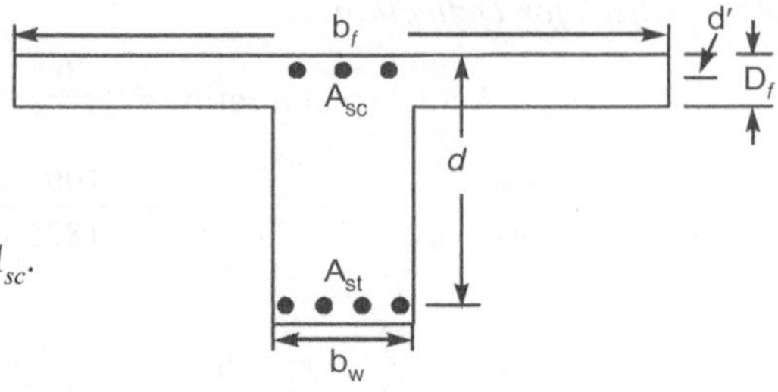

Fig. 6.14.

where, $$A_{st1} = \frac{M_1}{\sigma_{st} \cdot z}$$

$$A_{st2} = \frac{M_2}{\sigma_{st}.(d-d')}$$

4. Determine the area of compression steel, A_{sc}.

$$A_{sc} = \frac{M_2}{(1.5m-1)\sigma_{cbc}} \times \frac{x_c}{(x_c-d')} \times \frac{1}{(d-d')} \quad ...(6.16)$$

where, x_c = depth of critical $N.A.$

The various symbols used are defined in Fig. 6.14

6.10 SOLVED EXAMPLES

Example. 6.6.(a). Calculate the moment of resistance of the T-beam section as shown in Fig. 6.15. Concrete M 20 and HYSD, Fe415 bars used.

(b) Also calculate the $u.d.l.$ inclusive of self weight which the beam of effective span 12 m can carry.

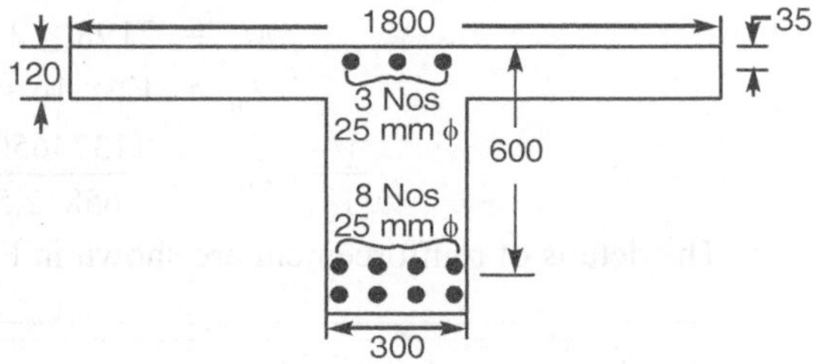

Fig. 6.15.

Sol. (a) $$m = \frac{280}{3 \times 7} = \frac{280}{21} = 13.33$$

Let x = depth of N.A. assumed to be in the rib.

Taking moments of reinforcement areas about $N.A.$ Equivalent area of compression steel (1.5m – 1) A_{sc}

$$1800 \times 120\left(x - \frac{120}{2}\right) + (1.5 \times 13.33 - 1)3 \times 491(x-35)$$
$$= 13.33 \times 8 \times 491\ (600 - x)$$
$$x = 153.05 \text{ mm}$$
$$x_c = 0.29 \times d = 0.29 \times 600 = 174 \text{ mm}, x < x_c$$

Section is under reinforced, steel reaches maximum stress first.

$$M = A_{st}.\sigma_{st}.z$$
$$= 8 \times 491 \times 230\left(600 - \frac{153.05}{3}\right) \qquad \sigma_{st} = 230 \text{ N/mm}^2$$
$$= 4.9597 \times 10^8 \text{ N. mm} = 495970 \text{ N.m}$$

(b) Maximum Bending Moment $= \dfrac{wl^2}{8}$

or $$w \times \frac{12 \times 12}{8} = 495970$$

or $$w = 27554 \text{ N/m}$$

Maximum $u.d.l.$ including self weight of T-beam which the section can carry = 27554 N/m

Example 6.7. The T-beam shown in Fig. 6.16 simply supported over a span of 10m, carries a load of 40,000 N/m (4000 kg/m) including the self weight of the beam. Calculate the area of steel in tension and

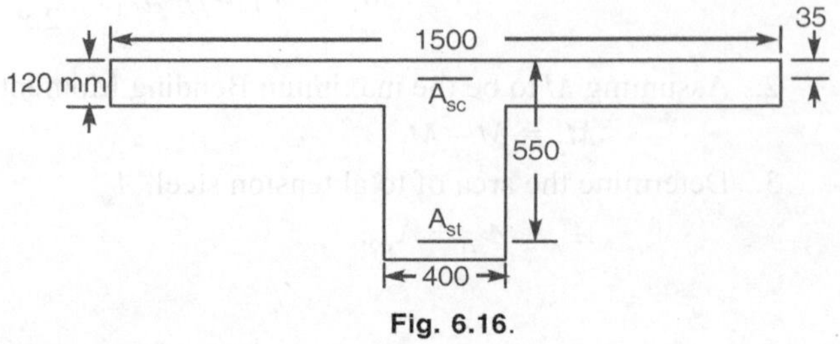

Fig. 6.16.

compression in the section. M20 concrete and HYSD, Fe 415 reinforcement is used.

Sol. Assuming the section to be balanced.

$$x_c = 0.29\,d = 0.29 \times 550 = 159.5 \text{ mm}$$

Moment of resistance of T-beam

(*i*)
$$M_1 = b_f.D_f.\sigma_{cbc}\left(\frac{2x_c - D_f}{2x_c}\right)\left\{d - \frac{D_f}{3}\left(\frac{3x_c - 2D_f}{2x_c - D_f}\right)\right\}$$

$$= 1500\times120\times7\left(\frac{2\times159.5-120}{2\times159.5}\right)\left\{550-\frac{120}{3}\left(\frac{3\times159.5-2\times120}{2\times159.5-120}\right)\right\}$$

$$= 3.9 \times 10^8 \text{ N.mm}$$

$$= 3.9\times10^5 \text{ N.m} = 390{,}000 \text{ N.m}$$

(*ii*)
$$M = M_{ax}\ B.M. = \frac{wl^2}{8} = \frac{40{,}000\times10^2}{8} = 500{,}000 \text{ N.m}$$

$$M_2 = 500{,}000 - 390{,}000 = 110000$$

(*iii*) Assuming the bar dia. to be more than 20 mm, $\sigma_{st} = 230$ N/mm²

$$A_{st1} = \frac{M_1}{\sigma_{st}.z} = \frac{390000\times1000}{230\left(550-\frac{120}{3}\right)} = 3324$$

$$A_{st2} = \frac{M_2}{\sigma_{st}.(d-d')} = \frac{110{,}000\times1000}{230(550-35)} = 928 \text{ mm}^2$$

Total tension steel,

$$A_{st} = A_{st1} + A_{st2} = 3324 + 928$$

$$= 4252 \text{ mm}^2$$

Provide 28 mm ϕ 4 Nos + 4 Nos. 25 mm ϕ, A_{st} provided = 4426 mm² in two rows.

(*iv*) Area of compression steel, A_{sc}.

$$A_{sc} = \frac{M_2}{(1.5m-1)\sigma_{cbc}}\times\frac{x_c}{(x_c-d')}\times\frac{1}{(d-d')}$$

$$= \frac{110{,}000\times1000}{\left(\frac{1.5\times280-1}{3\times7}\right)7}\times\frac{159.5}{(159.5-35)}\times\frac{1}{(550-35)}$$

$$= 827 \text{ mm}^2$$

Provide 5 Nos. 16 mm ϕ, A_{sc} = 1005 mm²

Example 6.8. The reinforced concrete *T*-beam section is shown in Fig. 6.17. It is subjected to a maximum Bending Moment of 249850 N.m. Calculate the maximum stress developed in concrete, tension steel and compression steel. Concrete used is M 20.

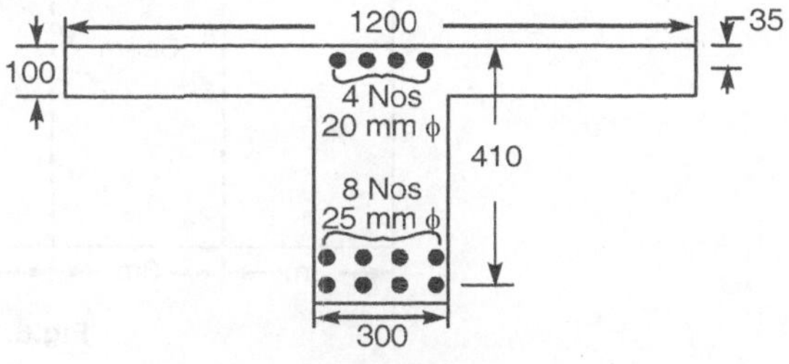

Fig. 6.17.

Sol. $m = \frac{280}{3\times7} = 13.33$

Let x = the depth of *N.A.* assumed to lie in rib. Taking moments of equivalent areas about *N.A.*

$$1200 \times 100\,(x - 100/2) + (1.5 \times 13.33 - 1) \times 4 \times 314\,(x - 35)$$
$$= 13.33\,(8 \times 491)\,(410 - x)$$
$$x = 144.23 \text{ mm}$$

Assuming it to be balanced section

$$M = b_f . D_f . \sigma_{cbc} \left(\frac{2x - D_f}{2x} \right) \left\{ d - \frac{D_f}{3} \left(\frac{3x - 2D_f}{2x - D_f} \right) \right\}$$
$$+ (1.5\,m - 1)\,A_{sc} . \sigma_{cbc}\,(x - d')\,(d - d')/x$$
$$= 1200 \times 100 \times \sigma_{cbc} \frac{2 \times 144.24 - 100}{2 \times 144.23} \left\{ 410 - \frac{100}{3} \left(\frac{3 \times 144.23 - 2 \times 100}{2 \times 144.23 - 100} \right) \right\}$$
$$+ (1.5 \times 13.33 - 1) 4 \times 314 \times \sigma_{cbc} \left(\frac{144.23 - 35}{144.23} \right) (410 - 35)$$
$$= 28917288\ \sigma_{cbc} + 6775580\ \sigma_{cbc}$$
$$= 35692869\ \sigma_{cbc} \text{ N.mm}$$
$$M = 249850 \text{ N.m}$$
$$= 249850000 \text{ N.mm}$$

Streess in concrete

$$\sigma_{cbc} = \frac{249{,}850000}{25{,}692869} = 7.00 \text{ N/mm}^2$$

Stress in compressive steel

$$= 1.5 \text{ m } \sigma_{cbc} . \frac{x - d'}{x}$$
$$= 1.5 \times 13.33 \times 7.0 \times \frac{144.23 - 35}{144.23} = 106.0 \text{ N/mm}^2$$

Stress in tensile steel $= m . \sigma_{cbc} \dfrac{d - x}{x}$

$$= 13.33 \times 7.00 \times \frac{410 - 144.23}{144{,}23} = 171.94 \text{ N/mm}^2$$

Example 6.9. Design the beam in a room as shown in Fig. 6.18 if the loading on the roof slab of the beam is 8000N/m² (800 kg/m²) including self weight of the slab. Use M20 concrete and HYSD bars of Fe415 grade.

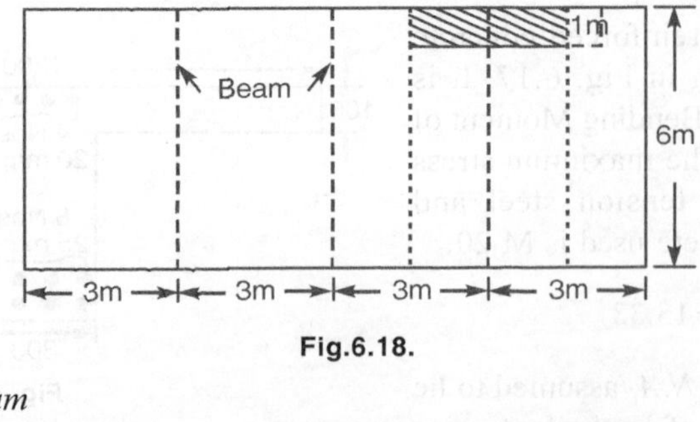

Fig.6.18.

Sol. *Loads on Beam*

u.d.l. from slab $= 8000 \times 3 = 24000$ N/m

Assume the size of beam to be 300 mm × 300 mm (Hang)

$$\text{Self weight of the beam} = 0.3 \times 0.3 \times 25000 = 2250 \text{ N/m}$$

$$\text{Total design } u.d.l.\ w = 26250 \text{ N/m}$$

$$\text{Maximum Bending Moment} = \frac{wl^2}{8}$$

$$= \frac{26250 \times 6 \times 6}{8}$$

$$= 118125 \text{ N.m}$$

$$\text{Maximum } S.F. = \frac{wl}{2}$$

$$= \frac{26250 \times 3}{2} = 78750 \text{ N}$$

Breadth of Flange (Fig. 6.19)

$$b_f = \frac{L_0}{6} + b_w + 6\,D_f$$

$$= \frac{6000}{6} + 300 + 6 \times 125$$

$$= 2050 \text{ mm}$$

Assuming 40 mm as effective cover to tension steel.

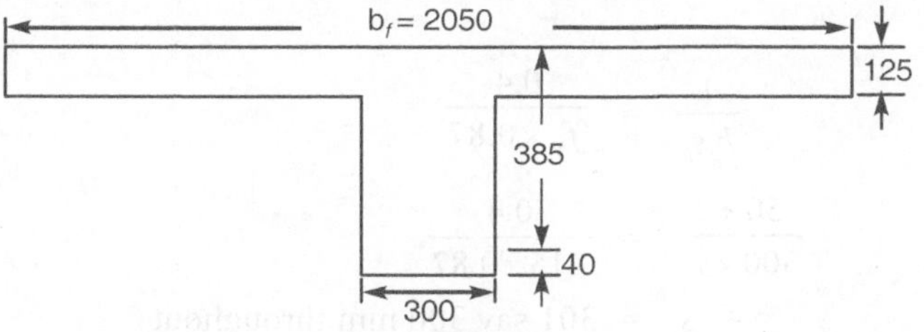

Fig.6.19.

$$x_c = 0.29\,d = 0.4 \times 385 = 112 \text{ mm (for a balanced section)}$$

$$M = b_f.D_f\ \sigma_{cbc}\left(\frac{2x_c - D_f}{2x_c}\right)\left\{d - \frac{D_f}{3}\left(\frac{3x_c - 2\,D_f}{2x_c - D_f}\right)\right\}$$

$$= 27.75 \times 10^6 \text{ N.mm}$$

$$= 277500 \text{ N. m} > B.M. = 118125 \text{ N.m}$$

The section is to be designed as siggly reinforced section.

$$\text{Area of tensile steel required} = A_{st} = \frac{M}{\sigma_{st}.jd}$$

$$A_{st} = \frac{118125 \times 10^3}{230 \times 0.90 \times 385}$$

$$= 1465.9$$

Provide 20 mm φ 6 Nos. thus providing

$$A_{st} = 1885 \text{ mm}^2$$

Check for Shear

$$\text{Maximum } S.F. \text{ '}V\text{'} = 78750 \text{ N}$$

$$\tau_v = \frac{V}{bd}$$

$$= \frac{78750}{300 \times 385}$$

$$= 0.682 \text{ N/mm}^2$$

$$\frac{100\,A_s}{bd} = \frac{100 \times 1885}{300 \times 385} = 1.63$$

$$\tau_c = 0.46 \text{ N/mm}^2 < \tau_v$$

$$\text{Design shear '}V_s\text{'} = V - \tau_c\, bd$$

$$= 78750 - 0.46 \times 300 \times 385$$

$$= 25620 \text{ N}$$

Assume that 2 Nos. bars out of 6 bars of 20 mm ϕ provided in tension zone, are bent up at 45° to the horizontal.

Shear carried by 2 Nos, bent up bars

$$= A_{sv} \cdot \sigma_{sv} \cdot \text{Sin}\phi$$

$$= 314 \times 2 \times 230 \times \text{Sin } 45°$$

$$= 102134 \text{ N} > \text{design shear } V_s$$

Minimum shear reinforcement in the form of vertical stirrups are required. Provide 8 mm ϕ 2 legged stirrups.

$$\frac{A_s}{b.s_t} = \frac{0.4}{f_y \times 0.87}$$

$$\frac{50 \times 2}{300 \times s_v} = \frac{0.4}{415 \times 0.87}$$

$$s_v = 301 \text{ say } 300 \text{ mm throughout}$$

Check for Deflection

$$\text{Actual span}/d \text{ ration} = \frac{6}{0.385} = 15.58$$

$$\frac{b_w}{b_f} = \frac{300}{2050} = 0.146$$

$$p = \frac{A_{st} \times 100}{b_f d} = \frac{100 \times 1885}{2050 \times 385} = 0.239$$

Modification factor for tension steel = 2.00 from Fig. 5.1 where

$$f_s = 0.58 \times 415 \times 1465.9/1995 = 187.2 \text{ N/mm}^2$$

Reduction factor for flanged beam for $\dfrac{b_w}{b_f = 0.146,}$ is 0.8 from Fig. 5.3

Allowable span/d ratio for deflection to be within permissible limits

$$= 20 \times 2.00 \times 0.8 = 32 > 15.58 \qquad \text{O.K.}$$

Check for End Anchorage

$$L_d \not> \frac{M_1}{V} + L_0$$

$$L_d = \frac{\sigma_s . \phi}{4\,\tau_{bd}} = \frac{230 \times 20}{4 \times 1.28} = 898.4$$

$$M_1 = 4 \times 314 \times 230 \times 0.90 \times 385 = 101209108 \text{ N.mm}$$

$$V = 78750 \text{ N}$$

$$L_0 = 16\phi \text{ or } d_{eff} \text{ is standard hook is provided } L_0 = 16\phi$$

$$= 16 \times 20 \text{ or } 385$$

$$= 385 \text{ mm}$$

$$\frac{M_1}{V} + L_0 = \frac{101209108}{78750} + 385$$

$$= 1670 \text{ mm} > L_d = 898.4 \text{ mm} \;\therefore\; \text{O.K.}$$

The reinforcement details are given in Fig. 6.20

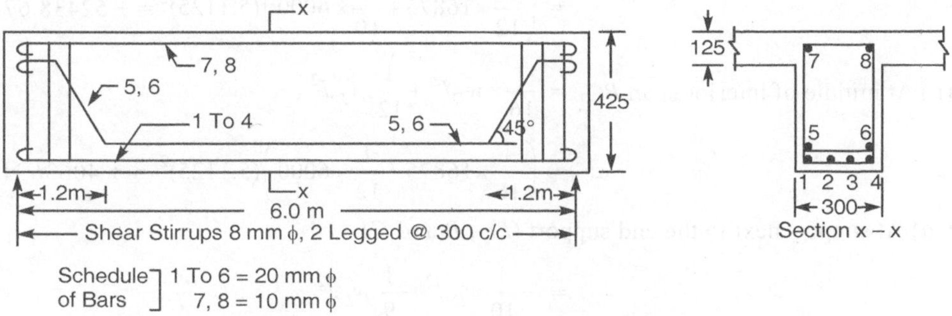

Fig. 6.20.

Example 6.10. Design the continuous beam to be provided over a hall of 15 m × 9 m with columns as shown in the Fig. 6.21. Thickness of slab = 125 mm. Superimposed *L. L.* on slab = 2000 N/m² (200 kg/m²) and the weight of finishes = 2500 N/m² (250 kg/m²). Use M20 concrete and HYSD, Fe415 Bars.

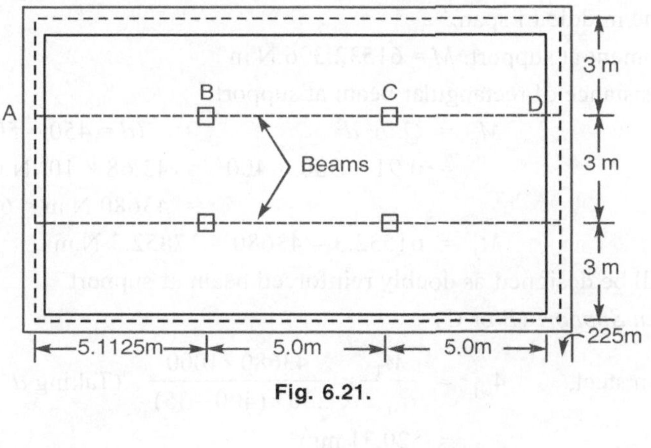

Fig. 6.21.

The size of beam is restricted to 300 mm wide × 450 mm deep (overall including slab thickness).

Sol. The effective end spans are $5+\frac{0.225}{2} = 5.1125$ m

$$\begin{aligned} D.L. \text{ of slab} &= 0.125 \times 1 \times 1 \times 25000 \\ &= 3125 \text{ N/m}^2 \\ \text{Finishes} &= 2500 \text{ N/m}^2 \\ \text{Total } D.L. &= 3125 + 2500 = 5625 \text{ N/m}^2 \\ D.L/m &= 5625 \times 3 = 16875 \text{ N/m} \\ L.L/m &= 2000 \times 3 = 6000 \text{ N/m} \end{aligned}$$

Bending Moment (*Maximum*) : The span do not differ by more than 15% of the longest : Calculate moment with longest span.

(*i*) Near middle of end spans *AB* and *CD*

$$= +\frac{1}{12} w_D . l^2 + \frac{1}{10} w_L . l^2$$

$$= \left(\frac{1}{12} \times 16875 + \frac{1}{10} \times 6000\right)(5.1125)^2 = +\ 52438.67 \text{ N.m}$$

(*ii*) At middle of interior span *BC*

$$= \frac{1}{16} \times w_D l^2 + \frac{1}{12} . w_L . l^2$$

$$= \left(\frac{1}{16} \times 16875 + \frac{1}{12} \times 6000\right)(5.1125)^2 = +\ 40636 \text{ N.m.}$$

(*iii*) At support next to the end support (*i.e.*, *B* and *C*)

$$= -\frac{1}{10} w_D . l^2 - \frac{1}{9} . w_L . l^2$$

$$= \left(-\frac{1}{10} \times 16875 - \frac{1}{9} \times 6000\right)(5.1125)^2$$

$$= -\ 61532.396 \text{ N.M}$$

The maximum moment occurs at the support and is negative. This show, that the tension is at top and compression at bottom face of the beam . The beam acts as rectangular beam at support and as T-beam in the middle of span.

Maximum Moment at support, $M = 61532.396$ N.m

Moment of resistance of rectangular beam at support

$$M_1 = Q.\ b.\ d^2 \qquad (d = 450 - 50 = 400 \text{ mm})$$

$$= 0.91 \times 300 \times 400^2 = 43.68 \times 10^6 \text{ N.mm}$$

$$= 43680 \text{ N.m} < 61532.396 \text{ N.m}$$

$$M_2 = 61532.3 - 43680 = 17852.3 \text{ N.m}$$

The beam shall be designed as doubly reinforced beam at support.

Area of Steel at Support (*B* or *C*)

Area of tension steel, $4_{st1} = \frac{M_1}{\sigma_{st} . z} = \frac{43680 \times 1000}{230 \times (400 - 35)}$ (Taking $d' = 35$ mm)

$$= 520.31 \text{ mm}^2$$

Additional tensile steel for M_2 = $A_{st2} = \dfrac{M_2}{\sigma_{st}.z} = \dfrac{17852.3\times10^3}{230\times(400-35)}$

$= 212.65$

Total A_{st} = $A_{st1} + A_{st2} = 520.31 + 212.65 = 732.46\ \text{mm}^2$

Provide 12 mm ϕ 7 Nos. (A_{st} provided = $7 \times 113 = 791\ \text{mm}^2$)

Area of compressive steel, A_{sc} = $\dfrac{M_2}{(1.5\,m-1)\,\sigma_{cbc}(d-d')}$

$= \dfrac{17852.3\times10^3}{(1.5\times13-1)\times7\times(400-35)} = 367.75\ \text{mm}^2$

Provide 12 mm ϕ, 5 Nos.

The beam acts as *T*-beam near middle of span.

Moment of Resistance of T-Beam : Assuming it to balanced section, *M.O.R.* can be found it flange width is known.

Flange width is least of the following:

(*i*) $b_f = \dfrac{l_0}{6} + b_w + 6.D_f$

$= \dfrac{5112.5\times-0.7}{6} + 300 + 6\times125 = 1646\ \text{mm}$

(*ii*) $b_f = \dfrac{1}{2}(3000+3000) = 3000\ \text{mm}$

$b_f = 1646\ \text{mm}$

$x_c = 0.29\,d = 0.29 \times 400$

$= 116\ \text{mm}$

$$M = b_f.D_f\sigma_{cbc}\left(\frac{2x_c - D_f}{2x_c}\right)\left\{d - \frac{D_f}{3}\left(\frac{3x_c - 2D_f}{2x_c - D_f}\right)\right\}$$

$$= 1646\times125\times5\left(\frac{2\times116-125}{2\times116}\right)\left\{400 - \frac{125}{3}\left(\frac{3\times116-2\times125}{2\times116-125}\right)\right\}$$

$= 1.716 \times 10^8\ \text{N.mm}$

$= 171,00\ \text{N.m}$

$> 52,438.67\ \text{N.m}$

Resistance as *T*-beam is approximately five times the maximum moment at middle of span.

Area of Steel near middle of AB and CD.

$$A_{st} = \frac{52438.67\times1000}{230\times\left(400-\frac{116}{3}\right)} = 631.4\ \text{mm}$$

Provide 6 Nos. 12 mm ϕ

Area provided = $678\ \text{mm}^2$

Area of Steel near middle of BC

$$A_{st} = \frac{40635.887 \times 1000}{230\left(400 - \frac{116}{3}\right)} = 489 \text{ mm}^2$$

Use 4 Nos. 12 mm ϕ

Area = 452 mm²

Shear Force.

(*i*) At end support (i.e., *A* and *D*) $= 0.4\,.w_D.\,l + 0.45\,w_L.l$

$= 0.4 \times 16875 \times 5.1125 + 0.45 \times 6000 \times 5.1125$

$= 48313$ N

(*ii*) At support next to end support (i.e., *B* and *C*)

(a) Outer side $= 0.6 \times w_D.1 + 0.6 \times w_L.l$

$= 0.6\ 1\ (w_D + w_L)$

$= 0.6 \times 5.1125\ (16875 + 6000)$

$= 70169$ N

(b) Inner side $= 0.55 \times w_D.\ l + 0.6\ w_L.l$

$= (0.55 \times 16875 + 0.6 \times 6000)\ 5.1125$

$= 65855.39$ N

Maximum *S.F.* = 70169 *N* at *B* or *C*

$$p = \frac{100 \times A_s}{b.d} = \frac{100 \times 6 \times 113}{300 \times 400} = 0.565$$

$\tau_c = 0.317$ N/mm²

Shear resistance of concrete $= 0.317 \times 300 \times 400$

$= 38040$ N

$V_s = 70169 - 38040 = 32129$ N

Use Vertical Stirrups ; 8 mm ϕ 2 legged stirrups at a spacing which is least of the following :

(*i*) Spacing $= \dfrac{A_{sv}.\sigma_{sv}.d}{V_s}$

$$= \frac{2 \times 50 \times 230 \times 400}{32129} = 285 \text{ mm}$$

(*ii*) Maximum spacing as per min. shear reinforcement

requirement, $S_v = \dfrac{A_{sv}.f_y 0.87}{0.4\,b} = \dfrac{2 \times 50 \times 415 \times 0.87}{0.4 \times 300}$

$= 301$ mm

(*iii*) $0.75\ d = 0.75 \times 400 = 300.00$ mm

provide 8 mm ϕ 2 legged stirrups @ 280 mm *c/c* throughout.

Check for Deflection

Actual span/*d* ration $= \dfrac{3.1125}{0.4} = 12.78$

$$\frac{b_w}{b_f} = \frac{300}{1902} = 0.156, \text{ Reduction factor} = 0.8 \qquad \text{(Fig. 5.3)}$$

$$p = \frac{100 \times 6 \times 113}{i\,902 \times 400} = 0.09 \text{ Modification factor} = 2 \qquad \text{(Fig. 5.1)}$$

where $f_s = 0.58 \times 415 \times 631/678 = 224 \text{ N/mm}^2$

Permissible span/d ration $= 26 \times 2 \times 0.8 = 41.6 \quad > 12.78$ O.K.

Check for End Anchorage

$$L_d \ngtr \frac{M_1}{V} + L_0$$

$$L_d = \frac{\sigma_s.\phi}{4\tau_{bd}} = \frac{230 \times 12}{4 \times 1.28} = 539 \text{ mm}$$

Assuming 2 Nos. 12 mm ϕ bars bent up at ends A and D

$$M_1 = 4 \times 113 \times 230 \times 0.90 \times 400 = 18{,}712{,}800 \text{ N.mm}$$

$$V = 48313 \text{ N}$$

$$L_0 = 16\,\phi = 16 \times 16 = 256 \text{ mm}$$

$$\frac{M_1}{V} + L_0 = \frac{18{,}712{,}800}{48313} + 256 = 643 \text{ mm} > L_d \qquad \text{O.K.}$$

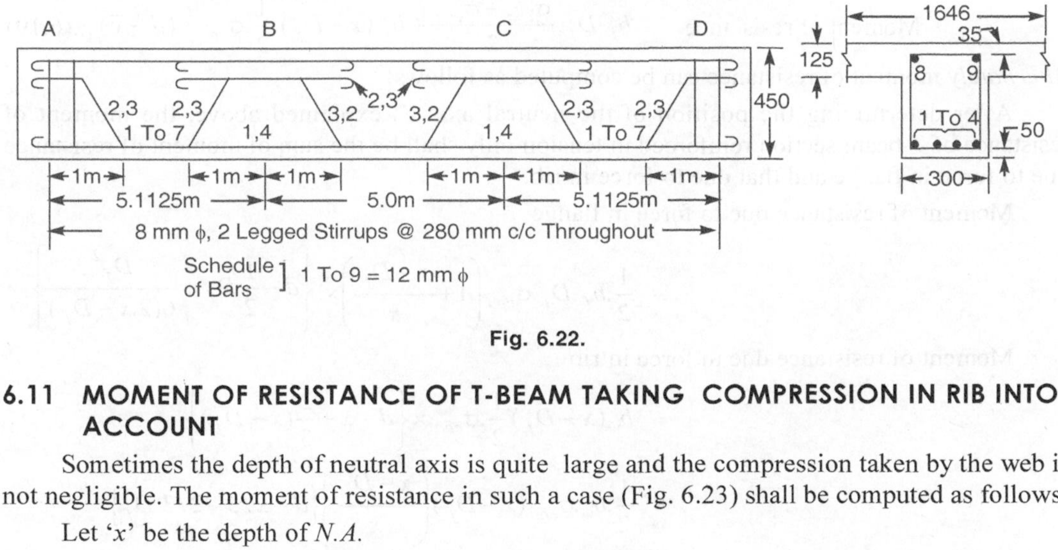

Fig. 6.22.

6.11 MOMENT OF RESISTANCE OF T-BEAM TAKING COMPRESSION IN RIB INTO ACCOUNT

Sometimes the depth of neutral axis is quite large and the compression taken by the web is not negligible. The moment of resistance in such a case (Fig. 6.23) shall be computed as follows:

Let 'x' be the depth of $N.A.$

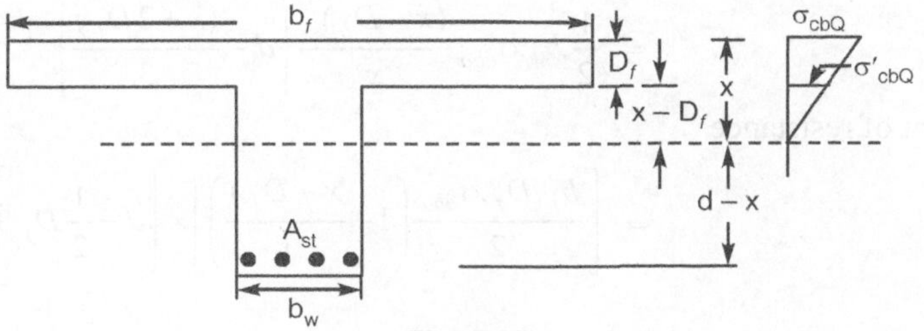

Fig. 6.23.

Taking moments of equivalent areas about *N.A.*

$$b_f.D_f(x-\frac{1}{2}D_f)+b_w.\frac{(x-D_f)^2}{2}=m.\,A_{st}.(d-x) \quad ...(6.17)$$

From Eq. 6.17, the value '*x*' can be determined. Knowing the position of the neutral axis, the moment of resistance of the section can be found by taking moments of compressive forces about centre of tension steel. Alternatively the centre of gravity of compressive forces can be found and the lever arm will be the distance between centre of gravity of compressive forces and tensile forces. The total compressive force multiplied by the lever arm gives moment of resistance.

Let $\bar{y}$ be the distance of c.g. of compressive force from top. Taking moments about top.

$$\bar{y}=\frac{b_f.D_f.\frac{1}{2}(\sigma_{bca}+\sigma'_{cba}).\frac{2\sigma'_{cba}+\sigma_{cba}}{\sigma_{cba}+\sigma'_{cba}}.\frac{D_f}{3}+b_w(x-D_f)\frac{1}{2}.\sigma'_{cba}}{b_f.D_f.\frac{1}{2}(\sigma_{cba}+\sigma'_{cba})+b_w(x-D_f)\sigma'_{cba}/2}\times\left[D_f+\frac{(x-D_f)}{3}\right] \quad ...(6.18)$$

where σ'_{cba} is stress at bottom of flange

$$\sigma'_{cba}=\sigma_{cba}.\frac{x-D_f}{x}$$

$$\text{Lever arm}=z=d-\bar{y}$$

$$\text{Moment of resistance}=b_f.D_f.\frac{\sigma'_{cba}+\sigma_{cba}}{2}+b_w.(x-D_f).\frac{1}{2}.\sigma'_{cba}\times(d-\bar{y}) \quad ...(6.19)$$

Alternately moment of resistance can be computed as follows:

After determining the position of the neutral axis as explained above, the moment of resistance of *T*-beam section reinforced in tension only shall be the sum of moment of resistance due to force in flange and that due to force in rib.

Moment of resistance due to force in flange

$$=\frac{1}{2}.b_f.D_f.\sigma_{cba}\left(1+\frac{x-D_f}{x}\right)\times\left\{d-\frac{1}{2}D_f+\frac{D_f^{\,2}}{6(2.x-D_f)}\right\}$$

Moment of resistance due to force in rib

$$=b_w(x-D_f)\frac{1}{2}.\sigma_{cba}\times\left\{d-x+\frac{2}{3}(x-D_f)\right\}$$

$$=\frac{1}{2}.b_w.\sigma_{cba}(x-D_f)\left(\frac{x-D_f}{x}\right)\{d-x/3-2/3\,D_f\}$$

$$=\frac{1}{2}.b_w.\sigma_{cba}\frac{(x-D_f)^2}{x}\left\{d-\frac{(x+2D_f)}{3}\right\}$$

Total moment of resistance

$$=\left[\frac{b_f.D_f.\sigma_{cba}}{2}\left(1+\frac{x-D_f)}{x}\right)\right]\times\left\{d-\frac{1}{2}D_f+\frac{D_f}{6(2.x-D_f)}\right\}$$

$$+\frac{b_w.\sigma_{cba}}{2}\frac{(x-D_f)^2}{x}\left\{d-\frac{(x+2D_f)}{3}\right\} \quad ...(6.20)$$

Example 6.11. Determine the moment of resistance of the *T*-beam section as shown in Fig. 6.7 (Ex.6.2) considering the compression into rib into account.

Sol. Let x be the depth of N.A.

Taking moment about N.A.

$$b_f.D_f(x-\frac{1}{2}D_f)+\frac{1}{2}.b_w(x-D_f)^2 = m.A_{st}.(D-x)$$

$$1200\times100\left(x-\frac{100}{2}\right)+\frac{300}{2}(x-100)^2 = 13.33\ (6\times491)\ (360-x)$$

$$\therefore\ x = 125.8$$

$$x_c = 0.29\times360 = 104.4$$

As $x > x_c$

$\therefore$ The section is over reinforced.

Moment of resistance

$$M = \frac{b_f.D_f.\sigma_{cbc}}{2}\left(1+\frac{x-D_f}{x}\right)\left\{d-\frac{D_f}{3}\left(\frac{3.x-2D_f}{2.x-D_f}\right)\right\}$$

$$+\frac{b_w.\sigma_{abc}}{2}.\frac{(x-D_f)^2}{x}\left\{d-x+\frac{2}{3}(x-D_f)\right\}$$

$$M = \frac{1200\times100\times7}{2}\left(1+\frac{125.8-100}{125.8}\right)\left\{360-\frac{100}{3}\left(\frac{3\times125.8-2\times100}{2\times125.8-100}\right)\right\}$$

$$+\frac{300\times7}{2}\frac{(125.8-100)^2}{125.8}\left\{360-125.8+\frac{2}{3}(125.8-100)\right\}$$

$$= 162{,}466{,}840 + 1396732.7$$

$$= 163{,}863.57\ \text{N.m.}$$

$$= 163{,}863{,}572.7\ \text{N.mm}$$

It is evident from this example when compared to Ex. 6.2 that the moment of resistance of *T*-beam section with same reinforcement and dimensions is more when rib contribution is considered than that when rib contribution is neglected.

6.12 L-BEAMS

The end beams which have slab on one side only, act as *L*-beams. Since these *L*-beams receive their load from one side only, they are subjected to considerable amount of torsional moment. The computation of effective flange width in these types of beams, is explained in article 6.2 of this chapter. A *L*-beam section is, in general, subjected to bending moment, shear force and torsional moment.

The design principles for L-beams are same as for *T*-beams except that effect of torsional moments are to be added to bending moments and shear forces.

Example 6.12. Design the canopy slab and the canopy beam as shown in Fig. 6.24. Take live load as 750 N/m^2 and no finishes. Use M20 concrete and Fe415 bars.

Sol. Take the thickness of slab varying from 150 mm at junction of beam to 75 mm at free end.

Average thickness = $\frac{1}{2}(150+75) = 112.5$ mm

Dead Load = $0.1125 \times 1 \times 1 \times 25000$

= 2812.5 N/m²

Live load = 750 N/m²

Total load (*u.d.l.*) = 3562.5 N/m²

Considering 1 m width of canopy slab.

Design of Slab

The design of cantilever (canopy) slab is to be carried out in the same way as for rectangular beams but the width of such a beam is taken as 1 m.

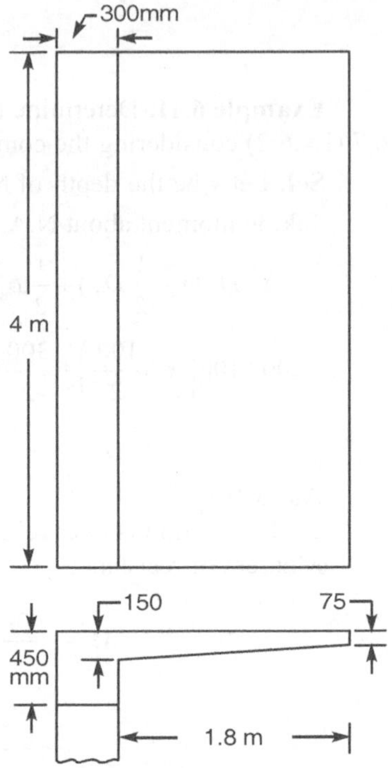

Fig. 6.24.

$$\text{Max. } B.M. = \frac{w.L^2}{2} = \frac{3562.5 \times 1.8^2}{2} = 5771.25 \text{ N.m.}$$

$$d = \sqrt{\left(\frac{5771.25 \times 1000}{0.91 \times 1000}\right)} = 79.64 \text{ mm}$$

Provide $D = 150$ mm $\qquad d = 130$ mm

$$A_{st} = \frac{5771.25 \times 1000}{230 \times 0.90 \times 130} = 214.46 \text{ mm}^2$$

Use 8 mm φ bars @ $\frac{50 \times 1000}{214.46} = 233$ mm

say 200 mm *c/c*.

In slabs, distribution steel at the rate of 0.12% of gross sectional of slab is provided in transverse direction to distribute the loads and to take care of temperature and shrinkage stresses.

$$\text{Distribution steel} = \frac{0.12 \times 112.5 \times 1000}{100} = 135 \text{ mm}^2$$

Provide 8 mm φ @ 300 mm *c/c*.

Design of Canopy Beam

the beam will be designed as *L*-beam.

Width of Flange

Width of flange shall be least of the following:

(*i*) Actual width = 300 + 1800 = 2100 mm

(*ii*) $\frac{0.5\, l_{0'}}{l_{0'}/b+4} + b_w$ (for isolated beams)

$$\frac{0.5 \times 4000}{4000/2100+4} + 300 = 638.7 \text{ mm}$$

$\therefore \qquad b_f = 638.7$ mm

Moment of resistance of *L*-beam as balanced section

$$M = \frac{b_f . D_f . \sigma_{cbc}}{2}\left(1+\frac{x-D_f}{x}\right)\left\{d - \frac{D_f}{3}\left(\frac{3.x-2D_f}{2.x-D_f}\right)\right\}$$

$$= \frac{638.7 \times 143^* \times 7}{2 \times 1000}\left(1 + \frac{0.29 \times 400 - 143}{0.29 \times 400}\right)\left\{400 - \frac{143}{3}\left(\frac{3 \times 0.29 \times 400 - 2 \times 143}{2 \times 0.29 \times 400 - 143}\right)\right\}$$

$$= 89961.2$$

*D_f has been calculated as 143 mm from Fig. 6.25

Average $D_f = \dfrac{136 + 150}{2}$

$= 143$ mm

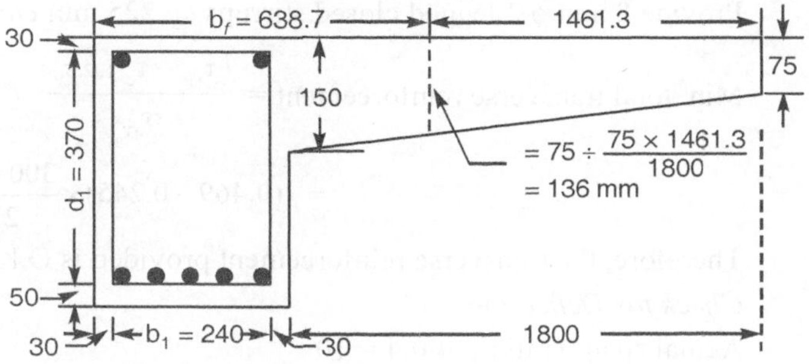

Fig. 6.25.

Bending Moment

$w = 0.45 \times 0.3 \times 1 \times 25000 + 3562.5 \times 1.8 = 10162.5$ N/m

Taking partial fixity at columns

$$\text{Max. } B.M. = \frac{w.l^2}{10} = \frac{10162.5 \times 4^2}{10} = 16260 \text{ N.m}$$

$$\text{Torsional moment} = T = 3562.5 \times 1.8\left(\frac{1}{2} \times 1.8 + \frac{1}{2} \times 0.3\right) = 6733 \text{ N.m.}$$

$$M_t = \frac{T(1 + D/b)}{1.7} = 6733\frac{(1 + 450/300)}{1.7} = 9901.5 \text{ N.m.}$$

$$\text{Design } B.M. = M + M_t$$

$$= 16260 + 9901.5 = 26161.5 \text{ N.m. (less than 89961.2 N.m} = M)$$

$$A_{st} = \frac{M}{\sigma_{st}.j.d} = \frac{26161.5 \times 1000}{230 \times 0.90 \times 400} = 316 \text{ mm}^2$$

Provide 5 Nos. 10 mm ϕ ($A = 392.5$ mm^2)

Shear Force

$$V = \frac{w.L}{2} = \frac{10162.5 \times 4}{2} = 20325 \text{ N}$$

$$V_c = V + 1.6\ T/b$$

$$= 20325 + 1.6 \times \frac{6733}{0.3} = 562334 \text{ N}$$

$$\tau_v = \frac{V}{b.d} = \frac{56234}{300 \times 400} = 0.469 \text{ N/mm}^2$$

$$p = \frac{392.5 \times 100}{300 \times 400} = 0.327, \quad \tau_c = 0.245 \text{ N/mm}^2 \text{ less than } \tau_v$$

$\therefore$ Shear reinforcement is required.

$$b_1 = 300 - 30 - 30 = 240 \text{ mm}$$

$$d_1 = 450 - 30 - 50 = 370 \text{ mm}$$

$$A_{sv} = \frac{T.s_v}{b_1.d_1.\sigma_{sv}} + \frac{V.s_v}{2.5\ d_1.\sigma_{sv}}$$

Using 8 mm ϕ 2 legged closed stirrups

$$S_v = \frac{2\times 50}{\dfrac{6733\times 1000}{240\times 370\times 230}+\dfrac{20325}{2.5\times 370\times 230}} = 235 \text{ say } 225 \text{ mm}$$

Provide 8 mm ϕ 2 legged closed stirrups @ 225 mm *c/c* throughout.

$$\text{Min. total transverse reinforcement} = \frac{(\tau_{ve}-\tau_c).b.s_v}{\sigma_{sv}}$$

$$= (0.469-0.245)\times\frac{300\times 225}{230} = 65.74 \text{ mm}^2$$

Therefore, the transverse reinforcement provided is O.K.

Check for Deflection

Actual span/d ratio = 4/0.4 = 10

$$p = \frac{A_{st}\times 100}{b_f.d} = \frac{5\times 78.5\times 100}{638.7\times 400} = 0.154,\ K_1 = 2.0 \text{ from Fig. 5.1}$$

$$\frac{b_w}{b_f} = \frac{300}{638.7} - 0.47 \qquad K_2 = 0.85$$

Permissible span/d ratio for control of deflection = 20 K1, K2

= 20 × 2.0 ×0.85 = 34

> 10, O.K.

The reinforcement details are shown in Fig. 6.26

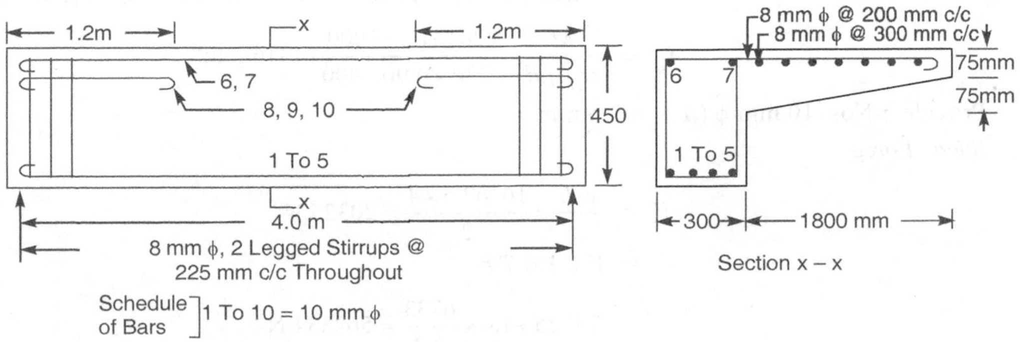

Fig. 6.26.

6.13 INVERTED FLANGED BEAMS

Beams cast monolithically with the slab having their web, turned upwards, are called 'Inverted flanged Beams'. Such beams act as *T*-beam or *L*-beam, though inverted, in the case of overhanging or cantilever spans and as simple rectangular beams in the case of simply supported spans because here the bottom side (slab side) is in tension. However it may be noted that in the case of simply supported spans, overall depth of the beam will include depth of slab also.

EXERCISE 6

1. Determine the maximum stresses developed in steel (Fe 500) and concrete (m = 11) in a R.C. section shown in Fig. 6.27, subjected to a B.M. of 150 KN.m.

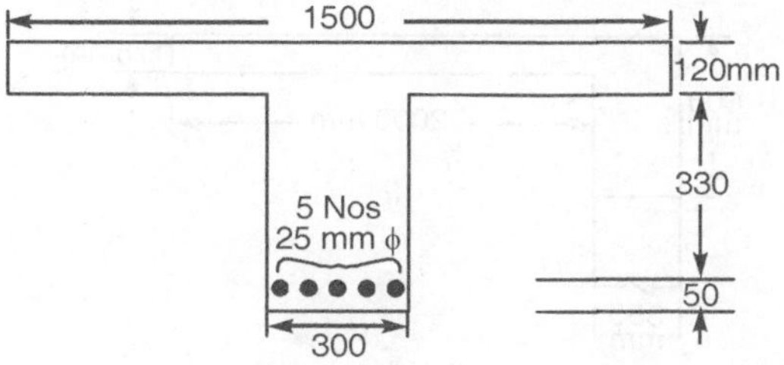

Fig. 6.27.

2. Determine the moment of resistance of a *T*-beam section reinforced with Fe 415 on tension side only as shown in Fig. 6.28. Concrete used is M 20. m = 13.

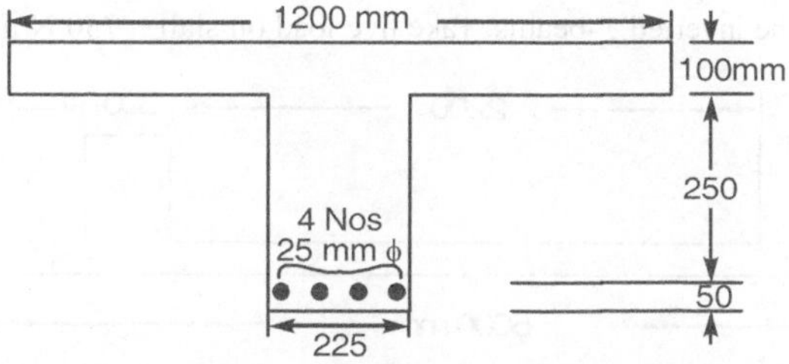

Fig. 6.28.

3. Determine the moment of resistance of a *T*-beam section reinforced on tension and compression side as shown in Fig. 6.29. M20, Fe500.

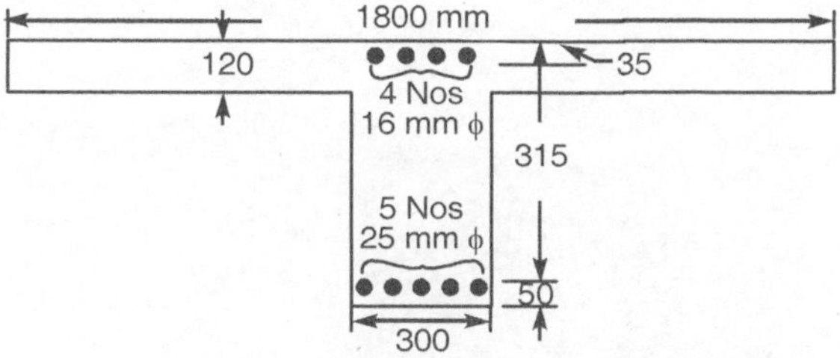

Fig. 6.29.

4. Determine the moment of resistance of a *T*-beam section placed midspan in a room 5 m × 10 m (centre line dimensions) as shown in Fig. 6.30, cast monolithically with a slab 100 mm thick. Take load = 2500 N/m m^2 and finishes = 3000 N/m^2. Use M25 concrete and Fe 500. Take width of beam as 225 mm and hand of beam downwards as 225 mm.

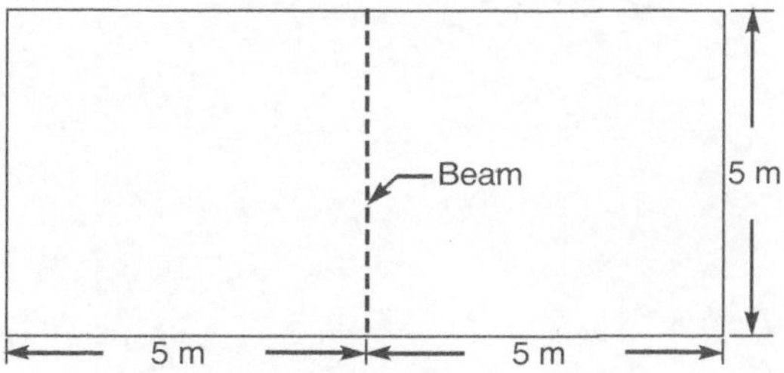

Fig. 6.30.

5. Design the *L*-beam section of the porch as shown in Fig. 6.31. The beam is rigidly connected to R.C.C. columns 350 mm × 350 mm and has a span of 3 m. Take $L.L$ = 750 N/m², finishes = 250N/m². use M20 concrete and Fe415 steel.

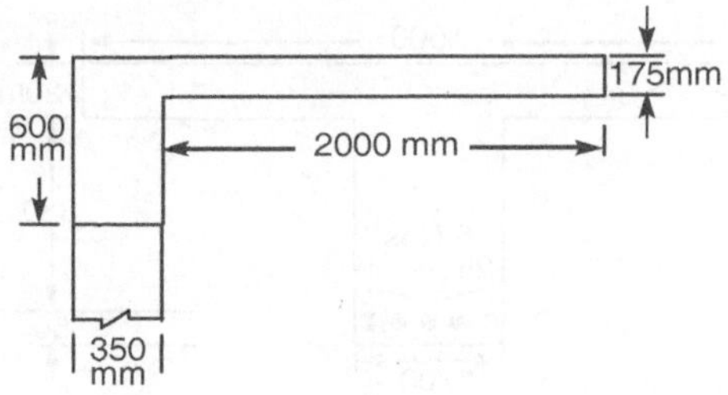

Fig. 6.31.

6. A R.C.C. cantilever car porch with 3 effective obverhand and 5 m wide, is supported on two beams cantilevering out of R.C.C. columns as shown in Fig. 6.32. the beams 300 mm wide and 500 mm deep including slab, are cast monolithically with the slab. Using M20 concrete and HYSD, F415 reinforcement, design the inverted *T*-beams. Take live load on slab = 750 N/m² and finishes = 250 N/m².

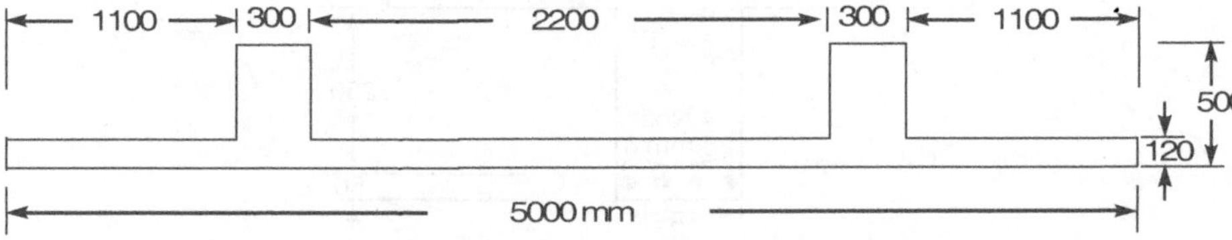

Fig. 6.32.

Slabs

CHAPTER 7

7.1 INTRODUCTION

Slabs are plane structural members whose thickness is quite small as compared to its other dimensions. Slabs are most frequently used as roof coverings and floors in various shapes such as square, rectangular, circular and triangular in buildings, tanks, etc. Some simple forms of slabs are show in Fig. 7.1.

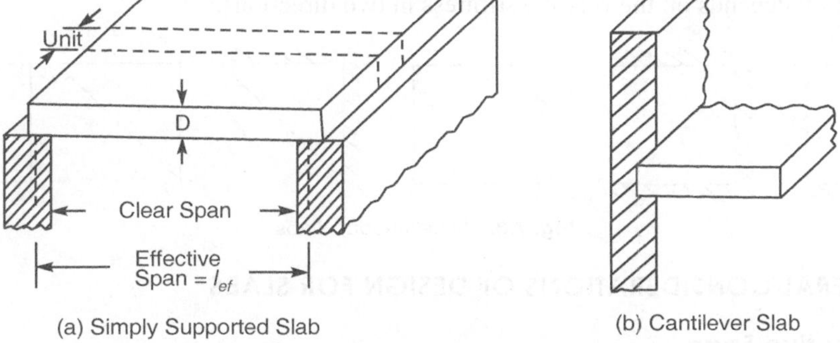

Fig. 7.1. Slabs

The slabs are designed just like beams keeping the breadth of such a beam as unity depending upon system of units, *e.g.*, 1m in M.K.S. Units, 1000 mm in SI units and 1 foot in Imperial units. Thus the total slab is assumed to be consisting of a number of strips, though the slab is cast in one piece and not in strips of unit width. For designing the slabs, the bending moments and shear forces are computed in the same way as is done for the design of rectangular beams. The load per unit area will be considered as load per unit run of the imaginary beam having unit width and span *l*.

Slabs are, in general, divided into two categories depending upon the ratio of long span to short span. When this ratio is greater than or equal to 2, the slab is to be designed as slab spanning in one direction (often called one-way slab) but if this ratio is less than 2, the slab is to be designed

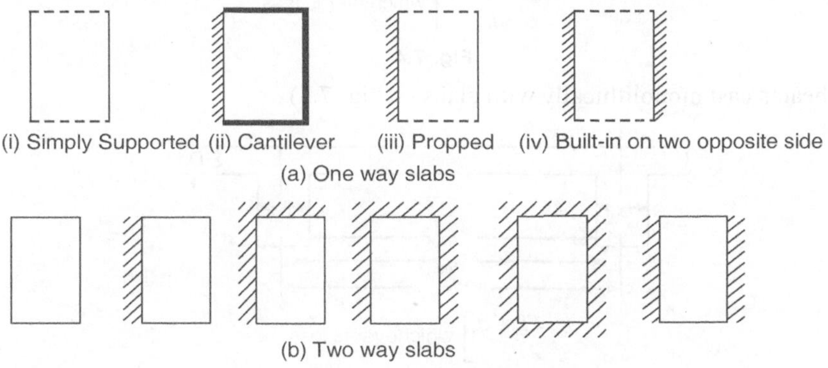

Fig. 7.2. Types of supports

as slab spanning in two directions (often called two-way slab), provided in both the cases the slab is supported on all the four sides. However, if the slab is supported over two opposite edges only as in case of verandahs, corridors, foyers in cinema houses etc., such slabs are designed as one-way slabs. Whether the slab is to be considered as one-way or two-way slab, depends upon the supporting conditions of the slab. Fig. 7.2 shows some of the types of boundary conditions making the slab as one-way or two way as the case may be.

(*i*) Free edge - - - - - - - -

(*ii*) Simply supported △

(*iii*) Built-in or Encastered edge / / / / / / /

Symbols representing types of support

Some of the typical cases of slabs with different boundary conditions shown in Fig. 7.3 can be designed either one-way or two-way slab depending upon designer's will or they may be designed as one-way slab in each direction depending upon the flow of load in each direction. The flow of load in any direction depends on the relative stiffness in two directions.

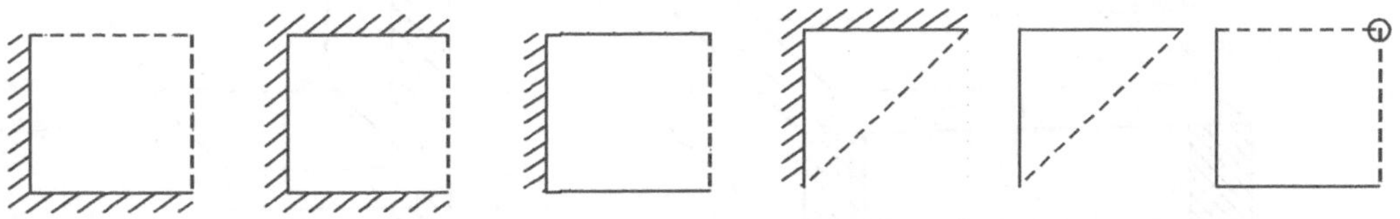

Fig. 7.3. Miscellaneous slabs

7.2 GENERAL CONSIDERATIONS OF DESIGN FOR SLABS

7.2.1 Effective Span

Proper assessment of the effective span is very important as the bending moment varies with the square of effective span. The effective spans for slabs supported in different ways are as follows :

(*a*) On brick walls or Rolled Steel Joists. (Fig. 7.4)

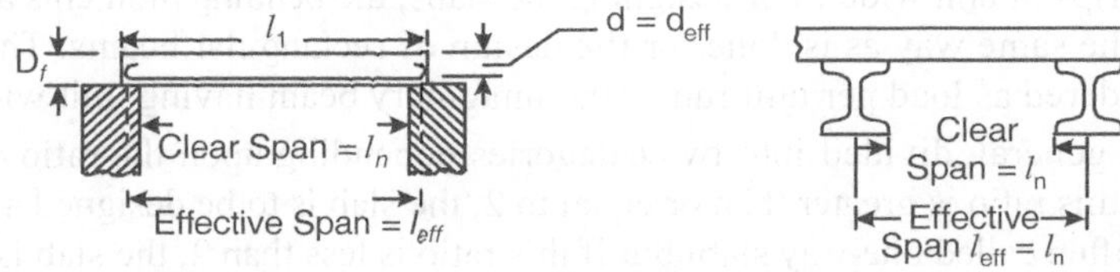

$$\left.\begin{aligned} l_{eff} &= l_n + d \\ &= l_1 \end{aligned}\right] \text{ whichever is less}$$

Fig. 7.4

(*b*) On beams cast monolithically with slabs. (Fig. 7.5)

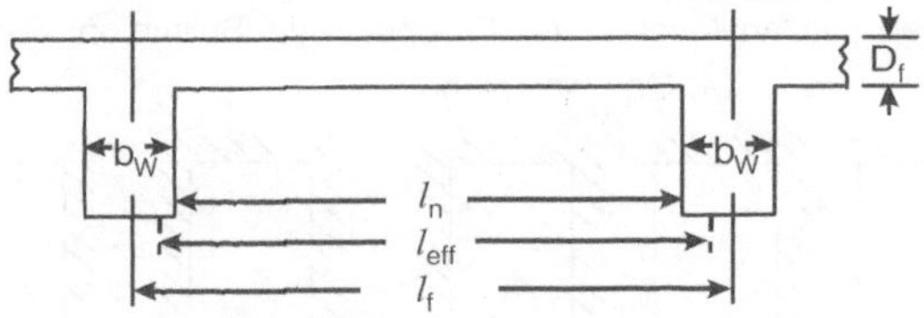

$$\left.\begin{aligned} l_{eff} &= l_n + d \\ &= l_1 \end{aligned}\right] \text{ whichever is less}$$

Fig. 7.5

$b_w \le \frac{l_n}{12}$ or 600 mm which ever is less effective span to be determined as per (*a*)

$$b_u \ge \frac{l_n}{12}$$

(*i*) For end span with one end fixed and the other continuous or for intermediate spans

$$l_{eff} = l_n$$

(*ii*) For end span with one end free and the other continuous

$$l_{eff} = l_n + d/2$$

$$= l_n + \text{half the width of discontinuous support.}$$

7.2.2 Control of Deflection

The provisions given in section 5.2-4 of chapter 5 for beams shall also apply to slabs,

For slabs spanning in two directions, the shorter of the two spans shall be used for calculating effective depth ratios.

For two-way slabs of small spans (upto 3.5 m) with mild steel reinforcement, the span to overall depth ratios given below may generally be assumed to satisfy the vertical deflection limits for loading class upto 3000 N/mm^2 (300 kg/m^2).

Simply supported slabs	35
Continuous slabs	40

For H.S. deformed bars of grade F_e 415, the values given above should be multiplied by 0.8.

7.2.3 Slabs Continuous over Supports

The slabs continuous over supports shall be designed according to the provisions applicable to continuous beams.

7.2.4 Slabs Monolithic with Supports

Bending moments in slabs constructed monolithically with the supports shall be calculated by taking such slabs either as continuous over supports and capable of free rotation, or as members of a continuous framework with the supports, taking into account the stiffness of such supports. If such supports are formed due to beams which justify fixity at the support of slabs, then the effects on such supporting beam, such as the bending of the web in the transverse direction of the beam and the torsion in the longitudinal direction of the beams, wherever applicable, shall also be considered in the design of the beam.

For the purpose of calculation of moments in slabs in a monolithic structure; it will generally be sufficiently accurate to assume that members connected to the ends of such slabs are fixed in position and direction at the ends remote from their connections with the slabs.

7.2.5 Slabs Carrying Concentrated Load

If a solid slab supported on two opposite edges, carries concentrated loads the maximum bending moment caused by the concentrated loads shall be assumed to be resisted by an effective width of slab (measured parallel to the supporting edges) as follows:

(*a*) For a single concentrated load, the effective width shall be calculated in accordance with the following equation provided that it shall not exceed the actual width of slab.

$$b_{ef} = k \cdot x \left(1 - \frac{x}{l_{ef}}\right) + a$$

where b_{ef} = effective width of slab.

K = constant having the values given in Table 7.1 depending upon the ratio of the width of the slab (l) to the effective span l_{eff}

x = distance of centroid of the concentrated load from nearer support.

l_{eff} = effective span, and

a = width of contact area of the concentrated load measured parallel to the supported edge.

And provided further that in case of a load near the unsupported edge of a slab, the effective width shall not exceed the above value plus the distance of the load from the unsupported edge.

(*b*) For two or more concentrated loads placed in a line in the direction of the span, the bending moment per meter width of slab shall be calculated separately for each load according to its appropriate effective width of slab calculated as in (*a*) above and added together for design calculations.

Table 7.1. Values of K for Simply Supported and Continuous Slabs

l'/l_{eff}	K for simply supported slabs	K for continuous slabs
0.1	0.4	0.4
0.2	0.8	0.8
0.3	1.16	1.16
0.4	1.48	1.44
0.5	1.72	1.68
0.6	1.96	1.84
0.7	2.12	1.96
0.8	2.24	2.08
0.9	2.36	2.16
1.0 or more	2.48	2.24

(*c*) For two or more loads not in a line in the direction of the span if the effective width of slab for one load does not overlap the effective width of slab for another load, both calculated as in (*a*) above, then the slab for each load can be designed separately. If the effective width of slab for one load overlaps the effective width of slab for an adjacent load, the overlapping portion of the slab shall be designed for the combined effect of the two loads.

(*d*) For cantilever solid slabs, the effective width shall be calculated in accordance with the following equation.

$$b_{ef} = 1.2\, a_1 + a$$

where b_{ef} = effective width

a_1 = distance of the concentrated load from the face of the cantilever support.

a = width of contact area of the concentrated load measured parallel to the supporting edge

Provided that the effective width of the cantilever slab shall not exceed one third the length of the cantilever slab measured parallel to the fixed edge and provided further that when the concentrated load is placed near the extreme ends of the length of the cantilever slab in the direction parallel to the fixed edge, the effective width shall not exceed the above value, nor shall it exceed half the above value plus the distance of the concentrated load from the extreme and measured in the direction parallel to the fixed edge.

For slabs other than solid slabs, the effective width shall depend on the ratio of the transverse and longitudinal flexural rigidities of the slab. Where this ratio is one, that is, where the transverse and longitudinal flexural rigidities are approximately equal the value of the effective width as found for solid slabs may be used. But as the ratio decreases, proportionately smaller value shall be taken.

7.2.6 Minimum Reinforcement

The mild steel reinforcement in either direction in slabs shall not be less than 0.15% of the total area of cross-section. However, this value can be reduced to 0.12% when H.S. deformed bars or welded wire fabrics are used. According to ACI-318 Code, minimum inforcement in either direction shall not be less than 0.2% of total area of cross-section.

7.2.7 Maximum Diameter

The diameter of reinforcing bar shall not exceed the one eighth of the total thickness of the slab.

7.2.8 Maximum Distance between Bars

(*i*) The horizontal distance between parallel main reinforcement shall not be more than 3 times the effective depth of a solid slab or 450 mm, whichever is smaller.

(*ii*) The horizontal distance between parallel reinforcing bars provided against shrinkage, temperature and uniform distribution of loads (such steel is called distribution steel) shall not be more than 5 times the effective depth of slab or 450 mm whichever is less. The area of such reinforcement shall not be less than as indicated in 7.2.6.

7.2.9 Cover to Reinforcement

Reinforcement shall have concrete and the thickness of such cover (exclusive of plaster or other decorative finish) shall be as follows :

(*a*) The reinforcement shall have a minimum clear cover of 20 mm or dia. of such bar whichever is more.

(*b*) At each end of reinforcing bar not less than 25 mm, nor less than twice the diameter of such bar.

(*c*) Increased cover thickness may be provided when surface of concrete members is exposed to the action of harmful chemicals (as in the case of concrete in contact with earth faces contaminated with such chemical) acid, vapour, saline atmosphere, sulphurous smoke (as in the case of steam operated railways) etc, and such increase may be between the cover should not exceed 75 mm.

7.2.10 Permissible Shear Stress in Slabs

For solid slabs, the permissible shear stress in concrete shall be $K.T_c$ where K has the value given below

Overall depth of slab	300 or more	275	250	225	200	175	150 or less
K	1.00	1.05	1.10	1.15	1.2	1.25	1.30

where τ_c = permissible shear stress in concrete depending upon $\dfrac{A_0 \times 100}{bd}$ and grade of concrete.

For slabs, τ_v shall not exceed half the value of $T_{c\,max}$ (given in Table 4.1).

7.3 RECTANGULAR SLABS SPANNING IN ONE DIRECTION (ONE-WAY SLABS)

One-way slabs may be

(*a*) simply supported or cantilever.

(*b*) continuous over two or more supports.

Following steps may be followed for the design of one-way slabs.

1. Assume some suitable thickness of slab, Table 7.2 may be useful.

Table 7.2. Guide to Assume Thickness of Slabs

Cantilever slab		*Simply supported slabs*		
Span (meters)	*Thickness (mm)*	*Span (meters)*	*Thickness (mm)*	
			One-way	*Two-way*
Upto 0.8	75	upto 1.5	75	—
0.8 to 1.2	100	1.5 to 2.5	100	75
1.2 to 1.5	125	2.5 to 4.0	125	100
1.5 to 2.5	150	4.0 to 5.0	150	125
		5.0 to 6.0	—	150

2. Compute maximum B.M. and maximum *S.F.*

For simply supported slabs loaded with *u. d. l*, the maximum bending moment is given by

$$M = \frac{WL}{8} = \frac{wL^2}{8}$$

where W = Total load on eff. span $L = wL$

w = Uniformly distributed load

L = Effective span.

For cantilever slabs, the maximum bending moment is given by :

$$M = \frac{WL}{2} = \frac{wl^2}{2}$$

The main reinforcement is provided at the bottom face in the case of simply supported spans and to allow for partial restraint for slabs built into brick walls, it is usual to provide top reinforcement at the face of the support for a moment of $WL/24$ or such other restraining moment as may be shown to be applicable. For such a condition shear coefficient given in Table at the end support may be increased by 0.05.

For continuous slabs, coefficients for computing bending moments and shear forces are same as for beams given in Tables 5.2 and 5.3.

For other loadings, conventional methods may be used to compute maximum Bending Moment and *S.F.*

3. Obtain effective depth from the relation

$$d = \sqrt{\left(\frac{M}{bQ}\right)}$$

where b = may be taken as unity depending upon the units adopted in design.

= 100 cm in *MKS*

= 1000 mm in *SI*

= 12″ in *FPS*

4. Overall depth shall be obtained as :

$$D_f = d + 25 \text{ mm (clear cover)} + \text{half the proposed dia. of bar.}$$

5. This overall depth should be in close approximately to assumed depth for assessment of self weight, otherwise modify the size and repeat steps 1 to 4.

6. Calculate main reinforcement on tension side by the relation

$$A_{st} = \frac{M}{\sigma_{st \cdot j} . d}$$

Provide suitable dia. of bars with proper spacing so that area provided is equal to or more than that required.

7. In one-way slabs, provide distribution steel as 0.15% of gross cross sectional area of concrete in case of *M.S.* bars and 0.12% of gross cross sectional area of concrete in case of twisted deformed bars.

8. Check the slab for shear. Mostly slabs are safe in shear.

9. Check for deflection as in case of beams.

7.4 SOLVED EXAMPLES

Example 7.1. Design a verandah slab to cover a clear span of 2.5 metres for concrete grade M 20 and to be used as steel reinforcement. It carries a superimposed load of 2500 N/m² (250 kg/m²) including finishes. The thickness of walls on two opposite supporting edges may be taken as 225 mm.

Solution. (Fig. 7.6)

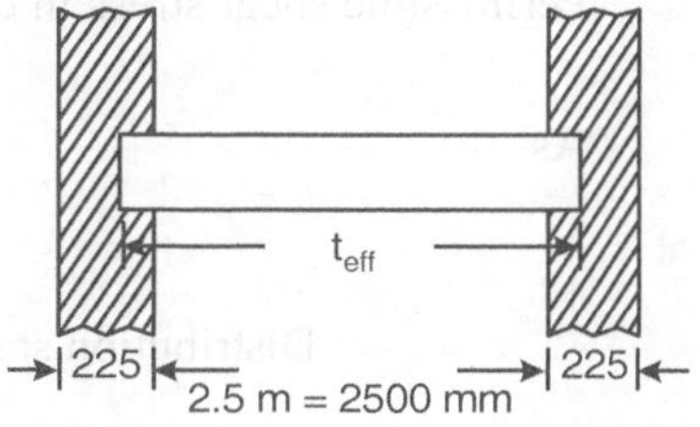

Assume the thickness of slab as 100 mm

Loads

$$\text{Self weight} = 0.100 \times 1 \times 25000$$

$$= 2500 \text{ N/m}^2 = 250 \text{ kg/m}^2$$

$$\text{Superimposed load} = 2500 \text{ N/m}^2 = 250 \text{ kg/m}^2$$

$$\text{Total load} = 5000 \text{ N/m}^2$$

$$\text{Effective span} \approx 2.5 + 0.1 \approx 2.6 \text{ m}$$

$$\text{Maximum } B.M. = M = \frac{5000 \times 2.6^2}{8} = 4225 \text{ N.m}$$

For a balanced design $d = \sqrt{\left(\frac{M}{Qb}\right)}$

$$d = \sqrt{\left(\frac{4225 \times 1000}{0.91 \times 1000}\right)} = 68.5 \text{ mm}$$

Provide 20 mm effective cover to the centre line of steel

Overall depth = 100 mm $d_{eff} = 80$ mm

Assumed overall thickness of slab as 100 mm is O.K.

Area of steel required

$$A_{st} = \frac{M}{\sigma_{st} \, j d}$$

$$A_{st} = \frac{4225 \times 1000}{230 \times 0.90} = 255 \text{ mm}^2$$

Provide 8 mm dia @ 150 mm *c/c*, thus providing an area of steel = 333 mm². It is less than maximum permissible spacing of $3d = 3 \times 80 = 240$ mm

$$\text{Max. } S.F. = V = \frac{wl}{2} = \frac{5000 \times 2.5}{2}$$

$$= 6250 \text{ N}$$

$$\text{Nominal shear stress} = \tau_v = \frac{V_s}{bd} = \frac{6250}{1000 \times 80}$$

$$= 0.078 \text{ N/mm}^2$$

$$p = \frac{A_{st} \times 100}{bd} = \frac{333 \times 100}{1000 \times 80}$$

$$= 0.416$$

For M 20

Permissible shear stress in concrete

$$\tau_c = 0.27 \times 1.3 = 0.35$$

As $\tau_v < \tau_c$

Safe

$$\text{Distribution steel} = \frac{0.12}{100} \times bD$$

$$= 0.0012 \times 1000 \times 100 = 120 \text{ mm}^2$$

Provide 8 mm dia. @ *c/c*, $5d = 5 \times 80 = 400$ mm

Thus providing an area of steel as 168 mm².

Check for Vertical deflection

$$f_s = 0.58 f_y \cdot \frac{\text{Area of Cross-section of steel required}}{\text{Area of cum-section of steel provided}}$$

$$= 0.55 \times 415 \times \frac{255}{333} = 184$$

Modification factor for tension steel from Fig. 5.1 = 1.7

$$\therefore \quad \frac{L_{ef}}{d} = \frac{2600}{80} = 32.5$$

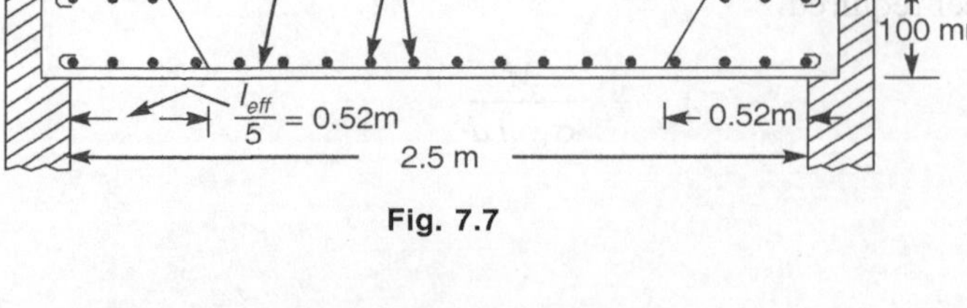

Fig. 7.7

$$\text{Permissible } \frac{L_{ef}}{d} = 20 \times 1.7 = 34 > 32.5$$

Hence vertical deflection limit is satisfied.

Main Steel

$$a + b + c = 8 \text{ mm } \phi \text{ @ } 150 \; c/c$$
$$a = 8 \text{ mm } \phi \text{ @ } 450 \; c/c$$
$$b = 8 \text{ mm } \phi \text{ @ } 450 \; c/c$$
$$c = 8 \text{ mm } \phi \text{ @ } 450 \; c/c$$

Check for End Anchorage/Development Length

If alternate bars of main steel 8 mm ϕ @ 150 mm c/c are bent-up

$$L_d \ngtr \frac{M_1}{V} + L_0$$

$$L_d = \frac{\phi \cdot \sigma_s}{4 \cdot T_{bd}} = \frac{8 \times 230}{4 \times 1.28} = 359 \text{ mm}$$

M_1 = Moment of resistance of the section assuming all reinforcement to be stressed to σ_{st},

$$= \left(\frac{50 \times 1000}{300}\right) 230 \times 0.9 \times 80$$
$$= 2760000 \text{ N.mm}$$
$$V = 6250 \text{ N}$$
$$L_0 = 12\,\phi \text{ or eff. depth whichever is greater.}$$
$$= 12 \times 8 \text{ or } 80 = 96 \text{ or } 80 \text{ mm}$$

Take $L_0 = 96$ mm

$$\frac{M_1}{V} + V_0 = \frac{2760000}{6250} + 96$$
$$= 442 + 96 = 538 \text{ mm} > 359.$$
$$L_d > \frac{M_1}{V} + L_0$$

It is suggested to bend every third bar instead of alternate bars.

$$M_1 = \left(\frac{2}{3} \times \frac{50 \times 1000}{110}\right) 140 \times 0.87 \times 80$$
$$= 2952727.2 \text{ N.mm}$$
$$\frac{M_1}{V} + L_0 = \frac{2952727.2}{6250} + 96 = 472.4 + 96$$
$$= 568.4 \text{ mm} > 467 \text{ mm}$$

Now $L_d < \frac{M_1}{V} + L_0$

Therefore, bending every third bar and thus leaving two-third bars straight makes the design for end anchorage safe.

The placing and details of reinforcement for this are shown in Fig. 7.7.

Example 7.2. Design a cantilever slab of 2.0 m span and carrying a superimposed load of 2500 N/m² (250 kg/m²) including finishes. Use concrete of grade M20 and HYSD (Fe 415) bars as reinforcement.

Solution. Assume the thickness of slab 150 mm

Loads

Self weight of slab $= 0.15 \times 1 \times 25000 = 3750$ N/m²

Superimposed load on slab $= 2500$ N/m²

Total *u. d. l.* $= w = 6250$ N/m²

Effective span $= l_{ef} = 2$ m

$$\text{Maximum B. M. at built-in end} = \frac{wl^2}{2} = \frac{6250 \times 2^2}{2}$$

$$= 12500 \text{ N.n}$$

For a balanced design

$$d = \sqrt{\left(\frac{M}{Qb}\right)}$$

$$= \sqrt{\left(\frac{12500 \times 1000}{0.91 \times 1000}\right)} = 117 \text{ mm}$$

Providing 25 mm cover to the centre of steel bars.

Let Overall depth $= 150$ mm

$d = 150 - 25 = 125$ mm

Assumed 150 mm thickness is O.K.

$$\text{Area of steel required} = \frac{M}{\sigma_{st}\, jd} = \frac{12500 \times 1000}{230 \times 0.90 \times 125}$$

$$= 483 \text{ mm}^2$$

Providing 10 mm ϕ @ 150 mm *c/c* (thus providing 523 mm²)

The spacing of bars is $< 3 \times 125 = 375$ mm

Max. *S.F.* $= V = wl$

$= 6250 \times 2 = 12500$ N

$$\text{Nominal shear stress} = \tau_v = \frac{V}{bd} = \frac{12500}{1000 \times 125}$$

$$= 0.10 \text{ N/mm}^2$$

$$p = \frac{A_{st} \times 100}{bd} = \frac{523 \times 100}{1000 \times 125} = 0.418$$

For M20 concrete

Permissible shear stress $\tau_c = 0.18 \times 1.3 = 0.234$ N/mm²

$\tau_v < \tau_c$

$\therefore$ Safe in shear.

Distribution Steel $A_{st} = \frac{0.12}{100} \times 1000 \times 150$

$= 180 \text{ mm}^2$

Provide 8 mm ϕ @ 220 *c/c*,

(thus providing area of steel = 229 mm²) < 5, $d = 5 \times 130 = 650$ mm

Check for Deflection

$$\text{Actual } \frac{l_{eff}}{d} = \frac{2000}{125} = 16$$

For $p = 0.41$ for Fe 415, Modification factor = 1.5

$$\text{Permissible } \frac{l_{eff}}{d} = 7 \times 1.5 = 10.5 < 15.4$$

As Deflection criteria is *not satisfied,*

Increase the overall depth of slab to 180 mm and $d = 150$ mm

$$\text{Self weight of slab} = 0.18 \times 1 \times 25000 = 4500 \text{ N/m}^2$$

$$\text{Actual } \frac{l_{eff}}{d} = \frac{2000}{150} = 13.33$$

$$\text{Total weight on slab } w = 2500 + 4500 = 7000 \text{ N/m}^2$$

$$\text{Maximum } B.M = M = \frac{7000 \times 2^2}{2} = 14000 \text{ N.m.}$$

$$d = \sqrt{\left(\frac{14000 \times 1000}{1000 \times 0.91}\right)} = 125 \text{ mm} < 150 \text{ mm}$$

Take $d = 150$ m $D_f = 180$ mm

$$\text{Self weight} = 0.18 \times 1 \times 25000 = 4500 \text{ N/m}^2$$

$$A_{st} = \frac{M}{\sigma_{st} \cdot j \cdot d} = \frac{14000 \times 1000}{230 \times 0.90 \times 150}$$

$$= 450.8 \text{ mm}^2$$

Provide 10 mm ϕ @ 100 *c/c*

Thus providing $A_{st} = 785 \text{ mm}^2$

Check for Deflection

$$p = \frac{A_{st} \times 100}{bd} = \frac{785 \times 100}{1000 \times 150} = 0.523$$

For HYSD Bars.

$$\text{Modification factor} = 2.0 \text{ from Fig. 5.1 where } f_s = 0.58 \times 415 \times \frac{450.8}{785}$$

$$= 138 \text{ N/mm}^2$$

$$\frac{l_{ef}}{d} = 7 \times 2 = 14 > 13.33$$

O.K.

Dist. steel $A_{st} = 0.0012 \times 180 \times 1000 = 216 \text{ mm}^2$

Provide 8 mm ϕ @ 200 *c/c*, thus providing $A_{st} = 250 \text{ mm}^2$

Check for Shear

For $p = 0.348$ and M 20 concrete

$\tau_c = 0.28 \text{ N/mm}^2$

Max $S.F. = 7000 \times 2 = 14000$ N

$$T_v = \frac{14000}{1000 \times 150} = 0.093 < 0.28 \text{ N/mm}^2 \qquad \text{Safe}$$

Check for End Anchorages/Development Length

$$L_d \ngtr \frac{M_1}{V} + L_0$$

$$L_d = \frac{\sigma_s \cdot \phi}{4_{T_{bd}}} = \frac{230 \times 10}{4 \times 1.28} = 449 \text{ mm}$$

$$M_1 = \left(\frac{78.5 \times 1000}{150}\right) \times 230 \times 0.90 \times 150 = 16249500 \text{ N.mm}$$

$$V = 14000 \text{ N}$$

$$L_0 = 12\,\phi \text{ or } d_{eff} \text{ whichever is more}$$

$$= 12 \times 10 \quad \text{or} \quad 150 \text{ take } L_0 = 150 \text{ mm}$$

$$\frac{M_1}{V} + L_0 = \frac{16249500}{14000} + 150 = 1310 \text{ mm} > 449 \text{ mm}$$

The reinforcement details are shown in Fig. 7.8

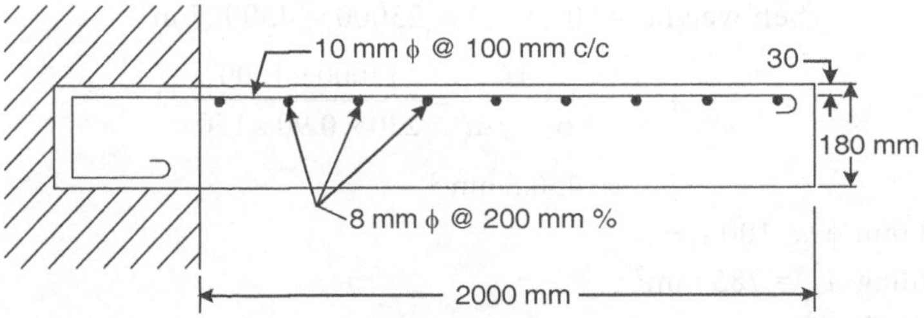

Fig. 7.8

Example 7.3. Design a continuous slab over a room 10 m × 15 m in plan. Four beams 300 mm × 800 mm are provided to cover 10 m span. The slab carries a superimposed live load of 2500 N/m² (250 kg/m²). Use M 20 concrete and HYSD (Fe 415) bars as reinforcement.

Solution. (Fig. 7.9)

Assume thickness of slab = 120 mm

Loads

Self weight of slab $= 0.12 \times 1 \times 1 \times 25000 = 3000 \text{ N/m}^2$

Superimposed live load $= 2500 \text{ N/m}^2$

Finishes $= 2000 \text{ N/m}^2$

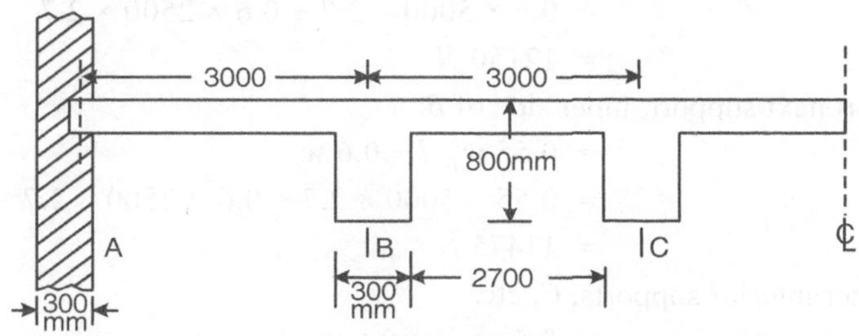

Fig. 7.9

Total Dead load $w_D = 3000 + 2000 = 5000 \text{ N/m}^2$

Total Lived Load $w_l = 2500 \text{ N/m}^2$

$l_{ef} = 3 \text{ m}$ or $2.7 + 0.12 = 2.82$ m whichever is less.

$l_{ef} = 2.82$ m

Computation of Bending Moment at Various Points

$$B.M. \text{ at middle of end span } AB = \frac{w_d\, l_{ef^2}}{12} + \frac{w_l\, l_{ef^2}}{10}$$

$$= \frac{5000 \times 2.82^2}{12} + \frac{2500 \times 2.82^2}{10}$$

$$= 3312 + 1988 = 5300 \text{ Nm}$$

$$B.M. \text{ at middle of interior spans } BC, \text{ etc.} = \frac{w_d\, l_{ef^2}}{16} + \frac{w_l\, l_{ef^2}}{12}$$

$$= \frac{5000 \times 2.82^2}{16} + \frac{2500 \times 2.82^2}{12}$$

$$= 2485 + 1656.75 = 41461.75 \text{ N.m}$$

$$B.M. \text{ at support next to end support, } i.e., B = -\frac{w_d\, l_{ef^2}}{10} - \frac{w_l \,.\, l_{ef^2}}{9}$$

$$= -\frac{5000 \times 2.82^2}{10} - \frac{2500 \times 2.82^2}{9}$$

$$= -3975 - 2210 = -6185 \text{ N.m}$$

$$B.M. \text{ at interior supports } C, \text{ etc.} = -\frac{w_d \,.\, l_{ef^2}}{12} - \frac{w_l \,.\, l_{ef^2}}{9}$$

$$= -\frac{5000 \times 2.82^2}{12} - \frac{2500 \times 2.82^2}{9}$$

$$= -3312 - 2210 = -5522 \text{ N·m}$$

Computation of Shear Forces

$S.F.$ at end support $A = w_d \cdot l_{ex} \times 4 + w_d\, l \times 0.45$

$= 5000 \times 2.7 \times 0.4 + 2500 \times 2.7 \times 0.45 = 8437.5$ N

$S.F.$ at support next to end support, outer side of B

$= 0.6\, w_d\, l + 0.6\, w_l\, l$

$$= 0.6 \times 5000 \times 2.7 + 0.6 \times 2500 \times 2.7$$
$$= 12150 \text{ N}$$

S.F. at support next support, inner side of *B*

$$= 0.55\, w_d \cdot l + 0.6\, w_l \cdot l$$
$$= 0.55 \times 5000 \times 2.7 + 0.6 \times 2500 \times 2.7$$
$$= 11475 \text{ N}$$

S.F. at all other interior supports, *C*, etc.

$$= 0.5\, w_d \cdot l + 0.6\, w_l \cdot l$$
$$= 0.5 \times 5000 \times 2.7 + 0.6 \times 2500 \times 2.7$$
$$= 10800 \text{ N}$$

Maximum $B.M. = 6185$ N.m

For a balanced design $d = \sqrt{\left(\dfrac{M}{Qb}\right)} = \sqrt{\left(\dfrac{6185 \times 1000}{0.91 \times 1000}\right)} = 83$ mm

Providing overall depth $D_f = 120$ mm

$$d = D_f - \text{eff. cover}$$
$$d = 120 - 25 = 95 \text{ mm}$$

Areas of Steel Required at Various Points of Interest

A_{st} at middle of $AB = \dfrac{5300 \times 1000}{230 \times 0.90 \times 95} = 270 \text{ mm}^2$

Provide 8 mm ϕ @ 150 *c/c*, thus providing $A_{st} = 335 \text{ mm}^2$

A_{st} at middle of $BC = \dfrac{3313.5 \times 1000}{230 \times 0.90 \times 95} = 169 \text{ mm}^2$

Provide 8 mm ϕ @ 190 *c/c*, thus providing $A_{st} = 335 \text{ mm}^2$

A_{st} at support $B = \dfrac{6185 \times 1000}{230 \times 0.90 \times 95} = 315 \text{ mm}^2$

Provide + 8 mm @ 150 *c/c*. Thus providing $A_{st} = 335 \text{ mm}^2$

A_{st} at support $C = \dfrac{5522 \times 1000}{230 \times 0.90 \times 95} = 280 \text{ mm}^2$

Provide 8 mm ϕ @ 150 *c/c*. Thus providing $A_{st} = 335 \text{ mm}^2$

Check for Shear

Maximum $S.F. = V = 12150$ N

Nominal shear stress $\tau_v = \dfrac{V}{bd}$

$$\tau_v = \frac{12150}{1000 \times 95} = 0.128$$

$$p = \frac{335 \times 100}{1000 \times 95} = 0.35$$

Permissible shear stress

$$\tau_c \text{ for M 20 concrete} = 0.25 \times 1.3 = 0.375 \text{ N/mm}^2$$

Safe

The distribution steel

$$A_{st} = 0.0012\, bD$$
$$= 0.0012 \times 1000 \times 120 = 144 \text{ mm}^2$$

Provide 8 mm ϕ @ 300 *c/c*, thus providing

$$A_{st} = 168 \text{ mm}^2$$

Arrangement of Reinforcement

Reinforcement will have to be provided near top face for negative moments over supports, and at bottom face for the positive moment in middle of spans. Bars which are no longer required for positive moments, have to be bent up from a distance 0.15 *l* from face of supports and these bars must extend 0.3 *l* from other face of support on top face of slab. Details of reinforcement are shown in Fig. 7.10.

Check for Deflection

$$\frac{L_{eff}}{d} = \frac{2.82}{0.095} = 29.7$$

$$p \text{ at middle of interior span } BC = \frac{335 \times 100}{1000 \times 95} = 0.352$$

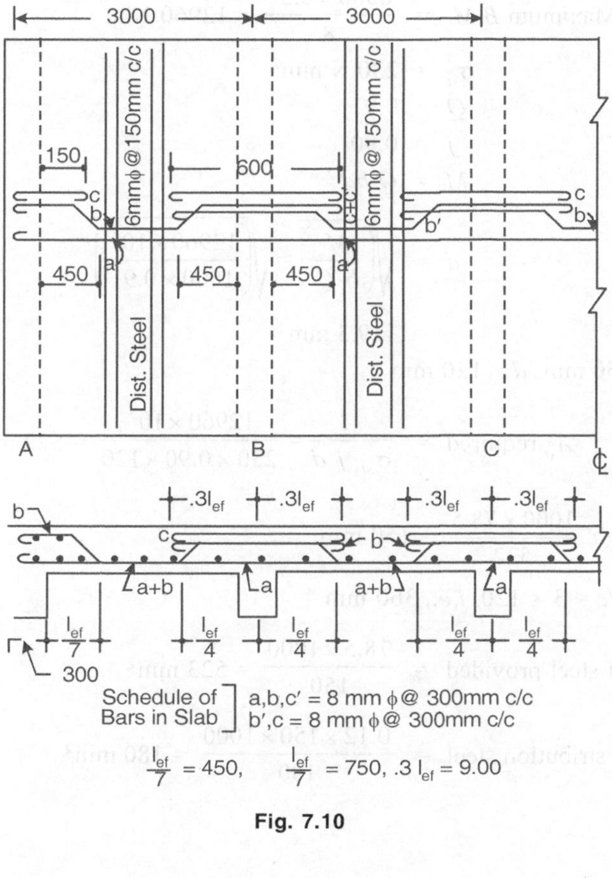

Fig. 7.10

Allowable L_{eff} / d for deflection to be within limits

$$= 26 \times 1.8 = 46.8 > 29.7$$

$\therefore$ Safe

Check for End Anchorage

Max. *S.F.* is at support *B* (next to the end support) = 12150 N.

$$\frac{M_1}{V} + L_0 = \frac{335 \times 230 \times 0.9 \times 95}{12150} + 95 = 635 \text{ mm}$$

$$L_d = \frac{\sigma_s \phi}{4.\tau_{bd}} = \frac{8 \times 230}{4 \times 1.28} = 359 \text{ mm}$$

$$L_d \not> \frac{M_1}{V} + L_0$$

Safe

Example 7.4. Design a reinforced concrete slab for an office floor to carry a load of 8000 N/m^2 (800 kg/m^2) inclusive of its own weight over in effective span of 3.5 m simply supported at its ends. Use M20 concrete and HYSD (Fe 415) steel with σ_{st} = 230 N/mm^2.

Solution. Consider 1 m width

u. d. l. over span, $w = 8000 \times 1 = 8000$ N/m

$$\text{Maximum } B.M. = \frac{8000 \times 3.5^2}{8} = 12960 \text{ N.m}$$

$$\sigma_{st} = 230 \text{ N/mm}^2$$

$$Q = 0.91$$

$$j = 0.90$$

$$M = Q\, bd^2$$

$$d = \sqrt{\left(\frac{M}{b \cdot Q}\right)} = \sqrt{\left(\frac{12960 \times 10^3}{1000 \times 0.91}\right)}$$

$$= 119.3 \text{ mm}$$

Provide $D = 150$ mm, $d = 120$ mm

$$A_{st} \text{ required} = \frac{M}{\sigma_{st} \cdot j.d} = \frac{12960 \times 10^3}{230 \times 0.90 \times 120} = 522 \text{ mm}^2$$

Use 10 mm Φ @ $\dfrac{1000 \times 78.5}{522} = 150$ mm

Say 150 mm *c/c* < 3 × 120, *i.e.*, 360 mm

$$\text{Area of steel provided} = \frac{78.5 \times 1000}{150} = 523 \text{ mm}^2$$

$$\text{Distribution steel} = \frac{0.12 \times 150 \times 1000}{100} = 180 \text{ mm}^2$$

Use 8 mm Φ @ $\dfrac{50 \times 1000}{180}$ = 272 mm say 275 mm *c/c*.

less than 5 × 120 = 600 mm or 450 mm

$\therefore$ O. K.

Check for Shear

$$S.F. = \frac{8000 \times 3.5}{2} = 14000 \text{ N}$$

$$\tau_t = \frac{14000}{1000 \times 120} = 0.116 \text{ N/mm}^2$$

$$\frac{100 \times A_s}{b\ d} = \frac{100 \times 523}{1000 \times 120} = 0.435$$

$$\tau_c = 0.28 \times 1.3 = 0.364 \text{ N/mm}^2 > 0.116 \text{ N/mm}^2, \text{ Safe}$$

Check for Deflection

$$\text{Actual span}/d = \frac{3500}{120} = 29.2$$

$$p = \frac{523 \times 100}{1000 \times 120} = 0.435$$

$$\text{Modification factor} = 1.35 \text{ from Fig. 7.1 where } f_s = 0.58 \times 415 \times \frac{522}{523}$$

$$= 240.7 \text{ N/mm}^2$$

$$\text{Allowable span}/d \text{ ratio} = 20 \times 1.35 = 27 > 29.2$$

$$\text{Actual deflection} = \frac{5}{384} \times \frac{8000 \times 3.5^4 \times 10^9 \times 12}{223 \times 10^4 \times 1000 \times 150^3} = 2.5 \text{ mm} \quad \text{(very small)}$$

The details of slab reinforcement are shown in Fig. 7.11.

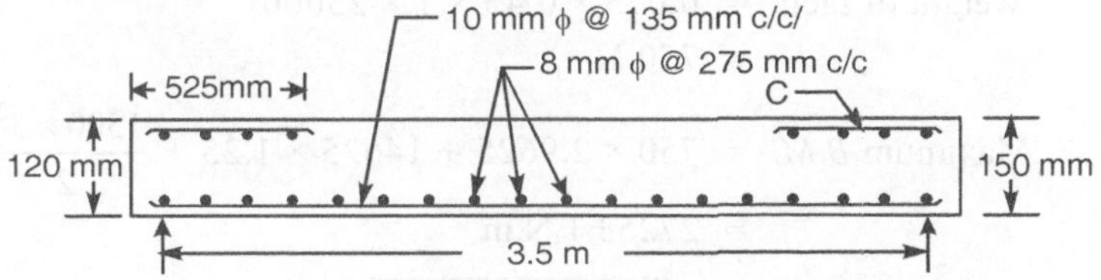

Fig. 7.11

Example 7.5. The roof of a car parking place consists of a reinforced concrete slab cantilevering 3 m from *R. C. C.* Beam which is well supported on columns. Design the slab assuming the live load on the roof to be 1500 N/m² with no finishes. Use M 20 concrete and ribbed tor steel with σ_{st} = 230 N/mm².

Solution. (Fig. 7.12)

To find the position of *N. A.*

$$x = k\,d$$

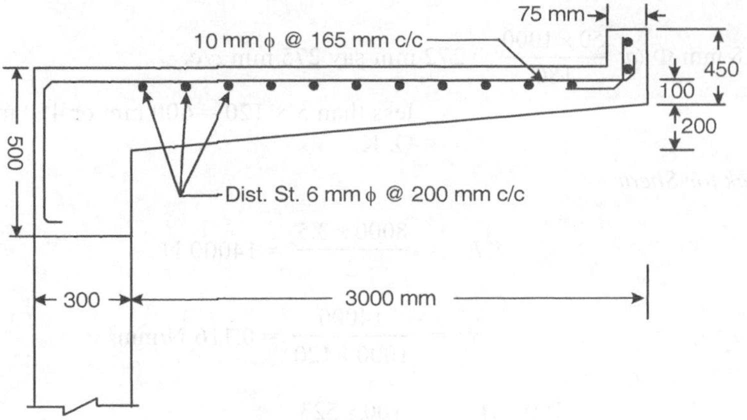

Fig. 7.12

where
$$k = \frac{1}{1+\dfrac{\sigma_{st}}{m\cdot\sigma_{cbc}}} = \frac{1}{1+\dfrac{230}{13.3.\times 7}} = 0.288$$

$$j = 1-\frac{k}{3} = 1-\frac{0.288}{3} = 0.904$$

$$Q = \frac{1}{2}\sigma_{cbc}.\, j\cdot k = \frac{1}{2} + 7 \times 0.904 \times 0.288$$

$$= 0.91$$

Assuming the overall depth of slab = 300 mm tapered to 100 mm at free end.

$$\text{Average thickness} = \frac{100+300}{2} = 200.0 \text{ mm}$$

$$\text{Self weight} = 0.2000 \times (3 - 0.075) \times 1 \times 25000$$

$$= 14625 \text{ N acting at 1.25 m from fixed end.}$$

$$\text{Weight of facia} = 0.075 \times 0.45 \times 1 \times 25000$$

$$= 750 \text{ N}$$

$$\text{Maximum } B.M. = 750 \times 2.9625 + 14625 \times 1.25 + \frac{1500 \times 3^2}{2}$$

$$= 27253.1 \text{ N.m}$$

$$d = \sqrt{\left(\frac{27253.1\times 1000}{0.91\times 1000}\right)} = 173 \text{ mm}$$

$$D = 300 \text{ mm is O. K.}, d = 270 \text{ mm}$$

$$A_{st} = \frac{27253.1\times 1000}{230\times 0.904\times 270} = 487 \text{ mm}^2$$

Use 10 mm Φ (tor) bars @ 100 mm *c/c*

$$A_{st} \text{ provided} = 785 \text{ mm}^2$$

Check for Shear

Maximum *S.F.* at fixed end, $V = 750 + 14625 + 1500 \times 3$

$= 19875$ N

Nominal shear stress, $\tau_v = \dfrac{V}{bd} = \dfrac{19875}{1000 \times 280}$

$= 0.071$ N/mm^2

$$\frac{A_{st} \times 100}{b\,d} = \frac{475 \times 100}{1000 \times 280} = 0.17$$

$\therefore \quad \tau_c = 0.22 \times 1.0 = 0.22 \text{ N/mm}^2 > 0.071 \text{ N/mm}^2$

$\therefore$ Safe

Check for Deflection

$$\text{Actual span}/d = \frac{300}{270} = 11.1$$

$$\frac{A_s \times 100}{b \cdot d} = \frac{785 \times 100}{1000 \times 270} = 0.29$$

$K_1 = 2.00$ from Fig. 5.1 where $f_s = 0.58 \times 415 \times \dfrac{487}{785}$

$= 150$ N/mm^2

Permissible span/*d* ratio for a cantilever

$= 7.0 \times K_1 = 7.0 \times 2.0 = 14.0 > 10.7$

$\therefore$ Safe

Distribution Steel

Area of distribution steel $= 0.15\%$ of cross-sectional area for M. S.

$= 0.8 \times 0.15$, *i.e.*, 0.12% for Tor Steel

$$= \frac{0.12}{100} \times \left(\frac{100 + 300}{2}\right) \times 1000 = 240 \text{ mm}^2$$

Provide 8 mm Φ (tor) @ 200 mm *c/c*, Area provided = 252 mm^2

The reinforcement details are shown in Fig. 7.12.

7.5 RECTANGULAR SLABS SPANNING IN TWO DIRECTIONS (TWO-WAY SLABS)

A slab supported on all its edges with ratio of longer side to shorter side smaller than or equal to two, is termed as two-way slab. A two-way designed slab is smaller in thickness than a one-way slab for similar loading and supporting conditions and so it is more economical. The load which flows or gets support in one direction in one-way slabs, gets distributed in two directions at right angles in two-way slabs. For two-way slabs, steel in each direction is considered as main steel. The computation of bending moments in the two directions is not very easy if all conditions of elasticity have to be satisfied. Therefore, some empirical formulae and approximate theories have been formulated. The method, most suited to a particular situation depends upon the support conditions and the type of loading. Bending Moments, in two directions can be computed by any one of the following methods, most frequently used.

7.6 METHODS OF ANALYSIS

The most commonly used elastic methods are based on Pigeaud Grashof, Rankine or Marcus method. Westergaard coefficient method as presented in Appendix *C* of IS : 456-1978 (I.S. Code Method) is given below.

7.6.1 I. S. Code Method

7.6.1.1 Restrained Slabs

This method is based on the mathematical analysis carried out by Westergaard in America and supplemented by experimental data. It is applicable for slabs loaded with *u. d. l.* restrained on all sides, the corners of which are prevented from lifting and adequate provision for torsion is made.

(*a*) Slabs are considered as divided in each direction into middle strips and edge strips as shown in Fig. 7.13. The middle strips being three quarters of the width and edge strip one-eighth of the width of slab.

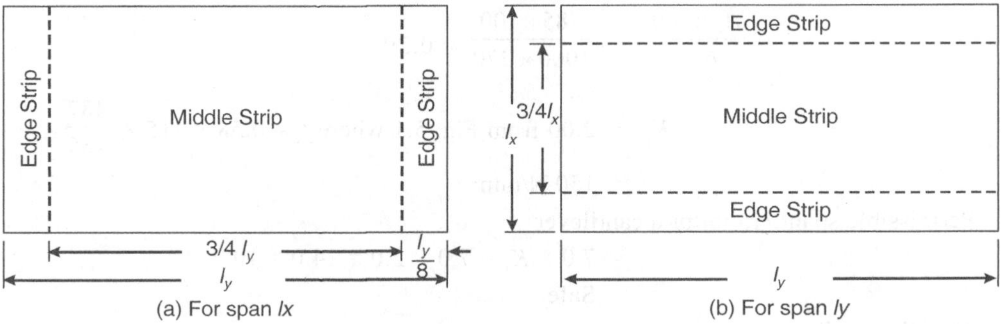

Fig. 7.13

(*b*) The maximum bending moments per unit width of slab are given by the following equations.

$$M_x = \propto_x \cdot w \cdot l_x^2$$
$$M_y = \propto_y \cdot w \cdot l_x^2$$

where $\propto_x$ and $\propto_y$ are coefficients given in Table 7.3.

w = total design load per unit area

M_x, M_y = moments on strips of unit width spanning l_x and l_y respectively, and

l_x and l_y = lengths of the shorter span and longer span respectively.

(*c*) The maximum moments calculated as above in (*b*) apply only to the middle strip and no redistribution shall be made.

(*d*) Tension reinforcement provided at mid span in the middle strip shall extend in the lower part of the slab to within 0.25 *l* of a continuous edge or 0.15 *l* of a discontinuous edge.

(*e*) Over the continuous edges of a middle strip the tension reinforcement shall extend in the upper part of the slab a distance of 0.15 *l* from the support, and at least 50 per cent shall extend a distance 0.3*l*.

(*f*) At a discontinuous edge, negative moments may arise. They depend on the degree of fixity at the edge of the slab but, in general, tension reinforcement equal to 50 per cent of that provided at mid span extending 0.1*l* into the span will be sufficient.

Table 7.3. Bending Moment Coefficients for Rectangular Panels Supported on Four Sides with Provision for Torsion at Corners

Case No	*Type of Panel and Moments Considered*	*Short Span Coefficients* α_x *(Values of* l_y/l_x*)*								*Long Span Coefficients* α_y *for all Values of* l_y/l_x
		1.0	1.1	1.2	1.3	1.4	1.5	1.75	2.0	
1.	INTERIOR PANELS									
	Negative B.M. at cont. edge	0.032	.037	.043	.047	.051	.053	.060	.065	0.032
	Positive B.M. at mid-span .	0.024	.028	.032	.036	.039	.041	.045	.049	0.024
2.	One Short Edge Discontinuous									
	Negative B.M. at cont. edge	0.037	.043	.048	.051	.055	.057	.064	.068	0.037
	Positive B.M. at mid-span .	0.028	.032	.036	.039	.041	.044	.048	.052	0.028
3.	One Long Edge Discontinuous									
	Negative B.M. at cont. edge	0.037	.044	.052	.057	.063	.067	.077	.085	0.037
	Positive B.M. at mid-span .	0.028	.033	.039	.044	.047	.051	.059	.065	0.028
4.	Two Adjacent Edges Discontinuous									
	Negative B.M. at cont. edge	0.047	.053	.060	.065	0.071	.075	.084	.091	0.047
	Positive B.M. at mid-span	0.035	.040	.045	.049	.053	.056	.063	.069	0.035
5.	Two Short Edges Discontinuous									
	Negative B.M. at cont. edge	0.045	.049	.052	.056	.059	.060	.065	.069	
	Positive B.M. at mid-span .	0.035	.037	.040	.043	.044	.045	.049	.052	0.035
6.	Two Long Edges Discontinuous									
	Negative B.M. at cont. edge	–	–	–	–	–	–	–	–	0.045
	Positive B.M. at mid-span .	0.035	.043	.051	.057	.063	.068	.080	.088	0.035
7.	Three Edges Discontinuous (One Long Edge Continuous)									
	Negative B.M. at cont. edge	0.057	.064	.071	.076	.080	.084	.091	.097	–
	Positive B.M. at mid-span .	0.043	.048	.053	.057	.060	.064	.069	.073	0.043
8.	Three Edges Discontinuous (One Short Edge Continuous)									
	Negative B.M. at cont. edge	–	–	–	–	–	–	–	–	0.057
	Positive B.M. at mid-span .	0.043	.051	.059	.063	.071	.076	.087	.096	0.043
9.	Four Edges Discontinuous									
	Positive B.M. at mid-span .	0.056	.064	.072	0.079	.085	.089	.100	.107	0.056

(*g*) Reinforcement in edge strip, parallel to that edge, shall comply with the minimum tension steel requirement that the minimum area of tension reinforcement shall not be less than that given by 7.2.6

The edge strip shall also comply with minimum requirements of torsion given subsequently.

(*h*) Reinforcement for torsion at corners. When loads are applied to a rectangular slab simply supported on the four sides, there is a tendency for the slab to lift at corners, torsional moments are induced as they are held down. The problem of lifting does not arise, if the slab is fixed at edges or is continuing. Special attention becomes necessary only when the slabs have discontinuous edges. Exact calculation of moments in a slab in which the corners can lift, is extremely difficult. But at the same time neglect of torsion makes it desirable to use increased bending moments in the span.

Torsion reinforcement shall be provided at any corner, where the slab is simply supported on both edges meeting at that corner. It shall consist of top and bottom reinforcement, each with layers of bars placed parallel to the sides of the slab and extending from the edges a minimum distance of one-fifth of the shorter span. The area of reinforcement in each of these four layers shall be three-quarters of the area required for the maximum mid-span moment in slab.

Torsion reinforcement equal to half of that prescribed in the previous paragraphs, shall be provided at a corner contained by edges over only one of which the slab is continuous.

Torsion reinforcement need not be provided at any corner contained by edges over both of which the slab is continuous.

7.6.1.2 Simply Supported Slabs

When simply supported slabs do not have adequate provisions to resist torsion at corners and to prevent the corners from lifting, the maximum moment per unit width are given by the following equation.

$$Mx = \propto_{x'} w \cdot l_x^2$$

$$M_y = \propto_{y'} w \cdot l_x^2$$

Where M_x, M_y, w, l_x and l_y are same as above but

$\propto_{x'}$ and $\propto_y$ are moment coefficients given in Table 7.4.

At least 50% of the tension reinforcement provided at mid span should extend to support. The remaining 50% should extend to within 0.1 l_x or 0.1 l_y of the support as appropriate.

The following specific example problem will illustrate the use of the tables.

Table 7.4. Bending Moment Coefficients for Slabs Spanning in Two Directions at Right Angles, Simply Supported on Four Edges

L_y/L_x	1.0	1.1	1.2	1.3	1.4	1.5	1.75	2.0	2.5	3.0
$\propto_x$	.062	.074	.084	.093	.099	.104	.113	.118	.122	.121
$\propto_y$	.062	.061	.059	.055	.051	.046	.037	.029	.020	.01

The maximum Bending Moment per unit width in a slab is given by

$$Mx = \propto_{x'} w \cdot l_x^2$$

$$M_y = \propto_{y'} w \cdot l_x^2$$

Note: Where L_y/L_x is greater than 2, the slabs shall be designed as spanning one way.

7.7 SOLVED EXAMPLES

Example 7.6. Design a two-way slab for a room having clear dimensions of 400 mm × 5000 mm (4m × 5m) Take live load as 2000 N/m² (200 Kg/m²) and finishes as 500 N/m² (50 kg/m²). Use M 20 mix and HYSD (Fe 415) bar as reinforcement.

Solution. Assume the thickness of slab as 120 mm and effective depth as = 100 mm.

Effective span in two directions will be :

$$l_x = 4000 + 100 = 4100 \text{ mm} = 4.1 \text{ m}$$

$$l_y = 5000 + 100 = 5100 \text{ mm} = 5.1 \text{ m}$$

Load per square meter

Self weight of slab	= 0.12 × 1 × 1 × 25000	=	3000 N
Live load		=	2000 N
Finishes		=	500 N
Total load w		=	5500 N

$$K = \frac{l_y}{i_x} = \frac{5.1}{4.1} = 1.244$$

Ref. Table 7.4

$$\propto_x = 0.08796 \quad \text{By interpolation}$$

$$\propto_y = 0.05724$$

The maximum bending moments per unit width of slabs are given by

$$M_x = 0.08796 \times 5500 \times 4.1^2 = 8132.34 \text{ N.m}$$

$$M_y = 0.05724 \times 5500 \times 4.1^2 = 5292.12 \text{ N.m}$$

$$d = \sqrt{\left(\frac{M}{Qb}\right)} = \sqrt{\left(\frac{8132.34 \times 1000}{0.91 \times 1000}\right)} = 94.5 \text{ mm}$$

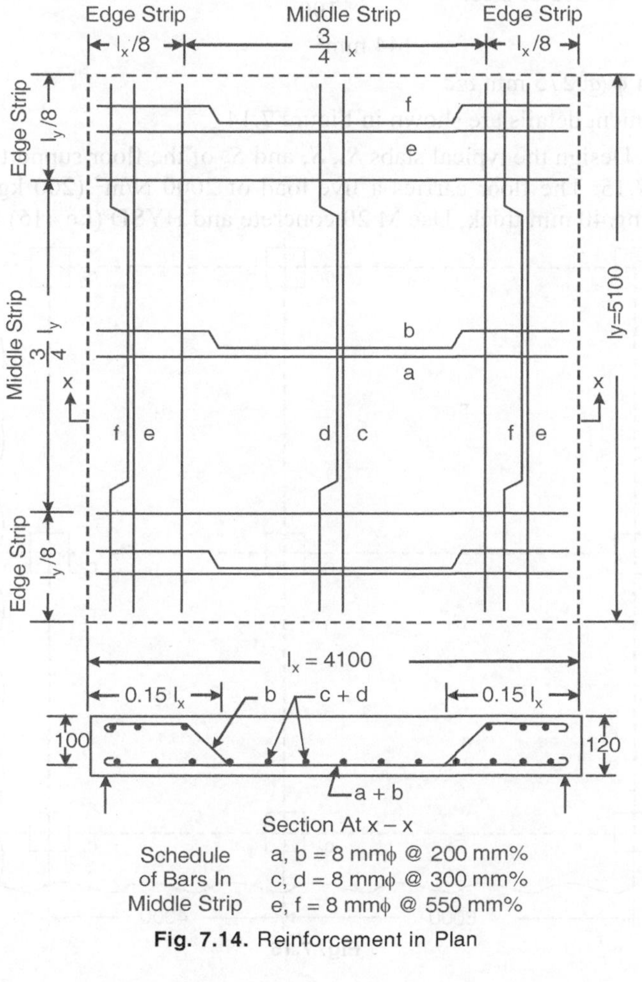

Fig. 7.14. Reinforcement in Plan

Provide overall depth of 120 mm and $d = 95$ mm

Reinforcement along short span in middle strip

$$A_{st} = \frac{8132.34 \times 1000}{230 \times 0.90 \times 95} = 413 \text{ mm}^2$$

Provide 8 mm ϕ bars @ 100 mm *c/c*

Thus providing $A_{st} = 500 \text{ mm}^2$

Bend half bars at 0.15 of span from ends

Reinforcement along long span in middle strip

$$A_{st} = \frac{5292.12 \times 1000}{230 \times 0.90 \times (95 - 8)} = 294 \text{ mm}^2$$

Provide 8 m bars @ 150 *c/c*

Thus providing $A_{st} = 335 \text{ mm}^2$

Bend half bars at 0.15 of span from ends.

Reinforcement in edge strip

$$\text{Area of steel} = \frac{0.12 \times 1000 \times 120}{100} = 144 \text{ mm}^2$$

Provide 8 mm ϕ @ 275 mm *c/c*

The reinforcement details are shown in Figure 7.14

Example 7.7. Design the typical slabs S_1, S_2 and S_3 of the floor supported on *R.C.* columns as shown in Fig. 7.15. The floor carries a live load of 2000 N/m² (200 kg/m²); 60 mm plain concrete and flooring 40 mm thick, Use M 20 concrete and HYSD (Fe 415) bars.

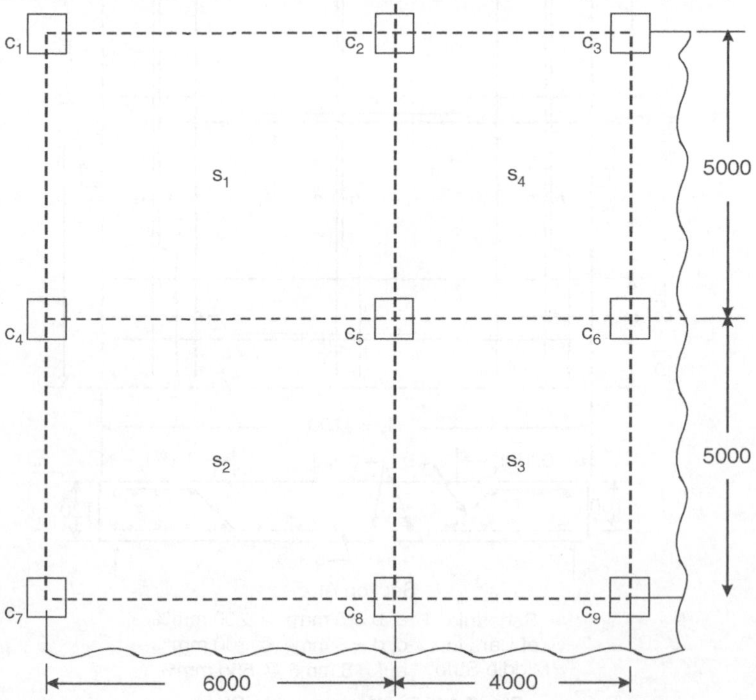

Fig. 7.15

Solution.

(*i*) *Design of Slab S_1*

$$l_x = 5\text{m} \quad l_y = 6\text{m}$$

Edge conditions are : two adjacent edges discontinuous.

$$k = 6/5 = 1.2$$

Referring Table 7.3, bending moment coefficients are

$$\propto_x = -0.06 \qquad \propto_y = -0.047$$

$$\propto_x = +0.045 \qquad \propto_y = +0.035$$

Assume slab thickness as 150 mm

Load per square meter

$$\text{Slab load} = 0.15 \times 1 \times 2500 = 37540 \text{ N/m}^2$$

$$\text{live load} = 2000 \text{ N/m}^2$$

$$\text{Plain concrete} = 0.06 \times 1 \times 1 \times 24000 = 1440 \text{ N/m}^2$$

$$\text{Flooring} = 0.04 \times 1 \times 1 \times 24000 = 960 \text{ N/m}^2$$

$$\text{Total load} = 8150 \text{ N/m}^2$$

Maximum bending moments per unit width of slab are given by

Short Span

$$-M_x = 0.06 \times 8150 \times 5^2 = 12225 \text{ N.m}$$

$$+M_x = 0.045 \times 8150 \times 5^2 = 9168.75 \text{ N.m}$$

Long Span

$$-M_y = 0.047 \times 8150 \times 5^2 = 9576.25 \text{ N.m}$$

$$+M_y = 0.035 \times 8150 \times 5^2 = 7131.25 \text{ N.m}$$

$$d = \sqrt{\left(\frac{12225 \times 1000}{0.91 \times 1000}\right)} = 116.6 \text{ mm}$$

Provide overall depth of slab as 150 mm

$$d = 150 - \frac{10}{2} - 30 = 115 \text{ mm}$$

Reinforcement along short span

$$A_{st} = \frac{12225 \times 1000}{230 \times 0.90 \times 115} = 513 \text{ mm}^2$$

Provide 10 mm ϕ @ 300 *c*/*c* + 12 mm ϕ @ 300 *c*/*c*

Thus providing $A_{st} = 262 + 377 = 639 \text{ mm}^2$

Reinforcement along short span in the middle strip

$$+A_{st} = \frac{9168.75 \times 1000}{230 \times 0.90 \times (115)} = 385 \text{ mm}^2$$

Provide 10 mm ϕ @ 150 mm *c*/*c*

Thus providing $A_{st} = 523 \text{ mm}^2$

Reinforcement along long span, $d = 115 - 10 = 105$ mm

$$-A_{st} = \frac{9576.25 \times 1000}{230 \times 0.90 \times (115 - 10)} = 441 \text{ mm}^2$$

Provide 10 mm ϕ @ 400 *c/c* + 12 mm @ 400 *c/c*

Thus providing 478 mm²

$$+A_{st} = \frac{7131.25 \times 1000}{230 \times 0.90 \times (105)} = 328 \text{ mm}^2$$

Provide 10 mm ϕ @ 200 *c/c*

Thus providing 392 mm²

Reinforcement in edge strips

$$= \frac{0.12 \times 1000 \times 150}{100} = 180 \text{ mm}^2$$

Provide 8 mm ϕ @ 200 *c/c*

Thus providing 251 mm²

Torsion reinforcement at corner C_1

$$= \frac{3}{4} \times 385 = 290 \text{ mm}^2$$

Provide 10 mm ϕ @ 200 *c/c* at top and bottom in 1m × 1m distances in the two directions

(*ii*) *Design of Slab S_2*

$$l_x = 5 \text{ m} \quad l_y = 6\text{m}$$

$$k = \frac{l_y}{l_x} = 1.2$$

Edge conditions are : One short edge discontinuous

Short span	*Long span*
$-\propto_x = 0.048$	$-\propto_y = 0.037$
$+\propto_x = 0.036$	$+\propto_y = 0.028$

Maximum bending moments per unit width

Short Span

Negative Moment at continuous $= 0.048 \times 8150 \times 5^2$

$= 9780$ N.m

Positive moment at mid span $= 0.036 \times 8150 \times 5^2$

$= 7335$ N.m

Long Span

Negative Moment at continuous edge

$= 0.037 \times 8150 \times 5^2$

$= 7538.75$ N.m

$$d = \sqrt{\left(\frac{9780 \times 1000}{0.91 \times 1000}\right)} = 105 \text{ mm}$$

$$D = 150 \text{ mm} \qquad d = 115 \text{ mm}$$

Areas of Steel

Short Span

$$-A_{st} = \frac{9780 \times 1000}{230 \times 0.90 \times 115} = 410 \text{ mm}^2,$$

Provide 10 mm ϕ @ 300 *c/c* + 12 mm ϕ @ 300 *c/c*

$$+A_{st} = \frac{7335\times1000}{230\times0.90\times115} = 308 \text{ mm}^2$$

10 mm ϕ @ 150 mm *c/c*

Long Span

$$-A_{st} = \frac{7538.75\times1000}{230\times0.90\times105} = 346 \text{ mm}^2$$

Provide 10 mm ϕ @ 400 *c/c* + 12 mm ϕ @ 400 *c/c*

Thus providing $Ast = 478 \text{ mm}^2$

$$+A_{st} = \frac{5705\times1000}{230\times0.90\times105} = 263 \text{ mm}^2$$

Provide 10 mm ϕ 200 mm *c/c*

Thus providing $+A_{st} = 392 \text{ mm}^2$

Distribution steel in edge

$$\text{strips} = \frac{0.12\times1000\times150}{100} = 180 \text{ mm}^2$$

Provide 8 mm ϕ @ 300 mm *c/c*

Thus providing 277 mm²

Torsion reinforcement = $\frac{3}{4}\times308 = 231 \text{ mm}^2$

Provide 8 mm ϕ @ 200 mm *c/c* in the required directions.

(iii) Design of Slabs S_3

Take this slab as interior slab, $l_x = 4$ m $l_y = 5$ m

$$K = \frac{5}{4} = 1.25$$

Maximum bending moments per unit width

Short Span

Negative moment at continuous edge = $0.045 \times 8150 \times 4^2 = 5868$ N.m

Positive moment at mid span = $0.034 \times 8150 \times 4^2 = 4434$ N.m

Long Span

Negative moment at continuous edge = $0.032 \times 8150 \times 4^2 = 4173$ N.m

Positive moment at mid span = $0.024 \times 8150 \times 4^2 = 3130$ N.m

$$d = \sqrt{\left(\frac{5868\times1000}{0.91\times1000}\right)} = 81 \text{ mm}$$

$$D = 150 \text{ m} \qquad d = 115 \text{ mm}$$

Areas of Steel

Short Span

$$-A_{st} = \frac{5868\times1000}{230\times0.90\times115} = 246 \text{ mm}^2, \text{ (8 mm } \phi \text{ @ 400 mm } c/c$$

+ 12 mm ϕ @ 400 mm *c/c*

$-A_{st} = 408 \text{ mm}^2$)

$$+A_{st} = \frac{4434 \times 1000}{230 \times 0.90 \times 115} = 186 \text{ mm}^2 \text{ (8 mm } \phi \text{ @ 400 } c/c + 10 \text{ mm } \phi \text{ @ 400 } c/c, A_{st} \text{ Provide} = 321 \text{ mm}^2)$$

Long Span

$$-A_{st} = \frac{4173 \times 1000}{230 \times 0.90 \times 105} = 192 \text{ mm}^2 \text{ (8 mm } \phi \text{ @ 400 mm } c/c + 10 \text{ mm } \phi \text{ @ 400 } c/c, A_{st} \text{ Provide} = 321 \text{ mm}^2)$$

$$+A_{st} = \frac{3130 \times 1000}{230 \times 0.90 \times 105} = 144 \text{ mm}^2 \text{ (8 mm } \phi \text{ @ 200 mm } c/c, A_{st} = 250 \text{ mm}^2)$$

The reinforcement details of slabs, S_1, S_2, S_3 are shown in Fig. 7.16

Schedule of Bars in Slab (Bottom face – B, Crank – C, Top face – T)

a = 10 mm ϕ @ 300 mm c/c (B)
b = 10 mm ϕ @ 300 mm c/c (C)
c = 12 mm ϕ @ 300 mm c/c (T)
d = 10 mm ϕ @ 400 mm c/c (B)
e = 10 mm ϕ @ 400 mm c/c (C)
f = 12 mm ϕ @ 400 mm c/c (T)
h = 10 mm ϕ @ 400 mm c/c (B)

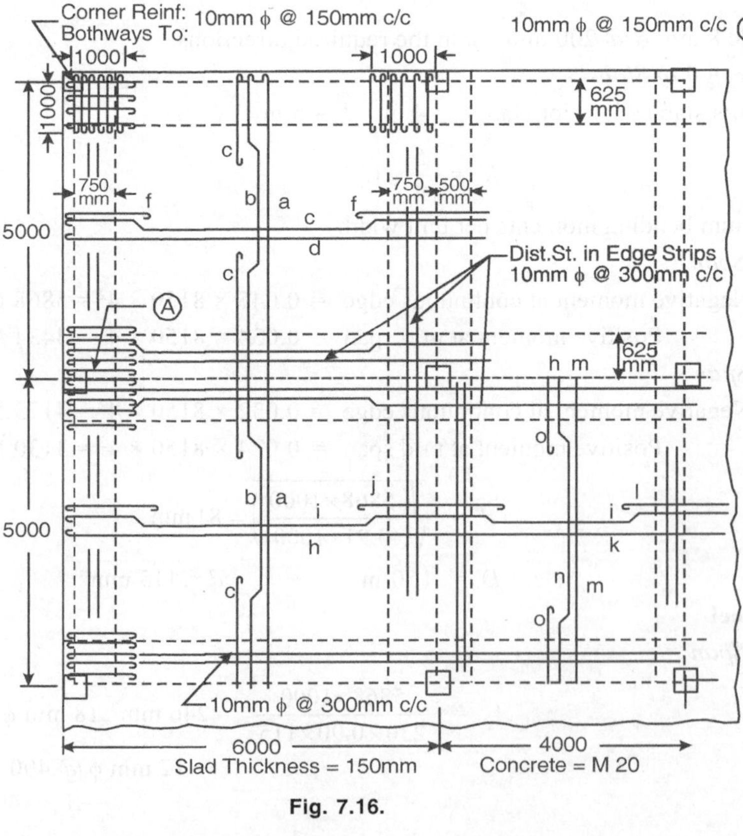

Fig. 7.16.

i = 10 mm φ @ 400 mm c/c (C)
j = 10 mm φ @ 400 mm c/c (T)
k = 08 mm φ @ 400 mm c/c (B)
l = 10 mm φ @ 400 mm c/c (T)
m = 08 mm φ @ 400 mm c/c (B)
n = 08 mm φ @ 400 mm c/c (B)
0 = 10 mm φ @ 400 mm c/c (T)

7.8 GRASHOF AND RANKINE METHOD

Grashof and Rankine Method for slabs simply supported on four edges will corners not held down and carrying uniformly distributed loads.

Consider the two strips shaded as shown in Fig. 71.7 of the slab simply supported on four edges and under *u.d.l.* load.

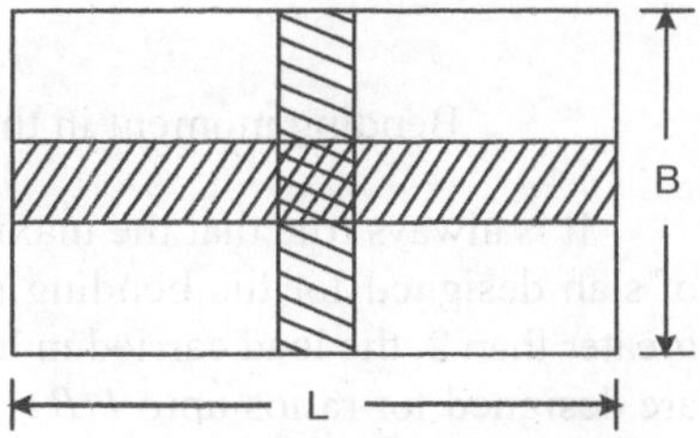

Fig. 7.17.

The deflection of each strip in longer and shorter directions at centre will be same.

$$\text{Deflection of strip in direction } L \text{ at mid span} = \frac{5\, w_L . L^4}{384\, EI}$$

$$\text{Deflection of strip in direction } B \text{ at mid span} = \frac{5\, w_B . B^4}{384\, EI}$$

Equating the two expressions

$$\frac{5\, w_L . L^4}{384\, EI} = \frac{5\, w_B . B^4}{384\, EI}$$

where

L = Length of slab
B = Breadth of slab
w_L = Load carried by slab in L-direction
w_B = Load carried by slab in B-direction

Let total *u.d.l.* on slab be w

$$w = w_L + w_B$$

$$w = w_B . (B/L)^4 + w_B = w_B \left(1 + \frac{B^4}{L^4}\right)$$

$$w_B = w . \frac{L^4}{L^4 + B^4}$$

$$= w \times \frac{1}{1 + (B/L)^4}$$

and

$$w_L = w . \frac{1}{1 + (L/B)^4}$$

Let $\quad L/B = K$

$$w_B = w.\frac{K^4}{1+K^4} = K_L.w$$

$$w_L = w.\frac{1}{1+K^4} = K_B.w$$

The values of K_L and K_B depend on ratio L/B and their values for various L/B ratios are tabulated in Table 7.5

$$\text{Bending moment in the direction } B = M_B = K_B.\frac{w\ B^2}{8}$$

$$\text{Bending moment in the direction } L = M_L = K_L.\frac{w\ L^2}{8}$$

It is always true that the maximum bending moment occurs along shorter span and the depth of slab designed for his bending moment. It may be observed from Table 7.5 that when L/B is greater than 2, the load carried in long direction is very small. This is the reason that two way slabs are designed for ratios upto $L/B = 2$ and when L/B exceeds 2, the slab is assumed to be spanning in short direction only and the maximum bending moment is taken as $w\,B^2/8$. It may be noted that ratio of load carried by slab in two directions as given by above Table 7.5 is not same at all points of slab. It is true only at centre of slab.

Table 7.5.

Ratio L/B	1.0	1.05	1.1	1.15	1.20	1.25	1.3	1.4	1.5	1.6	1.75	2.0	2.5	3.0
K_B	0.50	0.55	0.59	0.64	0.68	0.71	0.74	0.79	0.84	0.87	0.9	0.94	0.97	0.98
K_L	0.50	0.45	0.41	0.36	0.32	0.29	0.26	0.21	0.16	0.13	0.1	0.06	0.03	0.02

Slabs simply supported on four edges with corners held down and carrying u.d.l. loads.

Slabs are very frequently, monolithic with thin beams or build into thin walls. The bending moments in such slabs can be found by either of the two methods given below :

(*a*) The bending moments in such slabs in directions B and L are

$$M_B = K_B'\frac{w\ B^2}{8} \text{ and}$$

$$M_L = K_L'\frac{w\ L^2}{8}$$

The coefficients K'_B and K'_L in this case are given in Table 7.6

(*b*) *Dr. Marcus Method* :

Dr. Marcus has presented an approximate method for computing bending moments in slabs supported on four edges with corners prevented from lifting. He has modified Grashof and Rankine's method by adopting multiplication factor C which is less than unity as the positive bending moments are reduced due to corner restraints. The factor C is given by :

$$C = 1-\frac{5}{6}\left(\frac{K^2}{1+K^4}\right)$$

where $\quad K = L/B$

Bending Moment in direction of B = $M_B = C.K_B.\frac{wB^2}{8}$

Bending Moment in direction of L = $M_L = C.K_L.\frac{wL}{8}$

The values of C for different ratios of L/B can be obtained from Table 7.7. For slabs fixed on all four edges carrying *u.d.l.* loads, the bending moments may be calculated by Dr. Marcus method.

Short Span

$$-\text{Moment} = M'_B = -K_B \frac{wB^2}{12}$$

$$+\text{Moment} = M_B = -C'.K_B \frac{wB^2}{24}$$

Long Span

$$-\text{Moment } M'_L = K_L.\frac{wL^2}{12}$$

$$+\text{Moment } M_L = C'.K_L.\frac{wL^2}{24}$$

where

$$C' = 1 - \frac{5}{18}\frac{K^2}{1+K^4}$$

Table 7.6

Ratio L/B	1.0	1.05	1.1	1.15	1.20	1.25	1.3	1.4	1.5	1.6	1.75	2.0	2.5	3.0
K'_B	0.30	0.33	0.36	0.39	0.42	0.45	0.48	0.53	0.58	0.63	0.68	0.76	0.87	0.94
K'_L	0.30	0.27	0.24	0.22	0.19	0.17	0.15	0.13	0.11	0.09	0.07	0.05	0.03	0.02

Table 7.7

L/B	1.0	1.05	1.1	1.15	1.20	1.25	1.3	1.4	1.5	1.6	1.75	2.0	2.5	3.0
C	0.583	0.586	0.591	0.589	0.610	0.622	0.635	0.663	0.691	0.718	0.755	0.804	0.870	0.970
C'	0.861	0.862	0.864	0.866	0.970	0.874	0.879	0.888	0.897	0.907	0.919	0.935	0.957	0.970

The values of C' for different values of L/B are given in Table 7.7.

The bending moment formulae for various end conditions of slabs under *u.d.l.* load are given in Table 7.8.

7.9 SOLVED EXAMPLES

Example 7.8. Design a two-way slab for a room having clear dimensions 5m × 4m. The superimposed live load is 2000 N/m^2 and finishes (plain concrete known as screeding and flooring) as 2500 N/m^2. Use M 20 mix and HYSD (Fe 415) bars.

Solution. Assume overall depth of slab as 140 mm. Effective spans in two directions shall be 5.12 m and 4.12 m (taking d = 120 mm).

Table 7.8.

Conditions at support	$\frac{w_B}{w}$	$\frac{w_L}{w}$	Span Moment M_B	Span Moment M_L	Multiplying factor for span moment: For M_B	Multiplying factor for span moment: For M_L	Support Moment M'_B	Support Moment M'_L
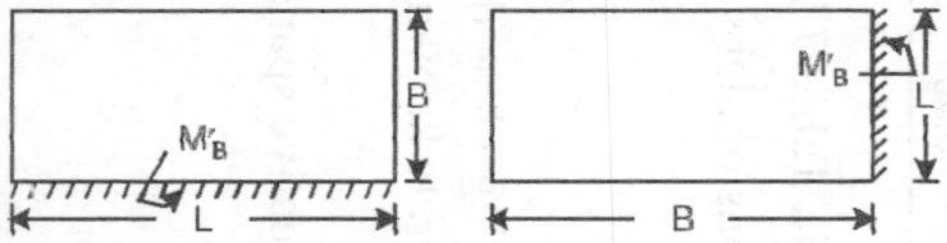	$\frac{5K^4}{2+5K^4}$	$\frac{2}{2+5K^4}$	$\frac{9}{128}\cdot w_B\cdot B^2$	$\frac{1}{8}\cdot w_L\cdot L^2$	$1-\frac{75}{32}\frac{K^2}{2+5K^4}$	$1-\frac{5}{3}\frac{K^2}{2+5K^4}$	$\frac{w_B\cdot B^2}{8}$	
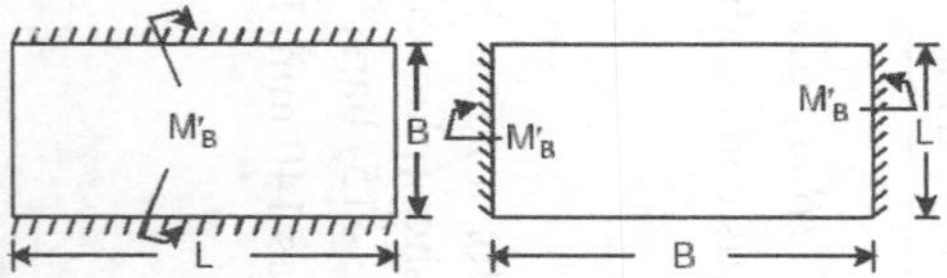	$\frac{5K^4}{1+5K^4}$	$\frac{2}{1+5K^4}$	$\frac{1}{24}\cdot w_B\cdot B^2$	$\frac{1}{8}\cdot w_L\cdot L^2$	$1-\frac{25}{18}\frac{K^2}{2+5K^4}$	$1-\frac{5}{6}\frac{K^2}{2+5K^4}$	$\frac{w_B\cdot B^2}{12}$	
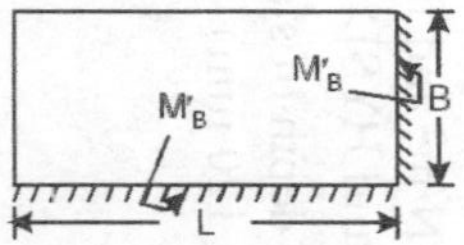	$\frac{K^2}{1+K^4}$	$\frac{1}{1+K^4}$	$\frac{9}{128}\cdot w_B\cdot B^2$	$\frac{1}{128}\cdot w_L\cdot L^2$	$1-\frac{15}{32}\frac{K^2}{1+K^4}$	$1-\frac{5}{32}\frac{K^2}{1+K^4}$	$\frac{w_B\cdot B^2}{8}$	$\frac{w_L\cdot L^2}{8}$
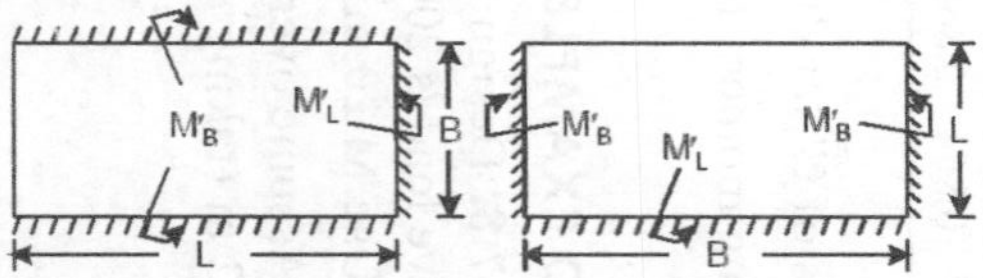	$\frac{2K^2}{1+2K^4}$	$\frac{1}{1+2K^4}$	$\frac{1}{24}\cdot w_B\cdot B^2$	$\frac{9}{128}\cdot w_L\cdot L^2$	$1-\frac{5}{9}\frac{K^2}{1+2K^4}$	$1-\frac{15}{32}\frac{K^2}{1+5K^4}$	$\frac{w_B\cdot B^2}{12}$	$\frac{w_L\cdot L^2}{8}$

M'_L = Maximum Bending Moment at support span L

M'_B = Maximum Bending Moment at support span B

Loading

Self weight of slab	$= 0.14 \times 1 \times 1 \times 1 \times 25000$	$= 3500 \text{ N/m}^2$
Superimposed *L.L.*		$= 2000 \text{ N/m}^2$
Finishes (screeding and flooring)		$= 2500 \text{ N/m}^2$
Total load w		$= 8000 \text{ N/m}^2$

$$k = \frac{512}{4.12} = 1.243$$

$$w_B = w.\frac{k^4}{1+k^4}$$

$$= 8000 \times \frac{1.243^4}{1+1.243^4} = 5638 \text{ N/m}^2$$

$$w_L = 8000 - 5683 = 2362 \text{ N/m}^2$$

(*i*) *Taking Corners of slab not held down*

$$M_B = \frac{w_B.B^2}{8}$$

$$= \frac{5638 \times 4.12^2}{8} = 11962.7 \text{ N.m}$$

$$M_L = \frac{w_L.L^2}{8}$$

$$= \frac{2362 \times 5.12^2}{8} = 7740 \text{ N.m}$$

$$d = \sqrt{\left(\frac{M}{Q.b}\right)} = \sqrt{\left(\frac{11962.7 \times 1000}{0.91 \times 1000}\right)} = 114.65 \text{ mm}$$

Provide D = 140 mm d = 120 mm

$$A_{st} \text{ required along short span} = \frac{11962.7 \times 1000}{230 \times 0.90 \times 129} = 481.6 \text{ mm}^2$$

Provide 10 mm ϕ @ 150 mm *c/c* in short span direction (A_{st} = 524 mm²)

$$A_{st} \text{ required along long span} = \frac{7740 \times 1000}{230 \times 0.90 \times (120 - 10)} = 340 \text{ mm}^2$$

Provide 10 mm ϕ @ 200 mm *c/c* along long span direction. (A_{st} = 393 mm²)

(*ii*) *Taking Corners held down* : (Using Marcus Method)

$$M_B = \frac{w_B.B^2}{8}\left(1 - \frac{5}{6}.\frac{k^2}{1+k^4}\right)$$

$$= \frac{5638 \times 4.12^2}{8}\left(1 - \frac{5}{6}.\frac{1.243^2}{1+1.243^4}\right)$$

$$= 7417 \text{ N.m}$$

$$M_L = \frac{2362 \times 5.12^2}{8}\left(1 - \frac{5}{6}.\frac{1.243^2}{1+1.243^4}\right)$$

$$= 4798.5 \text{ N.m}$$

$$d = \sqrt{\left(\frac{7417\times1000}{0.91\times1000}\right)} = 90.3 \text{ mm}$$

$$D = 140 \text{ mm} \quad d = 120 \text{ mm}$$

$$A_{st} = \text{required along short span} = \frac{7417\times1000}{230\times0.90\times120} = 299 \text{ mm}^2$$

Provide 10 mm ϕ @ 150 mm c/c (A_{st} = 529 mm²)
(Bend alternate bars at 0.15 × B = 0.62 m)

$$A_{st} \text{ required along long span} = \frac{4798.5\times1000}{230\times0.90\times110} = 211 \text{ mm}^2$$

Provide 8 mm ϕ 330 mm c/c
+10 mm ϕ 330 mm c/c $\quad A_{st}$ provided = 389.4 mm²

Torsion Reinforcement

Torsional reinforcement of 3/4 of area of maximum main reinforcement at mid span is provided in both directions parallel to sides at top and bottom faces upto a distance of one-fifth of the shorter span.

$$A_s = 3.4 \times 299 = 224 \text{ mm}^2$$

Provide at top and bottom $\left\{\begin{array}{l} 8 \text{ mm } \phi\ 300 \text{ mm } c/c \\ +10 \text{ mm } \phi\ 300 \text{ mm } c/c \end{array}\right\}$ in short direction

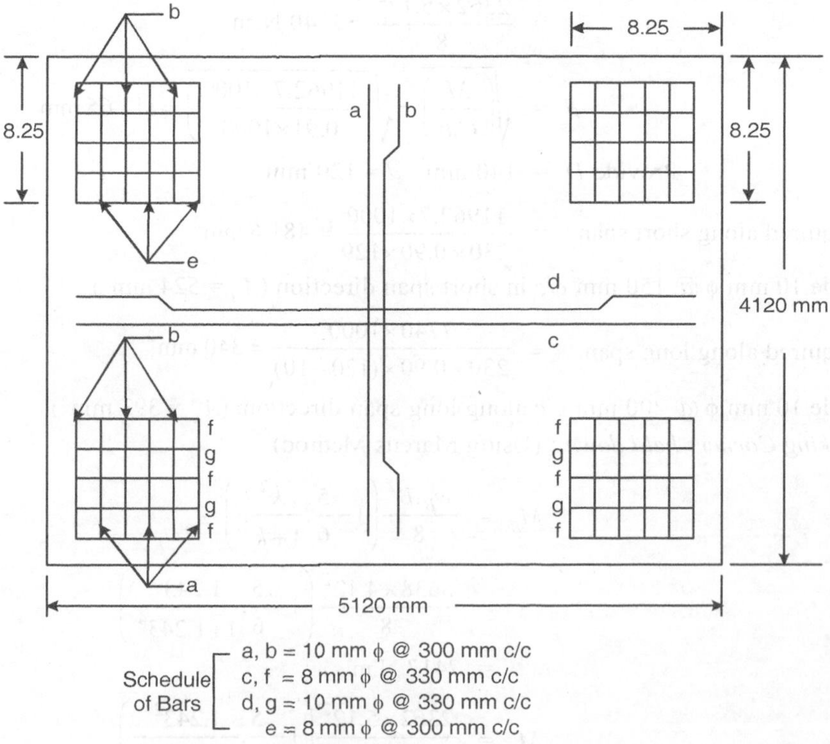

Fig.7.19. Corners held down

$$\begin{Bmatrix} 8 \text{ mm } \phi \text{ } 330 \text{ mm } c/c \\ +10 \text{ mm } \phi \text{ } 330 \text{ mm } c/c \end{Bmatrix} \text{ in long direction}$$

For a distance of 4.12/5 = 0.825 m

The details of reinforcement in slab for corners held down, is shown in Fig. 7.18.

7.10 NON-RECTANGULAR SLABS

Quite often, non-rectangular slabs such as trapezoidal, triangular, polygonal or circular, etc., are encountered in Civil Engineering structures. The bending moments in such slabs can be obtained by taking slab as a rectangular two-way slab or circular slab depending upon the shape of the actual panel of slab.

7.10.1 Trapezoidal Slabs

If the slab is trapezoidal, it may be treated as rectangular slab having the same area and the ratio of its sides the same as the ratio of the mean axes of the trapezium as shown in Fig. 7.19.

7.10.2 Circular Slabs

Circular slabs are generally used in water tanks as roof coverings or base slabs. The slab when loaded deflects in the form of a saucer thereby creating radial and circumferential moments. At the centre of slab, cracks are always observed circumferential as well as radially requiring reinforcement in both directions. It will be ideal to place reinforcement in radial and circumferential at the tension face. However, reinforcement in the form of rectangular mesh is also sometimes provided for easier working. The following six cases of circular slabs are more commonly encountered in practice.

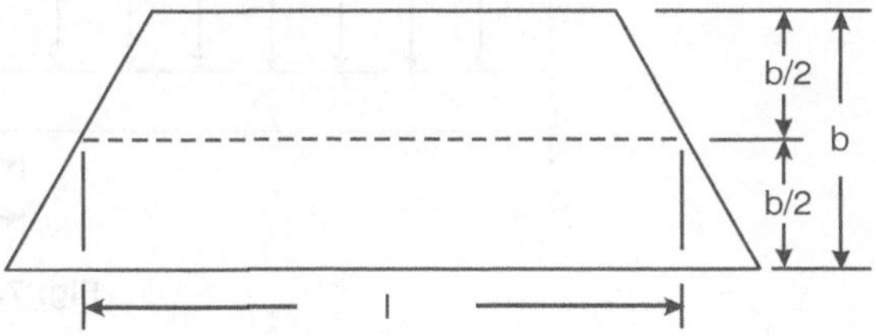

Fig. 7.19

Case I : Slabs freely supported at edges and loaded uniformly (Fig. 7.20)

Let p = load per unit area

a = radius of slab

M_r = radial bending moment per unit width (parabolic variation)

M_θ = circumferential bending moment per unit width (parabolic variation)

At centre of slab

$$M_r = M_\theta = \frac{3}{16}.p.a^2$$

At free end of slab

$$M_r = 0$$

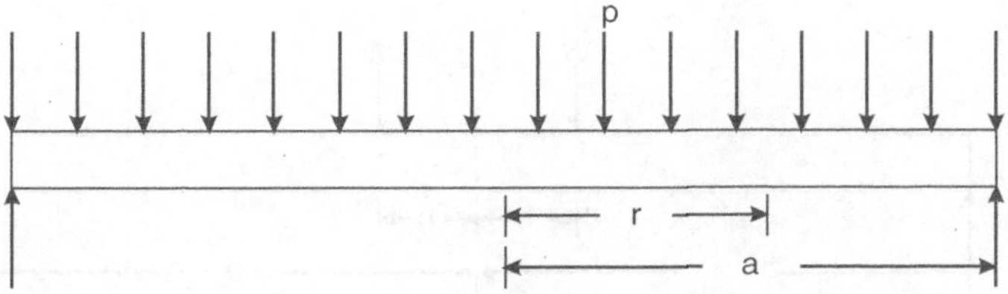

Fig. 7.20

$$M_\theta = \frac{2}{16}.p.a^2$$

Radial *S.F.* at any radius $r = \frac{1}{2}$ $p.r$ per unit width

There will be no circumferential *S.F.*

Case II : Slabs fixed at edges and loaded uniformly (Fig. 7.21)

At centre of slab

$$M_r = M_\theta = \frac{1}{16}.p.a^2$$

At edges

$$M_r = \frac{2}{16}.p.a^2$$

$$M_\theta = 0$$

$$M_r = 0 \text{ at } r = \frac{a}{3}$$

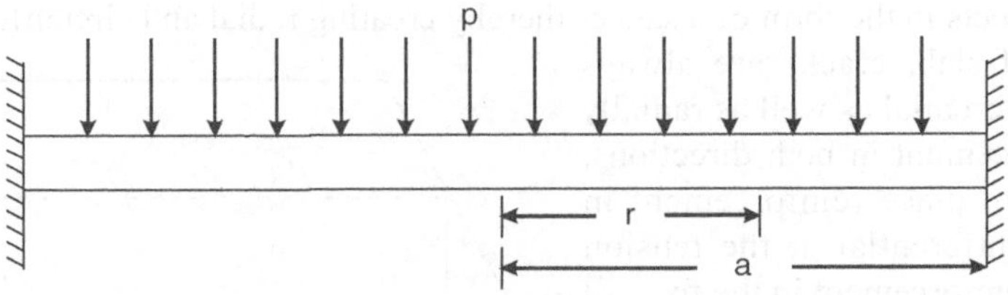

Fig. 7.21.

S.F. at any radius $r = \frac{a}{3}$ $p.r$

Case III : Slabs simply supported at edges with load *W* distributed along the circumference of a concentric circle (Fig. 7.22)

If a = radius of slab

b = radius of the load circle

For $r < b$

$$M_r = M_\theta = \frac{W}{8.\pi}\left[2\log_e(a/b)+1-(b/a)^2\right]$$

$$S.F. = 0$$

For $r > b$

$$M_r = \frac{W}{8.\pi}\left[2\log_e(a/r)-(b/a)^2+(b/r)^2\right]$$

$$M_\theta = \frac{W}{8.\pi}\left[2\log_e(a/r)-(b/r)^2+2+(b/a)^2\right]$$

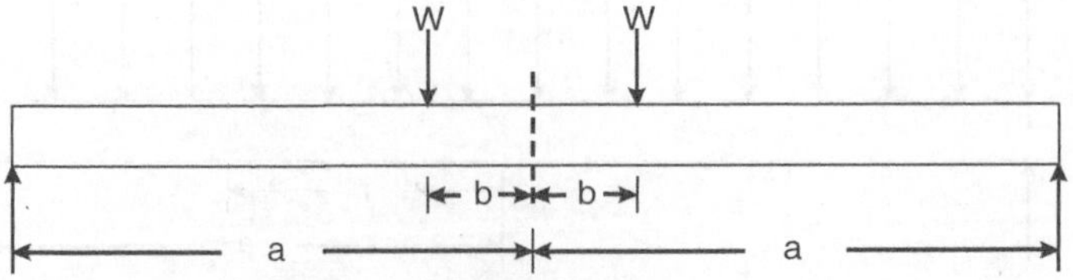

Fig. 7.22

$$S.F. = \frac{W}{2\pi r}$$

Case IV : Slabs simply supported at edges with load uniformly distributed inside a concentric circle (Fig. 7.23)

If a = radius of slab
b = radius of load circle
p = load per unit area

For $r < b$

$$M_r = -\frac{3}{16}.p.r^2 + \frac{1}{2}p.b^2\left[1-\log_c(b/a)-\frac{b^2}{4.a^2}\right]$$

$$M_\theta = -\frac{1}{16}.p.r^2 + \frac{1}{4}.p.b^2\left[1-\log_e(b/a)-\frac{b^2}{4.a^2}\right]$$

$$S.F. = p.r/2$$

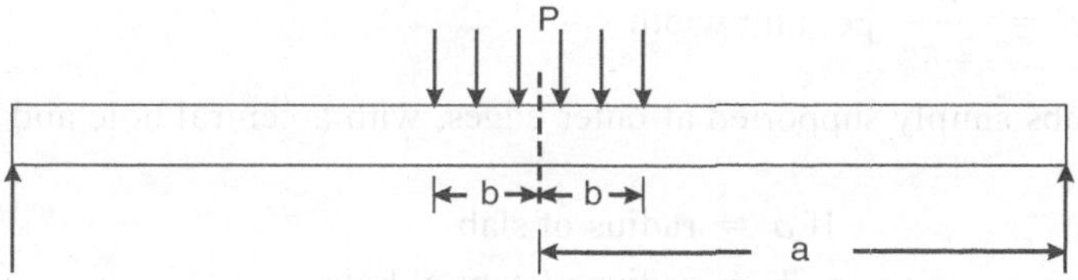

Fig. 7.23

For $r > b$

$$M_r = -p.b^2\left[\frac{1}{4}\log_e(r/a)+\frac{b^2}{16}(1/a^2-1/r^2)\right]$$

$$M_\theta = -p.b^2\left[\frac{1}{4}\log_e(r/a)-\frac{1}{4}+\frac{b^2}{16}(1/a^2+1/r^2)\right]$$

$$S.F. = \frac{1}{2}.p.b^2/r$$

Case V : Slabs simply supported at outer edges, with a central hole and loaded with W distributed along the circumference of a concentric circle (Fig. 7.24).

If a = radius of slab
b = radius of central hole
c = radius of load circle

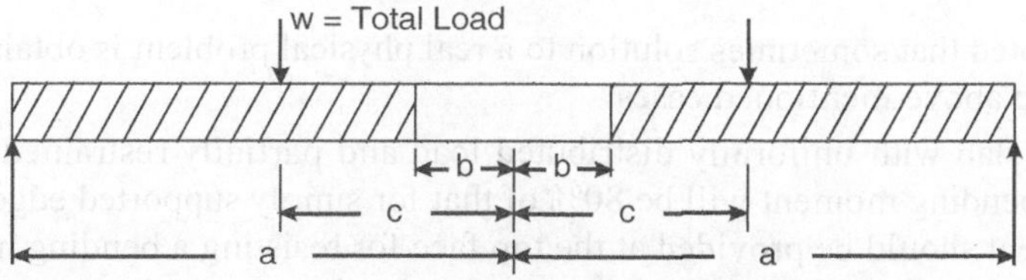

Fig. 7.24

For $r < c$

$$M_r = \frac{W}{4.\pi}\cdot\frac{a^2}{a^2-b^2}\cdot\left[(1-b^2/r^2).\left(\log_e a/c+\frac{1}{2}-\frac{c^2}{2.a^2}\right)\right]$$

$$M_\theta = \frac{W}{4.\pi}\cdot\frac{a^2}{a^2-b^2}\cdot\left[(1+b^2/r^2).\left(\log_e a/c+\frac{1}{2}-\frac{c^2}{2.a^2}\right)\right]$$

$$S.F. = 0$$

For $r > c$

$$M_r = \frac{W}{4.\pi}\left[\log_e c/r-\frac{1}{2}+\frac{a^2}{a^2-b^2}\cdot\left(\frac{r^2-b^2}{r^2}\right)\left(\log_e a/c+\frac{1}{2}-\frac{c^2}{2.a^2}\right)+\frac{c^2}{2r^2}\right]$$

$$M_\theta = \frac{W}{4.\pi}\cdot\left[\log_e c/r+\frac{1}{2}+\frac{a^2}{a^2-b^2}\cdot\left(\frac{r^2+b^2}{r^2}\right)\left(\log_e a/c+\frac{1}{2}-\frac{c^2}{2.a^2}\right)-\frac{c^2}{2.r^2}\right]$$

$$S.F. = \frac{W}{2.\pi r} \text{ per unit width}$$

Case VI : Slabs simply supported at outer edges, with a central hole and loaded uniformly (Fig. 7.25).

If a = radius of slab
b = radius of central hole
p = load per unit area

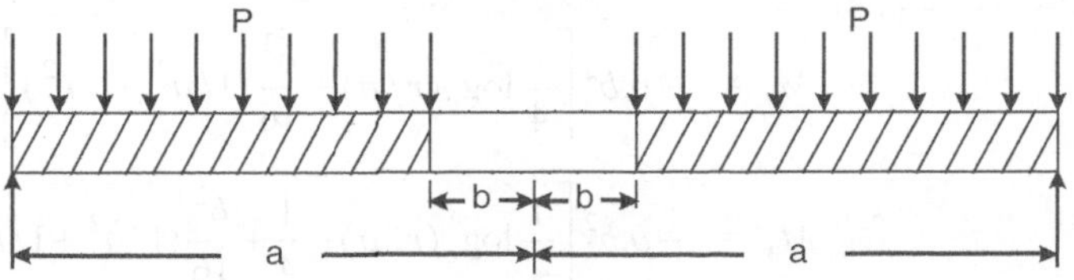

Fig. 7.25

$$M_r = -\frac{3}{16}.p.r^2+\frac{p.b^2}{4}\left[\log_e r/a+3/4(1+a^2/b^2-a^2/r^2)+\frac{a^2-r^2}{r^2}\cdot\frac{b^2}{a^2-b^2}.\log_e a/b\right]$$

$$M_\theta = -\frac{1}{16}.p.r^2+\frac{p.b^2}{4}\left[\log_e r/a+3/4(-1/3)+a^2/b^2+a^2/r^2)-\frac{(a^2+r^2)}{r^2}\cdot\frac{b^2}{(a^2-b^2)}.\log_e a/b\right]$$

$$S.F. = \frac{1}{2}.p.r.-\frac{1}{2}.p.b^2/r$$

It may be noted that sometimes solution to a real physical problem is obtained by combining two or more of the above mentioned cases.

For circular slab with uniformly distributed load and partially restrained at the edges, the maximum radial bending moment will be 80% of that for simply supported edges. Near the edges radial reinforcement should be provided at the top face for resisting a bending moment of at least one-third the maximum positive bending moment for simply supported edges.

7.10.3 Triangular and Polygonal Slabs

In the case of triangular slabs or slabs of any regular polygonal shape, these may be treated as circular slabs having diameter equal to the mean of the circumscribed and incribed circles as shown in Fig. 7.26.

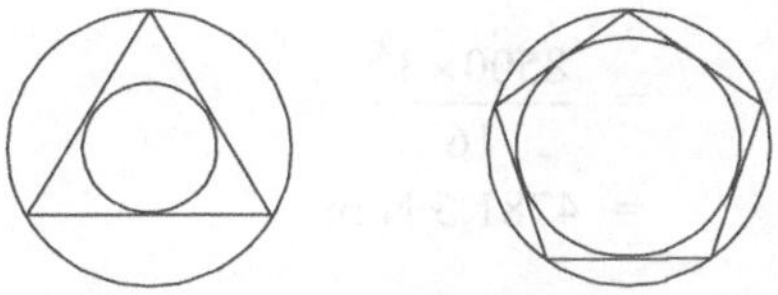

Fig. 7.26

Example 7.9. Design a circular slab for a circular room 6 m in diameter. Take live load on slab = 2000 N/m², and finishes (earth plus titles) = 2750 N/m². The edges of the slab are partially restrained. Take M 20 concrete for slab and use HYSD (Fe 415) bars. m = 13.

Sol. Let the slab thickness be = 150 mm

Self weight of slab = 0.15 × 1 × 1 × 25000 = 3750 N/m²

L.L. = 2000 N/m²

Finishes = 2750 N/m²

Total *p* = 8500 N/m²

Maximum positive radial moment = 80% of *B.M.* for simply supported case and Circumferential moment.

$$= 0.8 \times \frac{3}{16} p.a^2$$

$$= 0.8 \times 8500 \times 3^2 \times \frac{3}{16}$$

$$= 11475 \text{ N.m}$$

$$d = \sqrt{\left(\frac{M}{Q.b}\right)} = \sqrt{\left(\frac{11475 \times 1000}{0.91 \times 1000}\right)} = 112.30 \text{ mm}$$

∴ $D = 150$ mm $\quad d = 150 - 20$ mm cover = 130 mm

$$A_{st} \text{ required} = \frac{M}{\sigma_{st} \cdot j.d}$$

$$= \frac{11475 \times 1000}{230 \times 0.90 \times 130}$$

$$= 426 \text{ mm}^2$$

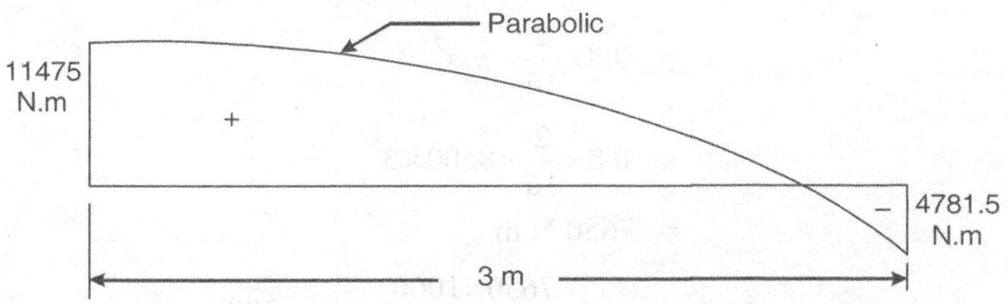

Fig 7.27.

Provide 8 mm ϕ @ 100 mm *c/c* both ways (mesh) in 2m central diameter slab. Negative *B.M.* at support = 1/3 of Max. positive *B.M.* at centre for simply supported edges.

$$= \frac{p.a^2}{16}$$

$$= \frac{8500 \times 3^2}{16}$$

$$= 4781.5 \text{ N.m}$$

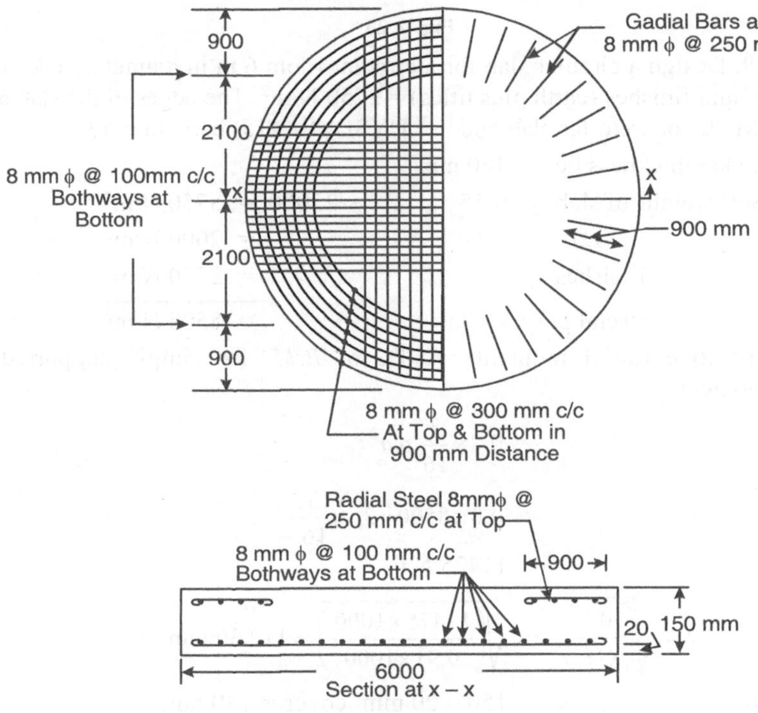

Fig. 7.28.

The *B.M.* diagram is parabolic as shown in Fig. 7.27.

Area of –ve steel required at edges $= \dfrac{4781.5 \times 1000}{230 \times 0.90 \times 130} = 178 \text{ mm}^2$

Provide 8 mm ϕ @ 250 mm *c/c* at top face in radial direction in a distance of $0.15 \times 6000 = 900$ mm from edges.

Circumferential *B.M.* at edge = 80% of that for simply supported slab

$$= 0.8 \times \frac{2}{16} \times p.a^2$$

$$= 0.8 \times \frac{2}{16} \times 8500 \times 3^2$$

$$= 7650 \text{ N.m}$$

$$A_{st} \text{ required} = \frac{7650 \times 1000}{230 \times 0.90 \times (130 - 10)} = 308 \text{ mm}^2$$

Provide 8 mm ϕ @ 150 mm *c/c*, ie., half of this steel at top edge and half at bottom face of slab in 900 mm distance from edge.

The reinforcement details of circular slab, are shown in Fig. 7.28.

7.11 FLAT SLABS

A flat slab means a reinforced concrete slab with or without drops supported, generally without beams, by columns with or without flared column heads as shown in Fig. 7.29. A flat slab may be solid slab or may have recesses formed on the soffit so that the soffit comprises of a series of ribs in two directions. The recesses may be formed by removable or permanent filler blocks. The reinforcement is provided in two or more directions as per the requirements. Generally the slab is divided into column strips and middle strips, the former acting as supports to the latter.

Definitions :

1. *Drop* : The thickened part of the slab over the column is called drop.
2. *Panel* : Panel means that part of a slab bounded on each of its four sides by the centre-line of a column or centre lines of adjacent spans.
3. *Column Head* : The widened area at the top of column to provide more support aea to slab is called column-head or column capital.
4. *Column Strip* : Column strip means a design strip having a width of 0.25 l_2, but not greater than 0.25 l_1 on each side of the column centre-line where l_1 is the span in the direction moments are being determined, measured centre to centre of supports and l_2 is the span transverse to l_1, measured centre to centre of supports. (Fig. 7.29).
5. *Middle Strips* : Middle strip means a design strip bounded on each of its opposite sides by the column strip.

7.11.1 Advantages of Flat Slabs

The various advantages of flat slab construction are given as under :

1. It has plain ceiling which gives an attractive appearance and better illumination to the room.
2. Flat slab floors require less thickness thus allowing for reduction in storey height.
3. Since no beams are provided, form-work required is simple.
4. Reduced loads on foundation because of less thickness and less height of structure.
5. Curing is easy because of flat surface.

7.11.2 General Notes on Design of Flat Slabs (IS : 456-2000)

7.11.2.1 Proportioning

(*a*) *Thickness of Flat Slab* : The thickness of flat slab shall be generally controlled by considerations of span to effective depth ratios given in article 5.2. For slabs with drops conforming to (*b*) span to effective depth ratios given in article 5.2 shall be applied directly, otherwise this ratio shall be multiplied by 0.9. For this purpose, the longer span shall be considered. The minimum thickness of flat slab shall be 125 mm.

(*b*) *Drops* : The drops when provided shall be rectangular (or square) in plan, and have a length in each direction not less than one-third of the panel length in that direction. For exterior panels, the width of drops at right angles to the non-continuous edge and measured from the centre line of the columns shall be equal to one-half the width of drops for interior panels. (See Fig. 7.29).

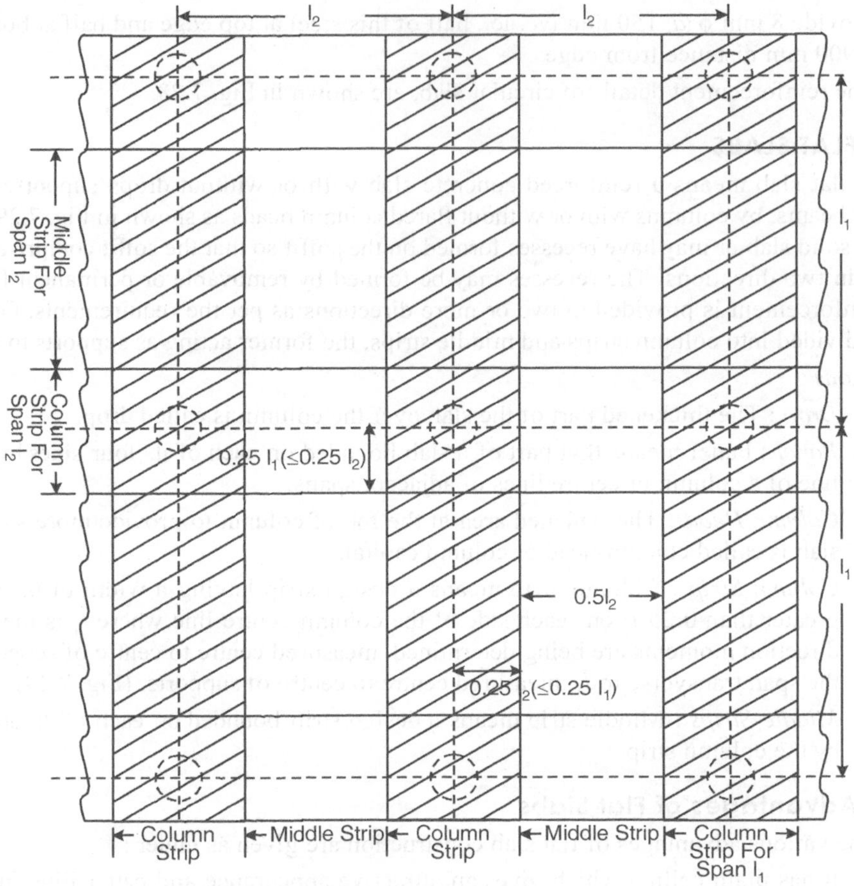

Fig. 7.29. Dimension of Flat-slab panel into strips

(*c*) *Column Heads* : Where column heads are provided, that portion of a column head which lies within the largest right vertex angle of 90° and can be included entirely within the outlines of the column and the column heads, shall be considered for design purposes (See Fig. 7.29)

7.11.2.2 Determination of Bending Moment

(*A*) *Methods of Analysis and Design* : It shall be permissible to design the flat slab by one of the following methods :

(*i*) The direct Design Method.

(*ii*) The Equivalent Frame Method.

(*i*) *Direct Design Method* :

Limitations : Flat slabs designed by this method shall fulfil the following conditions :

(a) There will be minimum of three continuous spans in each direction.

(b) The panel shall be rectangular, and the ratio of the longer span to the shorter span within a panel shall not be greater than 2.0.

(c) It shall be permissible to offset columns to a maximum of 10 per cent of the span in the direction of the off set notwithstanding the provisions in (*b*).

(d) The successive span lengths in each direction shall not differ by more than one-third of the longer span. The end spans may be shorter but not longer than the interior spans.

(e) The design live load shall not exceed three times the design dead load.

Total Design Moment for a Span : In the direct design method, the total design moment for a span shall be determined for a strip bounded laterally by the centre line of the panel on each side of the centre line of the supports.

The absolute sum of the positive and average negative bending moments in each direction shall be taken as :

$$M_0 = \frac{W . l_n}{8}$$

where M_0 = Total moment,

W = Design load on an area $l_2 \, l_n$;

l_n = Clear span extending from face to face of columns, capitals, brackets or walls, but not less than 0.65 l_1

l_1 = length of span in the direction of M_0

l_2 = length of span transverse to l_1

Circular supports shall be treated as square supports having the same area.

When the transverse span of the panels on either side of the centre line of supports varies, l_2 shall be taken as the average of the transverse spans.

When the span adjacent and parallel to an edge is being considered, the distance from the edge to the centre line of the panel shall be substituted for l_2.

Negative and Positive Design Moments : The negative design moment shall be located at the face of rectangular supports, circular supports being treated as square supports having the same area. The distribution of moments between exterior and interior span is shown in Fig. 7.30.

In an interior span, the total design moment M_0 in a direction shall be distributed in the following proportions:

M_4 = Negative design moment . . . 0.65

M_5 = Positive design moment . . . 0.35

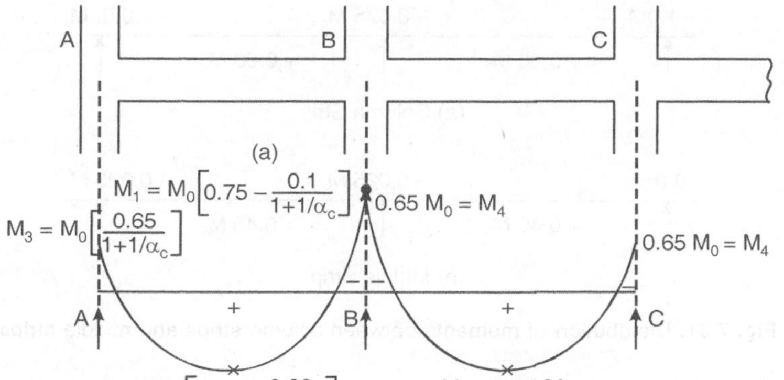

Fig. 7.30. Distrbution of moments between exterior and interior spans of Flat slabs

In the end span, the total design moment M_0 shall be distributed in the following proportions:

$$M_1 = \text{Interior negative design moment} = 0.75 - \frac{0.10}{1+\frac{1}{\infty_c}}$$

$$M_2 = \text{Positive design moment} = 0.63 - \frac{0.28}{1+\frac{1}{\infty_c}}$$

$$M_3 = \text{Exterior negative design moment} = \frac{0.65}{1+\frac{1}{x_c}}$$

where

$$\propto_c = \frac{\sum K_c}{K_s}$$

ΣK_c = sum of the flexural stiffneses, of the columns meeting at the joint.

K_s = Flexural stiffness of the slab, expressed as moment for unit rotation.

It shall be permissible to modify these design moments by upto 10 per cent, so long as the total design moment M_0 for the panel in the direction considered is not less than

$$M_0 = \frac{W \cdot l_n}{8}$$

The negative moment section shall be designed to resist the larger of the two interior negative design moments determined for the spans framing into a common support unless an analysis is made to distribute the unbalanced moment in accordance with the stiffness of the adjoining parts.

Distribution of Bending Moments Across the Panel Width : Bending moments at critical cross section shall be distributed to the column strips and middle strips as under (Fig. 7.31).

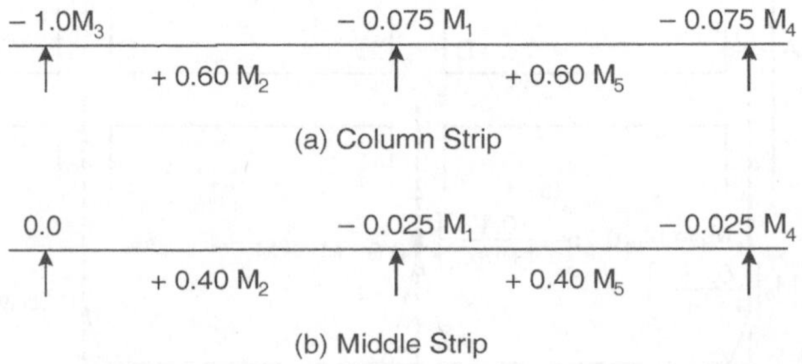

Fig. 7.31. Distribution of moments between column strips and middle strips

Column Strips

Negative Moment at an Interior Support : At an interior support, the column strip shall be designed to resist 75 per cent of the total negative moment in the panel at that support.

Negative Moment at an Exterior Support : At an exterior support the column strip shall be designed to resist the total negative moment in the panel at the support.

When the exterior support consists of column or a wall extending for a distance equal to or greater than the three quarters of the value of l_2, the length of span transverse to the direction moments shall be considered to be uniformly distributed across the length l_2.

Positive Moment for Each Span : For each span, the column strip shall be designed to resist 60 per cent of the total positive moment in the panel,

Middle Strip : The middle strip shall be designed on the following basis :

The portion of the design moment not resisted by the column strip shall be assigned to the adjacent middle strips.

Each middle strip shall be proportioned to resist the sum of the moment assigned to its two half middle strips.

The middle strip adjacent and parallel to an edge supported by a wall shall be proportioned to resist twice the moment assigned to half the middle strip corresponding to the first row of interior columns.

Moments in Columns : Columns built integrally with the slab system shall be designed to resist moments arising from loads on the slab system.

At an interior support, the supporting members above and below shall be designed to resist the moment M given by the following equation, in direct proportion to their stiffnesses unless a general analysis is made.

$$M = 0.08\frac{(w_d + 0.5\,w_l)l_2\,l_n^2 - w'_d\,l'_2\,l'^2_n}{1+\dfrac{1}{\alpha_c}}$$

where w_d, w_l = design dead and live loads respectively per unit area.

l_2 = length of span transverse to the direction of M.

l_n = length of the clear span in the direction of M, measured face to face of supports.

$$\alpha_c = \frac{\sum K_c}{\sum K_s}$$ w', l'_2, l'_c refer to the shorter span.

Effects of Pattern Loading: In the direct design method, when the ratio of live load to dead load exceeds 0.5,

(*a*) ΣK_c shal be such that α_c is not less than the appropriate minimum value $\alpha_{c,\,min}$ specified in Table 7.9.

(*b*) If ΣK_c does not satisfy (*a*), the positive design moments for the panel shall be multiplied by the coefficient β_s given by :

$$\beta_s = 1 = \left(\frac{2-\dfrac{w_d}{w_l}}{4+\dfrac{w_d}{w_l}}\right)\left(1-\frac{\alpha_c}{\alpha_{c,min}}\right)$$

where $\alpha_c = \dfrac{\sum K_c}{\sum K_s}$ at a joint taken in the direction moments are being determined.

(*ii*) *Equivalent Frame Method*

The bending moments and shear forces may be determined by an analysis of the structure as a continuous frame by any of the well known methods namely Moment Distribution method,

Table 7.9. Minimum Permissible Values of α_c

Live load/Dead Load	Ratio $\frac{l_2}{l_1}$	Values of a_c min
0.5	0.5 to 2.0	0
1.0	0.5	0.6
1.0	0.8	0.7
1.0	1.0	0.7
1.0	1.25	0.8
1.0	2.0	1.2
2.0	0.5	1.3
2.0	0.8	1.5
2.0	1.0	1.6
2.0	1.25	1.9
2.0	2.0	4.9
3.0	0.5	1.8
3.0	0.8	2.0
3.0	1.0	2.3
3.0	1.25	2.8
3.0	2.0	3.0

Kani's method or stiffness method, the design of flat slabs by direct design method is more common and easy. However, where computations can be made on Electronic digital computers, equivalent frame method is superior and quicker. The following assumptions are made in Equivalent Frame method.

(*a*) The structure shall be considered to be made up of equivalent frames on column lines taken longitudinally and transversely through the building. Each frame consists of a row of equivalent columns or supports and beams equivalent to slab width bounded by the centre line of the panel on each side of the centre line of the columns or supports. Frames adjacent and parallel to an edge shall be bounded by the edge and the centre line of the adjacent panel.

(*b*) Each such frame may be analysed in its entirety, or for vertical loading, each floor thereof and the roof may be analysed separately with its columns being assumed fixed at their remote ends where slabs are thus analysed separately, it may be assumed in determining the bending moment at a given support that the slab is fixed at any support that two panels distant therefrom provided the slab continues beyond that point.

(*c*) For the purpose of determining relative stiffness of members, the moment of inertial of any slab or column may be assumed to be that of the gross cross section of the concrete alone.

(*d*) Variations of moment of interia along the axis of the slab on account of provision of drops shall be taken into account. In the case of recess or coffered slab which is made solid in the region of the columns, the stiffening effect may be ignored provided the solid part of the slab does not extend more than 0.15 l_{ef}, into the span measured from the centre line of the columns. The stiffening effect of flared column heads may be ignored.

Loading Pattern

When the loading pattern is known, the structure shall be analysed for the load concerned.

When the live load is variable but does not exceed three-quarters of the dead load or the nature of the live load is such that tall panels will be loaded simultaneously, the maximum moments may be assumed to occur at all sections when full design live load is on the entire slab system.

For other conditions of live load/dead load ratio and when all panels are not loaded simultaneously :

(*a*) *Maximum Positive Moment* : Near mid span of a panel may be assumed to occur when three-quarters of the full design live load is one the panel and on alternate panels; and

(*b*) *Maximum Negative Moment* : In the slab at a support may be assumed to occur when three-quarters of the full design live load is one the adjacents panels only.

Minimum Design Moment : In no case shall design moments be taken to be less than those occurring with full design live load on all panels.

Negative Design Moment

At interior supports, the critical section for negative moment, in both the column strip and middle strip, shall be taken at the face of rectilinear supports, but in no case at a distance greater than 0.175 l_1 from the centre of the column.

At exterior supports provided with brackets or capitals, the critical section for negative moment in the direction perpendicular to the edge shall be taken at a distance from the face of the supporting element not greater than one-half of the projection of the bracket or capital beyond the face of the supporting element.

Circular or regular polygon shaped supports shall be treated as square supports having the same area.

Modification of Maximum Moment

Moments determined by means of the equivalent frame method, for slabs which fulfil the limitations of direct design method may be reduced in such proportions that the numerical sum of the positive and average negative moments is not less than the value of total design moment

$$M_0 = \left(\frac{w \cdot l_n}{8}\right)$$

Distribution of Bending Moment Across the panel width

This is the same as discussed in the direct design method.

(*B*) *Bending Moments in Panels with Marginal Beams or Walls*

Where the slab is supported by a marginal beam with a depth greater than 1.5 times the thickness of the slab, or by a wall, then:

(*a*) the total load to be carried by the beam or wall shall comprise those loads directly on the wall or beam plus a uniformly distributed load equal to one-quarter of the load on the slab, and

(*b*) the bending moments on the half-column strip adjacent to the beam or wall shall be one-quarter of the bending moments for the first interior column strip.

(*C*) *Transfer of Bending Moments to Columns*

When unbalanced gravity load, wind, earthquake or other lateral loads cause transfer of bending moment between slab and column, the flexural stresses shall be investigated using a fraction, $\propto$ of the column given by :

$$\propto = \frac{1}{1 \pm \frac{2}{3}\sqrt{\left(\frac{a_1}{a_2}\right)}}$$

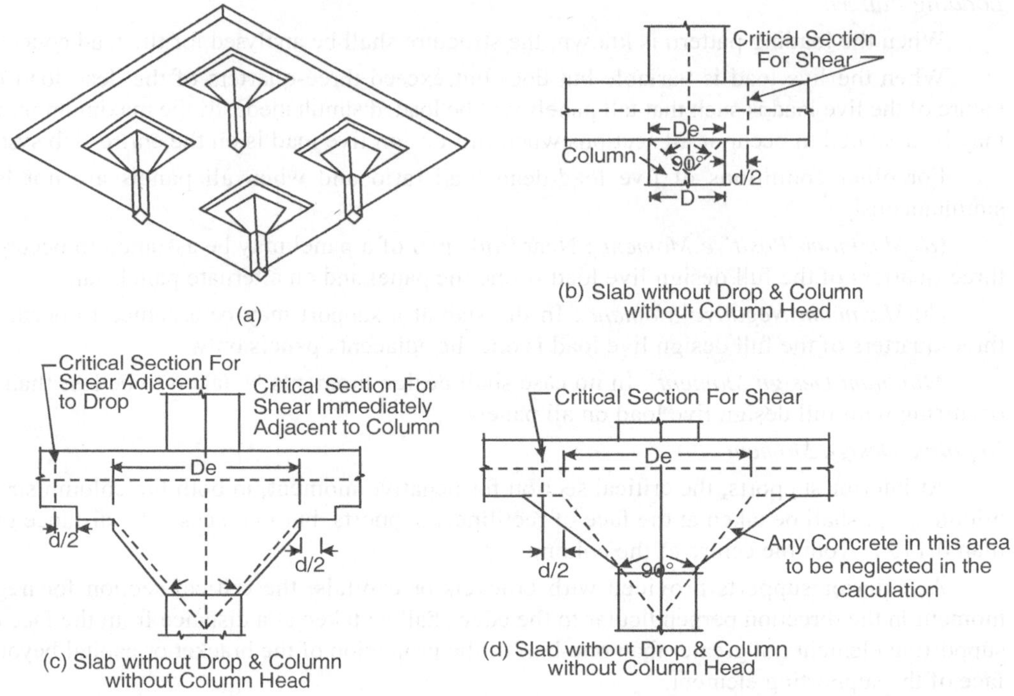

Fig. 7.32

where a_1 = overall dimension of the critical section for shear in the direction in which moment acts, and

a_2 = overall dimension of the critical section for shear transverse to the direction in which moment acts.

A slab width between lines that are one and one-half slab or drop panel thickness. 1. 5 D on each side of the column or capital may be considered effective, D being the size of the column.

Concentration of reinforcement over column head by closer spacing or additional reinforcement may be used to resist the moment on this section.

7.11.2.3 Shear in Flat Slabs

The critical section for shear shall be at a distance $d/2$ from the periphery of the column/capital/drop panel, perpendicular to the plane of the slab where d is the effective depth of the section (Fig. 7.32). the shape in plan is geometrically similar to the support immediately below the slab as shown in Fig. 7.32 (*a*) and (*b*). For column, sections with re-entrant angles, the critical section shall be taken as indicated in Fig. 7.32 (*c*) and (*d*)

In case of columns near the free edge of a slab, the critical section shall be taken as shown in Fig. 7.33.

When the openings in flat slabs are located at a distance less than 10 times the thickness of the slab from a concentrated reaction or when the openings are located within the column strips, the critical sections specified above shall be modified so that the part of the periphery of the critical section which is enclosed by radial projections of the openings to the centroid of the reaction area shall be considered ineffective and openings shall not encroach upon column head.

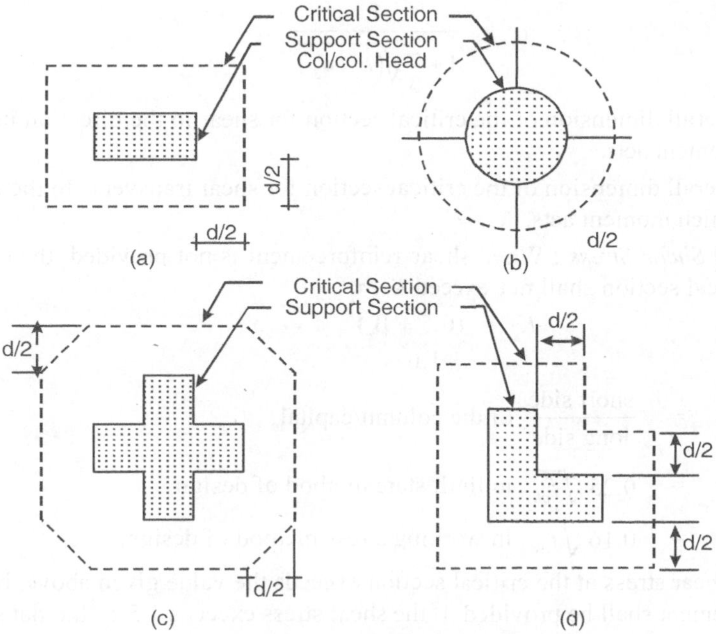

Critical sections in plan for shear in flat slab

Fig. 7.32

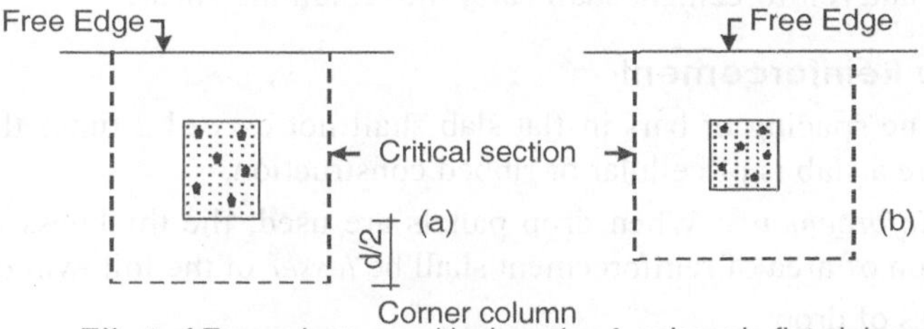

Corner column

Effect of Free edges on critical section for shear in flat slab

Fig. 7.33

Calculation of Shear Stress : The shear stress τ_v shall be the sum of the following two vlaues. The nominal shear stress in flat slab shall be taken as :

$$\tau_v = \frac{V}{b_0 . d}$$

where

V = shear force due to design load

b_0 = periphery of the critical section

d = effective depth.

When unbalanced gravity load, wind, earthquake, or other forces caused transfer of bending moment between slab and column, a fraction $(1 - \propto)$ of the moment shall be considered transferred by eccentricity of the shear about the centroid of the critical section. Shear stresses shall be taken as varying linearly about the centroid of the critical section. The value of $\propto$ shall be obtained from equation.

$$\alpha = \frac{1}{1+\frac{2}{3}\sqrt{(a_1/a_2)}}$$

where a_1 = overall dimension of the critical section for shear in the direction in which moment acts.

a_2 = overall dimension of the critical section for shear transverse to the direction in which moment acts.

Permissible Shear Stress : When shear reinforcement is not provided, the calculated shear stress at the critical section shall not exceed K_2, τ_c

where $K_s = (0.5 + \beta_c)$

$\not< 1.0$

$$\beta_c = \frac{\text{short side}}{\text{long side}} \text{ of the column/capital}$$

$$\tau_c = 0.25\sqrt{f_{ck}} \text{ in limit state method of design}$$

$$= 0.16\sqrt{f_{ck}} \text{ in working stress method of design.}$$

When the shear stress at the critical section exceeds the value given above, but less than 1.5 τ_c shear reinforcement shall be provided. If the shear stress exceeds 1.5 τ_c, the flat slab shall be re-designed. Shear stresses shall be investigated at successive sections more distant form the support and shear reinforcement shall be provided upto a section where the shear stress does not exceed 0.5 τ_c. While designing the shear reinforcement, the shear stress carried by the concrete shall be assumed to be 0.5 τ_c and reinforcement shall carry the remaining shear.

7.11.2.4 Flat Slab Reinforcement

(*i*) *Spacing* : The spacing of bars in flat slab shall not exceed 2 times the slab thickness, except where a slab is of cellular or ribbed construction.

(*ii*) *Area of Reinforcement* : When drop panels are used, the thickness of drop panel for determination of area of reinforcement shall be *lesser* of the following:

(*a*) thickness of drop

(*b*) thickness of slab plus one quarter the distance between edge of drop and edge of capital

(*iii*) *Minimum Length of Reinforcement* : (a) Reinforcement in flat slab shall have the minimum lengths specified in Fig. 7.34. Larger lengths of reinforcement shall be provided when required by analysis.

(b) Where adjacent spans are unequal, the extension of negative reinforcement beyond each face of the common column shall be based on the longer span.

(*iv*) *Anchoring Reinforcement* : (a) All slab reinforcement perpendicular to a discontinuous edge shall have an anchorage (straight, bent or otherwise anchored) past the internal face of the spandrel beam, wall or column, of an amount:

(1) *For Positive Reinforcement* : not less than 150 mm except that with fabric reinforcement having a fully welded transverse we directly over the support it shall be permissible to reduce this length to one half of the width of the support or 50 mm, whichever is greater; and

(2) *For Negative Reinforcement* : such that the design stress is developed at the internal face, in accordance with the given in Chapter 5 of the book.

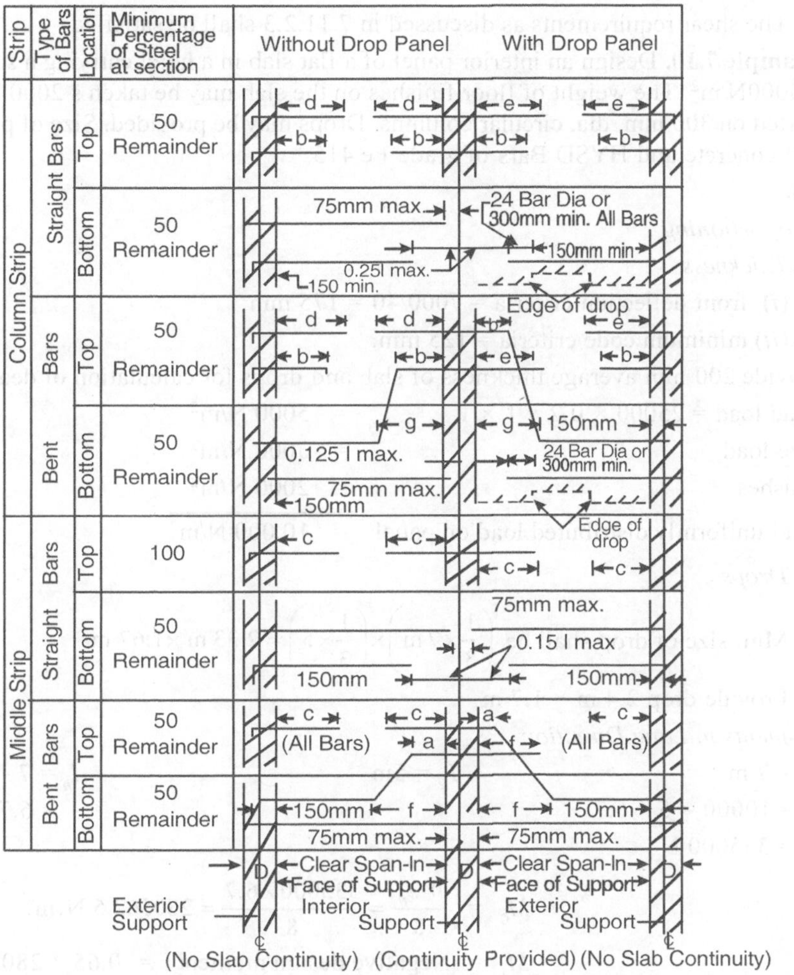

Fig. 7.34. Minimum Bend Joint Locations and Extensions for Reinforcement in Flat Slabs

(3) Where the slab is not supported by a spandrel beam or wall, or where the slab cantilevers beyond the support, the achorage shall be obtained within the slab.

7.11.2.5 Opening in Flat Slabs

Openings of any size may be provided in the flat slab if it is shown by analysis that the requirements of strength and serviceability are met. However, for openings conforming to the following, no special analysis is required:

(*a*) Openings of any size may be placed within the middle half of the span in each direction, provided the total amount of reinforcement required for the panel without the opening is maintained.

(*b*) In the area common to two column strips, not more than one-eighth of the width of strip in either shall be interrupted by the openings.

(*c*) In the area common to one column strip an done middle strip, not more than one quarter of the reinforcement in either strip shall be interrupted by the openings. The equivalent of reinforcement interrupted shall be added on all sides of the openings.

(*d*) The shear requirements as discussed in 7.11.2.3 shall be satisfied.

Example 7.10. Design an interior panel of a flat slab in a hotel carrying a superimposed live load of 3000N/m². The weight of floor finishes on the slab may be taken s 2000 N/m². The panel is supported on 300 mm. dia. circular columns. Drops may be provided. Size of panel is 5m × 7m. Use M20 concrete and HYSD Bars of grade Fe 415.

Sol.

(*A*) *Proportioning*

(1) *Thickness.*

(*i*) from deflection criteria = 7000/40 = 175 mm

(*ii*) minimum code criteria = 125 mm.

Provide 200 mm average thickness of slab and drops for calculation of dead load.

Dead load = 25000 × 0.2 × 1 × 1	=	5000 N/m²
Live load	=	3000 N/m²
Finishes	=	2000 N/m²
Total uniformly distributed load on panel	=	10,000 N/m²

(2) *Drops*

Min. size of drop shall be $\left(\frac{1}{3}\times 7\,\text{m}\right)\times\left(\frac{1}{3}\times 5\right) = 2.33\text{ m}\times 1.67\text{ m}$

Provide drop 2.4 m × 1.7 m

(*B*) *Moments in Long Direction*

$l_1 = 7$ m, $l_2 = 5$m $l_n = 7 - 0.3 = 6.7$ m

$W = 10000 \times 5 \times 6.7$

$= 335000$ N

$$M_0 = \frac{W.l_n}{8} = \frac{335000\times 6.7}{8} = 280562.5\text{ N.m.}$$

M_4 = Negative design moment = 0.65 × 280562.5

= 182365. 6 N.m

M_5 = Positive design moment = 0.35 × 280562.5

= 98196.9 N.m.

Width of column strip in long direction

= 0.25 × 5 = 1.25 m on each side of column centre line

≯ 0.25 × 7 = 1.75 m

Adopt 1.25 × 2 = 2 = 2.5 m

Width of middle strip = 5 – (1.25 + 1.25) = 2.5

B.M. for column Strip :

Negative moment = 75% of negative design moments = 0.75 M_4

= 0.75 × 182365.6

= 136775 N.m.

Positive moment = 60% of positive design moment M_5

= 0.60 × 98196.9

= 58,918 N.m

B.M. for Middle Strip:

Negative moment = 25% of negative design moment M_4
= 0.25 × 182365.6
= 45,591.4 N.m

Positive moment = 40% of positive design moment M_5
= 0.40 × 98196.19
= 39278.8 N.m

(*C*) *Moments in Short Direction* :

l_1 = 5 m, l_2 = 7m l_n = 5 – 0.3 = 4.7 m

$$W = 10000 \times 7 \times 4.7 = 329000 \text{ N}$$

$$M_0 = \frac{W.l_n}{8} = \frac{329000 \times 4.7}{8} = 193287.5 \text{ N.m}$$

M_4 = Negative design moment = 0.65 × 193287.5 = 125636.9 N.m

M_5 = Positive design moment = 0.35 × 193287.5 = 67650.6 N.m

Width of column strip in short direction = 0.25 × l_2 = 0.25 × 7 = 1.75 m
≯ 0.25 l1 = 0.25 × 5 = 1.25
Adopt 1.25 × 2 = 2.5 m

Width of the middle strip in short direction = 7 – 2 × 1.25 = 4.5 m

B.M. for Colum Strip :

Negative moment = 75% of M_4
= 0.75 × 125636.9
= 94227.7 N.m

Positive moment = 60% of M_5
= 0.60 × 67650.6
= 40,590.4 N.m

B.M. for Middle Strip :

Negative moment = 25% of M_4
= 0.25 × 125636.9
= 31409.2 N.m

Positive moment = 40% of M_5
= 0.40 × 67650.5
= 27060.2 N.m

(*D*) *Design Moments for Thickness* :

(*i*) *Drop* : The design moment for computing drop thickness shall be taken as the negative moment in the column strip of the long span, *i.e.*, 1,36,775. N.m. Width of drop is 1.7 m.

(*ii*) *Flat Slab* : The design moment for computing thickness of slab shall be taken as positive in the column strip of the long span, i.e., 58,918 N.m. Whose width is 2.5 m.

(*E*) *Thickenss*

(*i*) *Drop*

$$\text{Effective depth} = \sqrt{\left(\frac{136775\times1000}{0.91\times1.7\times1000}\right)}$$

$$= 297.3 \text{ mm}$$

Adopt D = 330 mm, d = 305 mm

(*ii*) *Slab*

$$\text{Effective depth} = \sqrt{\left(\frac{58918\times1000}{0.91\times2.5\times1000}\right)}$$

$$= 164.6 \text{ mm}$$

Adopt D = 185 mm, d = 165 mm

(*F*) *Areas of Steel* :

Column strip (long direction) 2.5 m wide

$$\text{Steel for the +ve } B.M. = \frac{58918\times1000}{230\times0.90\times165} = 1706 \text{ mm}^2/2.5 \text{ m}$$

$$= 682 \text{ mm}^2/\text{m}$$

Provide 10 mm dia. bars @ 100 mm *c*/*c* (A_s provided = 785 mm²)

$$\text{Steel for -- ve } B.M. = \frac{136775\times1000}{230\times0.90\times305} = 2142 /\text{mm}^2/2.5 \text{ m}$$

$$\text{–ve steel/m} = 857 \text{ mm}^2$$

Provide 10 mm ϕ @ 200 *c*/*c* and 12 m ϕ @ 200 mm *c*/*c* thus providing = 958 mm²

Middle Strip (*Long Direction*) 2.5m wide

$$\text{Steel for +ve } B.M. = \frac{39278.8\times1000}{230\times0.90\times165} = 1137.4 \text{ mm}^2/2.5 \text{ m}$$

$$= 455 \text{ mm}^2/\text{m}$$

Provide 10 mm ϕ @ 150 mm *c*/*c*

$$\text{Steel for –ve } B.M. = \frac{45591.4\times1000}{230\times0.90\times165} = 1320.2 \text{ mm}^2/2.5 \text{ m}$$

Provide 10 mm ϕ @ 300 (A = 639mm² mm *c*/*c*)
+12 mm ϕ @ 300 (A = 639mm² mm *c*/*c*)

Column Strip (*Short Direction*) 2.5 m wide

$$\text{Steel for +ve } B.M. = \frac{40590.4\times1000}{230\times0.90\times165} = 1175.4 \text{ mm}^2/2.5 \text{ m}$$

$$= 470 \text{ mm}^2/\text{m}$$

Provide 10 mm ϕ 150, A = 524 mm²

$$\text{Steel for –ve } B.M. = \frac{94227.7\times1000}{230\times0.90\times305} = 1476 \text{ mm}^2/2.5 \text{ m}$$

$$= 590.4 \text{ mm}^2/\text{m}$$

Provide $10 \text{ mm } \phi @ 300 \quad (A = 639 \text{ mm}^2)$

$+12$ mm f @ 300 $A = 639 \text{ mm}^2$

Middle Strip Short Direction : 4.5 m wide

$$\text{Steel for +ve } B.M. = \frac{27060.2 \times 1000}{230 \times 0.90 \times 165} = 783.6 \text{ mm}^2/4.5 \text{ m}$$

$$= 174.1 \text{ mm}^2/\text{m}$$

$8 \text{ mm } \phi @ 165 \text{ mm } c/c, A = 305 \text{ mm}^2$

$$\text{Steel of –ve } B.M. = \frac{31409.2 \times 1000}{230 \times 0.90 \times 165} = 909.5 \text{ mm}^2/4.5 \text{ m}$$

$$= 202.1 \text{ mm}^2/\text{m}$$

$8 \text{ mm } \phi @ 330 \text{ mm } c/c$ + $10 \text{ mm } \phi @ 330 \text{ mm } c/c$ $\Big] A = 151.5 + 237.9 = 389.4 \text{ mm}^2$

Check for Shear Stresses in drops and slab.

(a) *Critical Section in Drop*:

The critical section for shear is to be taken at a distance of half the effective depth of drop i.*e.*, 305/2 mm from face to column as shown in Fig. 7.35. Dia. of the critical circle

$$= 300 + 305/2 + 305/2 = 605\text{mm}$$

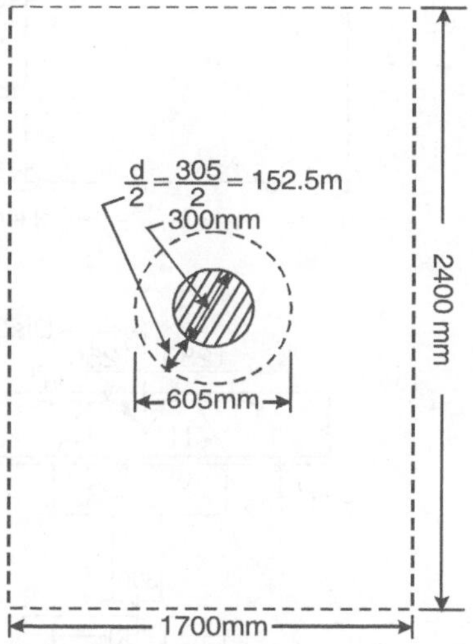

Fig. 7.35.

Load on One Column i.e., S.F. at

$$\text{Load from the slab} = 10000 \times \left((7 \times 5) - \left(\frac{\pi}{4} \times 0.605^2\right)\right)$$

$$= 347127 \text{ N}$$

Nominal shear stress 'τ_v'

$$= \frac{347127}{(\pi \times 605) \times 305} = 0.599 \text{ N/mm}^2$$

$$\text{Permissible shear stress'}\tau_v\text{'} = 0.16\sqrt{(f_{ck})}$$

$$= 0.16\sqrt{20}$$

$$= 0.7155 \text{ N/mm}^2$$

$$\tau_v < \tau_c$$

Safe in shear.

(b) *Critical Section in slab.*

This critical section is to be taken at a distance of half the effective depth of slab *i.e.* $\frac{1625}{2}$ mm from the edge of the drop as shown in Fig. 7.36.

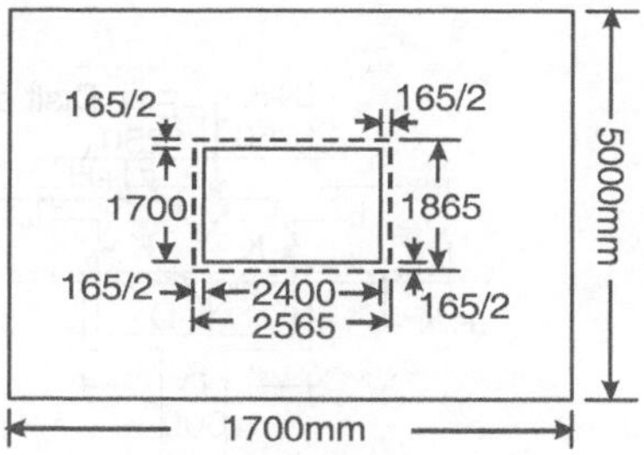

Fig. 7.36.

Total perimeter of the critical section

$$= 2(2565 + 1865) = 8860 \text{ mm}$$

Total shear on the critical section

$$= 10000 \times (7 \times 5 - 2.565 \times 1.865) = 302163 \text{ N}$$

$$\text{Nominal shear stress '}\tau_v\text{'} = \frac{302163}{(8860 \times 165)} = 0.207 \text{ N/mm}^2 < \tau_c = 0.7155 \text{ N/m}^2$$

Safe in shear.

The reinforcement details are shown in Fig. 7.37 (a) to (e).

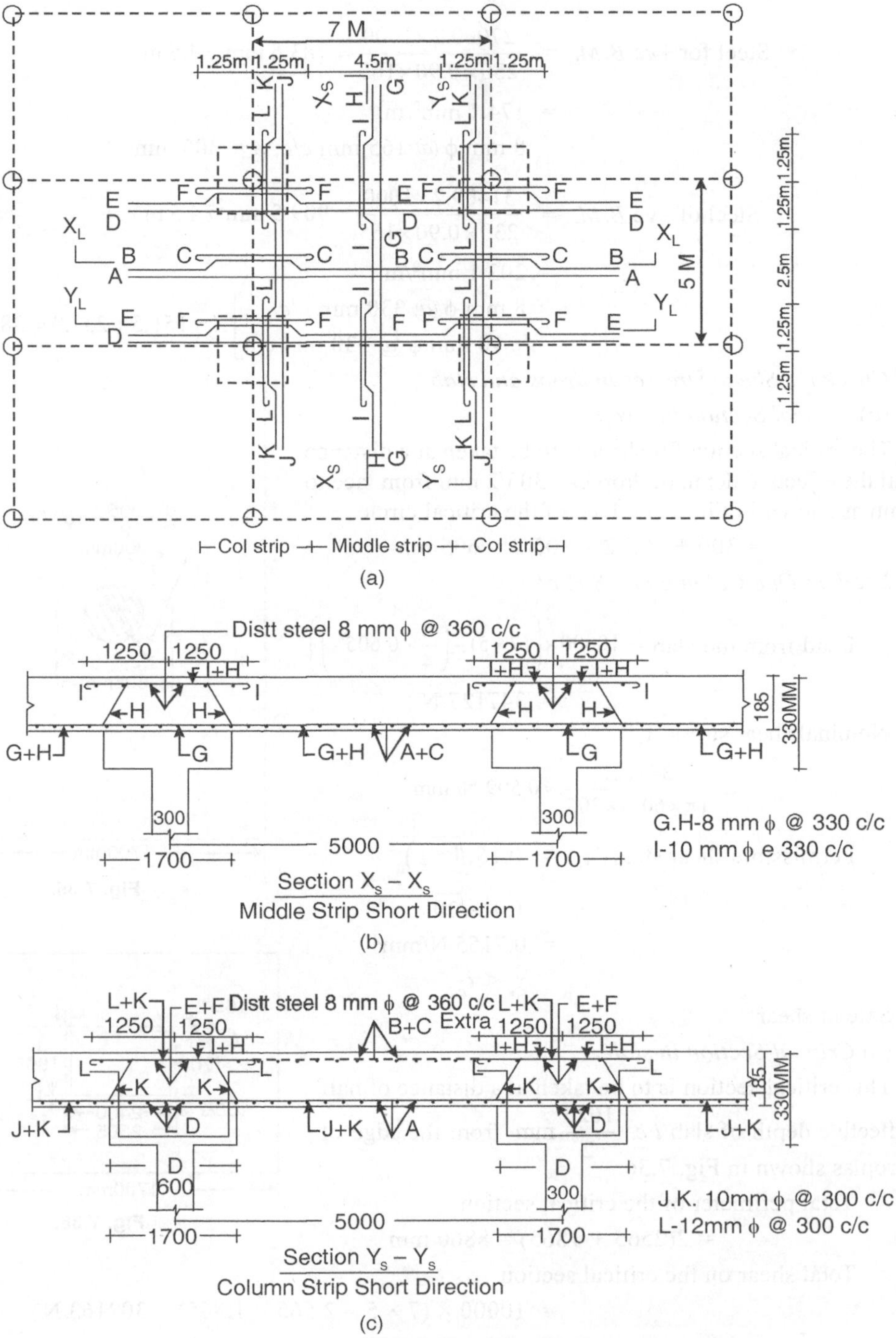

Fig. 7.37.

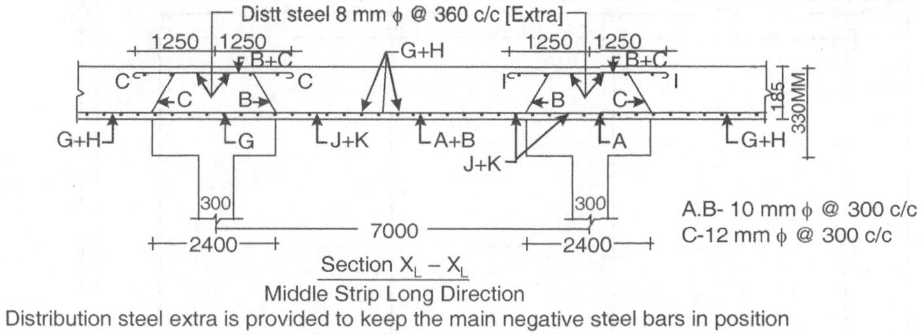

Fig. 7.37d

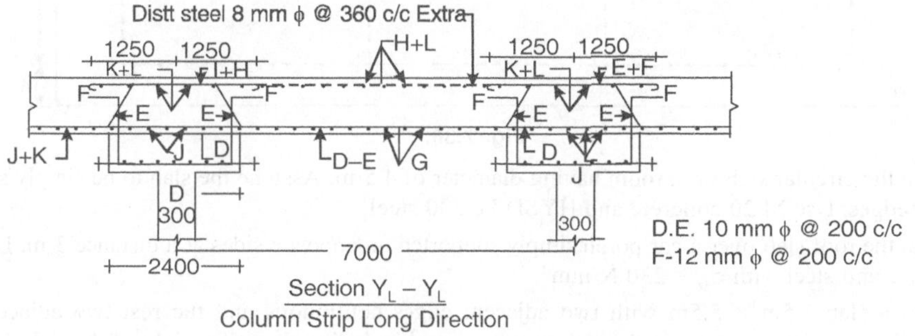

Fig. 7.37e

7.12 OPENINGS IN SLABS

Quite often openings are provided in slab panels to get natural light from the top when it is not possible to get it from any of the sides. If the opening size is small as compared to overall dimensions of the slab panel, no special beams are necessary but thickening of edges of the opening and extra reinforcement in thickened portion are required to take care of stress concentration. This extra reinforcement normally equals the amount of steel saved because of the opening. However, if the openings are quite large, special beam arrangement is to be provided. Sometimes these beams are concealed in the floor thickness for aesthetic reasons. Concealed beams are mostly designed on the basis of Steel Beam theory.

EXERCISE 7

1. The clear dimensions of a room with 225 mm thick masonry walls on all four sides are 4m × 5m. The superimposed live load, earth load and tiles load may be taken as 750 N/m², 1500 N/m² and 1000 N/m² respectively. Design the roof slab which is simply supported on all sides. Use M15 concrete and HYSD (Fe 415) Steel with σ_{st} = 230 N/mm².
2. Design a roof slab for verandah with effective span 2.2 m. The superimposed load including live load and other finishes may be taken 3500 N/m². Use M 20 concrete and HYSD (Fe 415) bars.
3. Design an interior panel slab at first floor level with effective spans in two directions as 4m × 6m. The superimposed live load and weight of finishes may be taken as 3000 N/m². Us M 20 concrete and HYSD (Fe 500) bars.
4. Design a cantilever balcony projecting out of wall by a distance 1.5 m with a facia 75 mm thick and 450 mm height at its free end. Take live load and finishes weight as 3000 N/m². Use M 20 concrete and HYSD (Fe 500) bars.
5. Fig. 7.38 shows the plan of a floor slab of an industrial building. The live load including the weight of machinery may be taken as 500 N/m². Take M 20 mix and HYSD (Fe 500) bars.

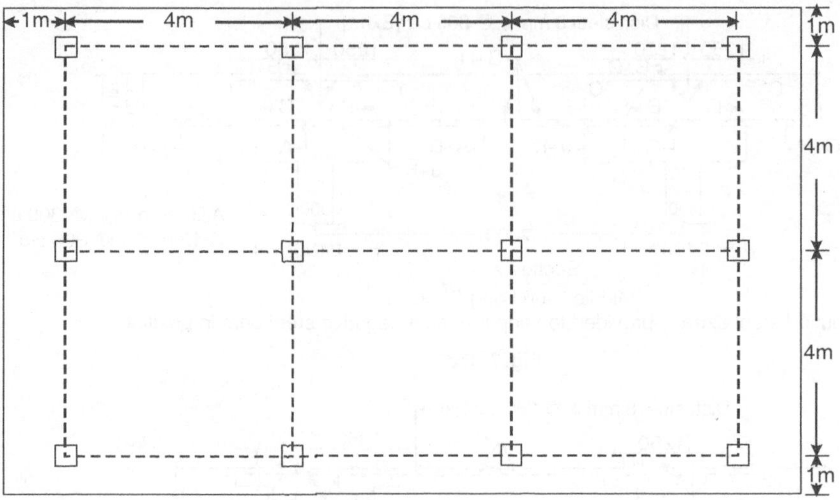

Fig. 7.38.

6. Design the circular slab for a room having diameter of 4.5 m. Assume the slab to be simply supported on its edges. Use M 20 concrete and HYSD Fe 500 steel.
7. Design the roof slab over a car porch simply supported on opposite sides at a distance 3 m. Use M 20 concrete and steel with σ_{st} = 230 N/mm^2.
8. Design a slab 3.5m × 5.5m with two adjacent edges continuous and the rest two adjacent edges discontinuous. Take loading on the slab as 8000 N/m^2 including the self weight of the slab. Use M 20 concrete and HYSD (Fe 500).

CHAPTER

8

Axially Loaded Columns

8.1 INTRODUCTION

Columns are vertical compression members used to transfer the loads of the structures such as buildings, factory floors, Cinema balconies, Auditorium halls, floors of framed buildings, etc., to the foundation below. The transfer of load may be :

(*i*) direct from the roof or floor slabs through the columns to the foundation.

(*ii*) indirect through a beam to the columns and then to the foundation.

Columns form the most important part of a structure. The failure of any other part (beam or slab) of a structure may not cause so serious a damage as that caused by the failure of a column. It can endanger even the whole structure. As such the columns must be designed very carefully.

All vertical members may not be termed as columns. Only those members whose effective length is more than three times the least lateral dimension are called **columns** and those members whose effective length is less than 3 times the least lateral dimension are called "**pedastals**". Concrete is a very strong material in compression and weak in tension. Thus the best utility of this property of concrete can be made use of in designing a compression member of concrete, i.e., a column. Though column is a member in compression yet the possibility of tension in some part of it cannot be ruled out. This tension may occur due to :

(*i*) accidental eccentricity of loading

(*ii*) bending in columns.

To take care of this tension, some amount of steel has to be provided on tension face in the longitudinal direction. The amount of steel should be just sufficient to take care of the tensile stresses only. The longitudinal steel is kept its position with the help of binders (steel wires of small diameter) or ties which also prevent the buckling of vertical bars.

Axially loaded columns are those in which the line of action of resultant thrust of the loads supported by a column coincides with the centre of gravity of the column cross section. As per IS:456-2000, there is no column which is truly axially loaded. It says that all columns shall be designed for minimum eccentricity equal to the unsupported length of column/500 plus lateral dimension/30, subject to a minimum of 20 mm.

8.2 TYPES OF COLUMNS

The various types of R.C. columns are :

(*i*) Columns with longitudinal steel and with lateral ties or spirals.

(*ii*) Composite columns with structural Rolled Steel section encased in concrete.

(*iii*) Concrete filled steel tubular columns in which steel tube is filled with concrete inside it.

The column may be of any shape; rectangular, square, circular, hexagonal, etc. The steel should be placed near each sharp corner in case the section is rectangular, square or hexagonal and well distributed if the section is circular as shown in Fig. 8.1.

Column section shown in Fig. 8.1 (a) to (d) are simple R.C.C. columns whereas columns shown in Fig. 8.1 (e) and (f) are termed as "composite columns": in which either a Rolled Steel section (R.S.J., I-beam or channel) is encased in concrete or concrete is filled in steel tube.

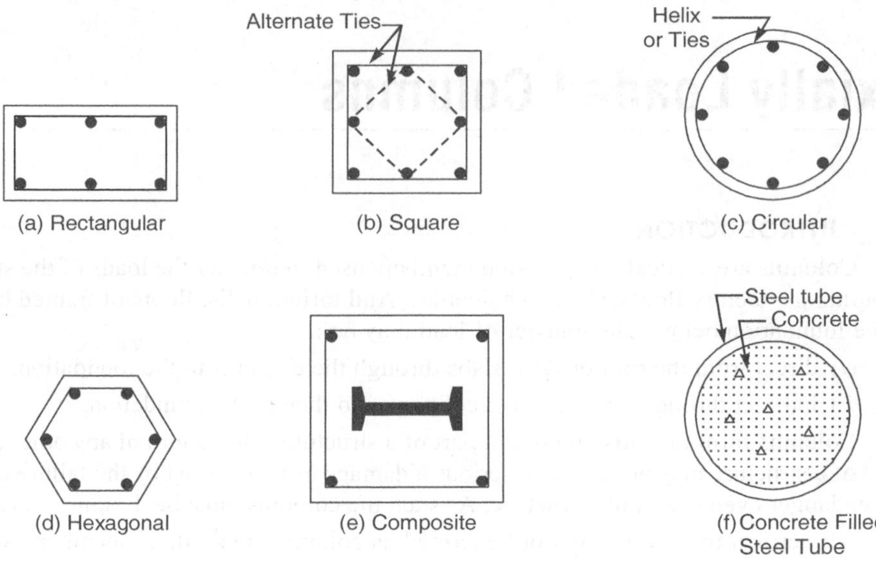

Fig. 8.1. Shapes of Columns

The helical ties, Fig. 8.1 (c) provide lateral confinement to the concrete. This also prevents buckling of longitudinal bars due to compression. The extreme case is when the hoop steel in the form of steel tube is provided. This type of column may be termed as concrete Filled Steel Tubular Column.

8.3 SHORT AND SLENDER COLUMNS

A compression member may be considered as short when both the slenderness ratios $\frac{l_{ex}}{D}$ and $\frac{l_{ey}}{b}$ are less than 12,

where l_{ex}, l_{ey} = effective length in respect of the major axis and minor axis respectively.

D = Depth in respect of the major axis.

b = Width of the member.

If the above slenderness ratios are greater than 12, then it shall be considered as slender compression member or a long column.

8.4 PERMISSIBLE LOADS IN COMPRESSION MEMBERS

The load carrying capacity of different types of columns is given below:

(*a*) *Pedestals and Short Columns with Lateral Ties* : The axial load P permissible on a short coloumn reinforced with logitudinal bars and lateral ties shall not exceed that given by the following equations.

where $$P = \sigma_{cc} \cdot A_c + \sigma_{sc} \cdot A_{sc} \quad ...(8.1)$$

σ_{cc} = permissible stress in concrete in direct compression.

A_c = cross-sectional area of concrete excluding any finishing material and reinforcing steel,

σ_{sc} = permissible compressive stress for column bars, and

A_{sc} = cross-sectional area of the longitudinal steel.

Note : The minimum eccentricity of 20 mm mentioned in article 81. may be deemed to be incorporated in the above equation.

(*b*) *Short Columns with Helical Reinforcement*: The permissible load for columns with helical reinforcement shall be 1.05 times the permissible load for similar member with lateral ties or rings, *i.e.*,

$$P = 1.05\,(\sigma_{cc} \,.\, A_c + \sigma_{sc} \,.\, A_{sc}) \qquad ...(8.2)$$

provided such columns satisfy the following condition:

The ratio of the volume of helical reinforcement to the volume of the core shall not be less than

$$0.36\left(\frac{A_g}{A_c} - 1\right).\frac{f_{ck}}{f_y}$$

where in this equation only

A_g = gross area of the section

A_c = area of the core of the helically reinforced column measured to the outside diameter of the helix.

f_{ck} = characteristic compressive strength of the concrete, and

f_y = characteristic strength of helical reinforcement but not exceeding 415 N/mm^2.

(*c*) *Long Columns* : The maximum permissible stress in a reinforced concrete column or part thereof having a ratio of effective column length to least lateral dimension above 12 shall not exceed that which results from the multiplication of the appropriate maximum permissible stress in concrete and steel by the coefficient C_r given by the following formula :

$$C_r = 1.25 - \frac{l_{ef}}{48\,b}$$

where C_r = reduction coefficient

l_{ef} = effective length of column; and

b = least lateral dimension of column; and is equal to diameter of the core in case of column with helical reinforcement.

For more exact calculations, the maximum permissible stresses in a reinforced concrete column of part thereof having a ratio of effective column length to least lateral radius of gyration above 40 shall not exceed those which result from the multiplication of the appropriate maximum permissible stresses in concrete and steel by the coefficient C_r given by the following formula :

$$C_r = 1.25 - \frac{l_{ef}}{160\,i_{\min}}$$

where i_{min} = least radius of gyration.

(*d*) *Composite Columns* ; The allowable axial load P on a composite column consisting of structural steel or cast-iron column thoroughly encased in concrete reinforced with both longitudinal and spiral reinforcement, shall not exceed that given by the following formula :

$$P = \sigma_{cc} \,.\, A_c + \sigma_{sc} \,.\, A_{sc} + \sigma_{mc} \,.\, A_m$$

where

A_c = net area of concrete section; which is equal to the gross area of the concrete section minus A_{sc} minus A_m;

A_{sc} = cross-sectional area of longitudinal bar reinforcement,

σ_{mc} = allowable unit stress in metal core, not to exceed 125 N/mm² for a steel core, or 70 N/mm² for a cast-iron core; and

A_m = the cross-sectional area of the steel or cast iron core.

(*e*) *Concrete Filled Steel Tubular Columns* : Concrete filled tubes have many advantages over the conventional reinforced concrete coloumns which are given below :

1. The cross-section of column is reduced.
2. It increases the ultimate load carrying capacity of the column many times.
3. It prevents bucking of the wall of the tube because of concrete filling inside it.
4. The steel tube serves as form-work. So no extra form-work is required, thus saving in its cost.
5. It looks beautiful architecturally.
6. Provides more room space because of its reduced size as compared to that with R.C. coloumn.

The allowable axial load P on a column consisting of mild steel pipe filled with concrete shall be as given by equation.

$$P = A_c.\sigma_{cc}\left(1 - 0.000025\frac{L^2}{K_c^2}\right) + \sigma_{sc}.A_{sc}$$

where L = unsupported length of the column.

k_c = radius of gyration of concrete alone.

$$\sigma_{sc} = \left(1200 - 0.034\frac{L^2}{K_s^2}\right) \text{kg/cm}^2,$$

Provided yield strength of the pipe is not less than 2400 kg/cm² and L/k_s is not more than 120.

A_{sc} = cross-sectional area of the shell of the steel pipe

and k_s = radius of gyration of steel pipe alone.

8.5 IS: CODE REQUIREMENTS FOR DESIGN OF COLUMNS

1. *Unsupported Length*: The unsupported length, l, of a compression member shall be taken as the clear distance between end restraints except that :

(*a*) in flat slab construction it shall be clear distance between the floor and the lower extremity of the capital, the drop panel or slab whichever is the least.

(*b*) in beam and slab construction, it shall be clear distance between the floor and the underside of the shallower beam framing into the columns in each direction at the next higher floor lever.

(*c*) in columns restrained laterally by struts, it shall be the clear distance between consecutive struts in each vertical plan provided that to be an adequate support, tow such struts shall meet the columns at approximately the same level and the angle between vertical planes through the struts shall not vary more than 30° from a right angle. Such struts shall be of adequate dimensions and shall have sufficient anchorage to restrain the member against lateral deflection.

(*d*) in columns restrained laterally by struts of beams, with brackets used at the junction, it shall be the clear distance between the floor and the lower edge of the bracket, provided that the bracket width equals that of the beam strut and is at least half that of the column.

2. *Effective Length*: In the absence of more exact analysis, the effective length of columns in framed structures may be obtained from the ratio of the effective length to unsupported length l_{ef}/l given in Fig. 82. when relative displacement of the ends of column is prevented and in Fig. 8.3 when relative lateral displacement of the ends is not prevented. In the latter case, it is recommended that the effective length ratio $1_{ef}/l$ may not be taken to be less than 1.2.

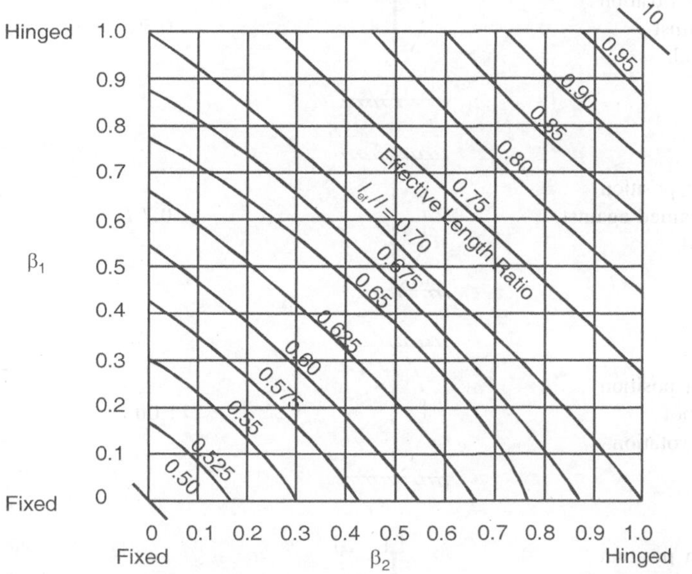

Fig. 8.2. Effective Length Ratios for a Column in a Frame with no Sway

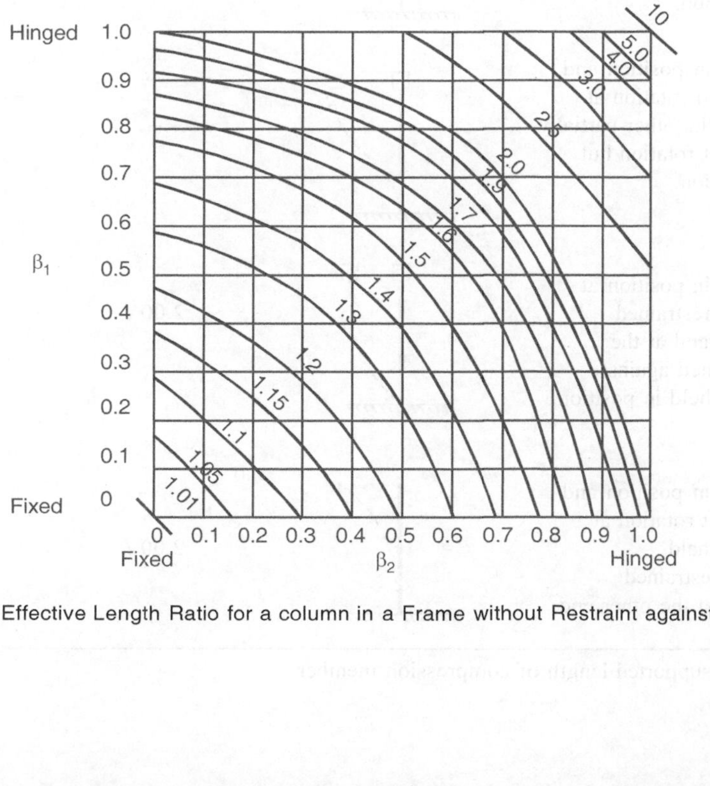

Fig. 8.3. Effective Length Ratio for a column in a Frame without Restraint against sway

Table 8.1. Effective Length of Compression Members

Degree of end restraint of compression member	*Symbol*	*Theoretical value of effective length*	*Recommended value of effective length*
Effectively held in Position and restrained against rotation at both end.		0.5 *l*	0.65 *l*
Effectively held in position at both ends, restrained against rotation at one end.	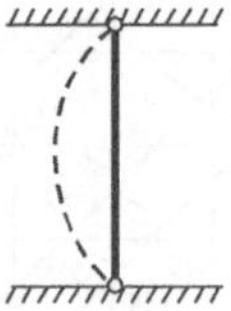	0.7 *l*	0.80 *l*
Effectively held in position at both ends, but not restrained against rotation.	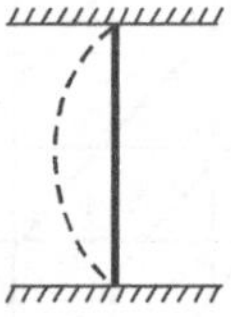	1.00 *l*	1.00 *l*
Effectively held in position and restrained against rotation at one end, and at the other restrained against rotation, but not held in position.	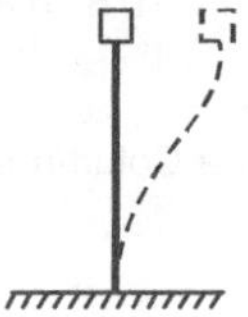	1.00 *l*	1.20 *l*
Effectively held in position and restrained against rotation at one end, and at the other partially restrained against rotation but not held in position.	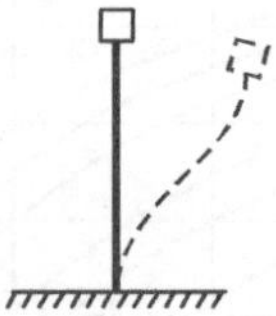	–	1.50 *l*
Effectively held in position at one end but not restrained against rotation, and at the other end restrained against rotation but not held in position.	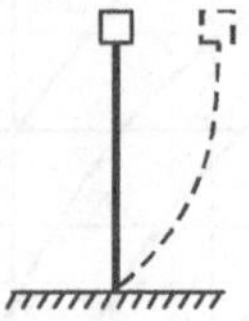	2.00 *l*	2.00 *l*
Effectively held in position and restrained against rotation at one end but not held in position nor restrained against rotation at the other end.	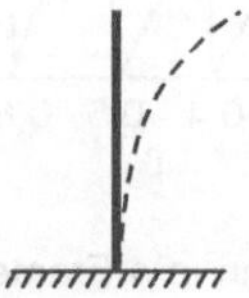	2.00 *l*	2.00 *l*

Note: *l* is the unsupported length of compression member.

In Fig. 8.2 and 8.3 β_1 and β_2 are equal to $\dfrac{\sum k_c}{\sum k_c + \sum k_b}$ where the summation is to be done for the members framing into a joint at top and bottom respectively; K_c and K_b being the flexural stiffness for column and beam respectively.

To determine whether a column is a no sway or a sway column, stability index Q may be computed as given below:

$$Q = \frac{\sum P_u \Delta_u}{H_u h_s}$$

where ΣP_u = sum of axial loads *ou* all columns in the storey

Δ_u = elastically computed first order lateral deflection

H_u = total lateral force acting with in the storey,

and h_s = height of the storey.

If $Q \leq 0.04$, then the column in the frame may be taken as no sway column, otherwise the column will be considered as sway column.

For normal usage assuming idealized conditions, the effective length l_{ef} in a given plane may be assessed on the basis of Table 8.1.

3. *Slenderness Limits* :

(*a*) the unsupported length between end restrained shall not exceed 60 times the least lateral dimension of a column.

(*b*) If, in any given plane, one end of a column is restrained its unsupported length, l, shall not exceed $100\dfrac{b^2}{D}$

where b = width of that cross-section, and

D = depth of the cross-section measured in the plane under consideration.

4. *Minimum Eccentricity*

All columns shall be designed for minimum eccentricity, equal to the unsupported length of column/500 plus lateral dimesnions / 30, subject to a minimum of 20 mm. Where bi-axial bending is considered, it is sufficient to ensure that eccentricity exceeds the minimum about one axis at a time.

5. *Longitudinal Reinforcement* :

(*a*) The cross-sectional area of longitudinal reinforcement, shall be not less than 0.8% nor more than 6% of the gross cross-sectional area of the column.

The use of 6% reinforcement may involve practical difficulties in placing and compacting of concrete; hence lower percentage is recommended. Where bars from the columns below have to be lapped with those in the column under consideration, the percentage of steel shall usually not exceed 4%.

(*b*) In any column that has a cross-sectional area than that required to support the load, the minimum percentage of steel shall be based upon the area of concrete required to resist the direct stress and not upon the actual area.

(*c*) The minimum number of longitudinal bars provided in a column shall be 4 in rectangular and 6 in circular columns.

(*d*) The bars shall not be less than 12 mm in diameter.

(*e*) A reinforced column having helical reinforcement shall have at least 6 bars of longitudinal reinforcement within the helical reinforcement.

(*f*) In a helically reinforced column, the longitudinal bars shall be in contact with the helical reinforcement and equidistant around its inner circumference.

(*g*) Spacing of longitudinal bars measured along the periphery of the column shall not exceed 300 mm.

(*h*) In case of pedestals in which longitudinal reinforcement is not taken into account in strength calculations, nominal longitudinal reinforcement not less than 0.15% of the cross-sectional area shall be provided. Pedestal is a compression member, the effective length of which does not exceed three times the least lateral dimension.

6. *Transverse Reinforcement* :

(*a*) *General*-A reinforced concrete compression member shall have transverse or helical reinforcement so disposed that every logitudinal bar nearest to the compression face has effective lateral support against buckling subject to provision given in (*c*) below. The effective lateral support is given by transverse reinforcement either in the form of circular rings capable of taking up circumferential tension or by polygonal links (lateral ties) with internal angles not exceeding 135°. The ends of transverse reinforcement shall be properly anchored as given in (*b*) which follows.

(*b*) Anchoring the transverse reinforcement or secondary reinforcement, such as stirrups and transverse ties, complete development lengths and anchorage shall be deemed to have been provided when the bar is bent through an angle of at least 90° round a bar of atleast its own diameter and is continued beyond the end of the curve for a length of atleast eight diameters or when the bar is bent through an angle of 135° and is continued beyond the end of the curve for a length of at least six bar diameters or when the bar is bent through an angle of 180 and is continued beyond the end of curve for a length of at least 4 times bar diameters.

(*c*) *Arrangement of Transverse Reinforcement*

(*i*) If the longitudinal bars are not spaced more than 75 mm on either side, transverse reinforcement need only to go round corner and alternate bars for the purpose of providing effective lateral supports (Fig. 8.4).

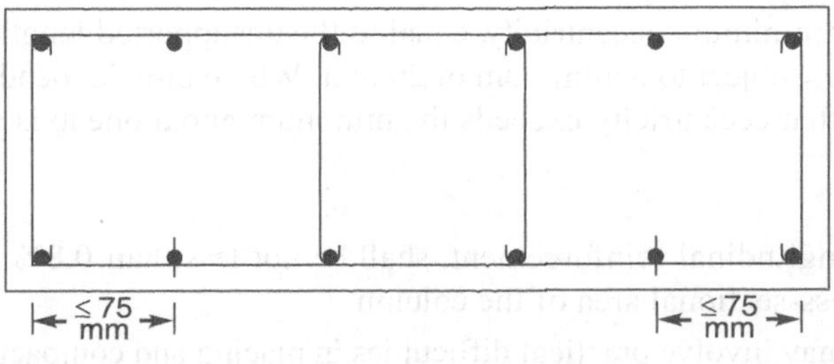

Fig. 8.4

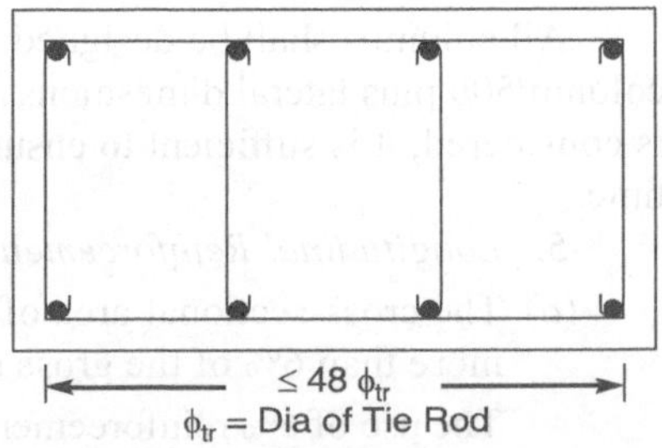

Fig. 8.5

(*ii*) If the longitudinal bars spaced at a distance of not exceeding 48 times the diameter of the tie are effectively tied in two directions, additional longitudinal bars in between these bars need to be tied in one direction by open ties (Fig. 8.5).

(*iii*) Where the longitudinal reinforcing bars in a compression member are placed in more than one row, effective lateral support to the longitudinal bars in the inner rows may be assumed to have been provided, if

(*a*) Transverse reinforcement is provided for the outer most row in accordance with (6) and

(*b*) No bar of the inner row is closer to the nearest compression face than three times the diameter of the largest bar in the inner row (Fig. 8.6).

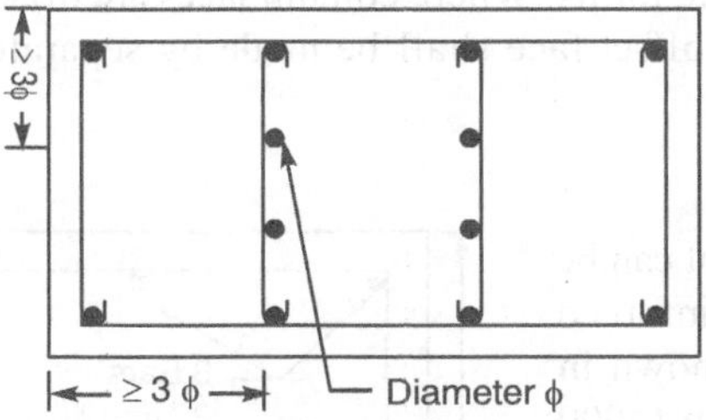

Fig. 8.6.

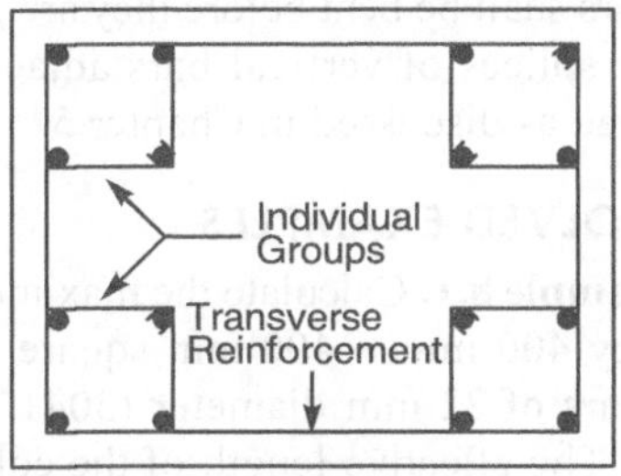

Fig. 8.7.

(*iv*) Where the logitudinal bars in a compression member are grouped (not in contact) and each group adequately tied with transverse reinforcement in accordance with (6), the transverse reinforcement for the compression member as a whole may be provided on the assumption that each group is a single logitudinal bar for purpose of determining the pitch and diameter of the transverse reinforcement in accordance with (6). The diameter of such transverse reinforcement need not, however, exceed 20 mm (Fig. 8.7)

(*d*) *Pitch and Diameter of Lateral Ties*

(*i*) Pitch : The pitch of transverse reinforcement shall be not more than the least of the following distances.

(A) The least lateral dimension of the compression member;

(B) 16 times the smallest diameter of the longitudinal reinforcement bar to be tied:

(C) 48 times the diameter of the transverse reinforcement.

(*ii*) Diameter: The diameter of the polygonal links or laterial ties shall be not less than 1/4 of the diameter of the largest bar and in no case less than 5 mm.

(*e*) *Helical Reinforcement*

(*i*) Pitch : Helical reinforcement shall be of regular formation with the turns of the helix spaced evenly and its ends shall be anchored properly by providing one and a half extra turns of the spiral bar. Where an increased load on the column on the strength of the helical reinforcement is allowed for, the pitch of the helical turns shall be,

(*i*) not mor than 75 mm,

(*ii*) not more than one-sixth of the core diameter of the column :

(*iii*) not less than 25 mm,

(*iv*) not less than 3 times the diameter of the steel bar forming the helix.

In other cases, the requirements of (6) shall be complied with.

(*ii*) Diameter: The diameter of the helical reinforcement shall be in accordance with that of lateral ties "*d* (*ii*)".

(6) In columns where longitudinal bars are offset at a splice, the slope of the inclined portion of the bar with the axis of the column shall not exceed in 6, and the portions of the bar above and below the offset shall be parallel to the axis of the column. Adequate horizontal support at the offset bends shall be treated as a matter of design, and shall be provided by metal ties, spirals, or parts of the floor construction. Metal ties or spirals so designed shall be placed near (not more than 8 bear diameter from) the point of bend. The horizontal thrust to be resisted shall be assumed as $1\frac{1}{2}$ times the horizontal components of the nominal stress in the inclined portion of the bar.

Offset bars shall be bent before they are placed in the forms. Where column faces are offset 75 mm or more, splices of vertical bars adjacent to the offset face shall be made by separate dowels overlapped as discussed in Chapter 5.

8.6 SOLVED EXAMPLES

Example 8.1. Calculate the maximum load that can be carried by 400 mm × 400 mm square column reinforced with 8 bars of 22 mm diameter (3041 mm²) as shown in Fig. 8.8. The effective length of the column is 4 m (4000 mm).

$$\sigma_{cc} = 5 \text{ N/mm}^2 \ (50 \text{ kg/cm}^2)$$
$$\sigma_{sc} = 190 \text{ N/mm}^2 \ (1900 \text{ kg/cm}^2)$$

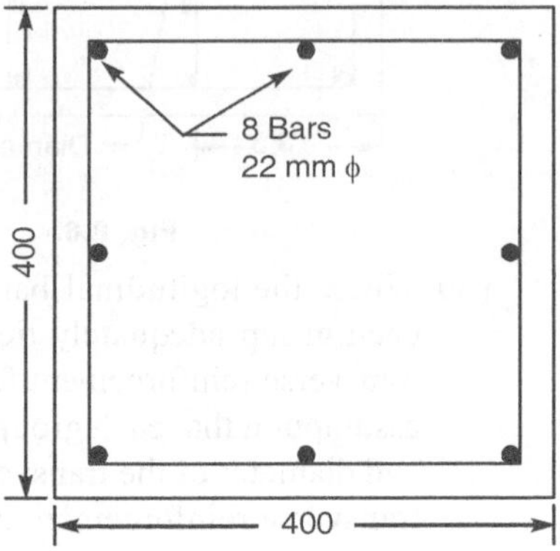

Fig. 8.8.

Sol. $A_{sc} = 8 \times \frac{\pi}{4} \times 22 \times 22 = 3041 \text{ mm}^2$

Effective length of the column = 4000 mm

Least lateral dimension

$b = 400$ mm, $l/b = 4000/400 = 10 < 12$

It is a short column. Maximum permissible load,

$$\begin{aligned} P &= \sigma_{cc}.\, A_c + \sigma_{sc}.\, A_{sc} \\ &= 5(400 \times 400 - 3041) + 190 \times 3041 \\ &= 5 \times 156959 + 190 \times 3041 \\ &= 784795 + 577790 \\ &= 1362585 \text{ N} \\ &= 136258.5 \text{ Kg} \\ &= 136.2585 \text{ tonnes} \end{aligned}$$

Example 8.2. Solve Example 8.1. if the column is 400 mm diameter (Fig. 8.9) with (*a*) 10 mm dia. spirals at 60 mm pitch, (*b*) 8 mm dia. lateral ties @ 100 mm *c/c*. The clear cover upto the main reinforcement is 50 mm.

Sol. (a) : Permissible load neglecting reinforcement for short column since

$$\begin{aligned} l/b &= 10 < 12 \\ P &= \sigma_{cc}.\, A_c + \sigma_{sc}.\, A_{sc} \\ &= 5\left[\frac{\pi}{4} \times (400)^2 - 3041\right] + 190 \times 3041 \\ &= 612795 + 577790 \\ &= 1190585 \text{ N} \\ &= 1190.585 \text{ KN} \end{aligned}$$

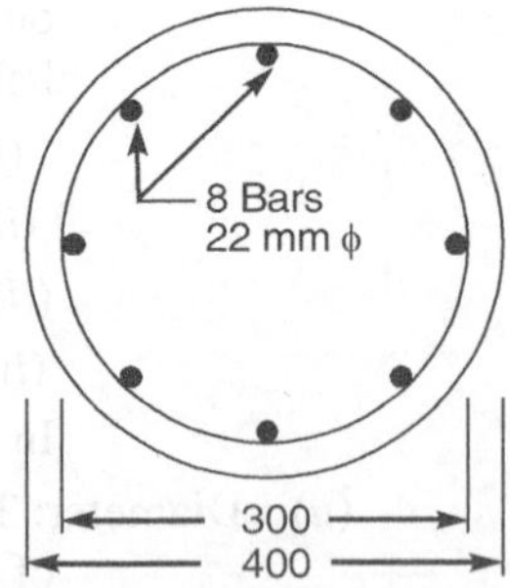

Fig. 8.9.

Permissible load with helical reinforcement

$$P = 1.05\,(\sigma_{cc}.\, A_c + \sigma_{sc}\ A_{sc})$$

$$\text{if } \frac{\text{volume of helical reinforcement}}{\text{volume of core}} \not> 0.36\left(\frac{A_g}{A_c} - 1\right)\frac{f_{ck}}{f_y}$$

$$\begin{aligned} \text{Length of spiral} &= \sqrt{((\pi D)^2 + p^2)} \\ &= \sqrt{((\pi \times 310)^2 + 60^2)} = 975.74 \text{ mm} \end{aligned}$$

$$\text{Volume of spiral per mm height} = \frac{\frac{\pi}{4} \times 10^2 \times 975.74}{60}$$

$$= 1277 \text{ mm}^3$$

$$\text{Volume of core per mm height} = \frac{\pi}{4} \times (320)^2 \times 1 = 80425 \text{ mm}^3$$

$$L.H.S. = \frac{\text{Volume of helical reinforcement}}{\text{Volume of core}}$$

$$= \frac{1277}{80425} = 0.016$$

$$f_{ck} = 20 \text{ N/mm}^2 \qquad f_y = 500 \text{ N/mm}^2$$

$$R.H.S. = 0.36\left(\frac{A_g}{A_c} - 1\right).\frac{f_{ck}}{f_y}$$

A_g = gross area of the section

$$= \frac{\pi}{4}(400)^2 = 125664 \text{ mm}^2$$

A_c = area of core of the helically reinforced column measured to the outside diameter of helix.

$$= \frac{\pi}{4}(320)^2 = 80425 \text{ mm}^2$$

$$\therefore \quad R.H.S. = 0.36\left(\frac{A_g}{A_c} - 1\right)\frac{f_{ck}}{f_y}$$

$$= 0.36\left(\frac{125664}{80425} - 1\right)\frac{20}{500}$$

$$= 0.0081$$

$$L.H.S. \nless R.H.S$$

Permissible load with helical reinforcement

$$= 1.05 \times P$$

$$= 1.05 \times 1190.585$$

$$= 1250.1142 \text{ KN}$$

(*b*) Permissible load which the column can safely carry with lateral ties.

$$P = \sigma_{cc}.\, A_c + \sigma_{sc}.\, A_{sc}$$

$$= 5\left[\frac{\pi}{4}(400)^2 - 3041\right] + 190 \times 3041$$

$$= 612795 + 577790$$

$$= 1190585 \text{ N}$$

$$= 1190.585 \text{ KN}$$

Example 8.3. Design a short column of reinforced concrete to carry an axial load of 11,00,000 N (110,000 kg or 110 *t*). Use M 20 concrete and σ_{sc} = 190 N/mm^2.

Sol. $\sigma_{cc} = 5.0 \text{ N/mm}^2$ for M 20

Longitudinal steel in column varies from 0.8 to 6%. Using minimum of 0.8%

$$A_{sc} = 0.008\, A_g$$
$$A_c = A_g - 0.008\, A_g$$
$$= 0.992\, A_g$$
$$P_{short} = \sigma_{cc}.\, A_c + \sigma_{sc}.\, A_{sc}$$
$$1{,}100{,}000 = 5 \times 0.992\, A_g + 190 \times 0.008 \times A_g$$
$$A_g = 169753.08 \text{ mm}^2$$

Any of the following sections can be used.

(A)Square Section :

Side of the square required $= \sqrt{(169753.08)}$

$= 412.01$ mm

Use 450 × 450 mm Square Column

$$P = 1{,}100{,}00 = 5\,(450 \times 450 - A_{sc}) + 190\, A_{sc}$$
$$A_{sc} = 473 \text{ mm}^2$$

Minimum diameter of the bar which can be used in a column is 12 mm. If 4 Nos. 12 mm ϕ are used, then $A_{sc} = 4 \times 113 = 452 < 473$,

$\therefore$ Use 8 Nos, 12 mm ϕ, Asc provided = 905 mm²

or 4 bars can also be used, then the area of one bar would be

$= 473/4$

$= 118 \text{ mm}^2$

Use 4 Nos. 16 mm ϕ

$$A_{sc} = 804 \text{ mm}^2 > 473 \text{ mm}^2$$

It is clear that using 4 Nos. 16 mm dia. bars is economical than using 8 nos. 12 mm but use 8 nos. 12 mm ϕ bars.

Transverse Reinforcement.

(*a*) *Dia. of Lateral Ties,* shall be not less than the greater of the following :

(*i*) 1/4th. dia. of largest longitudinal bar, i.e., 12/4 = 3 mm

(*ii*) 5 mm.

(*iii*) Minimum available in market = 8 mm

Adopt 8 mm dia.

(*b*) *Pitch of Lateral Ties,* shall not be more than the least of the following:

(*i*) least lateral dimension of the compression member, i.e., 450 mm

(*ii*) 16 × dia. main bar, i.e., 16 × 12 = 192 mm

(*iii*) 48 × dia. of lateral ties, ie., 48 × 8 = 384 mm

So the pitch has to be less than 192 mm, adopt a pitch of 190 mm,

Provide 8 mm ϕ lateral ties @ 190 mm *c/c*

For design details see, Fig. 8.10

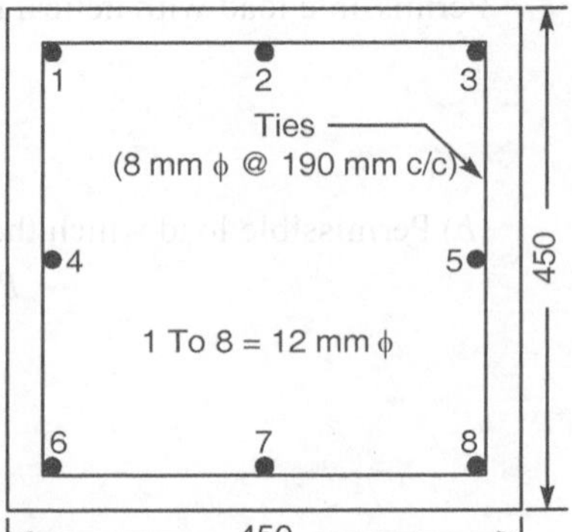

Fig. 8.10.

1 to 8 = 12 mm ϕ

Ties = 8 mm ϕ @ 190 mm *c/c*

(B) Circular Section :

$$A_g = 169753.08 \text{ mm}^2 = \frac{\pi}{4} d^2$$

$$d = \sqrt{\left(\frac{4}{\pi} \times 169753.08\right)}$$

$$= 465 \text{ mm}$$

Use circular column 450 mm diameter

$$A_g = \frac{4}{\pi}(450)^2 = 158962.5 \text{ mm}^2$$

$$P = 11{,}00{,}000 = 5(158962.5\text{-}A_{sc}) + 190 \times A_{sc}, \; A_{sc} = 1650 \text{ mm}^2$$

Minimum bars to be used in circular column is six, hence provide 6 bars

$$\text{Area per bar} = \frac{1650}{6} = 275 \text{ mm}^2$$

Use 6 Nos. 20 mm ϕ, A_{sc} 1885 mm^2

Lateral Ties

(*a*) *Diameter,* shall not be less than the greater of the following :

(*i*) 1/4th. dia. of largest long. bar = 20/4 = 5 mm

(*ii*) 5 mm

(*iii*) minimum available in market = 8 mm

Adopt 8 mm ties.

(*b*) *Pitch* of lateral ties shall not be more than the least of the following :

(*i*) least, lateral dimension of the compression member, i.e., 450 mm

(*ii*) 16 × dia. of main bar, i.e., 16 × 20 = 320 mm

(*iii*) 48 × dia. of tie, i.e., 48 × 8 = 384 mm

Pitch 320 mm

Adopt 8 mm ϕ ties @ 300 mm *c/c*

It may be observed from the above example that it is always economical to provide a square/rectangular section. Steel area provided in the circular section for carrying same load is 2 times that required for square section though saving in concrete is only of the order of 20% compared to steel.

For design details, see Fig. 8.11

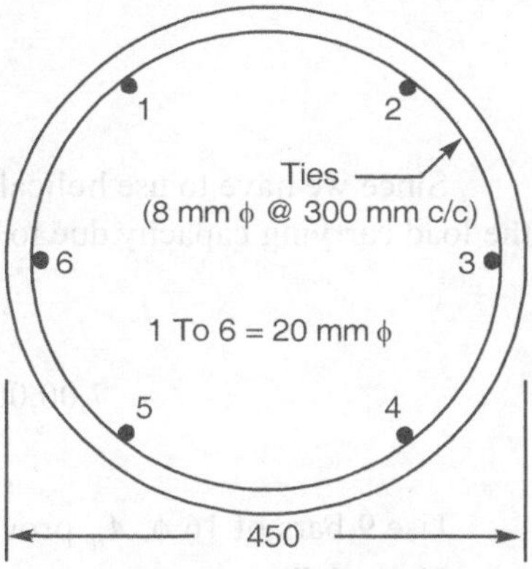

1 to 6 = 20 mm ϕ

Ties = 8 mm ϕ @ 300 mm *c/c*

Fig. 8.11.

Example 8.4. Design a R.C. column to carry an axial load of 2000 KN (200,000 kg.). The size of the column is restricted to 600 mm square. The effective height of column is 9 m. Use M 20 concrete and σ_{sc} = 190 N/mm^2

Sol. Least lateral dimension of the column = 600 mm

Effective length of column = 9000 mm

$$l/b = 9000/600 = 15 > 12$$

It is a slender (long) column.

$$\text{Reduction coefficient, } C_r = 1.25 - \frac{l_{ef}}{48.b}$$

$$= 1.25 - \frac{15}{48}$$

$$= 1.2500 - 0.3125$$

$$= 0.9375$$

Permissible load, $P = C_r (\sigma_{cc} \cdot A_c + \sigma_{sc} \cdot A_{sc})$

$$20,00,000 = 0.9375 [5 (600 \times 600 - A_{sc}) + 190 A_{sc}] \text{ or } A_{sc} = 1801.8 \text{ mm}^2$$

Provide 4 Nos. 25 mm ϕ, A_{sc} = 1963 mm² with 50 mm cover to the reinforcement

Lateral Ties

(*a*) Diameter, greater of the following :

(*i*) 1/4 the dia. of longitudinal bar, i.e., 25/4 = 6.25

(*ii*) 5 mm

Adopt 8 mm ϕ ties (minimum available in market).

(*b*) *Pitch*, the least of the following :

(*i*) least lateral dimension of the compression member, i.e., 600 mm

(*ii*) 16 × dia. of main bar, i.e., 16 × 25 = 400 mm

(*iii*) 48 × dia. of tie, i.e., 48 × 8 = 384 mm

Provide 8 mm ϕ ties @ 380 mm *c*/*c*

For design details, see Fig. 8.12

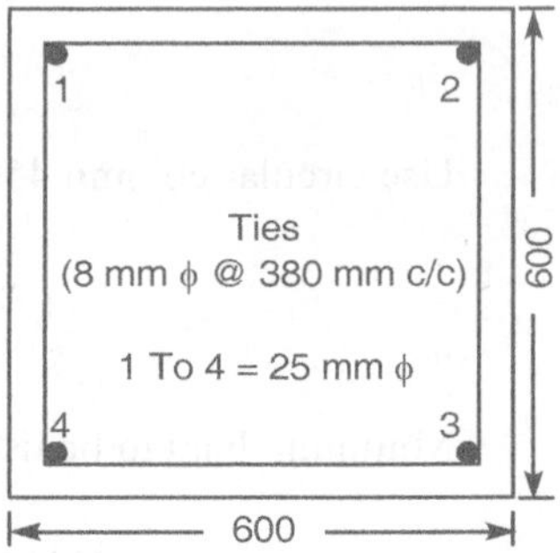

1 to 4 = 25 mm ϕ

Ties = 8 mm ϕ @ 380 mm *c*/*c*

Fig. 8.12.

Example 8.5. Design a circular column 8 m effective height and 400 mm diameter to carry an axial load of 700 KN (70 *t*). σ_{cc} = 5N/mm², σ_{sc} = 190 N/mm². Use helical ties.

Sol.

$$\frac{l_{ef}}{D} = \frac{8000}{400} = 20 > 12$$

It is a long column.

Reduction factor,

$$C_r = 1.25 - \frac{l_{ef}}{48\,D} = 1.25 - \frac{20}{48} = 1.25 - 0.417 = 0.833$$

Since we have to use helical reinforcement, the permissible load, neglecting any increase in the load carrying capacity due to helical reinforcement

$$P = C_r(\sigma_{cc} \cdot A_c + \sigma_{sc} \cdot A_{sc})$$

$$7,00,000 = 0.833 \left[5\left(\frac{\pi}{4} \times 400^2 - A_{sc}\right) + 190 \times A_{sc} \right]$$

$$A_{sc} = 1148 \text{ mm}^2$$

Use 9 bars of 16 ϕ, A_{sc} provided = 1809 mm²

Helical Ties

(*a*) Diameter : It should not be less than the greater of the following:

(*i*) $\frac{1}{4}$ diameter of the longitudinal bar, i.e., $\frac{1}{4}$ of 16 = 4 mm.

(*ii*) 5 mm

Use 8 mm ϕ helical reinforcement as bars below 8 mm ϕ are not available in the market at present.

(*b*) *Pitch* : Since the increase in load on the column due to helical reinforcement is not considered, the pitch shall be decided as that for lateral ties, i.e.,

Pitch $\ngtr$ the least of the following:

(*i*) Least lateral dimension of the compression member, *i.e.*, 400 mm.

(*ii*) 16 times the dia. of the main bar, i.e., $16 \times 16 = 256$ mm

(*iii*) 48 times the dia. of the helical bar., i.e. $48 \times 8 = 384$ mm

Adopt a pitch of 250 mm *c/c*

For reinforcement, see Fig. 8.13.

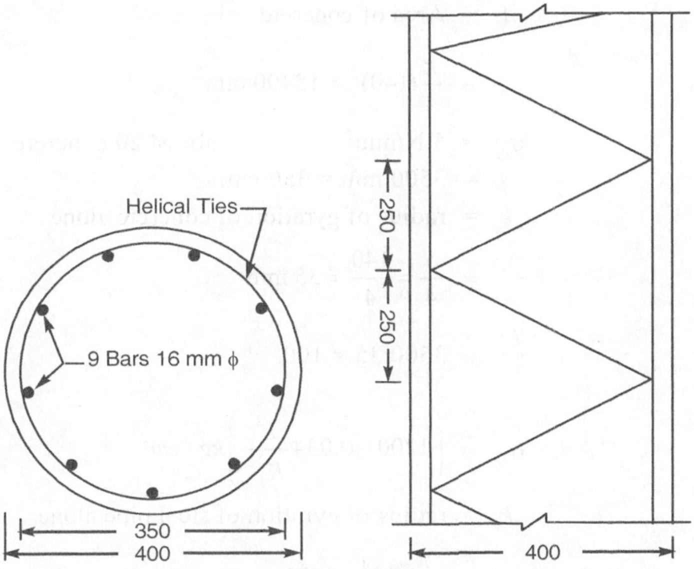

Fig. 8.13.

Example 8.6. Determine the safe axial load carrying capacity for the column shown in Fig. 8.14. The rolled steel section used is ISHB 225 @ 43.1 kg/m. Area of cross section of ISHB is 5494 mm². The effective height of the column is 5 m.

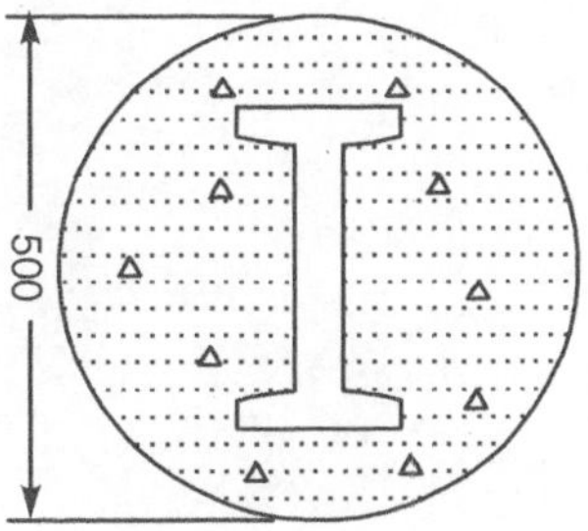

Fig. 8.14.

Use M 20 concrete, $\sigma_{mc} = 125 \text{ N/mm}^2$

Solution. $L_{ef} = 5000 \text{ mm}, D = 500 \text{ mm}$

$$L_{ef} = \frac{5000}{500} = 10 < 12$$

$$\sigma_{cc} = 5 \text{ N/mm}^2 \text{ for M 20}$$

It is a short column Permissible load is :

$$P = (\sigma_{cc} A_c + \sigma_{sc}.A_{sc} + \sigma_{mc} A_m)$$

$$= 5\left\{\frac{\pi}{4}(500)^2 - A_{sc} - A_m\right\} + \sigma_{sc}.A_{sc} + \sigma_{mc}.A_m$$

$$A_{sc} = 0, A_m = 5494 \text{ mm}^2 \text{ for ISHB 225}$$

$$P = 5\left\{\frac{\pi}{4}\times 2,50,000 - 5494\right\} + 0 + 125\times 5494$$

$$= 1640530 \text{ N}$$

$$= 1640.530 \text{ KN}$$

Example 8.7. Determine the safe axial load capacity of the concrete filled pipe column shown in Fig. 8.15. The external diameter of column is 150 mm and internal diameter is 140 mm and is 3500 mm (3.5m) long. Use M 20 concrete.

Sol. Permissible load

$$P = A_c.\sigma_{cc}\left(1 - 0.000025\frac{L^2}{K_c{}^2}\right) + \sigma_{sc}.A_{sc}$$

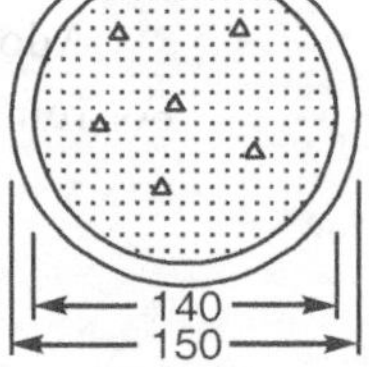

Fig. 8.15.

A_c = Area of concrete

$$= \frac{\pi}{4}(140)^2 = 15400 \text{ mm}^2$$

σ_{cc} = 5 N/mm² for M 20 concrete

L = 3500 mm = 350 cm.

k_c = radius of gyration of concrete alone

$$= \frac{d}{4} = \frac{140}{4} = 35 \text{ mm}$$

$$\frac{L}{k_c} = 3500/35 = 100$$

$$\sigma_{sc} = \left(1200 - 0.034\frac{L^2}{K_s{}^2}\right) kg/cm^2$$

k_s = radius of gyration of steel pipe alone

$$k_s{}^2 = \frac{\frac{\pi}{64}(15^4 - 14^4)}{\frac{\pi}{4}(15^2 - 14^2)} = \frac{1}{16}(15^2 + 14^2) = \frac{225+196}{16}$$

$$= \frac{421}{16} = 26.3125 \text{ cm}.$$

$$\sigma_{sc} = 1200 - 0.034\frac{(350)^2}{26.3125}$$

$$= 1200 - 158.29$$

$$= 1041.71 \text{ kg/cm}^2$$

$$= 104.171 \text{ N/mm}^2$$

A_{sc} = cross-sectional area of the shell of the steel pipe

$$= \frac{\pi}{4}(150^2 - 140^2)$$

$$= 2280 \text{ mm}^2$$

$$P = 15400 \times 5\,[1 - 0.000025 \times (100)^2] + 104.171 \times 2280$$

$= 309700$ N

$= 309.7$ KN

Example 8.8. Determine the safe axial load capacity for the column shown in Fig. 8.16. The rolled steel section used is ISHB 300 @ 63.0 kg/m, (Area = 8025 mm^2). In addition 4 bars of 20 mm ϕ are used. Use M 20 concrete, $\sigma_{sc} = 190$ N/mm^2

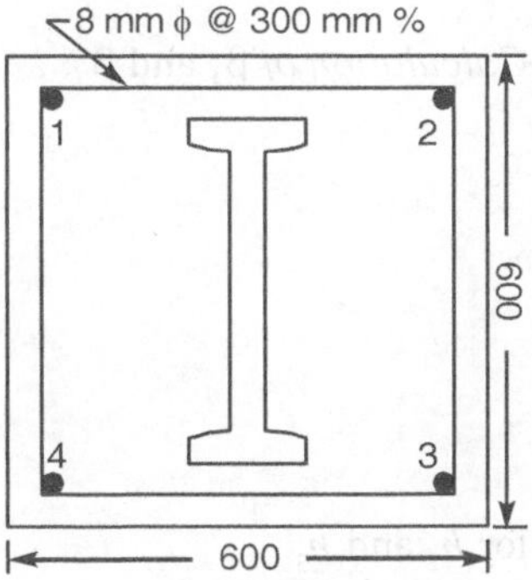

Fig. 8.16.

Sol. Permissible load

$$P = \sigma_{cc} . A_c + \sigma_{sc} . A_{sc} + \sigma_{mc} . A_m$$

$$\sigma_{cc} = 5.0 \text{ N/mm}^2 \text{ for M20 concrete}$$

$$A_g = \text{gross area}$$

$$= 600 \times 600 = 360000 \text{ mm}^2$$

$$A_{sc} = 4 \times 314 = 1256 \text{ mm}^2$$

$$A_m = 8025 \text{ mm}^2$$

$$A_c = A_c - A_{sc} - A_m$$

$$= 360000 - 1256 - 8025$$

$$= 350{,}719 \text{ mm}^2$$

$$P = 5 \times 350{,}719 + 190 \times 1256 + 125 \times 8025$$

$$= 1753595 + 238{,}640 + 1{,}003{,}125$$

$$= 2995360 \text{ N}$$

$$= 2995.360 \text{ KN}$$

Example 8.9. Find the effective length of a typical interior column (Fig. 8.17.) for a multistoreyed building assuming :

(*a*) The column with no sway

(*b*) The column without restraint against sway

Fig. 8.17.

Sol.

Stiffnesses

(*i*) Column 300 × 300

$$I_1 = \frac{300 \times 300^3}{12}$$

$$= 6.75 \times 10^8 \text{ mm}^4$$

$$\frac{I_1}{h_1} = \frac{6.75 \times 10^8}{3000} = 225{,}000 \text{ mm}^3$$

(*ii*) Column 450 × 450

$$I_2 = \frac{450 \times 450^3}{12}$$

$$= 3.417 \times 10^9 \text{ mm}^4$$

$$\frac{I_2}{h_2} = \frac{3.417 \times 10^9}{3000}$$

$$= 1139062.5 \text{ mm}^3$$

(*iii*) Beam 450 × 600 (2 Nos. l_1 and l_2 or l_3 and l_4)

$$I_3 = \frac{450 \times 600^3}{12}$$

$$= 8.1 \times 10^9 \text{ mm}^4$$

$$\frac{I_3}{l_3} = \frac{8.1\times10^9}{6000}$$

$$= 1350000 \text{ mm}^3$$

Calculation of β_1 and β_2

$$\beta_1 = \frac{\sum k_c}{\sum k_c + \sum k_b}$$

$$= \frac{225000+1139062.5}{(225500+1139062.5)+(1350{,}000+1350000)}$$

$$= 0.3356$$

for h_2 and h_3

$$\beta_2 = \frac{1139062.5+1139062.5}{(1139062.5+1139062.5)+(1350{,}000+1350{,}000)}$$

$$= 0.4577$$

(*a*) Column in a frame with no side sway

$$\frac{l_{ef}}{l} = 0.635 \text{ from Fig. 8.2}$$

$$l_{ef} = 0.635 \times 3 = 1.905 \text{ m}$$

(*b*) Column in a frame without restraint against sway

$$\frac{l_{ef}}{l} = 1.32 \text{ from Fig. 8.3}$$

$$l_{ef} = 1.32 \times 3 = 3.96 \text{ m}$$

EXERCISE 8

1. A short column 300 mm × 300 mm is reinforced with 4 bars of 20 mm diameter. Determine the safe working load on the column. Concrete used M 20.
2. A rectangular column 400 mm × 300 mm of 5 metres effective height has to support an axial load of 600 KN. Design the column reinforcement showing reinforcement details. Concrete M 20.
3. A short column 300 mm diameter has a core of 220 mm and is reinforced with 6 Nos. 16 mm ϕ. The spiral ties are 8 mm ϕ at 75 mm pitch. Determine the safe load the column can carry. Concrete used M 20.
4. Design and sketch a circular column to carry an axial load of 1000 KN. The effective height is 8 m. Concrete used M 20.
5. A short composite column has to support a load of 2000 KN. The size of the column is to be 500 mm square. Design the column section. Concrete used M 20 and steel has $\sigma_{sc} = 190$ N/mm^2
6. A concrete filled column consists of steel pipe with external diameter 300 mm and internal diameter 250 mm. The column is 15 m long. Find the safe working load on the column. Use M 20.
7. A short composite column has to support a load of 2000 KN. The size of the column is to be 400 mm × 400 mm. Design and sketch the column. Use M 20 and steel having $\sigma_{sc} = 190$N/mm^2

CHAPTER 9

Combined Bending and Direct Stresses

9.1 GENERAL

In most of the concrete structures we encounter some such members which are subjected to bending moments in addition to direct compressive loads, e.g., columns rigidly connected to beams, columns of multistoreyed buildings and water towers, arches, tank walls, chimneys, slios and bins, electric towers, etc. Such conditions arise whenever there are lateral loads acting in addition to vertical loads or there are unsymmetrical loading conditions at the same end of the column on either side. If on a column, some axial force P and a moment M is acting as shown in Fig. 9.1, it is always possible to have its equivalent in the form of the axial load P and some eccentricity 'e' such that $M = P \times e$.

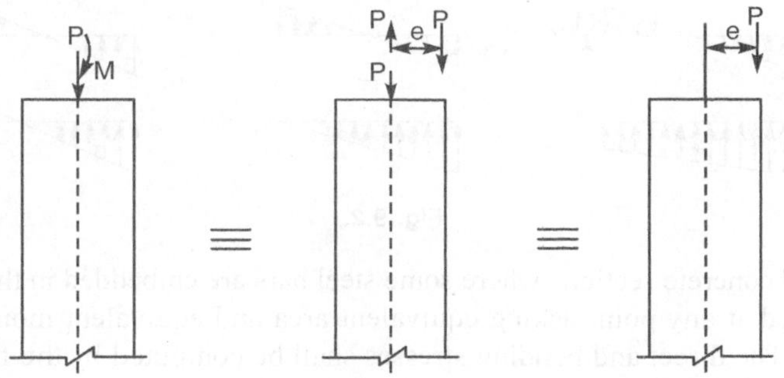

Fig. 9.1.

The combined stress at any point in rectangular column cross-section of homogenous material is the algebraic sum of direct stress due to $P(= P/A)$ and bending stress due to $M\left(= \pm\frac{M}{I}y\right)$ and is given by the following relation :

$$\sigma = \frac{P}{A} \pm \frac{M.y}{I} \qquad ...(9.1)$$

where
- σ = combined stress at any point in the cross-section,
- P = axial load,
- A = area of cross-section of the column,
- M = bending moment,
- y = distance from the N.A. to the point at which the stress is required,
- I = moment of inertia of the section

In some cases, the direct stresses may be predominant as compared to bending stresses, this will result in complete compressive stresses over whole of the section while in some other cases, bending stresses may dominate resulting in some tensile stresses development in part of the section as shown in Fig. 9.2.

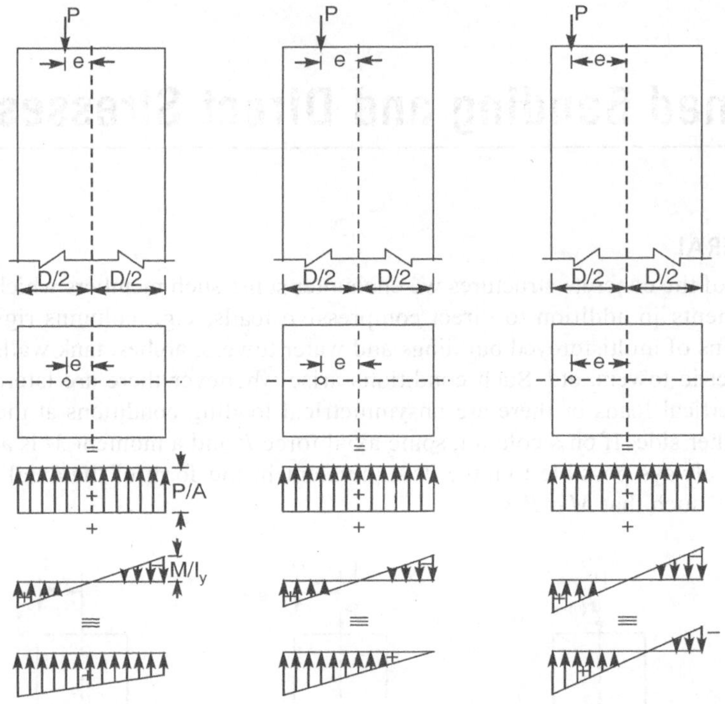

Fig. 9.2.

For reinforced concrete sections where some steel bars are embedded in the concrete, stresses have to be computed at any point asking equivalent area and equivalent moment of inertia of a composite section. The direct and bending stresses shall be computed by the following relations:

$$\text{Direct stress} = \frac{P}{A_c}$$

where $A_e = (A_c + 1.5 \text{ m } A_{sc})$ for sections completely under compressive stressess

$A_c = (A_c + 1.5\ m\ .\ A_{sc} + m.A_{st})$

for sections which are partly under tension

A_c = gross cross-sectional area of concrete

m = modular ratio = $\dfrac{280}{3.\sigma_{cbc}}$

A_{sc} = area of steel under compression

A_{st} = area of steel under tension

$$\text{Bending stress} = \frac{M}{I_e} \cdot y$$

where I_e = equivalent moment of intertia of the composite section

= moment of inertia of the concrete cross-section

+ (1.5 m A_{sc} × distance^2)

Note: Take $m\ A_{st}$ instead of 1.5 $m\ A_{sc}$ for bars in tension

It is important to note that in concrete columns, compressive forces are dominating (concrete in tension being neglected), compressive stresses are taken positive.

9.2. GENERAL NOTES ON DESIGN OF COLUMNS SUBJECTED TO COMBINED BENDING AND DIRECT STRESSES

9.2.1 Design Based on Uncracked Section

A member subjected to axial load and bending (due to eccentricity of load, monolithic construction, lateral forces, etc.) shall be considered safe provided following conditions are satisfied:

$$(a)\quad \frac{\sigma_{cc,cal}}{\sigma_{cc}} + \frac{\sigma_{cbc,cal}}{\sigma_{cbc}} \le 1 \qquad \text{...(9.2)}$$

where $\sigma_{cc,\ cal}$ = calculated direct compressive stress in concrete,

σ_{cc} = permissible axial compressive stress in concrete,

$\sigma_{cbc,\ cal}$ = calculated bending compressive stress in concrete,

σ_{cbc} = permissible bending compressive stress in concrete,

(*b*) The resultant tension in concrete is not greater than 35% and 25% of the resultant compression for biaxial and uni axial-bending respectively, or exceeds 3/4th the 7-day modulus of rupture of concrete.

Note 1. $\sigma_{cc,\ cal} = \dfrac{P}{A_c + 1.5\, m\, A_{sc}}$ for columns with ties

where

P = axial load permissible on column

A_c = cross-sectional area of concrete excluding any finishing material and reinforcing steel

A_{sc} = cross-sectional area of the longitudinal steel

m = modular ratio

Note 2. $\sigma_{cbc,\ cal} = \dfrac{M}{Z}$

where M = moment

Z = section modulus

In the case of sections subjected to moments in two directions, the stress shall be calculated separately and added algebraically.

9.2.2 Design Based on Cracked Section

If the requirements specified in 9.2.1 above are not satisfied the stresses in concrete and steel shall be calculated by the theory of cracked section in which the tensile resistance of concrete is ignored. If the calculated stresses are within the permissible limits, the section may be assumed to be safe.

Note : The maximum stress in concrete and steel may be found from tables and charts based on the cracked section theory or directly by determining the no-stress line which should satisfy the following requirements.

(*a*) The direct load should be equal to the algebraic sum of the forces on concrete and steel

(*b*) The moment of the external loads about any reference lines should be equal to the algebraic sum of the moment of the forces in concrete (ignoring the tensile forces in concrete) and steel about the same line,

(*c*) The moment of the external loads about any other reference line should be equal to the algebraic sum of the moment of the forces in concrete (ignoring the tensile force in concrete) and steel about the same line.

9.2.3 Members Subjected to Combined Direct Load and Flexure

Members subjected to combined direct load and flexure are designed by the methods based on elastic theory should be further checked for their strength under ultimate load conditions to ensure the desired margin of safety; this check is specially necessary when the bending moment is due to horizontal loads or by seismic loads.

9.3 SOLVED EXAMPLES

Example 9.1. Design a R.C.C. column carrying a direct axial load 500 KN and a bending moment of 6 KN.m. Use M25 concrete and Fe415 steel.

Sol. Try 300 × 300 mm section reinforced with 8 bars of 16 mm ϕ as shown in Fig. 9.3.

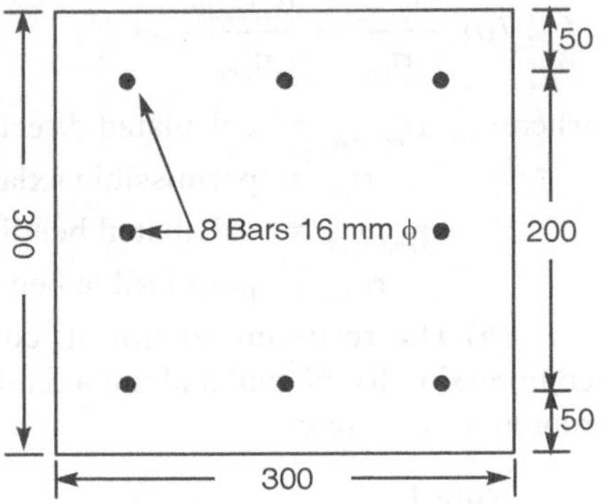

Fig. 9.3.

Equivalent area

$$A_c = A_c + 1.5\, m\, A_{sc}$$
$$= (300 \times 300 - A_{sc}) + 1.5\, m\, A_{sc}$$
$$= 300 \times 300 + A_{sc}\,(1.5\, m - 1)$$
$$= 90000 + 8 \times 201\left(1.5 \times \frac{280}{3 \times 8.5} - 1\right)$$
$$= 90000 + 24877$$
$$= 114877 \text{ mm}^2$$

Equivalent moment of Inertia

$$= \frac{300^4}{12} + \left(1.5 \times \frac{280}{3 \times 8.5} - 1\right) \times 6 \times 201 \times 100^2$$
$$= 6.75 \times 10^8 + 18657.6 \times 10^4$$
$$= 8.616 \times 10^8 \text{ mm}^4$$

Direct stress, $\sigma'_{cc_{cal}} = P / A_c$

$$= \frac{500 \times 1000}{114877} = 4.352 \text{ N/mm}^2$$

Bending stress, $\sigma_{cbc,\, cal}$ $= \frac{M}{Z} = \frac{6000 \times 1000}{8.616 \times 10^8} \times 150 = 1.044 \text{ N/ mm}^2$

$$\therefore \quad \frac{\sigma_{cc,cal}}{\sigma_{cc}} + \frac{\sigma_{cbc,\, cal}}{\sigma_{cbc}} \le 1$$
$$= \frac{4.352}{6} + \frac{1.044}{8.5} \le 1$$

$= 0.725 + 0.123 = 0.848 < 1$. Hence section is safe.

Example 9.2. A column 350 mm × 350 mm is reinforced with 4 bars of 25 mm dia. (Fig. 9.4). Check the safety of the section under a load of 300 KN (30 t) acting at an eccentricity of 75 mm about one of the axis. Also determine the maximum and minimum stresses as developed in the section. Take the allowable stressed as $\sigma_{cc} = 5$ N/mm², $\sigma_{cbc} = 7$ N/mm², $m = 13.33$

Sol. Equivalent area of column

$$A_c = A_c + 1.5\, m\, A_{sc}$$
$$= (A_g - A_{sc}) + 1.5\, m\, A_{sc}$$

$$= 350\times350+(1.5\times\frac{280}{3\times7}-1)4\times491$$

$= 122500 + 37316$

$= 159816 \text{ mm}^2$

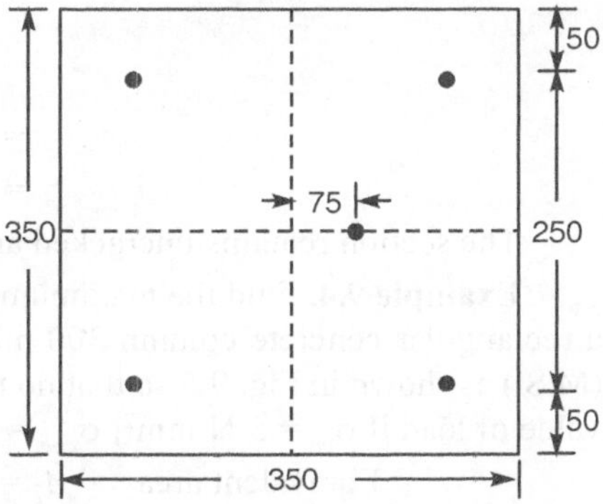

Fig. 9.4.

Equivalent moment of inertia

$$I_e = \frac{350^4}{12}+(1.5\times\frac{280}{3\times7}-1)\times4\times491\times\left(\frac{250}{2}\right)^2$$

$= 10^8\ (12.505 + 5.831)$

$= 18.335 \times 10^8 \text{ mm}^4$

Direct stress, $\sigma_{cc,\ cal} = P/A_c$

$$= \frac{300\times1000}{159816} = 1.88 \text{ N/mm}^2$$

$B.M. = P \times e = 300 \times 1000 \times 75 = 225 \times 10^5$ N – mm

Bending stresses at extreme fibre $= \frac{M}{I_e}.y$

$$\sigma_{cbc,\ cal} = \frac{225\times10^5}{18.335\times10^8}\times175 = 2.147 \text{ N/mm}^2$$

$$\text{Checking of Section} = \frac{\sigma_{cc,cal}}{\sigma_{ce}}+\frac{\sigma_{cbc,cal}}{\sigma_{cbc}} \le 1$$

$$= \frac{1.88}{5}+\frac{2.147}{7} < 1$$

$= 0.3754 + 0.3068 = 0.6822 < 1$

Hence section is safe under the eccentric loading.

Stresses Developed in the Section

Maximum stress $= 1.88 + 2.147 = 4.027 = \text{N/mm}^2$ (Compressive)

Minimum stress $= 1.88 - 2.147 = -0.27 \text{ N/mm}^2$ (tensile)

Example 9.3. In example 9.2 find the maximum value of eccentricity so that no tension is developed in the section.

Sol. Direct stress $= \sigma_{cc,\ cal} = 1.88 \text{ N/mm}^2$

$P = 300 \text{ KN} = 3 \times 10^5$ N ; Let e = the eccentricity

∴ Bending stress, $\sigma_{cc,\ cal}$

$$= \frac{P.e}{I_e}.y = \frac{3\times10^5\times e}{18.335\times10^8}\times175 = 0.0286e$$

For no tension

Direct stress = Bending stress

$1.88 = 0.0286e \quad \therefore e = 65.73$ mm

For uncracked section

$$\frac{\sigma_{cc,cal}}{\sigma_{cc}}+\frac{\sigma_{cbc,\ cal}}{\sigma_{cbc}} \le 1$$

$$0.3754 + \frac{(3\times10^5\times65.73)175}{7\times8.335\times10^8} \le 1$$

$$= 0.3754 + 0.2689$$

$$= 0.6443 < 1$$

The section remains uncracked and no tension developed.

Example 9.4. Find the maximum eccentricity about y-axis at which a load can be applied on a rectangular concrete column 300 mm × 600 mm reinforced with 12 bars of 20 mm diameter (M.S.) as shown in Fig. 9.5 so that no tension is developed in the section. Also find the maximum value of load if σ_{cc} = 5 N/mm², σ_{cbc} = 7N/mm², m = 13.33.

$$\text{Equivalent area} = A_c = A_g + (1.5\text{ m} - 1)\,A_{sc}$$

$$Ac = 300\times600 + (1.5\times13.33 - 1)\times12\times314$$

$$= 180000 + 71.592$$

$$= 251592 \text{ mm}^2$$

Equivalent moment of inertia about $y - y$ axis

$$I_e = \frac{300\times600^3}{12} + (1.5\times13.33-1)(8\times314\times250^2 + 4\times314\times100^2)$$

$$= 5.4\times10^9 + 3.222\times10^9$$

$$= 8.622\times10^9 \text{ mm}^4$$

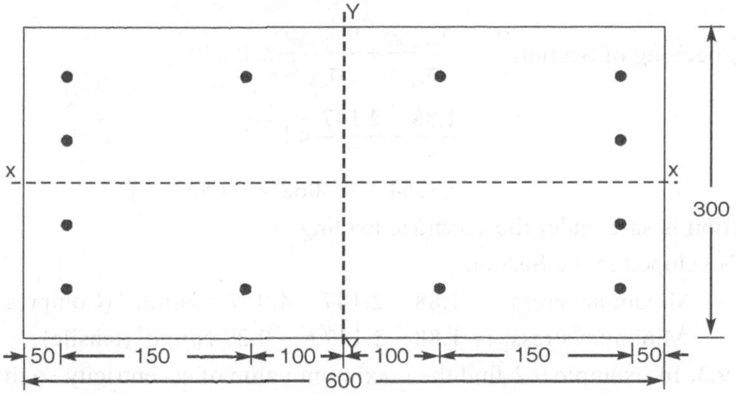

Fig. 9.5.

For no tension to develop in the section

$$\text{Direct stress} = \text{Bending stress}$$

$$\frac{P}{A_e} = \frac{P\times e.B/2}{I_e} \qquad e = \frac{2I_e}{A_e.B}$$

$$\therefore \quad e = \frac{2\times8.622\times10^9}{251592\times600} = 114.23 \text{ mm}$$

For Uncracked section

$$\frac{\sigma_{cc,cal}}{\sigma_{cc}} + \frac{\sigma_{cbc,cal}}{\sigma_{cbc}} \le 1$$

If P = the maximum load on the section, then

$$\sigma_{cc,\ cal} = \frac{P}{A_c} = \frac{P}{251592}$$

$$\sigma_{cbc,\ cal} = \frac{P \times e \times B/2}{I_e}$$

$$= \frac{P \times 114.23 \times 300}{8.622 \times 10^9} = \frac{9}{251592}$$

$$\therefore \quad P = \frac{\left(\frac{P}{251592}\right)}{5} + \frac{\left(\frac{P}{251592}\right)}{7} = 1$$

$$\therefore \quad P = 251592\left(\frac{35}{7+5}\right)$$

$$= 737810 \text{ N}$$

$$= 733.810 \text{ KN}$$

Example 9.5. The crown section of an arch rib is 250 mm wide and 400 mm deep reinforced with 3 bars of 20 mm ϕ at top and bottom as shown in Fig. 9.6 The effective cover to steel is 25 mm. Determine the maximum normal thrust which the section can withstand while it acts at a distance of 40 mm from the centre line. The compressive stress over the section is not to exceed 7 N/mm². Take m = 13.33.

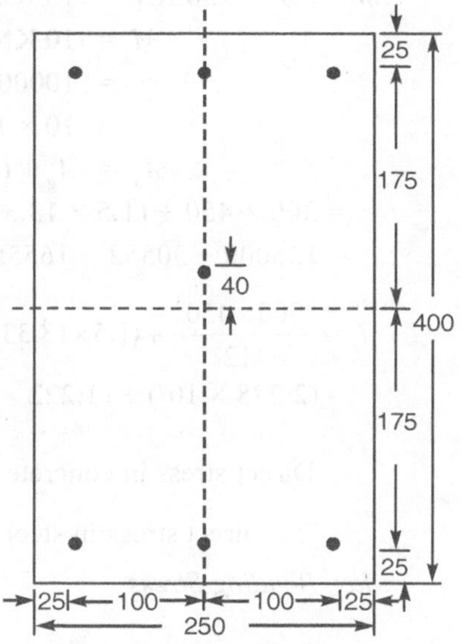

Fig. 9.6.

Sol. $e = 40$ mm

$$\frac{D}{6} = \frac{400}{6} = 66.7 \text{ mm}$$

$$e < \frac{D}{6}$$

∴ The eccentricity is small and no tension will develop at the section Equivalent area of section,

$$A_c = A_g + (1.5\,m - 1)\,A_{sc}$$

$$= 250 \times 400 + (1.5 \times 13.33 - 1)\,6 \times 314$$

$$= 100000 + 35796$$

$$= 135796 \text{ mm}^2$$

Equivalent moment of inertia

$$I_e = \frac{250 \times 400^3}{12} + (1.5 \times 13.33 - 1) \times 6 \times 314 \times 175^2$$

$$= (1.333 \times 10^9) + (1.096 \times 10^9)$$

$$= 2.429 \times 10^9 \text{ mm}^4$$

If P = max. normal thrust acting at 40 mm from the centre line, then M = the bending moment at the section = $P \times 40 = 40\,P$

$$\text{Direct stress} = \sigma_{cc,\ cal} = \frac{P}{A_e}$$

$$\text{Bending stress} = \sigma_{cbc,\ cal} = \frac{40P}{I_e} \times y_{max}$$

Max. compressive stress at the section

$$= \text{Direct stress} + \text{Bending stress}$$

$$\therefore \quad 7 = \frac{P}{135796} + \frac{P \times 40 \times 200}{2.429 \times 10^9} = \frac{P}{135796} + \frac{P}{303666}$$

$$P = 656839 \quad N = 656.839 \text{ KN}$$

Example 9.6. The crown section of a two hinged arch is 300 mm × 450 reinforced with 4 nos. 16 mm ϕ at both top and bottom faces as shown in Fig. 9.7 with effective cover of 25 mm. The horizontal thrust of the arch is found to be 150 *KN* and the bending moment on the section is 10 KN. m. Calculate the max. and min. stresses in concrete and steel. Take m = 13.33.

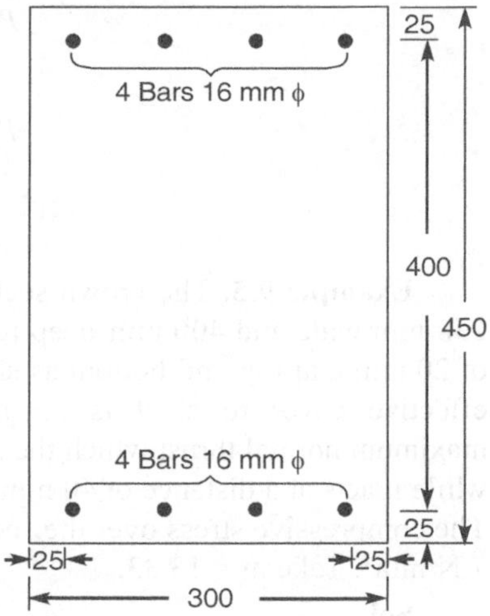

Fig. 9.7.

Sol. $P = 150\ KN = 15 \times 10^4$ N

$M = 10$ KN.m

$= 10000$ N.m.

$= 10 \times 10^6$ N.mm

$A_e = A_g + (1.5\text{ m} - 1)\,A_{sc}$

$= 300 \times 450 + (1.5 \times 13.33 - 1) \times 8 \times 201$

$= 135000 + 30552 = 165552\ \text{mm}^2$

$$I_e = \frac{300 \times 450^3}{12} + (1.5 \times 13.33 - 1) \times 8 \times 201 \times 200^2$$

$= (2.278 \times 10^9) + (1.222 \times 10^9) = 3.5 \times 10^9\ \text{mm}^4$

$$\text{Direct stress in concrete} = \frac{15 \times 10^4}{165552} = 0.906\ \text{N/mm}^2$$

$$\text{Direct stress in steel} = 13.33 \times 0.906 = 12.08\ \text{N/mm}^2$$

Max. Bending Stress

(*i*) in extreme concrete fibre $= \dfrac{10 \times 10^6 \times 225}{3.50 \times 10^9} = 0.643\ \text{N/mm}^2$

(*ii*) in steel $= \dfrac{13.33 \times 10 \times 10^6 \times 200}{3.50 \times 10^9} = 7.62\ \text{N/mm}^2$

Max. Stress

(*i*) in concrete = 0.906 + 0.643 = 1.549 N/mm² (Comp.)

(*ii*) in steel = 12.08 + 7.62 = 19.70 N/mm² (Comp.)

Min. Stress

(*i*) In concrete = 0.906 – 0.643 = 0.263 N/mm² (Comp.)

(*ii*) in steel = 12.08 – 7.62 = 4.46 N/mm² (Comp.)

Example 9.7. Check the safety of a circular R.C.C. column 300 mm dia. reinforced with 6 nos. 16 mm ϕ bars as shown in Fig. 9.8. The effective cover is 25 mm. The column is subjected to a load of 200 KN and a B.M. of 5 KN.m $\sigma_{cc} = 5\ \text{N/mm}^2$, $\sigma_{cbc} = 7\ \text{N/mm}^2$, m = 13.33.

Sol. $R = 150$ mm

$$r = \frac{250}{2} = 125 \text{ mm}$$

$$r_y = r \sin 60 = \frac{r\sqrt{3}}{2} = 125\frac{\sqrt{3}}{2}$$

$$= 108.25 \text{ mm}$$

$$A_e = Ag + (1.5\,m - 1)\,Asc$$

$$= \frac{\pi}{4}(300 \times 300) + (1.5 \times 13.33 - 1) \times 6 \times 201$$

$$= 70650 + 22914 = 93564 \text{ mm}^2$$

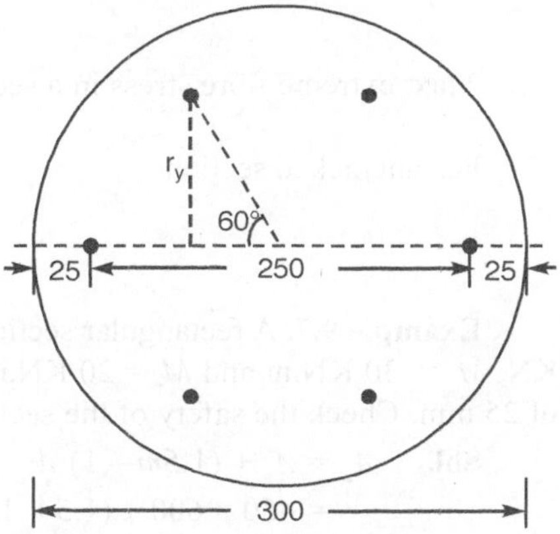

Fig. 9.8.

$$I_e = \frac{\pi}{64} \times (300)^4 + (1.5 \times 13.33 - 1) \times 4 \times 201 \times 108.25^2$$

$$= (3.974 \times 10^8) + (1.790 \times 10^8) = 5.764 \times 10^8 \text{ mm}^4$$

$$\sigma_{cc,\ cal} = \frac{200 \times 1000}{93564} = 2.138$$

$$\sigma_{cbc,\ cal} = \frac{5 \times 1000 \times 1000}{5.764 \times 10^8} \times 150 = 1.301$$

$$\frac{\sigma_{cc,cal}}{\sigma_{cc}} + \frac{\sigma_{cbc,cal}}{\sigma_{cbc}} = \frac{2.138}{5} + \frac{1.301}{7} = 0.428 + 0.186 = 0.614 < 1.0 \quad \text{Hence safe}$$

9.4 COLUMNS SUBJECTED TO BIAXIAL BENDING ABOUT TWO AXES SO THAT NO TENSION DEVELOPS IN THE SECTION

Bending about two axes may take place in different situations such as

(*i*) Corner columns in multistorey buildings.

(*ii*) In any intermediate column where there is unsymmetrical loading on two sides on both axes of the column cross section.

For design of such columns, the bending stresses are calculated about both the axes by considering the moment of inertias about corresponding axes. For a rectangular reinforced section (Fig. 9.9)

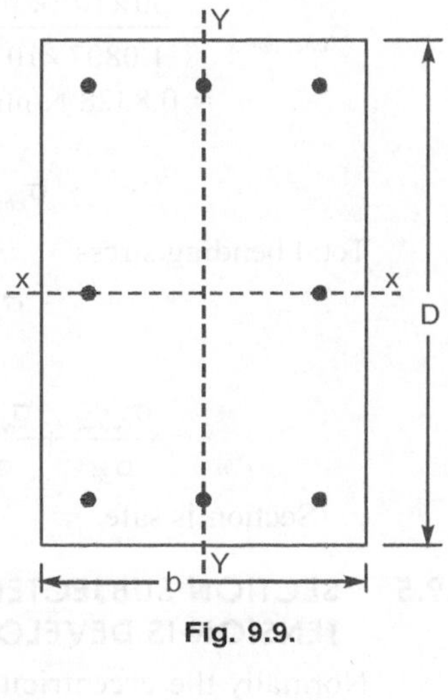

Fig. 9.9.

$$\sigma_{cc,\ cal} = \text{Direct stress} = \frac{P}{A_e}$$

$$\sigma_{cbc,\ cal,\ x} = \text{Bending stress about } x\text{-axis} = \frac{M_x}{I_x}.y$$

$$\sigma_{cbc,\ cal,\ y} = \text{Bending stress about } y\text{-axis} = \frac{M_y}{I_y}.x$$

$$\therefore \quad \text{Total stress at any fibre} = \frac{P}{A_e} \pm \frac{M_x.y}{I_x} \pm \frac{M_y.x}{I_y}$$

Max. extreme fibre stress in a section $= \dfrac{P}{A_e} \pm \dfrac{M_x \cdot \frac{1}{2} D}{I_x} + \dfrac{M_y \cdot \frac{1}{2} B}{I_y}$ ...(9.3)

For uncracked section

$$\frac{\sigma_{cc,c.l}}{\sigma_{cc}} + \frac{\sigma_{cbc,\,cal}}{\sigma_{cbc}} \le 1 \quad ...(9.4)$$

Example 9.7. A rectangular section 450 mm × 600 mm is acted upon by an axial load of 500 KN, M_x = 30 KN.m and M_y = 20 KN.m. The steel is 8 nos. 20 mm ϕ bars with an effective cover of 25 mm. Check the safety of the section m = 13.33. Take M 20. (Fig. 9.10).

Sol. $A_e = A_g + (1.5m - 1)\, A_{sc}$

$= 450 \times 600 + (1.5 \times 13.33 - 1) \times 8 \times 314$

$= 270000 + 47728 = 317728 \text{ mm}^2$

$$I_{e,x} = \frac{450 \times 600^3}{12} + (1.5 \times 13.33 - 1) \times 6 \times 314 \times 275^2$$

$= (0.810 \times 10^{10}) + (0.2707 \times 10^{10})$

$= 1.0807 \times 10^{10} \text{ mm}^4$

$$I_{e,y} = \frac{600 \times 450^3}{12} + (1.5 \times 13.33 - 1) \times 6 \times 314 \times 200^2$$

$= (4.556 \times 10^9) + (1.432 \times 10^9)$

$= 5.988 \times 10^9 \text{ mm}^4$

$$\sigma_{cc,\,cal} = \frac{500,000}{317728} = 1.57 \text{ N/mm}^2$$

$$\sigma_{cbc,\,cal,\,x} = \frac{30 \times 10^6 \times 300}{1.0807 \times 10^{10}}$$

$= 0.8328 \text{ N/mm}^2$

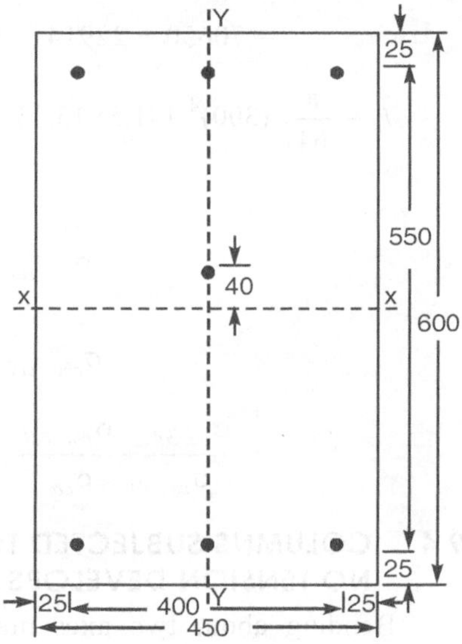

Fig. 9.10.

$$\sigma_{cbc,\,cal,\,y} = \frac{20 \times 10^6 \times 225}{5.988 \times 10^9} = 0.7515 \text{ N/mm}^2$$

Total bending stress

$$\sigma_{cbc,\,cal} = \sigma_{cbc,\,cal,\,x} + \sigma_{cbc,\,cal,\,y}$$

$= 0.8328 + 0.7515 = 1.5843 \text{ N/mm}^2$

$$\frac{\sigma_{cc,cal}}{\sigma_{cc}} + \frac{\sigma_{cbc,cal}}{\sigma_{cbc}} = \frac{1.57}{5} + \frac{1.5843}{7} = 0.315 + 0.2263 = 0.5413 < 1$$

∴ Section is safe.

9.5 SECTION SUBJECTED TO LARGE ECCENTRICITY SO THAT TENSION IS DEVELOPED

Normally the eccentricity is said to be large if it is more than 1.5 times the depth of the section. Under such conditions, tensile stresses are developed in concrete below the neutral axis resulting in the development of cracks in this portion. As we know that concrete is quite weak in tension, we neglect the resistance to loading of this part. So all the stresses are resisted by the concrete in the compression zone, i.e., above the neutral axis. However, the steel in both the compression and tension zones shall carry the respective stresses.

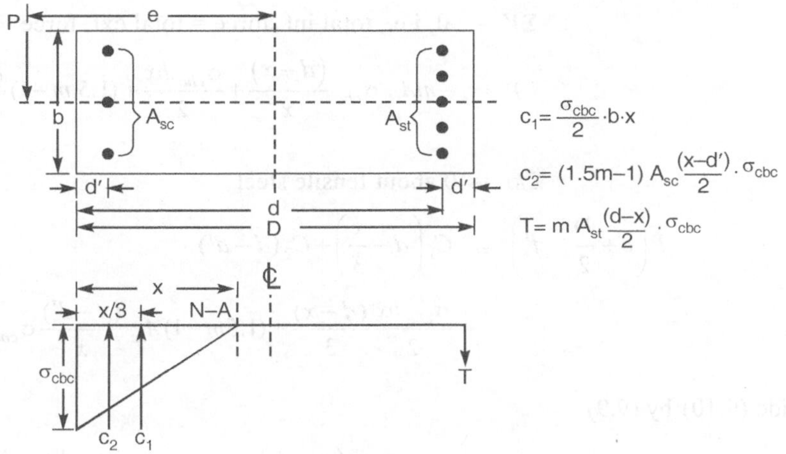

Fig. 9.11.

To Find the Position of Neutral Axis

P = load acting at an eccentricity from the centre of the section

x = depth of neutral axis

σ_{cbc} = max. stress in bending in concrete

σ_{st} = stress in tensile steel

σ_{sc} = stress in compressive steel

d' = effective cover to the centre of compressive reinforcement

d = effective depth to centre of tension reinforcement

A_{st} = Area of tensile steel

A_{sc} = Area of compressive steel

C_1 = Force on the concrete in compression

C_2 = Force on the compression steel

T = Force on the tensile steel

$$\text{Stress in compressive steel} = 1.5\text{ m } \sigma_{cbc} \frac{(x-d')}{x} \qquad ...(9.5)$$

$$\text{Stress in tensile steel} = m.\sigma_{cbc} \frac{(d-x)}{x} \qquad ...(9.6)$$

$$\therefore \quad C_1 = \text{stress} \times \text{comp. area}$$

$$= \frac{\sigma_{clc}}{2} \times (b \times x)$$

$$C_2 = (1.5\, m - 1)\sigma_{cbc} \cdot \frac{(x-d')}{x} . A_{sc}$$

Total compressive force

$$= C_1 + C_2$$

$$= \frac{\sigma_{cbc}}{2}(b.x) + (1.5\, m - 1)\sigma_{cbc} \frac{(x-d')}{x} A_{sc} \qquad ...(9.7)$$

Total tensile force,

$$T = m\, A_{st} \frac{(d-x)}{x} \sigma_{cbc} \qquad ...(9.8)$$

$$\Sigma V = 0, \text{ i.e., total int. force} = \text{total ext. force}$$

$$P = -mA_{st}\,\sigma_{cbc}\frac{(d-x)}{x} + \frac{\sigma_{cbc}.bx}{2} + (1.5\,m-1)\frac{(x-d')}{x}.A_{sc}\sigma_{cbc} \quad ...(9.9)$$

$$\Sigma M = 0 \text{ about tensile steel}$$

$$P\left(e+\frac{D}{2}-d'\right) = C_1\left(d-\frac{x}{3}\right)+C_2(d-d')$$

$$= \frac{\sigma_{cbc.}bx}{2}\frac{(d-x)}{3} + (1.5m-1)A_{sc}\frac{(x-d')}{x}\sigma_{cbc}\times(d-d') \quad ...(9.10)$$

Devide (9.10) by (9.9)

$$\left(e+\frac{D}{2}-d'\right) = \frac{b.x.\frac{\sigma_{cbc}}{2}\left(d-\frac{x}{3}\right)+(1.5m-1)\frac{(x-d')}{x}(d-d')A_{sc}.\sigma_{cbc}}{bx.\frac{\sigma_{cbc}}{2}+(1.5m-1)\frac{(x-d')}{x}A_{sc}\,\sigma_{cbc}-mA_{st}\sigma_{cbc}\frac{(d-x)}{x}}$$

$$= \frac{\frac{bx}{2}\left(d-\frac{x}{3}\right)+(1.5m-1)\frac{(x-d')}{x}(d-d')A_{sc}}{\frac{bx}{2}+(1.5m-1)\frac{(x-d')}{x}A_{sc}-A_{st}\frac{(d-x)}{x}} \quad ...(9.11)$$

The value of x can be found from Eq. 9.11 which is a cubic in x, knowing x, σ_{cbc} can be found from Eq. 9.9.

Knowing σ_{cbc}, stress in compressive and tensile steel can be calculated from Eq. 9.5 Eq. 9.6.

Alternatc Method of Finding the Position of Neutral Axis

Considering the effective section only, (Fig. 9.12)

Let d = distance of N.A. below max. compressed fibre

Stress at any point is proportional to its distance from the neutral axis. Consider a small element of thickness dy at y from N.A.

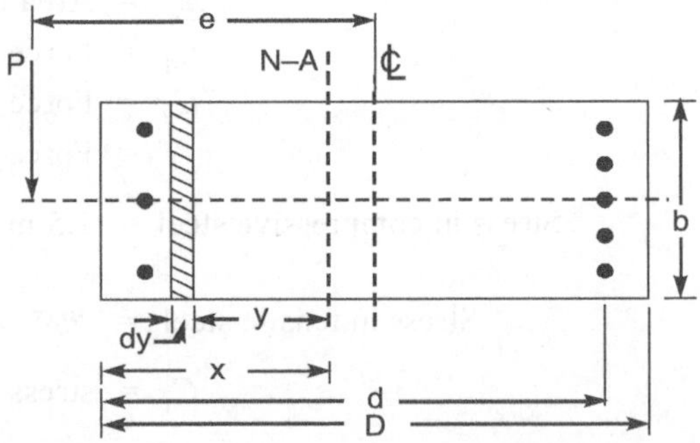

Fig. 9.12.

$$dA = b.dy$$

$$d\sigma \propto y$$

$$= K.y$$

where K = constant of proportionality

$$dF = \text{Force on element}$$

$$= d\sigma.dA$$

$$= K.y.b.dy.$$

∴ The internal force = total external force

i.e., $\Sigma dF = P$

$$\sum_A K.y.dA = P$$

or $$K\sum_A y.dA = P \qquad ...(9.12)$$

Also moment of force on element dA about neutral axis should be equal to the moment of P about the neutral axis

$$\therefore \quad \sum_A dF.y = P\left(e-\frac{D}{2}+x\right)$$

$$\sum_A K.y.dA.y = P\left(e-\frac{D}{2}+x\right)$$

or $$K\sum_A y^2.dA = P\left(e-\frac{D}{2}+x\right) \qquad ...(9.13)$$

Divide (9.13) by (9.12), we get

$$e-\frac{D}{2}+x = \frac{\sum_A y^2 \cdot dA}{\sum_A y.dA}$$

$$\text{Distance of } P \text{ from n.a.} = \frac{\text{Moment of inertia of the eff.section about the n.a}}{\text{Moment of the area of eff.section about n.a.}}$$

Rule : The eccentricity of load P from the neutral axis is equal to the ratio of the moment of inertia of the effective about the neutral axis divided by the moment of the area of the effective section about the neutral axis.

Example 9.8. The section of a column shown in Fig. 9.13 is to carry an axial load of 300 KN and a B.M of 150 KN.m. Find the maximum stresses developed in steel and concrete. m = 13.33

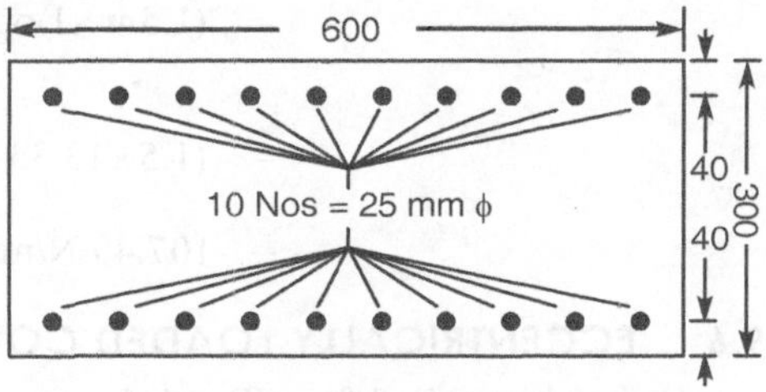

Fig. 9.13.

Sol. Eccentricity 'e' = 150/300 = 0.5 m

e/D = 0.5/0.3 = 1.67 > 1.5

It is a case of large eccentricity. Let 'x' be the depth of the N.A.

Using Eq. (9.11)

$$\left(e+\frac{D}{2}-d'\right) = \frac{\frac{bx}{2}\left(d-\frac{x}{3}\right)+(1.5m-1)\frac{(x-d')}{x}(d-d')A_{sc}}{\frac{bx}{2}+(1.5m-1)\frac{(x-d')}{x}A_{sc}-A_{st}\frac{(d-x)}{x}}$$

$$\left(500+\frac{300}{2}-40\right)$$

$$= \frac{\frac{600x}{2}\times(260-x/3)+(1.5\times13.33-1)\left(\frac{x-40}{x}\right)\times(260-40)10\times491}{\frac{600x}{2}+(1.5\times13.33-1)\left(\frac{x-40}{x}\right)\times10\times491-13.33\times10\times491\left(\frac{260-x}{x}\right)}$$

i.e., $$= x^3 + 1050\,x^2 + 763078\,x - 118{,}357{,}410 = 0$$

Solving the cubic equation $x = 129$ mm

$$\text{Total force} = 300{,}000 = \frac{bx.\sigma_{cbc}}{2} + (1.5m-1)A_{sc}.$$

$$\sigma_{cbc}\left(\frac{x-d'}{x}\right) - m\,A_{sc}\cdot\sigma_{cbc}\left(\frac{d-x}{x}\right)$$

$$300{,}000 = \frac{600\times129\,\sigma_{cbc}}{2} + (1.5\times13.33-1)\times10\times491\times\sigma_{cbc}\left(\frac{129-40}{129}\right)$$

$$-13.3\times10\times491\times\sigma_{cbc}\left(\frac{260-129}{129}\right)$$

$$= \sigma_{cbc}\,(38700 + 64363 - 66465)$$

$$= 36598\,\sigma_{cbc}$$

$$\sigma_{cbc} = \frac{300{,}000}{36598} = 8.20\text{ N/mm}^2$$

$$\text{Stress in tensile steel} = m\,\sigma_{cbc}\left(\frac{d-x}{x}\right)$$

$$= 13.33\times8.20\left(\frac{260-129}{129}\right)$$

$$= 111\text{ N/mm}^2$$

Stress in Compressive steel

$$= (1.5m-1)\sigma_{cbc}\left(\frac{x-d'}{x}\right)$$

$$= (1.5\times13.33-1)8.20\left(\frac{129-40}{129}\right)$$

$$= 107.49\text{ N/mm}^2$$

9.6 ECCENTRICALLY LOADED COLUMNS BASED ON AVERAGE STRESS

It can be applied for $e/D < 1.5$

P = percentage of steel in cross-section

$$A_{sc} = p\times\frac{bD}{100}$$

$$A_{st} = p\times\frac{bD}{100}$$

From Equation 9.9 it can be seen that external load P is :

$$p = b.x.\frac{\sigma_{cbc}}{2} + (1.5m-1)\left(\frac{x-d'}{x}\right)\frac{pbD}{100}.\sigma_{cbc} - m\frac{pbD}{100}.\sigma_{cbc}\left(\frac{d-x}{x}\right)$$

$$x = KD, d' = K_1\,D, d = K_1 D = D - K_1 D = D(1-K_1), \therefore K_2 = 1-K_1$$

where K is a fraction, less than 1.0

For M20 Concrete, $m = \dfrac{280}{3\cdot\sigma_{cbc}} = \dfrac{280}{3\times7} = \dfrac{40}{3}$

$$\therefore \quad P = \sigma_{cbc}.bD\left[\frac{K}{2}+27\left(\frac{KD-K_1D}{KD}\right)\frac{p}{100}-\frac{40}{3}.\frac{p}{100}\left(\frac{K_2D-KD}{KD}\right)\right]$$

$$= \sigma_{cbc}.bD\left[\frac{K}{2}+\frac{27p}{100}\left(\frac{K-K_1}{K}\right)-\frac{40p}{300}\left(\frac{K_2-K}{K}\right)\right]$$

$$= \frac{\sigma_{cbc}.bD}{100}\left[50K+p\frac{(137K-25K_1-40)}{3K}\right]$$

$$\frac{P}{bD} = \frac{\sigma_{cbc}}{100}\left[50K+p\left(\frac{137}{3}-\frac{25K_1}{3K}-\frac{40}{3K}\right)\right] \qquad ...(9.14)$$

$$\frac{P}{bD} = \text{Average stress on the section}$$

Therefore Eq. 9.14 gives the average stress on the section. Taking moments about the centre of tensile steel

$$P\left(e+\frac{D}{2}-k_1D\right) = b.x\frac{\sigma_{cbc}}{2}\left(D-K_1D-\frac{x}{3}\right)+(1.5m-1)A_{sc}\left(\frac{x-d'}{x}\right)\sigma_{cbc}(D-2K_1D)$$

$$= bD^2\sigma_{cbc}\left[\frac{K}{2}\left(1-K_1-\frac{K}{3}\right)+\frac{27p}{300}\left(\frac{K-K_1}{K}\right)(1-2K_1)\right]$$

$$P\frac{(e+D/2-K_1D)}{b.D^2} = \frac{\sigma_{cbc}}{100}\left[50K\left(1-K_1-\frac{K}{3}\right)+27p\left(\frac{K-K_1}{k}\right)(1-2K_1)\right]$$

$$\frac{P}{bD}\left[\frac{e}{D}+\frac{1}{2}-K_1\right] = \frac{\sigma_{cbc}}{100}\left[50K\left(1-K_1-\frac{K}{3}\right)+27p\left(\frac{K-K_1}{K}\right)(1-2K_1)\right] \qquad ...(9.15)$$

Substitute $\frac{p}{bD}$ from Eq. (9.14)

$$\frac{e}{D}+0.5-K_1 = \frac{50K\left(1-K_1-\frac{K}{3}\right)+27p\frac{(K-K_1\_(1-2K_1)}{K}}{50K+p\left(\frac{137}{3}-\frac{25}{3}\frac{K_1}{K}-\frac{40}{3K}\right)}$$

$$\frac{e}{D} = K_1-0.5+\frac{50K\left(1-K_1-\frac{K}{3}\right)+27p\frac{(K-K_1)(1-2K_1)}{K}}{50K+p\left(\frac{137}{3}-\frac{25}{3}\frac{K_1}{K}-\frac{40}{3K}\right)} \qquad ..(9.16)$$

Example 9.9. Design a column to take a load of 180 KN and a bending moment 36 K.N.m. Use M 20 concrete.

Sol. Assuming steel to be provided as 1.2% of the cross-sectional area and the cover to the centre of steel as $D/6$.

$$e = 36/180 = 0.2 \text{ m} = 200 \text{ mm}$$

Assume depth $D = 400$ mm $e/D = 200/400 = 0.5 < 1.5$

Using Eq. 9.16 to find the position of neutral axis

$$\frac{e}{D} = K_1 - 0.5 + \frac{50K(1-K_1-K/3) + 27\frac{p}{k}(K-K_1)(1-2K_1)}{50K + p\left(\frac{137}{3} - \frac{25K_1}{3K} - \frac{40}{3K}\right)}$$

$$K_1 = 1.6, p = 1.2\% = 0.012$$

$$0.5 = \frac{1}{6} - 0.5 + \frac{50K\left(1-\frac{1}{6}-\frac{K}{3}\right) - 27\times\frac{0.012}{K}\left(K-\frac{1}{6}\right)\left(1-2\times\frac{1}{6}\right)}{50K + 0.012\left(\frac{137}{3} - \frac{25}{3}.\frac{1}{6}.\frac{1}{K} - \frac{40}{3K}\right)}$$

$$16.66K^3 - 41.67K_2 + 8.574K - 0.164 = 0$$

Solving by trial and error.

$K = 0.205$, Using Eq. 9.14

$$\frac{P}{bD} = \frac{\sigma_{cbc}}{100}\left[50\,K + p\left(\frac{137}{3} - \frac{25}{3}.\frac{K_1}{K} - \frac{40}{3K}\right)\right]$$

$$= \frac{7}{100}\left(50\times 0.205 + 0.012\left(\frac{137}{3} - \frac{25}{3}\times\frac{1}{6\times 0.205} - \frac{40}{3\times 0.205}\right)\right)$$

$$= \frac{7}{100}(10.25 - 0.313) = 0.695$$

$$b = \frac{P}{0.695}$$

$$= \frac{180\times 1000}{0.695\times 400} = 647 \text{ mm}$$

Use column section 668 × 400 mm

$$A_s = 668\times 400\times\frac{1.2}{100} = 3206 \text{ mm}^2$$

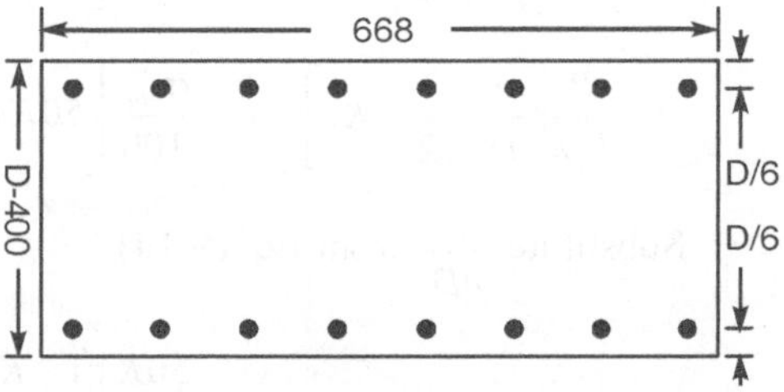

Fig. 9.14.

Use 16 bars 16 mm ϕ, 8 Nos. on each side as shown in Fig. 9.14.

***Section Subjected to Loads with Very Large Eccentricities. Design based on Average Stress*:**

If $e/D > 1.5$, then the effect of bending stress is predominant, and the process of superimposing bending and direct stresses can be adopted with fair accuracy. In such a case, the section may be designed as a beam reinforced on one side for bending moment M and extra reinforcement provided to take up direct force P.

Let *a* rectangular section shown in Fig. 9.15 be subjected to a moment M only. xx = neutral axis for pure bending.

1. Calculate the bending stresses from $M/I = \sigma_{cbc}/y$ as usual. The concrete will have the stresses represented by. ABC in the stress diagram. The portion below xx will be cracked.

2. Superimpose the direct stress (= P/A) as shown by AD on stress diagram, the new distribution of stresses is given by BDE. Neutral axis shifts from xx to $x'x'$.

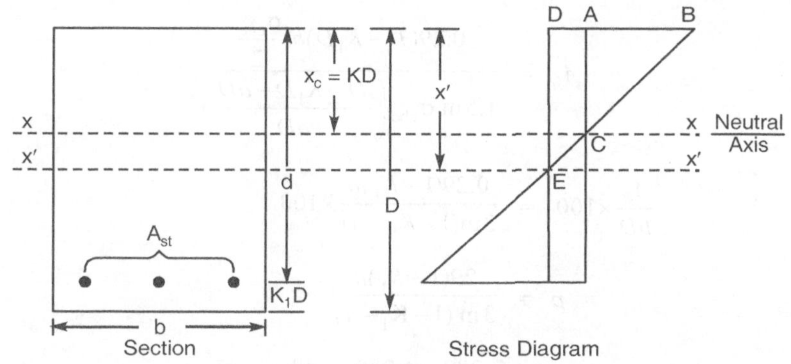

Fig. 9.15.

As the new neutral axis shifts downwards, the effective area changes and this will affect the bending stresses calculated above. The smaller is the value of direct stress, the smaller is the shifting of the neutral axis and less is the inaccuracy in the bending stresses calculated before as for pure bending only. So the stresses can be analysed as an ordinary R.C. beam for bending moment and the direct stress superimposed without appereciable error.

For M20 concrete, $M = 0.91\ bd^2$

Effective depth d can be calculated if we assume b, thus fix overall depth D.

x' = new position of neutral axis after considering the direct stress due to axial force P.

$$x_c = \text{critical neutral axis}$$

$$= 0.29\ d \qquad = 0.29\ (D - K_1D) \qquad = 0.29\ D(1 - K_1)$$

Let $x' = a\ .\ D$ Then $P = (x' - x_c)\dfrac{\sigma_{cbc}}{2}.b$

Since the neutral axis shifts from xx to $x'x'$ due to P only, therefore the added area on compression side gives the load P.

$$P = [a - 0.29(1 - K_1)]\,\frac{bD.\sigma_{cbc}}{2}$$

or

$$\frac{p}{bD} = [0.5\,a - 0.145(1 - K_1)]\sigma_{cbc} \qquad ...(9.17)$$

This equation gives average stress.

Tensile force = comp. force under critical conditions

$$= 0.29(1 - K_1)\frac{\sigma_{cbc}}{2}.b$$

For the new position of neutral axis, stress in tensile steel

$$= \left(d - x'\right)\frac{1.5\ \text{m}\ \sigma_{cbc}}{x'}$$

$$= 1.5\ \text{m}\ \sigma_{\text{cbc}}\left(\frac{D - K_1D - aD}{aD}\right)$$

$$\therefore \quad \text{Area of steel required} = \frac{\text{Tensile force}}{\text{stress in steel}}$$

$$A_{st} = \frac{0.29(D - K_1 D)b.\frac{\sigma_{cbc}}{2}}{1.5\,\text{m}\,\sigma_{cbc}\left(\frac{D - K_1 D - aD}{aD}\right)}$$

$$\frac{A_{st}}{bD}\times 100 = \frac{0.29(1 - K_1)a}{3\text{m}(1 - K_1 - a)}\times 100$$

or

$$p = \frac{29(1 - K_1)a}{3\,\text{m}\,(1 - \text{K}_1 - a)}$$

$$\text{For M 20, m} = \frac{280}{3\,\sigma_{cbc}} = \frac{280}{3\times 7} = \frac{40}{3}$$

$$\therefore \quad p = \frac{29(1 - K_1)a}{40(1 - K_1 - a)}$$

$$a = \frac{40p\,(1 - K_1)}{(29 - 29K_1 + 40p)} \qquad \text{...(9.18)}$$

From this equation, the position of new neutral axis can be found out. From 9.17,

$$\frac{P}{bD} = [0.5\,a - 0.145\,(1 - K_1)\,\sigma_{cbc}$$

$$M = 0.91\,bd^2$$

$$\therefore \quad \frac{Mb}{P^2} = \frac{0.91\,b^2(D - K_1 D)^2}{P^2} = \frac{0.91(1 - K_1)^2}{\left(\frac{P}{bD}\right)^2} \qquad \text{...(9.19)}$$

Example 9.10. An R.C.C. Column section is subjected to the axial load of 60 KN and bending moment of 40 KN.m Design the section. Use M 20 concrete.

Sol. $e = 40/60 = 0.67$ m

As the eccentricity is quite large, the member will behave more like a flexural member than a compression member. Therefore, the section shall be designed as a beam and reinforcement shall be provided on tension side only.

Take width, $b = 300$ mm

$$M = Q.bd^2$$

$$40 \times 1000 \times 1000 = 0.91 \times 300 \times d^2$$

$$d = \sqrt{\left(\frac{40\times 1000\times 1000}{0.91\times 300}\right)} = 382.8 \text{ mm}$$

$$D = 450 \text{ mm}, \quad d = 400 \text{ mm}$$

X_c = Depth of critical neutral axis,

$$= 0.29 \times 400 = 116 \text{ mm}$$

To compute the actual position of neutral axis, x' (Fig. 9.16)

$$P = (x' - x_c)\frac{b\,.\,\sigma_{cbc}}{2}$$

$$60 \times 1000 = (x' - 116) \times \frac{300 \times 7}{2}$$

$$x' = 173.14$$

$$a = \frac{x'}{D} = 173.14/450 = 0.385$$

$$p = \frac{29(1 - K_1)a}{40(1 - K_1 - a)}$$

$$= \frac{29(1 - 0.1)0.385}{40(1 - 0.1 - 0.385)}$$

(Taking $K_1 = 1/10$)

$$= 0.4878\ \%$$

$$A_{st} = \frac{0.4878\%}{100} \times 300 \times 450$$

$$= 658.5\ \text{mm}^2$$

Use 6 bars 12 mm ϕ, A_{st} provided = 678 mm²

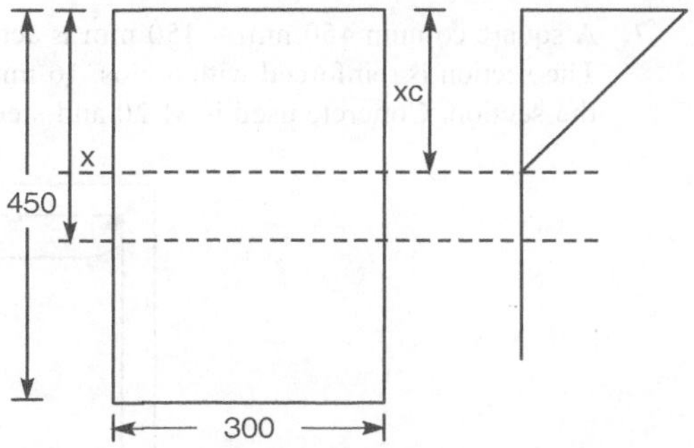

Fig. 9.16.

EXERCISE 9

1. Explain briefly the stress distribution pattern for rectangular section under eccentric loads.
2. A square column of side 500 mm is reinforced with 8 bars of 16 mm diameter as shown in Fig. 9.17. The section is subjected to a load of 300 KN at an eccentricity of 40 mm about one of the axis. Determine the maximum and minimum stresses. Check whether the section is safe, if the permissible stressess are σ_{cc} =5 N/mm² and σ_{cbc} = 7 N/mm² and m = 13.33

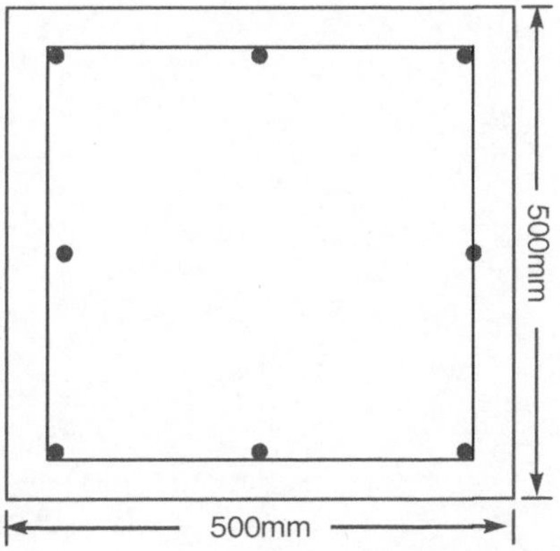

Fig. 9.17.

3. Design a R.C. column carrying an axial load of 300 KN and bending moment of 4500 N.m. The allowable stresses in concrete and steel are : σ_{cc} = 5 N/mm², σ_{cbc} = 7 N/mm², m = 13.33.
4. The cross- section of a curved member is reinforced as shown in Fig. 9.18. Determine the load and the bending moment on the section so that no tension develops in the section and the section is safe as uncracked section. Given σ_{cc} = 5 N/mm², σ_{cbc} = 7 N/mm² and m = 13.33.
5. Check the safety of a circular R.C.C. column 400 mm diameter reinforced with 8 Nos. 16 mm ϕ bars. The column is subjected to a load of 250 KN and a B.M. = 4 KN. m. The effective cover is 30 mm. Concrete M 20 and Fe500 Steel are used.
6. Explain briefly the procedural steps to design a biaxially loaded R.C. rectangular column.

7. A square column 450 mm × 450 mm is acted upon by 300 KN and M_x = 30 KN.m, Mg = 15 K Nm. The section is reinforced with 6 Nos. 16 mm ϕ with an effective cover of 30 mm. Check the safety of the section. Concrete used is M 20 and steel is Fe 415 grade.

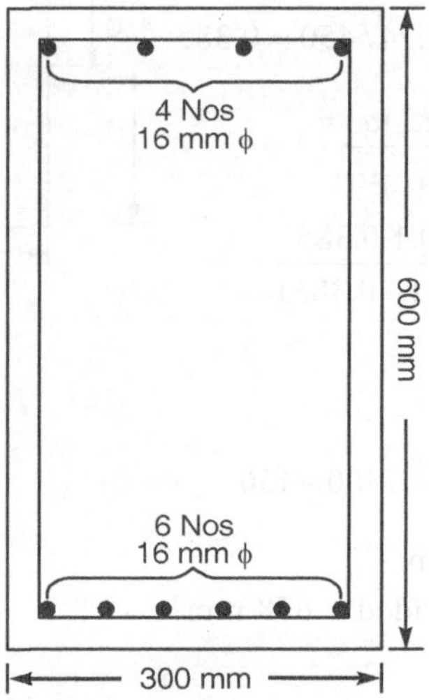

Fig. 9.18.

Foundations

CHAPTER **10**

10.1 INTRODUCTION

Foundation is that part of the building whose function is to distribute the loads of the structure to the soil supporting the structure. The foundation distributes the load over a larger area at a uniform rate so that the pressure on the soil does not exceed its allowable bearing capacity. Foundation increases the stability of the structure. It restricts the settlement of the structure within reasonable limits. The settement depends upon the intensity of load, type of soil and depth at which the load of the structure is transferred to the soil. To avoid unequal settlement it is necessary that the load per unit area on the soil should be same under different footings. If this load is different, as is usual, under different footings, unequal settlement may take place. It is customary to design a footing for the maximum loaded part and provide the same footing for other parts which are comparatively less loaded.

10.2 DISTRIBUTION OF PRESSURE BELOW FOUNDATIONS

Settlement of a foundation occurs mainly due to the excessive vertical pressures being transmitted from the base of the foundation to the soil below. It has been observed that the

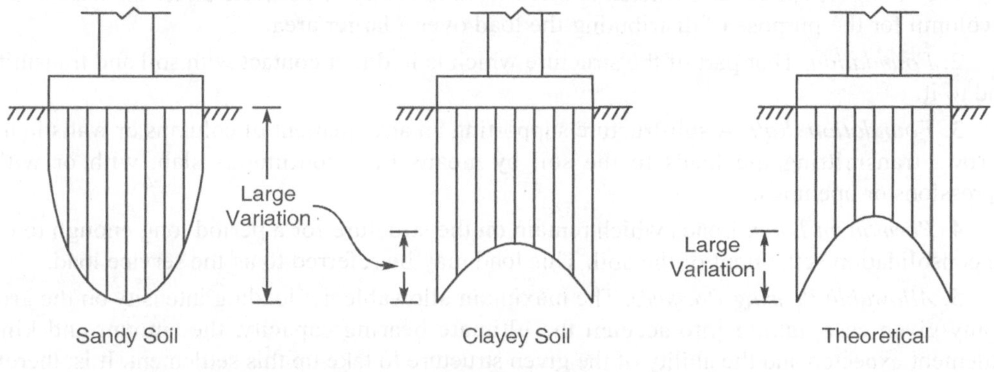

Fig. 10.1. Pressure Variation in Shallow Foundation

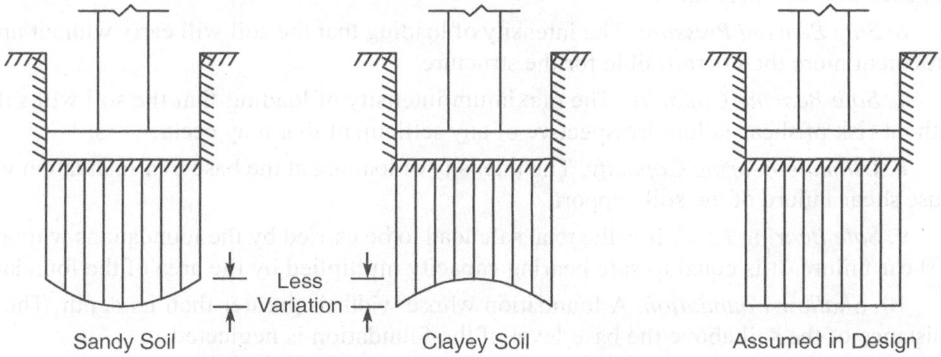

Fig. 10.2. Pressure Variation in Deep Foundation

variation of pressure changes from the centre of the footing to the edges. This variation depends upon (*i*) the rigidity of the footing base (*ii*) the type of soil underneath and (*iii*) the depth of excavation at which the footing is laid.

For a rigid base, the variation of pressure, along the width of footing is more in case of shallow footing and gets increased as the depth of excavation increases as shown in Fig. 10.1 and Fig. 10.2.

10.3 CAUSES OF FAILURE OF FOUNDATIONS

Any one or more of the following may cause failure of foundation.

1. Unequal settlement of the sub-soil.
2. Horizontal movement of the soil adjoining the structure.
3. Shrinkage due to withdrawal of moisture from the soil below the foundation.
4. Lateral pressure tending to over-turn the structure.
5. Action of atmosphere.

10.4 ASSUMPTIONS IN THE DESIGN OF FOUNDATIONS

The following assumptions are made :

1. Foundation is rigid.
2. The distribution of pressure from soil on the base is uniform.

10.5 TERMINOLOGY

1. *Footing.* A spread constructed in brick work masonry or concrete under the base of a wall or column for the purpose of distributing the load over a larger area.

2. *Foundation.* That part of the structure which is in direct contact with soil and transmitting load to it.

3. *Foundation, Raft.* A substructure supporting an arrangement of columns or walls in a row or rows transmitting the loads to the soil by means of a continuous slab, with or without depressions or openings.

4. *Permanent Load.* Loads which remain on the structure for a period long enough to cause the consolidation settlement of the soil. This load may be referred to as the service load.

5. *Allowable Bearing Pressure.* The maximum allowable net loading intensity on the ground in any given case, taking into account the ultimate bearing capacity, the amount and kind of settlement expected and the ability of the given structure to take up this settlement. It is, therefore, a combined function of both the site conditions and characteristics of the particular structure it is proposed to erect thereon.

6. *Safe Bearing Pressure.* The intensity of loading that the soil will carry without undergoing settlement more than permissible for the structure.

7. *Safe Bearing Capacity.* The maximum intensity of loading that the soil will safely carry without risk of shear failure irrespective of any settlement that may occur.

8. *Ultimate Bearing Capacity.* The intensity of loading at the base of a foundation which will cause shear failure of the soil support.

9. *Safe Bearing Load.* It is the total safe load to be carried by the foundations without the risk of shear failure. It is equal to safe bearing capacity multiplied by the area of the foundation.

10. *Shallow Foundation.* A foundation whose width is greater than its depth. The shearing resistance of the soil above the base level of the foundation is neglected.

10.6 DEPTH OF FOUNDATIONS

The depth to which foundations should be carried depends upon four principal factors :

(*a*) Securing of adequate bearing capacity.

(*b*) In the case of clay soils, penetration below the zone where shrinkage and swelling due to seasonal weather changes are likely to cause appreciable movements.

(*c*) In fine sands and silts, penetration below the zone in which trouble may be expected from frost.

(*d*) The maximum depth of scour wherever relevant should also be considered and the foundation should be located sufficiently below this depth.

Minimum Depth of Foundation. All foundations shall extend to a depth of at least 0.50 m below natural ground level. On rock or such other weather-resisting natural ground, removal of the top soil may be all that is required. In such cases the surface shall be cleaned and, if necessary, stepped or otherwise prepared so as to provide a suitable bearing and thus prevent slipping or other unwanted movements.

Where there is excavation, ditch, pond, water course, filled up ground or similar condition adjoining or adjacent to the subsoil on which the building is to be erected which is likely to impair the stability of a building, either the foundation of such building shall be carried down to a depth beyond the detrimental influence of such conditions, or retaining walls or similar works shall be constructed for the purpose of shielding from their effects.

A foundation in any type of soil shall be below the zone significantly weakened by root holes or cavities produced by burrowing animals or worms. The depth shall also be enough to prevent the rain water scouring below the footings.

In soils which are affected by (*i*) drying, shrinkage and cracking in dry and hot weather and by (*ii*) swelling in the following wet weather, it is necessary to place the foundation bearing at such a depth where the effects of seasonal changes are not important or to make foundation capable of eliminating the undesirable effects due to relative movement by providing flexible type of construction or rigid foundations as the can may be.

10.7 TYPES OF FOUNDATIONS

The various types of foundations are given below :

1. Spread Footings
2. Combined Footings
3. Eccentrically Loaded Footings
4. Raft or Mat Foundation
5. Pile Foundations

10.7.1 Spread Footings

It distributes the load over a wider area because the base of the member transmitting load to the soil is made wider. The various types of footings under this head are :

(*a*) Wall Footings

(*b*) Column Footings

(*a*) *Wall Footings.* When the walls are subjected to relatively heavy loads and the bearing capacity of the soil on which it rests is very low, then it is desirable to provide R.C.C. continuous footing below the wall [Fig. 10.3(*a*)]

(*b*) *Column Footings.* These footings when provided for independent columns are called isolated or independent footings. It is provided to distribute the concentrated loads in the form of uniformly distributed load on the soil below. The footings may be circular, square or rectangular in plan [Fig. 10.3(*b*) to (*e*)].

10.7.2 Combined Footings

When a single footing is provided for two or more columns, it is known as combined footing. The footing is proportioned in such a way that the C.G. of the column loads is in line with the C.G. of the footing. A combined footing can be of two shapes in Plan :

(*a*) *Rectangular Shape*. When the two column loads are equal in magnitude [Fig. 10.3(*g*)].

(*b*) *Trapezoidal Shape*. When the two column loads are unequal in magnitude [Fig. 10.3(*f*)] or under any condition of loading.

The combined footings can be provided under the following conditions :

(*i*) When isolated footings for individual columns are practically touching one another or overlapping.

(*ii*) When the columns are located on the property boundary line.

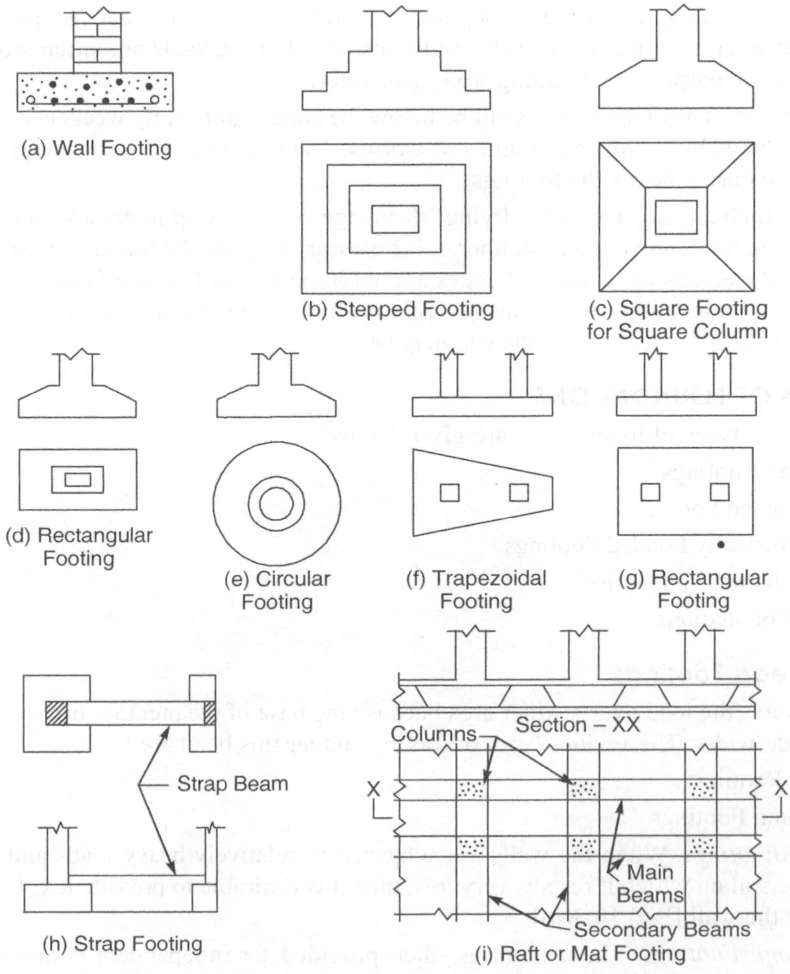

Fig. 10.3. Types of Foundations

10.7.3 Eccentrically Loaded Footings

Such footings are used for columns provided at property boundary to avoid any encroachment of the adjoining property. In such cases a strap footing in which load from the outer column is balanced by the load from the inner column through a cantilever beam acting about a fulcrum is provided. In case, the interior column cannot be provided, a suitable concrete block may be built to provide the necessary anchorage [Fig. 10.3(*h*)].

10.7.4 Raft or Mat Foundation

A mat or thick reinforced concrete slab is sometimes provided to cover the entire foundation area under a building or under a number of columns in more than one row [Fig. 10.3(*i*)]. It is provided under such site conditions where (*i*) the bearing power of the soil is so low that independent column footings are impracticable (*ii*) pile foundations cannot be used advantageously.

10.7.5 Pile Foundations

The load is transmitted to the soil by driving long vertical members of concrete called piles. The piles can be named depending upon :

(*a*) *Method of Transfer of Loads*

(*i*) *Bearing Piles*. The piles which are driven to the very hard stratum thus transfer of load is direct to a strong strata [Fig. 10.4(*a*)].

(*ii*) *Friction Piles*. The transfer of load takes place through friction between the pile surface and the surrounding soil [Fig. 10.4(*b*)].

(*iii*) *Bearing and Friction Piles*. When the transfer of load takes place partly by bearing and partly by friction [Fig. 10.4(*c*)]

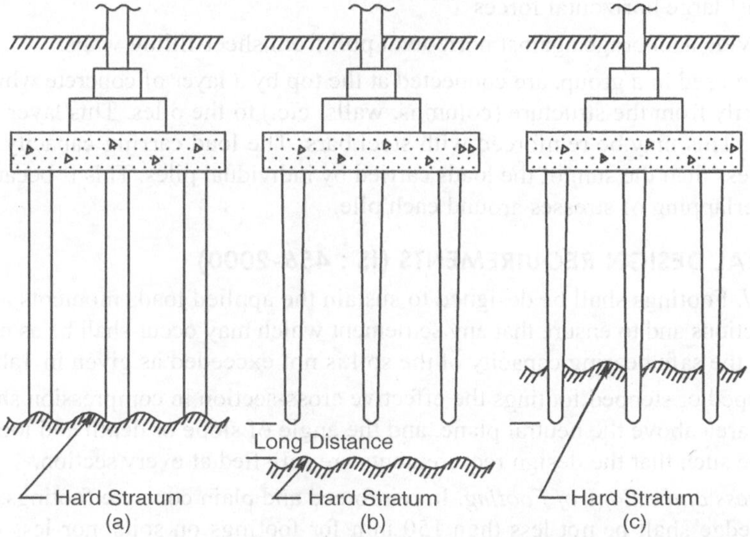

Fig. 10.4. (a) Bearing Pile (b) Friction Pile (c) Bearing and Friction Pile

(*b*) *Method of Construction of Piles*

(*i*) *Precast Concrete Piles*. These piles are cast in workshop, cured and then driven into the ground. These are square in section with chamfered corners. A solid shoe is used at the bottom of the pile to facilitate driving (Fig. 10.5). The length varies from 5 to

30 metre and dia./side varies from 250 mm to 600 mm depending upon the loads to be transferred.

(*ii*) *Cast-in-situ Concrete Piles.* It is a pile which is cast in its permanent location within a hole made for this purpose. Such piles are not subjected to any type of loading till these attain sufficient strength.

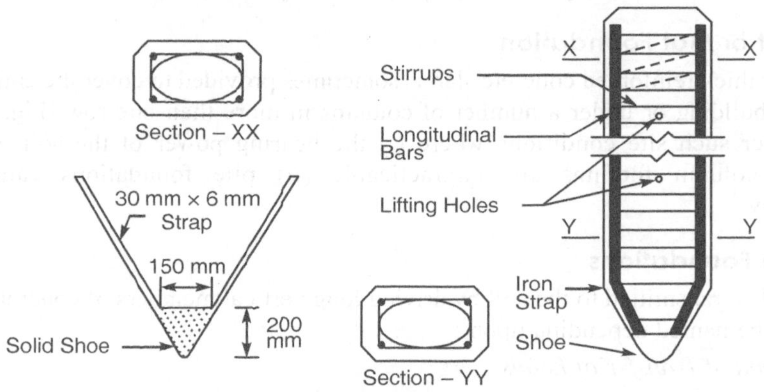

Fig. 10.5. Precast Pile

Piles are used under the following conditions :

1. Load is to be transferred through water or soft soil
2. To prevent excessive settlement
3. To carry the foundation below the scour depth
4. To resist large horizontal forces
5. To provide anchorage against horizontal pull from sheet-filling walls.

Piles, when used in a group, are connected at the top by a layer of concrete which distributes the load uniformly from the structure (columns, walls, etc.) to the piles. This layer of concrete is called Pile Cap. This may be reinforced with steel bars. The load carring capacity of a group of piles is always less than the sum of the loads carried by individual piles. This is because of the fact that there is overlapping of stresses around each pile.

10.8 GENERAL DESIGN REQUIREMENTS (IS : 456-2000)

1. *General.* Footings shall be designed to sustain the applied loads moments and forces and the induced reactions and to ensure that any settlement which may occur shall be as nearly uniform as possible and the safe bearing capacity of the soil is not exceeded as given in Table 10.1.

(*a*) In slopped or stepped footings the effective cross-section in compression shall be limited by the area above the neutral plane, and the angle of slope or depth and location of steps shall be such that the design requirements are satisfied at every section.

(*b*) *Thickness at the Edge of Footing.* In reinforced and plain concrete footings, the thickness at the edge shall be not less than 150 mm for footings on soils, nor less than 300 mm above the tops of piles for footing on piles (Fig. 10.6).

(*c*) In the case of plain concrete pedestals, the angle α between the plane passing through the bottom edge of the pedestal and the corresponding junction edge of the column with pedestal and the horizontal plane (Fig. 10.7) shall be governed by the expression :

Table 10.1. Safe Bearing Capacity

Sl. No.	*Type of Rocks/Soils*	*Safe Bearing Capacity* kN/m^2 (kgf/cm^2)	*Remarks*
(*1*)	(*2*)	(*3*)	(*4*)
(*a*)	*Rocks*		
(*i*)	Rocks-hard without lamination and defects, for example, granite, trap and diorite	3240(33)	—
(*ii*)	Laminated rocks, for example, sand-stone and limestone in sound condition	1620(16.5)	—
(*iii*)	Residual deposits of shattered and broken bed rock and hard shale, cemented material	880(9.0)	—
(*iv*)	Soft rock	440(4.5)	—
(*b*)	*Non-Cohesive Soils*		
(*v*)	Gravel, sand and gravel, compact and offering high resistance to penetration when excavated by tools	440(4.5)	(see Note 2)
(*vi*)	Coarse sand, compact and dry	440(4.5)	Dry means that the ground water level is at a depth not less than the width of foundation below the base of the foundation
(*vii*)	Medium sand, compact and dry	245(2.5)	—
(*viii*)	Fine sand, silt (dry lumps easily pulverized by the fingers)	150(1.5)	—
(*ix*)	Loose gravel or sand gravel mixture loose to medium sand, dry	245(2.5)	(see Note 2)
(*x*)	Fine sand, loose and dry	100(1.0)	
(*c*)	Cohesive Soils		
(*xi*)	Soft shale, hard or stiff clay in deep bed, dry	440(4.5)	This group is susceptible to long term consolidation settlement
(*xii*)	Medium clay, readily indented with a thumb nail	245(2.5)	—
(*xiii*)	Moist clay and sand clay mixture which can be indented with strong thumb pressure	150(1.5)	—
(*xiv*)	Soft clay indented with moderate thumb pressure	100(1.0)	—
(*xv*)	Very soft clay which can be penetrated several inches with the thumb	50(0.5)	—
(*xvi*)	Black cotton soil or other shrinkable or expansive clay in dry condition (50 per cent saturation)	—	See Note 3. To be determined after investigation
(*d*)	*Peat*		
(*xvii*)	Peat	—	See Note 3 and Note 4. To be determined after investigation

(*e*)	*Made-Up Ground*		
(*xviii*)	Fills or made-up ground	—	See Note 2 and Note 4. To be determined after investigation.

Note: 1. Values listed in the Table are from shear consideration only.

2. Values are very much rough due to the following reasons :
   (*a*) Effect of characteristics of foundations (that is, effect of depth, width, shape, roughness, etc.) has not been considered.
   (*b*) Effect of range of soil properties (that is, angle of frictional resistance, cohesion, water table, density, etc.) has not been considered.
   (*c*) Effect of eccentricity and indication of loads has not been considered.
3. For non-cohesive soils, the values listed in the table shall be reduced by 50 per cent if the water tables is above or near the base of footing.
4. Compactness or looseness of non-cohesive soils may be determined by driving the cone of 65 mm dia. and 60 apex angle by a hammer of 65 kg falling from 75 cm. If corrected number of blows (N) (see IS : 6403-1971*) for 30 cm penetrations are less than 10, the soil is called loose, if N lies between 10 and 30, it is medium, if more than 30, the soil is called as dense.

* Code of practice for determination of allowable bearing pressure on shallow foundations.

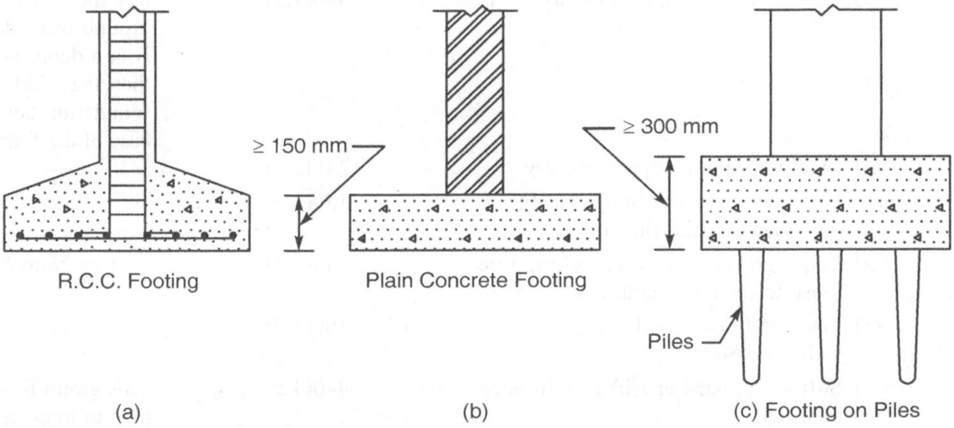

Fig. 10.6. Thickness at the Edge of Footings

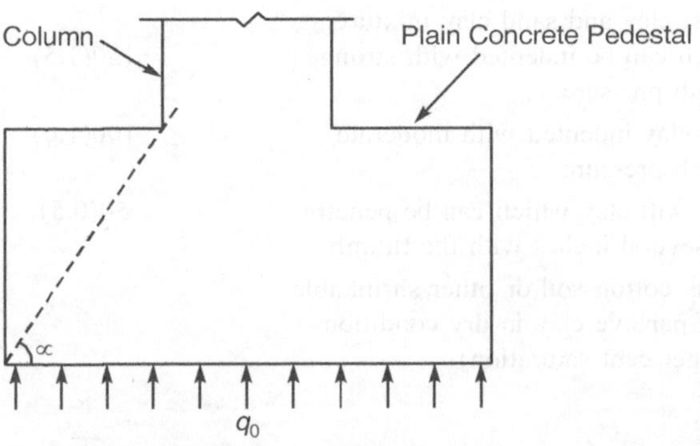

Fig. 10.7. Plain Concrete Pedestal

$$\tan \propto \not< 0.9\sqrt{\left(\frac{100q_0}{f_{ck}}\right)+1}$$

where q_0 = calculated maximum bearing pressure at the base of the pedestal in N/mm²
f_{ck} = characteristic strength of concrete at 28 days in N/mm²

2. *Moments and Forces*

(*a*) In case of footings on piles, computation for moments and shears may be based on the assumption that the reaction from any pile is concentrated at the centre of the pile.

(*b*) For the purpose of computing stresses in footings which support a round or octagonal concrete column or pedestal, the face of the column or pedestal shall be taken as the side of a square inscribed within the perimeter of the round or octagonal column (Fig. 10.8) or pedestal.

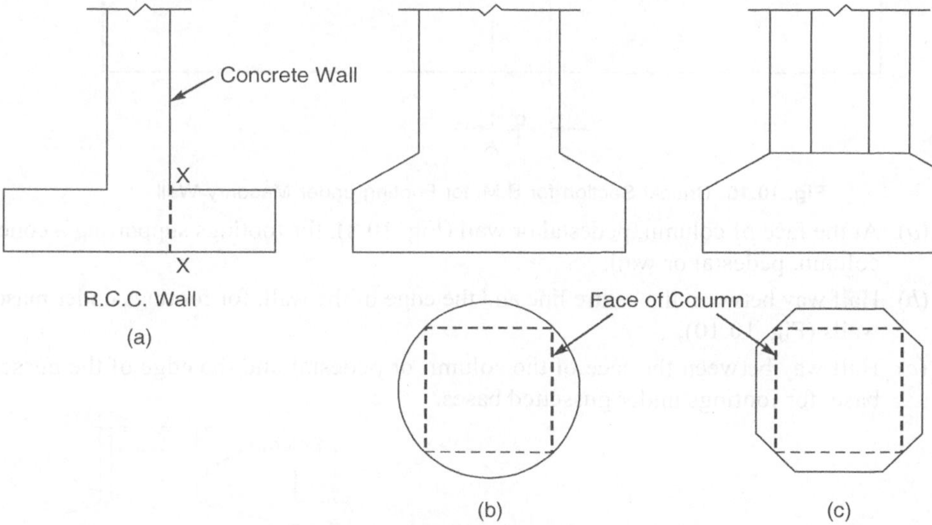

Fig. 10.8. Face of Column (Reinforced Concrete Wall)

(*c*) *Bending Moment*

(*i*) The bending moment at any section shall be determined by passing through the section a vertical plane which extends completely across the footing, and computing the moment of the forces acting over the entire area of the footing on one side of the said plane (Fig. 10.9).

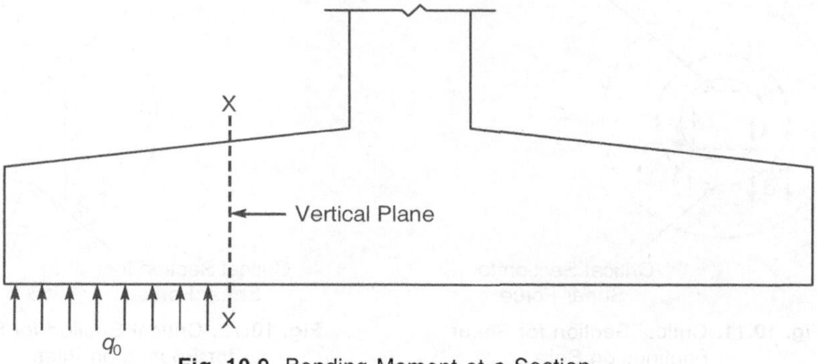

Fig. 10.9. Bending Moment at a Section

(*ii*) The greatest bending moment to be used in the design of an isolated concrete footing which supports a column, pedestal or wall, shall be the moment computed in the manner prescribed in (*i*) at sections located as follows :

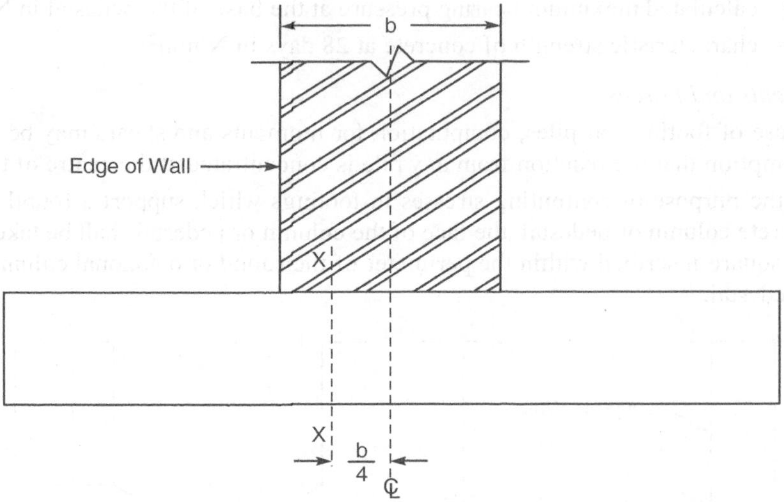

Fig. 10.10. Critical Section for B.M. for Footing under Masonry Wall

(*a*) At the face of column, pedestal or wall (Fig. 10.8), for footings supporting a concrete column, pedestal or wall;

(*b*) Half way between the centre line and the edge of the wall, for footings under masonry walls (Fig. 10.10);

(*c*) Half way between the face of the column or pedestal and the edge of the gussetted base, for footings under gussetted bases.

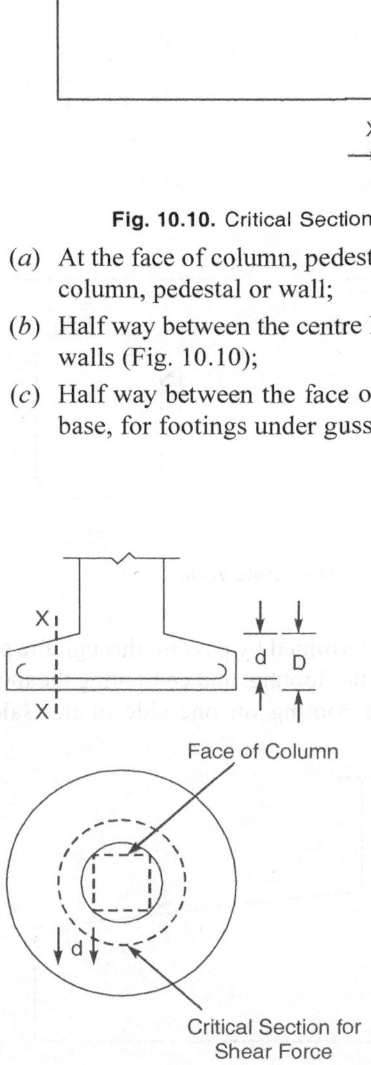

Fig. 10.11. Critical Section for Shear for Footings on Soils

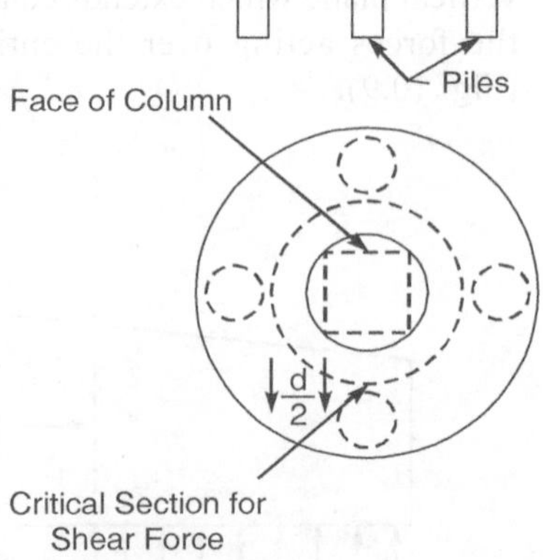

Fig. 10.12. Critical Section for Shear for Footing on Piles

(*d*) *Shear and Bond*

(*i*) The shear strength of footings is governed by the more severe of the following two conditions :

(*i*) The footing acting essentially as a wide beam, with a potential diagonal crack extending in a plane across the entire width; the critical section for this condition shall be assumed as a vertical section located from the face of the column, pedestal or wall at a distance equal to the effective depth of the footing in case of footings on soils (Fig. 10.11) which is most common use, and a distance equal to half the effective depth of footing for footing on piles (Fig. 10.12).

(*ii*) Two-way action of the footing, with potential diagonal cracking along the surface of truncated cone or pyramid around the concentrated load; in this case, the footing shall be designed for shear in accordance with provisions for flat slabs as follows :

(*a*) The critical section for punching shear shall be at a distance d/2 from the periphery of the column/capital/drop panel, perpendicular to the plane of the slab where d is the effective depth of the section (Fig. 10.13). The shape in plan is shown in Fig. 10.14.

(*b*) *Calculation of Shear Stress*. The shear stress τ_v shall be the sum of the values calculated below :

(*i*) The nominal shear stress in footing shall be taken as $\frac{V}{b_0 \cdot d}$ where V is the shear force due to design load, b_0 is the periphery of the critical section and d is the effective depth.

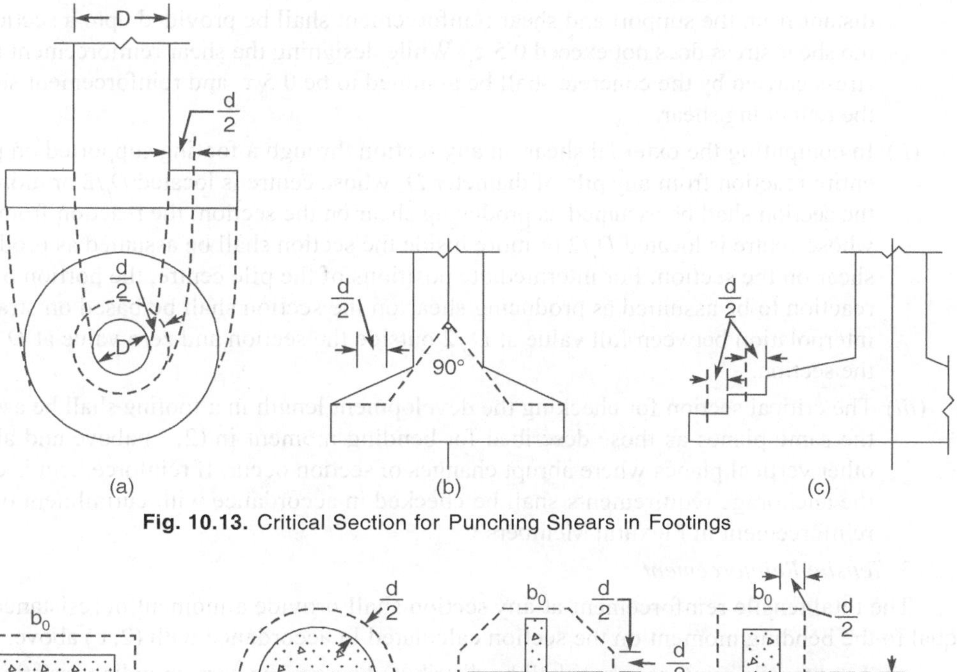

Fig. 10.13. Critical Section for Punching Shears in Footings

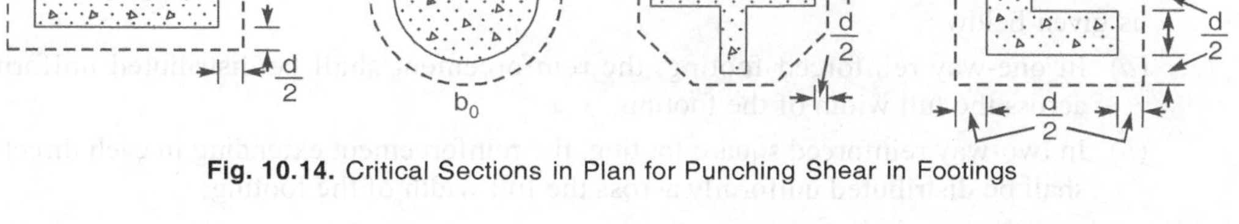

Fig. 10.14. Critical Sections in Plan for Punching Shear in Footings

(*ii*) When unbalanced loads, earthquake or other forces cause transfer of bending moment between footing and column a fraction (1 – ∝) of the moment shall be considered transferred by eccentricity of the shear about the centroid of the critical section. Shear stresses shall be taken as varying linearly about the centroid of the critical section. The value of ∝ is given by

$$\propto = \frac{1}{1 + 2/3\sqrt{(a_1/a_2)}}$$

where a_1 = overall dimension of the critical section for shear in the direction in which moment acts.

a_2 = overall dimension of the critical section for shear transverse to the direction in which moment acts.

(*c*) *Permissible Shear Stress*. When shear reinforcement is not provided, the calculated shear stress at the critical section shall not exceed $k_s \cdot \tau_c$,

where $k_s = (0.5 + \beta_c) \ngtr 1$

β_c = the ratio of short side to long side of the column/capital

$\tau_c = 0.16\sqrt{f_{ck}}$ in working stress method of design, $0.25\sqrt{f_{ck}}$ in case of limit stata design

(*d*) When shear stress at the critical section exceeds the value given in (*c*) above, but less than 1.5 τ_c shear reinforcement shall be provided. If the shear-stress exceeds 1.5 τ_c, the footing shall be redesigned. Shear stresses shall be investigated at successive sections more distant from the support and shear reinforcement shall be provided upto a section where the shear stress does not exceed 0.5 τ_c. While designing the shear reinforcement the shear stress carried by the concrete shall be assumed to be 0.5 τ_c and reinforcement shall carry the remaining shear.

(*ii*) In computing the external shear on any section through a footing supported on piles, the entire reaction from any pile of diameter D_p whose centre is located $D_p/2$ or more outside the section shall be assumed as producing shear on the section; the reaction from any pile whose centre is located $D_p/2$ or more inside the section shall be assumed as producing no shear on the section. For intermediate positions of the pile centre, the portion of the pile reaction to be assumed as producing shear on the section shall be based on straight line interpolation between full value at $D_p/2$ outside the section and zero value at $D_p/2$ inside the section.

(*iii*) The critical section for checking the development length in a footing shall be assumed at the same planes as those described for bending moment in (2, *c*) above and also at all other vertical planes where abrupt changes of section occur. If reinforcement is curtailed, the anchorage requirements shall be checked in accordance with curtailment of tension reinforcement in Flexural Members.

3. *Tensile Reinforcement*

The total tensile reinforcement at any section shall provide a moment of resistance at least equal to the bending moment on the section calculated in accordance with (2, *c*) above.

(*i*) Total tensile reinforcement shall be distributed across the corresponding resisting section as given below :

(*a*) In one-way reinforced footing, the reinforcement shall be distributed uniformly across the full width of the footing;

(*b*) In two-way reinforced square footing, the reinforcement extending in each direction shall be distributed uniformly across the full width of the footing;

(*c*) In two-way reinforced rectangular footing, the reinforcement in the long direction shall be distributed uniformly across the full width of the footing. For reinforcement in the short direction, a central band equal to the width of the footing shall be marked along the length of the footing and portion of the reinforcement determined in accordance with the equation given below shall be uniformly distributed across the central band :

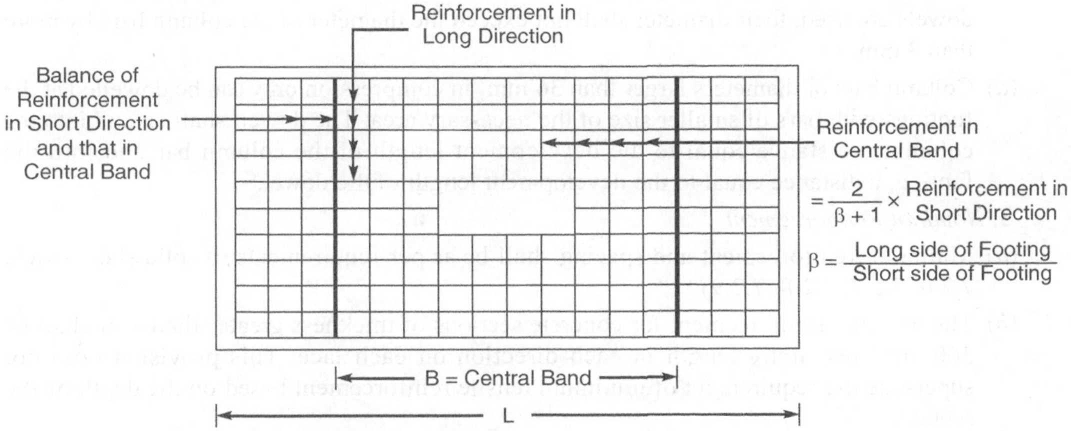

Fig. 10.15. Two-Way Reinforced Rectangular Footing

$$\frac{\text{Reinforcement in central band width}}{\text{Total reinforcement in short direction}} = \frac{2}{\beta + 1}$$

where β = the ratio of the long side to the short side of the footing

$$= \frac{\text{long side of footing}}{\text{short side of footing}}$$

The remainder of the reinforcement shall be uniformly distributed in the outer portions of the footing.

4. *Transfer of Load at the Base of Column*

The compressive stress in concrete at the base of a column or pedestal shall be considered as being transferred by bearing to the top of the supporting pedestal or footing. The bearing pressure on the loaded area shall not exceed the permissible bearing stress in direct compression multiplied by a value equal to $\sqrt{\left(\frac{A_1}{A_2}\right)}$ but not greater than 2;

where A_1 = supporting area for bearing of footing, which is slopped or stepped footing may be taken as the area of the lower base of the largest frustrum of a pyramid or cone contained wholly within the footing and having for its upper base, the area actually loaded and having side slope of one vertical to two horizontal; and

A_2 = loaded area at the column base

For working stress method of design the permissible bearing stress on full area of concrete shall be taken as $0.25 f_{ck}$; for limit state method of design the permissible bearing stress shall be $0.45 f_{ck}$.

(*a*) Where the permissible bearing stress on the concrete in the supporting or supported member would be exceeded, reinforcement shall be provided for developing the excess force, either by extending the longitudinal bars into the supporting member, or by dowels as given in (*c*) below.

(*b*) Where transfer of force is accomplished by reinforcement, the development length of the reinforcement shall be sufficient to transfer the compression or tension to the supporting member in accordance with Development of Stress in reinforcement.

(*c*) Extended longitudinal reinforcement or dowels of atleast 0.5% of the cross-sectional area of the supported column or pedestal and a minimum of 4 bars shall be provided. Where dowels are used, their diameter shall not exceed the diameter of the column bars by more than 3 mm.

(*d*) Column bars of diameters larger than 36 mm, in compression only can be dowelled at the footings with bars of smaller size of the necessary area. The dowell shall extend into the column, a distance equal to the development length of the column bar and into the footing, a distance equal to the development length of the dowel.

5. *Nominal Reinforcement*

(*a*) Minimum reinforcement and spacing shall be as per requirements of solid slab (article 7.2.6, 7.2.7, 7.2.8, 7.2.9).

(*b*) The nominal reinforcement for concrete sections of thickness greater than 1 m shall be 360 mm² per metre length in each direction on each face. This provision does not supersede the requirement of minimum tensile reinforcement based on the depth of the section.

10.9 MINIMUM THICKNESS

The minimum thickness of foundation should be not less than 150 mm. The thickness, in working stress method, is invariably governed by bending criteria while it is governed by shear considerations in limit state method of design.

10.10 SPREAD FOOTING FOR WALLS

10.10.1 Reinforced Concrete Wall

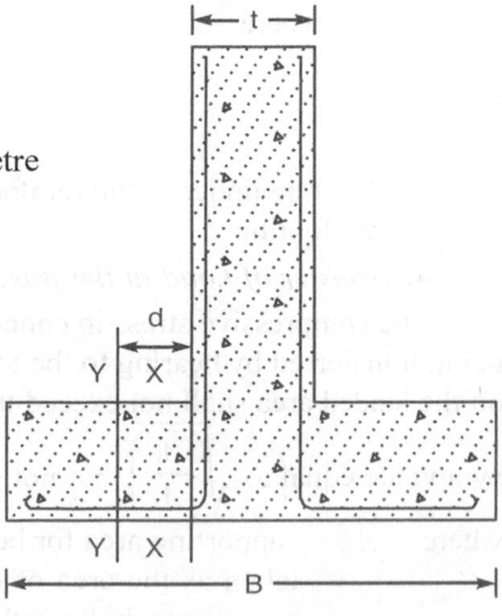

Fig. 10.16.

Let W = (Load on wall + self weight of wall) in N/metre

W_1 = Self weight of footing generally 10-20% of W (N/metre)

p_0 = bearing capacity of soil N/m²

B = Width of footing

$$B = \frac{W + W_1}{p_0} \text{ in metres}$$

q_0 = Calculated max. bearing pressure of soil

$$= \frac{W}{B \times 1} \text{ N/m}^2$$

t = thickness of wall (metres)

Critical section *XX* for max. B.M. is at the face of wall. Actual max. B.M. per metre length,

$$M = q_0 \times 1 \left(\frac{B-t}{2}\right)\left(\frac{B-t}{4}\right) \text{ N/m}$$

$$= \frac{q_0}{8}(B-t)^2 \times 1000 \text{ N/mm} \qquad (10.1)$$

Theoretical Max. B.M., $M = Qbd^2 = 0.91 \times 1000 \times d^2$ $(Q = 0.91 \text{ for M 20 conc.})$

$$\therefore \quad 0.91 \times 1000 \times d^2 = \frac{q_0}{8}(B-t)^2 \times 1000$$

$$d = \sqrt{\left(\frac{q_0 (B-t)^2}{8 \times 0.91}\right)}$$

∴ Overall depth, D = effective depth + Cover = d + cover

Area of steel, $$A_{st} = \frac{M}{\sigma_{st} \times 0.90 \times d}$$

σ_{st} = 140 N/mm² for < 20 mmϕ mild steel bars

= 130 N/mm² for > 20 mmϕ mild steel bars

σ_{st} = 275 N/mm² for Fe 500 grade bars of all diameters

= 230 N/mm² for High strength Deformed bars of all diameters of grade Fe 415

Check for Shear (Fig. 10.17)

V = Maximum shear force at critical section YY at distance d from the face of wall or $(B/2 - t/2 - d)$ metres from the edge of footing

= $q_0 \times 1 \times (B/2 - t/2 - d)$ Newtons

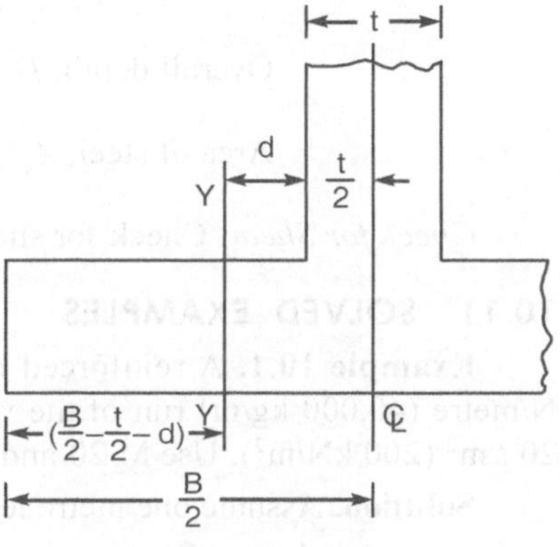

Fig. 10.17.

Nominal Shear Stress

$$\tau_v = \frac{V}{bd}$$

$$= \frac{q_0 (B/2 - t/2 - d)}{1000 \times d} \text{ N/mm}^2 \quad (10.2)$$

$k\tau_c$ = Allowable shear stress depending upon A_{st} provided (N/mm²)

If, (*i*) $\tau_v < k\tau_c$, safe in shear

(*ii*) $\tau_v > k\tau_c$, not safe in shear. Either increase the effective depth d or A_{st} to make it safe in shear.

10.10.2 Masonry Walls

Let W = (Load on wall + self weight of wall) in N/metre

W_1 = Self weight of the footing

p_0 = Bearing capacity of soil (N/m²)

B = Width of footing

$$= \frac{W + W_1}{p_0} \text{ in metres}$$

q_0 = Calculated maximum bearing pressure of soil

$$= \frac{W}{B \times 1} \text{ (N/m}^2\text{)}$$

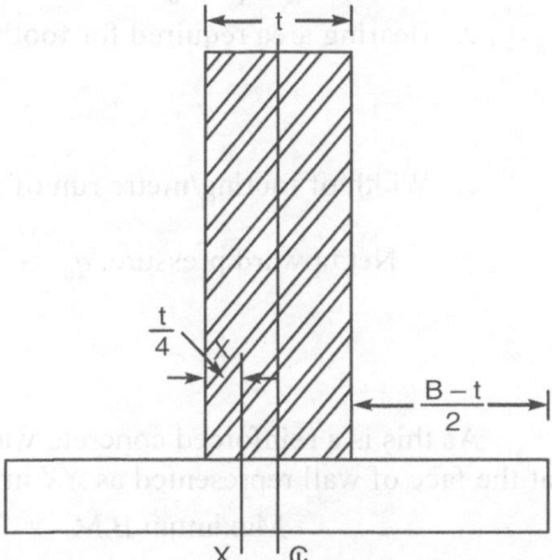

Fig. 10.18.

Critical section XX for maximum B.M. in Reinforced Concrete foundation for masonry wall is located half way between the centre line and the edge of the wall (Fig. 10.18).

Actual maximum B.M. per metre length of footing,

$$M = q_0 \times 1 \times \frac{\left(\frac{B-t}{2}+\frac{t}{4}\right)^2}{2} - \left(\frac{W}{t \times 1}\right)\left(1 \times \frac{t}{4}\right)\left(\frac{t}{8}\right)$$

$$= \frac{W}{2B \times 1}\left(\frac{2B-2t+t}{4}\right)^2 - \frac{Wt}{32}$$

$$= \frac{q_0}{32}(4B-t)(B-t) \text{ N-metre} \qquad (10.3)$$

$$\therefore \quad M = \frac{q_0}{8}\left(B-\frac{t}{4}\right)(B-t) \times 1000 \text{ N/mm}$$

$$\text{Effective depth, } d = \sqrt{\left(\frac{M}{0.91 \times 1000}\right)} \text{ mm}$$

$$\text{Overall depth, } D = d + \text{cover}$$

$$\text{Area of steel, } A_{st} = \frac{M}{\sigma_{st} \times 0.90 \times d}$$

Check for Shear. Check for shear is same as that for Reinforced Concrete wall.

10.11 SOLVED EXAMPLES

Example 10.1. A reinforced concrete wall 150 mm thick is to carry a load of 400,000 N/metre (40,000 kg/m) run of the wall. Design the footing if the bearing capacity of the soil is 20 t/m^2 (200 kN/m^2). Use M 20 and Fe 500 grade bars.

Solution. Assume one metre length of wall.

Load on wall/metre run = 400,000 N

Assume self weight of footing @ 10% of load on wall = 40,000 N

Total load = 440,000 N

Bearing capacity of soil = 200 kN/m^2 = 200,000 N/m^2

∴ Bearing area required for footing

$$= \frac{\text{Load}}{\text{Bearing capacity}} = \frac{440,000}{200,000} = 2.2 \text{ m}^2$$

∴ Width of footing/metre run of wall = 2.2 m.

$$\text{Net upward pressure, } q_0 = \frac{\text{Load on wall (exclude self weight of footing)}}{\text{Bearing area provided}}$$

$$= \frac{400,000}{2.2 \times 1} = 181818.18 \text{ N/m}^2$$

As this is a reinforced concrete wall, the critical section for maximum bending moment lies at the face of wall represented as *XX* in Fig. 10.19.

Maximum B.M. = cantilever moment about XX

$$= q_0 \times 1.025 \times \frac{1.025}{2}$$

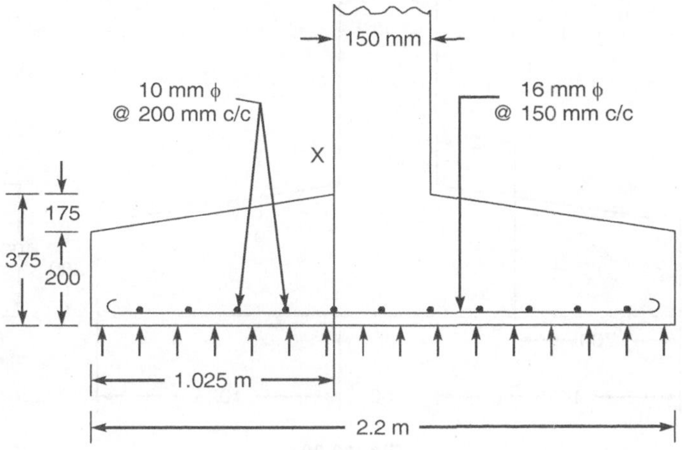

Fig. 10.19.

$$= 181818.18 \times \frac{1.025^2}{2}$$

$$= 95511.36 \text{ N/m}$$

$$d = \sqrt{\left(\frac{M}{Q \cdot b}\right)} = \sqrt{\left(\frac{95511.36 \times 1000}{0.812 \times 1000}\right)}$$

$$= 343 \text{ mm} = 34.3 \text{ cm}$$

Provide overall depth of footing, $D = 400$ mm reduced to 200 mm at the edge of footing, effective depth, $d = 325$ mm

$$A_{st} = \frac{95511.36 \times 1000}{275 \times 0.9155 \times 325} = 1167.3 \text{ mm}^2$$

Use 16 mm ϕ bars @ 150 mm *c/c*

Area provided = 1340 mm^2

Longitudinal Reinforcement

Longitudinal Reinforcement = 0.12% of the sectional area

$$= \frac{0.12}{100} \times \frac{200 + 400}{2} \times 1000$$

$$= 360 \text{ mm}^2$$

Use 10 mm ϕ @ 200 mm *c/c*. ($A = 393$ mm^2)

Check for Shear. The footing will be acting as a wide beam, therefore, critical section for maximum shear will be at a distance equal to the effective depth (325 mm) of the footing from the face of the wall or at 700 mm from the edge of the footing as shown in Fig. 10.20.

Max. shear force per metre length $= q_0 (700 \times 1000)$

$$= \frac{181818.18}{10^6} \times 700 \times 1000$$

$$= 127272.72 \text{ N}$$

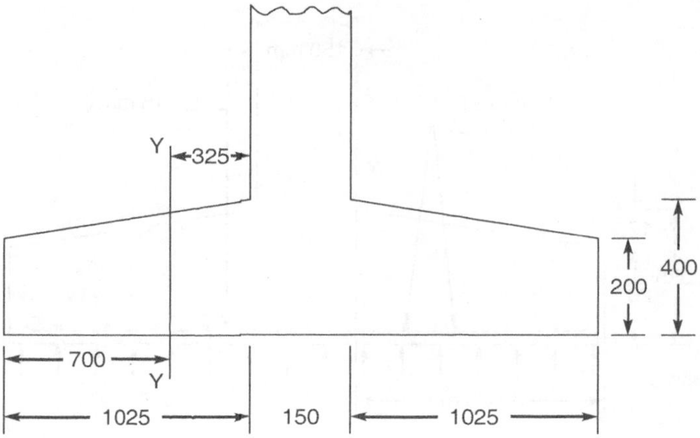

Fig. 10.20.

$$\therefore \quad \text{Nominal shear stress, } \tau_c = \frac{V}{bd}$$

$$= \frac{127272.72}{1000 \times d_{yy}}$$

$$d_{yy} = \left(200 + \frac{400 - 200}{1025} \times 700\right) - 75$$

$$= 261.58 \text{ mm}$$

$$\therefore \quad \tau_v = \frac{127272.72}{1000 \times 261.58}$$

$$= 0.4865 \text{ N/mm}^2 > 0.285 \text{ N/mm}^2 < 0 \text{ N/mm}^2 \text{ for M20}$$

Possibilities

(*i*) Increase the effective depth from 261.58 mm to 325 mm

$$\tau_v = \frac{127272.72}{1000 \times 325} = 0.3916 \text{ N/mm}^2$$

$$\frac{100 \times A_s}{b \cdot d} = \frac{1340 \times 100}{1000 \times 325} = 0.412$$

$$\tau_c = 0.285 \text{ N/mm}^2$$

Permissible shear stress = $k \times \tau_c = 1.0 \times 0.285 = 0.285$ N/mm^2, where $k = 1.0$ for slab thickness of 375 mm.

$$\tau_v > \tau_c \quad \therefore \text{Not safe in shear}$$

(*ii*) Keep the effective depth at *yy* as 325 mm and provide area of steel A_{st} for shear requirement as shown in Fig. 10.21.

Nominal shear stress, $\tau_v = 0.3916$ N/mm^2

To make $\tau_v = \tau_c$,

$$\frac{100 A_s}{b \cdot d} = 1.25 - \frac{1.25 - 1.00}{0.40 - 0.39} \times (0.42 - 0.3916) = 1.013$$

$$\therefore \quad A_s = \frac{1.013 \times 1000 \times 325}{100} = 3293 \text{ mm}^2$$

Provide 20 mm ϕ @ 90 mm c/c

$$A_s - \text{provided} = 3489 \text{ mm}^2$$

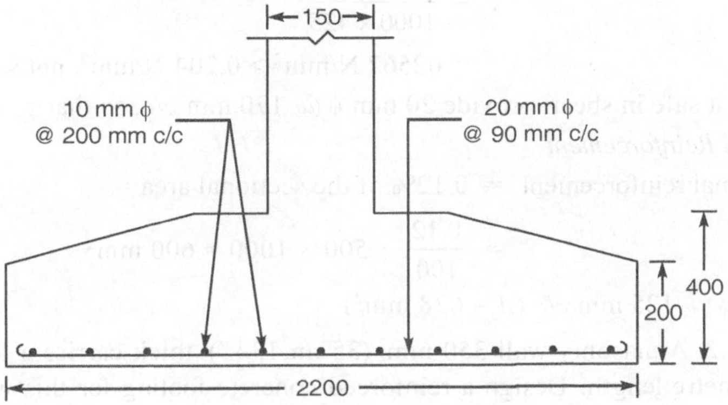

Fig. 10.21.

(*iii*) Increase the depth to 500 mm (Fig. 10.22)

$D = 500$ mm; $d = 425$ mm as shown in Fig. 10.22.

$$\therefore \quad A_{st} = \frac{9551.36 \times 1000}{275 \times 0.9155 \times 425} = 892.64 \text{ mm2}$$

Use 12 mm ϕ @ 125 mm c/c ($A = 905 \text{ mm}^2$)

$$\frac{100 \times A_s}{b \cdot d} = \frac{100 \times 905}{1000 \times 425} = 0.213$$

$$\tau_c = 0.204 \text{ N/mm}^2$$

$$\tau_v = \frac{V}{b \cdot d}$$

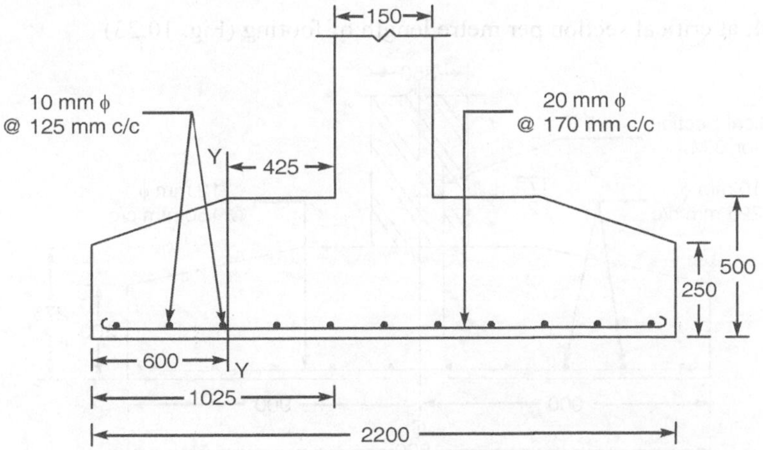

Fig. 10.22.

$$V = \frac{181818.18}{10^4} \times 600 \times 1000$$
$$= 109090.9 \text{ N}$$
$$\tau_v = \frac{109090.9}{1000 \times 425}$$
$$= 02567 \text{ N/mm}^2 > 0.204 \text{ N/mm}^2, \text{ not safe in shear.}$$

To make it safe in shear, provide 20 mm ϕ @ 170 mm *c/c*, so that $\tau_c = 0.2751$ N/mm².

Longitudinal Reinforcement

Longitudinal reinforcement = 0.12% of the sectional area

$$= \frac{0.12}{100} \times 500 \times 1000 = 600 \text{ mm}^2$$

Use 10 mm ϕ @ 125 mm *c/c* (A = 628 mm²)

Example 10.2. A masonry wall 350 mm (35 cm, $13\frac{1}{2}''$) thick carries a load of 180,000 N (18,000 kg) per metre length. Design a reinforced concrete footing for this wall if the bearing capacity of the soil over which it rests is 110 kN/m² (0.11 N/mm²). Use M 20 and Fe 415 bars.

Solution. Assume 1 m length of wall

Load on 1 m length of wall = 180,000 N

Assume self wt. of footing = 10% of load on wall

= 18,000 N

Total load = 198,000 N

$$\text{Bearing area required for footing} = \frac{198.000}{0.11} = 1800000 \text{ mm}^2$$

$$\text{Width of footing} = \frac{1,800,000}{1000} = 1800 \text{ mm per metre length}$$

$$\text{Net upward pressure, } q_0 = \frac{\text{Load on wall}}{\text{Area provided}}$$

$$= \frac{180,000}{1800 \times 1000} = 0.1 \text{ N/mm}^2$$

Max. B.M. at critical section per metre length of footing (Fig. 10.23)

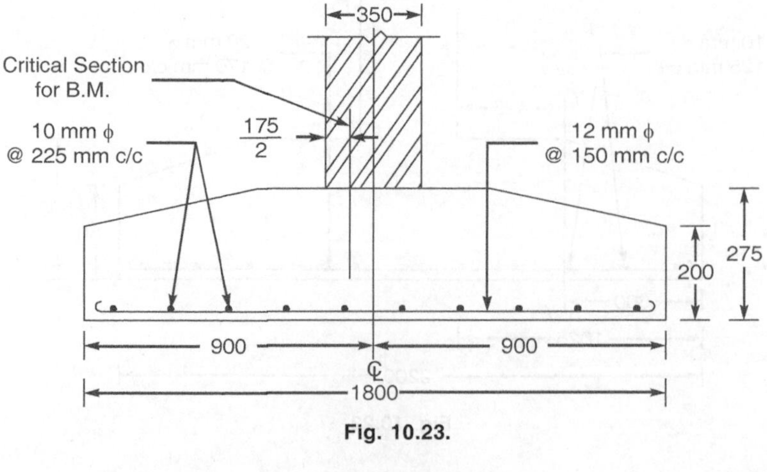

Fig. 10.23.

$$= (0.1 \times 812.5 \times 1000)\,\frac{812.5}{2} - \frac{180,000}{1000 \times 350} \times \left(1000 \times \frac{175}{2}\right) \times \left(\frac{175}{2} \times \frac{1}{2}\right)$$

$= 31039062$ N/mm

$$d = \sqrt{\left(\frac{31039062}{0.91 \times 1000}\right)} = 184.7 \text{ mm}$$

Provide $D = 275$ mm; $d = 200$ mm

$$A_{st} = \frac{31039062}{230 \times 0.90 \times 200} = 749.74 \text{ mm}^2$$

Provide 12 mm ϕ @ 150 mm c/c

A_{st} provided $= 754$ mm^2

Distribution steel $=$ 0.12% of sectional area

$$= \frac{0.12}{100} \times 1000 \times 275 = 330 \text{ mm}^2$$

Provide 10 mm ϕ bars at 225 mm c/c (Area provided = 349 mm^2)

Check for Shear

The footing will be acting as a wide beam, therefore, critical section *xx* for maximum shear will be at a distance equal to the effective depth (200 mm) of the footing from the face of the wall or 525 mm from the edge of the footing as shown in Fig. 10.24.

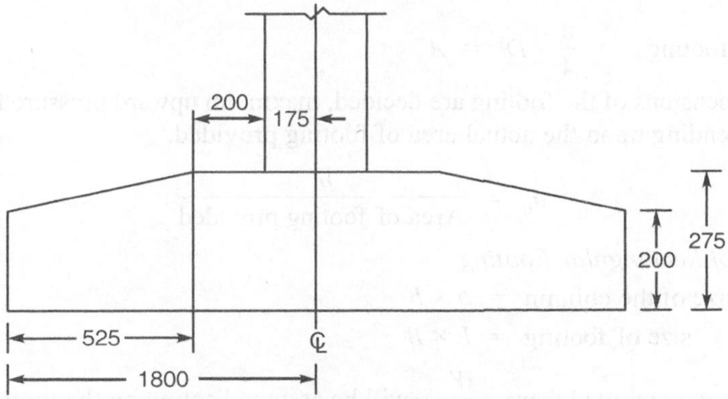

Fig. 10.24.

Max. shear force per metre,

length $= q_0 \times (525 \times 1000)$

$= 0.1 \times 525 \times 1000 = 52,500$ N

$$\tau_v = \frac{V}{b \cdot d}$$

$$= \frac{52500}{1000 \times 200} = 0.2625 \text{ N/mm}^2$$

$$\frac{100\,A_s}{b \cdot d} = \frac{100 \times 754}{1000 \times 200} = 0.377$$

$\therefore \quad \tau_c = 0.26$ N/mm^2

Permissible shear stress $= k \cdot \tau_c = 1.05 \times 0.26 = 0.273$ N/mm^2

where $k = 1.05$ for 275 mm thickness of slab.

$\therefore$ $\tau_v <$ Permissible, safe in shear.

The reinforcement details are shown in Fig. 10.23.

10.12 COLUMN FOOTINGS

Individual footing provided to distribute the load of a column to the soil, is called 'Isolated Footing'. This footing can be :

(*i*) Square for a square or circular column.

(*ii*) Rectangular for a rectangular column.

(*iii*) Circular for circular columns.

Let W = load on the column footing from column,

W_1 = Self weight of footing, generally taken as 10 to 20% of W,

p_0 = Bearing capacity of soil.

Area of footing rquired can be computed from the relation

$$A = \frac{W + W_1}{p_0}$$

Provide this much area in a form depending upon the type of the column.

(*i*) Square footing $B = \sqrt{A}$

(*ii*) Rectangular footing $B \times L = A$

(*iii*) Circular footing $\frac{\pi}{4} \cdot D^2 = A$

Once the dimensions of the footing are decided, maximum upward pressure from soil 'q_0' can be calculated depending upon the actual area of footing provided.

$$q_0 = \frac{W}{\text{Area of footing provided}}$$

(A) *Design of Rectangular Footing*

Let size of the column $= a \times b$

size of footing $= L \times B$

The pressure q_0 computed from $\frac{W}{L \times B}$ will be assumed acting on the footing upwards from soil on trapezoids shown in Fig. 10.25. The critical sections for bending moment shall be located along each face of the column marked $X - X$ and $Y - Y$. The bending moments at the critical sections are computed as follows :

$$M_{x-x} = q_0 \left(\frac{a+L}{2} \cdot \frac{B-b}{2}\right) \cdot \left(\frac{2L+a}{L+a} \cdot \frac{1}{3} \cdot \frac{B-b}{2}\right)$$

$$= \frac{q_0}{24} (2L + a)(B - b)^2 \qquad (10.4)$$

$$M_{y-y} = q_0 \left(\frac{B+b}{2} \cdot \frac{L-a}{2}\right) \cdot \left(\frac{2B+b}{B+b} \cdot \frac{1}{3} \cdot \frac{L-a}{2}\right)$$

$$= \frac{q_0}{24} (2B + b)(L - a)^2 \qquad (10.5)$$

The effective depth for the footing shall be larger of the two calculated as follows :

$$d_{x-x} = \sqrt{\left(\frac{M_{x-x}}{Q \cdot a}\right)}$$

$$d_{y-y} = \sqrt{\left(\frac{M_{y-y}}{Q \cdot b}\right)}$$

As an alternative to the above, a simple but approximate method of calculation of M_{x-x} and M_{y-y} is explained in Fig. 10.26.

By this approximation method, higher values of M_{x-x} and M_{y-y} are computed because some extra portion of the area of footing is also considered (shown double hatched in Fig. 10.26).

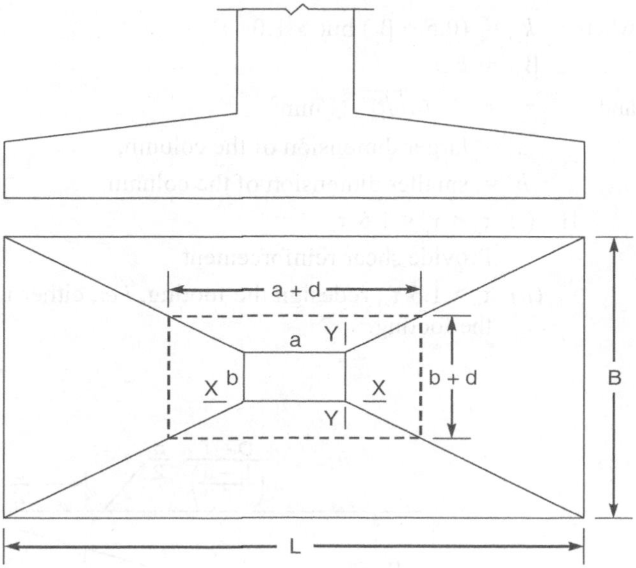

Fig. 10.25.

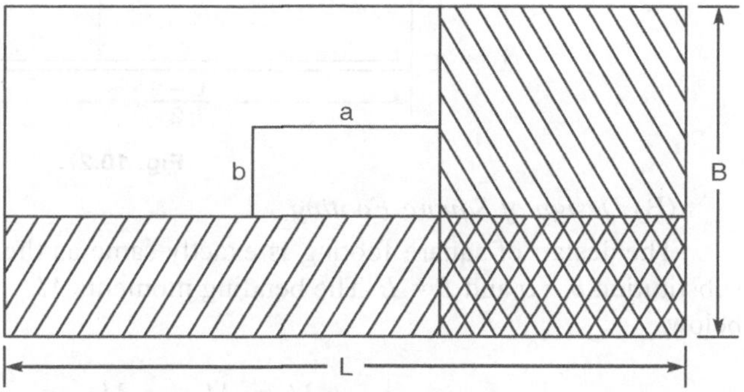

Fig. 10.26.

$$M_{x-x} = q_0 \left(\frac{B-b}{2}\right) \cdot L \cdot \frac{1}{2} \cdot \frac{B-b}{2}$$

$$= q_0 \cdot \frac{L}{8} (B-b)^2 \qquad (10.6)$$

$$M_{y-y} = q_0 \left(B \times \frac{L-a}{2}\right) \cdot \left(\frac{1}{2} \cdot \frac{L-a}{2}\right)$$

$$= q_0 \cdot \frac{B}{8} (L-a)^2 \qquad (10.7)$$

The effective depth of the footing can be calculated by the expressions d_{x-x} and d_{y-y} as explained above and the area of steel can be provided in the two directions according to the values of M_{x-x} and M_{y-y} respectively.

Check for Shear

The critical section for shear shall be located at a distance $d/2$ from the periphery of the column as shown in Fig. 10.25.

$$\text{Shear force } V = q_0 \cdot [L \cdot B - (a+d)(b+d)]$$

$$\tau_v = \frac{V}{2\,[(a+d)+(b+d)] \cdot d_{cc}} \qquad (10.8)$$

where d_{cc} = depth at critical section (Fig. 10.27)

$$= d - \frac{D-T}{\frac{1}{2}(L-a)} \cdot \tfrac{1}{2} d$$

τ_v should be less than $k_s \cdot \tau_c$

where $k_s = (0.5 + \beta_c)$ but $\ngtr 1.0$

$\beta_c = b/a$

and $\tau_c = 0.16\sqrt{f_{ck}}$ N/mm^2

a = larger dimension of the column,

b = smaller dimension of the column.

If (*i*) $\tau_c < \tau_v < 1.5\ \tau_c$

Provide shear reinforcement

(*ii*) $\tau_v > 1.5\ \tau_c$, redesign the footing, *i.e.*, either increase the depth or change the steel in the footing.

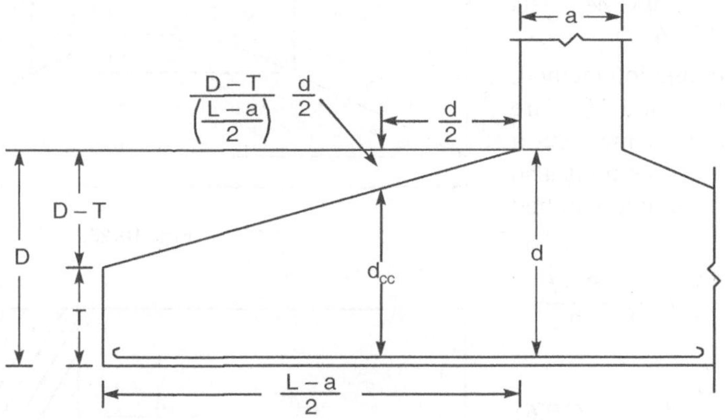

Fig. 10.27.

(B) *Design of Square Footing*

The design of square footing is exactly same as the design of a rectangular footing with a substitution $b = a$ and $B = L$. The bending moments M_{x-x} and M_{y-y} shall be equal and are given below.

$$M = M_{x-x} = M_{y-y} = \frac{q_0}{24}\left[(2L + a)(L - a)^2\right] \qquad (10.9)$$

Square footing can also be provided under circular columns treating it as a square column of side of inscribed square.

(C) *Design of Circular Footing*

Let r = radius of column

R = radius of the circular footing

Maximum pressure from soil on footing $q_0 = \dfrac{W}{\pi \cdot R^2}$

The critical section for maximum bending moment is located at the face of the column at $\dfrac{r}{\sqrt{2}}$ from the centre of the column as shown in Fig. 10.28.

Maximum bending moment in the case of circular footing with circular column over it, can be calculated with the help of the Fig. 10.29.

Considering a quadrant of the column and the footing.

$$OG_1 = 0.6r \quad a_1 = \text{Area of quadrant} = \frac{\pi}{4} \cdot r^2$$

$$OG_2 = 0.6R \quad a_2 = \frac{\pi}{4} \cdot R^2$$

X' = distance of C.G. of the shaded area from the centre of the column

$$= \frac{\frac{\pi}{4} \cdot R^2 \times 0.6R - \frac{\pi}{4} \cdot r^2 \times 0.6r}{\frac{\pi}{4} \cdot R^2 - \frac{\pi}{4} \cdot r^2}$$

$$= \frac{0.6\,(R^2 + R \cdot r + r^2)}{R + r}$$

$\bar{x}$ = distance of C.G. (G_3) from face of the column

$$= X' - \frac{r}{\sqrt{2}}$$

$$= 0.107\,\frac{(5.6R^2 - R \cdot r - r^2)}{R + r} \qquad (10.10)$$

Approximate value of $\bar{x}$ can be evaluated from the periphery of the column instead of the face of the column.

$$\bar{x} = 0.6\,\frac{(R^2 + R \cdot r + r^2)}{R + r} - r$$

$$= \frac{0.2}{R + r}\,(3R^2 - 2 \cdot R \cdot r - 2 \cdot r^2) \qquad (10.11)$$

Upward soil pressure on the shaded area $= \frac{\pi}{4}\,(R^2 - r^2) \cdot q_0$

B.M. at the face of the column M

$$= \frac{\pi}{4}\,(R^2 - r^2) \cdot q_0 \times \frac{0.107}{R + r}\,(5.6R^2 - R \cdot r - r^2)$$

$$= 0.084 q_0\,(R - r)\,(5.6R^2 - R \cdot r - r^2) \qquad (10.12)$$

Breadth at the face of the column $= \frac{D}{\sqrt{2}} = r \cdot \sqrt{2}$.

Effective depth can now be calculated as $d = \sqrt{\left(\frac{M}{Qr\sqrt{2}}\right)}$

Area of steel can be calculated as $A_{st} = \frac{M}{\sigma_{st} \cdot j \cdot d}$.

Check for Shear

The critical section for shear is at a distance $d/2$ from the periphery of the column as shown dotted in Fig. 10.28.

$$\text{Shear force } V = q_0 \cdot \pi\,[R^2 - (r + \tfrac{1}{2}d)^2] \qquad (10.13)$$

The periphery of the critical section

$$b_0 = 2 \cdot \pi\,(r + \tfrac{1}{2} \cdot d)$$

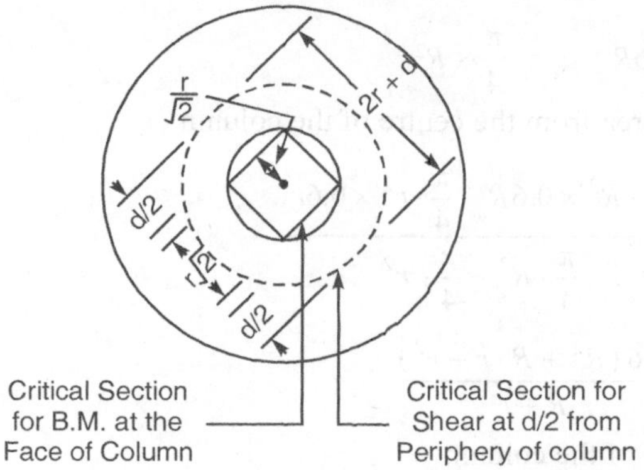

Fig. 10.28.

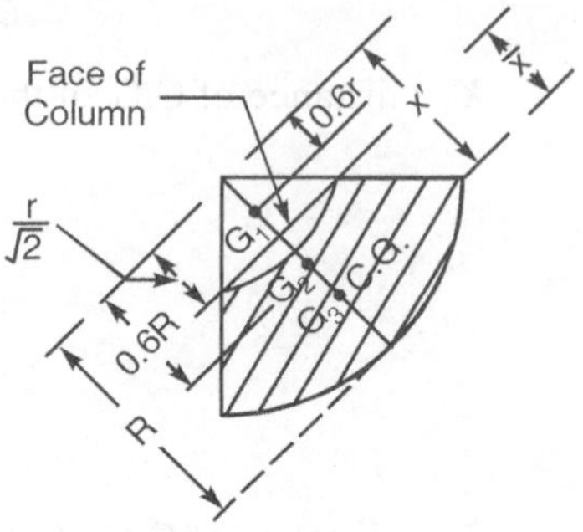

Fig. 10.29.

The nominal shear stress $\tau_c = \dfrac{V}{b_0 \cdot d}$ (10.14)

Check for shear stresses shall be applied in a similar manner as for rectangular or square footings.

10.13 SOLVED EXAMPLES

Example 10.3. Design the footing for a reinforced concrete column 225 × 450 mm carrying an axial load of 1075 kN. The bearing capacity of the soil is 100 kN/m². Use M 20 concrete and Fe 500 grade steel as reinforcement.

Solution. Load on the column = 1075000 N

Assume self weight of footing @ 15% = 0.15 × 1075000 = 161250 N

Total Load = 1,236,250 N

$$\text{Area of footing required} = \frac{\text{Total load}}{\text{Bearing capacity of soil}}$$

$$= \frac{123625}{100} = 12.36 \text{ m}^2$$

Provide 3 m × 4.5 m footing

$$\text{Bearing pressure } q_0 = \frac{1075000}{3 \times 4.5} = 79630 \text{ N/m}^2$$

$$M_{x-x} = \frac{q_0}{24}\,[(2L + a) \cdot (B - b)^2]$$

$$= \frac{79630}{24}\,[(2 \times 4.5 + 0.45) \cdot (3 - 0.225)^2]$$

$$= 241{,}447 \text{ N/m}$$

$$M_{y-y} = \frac{q_0}{24}\,[(2B + b) \cdot (L - a)^2]$$

$$= \frac{79630}{24} . [(2 \times 3 + 0.225) \cdot (4.5 - 0.45)^2]$$

$$= 338{,}777 \text{ N/m}$$

The effective depths in two directions can be computed from :

$$d_{x-x} = \sqrt{\left(\frac{241447 \times 10^3}{0.812 \times 450}\right)} = 812.9 \text{ mm}$$

$$d_{y-y} = \sqrt{\left(\frac{338777 \times 10^3}{0.812 \times 225}\right)} = 1361.72 \text{ mm}$$

$$\therefore \quad D = 1400,\ d = 1400 - 75 - 1325 \text{ mm}$$

$$A_{st\,(x-x)} = \frac{241447 \times 1000}{275 \times 0.9155 \times 1325} = 723.8 \text{ mm}^2$$

$$A_{st\,(y-y)} = \frac{338777 \times 1000}{275 \times 0.9155 \times 1325} = 1015.56 \text{ mm}^2$$

Provide 10 mm ϕ 15 Nos. bars in the long direction (A = 11785 mm^2).

Reinforcement in the Short Span

$$\beta = b/a = 4.5/3 = 1.5$$

The reinforcement in central band which is equal to short span

$$= \frac{2}{\beta + 1} \cdot A_{st\,(x-x)}$$

$$= \frac{2}{1.5 + 1} \times 723.8 = 579.04 \text{ mm}^2$$

$$\text{Balance steel} = 723.8 - 579.04 = 144.76 \text{ mm}^2$$

Provide 8 mm ϕ 17 Nos. (A = 854 mm^2) in the central band of 3 m distance and the balance of steel, *i.e.*, 144.76 mm^2 required shall be provided in the form of 8 mm ϕ 4 Nos. on each side of the central band.

Check for Shear

$$V = q_0\,[L \times B - (a + d)\,(b + d)]$$

$$= 79630\,[3 \times 4.5 - (0.45 + 1.325)\,(0.225 + 1.325)]$$

$$= 8{,}55{,}923 \text{ N}$$

$$\tau_v = \frac{V}{2\,[(a+d)+(b+d)] \cdot d_{cc}}$$

$$= \frac{855923}{2\,[(450+1325)+(225+1325)] \times 965}$$

$$= 0.136 \text{ N/mm}^2$$

$$\text{Permissible shear stress } \tau_c = k_s \times \tau_c$$

$$= \left(0.5 + \frac{225}{450}\right) \cdot 0.16\sqrt{20}$$

$$= 0.7155 \text{ N/mm}^2$$

$\therefore$ No shear reinforcement is required.

The reinforcement details in the footing are shown in Fig. 10.30.

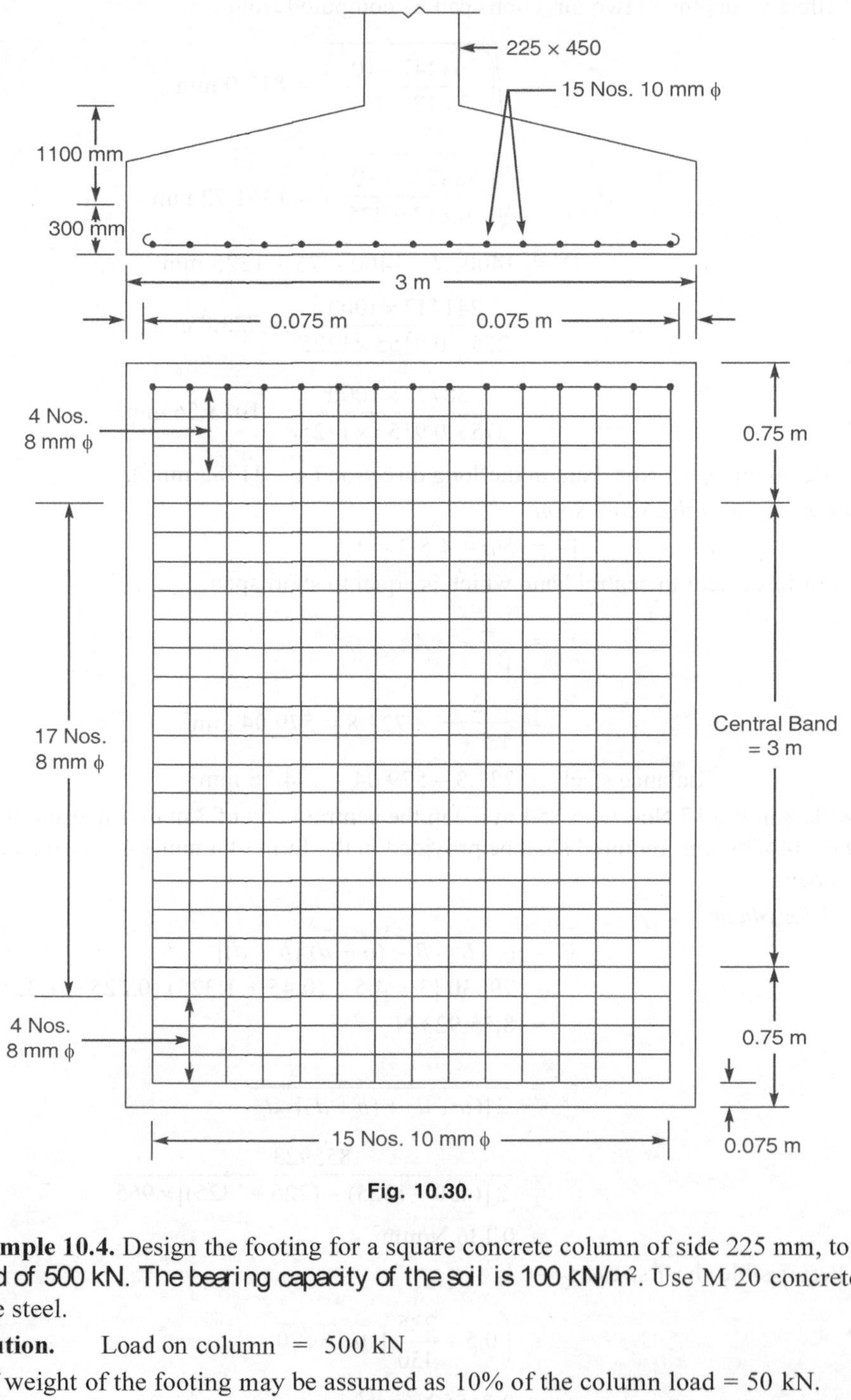

Fig. 10.30.

Example 10.4. Design the footing for a square concrete column of side 225 mm, to carry an axial load of 500 kN. The bearing capacity of the soil is 100 kN/m². Use M 20 concrete and Fe 500 grade steel.

Solution. Load on column = 500 kN

Self weight of the footing may be assumed as 10% of the column load = 50 kN.

$$\text{Total load} = 550 \text{ kN}$$

$$\text{Area of footing required} = \frac{550}{100} = 5.5 \text{ m}^2$$

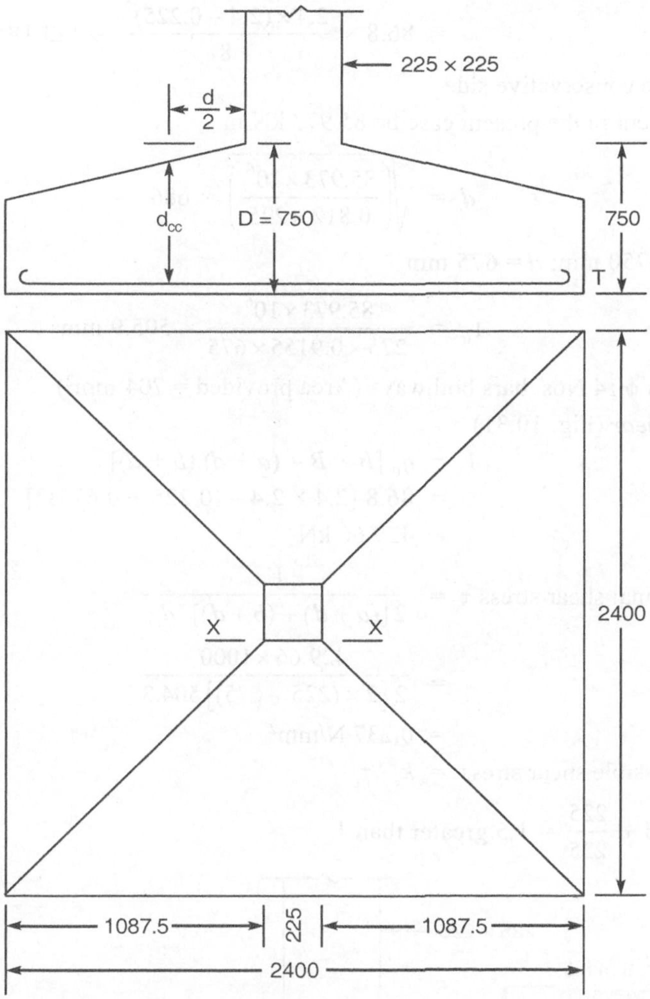

Fig. 10.31.

Side of square footing required = $\sqrt{(5.5)}$ = 2.345 m

Provide square footing of 2.4 × 2.4 m

The actual bearing pressure = $\dfrac{500}{2.4 \times 2.4}$ = 86.8 kN/m²

Refer Fig. 10.31,

$$M_{x-x} = 86.8 \times \left(\frac{2.4 + 0.225}{2} \times 1.0875\right) \times \left(\frac{0.225 + 2 \times 2.4}{0.225 + 2.4} \times \frac{1.0875}{3}\right)$$

$$= 85.973 \text{ kN/m}$$

It may be noted that with approximate method, the maximum bending moment

$$M_x = q_0 \times \frac{B\,(B - b)^2}{8}$$

$$= 86.8 \times \frac{2.4 \times (2.4 - 0.225)^2}{8} = 123.185 \text{ kN/m}$$

This is on the conservative side.

Design moment in the present case be 85.973 kN/m

$$d = \sqrt{\left(\frac{85.973 \times 10^6}{0.812 \times 225}\right)} = 686$$

Provide $D = 750$ mm; $d = 675$ mm

$$A_{st} = \frac{85.973 \times 10^6}{275 \times 0.9155 \times 675} = 505.9 \text{ mm}^2$$

Provie 8 mm ϕ 14 Nos. bars bothways (Area provided = 704 mm²)

Check for Shear (Fig. 10.31)

$$V = q_0 [B \times B - (a + d)(b + d)]$$
$$= 86.8 [2.4 \times 2.4 - (0.225 + 0.675)2]$$
$$= 429.66 \text{ kN}$$

$$\text{The nominal shear stress } \tau_v = \frac{V}{2[(a+d)+(b+d)] \cdot d_{cc}}$$
$$= \frac{429.66 \times 1000}{2[2 \times (225 + 675)] 504.3}$$
$$= 0.237 \text{ N/mm}^2$$

$$\text{Permissible shear stress} = k_s \cdot \tau_c$$

where $k_s = 0.5 + \frac{225}{225} = 1.5$ greater than 1

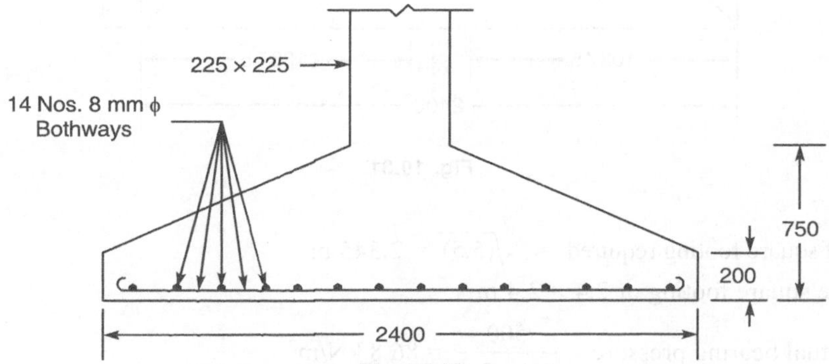

Fig. 10.32.

$\therefore$ $k_s = 1.0$

$\therefore$ Permissible shear stress $= 1 \times 0.16\sqrt{20}$

$= 0.7155 \text{ N/mm}^2$

Hence safe in shear.

The reinforcement details are shown in Fig. 10.32.

Example 10.5. Design a square footing for a circular R.C.C. column 600 mm diameter to carry an axial load of 1400 kN (140 tonnes) including the self weight of the column. The bearing capacity of the soil is 200 kN/m². Use 25 concrete, and Fe 500 grade bars.

Solution. $Q = 0.986, j = 0.9155$

$$\text{Load on column} = 1400 \text{ kN}$$

$$\text{Weight of footing @ 10\% of column load} = 140 \text{ kN}$$

$$\text{Total load on soil} = 1540 \text{ kN}$$

$$\text{Area of footing required} = \frac{1540}{200} = 7.7 \text{ m}^2$$

Side of square footing required = 2.775 m

Provide a R.C.C. footing of 2.8 m × 2.8 m

$$\text{Bearing pressure } q_0 = \frac{1400 \times 1000}{2.8 \times 2.8} = 178600 \text{ N/m}^2$$

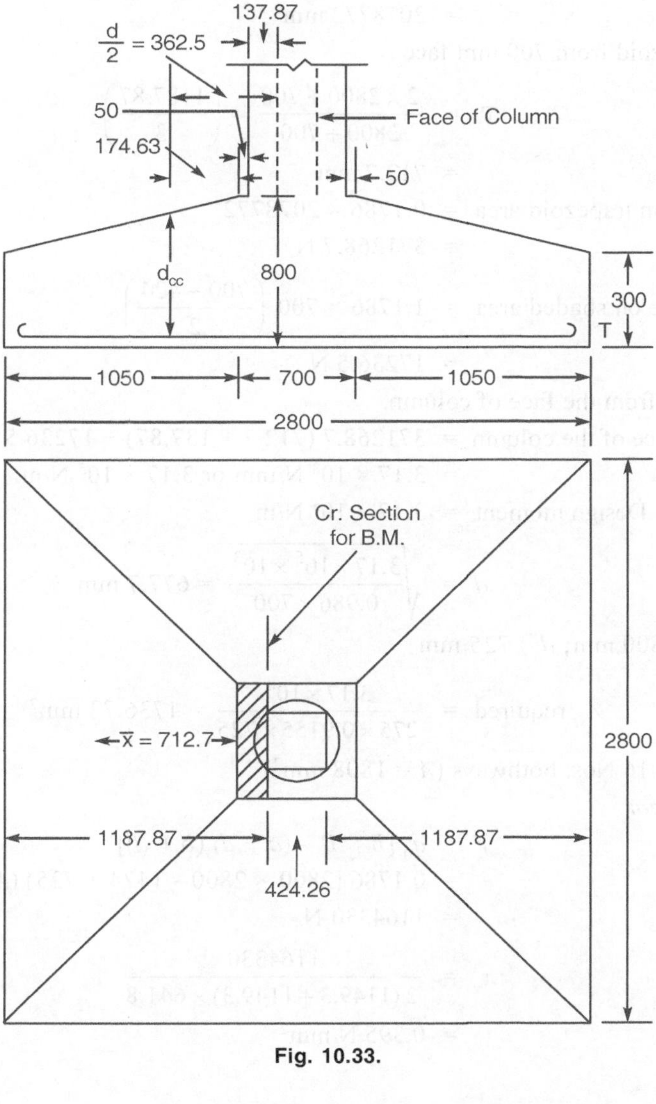

Fig. 10.33.

Side of the inscribed square in the circular column

$$= \frac{0.6}{\sqrt{2}} = 0.424 \text{ m}$$

$$\text{Maximum B.M.} = q_0 \cdot B \, \frac{(B-a)^2}{8}$$

Note: Max. B.M. is calculated by considering rectangular area

$$= \frac{178600 \times 2.8\,(2.8 - 0.424)^2}{8}$$

$$= 3.528 \times 10^5 \text{ N/m}$$

Alternatively (Fig. 10.33)

$$\text{Area of Trapezoid} = \frac{700 + 2800}{2} \times 1187.87$$

$$= 2078772 \text{ mm}^2$$

C.G. of trapezoid from 700 mm face

$$\bar{x} = \frac{2 \times 2800 + 700}{2800 + 700} = \left(\frac{1187.87}{3}\right)$$

$$= 712.7 \text{ mm}$$

$$\text{Force on trapezoid area} = 0.1786 \times 2078772$$

$$= 371268.7 \text{ N}$$

$$\text{Force on shaded area} = 1.1786 \times 700 \left(\frac{700 - 424}{2}\right)$$

$$= 17236.5 \text{ N}$$

acting at 68.9 mm from the face of column.

$$\text{Moment at face of the column} = 371268.7\,(712.7 + 137.87) + 17236.5 \times 68.935$$

$$= 3.17 \times 10^8 \text{ N/mm or } 3.17 \times 10^5 \text{ N/m}$$

$$\therefore \quad \text{Design moment} = 3.17 \times 10^5 \text{ N/m}$$

$$d = \sqrt{\frac{3.17 \times 10^5 \times 10^3}{0.986 \times 700}} = 677.7 \text{ mm}$$

Provide $D = 800$ mm; $d = 725$ mm

$$A_{st} \text{ required} = \frac{3.17 \times 10^8}{275 \times 0.9155 \times 725} = 1736.72 \text{ mm}^2$$

Use 12 mm ϕ 16 Nos. bothways ($A = 1808$ mm^2)

Check for Shear

$$V = q_0\,[B \times B - (a + d)\,(b + d)]$$

$$= 0.1786\,[2800 \times 2800 - (424 + 725)\,(424 + 725)]$$

$$= 1164330 \text{ N}$$

$$\tau_v = \frac{1164330}{2\,(1149.3 + 1149.3) \times 641.8}$$

$$= 0.395 \text{ N/mm}^2$$

$$\text{Permissible shear stress} = k_s \cdot \tau_c \qquad \text{where } k_s = 0.5 + \frac{700}{700}$$

$$= 0.16\sqrt{25} \qquad \text{Take } k_s = 1$$

$$= 0.80 \text{ N/mm}^2$$

Safe in shear.

The reinforcement details are shown in Fig. 10.34.

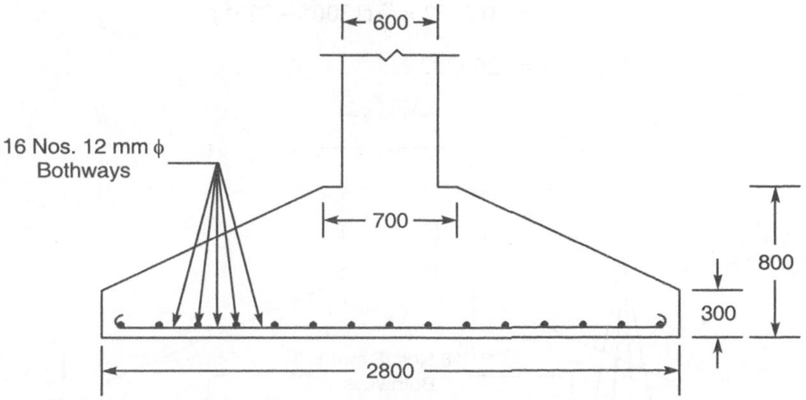

Fig. 10.34.

Example 10.6. Design a circular footing for a circular column of 450 mm diameter carrying an axial load of 1100 kN including the self weight of the column. The bearing capacity of the soil is 250 kN/m² (25 *t*/m²). Use M 25 concrete and Fe 500 grade steel.

Solution. $Q = 0.986$, $j = 0.9155$

$$\text{Load on the column including self weight } W = 1100 \text{ kN}$$

$$\text{Let self weight of footing be 15\% of column load} = 165 \text{ kN}$$

$$\text{Total load on soil} = 1265 \text{ kN}$$

$$\text{Area of footing required} = \frac{1265}{250}$$

$$= 5.06 \text{ m}^2$$

Provide a circular footing of radius $R = 1.3$ m, thus providing area of 5.3 m².

$$q_c = \frac{1100}{5.3}$$

$$= 207.54 \text{ kN/m}^2$$

$$\text{Radius of the column } r = 225 \text{ mm}$$

$$\text{Side of the inscribed square} = 450\sqrt{2} = 318 \text{ mm}$$

Critical section for maximum B.M. shall be at edge of side face of the inscribed square at $r/\sqrt{2} = 159$ mm from centre of the column as shown in Fig. 10.28.

$$\bar{x} = \text{Distance of C.G. of the shaded area from face of the column}$$

$$= \frac{0.107}{R + r}(5.6R^2 - R \cdot r - r^2) \tag{10.10}$$

$$= 0.107 \; \frac{(5.6 \times 1300^2 - 1300 \times 225 - 225^2)}{1300 + 225}$$

$$= 640 \text{ mm}$$

Upward soil pressure on shaded area

$$= q_0 \cdot \frac{\pi}{4}(R^2 - r^2)$$

$$= 0.2072 \times \frac{\pi}{4}(1300^2 - 225^2)$$

$$= 266782 \text{ N}$$

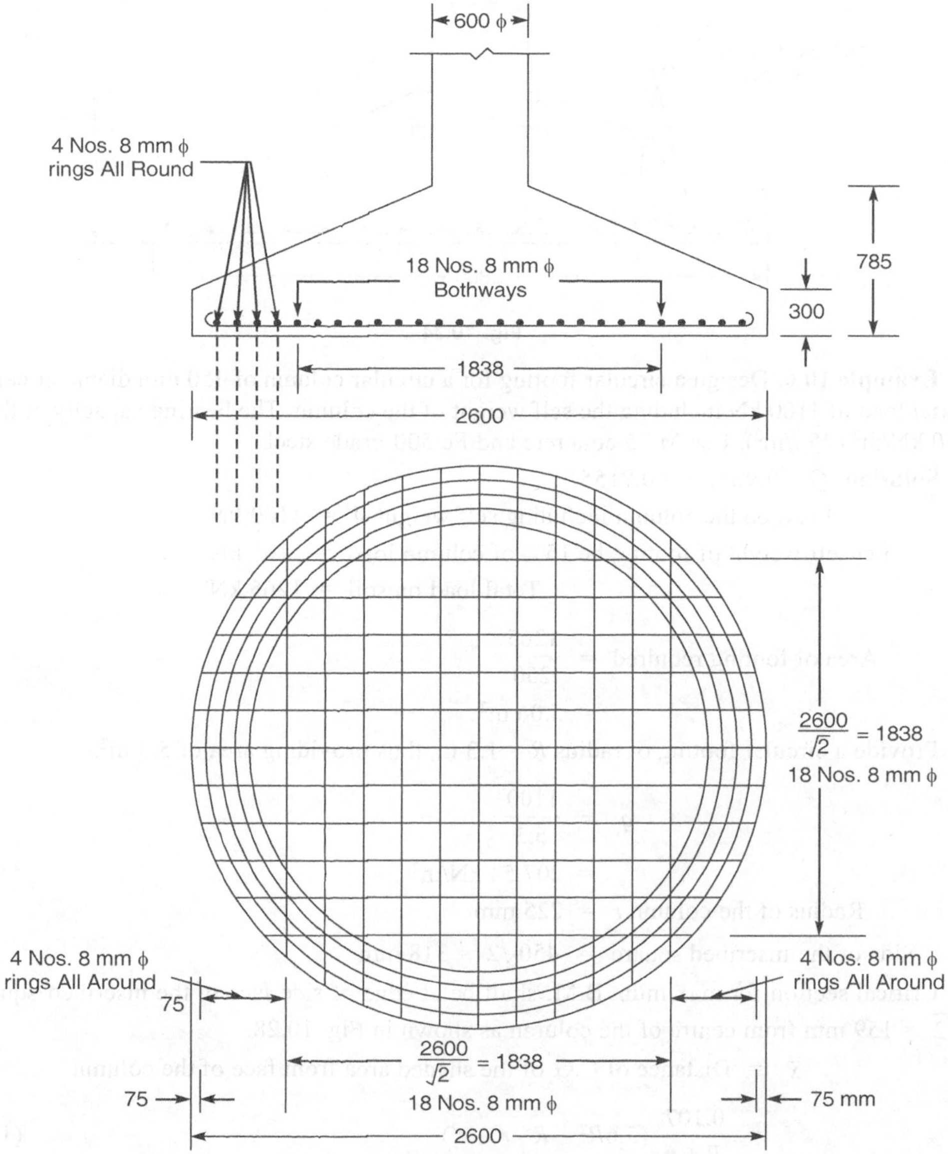

Fig. 10.35.

Maximum B.M. at the face of the column

$$= 266782 \times 640$$
$$= 1.7073 \times 10^8 \text{ N/mm}$$
$$= 170.73 \text{ kN/m}$$

Breadth of the face of the column = $D/\sqrt{2} = 318.2$ mm

$$d = \sqrt{\left(\frac{1.7073 \times 10^8}{0.986 \times 318.2}\right)} = 737.68 \text{ mm}$$

Adopt $D = 875$ mm $d = 800$ mm

$$A_{st} \text{ required} = \frac{1.7073 \times 10^8}{275 \times 0.9155 \times 800} = 847.67 \text{ mm}^2$$

Provide 18 nos. bars of 8 mm φ in two directions (18 nos. each direction) at right angles to each other. These bars are placed in the inscribed square as shown in Fig. 10.35. In addition, a 8 mm φ bar is provided circumferentially at a minimum distance of 75 mm from edge. Also provide 4 nos. 8 mm φ on each side on outer side of the circumscribed square.

Check for Shear

Shear force at the critical section at a distance $d/2$ from the periphery of the column

$$= q_0 \cdot [R^2 - (r + \tfrac{1}{2}d)^2] \quad (10.13)$$
$$= 0.2072 \times [1300^2 - (225 + 400)^2]$$
$$= 84581.5 \text{ N}$$

b_0 = periphery of critical section

$$= 2\pi\,(r + \tfrac{1}{2}d) = 2\pi\,(225 + 400) = 3927 \text{ mm}$$

$$\text{Nominal shear stress } \tau_v = \frac{v}{b_0 \cdot d}$$

$$= \frac{84581.5}{3927 \times 800}$$

$$= 0.2692 \text{ N/mm}^2$$

Perimissible shear stress $= 0.16\sqrt{(f_{ck})}$

$= 0.8$ N/mm² for M25

∴ Safe in shear.

Hence no shear reinforcement is required.

10.14 INDIVIDUAL FOOTINGS IN PRACTICE

It is important to note that top of the footing is normally provided below the natural ground level a distance which normally varies from 150 mm to 450 mm depending upon the type of the structure above and the soil conditions. It is also customary to provide lean mix base (called Gatka) of thickness varying from 75 mm to 300 mm below the R.C.C. footing. The lean mix base generally has some projections on all sides beyond the R.C.C. footing ranging from zero to 150 mm. In case R.C.C. footing base area provided is equal to or greater than area required to transmit load to soil, projections are not necessary. However, if lean mix base area is provided equal to that required and the R.C.C. footing is kept a little less than lean mix base, the pressure on R.C.C. base may exceed the bearing capacity of the soil. This is acceptable under such circumstances. The lean mix concrete base is generally of grade 1 : 8 : 16 (Cement : Sand : Gravel) for residential and office buildings, 1 : 6 : 12, 1 : 5 : 10 or 1 : 4 : 8 mix for Industrial buildings, Water reservoirs. Richer lean mix is used where the structure is likely to be subjected to vibrations. A column with footing in practice is normally represented in the form shown in Fig. 10.36.

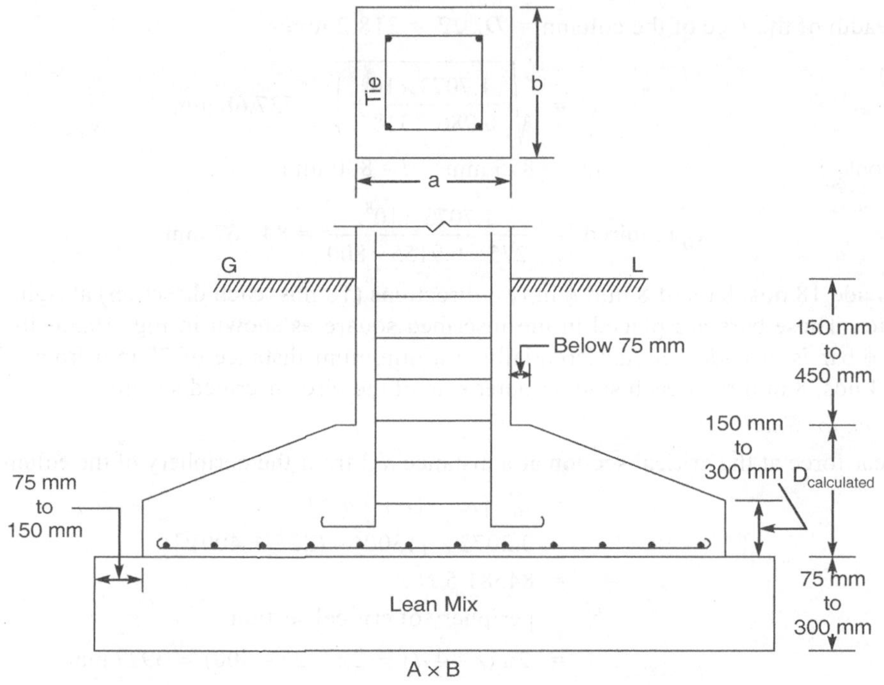

Fig. 10.36.

10.15 STRAP FOOTING

Strap footing are quite often provided where an independent footing cannot be extended into the adjoining land. Strap beams are provided to connect two individual column footings as shown in Fig. 10.37. It is assumed that the load from columns is transferred through the footings only and the strap beam does not transfer any load to the soil directly. The areas of the independent footings are so chosen that the pressure on the footings is uniform and the resultant soil pressure on the footings areas coincide with the C.G. of columns load. This type of footing is economical than combined footings where distance between the columns is large.

Assumption :

1. The strap beam is rigid.
2. The strap beam transfers the load from columns to the footings and not to the soil directly.
3. The loads are transferred to the soil through the independent footings only.

Size of Straps Beam. The width of strap beam is generally equal to or greater than the column side at right angles to strap. The depth of the strap is decided (*i*) on bending moment/S.F. considerations as the case may be; (*ii*) such that it does not have direct bearing on the soil because of assumption No. 2.

Design Steps. The complete design of strap footing involves the following steps.

1. The area A_1, A_2 of footing under each column C_1, C_2 is calculated.

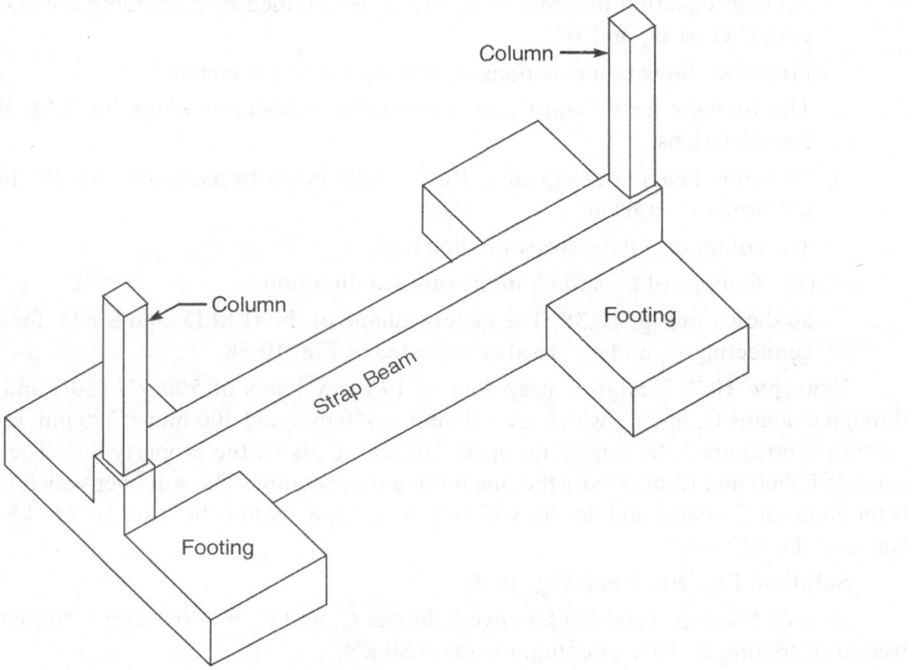

Fig. 10.37.

$$\text{Area required} = \frac{W_1 + W_2 + W_3}{p_0}$$

where W_1, W_2 = loads on columns C_1 and C_2 respectively.

W_3 = self weight of footing.

p_0 = bearing capacity of soil

$$A_1 + A_2 = \frac{W_1 + W_2 + W_3}{p_0} \quad (10.15)$$

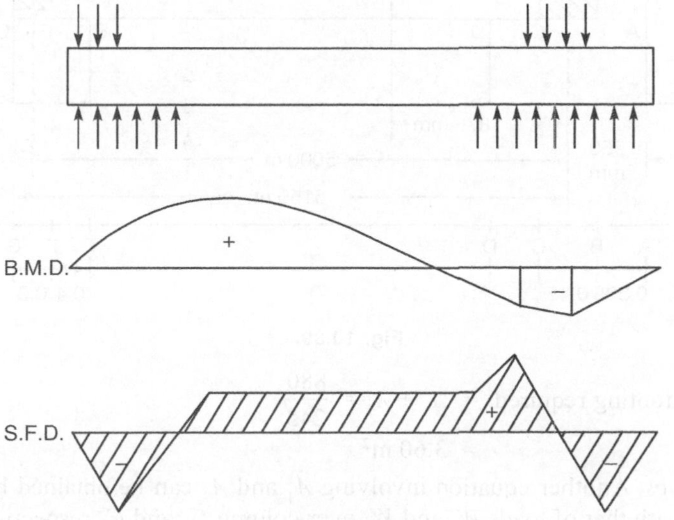

Fig. 10.38.

Another equation in terms of A_1 and A_2 is obtained by coinciding the C.G. of A_1 and A_2 with C.G. of W_1 and W_2.

From the above two equations A_1 and A_2 can be evaluated.

2. The footings for C_1 and C_2 are designed as isolated footings for B.M. and shear force considerations.
3. The strap beam is designed as Rectangular beam by assuming that the loads are acting uniformly on it from
   (*i*) columns in the downward direction
   (*ii*) footings of C_1 and C_2 in the upward direction,
   as shown in Fig. 10.38. The general shape of the B.M.D. and S.F.D. for the strap beam connecting C_1 and C_2 are also indicated in Fig. 10.38.

Example 10.7. Design a strap footing to carry loads of 500 kN (50*t*) and 300 kN (30*t*) through columns C_1 and C_2 which are 450 mm × 450 mm and 300 mm × 300 mm respectively. The column centres are 5000 mm (5 m) apart. Column C_2 is on the property line. The footing can be extended 2000 mm (2 m) across the line joining the columns. The soil over which the footing rests is medium sand, coarse and dry for which bearing capacity may be taken as 245 kN/m². Use M 20 concrete, Fe 415 steel.

Solution. Fig. 10.39 and Fig. 10.43

Area of Footing. Total load on two columns C_1 and C_2 = 300 + 500 = 800 kN. Assume self weight of footing @ 10% of column loads = 80 kN.

$$\text{Total load on soil} = 880 \text{ kN}$$

Let A_1 and A_2 be the areas of the individual footings under C_1 and C_2. Assume that the strap beam is not touching the soil.

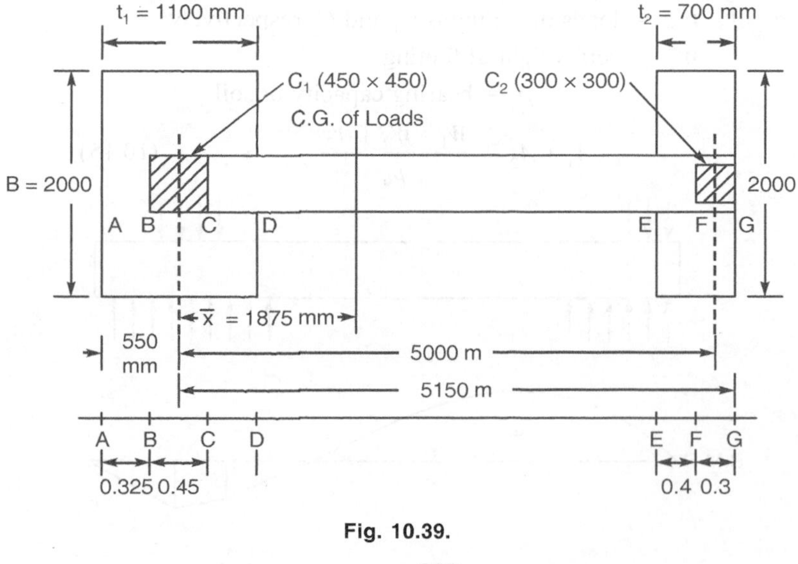

Fig. 10.39.

$$\text{Area of footing required} = A_1 + A_2 = \frac{880}{245}$$

$$= 3.60 \text{ m}^2$$

Size of Footings. Another equation involving A_1 and A_2 can be obtained by coinciding the C.G. of A_1 and A_2 with that of loads W_1 and W_2 over column C_1 and C_2 respectively.

Let us assume that the footings A_1 and A_2 extend to the full width available (*i.e.*, $B = 2$ m given) and the length of each be l_1 and l_2 respectively. Without fixing these, it may be quite difficult to calculate the C.G. of A_1 and A_2.

Let x = distance of C.G. of loads from C_1

$$= \frac{W_2 \cdot l}{W_1 + W_2} = \frac{300 \times 5.0}{500 + 300} = 1.875 \text{ m}$$

Take moments of areas A_1 and A_2 about C_1

$$(A_1 + A_2)\,x = Bl_2\,(5.150 - l_2/2)$$

$$x = \frac{Bl_2\,(5.150 - l_2/2)}{A_1 + A_2}$$

$$1.875 = \frac{2.0 l_2\,(5.150 - l_2/2)}{3.6}$$

$$\therefore \quad l_2 = 0.7 \text{ m}$$

$$A_1 + A_2 = B\,(l_1 + l_2) = 2\,(l_1 + 0.7) = 3.6$$

$$l_1 = 3.6/2 - 0.7$$

$$= 1.1 \text{ m}$$

The size of footings are fixed as

$$l_1 = 1.1 \text{ m}, \quad l_2 = 0.7, \quad B = 2.0 \text{ m}$$

$$\text{Area provided} = 2\,(1.1 + 0.7) = 3.6 \text{ m}^2 \quad \text{O.K.}$$

Design of Independent Footings: (Fig. 10.40)

q_0 = net upward soil pressure

$$= \frac{W_1 + W_2}{\text{area provided}} = \frac{500 + 300}{3.6}$$

$$= 222.222 \text{ kN/m}^2$$

Assume the width of strap beam = 450 mm (0.45 m).

Projection of the footing beyond strap beam = $p = \dfrac{2 - 0.45}{2} = 0.775$ m.

Consider 1 m width of footing

$$\text{Max. B.M.} = q_0 \times \frac{p^2}{2}$$

$$= 222.2 \times 0.775^2$$

$$= 66.736 \text{ kN/m}$$

$$d = \sqrt{\left(\frac{66.736 \times 1000 \times 1000}{0.91 \times 1000}\right)} = 270.8 \text{ mm}$$

Provide $d = 300$ mm, $D = 375$ mm

$$A_{st} = \frac{66736 \times 1000}{230 \times 0.90 \times 300} = 1074.65 \text{ mm}^2$$

Use 12 mm dia. bars at spacing = $\dfrac{113 \times 1000}{1074.65} = 105.15$ mm

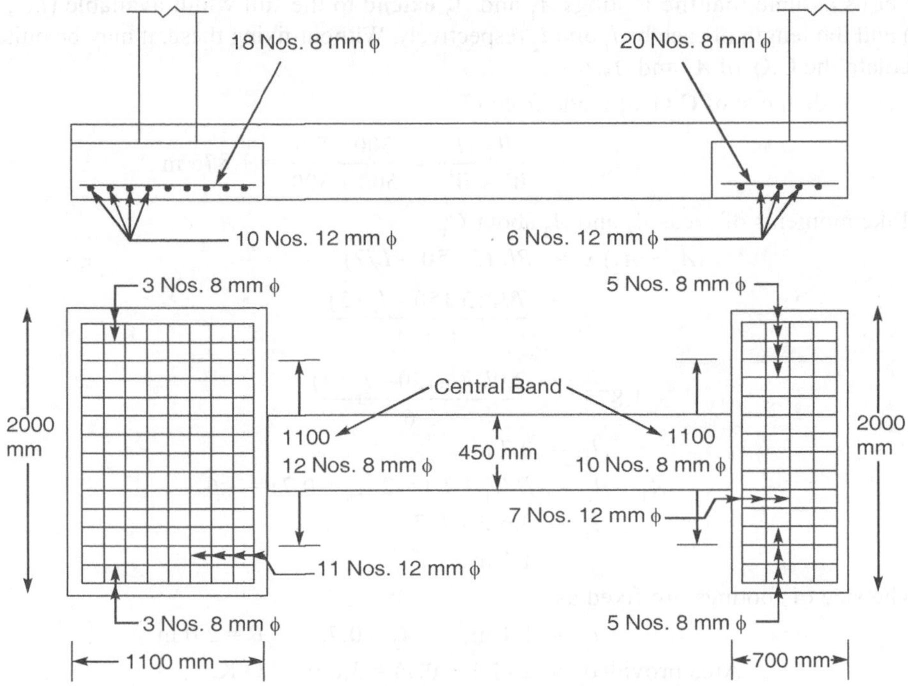

Fig. 10.40.

Provide 12 mm ϕ bars @ 100 mm *c/c* at bottom of the footing along 2 m direction.

$$\text{Distribution steel} = 0.12\% \text{ of gross} \times \text{section area}$$

$$= \frac{0.12 \times 375 \times 1000}{100}$$

$$= 450 \text{ mm}^2$$

$$\text{Use 8 mm } \phi \text{ bars at spacing} = \frac{50 \times 1000}{450} = 111.11 \text{ mm}$$

Provide 8 mm ϕ bars @ 110 mm *c/c* at right angles to main steel.

Reinforcement in Footings (Fig. 10.40)

(*i*) *Under Column* C_1 :

$$\text{Main steel, No. of bars} = \frac{1100 - 75 - 75}{100}$$

$$= 9.5 \text{ say } 10$$

Use 10 Nos. 12 mm ϕ in long direction.

Distribution Steel : No. of bars $= \dfrac{2000 - 75 - 75}{110}$

$$= 16.8 \text{ say } 17.$$

$$\text{in Central band} = \frac{2}{\frac{2}{1.1} + 1} \times (17 \times 50) = 603.2 \text{ mm}^2, \text{ use 12 Nos. 8 mm } \phi.$$

Use 3 Nos. 8 mm ϕ in short direction on each side of central band.

(ii) Under Column C_2 :

$$\text{Main Steel, No. of bars} = \frac{700 - 75 - 75}{100} = 5.5 \approx 6$$

Use 6 Nos. 12 mm ϕ in long direction.

Distribution Steel

$$\text{Central band} = \frac{2}{\frac{2}{0.7}+1} \times (17 \times 50) = 440.74 \text{ mm}^2$$

Use 10 Nos. 8 mm ϕ and 5 Nos. 8 mm ϕ on one side and 5 Nos. 8 mm ϕ on other side of the central band.

Design of Strap Beam. The soil pressure does not act on the strap beam directly and it transfers the pressure from one footing to the other footings.

Loading on the Beam. Soil pressure on the footing per metre length of beam

$$= q_0 \times B \times 1$$
$$= 222.222 \times 2 \times 1$$
$$= 444.444 \text{ kN/m}$$

S.F.D. (Fig. 10.43)

$$V_A = \text{S.F. at } A = 0$$
$$V_B = \text{S.F. at } B = 444.444 \times 0.325 = 144.444 \text{ kN}$$
$$V_C = \text{S.F. at } C = 444.444\,(0.325 + 0.450) - 500 = -155.556$$
$$V_D = \text{S.F. at } D = 444.444\,(1.10) - 500 = -11.112$$
$$V_E = \text{S.F. at } E = 444.444\,(1.10) - 500 = -11.112$$
$$V_F = \text{S.F. at } F = 300 - 444.444 \times 0.30 = 166.667$$
$$V_G = \text{S.F. at } G = 0$$

Points on Zero Shear

(1) $x_1 = AH = AB + BH$

$$AB = 0.325 \text{ m (Fig. 10.40)}$$

$$\frac{BH}{144.444} = \frac{HC}{155.556}$$

$$BH = \frac{HC \times 144.444}{155.556}$$

$$155.556\,BH = (0.45 - BH) \times 144.444$$

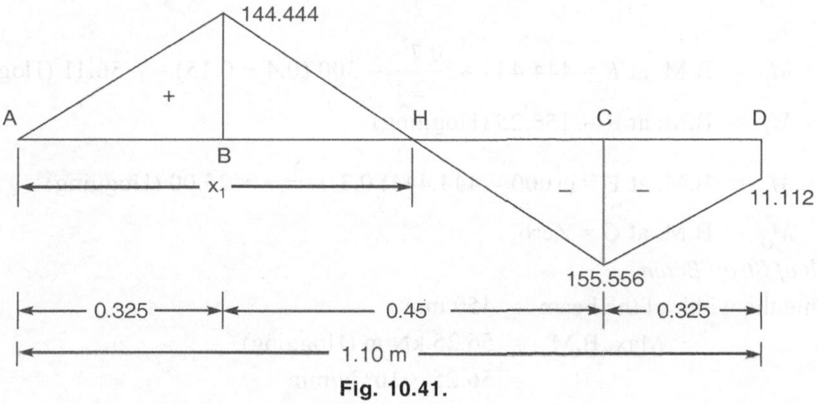

Fig. 10.41.

$$BH = 0.217 \text{ m}$$

$$x_1 = 0.325 + 0.217 = 0.542 \text{ m from } A$$

(2) $x_2 = FG = FK$

$$FG = 0.3 \text{ m (Fig. 10.41)}$$

$$\frac{FK}{166.667} = \frac{EK}{11.112} = \frac{(0.4 - FK)}{11.112}$$

$$= \frac{(0.4 - FK)}{11.112}$$

$$11.112\, FK = 166.667\,(0.4 - FK)$$

$$FK = 0.375 \text{ m}$$

$$X_2 = 0.3 + 0.375 = 0.675 \text{ from } G$$

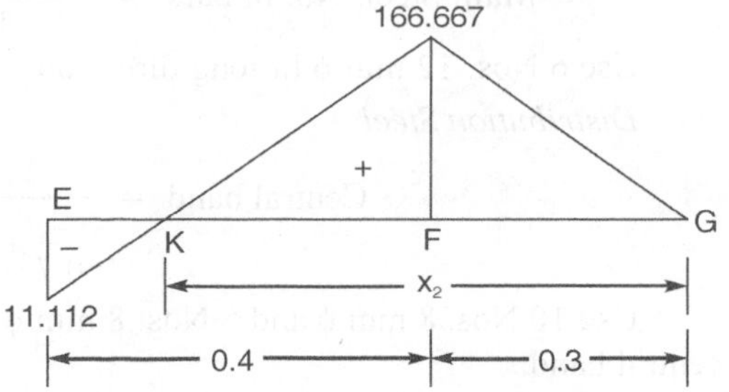

Fig. 10.42.

The moment shall be maximum either at H or K.

B.M.D. (Fig. 10.43)

$$\text{B.M. at } H = 444.444 \times \frac{0.542^2}{2} - 1111.11 \times \frac{(0.542 - 0.325)^2}{2}$$

$$= 65.28 - 26.16$$

$$= 39.12 \text{ kN/m (Sagging)}$$

$$\text{B.M. at } K = 444.444 \times \frac{0.675^2}{2} - 300\left(0.675 - \frac{0.3}{2}\right)$$

$$= 101.25 - 157.50$$

$$= -56.25 \text{ kN/m (Sagging)}$$

$$= 56.25 \text{ kN/m (Hogging)}$$

M_A = B.M. at A = zero

M_B = B.M. at $B = 444.444 \times \frac{0.325^2}{2} = 23.47$ (Sagging)

M_H = B.M. at $H = 39.12$ kN/m (Sagging)

M_C = B.M. at $C = 444.444 \times \frac{0.775^2}{2} - 500 \times 0.225 = 20.97$ (Sagging)

M_D = B.M. at $D = 444.444 \times \frac{1.1^2}{2} - 500\,(0.225 + 0.325) = -6.11$ (Hoggging)

M_E = B.M. at $E = 444.444 \times \frac{0.7^2}{2} - 300\,(0.4 + 0.15) = -56.11$ (Hogging)

M_K = B.M. at K = 156.25 (Hogging)

M_F = B.M. at F = $(1000 - 444.444)\ 0.3 \times \frac{0.3}{2} = 25.00$ (Hogging)

M_G = B.M. at G = Zero.

Depth of Strap Beam

Assume the width of the beam = 450 mm

$$\text{Max. B.M.} = 56.25 \text{ kN/m (Hogging)}$$

$$= 56.25 \times 10^6 \text{ N/mm}$$

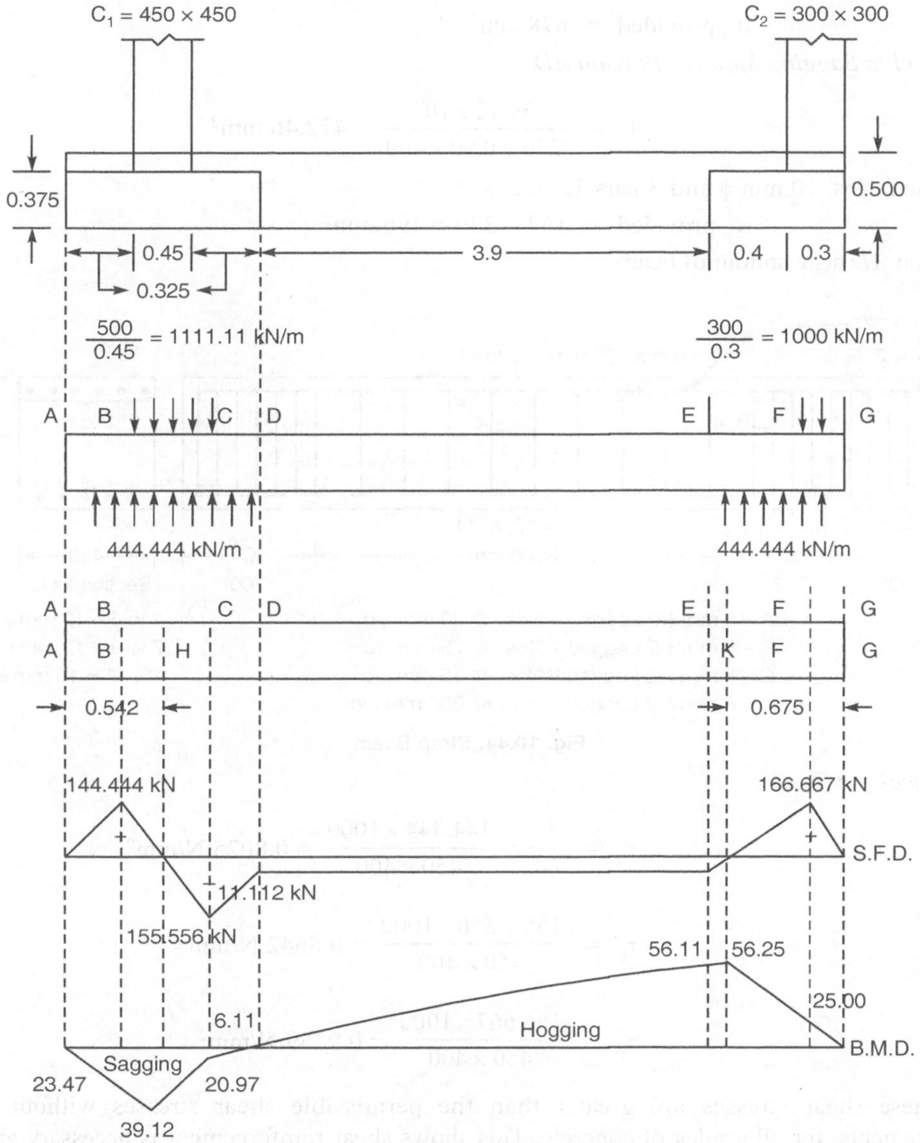

Fig. 10.43.

$$d = \sqrt{\left(\frac{56.25 \times 10^6}{0.91 \times 450}\right)} = 370.62 \text{ mm}$$

Use $d = 400$ mm, $D = 500$ mm

Reinforcement in Strap Beam

$$A_{st} = \frac{56.25 \times 10^6}{230 \times 0.9 \times 400} = 679.35 \text{ mm}^2$$

Provide 6 Nos. 12 mm ϕ near the top face of the beam in the portion *CG* of hogging moment (Fig. 10.44).

$$A_{st} \text{ provided} = 678 \text{ mm}^2$$

Steel in Sagging Moment Portion AD

$$A_{st} = \frac{39.12 \times 10^6}{230 \times 0.90 \times 400} = 472.46 \text{ mm}^2$$

Use 2 Nos. 10 mm ϕ and 3 bars 12 mm ϕ

$$A_{st} \text{ provided} = 157 + 339 = 496 \text{ mm}^2$$

in portion *AD* near bottom of beam.

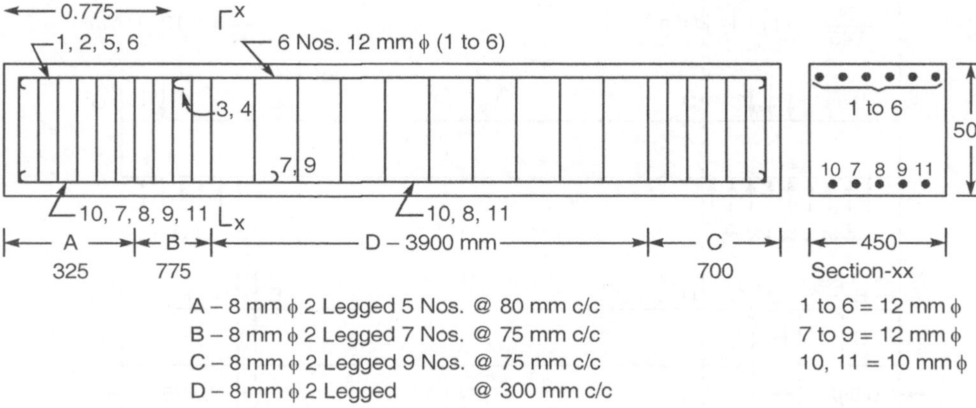

Fig. 10.44. Strap Beam

Check for Shear

$$\tau_B = \frac{V}{bd} = \frac{144.444 \times 1000}{450 \times 400} = 0.8025 \text{ N/mm}^2$$

$$\tau_C = \frac{155 - 556 \times 1000}{450 \times 400} = 0.8642 \text{ N/mm}^2$$

$$\tau_F = \frac{166.667 \times 1000}{450 \times 400} = 0.9259 \text{ N/mm}^2$$

These shear stresses are greater than the permissible shear stresses without shear reinforcements, for all grades of concrete. This shows shear reinforcement is necessary and the permissible values of shear stresses are given by Table 2.2.

$$\tau_{c\,max} \text{ for M 20 concrete} = 1.8 \text{ N/mm}^2$$

Use 8 mm ϕ 2 legged stirrups as shear reinforcement.

Shear Reinforcement

At *B*

$$\frac{100\,A_s}{bd} = \frac{100 \times 496}{450 \times 400} = 0.278$$

$$\tau_c = 0.22 + \frac{(0.30 - 0.22)}{0.5 - 0.25}\,(0.278 - 0.25)$$

$$= 0.228 \text{ N/mm}^2$$

$$\text{Shear strength of concrete} = \tau_c \cdot bd$$
$$= 0.22 \times 450 \times 400 = 39600 \text{ N}$$
$$= 39.60 \text{ kN}$$
$$V_s = \text{shear to be catered by shear reinforcement}$$
$$= 144.444 - 39.60$$
$$= 104.844$$
$$S_v = \text{spacing of shear reinforcement}$$
$$= \frac{\sigma_{st} \cdot A_{sv} \cdot d}{V_s}$$
$$= \frac{230 \times 2 \times 50 \times 400}{104.844 \times 1000}$$
$$= 87.75 \text{ mm}$$

Use 8 mm ϕ 2 legged, 5 Nos. @ 80 mm *c/c* under *B* from *A* to *H*.

At *C*

$$V_s = 155.556 - 39.600$$
$$= 115.956 \text{ kN}$$
$$S_v = \frac{230 \times 2 \times 50 \times 400}{115.956 \times 1000}$$
$$= 79.34 \text{ mm}$$

Use 8 mm ϕ 2 legged stirrups, 7 Nos. @ 75 mm *c/c* at *C* from *H* to *D*.

At *F*

$$\frac{100\,A_s}{bd} = \frac{100 \times 678}{450 \times 400} = 0.377$$
$$\tau_c = 0.26 \text{ N/mm}^2$$
$$\text{Shear strength of concrete} = \tau_c \cdot bd$$
$$= 0.26 \times 450 \times 400$$
$$= 46800 \text{ N}$$
$$= 46.8 \text{ kN}$$
$$V_1 = \text{Shear force to be catered by shear reinforcement}$$
$$= 166.667 - 46.8$$
$$= 119.867 \text{ kN}$$

8 mm ϕ 2 legged stirrups

$$\text{spacing } S_v = \frac{230 \times 2 \times 50 \times 400}{119.867 \times 1000}$$
$$= 76.75 \text{ mm}$$

Use 8 mm ϕ 2 legged stirrups, 9 Nos. @ 75 mm *c/c* at *F* from *K* to *G*.

Between D and E

$$\tau_v = \frac{11.112 \times 1000}{450 \times 400} = 0.0617 \text{ N/mm}^2$$
$$\frac{100\,A_s}{bd} = \frac{100 \times 678}{450 \times 400} = 0.377$$
$$\tau_c = 0.26 \text{ N/mm}^2$$

$$\tau_v < \frac{\tau_c}{2} \text{ no shear reinforcement required.}$$

Provide 8 mm ϕ 2 legged max. spacing 300 mm *c/c*.

Reinforcement details are shown in Fig. 10.44.

10.16 COMBINED FOOTING (RECTANGULAR)

Let W_1 and W_2 be loads from columns C_1 and C_2 and l be the distance between centres of columns. The loads W_1 and W_2 are inclusive of self weight of the columns C_1 and C_2.

Let W_3 be the weight of the footing which is taken as varying between 10% to 20% of the columns loads $(W_1 + W_2)$.

$$\text{Total load on soil} = W_1 + W_2 + W_3$$

$$\text{Area of footing required} = \frac{W_1 + W_2 + W_3}{\text{Bearing capacity of soil}}$$

$$= \frac{W_1 + W_2 + W_3}{p_0}$$

Size $(L \times B)$ of footing is selected.

$$\text{Net upward pressure on footing } q_0 = \frac{W_1 + W_2}{L \times B}.$$

The projections p_1 and p_2 of a footing beyond by the centres of columns, C_1 and C_2 respectively are selected in such a fashion that the resultant of loads W_1 and W_2 passes through the C.G. of area of the footing.

Let X be the distance of the C.G. of columns loads W_1 and W_2 from C_1. Taking moments about centre of C_1.

$$(W_1 + W_2)\,X = W_2 \cdot l$$

$$X = \frac{W_2 \cdot l}{W_1 + W_2}$$

$$p_1 + X = \frac{L}{2} = p_2 + l - X$$

From this equation, the values of projections p_1 and p_2 are decided. The B.M.D. and S.F.D. are shown in Fig. 10.45.

The depth of the footing is designed for max. B.M. which occurs at a point where S.F. is zero, *i.e.*, at a distance of

$$x = \frac{W_1}{q_0\, B} \text{ from the left edge of the footing.}$$

$$\text{Max. B.M.} = (q_0 \cdot B)\,\frac{x^2}{2} - W_1\,(x - p_1)$$

$$= \frac{q_0 \cdot B}{2}\,\frac{W_1^2}{(q_0 \cdot B)^2} - W_1\left(\frac{W_1}{q_0\, B} - p_1\right)$$

$$= \frac{W_1^2}{2 q_0\, B} - \frac{W_1^2}{q_0\, B} + W_1\, p_1$$

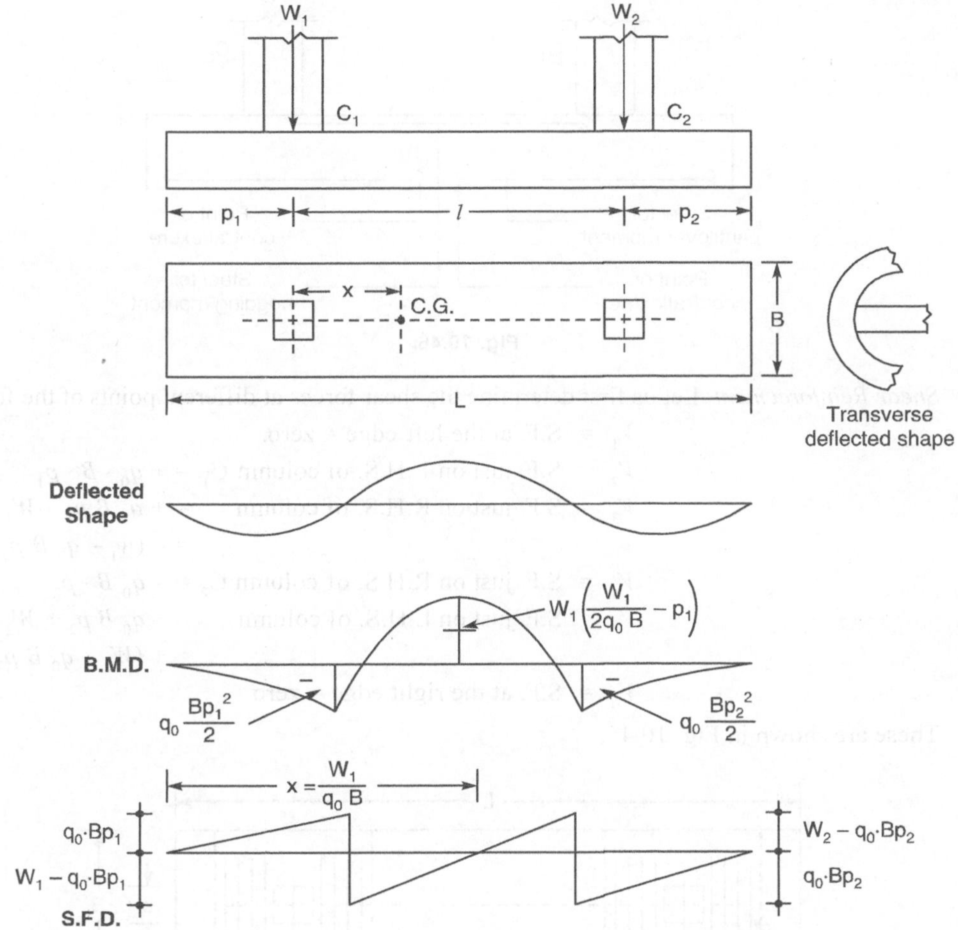

Fig. 10.45.

$$= W_1\, p_1 - \frac{W_1^2}{2q_0\, B} = W_1\left(p_1 - \frac{W_1}{2q_0\, B}\right) \text{ Sagging}$$

$$= W_1\left(\frac{W_1}{2q_0\, B} - p_1\right) \text{ Hogging}$$

As the pressure from soil is acting upwards, the two columns act as supports. Therefore, the portion of the footing between columns has tension at top, while in cantilever portion tension at bottom. So, the main steel for cantilever moments at C_1 and C_2 will be placed near the lower face of the footing, and that for central part moment between C_1 and C_2, the main steel will be placed near the upper face of the footing as shown in Fig. 10.46.

The point of max. B.M. lies in between the columns.

$$\text{Max. B.M.} = W_1\left(\frac{W_1}{2q_0\, B} - p_1\right)$$

$$d = \sqrt{\left(\frac{\text{Max. B.M.}}{\text{Q.B.}}\right)}$$

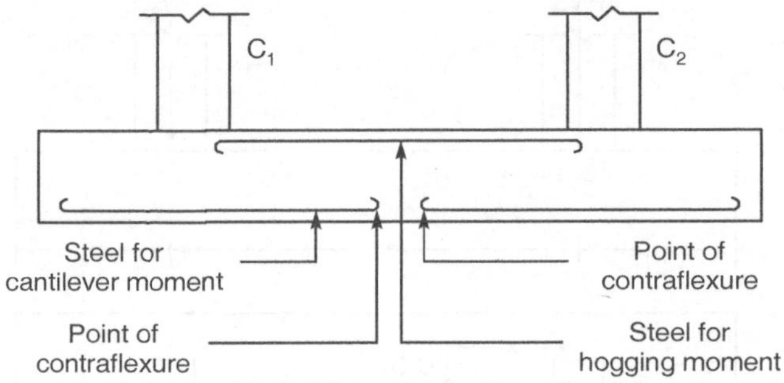

Fig. 10.46.

Shear Reinforcement. Let us first determine the shear forces at different points of the footing.

V_1 = S.F. at the left edge = zero.

V_2 = S.F. just on L.H.S. of column C_1 $= + q_0 \cdot B \cdot p_1$

V_3 = S.F. just on R.H.S. of column $= + q_0 B p_1 - W_1$

$= -(w_1 - q_0 B p_1)$

V_4 = S.F. just on R.H.S. of column C_2 $= - q_0 B \cdot p_2$

V_5 = S.F. just on L.H.S. of column $= - q_0 B p_2 + W_2$

$= +(W_2 - q_0 B p_2)$

V_6 = S.F. at the right edge = zero

These are shown in Fig. 10.45.

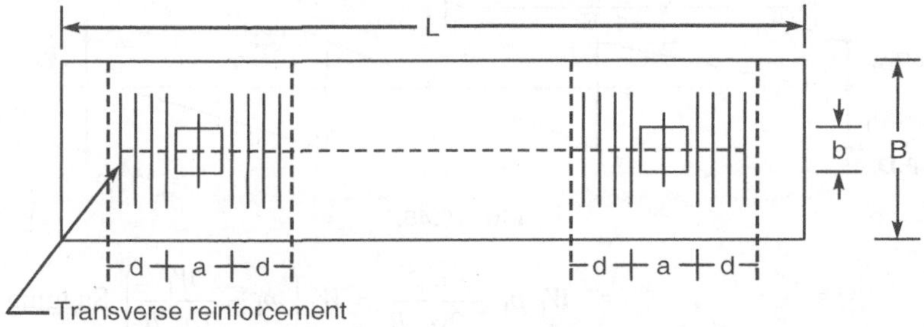

Fig. 10.47.

Transverse Bending of Footing Slab. Near the columns bending takes place both in longitudinal as well as transverse direction. Transverse steel has to be provided, to take care of transverse B.M. in a width = (width of column + 2 × effective depth of footing) as shown in Fig. 10.47.

$$\text{Transverse B.M.} = q_0 \left(\frac{B-b}{2}\right)\left(\frac{B-b}{4}\right)$$

at the face of column

$$= q_0 \left(\frac{B-b}{8}\right)^2$$

Example 10.8. Design a rectangular combined footing for two columns 450 mm × 450 mm and 600 mm × 600 mm carrying 800 kN (80 *t*) and 1000 kN (100 *t*) respectively. The columns are located 4.0 m apart. The safe bearing capacity of the soil is 200 kN/m² (0.2 N/mm²). Use M 20 concrete and Fe 415 grade steel.

Solution. For M20 concrete

$$\sigma_{cbc} = 7 \text{ N/mm}^2, \qquad m = 13.33$$

$$Q = 0.91, \qquad j = 0.90$$

$$\sigma_{st} = 230 \text{ N/mm}^2$$

$$\text{Load on columns} = 800 + 1000 = 1800 \text{ kN}$$

Assume self weight of footing @ 10% = 180 kN

$$\text{Total load} = 1980 \text{ kN}$$

$$\text{Area of footing required} = \frac{1980}{200} = 9.9 \text{ m}^2$$

Adopt 1.9 m × 5.5 m footing size.

$$q_0 = \frac{1800}{1.9 \times 5.5} = 172.25 \text{ kN/m}^2$$

$$X = \text{C.G. of loads}$$

$$= \frac{W_2 \cdot l}{W_1 + W_2} = \frac{1000 \times 4.0}{800 + 1000} = 2.22 \text{ m}$$

$$p_1 + X = \frac{L}{2}$$

$$p_1 = \frac{5.5}{2} - 2.22 = 2.750 - 2.22 = 0.530 \text{ m}$$

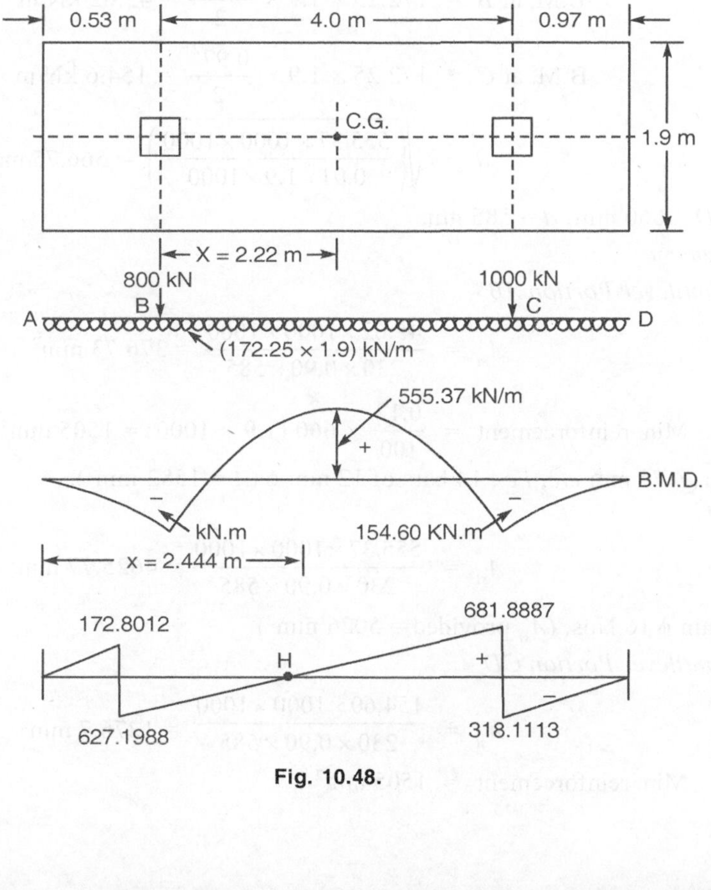

Fig. 10.48.

$$p_2 + l - X = \frac{L}{2}$$

$$p_2 = \frac{5.5}{2} - 4 + 2.22 = 2.75 + 2.22 - 4.0 = 0.97 \text{ m}$$

Therefore, projections of 0.53 m and 0.97 m are provided beyond columns C_1 and C_2 respectively so that C.G. of loads coincides with C.G. of footing.

S.F.D. (Fig. 10.48)

$$q_0 \, B \, p_1 = 172.25 \times 1.9 \times 0.530 = 172.8012 \text{ N}$$

$$W_1 - q_0 \, B \, p_1 = 800 - 172.8012 = 627.1988 \text{ N}$$

$$q_0 \, B \, p_2 = 172.25 \times 1.9 \times 0.97 = 318.1113 \text{ N}$$

$$W_2 - q_0 \, B \, p_2 = 1000 - 318.1113 = 681.8887 \text{ N}$$

x = distance of zero shear from left edge

$$= \frac{W_1}{q_0 \, B} = \frac{800}{172.25 \times 1.9} = 2.444 \text{ m}$$

$$\text{Max. B.M. at this point} = W_1\left(\frac{W_1}{2q_0 \, B} - p_1\right)$$

$$= 800\left(\frac{800}{2 \times 172.25 \times 1.9} - 0.53\right)$$

$$= 555.37 \text{ kN/m}$$

$$\text{B.M. at } B = 172.25 \times 1.9 \times \frac{0.53^2}{2} = 45.62 \text{ kN/m}$$

$$\text{B.M. at } C = 172.25 \times 1.9 \times \frac{0.97^2}{2} = 154.6 \text{ kN/m}$$

$$d = \sqrt{\left(\frac{555.37 \times 1000 \times 1000}{0.91 \times 1.9 \times 1000}\right)} = 566.75 \text{ mm}$$

Provide $D = 660$ mm; $d = 585$ mm

Reinforcement

(*a*) *In Cantilever Portion AB*

$$A_{st} = \frac{45.62 \times 1000 \times 1000}{230 \times 0.90 \times 585} = 376.73 \text{ mm}^2$$

$$\text{Min. reinforcement} = \frac{0.12}{100} \times 660\,(1.9 \times 1000) = 1505 \text{ mm}^2$$

or 12 mm ϕ @ 75 mm *c/c*, *i.e.*, 14 bars of 12 mm ϕ ($A = 1582$ mm^2).

(*b*) *In B.C.*

$$A_{st} = \frac{555.37 \times 1000 \times 1000}{230 \times 0.90 \times 585} = 4625.77 \text{ mm}^2$$

Use 20 mm ϕ 16 Nos. (A_{st} provided = 5026 mm^2)

(*c*) *In Cantilever Portion CD*

$$A_{st} = \frac{154.60 \times 1000 \times 1000}{230 \times 0.90 \times 585} = 1276.7 \text{ mm}^2$$

$$\text{Min. reinforcement} = 1505 \text{ mm}^2$$

Provide 12 mm ϕ 14 Nos. (A_{st} provided = 1582 mm^2)

Transverse Reinforcement

Transverse B.M. at

(*i*) the face of column in C_1 $= q_0 \dfrac{(B-b)^2}{8}$

$$= \frac{172.25}{8}(1.9-0.45)^2$$

$$= 45.27 \text{ kN-m}$$

$$A_{st} = \frac{45.27 \times 1000 \times 1000}{230 \times 0.90 \times 585} = 373.84 \text{ mm}^2$$

Min. reinforcement in 1620 mm

$$= \frac{0.12}{100}[1620 \times 660] = 1283.04 \text{ mm}^2$$

Use 12 mm ϕ @ 135 mm *c/c* in length = 450 + 2 × 585 = 1620 mm (12 bars).

A_{st} provided = 1356 mm^2

(*ii*) the face of column C_2 $= \dfrac{172.25}{8}(1.9-0.6)^2$

$$= 36.39 \text{ kN-m}$$

$$A_{st} = \frac{36.39 \times 10^6}{230 \times 0.90 \times 585} = 300.51 \text{ mm}$$

Min. reinforcement in 1770 mm

$$= \frac{0.12}{100}[1770 \times 660] = 1401.84 \text{ mm}^2$$

Use 12 mm ϕ @ 140 mm *c/c* in length = 600 + 2 × 585 = 1770 mm (13 bars).

A_{st} provided = 1469 mm^2

Check for Shear

Since the footing behaviour is under two-way action, allowable shear stress τ_c shall be

$$\tau_c = 0.16\sqrt{(f_{ck})}$$

$$= 0.16\sqrt{(20)}$$

$$= 0.7155 \text{ N/mm}^2$$

Actual shear stress

(*i*) To left of *B*

$$\tau_v = \frac{172.8012 \times 1000}{1900 \times 585} = 0.1550 \text{ N/mm}^2 < \frac{0.7155}{2} \text{ N/mm}^2$$

No. shear reinforcement is required.

(*ii*) To right of *B*

$$\tau_v = \frac{627.1988 \times 1000}{1900 \times 585} = 0.564 \text{ N/mm}^2 < 0.7155 \text{ N/mm}^2$$

(*iii*) To the left of *C*

$$\tau_v = \frac{681.887 \times 1000}{1900 \times 585} = 0.613 \text{ N/mm}^2$$

(*iv*) To the right of *C* $\tau_v = \dfrac{318.1113 \times 1000}{1900 \times 585} = 0.286 \text{ N/mm}^2 < \dfrac{0.7155}{2} \text{ N/mm}^2$

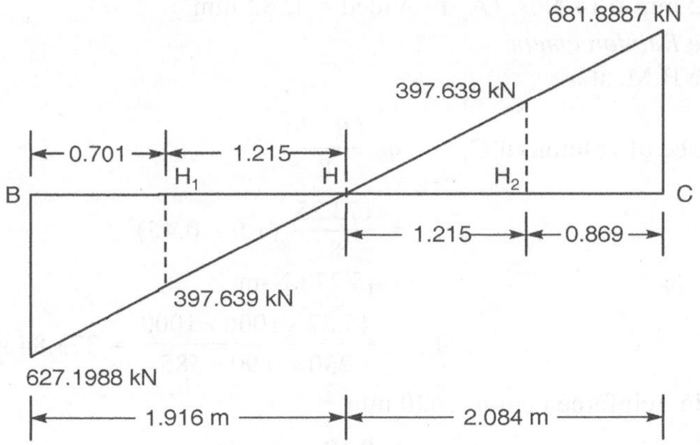

Fig. 10.49.

Safe in shear.

However, min. shear reinforcement shall be provided. It may be noted that even minimum reinforcement shall not be provided in the zone $H_1 H_2$ (Ref. Fig. 10.49).

Shear resistance of concrete section = $0.5\ \tau_c$ (b.d)

$$= (0.5 \times 0.7155) \times 1900 \times 585$$

$$= 397639 \text{ N}$$

$$H H_1 = \frac{397.639}{627.1988} \times 1.916 = 1.215 \text{ m}$$

$$H H_2 = \frac{397.639}{681.8877} \times 2.084 = 1.215 \text{ m}$$

Min. shear reinforcement.

Use 8 mm ϕ 10 legged

$$S_v = \frac{A_{sv} \cdot f_y \times 0.87}{b \times 0.4}$$

$$= \frac{(50 \times 10)\ 415 \times 0.87}{1900 \times 0.4} = 237.5 \text{ mm}$$

Use 8 mm ϕ 10 legged @ 225 mm *c/c* in the zones BH_1 and H_2C.

Distribution and temperature steel

$$= \frac{0.12 \times 660 \times 1000}{100}$$

$$= 792 \text{ mm}^2$$

Provide 10 mm ϕ @ 100 mm *c/c* at top and bottom where main steel is not placed.

The reinforcement details are shown in Fig. 10.50.

10.17 COMBINED FOOTING (TRAPEZOIDAL)

This type of footing is provided when

(*i*) the two column loads are unequal.

(*ii*) the space available is restricted on both sides of the building due to neighbouring structure.

In such cases, the distance available is known which becomes the length of the footing. To decide the width of footing at two ends, the principle "that C.G. of column loads should coincide with C.G. of footing" is utilized.

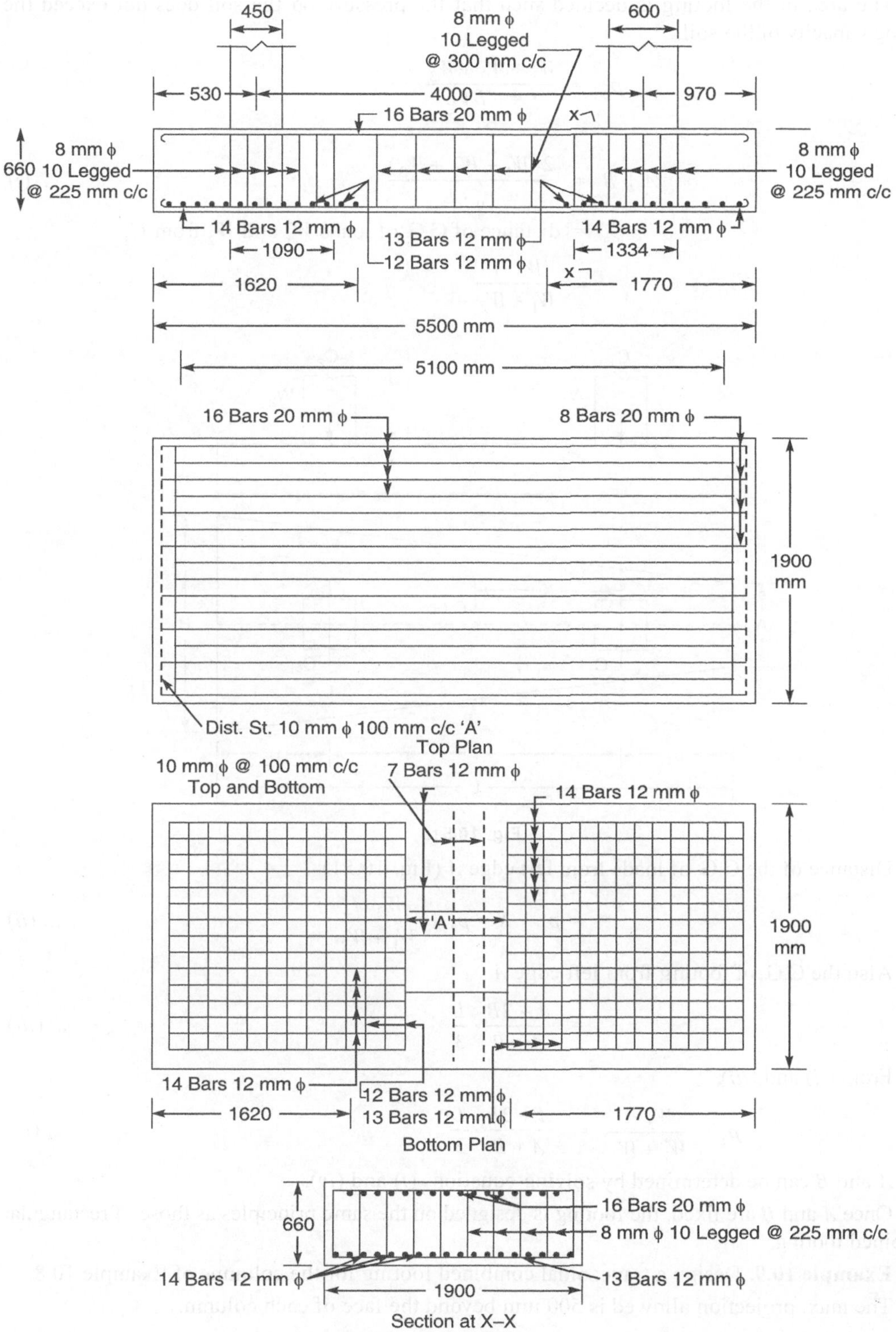
450
8 mm ϕ
10 Legged
@ 300 mm c/c
600
530
4000
970
16 Bars 20 mm ϕ
X
8 mm ϕ
660 10 Legged
@ 225 mm c/c
8 mm ϕ
10 Legged
@ 225 mm c/c
14 Bars 12 mm ϕ
14 Bars 12 mm ϕ
13 Bars 12 mm ϕ
1090
1334
12 Bars 12 mm ϕ
1620
X
1770
5500 mm
5100 mm
16 Bars 20 mm ϕ
8 Bars 20 mm ϕ
1900 mm
Dist. St. 10 mm ϕ 100 mm c/c 'A'
Top Plan
10 mm ϕ @ 100 mm c/c
Top and Bottom
7 Bars 12 mm ϕ
14 Bars 12 mm ϕ
'A'
1900 mm
14 Bars 12 mm ϕ
12 Bars 12 mm ϕ
13 Bars 12 mm ϕ
1620
1770
Bottom Plan
660
16 Bars 20 mm ϕ
8 mm ϕ 10 Legged @ 225 mm c/c
14 Bars 12 mm ϕ
13 Bars 12 mm ϕ
1900
Section at X–X

Fig. 10.50.

The area of the footing is decided such that the pressure on the soil does not exceed the bearing capacity of the soil.

$$p_0 = \frac{W_1 + W_2 + W_3}{\frac{(A+B)\,L}{2}}$$

$$A + B = \frac{2\,(W_1 + W_2 + W_3)}{p_0 \cdot L} \qquad ...(i)$$

X = distance of C.G. of loads W_1 and W_2 from C_1

$$= \frac{W_2 \cdot l}{W_1 + W_2}$$

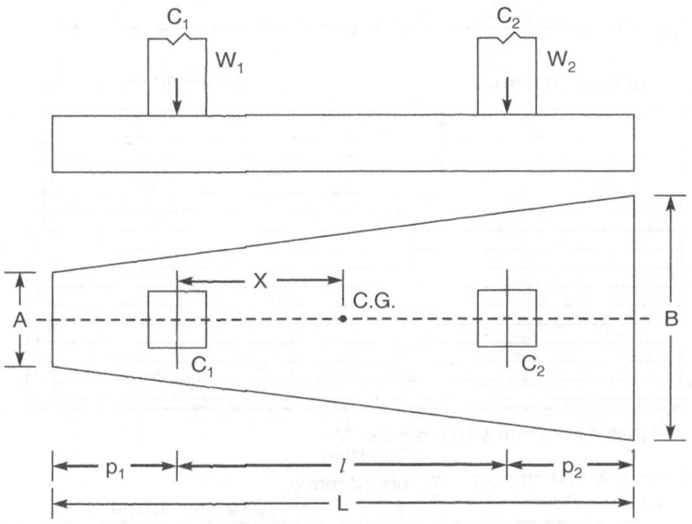

Fig. 10.51.

Distance of the C.G. of loads from left edge A (Fig. 10.51)

$$= p_1 + X = p_1 + \frac{W_2 \cdot l}{W_1 + W_2} \qquad ...(ii)$$

Also the C.G. of footing from left edge A

$$= \frac{A + 2B}{A + B} \cdot \frac{L}{3} \qquad ...(iii)$$

From (*ii*) and (*iii*),

$$p_1 + \frac{W_2 \cdot l}{W_1 + W_2} = \frac{A + 2B}{A + B} \cdot \frac{L}{3} \qquad ...(iv)$$

A and B can be determined by solving equations (*i*) and (*iv*).

Once A and B are fixed, the footing is designed on the same principles as those of rectangular combined footing.

Example 10.9. Design a trapezoidal combined footing for the columns of Example 10.8. The max. projection allowed is 500 mm beyond the face of each column.

Solution. (Fig. 10.52)

$$p_1 = 500 + 225$$
$$= 725 \text{ mm} = 72.5 \text{ cm}$$
$$p_2 = 500 + 300$$
$$= 800 \text{ mm} = 80 \text{ cm}$$
$$L = 72.5 + 400 + 80$$
$$= 552.5 \text{ cm} = 5525 \text{ mm}$$
$$l = 4 \text{ m}$$

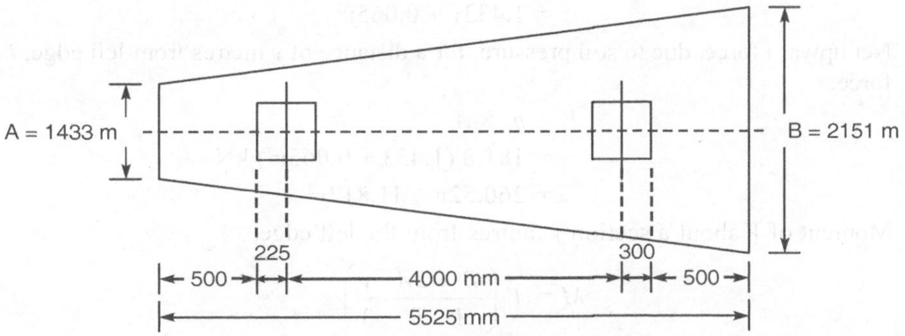

Fig. 10.52.

$$X = \text{C.G. of loads from centre of } C_1$$

$$X = \frac{W_2 \cdot l}{W_1 + W_2} = \frac{1000 \times 4.0}{800 + 1000} = 2.222 \text{ m}$$

From equation (*i*),

$$A + B = \frac{2\,(W_1 + W_2 + W_3)}{p_0 \cdot L}$$

$$= \frac{2\,(800 + 1000 + 180)}{200 \times 5.525} = 3.584 \text{ m}$$

From equation (*iv*),

$$\frac{2B + A}{A + B} \cdot \frac{L}{3} = p_1 + X$$

$$B = \frac{3\,(A + B)\,(p_1 + X)}{L} - (A + B)$$

$$= \frac{3 \times 3.584\,(0.725 + 2.222)}{5.525} - 3.584 = 2.151 \text{ m}$$

$$A = 3.584 - 2.151 = 1.433 \text{ m}$$

$$q_0 = \frac{W_1 + W_2}{\dfrac{(A + B)\,L}{2}} = \frac{800 + 1000}{\dfrac{(1.433 + 2.151) \times 5.525}{2}}$$

$$= 181.8 \text{ kN/m}^2$$
$$= (181.8 \times 10^{-3}) \text{ N/mm}^2$$

(*i*) Width of the footing at any distance *y* metres from the left edge.

$$B' = 1.433 + (2.151 - 1.433)\ \frac{y}{5.525}$$
$$= 1.433 + 0.13y$$

(*ii*) Area of the footing for a distance *y* metres from the left edge.

$$A = \int_0^y (1.433 + 0.13y)\, dy$$
$$= 1.433y + 0.065y^2$$

(*iii*) Net upward force, due to soil pressure, for a distance of *y* metres from left edge, *i.e.*, shear force.

$$V = q_0 \times A$$
$$= 181.8\ (1.433 + 0.065y^2)\ \text{kN}$$
$$= 260.52y + 11.817y^2$$

(*iv*) Moment of *V* about a section *y* metres from the left edge,

$$M = V\left(\frac{B' + 2A}{A + B'} \cdot \frac{y}{3}\right)$$
$$= 181.8\ (1.433y + 0.065y^2)\ \frac{y}{3}\left[\frac{1.433 + 0.13y + 2 \times 1.433}{1.433 + (1.433 + 0.13y)}\right]$$
$$= (130.26y^2 + 3.393y^2)\ \text{kN/m}$$

Alternatively

$$V = \frac{dM}{dy}$$
$$M = \int_0^y V \cdot dy$$
$$= \int_0^y 181.8\ (1.433y + 0.065y^2)\ dy$$
$$= (130.26y^2 + 3.939y^3)\ \text{kN.m}$$

Max. B.M. shall occur where S.F. is zero.

$$260.52y + 11.817y^2 = 800$$
$$y = 2.732\ \text{m},\quad 24.778\ \text{m}$$

Neglect second term as $y \not> 5.525$ m.

$$\therefore \quad y = 2.732\ \text{m}$$

B.M. Under Column C_1

Put

$$y = 500 + 225 = 725\ \text{mm} = 0.725\ \text{m}$$
$$M = 130.26y^2 + 3.939y^3$$
$$= 69.969\ \text{kN.m} \quad \text{(Sagging)}$$

$$\text{B.M. at } y = 2.732\ \text{m}$$
$$M = 800\ (y - 0.725) - (130.26y^2 + 3.939y^3)$$
$$= 553.041\ \text{kN.m} \quad \text{(Hogging)}$$

B.M. Under Column C_2

Put $$y = 5.525 - 0.8 = 4.725 \text{ m}$$

$$M = (2908.136 + 415.520) - 800\,(x - 0.725)$$
$$= 123.656 \text{ kN/m} \quad \text{Sagging}$$

B.M.D is shown in Fig. 10.53.

Depth of Footing

$$\text{Width of footing at } y = 2.732 \text{ m}$$
$$B' = 1.433 + 0.13y = 1.788 \text{ m}$$
$$d = \sqrt{\frac{553.041 \times 10^6}{0.91 \times 1788}} = 583.06 \text{ mm}$$

Adopt $D = 675$ mm, $d = 600$ mm

Reinforcement

(*i*) In cantilever position *AB* (at bottom)

$$A_{st} = \frac{69.969 \times 10^6}{230 \times 0.90 \times 600} = 563.36 \text{ mm}^2$$

$$\text{Min. Steel} = 0.12\% \text{ of cross sectional area}$$
$$= \frac{0.12}{100} \times 675\,(1.433 + 0.13 \times 0.725)\,1000$$
$$= 1237 \text{ mm}^2$$
$$\text{No. of Bars} = \frac{1237}{113} = 10.95 \text{ say 11 of 12 mm } \phi$$

Use 11 bars of 12 mm ϕ at bottom in *AB*.

(*ii*) In *BC* (at top)

$$A_{st} = \frac{553.041 \times 10^6}{230 \times 0.90 \times 600} = 4452.83 \text{ mm}^2$$

Use 17 bars of 20 mm ϕ at top in *BC*.

(*iii*) In cantilever position *CD* (at bottom)

$$A_{st} = \frac{123.656 \times 10^6}{230 \times 0.90 \times 600} = 995.62 \text{ mm}^2$$

$$\text{Min. steel} = \frac{0.12}{100} \times 675\,(1.433 + 0.13 \times 4.725)\,1000$$
$$= 1658.24 \text{ mm}^2$$
$$\text{No. of bars} = 1658.24 = \frac{1658.24}{113} = 14.67 \text{ say 15 of 12 mm } \phi$$

Use 15 bars of 12 mm ϕ at bottom in *CD*.

Transverse Reinforcement

(*i*) Near Column C_1

$$\text{Width of footing at } C_1 = 1.433 + 0.13 \times 0.725 = 1.527 \text{ m}$$

$$\text{Projection from face of column} = \frac{1.527 - 0.45}{2}$$

$$= 0.539 \text{ m}$$
$$= 539 \text{ mm}$$

$$\text{B.M. per mm width} = q_0 \times \frac{(539)^2}{2}$$
$$= (181.8 \times 10^{-3})\ \frac{(539)^2}{2}$$
$$= 26408 \text{ N/mm}$$

$$A_{st} = \frac{26408}{230 \times 0.90 \times (600 - 16)} = 0.22 \text{ mm}^2\text{/mm}$$

$$\text{Min. steel} = \frac{0.12}{100} \times 1 \times 675 = 0.81 \text{ mm}^2$$

Use 12 mm ϕ bars, spacing $= \dfrac{113}{0.81} = 139.5$ mm

Width of strip in which reinforcement is to be provided

$$= 450 + 2 \times 600$$
$$= 1650 \text{ mm}$$

$$\text{No. of bars} = \frac{1650}{139.5} = 11.83$$

Use 12 bars of 12 mm ϕ in a width = 1.65 m.

(*ii*) *Near Column* C_2

$$\text{Width of footing at } C_2 = 1.433 + 0.13 \times 4.725 = 2.047 \text{ m}$$

$$\text{Projection from face of column} = \frac{2.047 - 0.6}{2}$$
$$= 0.7235 \text{ m}$$
$$= 723.5 \text{ mm}$$

$$\text{B.M. per mm width} = (181.8 \times 10^{-3})\ \frac{(723.5)^2}{2}$$
$$= 47582 \text{ N/mm}$$

$$A_{st} = \frac{47582}{230 \times 0.90 \times (600 - 16)} = 0.394 \text{ mm}^2$$

$$\text{Min. steel} = 0.81 \text{ mm}^2$$

$$\text{Spacing of 12 mm } \phi \text{ bar} = \frac{113}{0.81} = 139.5 \text{ mm}$$

$$\text{Width} = 600 + 2 \times 600 = 1800 \text{ mm}$$

$$\text{No. of bars} = \frac{1800}{139.5} = 12.9$$

Use 13 bars of 12 mm ϕ in a width of 1.8 m.

Shear Reinforcement

First of all calculate the shear forces at critical points of footing.

$$V = 260.52y + 11.817y^2 \text{ kN upto } C_1$$
$$V_1 \uparrow = \text{S.F. just to left of } B = 188.877 + 6.211$$
$$= 195.088 \text{ kN } (y = 0.725 \text{ m})$$

$$V_2 \downarrow = \text{S.F. just to right of } B = 800 - 195.088 = 604.912 \text{ kN}$$

$$V_3 \uparrow = \text{S.F. just to left of } C = 800 - (260.52y + 11.817y^2)$$

$$(y = 4.725 \text{ m}) = 694.779 \text{ kN}$$

$$V_4 \uparrow = \text{S.F. just to right of } C = (260.52y + 11.817y^2) - 1800$$

$$(y = 4.725 \text{ m}) = 305.221 \text{ kN}$$

S.F.D. and B.M.D. are shown in Fig. 10.53.

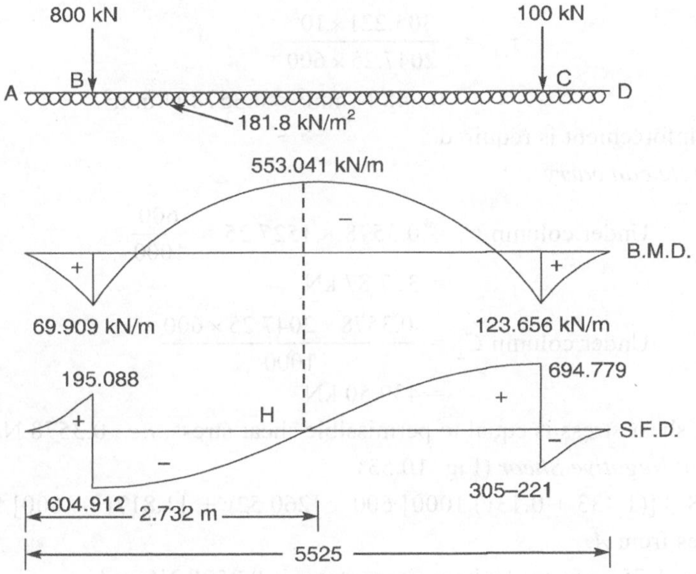

Fig. 10.53.

Width of footing under column C_1,

$$B' = 1.433 + 0.13 \times 0.725$$
$$= 1.52725 \text{ m}$$
$$= 1527.25 \text{ mm}$$

Width of footing under column C_2,

$$B' = 1.433 + 0.13 \times 4.725$$
$$= 2.04725 \text{ m}$$
$$= 2047.25 \text{ mm}$$

$$\text{Permissible shear stress, } \tau_c = 0.7155 \text{ N/mm}^2$$

$$\text{Nominal shear stress, } \tau_v = \frac{V}{B'd}$$

$$\text{Shear stress due to } V_1,\ \tau_{v1} = \frac{195.088 \times 10^3}{1527.25 \times 600}$$

$$= 0.213 \text{ N/mm}^2 < (\tfrac{1}{2} \times 0.7155)\text{, } i.e.\text{, } 0.3578 \text{ N/mm}^2$$

No shear reinforcement is required.

$$\text{Due to } V_2,\ \tau_{v2} = \frac{604.912 \times 10^3}{1527.25 \times 600} \text{ N/mm}^2$$

$= 0.66 \text{ N/mm}^2 < 0.7155 \text{ N/mm}^2$

Shear reinforcement is not required.

$$\text{Due to } V_3, \tau_{v3} = \frac{694.779 \times 10^3}{2047.25 \times 600}$$

$$= 0.566 \text{ N/mm}^2 < 0.7155 \text{ N/mm}^2 > 0.3578 \text{ N/mm}^2$$

Minimum shear reinforcement is required.

$$\text{Due to } V_4, \quad \tau_{v4} = \frac{305.221 \times 10^3}{2047.25 \times 600}$$

$$= 0.248 \text{ N/mm}^2 < 0.3578 \text{ N/mm}^2$$

No shear reinforcement is required.

Shear concrete can carry

(*i*) Under column $C_1 = 0.3578 \times 1527.25 \times \frac{600}{1000}$

$= 327.87$ kN

(*ii*) Under column $C_2 = \frac{0.3578 \times 2047.25 \times 600}{1000}$

$= 439.50$ kN

Zone where shear stress is equal to permissible shear stress, *i.e.*, 0.3578 N/mm².

(*i*) *In Zone of Negative Shear* (Fig. 10.53)

$-0.3578 \times [(1.433 + 0.13y)\ 1000]\ 600 = [260.52y + 11.817y^2 - 800]\ 1000$

y is in metres from A.

$AH_1 = y = 1.75$ m from A where shear stress is 0.3578 N/mm²

$BH_1 = 1.75 - 0.7250 = 1.025$ m from B

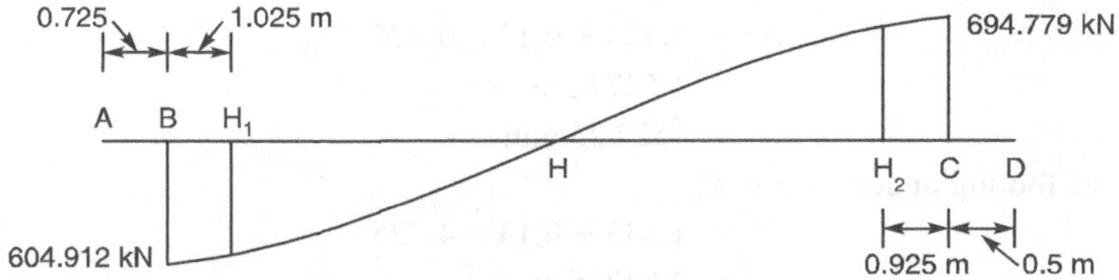

Fig. 10.54.

(*ii*) *In Zone of Positive Shear* (Fig. 10.54)

$+ 0.3578 [(1.433 + 0.13y)\ 1000]\ 600 = [260.52y + 11.817y^2 - 800]\ 1000$

$AH_2 = y = 3.8$ m from A

or $CH_2 = 4.725 - 3.8 = 0.925$ m = from C

Stirrups : 8 mm ϕ 8 legged.

Stirrups are required in distance $BH_1 = 1.025$ m and $CH_2 = 0.925$ m.

V_{s1} = Shear force for which shear reinforcement is required under column C_1

$= 604.912 - 284.07 = 320.842$ kN

V_{s2} = Shear force for which shear reinforcement is required under column C_2

$= 694.779 - 380.79 = 313.99$ kN

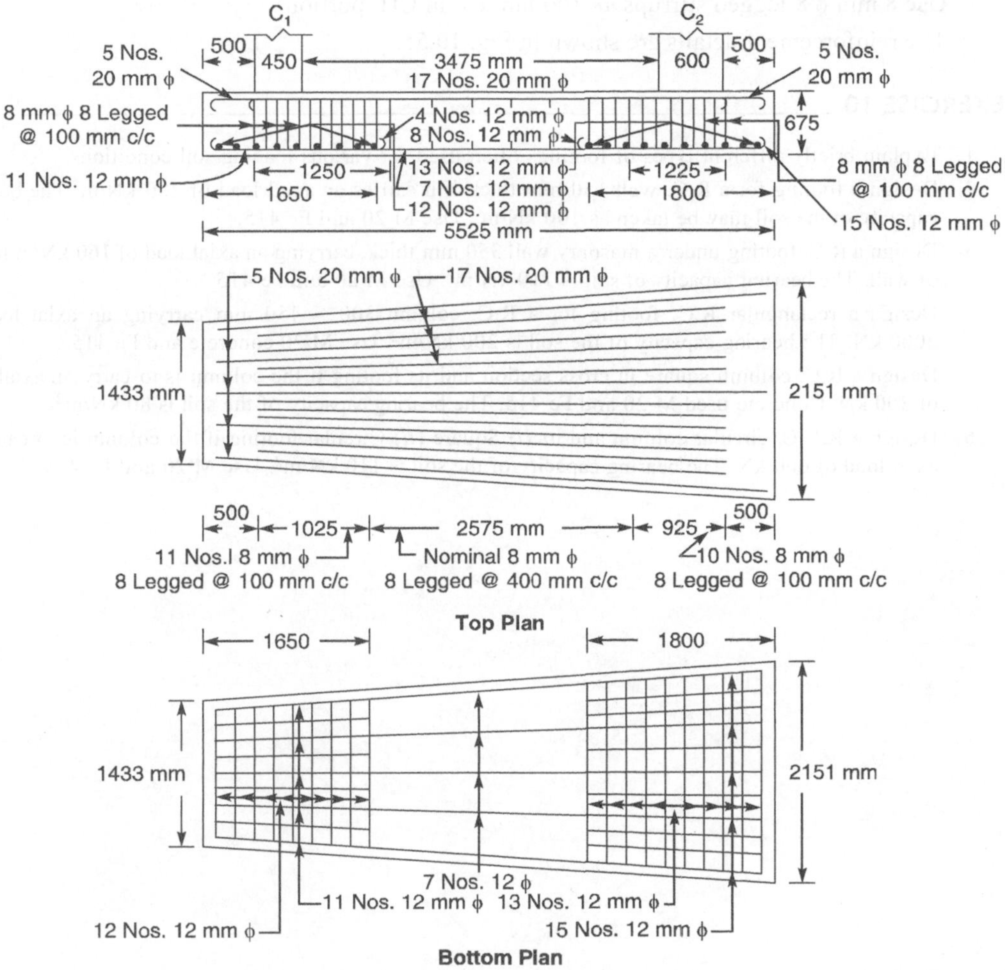

Fig. 10.55.

Under Column C_1

$$\text{Spacing} = V = \frac{\sigma_{st} \cdot A_{sv} \cdot d}{V_s}$$

$$= \frac{230 \times (50 \times 8)\,600}{320.842 \times 1000}$$

$$= 172 \text{ mm}$$

Use 8 mm ϕ 8 legged stirrups @ 100 mm *c/c* in distance 1.025 mm from C_1 towards right, *i.e.*, only 11 stirrups.

Under Column C_2

$$\text{Spacing } S_v = \frac{230 \times (50 \times 8)\,600}{313.99 \times 1000} = 175.8 \text{ mm}$$

$$\text{Max. spacing allowed} = 0.75d$$

$$= 0.75 \times 600 = 450 \text{ mm}$$

Use 8 mm ϕ 8 legged stirrups @ 100 mm *c/c* in CH_2 portion.

The reinforcement details are shown in Fig. 10.55.

EXERCISE 10

1. Explain briefly different types of footings often used for various loading soil conditions.
2. Design a footing for a R.C. wall 150 mm thick if it carries an axial load of 200 kN/m. The bearing capacity of the soil may be taken as 100 kN/m^2. Use M 20 and Fe 415.
3. Design a R.C. footing under a masonry wall 350 mm thick, carrying an axial load of 160 kN/m length of wall. The bearing capacity of soil is 110 kN/m^2. Use M 20 and Fe 415.
4. Design a rectangular R.C. footing for a R.C. column 300 × 450 mm carrying an axial load of 2000 kN. The bearing capacity of the soil is 200 kN/m^2. Use M 20 concrete and Fe 415.
5. Design a R.C. column square in cross section and its footing if the column is to carry an axial load of 300 kN. Concrete used M 20 and Fe 415. The bearing capacity of the soil is 80 kN/m^2.
6. Design a R.C.C. circular column and its (*i*) Square (*ii*) Circular footing if the column is to carry an axial load of 600 kN. The bearing capacity of the soil is 110 kN/m^2. Use M 20 and Fe 415.

Stairs

CHAPTER 11

11.1 GENERAL

Stairs are provided in a building to afford a means of communication between the various floors. These are steps arranged in series and placed in an enclosure called 'Staircase'. Since they have to perform a very important function, the slab over which the steps rest, should be designed properly to provide maximum comfort, ease and safety.

The most important aspect in providing stairs is its location. The location of stairs should be such as to provide an easy access so that in case of any casuality, e.g., fire break-out., earthquake, floods, etc., occupants should be able to use it to reach safer places. The location depends upon the position of rooms and the type of approach needed. In residential buildings, it should be placed centrally so as to (*i*) provide easy access from all the rooms (*ii*) maintain privacy. In public buildings, the stairs should be located near the main entrance.

Terms. The most common terms required by the construction/design Engineer, are summarized below, and are shown in Fig. 11.1.

1. *Tread* : Horizontal upper portion of the step on which the foot is placed for ascending and descending.
2. *Riser* : Vertical portion of a step which provides support to the tread.
3. *Rise* : Vertical distance between the upper horizontal portions of any two consecutive steps.
4. *Landing* : A flat platform provided between two flights.
5. *Flight* : A series of steps without an intermediate platform, break or landing. Two flights are normally separated by a landing.

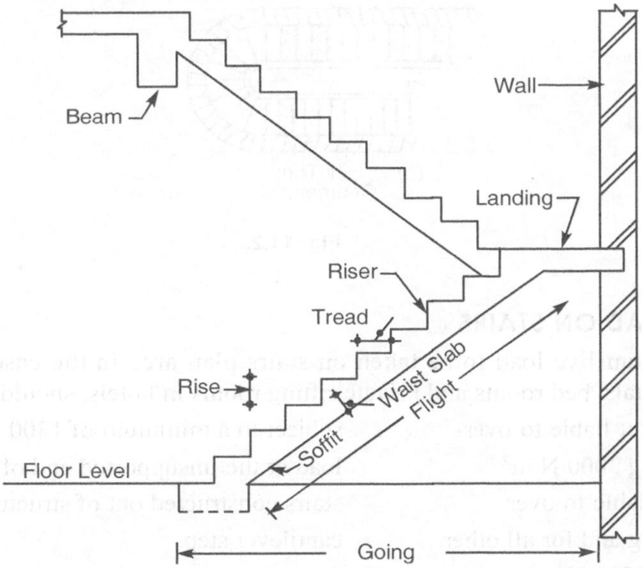

Fig. 11.1.

6. *Winders* : Radiating or angular or tapering steps which give a change of direction to the stair.
7. *Going* : The horizontal projection of the flight.
8. *Soffit* : The under surface of a stair-slab.

11.2 TYPES OF STAIRS

The different types of stairs in common use are :

(*i*) Straight Flight Stair (Fig. 11.2 *a*)
(*ii*) Quarter Turn Stair (Fig. 11.2 *b*)
(*iii*) Half Turn Stair or dog legged (Fig. 11.2 *c*)
(*iv*) Three Quarter Turn Stair (Fig. 11.2 *d, e*)
(*v*) Geometrical Stairs (Fig. 11.2 *f, g*)

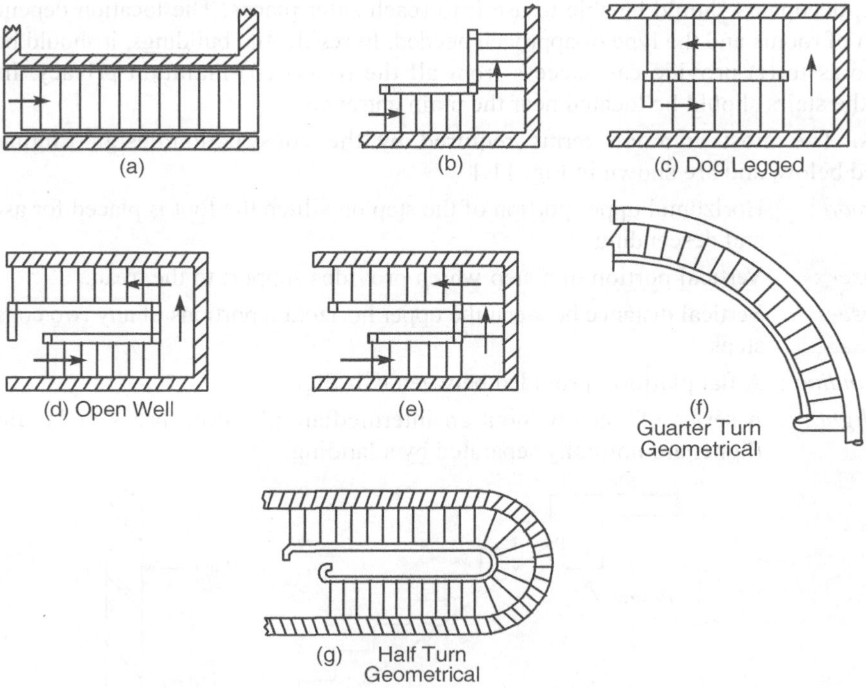

Fig. 11.2.

11.3 LIVE LOAD ON STAIRS

The minimum live load to be taken on stairs plan area in the case of dwelling houses, tenaments, hospitals, bed rooms and private sitting rooms in hotels, should be taken as:

(*i*) Places not liable to over-crowding 3000 N/m^2

(*ii*) Places liable to over-crowding and for all other classes 5000 N/m^2

subject to a minimum of 1300 N concentrated load at the unsupported end of each step for stairs constructed out of structurally independent cantilever steps.

11.4 GENERAL NOTES ON DESIGN OF STAIRS (IS: 456-2000)

(1) Effective Span of Stairs: The effective span of stairs without stringer beam shall be taken as the following horizontal distances:

(*a*) Where supported at top and bottom risers by beams standing parallel with the risers, the distance centre to centre of beams.

(*b*) Where spanning on to the edge of a landing slab, which spans parallel with the risers (Fig. 11.3), a distance equal to the going of the stairs plus at each either half the width of landing or 1 m, whichever is smaller:

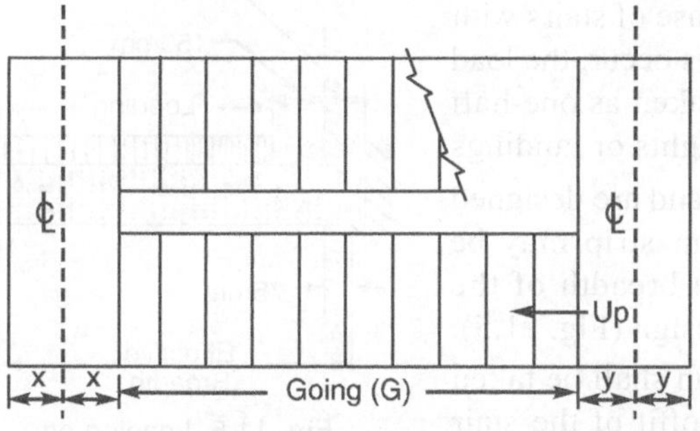

X	Y	Span
		In Metres
<1	<1	G + X + Y
<1	≥1	G + X + 1
≥1	<1	G + Y + 1
≥1	≥1	G + 1 + 1

Fig. 11.3. Effective Span for Stairs Supported at each End by Landings Spanning Parallel with the Risers

(*c*) Where the landing slab spans in the same direction as the stairs, they shall be considered as acting together to form a single slab and the span determined as the distance centre to centre of the supporting beams or walls, the going being measured horizontally.

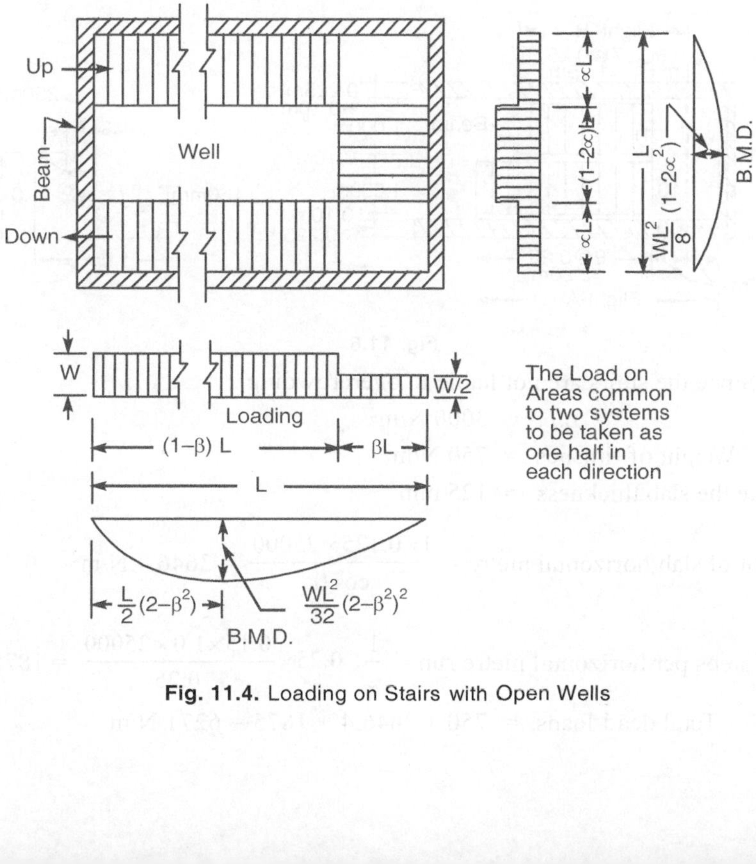

Fig. 11.4. Loading on Stairs with Open Wells

(2) *Size of Rise and Tread*:

	Rise (mm)	Tread (mm)
1. Public Building	150	300
2. Residential Buildings	150 – 200	200 – 300

Rules (Dimensions in mm)

(*i*) 400 < (Rise + Tread) < 450

(*ii*) 580 < (Rise + Tread) < 630

(*iii*) 4000 < (Rise × Tread) < 5000

(3) *Distribution of Loading on Stairs*: In the case of stairs with open well where spans partly crossing at right angles occur, the load on areas common to any two such spans may be taken as one-half in each direction as shown in Fig. 11.4. Where flights or landings are embedded into walls for a length of $\not<$ 110 mm and are designed to span in the direction of the flight, a 150 mm strip may be deducted from the loaded area and the effective breadth of the section increased by 75 mm for the purposes of design (Fig. 11.5).

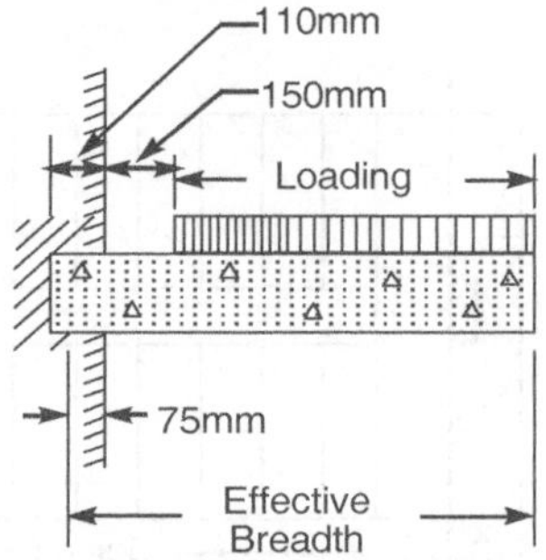

Fig. 11.5. Loading on Stairs built into Walls

(4) *Depth of Section*: The depth of the section shall be taken as the minimum thickness perpendicular to the soffit of the stair case.

11.5 SOLVED EXAMPLES

Example 11.1. Design the arrangement of stair case shown in Fig. 11.6. The building is not liable to overcrowding. Rise of step = 150 mm, Tread = 250 mm and width of stairs = 1m. The weight of finishes = 750 N/m². Use M 20 concrete and Fe 415 bars.

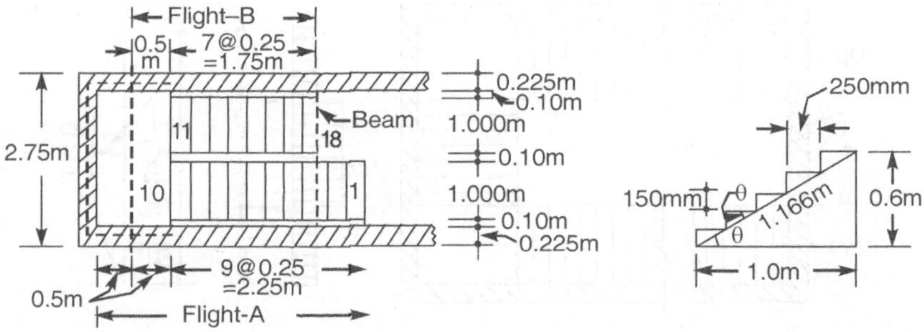

Fig. 11.6.

Solution. Since the stairs are not liable to overcrowding,

$$L.L. = 3000 \text{ N/m}^2$$

$$\text{Weight of finishes} = 750 \text{ N/m}^2$$

$$\text{Assume the slab thickness} = 125 \text{ mm}$$

$$\text{Soft weight of slab/horizontal metre} = \frac{1 \times 0.125 \times 25000}{\cos\theta} = 3646.4 \text{ N/m}^2$$

$$\text{Weight of steps per horizontal metre run} = \frac{1}{2} \times 0.25 \times \frac{0.15 \times 1.0 \times 25000}{0.25} = 1875 \text{ N/m}^2$$

$$\text{Total dead loads} = 750 + 3646.4 + 1875 = 6271 \text{ N/m}$$

Live Load = 3000 N/m

Total Load = 9271 say 9300 N/m

Flight A

Flight *A* comprises of step 1 to step 10. It starts from Ground floor to landing as shown in Fig. 11.6.

Effective span = 9 × 0.25 + 0.5 = 2.75 m

$$\text{Maximum } B.M. = \frac{w \cdot L^2}{8}$$

$$= \frac{9300 \times 2.75^2}{8} = 8791.4 \text{ N.m}$$

$$d = \sqrt{\left(\frac{8791.4 \times 1000}{0.91 \times 1000}\right)} = 98.3 \text{ mm}$$

∴ Assumed overall depth of 125 mm is O.K.

∴ d = 100 mm

$$A_{st} = \frac{8791.4 \times 1000}{230 \times 0.90 \times 100} = 425 \text{ mm}^2$$

Use 8 mm ϕ bars at spacing $\frac{1000 \times 50}{425}$ = 118 mm say 120 mm c/c, i.e., in one metre width of stair provide 8 Nos. of 8 mm ϕ.

$$\text{Minimum distribution steel} = \frac{0.12 \times 125 \times 1000}{100} = 150.0 \text{ mm}^2$$

Use 8 mm ϕ @ 250 mm *c/c*. (A_{st} = 201 mm²)

Landing Slab. Effective span = c/c of wall or (clear span of landing + Eff. depth of slab) which ever is less

= 2.525 m or (2.3 + 0.105)

I_{ef} = 2.405 m

Assume thickness of landing slab = 125 mm

∴ Self weight of landing slab/m = 0.125 × 1 × 1 × 25000 = 3125 N/m

Live load = 3000 N/m

Finsihes = 750 N/m

Total w = 6875 N/m

$$\text{Max. } B.M. = \frac{6875 \times 2.405^2}{8} = 4970.6 \text{ N.m}$$

$$d = \sqrt{\left(\frac{4970.6 \times 1000}{0.91 \times 1000}\right)} = 73.9 \text{ mm}$$

Assumed depth of 125 mm is O.K.

d = 100 mm

$$A_{st} = \frac{4970.6 \times 1000}{230 \times 0.90 \times 100} = 240 \text{ mm}^2$$

Provide 8 mm φ @ 200 mm c/c (A_{st} provided = 251 mm²)

Minimum distribution steel of 8 mm φ @ 250 mm c/c may be provided on upper face of landing to give support to the main bars.

Flight B

$$\text{Effective span} = \frac{1}{2} \text{ width of landing} + \text{Going}$$

$$= \frac{1}{2} \times 1 + 7 \times 0.25 = 2.25 \text{ m}$$

Loading is same as for Flight *A*.

$$\text{Max. } B.M. = \frac{9300 \times 2.25^2}{8} = 5885.1 \text{ N.m}$$

$$d = \sqrt{\left(\frac{5885.1 \times 1000}{0.91 \times 1000}\right)} = 80.42 \text{ mm}$$

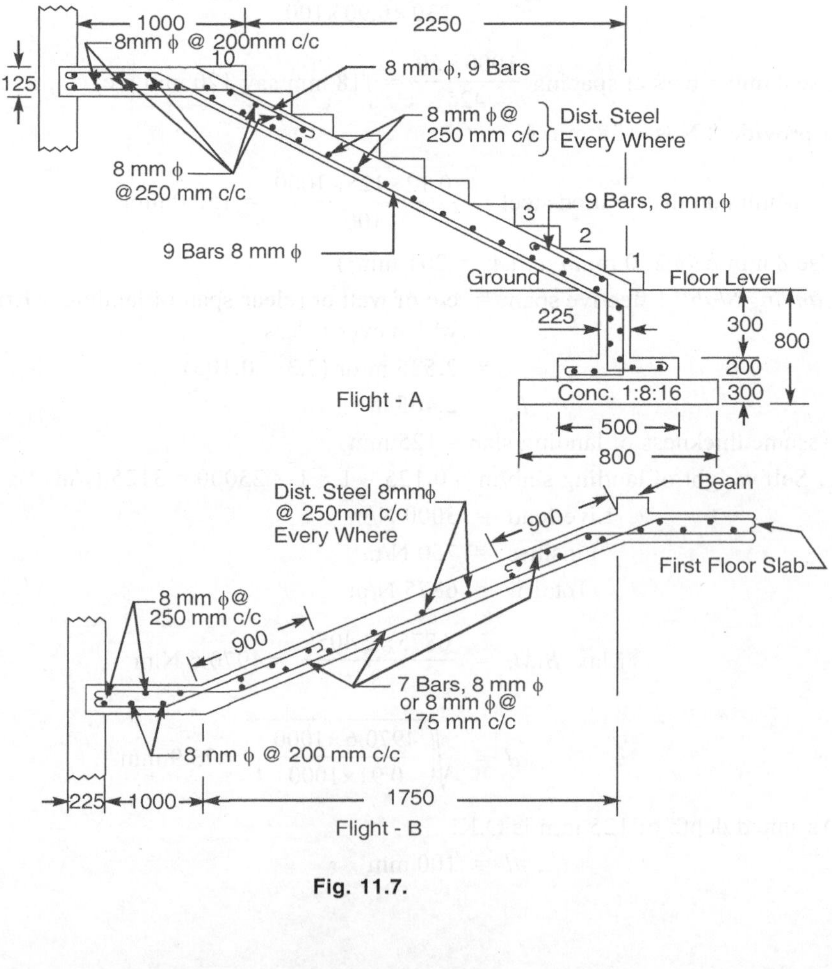

Fig. 11.7.

$$D = 125 \text{ mm} \qquad d = 100 \text{ mm}$$

$$A_{st} = \frac{5885.1 \times 1000}{230 \times 0.90 \times 100} = 284 \text{ mm}$$

Provide 8 mm ϕ @ 175 mm *c/c i.e.*, 6 bars of 8 mm ϕ in the width of Flight *B*. Distribution steel 8 mm ϕ @ 250 mm *c/c*

The reinforcement details of Flights *A* and *B* are shown in Fig. 11.7.

Example 11.2. Design a single fligth stair (Fig. 11.8) to climb a roof slab 3 m high. The rise and tread of steps may be taken as 200 mm and 250 mm respectively. Width of stair is 0.9 m. Take Live load = 3000 N/m² finishes = 750 N/m². Use M 20 concrete and HYSD (Fe 415) bars.

Solution. No. of setps = Height/Rise = 3×1000 / 200 =15

∴ Effective span = 15 × 250 = 3750 mm = 3.75 m

Assume waist slab thickness = 150 mm

Self weight of slab/hor.m

$$= \frac{1 \times 0.15 \times 25000 \times \sqrt{(25^2 + 20^2)}}{25}$$

$$= 4802.4 \text{ N/m}^2$$

Weight of steps/hor.m

$$= \frac{1/2 \times 0.25 \times 0.2 \times 1 \times 25000}{0.25}$$

$$= 2500 \text{ N/m}^2$$

Live load = 3000 N/m²

Finishes = 750 N/m²

Total load = 11052.4 N/m²

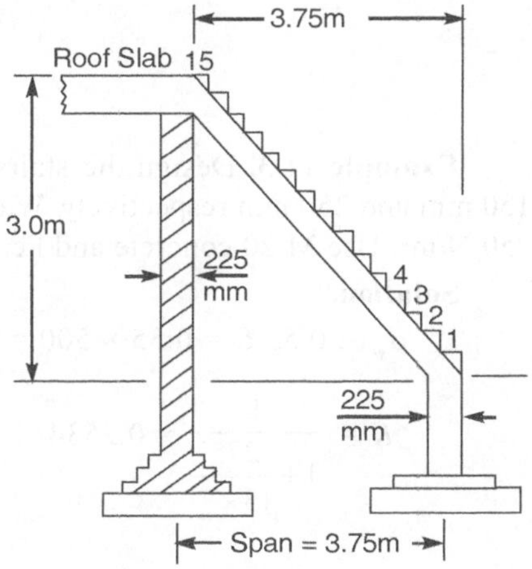

Fig. 11.8.

Load/m of waist slab 0.9 m wide = 11052.4 × 0.9 = 9947.16 N/m say 10,000 N/m

$$\text{Maximum B.M.} = \frac{10000 \times 3.75^2}{8} = 17578 \text{ N.m}$$

$$d = \sqrt{\left(\frac{17578 \times 1000}{0.91 \times 900}\right)} = 146.5 \text{ mm}$$

$$D = 150 \text{ mm} \qquad d = 125 \text{ mm}$$

$$A_{st} = \frac{17578 \times 1000}{230 \times 0.90 \times 125} = 679.34 \text{ mm}^2$$

Provide 10 mm ϕ @ 100 mm *c/c*, *i.e.*, 10 bars.

$$\text{Distribution steel} = \frac{0.12 \times 900 \times 150}{100} = 162 \text{ mm}^2$$

Provide 8 mm ϕ @ 300 mm *c/c*.

The reinforcement details are shown in Fig. 11.9.

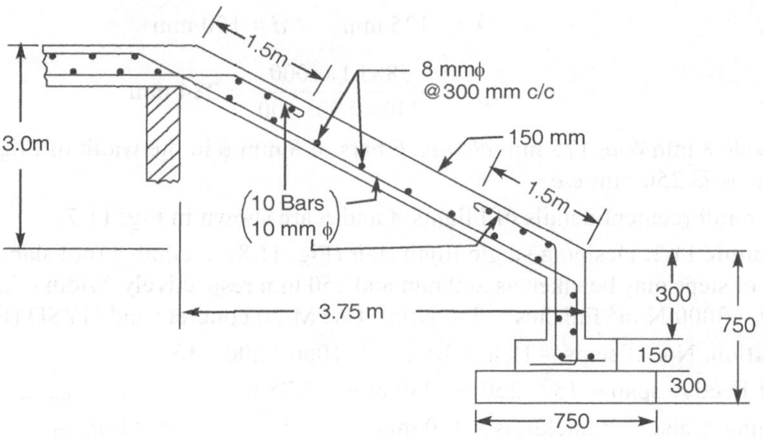

Fig. 11.9.

Example 11.3. Design the stairs as shown in Fig. 11.10. Take rise and tread of steps as 150 mm and 250 mm respectively. Width of stairs = 1.5 m. Take L.L = 5000 N/m² and finishes = 750 N/m². Use M 20 concrete and Fe 500 Steel.

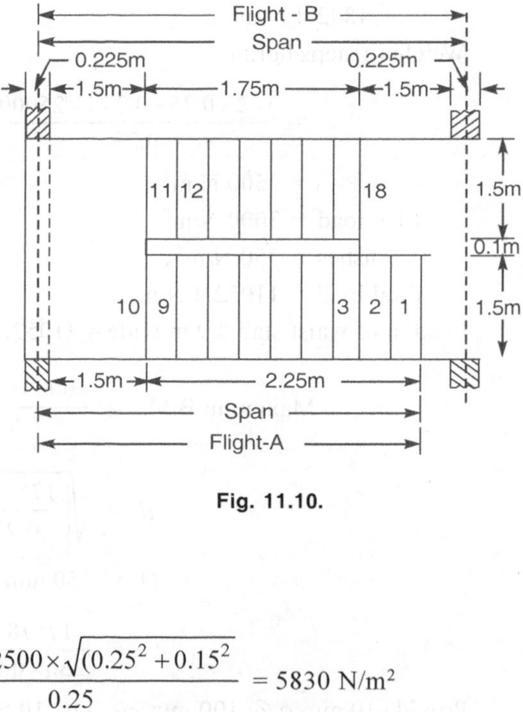

Fig. 11.10.

Solution.

$$\sigma_{st} = 0.55 fy = 0.55 \times 500 = 275 \text{ N/m}^2$$

$$K = \frac{1}{1+\frac{275}{13\times 7}} = 0.2534$$

$$j = 1 - \frac{K}{3} = 0.9155$$

$$Q = \frac{1}{2}.k.j.\sigma_{cbc} = 0.812$$

Flight A

Effective span = 2.25 + 1.5 + 1/2 × 0.225

= 3.8625 m

Assume waist and landing

slab thickness = 200 mm

Loads per meter width of Waist Slab

$$\text{Self weight of slab} = \frac{0.2\times 1\times 2500\times\sqrt{(0.25^2+0.15^2}}{0.25} = 5830 \text{ N/m}^2$$

$$\text{Weight of steps} = \frac{1/2\times 0.25\times 0.15\times 1\times 25000}{0.25} = 1875 \text{ N/m}^2$$

Live load = 5000 N/m²

Finishes = 750 N/m²

Total load = 13455 N/m² say 14000 N/m²

$$\text{Maximum } B.\ M. = \frac{14000 \times 3.8625^2}{8} = 26108 \text{ N.m}$$

$$d = \sqrt{\frac{26108 \times 1000}{0.812 \times 1000}} = 179.31 \text{ mm}$$

$$D = 200 \text{ mm} \qquad d = 170 \text{ mm}$$

$$A_{st} = \frac{26108 \times 1000}{275 \times 0.955 \times 170} = 610 \text{ mm}^2$$

Provide 10 mm φ @ 125 mm *c/c* or 13 Nos. in 1.5 m width of waist slab.

$$\text{Distribution steel} = \frac{0.12 \times 1000 \times 200}{100} = 240 \text{ mm}^2$$

Provide 8 mm φ @ 200 *c/c* (A_{st} provided = 251 mm²)

Flight B

Assume the thickness of waist slab of Flight *B* as 200 mm.

$$\text{Effective span} = \frac{0.225}{2} + 1.5 + 1.75 + 1.5 + \frac{0.225}{2} = 4.975 \text{ m}$$

$$\text{Maximum } B.\ M. = \frac{w.L^2}{8}$$

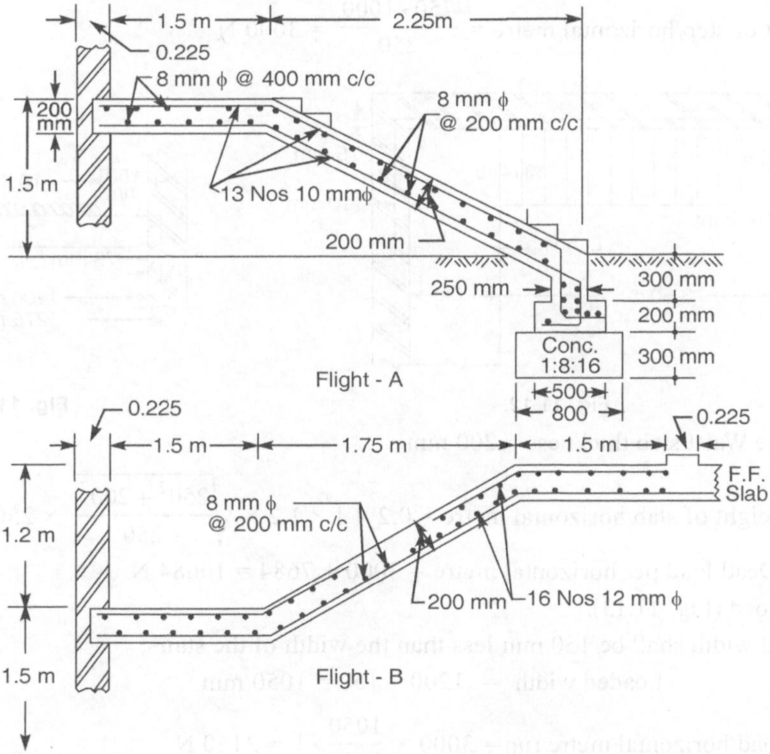

Fig. 11.11.

$$= \frac{14000 \times 4.975^2}{8} = 43313.6 \text{ N.m}$$

$$d = \sqrt{\frac{43313.6 \times 1000}{0.812 \times 1000}} = 231 \text{ mm}$$

$$D = 200 \text{ mm} \qquad d = 170 \text{ mm}$$

Note: The effective depth provided is 18 mm against 233 mm. This will be adjusted by providing more steel than required for a normal balanced section.

$$A_{st} = \frac{43313.6 \times 1000}{275 \times 0.9155 \times 170} = 1012 \text{ mm}^2 \qquad (A_{st} = 1131 \text{ mm}^2)$$

Provide 12 mm ϕ bars @ 100 mm *c/c*, *i.e.*, 12 mm ϕ 16 Nos.

Distribution steel is exactly same as provided in flight *A*.

The reinforcement details are shown in Fig. 11.11.

Example 11.4. Design the arrangement of staircase shown in Fig. 11.12. The stairs are built in the wall with a bearing of 110 mm along the flight. Tread = 250 mm, Rise = 200 mm, Live load = 3000 N/m², Weight of finishes = 500 N/m². Use Fe 500 and M 20 concrete.

Solution. Live load = 3000 N/m²

Finishes = 500 N/m²

$$\text{Self weight of each step} = \frac{1.2 \times 0.25 \times 0.2 \times 25000}{2} = 750 \text{ N}$$

$$\text{Weight of step/horizontal metre} = \frac{750 \times 1000}{250} = 3000 \text{ N}$$

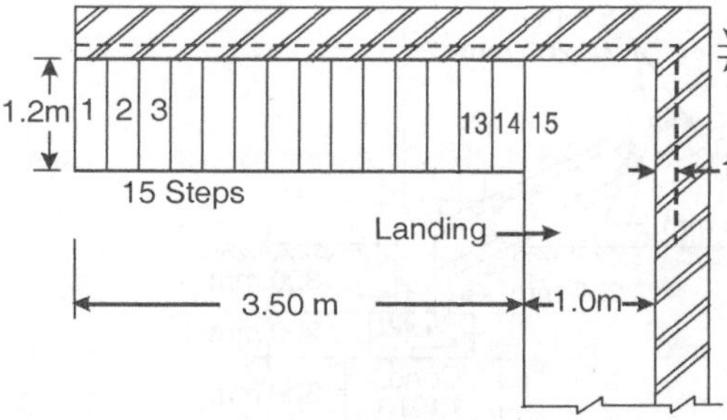

Fig. 11.12.

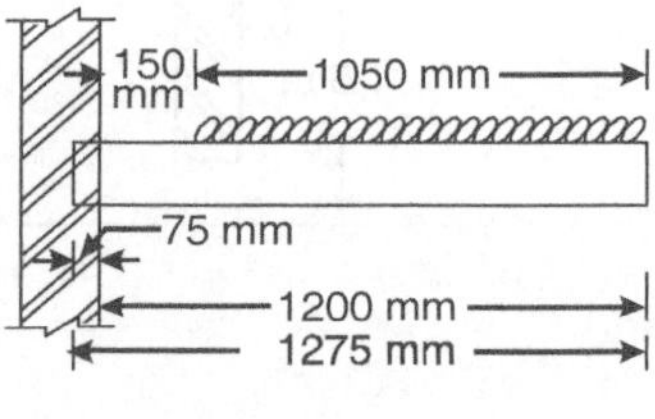

Fig. 11.13.

Assume Waist slab thickness = 200 mm

$$\text{Self weight of slab/horizontal metre} = 0.2 \times 1 \times 1.2 \times \sqrt{\frac{250^2 + 200^2}{250}} \times 25000 = 7684 \text{ N}$$

Total Dead load per horizontal metre = 3000 + 7684 = 10684 N

Live Load (Fig. 11.13).

Loaded width shall be 150 mm less than the width of the stairs.

∴ Loaded width = 1200 – 150 = 1050 mm

$$\text{Live load/horizontal metre run} = 3000 \times \frac{1050}{1000} \times 1 = 3150 \text{ N}$$

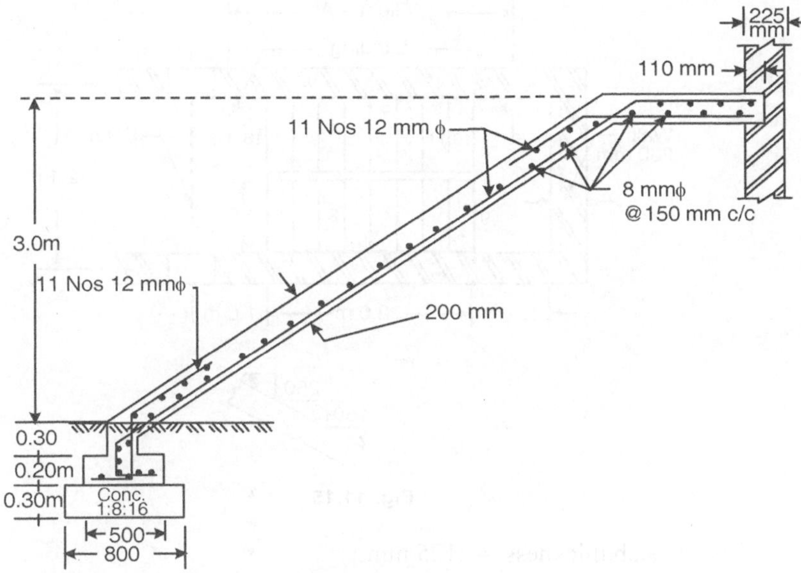

Fig. 11.14.

Total load per horizontal metre = 10684 + 3150 = 13834 N say 14000 N

In this case the flight shall be assumed to be simply supported in the direction of the flight. The effective span will be (3.5 + 1.0 + 0.11) 4.61 m.

$$\therefore \quad \text{Max } B.\,M. = \frac{w.L^2}{8}$$

$$= \frac{14000 \times 4.61^2}{8}$$

$$= 37191 \text{ N. m.}$$

$$\text{Breadth of slab for design} = 1200 + 75 = 1275 \text{ mm}$$

$$d = \sqrt{\frac{37191 \times 1000}{0.812 \times 1275}} = 189.53$$

$$D = 200 \text{ mm} \qquad d = 170 \text{ mm}$$

$$A_{st} = \frac{37191 \times 1000}{275 \times 0.9155 \times 170} = 869 \text{ mm}^2$$

Profide 12 mm ϕ bar @ 125 mm *c/c*., *i.e.*, 11 bars in 1.275 m width.

Distribution steel = 0.12% of 1275 × 200 = 306 mm²

Provide 8 mm ϕ @ 150 mm *c/c*.

The reinforcement details are shown in Fig. 11.14

Example 11.5. Design the staircase as shown in Fig. 11.15. The stair is not liable to over-crowding. Rise of step is 150 mm and the tread is 250 mm. Finishes may be taken as 750 N/m². Concrete M 20 and Fe 500.

Solution. Live load = 3000 N/m²

Finishes = 750 N/m²

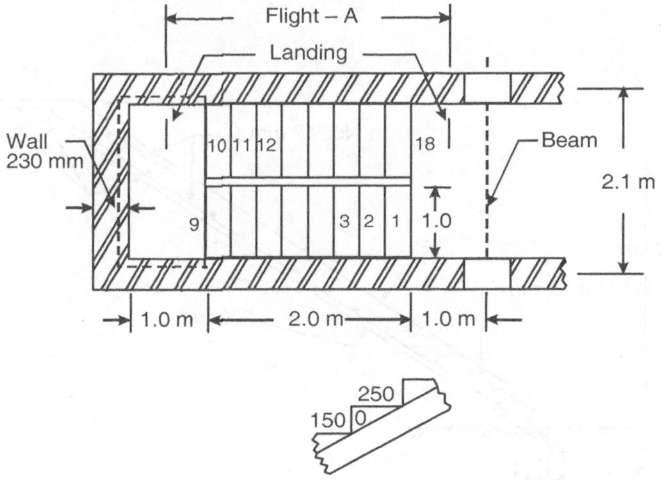

Fig. 11.15.

Assume slab thickness = 125 mm

Self weight of slab per horizontal metre

$$= 1.0 \times 0.125 \frac{\sqrt{250^2 + 150^2}}{250} \times 25000 = 3645 \text{ N/m}^2$$

Weight of step per horizontal metre run

$$= \frac{1}{2} \times 0.25 \times 0.15 \times 1.0 \times 25000 \times \frac{1000}{250} = 1875 \text{ N}$$

Loads per horizontal metre run

Dead loads = 1875 + 3645 + 750 = 6270 N/m

Live load = 3000 N/m

Total load = 9270 N/m say 9300 N/m

Span (centre to centre of landings) = 2.0 + 0.5 + 0.5 = 3.0 m

$$\text{Max. } B.M. = \frac{w.L^2}{8} = \frac{9300 \times 3^2}{8} = 10462.5 \text{ N.m}$$

Width of slab = 1 m

$$d = \sqrt{\frac{10462.5 \times 1000}{0.812 \times 1000}} = 113.51 \text{ mm}$$

D = 125 mm $\qquad d$ = 100 mm

$$A_{st} = \frac{10462.5 \times 1000}{275 \times 0.9155 \times 100} = 415.8 \text{ mm}^2$$, Provide 8 mm ϕ@ 125 mm *c/c*

Distribution steel 0.9155 = 0.12% of 1000 × 125 = 150 mm^2; Provide 8 mm ϕ @ 300 mm *c/c*.

Design of Landing Slab

Span = 2.1 m

Assume the thickness of slab = 125 mm

Self weight of slab = 0.125 × 1 × 1 × 25000 = 3125 N/m

Live load = 3000 N/m

Finishes = 750 N/m^2

Total load = 6875 N/m

$$\text{Max. } B.\,M. = \frac{6875 \times 2.1^2}{8} = 3790 \text{ N.mm}$$

$$d = \sqrt{\frac{3790 \times 1000}{0.812 \times 1000}} = 68.32 \text{ mm}$$

$$D = 125 \text{ mm} \qquad d = 100 \text{ mm}$$

$$A_{st} = \frac{3790 \times 1000}{275 \times 0.9155 \times 100} = 150 \text{ Provide 8 mm } \phi \text{ @ 300 mm } c/c$$

Distribution Steel : 8 mm ϕ @ 300 mm *c*/*c*.

The reinforcement details are shown in Fig. 11.16.

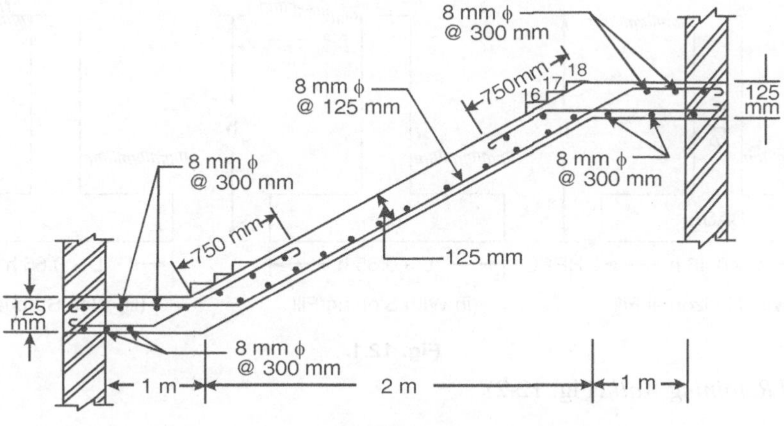

Fig. 11.16.

EXERCISE 11

1. Define the following terms:
   (*a*) Tread (*b*) Rise (*c*) Landing (*d*) Flight (*e*) Going
2. Discuss various types of staircases used in different types of buildings.
3. Design a single flight staircase to cover a horizontal span of 4.5 if the total vertical rise is 3.6. There are total 18 steps to rise. The tread is 250 mm. Take live load as 3000 N/m^2. M25, Fe 415.
4. Design a U-turn staircase, supported on three adjacent 225 mm wlass and having a 0.9 m landing, in a school building where live load can be taken as 4000 N/m^2. The span in a flight is 3.2 m including landing part. The second flight is hinged to a floor beam. The width of waist slab is 0.9 m. Weight of finishes may be taken as 400 N/m^2. M 20, Fe 500
5. Design a staircase flight spanning 3 m if last 0.9 m is a landing. The staircase starts from ground and the flight rests on a 225 mm masonry wall. Take tread = 250 mm, Riser = 175 mm. Live load may be taken as 3000 N/m^2. M25, Fe 500.

CHAPTER

12

Retaining Walls

12.1 INTRODUCTION

The wall, used to retain some material on one or both sides of it, can be termed as "retaining wall". In case of civil engineering structures, it is used to retain soil, usually at two levels on either side. The materials to be retained on either side may be different, e.g., the wall of a swimming pool retains soil on one side and water on the other side. Retaining walls are mostly used for roads in hilly areas, swimming pools, at the ends of the bridges in the form of abutments, constructing a building on a site where filling is required and under-ground water tanks.

12.2 TYPES OF RETAINING WALLS

The various types of retaining walls are described below:

Fig. (12.1) (*a*) *T-Shaped Cantilever Wall*

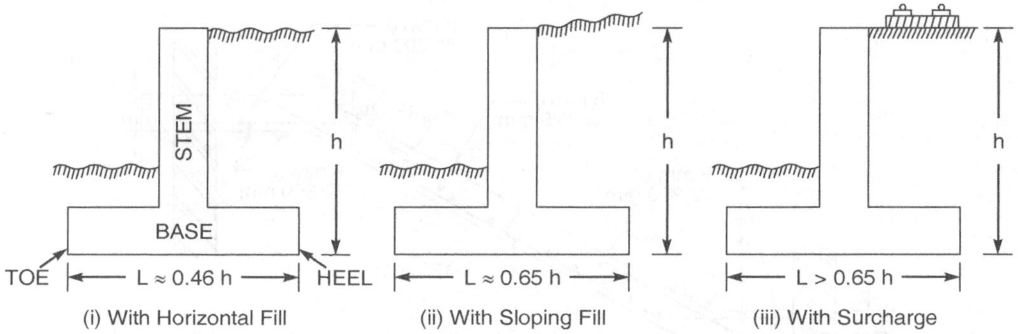

Fig. 12.1.

(*b*) *L-Shaped Retaining Wall* (Fig. 12.2)

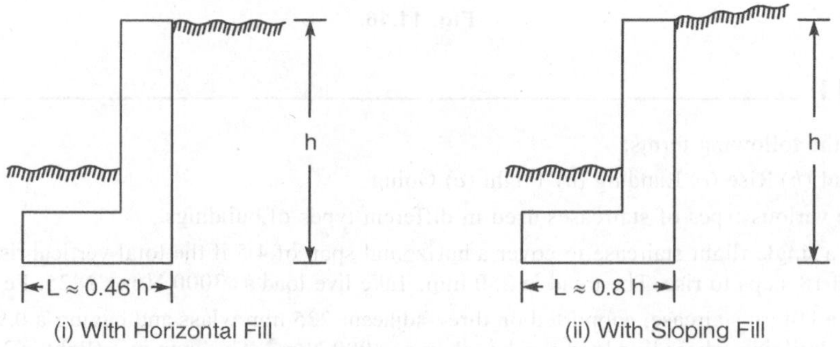

Fig. 12.2.

(*c*) *Reversed L-Shaped Retaining Wall* (Fig. 12.3)

(*d*) *Basement Wall* (Fig. 12.4)

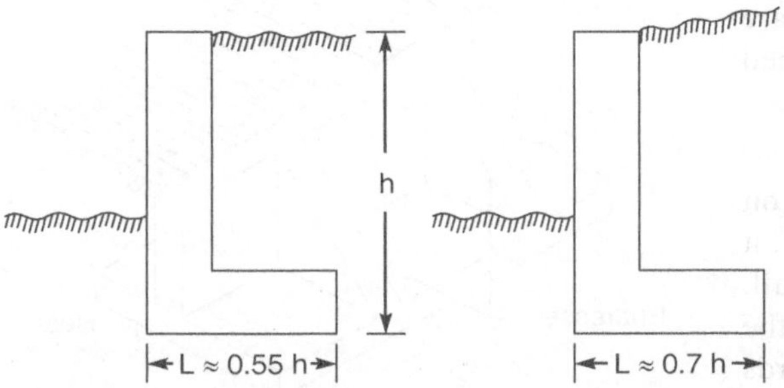

(i) Horizontal Fill (ii) Sloping Fill

Fig. 12.3.

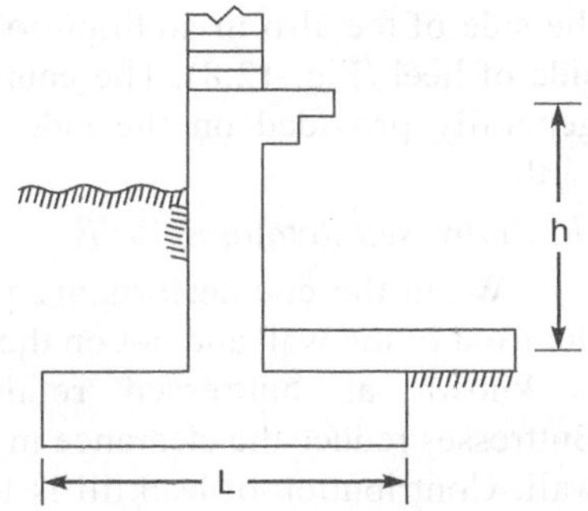

Fig. 12.4.

(*e*) *Gravity Retaining Wall* (Fig. 12.5)

It consists of mass of concrete or masonry. Its dead weight is more to give stability against thrust of retained earth. Its base is so proportioned that there are no tensile stresses developed. Little reinforcement is required in this type of wall.

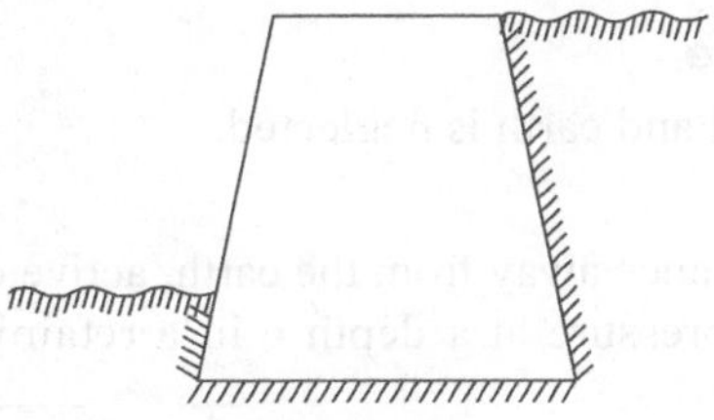

Fig. 12.5.

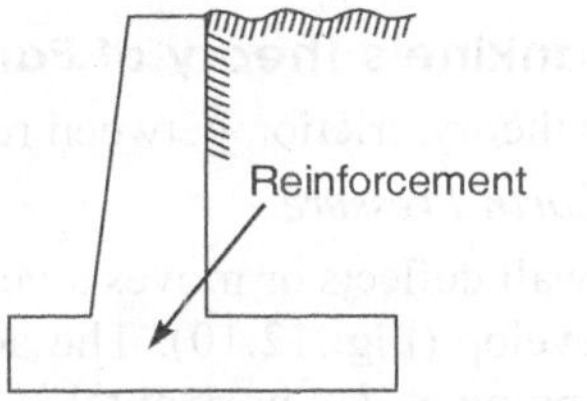

Fig. 12.6.

(*f*) *Semi-Gravity Retaining Walls*

The section is lighter than gravity retaining wall and its stability is increased by providing reinforcement in the stem as well as toe as shown in Fig. 12.6.

(g) *Counterfort Retaining Walls*

When the height of retaining wall is more than 6 m, walls are also provided perpendicular to stem wall. In such a case, it is known as counterfort retaining wall (Fig. 12.7).

The counterforts act as supports to stem and heel slab. Therefore, stem and heel slabs are designed as continuous slabs supported over counterforts. The sizes of concrete components and the steel reinforcement shall be reduced as the bending moments are less than that in case of cantilever retaining wall. Extra cost of formwork to make counterforts is involved.

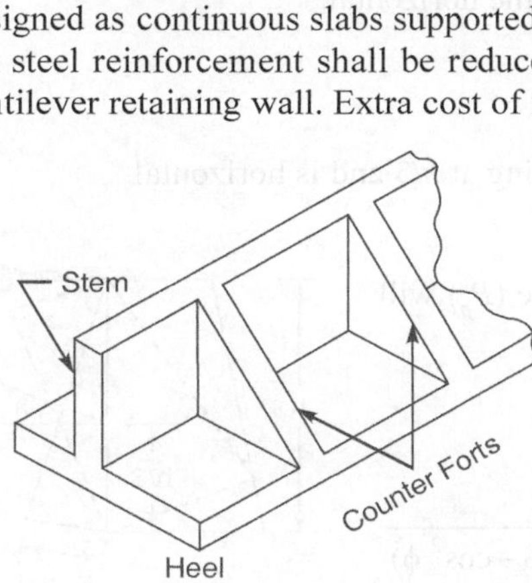

Fig. 12.7.

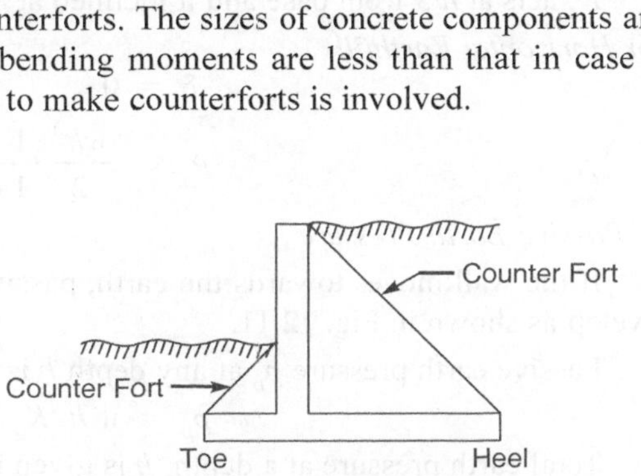

Fig. 12.8.

Sometimes counterforts may be used on the side of toe also in addition to that on the side of heel (Fig. 12.8). The counterforts are generally provided on the side of retained earth.

(*h*) *Buttressed Retaining Walls*

When the counterforts are provdied on the front of the wall and not on the soil side, it is known as buttressed retaining wall. Buttresses reduce the clearance in front of the wall. Contribution of backfill is less towards the stability of the wall because heel projection is small. It is shown in Fig. 12.9.

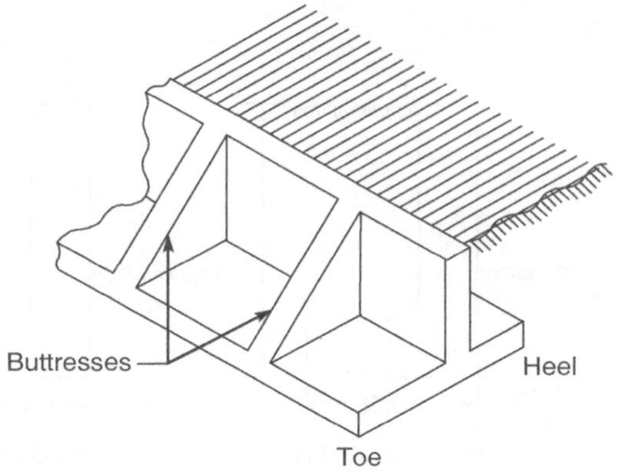

Fig. 12.9.

12.3 THEORIES OF EARTH PRESSURE

There are two theories for calculating earth pressure on retaining walls.

1. Rankine's Theory
2. Coulumb's Theory

12.3.1 Rankine's Theory of Earth Pressure

In this theory, friction between retaining wall and earth is neglected.

(*a*) *Active Earth Pressure*

If the wall deflects or moves a very small distance away from the earth, active earth pressure (P_a) will develop (Fig. 12.10). The active earth pressure at a depth *h* in a retaining wall with earthfill slopping at *d* to horizontal is given by

$$p_a = wh.\cos\delta\,\frac{\cos\delta - \sqrt{(\cos^2\delta - \cos^2\phi)}}{\cos\delta + \sqrt{(\cos^2\delta - \cos^2\phi)}}$$

Total pressure P_A at a depth '*h*' is given by

$$P_A = \frac{wh^2}{2}\cdot\cos\delta\,\frac{\cos\delta - \sqrt{(\cos^2\delta - \cos^2\phi)}}{\cos\delta + \sqrt{(\cos^2\delta - \cos^2\phi)}}$$

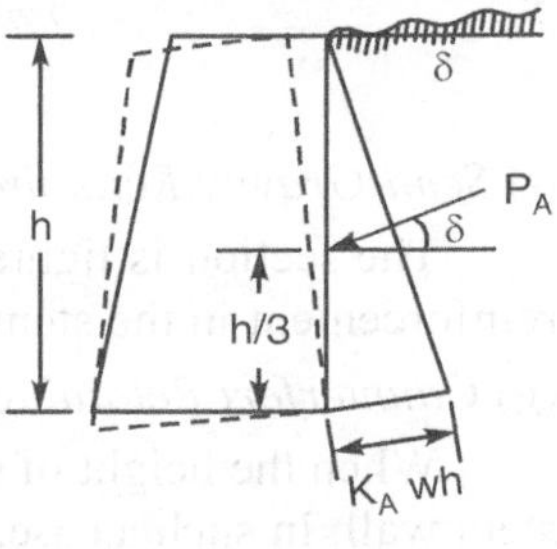

Fig. 12.10.

P_A acts at $h/3$ from base and it inclined at angle δ to the horizontal.

For Horizontal Earthfill

$$\delta = 0$$

$$\therefore \quad P_A = \frac{wh^2}{2}\cdot\frac{1-\sin\phi}{1+\sin\phi}$$ acting at $h/3$ and is horizontal.

(*b*) *Passive Earth Pressure*

If the wall moves towards the earth, passive pressure (P_p) will develop as shown in Fig. 12.11.

Passive earth pressure p_p at any depth *h* is given by

$$p_p = w\,h.\,K_p$$

Total earth pressure at a depth *h* is given by,

$$P_p = \frac{wh^2}{2}\cdot Kp,\ \text{where } K_p = \cos\delta.\,\frac{\cos\delta + \sqrt{(\cos^2\delta - \cos^2\phi)}}{\cos\delta - \sqrt{(\cos^2\delta - \cos^2\phi)}}$$

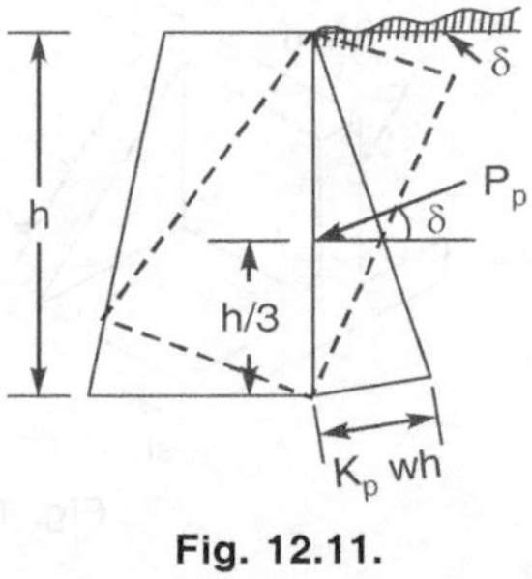

Fig. 12.11.

This pressure acts at $h/3$ from base and is inclined at an angle δ with horizontal.

For Horizontal Earthfill

$$\delta = 0$$

$$P_p = \frac{wh^2}{2} \cdot \frac{1+\sin\phi}{1-\sin\phi} \text{ acting at } h/3 \text{ and is horizontal.}$$

12.3.2 Columb's Theory of Earth Pressure

In this case, the friction between the retaining wall and earth filling is considered, because the back of the retaining wall is never so smooth to neglect the friction.

(a) Active Earth Pressure

If the wall moves outwards, then active earth pressure (P_a) is developed. The intensity of active earth pressure at a depth h is given by,

$$Pa = K_A .\, wh. \cos\theta, \text{ where } K_A = \frac{\cos^2(\phi-\theta)}{\cos^2\theta\cos(\beta+\theta)\left[1+\sqrt{\left(\dfrac{\sin(\beta+\phi)\sin(\phi-\delta)}{\cos(\beta+\theta)\cos(\theta-1)}\right)}\right]^2}$$

and Total pressure, $P_a = K_A . \dfrac{wh^2}{2}.\cos\theta$

where θ = angle made by back of wall with vertical.

ϕ = angle of internal friction of soil

β = angle of friction between retaining wall and earth.

The resultant earth pressure acts at $h/3$ from base and is inclined at angle β to the normal to the face of the wall as shown in Fig. 12.12.

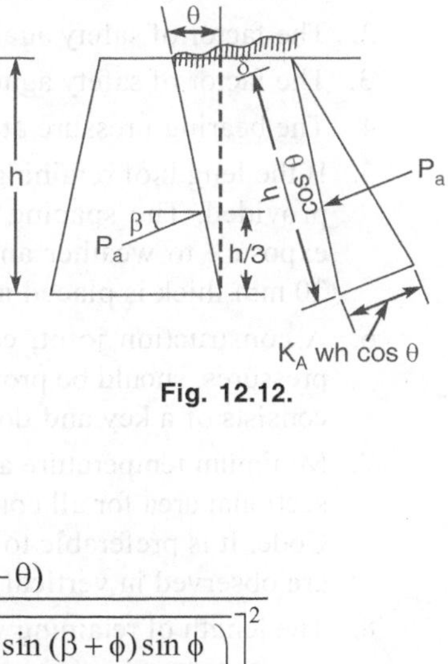

Fig. 12.12.

For Horizontal Earthfill, $\delta = 0$

$$\therefore \quad P_a = K_A . \frac{wh^2}{2}.\cos\theta$$

where
$$K_A = \frac{\cos^2(\phi-\theta)}{\cos^2\theta\cos(\beta+\theta)\left[1+\sqrt{\left(\dfrac{\sin(\beta+\phi)\sin\phi}{\cos(\beta+\theta)\cos\theta}\right)}\right]^2}$$

(b) Passive Earth Pressure

If the wall moves inwards, then passive earth pressure (P_p) is developed. The intensity of passive earth pressure at a depth h is given by $p_p = K_p .\, wh \cos\theta$

where
$$K_p = \frac{\cos^2(\phi-\theta)}{\cos^2\theta.\cos(\theta+\beta)\left[1-\sqrt{\left(\dfrac{\sin(\phi+\beta)\sin(\phi+\delta)}{\cos(\theta-\beta)\cos(\theta-\delta)}\right)}\right]^2}$$

and total pressure, $p_p = K_p . \dfrac{wh^2}{2}\cdot\cos\theta$

This pressure acts at $h/3$ from base and it inclined at an angle β to the face of the wall as shown in Fig. 12.13.

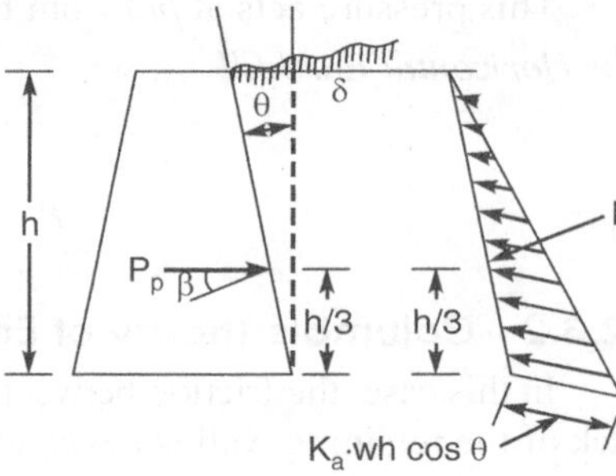

Fig. 12.13.

For Horizontal Earthfill, $\delta = 0$

$$P_p = K_p \cdot \frac{wh^2}{2} \cdot \cos\theta$$

where $$K_p = \frac{\cos^2(\phi+\theta)}{\cos^2\theta\cos(\theta+\beta)\left[1-\sqrt{\left(\frac{\sin(\phi+\beta)\sin\phi}{\cos(\phi-\beta)\cos\theta}\right)}\right]^2}$$

acting at $h/3$ from base and is inclined at an angle β to the normal to the face of the wall.

12.4 GENERAL DESIGN REQUIREMENTS

The following requirements should be satisfied by a retaining wall in order to avoid its failure.

1. The thickness of different components of the retaining wall, *i.e.*, stem, heel and toe should be sufficient to resist the moments, shear forces acting on the wall.
2. The factor of safety against sliding should be at least 1.5.
3. The factor of safety against overturning should be at least 1.5.
4. The bearing pressure at toe should be less than the bearing capacity of the soil.
5. If the length of retaining the wall exceeds 45 metres, then the expansion joint should be provided. The spacing of such joints would depend upon the temperature variation, exposure to weather and the season of laying concrete. Flexible joint filler 12 mm to 20 mm thick is placed in the staggard form.
6. A construction joint, capable of resisting the shear force caused by the lateral earth pressures, should be provided between the base slab and stem of retaining wall. This joint consists of a key and dowel bars. The various types are shown in Fig. 12.14.
7. Minimum temperature and distribution reinforcement should be 0.12% of the gross cross-sectional area for all componenets of retaining wall as for IS code and 0.20% as per ACI Code. It is preferable to provide Min 0.2% reinforcement in a direction otherwise cracks are observed in vertical walls of thickeness 300 mm or more.
8. The length of retaining wall to be cast in one go should not exceed 10 m otherwise cracks may appear.

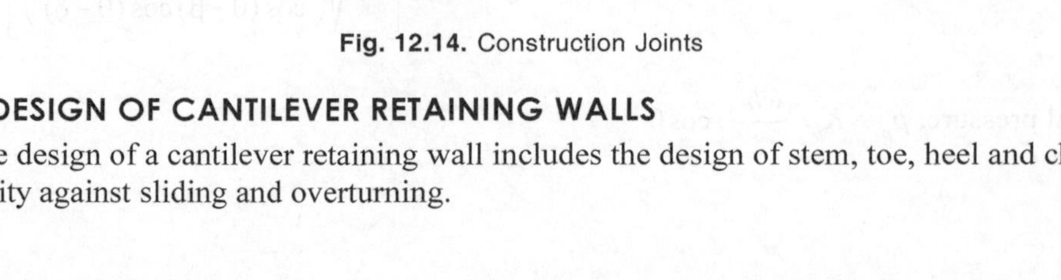

Fig. 12.14. Construction Joints

12.5 DESIGN OF CANTILEVER RETAINING WALLS

The design of a cantilever retaining wall includes the design of stem, toe, heel and checking its stability against sliding and overturning.

12.5.1 Design of Stem

Stem acts as a cantilever and is subjected to lateral pressures which cause maximum *B. M.* and *S. F.* at the junction of stem with base slab. Generally thickness of 15 cm is provided at the top of stem. The thickness at the junction and the reinforcement are designed for the maximum *B. M.* The vertical reinforcement is curtalied in the top portion either at mid-height or at one-third and two-third points from junction. Temperature reinforcement at the rate of 0.12% of gross cross sectional area is provided on the face of stem and is distributed equally in vertical and horizontal directions.

12.5.2 Design of Base Slab

The base width should be such that the resultant of the various forces acting on wall should cut the base in the middle third so that no tension is developed at the base. Various forces acting on wall are earth pressure, weight of earth supported on heel, weight of stem and the weight of base slab, weight of earth supported on toe, if any, is neglected. The maximum and minimum pressures at the base are gien by the following equation.

$$\text{Pressure} = \frac{W}{B}\left(1 \pm \frac{6e}{B}\right)$$

where W is the total vertical weight, e is the eccentricity of the resultant force acting on the base and B is the base width. Maximum pressure should not exceed the allowable bearing capacity of the soil.

12.5.3 Design of Toe

The toe acts as a cantilever. It is subjected to upward pressure from the soil and the self weight acting downwards. Maximum B.M. and S.F. occurs at the face of stem for which adequate thickness and steel are provided. Maximum B.M. is due to upward pressure of soil. Steel is placed near the lower face.

12.5.4 Design of Heel

The heel acts as a cantilever. It is subjected to upward pressure from the soil and the downward loads due to self weight and earth above the heel. B. M. due to downward loads is more than that due to the upward loads. Maximum B.M. and S.F. occurs at the junction of heel with stem for which adequate thickness and steel are provided. Steel is placed near the upper face.

12.5.5 Check against Sliding

The wall has a tendency to slide, at the junction of base with soil, due to the lateral pressures. This tendency is counter-balanced by the friction between the base and the soil. Resisting force is the frictional force (μW) which should be at least 1.5 times the sliding force (the horizontal component of earth pressures). μ is the coefficient of friction. In case the frictional resistance is not sufficient, either increase the value of W by increasing the width of heel or provide a key below the base. The key may be located anywhere in the base width but it is preferred if provided below the stem.

12.6 APROXIMATE FORMULA FOR BASE WIDTH

The approximate formula can be used for level ground surface. It is obtained on the assumption that the density of earth retained and concrete is same and that the weight of toe and earth above it is neglected.

$$B = H\sqrt{\frac{\left(\dfrac{1-\sin\phi}{1+\sin\phi}\right)}{(1-K)\,(3K+1)}}$$

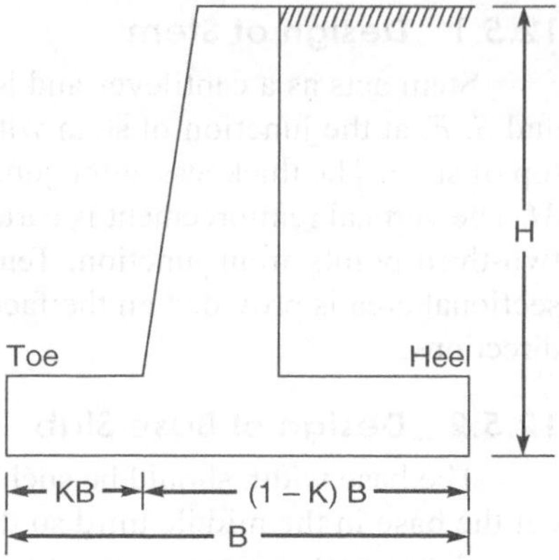

Fig. 12.15.

where $K = 1 - \dfrac{p_0}{2H.w}$

p_0 = allowable bearing capacity of soil

w = density of soil

H = height of retaining wall

ϕ = angle of repose of soil

For surcharge of earth at an angle δ with horizontal

$$B = H\sqrt{\left(\frac{C_p \cdot \cos\delta}{(1-K)\,(1+3K)}\right)}$$

where $C_p = \cos\delta \cdot \dfrac{\cos\delta + \sqrt{(\cos^2\delta - \cos^2\phi)}}{\cos\delta - \sqrt{(\cos^2\delta - \cos^2\phi)}}$

$$K = 1 - \frac{3}{8\beta}$$

$$\beta = \frac{w \cdot h}{p_0}$$

For no surcharge of earth

$$B = 0.95\sqrt{\left(\frac{C_p}{(1-K)(1+3K)}\right)}$$

where $K = 1 - \dfrac{4}{9\beta}$

12.7 DESIGN OF COUNTERFORT RETAINING WALL

The design of counterfort retaining wall includes the design of vertical slab, toe slab, heel slab and counterforts. Counterfort retaining walls are generally economical when its height exceeds 6 m.

12.7.1 Spacing of Counterforts

The spacing of counterforts depends upon the height of the wall, cost of steel and concrete, allowable bearing capacity of soil and the cost of formwork. The thickness of vertical slab and heel slab can be reduced by keeping the counterforts closure. But the reduction in cost of concrete and steel due to less thickness of vertical and heel slabs is offset by the quantity of concrete and steel required for large number of counterforts which are required. The cost of formwork required for large number of counterforts also offsets the above factor. The best spacing is the one which makes the design economical. The most economical spacing is one-third to one-half the height of retaining wall.

12.7.2 Thickness of Counterforts

The thickness of counterforts is governed by the space required to accommodate the number of steel bars.

12.7.3 Design of Counterforts

The counterforts act as a cantilever beam. They are designed for a load from stem portion betwen the adjacent counterforts. The effective depth of counterfort is the distance between its junction with stem and base slab. The stem will have a tendency to move itself and the counterforts away under the action of earth pressure, thus creating the tension on inclined side of the counterfort. Therefore, main steel is povided on the inclined side. The section is checked for shear and end anchorage. Two legged stirrups are provided vertically and horizontally to connect the counterfort with the base and stem respectively. These stirrups are in tension. The front counterfort, if any, is designed as a cantilever beam subjected to self load acting downwards and net earth pressure on the toe portion between two counterforts acting upwards. The section is checked for shear and end anchorage.

12.7.4 Design of Stem (*i.e.*, Vertical Slab)

The stem is designed as a continuous slab with span equal to the spacing of counterforts. The spacing of counterforts may vary from 2.5 m to 4 m. Maximum load on stem is at its lowest portion due to maximum horizontal earth pressure. Consider one metre height of vertical slab and design it for maximum moments. The maximum negative moment at the end supports may be taken as $\frac{wl^2}{10}$ and that at intermediate supports as $\frac{wl^2}{12}$, where l is the span and w is the earth pressure intensity at the lowest portion of vertical slab. The reinforcement is curtailed towards top. The section is checked for shear and end anchorage.

12.7.5 Design of Toe Slab

The design of toe slab depends upon whether the front counterforts are provided or not. If provided, then toe slab is designed as a continuous slab supported on counterforts; and if not provided, then toe slab is designed as a cantilever. In both the cases, the slab is subjected to upward earth pressure and downward self load of slab, the downward load of earth on toe slab being neglected.

12.7.6 Design of Heel Slab

The heel slab is designed as a continuous slab with span equal to approximate depth of heel slab can be fixed by (*i*) $d = 40\ l\ \sqrt{H}$ from B.M. consideration and (*ii*) $d = 40/3\ Hl$ from S.F. consideration for M20 concrete where d is the depth of heel slab in mm, l is the clear distance between counterforts in metre and H is the overall height of the retaining wall in metres.

12.8 SOLVED EXAMPLES

Example 12.1. Design a cantilever retaining wall to retain a lever earthfill to 4.5 m above ground level. The surcharge on the earth fill is 15 KN/m^2. The angle of repose of soil is 30°, unit weight of soil is 16KN/m^3, coefficient of friction between soil and concrete is 0.6 and the safe bearing capacity of soil is 150 KN/m^2. Use M 20 concrete and Fe 500 steel.

Solution.

$$K_A = \frac{1-\sin\phi}{1+\sin\phi} = \frac{1-\sin 30°}{1+\sin 30°}$$

$$\text{Depth of foundation below ground level} = \frac{P_0}{w}\left(\frac{1-\sin\phi}{1+\sin\phi}\right)^2$$

$$= \frac{150}{16}\left(\frac{1-\sin 30°}{1+\sin 30°}\right)^2 = 1.05 \text{ m}$$

Total height of wall above foundation level = 4.5 + 1.05 = 5.55 m.

Approximate Base Width

$$B = H\cdot\sqrt{\left(\frac{(1-\sin\phi)/(1+\sin\phi)}{(1-K)(1+3.K)}\right)} \quad \text{where } K = 1-\frac{P_0}{2.H.w}$$

$$= 5.55\sqrt{\left(\frac{1/3}{(1-0.155)(1+3\times 0.155)}\right)} \quad = 1-\frac{150}{2\times 5.55\times 16}$$

$$= 2.9 \text{ m} \quad = 0.155$$

Generally the base width of the earth retaining wall lies near to 0.6 H

$$B = 0.6 \times 5.55 = 3.33 \text{ say } 3.4 \text{ m}$$

Take B = 3.4 m

Converting surcharge into equivalent depth of earth (h_1)

$$w.h_1 = \text{surcharge} = 15, \therefore h_1 = \frac{15}{16} = 0.9375 \text{ m}$$

$$K_A.w.\ h_1 = \frac{1}{3}(16)\times\left(\frac{15}{16}\right) = 5\text{KN/m}^2$$

$$K_A.\ w(H+h_1) = 1/3 \times 16 \times (5.55 + 0.9375) = 34.6 \text{ KN/m}^2$$

Resultant active earth pressure

$$P = \frac{5+34.6}{2}\times 5.55 = 109.89 \text{ KN}$$

$$\text{lever arm} = \frac{34.6+2\times 5}{34.6+5}\times\frac{5.55}{3} = 2.08 \text{ m}$$

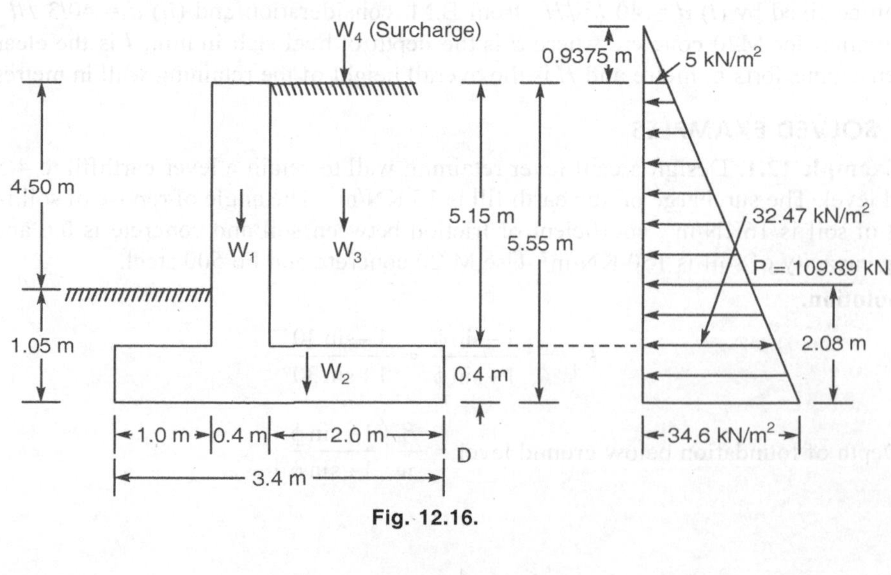

Fig. 12.16.

Check for Base Width

Assume the average thickness of stem, heel and toe as 400 mm.

Description	*(N)*	*C.G. from D* (m)	*Moment about D* (N.m)
$W_1 = 5.15 \times 0.4 \times 1 \times 25000$	= 51500	2.2	113300
$W_2 = 3.4 \times 0.4 \times 1 \times 25000$	= 34000	1.7	57800
$W_3 = 2.0 \times 5.15 \times 1 \times 16000$	= 164800	1.00	164800
W4 = 0.9375 × 2 × 1 × 16000	= 30000	1.00	30000
Total	280300		365900
P = 109890		2.08	228571
			M = 594471

Resultant strikes the base at a distance Z from D $\dfrac{594471}{280300} = 2.12$ m.

$$\frac{2}{3}\times 3.4 = 2.267 \text{ m greater than } 2.12 \text{ m.}$$

Therefore no tension develops at the base.

$$e = 2.12 - \frac{1}{2}\times 3.4 = 0.42 \text{ m}$$

Pressure Distribution at Base

$$\text{Pressure} = \frac{W}{B}\left(1 \pm \frac{6.e}{B}\right)$$

$$P_{max} = \frac{280300}{3.4}\left(1+\frac{6\times 0.42}{3.4}\right) = 143550 \text{ N/m}^2 \text{ less than } 150 \text{ KN/m}^2$$

$$P_{min} = \frac{280300}{3.4}\left(1-\frac{6\times 0.42}{3.4}\right) = 21340 \text{ N/m}^2$$

Pressure diagram is plotted as shown in Fig. 12.17.

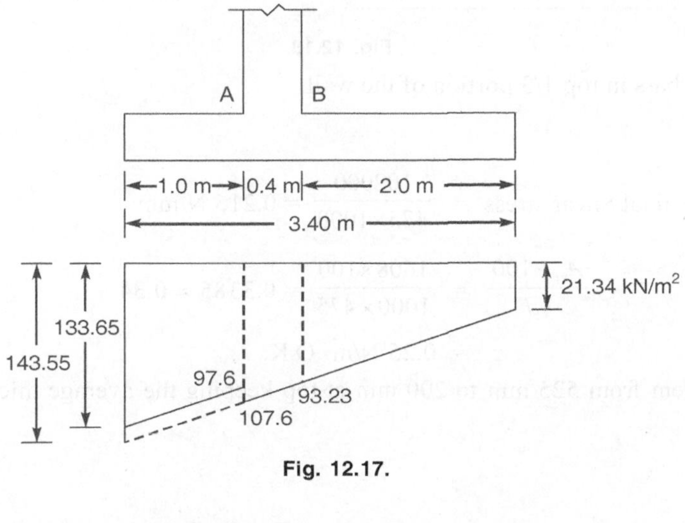

Fig. 12.17.

$$\text{Pressure at } A = 21.34 + \frac{143.55 - 21.34}{3.4} \times 2.4 = 107.6 \text{ KN/m}^2$$

$$\text{Pressure at } B = 21.34 + \frac{143.55 - 21.34}{3.4} \times 2 = 93.23 \text{ KN/m}^2$$

Design of Stem

Max. *B.M.* at *AB* level

$$= \left(\frac{5 + 32.47}{2} \times 5.15\right)\left[\frac{32.47 + 2 \times 5}{32.47 + 5} \times \frac{5.15}{3}\right] = 187.735 \text{ KN.m}$$

$$\text{Max. S.F.} = (5 + 34.6) \times \frac{1}{2} \times 5.15 = 102 \text{ KN}$$

$$d = \sqrt{\left(\frac{187.735 \times 1000 \times 1000}{0.812 \times 1000}\right)} = 480.83 \text{ mm}$$

Provide D = 525 mm, d = 475 mm.

$$A_{st} = \frac{187.735 \times 10^6}{275 \times 0.9155 \times 475} = 1570 \text{ mm}^2$$

Provide 16 mm ϕ @ 125 mm *c/c*, (A_{st} provided = 1608 mm²).

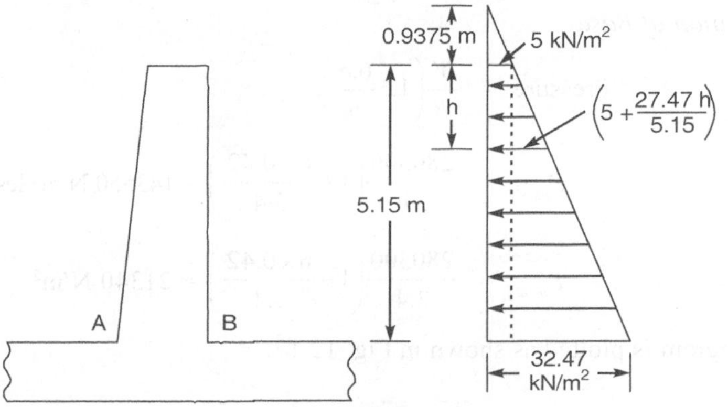

Fig. 12.18.

Curtail half bars in top 1/3 portion of the wall.

Check of Shear

$$\text{Nominal Shear stress} = \frac{102000}{475 \times 1000} = 0.215 \text{ N/mm}^2$$

$$\frac{A_{st} \times 100}{bd} = \frac{1608 \times 100}{1000 \times 475} = 0.3385 \approx 0.34$$

$$\tau_c = 0.25 \text{ N/m}^2 \text{ O.K.}$$

Taper the stem from 525 mm to 200 mm at top keeping the average thickness as 400 mm approximately.

Design of Toe

Design it as a cantilever slab under the action of self weight und upward pressure from soil.

$$\text{Weight of toe slab} = 0.4 \times 1 \times 1 \times 25 = 10 \text{ KN/m}^2$$

$$\text{Average upward pressure} = \frac{133.55 + 97.6}{2} = 11.66 \text{ KN/m}^2$$

$$B.M. \text{ at } A = 116.6 \times 1 \times \frac{107.6 + 2 \times 143.55}{107.6 + 143.55} \times 1/3 = 61.08 \text{ KN.m}$$

$$d = \sqrt{\left(\frac{61.08 \times 1000 \times 1000}{0.812 \times 1000}\right)} = 274.3 \text{ mm. Provide } D = 400 \text{ mm, } d = 350 \text{ mm.}$$

$$A_{st} \text{ required} = \frac{61.08 \times 1000 \times 1000}{275 \times 0.9155 \times 350} = 693.2 \text{ mm}^2$$

Extend tension steel of stem in toe (A_{st} provided = 1608 mm^2)

Check for Shear

$$\text{Shear force} = 116.6 \text{ KN,}$$

$$\text{Nominal shear stress, } \tau_v = \frac{116.6 \times 1000}{1000 \times 350} = 0.33 \text{ N/mm}^2$$

$$< 0.16\sqrt{(f_{ck})} = 0.7155 \text{ N/mm}^2$$

$$\frac{A_s \times 100}{bd} = \frac{1608 \times 100}{1000 \times 350} = 0.46$$

$$\therefore \quad \tau_c = 0.29 \text{ N/mm}^2$$

Since $\tau_v > \tau_c$, shear reinforced is required.

$$\text{Shear concrete can carry} = \tau_c . b.d = 0.29 \times 1000 \times 350 = 101500 \text{ N}$$

$$V_s = \text{Shear force reinforcement} = 116600 - 101500 = 15100 \text{ N}$$

$$8 \text{ mm } \phi, 2 \text{ legged at spacing} = A_{sv} . \sigma_{sv} / d/v_s$$

$$= 2 \times 50 \times 230 \times 350/15100$$

$$= 533 \text{ mm } c/c.$$

$$\text{Max spacing} = 0.75\, d = 0.75 \times 350 = 262.5 \text{ mm}$$

Use 8 mm ϕ, 2 legged shear stirrups @ 250 mm *c/c*.

Design of Heel

The downward weights acting on heel are W_3, W_4 and self weight of heel and the upward pressure of soil.

$$\text{Self weight of heel} = 0.4 \times 2 \times 1 \times 25000 = 20000 \text{ N}$$

$$\text{Upward pressure of soil} = \frac{93.23 + 21.34}{2} \times 2.0 = 114.57 \text{ KN}$$

$$\text{Acting at } \frac{93.23 + 2 \times 21.34}{93.23 + 21.34} \times 2/3 = 0.79 \text{ m from } B.$$

$$B.M. \text{ at } B = 164800 \times 1 + 20000 \times 1.0 - 114570 \times 0.79 = 124290 \text{ N.m.}$$

$$d = \sqrt{\left(\frac{124290 \times 1000}{0.812 \times 1000}\right)} = 391.24 \text{ mm}$$

Provide $D = 400$ mm, $d = 350$ mm

$$A_{st} \text{ required} = \frac{124290 \times 1000}{275 \times 0.9155 \times 350} = 1410.5 \text{ mm}^2$$

Provide 16 mm ϕ @ 125 mm *c*/*c*. A_{st} provided = 1608 mm^2.

Check for Shear

$$\text{Shear force} = 164800 + 30000 + 20000 - 114570 = 100230 \text{ N}$$

$$\text{Nominal shear stress} = \frac{100230}{1000 \times 350} = 0.286 \text{ N/mm}^2 \text{ O.K.}$$

Bars shall be extended 58 × 20 = 1160 mm from *B*.

Check against Sliding

$$\text{Sliding force} = \text{Earth pressure} = 109890 \text{ N}$$

$$\text{Frictional force} = \mu.\ W = 0.6 \times 280300 = 168180 \text{ N}$$

$$\text{Factor of safety against sliding Safe} = \frac{168180}{109890} = 1.53 > 1.5$$

No key is requried.

In case the factor of safety in sliding works out to be less than safe value of 1.5, a key may be provided below base. The design of key in that case is explained below.

A key 0.65 m deep and 0.5 m wide may be considered to be provided to increate resistance against sliding.

Passive pressure $p_p = K_p.\ w.\ h. = 3 \times 16 \times 1.05 = 50.4$ KN/m^2. Active pressure $P_a = K_A.\ w.\ (H + h_1) = \frac{1}{3} \times 16 \times (6.2 + 0.9375) = 38.07$ KN/m^2. The section is checked at bottom of key (at

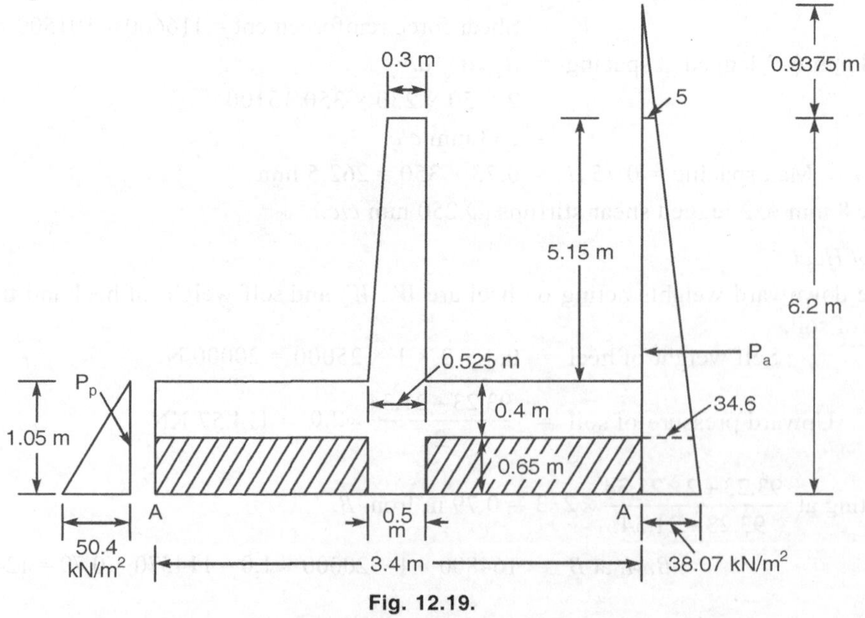

Fig. 12.19.

A-A in Fig. 12.19). The weight of key and earth shown hatched will be additional weight coming at A-A. There will be passive pressure in front of the wall as calculated above.

$$\text{Total weight} = 280300 + (3.4 + 0.5) \times 0.65 \times 16 \times 1000 + 0.65 \times 0.5 \times 25000$$

$$= 318586 \text{ N}$$

$$P_a = \frac{5+38.07}{2} \times 6.2 = 133.517 \text{ KN}$$

$$P_p = \frac{1}{2} \times 50.4 \times 1.05 = 26.46 \text{ KN}$$

Net horizontal force = 133.517 – 26.46 = 107.057 KN = 107057 N

$$\text{Factory of safety} = \frac{0.6 \times 318586}{107057} = 1.78 > 1.5. \text{ Safe}$$

Take all the bars from stem inside the key.

Check for Shear

Max. S.F. at junction of key with base = $(P_a - P_p)$ at base level

$$= 109890 - \frac{16000 \times 0.4^2}{2} \times 3 = 106050 \text{ N}$$

$$\text{Nominal shear stress} = \frac{106050}{1000 \times 450} = 0.236 \text{ N/mm}^2$$

$$\frac{100 \times 1608}{1000 \times 450} = 0.357, \text{ Permissible shear stress} = 0.259 \text{ N/mm}^2 \text{ safe.}$$

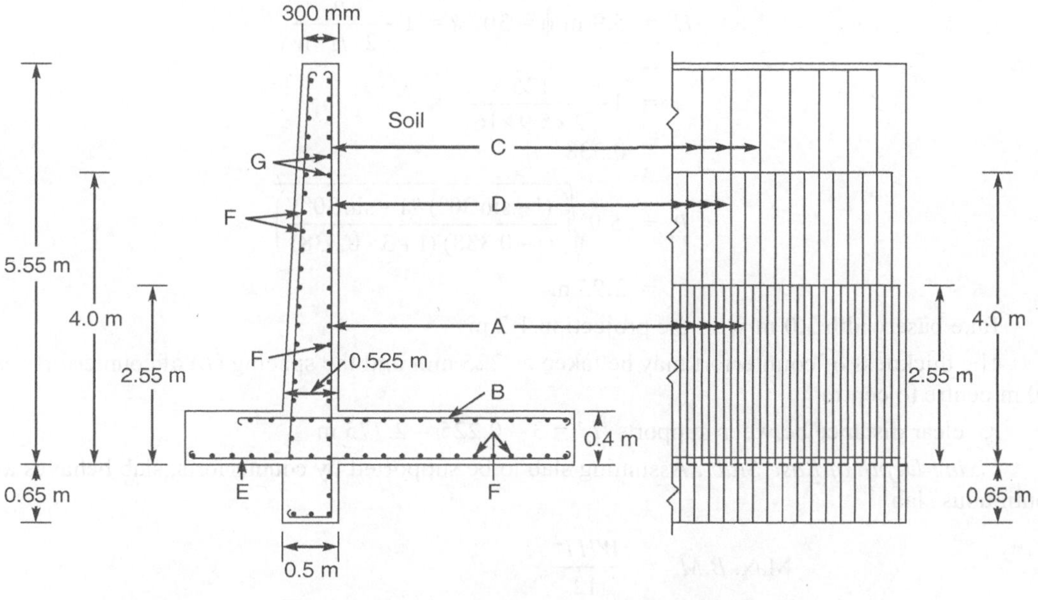

SCHEDULE OF SCHEDULE
A = 16 mm ϕ @ 125 mm c/c
B = 16 mm ϕ @ 125 mm c/c
C = 16 mm ϕ @ 250 mm c/c
D = 16 mm ϕ @ 125 mm c/c
E = 16 mm ϕ @ 125 mm c/c
F, G = 8 mm ϕ @ 200 mm c/c

Fig. 12.20.

The reinforcement details are shown in Fig. 12.20.

$$\text{Average thickness} = 400 \text{ mm}$$

$$\text{Minimum Distribution steel} = 0.12\% \times 400 \times 1000 = 480 \text{ mm}^2$$

$$\text{Distribution steel per face} = \frac{1}{2}\times 180 = 240 \text{ mm}^2$$

provide 8 mm ϕ @ 200, mm c/c $A_{st} = 251 \text{ mm}^2$

Example 12.2. Design a counterfort retaining wall to retain earth 5.0 m above the basement level. The density of the earth is 16000 N/m^3 and the angle of repose of soil is 30°. The bearing capacity of the soil is 125 KN/m^2. Use M 20 and Fe 500.

Solution. The depth of base below basement level

$$= \frac{p_0}{w}\left(\frac{1-\sin\phi}{1+\sin\phi}\right)^2$$

$$\approx \frac{125}{16}\left(\frac{1-\sin 30°}{1+\sin 30°}\right)^2$$

$$= 0.868$$

$$= 0.90 \text{ m say}$$

$$\therefore \quad H = 5.0 + 0.9 = 5.9 \text{ m}$$

Approximate Base width

$$B = H\sqrt{\left(\frac{(1-\sin\phi)/(1+\sin\phi)}{(1-k)\cdot(1+3k)}\right)}$$

$$H = 5.9 \text{ m} \quad \phi = 30° \quad k = 1 - \frac{p_0}{2\cdot H\cdot w}$$

$$= 1 - \frac{125}{2\times 5.9\times 16}$$

$$\approx 0.338$$

$$\therefore \quad B = 5.9\sqrt{\left(\frac{(1-\sin 30°)/(1+\sin 30°)}{(1-0.338)\,(1+3\times 0.338)}\right)}$$

$$= 2.95 \text{ m}$$

Take base width 3.0 m with the projection 1.2 m

The thickness of counterfort may be taken as 225 mm and the spacing (L) of counterforts as 3.0 m centre to centre.

$\therefore$ clear distance between supports 'l_c' = 3 – 0.225 = 2.775 m

Fixing Depth of Base Slab : Assuming slab to be supported by counterforts, slab behaves as continuous slab

$$\text{Max. } B.M. = \frac{WHl^2}{12}$$

$$= \frac{16000\times 5.9\times 3.0^2}{12} = 70800 \text{ N.m}$$

$$\text{Effective depth} = \sqrt{\left(\frac{70800 \times 1000}{0.812 \times 1000}\right)} = 295.3 \text{ mm}$$

$$\text{Provide base thickness} = 300 \text{ mm}$$

The arrangement of counterforts is shown in Fig. 12.21.

Design of Vertical Slab

Consider 1 m height of slab near base

$$\text{Earth pressure} = wh\frac{1-\sin\phi}{1+\sin\phi}$$

$$q = 16000 \times 5.6\left(\frac{1-\sin 30°}{1+\sin 30°}\right)$$

$$= 29866.6 \text{ N}$$

Max. *B. M.* due to

$$\text{Eart pressure} = \frac{q\,l^2}{12} = \frac{29866 \times 2.775^2}{12}$$

$$= 19166 \text{ NM}$$

$$\text{Effective depth} = \sqrt{\left(\frac{19166 \times 1000}{0.812 \times 1000}\right)}$$

$$= 153.6 \text{ mm}$$

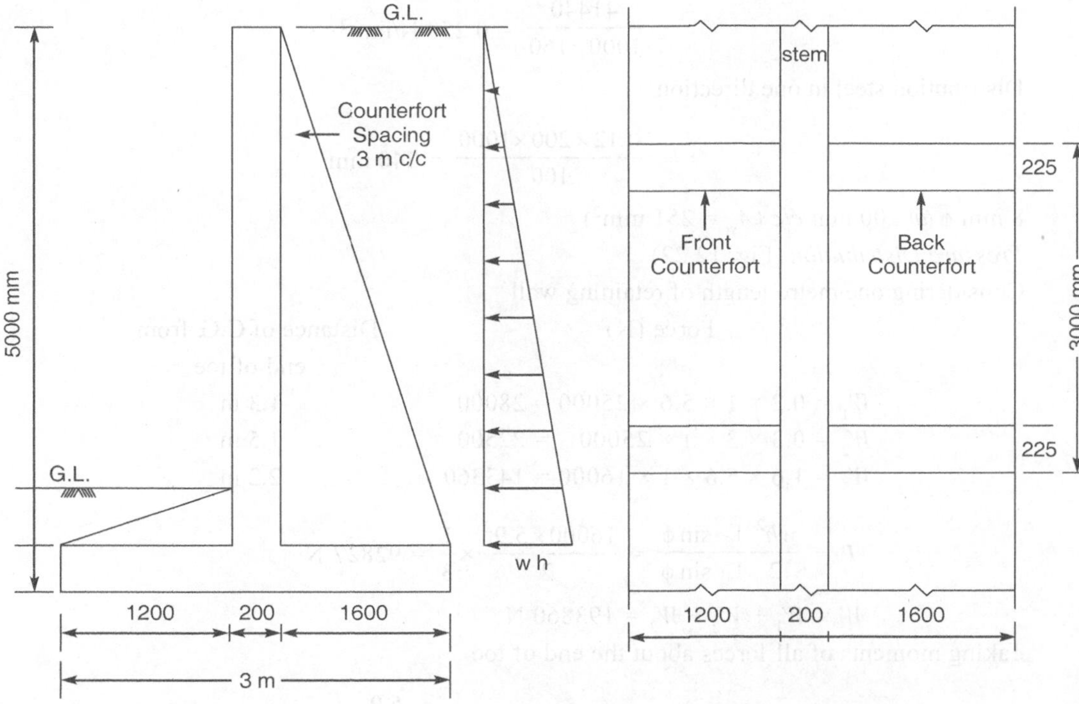

Fig. 12.21.

$$\text{Provide overall depth } `D' = 200 \text{ mm}$$
$$\text{and } d = 160 \text{ mm}$$
$$\text{Effective cover} = 40 \text{ mm}$$

Provide same thickness throughout height.

$$\text{Area of steel required} = \frac{M}{\sigma_{st} jd}$$

$$= \frac{19166 \times 1000}{275 \times 0.9155 \times 160} = 475.8 \text{ mm}^2$$

Provide 8 mm ϕ bars @ 100 mm *c*/*c* (A_{st} provided = 503 mm^2)

Since the earth pressure decreases towards top, B.M. will decrease linearly towards top. The spacing of bars can be increased to 400 mm *c*/*c* near top gradually.

Check for Shear

$$\text{Max. shear force} = wh\frac{1-\sin\phi}{1+\sin\phi} \times \frac{l_c}{2}$$

$$V = 16000 \times 5.6 \times \frac{1}{3} \times \frac{2.775}{2}$$

$$= 41440 \text{ N}$$

$$\tau_v = \frac{V}{bd}$$

$$= \frac{41440}{1000 \times 160} = 0.259 \text{ N/mm}^2$$

Distribution steel in one direction

$$= \frac{0.12 \times 200 \times 1000}{100} = 240 \text{ mm}^2$$

8 mm ϕ @ 200 mm *c*/*c* (A_{st} = 251 mm^2)

Pressure Distribution (Fig. 12.22)

Considering one metre length of retaining wall

Force (N)	Distance of C.G. from end of toe
$W_1 = 0.2 \times 1 \times 5.6 \times 25000 = 28000$	1.3 m
$W_2 = 0.3 \times 3 \times 1 \times 25000 = 22500$	1.5 m
$W_3 = 1.6 \times 5.6 \times 1 \times 16000 = 143360$	2.2 m

$$P = \frac{wh^2}{2}\frac{1-\sin\phi}{1+\sin\phi} = \frac{16000 \times 5.9^2}{2} \times \frac{1}{3} = 92827 \text{ N}$$

$$W = W_1 + W_2 + W_3 = 193860 \text{ N}$$

Taking moments of all forces about the end of toe

$$28000 \times 1.3 + 22500 \times 1.5 + 143360 \times 2.2 = 92827 \times \frac{5.9}{3} + W\bar{x}$$

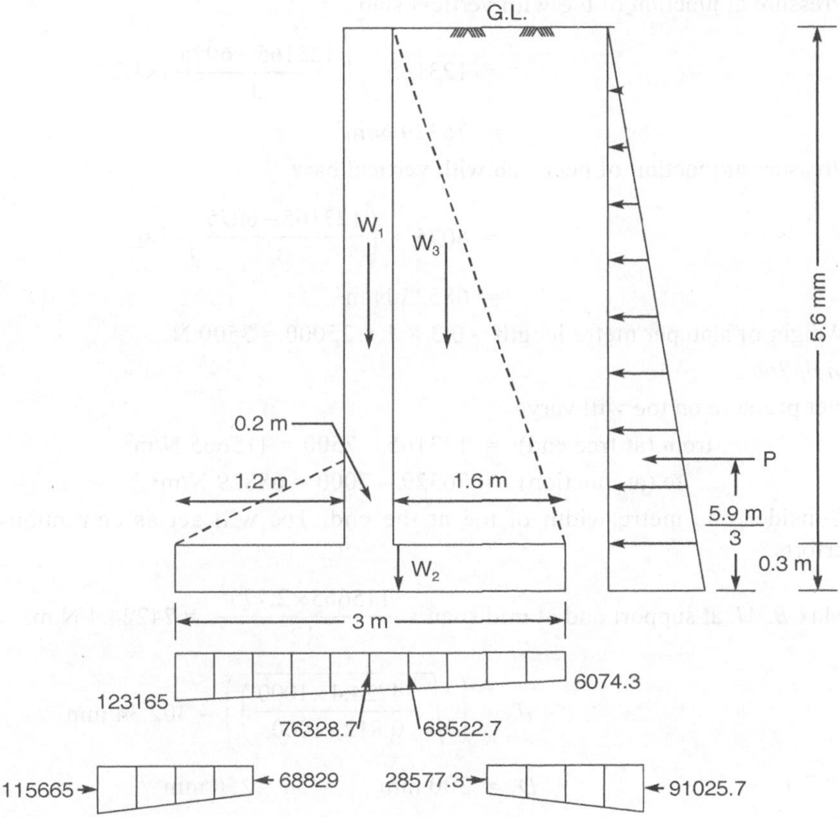

Fig. 12.22

$$\bar{x} = \frac{36400 + 33750 + 315392 - 182560}{193860}$$

$$= 1.047 \text{ m}$$

∴ Resultant lies within middle third of base.

$$\text{Eccentricity} = \frac{3}{2} - 1.047 = 0.453 \text{ m}$$

Pressure distribution at base is given by

$$P = \frac{W}{b}\left(1 \pm \frac{6e}{b}\right)$$

$$= \frac{193860}{3}\left(1 \pm \frac{6 \times 0.453}{3}\right)$$

$$= 64620\ (1 \pm 0.906)$$

P_{max} = 123165 N/m² < 125000 N/m² ∴ O.K.

P_{min} = 6075 N/m²

Pressure at junction of toe with vertical slab

$$= 123165 - \left(\frac{123165 - 6075}{3}\right) \times 1.2$$

$$= 76329 \text{ N/m}^2$$

Pressure at junction of heel slab with vertical base

$$= 6075 + \left(\frac{123165 - 6075}{3}\right) \times 1.6$$

$$= 68523 \text{ N/m}^2$$

Weight of slab per metre length = 0.3 × 1 × 25000 = 7500 N

Design of Toe

Net pressure on toe will vary

$$\text{from (at free end)} = 123165 - 7500 = 115665 \text{ N/m}^2$$

$$\text{to (at junction)} = 76329 - 7000 = 68829 \text{ N/m}^2$$

Consider one metre width of toe at the end. Toe will act as continuous slab between counterforts.

$$\text{Max } B.\ M. \text{ at support and at mid span} = \frac{115665 \times 2.775^2}{12} = 74224.4 \text{ N.m.}$$

$$d = \left(\sqrt{\left(\frac{74224.4 \times 1000}{0.812 \times 1000}\right)}\right) = 302.34 \text{ mm}$$

$$D = 300 \text{ mm} \qquad d = 250 \text{ mm}$$

$$A_{st} \text{ required} = \frac{74224.4 \times 1000}{275 \times 0.9155 \times 250} = 1179.3 \text{ mm}^2$$

Provide 12 mm ϕ @ 80 mm *c/c* at top and 12 mm ϕ @ 160 mm *c/c* at bottom throughout. Bend down alternate bars at 0.6 m from fact of supports. A_{st} provided = 1413 mm² (Fig. 12.25 (*a*))

Dist steel

$$= \frac{0.12}{100} \times 300 \times 1000$$

$$= 420 \text{ mm}^2$$

Provide 8 mm ϕ @ 100 *c/c* on each face (A_{st} = 503 mm²)

Check for Shear

$$\text{Max. } S.\ F. = \frac{115665 \times 2.775}{2} = 160485 \text{ N}$$

$$\text{Nominal shear stress } \tau_v = \frac{160485}{1000 \times 250} = 0.642 \text{ N/mm}^2$$

$$\frac{A_s \times 100}{bd} = \frac{1413 \times 100}{1000 \times 250} = 0.565$$

$$\tau_c = 0.312 \text{ N/mm}^2$$

Shear reinforcement is necessary

$$\text{Design shear } V_s = (0.642 - 0.312) \times 1000 \times 250$$
$$= 82500 \text{ N}$$

Providing 10 mm ϕ 4 legged in one metre width at

$$\text{Spacing} = \frac{230 \times 4 \times 78.5 \times 250}{82500} = 218.85$$

Say 200 mm c/c

Since the width of toe slab is 1.2 m, it requires 4 × 1.2 = 4.8 sy 5 legged stirrups. These stirrups may be provided in 0.7 m distance from face of counterfort

For Min. steel in the form of shear stirrups, the spacing shall be

$$S_v = \frac{A_{sv} \times f_y \times 0.87}{0.4b}$$
$$= \frac{5 \times 78.5 \times 415 \times 0.87}{0.4 \times 1200}$$
$$= 295 \text{ mm say } 280 \text{ mm},$$

∴ Provide 10 mm ϕ 5 legged in width 1.2 m @ 200 mm c/c upto 0.7 m from each face of counterfort and in rest of proportion, (i.e., 2.775 – 2 × 0.7 = 1.375 m) @ 280 mm c/c

Design of Heel Slab

$$\text{Weight of earth per metre} = 5.6 \times 1 \times 1 \times 16000$$
$$= 89600 \text{ N/m}$$
$$\text{Weight of heel slab} = 7500 \text{ N/m}$$

Pressure on heel will vary from

$$\text{at free end} = 89600 + 7500 - 6075$$
$$= 91025 \text{ N/m}^2$$
$$\text{to at junction} = 89600 + 7500 - 68523$$
$$= 28577 \text{ N/m}^2$$

Consider 1 m width of slab near and

$$\text{Max. } B.\,M \text{ at support and mid span} = \frac{91025 \times 2.775^2}{12}$$
$$= 58412.4 \text{ N.m}$$

$$d = \sqrt{\left(\frac{58412.4 \times 1000}{0.812 \times 1000}\right)} = 268.2 \text{ mm}$$

$$D = 300 \quad d = 250$$

$$A_{st} \text{ required} = \frac{58412.9 \times 1000}{275 \times 0.9155 \times 250} = 928 \text{ mm}^2$$

Provide 12 mm ϕ @ 100 mm c/c at bottom face and 12 mm ϕ @ 200 mm c/c at top face throughout. Bend up bottom alternate bars at 0.6 m from face of counterfort supports. A_{st} provided = 1131 mm² [Fig. 12.25 (*a*)]

$$\text{Average intensity of pressure} = \frac{91025 + 28577.3}{2} = 59801 \text{ N/m}^2$$

$$\text{Max. } S.\ F. \text{ per metre width} = \frac{59801 \times 2.775}{2}$$
$$= 82974.6 \text{ N}$$
$$\tau_v = \frac{82974}{1000 \times 250} = 0.332 \text{ N/mm}^2$$
$$\frac{A_s \times 100}{bd} = \frac{1131 \times 100}{1000 \times 250} = 0.45$$
$$\tau_c = 0.285 \text{ N/mm}^2 < \tau_v; \therefore \text{ Unsafe}$$

Since $\tau_v > \tau_c$, shear reinforcement is required.

$$V_s = (0.332 - 0.285)\ 1000 \times 250 = 11750 \text{ N}$$

10 mm ϕ, 7 legged @ spacing $= A_{sv} \cdot \sigma_{sv} \cdot d/v_s$

$$= (7 \times 78.5)\ 230 \times 250/11750$$
$$= 2689 \text{ mm}$$
$$\text{Max. spacing} = 0.75 \text{ d} = 0.75 \times 250 = 187.5 \text{ mm}$$

Spacing for minimum shear reinforcement

$$S_v = \frac{A_{sv}(0.87 f_y)}{0.4b}$$
$$= \frac{(7 \times 78.5)(0.87 \times 415)}{0.4 \times 1600}$$
$$= 310 \text{ mm}$$

Provide minimum shear reinforcement, *i.e.*, 10 mm ϕ, 4 × 1.6 = 6.4 say 7 legged @ 175 mm *c/c* upto 0.7 m from each face of supports and thereafter no shear reinforcement is to be provided since the actual shear stress shall be less than half the allowable value of t_c.

Design of Counterforts at Back (Fig. 12.23)

Let θ be the inclination of counterfort with vertical

$$\tan \theta = \frac{1.6}{5.6} = 0.2857, \qquad \theta = 15.945°$$

The section of counterfort shall be designed at 5 m below top as below this front counterfort is also provided.

$$\text{Effective depth} = 5000 \sin 15.945° - 75$$
$$= 1298.6 \text{ mm}$$
$$\text{Max. } B.\ M. = \frac{wh^3}{6}\left(\frac{1-\sin\phi}{1+\sin\phi}\right) + 3$$
$$= \frac{16000 \times 5^3}{6}\left(\frac{1-\sin 30°}{1+\sin 30°}\right) \times 3$$
$$= 333333 \text{ N.m}$$
$$d = \sqrt{\left(\frac{333333 \times 1000}{0.812 \times 225}\right)} = 1350.7 \text{ mm}$$

d provided = 1298.6 mm.

$$A_{st} = \frac{333333 \times 1000}{275 \times 0.9155 \times 1298.6} = 1019.6 \text{ mm}^2$$

Provide 7 Nos. of 16 mm ϕ in two layers (A_{st} provided = 1407 mm²) as shown in Fig. 12.25 (*b*). 3 bars are curtailed at 2.5 m from top and 4 bars continue on the top.

Check for Shear

Shear force due to horizontal thrust at the level of front counterfort

$$= \frac{wh^2}{2}\left(\frac{1-\sin\phi}{1+\sin\phi}\right) \times \text{Spacing of counterforts}$$

$$= \frac{16000 \times 5^2}{2}\left(\frac{1-\sin 30°}{1+\sin 30°}\right) \times 3$$

$$= 200{,}000 \text{ N}$$

$$\text{Net shear force} = F - \frac{M}{d}\sin\theta$$

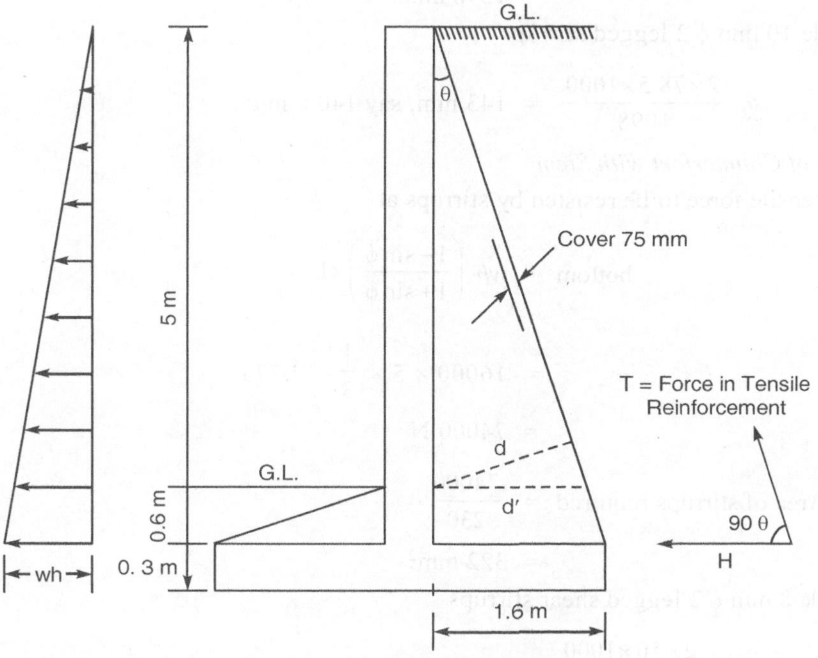

Fig. 12.23.

$$= 200{,}000 - \frac{333333}{1.2986}\sin 15.945°$$

$$= 129483 \text{ N}$$

$$d' = \frac{d}{\cos\theta} = \frac{1298.6}{\cos 15.945°} = 1350 \text{ mm}$$

$$\text{Shear stress } \tau_v = \frac{129483}{225 \times 1350} = 0.426 \text{ N/mm}^2$$

$$\frac{A_s \times 100}{bd} = \frac{7 \times 201 \times 100}{225 \times 1350} = 0.463$$

$$\tau_c = 0.29 \text{ N/mm}^2$$

$$\text{Design shear} = (0.426 - 0.29) \times 225 \times 1350 = 41310 \text{ N}$$

Provide 8 mm ϕ shear stirrups 2 legged

$$@ \frac{50 \times 2 \times 230 \times 1350 \times \cos 15.945^\circ}{41310} = 722.5 \text{ mm}$$

say @ 300 mm *c*/*c*

Connection of Counterfort with Base:

Consider 1 m width of base slab near the end of heel

$$\text{Total downward load} = 91025.7 \times 2.775 = 252596 \text{ N}$$

$$\text{Area of vertical stirrups requried} = \frac{252596}{230}$$

$$= 1098 \text{ mm}^2$$

Provide 10 mm ϕ 2 legged stirrups

$$@ \frac{2 \times 78.5 \times 1000}{1098} = 143 \text{ mm, say 140 mm } c/c$$

Connection of Counterfort with Stem

Max. tensile force to be resisted by stirrups at

$$\text{bottom} = wh \left(\frac{1 - \sin\phi}{1 + \sin\phi}\right) \times 1$$

$$= 16000 \times 5 \times \frac{1}{3} \times 2.775$$

$$= 74000 \text{ N}$$

$$\text{Area of stirrups required} = \frac{74000}{230}$$

$$= 322 \text{ mm}^2$$

Provide 8 mm ϕ 2 legged shear stirrups

$$@ \frac{2 \times 50 \times 1000}{322} = 310 \text{ say 300 mm } c/c$$

As the pressure reduces towards the top of stem gradually and vanishes at top, the spacing may be increased to 450 mm *c*/*c*

Anchorage length of 10 mm ϕ bar required

$$= \frac{\phi\, \sigma_s}{4\, \tau_{bd}}$$

$$= \frac{10\times140}{4\times0.6} \quad \text{say } 600 \text{ mm}$$

Front Counterfort (Fig. 12.24)

Neglecting effect of the self weight of counterfort

Max. *B. M.*

$$= \frac{68829\times1.2^2\times3}{2}+\left(\frac{115665-68829}{2}\right)\times1.2\times3\times\frac{2}{3}\times1.2$$

$$= 216114.5 \text{ N.m}$$

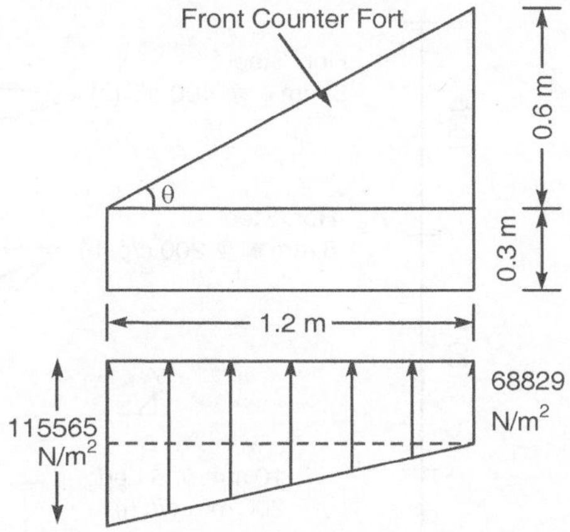

Fig. 12.24.

$$\text{Effect depth required} = \sqrt{\left(\frac{216114.50\times10000}{0.812\times225}\right)}$$

$$= 1087.6 \text{ mm}$$

D provided = 900 mm, d = 825 mm

Moment of resistance as balanced section

$$= \frac{0.812\times225\times825^2}{1000} = 124350 \text{ N.m}$$

$$\text{Unbalanced balanced moment} = 216114 - 124350$$

$$= 91764 \text{ N.m}$$

$$A_{st_1} = \frac{124350\times1000}{275\times0.9155\times825} = 599 \text{ mm}^2$$

$$A_{st_2} = \frac{91764\times1000}{275\times(825-50)} = 430 \text{ mm}^2$$

$$\text{Total } A_{st} = 599 + 430 = 1029 \text{ mm}^2$$

Provide 16 mm f 6 Nos. (A_{st} = 1206 mm²) in two layers are shown in Fig. 12.25 (*b*)

$$x_c = 0.4\times825 = 330 \text{ mm}$$

$$A_{sc} = \frac{91764\times1000}{7\times(1.5\times13.33-1)\dfrac{330-50}{330}(825-50)}$$

$$= 1049.24 \text{ mm}^2$$

Provide 20 mm ϕ, 4 Nos. (A_s = 1257 mm²)

Check for Shear

$$\text{Shear force } V = \left(\frac{115665-68829}{2}\right)\times1.2\times2.775$$

$$= 307183 \text{ N}$$

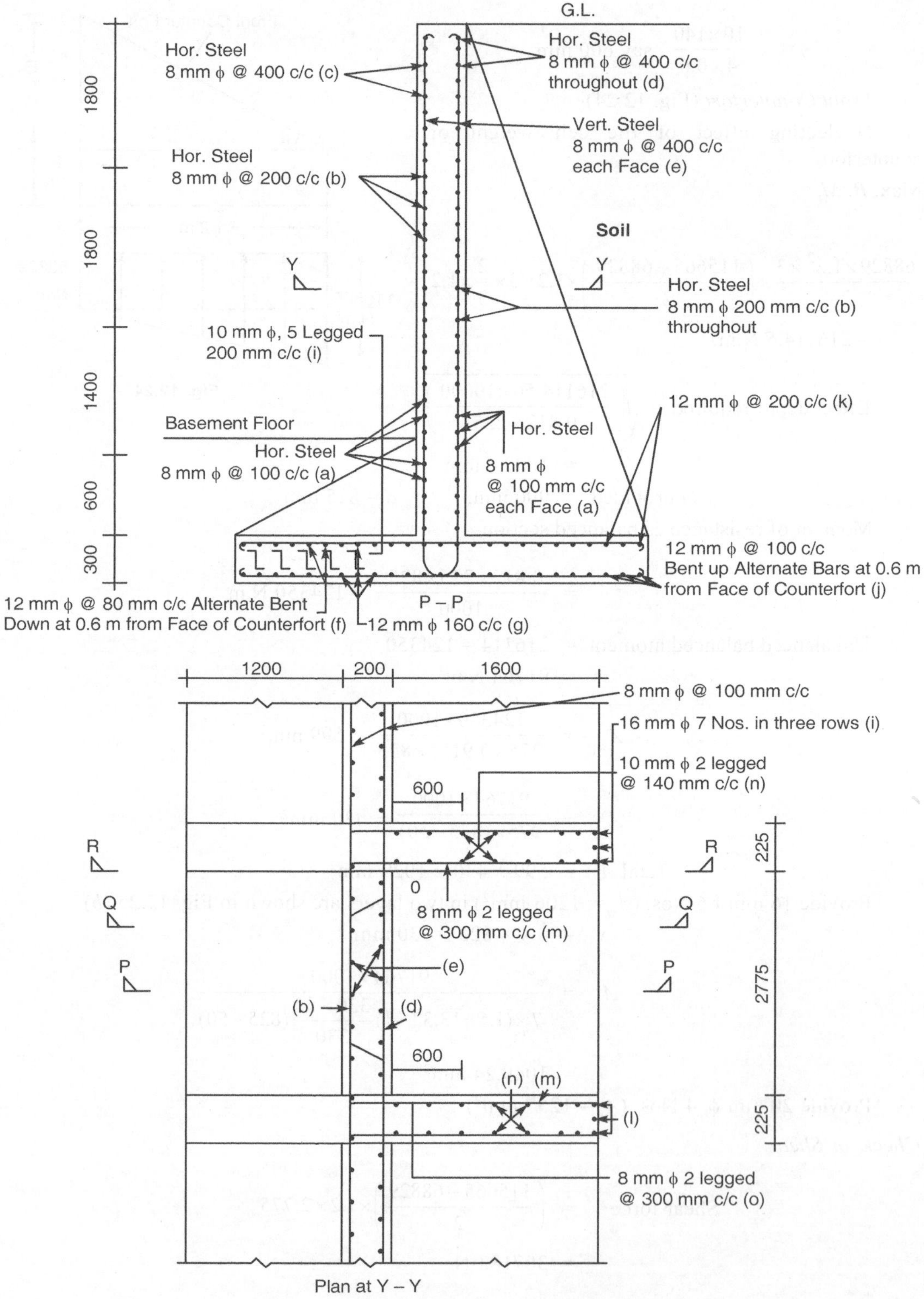

Fig. 12.25. (*a*) Reinforcement Details of Counterfort Retaining Wall

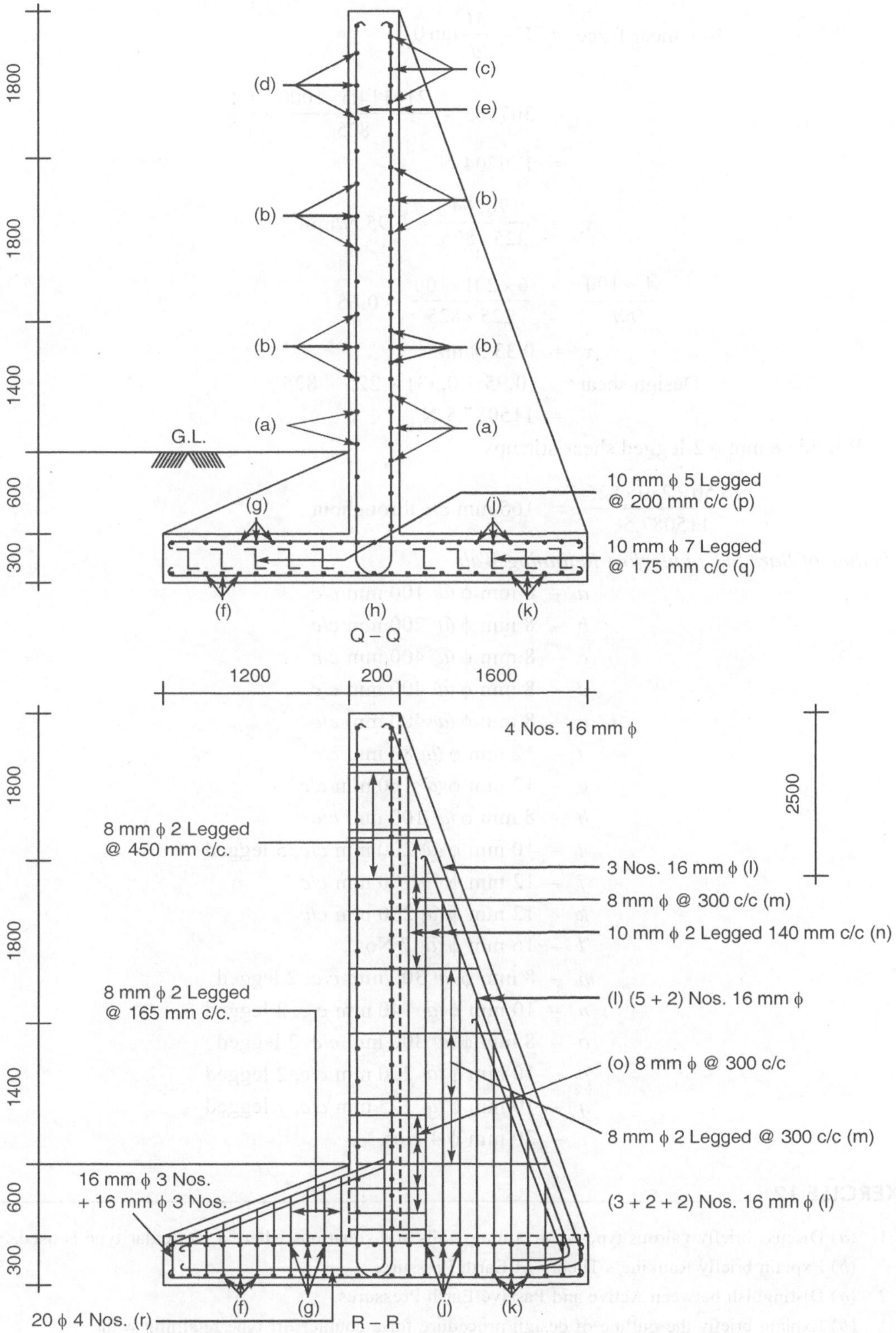

Fig. 12.25. (*b*) Reinforcement Details of Counterfort Retaining Wall

$$\text{Net shear force} = V - \frac{M}{d}\tan\theta$$

$$= 307183 - \frac{216114.5 \times 1000}{825} \times \frac{0.6}{1.2}$$

$$= 176204 \text{ N}$$

$$\tau_v = \frac{176204}{225 \times 825} = 0.95 \text{ N/mm}^2$$

$$\frac{A_s \times 100}{bd} = \frac{6 \times 201 \times 100}{225 \times 825} = 0.65$$

$$\tau_c = 0.33 \text{ N/mm}^2$$

$$\text{Design shear} = (0.95 - 0.33) \times 225 \times 825$$

$$= 115087.5 \text{ N}$$

Provide 8 mm ϕ 2 legged shear stirrups

$$@ \frac{2 \times 50 \times 230 \times 825}{115087.5} = 165 \text{ mm } c/c \text{ throughout}$$

Schedule of Bars in Counterfort Retaining Wall

a – 8 mm ϕ @ 100 mm *c*/*c*
b – 8 mm ϕ @ 200 mm *c*/*c*
c – 8 mm ϕ @ 400 mm *c*/*c*
d – 8 mm ϕ @ 400 mm *c*/*c*
e – 8 mm ϕ @ 400 mm *c*/*c*
f – 12 mm ϕ @ 80 mm *c*/*c*
g – 12 mm ϕ @ 160 mm *c*/*c*
h – 8 mm ϕ @ 100 mm *c*/*c*
i – 10 mm ϕ @ 200 mm *c*/*c*, 5 legged
j – 12 mm ϕ @ 100 mm *c*/*c*
k – 12 mm ϕ @ 200 mm *c*/*c*
l – 16 mm ϕ @ 7 Nos.
m – 8 mm ϕ @ 300 mm *c*/*c*, 2 legged
n – 10 mm ϕ @ 140 mm *c*/*c*, 2 legged
o – 8 mm ϕ @ 300 mm *c*/*c*, 2 legged
p – 10 mm ϕ @ 200 mm *c*/*c*, 2 legged
q – 10 mm ϕ @ 175 mm *c*/*c*, 7 legged
r – 20 mm ϕ 4 Nos.

EXERCISE 12

1. (*a*) Discuss briefly vairous types of retaining walls and situations where a particular type is used.
   (*b*) Explain briefly Rankine's Theory of Earth Pressure.
2. (*a*) Distinguish between Active and Passive Earth Pressures.
   (*b*) Explain briefly the outline of design procedure for a counterfort type retaining wall.

3. Design a cantilever retaining wall to retain a level earth of 3 m above Basement level. The angle of repose of soil is 30°. The unit weight of soil is 16 KN/m^3. The coefficient of friction between soil and concrete is 0.6. The bearing capacity of soil is 100 KN/Sq. m. Use M 20 concrete and Fe 500 steel.
4. Design a retaining wall to retain an earthfill of 3.5 m above basement level. The soil has the following characteristics.

   Angle of internal friction = 30°

   Unit weight of soil = 16 KN/m^2

   Coefficient of friction between concrete and soil = 0.6
5. Design a R. C. retaining wall to retain water on one side or soil on the other side of 3 m height. The soil characteristics are given below.

   Unit weight of soil = 16 KN/m^2

   Angle of repose of soil = 30°

   The bearing capacity of soil = 100 KN/m^2.

   Use M 25 and Fe 500 steel.

Domes

CHAPTER 13

13.1 GENERAL

A dome is a thin curved surface obtained by revolution of curved surface about a vertical axis. Domes are used to cover circular roof openings, e.g., top of Temples, Gurudwaras. Mosques, etc. However, these are mostly provided over large spans as in the case of auditoriums, assembly halls and the like. The thickness of dome is very small as compared to its other dimensions. It varies from 75 mm to 150 mm. These are, therefore, more economical materially but their cost is offset by the cost of curved shuttering. Domes prove very economical if the use of shuttering is repetitive. The material in domes, is best used because domes are generally subjected to compressive stresses only.

Domes can be circular or conical and these can be obtained by revolution of a segment of a circle or a slanting line about vertical axis. They are subjected to membrane stresses, bending stresses being negligible. Bending is significant when the rise of the dome is very small. The analysis of domes under vertical loads is discussed subsequently. Analysis for wind forces and other effects is quite cumbersome and allowance for such effects is made by stressing the dome to 2/5th of allowable stresses under vertical loads.

A dome can be subjected to membranae compressive or tensile stresses if (i) the support to the dome is at the same level (ii) the loading is symmetrical about the axis of revolution.

Any fibre in a dome is subjected to meridinal thrust and hoop stress in the direction of meridians and latitudes respectively as shown in Fig. 13.1 At the free edge of the dome, meridinal stresses have a large horizontal component which is taken care of by providing a ring beam there. This ring beam is subjected to hoop tension.

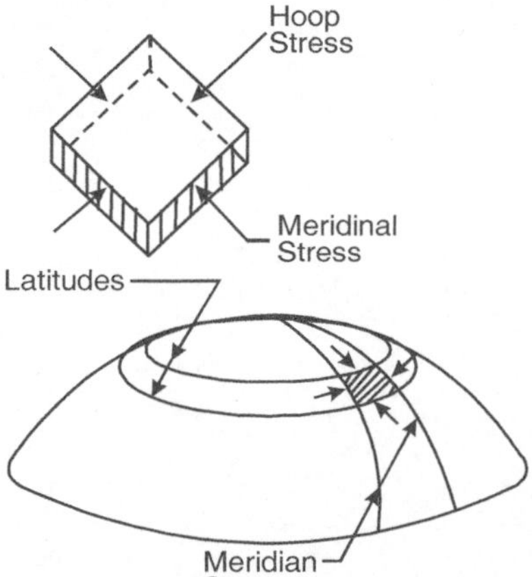

Fig. 13.1.

Steel is generally placed at the centre of the thickness of the dome along meridians and latitudes. The percentage of steel is generally 0.3% for mild steel and 0.24% for ribbed tor steel of the gross-sectional area of the dome section. Concrete to be used in shells/domes as recommended by IS : 2210-1962, is M 20.

13.2 TYPES OF DOMES

(1) Spherical Domes of Uniform Thickness

(2) Conical Domes of Uniform Thickness

13.2.1 Analysis of Spherical Domes of Uniform Thickness

The analysis consists in determining the meridinal thrust and hoop stress (tensile or compressive) under different loading conditions such as *u.d.l.* or concentrated load.

(*a*) *Under u.d.l. of w Per Unit Surface Area*

Consider a meridinal circle of the dome. Considering the equilibrium of a ring between two horizontal planes *A B* and *C D* making angle 2 ϕ and (ϕ : $d\phi$) at the centre respectively as shown in Fig. 13.2.

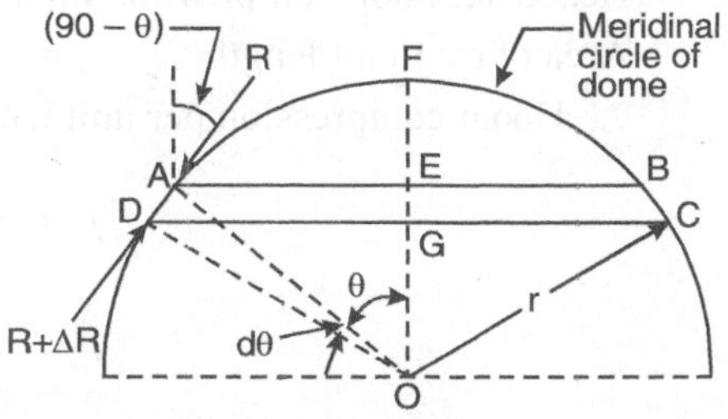

Fig. 13.2.

The forces acting on the ring are :

(*i*) Weight of the ring acting downwards

(*ii*) Thrust *R* per unit length of circle of latitude *AB* acting at right angles to *OA*.

(*iii*) Thrust $R + \Delta R$ per unit length of circle of latitude *CD* acting upwards at right angles to *OD*.

The two thrusts *R* and $(R + \Delta R)$ cannot be equal because of the weight of the ring *AD* acting vertically downwards.

W = Weight of the dome surface *ABF* above ring *AB*

= Area of dome surface × intensity of load per unit surface area

$= 2p\, r.\ EF\ .\ w$ ...(*a*)

$= 2\pi r^2 (1 - \cos\theta)\, w$

R = thrust/unit length of latitude circle *AB*.

Sum of vertical components of the thrusts *R* acting along the circumference of circle *AB* must be equal to the total weight above it.

Total vertical component of the thrust, *R* on the ring *AB*

$= R \operatorname{Sin}\theta \times 2\pi\ .\ AE$

$= R \operatorname{Sin}\theta \times 2\pi\ .\ r \operatorname{Sin}\theta$

$= R \times 2\pi\ .\ r \operatorname{Sin}^2\theta$ ...(*b*)

$\therefore \quad R \times 2\pi r \operatorname{Sin}^2\theta = 2\pi r^2 w (1 - \operatorname{Cos}\theta)$

$$\therefore \quad R = \frac{w\,r\,(1-\operatorname{Cos}\theta)}{\operatorname{Sin}^2\theta}$$

$$= \frac{w\,.\,r}{1+\operatorname{Cos}\theta} \qquad ...(13.1)$$

At Crown $\quad \theta = 0° \quad \therefore \quad R = \dfrac{w\,r}{2}$

At $\theta = 90°$, $R = wr$

Thus Meridinal thrust is maximum at the springings for hemispherical dome for which a ring beam has to be provided.

Horizontal component of *R* acting radially outwards = $R.\cos\theta$

Hoop tension due to $R\cos\theta$, $H = R\cos\theta . \dfrac{Dia}{2}$

$$= \frac{w\,r^2}{1+\cos\theta} . \sin\theta\cos\theta \qquad ...(c)$$

Horizontal component of $R + \Delta R$ acting radially inwards = $(R + \Delta R)\cos(\theta + d\theta)$ Hoop compression due to $(R + \Delta R).\ \cos(\theta + d\theta)$,

$$H + dH = (R + \Delta R) \cos(\theta + d\theta).\ DG$$
$$= (R + \Delta R) \cos(\theta + d\theta).\ r \sin(\theta + d\theta)$$

Hence net hoop compression on the ring = dH

Area of ring/unit length $= r\ .\ d\theta \times 1$

$\therefore$ Hoom compression per unit length,

$$f = \frac{dH}{r\ .\ d\theta}$$
$$= \frac{1}{r}\frac{dH}{d\theta} \quad \text{...}(d)$$

From (*c*), $$H = \frac{w r^2}{1+\cos\theta}\ .\ \sin\theta\cos\theta$$

$$\therefore \frac{dH}{d\theta} = wr^2 \frac{(1+\cos\theta)[\sin\theta.(-\sin\theta)+\cos\theta.\cos\theta] - \sin\theta\cos\theta(-\sin\theta)}{(1+\cos\theta)^2}$$

$$= w r^2 \left(\cos\theta - \frac{1}{1+\cos\theta}\right)$$

$\therefore$ Hoom compression per unit length, $f = \pm \frac{1}{r}\frac{dH}{d\theta}$

$$= \pm w r \left(\cos\theta - \frac{1}{1+\cos\theta}\right) \quad \text{...(13.2)}$$

+ ve for hoom compression

– ve for hoop tension

Eq. (13.2) gives value of hoop stress at any section depending upon the value of θ.

At Crown, $\theta = 0$

$$\therefore \quad f = +\frac{w r}{2} \text{ (Maximum compression)}$$

If t = thickness of dome shell then intensity of Hoop stress

$$= +\frac{w r}{2} \text{ (Compression).}$$

At *Base*, $\theta = 90°$ for semi-spherical dome

Therefore $f = w r$ (maximum tension).

This shows that there is certain value of θ for which hoop stress will change from compression to tension (i.e., zero).

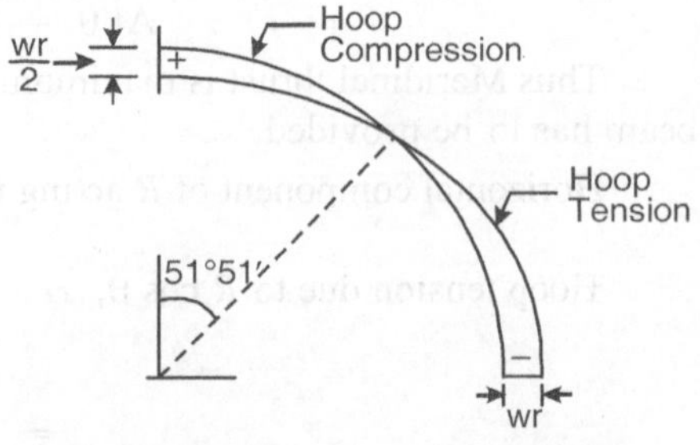

Fig.13.3. Variation of Hoop Stress

For $$f = 0, \quad w r \left(\cos\theta - \frac{1}{1+\cos\theta}\right) = 0$$

$$\therefore \quad \cos^2\theta + \cos\theta - 1 = 0$$

$$\theta = 51°\ 51'$$

From the above it is clear that maximum meridinal thrust (= $w\ r$) occurs at $\theta = 90°$, thus maximum meridinal stress $\frac{w r}{t}$. and minimum at the crown $\left(\frac{w r}{2 t}\right)$.

Also Hoop stress is compressive $\left(=\frac{wr}{2}\right)$ at crown and is tensile at $\theta = 90°$ (*i.e.*, $= w\,r$) having a zero value at $\theta = 51°\,51'$ as shown in Fig. 13.3.

(*b*) *Under Concentrated Load W at the Crown of the Dome* (Fig. 13.4).

W = Concentrated load at crown

Vert. load/unit length of ring *AB*

$$= \frac{W}{\text{Circumference of ring } AB}$$

$$= \frac{W}{2\,\pi\,.\,AE}$$

$$= \frac{W}{2\,\pi\,.\,r\,.\sin\theta}$$

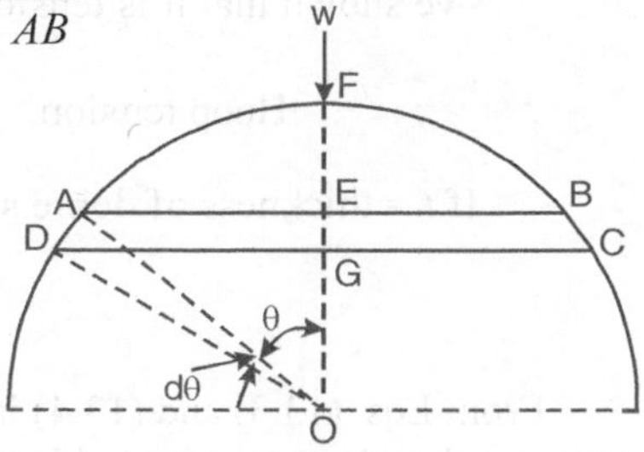

Fig. 13.4.

Vertical component of meridinal thrust $R = R.\ \sin\theta$

$$\therefore \qquad R.\sin\theta = \frac{W}{2\,\pi\,.\,r\,.\sin\theta}$$

$$R = \frac{W}{2\,\pi\,.\,r\,.\sin^2\theta} \qquad \text{...(13.3)}$$

Horizontal component of R (= $R\cos\theta$) will procedure hoop tension of

$$H = R\cos\theta.\frac{Dia}{2}$$

$$= R\cos\theta,\ r\sin\theta$$

$$= \frac{W}{2\pi r\sin^2\theta}.r\sin\theta\cos\theta$$

$$= \frac{W}{2\pi}.\cot\theta \qquad \text{...(e)}$$

Similarly horizontal component of $R + \Delta R$ [$=(R + \Delta R)\cos(\theta + d\theta)$] will produce hoop compression at the lower face of ring and its magnitude is

$$H + dH = (R + \Delta R)\cos(\theta + d\,\theta).\ r\sin(\theta + d\theta)$$

Hence net hoop compression on the ring $= H + dH - H = dH$

Area of ring/ unit length $= r\,.\,\delta\,\theta \times 1$

$\therefore$ Hoop compression per unit length

$$f = \frac{dH}{r\,.\,d\theta} = \frac{1}{r}.\frac{dH}{d\,\theta}$$

From Eq. (*e*)
$$H = \frac{W}{2\,\pi}\cot\theta = \frac{W}{2\,\pi}.\frac{\cos\theta}{\sin\theta}$$

$$\frac{dH}{d\,\theta} = \frac{W}{2\,\pi}.\frac{\sin\theta(-\sin\theta) - \cos\theta.\cos\theta}{\sin^2\theta}$$

$$= \frac{W}{2\,\pi}.\frac{\sin^2\theta + \cos^2\theta}{\sin^2\theta}$$

$$= \frac{W}{2\pi} \cdot \frac{1}{\sin^2 \theta}$$

$$= \frac{W}{2\pi} \cdot \text{cosec}^2 \theta$$

$\therefore$ Hoom compression, $f = \frac{W}{2\pi} \cdot \text{cosec}^2 \theta$

–ve shown that it is tension and not compression.

$\therefore$ Hoop tension, $f = +\frac{W}{2\pi r} \cdot \text{cosec}^2 \theta$

If t = thickness of dome shell then Hoop stress,

$$= \frac{W}{2\pi r t \,\text{Sin}^2 \theta} \quad \text{(tensile)} \qquad ...(13.4)$$

From Eqs. (13.3) and (13.4) it is clear that both R and f are infinity at crown (i.e., $\theta = 0$). This indicates that the concentrated load should never be placed at the crown but must be spread over sufficient area so as not to cause excessive stresses. In order to achieve this aim, either the top of dome should be thickened or an opening may be provided at the crown.

The meridinal thrust is always compressive while hoop stress will be compressive if f is positive and tensile if f is negative.

Since the meridinal thrust and Hoop stress are in terms of θ, their values are given in Table 13.1 for various values of θ, for both the u.d.l. as well as concentrated load at the crown for a dome shell supported on the circle of latitude. In case a dome is to support a concentrated load in addition to its own weights, the above two cases can be algebraically added.

Table 13.1. Coefficients for Hoop Stress and Meridinal Thrust

θ	*u.d.l.*		*Concentrated load*	
	R	f	R	f
	$\frac{1}{1+\cos\theta}$	$\cos\theta - \frac{1}{1+\cos\theta}$	$\frac{\text{cosec}^2\theta}{2\pi}$	$\frac{\text{cosec}^2\theta}{2\pi}$
0	0.5	+ 0.5	∞	∞
5	0.5	+ 0.496	21.0	21.0
10	0.505	+ 0.48	5.3	5.3
20	0.516	+ 0.425	1.37	1.37
30	0.537	+ 0.33	0.64	0.64
40	0.566	+ 0.20	0.38	0.38
50	0.608	+ 0.034	0.27	0.27
51° 51′	0.618	+ 0.0	0.26	0.26
60	0.667	–0.167	0.21	0.21
70	0.747	–0.402	0.18	0.18
80	0.838	–0.68	0.16	0.16
90	1.000	–1.00	0.16	0.16

Example 13.1. Design a spherical dome for a temple hall of 8 m dia. The rise of the dome is to be 1.60 m. The live load may be taken as 750 N/m^2 and finishes as 250 N/m^2. The dome carries a lantern load of 10 KN concentrated at the crown.

Solution.

Assume the thickness of dome = 75 mm

$$\text{Self weight of dome} = \frac{75}{1000} \times 25000 = 1875 \text{ N/m}^2$$

L.L. = 1000 N/m²

Total = 2875 N/m²

Let R = radius of dome in meters.

$\therefore$ 1.6 (2R – 1.6) = 4 × 4, R = 5.8 m

sin θ = 4/5.8 = 0.6896

$\therefore$ θ = 43° 36'; cos θ = 0.7242

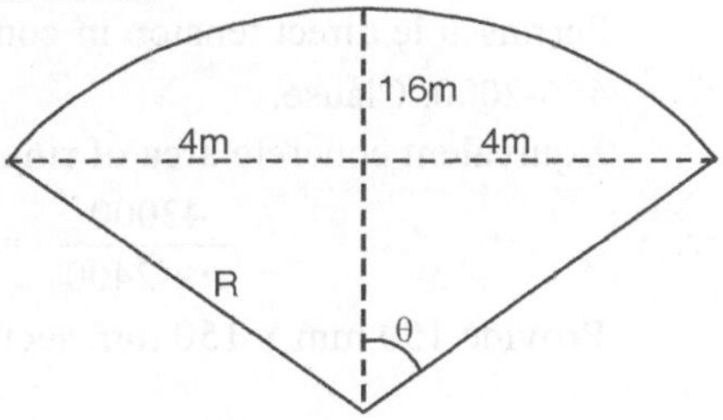

Fig. 13.5.

Meridinal thrust N/m run at springing,

$$R = \frac{wr}{1+\cos\theta} + \frac{W}{2\pi r \sin^2\theta}$$

$$= \frac{2875 \times 5.8}{1+0.7242} + \frac{10 \times 1000}{2\,\pi \times 5.8 \times (0.6896)^2} = 10227 \text{ N}$$

Compressive stress due to meridinal thrust

$$= \frac{10227}{75 \times 1000} = 0.1365 \text{ N/mm}^2$$

Hoop stress at the springing level

$$f = \frac{wr}{t}\left[\cos\theta - \frac{1}{1+\cos\theta}\right] - \frac{W}{2\,\pi r t \sin^2\theta}$$

$$= \frac{2875 \times 5.8}{.075}\left[0.7242 - \frac{1}{1.7242}\right]$$

$$= \frac{10 \times 1000}{2\,\pi 5.8 (0.6896)^2}$$

$$= 150273 \text{ N/m}^2$$

$$= 0.150273 \text{ N/mm}^2$$

Similarly the stresses can be calculated at other levels also.

Since the stresses are very small, a minimum of 0.3% steel may be provided in each direction.

$$\text{Min. steel required} = \frac{0.3}{100} \times 1000 \times 75 = 225 \text{ mm}^2$$

$$\text{Spacing of 8 mm } \phi \text{ bars} = \frac{50 \times 1000}{225} = 222.2 \text{ mm}$$

Provide 8 mm ϕ bars @ 220 mm *c/c* bothways.

Ring Beam

Horizontal component of meridinal thrust = R cos θ

= 10227 × 0.7242 = 7400 N/m

Hoop tension in the ring beam = R cos θ.r = 7400 × 5.8 = 43000 N

Stress in steel = 230 N/mm²

$\therefore$ Area of steel required, $A_{st} = \dfrac{43000}{230} = 187 \text{ mm}^2$

Provide 4 bars of 8 mm ϕ

Area of steel provided = 200 mm^2.

Size of Ring Beam

If A is the minimum area of ring beam section in mm^2 using M 20 concrete, m = 13

Permissible direct tension in concrete is 2.8 N/mm^2 as per *IS* : 456-2000, Clause.

$\therefore$ Equivalent concrete area of ring beam = $A + (13 - 1)\,200 = A + 2400$

$$\therefore \quad \frac{43000}{A + 2400} = 2.8 \qquad A = 12957 \text{ mm}^2$$

Provide 150 mm × 150 mm section. Reinforcement details are shown in Fig. 13.6.

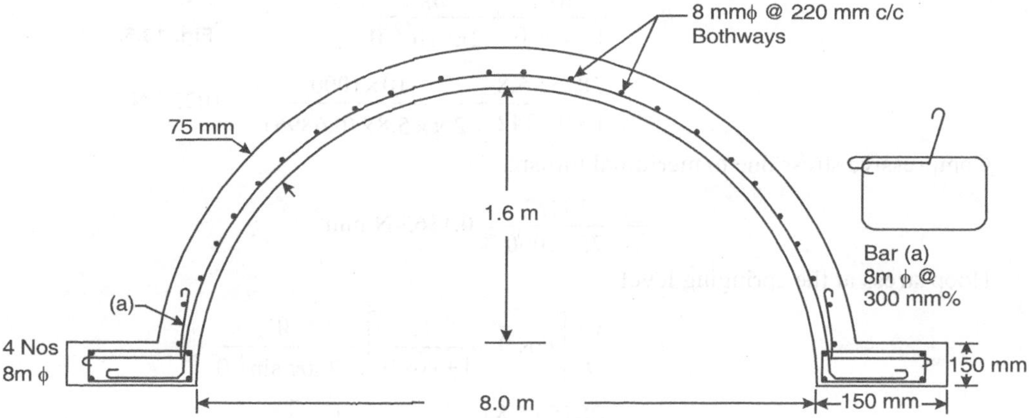

Fig. 13.6.

Example 13.2. Design a spherical dome for a hall of 8 m diameter. The rise of the dome is to be 2.0 m. The dome carries a lantern load of 10 KN at the crown acting on an area of 1.5 m diameter. Live load including finishes may be taken as 1000 N/m^2.

Sol. *Loading*

Assume thickness of dome = 100 mm

$$\text{Self weight of dome/m}^2 = \frac{100}{1000} \times 25000$$

$$= 2500 \text{ N}$$

Live load on dome/m^2 = 1000 N

Total = 3500 N/m^2

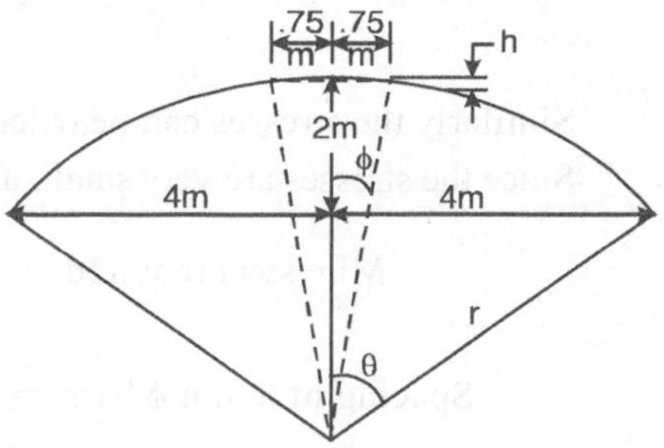

Fig. 13.7.

Radius of Dome (r)

$2.0\,(2r - 2) = 4 \times 4, \qquad r = 5$ m.

$\text{Sin } \theta = 4/5 = 0.8, \qquad \theta = 53°8'. \qquad \cos\theta = 3/56 = 0.6$

$h(2r - h) = .75 \times .75$

$h = 0.06$ m.

Concentrated load

$= 10000 + 2\pi\, r\, h \times 3500$

$$= 16600 \text{ N}$$

$$\text{Sin } \phi = \frac{1.5/2}{5} = \frac{0.75}{5} = 0.15$$

$$\phi = 8°38', \cos \phi = 0.9889$$

Stresses Between 0° *and* 8°38′

Load per unit area = self weight of dome + conc. load distributed over 1. 5 m diameter

$$= 2500 + \frac{16.6 \times 1000}{\frac{\pi}{4} \times (1.5)^2}$$

$$= 11900 \text{ N/m}^2$$

$$\text{Meridinal thrust at } 8°38' = \frac{wr}{1+\cos\phi}$$

$$= \frac{11900 \times 5}{1.9889} = 29900 \text{ N/m}$$

$$\text{Meridinal stress} = \frac{29900}{100 \times 1000} = 0.299 \text{ N/mm}^2$$

$$\text{Hoop force at } 8°38' = wr\left(\cos\phi - \frac{1}{1+\cos\phi}\right) = 11900 \times 5\left(0.9889 - \frac{1}{1.9889}\right)$$

$$= 28950 \text{ N}$$

$$\text{Hoop stress} = \frac{28950}{100 \times 1000}.0.2895 \text{ N/mm}^2$$

$$\text{Meridinal force at } 0° = \frac{wr}{2} = \frac{11900 \times 5}{2}$$

$$= 29750 \text{ N}$$

$$\text{Meridinal stress at } 0° = \frac{29750}{100 \times 1000} = 0.2975 \text{ N/mm}^2$$

$$\text{Hoop force at } 0° = \frac{wr}{2} = 29750 \text{ N}$$

$$\text{Hoop stress at } 0° = \frac{29750}{100 \times 1000} = 0.2975 \text{ N/mm}^2$$

The stresses are within permissible limits.

Stresses Between 8°38′ *and* 53°8′

(*a*) *With Live Load*

w = 3500 N/m², W = 16.6 KN including lantern load,

D.L. of dome shell and Live Load, t = 100 mm, r = 5 m.

$$\text{Meridinal stress} = \frac{wr}{t\,(1+\cos\theta)} + \frac{W}{2\,\pi\, rt \sin^2\theta}$$

$$= \frac{3500 \times 5}{0.10(1+\cos\theta)} \times 10^{-6} + \frac{16.6 \times 1000}{2\pi \times 5 \times 0.1 \times \sin^2\theta} \times 10^{-6}$$

$$= \left(\frac{0.175}{1+\cos\theta)} + \frac{0.0332}{2\pi \sin^2\theta} \right). \text{N/mm}^2$$

$$\text{Hoop stress} = \frac{wr}{t}\left(\cos\theta - \frac{1}{1+\cos\theta} \right) - \frac{W}{2\pi rt \sin^2\theta}$$

$$= \left[\frac{3500\times 5\times 10^{-6}}{0.1}\left(\cos\theta - \frac{1}{1+\cos\theta} \right) - \frac{0.0332}{2\pi \sin^2\theta} \right] \text{N/mm}^2$$

$$= \left[0.175000\left(\cos\theta - \frac{1}{1+\cos\theta} \right) = \frac{0.0332}{2\pi \sin^2\theta} \right] \text{N/mm}^2$$

Using Table 13.1 prepare Table 13.2 for meridinal and Hoop stresses for various values of θ.

(*b*) *Without Live Load*

$w = 2500 \text{ N/m}^2$ $\quad W = 10000 + 2\pi \times 5 \times 0.06 \times 2500$

$= 14710$ N without *L.L.*

$t = 100 \text{ mm} = 0.10 \text{ m}$ $\quad r = 5 \text{ m}$

$$\text{Meridinal stress} = \frac{2500\times 5}{0.10(1+\cos\theta)}\times 10^{-6} + \frac{14710}{2\pi\times 5\times 0.1\sin^2\theta}\times 10^{-6}$$

$$= \left(\frac{0.125}{1+\cos\theta} + \frac{0.02942}{2\pi \sin^2\theta} \right) \text{N/mm}^2$$

$$\text{Hoop stress} = \frac{2500\times 5}{0.10}\left(\cos\theta - \frac{1}{1+\cos\theta} \right)\times 10^{-6}$$

$$- \frac{14710}{2\pi\times 5\times 0.1\sin^2\theta}\times 10^{-6}$$

Table 13.2. (Including *L.L.*)

θ	*Meridinal stress N/mm²*	*Hoop stress N/mm²*
10°	0.175 × 0.505 + 0.0332 × 5.3 = 0.2645	0.175 × 0.48 – 0.0332 × 5.3 = – 0.092
20°	0.175 × 0.516 + 0.0332 × 1.37 = 0.1359	0.175 × 0.425 – 0.0332 × 1.37 = + 0.028
30°	0.175 × 0.537 + 0.0332 × 0.64 = 0.1152	0.175 × 0.33 – 0.0332 × 0.64 = + 0.0365
40°	0.175 × 0.566 + 0.0332 × 0.38 = 0.1116	0.175 × 0.20 – 0.0332 × 0.38 = + 0.0224
50°	0.175 × 0.608 + 0.0332 × 0.27 = 0.1153	0.175 × 0.034 – 0.0332 × 0.27 = – 0.0030
53°8'	0.175 × 0.625 + 0.0332 × 0.248 = 0.1175	0.175 × (–0..025) – 0.0332 × 0.248 = –0.0126

Table 13.3. (Without *L.L.*) + compression, – tension

θ	*Meridinal stress N/mm²*	*Hoop stress N/mm²*
10°	0.125 × 0.505 + 0.02942 × 5.3 = 0.2191	0.125 × 0.48 – 0.02942 × 5.3 = – 0.096
20°	0.125 × 0.516 + 0.02942 × 1.37 = 0.1049	0.125 × 0.425 – 0.02942 × 1.37 = + 0.0127
30°	0.125 × 0.537 + 0.02942 × 0.64 = 0.0859	0.125 × 0.33 – 0.02942 × 0.64 = + 0.0224
40°	0.125 × 0.566 + 0.02942 × 0.38 = 0.0820	0.125 × 0.20 – 0.02942 × 0.38 = + 0.0138
50°	0.125 × 0.608 + 0.02942 × 0.27 = 0.0840	0.125 × 0.034 – 0.2942 × 0.27 = –0.0037
53°8'	0.125 × 0.625 + 0.02942 × 0.248 = 0.0854	0.125 × (–0.025) – 0.02942 × 0.248 = – 0.00418

$$= \left[0.125\left(\cos\theta - \frac{1}{1+\cos\theta}\right) - \frac{0.02942}{2\pi\sin^2\theta}\right] \text{N/mm}^2$$

Using table 13.1, prepare Table 13.3 for meridinal stress and Hoop stress for various values of θ.

Maximum hoop compressive stress = 0.0365 N/mm²

Maximum meridinal stress = 0.2191 N/mm² (compressive)

Maximum hoop tensile stress = 0.096 N/mm²

The stresses are very small and hence safe.

Steel Reinforcement

Provide a minimum of 0.3% of cross-sectional area in both meridinal as well as latitude direction.

$$\therefore \quad \text{Area of steel} = \frac{0.3}{100} \times 100 \times 1000 = 300 \text{ mm}^2$$

$$\text{Spacing of 8 mm } \phi \text{ bars} = \frac{50 \times 1000}{300} 167 \text{ mm}$$

Provide 8 mm ϕ bars @ 100 mm *c/c* along latitude and meridinal circles.

Ring Beam at Bottom Edge

Meridinal thrust per mm length = 0.1175 × 100 = 11.75 N

Horizontal component of thrust = 11.75 cos 53°8′

Hoop tension = (11.75 × 0.6) (5 × 1000) = 35250 N

$$\text{Area of steel required} = \frac{35250}{230} = 153 \text{ mm}^2$$

Provide 4 bars of 8 mm ϕ. A_{st} = 200 mm²

Size of Ring Beam

If A is the minimum area of ring beam section in mm²; Use M 20 concrete, m = 13.33 and permissible direct tension in concrete is 2.8 N/mm²

∴ Equivalent concrete area of ring beam = A + (13.33 – 1) 200 = A + 2466

$$\therefore \quad \frac{35250}{A + 2466} = 2.8, \quad A = 10123 \text{ mm}^2$$

Provide 200 mm × 150 mm section.

Ring Beam at the Lantern Opening

The meridinal thrust at the lantern opening (at 8°38′) is 29900 N/m. A ring beam will be provided to resist its horizontal component. This may be provided by thickening the dome shell at this point. Limiting the direct compressive stress σ_{cc} to 4 N/mm² in M 15.

$$\text{Area required in compression/m} = \frac{29900}{4} = 7475 \text{ mm}^2$$

$$\text{Thickness required} = \frac{7475}{1000} = 7.475 \text{ mm}$$

Thickness of dome shell = 100 mm

∴ No need of thickening the dome shell at the lantern opening. The reinforcement details are shown in Fig. 13.8.

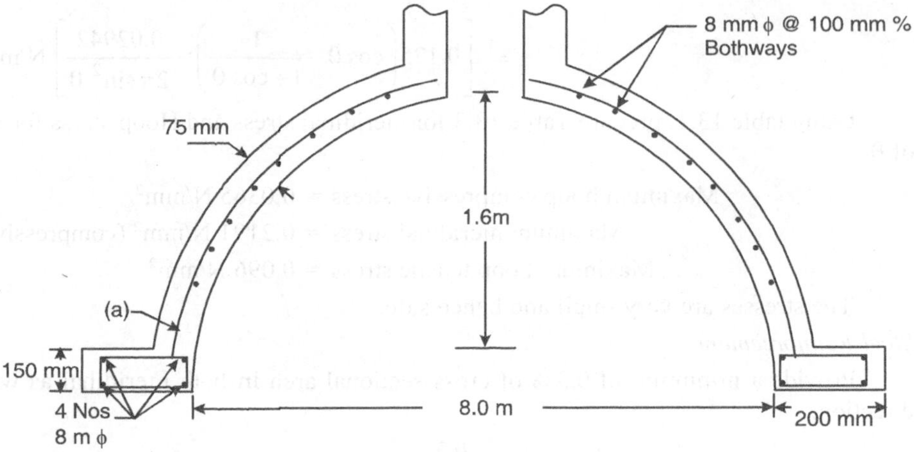

Fig. 13.8.

13.2.2 Analysis of Conical Domes of Uniform Thickness

(a) *Under u.d. l of w Per Unit Area*

Consider a conical dome of uniform thickness '*t*'. Let 2 θ = the angle at the apex of the dome.

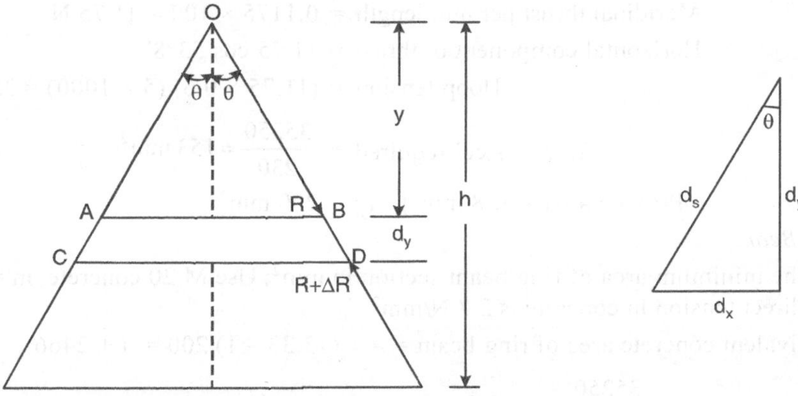

Fig. 13.9.

Consider the equilibrium of a ring between two horizontal planes *AB* and *CD* at distance of *y* and *y* + *dy* from 0 as shown in Fig. 13.9.

The forces acting on the ring are :

(*i*) Weight of the ring acting downwards.

(*ii*) Thrust *R* per unit length of the ring *AB* acting downwards along *OB* as shown.

(*iii*) Thrust *R* + Δ*R* per unit length of the ring *CD* acting upwards along *DO* as shown.

The two thrust *R* and *R* + Δ*R* cannot be equal because of the weight of the ring *BD* acting vertically downwards.

W = Weight of the dome surface *ABO* above ring *AB*.

= Area of dome surface × intensity of load per unit surface area

$$= (2\pi\, y \tan\theta)\frac{y}{\cos\theta}.\frac{w}{2} \qquad \ldots(a)$$

This should be equal to the sum of vertical components of total meridinal thrust R at level AB which is equal to

$$(R \cos \theta)(2\ \pi y \tan \theta) \quad \text{...(b)}$$

$$\therefore \quad (R \cos \theta)(2\ \pi y \tan \theta) = (2\pi\ y \tan \theta)\frac{y}{\cos \theta}.\frac{w}{2}$$

$$R = \frac{w.y}{2 \cos^2 \theta} \quad \text{...(13.7)}$$

At Crown, $y = O$, $\therefore R = O$

At Base $y = h$, $R = \dfrac{wh}{2 \operatorname{Cos}^2 \theta}$ $\therefore$ Maximum at base.

Horizontal component of R acting radially outwards = $R \sin \theta$ Hoop tension due to R. $\sin \theta$.

$$H = R \sin \theta \times \text{radius of ring } AB$$
$$= R \sin \theta.\ y \tan \theta$$
$$= R.y.\frac{\operatorname{Sin}^2 \theta}{\operatorname{Cos} \theta} \quad \text{...(c)}$$

Horizontal component of $R + \Delta R$ acting radially inwards = $(R \times \Delta R)$ Sin θ.

Hoop compression due to $[(R + \Delta R) \sin \theta]$,

$$H + dH = (R + \Delta R) \sin \theta.\ \text{radius of ring } CD$$
$$= (R + \Delta R) \sin \theta.\ (y + dy) \tan \theta$$
$$= (R + \Delta R)(y + dy).\ \frac{\sin^2 \theta}{\cos \theta}$$

Hence net hoop compression on the ring = dH

Area of ring/unit length = (inclined length AC) × 1

$$= ds \times 1$$
$$= \frac{dy}{\cos \theta}$$

$\therefore$ Hoop compression per unit length,

$$f = \frac{dH}{(dy / \cos \theta)} = \frac{dH}{dy}.\ \cos\theta \quad \text{...(d)}$$

$$\text{From } (c)\ H = R.y.\frac{\sin^2 \theta}{\cos\theta}$$
$$= \frac{w.y}{2 \cos^2 \theta}.y\frac{\sin^2 \theta}{\cos\theta}$$
$$= \frac{w}{2}y^2.\frac{\sin^2 \theta}{\cos^2 \theta}$$
$$\frac{dH}{dy} = wy\frac{\sin^2 \theta}{\cos^3 \theta}$$

From (*d*), $f = \dfrac{dH}{dy}.\cos\theta$

$$= wy.\frac{\sin^2\theta}{\cos^3\theta}.\cos\theta$$

$$= wy.\frac{\sin^2\theta}{\cos^2\theta}$$

$$= wy\tan^2\theta \qquad ...(13.8)$$

Dome is under Hoop compression at all levels since θ will always be less than 90° and hence tan θ = positive.

Intensity of Hoop compression = $\dfrac{wy}{t}.\tan^2\theta$

(*b*) *Under Concentrated Load W at the Crown of the Dome*

$$W = \text{Concentrated load at crown.}$$

$$\text{Vertical load.unit length} = \frac{W}{\text{Circumference of ring } AB}$$

$$= \frac{W}{2\pi.\,y\tan\theta}$$

Vertical component of meridinal thrust

$$R = R.\cos\theta$$

$$\therefore \quad R\cos\theta = \frac{W}{2\pi.y\tan\theta}$$

$$\therefore \quad R = \frac{W}{2\pi.y\dfrac{\sin\theta}{\cos\theta}\cos\theta}$$

$$= \frac{W}{2\pi\,y\sin\theta} \qquad ...(13.9)$$

Horizontal component of $R(= R\sin\theta)$ will produce hoop tension. *H*.

$$H = R\sin\theta\,(y.\tan\theta)$$

$$= R.y.\frac{\sin^2\theta}{\cos\theta} = \frac{W}{2\pi\,y\sin\theta}.y.\frac{\sin^2\theta}{\cos\theta} = \frac{W}{2\pi}.\tan\theta$$

Similarly Horizontal component of $R + \Delta R = [(R + \Delta R).\sin\theta]$ will produce hoop compression at the lower face of ring and its magnitude is

$$H + dH = (R+\Delta R)\sin\theta.(y+dy)\tan\theta = (R+\Delta R)\,(y+dy)\frac{\sin^2\theta}{\cos\theta}$$

Bence Net hoop compression on ring = $H + dH - H = dH$

$$\text{Area of ring/unit length} = (d_s \times 1) = \frac{dy}{\cos\theta}$$

$$\therefore \quad \text{Hoop compression/unit length } f = \frac{dH}{dy/\cos\theta} = \frac{dH}{dy}.\cos\theta$$

Now $$H = \frac{W}{2\pi} \tan\theta \quad \text{...(e)}$$

It is clear from the eq.(*e*) that H is a constant quantity as W, π and θ are constant for a particular dome.

$$\therefore \quad \frac{dH}{dy} = 0$$

$dH = 0$ shows that the thrust on the bottom of the ring will also produce a hoop compression equal to H itself.

Hence net hoop load on section = 0

Example 13.3. Design a conical dome roof for a room with base diameter as 12 m. The live load due to wind, snow, etc., may be taken as 1000 N/m^2. The height of the roof is 4.0 m. Use M20 concrete and Fe 415 steel reinforcement.

Sol. Assume the thickness of dome = 100 mm

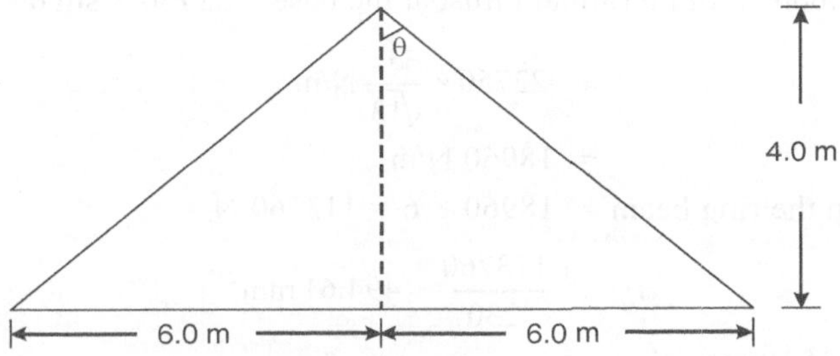

Fig. 13.10.

Loads

(*i*) Self weight of dome/m^2 $= \frac{100}{1000} \times 25000 = 2500$ N

(*ii*) Live load on dome/m^2 $= 1000$ N

Total load $= 3500$ N/m^2

$$\tan\theta = \frac{6}{4} = 1.5 \text{ (Fig. 13.10)}$$

$$\sin\theta = \frac{6}{2\sqrt{13}} = 3/\sqrt{13}$$

$$\cos\theta = \frac{4}{2\sqrt{13}} = 2/\sqrt{13}$$

$$\text{Meridinal thrust, } R = \frac{w.y}{2\cos^2\theta}$$

R is max. at $y = 4$, $\therefore$ $$R_{max} = \frac{3500 \times 4 \times 13}{2 \times 4} = 22750 \text{ N}$$

$$\text{Hoop Compression, } f = w.y.\tan^2\theta$$

$$f_{max} = 3500 \times 4 \times \frac{36}{16} = 31500 \text{ N}$$

$$\text{Compressive stress due to meridinal thrust} = \frac{22750}{100 \times 1000}$$

$$= 0.2275 \text{ N/mm}^2$$

The stresses are very small. Therefore, only minimum reinforcement of 0.3% shall be provided.

$$\text{Area of steel} = \frac{0.3}{100} \times 100 \times 1000 = 300 \text{ mm}^2 \text{ in 1 m width.}$$

$$\text{Spacing of 10 mm } \phi \text{ bars} = \frac{78.5 \times 1000}{300} = 261.7 \text{ mm}$$

∴ Provide 10 mm ϕ bars at 260 mm *c/c* both ways.

Ring Beam at Bottom

Horizontal component of meridinal thrust at the base = 22750 × sin θ

$$= 22750 \times \frac{3}{\sqrt{13}} \text{ N/m}$$

$$= 18960 \text{ N/m}$$

$$\text{Hoop tension in the ring beam} = 18960 \times 6 = 113760 \text{ N}$$

$$\sigma_{st} = \frac{113760}{230} = 494.61 \text{ mm}^2$$

Provide 8 bars of 10 mm dia.

Area of steel provided = 628 mm²

Size of Ring Beam

Let A = area of ring beam

m = 13 for M 20 concrete

σ_{ct} = permissible direct tension in concrete = 2.8 N/mm²

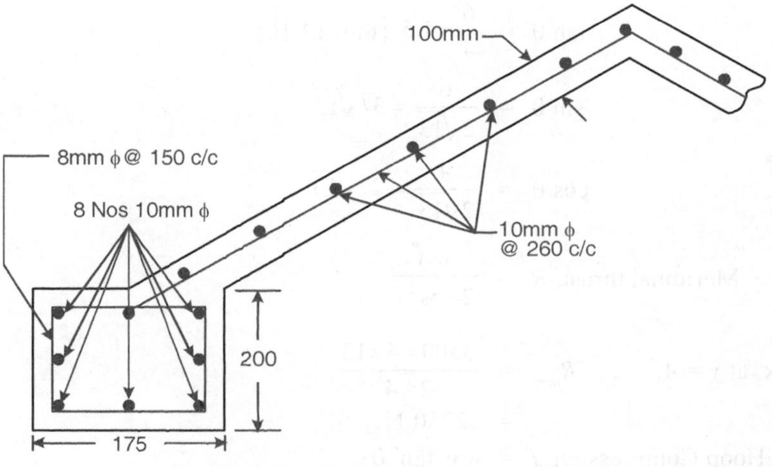

Fig. 13.11.

∴ Equivalent concrete area of ring beam

$$= A + (13 - 1)\,628 = A + 7536$$

$$\therefore \quad \frac{113760}{A + 7536} = 2.8$$

$$A = 33093 \text{ mm}^2$$

Provide a ring beam of size 175 mm × 200 mm

$$\text{Area provided} = 35000 \text{ mm}^2$$

O.K.

Reinforcement details are shown in Fig. 13.11

EXERCISE 13

1. What is a dome? Obtain the expressions for circumferential and meridinal stress resultants for a hemispherical dome under surface loads,
2. Derive membrance stress resultants for a dome loaded with a concentrated load at its apex.
3. Design a spherical dome over a hall of 6 m in diameter. The rise of the dome is 1.8 m. The live load and the weight of finishes may be taken as 750 N/m^2 and 200 N/m^2. The dome carries a concentrated load of 12 KN at crown. Design the ring beam at base also. Use M20 and Fe 500 bars.
4. Derive expressions for membrane stress resultants for a conical dome of uniform thickness under its self weight.
5. Design a conical dome roof for a hall of 10 m in diameter and rise 4 m. The live load and finishes may be taken as 750 KN/m^2 and 250 KN/m^2 respectively. Use M 20 and Fe 415 bars.

CHAPTER

14

Beams Curved in Plan

14.1 GENERAL

Curved beams are most frequently encountered in circular reservoirs, curved balconies, curved ramps in Hospitals, Schools, etc. Curved beams are subjected to torsional moments in addition to shear and bending because the centre of gravity of loads does not coincide with the centre line axis of the member. The torsional moments cause overturning of the beams also unless the ends of the beams are properly restrained. Hence such beams have to be designed for bending moment, shear force and torsional moment. The torsional moment is converted into equivalent bending and equivalent shear by the procedure as discussed earlier.

In case of circular beams supported over evenly spaced columns, the torsion at the centre of beam between any two adjacent supports shall be zero and also there will be no such moments at support itself. Thus, the support section of the beam shall be designed for bending moment and shear force only.

14.2 CIRCULAR BEAMS LOADED UNIFORMLY AND SUPPORTED ON SYMMETRICALLY PLACED COLUMNS

Consider a portion *AB* of a circular beam between two consecutive supports, loaded with *u.d.l.* of intensity '*w*' as shown in Fig. 14.1. Let the angle subtended at the centre by these two consecutive supports *A* and *B* be θ.

The load on the portion *AB* of the beam = $w.R.\theta$

The centre of gravity of the load will lie at a distance.

$$\frac{R \sin \theta/2}{\theta/2} \text{ from the centre}$$

Let M_0 be the *B.M.* and V_0 be the *S.F.* at the supports.

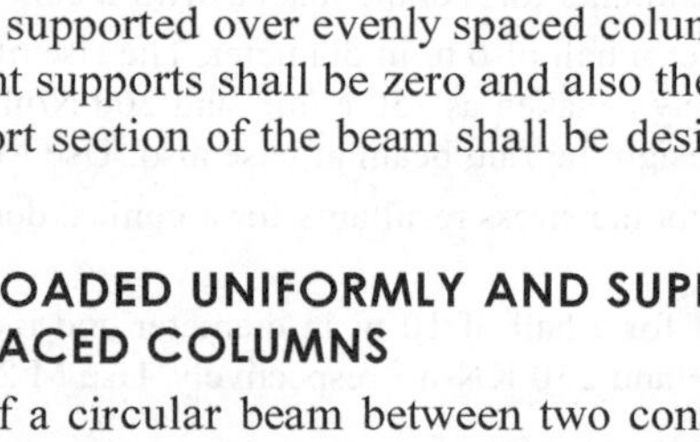

Fig. 14.1.

$$\text{From symmetry } V_0 = \frac{WR\theta}{2} \qquad ...(14.1)$$

The moment M_0 at support can be resolved along the chord *AB* and at right angles to it. The two components will be $M_0 \sin \theta/2$ and $M_0.\cos \theta/2$ respectively.

Taking moment of all forces about line *AB*.

$$2M_0 . \sin \theta/2 = wR\,\theta\left[\frac{R \sin \theta/2}{\theta/2} - R \cos \theta/2\right]$$

$$M_0 = wR^2\left[1 - \frac{\theta \cot \theta/2}{2}\right] \qquad ...(14.2)$$

To find *S.F.*, *B.M.* and torsional moment at a point *P*, at angle ϕ from one support (Fig. 14.2).

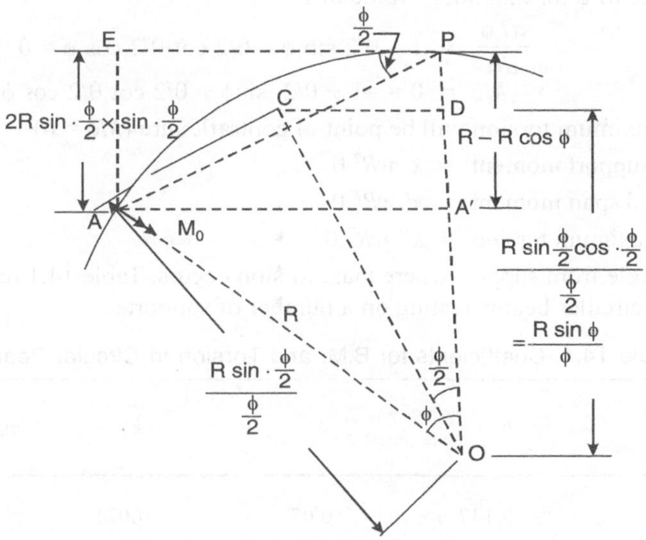

Fig. 14.2.

Load between $AP = wR\,\phi$

$$S.F. \text{ at } P = V_p = \frac{wR\,\phi}{2} - WR\,\phi \qquad ...(14.3)$$

Distance of *C.G.* of load between *AP* from centre

$$= \frac{R \sin \phi/2}{\phi/2}$$

The vector representing *B.M.* at *P* shall be acting along *PO* and vector representing torsion at *P* will be acting at right angles to *PO*.

$$M\phi = V_0 . R \sin \phi - M_0 \cos \phi - wR\,\phi \frac{R \sin \phi/2}{\phi/2} . \sin \phi/2$$

$$= \frac{wR\theta}{2} . R \sin \phi - wR^2\left(1 - \theta . \frac{\cot \theta/2}{2}\right)\cos \phi - wR^2 \times 2 \sin^2 \phi/2$$

$$= wR^2[\theta/2 . \sin \phi - \cos \phi + \theta/2 \cot \theta/2 \cos \phi - 2 \sin^2 \phi/2]$$

$$= wR^2[-1 + \theta/2 \sin \phi + \theta/2 \cot + \theta/2 \cos \phi] \qquad ...(14.4)$$

$$(\therefore -\cos \phi - 2 \sin^2 \phi/2 = -1)$$

$$T\phi = M_0 \sin \phi - wR\phi/2 \times 2R \sin^2 \phi/2 + wR\phi\left(R - R\frac{\sin \phi/2}{\phi/2} . \cos \phi/2\right)$$

$$= wR^2\left(1-\theta.\frac{\cot\theta/2}{2}\right)\sin\phi - wR^2\ \theta.\sin^2\theta/2 + wR^2(\phi - 2\sin\phi/2\cos\phi/2)$$

$$= wR^2[\phi - \theta/2 + \theta/2\cos\phi - \theta/2\cot\theta/2\sin\phi] \qquad ...(14.5)$$

The obtain maximum value of torsional moment, $\frac{dT\phi}{d\phi} = 0.$

The gives value of ϕ for maximum value of $T.$

$$\frac{dT\phi}{d\phi} = 1 - \theta/2\ \sin\phi - \theta/2\ \cot\theta/2\ \cos\phi = 0$$

$$M\phi = 0 = -1 + \theta/2.\ \sin\phi + \theta/2\ \cot\theta/2\ \cos\phi$$

$\therefore$ Point of maximum torsion will be point of contraflexure ($M\phi = 0$)

$$\text{Let support moment} = k.\ wR^2\ \theta \qquad ...(14.6)$$

$$\text{Mid span moment} = k'\ wR^2\ \theta \qquad ...(14.7)$$

$$\text{Maximum torsion} = k''\ wR^2\ \theta \qquad ...(14.8)$$

Let ϕ be the angle from support where max. torsion occurs. Table 14.1 represents the values of k, k', k'' an ϕ for circular beams resting on a number of supports .

Table 14.1. Coefficients for B.M. and Torsion in Circular Beams

Number of support	θ	k	k	k''	*Value of f for maximum twisting moment*
4	90	0.137	0.07	0.021	$19\frac{1}{4}°$
5	72°	0.108	0.054	0.014	$15\frac{3}{4}°$
6	60°	0.089	0.045	0.009	$12\frac{3}{4}°$
7	$51\frac{3}{7}°$	0.077	0.037	0.007	$10\frac{3}{4}°$
8	45°	0.066	0.03	0.005	$9\frac{1}{2}°$
9	40°	0.060	0.027	0.004	$8\frac{1}{2}°$
10	36°	0.054	0.023	0.003	$7\frac{1}{2}°$
12	30°	0.045	0.017	0.002	$6\frac{1}{4}°$

14.3 SEMI-CIRCULAR BEAM SIMPLY SUPPORTED ON THREE EQUALLY SPACED SUPPORTS

Let semi-circular beam ABC be supported on three evenly spaced supports A, B and C, loaded with w per unit length. Let R be the radius of centre line of the beam as shown in Fig. 14.3.

Total load on semi-curcular

$$\text{beam} = w\,R\,\pi,\ (\phi = \pi)$$

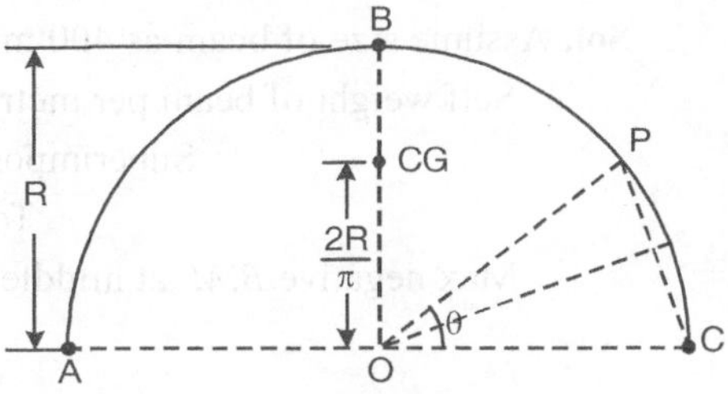

Fig. 14.3

Distance of *C.G.* of load from Centre 0

$$= \frac{R\sin\phi/2}{\phi/2} = \frac{R\sin\pi/2}{\pi/2} = \frac{2R}{\pi}$$

Let V_1 be the reaction at supports *A* and *C* and V_2 be the reaction at *B*.

Taking moments of reactions about line passing through *B* and parallel to *AC*.

$$2V_1R = wR\pi(R - 2R/\pi),\ V_1 = wR/2(\pi - 2)$$

$$V_2 = wR\,\pi - 2\times wR/2(\pi-2)$$

$$= 2wR$$

B.M. at a point *P* at angle θ from one of supports say *C*.

$$\text{Load on portion } CP = WR\,\theta$$

$$\text{Distance of } C.G. \text{ of load from } O = \frac{R.\sin\theta/2}{\theta/2}$$

$$M_0 = V_1 \times R\sin\theta - wR\theta\left[\frac{R\sin\theta/2}{\theta/2}\times\sin\theta/2\right]$$

$$= \frac{wR}{2}.(\pi-2)\,R\sin\theta - 2wR^2\sin^2\theta/2$$

$$= wR^2\left[\frac{\pi-2}{2}\sin\theta - 2\sin^2\theta/2\right]$$

$$\text{At } \theta = \pi/2$$

$$M_b = 0.429\ wR^2$$

For maximum +ve *B.M.*, putting

$$\frac{dM\theta}{d\theta} = 0, \text{ we get}$$

$$\theta = 29°\ 44'$$

$$\text{Mmax} = 0.152\ wR^2 \quad ...(14.9)$$

$$T_\theta = V_1\,(R - R\cos\theta) - wR\theta\left[R - \frac{R\sin\theta/2}{\theta/2}.\cos\theta/2\right]$$

$$= wR^2\left[\frac{\pi-2}{2} - \frac{\pi-2}{2}\cos\theta - \theta + \sin\theta\right]$$

For maximum value of torsional moment, put $\frac{dT\theta}{d\theta} = 0$

$$\therefore \quad \theta = 59°\ 29'$$

$$T_{max} = 0.103\ wR^2 \quad ...(14.10)$$

14.4 SOLVED EXAMPLE

Example 14.1. Design a semi-circular beam supported on three equally spaced columns. The centres of columns are on a circular curve of diameter 8 m. the superimposed load on beam is 16 kN/m. Use M 20 and Fe 415 steel as reinforcement.

Sol. Assume size of beam as 400 mm × 750 mm.

Self weight of beam per metre length = $0.4 \times 0.75 \times 25000 = 7500$ N/m

Superimposed load = 16000 N/m

Total load = 23500 N/m

Max negative *B.M.* at middle support $= -0.429\, wR^2$

$= 0.429 \times 23500 \times 4^2$

$= -169952$ N.m.

Maximum positive *B.M.* at 29° 44′ $= 0.152\, wR^2$

$= 0.152 \times 23500 \times 4^2$

$= 57152$ N.m.

Maximum Torsional moment at 59° 29′ $= 0.103\, wR^2$

$= 0.103 \times 23500 \times 4^2$

$= 38728$ N.m.

$$\text{Equivalent moment from torsion} = \frac{T(1 + D/b)}{1.7}$$

$$= \frac{387828 \times (1 + 750/400)}{1.7}$$

$$= 65496 \text{ N.m.}$$

$$B.M. \text{ at } 59^\circ\ 29' = wR^2\left[\frac{\pi - 2}{2} \times \sin\theta - 2\sin^2\theta/2\right]$$

$$= 23500 \times 4^2\left[\frac{\pi - 2}{2} \times 0.8615 - 2 \times 0.2467\right]$$

Max. *B.M.* including effect of torsion at 59°29′ = – 623.77 – 65496 = – 66119.77 N.m.

For Mix. M 20

$$d = \sqrt{\left(\frac{169952 \times 1000}{0.91 \times 400}\right)} = 683.3 \text{ mm}$$

$$D = 750 \text{ mm} \qquad d = 750 - 50 = 700 \text{ m}$$

$$A_{st} \text{ required for +ve } B.M. \text{ at } 29^\circ\ 44' = \frac{57152 \times 1000}{230 \times 0.90 \times 700} = 394.4 \text{ mm}^2$$

Provide 4 Nos. 12 mm ϕ throughout.

$$A_{st} \text{ required for –ve } B.M. \text{ at } 59^\circ\ 29' = \frac{65935.3 \times 1000}{230 \times 0.90 \times 700} = 455 \text{ mm}^2$$

Provide 16 mm ϕ 3 Nos.

$$A_{st} \text{ required at middle support} = \frac{169952 \times 1000}{230 \times 0.90 \times 700} = 1172.9 \text{ mm}^2$$

Provide 16 mm ϕ, 7 Nos.

$$\text{Max. } S.F. \text{ due to } u.d.l. \text{ at end column} = \frac{wR}{2}(\pi - 2)$$

$$= \frac{23500 \times 4}{2} \times (\pi - 2)$$

$$= 53580 \text{ N}$$

$$\text{Equivalent S.F. due to torsion} = 1.6\ T/b$$

$$= 1.6 \times 38728/0.4 = 154912 \text{ N.}$$

$$S.F. \text{ at } 59° 29' \text{ due to } u.d.l. \text{ on beam} = 53580 - 23500 \times 2/3 \times \pi/2 \times 4 = 44806 \text{ N.}$$

$$\text{Total } S.F. \text{ at } 59° 29' = 154912 + 44806 = 199718 \text{ N.}$$

$$\frac{A_s \times 100}{bd} = \frac{3 \times 201 \times 100}{400 \times 700} = 0.215$$

$$\tau_c = 0.205 \text{ N/mm}^2$$

$$\text{Shear strength of section} = 0.205 \times 400 \times 700 = 57400 \text{ N}$$

$$V_s = 199718 - 57400 = 142318 \text{ N}$$

Using 10 mm ϕ 2 legged closed stirrups,

$$@ \frac{230 \times 700 \times 78.5 \times 2}{142318} = 177.61 \text{ mm}$$

Say 175 mm throughout

The reinforcement detail is shown in Fig. 14.4

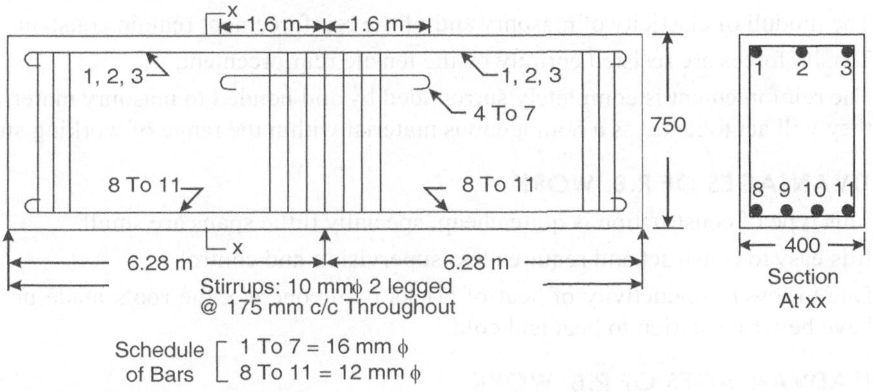

Fig. 14.4.

EXERCISE 14

1. Under what circumstances the beams curved in plan are used?
2. Derive expressions for maximum bending moment and torsional moments for a semi-circular beam siply supported on three equally spaced supports.
3. The centres of supports are on a circular curve of diameter 6 m. The loading on the beam is 20 KN/m. Design the semi-circular beam supported on three equally spaced column supprts. Use M 20 Concrete and Fe 415 grade rinforcement.
4. Design a circular ring beam 500 mm × 1000 mm supported on six columns, carrying a *u.d.l.* of 120 KN/m including self weight. The six columns are equally spaced and placed on a circle of diameter 8.5 m. Use concrete M 20 and Fe 415 grade reinforcement.
5. Design a circular beam 500 × 1000 mm supported on 8 columns equally spaced on a circle of 11.5 m diameter and carrying a *u.d.l.* of 150 KN/m. Use M 25 concrete and Fe 500 grade reinforcement.

CHAPTER 15

Reinforced Brick Work

15.1 GENERAL

Brick masonry is quite strong in compression but extremely weak in diagonal tension. Sometimes because of lighter loads coming on the members and also because of economy considerations, brick masonry reinforced with steel bars, well known as "Reinforced Brick Work" is used. This type of construction is often adopted in small span beams and slabs in residential buildings. The limited scope of its use is due to the reason that brick masonry is far more heterogenous material and the applicability of elastic theory of bending is not fully justified :

15.2 ASSUMPTIONS

While designing a reinforced masonry structure, following assumptions are made :

(*a*) A section that is plane before bending remains plain after bending.

(*b*) The moduli of elasticity of masonry and of the reinforcement remain constant.

(*c*) Tensile forces are resisted entirely by the tensile reinforcement.

(*d*) The reinforcement is completely surrounded by and bonded to masonry material so that they will act together as a homogenous material within the range of working stresses.

15.3 ADVANTAGES OF R.B. WORK

(1) This type of construction is quite cheap, specially if the spans are small.

(2) It is easy to construct and requires less supervision and control.

(3) Due to lower conductivity of heat of bricks than concrete, the roofs made up of bricks have better insulation to heat and cold.

15.4 DISADVANTAGES OF R.B. WORK

Structures in brick masonry need more protection from moisture which can easily penetrate through the bricks, thereby causing leakage and ultimately corrode steel.

15.5 DESIGN CONSIDERATIONS

1. *Effective Span*

The effective span 'l' of a freely supported member shall be clear span plus the depth of the member, but should not exceed the distance between centres of supports.

2. *Effective Depth*

The effective depth 'd' of the member should be taken as the distance from the centroid of the tensile reinforcement to the compression face, except for deep beams.

3. *Effective Width*

In beams the effective width 'b' should not be more than eight times the effective depth nor one-fourth of the span length nor the centre to centre distance between beams.

4. *Lateral Support*

The effects of lateral eccentricity of load should be taken into account in determining the spacing of lateral supports for a beam which should not exceed 32 times the least width b of the compression flange or face.

5. *Moments and Shears*

Where larger of the two adjacent spans does not exceed the shorter by more than 20% and loads are uniformly distributed and the live load does not exceed three times the dead load, the following moments and shears for beams may be used.

(A) Positive moment :

End spans

If discontinuous end is unrestrained ... $w.l^2/11$

If discontinuous end is integral with support ... $w.l^2/14$

Interior spans ... $w.l^2/16$

(B) Negative Moment at the exterior face of first interior support

Two spans ... $w\,l^2/9$

More than two spans ... $w\,l^2/10$

(C) Negative Moment at face of other interior supports ... $w\,l^2/11$

(D) Shear in end members at first interior support ... $1.15\,w.l/2$

(E) Shear at other supports where l is effective span. ... $wl/2$

6. *Allowable Stresses*

The following values of safe working stresses are recommended :

(*a*) Permissible compressive stress in brick work σ_{cb} = 2.1 N/mm² to 3.5 N/mm² increasing with coarseness of sand in the mortars and consequent decrease in *w/c* ratio. It also increases with better quality of works.

(*b*) Permissible tensile stress in plain Mild steel = 140 N/mm²

(*c*) Permissible tensile stress in Tor-steel = 230 N/mm²

(*d*) Modular ratio = m = 40

(*e*) Permissible diagonal tension τ_c = 0.07 N/mm² to 0.14 N/mm² τ_c max = 0.28 N/mm² to 0.35 N/mm²

(*f*) Permissible bond stress τ_b = 0.6 N/mm² average

7. *Some General Precautions*

Only sound bricks of good quality should be used and the mortar of 1 part of cement with 3 parts of clean coarse sand should be used. As small a quantity of water should be used in mixing the mortar as is necessary to make it workable in the joints. The bricks must be soaked in water before use, otherwise, they will absorb water from the mortar and interfere in its curing. The diameter of the steel bars should be kept as small as possible and the joints should be sufficiently wide so that mortar may be filled all around the bars. If the bars come in contact with bricks at any point due to lack of mortar, they are liable to rust. The bricks near the edges should be half on the wall and half projecting out.

It is preferable to use two layers of bricks, so that the mortar between two layers prevents the leakage. The thickness of slab has to be a multiple of brick dimensions. The spacing of steel bars depends upon the distance between joints.

8. *Shear Stress*

The nominal shear stress, as a measure of diagonal tension, in reinforcement concrete masonry members should be computed by

$$\tau_v = \frac{V}{bd}$$

For a beam with *T* section, flange width should be substituted for *b*.

Web reinforcement should be provided to carry the entire shear in any of the following manner.

(*a*) Where the computed shear stress exceeds that allowed for masonry with no shear reinforcement.

(*b*) Where the column is subject to reversal of stresses involving cracked sections.

(*c*) In the portion of the length of a member where negative reinforcement is required and for a minimum distance of 1/16th clear span beyond the inflexion point.

(*d*) In flexural members where jongitudinal reinforcement is designed to act in compression.

The area of steel required in stirrups, placed perpendicular to the longitudinal reinforcement, should be computed by:

$$A_s = \frac{V.s}{\sigma_{sv}.d}$$

Where web reinforcement is required, it should be so spaced that every 45 degree line (representing a potential diagonal crack) extending towards the reaction from mid depth (*d*/2) of the member to the longitudinal tension bars should be crossed by at least one line of web reinforcement.

Where web reinforcement is required, its area should not be less than 0.10% of the area *b.s,* computed as the product of the width of the web and the spacing of the web reinforcement along the longitudinal axis of the member.

9. *Computation of Bond*

In flexural members in which the tension reinforcement is parallel to the compression face, the flexural bond stress at any cross-section should be computed by

$$\tau_b = \frac{V}{\sum 0.j\,d}$$

Bent up bars that are not more than *d*/3 from the level of the main longitudinal reinforcement may be induced. Critical sections occur at the face of the support, at each point where tension bars terminate within a span, and at a point of inflection.

10. *General Acnhorage Requirements*

The calculated tension or compression in any bar at any section should be developed on each side of that section by proper embedment length, end anchorage or hooks. A tension bar may be anchored by bending it across the web at an angle of not less than 15 degrees with the logitudinal protion of the bar and making it continuous with the reinforcement on the opposite face of the member.

Except at supports, every reinforcing bar should be extended beyond the point at which it is no longer required to resist flexural stress, for a distance equal to the effective depth of the member or 12 bar diameters, whichever is greater.

At least one-third the positive moment reinforcement in simple beams and one-fourth the positive moment reinforcement in continuous beams should extend along the same face of the beams into the support at least 150 mm.

Tensile negative reinforcement in any span of continuous, restrained or cantilever beam, or in any member of a rigid frame should be adequately anchored by bond, hooks or mechanical anchors in and through the member. At least one-third of the total reinforcement provided for negative moment at the support should be extended beyond the extreme position of the point of inflexion a distance not less than 1/16 of the clear span, or the effective depth of a member, whichever is greater.

11. *Beam Reinforcement*

The ratio $p = \frac{A_s}{bd} \times 100$ should not be less than $\frac{100}{\sigma_y}$ where σ_y is the yield stress of steel.

15.6 SOLVED EXAMPLE

Example 15.1. Design the *R.B.* roof slab of a corridor 2.4 m wide if the live load is 1500 N/m² and there is 115 mm lime concrete terracing over it. Using Fe 415 grade reinforcing bars.

Sol. Assume 150 mm thick *R.B.* slab

$$\text{Effective span} = \text{Clear span} + \text{depth of slab} = 2.4 + 0.15 = 2.55 \text{ m}$$

$$\text{Self weight of slab} = 0.15 \times 1 \times 1 \times 20{,}000 = 300 \text{ N/m}^2$$

$$\text{Load of terracing} = 0.115 \times 1 \times 1 \times 20{,}000 = 2300 \text{ N/m}^2$$

$$\text{Live load} = \underline{1500 \text{ N/m}^2}$$

$$\text{Total} = 6800 \text{ N/m}^2$$

$$\therefore \quad \text{Max. } B.M. = \frac{6800 \times 2.55^2}{8} = 5527 \text{ N.m.}$$

$$\therefore \quad \text{Max. } S.F. = \quad = 6800 \times 1.2 = 8160 \text{ N}$$

Taking $\sigma_{cb} = 2.1 \text{ N/mm}^2$, m = 40

$$x = k.d.$$

$$= \frac{1}{1+\frac{230}{40 \times 2.1}} = 0.268\, d \qquad \therefore \; k = 0.268$$

$$j = 1 = k/3 = 1 - \frac{0.268}{3} = 0.911$$

$$Q = 1/2k.\sigma_{cb}.j = 1/2 \times 0.268 \times 2.1 \times 0.911 = 0.256 \text{ N/mm}^2$$

$$d = \sqrt{\left(\frac{5527 \times 1000}{0.256 \times 1000}\right)} = 146.9 \text{ mm}$$

$$D = 150 \text{ mm.}$$

Use one Tile 200 mm × 100 mm × 50 mm over a brick 200 mm × 100 mm × 100 mm.

$$d = 150 - 20 = 130 \text{ mm.}$$

$$A_{st} = \frac{M}{\sigma_{st}.j.d} = \frac{5527 \times 1000}{230 \times 0.911 \times 130} = 202.91 \text{ mm}^2$$

Since spacing of bars can only be 120 mm or 220 mm taking width of joints 30 mm. Use 8 mm ϕ - bars @ 120 mm *c/c*.

$$\tau_v = \frac{V}{bd}$$

$$= \frac{8160}{1000 \times 130} = 0.0627 \text{ N/mm}^3 \qquad < 0.07 \text{ N/mm}^2$$

$\therefore$ O.K.

Bend alternate bars at 1/7th span from the ends.

$$\text{Distribution steel} = \frac{0.12 \times 150 \times 1000}{100} = 180 \text{ mm}^2$$

Provide 8 mm ϕ @ 220 mm *c/c* assuming 30 mm wide joints.

The reinforcement is shown in Fig. 15.1

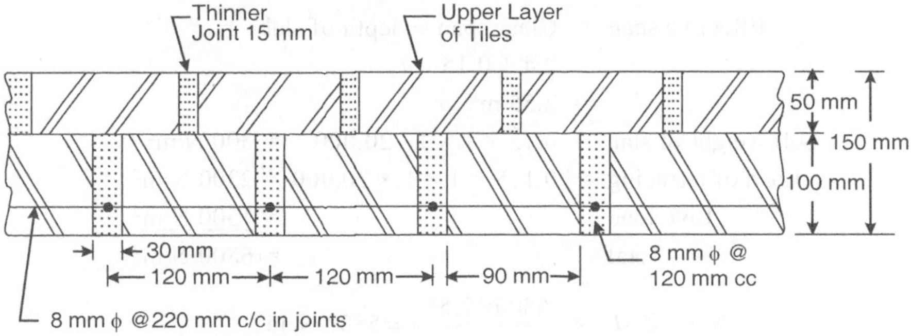

Fig. 15.1. Section of Slab Along Length

$$\text{Bond stress} = \frac{V}{jd\sum 0}$$

$$= \frac{8160}{0.911 \times 130 \times \frac{1000}{120} \times 8\pi}$$

$$= 0.329 \text{ N/mm}^2 \qquad \text{Safe}$$

15.7 REINFORCED MASONRY COLUMNS

(*a*) *Minimum Dimensions*

Such columns should have a minimum dimension of 300 mm where they are the principal supports of a floor of roof. When they are stressed to less than one-half the allowable stress or when they are auxiliary supports the minimum dimensions may be reduced to 200 mm.

(*b*) *Maximum Height*

The maximum unsupported height of a column should not be more than 25 times the least lateral dimension of the column. The unsupported height of a column should be taken as the clear distance between supports.

(*c*) *Minimum Reinforcement*

The vertical reinforcement for columns should not be less than 1/2 per cent nor more than 4 per cent of the gross cross-sectional area.

The minimum size of bars should be 12 mm. The minimum number of bars should be 4.

All vertical reinforcement for columns should be enclosed by lateral ties spaced apart not over 16 bar diameters, 48 tie diameters or the least dimension of the column. Where bars are located around the periphery of a circle, a complete circular tie may be used.

(d) *Allowable Load and Stresses*

The maximum allowable axial load on reinforced masonry columns should be given by :

$$P_a = Ag.\ (0.20\ \sigma'_m + 0.65\ p_g.\sigma_s)\left[1-\left(\frac{h}{40t}\right)^3\right]$$

where

P_a = Allowable axial load
A_g = Gross cross-sectional area
σ'_m = Permissible ultimate axial compressive stress in masonry.
P_g = percentage steel provided
σ_s = stress in steel
h = effective height
t = least lateral dimension.

15.8 REINFORCED MASONRY WALLS

Walls should be designed for any lateral, vertical or other pressures to which they are subjected. Proper provisions should be made for all eccentric and concentrated loads.

(a) *Allowable Axial Stress*

The allowable compressive stress F_a for axial loads based on the net horizontal cross-section of the wall should be :

$$F_a = 0.225\,\sigma'_m\left[1-\left(\frac{4}{40t}\right)^3\right]$$

When the reinforcement in bearing walls is designed, placed and anchored in position as for columns, the allowable stress should be that for columns. Combined flexural and axial stresses should not exceed the value determined by the interaction formula.

$$\frac{f_a}{F_a}+\frac{f_m}{F_m} \not> 1.0$$

For concentrated loads, the length of the wall to be considered effective for each load should not exceed the centre to centre distance between loads, nor exceed the width of bearing plus four times the wall thickness.

Masonry walls should have a thickness of at least 1/36 of the unsupported height or width, whichever is shorter, and should be minimum 100 mm. In computing flexural stresses where reinforcement occurs in the wall, the effective width should not be greater than six times the wall thickness, where the masonry is laid in running bond nor three times the wall thickness where the masonry is laid in stacked bond.

The minimum area of reinforcement in either direction should be not less than 0.07 per cent of the gross cross-sectional area of the wall taken perpendicular to the steel considered. The sum of the percentages of horizontal and vertical reinforcement should be at least 0.2 per cent. Reinforcement should be spaced not more than 2.44 m on centre in both directions. The vertical spacing in metres times the horizontal spacing in metres should not exceed 3.0

EXERCISE 15

1. (*a*) List the assumptions made in Reinforced brick masonry construction.
   (*b*) Discuss briefly advantages and disadvantages of *R.B.* construction.

2. Design the following *R.B.* roof slab over rooms as shown in Fig. 15.2. Take 30 mm thick M 20 concrete over slab and in between joints. The live load may be taken as 1500 N/m^2. Use Fe 415 grade reinforcement.

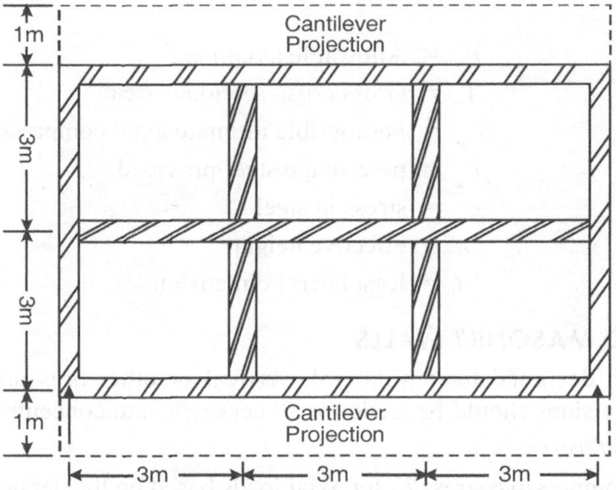

Fig. 15.2.

3. (*a*) Discuss briefly design concepts of *R.B.* Columns.

   (*b*) Determine the working load for a 350 mm × 350 mm square column of 4 m effective height and reinforced with 4 Nos. 21 mm bars of grade Fe 415.

CHAPTER

16

Formwork

16.1 FORMWORK

For concrete construction, moulds or forms are required to keep the concrete in position till it sets. These forms are also sometimes called "Falsework, Centring and shuttering". Timber and steel are the most commonly used materials for formwork. Formwork is an important feature in the concrete structures, as it forms, generally, upto 25% of the cost of the structure. Therefore, it is essential to design it properly to effect economy without sacrificing strength and efficiency. For natural face-works steel shuttering is mostly used.

16.2 ESSENTIAL REQUIREMENTS OF FORMWORK

(1) It should be strong enough to take the dead and live loads during construction.

(2) The joints in the formwork should be rigid so that building, twisting or sagging due to dead and live loads is minimum. Excessive deformations may disfigure the surface of the concrete.

(3) All lines in the formwork should be true and the surface plane so that the cost of finishing the surface of the concrete on removing the shuttering is minimum.

(4) The formwork should be water tight.

(5) It should be easily removable without damage so that it can be used repeatedly.

(6) Shuttering recessis must be bevelled for free withdrawal.

16.3 PRECAUTIONS IN FORMWORK

Concrete is liable to stick to forms if no precautions are taken. Since timber is mostly used as formwork, therefore some special precautions are necessary.

(*i*) Interior surfaces of the forms should be coated with a thin layer of oil, soft soap or white washed. Sometimes, lining of oil paper of canvas, etc., are also introduced.

(*ii*) Stresses due to fluid concrete should be analysed so that minimum size of formwork with maximum spacing of supporting should be kept.

(*iii*) For timber forms, timber should be ordered of lengths to avoid wastage in cutting.

(*iv*) All shutters after they are stripped should be immediately cleaned for stacked neatly for reuse.

(*v*) Small cleats, wedges, boils and other loose parts should be cleaned for stacking and oiled for reuse.

(*vi*) Timber should be soft but not green, as green timber would dry and shrink and at the same time not too dry as it would absorb water from concrete. Partially seasoned timber is the most suitable.

(*vii*) To allow for any sags in beams, the forms should be given a chamber at the rate of 1 cm in 5 metre length of beam.

(*viii*) Clamps and screws should be preferred to nails for facility in removing and reuse.

(*ix*) One carpenter should always be in attendance during actual pouring of concrete.

(*x*) The points mentioned in Fig. 16.1 may be taken care of while shuttering.

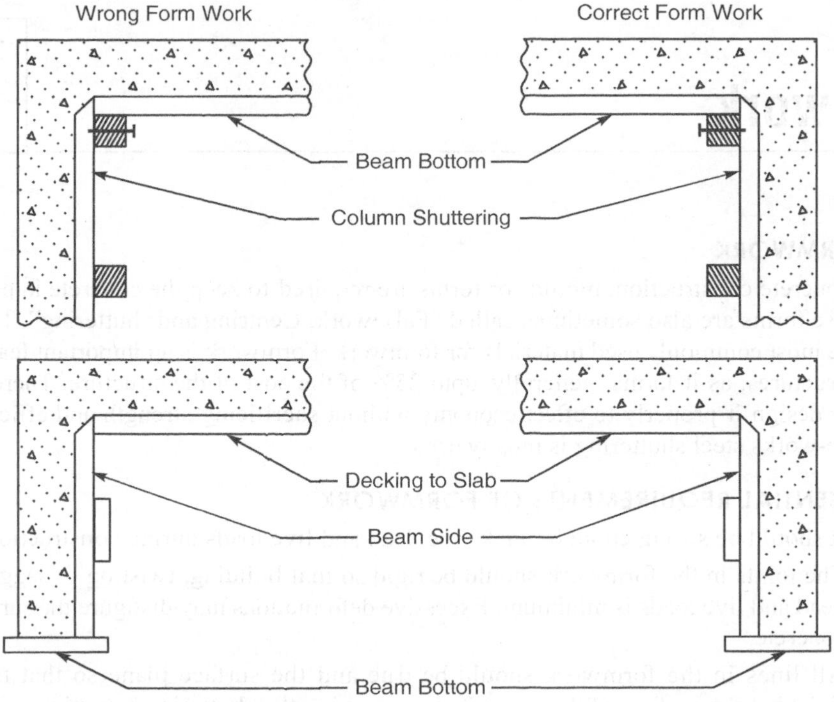

Fig. 16.1.

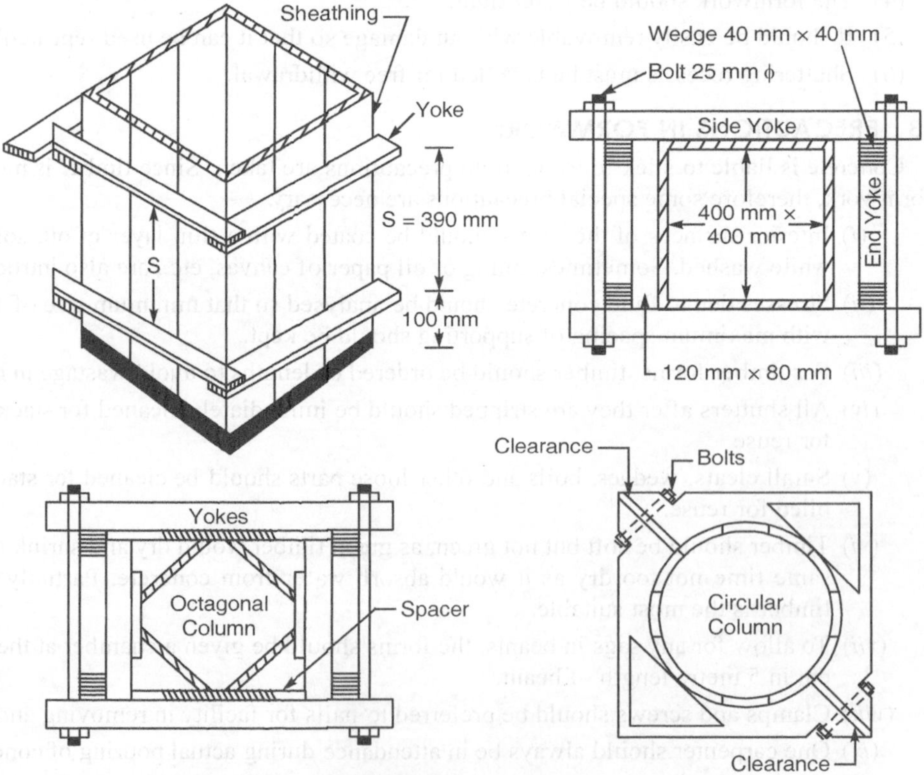

Fig. 16.2.

16.4 QUALITIES OF TIMBER FORMWORK

(*i*) It should be readily available and economical.

(*ii*) It should nail easily without splitting.

(*iii*) It should be as light as possible to permit easy handling.

(*iv*) It should be stiff to avoid undue deflection.

(*v*) It should not be liable to warp when exposed to sun and rain or wetted during concreting.

16.5 SHEATHING AND DECKING

The boards used to form the facing to shutters to wall columns are called 'SHEATHING' and those used for formwork to concrete slabs are called 'DECKING'.

16.6 FORMWORK FOR DIFFERENT TYPES OF STRUCTURES

(1) *Formwork for Columns* (Fig. 16.2)

(2) *Formwork for Beam and Floor* (Fig. 16.3)

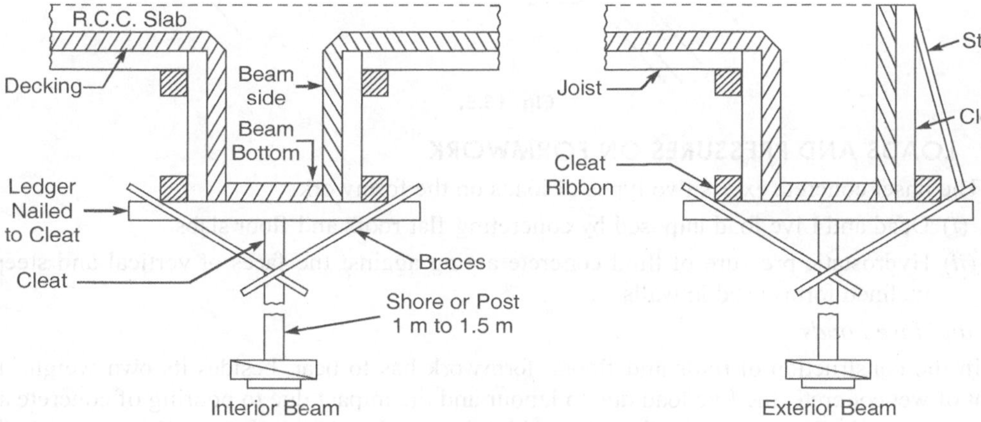

Fig. 16.3.

(3) *Formwork for Walls* (Fig. 16.4)

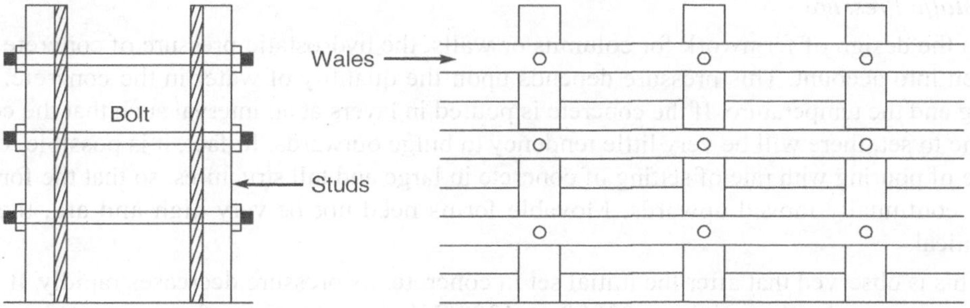

Fig. 16.4.

(4) *Form work of Stair Cases* (Fig. 16.5)

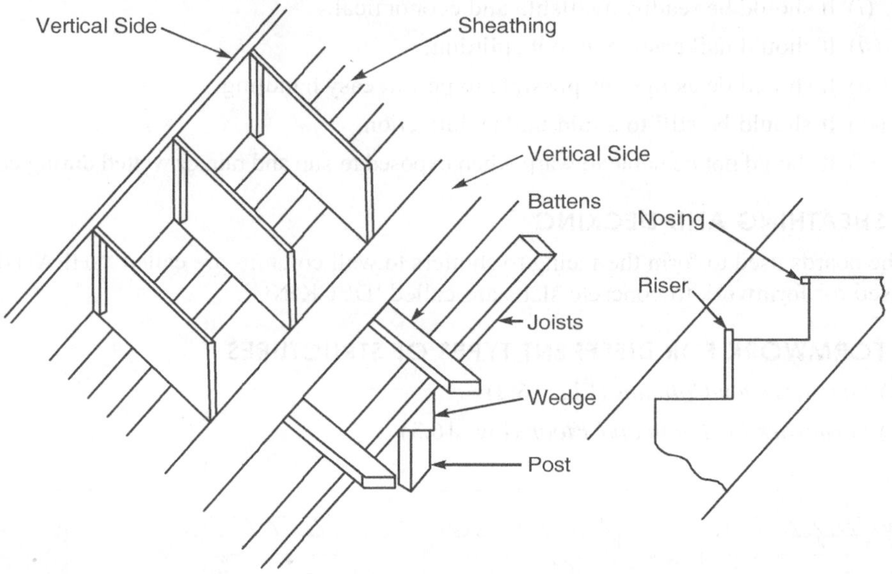

Fig. 16.5.

16.7 LOADS AND PRESSURES ON FORMWORK

The unset concrete exerts two types of loads on the formwork.

(*i*) Dead and Live load imposed by concreting flat roofs and floor slabs.

(*ii*) Hydrostatic pressure of fluid concrete acting against the faces of vertical and steeply inclined forms used in walls.

Dead and Live Loads

In the construction of roofs and floors, formwork has to bear, besides its own weight, the weight of wet concrete, the live load due to labour and the impact due to pouring of concrete and the workmen on it. For the design of planks and joists in the bending and shear, live load including impact may be taken as 3700 N/m^2. The sheathing with full live load of 3700N/m^2 should not deflect more than 0.25 cm and the joists with 2000N/sq. m. of live load should not deflect more than 0.25 cm. The dead load for R.C.C. may be taken as 25000 N/Cu.m.

Hydrostatic Pressure

In the design of formwork for columns or walls, the hydrostatic pressure of concrete should be taken into account. This pressure depends upon the quantity of water in the concrete, rate of pouring and the temperature. If the concrete is poured in layers at an interval such that the concrete has time to set, there will be very little tendency to bulge outwards. In fact, it is possible to adjust the rate of pouring with rate of setting of concrete in large and tall structures, so that the formwork can be continually moved upwards. Movable forms need not be very high and are, therefore, economical.

This is observed that after the initial set in concrete, its pressure decreases rapidly. It may be taken to act as a liquid weighing 23000 to 12000N/m^3 for heights of concrete 1.5 m to 6 m respectively.

It must be mentioned that as the design of it is based upon many assumptions the calculations usually made are only a rough guide. Members used for formwork must be strong enough to resist bending, shearing, bearing and deflection.

16.8 SOLVED EXAMPLE

Example 16.1. Design the formwork for a column 400 mm × 400 mm if its height is 3 m and it is proposed to pour entire concrete in one stage. For 3 m depth of concrete, the unit weight of fluid exerting pressure equivalent to that of concrete may be taken as 19090 N/cu.m.

Solution.

Design of Sheathing

Maximum load taken by yoke = 19090 × 3 × 0.4 × S

where S is spacing of Yokes in m.

Using Chir with thickness of sheathing 25 mm (f_b = 8.4 N/mm²)

$$\frac{M}{I} = \frac{f}{y} \qquad \therefore M = f.z$$

$$\therefore \quad 19090 \times 3 \times 0.4 \times S \times \frac{S}{10} = 8.4 \times \frac{0.4 \times 0.025^2}{6} \times 10^6$$

$$S = 0.391 \text{ m.}$$

(*ii*) *From Shear Criteria*

$$\text{Max. shear force} = 0.5 \times 19090 \times 3 \times 0.4 S = 11454 S$$

$$\text{Permissible shear stress } \sigma_s = 0.92 \text{ N/mm}^2$$

$$\frac{11454 S}{0.4 \times 0.025} = 0.92 \times 10^6, S = 0.8 \text{ m}$$

(*iii*) *From Deflection Criteria,*

$$\frac{5}{384}\left(\frac{19090 \times 3 \times 0.4 \times S^4}{98000 \times 10^5 \times 0.4 \times \dfrac{0.025^3}{12}}\right) = 0.0025$$

$$S = \sqrt[4]{\left(\frac{0.0025 \times 98000 \times 10^5 \times 0.4 \times 0.025^3 \times 384}{5 \times 12 \times 19090 \times 3 \times 0.4}\right)}$$

$$= 0.455 \text{ m}$$

Hence adopt spacing of yoke = 0.39 m.

Design of Yokes

The yokes will bend across two bolts due to hydrostatic pressure.

The pressure = 19090 × 3 × 0.4 × 0.39 = 8934 N

where 0.39 m is the spacing of yokes.

$$\text{Maximum bending moment} = \frac{W}{4}(L - b/2) = \frac{8934}{4}\left(720 - \frac{400}{2}\right) = 1161420 \text{ N.mm.}$$

(as the distance between the bolts in 720 mm.)

Direct pull in the yoke = 8934/2 = 4467 N

Let the section of the yoke be 120 mm × 80 mm.

$$\text{Max, fibre stress} = \frac{1161420 \times 6}{80 \times 120^2} = 6.05 \text{ N/mm}^2 < 8.4 \text{ N/mm}_2$$

∴ Safe.

Bolt

It is subjected to bending and tension. However, bending is more predominant.

Let the dia. of the bolt = 25 mm.

$$\text{Direct stress} = \frac{4467}{491} = 9.1 \text{ N/mm}^2$$

$$\text{Bending moment} = \frac{8934 \times 40}{2} = 178680 \text{ N.mm.}$$

Assuming 40 mm as the distance of the load from supports, load being transmitted through the wedges.

$$\text{Bending stress} = \frac{178680}{22/7 \cdot (25)^4/64} \times \frac{1}{2} \times (25) = 116.4 \text{ N/mm}^2$$

$$\text{Total stress} = 9.1 + 116.4$$

$$= 125.5 \text{ N/mm}^2 \quad < 140 \text{ N/mm}^2 \quad \therefore \text{Safe}$$

Washers may be used to avoid crushing.

16.9 TIMBER FOR FORMWORK

The timbers which are most commonly used for formwork, are Fir, Deodar, Kail and Chir. They are soft woods. The permissible stresses as per IS : 883 are given in Table 16.1.

16.10 LOADS ON DECKING

The bottom shuttering of slab in horizontal plane is called 'Decking'. The decking has to take the immediate impact of live loads which should be taken as 4000 N/m^2. Spacing of joist underneath the decking will depend upon the allowable deflection. The deflection should not exceed 3 mm. The deflection should be calculated from the following formula.

$$\Delta = \frac{3}{384} \frac{W.l^4}{E.I} \qquad ...(16.1)$$

where w = load on decking N/m^2

S = spacing of joists in metres

E = modulus of elasticity in N/m^2

I = moment of inertia = $b.d^3/12$

$$\therefore \quad \text{Spacing of joists } S = \sqrt{\left(\frac{12E.I.\Delta}{w}\right)}$$

where $l = S$, *i.e.*, span of deck slab = spacing of joists

D = deflection in metre

If 'L' metres is the span of the joists, the total load on the joists shall be $W = w.L.S.$

Considering a rectangular joist

$$\frac{w.s.L^2 \times 100}{8} = f.b.\frac{d^2}{6}$$

$$L = \sqrt{\left(\frac{f.b.d^2}{75\, w.s.}\right)}$$

Table 16.1.

Name	*Density* N/m^3	*Modulus of Elasticity* $K/N\ mm^2$	*Bending and tension* N/mm^2			*Shear*		*Compression parallel*			*Compression perpendicular*		
			Inside	*Outside*	*Wet*	*Horizontal*	*Along grain*	*Inside*	*Out-side*	*Wet*	*Inside*	*Outside*	*Wet*
Fit	4500	9.4	7.8	6.6	5.6	0.6	0.8	6.0	5.2	4.2	1.6	1.2	1.0
Deodar	5450	9.5	10.2	8.8	7.0	0.7	1.0	7.8	7.0	5.6	2.6	2.1	1.7
Kail	5150	6.8	6.6	5.6	5.0	0.6	0.8	5.2	4.6	3.8	1.7	1.3	1.0
Chir	5750	9.8	8.4	7.0	6.0	0.6	0.9	6.4	5.6	4.6	2.2	1.7	1.4

EXERCISE 16

1. (*a*) What is formwork? (*b*) List the essential requirement of formwork.
2. What precautions should be taken in formwork?
3. Explain with the help of sketches the forms which are used for various types of structures such as columns, beams, slabs and walls.
4. What lateral pressure should be taken for fresh concrete?
5. Design the formwork for a wall 2.5 m high and 225 mm thick. Take Deodar wood for formwork.

PART - III

LIMIT STATE DESIGN

In this method of design based on limit state concept, the structure shall be deisgned to withstand safely all loads liable to act on it throughout its life time; it shall also satisfy the serviceability requirements such as prevention of excessive deflection, excessive cracking and excessive vibrations. The acceptable limit for the safety and serviceability requirement before failure occurs is called "limit state". This method is modification of 'Ultimate Strength Design Procedure' where the structure is subjected to ultimate loads and the materials are allowed to develop ultimate stresses. In limit state design concept, design values are derived from characteristic values (ultimate values). Design loads are obtained by multiplying working loads with partial factors of safety for loads and in a similar manner, design strength of materials is obtained by dividing characteristics strength (ultimate strength) with respective partial factors of safety for materials. Members are, thus, proportioned to carry limit state design loads and materials are stressed to limitting design strengths.

This approach is more logical and has the merit of simplicity and economy. This is based on semi-probabilistic approach.

This Page has been intentionally left blank

CHAPTER 17

Introduction to Limit State Design

17.1 PURPOSE OF DESIGN

In the limit state of design, the structures shall be designed to with stand safely all loads liable to act on it throughout its life; it shall also satisfy the serviceability requirements such as limitations on excessive deflections, oscillations, cracking, local damage, vibrations, fire penetration, etc. The acceptable limit for safety and serviceability requirements before failure occurs, is called a 'Limit State'. The aim of the design is to achieve acceptable probabilities that the structure will not become unfit for the use for which it is intended, *i.e.*, it will not reach a limit state. In general, the structure shall be designed on the basis of the most critical state, *i.e.*, ultimate limit state and shall be checked for other limit states.

For ensuring the above objective, the design should be based on characteristic values for material strengths and applied loads, which take into account the variations in the material strengths and in the loads to be supported as well as the inadequacies in the methods of analysis, design and construction. The structure in all probabilities should not be allowed to reach a limit state, *i.e.*, probability of a limit state being attained is kept at an acceptably low value. However, exact probabilistic design is not possible at present because of lack of statistical data concerning the design variables. The design approach recommended in IS Code is, hence, 'semi-probabilistic.' in which the basic design variables are treated in a 'probabilistic manner' and other uncertainties in imposed loads, method of design, analysis, etc., are covered through 'partial safety factors'. The characteristic values should be based on statistical data if available, where such data is not available they should be based on experience. The design values are derived from the characteristic values through the use of partial safety factors, one for material strengths and the other for loads. The concept of limit state is not altogether new. It was embodied in the Russian Code as early as 1954. Later, the British Code also accepted it as method of design for reinforced concrete structures.

17.2 LIMIT STATE OF COLLAPSE

The limit state of collapse of the structure or part of the structure could be assessed from rupture of one or more critical sections and from buckling due to elastic or plastic instability or overturning or fatigue, etc. The resistance to bending, shear, torsion and axial loads at every section shall not be less than the appropriate value at that section produced by the probable most unfavourable combination of loads on the structure using the appropriate partial safety factors.

17.3 CHARACTERISTIC STRENGTH OF MATERIALS

The term 'Characteristic strength' means that value of the strength of the material below which not more than 5 per cent of the test results are expected to fall.

17.4 CHARACTERISTICS LOADS

The term "Charcteristic Load" means that value of load which has a 95 per cent probability of not being exceeded during the life of the structure.

17.5 PARTIAL SAFETY FACTORS AND DESIGN VALUES

The characteristic values allow for the inherent variations which occur in loads and material strengths. It is also necessary to consider other factors and causes of variations which can occur and to make allowances (up and down) for these uncertainties by introducing suitable partial safety factors. These uncertainties may be due to:

(*i*) Long term loading.

(*ii*) The material of the actual structure may be weaker than the characteristic value.

(*iii*) Constructional faults such as inadequate compaction or curing, impurities, bad weather, etc.

(*iv*) Repeated loading.

(*v*) Corrosion during life time of a structure.

(*vi*) The seriousness of the consequences that might follow if excessive loading or stress occurs and so on.

17.6 DESIGN VALUES

(*A*) *Materials*

The design strength of materials f_d is given by

$$f_d = \frac{f}{\gamma_m} \qquad \text{...(17.1)}$$

where

f = characteristic strength of material.

γ_m = partial safety factor appropriate to the material and limit state being considered.

γ_m = 1.5 for concrete

γ_m = 1.15 for steel

(*B*) *Load*

The design load, F_d is given by :

$$F_d = F.\gamma_f \qquad \text{...(17.2)}$$

where

F = Characteristic load,

γ_f = partial safety factor appropriate to the nature of loading and the limit state being considered.

Where the consequences of a structure attaining a limit state are of a serious nature such as huge loss of life and disruption of economy, higher values of γ_m and γ_f than those given subsequently may be applied.

(*C*) *Partial Safety Factors*

Partial safety factors γ_f for loads given in Table 17.1 shall normally be used.

Table 17.1. Values of Partial Safety Factor γ_f or Loads

Load Combination	*Limit State of Collapse*			*Limit States of Serviceability*		
	DL	*LL*	*WL*	*DL*	*LL*	*WL*
(1)	(2)	(3)	(4)	(5)	(6)	(7)
DL + *LL*	1.5		—	1.0	1.0	—
DL + *WL*	1.5 or 0.9*	—	1.5	1.0	–	1.0
DL + *LL* + *WL*	1.2			1.0	0.8	0.8

Notes: 1. While considering earthquake effects, substitute *EL* for *WL*.

2. For the limit states of serviceability, the values of γ_f given in this table are applicable for short effects. While assessing the long term effects due to creep, the dead load and that part of the live load likely to be permanent may only be considered.

* This value is to be considered when stability against overturning or stress reversal is critical.

EXERCISE 17

1. What do you understand by Limit State of Collapse?
2. Define the following :
   (*i*) Characteristic strength of mateirals.
   (*ii*) Characteristic loads.
   (*iii*) Partial safety factors and design values.
3. Why Limit State design is considered more rational than Working Stress design?

Limit State of Collapse : Flexure

CHAPTER

18

18.1 ASSUMPTIONS

In assessing the strength of any section at failure by limit analysis, the following assumptions are made:

(*a*) Plane sections normal to the axis remain plane after bending, *i.e.*, the distribution of strains across any section is linear. The strain at any point is proportional to its distance from the neutral axis.

(*b*) The maximum strain in concrete at the outermost compression fibre is taken as 0.0035 in bending.

(*c*) The relationship between the compressive stress distribution in concrete and the strain in concrete may be assumed to be rectangle, trapezoid, parabola or any other shape which results in prediction of strength in substantial agreeement with the results of tests. An acceptable stress-strain curve is given in Fig. 18.1. For the design purpose, the compressive strength of concrete in the structure shall be assumed to be 0.67 times the characteristic strength. The partial safety factor $\gamma_m = 1.5$ shall be applied to this.

For the stress-strain curve in Fig. 18.1, the design stress block parameters are as follows (Fig. 18.2).

$$\text{Area of stress block} = 0.36 f_{ck} \cdot x_u \qquad ...(18.1)$$

Depth of centre of compressive force from the extreme fibre in compression

$$= 0.42\, x_u \qquad ...(18.2)$$

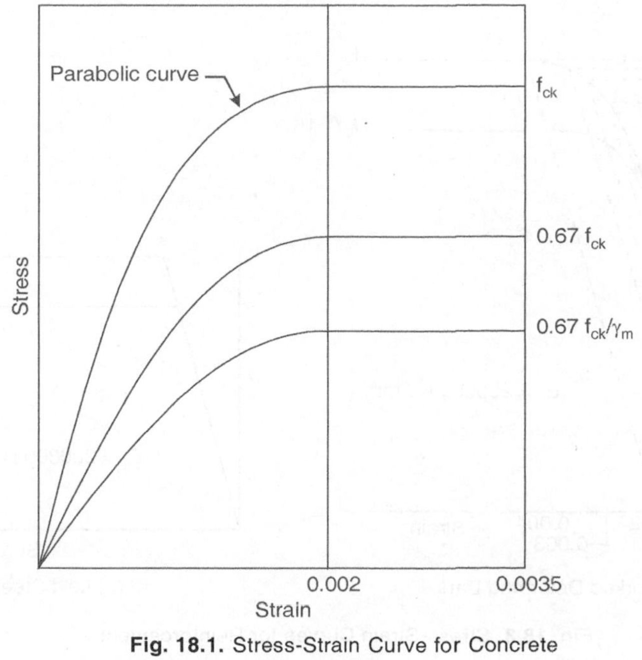

Fig. 18.1. Stress-Strain Curve for Concrete

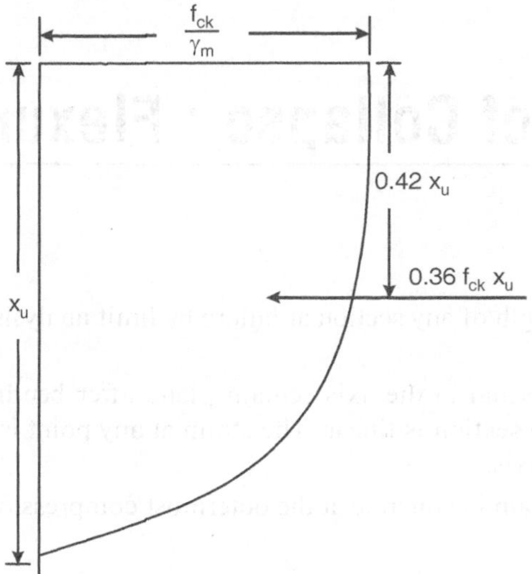

Fig. 18.2. Stress Block Parameters

where f_{ck} = characteristic compressive strength of concrete
x_u = depth of neutral axis.

(*d*) The tensile strength of the concrete is ignored.

(*e*) The stresses in reinforcement are derived from representative stress-strain curve for the type of steel used. Typical curves are given in Fig. 18.3. For design purposes, the partial safety factory γ_m equal to 1.15 shall be applied.

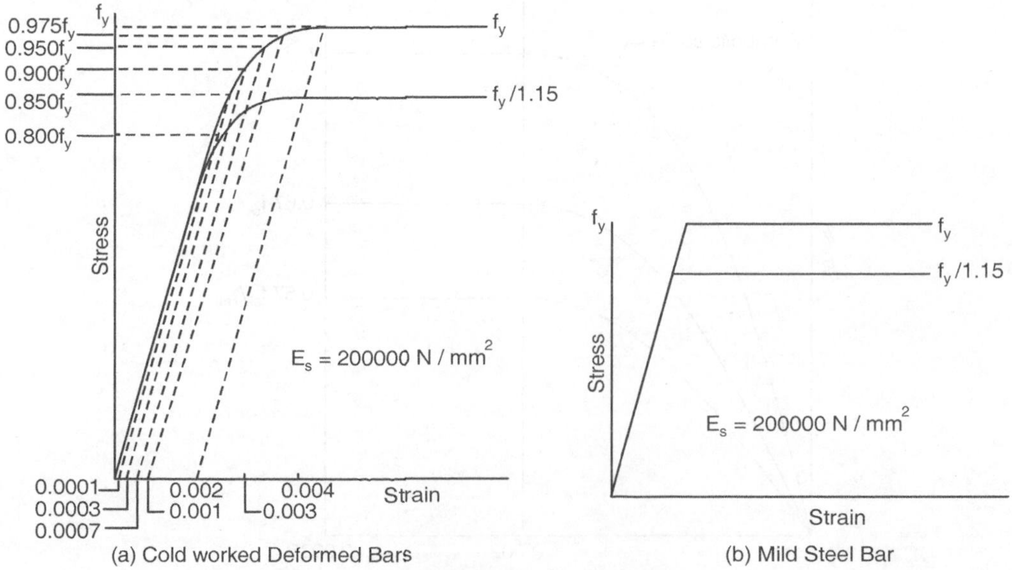

Fig. 18.3. Stress-Strain Curves for Reinforcement

(*f*) The maximum strain in the tension reinforcement in the section at failure shall not be less than:

$$\frac{f_y}{1.15E_s} + 0.002$$

where f_y = characteristic strength of steel, and
E_s = modulus of elasticity of steel.

18.2 MAXIMUM DEPTH OF NEUTRAL AXIS

Assumptions (*b*) and (*f*) govern the maximum depth of neutral axis in flexural members. The strain distribution across a member corresponding to those limiting conditions is shown in Fig. 18.4. The maximum depth of neutral axis, $x_{u,max}$, is obtained directly from the strain diagram considering similar triangles.

$$\frac{x_{u,max}}{d} = \frac{0.0035}{(0.0055+0.87f_y/E_s)}$$

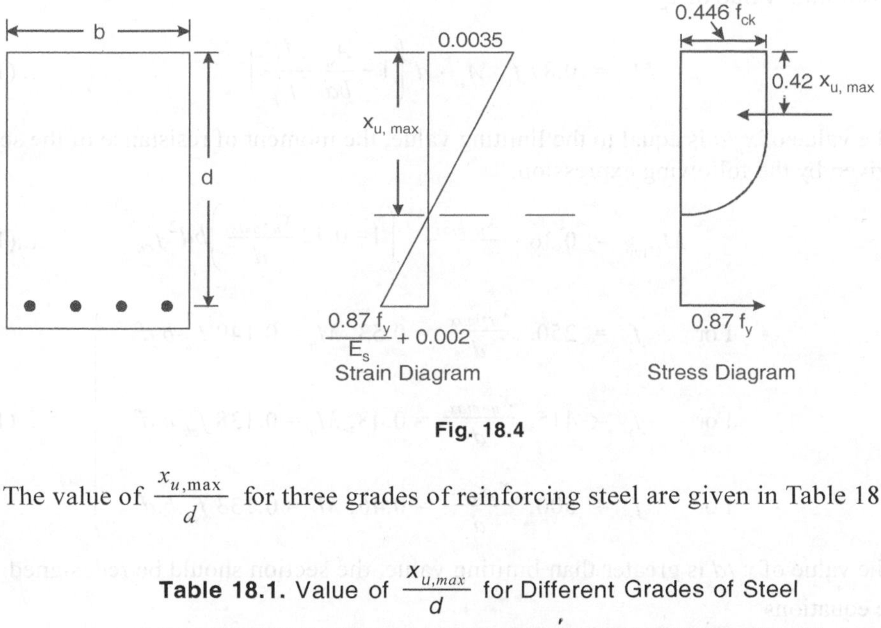

Fig. 18.4

The value of $\frac{x_{u,max}}{d}$ for three grades of reinforcing steel are given in Table 18.1.

Table 18.1. Value of $\frac{x_{u,max}}{d}$ for Different Grades of Steel

f_y N/mm²	$\frac{x_{u,max}}{d}$
250	0.53
415	0.48
500	0.46

18.3 DESIGN OF SECTIONS FOR FLEXURES AS PER IS : 456-2000

Expressions for obtaining moments of resistance of rectangular and *T*-beams are given as under.

18.3.1 Rectangular Sections without Compression Reinforcement

The moment of resistance of rectangular sections with tension reinforcement should be obtained as follows :

(*a*) Determine the depth of neutral axis by equating longitudinal forces (Fig. 18.2).

$$0.36\, x_u f_{ck} \cdot b = 0.87 f_y \cdot A_{st}$$

$$\therefore \quad x_u = \frac{0.87 f_y \cdot A_{st}}{0.36\, b \cdot f_{ck}} \quad \text{...(18.3)}$$

Dividing both sides by d

$$\frac{x_u}{d} = \frac{0.87 f_y \cdot A_{st}}{0.36 f_{ck} \cdot b.d}$$

$$= 2.42 \frac{f_y \cdot A_{st}}{b.d.f_{ck}} \quad \text{...(18.4)}$$

(*b*) If the value of x_u/d is less than the limiting value, calculate the moment of resistance by taking moments about concrete compressive face, $M_u = 0.87 f_y \cdot A_{st} \cdot (d - 0.42\, x_u)$
Substituting value of x_u

$$M_u = 0.87 f_y \cdot A_{st} \cdot d \left(1 - \frac{A_{st}}{bd} \cdot \frac{f_y}{f_{ck}}\right) \quad \text{...(18.5)}$$

(*c*) If the value of x_u/d is equal to the limiting value, the moment of resistance of the section is given by the following expression.

$$M_{u,\lim} = 0.36 \cdot \frac{x_{u,max}}{d} \cdot \left(1 - 0.42 \frac{x_{u,max}}{d}\right) bd^2 f_{ck} \quad \text{...(18.6)}$$

$$\left.\begin{array}{ll} \text{For} & f_y = 250, \ \frac{x_{u,max}}{d} = 0.53,\ M_u = 0.149 f_{ck}\, b.d^2 \\ \text{For} & f_y = 415, \ \frac{x_{u,max}}{d} = 0.48,\ M_u = 0.138 f_{ck}\, b.d^2 \\ \text{For} & f_y = 500, \ \frac{x_{u,max}}{d} = 0.46,\ M_u = 0.133 f_{ck}.b.d^2 \end{array}\right\} \quad \text{...(18.7)}$$

(*d*) If the value of x_u/d is greater than limiting value, the section should be redesigned.

In the above equations

x_u = depth of neutral axis ;
d = effective depth ;
f_y = characteristic strength of material ;
A_{st} = area of tension reinforcment ;
f_{ck} = characteristic compressive strength of concrete ;
b = width of section at compression face.
$M_{u,lim}$ = limiting moment of resistance of a section without compression reinforcement.
$x_{u,max}$ = limiting value of x_u.

Moment of resistance of concrete section is

$$M_u = 0.36 f_{ck}.b.x_u (d - 0.42\, x_u)$$

We can write the above equation as

$$\frac{M_u}{bd^2 f_{ck}} = 0.36 \frac{x_u}{d}\left(1 - 0.42\frac{x_u}{d}\right) \quad ...(18.8)$$

For any value of $\frac{M_u}{bd^2 f_{ck}}$, $\frac{x_u}{d}$ can be evaluated from equation (18.8) or from the scale shown in Fig. 18.5.

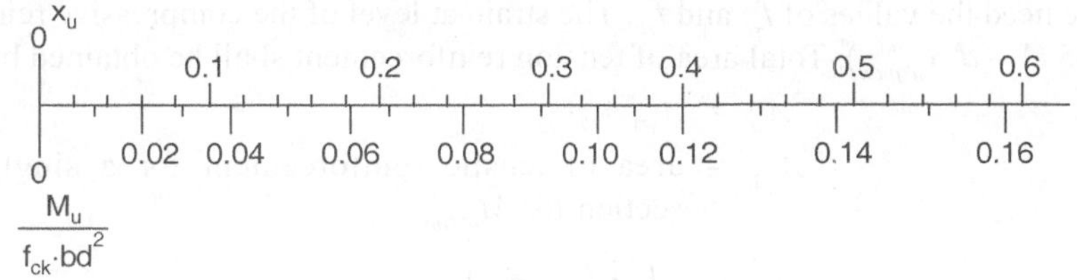

Fig. 18.5.

18.3.2 Doubly Reinforced Sections or Sections with Compression Reinforcement

Doubly reinforced sections are generally adopted when the dimensions of the beam have been predetermined from other considerations and the design moment exceeds the moment of resistance of a singly reinforced section. The additional moment of resistance needed, is obtained by providing compression reinforcement and additional tension reinforcement. The moment of resistance of a doubly reinforced section is thus the sum of the limiting moment of resistance $M_{u,lim}$ of a singly reinforced section and the additional moment of resistance M_{u2}. Given the values of M_u which is greater than $M_{u,lim}$, the value of M_{u2} can be calculated from Fig. 18.6.

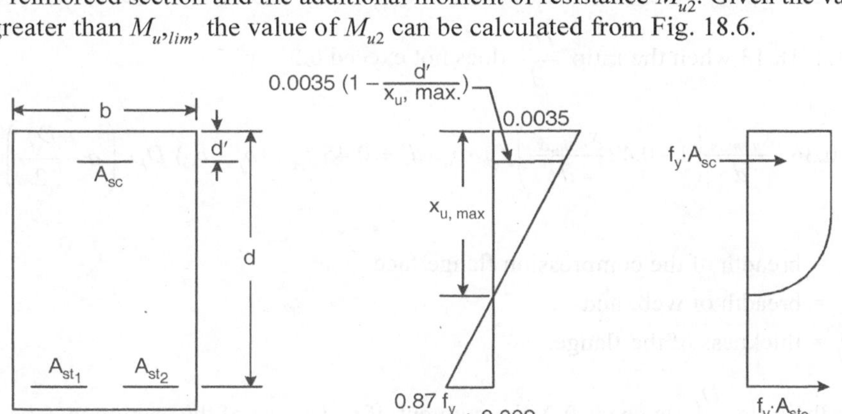

Fig. 18.6.

The lever arm for the additional moment of resistance is equal to the distance between centroids of tension reinforcement and compression reinforcement, that is $(d - d')$ where d' is the distance from the extreme compression fibre to the centroid of compression reinforcement. Therefore, considering the additional tensile reinforcement and the compressive reinforcement, we get the following :

$$M_{u2} = A_{st2}\,(0.87 f_y)\,(d - d') \quad ...(18.9)$$

also $$M_{u2} = A_{sc}(f_{sc} - f_{cc})(d - d') \quad ...(18.10)$$

where A_{st2} is the area of additional tensile reinforcement ;

A_{sc} is the area of compression reinforcement ;

f_{sc} is the stress in compression reinforcement ;

f_{cc} is the compressive stress in concrete at the level of the centroid of compresion reinforcement.

Since the additional tensile force is balanced by the additional compressive force.

$$A_{sc}(f_{sc} - f_{cc}) = A_{st2}(0.87 f_y) \quad ...(18.11)$$

Any two of the above three equations may be used for finding A_{st2} and A_{sc}. But before A_{sc} is determined, we need the values of f_{sc} and f_{cc}. The strain at level of the compressive reinforcement is equal to 0.0035 $(1 - d'/x_{u,max})$. Total area of tension reinforcement shall be obtained by

$$A_{st} = A_{st1} + A_{st2}$$

where A_{st1} = area of tensile reinforcement for a singly reinforced section for $M_{u,lim}$

$$A_{st2} = \frac{A_{sc}(f_{sc} - f_{cc})}{0.87 f_y}$$

$$\simeq \frac{A_{sc}(f_{sc})}{0.87 f_y} \quad ...(18.12)$$

18.3.3 Flanged Sections

For $x_u < D_f$, the moment of resistance is same as calculated for a rectangular section.

The limiting value of the moment of resistance of the section may be obtained by the following Eq. 18.13 when the ratio $\frac{D_f}{d}$ does not exceed 0.2.

$$M_u = 0.36\,\frac{x_{u,max}}{d}\left(1 - 0.42\frac{x_{u,max}}{d}\right) f_{ck} \cdot b_w \cdot d^2 + 0.45 f_{ck} \cdot (b_f - b_w)\, D_f \cdot \left(d - \frac{D_f}{2}\right) \quad ...(18.13)$$

where

b_f = breadth of the compression flange/face

b_w = breadth of web, and

D_f = thickness of the flange.

When the ratio $\frac{D_f}{d}$ exceeds 0.2, the moment of resistance of the section may be calculated by the following equation.

$$M_u = 0.36\,\frac{x_{u,max}}{d}\left(1 - 0.42\frac{x_{u,max}}{d}\right) f_{ck} \cdot b_w \cdot d^2 + 0.45 f_{ck} \cdot (b_f - b_w) y_f \cdot \left(d - \frac{y_f}{2}\right) \quad ...(18.14)$$

where $y_f = (0.15 x_u + 0.65\, D_f)$, but not greater than D_f.

For $x_{u,max} > x_u > D_f$, the moment of resistance may be calculated by the Eq. (18.13), when D_f/x_u does not exceed 0.43 and Eq. (18.14) when D_f/x_u exceeds 0.43; in both the cases substituting $x_{u,max}$ by x_u.

18.4 SOLVED EXAMPLES

Example 18.1. Calculate the area of steel of grade Fe 415 required for section of width 250 mm and overall depth 500 mm (effective depth 460 mm) in M20 if the limit state moment to be carried by the beam section is

(*i*) 100 KN.m (*ii*) 146 KN.m (*iii*) 200 KN.m

Solution. For steel $F_y = 415$ N/mm², $\dfrac{x_{u,max}}{d} = 0.48$

$$M_{u,lim} = 0.36\,\frac{x_{u,max}}{d}\left(1-0.42\frac{x_{u,max}}{d}\right)b\cdot d^2 f_{ck}$$

$$= 0.36 \times 0.48\,(1 - 0.42 \times 0.48) \times 250 \times 460^2 \times 20$$

$$= 146 \times 10^6 \text{ N.mm, } i.e., \text{ 146 KN.m}$$

(*i*) For $M_u = 100$ KN.m $< M_{u,lim} = 146$ KN.m

Area of steel required, A_{st}, is obtained from (Eq. 18.5)

$$M_u = 0.87\,f_y\,A_{st}\,d\left(1-\frac{A_{st}.f_y}{bd\,f_{ck}}\right)$$

$$100 \times 10^6 = 0.87 \times 415 \times A_{st} \times 460\left(1-\frac{A_{st}\times 415}{250\times 460\times 20}\right)$$

A_{st} (from quadratic equation) = 686 or 4850 mm²

Min.A_{st} required = 686 mm²

(*ii*) When $M_u = 146$ KN.m $= M_{u,lim}$

$$x_u = x_{u,max}$$

Area of tension reinforcement required

$$\frac{x_{u,max}}{d} = \frac{0.87\,f_y\,A_{st}}{0.36\,f_{ck}\,b\cdot d}$$

$$A_{st} = \frac{0.48\times 0.36\times 20\times 250\times 460}{0.87\times 415} = 1100 \text{ mm}^2$$

(*iii*) $M_u = 200$ KN.m $> M_{u,lim} = 146$ KN.m

Area of compression reinforcement is given by

$$M_u = M_{u,lim} = f_{sc}.A_{sc}\,(d - d')$$

where

f_{sc} is stress corresponding to strain of

$$\frac{0.0035(x_{u,max} - d')}{x_{u,max}}$$

$$= \frac{0.0035(0.48\times 460 - 40)}{0.48\times 460}$$

$$= 0.002866$$

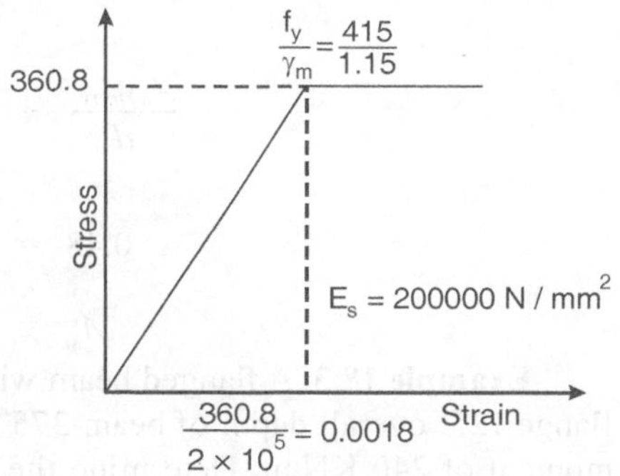

Fig. 18.7.

See Fig. 18.7

$$\therefore \quad f_{sc} = 360.8 \text{ N/mm}^2$$

$$(200 - 146) \times 10^6 = 360.8\, A_{sc}\, (460 - 40)$$

$$A_{sc} = 356 \text{ mm}^2$$

A_{st1} = Area of tension reinforcement corresponding to $M_{u'lim}$

$$146 \times 10^6 = 0.87 \times 460 \times 415\, A_{st1} \left(1 - \frac{A_{st1} \times 415}{250 \times 460 \times 20}\right)$$

$$A_{st1} = 1094 \text{ mm}^2 \text{ or } 4442 \text{ mm}^2 \text{ (neglected)}$$

$$A_{st2} = \frac{A_{sc} \cdot f_{sc}}{0.87\, f_y}$$

$$= \frac{356 \times 360.8}{0.87 \times 415} = 356 \text{ mm}^2$$

$$A_{st} = A_{st1} + A_{st2}$$

$$= 1094 + 356$$

$$= 1450 \text{ mm}^2$$

Example 18.2. Design a rectangular beam which carries a max. limiting bending moment of 65 KN.m. Use M20 concrete and Fe 415 as reinforcement.

Solution. At balanced failure condition

$$M_u = M_{u'lim}$$

$$M_{u'lim} = 0.36 \frac{x_{u'max}}{d} \left(1 - 0.42 \frac{x_{u\ max}}{d}\right) bd^2\, f_{ck}$$

$$= 0.36 \times 0.48 \times 20\, (1 - 0.42 \times 0.48)\, bd^2$$

$$= 2.759\, bd^2$$

$$d = \sqrt{\left(\frac{M_{u'lim}}{2.759\, b}\right)}$$

Assuming width of beam as 250 mm

$$d = \sqrt{\left(\frac{65 \times 10^6}{2.759 \times 250}\right)} = 307 \text{ mm}$$

$$\frac{x_{u'max}}{d} = \frac{0.87\, f_y \cdot A_{st}}{0.36\, f_{ck} \cdot bd}$$

$$0.48 = \frac{0.87 \times 415 \times A_{st}}{0.36 \times 20 \times 250 \times 307}$$

$$A_{st} = 734.66 \text{ mm}^2$$

Example 18.3. A flanged beam with flange width 960 mm, rib breadth 200 mm, thickness of flange 125, overall depth of beam 375 mm and effective depth 315 mm, is to support a factored moment of 240 KN.m. Determine the moment of resistance of section and also the amount of reinforcement if $f_{ck} = 20$ N/mm² and $f_y = 415$ N/mm².

Solution. (Fig. 18.8)

$$\frac{D_f}{d} = \frac{125}{315} = 0.3968 > 0.2$$

$$\text{For steel with } f_y = 415 \text{ N/mm}^2$$

$$\frac{x_{u,max}}{d} = 0.48$$

$$x_{u,max} = 0.48 \times 315 = 151.2 \text{ mm}$$

$$M_u = 0.36\,\frac{x_{u,max}}{d}\left(1-0.42\frac{x_{u,max}}{d}\right) f_{ck}\cdot b_w \cdot d^2 + 0.45\, f_{ck}\,(b_f - b_w)\, y_f \cdot\left(d-\frac{y_f}{2}\right)$$

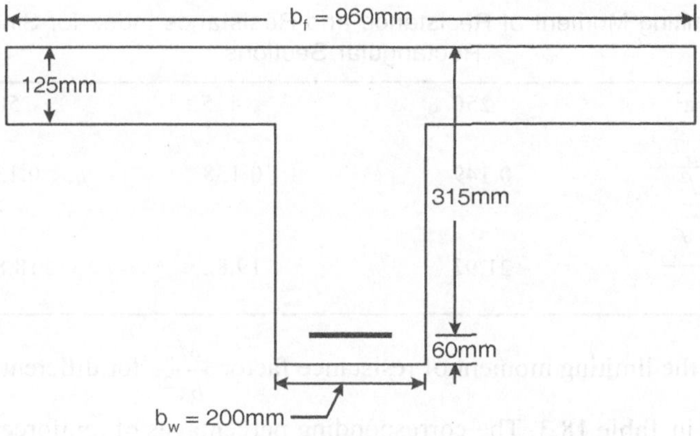

Fig. 18.8.

where

$$y_f = (0.15\,x_u + 0.65\,D_f)$$
$$= (0.15 \times 151.2 + 0.65 \times 125)$$
$$= 103.93 \text{ mm} < D_f = 125 \text{ mm}$$

$$M_u = 0.36 \times 0.48\,(1 - 0.42 \times 0.48) \times 20 \times 200 \times 315^2 + 0.45 \times 20 \times (960 - 200) \times 103.93 \times \left(315-\frac{103.93}{2}\right)$$
$$= 54.76 \times 10^6 + 186.98 \times 10^6$$
$$= 241.74 \times 10^6 \text{ N.mm}$$
$$= 241.74 \text{ KN.m}$$

$$\frac{x_{u,max}}{d} = \frac{0.87\, f_y . A_{st}}{0.36\, f_{ck} .bd}$$

$$A_{st} = \frac{0.48 \times 0.36 \times 20 \times 960 \times 315}{0.87 \times 415}$$
$$= 2894.60 \text{ mm}^2$$

18.5 USE OF SP 16–DESIGN AIDS FOR REINFORCED CONCRETE TO IS : 456-2000

Flexural Members

The limiting moment of resistance of a singly reinforced rectangle section is given by the following relation.

$$M_{u,lim} = 0.36 f_{ck}\, b \cdot x_{u,max}\,(d - 0.42\, x_{u,max}) \qquad ...(18.15)$$

Substituting for $x_{u,max}$ from Table 18.1 and transposing $f_{ck}\, bd^2$, we get the values of limiting moment of resistance factors for singly reinforced rectangular beams and slabs. These values are given in Table 18.2. The tensile reinforcement percentage, $p_{t,lim}$ corresponding to the limiting moment of resistance is obtained by equating the forces of tension and compression.

$$\frac{p_{t,lim} \times bd\,(0.87\, f_y)}{100} = 0.36\, f_{ck} \cdot b \cdot x_{u,max} \qquad ...(18.16)$$

Substituting the values of $x_{u,max}$ from Table 18.1, we get the values of $p_{t,lim} \cdot \dfrac{f_y}{f_{ck}}$ as given in Table 18.2.

Table 18.2. Limiting Moment of Resistance and Resistance Index for Singly Reinforced Rectangular Sections

f_y N/mm²	250	415	500
$\dfrac{M_{u,max}}{f_{ck} \cdot bd^2}$	0.149	0.138	0.133
$\dfrac{p_{t,lim}\, f_y}{f_{ck}}$	21.97	19.82	18.87

The values of the limiting moment of resistance factor $\dfrac{M_u}{bd^2}$ for different grades of concrete and steel are given in Table 18.3. The corresponding percentages of reinforcements are given in Table 18.4. These are the maximum permissible percentages for singly reinforced sections.

Table 18.3. Limiting Moment of Resistance Factor $\dfrac{M_{u,lim}}{bd^2}$, N/mm² for Singly Reinforced Rectangular Sections

f_{ck}	f_y *N/mm²*		
N/mm²	*250*	*415*	*500*
20	2.98	2.76	2.66
25	3.73	3.45	3.33
30	4.47	4.14	3.99

Table 18.4. Maximum Percentage of Tensile Reinforcement $p_{t,lim}$ for Singly Reinforced Rectangular Sections

f_{ck}	f_y N/mm²		
N/mm²	*250*	*415*	*500*
20	1.76	0.96	0.76
25	2.20	1.19	0.94
30	2.64	1.43	1.13

(A) Under-Reinforced Sections

Under-reinforced section means a singly reinforced section with reinforcement percentage not exceeding the appropriate value given in Table 18.4. For such sections, the depth of neutral axis x_u will be smaller than $x_{u,max}$. The strain in steel at the limit state of collapse will, therefore, be more than $\frac{0.87 f_y}{E_s} + 0.002$ and, the design stress in steel will be $0.87 f_y$. The depth of the neutral axis is obtained by equating forces of tension and compression.

$$\frac{p_t \cdot bd(0.87 f_y)}{100} = 0.36 f_{ck} \cdot bd \cdot x_u$$

$$\frac{x_u}{d} = \left(\frac{p_t}{100}\right)\frac{0.87 f_y}{0.36 f_{ck}} \qquad \text{...(18.17)}$$

The moment of resistance of the section is equal to the product of tensile force and lever arm.

$$M_u = \frac{p_t \cdot bd}{100}(0.87 f_y)(d - 0.42 x_u)$$

$$= 0.87 f_y \left(\frac{p_t}{100}\right)\left(1 - 0.42\frac{x_u}{d}\right) bd^2$$

Substituting for $\frac{x_u}{d}$, we get

$$M_u = 0.87 f_y \left(\frac{p_t}{100}\right)\left(1 - 1.005\frac{f_y}{f_{ck}}\frac{p_t}{100}\right) bd^2 \qquad \text{...(18.18)}$$

Charts 18.1 to 18.6 have been prepared by assigning different values of $\frac{M_u}{b}$ and plotting d versus p_t. The moment values in charts are in units of KN.m per metre width. Curves are given for two grades of steel and one grade of concrete namely M20, which is most commonly used for flexural members. Table 18.5 A, B and C cover a wider range, that is, five values of f_y and concretes M20, M25 and M30 respectively. In this Table, the values of percentage of reinforcement p_t have been tabulated against $\frac{M_u}{bd^2}$.

Example 18.4. Determine the main tensile reinforcement required for a rectangular beam with the following data.

Size of beam 30 cm × 50 cm

Concrete mix M20

Characteristic strength of Reinforcement = 250 N/mm²

Factored moment M_u = 130 KN.m

Solution.

Overall depth D = 50 cm

Assume effective depth = 46 cm.

From Table 18.3. for f_{ck} = 20 N/mm² and f_y = 250 N/mm²

$$\frac{M_{u,lim}}{bd^2} = 2.98 \text{ N/mm}^2 = 2.98 \times 10^3 \text{ KN/m}^2$$

$$M_{u'lim} = 2.98 \times 10^3\, bd^2$$
$$= 2.98 \times 10^3 \times 0.3 \times 0.46^2$$
$$= 189.18 \text{ KN.m}$$

Actual moment of 130 KN. m is less than $M_{u'lim}$ = 189.18 KN.m. The section is, therefore, to be designed as a singly reinforced (under reinforced section).

$$\frac{M_u}{b} = \frac{130}{0.3} = 433.33 \text{ KN.m per metre width}$$

Referring to Chart 18.2, corresponding to $\frac{M_u}{b}$ = 433.33 KN.m per metre and d = 46 cm.

$$\text{Percentage of steel, } p_t = \frac{100 A_s}{bd} = 1.08$$

$$A_s = \frac{1.08 \times 30 \times 46}{100} = 15.826 \text{ cm}^2$$

Method of Referring to Table 18.5A.

$$\frac{M_u}{bd^2} = \frac{130 \times 10^6}{30 \times 46^2 \times 10^3} = 2.048 \text{ N/mm}^2$$

From Table 18.5A for f_y = 250 N/mm²

$$p_t = 1.092$$

$$A_{st} = \frac{1.092 \times 30 \times 46}{100} = 15.066 \text{ cm}^2$$

Example 18.5. Determine the main reinforcement required for a slab subjected to Factored moment of 12.5 KN.m/metre width. Mix. of Concrete is M20. Assume the thickness of slab as 130 mm, and the characteristic strength of reinforcement as 415 N/mm².

Solution. D = 130 mm Assume d = 100 mm

For f_y = 415 N/mm²

From Table 18.3, $\frac{M_{u'lim}}{b.d^2} = 2.76 \text{ N/mm}^2$

$$M_{u'lim} = 2.76 \times 10^3 \times 1 \times \left(\frac{100}{1000}\right)^2 = 27.6 \text{ KN.m}$$

Actual bending moment of 12.5 KN.m is less than the limiting bending moment.

Referring to Chart 18.4, reinforcement percentage p_t = 0.38

8 mm ϕ (tor) @ 125 mm c/c or 10 mm ϕ (tor) @ 200 c/c.

(B) Doubly Reinforced Sections

In doubly reinforced sections

$$M_{u2} = M_u - M_{u'lim} \quad ...(18.19)$$
$$M_{u2} = A_{st2}\,(0.87 f_y)\,(d - d') \quad ...(18.20)$$

also

$$M_{u2} = A_{sc}\,(f_{sc} - f_{cc})\,(d - d') \quad ...(18.21)$$

Table 18.5A. Flexure–Reinforcement Percentage, p_t For Singly Reinforced Sections

f_{ck} = 20 N/mm²

M_u/bd^2,	f_y, N/mm²					M_u/bd^2,	f_y, N/mm²				
N/mm²	240	250	415	480	500	N/mm²	240	250	415	480	500
0.30	0.146	0.140	0.085	0.073	0.070	2.22	1.253	1.203	0.725	0.627	0.602
0.35	0.171	0.164	0.099	0.086	0.082	2.24	1.267	1.216	0.733	0.633	0.608
0.40	0.196	0.188	0.114	0.098	0.094	2.26	1.281	1.230	0.741	0.640	0.615
0.45	0.222	0.213	0.128	0.111	0.106	2.28	1.295	1.243	0.749	0.647	0.621
0.50	0.247	0.237	0.143	0.123	0.119	2.30	1.309	1.256	0.757	0.654	0.628
0.55	0.272	0.262	0.158	0.136	0.131	2.32	1.323	1.270	0.765	0.661	0.635
0.60	0.298	0.286	0.172	0.149	0.143	2.34	1.337	1.283	0.773	0.668	0.642
0.65	0.324	0.311	0.187	0.162	0.156	2.36	1.351	1.297	0.781	0.675	0.648
0.70	0.350	0.336	0.203	0.175	0.168	2.38	1.365	1.311	0.790	0.683	0.655
0.75	0.376	0.361	0.218	0.188	0.181	2.40	1.380	1.324	0.798	0.690	0.662
0.80	0.403	0.387	0.233	0.201	0.193	2.42	1.394	1.338	0.806	0.697	0.669
0.85	0.430	0.412	0.248	0.215	0.206	2.44	1.408	1.352	0.814	0.704	0.676
0.90	0.456	0.438	0.264	0.228	0.219	2.46	1.423	1.366	0.823	0.711	0.683
0.95	0.483	0.464	0.280	0.242	0.232	2.48	1.438	1.380	0.831	0.719	0.690
1.00	0.511	0.490	0.295	0.255	0.245	2.50	1.452	1.394	0.840	0.726	0.697
1.05	0.538	0.517	0.311	0.269	0.258	2.52	1.467	1.408	0.848	0.734	0.704
1.10	0.566	0.543	0.327	0.283	0.272	2.54	1.482	1.423	0.857	0.741	0.711
1.15	0.594	0.570	0.343	0.297	0.285	2.56	1.497	1.437	0.866	0.748	0.719
1.20	0.622	0.597	0.359	0.311	0.298	2.58	1.512	1.451	0.874	0.756	0.726
1.25	0.650	0.624	0.376	0.325	0.312	2.60	1.527	1.466	0.883	0.764	0.733
1.30	0.678	0.651	0.392	0.339	0.326	2.62	1.542	1.481	0.892	0.771	0.740
1.35	0.707	0.679	0.409	0.354	0.339	2.64	1.558	1.495	0.901	0.779	0.748
1.40	0.736	0.707	0.426	0.368	0.353	2.66	1.573	1.510	0.910	0.786	0.755
1.45	0.765	0.735	0.443	0.383	0.367	2.68	1.588	1.525	0.919	0.794	
1.50	0.795	0.763	0.460	0.397	0.382	2.70	1.604	1.540	0.928		
1.55	0.825	0.792	0.477	0.412	0.396	2.72	1.620	1.555	0.937		
1.60	0.855	0.821	0.494	0.427	0.410	2.74	1.636	1.570	0.946		
1.65	0.885	0.850	0.512	0.443	0.425	2.76	1.651	1.585	0.955		
1.70	0.916	0.879	0.530	0.458	0.440	2.78	1.667	1.601			
1.75	0.947	0.909	0.547	0.473	0.454	2.80	1.683	1.616			
1.80	0.978	0.939	0.565	0.489	0.469	2.82	1.700	1.632			
1.85	1.009	0.969	0.584	0.505	0.484	2.84	1.716	1.647			
1.90	1.041	1.000	0.602	0.521	0.500	2.86	1.732	1.663			
1.95	1.073	1.030	0.621	0.537	0.515	2.88	1.749	1.679			
2.00	1.106	1.062	0.640	0.553	0.531	2.90	1.766	1.695			
2.02	1.119	1.074	0.647	0.559	0.537	2.92	1.782	1.711			
2.04	1.132	1.087	0.655	0.566	0.543	2.94	1.799	1.727			
2.06	1.145	1.099	0.662	0.573	0.550	2.96	1.816	1.743			
2.08	1.159	1.112	0.670	0.579	0.556	2.98	1.833	1.760			
2.10	1.172	1.125	0.678	0.586	0.562						
2.12	1.185	1.138	0.685	0.593	0.569						
2.14	1.199	1.151	0.693	0.599	0.575						
2.16	1.212	1.164	0.701	0.606	0.582						
2.18	1.226	1.177	0.709	0.613	0.588						
2.20	1.239	1.190	0.717	0.620	0.595						

Note: Blanks indicate inadmissible reinforcement percentage (see Table *E*).

Table 18.5B. Flexure–Reinforcement Percentage, p_t For Singly Reinforced Sections

f_{ck} = 25 N/mm²

M_u/bd^2,	f_y, N/mm²					M_u/bd^2,	f_y, N/mm²				
N/mm²	240	250	415	480	500	N/mm²	240	250	415	480	500
0.30	0.146	0.140	0.084	0.073	0.070	2.55	1.415	1.358	0.818	0.708	0.679
0.35	0.171	0.164	0.099	0.085	0.082	2.60	1.448	1.390	0.837	0.724	0.695
0.40	0.195	0.188	0.113	0.098	0.094	2.65	1.482	1.422	0.857	0.741	0.711
0.45	0.220	0.211	0.127	0.110	0.106	2.70	1.515	1.455	0.876	0.758	0.727
0.50	0.245	0.236	0.142	0.123	0.118	2.75	1.549	1.487	0.896	0.775	0.744
0.55	0.271	0.260	0.156	0.135	0.130	2.80	1.584	1.520	0.916	0.792	0.760
0.60	0.296	0.284	0.171	0.148	0.142	2.85	1.618	1.554	0.936	0.809	0.777
0.65	0.321	0.309	0.186	0.161	0.154	2.90	1.653	1.587	0.956	0.827	0.794
0.70	0.347	0.333	0.201	0.174	0.167	2.95	1.689	1.621	0.977	0.844	0.811
0.75	0.373	0.358	0.216	0.186	0.179	3.00	1.724	1.655	0.997	0.862	0.828
0.80	0.399	0.383	0.231	0.199	0.191	3.05	1.760	1.690	1.018	0.880	0.845
0.85	0.425	0.408	0.246	0.212	0.204	3.10	1.797	1.725	1.039	0.898	0.863
0.90	0.451	0.458	0.261	0.225	0.216	3.15	1.834	1.760	1.061	0.917	0.880
0.95	0.477	0.458	0.276	0.239	0.229	3.20	1.871	1.796	1.082	0.936	0.898
1.00	0.504	0.483	0.291	0.252	0.242	3.25	1.909	1.832	1.104	0.954	0.916
1.05	0.530	0.509	0.307	0.265	0.255	3.30	1.947	1.869	1.126	0.973	0.935
1.10	0.557	0.535	0.322	0.279	0.267	3.32	1.962	1.884	1.135	0.981	0.942
1.15	0.584	0.561	0.338	0.292	0.280	3.34	1.978	1.899	1.144	0.989	
1.20	0.611	0.587	0.353	0.306	0.293	3.36	1.993	1.914	1.153		
1.25	0.638	0.613	0.369	0.319	0.306	3.38	2.009	1.929	1.162		
1.30	0.666	0.639	0.385	0.333	0.320	3.40	2.025	1.944	1.171		
1.35	0.693	0.666	0.401	0.347	0.333	3.42	2.040	1.959	1.180		
1.40	0.721	0.692	0.417	0.360	0.346	3.44	2.056	1.974	1.189		
1.45	0.749	0.719	0.433	0.374	0.359	3.46	2.072	1.989			
1.50	0.777	0.746	0.449	0.388	0.373	3.48	2.088	2.005			
1.55	0.805	0.773	0.466	0.406	0.387	3.50	2.104	2.020			
1.60	0.834	0.800	0.482	0.417	0.400	3.52	2.120	2.036			
1.65	0.862	0.856	0.499	0.431	0.414	3.54	2.137	2.051			
1.70	0.891	0.856	0.515	0.446	0.428	3.56	2.153	2.067			
1.75	0.920	0.883	0.532	0.460	0.442	3.58	2.170	2.083			
1.80	0.949	0.911	0.549	0.475	0.456	3.60	2.186	2.099			
1.85	0.979	0.940	0.566	0.489	0.470	3.62	2.203	2.115			
1.90	1.009	0.968	0.583	0.504	0.484	3.64	2.219	2.131			
1.95	1.038	0.997	0.601	0.519	0.498	3.66	2.236	2.147			
2.00	1.068	1.026	0.618	0.534	0.513	3.68	2.253	2.163			
2.05	1.099	1.055	0.635	0.549	0.527	3.70	2.270	2.179			
2.10	1.129	1.084	0.653	0.565	0.542	3.72	2.287	2.196			
2.15	1.160	1.114	0.671	0.580	0.557	3.74	2.304				
2.20	1.191	1.143	0.689	0.596	0.572						
2.25	1.222	1.173	0.707	0.611	0.587						
2.30	1.254	1.204	0.725	0.627	0.602						
2.35	1.285	1.234	0.743	0.643	0.617						
2.40	1.317	1.265	0.762	0.659	0.632						
2.45	1.350	1.296	0.781	0.675	0.648						
2.50	1.382	1.327	0.799	0.691	0.663						

Note: Blanks indicate inadmissible reinforcement percentage (*see* Table *E*).

Table 18.5C. Flexure–Reinforcement Percentage, p_t For Singly Reinforced Sections

$f_{ck} = 30$ N/mm²

M_u/bd^2,	f_y, N/mm²					M_u/bd^2,	f_y, N/mm²				
N/mm²	240	250	415	480	500	N/mm²	240	250	415	480	500
0.30	0.145	0.140	0.084	0.073	0.070	2.55	1.374	1.319	0.794	0.687	0.659
0.35	0.170	0.163	0.098	0.085	0.082	2.60	1.404	1.348	0.812	0.702	0.674
0.40	0.195	0.187	0.113	0.097	0.093	2.65	1.435	1.378	0.830	0.718	0.689
0.45	0.219	0.211	0.127	0.110	0.105	2.70	1.467	1.408	0.848	0.733	0.704
0.50	0.244	0.235	0.141	0.122	0.117	2.75	1.498	1.438	0.866	0.749	0.719
0.55	0.269	0.259	0.156	0.135	0.129	2.80	1.530	1.469	0.885	0.765	0.734
0.60	0.294	0.283	0.170	0.147	0.141	2.85	1.562	1.499	0.903	0.781	0.750
0.65	0.320	0.307	0.185	0.160	0.153	2.90	1.594	1.530	0.922	0.797	0.765
0.70	0.345	0.331	0.200	0.172	0.166	2.95	1.626	1.561	0.940	0.813	0.781
0.75	0.370	0.356	0.214	0.185	0.178	3.00	1.659	1.592	0.959	0.829	0.796
0.80	0.396	0.380	0.229	0.198	0.190	3.05	1.691	1.624	0.978	0.846	0.812
0.85	0.422	0.405	0.244	0.211	0.202	3.10	1.725	1.656	0.997	0.862	0.828
0.90	0.447	0.429	0.259	0.224	0.215	3.15	1.758	1.687	1.017	0.879	0.844
0.95	0.473	0.454	0.274	0.237	0.227	3.20	1.791	1.720	1.036	0.896	0.860
1.00	0.499	0.479	0.289	0.250	0.240	3.25	1.825	1.752	1.055	0.913	0.876
1.05	0.525	0.504	0.304	0.263	0.252	3.30	1.859	1.785	1.075	0.930	0.892
1.10	0.552	0.529	0.319	0.276	0.265	3.35	1.893	1.818	1.095	0.947	0.909
1.15	0.578	0.555	0.334	0.289	0.277	3.40	1.928	1.851	1.115	0.964	0.925
1.20	0.604	0.580	0.350	0.302	0.290	3.45	1.963	1.884	1.135	0.981	0.942
1.25	0.631	0.606	0.365	0.315	0.303	3.50	1.998	1.918	1.156	0.999	0.959
1.30	0.658	0.631	0.380	0.329	0.316	3.55	2.034	1.952	1.176	1.017	0.976
1.35	0.685	0.657	0.396	0.342	0.329	3.60	2.069	1.986	1.197	1.035	0.993
1.40	0.712	0.683	0.411	0.356	0.342	3.65	2.105	2.021	1.218	1.053	1.011
1.45	0.739	0.709	0.427	0.369	0.355	3.70	2.142	2.056	1.239	1.071	1.028
1.50	0.766	0.735	0.443	0.383	0.368	3.75	2.178	2.091	1.260	1.089	1.046
1.55	0.793	0.762	0.459	0.397	0.381	3.80	2.215	2.127	1.281	1.108	1.063
1.60	0.821	0.788	0.475	0.410	0.394	3.85	2.253	2.163	1.303	1.126	1.081
1.65	0.849	0.815	0.491	0.424	0.407	3.90	2.291	2.199	1.325	1.145	1.099
1.70	0.876	0.841	0.507	0.438	0.421	3.95	2.329	2.236	1.347	1.164	1.118
1.75	0.904	0.868	0.523	0.452	0.434	4.00	2.367	2.273	1.369	1.184	
1.80	0.932	0.895	0.539	0.466	0.448	4.05	2.406	2.310	1.391		
1.85	0.961	0.922	0.556	0.480	0.461	4.10	2.445	2.348	1.414		
1.90	1.989	0.950	0.572	0.495	0.475	4.15	2.485	2.386			
1.95	1.018	0.977	0.589	0.509	0.488	4.20	2.525	2.424			
2.00	1.046	1.005	0.605	0.523	0.502	4.25	2.566	2.463			
2.05	1.075	1.032	0.622	0.538	0.516	4.30	2.607	2.502			
2.10	1.104	1.060	0.639	0.552	0.530	4.35	2.648	2.542			
2.15	1.134	1.088	0.656	0.567	0.544	4.40	2.690	2.583			
2.20	1.163	1.116	0.673	0.581	0.558	4.45	2.733	2.623			
2.25	1.192	1.145	0.690	0.596	0.572						
2.30	1.222	1.173	0.707	0.611	0.587						
2.35	1.252	1.202	0.724	0.626	0.601						
2.40	1.282	1.231	0.742	0.641	0.615						
2.45	1.312	1.260	0.759	0.656	0.630						
2.50	1.343	1.280	0.777	0.671	0.645						

Note: Blanks indicate inadmissible reinforcement percentage

Since additional tensile force is balanced by the additional compressive force,

$$A_{sc}(f_{sc} - f_{cc}) = A_{st2}(0.87 f_y)$$

$$A_{st2} = \frac{A_{sc}(f_{sc} - f_{cc})}{0.87 f_y} \quad ...(18.22)$$

Total tensile reinforcement is given by

$$A_{st} = p_{t,lim} \frac{bd}{100} + A_{st2} \quad ...(18.23)$$

For values of d'/d upto 0.2, f_{cc} is equal to 0.446 f_{ck} and for mild steel reinforcement f_{sc} = 0.87 f_y. When the reinforcement is cold worked bars, the design stress in compressive reinforcement, f_{sc} for different values of d'/d upto 0.2 will be given in Table 18.6.

Table 18.6. Stress in Compression Reinforcement f_{sc}, N/mm² of Doubly Reinforced Beams with Cold Worked Bars

	d'/d			
f_y *N/mm²*	*0.05*	*0.10*	*0.15*	*0.20*
415	355	353	342	329
500	424	412	395	370

A_{st2} is plotted against $(d - d')$ for different values of M_{u2} in Charts 18.7 and 18.8. The charts have been prepared for f_s = 217.5 N/mm² and it is directly applicable for mild steel reinforcement with f_y = 250 N/mm². Values of A_{st2} for other grades of steel can be obtained by multiplying the value read from the chart by factors given in Table 18.7. The multiplying factors for A_{sc}, given in Table 18.7, are based on a value of f_{cc} corresponding to concrete grade M20, but it can be used for all other grades of concrete with little error.

Table 18.7. Multiplying Factors for Use with Charts 18.7 and 18.8

f_y *N/mm²*	*Factor for* A_{st2}	*Factor for* A_{sc} *for d'/d*			
		0.5	*0.10*	*0.15*	*0.2*
250	1.00	1.04	1.04	1.04	1.04
415	0.60	0.63	0.63	0.65	0.68
500	0.50	0.52	0.54	0.56	0.60

The expression for the moment of resistance of a doubly reinforced section may also be written as :

$$M_u = M_{u,lim} + \frac{p_{t2}.b.d}{100}(0.87 f_y)(d - d')$$

$$\frac{M_u}{b.d^2} = \frac{M_{u,lim}}{b.d^2} + \frac{p_{t2}}{100}(0.87 f_y)\cdot\left(1 - \frac{d'}{d}\right) \quad ...(18.24)$$

where p_{t2} is the additional percentage of tensile reinforcement,

$$p_t = p_{t,lim} + p_{t2}$$

$$p_c = p_{t2} \cdot \left[\frac{0.87 f_y}{f_{sc} - f_{cc}}\right]$$

Tables 18.8 A to F are presented for two grades steel (Fe 415 and Fe 500) and for three grades of concrete f_{ck} (20, 25 and 30) where from one get compute p_t (tension reinforcement) and p_c (compression force reinforcement) for different values of $\frac{M_u}{bd^2}$ and $\frac{d'}{d}$.

Tables 18.9 A and B are presented for two grades of steel (Fe 415 and Fe 500), to compute Limiting Moment of resistance factor $\frac{M_u, \text{lim}}{b_w \cdot d^2 \cdot f_{ck}}$ for singly reinforced T-beams for various values of $\frac{D_f}{d}$ and $\frac{b_f}{b_w}$.

where D_f = flange thickness
d = effective depth
b_f = flange width
b_w = width of web

Example 18.6. Determine the main reinforcement required for a rectangular beam section 300 mm × 500 mm of concrete mix M20, Characteristic strength of reinforcement is 415 N/mm². The section is subjected to Factored moment of 250 KN.m.

Solution. Assume effective depth of the section as 460 mm.

From Table 18.3, For f_y = 415 N/mm² and f_{ck} = 20 N/mm²

$$\frac{M_{u,lim}}{b.d^2} = 2.76 \text{ N/mm}^2 = 2.76 \times 10^3 \text{ KN/m}^2$$

$$M_{u,lim} = 2.76 \times 10^3 \cdot b.d^2$$

$$= 2.76 \times 10^3 \times 0.3 \times 0.46^2$$

$$= 175.2 \text{ KN.m}$$

Actual moment of 250 KN.m is greater than $M_{u,lim}$.

∴ The section is to be designed as a doubly reinforced section.

$$\frac{M_u}{b.d^2} = \frac{250 \times 10^6}{300 \times 460^2} = 3.94 \text{ N/mm}^2$$

$$d'/d = \frac{40}{460} = 0.087 \text{ (say 0.1)}$$

Referring to Table 18.8 corresponding

$$\frac{M_u}{b.d^2} = 3.94 \text{ and } \frac{d'}{d} = 0.1$$

$$p_t = 1.31 \text{ and } p_c = 0.369$$

$$A_{st} = 18.08 \text{ cm}^2 \text{ and } A_{sc} = 5.10 \text{ cm}^2$$

EXERCISE 18

1. State the assumptions made in Limit state of collapse : Flexure.
2. Explain maximum depth of neutral axis.
3. Obtain the expressions for depth of neutral axis and moment of resistance for (*i*) rectangular sections reinforced on tension face only (*ii*) rectangular sections reinforced on tension and compression faces both.
4. Determine the area of steel required for section 300mm × 500mm (overall) if the limit state moment carried by the beam section is :

   (*i*) 80 KN.m (*ii*) 160 KN.m
5. Design a rectangular section which carries a maximum limiting bending moment of 85 KN.m Use M 20 concrete and Fe 415 Steel.
6. Determine the main tensile reinforcement required for a rectangular beam 225 mm × 450 mm in M20 concrete and using Fe 415 Steel as reinforcement for a factored moment of 100 KN.m. use SP.16.

Table 18.8A. Flexure–Reinforcement Percentages For Doubly Reinforced Sections

f_{ck} = 20 N/mm²
f_y = 415 N/mm²

M_u/bd^2,	d'/d = 0.05		d'/d = 0.10		d'/d = 0.15		d'/d = 0.20	
N/mm²	P_t	P_c	P_t	P_c	P_t	P_c	P_t	P_c
2.77	0.958	0.002	0.958	0.002	0.959	0.003	0.959	0.003
2.80	0.967	0.011	0.968	0.012	0.968	0.013	0.969	0.015
2.90	0.996	0.042	0.998	0.045	1.001	0.049	1.004	0.054
3.00	1.025	0.072	1.029	0.077	1.034	0.084	1.038	0.093
3.10	1.055	0.103	1.060	0.109	1.066	0.119	1.073	0.132
3.20	1.084	0.133	1.091	0.142	1.099	0.154	1.108	0.171
3.30	1.113	0.164	1.122	0.174	1.131	0.190	1.142	0.210
3.40	1.142	0.194	1.152	0.207	1.164	0.225	1.177	0.249
3.50	1.171	0.224	1.183	0.239	1.197	0.260	1.212	0.288
3.60	1.200	0.255	1.214	0.271	1.229	0.295	1.246	0.327
3.70	1.230	0.285	1.245	0.304	1.262	0.331	1.281	0.366
3.80	1.259	0.316	1.276	0.336	1.294	0.366	1.315	0.405
3.90	1.288	0.346	1.306	0.369	1.327	0.401	1.350	0.444
4.00	1.317	0.376	1.337	0.401	1.360	0.437	1.385	0.483
4.10	1.346	0.407	1.368	0.433	1.392	0.472	1.419	0.522
4.20	1.375	0.437	1.399	0.466	1.425	0.507	1.454	0.561
4.30	1.405	0.468	1.429	0.498	1.457	0.542	1.489	0.600
4.40	1.434	0.498	1.460	0.530	1.490	0.578	1.523	0.640
4.50	1.463	0.528	1.491	0.563	1.523	0.613	1.558	0.679
4.60	1.492	0.559	1.522	0.595	1.555	0.648	1.593	0.718
4.70	1.521	0.589	1.553	0.628	1.588	0.683	1.627	0.757
4.80	1.550	0.620	1.583	0.660	1.620	0.719	1.662	0.796
4.90	1.558	0.650	1.614	0.692	1.653	0.754	1.696	0.835
5.00	1.609	0.680	1.645	0.725	1.686	0.789	1.731	0.874
5.10	1.638	0.711	1.676	0.757	1.718	0.825	1.766	0.913
5.20	1.667	0.741	1.707	0.790	1.751	0.860	1.800	0.952
5.30	1.696	0.772	1.737	0.822	1.783	0.895	1.835	0.991
5.40	1.725	0.802	1.768	0.854	1.816	0.930	1.870	1.030
5.50	1.755	0.832	1.799	0.887	1.849	0.966	1.904	1.069
5.60	1.784	0.863	1.830	0.919	1.881	1.001	1.939	1.108
5.70	1.813	0.893	1.861	0.952	1.914	1.036	1.974	1.147
5.80	1.842	0.924	1.891	0.984	1.946	1.071	2.008	1.186
5.90	1.871	0.954	1.922	1.016	1.979	1.107	2.043	1.225
6.00	1.900	0.985	1.953	1.049	2.012	1.142	2.078	1.264
6.10	1.930	1.015	1.984	1.081	2.044	1.177	2.112	1.303
6.20	1.959	1.045	2.014	1.114	2.077	1.213	2.147	1.342
6.30	1.988	1.076	2.045	1.146	2.109	1.248	2.181	1.381
6.40	2.017	1.106	2.076	1.178	2.142	1.283	2.216	1.421
6.50	2.046	1.137	2.107	1.211	2.175	1.318	2.251	1.460
6.60	2.075	1.167	2.138	1.243	2.207	1.354	2.285	1.499
6.70	2.105	1.197	2.168	1.276	2.240	1.389	2.320	1.538
6.80	2.134	1.228	2.199	1.308	2.272	1.424	2.355	1.577
6.90	2.163	1.258	2.230	1.340	2.305	1.459	2.389	1.616
7.00	2.192	1.289	2.261	1.373	2.338	1.495	2.424	1.655
7.10	2.221	1.319	2.292	1.405	2.370	1.530	2.459	1.694

Table 18.8B. Flexure–Reinforcement Percentages For Doubly Reinforced Sections

$f_{ck} = 25$ N/mm²
$f_y = 415$ N/mm²

M_u/bd^2,	$d'/d = 0.05$		$d'/d = 0.10$		$d'/d = 0.15$		$d'/d = 0.20$	
N/mm²	P_t	P_c	P_t	P_c	P_t	P_c	P_t	P_c
3.46	1.197	0.002	1.197	0.002	1.197	0.003	1.197	0.003
3.50	1.209	0.014	1.210	0.015	1.210	0.017	1.211	0.019
3.60	1.238	0.045	1.240	0.048	1.243	0.052	1.246	0.058
3.70	1.267	0.076	1.271	0.081	1.276	0.088	1.281	0.097
3.80	1.296	0.106	1.302	0.113	1.308	0.123	1.315	0.137
3.90	1.325	0.137	1.333	0.146	1.341	0.159	1.350	0.176
4.00	1.355	0.167	1.363	0.178	1.373	0.194	1.385	0.215
4.10	1.384	0.198	1.394	0.211	1.406	0.230	1.419	0.254
4.20	1.413	0.229	1.425	0.244	1.439	0.265	1.454	0.294
4.30	1.442	0.259	1.456	0.276	1.471	0.301	1.488	0.333
4.40	1.471	0.290	1.487	0.309	1.504	0.336	1.523	0.372
4.50	1.500	0.320	1.517	0.341	1.536	0.372	1.558	0.412
4.60	1.530	0.351	1.548	0.374	1.569	0.407	1.592	0.451
4.70	1.559	0.382	1.579	0.407	1.602	0.443	1.627	0.490
4.80	1.588	0.412	1.610	0.439	1.634	0.478	1.662	0.530
4.90	1.617	0.443	1.641	0.472	1.667	0.514	1.696	0.569
5.00	1.646	0.474	1.671	0.504	1.699	0.549	1.731	0.608
5.10	1.675	0.504	1.702	0.537	1.732	0.585	1.766	0.648
5.20	1.705	0.535	1.733	0.570	1.765	0.620	1.800	0.687
5.30	1.734	0.565	1.764	0.602	1.797	0.656	1.835	0.726
5.40	1.763	0.596	1.795	0.635	1.830	0.691	1.869	0.766
5.50	1.792	0.627	1.825	0.667	1.862	0.727	1.904	0.805
5.60	1.821	0.657	1.856	0.700	1.895	0.762	1.939	0.844
5.70	1.851	0.688	1.887	0.733	1.928	0.798	1.973	0.884
5.80	1.880	0.718	1.918	0.765	1.960	0.833	2.008	0.923
5.90	1.909	0.749	1.948	0.798	1.993	0.869	2.043	0.962
6.00	1.938	0.780	1.979	0.830	2.025	0.904	2.077	1.002
6.10	1.967	0.810	2.010	0.863	2.058	0.940	2.112	1.041
6.20	1.996	0.841	2.041	0.896	2.091	0.975	2.147	1.080
6.30	2.026	0.871	2.072	0.928	2.123	1.011	2.181	1.120
6.40	2.055	0.902	2.102	0.961	2.156	1.046	2.216	1.159
6.50	2.084	0.933	2.133	0.993	2.188	1.082	2.251	1.198
6.60	2.113	0.963	2.164	1.026	2.221	1.118	2.285	1.238
6.70	2.142	0.994	2.195	1.059	2.254	1.153	2.320	1.277
6.80	2.171	1.024	2.226	1.091	2.286	1.189	2.354	1.316
6.90	2.201	1.055	2.256	1.124	2.319	1.224	2.389	1.356
7.00	2.230	1.086	2.287	1.157	2.351	1.260	2.424	1.395
7.10	2.259	1.116	2.318	1.189	2.384	1.295	2.458	1.434
7.20	2.288	1.147	2.349	1.222	2.417	1.331	2.493	1.474
7.30	2.317	1.177	2.380	1.254	2.449	1.366	2.528	1.513
7.40	2.346	1.208	2.410	1.287	2.482	1.402	2.562	1.552
7.50	2.376	1.239	2.441	1.320	2.514	1.437	2.597	1.591
7.60	2.405	1.269	2.472	1.352	2.547	1.473	2.632	1.631
7.70	2.434	1.300	2.503	1.385	2.580	1.508	2.666	1.670
7.80	2.463	1.330	2.534	1.417	2.612	1.544	2.701	1.709

Table 18.8C. Flexure–Reinforcement Percentages For Doubly Reinforced Sections

$f_{ck} = 30$ N/mm^2
$f_y = 415$ N/mm^2

M_u/bd^2,	$d'/d = 0.05$		$d'/d = 0.10$		$d'/d = 0.15$		$d'/d = 0.20$	
N/mm^2	P_t	P_c	P_t	P_c	P_t	P_c	P_t	P_c
4.15	1.436	0.002	1.436	0.002	1.436	0.002	1.436	0.002
4.20	1.451	0.017	1.451	0.019	1.452	0.020	1.454	0.022
4.30	1.480	0.048	1.482	0.051	1.485	0.056	1.488	0.062
4.40	1.509	0.079	1.513	0.084	1.518	0.092	1.523	0.102
4.50	1.538	0.110	1.544	0.117	1550	0.127	1.558	0.141
4.60	1.567	0.141	1.575	0.150	1.583	0.163	1.592	0.181
4.70	1.596	0.171	1.605	0.183	1.615	0.199	1.627	0.220
4.80	1.626	0.202	1.636	0.215	1.648	0.235	1.661	0.260
4.90	1.655	0.233	1.667	0.248	1.681	0.270	1.696	0.300
5.00	1.684	0.264	1.698	0.281	1.713	0.306	1.731	0.339
5.10	1.713	0.295	1.729	0.314	1.746	0.342	1.765	0.379
5.20	1.742	0.325	1.759	0.347	1.778	0.378	1.800	0.418
5.30	1.771	0.356	1.790	0.380	1.811	0.413	1.835	0.458
5.40	1.801	0.387	1.821	0.412	1.844	0.449	1.869	0.498
5.50	1.830	0.418	1.852	0.445	1.876	0.485	1.904	0.537
5.60	1.859	0.449	1.883	0.478	1.909	0.521	1.939	0.577
5.70	1.888	0.479	1.913	0.511	1.941	0.556	1.973	0.616
5.80	1.917	0.510	1.944	0.544	1.974	0.592	2.008	0.656
5.90	1.946	0.541	1.975	0.576	2.007	0.628	2.042	0.696
6.00	1.976	0.572	2.006	0.609	2.039	0.664	2.077	0.735
6.10	2.005	0.603	2.036	0.642	2.072	0.699	2.112	0.775
6.20	2.034	0.634	2.067	0.675	2.104	0.735	2.146	0.814
6.30	2.063	0.664	2.098	0.708	2.137	0.771	2.181	0.854
6.40	2.092	0.695	2.129	0.741	2.170	0.807	2.216	0.894
6.50	2.121	0.726	2.160	0.773	2.202	0.842	2.250	0.933
6.60	2.151	0.757	2.190	0.806	2.235	0.878	2.285	0.973
6.70	2.180	0.788	2.221	0.839	2.267	0.914	2.320	1.012
6.80	2.209	0.818	2.252	0.872	2.300	0.950	2.354	1.052
6.90	2.238	0.849	2.283	0.905	2.333	0.985	2.389	1.092
7.00	2.267	0.880	2.314	0.937	2.365	1.021	2.424	1.131
7.10	2.296	0.911	2.344	0.970	2.398	1.057	2.458	1.171
7.20	2.326	0.942	2.375	1.003	2.431	1.093	2.493	1.210
7.30	2.355	0.972	2.406	1.036	2.463	1.128	2.527	1.250
7.40	2.384	1.003	2.437	1.069	2.496	1.164	2.562	1.290
7.50	2.413	1.034	2.468	1.102	2.528	1.200	2.597	1.329
7.60	2.442	1.065	2.498	1.134	2.561	1.236	2.631	1.369
7.70	2.471	1.096	2.529	1.167	2.594	1.271	2.666	1.408
7.80	2.501	1.126	2.560	1.200	2.626	1.307	2.701	1.448
7.90	2.530	1.157	2.591	1.233	2.659	1.343	2.735	1.488
8.00	2.559	1.188	2.621	1.266	2.691	1.379	2.770	1.527
8.10	2.588	1.219	2.652	1.299	2.724	1.414	2.805	1.567
8.20	2.617	1.250	2.683	1.331	2.757	1.450	2.839	1.606
8.30	2.646	1.280	2.714	1.364	2.789	1.486	2.874	1.646
8.40	2.676	1.311	2.745	1.397	2.822	1.522	2.908	1.686
8.50	2.705	1.342	2.775	1.430	2.854	1.557	2.943	1.725

Table 18.8D. Flexure–Reinforcement Percentages For Doubly Reinforced Sections

f_{ck} = 20 N/mm^2
f_y = 500 N/mm^2

M_u/bd^2,	d'/d = 0.05		d'/d = 0.10		d'/d = 0.15		d'/d = 0.20	
N/mm^2	P_t	P_c	P_t	P_c	P_t	P_c	P_t	P_c
2.67	0.758	0.002	0.758	0.003	0.758	0.003	0.758	0.003
2.70	0.765	0.010	0.765	0.011	0.766	0.012	0.767	0.014
2.80	0.789	0.035	0.791	0.038	0.793	0.042	0.795	0.048
2.90	0.813	0.061	0.816	0.066	0.820	0.073	0.824	0.083
3.00	0.837	0.086	0.842	0.094	0.847	0.103	0.853	0.117
3.10	0.862	0.111	0.868	0.121	0.874	0.134	0.882	0.152
3.20	0.886	0.137	0.893	0.149	0.901	0.164	0.910	0.186
3.30	0.910	0.162	0.919	0.176	0.928	0.195	0.939	0.221
3.40	0.934	0.188	0.944	0.204	0.955	0.225	0.968	0.255
3.50	0.958	0.213	0.970	0.232	0.982	0.256	0.997	0.290
3.60	0.983	0.238	0.995	0.259	1.009	0.286	1.025	0.324
3.70	1.007	0.264	1.021	0.287	1.036	0.316	1.054	0.359
3.80	1.031	0.289	1.046	0.314	1.064	0.347	1.083	0.394
3.90	1.055	0.314	1.072	0.342	1.091	0.377	1.112	0.428
4.00	1.080	0.340	1.098	0.369	1.118	0.408	1.140	0.463
4.10	1.104	0.365	1.123	0.397	1.145	0.438	1.169	0.497
4.20	1.128	0.391	1.149	0.425	1.172	0.469	1.198	0.532
4.30	1.152	0.416	1.174	0.452	1.199	0.499	1.227	0.566
4.40	1.176	0.441	1.200	0.480	1.226	0.530	1.255	0.601
4.50	1.201	0.467	1.225	0.507	1.253	0.560	1.284	0.635
4.60	1.225	0.492	1.251	0.535	1.280	0.591	1.313	0.670
4.70	1.249	0.517	1.276	0.563	1.307	0.621	1.342	0.704
4.80	1.273	0.543	1.302	0.590	1.334	0.651	1.370	0.739
4.90	1.297	0.568	1.328	0.618	1.361	0.682	1.399	0.773
5.00	1.322	0.593	1.353	0.645	1.388	0.712	1.428	0.808
5.10	1.346	0.619	1.379	0.673	1.415	0.743	1.457	0.843
5.20	1.370	0.644	1.404	0.701	1.442	0.773	1.485	0.877
5.30	1.394	0.670	1.430	0.728	1.469	0.804	1:514	0.912
5.40	1.418	0.695	1.455	0.756	1.496	0.834	1.543	0.946
5.50	1.443	0.720	1.481	0.783	1.524	0.865	1.572	0.981
5.60	1.467	0.746	1.506	0.811	1.551	0.895	1.600	1.015
5.70	1.491	0.771	1.532	0.839	1.578	0.925	1.629	1.050
5.80	1.515	0.796	1.558	0.866	1.605	0.956	1.658	1.084
5.90	1.540	0.822	1.583	0.894	1.632	0.986	1.687	1.119
6.00	1.564	0.847	1.609	0.921	1.659	1.017	1.715	1.153
6.10	1.588	0.873	1.634	0.949	1.686	1.047	1.744	1.188
6.20	1.612	0.898	1.660	0.976	1.713	1.078	1.773	1.223
6.30	1.636	0.923	1.685	1.004	1.740	1.108	1.802	1.257
6.40	1.661	0.949	1.711	1.032	1.767	1.139	1.830	1.292
6.50	1.685	0.974	1.736	1.059	1.794	1.169	1.859	1.326
6.60	1.709	0.999	1.762	1.087	1.821	1.200	1.888	1.361
6.70	1.733	1.025	1.788	1.114	1.848	1.230	1.917	1.395
6.80	1.757	1.050	1.813	1.142	1.875	1.260	1.945	1.430
6.90	1.782	1.076	1.839	1.170	1.902	1.291	1.974	1.464
7.00	1.806	1.101	1.864	1.197	1.929	1.321	2.003	1.499

Table 18.8E. Flexure–Reinforcement Percentages For Doubly Reinforced Sections

$f_{ck} = 25$ N/mm²
$f_y = 500$ N/mm²

M_u/bd^2, N/mm²	$d'/d = 0.05$		$d'/d = 0.10$		$d'/d = 0.15$		$d'/d = 0.20$	
	P_t	P_c	P_t	P_c	P_t	P_c	P_t	P_c
3.33	0.945	0.001	0.945	0.001	0.945	0.001	0.945	0.001
3.40	0.962	0.019	0.963	0.021	0.964	0.023	0.965	0.026
3.50	0.986	0.044	0.989	0.048	0.991	0.053	0.994	0.060
3.60	1.010	0.070	1.014	0.076	1.018	0.084	1.023	0.095
3.70	1.035	0.095	1.040	0.104	1.045	0.115	1.052	0.130
3.80	1.059	0.121	0.065	0.132	1.072	0.145	1.080	0.165
3.90	1.083	0.146	1.091	0.159	1.099	0.176	1.109	0.200
4.00	1.107	0.172	1.116	0.187	1.126	0.206	1.138	0.234
4.10	1.131	0.197	1.142	0.215	1.154	0.237	1.167	0.269
4.20	1.156	0.223	1.167	2.242	1.181	0.268	1.195	0.304
4.30	1.180	0.248	1.193	0.270	1.208	0.298	1.224	0.339
4.40	1.204	0.274	1.219	0.298	1.235	0.329	1.253	0.373
4.50	1.228	0.299	1.244	0.326	1.262	0.360	1.282	0.408
4.60	1.253	0.325	1.270	0.353	1.289	0.390	1.310	0.443
4.70	1.277	0.350	1.295	0.381	1.316	0.421	1.339	0.478
4.80	1.301	0.376	1.321	0.409	1.343	0.451	1.368	0.512
4.90	1.325	0.402	1.346	0.437	1.370	0.482	1.397	0.547
5.00	1.349	0.427	1.372	0.464	1.397	0.513	1.425	0.582
5.10	1.374	0.453	1.397	0.492	1.424	0.543	1.454	0.617
5.20	1.398	0.478	1.423	0.520	1.451	0.574	1.483	0.651
5.30	1.422	0.504	1.449	0.548	1.478	0.605	1.512	0.686
5.40	1.446	0.529	1.474	0.575	1.505	0.635	1.540	0.721
5.50	1.470	0.555	1.500	0.603	1.532	0.666	1.569	0.756
5.60	1.495	0.580	1.525	0.631	1.559	0.697	1.598	0.790
5.70	1.519	0.606	1.551	0.659	1.586	0.727	1.627	0.825
5.80	1.543	0.631	1.576	0.686	1.614	0.758	1.655	0.860
5.90	1.567	0.657	1.602	0.714	1.641	0.788	1.684	0.895
6.00	1.592	0.682	1.627	0.742	1.668	0.819	1.713	0.929
6.10	1.616	0.708	1.653	0.770	1.695	0.850	1.742	0.964
6.20	1.640	0.733	1.679	0.797	1.722	0.880	1.770	0.999
6.30	1.664	0.759	1.704	0.825	1.749	0.911	1.799	1.034
6.40	1.688	0.784	1.730	0.853	1.776	0.942	1.828	1.068
6.50	1.713	0.810	1.755	0.881	1.803	0.972	1.857	1.103
6.60	1.737	0.835	1.781	0.908	1.830	1.003	1.885	1.138
6.70	1.761	0.861	1.806	0.936	1.857	1.033	1.914	1.173
6.80	1.785	0.886	1.832	0.964	1.884	1.064	1.943	1.207
6.90	1.809	0.912	1.857	0.992	1.911	1.095	1.972	1.242
7.00	1.834	0.937	1.883	1.019	1.938	1.125	2.000	1.277
7.10	1.858	0.963	1.909	1.047	1.965	1.156	2.029	1.312
7.20	1.882	0.988	1.934	1.075	1.992	1.187	2.058	1.346
7.30	1.906	1.014	1.960	1.103	2.019	1.217	2.087	1.381
7.40	1.930	1.039	1.985	1.130	2.046	1.248	2.115	1.416
7.50	1.955	1.065	2.011	1.158	2.074	1.278	2.144	1.451
7.60	1.979	1.090	2.036	1.186	2.101	1.309	2.173	1.486
7.70	2.003	1.116	2.062	1.213	2.128	1.340	2.202	1.520

Table 18.8F. Flexure–Reinforcement Percentage For Doubly Reinforced Sections

$f_{ck} = 30$ N/mm²
$f_y = 500$ N/mm²

M_u/bd^2,	$d'/d = 0.05$		$d'/d = 0.10$		$d'/d = 0.15$		$d'/d = 0.20$	
N/mm²	P_t	P_c	P_t	P_c	P_t	P_c	P_t	P_c
4.00	1.135	0.002	1.135	0.002	1.135	0.003	1.135	0.003
4.10	1.159	0.028	1.161	0.030	1.162	0.034	1.164	0.038
4.20	1.183	0.054	1.186	0.058	1.189	0.064	1.193	0.073
4.30	1.208	0.079	1.212	0.086	1.216	0.095	1.222	0.108
4.40	1.232	0.105	1.237	0.114	1.244	0.126	1.250	0.143
4.50	1.256	0.130	1.263	0.142	1.271	0.157	1.279	0.178
4.60	1.280	0.156	1.289	0.170	1.298	0.188	1.308	0.213
4.70	1.305	0.182	1.314	0.198	1.325	0.218	1.337	0.248
4.80	1.329	0.207	1.340	0.226	1.352	0.249	1.365	0.283
4.90	1.353	0.233	1.365	0.254	1.379	0.280	1.394	0.318
5.00	1.377	0.259	1.391	0.281	1.406	0.311	1.423	0.353
5.10	1.401	0.284	1.416	0.309	1.433	0.342	1.452	0.388
5.20	1.426	0.310	1.442	0.337	1.460	0.372	1.480	0.423
5.30	1.450	0.336	1.467	0.365	1.487	0.403	1.509	0.458
5.40	1.474	0.361	1.493	0.393	1.514	0.434	1.538	0.493
5.50	1.498	0.387	1.519	0.421	1.541	0.465	1.567	0.528
5.60	1.522	0.413	1.544	0.449	1.568	0.496	1.595	0.563
5.70	1.547	0.438	1.570	0.477	1.595	0.526	1.624	0.598
5.80	1.571	0.464	1.595	0.505	1.622	0.557	1.653	0.633
5.90	1.595	0.490	1.621	0.533	1.649	0.588	1.682	0.688
6.00	1.619	0.515	1.646	0.560	1.676	0.619	1.710	0.703
6.10	1.643	0.541	1.672	0.588	1.704	0.650	1.739	0.738
6.20	1.668	0.566	1.697	0.616	1.731	0.680	1.768	0.773
6.30	1.692	0.592	1.723	0.644	1.758	0.711	1.797	0.807
6.40	1.716	0.618	1.749	0.672	1.785	0.742	1.825	0.842
6.50	1.740	0.643	1.774	0.700	1.812	0.773	1.854	0.877
6.60	1.765	0.669	1.800	0.728	1.839	0.804	1.883	0.912
6.70	1.789	0.695	1.825	0.756	1.866	0.835	1.912	0.947
6.80	1.813	0.720	1.851	0.784	1.893	0.865	1.940	0.982
6.90	1.837	0.746	1.876	0.812	1.920	0.896	1.969	1.017
7.00	1.861	0.772	1.902	0.839	1.947	0.927	1.998	1.052
7.10	1.886	0.797	1.927	0.867	1.974	0.958	2.027	1.087
7.20	1.910	0.823	1.953	0.895	2.001	0.989	2.055	1.122
7.30	1.934	0.849	1.979	0.923	2.028	1.019	2.084	1.157
7.40	1.958	0.874	2.004	0.951	2.055	1.050	2.113	1.192
7.50	1.982	0.900	2.030	0.979	2.082	1.081	2.142	1.227
7.60	2.007	0.926	2.055	1.007	2.109	1.112	2.170	1.262
7.70	2.031	0.951	2.081	1.035	2.136	1.143	2.199	1.297
7.80	2.055	0.977	2.106	1.063	2.164	1.173	2.228	1.332
7.90	0.079	1.002	2.132	1.091	2.191	1.204	2.257	1.367
8.00	2.103	1.028	2.157	1.118	2.218	1.235	2.285	1.402
8.10	2.128	1.054	2.183	1.146	2.245	1.266	2.314	1.437
8.20	2.152	1.079	2.209	1.174	2.272	1.297	2.343	1.472
8.30	2.176	1.105	2.234	1.202	2.299	1.327	2.372	1.507

Table 18.9A. Flexure–Limiting Moment of Resistance Factor, $M_{u,lim}/b_w d^2 f_{ck}$, For singly Reinforced T-Beams, N/mm²

$f_y = 415$ N/mm²

D_f/d	b_f/b_w									
	1.0	2.0	3.0	4.0	5.0	6.0	7.0	8.0	9.0	10.0
0.06	0.138	0.164	0.190	0.216	0.242	0.268	0.294	0.320	0.346	0.372
0.07	0.138	0.168	0.198	0.228	0.259	0.289	0.319	0.349	0.379	0.409
0.08	0.138	0.172	0.207	0.241	0.275	0.309	0.344	0.378	0.412	0.446
0.09	0.138	0.176	0.215	0.253	0.291	0.330	0.368	0.406	.445	0.483
0.10	0.138	0.180	0.223	0.265	0.308	0.350	0.392	0.435	0.477	0.519
0.11	0.138	0.184	0.231	0.277	0.324	0.370	0.416	0.463	0.509	0.555
0.12	0.138	0.188	0.239	0.289	0.339	0.390	0.440	0.490	0.541	0.591
0.13	0.138	0.192	0.247	0.301	0.355	0.409	0.463	0.518	0.572	0.626
0.14	0.138	0.196	0.254	0.312	0.370	0.428	0.487	0.545	0.603	0.661
0.15	0.138	0.200	0.262	0.324	0.386	0.448	0.509	0.571	0.633	0.695
0.16	0.138	0.204	0.269	0.335	0.401	0.466	0.532	0.598	0.663	0.729
0.17	0.138	0.207	0.277	0.346	0.416	0.485	0.554	0.624	0.693	0.762
0.18	0.138	0.211	0.284	0.357	0.430	0.503	0.576	0.649	0.723	0.796
0.19	0.138	0.215	0.291	0.368	0.445	0.522	0.598	0.675	0.752	0.828
0.20	0.138	0.218	0.299	0.379	0.459	0.540	0.620	0.700	0.780	0.861
0.21	0.138	0.221	0.305	0.388	0.471	0.554	0.638	0.721	0.804	0.887
0.22	0.138	0.224	0.309	0.395	0.480	0.566	0.651	0.737	0.822	0.908
0.23	0.138	0.226	0.314	0.402	0.489	0.577	0.665	0.753	0.841	0.928
0.24	0.138	0.228	0.318	0.408	0.498	0.588	0.678	0.768	0.859	0.949
0.25	0.138	0.230	0.323	0.415	0.507	0.600	0.692	0.784	0.876	0.969
0.26	0.138	0.233	0.327	0.422	0.516	0.611	0.705	0.800	0.894	0.989
0.27	0.138	0.235	0.331	0.428	0.525	0.622	0.718	0.815	0.912	1.008
0.28	0.138	0.237	0.336	0.435	0.534	0.632	0.731	0.830	0.929	1.028
0.29	0.138	0.239	0.340	0.441	0.542	0.643	0.744	0.845	0.946	1.047
0.30	0.138	0.241	0.344	0.448	0.551	0.654	0.757	0.860	0.963	1.066
0.31	0.138	0.243	0.349	0.454	0.559	0.664	0.770	0.875	0.980	1.085
0.32	0.138	0.245	0.353	0.460	0.568	0.675	0.782	0.890	0.997	1.104
0.33	0.138	0.248	0.357	0.466	0.576	0.685	0.795	0.904	1.014	1.123
0.34	0.138	0.250	0.361	0.473	0.584	0.696	0.807	0.919	1.030	1.142
0.35	0.138	0.252	0.365	0.479	0.592	0.706	0.819	0.933	1.160	1.046
0.36	0.138	0.254	0.369	0.485	0.600	0.716	0.831	0.947	1.063	1.178
0.37	0.138	0.256	0.373	0.491	0.608	0.726	0.843	0.961	1.079	1.196
0.38	0.138	0.258	0.377	0.497	0.616	0.736	0.855	0.975	0.094	1.214
0.39	0.138	0.260	0.381	0.503	0.624	0.746	0.867	0.989	1.110	1.232
0.40	0.138	0.262	0.385	0.508	0.632	0.755	0.879	1.002	1.126	1.249
0.41	0.138	0.263	0.389	0.514	0.640	0.765	0.890	1.016	1.141	1.267
0.42	0.138	0.265	0.393	0.520	0.647	0.775	0.902	1.029	1.156	1.284
0.43	0.138	0.267	0.396	0.526	0.655	0.784	0.913	1.042	1.172	1.301
0.44	0.138	0.269	0.400	0.531	0.662	0.793	0.924	1.055	1.187	1.318
0.45	0.138	0.271	0.444	0.537	0.670	0.803	0.936	1.068	1.201	1.334

Table 18.9B Flexure–Limiting Moment of Resistance Factor, $M_{u,lim}/b_w d^2 f_{ck}$, For singly Reinforced T-Beams, N/,mm²

f_y = 500 N/mm²

D_f/d	b_f/b_w									
	1.0	2.0	3.0	4.0	5.0	6.0	7.0	8.0	9.0	10.0
0.06	0.133	0.159	0.185	0.211	0.237	0.263	0.289	0.315	0.341	0.367
0.07	0.133	0.163	0.193	0.223	0.254	0.284	0.314	0.344	0.374	0.404
0.08	0.133	0.167	0.202	0.236	0.270	0.304	0.339	0.373	0.407	0.441
0.09	0.133	0.171	0.210	0.248	0.286	0.325	0.363	0.401	0.440	0.478
0.10	0.133	0.175	0.218	0.260	0.303	0.345	0.387	0.430	0.472	0.514
0.11	0.133	0.179	0.226	0.272	0.318	0.365	0.411	0.458	0.504	0.550
0.12	0.133	0.183	0.234	0.284	0.334	0.385	0.435	0.485	0.536	0.586
0.13	0.133	0.187	0.241	0.296	0.350	0.404	0.458	0.513	0.567	0.621
0.14	0.133	0.191	0.249	0.307	0.365	0.423	0.481	0.540	0.598	0.656
0.15	0.133	0.195	0.257	0.319	0.381	0.442	0.504	0.566	0.628	0.690
0.16	0.133	0.199	0.264	0.330	0.396	0.461	0.527	0.593	0.658	0.724
0.17	0.133	0.202	0.272	0.341	0.411	0.480	0.549	0.619	0.688	0.757
0.18	0.133	0.206	0.279	0.352	0.425	0.498	0.571	0.644	0.717	0.791
0.19	0.133	0.210	0.286	0.363	0.440	0.516	0.593	0.670	0.747	0.823
0.20	0.133	0.213	0.292	0.372	0.452	0.532	0.611	0.691	0.771	0.850
0.21	0.133	0.215	0.297	0.379	0.461	0.543	0.625	0.707	0.789	0.871
0.22	0.133	0.217	0.302	0.386	0.470	0.555	0.639	0.723	0.808	0.892
0.23	0.133	0.220	0.306	0.393	0.479	0.566	0.653	0.739	0.826	0.912
0.24	0.133	0.222	0.311	0.400	0.488	0.577	0.666	0.755	0.844	0.933
0.25	0.133	0.224	0.315	0.406	0.497	0.589	0.680	0.771	0.862	0.953
0.26	0.133	0.226	0.320	0.413	0.506	0.600	0.693	0.786	0.880	0.973
0.27	0.133	0.229	0.324	0.420	0.515	0.611	0.706	0.802	0.897	0.993
0.28	0.133	0.231	0.328	0.426	0.524	0.622	0.719	0.817	0.915	1.012
0.29	0.133	0.233	0.333	0.433	0.532	0.632	0.732	0.832	0.932	1.032
0.30	0.133	0.235	0.337	0.439	0.541	0.643	0.745	0.847	0.949	1.051
0.31	0.133	0.237	0.341	0.445	0.550	0.654	0.758	0.862	0.966	1.070
0.32	0.133	0.239	0.346	0.452	0.558	0.664	0.770	0.877	0.983	1.089
0.33	0.133	0.241	0.350	0.458	0.566	0.675	0.783	0.891	1.000	1.108
0.34	0.133	0.243	0.354	0.464	0.575	0.685	0.795	0.906	1.016	1.127
0.35	0.133	0.245	0.358	0.470	0.583	0.695	0.808	0.920	1.033	1.145
0.36	0.133	0.248	0.362	0.476	0.591	0.705	0.820	0.934	1.049	1.163
0.37	0.133	0.250	0.366	0.483	0.599	0.716	0.832	0.949	1.065	1.181
0.38	0.133	0.252	0.370	0.494	0.607	0.725	0.844	0.962	1.081	1.199
0.39	0.133	0.254	0.374	0.489	0.615	0.735	0.856	0.979	1.097	1.227
0.40	0.133	0.255	0.378	0.500	0.623	0.745	0868	0.990	1.112	1.235
0.41	0.133	0.257	0.382	0.506	0.630	0.755	0.879	1.004	1.128	1.252
0.42	0.133	0.259	0.386	0.512	0.638	0.764	0.891	1.017	1.143	1.270
0.43	0.133	0.261	0.389	0.518	0.646	0.774	0.902	1.030	1.158	1.287
0.44	0.133	0.263	0.393	0.523	0.653	0.783	0.913	1.043	1.174	1.304
0.45	0.133	0.265	0.397	0.529	0.661	0.793	0.925	1.056	1.188	1.320

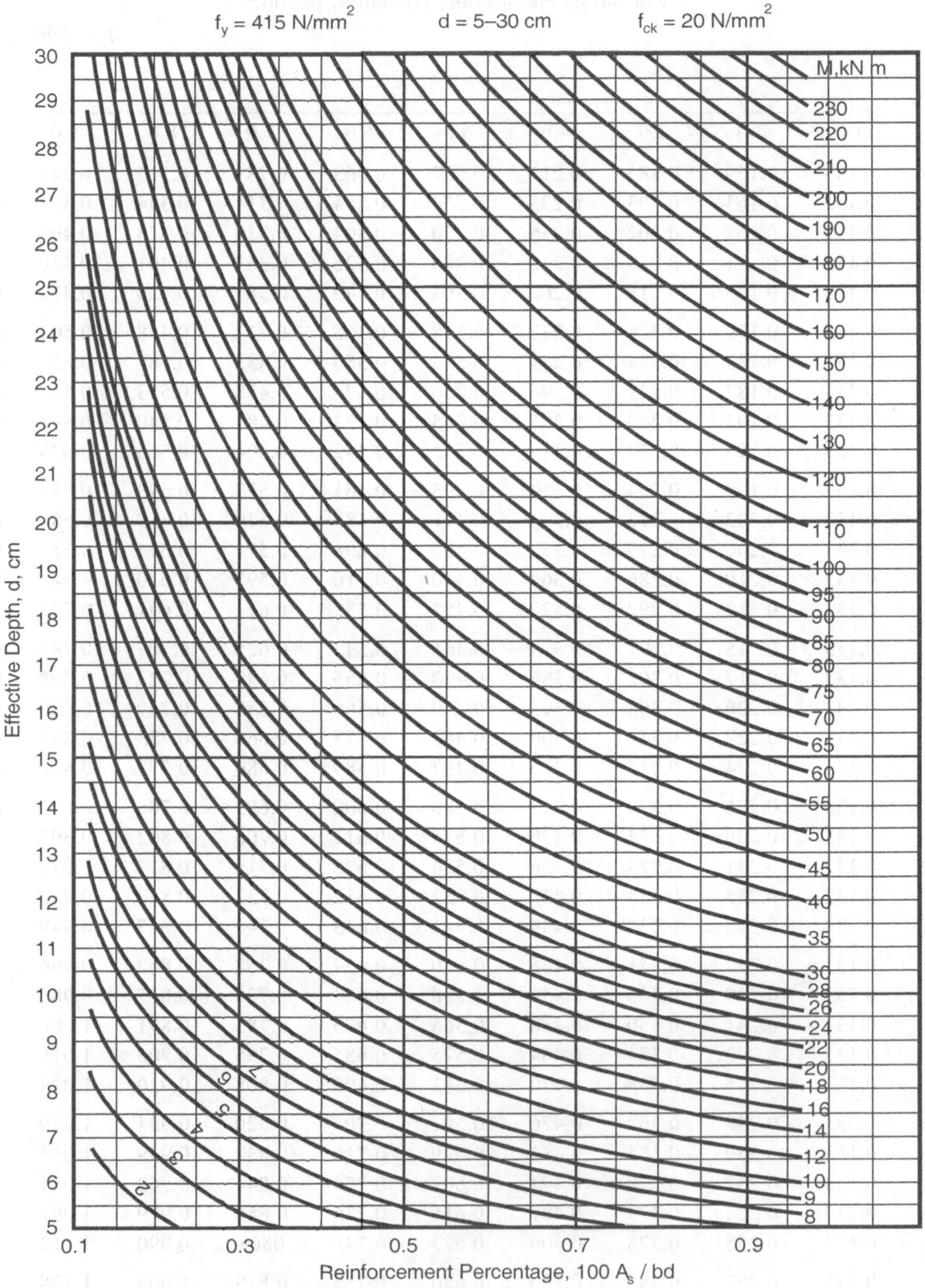

Chart 18.1. Flexure — Singly Reinforced Section
Moment of Resistance kN.m per Metre Width

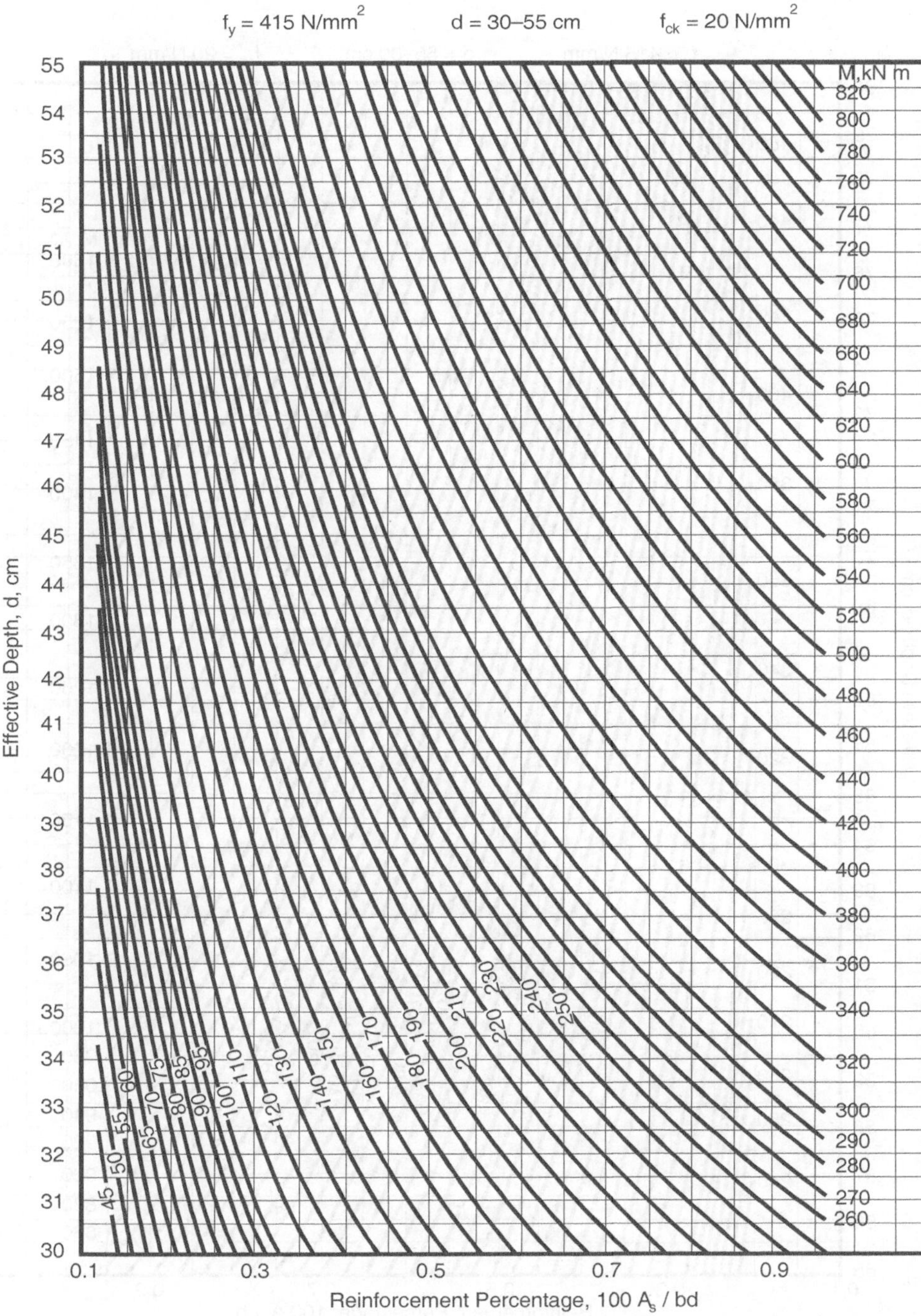

Chart 18.2. Flexure — Singly Reinforced Section
Moment of Resistance kN.m per Metre Width

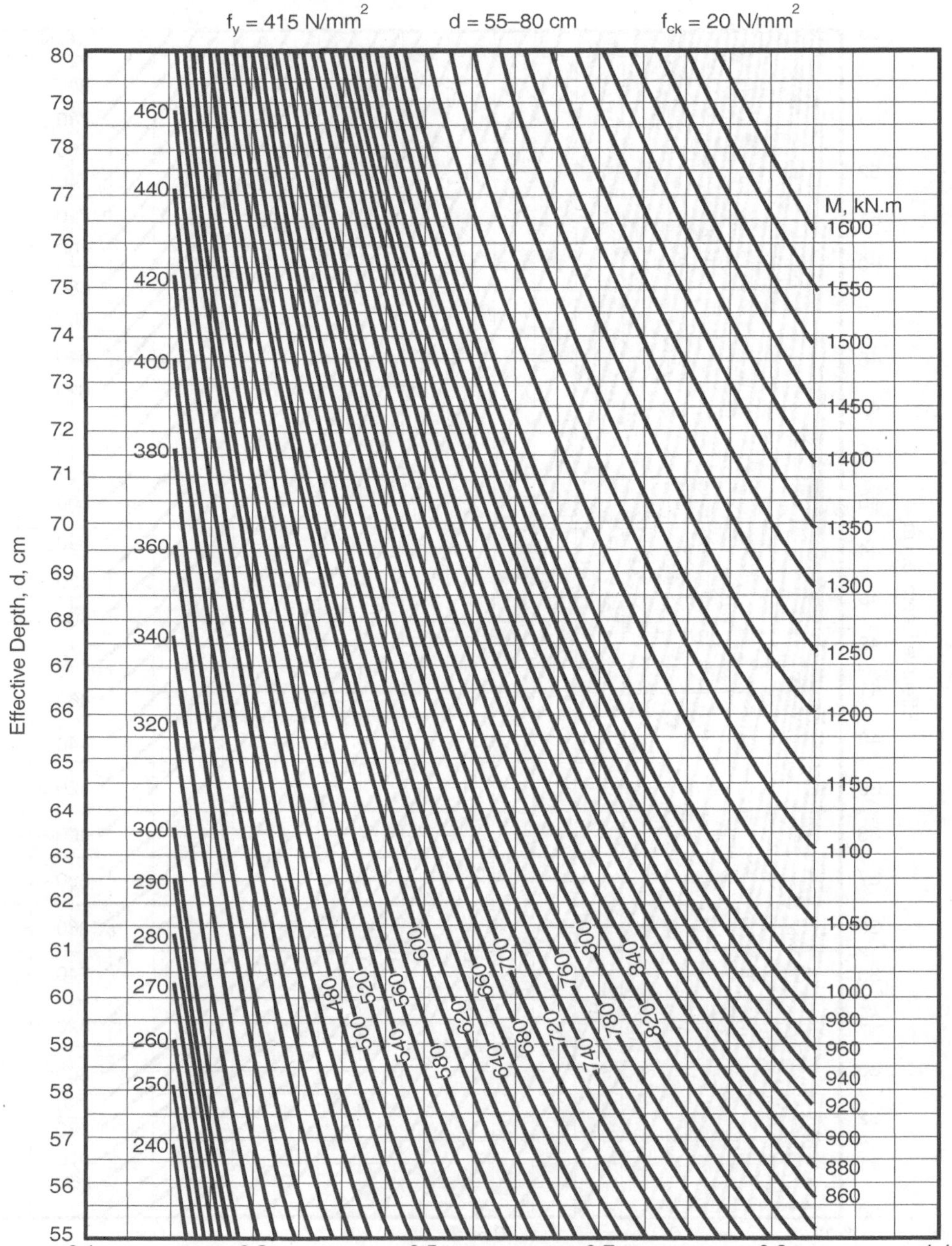

Chart 18.3. Flexure — Singly Reinforced Section
Moment of Resistance kN.m per Metre Width

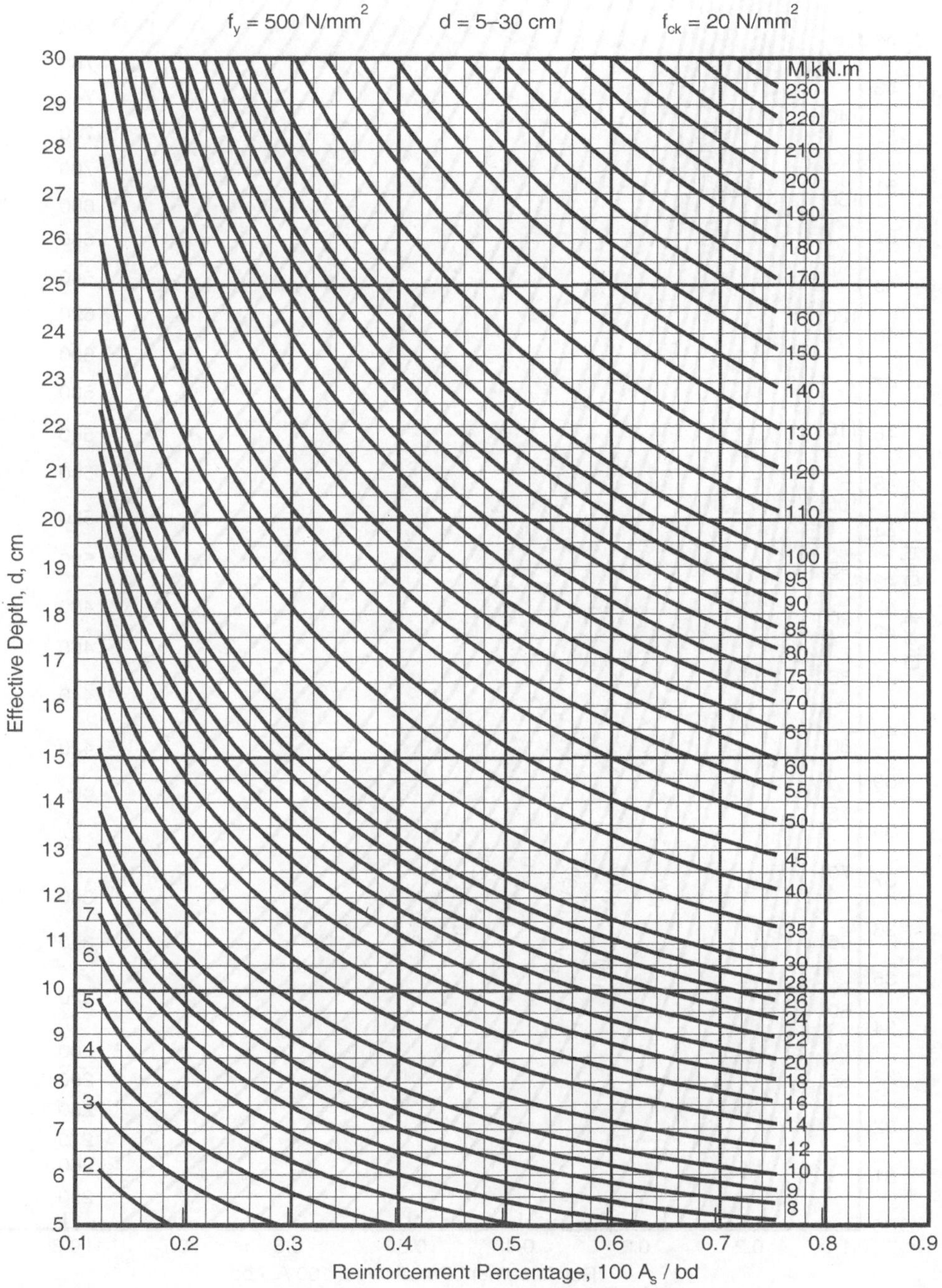

Chart 18.4. Flexure — Singly Reinforced Section
Moment of Resistance kN.m per Metre Width

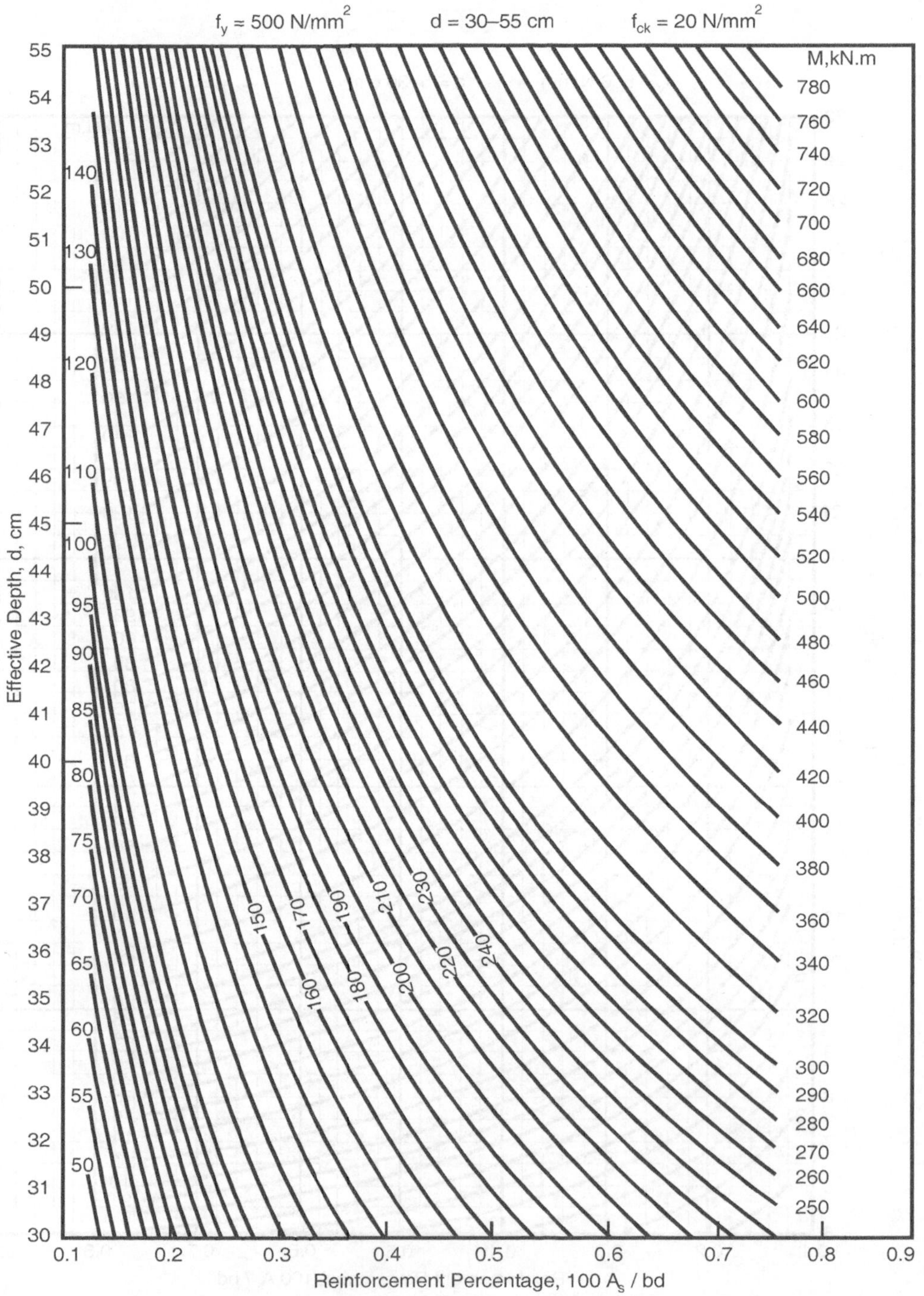

Chart 18.5. Flexure — Singly Reinforced Section

Moment of Resistance kN.m per Metre Width

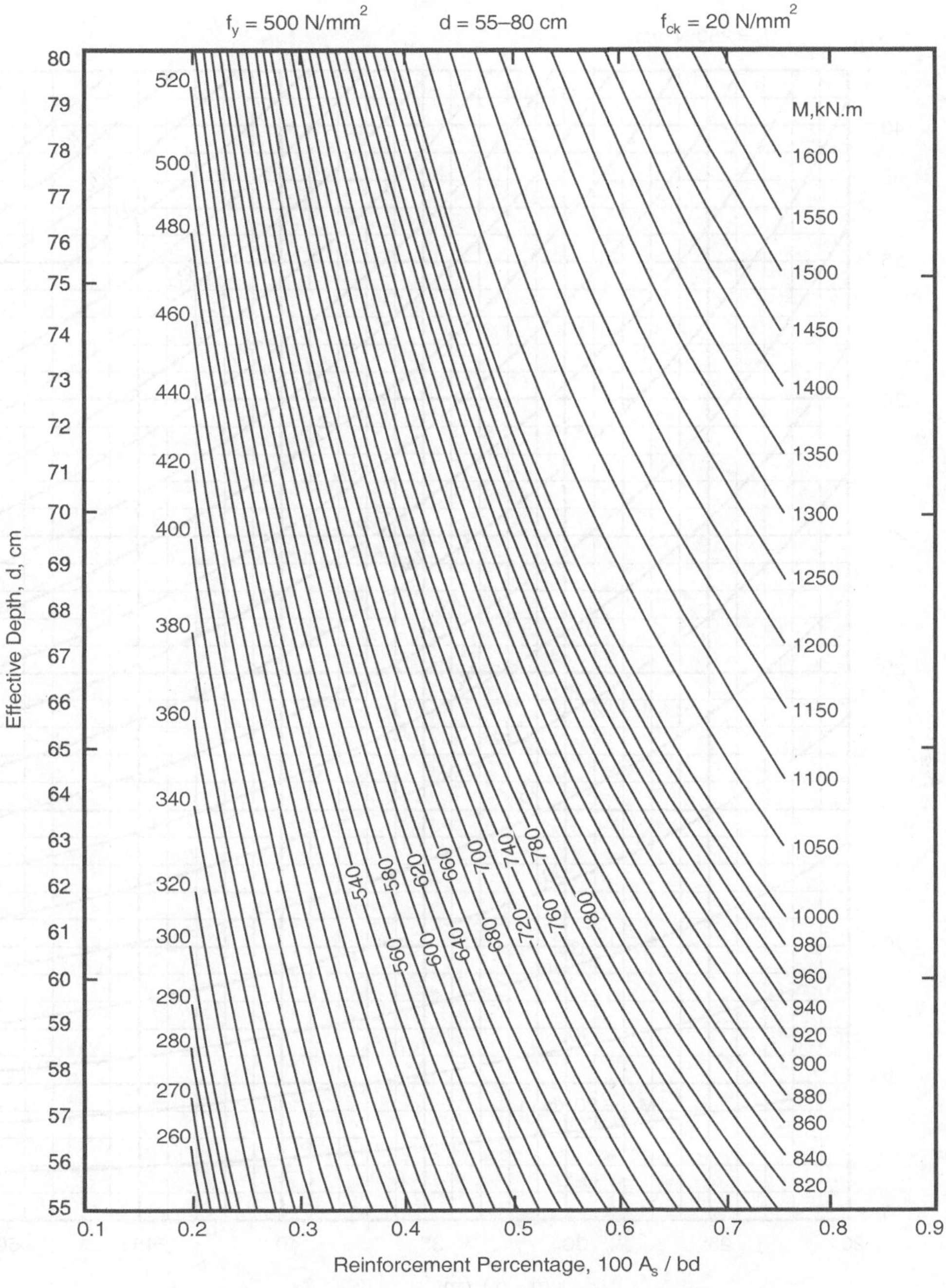

Chart 18.6. Flexure — Singly Reinforced Section
Moment of Resistance kN.m per Metre Width

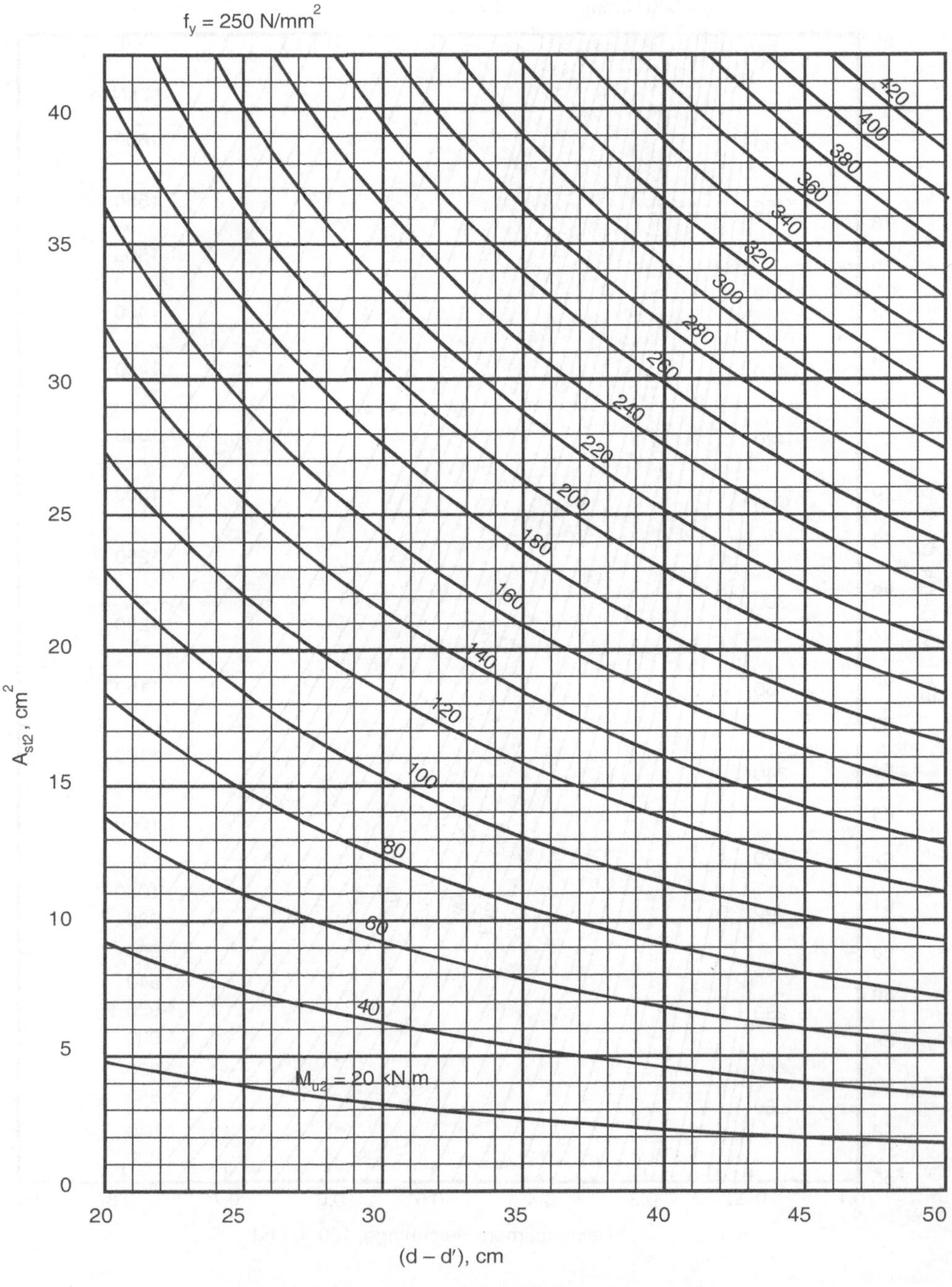

Chart 18.7. Flexure — Doubly Reinforced Section

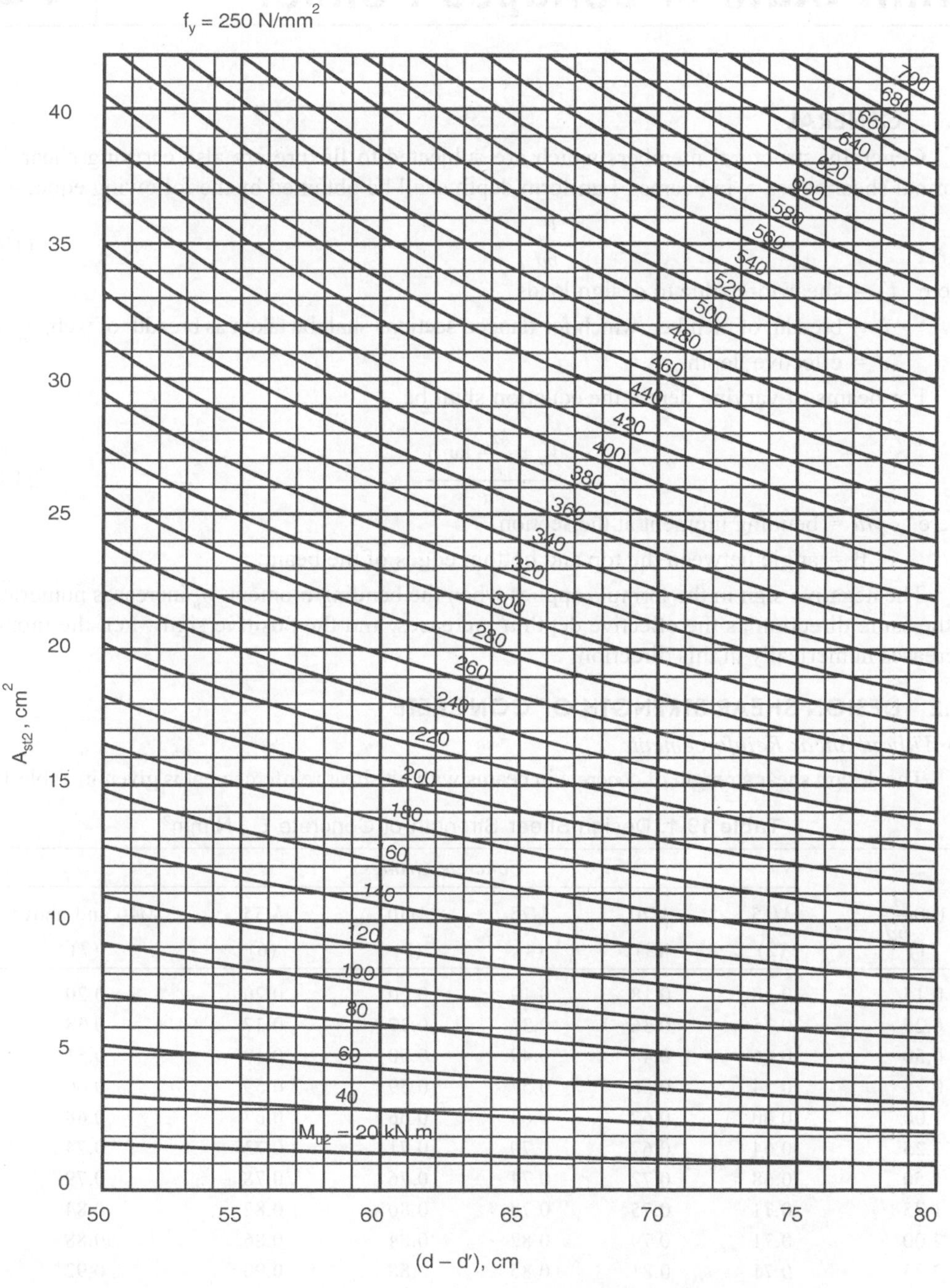

Chart 18.8. Flexure — Doubly Reinforced Section

Limit State of Collapse : Shear

CHAPTER **19**

19.1 GENERAL

Generally structural members which are subjected to flexure are also carrying shear. The nominal shear stress τ_v in beams of uniform depth shall be obtained by the following equation.

$$\tau_v = \frac{V_u}{bd} \qquad ...(19.1)$$

where V_u = shear force due to design loads

b = breadth of member, which for flanged sections shall be taken as breadth of web, b_w; and

d = effective depth.

For beams of varying depth, the equation shall be

$$\tau_v = \frac{V_u \pm \dfrac{M_u}{d} \tan\beta}{bd} \qquad ..(19.2)$$

where Mu = bending moment at the section

β = angle between the top and bottom edges of the beam.

The negative sign in the formula applies when the bending moment M_u increases numerically in the same direction as the effective depth d increases, and the positive sign when the moment decreases numerically in this direction.

19.2 DESIGN SHEAR STRENGTH OF CONCRETE

(*A*) *Without Shear Reinforcement*

The design shear strength of concrete in beams without shear reinforcement is given in Table 19.1.

Table 19.1. Design Shear Strength of Concrete τ_c, N/mm²

	Concrete Grades					
$100\frac{A_s}{bd}$ (1)	*M*15 (2)	*M*20 (3)	*M*25 (4)	*M*30 (5)	*M*35 (6)	*M*40 and above (7)
0.15	0.18	0.18	0.19	0.20	0.20	0.20
0.25	0.35	0.36	0.36	0.37	0.37	0.38
0.50	0.46	0.48	0.49	0.50	0.50	0.51
0.75	0.54	0.56	0.57	0.59	0.59	0.60
1.00	0.60	0.62	0.64	0.66	0.67	0.68
1.25	0.64	0.67	0.70	0.71	0.73	0.74
1.50	0.68	0.72	0.74	0.76	0.78	0.79
1.75	0.71	0.75	0.78	0.80	0.82	0.84
2.00	0.71	0.79	0.82	0.84	0.86	0.88
2.25	0.71	0.81	0.85	0.88	0.90	0.92
2.50	0.71	0.82	0.88	0.91	0.93	0.95
2.75	0.71	0.82	0.90	0.94	0.96	0.98
3.00	0.71	0.82	0.92	0.96	0.99	1.01

Note: The term A_s is the area of longitudinal tension reinforcement which continues at least one effective depth beyond the section being considered excpet at supports where the full area tension reinforcement may be used provided the detailing conforms to IS : 26.2.2 and 26.2.3.

For solid slabs, the design shaer strength for concrete shall be $K.\ \tau_c$ where K has the values given below.

Overall depth of slab, mm	300 or more	275	250	225	200	175	150 or less
K	1.00	1.05	1.1	1.15	1.20	1.25	1.30

For the design of flat slabs, the permissible shear stress is calculated from the following relation :

$$\tau_c = 0.25\sqrt{(f_{ck})} \qquad \text{...(19.3)}$$

(B) Minimum Shear Reinforcement

When τ_v is lessthan τ_c in Table 19.1, minimum shear reinforcement in the form of stirrups shall be provided such as

$$\frac{A_{sv}}{b.s_v} \geq \frac{0.4}{f_y \times 0.87} \qquad \text{...(19.4)}$$

where

A_{sv} = total cross-sectional areas of stirrup legs effective in shear.

s_v = stirrup spacing along the length of member

b = breadth of the beam or breadth of web of flanged beam, and

f_y = characteristic strength of the stirrup reinforcement in N/mm² which shall not be taken greater than 415 N/mm²

However, in members of minor structural importance such as lintels or where the max. shear stress calculated is less than half the permissible value, this provision need not be complied with.

Under no circumstances, even with shear reinforcement, shall the nominal shear stress in beams τ_v exceed τ_c max. given in Table 19.2 below.

Table 19.2. Maximum Shear Stress $\tau_{c\ max}$ N/mm²

Concrete Grade	M15	M20	M25	M30	M35	M40
$\tau_{cm\ max}$ N/mm²	2.5	2.8	3.1	3.5	3.7	4.0

(C) Design of Shear Reinforcement

In the region of the beam wehre τ_v exceeds τ_c given in Table 19.1, shear reinforcement shall be provided in any of the following forms:

(*a*) Vertical stirrups

(*b*) Bent-up bars along with stirrups

(*c*) Inclined stirrups.

When bent up bars are provided, their contribution towards shear resistance shall not be more than half that of the total shear reinforcement.

Shear reinforcement shall be provided to carry a shear 'V_{us}' equal to $V_u - \tau_c.\ b.d.$ The strength of shear reinforcement V_{us} shall be calculated as follows:

(*i*) For Vertical stirrups:

$$V_{us} = \frac{0.87\ f_y \cdot A_{sv} \cdot d}{s_v} \qquad \text{...(19.5)}$$

(*ii*) For inclined stirrups or a series of bars bent up at different cross sections:

$$V_{us} = \frac{0.87 f_y \cdot A_{sv} \cdot d}{s_v}(\sin\theta + \cos\theta) \quad ...(19.6)$$

(*iii*) For single bar or single group of parallel bars, all bent up at the same cross section:

$$V_{us} = 0.87 f_y \cdot A_{sv} \sin\theta \quad ...(19.7)$$

where

A_{sv} = total cross-sectional area of stirrup legs or bent-up bars within a distance s_v.

s_v = spacing of the stirrups or bent-up bars along the length of the member,

τ_v = nominal shear stress,

τ_c = design shear strength of the concrete.

b = breadth of the member which for flanged beams, shall be taken as the breadth of the web b_w.

f_y = characteristic strength of the stirrup or bent-up reinforcement which shall not be taken greater than 415 N/mm^2

θ = angle between the inclined stirrup or bent-up bar and the axis of the member, not less than 45° and

d = effective depth.

Where more than one type of shear reinforcement is used to reinforce the same portion of the beam, total shear resistance shall be computed as the sum of the resistances for the various types of reinforcements separately.

The spacing of the shear stirrups, shall be computed from the above formula, re-written here again.

$$s_v = \frac{0.87 f_y \cdot A_{sv} \cdot d}{V_{us}} \quad ...(19.8)$$

Further in the region upto the effective depth of section from the face of the support, the spacing of the stirrups 's_v' shall be kept constant.

It may, however, be noted that the maximum spacing of shear reinforcement measured along the axis of the member shall not exceed 0.75 *d* for vertical stirrups and *d* for inclined stirrups at 45° where *d* is the effective depth of the section under consideration. In no case shall the spacing exceed 450 mm.

(*D*) *Enhanced Shear strength of Sections close to supports-Refer article 4.6.8*

Example 19.1. A simply supported beam with clear span 6000 mm (6m) b = 400 mm, d = 560 mm, carries a limit state load of 175 KN/m (including self weight, dead load and live load). It is reinforced with 4 bars of 28 ϕ mm dia. tension steel (A_{st} = 2464 mm^2) which continue right into the support. Take f_{ck} = 20 N/mm^2, f_y = 250 N/mm^2. Design shear reinforcement.

Solution.

V_u at the face of support = 175 × 3 = 525 KN.

$$\tau_v = \frac{V_{u,\max}}{bd}$$

$$= \frac{525 \times 10^3}{400 \times 560}$$

$$= 2.343 \text{ N/mm}^2$$

For M20 concrete, from Table 19.2

$$\tau_{c\ max} = 2.8 \text{ N/mm}^2$$

τ_v is less than $\tau_{c,\ max}$

Hence section need not be revised.

$$\text{For} \quad \frac{100\ A_{st}}{bd} = \frac{100 \times 2464}{400 \times 560} = 1.1$$

From Table 19.1 for M20

$$\tau_c = 0.64 \text{ N/mm}^2$$

If the point of starting of nominal reinforcement be at a distance x from the face of either support (Fig. 19.1).

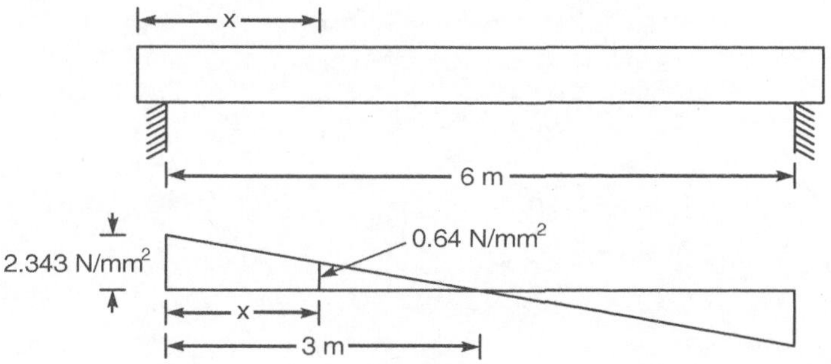

Fig. 19.1.

$$\frac{2.343}{3} = \frac{0.64}{3-x}$$

$$x = 2.18 \text{ m.}$$

In the region where τ_v is less than τ_c, the spacing of stirrups 8 mm ϕ 2 legged (say) shall be calculated as :

$$s_v = \frac{A_{sv} \cdot f_y}{b \times 0.4} = \frac{100 \times 250}{400 \times 0.4} = 156.25 \text{ mm}$$

In the region where τ_v exceeds τ_c, the spacing of stirrups 12 mm ϕ 2 legged shall be calculated as :

$$V_{us} = V_u - \tau_c \cdot bd = 525000 - 0.64 \times 400 \times 560$$

$$= 381640 \text{ N}$$

$$s_v = \frac{0.87\, f_y \cdot A_{sv} \cdot d}{V_{us}}$$

$$= \frac{0.87 \times 250 \times 226 \times 560}{381640}$$

$$= 72.1 \text{ mm}$$

Maximum spacing = 0.75 × d = 0.75 × 560 = 420 mm or 450 mm whichever is less.

Provide 12 mm ϕ 2 legged stirrups starting from face of supports @ 70 mm *c/c* gradually increased to 150 mm *c/c* upto distance 2.20 m and thereafter 8 mm dia. 2 legged stirrups @ 150 mm *c/c*.

EXERCISE 19

1. What do you understand by nominal shear stress? Write the formulas for uniform rectangular section and a rectangular section with varying depth.
2. Explain briefly the procedure for the deisgn of shear reinforcement.
3. A simply supported beam with clear span 5 m, width of beam = 300 mm and effective depth = 460 mm, carries a V_u = 300 KN. It is reinforced with 4 Nos. 25 mm ϕ dia. steel on tension face which continues right into the support. Take f_{ck} = 20 N/mm². f_y = 415 N/mm².
4. Is it always that we have to povide minimum shear reinforcement in a beam section? Comment.

CHAPTER 20

Limit State of Collapse : Torsion

20.1 GENERAL

Torsion in reinforced concrete members is caused due to monolithic construction and unequal loading on the two sides of the cross section of the member. Reinforced concrete members are quite often subjected to torsion combined with shear and bending (torsion alone exists very rarely) *e.g.*, in spandrel beams, edge beams, slabs, balcony beams with cantilever slabs, curved beams, spiral stair-cases, etc.

(*A*) *Critical Section*

Section located at a distance less than '*d*' from the face of support may be designed for the same torsion as computed at a distance '*d*' where *d* is the effective depth.

(*B*) *Shear and Torsion*

(*i*) *Equivalent shear*: The members subjected to torsion shall be designed for fictitious shear which is a function of the actual shear and torsion. This fictitious shear is called 'Equivalent Shear' and shall be calculated from the following formula:

$$V_e = V_u + 1.6\frac{T_u}{b} \quad ...(20.1)$$

where

V_e = equivalent shear

V_u = shear

T_u = torsional moment, and

b = breadth of beam,

The equivalent nominal shear stress τ_{ve} shall be calculated by :

$$\tau_{ve} = \frac{V_e}{bd} \quad ...(20.2)$$

which is obtained by substituting V_u by V_e in the expression for nominal shear stress. The values of τ_{ve} shall not exceed values of $\tau_{c, max}$, given in Table 19.2.

(*ii*) If the equivalent nominal shear stress τ_{ve} does not exceed τ_c, given in Table 19.1, minimum shear reinforcement shall be provided in the same way as discussed in shear articles (19.2)

(*iii*) If τ_{ve} exceeds τ_c given in Table 19.1, both longitudinal and transverse reinforcement shall be provided in the way described subsequently.

(*c*) *Reinforcement in Members Subjected to Torsion*

(*i*) *Longitudinal Reinforcement*: The longitudinal reinforcement shall be designed to resist an equivalent bending moment, M_{e1} given by :

$$M_{e1} = M_u + M_t$$

where Mu = Bending Moment at cross section and

$$M_t = T_u\left(\frac{1 + D/b}{1.7}\right)$$

where

T_u is the torsional moment, D is the overall depth of beam and b is the breadth of beam.

If the numerical value of M_t exceeds the numerical value of the moment M_u, the longitudinal reinforcement shall be provided on the flexural compression face, such that the beam can also withstand an equivalent M_{e2} given by $M_{e2} = M_t - M_u$, the moment M_{e2} being taken as acting in the opposite sense to the moment M_u.

(*ii*) *Transverse Reinforcement*: Two legged closed hoops, enclosing the corner logitudinal bars shall have an area of cross section A_{sv}, given by,

$$A_{sv} = \frac{T_u \cdot s_v}{b_1 d_1 (0.87 f_y)} + \frac{V_u \cdot s_v}{2.5\, d_1 (0.87 f_y)} \quad ...(20.4)$$

Transverse reinforcement shall not be less than

$$\frac{(\tau_{ve} - \tau_c)\, b \cdot s_v}{0.87\, f_y} \quad ...(20.5)$$

where

T_u = torsional moment,

V_u = shear force,

s_v = spacing of the stirrup reinforcement,

b_1 = centre to centre distance between corner bars in the direction of the width,

d_1 = centre to centre distance between corner bars in the direction of the depth,

b = breadth of the member,

f_y = characteristic strength of the stirrup reinforcement,

τ_{ve} = equivalent shear stress as specified in Eq. 20.2 and

τ_c = shear strength of the concrete as per table 19.1

20.2 SOLVED EXAMPLE

Example 20.1. A rectangular section with b = 200 mm and D = 350 mm is subjected to limit state shear of 80 KN. and BM of 25 KN.m. Design the torsional reinforcement for torsion moment of 6 KN.m. Given f_{ck} = 20 N/mm², f_y = 250 N/mm².

Solution. (Fig. 20.1)

Assuming 8 mm dia. stirrups are to be used with 20 mm clear side cover and main reinforcement as 20 mm dia. bars.

$$d = 350 - 40 - \frac{20}{2} = 300.0 \text{ mm}$$

Equivalent shear

$$V_e = V_u + 1.6 \frac{T_u}{b}$$

$$= 80{,}000 + 1.6 \times \frac{6 \times 10^6}{200}$$

$$= 1{,}28{,}000 \text{ N} = 128 \text{ KN}$$

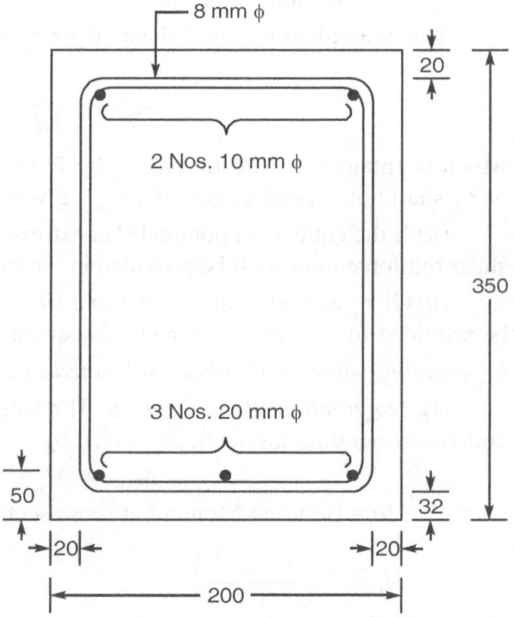

Fig. 20.1.

Equivalent nominal shear stress

$$\tau_{ve} = \frac{V_e}{bd}$$

$$= \frac{128 \times 10^3}{200 \times 297.5} = 2.15 \text{ N/mm}^2$$

From Table 19.2, for M 20 mix; $\tau_{c\ max} = 2.8$ N/mm²

Since $\tau_{ve} < \tau_{c\ max}$

∴ Section dimension need not be revised.

For an assumed value $p_t = 1.0$

$$\tau_c = 0.62 \text{ N/mm}^2 < \tau_{ve}$$

Hence longitudinal and transverse reinforcements are to be designed.

Equivalent bending moment

$$M_{e1} = M_u + M_t$$

$$= 25 + \frac{6}{1.7}\left(1 + \frac{350}{200}\right)$$

$$= 34.706 \text{ KN.m}$$

From Chart 18.1

$$M_{e1}/\text{metre} = \frac{34.706}{0.2} = 173.53 \text{ KN.m}$$

$$f_y = 415 \text{ N/mm}^2$$

$$d = 30.00 \text{ cm}$$

$$p_t = 0.65$$

$$A_{st} = 0.65 \times 20 \times \frac{30.00}{100} = 3.90 \text{ cm}^2 = 390 \text{ mm}^2$$

Provide 20 mm dia. 3 Nos.

($A_{st} = 314 \times 3 = 942$ mm²) on the flexural tensile face.

As M_t is less than M_u, it is not needed to consider M_{e2}. Therefore, provide only two bars of 10 mm dia. on compression face.

$$p_t = \frac{942 \times 100}{200 \times 300} = 1.571 \qquad \text{From Table 19.1, } \tau_c = 0.72 \text{ N/mm}^2$$

Side Face Reinforcement

As the depth of beam is not more than 450 mm, side face reinforcement is not required.

Transverse Reinforcement

Area of two legs of the stirrupps should satisfy the following :

$$A_{su} = \frac{T_u \cdot s_v}{b_1 \cdot d_1 (0.87 f_y)} + \frac{V_u \cdot s_v}{2.5 d_1 (0.87 f_y)}$$

where

b_1 = centre to centre distance of corner bars along width

= 200 – 76 = 124 mm

d_1 = centre to centre distance of corner bars along depth.

= 350 – 32 – 20 – 8 – 8 – 10 – 5 = 267 m (assuming 10 mm ϕ hanger bars)

$$\frac{A_{su}(0.87 f_y)}{s_v} = \frac{6\times10^6}{124\times267} + \frac{60\times10^3}{2.5\times267}$$

$$= 301.3 \text{ N/mm}$$

Area of all legs of the stirrups should satisfy the conditions that A_{su}/s_v should not be less than

$$\frac{(\tau_{ve} - \tau_c)\cdot b}{0.87 f_y}$$

$$\tau_c = 0.72 \text{ N/mm}^2 \text{ for } p_t = 1.571$$

$$\frac{A_{su}(0.87 f_y)}{s_v} = (\tau_{ve} - \tau_c)\cdot b = (2.15 - 0.72)\times 200$$

$$= 286 \text{ N/mm} < 301 \text{ N/mm}$$

∴ O.K.

Using 8 mm ϕ 2 legged shear stirrups at spacing

$$s_v = \frac{A_{su}0.87 f_y}{282} = \frac{2\times50\times0.87\times415}{286}$$

$$= 126.2 \text{ mm say 125 mm c/c throughout.}$$

Check for Spacing of Shear Stirrups

x_1 = shorter dimension of hoop stirrups = 200 – 40 – 8 = 152 mm

y_1 = longer dimension of the hoop stirrups = 350 – 32 – 20 = 290 mm

Spacing of hoop stirrups provided should not exceed the least of :

(*i*) x_1 = 152 mm

(*ii*) $\frac{x_1 + y_1}{4}$ = 110.5 mm

(*iii*) 300 mm

∴ Provide hoop stirrups 8 mm ϕ 2 legged @ 110 mm *c/c* throughout.

EXERCISE 20

1. (*a*) Define 'Equivalent Shear' as obtained from torsion.
   (b) Discuss briefly reinforcement in members subjected to torsion.
2. A rectangular section with width of section as 300 mm and depth as 500 mm is subjected to a limit state of shear of 80 KN and a *B. M.* of 30 KN.m. Design the torsional reinforcement if in addition to above forces, a torsional moment of 8 KN.m. is also acting on the member. Given f_{ck} = 20 N/mm², f_y = 415 N/mm².
3. A square cross section 300 mm × 300 mm is subjected to a limit state *S. F.* = 40 KN, *B.M.* = 20 KN.m and Torsional moment = 4 KN.m. Design the torsional reinforcement required. Use M20 concrete and HYSD (Fe 415) steel.

Limit State of Collapse : Compression

CHAPTER **21**

21.1 GENERAL

In addition to the assumptions made for flexural members, the following assumptions are made for members in compression.

(a) The maximum compression strain in concrete in axial compression is taken as 0.002.

(b) The maximum compressive strain at highly compressed extreme fibre in concrete subjected to axial compression and bending and when there is no tension on the section shall be 0.0035 minus 0.75 times the strain at the least compressed extreme fibre.

21.2 SHORT AXIALLY LOADED MEMBERS IN COMPRESSION

All the members in compression shall be designed for the minimum eccentricity equal to the unsupported length of column /500 plus lateral dimension /30, subject to a minimum of 20 mm, i.e, $e_{min} = \frac{l}{500} + \frac{D}{30}$, subject to a minimum of 20 mm. When the eccentricity does not exceed 0.05 times the lateral dimension, the members shall be designed by the following equaton.

$$P_u = 0.4 f_{ck} \cdot A_c + 0.67 f_y \cdot A_{sc} \qquad \text{...(21.1)}$$

where

P_u = axial load on the member

f_{ck} = characteristic compressive strength of the concrete

A_c = area of concrete

f_y = characteristic strength of compression reinforcement

A_{sc} = area of longitudinal reinforcement for columns.

The above equation can be written as :

$$P_u = 0.4 f_{ck}\left(A_g - \frac{pA_g}{100}\right) + 0.67\,\frac{p.A_g}{100}\,f_y \qquad \text{...(21.2)}$$

where

A_g = the gross area of cross-section

p = percentage of reinforcement

Dividing both sides by A_g

$$\frac{P_u}{A_g} = 0.4\, f_{ck}\left(1 - \frac{p}{100}\right) + 0.67\, f_y \cdot \frac{p}{100}$$

$$= 0.4\, f_{ck} + \frac{p}{100}(0.67\, f_y - 0.4\, f_{ck}) \qquad \text{...(21.3)}$$

Charts 21.1 to 21.2 can be used for designing short columns in accordance with the above equations. In the lower section of these charts, $\frac{P_u}{A_g}$ has been plotted against reinforcement percentage p for different grades of concrete. If the cross-section of the column is known, $\frac{P_u}{A_g}$ can be calculated, the reinforcement percentage can be read from the chart. In the upper section of the charts, $\frac{P_u}{A_g}$ is plotted against P_u for various values of A_g. The combined use of upper and the lower sections would eliminate the need for any calculation.

Example 21.1. Determine the cross-section and the reinforcement for an axially loaded column with the following data :

Factored load = 3000 KN
Concrete grade = M20
Characteristic strength of reinforcement = 415 N/mm²
Unsupported length of the column = 3 m

Sol.

The cross-sectional dimensions will depend on the percentage of reinforcement. Assuming 1.0% reinforcement, from the Equation

$$\frac{P_u}{A_g} = 0.4\, f_{ck} + \frac{p}{100}(0.67\, f_y - 0.4\, f_{ck})$$

$$\frac{3000{,}000}{A_g} = 0.4 \times 20 + \frac{1}{100}(0.67 \times 415 - 0.4 \times 20)$$

$$A_g = 280360.7 \text{ mm}^2 \qquad = 2803.607 \text{ cm}^2$$

Referring to chart 21.2

$$A_g = 2800 \text{ cm}^2$$

Provide a section 60 cm × 50 cm.

Area of reinforcement, $A_s = \frac{1 \times 60 \times 50}{100} = 30 \text{ cm}^2$

It is to be checked whether the minimum eccentricity to be considered is within 0.05 × least lateral dimension of the column. In the direction of the longer dimension,

$$e_{min} = \frac{l}{500} + \frac{D}{30}$$

$$= \frac{3000}{500} + \frac{600}{30} = 26 \text{ mm} = 2.6 \text{ cm}$$

$$\frac{e_{min}}{D} = \frac{2.6}{60} = 0.043 < 0.05$$

In the direction of the shorter dimension

$$e_{min} = \frac{3000}{500} + \frac{450}{30} = 21 \text{ mm} = 2.1 \text{ cm}$$

$$\frac{e_{min}}{b} = \frac{2.1}{45} = 0.047 < 0.05$$

Hence O.K.

21.3 COMPRESSION MEMBERS WITH HELICAL REINFORCEMENT

The strength of compression members with helical reinforcement satisfying the following para, shall be taken as 1.05 times the strength of similar member with lateral ties.

The ratio of the volume of helical reinforcement to the volume of the core, shall not be less than $0.36\,(A_g / A_c - 1)\dfrac{f_{ck}}{f_y}$

where

A_g = gross area of section

A_c = area of core of the helically reinforced column measured to the outside dia. of the helix.

f_{ck} = characteristic compression strength of the concrete and

f_y = characteristic strength of the helical reinforcement but not exceeding 415 N/mm²

21.4 COMBINED AXIAL LOAD AND UNIAXIAL BENDING

All compression members are subjected to some bending in addition to axial loads and these should be designed for minimum eccentricity of load. It should always be ensured that the section is designed for a moment which is not less than that due to the prescribed minimum eccentricity.

(A) Assumptions

1. Plane sections normal to the axis of the member remain plane after bending. This means that the strain at any point on the cross-section is directly proportional to the distance from the neutral axis.
2. The stress-strain relationship for concrete is taken as indicated in Fig. 18.1.
3. The tensile strength of concrete is ignored.
4. The design stresses in reinforcement are derived from the strains using the stress-strain relations used in Fig. 18.3.

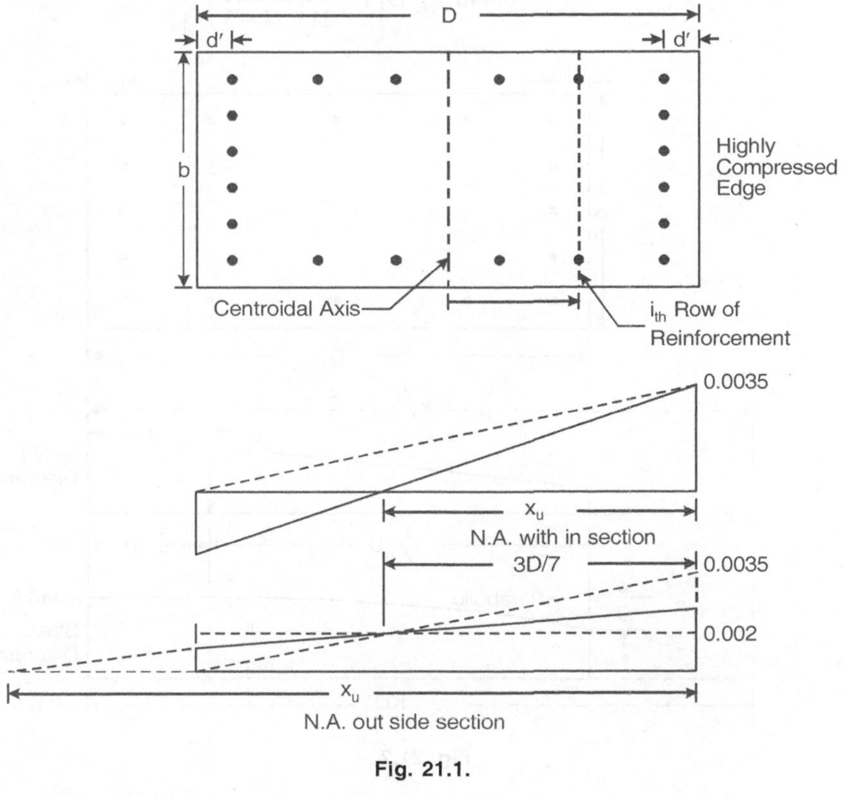

Fig. 21.1.

5. The maximum strain in concrete at the outermost compression fibre is 0.0035 when the neutral axis lies within the section and in the limiting case when the N.A. lies along the edge of the section; in the latter case, the strain varies from 0.0035 at the highly compressed edge to zero at the opposite edge.

For purely axial compression, the strain is assumed to be uniformly equal to 0.002 across the section. The strain distribution lines for these two cases intersect each other at a depth of 3D/7 from the highly compressed edge. This point is assumed to act as a fulcrum for the strain distribution line when the neutral axis lies outside the section (Fig. 21.1). This leads to the assumption that the strain at the highly compressed edge is 0.0035 minus 0.75 times the strain at the least compressed edge.

(B) Stress Block Parameters

(i) When the N.A. Lies Outside the Section

When the N.A. lies outside the section, the shape of the stress block will be as shown in Fig. 21.2. The stress-strain is uniformly 0.446 f_{ck} for a distance of 3D/7 from the highly compressed edge because the strain is more than 0.002 and thereafter the stress diagram is parabolic. Let $x_u = KD$ and let g be the difference between the stress at highly compressed edge and stress at least compressed edge.

Considering the geometric properties of a parabola

$$g = 0.446\, f_{ck}\left[\frac{4/7D}{kd - 3/7D}\right]^2 = 0.446\, f_{ck}\left(\frac{4}{7k-3}\right)^2 \qquad ...(21.4)$$

$$\text{Area of stress block} = 0.446 f_{ck}\, D - (g/3)\, 4/7D$$

$$= 0.446\, f_{ck}\, D\left[1 - \frac{4}{21}\left(\frac{4}{7k-3}\right)^2\right] \qquad ...(21.5)$$

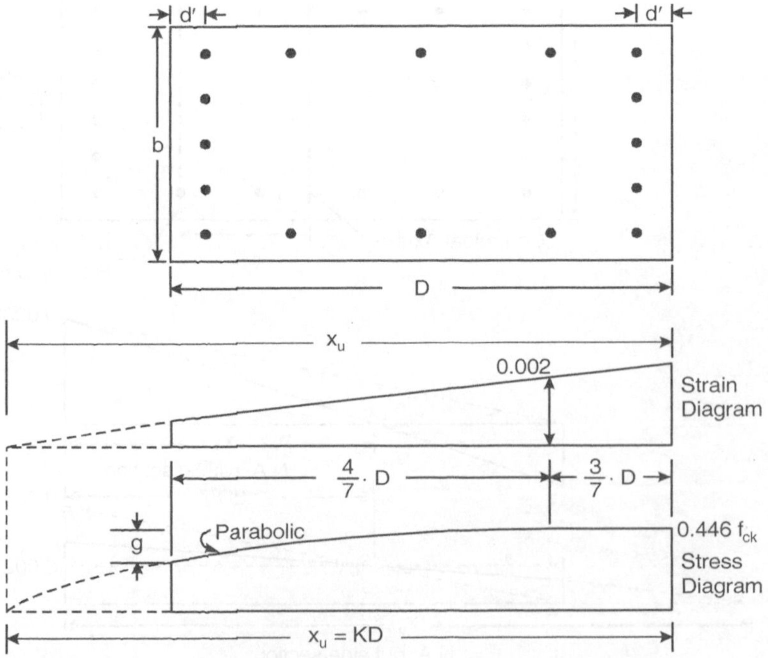

Fig. 21.2.

The centroid of the stress block will be found by taking moments about the compressed edge. Moments about the highly compressed edge.

$$= 0.446\, f_{ck}\, D(D/2) - \frac{4}{21} g\, D\left[\frac{3}{7}D + \frac{3}{4}(4/7D)\right]$$

$$= 0.446\, f_{ck}\, D^2/2 - \frac{8}{49} g\, D^2 \quad ...(21.6)$$

The position of the centroid is obtained by dividing the moment by the area. For different values of *k*, the area of stress block and the position of its centroid are given in Table 21.1.

Table 21.1. Stress Block Parameters when the Neutral Axis Lies Outside the Section

$k = \frac{x_u}{D}$	*Area of stress block*	*Distance of centroid from highly compressed edge*
1.00	$0.361\, f_{ck}\, D$	$0.416\, D$
1.05	$0.374\, f_{ck}\, D$	$0.432\, D$
1.10	$0.384\, f_{ck}\, D$	$0.443\, D$
1.20	$0.399\, f_{ck}\, D$	$0.458\, D$
1.30	$0.409\, f_{ck}\, D$	$0.468\, D$
1.40	$0.417\, f_{ck}\, D$	$0.475\, D$
1.50	$0.422\, f_{ck}\, D$	$0.480\, D$
2.00	$0.435\, f_{ck}\, D$	$0.491\, D$
2.50	$0.440\, f_{ck}\, D$	$0.495\, D$
3.00	$0.442\, f_{ck}\, D$	$0.497\, D$
4.00	$0.444\, f_{ck}\, D$	$0.499\, D$

Note: Values of stress block parameters have been tabulated for values of *k* up to 4.00 for information only. For construction of interaction diagrams it is generally adequate to consider values of *k* up to about 1.2.

Construction of Interaction Diagrams

Design charts for combined axial compression and bending are plotted in the form of interaction diagrams in which the curves for $\frac{P_u}{bDf_{ck}}$ Versus $\frac{M_u}{bD^2 f_{ck}}$ are plotted for different values of p/f_{ck} where *p* is the reinforcement percentage.

(a) For the case of purely axial compression, the points plotted on the *y*-axis of the charts are obtained as follows :

$$P_u = 0.446 f_{ck} bD + \frac{pbD}{100}(f_{sc} - 0.446 f_{ck}) \quad ...(21.7)$$

$$\frac{P_u}{f_{ck} bD} = 0.446 + \frac{p}{100 f_{ck}}(f_{sc} - 0.446 f_{ck}) \quad ...(21.8)$$

where

f_{sc} is the compression stress in steel corresponding to a strain of 0.002.

The 2nd term within parenthesis represents the deduction for the concrete replaced by the reinforcement bars.

(*b*) When bending moments are also acting in addition to axial load, the points for plotting the charts are obtained by assuming the different positions of N.A., strain distribution across the section and the stress block parameters. The stresses in the reinforcement are also calculated from the known strains. Thereafter the resultant axial force and the moment about the centroid of the section are calculated as follows :

(*i*) *When the Neutral Axis lies outside the Section*

$$P_u = C_1 f_{ck} bD + \sum_{i=1}^{n} \frac{p_i bD}{100}(f_{si} - f_{ci}) \qquad \text{...(21.9)}$$

where

C_1 = coefficient for area of stress block (Table 21.1)

$p_i = \frac{A_{si}}{bD}$ where A_{si} is the area of reinforcement in *i*th row

f_{si} = stress in *i*th row of reinforcement, compression being positive and tension being negative.

f_{ci} = stress in concrete at the level of the *i*th row of reinforcement.

n = number of rows of reinforcement.

The above expression can be written as :

$$\frac{P_u}{f_{ck} bD} = C_1 + \sum_{i=1}^{n} \frac{p_i}{100 f_{ck}}(f_{si} - f_{ci}) \qquad \text{...(21.10)}$$

Taking moments of the forces about the centroid of the section

$$M_u = C_1 f_{ck} bD\left(\frac{D}{2} - C_2 D\right) + \sum_{i=1}^{n} \frac{p_i bD}{100}(f_{si} - f_{ci}) y_i \qquad \text{...(21.11)}$$

where

C_2D = the distance of the centroid of concrete stress block, measured from the highly compressed edge ;

and y_i = the distance from the centroid of the section to the *i*th row of reinforcement ; positive towards highly compressed edge.

Dividing both sides by $f_{ck}\, bD^2$

$$\frac{M_u}{f_{ck} bD^2} = C_1(0.5 - C_2) + \sum_{i=1}^{n} \frac{p_i}{f_{ck} 100}(f_{si} - f_{ci})\frac{y_i}{D} \qquad \text{...(21.12)}$$

(*ii*) *When the N.A. Lies within the Section*

The following expressions are obtained when N.A. lies within the section.

$$\frac{P_u}{f_{ck} bD} = 0.36\, K + \sum_{i=1}^{n} \frac{p_i}{100 f_{ck}}(f_{si} - f_{ci}) \qquad \text{...(21.13)}$$

$$\frac{M_u}{f_{ck} bD^2} = 0.36k(0.5 - 0.416\, k) + \sum_{i=1}^{n} \frac{p_i}{f_{ck} 100}(f_{si} - f_{ci})\frac{y_i}{D} \qquad \text{...(21.14)}$$

where $K = \frac{\text{depth of N.A.}}{D}$

The above equation plotted in the form of charts 21.3 to 21.35 for axial compressive force combined with uniaxial bending for rectangular and circular sections for different grades of steel and concrete and for different values of *d*′ / *D*.

For the case of purely axial tension

$$P_u = \frac{pbD}{100}(0.87\,f_y) \qquad \text{...(21.15)}$$

$$\frac{P_u}{f_{ck}bD} = \frac{p}{100 f_{ck}}(0.87\,f_y) \qquad \text{...(21.16)}$$

Example 21.2. Determine the reinforcement to be provided in a square column subjected to uniaxial bending with the following data:

Size of the column = 450 mm × 450 mm

Concrete Mix = M 20

f_y of steel = 415 N/mm²

Factored load = 2025 KN

M_u = 200 KN.m

Arrangement of reinforcement

(a) On Two Sides

(b) On all Four Sides

Sol. Assume moment due to minimum eccentricity to be less than the actual moment.

Assume 25 mm dia. bars with 40 mm cover.

$$d' = 40 + 12.5 = 52.5 \text{ mm}$$

$$\frac{d'}{D} = 52.5/450 = 0.12$$

Chart 21.19 of d'/D = 0.15 will be used.

$$\frac{P_u}{f_{ck}.b.D} = \frac{2025\times10^3}{20\times450\times450} = 0.5$$

$$\frac{M_u}{f_{ck}.b.D^2} = \frac{200\times10^6}{20\times450\times450^2} = 0.1097$$

(a) Reinforcement on Two Sides

From Chart 21.8 $p = 0.11\,f_{ck} = 0.11 \times 20 = 2.2$

$$A_s = \frac{p.b.D}{100} = \frac{2.2\times450\times450}{100} = 4455 \text{ mm}^2$$

(b) Reinforcement on Four Sides

From Charts 21.19 $\quad p = 0.125 \times f_{ck} = 0.125 \times 20 = 2.5$

$$A_s = \frac{2.5\times450\times450}{100} = 5063 \text{ mm}^2$$

Example 21.3. Determine the reinforcement to be provided in a circular column with the following data :

Dia. of the column = 450 mm Grade of Concrete = M 20

f_y = 250 N/mm² for bars upto 20 mm dia. and 240 N/mm² for bars over 20 mm dia.

Factored load = P_u = 1500 KN. Factored Moment = 105 KN.m.

If the lateral reinforcement used is *(a)* hoop reinforcement in the form of ties *(b)* helical reinforcement.

Sol. Assume cover as 40 mm and diameter of main vertical bars as 20 mm.

$$d' = 40 + \frac{1}{2} \times 20 = 50 \text{ mm}$$

$d'/D = 50//450 = 0.11$ Using Charts for $d'/D = 0.1$

(*a*) *Column with Hoop Reinforcement in the Form of Ties*

$$\frac{P_u}{f_{ck} \cdot D^2} = \frac{1500 \times 10^3}{20 \times 450 \times 450} = 0.37, \quad \frac{M_u}{f_{ck} \cdot D^3} = \frac{105 \times 10^6}{20 \times 450^3} = 0.058$$

Referring to Chart 21.26 $p/f_{ck} = 0.13$

$\therefore$ $p = 0.13 \times 20 = 2.6$ $A_s = \frac{2.6 \times \pi \times 450^2}{400} = 4133 \text{ mm}^2$

Provide 20 mm ϕ 14 Nos.

(*b*) *Column with Helical Reinforcement*

The strength of a compression member with helical reinforcement is 1.05 times the strength of a similar member with lateral ties. Therefore, the given load and moment should be divided by 1.05 before referring to the chart.

$$\frac{P_u}{f_{ck} \cdot D^2 \times 1.05} = 0.37$$

$$\frac{M_u}{f_{ck} \cdot D^3 \times 1.05} = 0.055$$

Referring Chart 21.26

$$\frac{P}{f_{ck}} = 0.11$$

$$p = 0.11 \times 20 = 2.2$$

$$A_s = 2.2 \times \frac{\pi \times 450^2}{400} = 3497 \text{ mm}^2$$

Provide 20 mm ϕ 12 Nos.

Checking that the ratio of volume of helical reinforcement to the volume of Core shall not be less than $0.36\,(A_g/A_c - 1).\ \frac{f_{ck}}{f_y}$

Assuming 8 mm ϕ for helix

$$\text{Core dia.} = 450 - 2(40 - 8) = 386 \text{ mm}$$

$$\frac{A_g}{A_c} = \frac{450^2}{386^2} = 1.36$$

$$0.36\left(\frac{A_g}{A_c} - 1\right)\frac{f_{ck}}{f_y} = 0.36 \times (1.36 - 1)\ 20/250 = 0.0103$$

$$\frac{\text{Volume of helical reinforcement}}{\text{Volume of Core}} = \frac{A_{sh} . \pi(450 - 80 + 8)}{\frac{\pi}{4}(386)^2 . s_h}$$

$$= 0.0101 \frac{A_{sh}}{s_h}$$

where A_{sh} is the area of the bar forming the helix and s_h is the pitch of the helix. In order to satisfy the Codal requirement

$$0.0101 \frac{A_{sh}}{s_h} \rightarrow 0.0103$$

For 8 mm dia. bar, $A_{sh} = 50 \text{ mm}^2$

$$s_h = \frac{0.0101 \times 50}{0.0103}$$

$$= 49.03 \text{ mm}$$

Provide 8 mm ϕ at pitch of 49 mm

21.5 COMPRESSION MEMBERS SUBJECTED TO BIAXIAL BENDING

Compression members such as corner columns in a framed building, are subjected to biaxial bending in addition to direct loads. For such columns, exact design of members is extremely laborious. Therefore, the code permits the design of such members by the following equation :

$$\left(\frac{M_{ux}}{M_{ux1}}\right)^{\infty_n} + \left(\frac{M_{uy}}{M_{uy1}}\right)^{\infty_n} \leq 1.0 \qquad ...(21.17)$$

where

M_{ux}, M_{uy} are the moments about x and y axis respectively due to design load M_{ux1}, M_{uy1} are the maximum uniaxial moments capacities with an axial load P_u, bending about x and y axis respectively and ∞_n is an exponent whose value depends on

$$\frac{P_u}{P_{uz}} \text{ where } P_{u2} = 0.45 f_{ck} A_c + 0.75 f_y A_s$$

$\frac{P_u}{P_{uz}}$	∞_n
≤ 0.2	1.0
≥ 0.8	2.0

For intermediate values, linear interpolation may be done. Chart 21.36 can be used for evaluation of P_{uz}.

For different values of $\frac{P_u}{P_{uz}}$, the appropriate values of ∞_n has been taken and curves for the equation

$$\left(\frac{M_{ux}}{M_{ux1}}\right)^{\infty_n} + \left(\frac{M_{uy}}{M_{uy1}}\right)^{\infty_n} = 1.0 \qquad ...(21.18)$$

have been plotted in chart 21.37

Example 21.4. Determine the reinforcement to be provided in a short column subjected to biaxial bending with the following data.

Size of column = 300 mm × 600 mm

Concrete = M 20

f_y = 415 N/mm²

Factored Load P_u = 1200 KN

Factored moment acting parallel to larger dimension M_{ux} = 150 KN.m

Factored moment acting parallel to shorter dimension M_{uy} = 90 KN.m

Sol. Assuming the reinforcement to be distributed on all the four sides. As a first trial, assume the reinforcement percentage, $p = 1\%$

$$p/f_{ck} = 1/20 = 0.05$$

Uniaxial moment capacity of the section about $x - x$ axis.

Assume clear cover as 40 mm and dia of main bars as 25 mm

$$d'/D = \frac{52.5}{600} = 0.0875$$

$$\text{Chart for } d'/D = 0.1 \text{ will be used}$$

$$\frac{P_u}{f_{ck}bD} = \frac{1200 \times 10^3}{20 \times 300 \times 600} = 0.333$$

Referring to Chart 21.18, $\dfrac{M_u}{f_{ck}.bD^2} = 0.09$

where $M_{ux1} = 0.09 \times 20 \times 300 \times \dfrac{600^2}{10^6} = 194.4 \text{ KN.m}$

Uniaxial moment capacity of the section about $y - y$ axis.

$$d'/D = \frac{52.5}{300} = 0.175$$

$$\text{Chart 21.18 for } d'/D = 0.1, \quad \frac{M_u}{f_{ck}bD^2} = 0.09$$

$$\text{Chart 21.20 for } d'/D = 0.2, \quad \frac{M_u}{f_{ck}bD^2} = 0.075$$

$$\text{For } d'/D = 0.175 \frac{M_u}{f_{ck}bD^2} 0.0825$$

$$M_{uy1} = 0.0825 \times 20 \times 600 \times \frac{300^2}{10^6} = 158.5 \text{ KN.m}$$

Calculation of P_{uz}

Referring to Chart 21.36, $p = 1$ and $f_y = 415$ N/mm²

$$\frac{P_{uz}}{A_g} = 12 \qquad P_{uz} = 12 \times \frac{300 \times 600}{10^3} = 2160 \text{ KN}$$

$$\frac{P_u}{P_{uz}} = \frac{1200}{2160} = 0.5555$$

$$\frac{M_{ux}}{M_{ux1}} = \frac{150}{194.4} = 0.7716, \quad \frac{M_{uy}}{M_{uy1}} = \frac{90}{158.4} = 0.5682$$

$$\infty_n = 1.5916$$

$$\left(\frac{M_{ux}}{M_{ux1}}\right)^{\infty_n} + \left(\frac{M_{uy}}{M_{uy2}}\right)\infty_n \le 1$$

$$(0.7716)^{1.5916} + (0.5682)^{1.5916} = 0.6618 + 0.4 = 1.0618$$

Since 1.0618 is greater than 1.0, bigger value of p can be taken and the process repeated till interaction equation is satisfied.

21.6 SLENDER COMPRESSION MEMBERS

The design of compression members shall be based on the forces and moments determined from an analysis of the structure, including effect of deflections on moments and forces.

When the slenderness ratio $\frac{l_{ex}}{D}$ or $\frac{l_{ey}}{b}$ of a compression member exceeds 12, it is considered to be a slender compression member; l_{ex} and l_{ey}, being the effective lengths with respect to the major and minor axis respectively. When a compression member is slender with respect to the major axis, an additional moment M_{ax} given by the following equation should be taken into account in the design.

$$M_{ax} = \frac{P_u D}{2000}\left(\frac{l_{ex}}{D}\right)^2 \qquad ...(21.19)$$

where P_u = axial load on the member

D = depth of the cross-section at right angles to the major axis

b = width of member

Similarly, if the column is slender about the minor axis, an additional moment M_{ay} should be considered.

$$M_{ay} = \frac{P_u b}{2000}\left(\frac{l_{ex}}{b}\right)^2 \qquad ...(21.20)$$

The expression for the additional moments can be written in the form of eccentricities of load as follows:

$$M_{ax} = P_u . e_{ax} \qquad ...(21.21)$$

where $$e_{ox} = \frac{D}{2000}\left(\frac{l_{ex}}{D}\right)^2$$

$$\frac{e_{ox}}{D} = \frac{1}{2000}\left(\frac{l_{ex}}{D}\right)^2$$

Table 21.2 gives the values $\frac{e_{ax}}{D}$ or $\frac{e_{ay}}{b}$ for different values of slenderness ratio.

In accordance with 38.7.1.1 of the IS : 456-1978 Code, the additional moments may be reduced by the multiplication factor K given below :

$$k = \frac{P_{uz} - P_u}{P_{uz} - P_b} \le 1 \qquad ...(21.22)$$

where $$P_{uz} = 0.45 f_{ck} A_c + 0.75 f_y A_s$$

P_b is the axial load corresponding to the condition of maximum compressive strain of 0.0035 in concrete and tensile strain of 0.002 in outermost layer of tension steel.

Table 21.2. Additional Eccentricity for Slender Compression Members

l_{ex}/D or l_{ey}/D	e_{ax}/D or e_{ay}/b	l_{ex}/D or l_{ey}/b	e_{ax}/D or e_{ay}/b
12	0.072	25	0.313
13	0.085	30	0.450
14	0.098	35	0.613
15	0.113	40	0.800
16	0.128	45	1.013
17	0.145	50	1.250
18	0.162	55	1.513
19	0.181	60	1.800
20	0.200		

Though the modification is optional according to the code, it should always be taken advantage of, since the value of k should be substantially less than unity.

The value of P_b will depend on the arrangement of reinforcement and the cover ratio d'/D, in addition to the grades of concrete and steel. The values of the coefficients required for evaluating P_b for various cases are given in Table 21.3.

Table 21.3. Slender Compression Members – Values of P_b

Rectangular Section :

$$P_b/f_{ck}\,bD = k_1 + k_3\,p/f_{ck}$$

Circular section

$$P_b/f_{ck}\,D^2 = k_1 + k_2\,p/f_{ck}$$

Values of k_1

Section	d'/D 0.05	0.10	0.15	0.20
Rectangular	0.219	0.207	0.196	0.184
Circular	0.172	0.160	0.149	0.138

Values of k_2

Section	f_y N/mm²	d'/D 0.05	0.10	0.15	0.20
Rectangular : equal	250	–0.045	–0.045	–0.045	–0.045
reinforcement on	415	0.096	0.082	0.046	0.022
two opposite sides	500	0.213	0.173	0.104	–0.001
Rectangular : equal	250	0.215	0.146	0.061	–0.011
Reinforcement on	415	0.424	0.328	0.203	0.028
four sides	500	0.545	0.425	0.256	0.040
Circular	250	0.193	0.148	0.077	0.020
	415	0.410	0.323	0.201	0.036
	500	0.543	0.443	0.291	0.056

The expression for K can be written as follows :

$$K = \frac{1 - P_u / P_{uz}}{1 - P_b / P_{uz}} \leq 1$$

Example 21.5. Determine the reinforcement required for a column which is restrained against sway, with the following data :

Size of column = 400 × 300 mm

Concrete Grade = M20

F_y = 415 N/mm²

Effective length for bending parallel to larger dimension, l_{ex} = 6m

Effective length for bending parallel to shorter dimension, l_{ey} = 5 m

Unsupported length = 7 m

Factored load = 1000 KN

Factored moment in the direction of larger dimension = 36 KN.m at bottom

Factored moment in the direction of shorter dimensions = 25 KN.m

Sol. The column is bent in double curvature, Reinforcement will be distributed equally on four sides.

$$l_{ex} = \frac{6.0 \times 1000}{400} = 15 > 12$$

$$\frac{l_{ey}}{b} = \frac{5.0 \times 1000}{300} = 16.7 > 12$$

Therefore, the column is slender about both the axis. From Table 21.2.

For $\frac{l_{ex}}{D} = 15, \frac{e_x}{D} = 0.113$

For $\frac{l_{ey}}{b} = 16.7, \frac{e_y}{b} = 0.14$

Additional moments :

$$M_{cx} = P_u.e_x = 1000 \times 0.113 \times \frac{400}{1000} = 45.2 \text{ KN.m}$$

$$M_{ey} = P_u.e_y = 1000 \times 0.14 \times \frac{300}{1000} = 42 \text{ KN.m}$$

The above moments will have to be reduced in accordance with 39.7.1.1 of the IS Code, but multiplication factors can be evaluated only if the reinforcement is known. For first trial, assume p = 3% (equally distributed steel on all four sides).

$$A_g = 40 \times 30 = 1200 \text{ cm}^2$$

From Chart 21.36, $\frac{P_{uz}}{A_g}$ = 18 N/mm²

$$P_{uz} = 18 \times 1200 \times \frac{10^2}{10^3} = 2160 \text{ KN}.$$

Calculation of P_b

Assuming 25 mm dia. bars 40 mm cover

$$d'/D \text{ (about } x-x \text{ axis)} = 52.5/400 = 0.13$$

$$\text{Chart 21.19 of } d'/D = 0.15 \text{ will be used}$$

$$d'/D \text{ (about } y-y \text{ axis)} = 52.5/300 = 0.17$$

$$\text{Chart 21.20 for } d'/D = 0.2 \text{ will be used.}$$

$$P_{bx} \text{ (about } x-x \text{ axis)} = \left(k_1 + k_2\frac{p}{f_{ck}}\right) f_{ck}.bD$$

$$= \left(0.196 + 0.20 \times \frac{3}{20}\right)\frac{\times 20 \times 30 \times 40 \times 10^2}{10^3}$$

$$= 543.48 \text{ KN}$$

$$P_{by} \text{ (about } y-y \text{ axis)} = \left(0.184 + 0.028 \times \frac{3}{20}\right)\frac{\times 20 \times 30 \times 40 \times 10^2}{10^3}$$

$$= 451.68 \text{ KN}$$

$$k_x = \frac{P_{uz} - P_u}{P_{uz} - P_{bx}} = \frac{2160 - 1000}{2160 - 543.48} = 0.7176$$

$$k_y = \frac{P_{uz} - P_u}{P_{uz} - P_{by}} = \frac{2160 - 1000}{2160 - 451.68} = 0.679$$

The additional moments calculated above, will now be multiplied by the above values of k.

$$M_{ax} = 0.7176 \times 45.2 = 32.44 \text{ KN.m}$$

$$M_{ay} = 0.679 \times 42 = 28.52 \text{ KN.m}$$

The above moments should be compared with those obtained from minimum eccentricity.

$$e_x = \frac{l}{500} + \frac{D}{30} = \frac{700}{500} + \frac{40}{30} = 2.73 \text{ cm}$$

$$e_y = \frac{l}{500} + \frac{b}{30} = \frac{700}{500} + \frac{30}{30} = 2.4 \text{ cm}$$

Both e_x and e_y are greater than 2.0 cm. Moments due to minimum eccentricity

$$M_{ux} = 1000 \times \frac{2.73}{100} = 27.30 \text{ KN.m} < 32.44 \text{ KN.m}$$

$$M_{uy} = 1000 \times \frac{2.4}{100} = 24.00 \text{ KN.m} < 28.0 \text{ KN.m}$$

∴ Design moments are

$$M_{ax} = 32.44 \text{ KN.m}$$

$$M_{ay} = 28.52 \text{ KN.m}$$

The section is to be checked for biaxial bending

$$\frac{P_u}{f_{ck}.bD} = \frac{1000\times10^3}{20\times30\times40\times10^2} = 0.417$$

$$\frac{p}{f_{ck}} = \frac{3}{20} = 0.15$$

Referring to Chart 21.19 ($d'/D = 0.15$)

$$\frac{M_u}{f_{ck}bD^2} = 0.145$$

$$M_{ux} = 0.145\times20\times30\times40^2\times\frac{10^3}{10^6} = 139.2 \text{ KN.m}$$

Referring to Chart 21.20 ($d'/D = 0.2$)

$$\frac{M_u}{f_{ck}bD^2} = 0.13$$

$$M_{uy} = 0.13\times20\times30^2\times40\times\frac{10^3}{10^6} = 93.6 \text{ KN.m}$$

$$\frac{M_{ux}}{M_{ux1}} = \frac{32.44}{139.2} = 0.233$$

$$\frac{M_{uy}}{M_{uy1}} = \frac{28.52}{93.6} = 0.305$$

$$\frac{P_u}{P_{uz}} = \frac{1000}{2700} = 0.37$$

Referring to Chart 21.37

$$\frac{P_u}{P_{uz}} = 0.2<0.37$$

Assumed value of p = 3% is O.K.

$$A_s = \frac{pbD}{100} = \frac{3\times300\times400}{100} = 3600 \text{ mm}^2$$

Note 1: In the case of a braced column without any transverse loads occurring in its height, the additional moment shall be added to an initial moment equal to sum of 0.4 M_{u1} and 0.6 M_{u2} where M_{u2} is the larger end moment and M_{u1} is the smaller end moment (assumed negative if the column is bent in double curvature). In no case shall the initial moment be less than 0.4 M_{u2} ; nor the total moment including the initial moment be less than M_{u2}. For unbraced columns, the additional moment shall be added to the end moments.

Note 2: Unbraced compression members at any given level or storey, subject to lateral load are usually constrained to deflect equally. In such cases, slenderness ratio for each column may be taken as the average for all columns acting in the same direction.

Chart 21.1. Axial Compression

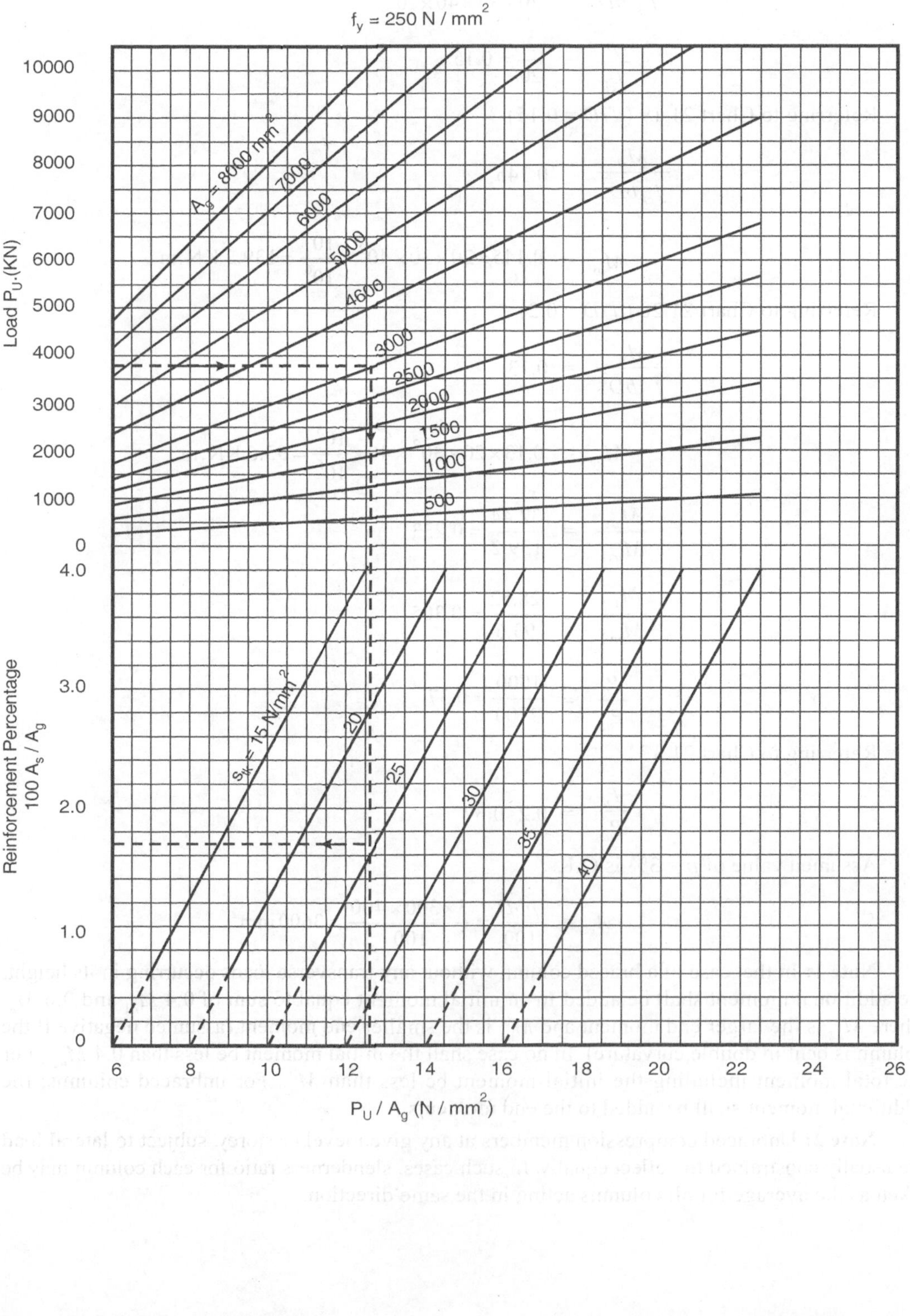

Chart 21.2. Axial Compression

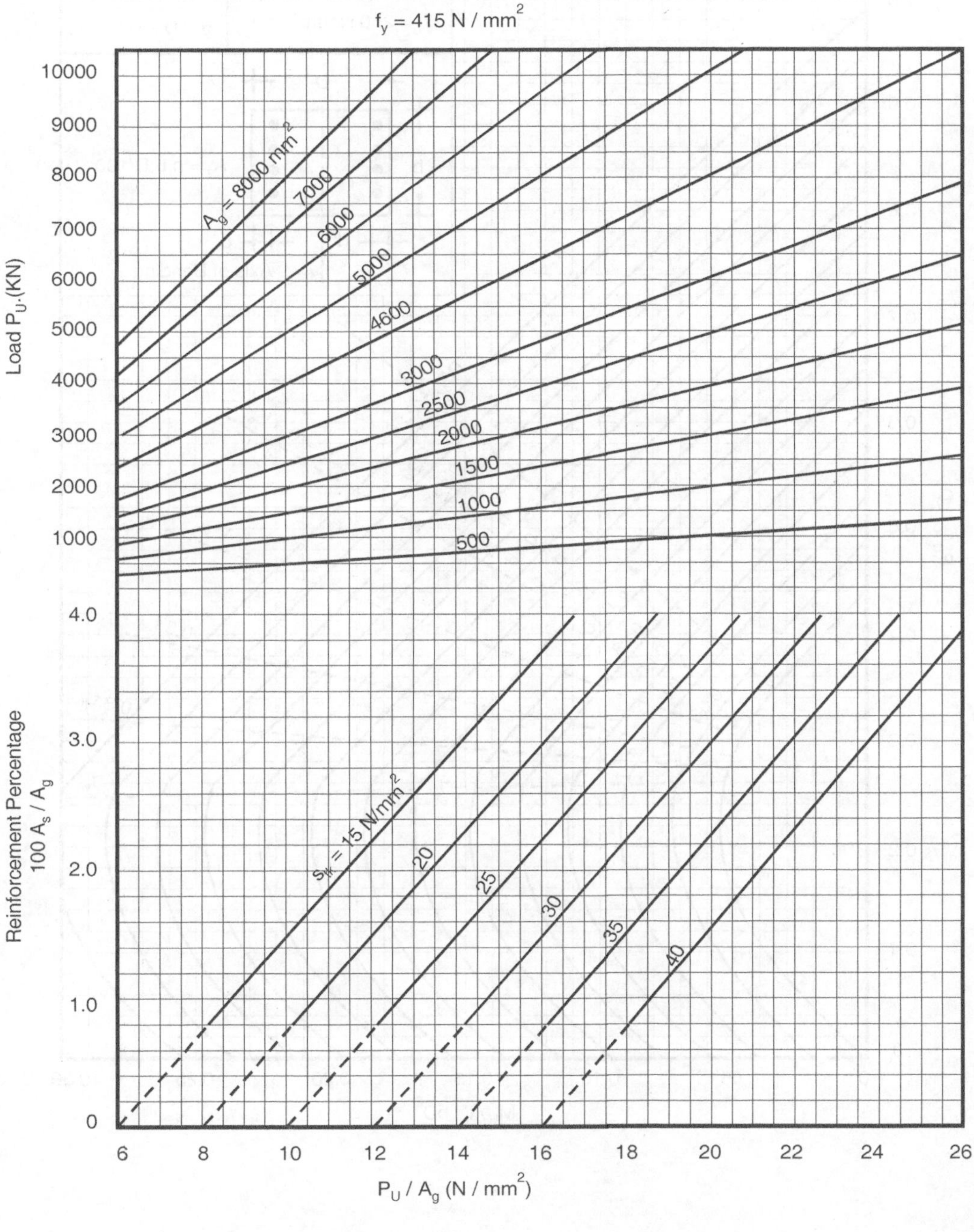

Chart 21.3. Compression with Bending – Rectangular Section – Reinforcement Distributed Equally on Two Sides

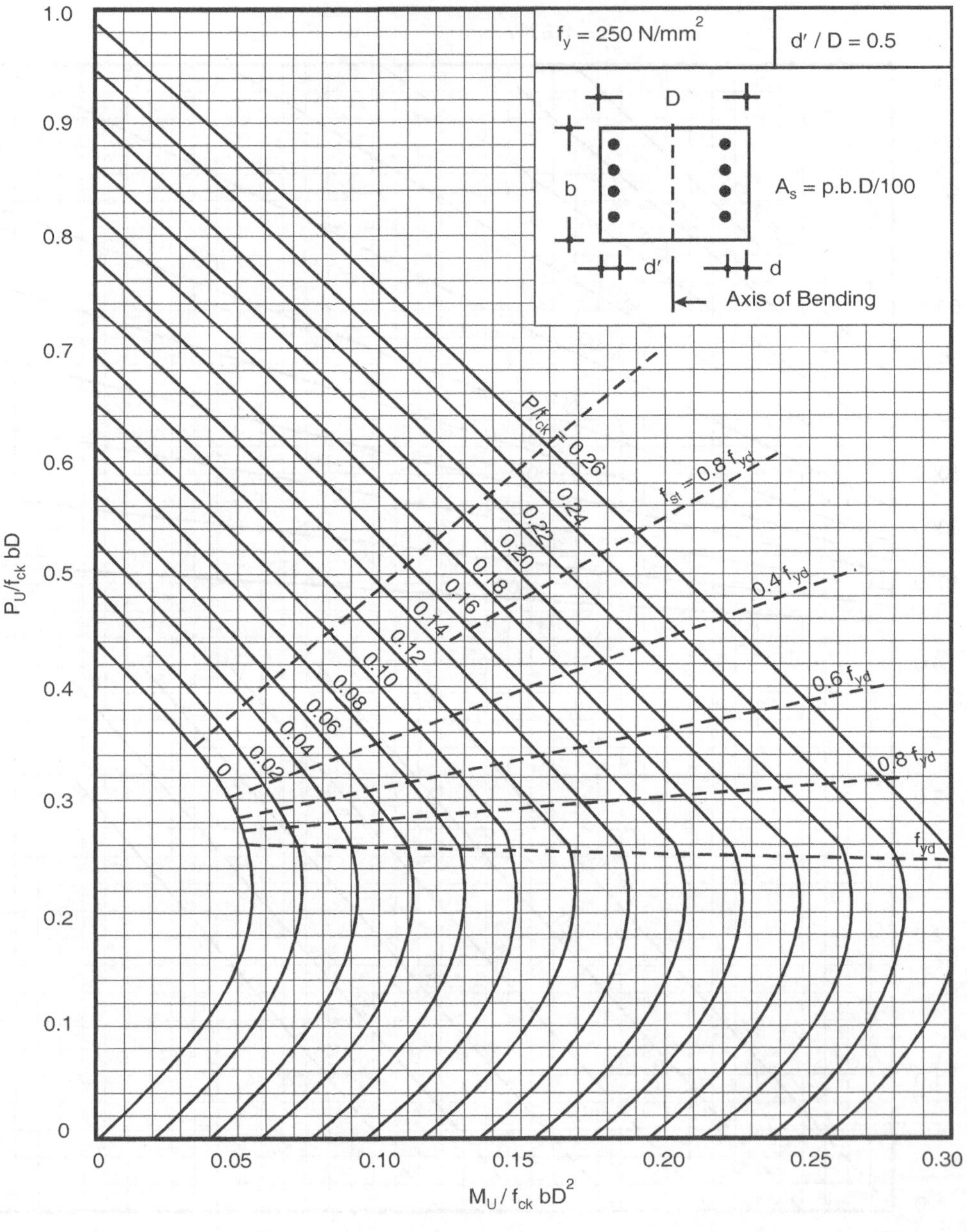

Chart 21.4. Compression with Bending – Rectangular Section – Reinforcement Distributed Equally on Two Sides

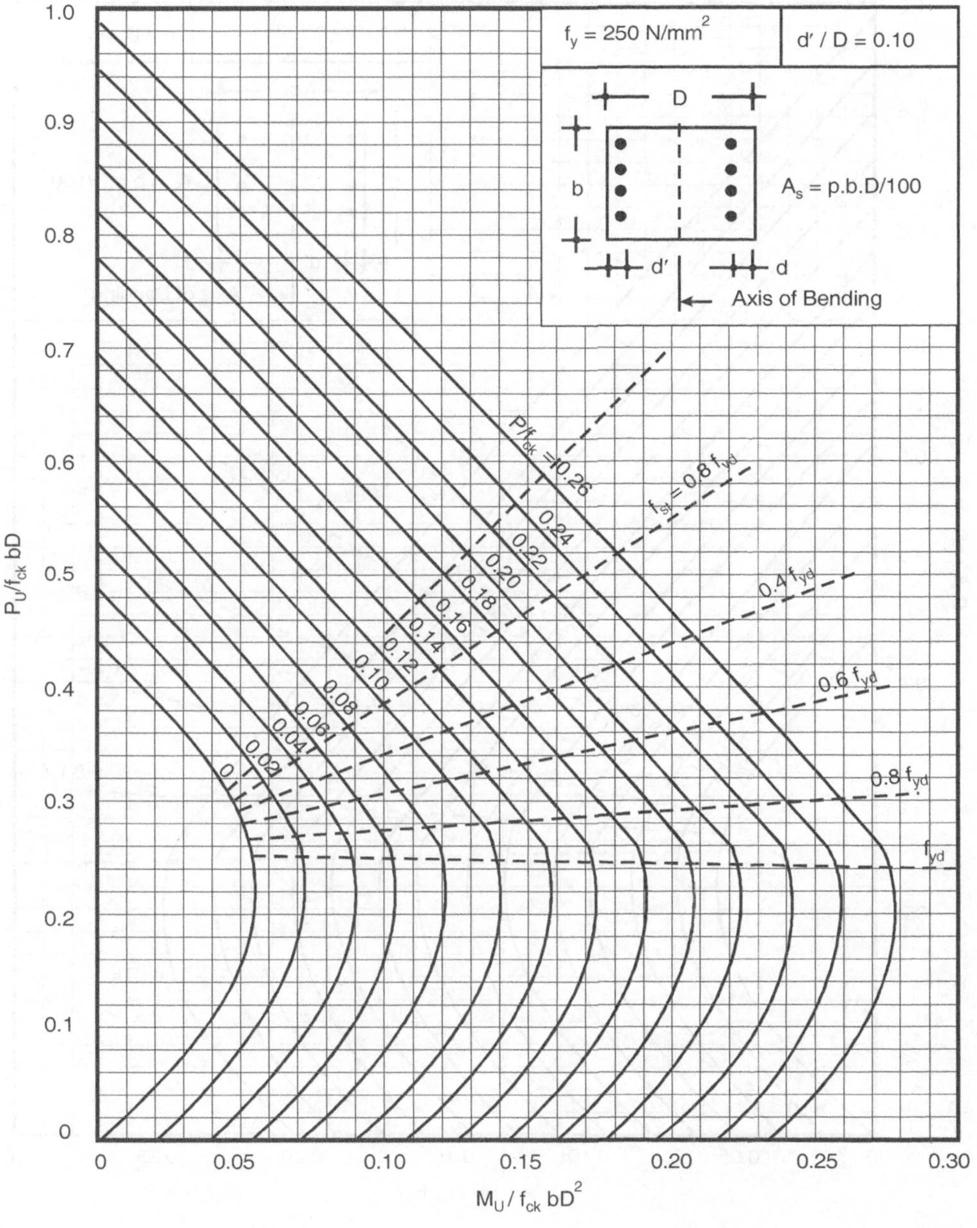

Chart 21.5. Compression with Bending – Rectangular Section – Reinforcement Distributed Equally on Two Sides

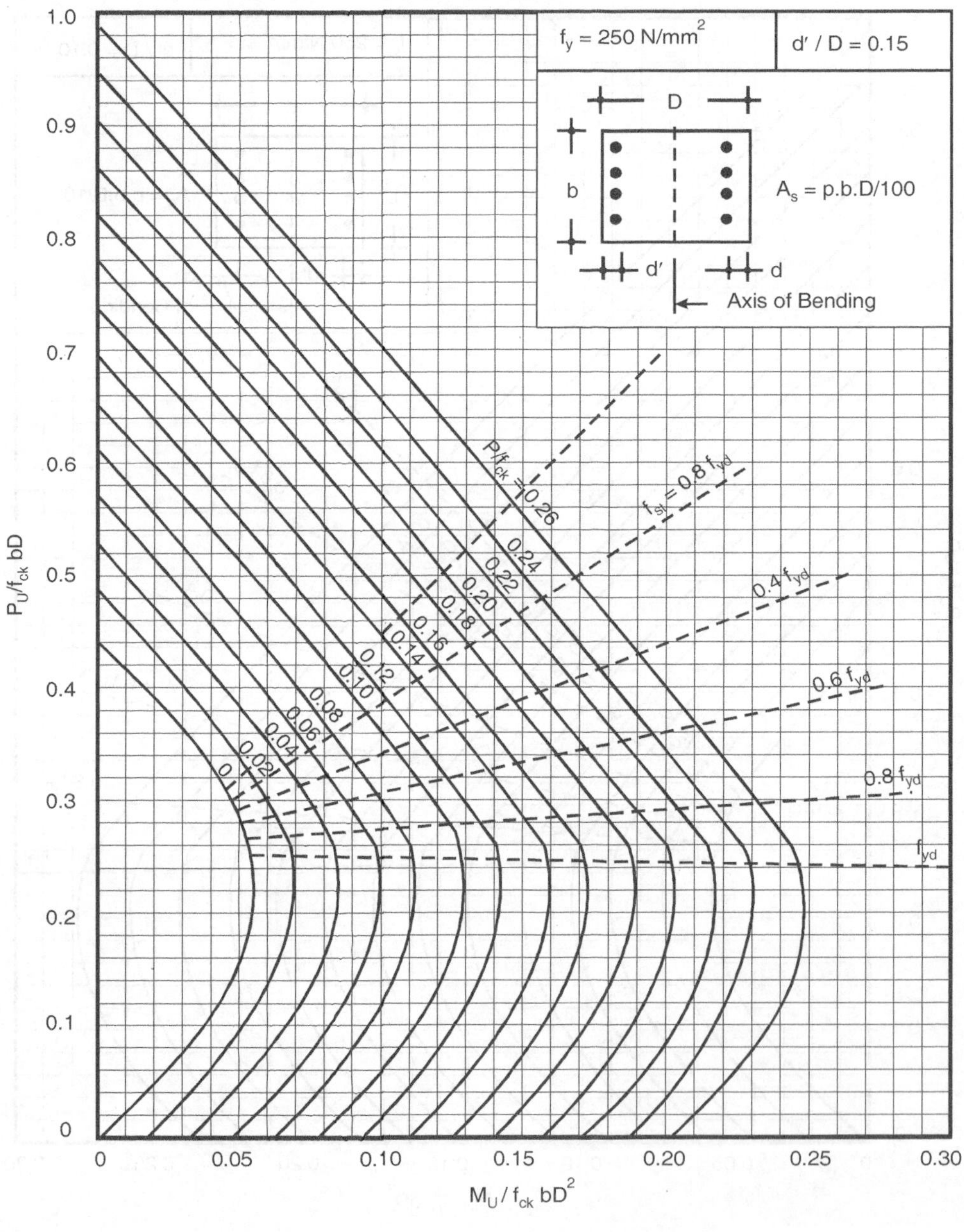

Chart 21.6. Compression with Bending – Rectangular Section – Reinforcement Distributed Equally on Two Sides

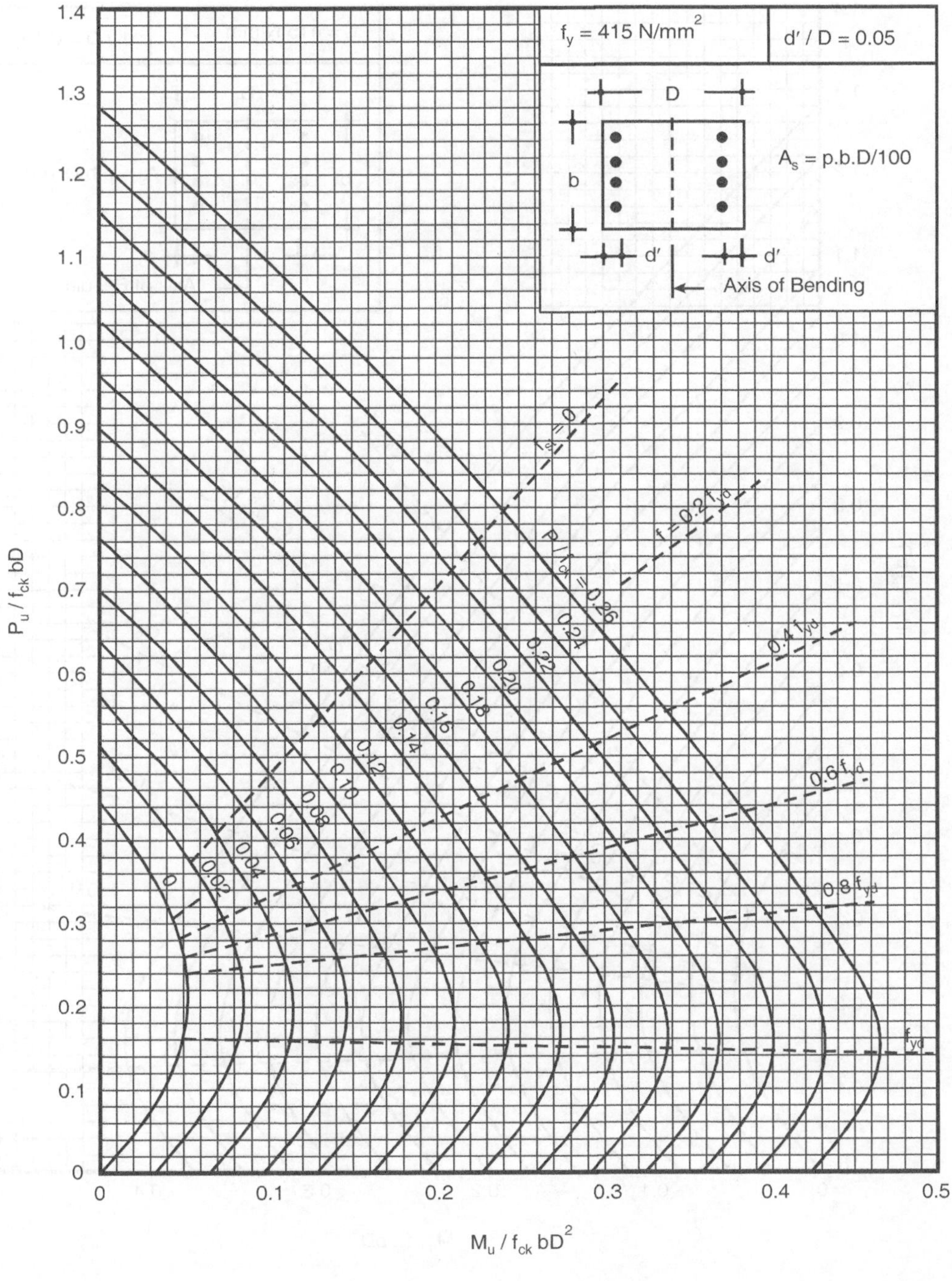

Chart 21.7. Compression with Bending – Rectangular Section – Reinforcement Distributed Equally on Two Sides

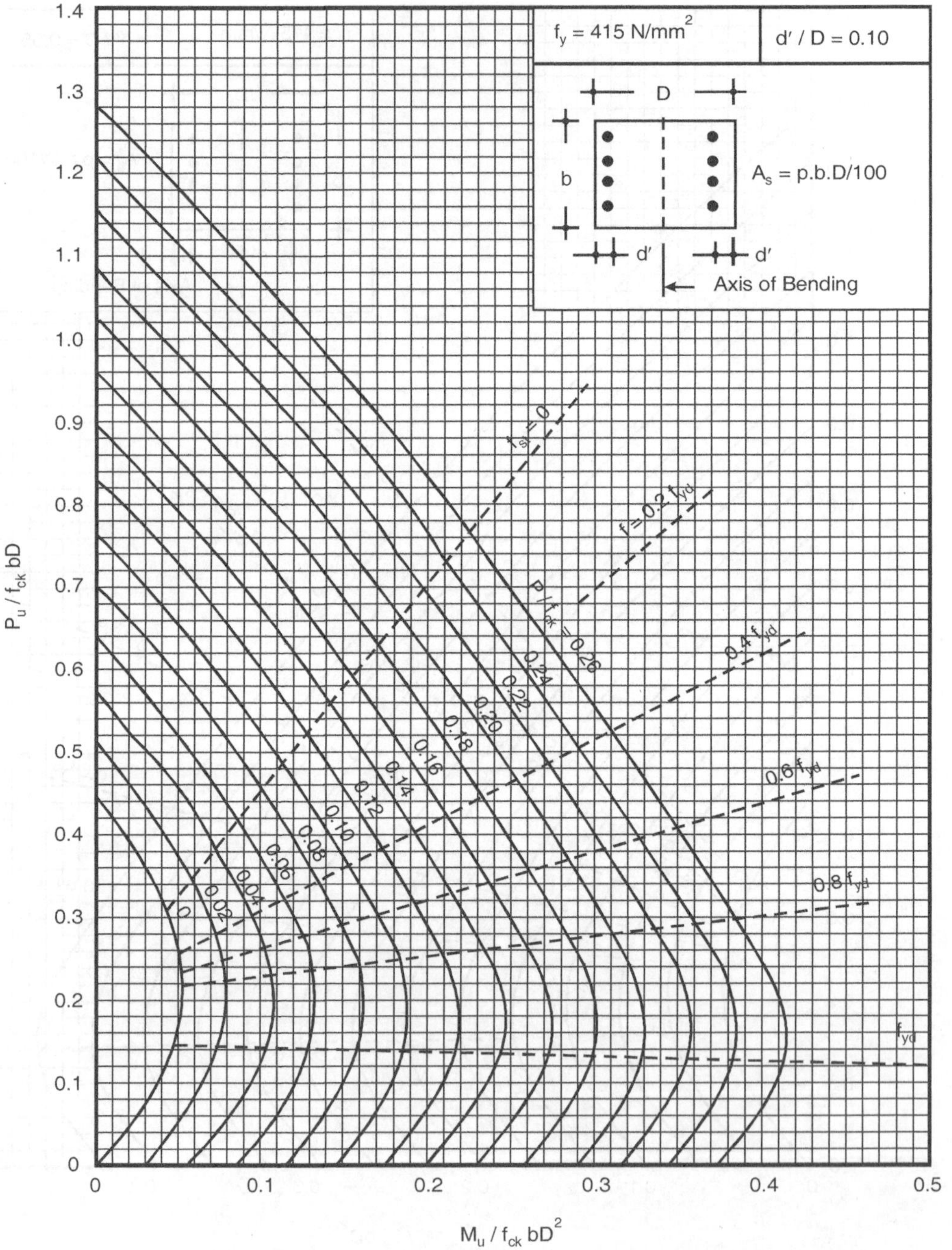

Chart 21.8. Compression with Bending – Rectangular Section – Reinforcement Distributed Equally on Two Sides

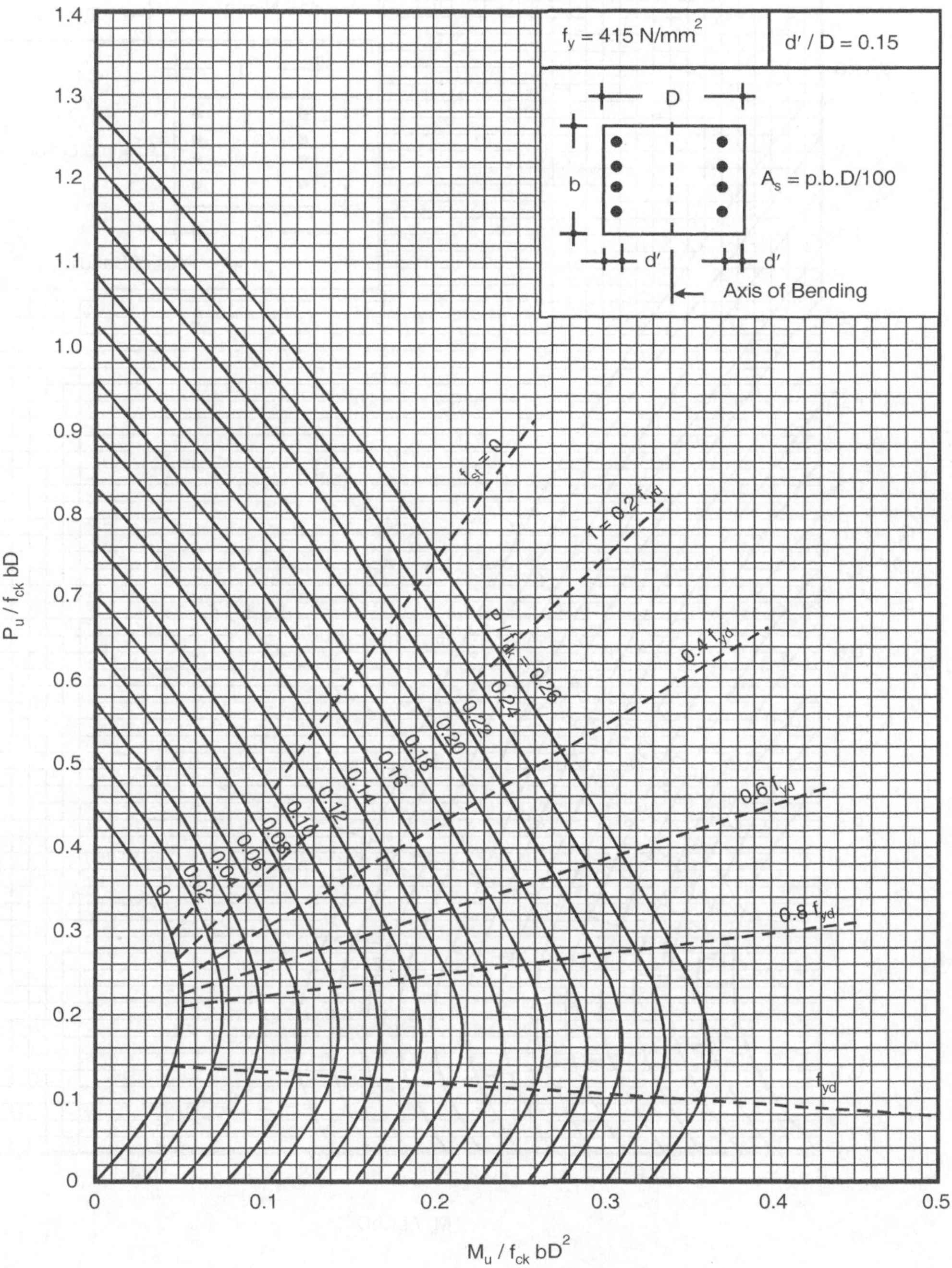

Chart 21.9. Compression with Bending – Rectangular Section – Reinforcement Distributed Equally on Two Sides

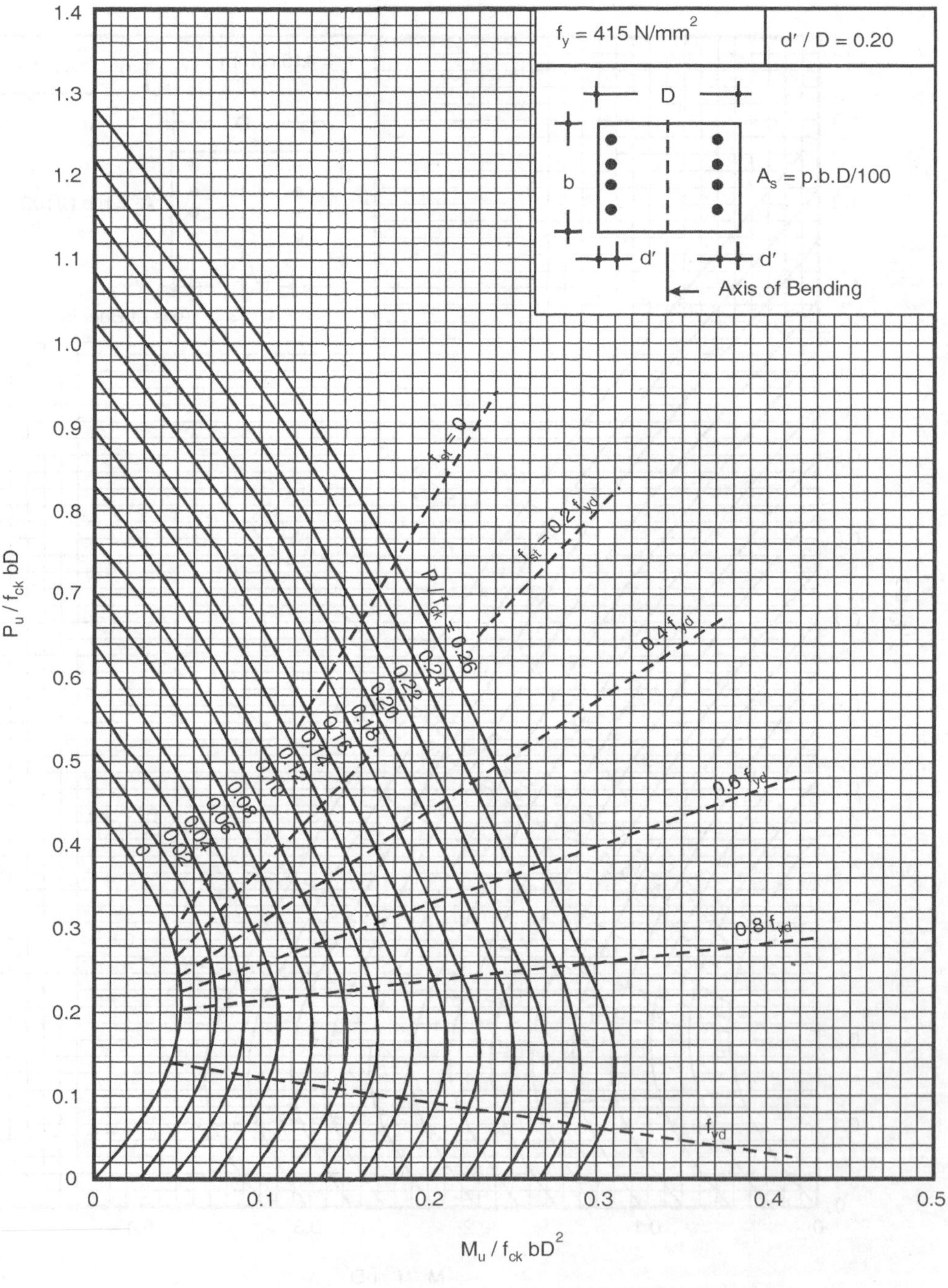

Chart 21.10. Compression with Bending – Rectangular Section – Reinforcement Distributed Equally on Two Sides

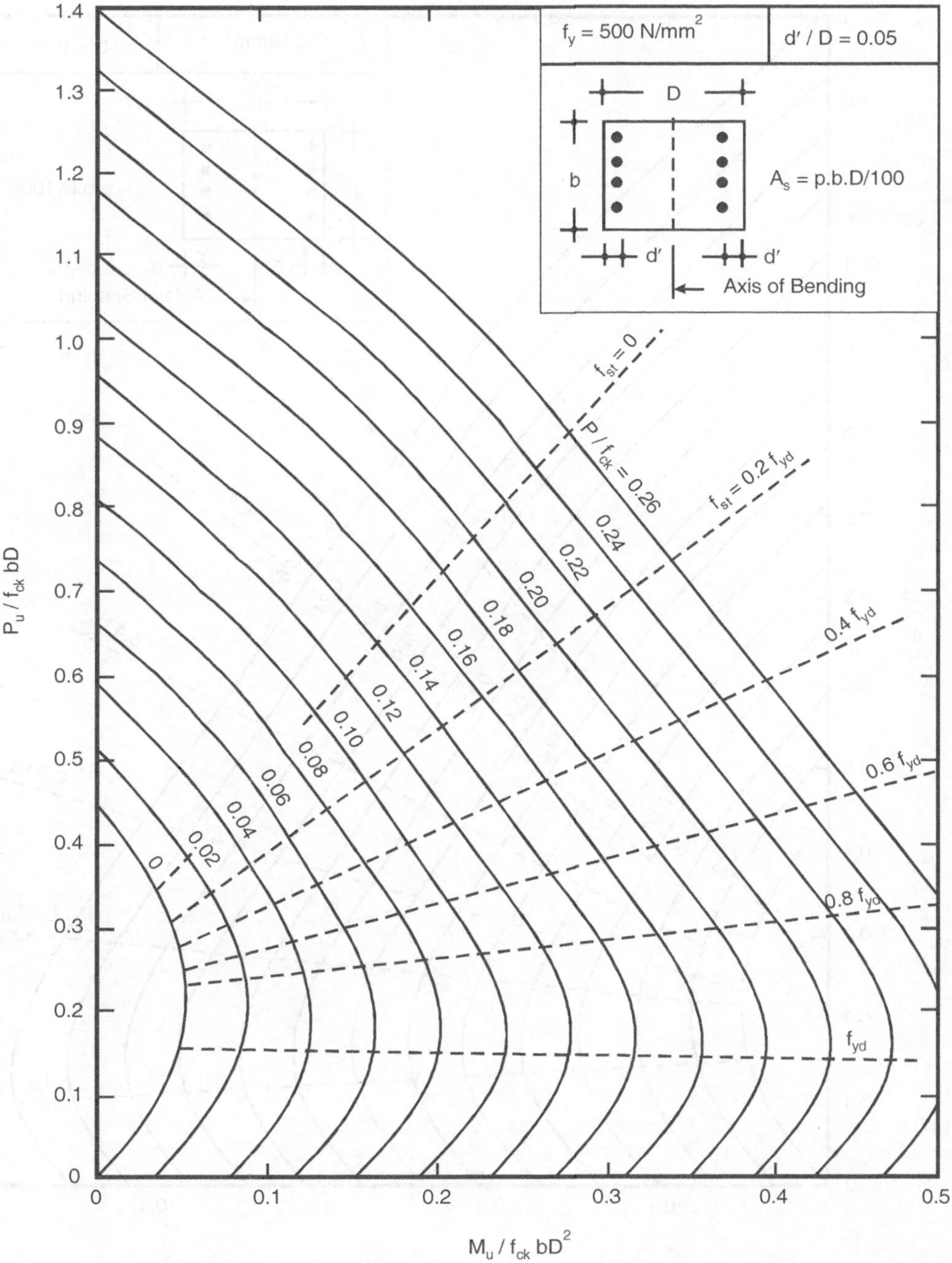

Chart 21.11. Compression with Bending – Rectangular Section – Reinforcement Distributed Equally on Two Sides

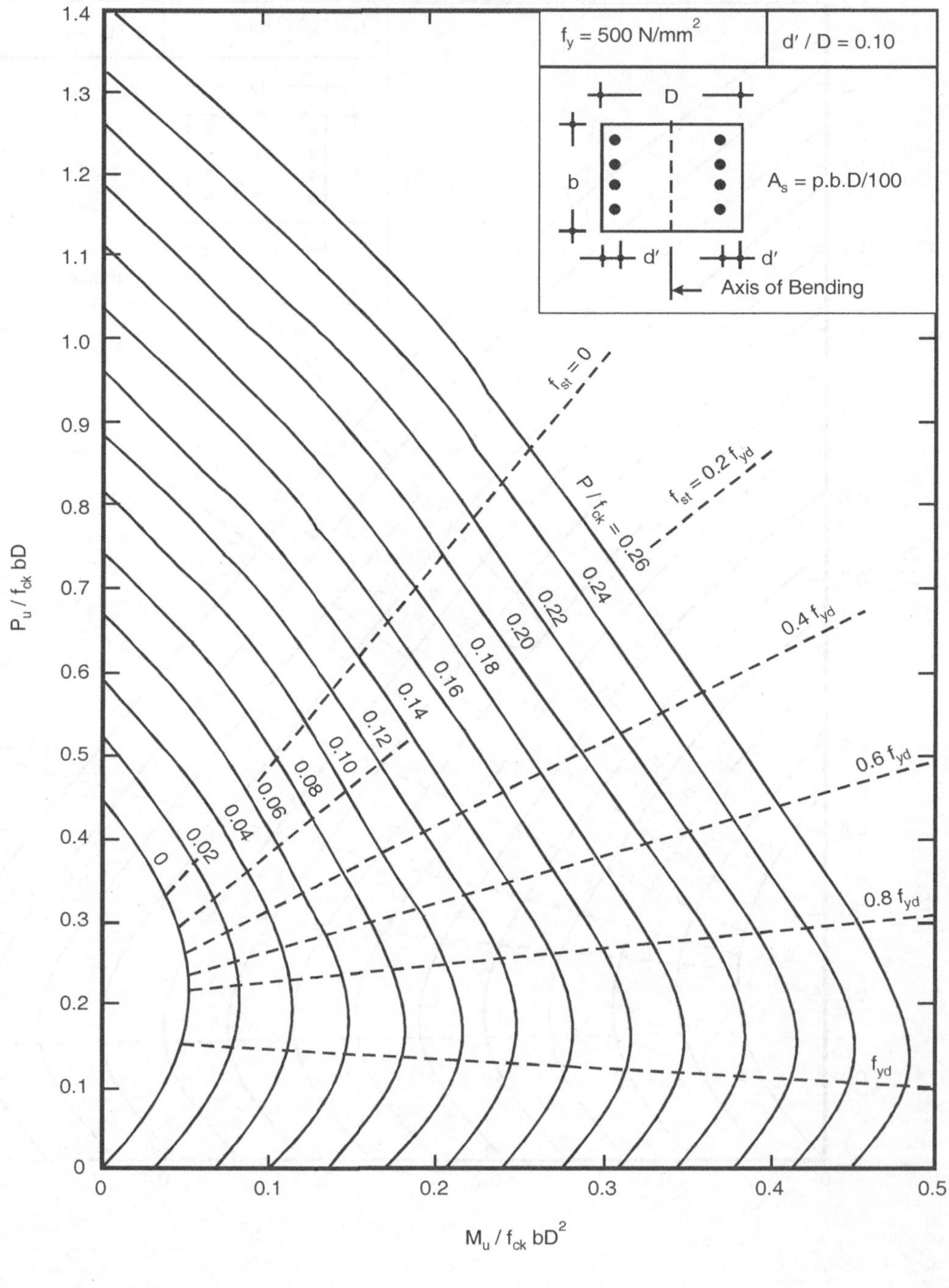

Chart 21.12. Compression with Bending – Rectangular Section – Reinforcement Distributed Equally on Two Sides

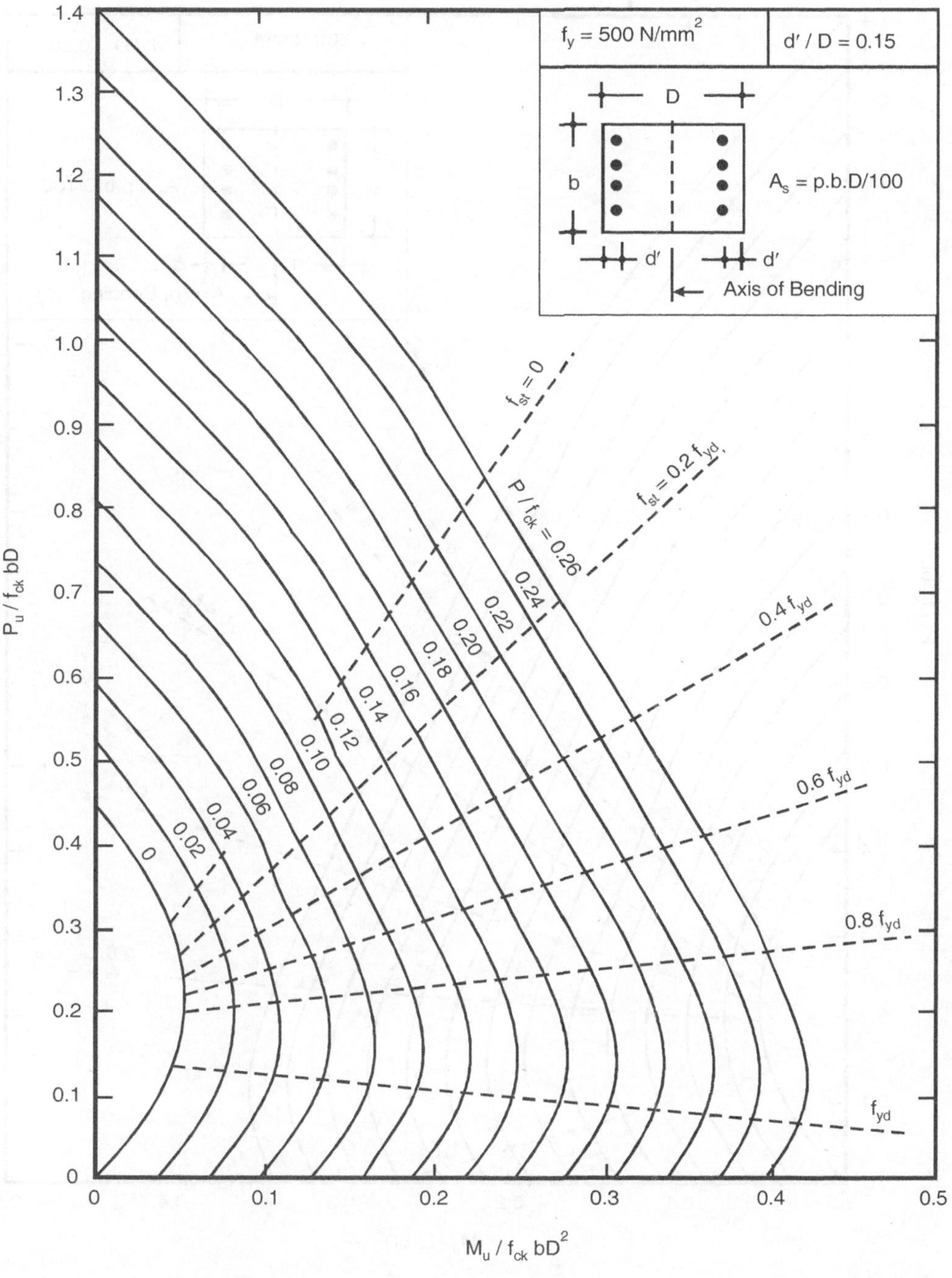

Chart 21.13. Compression with Bending – Rectangular Section – Reinforcement Distributed Equally on Two Sides

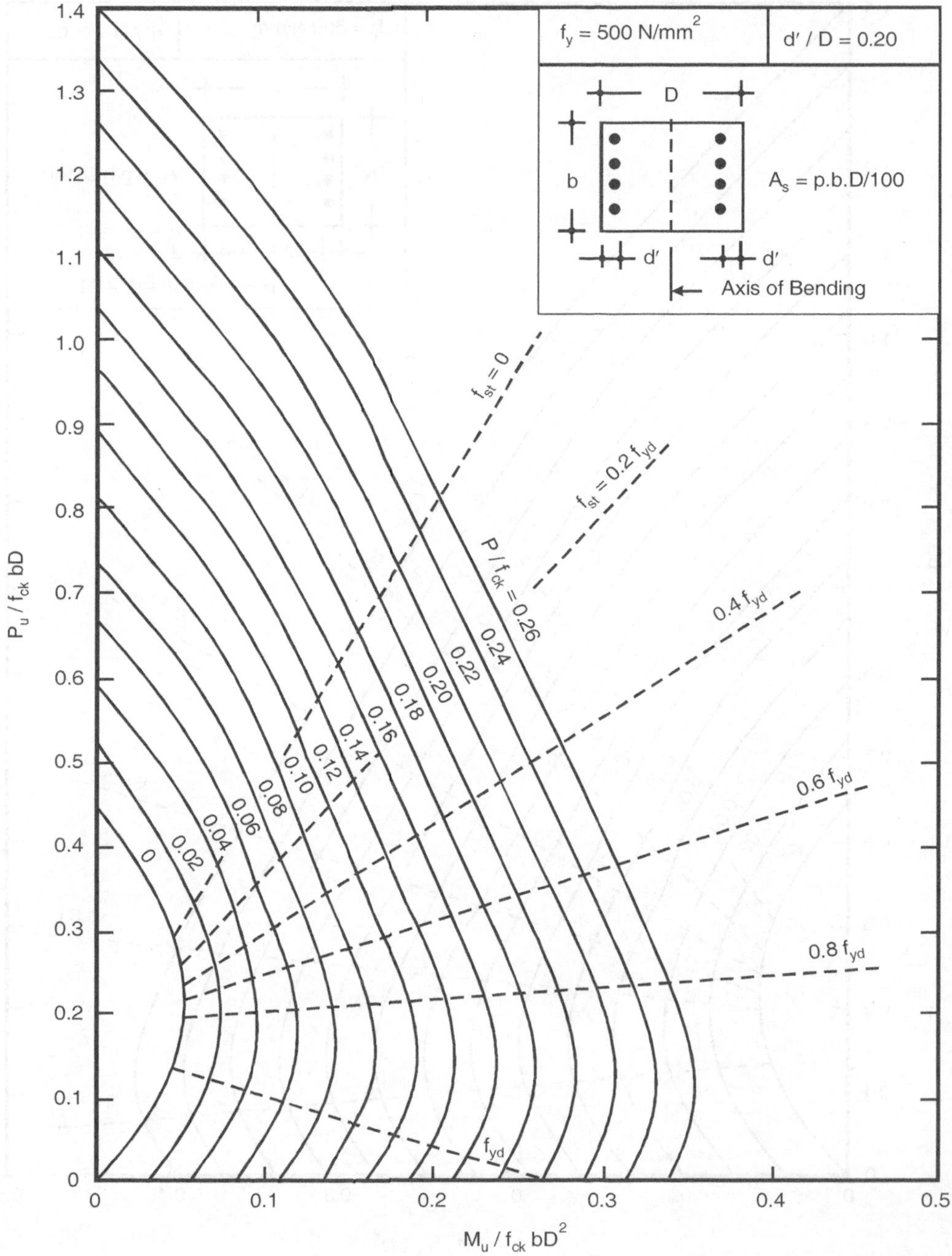

Chart 21.14. Compression with Bending – Rectangular Section – Reinforcement Distributed Equally on Four Sides

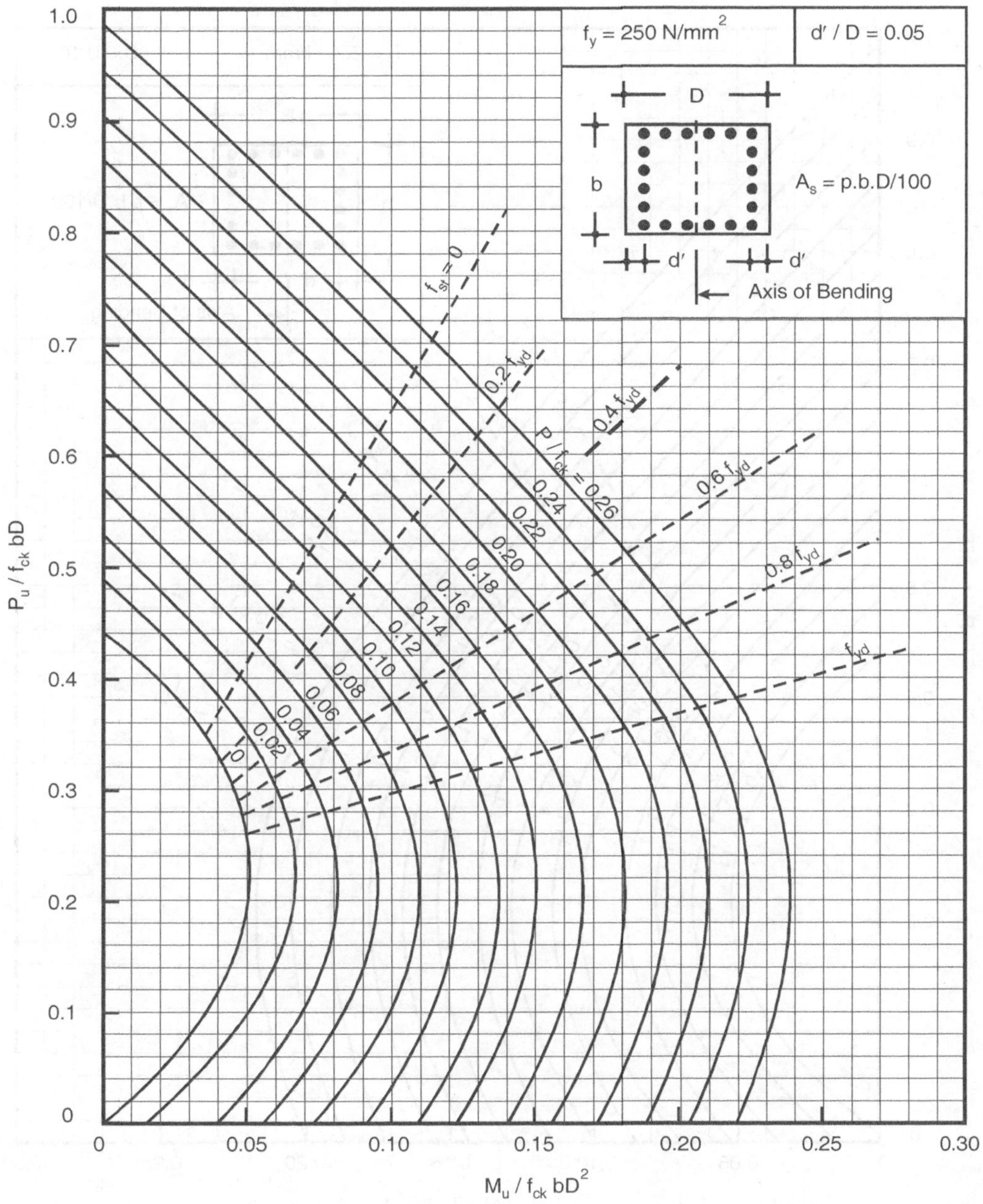

Chart 21.15. Compression with Bending – Rectangular Section – Reinforcement Distributed Equally on Four Sides

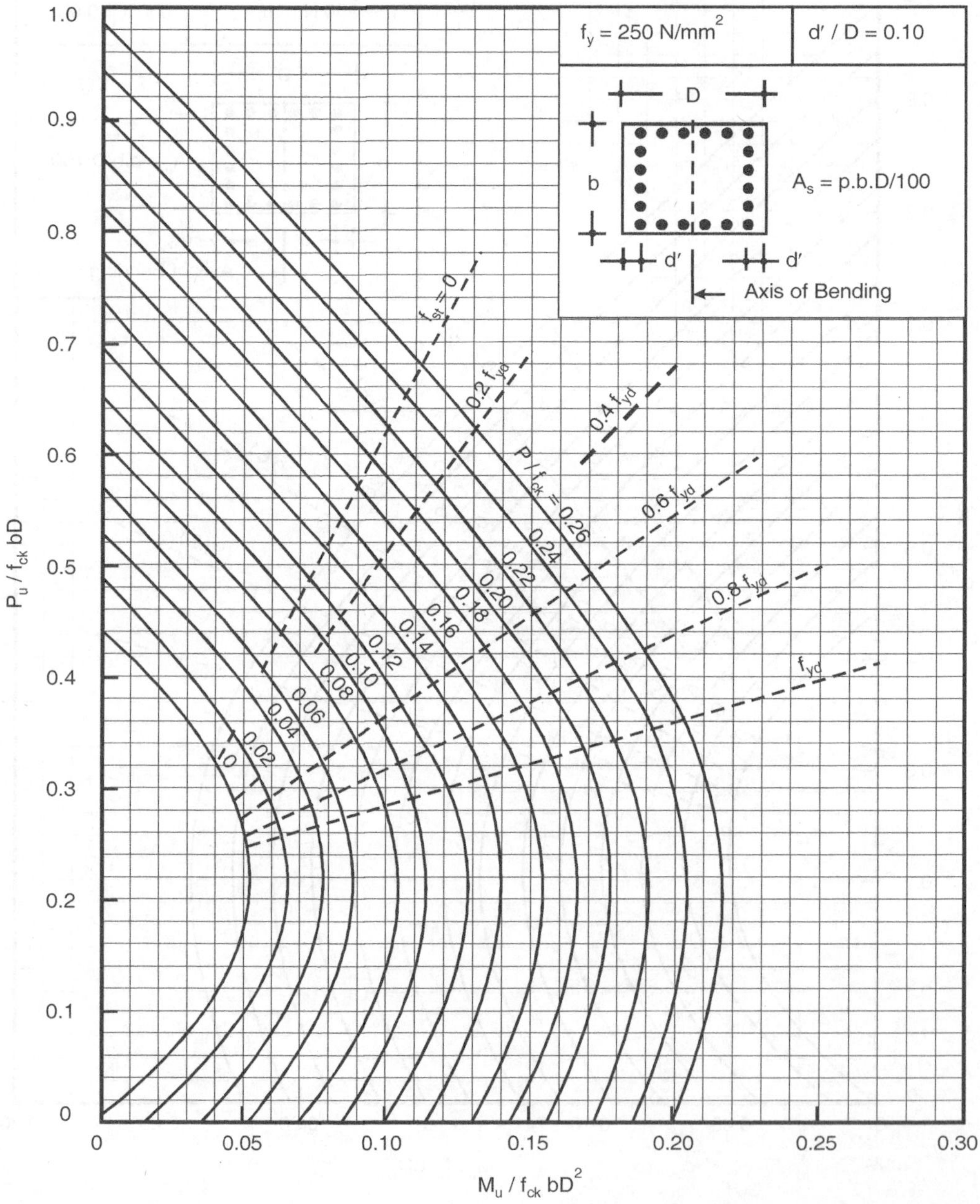

Chart 21.16. Compression with Bending – Rectangular Section – Reinforcement Distributed Equally on Four Sides

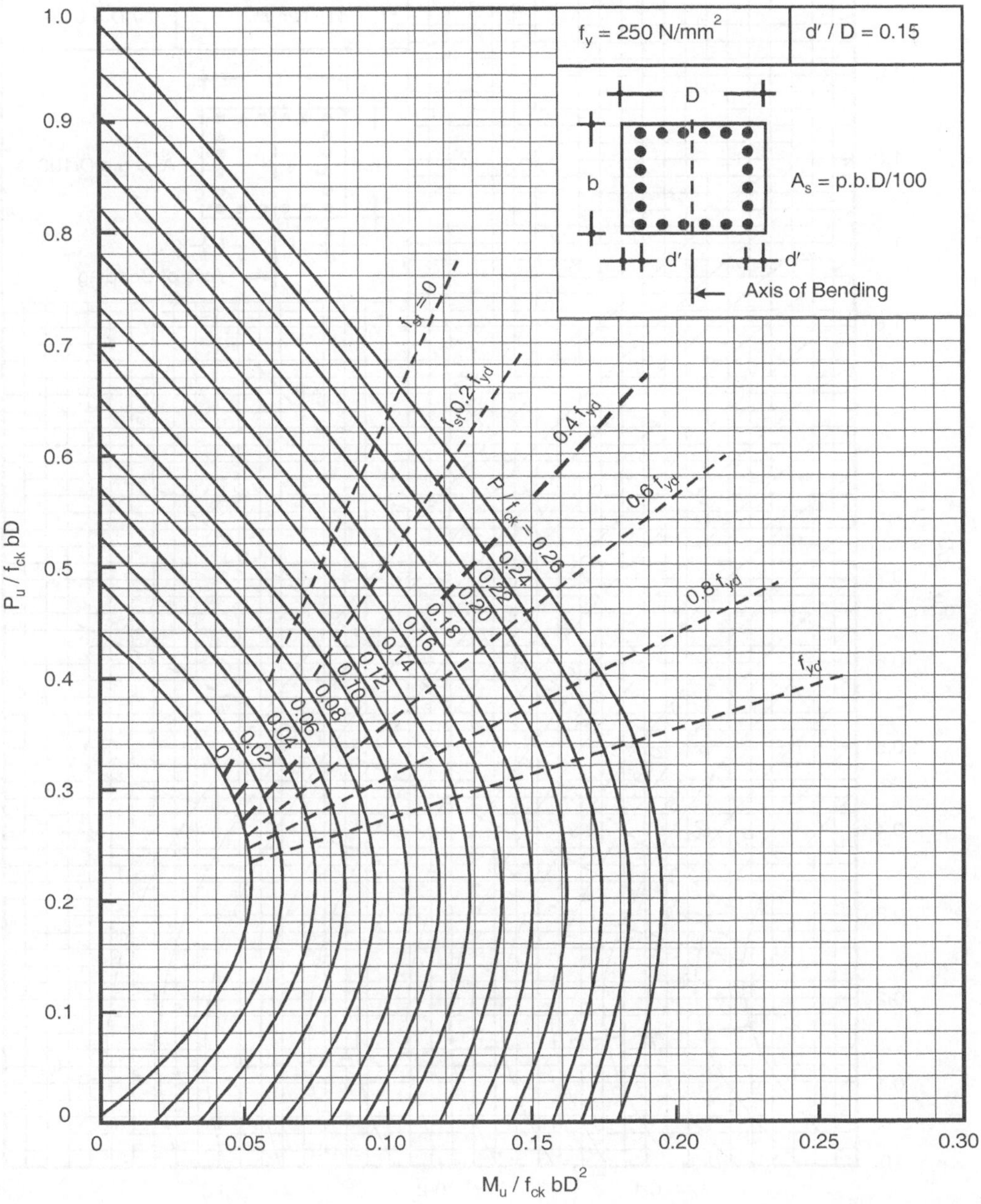

Chart 21.17. Compression with Bending – Rectangular Section – Reinforcement Distributed Equally on Four Sides

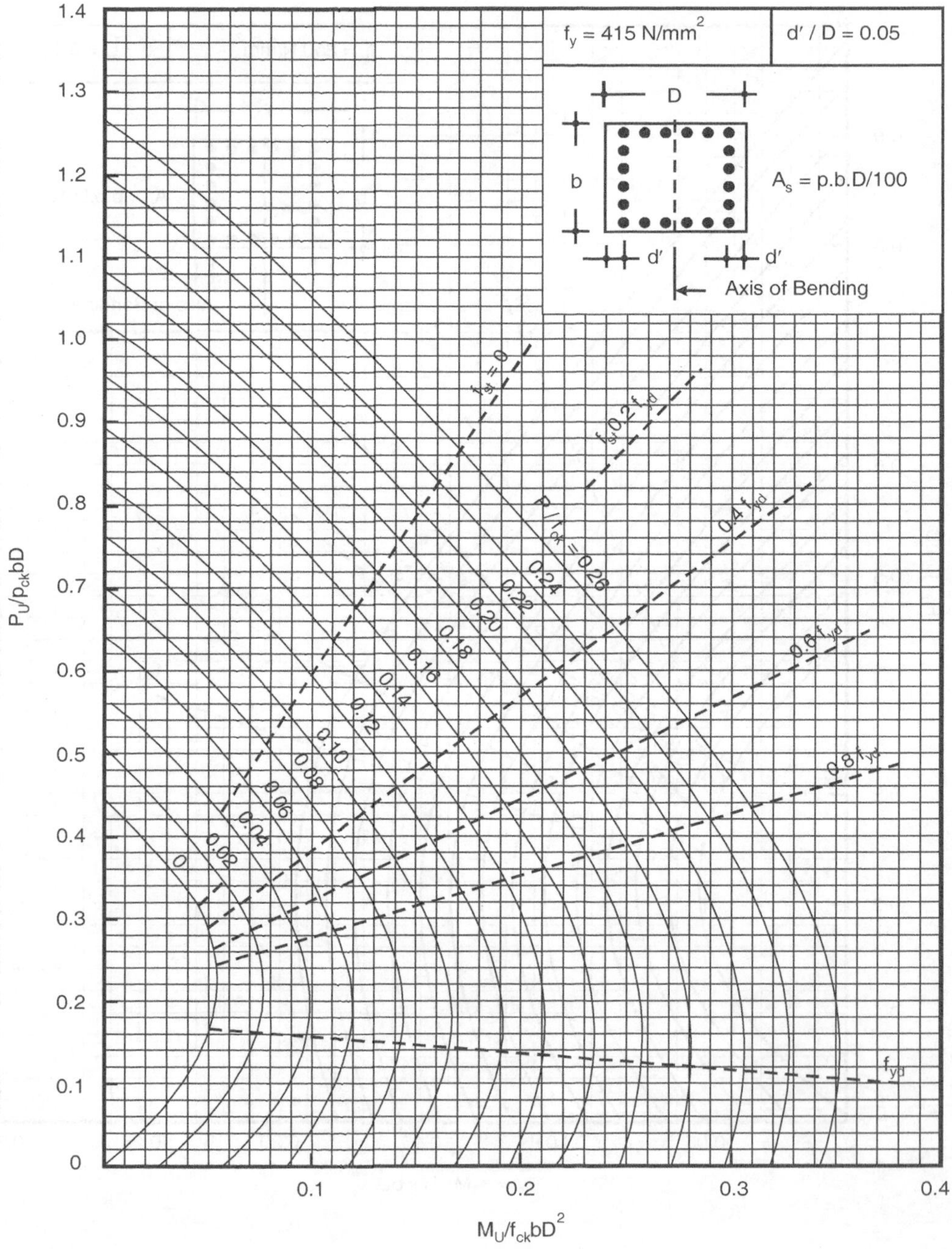

Chart 21.18. Compression with Bending – Rectangular Section – Reinforcement Distributed Equally on Four Sides

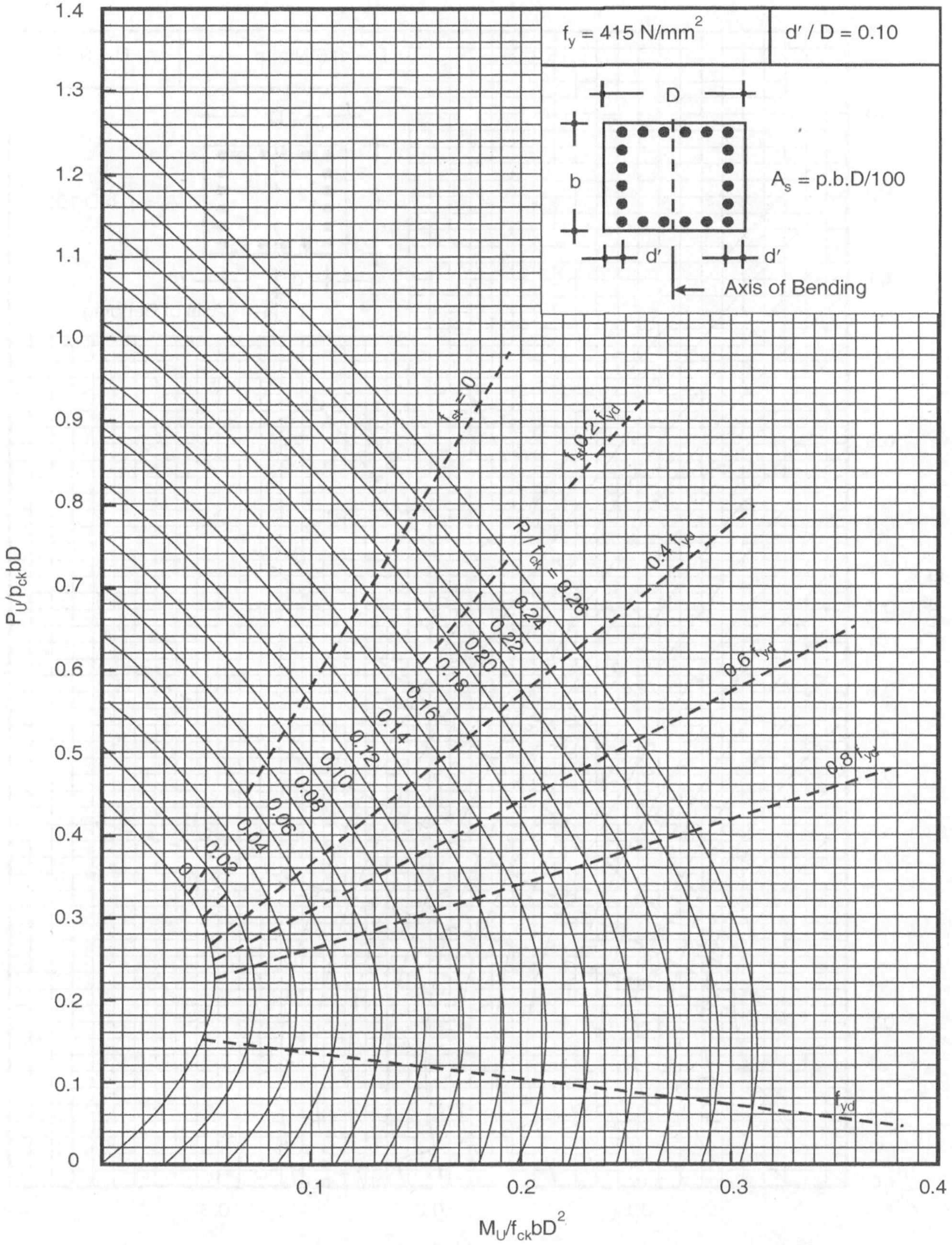

Chart 21.19. Compression with Bending – Rectangular Section – Reinforcement Distributed Equally on Four Sides

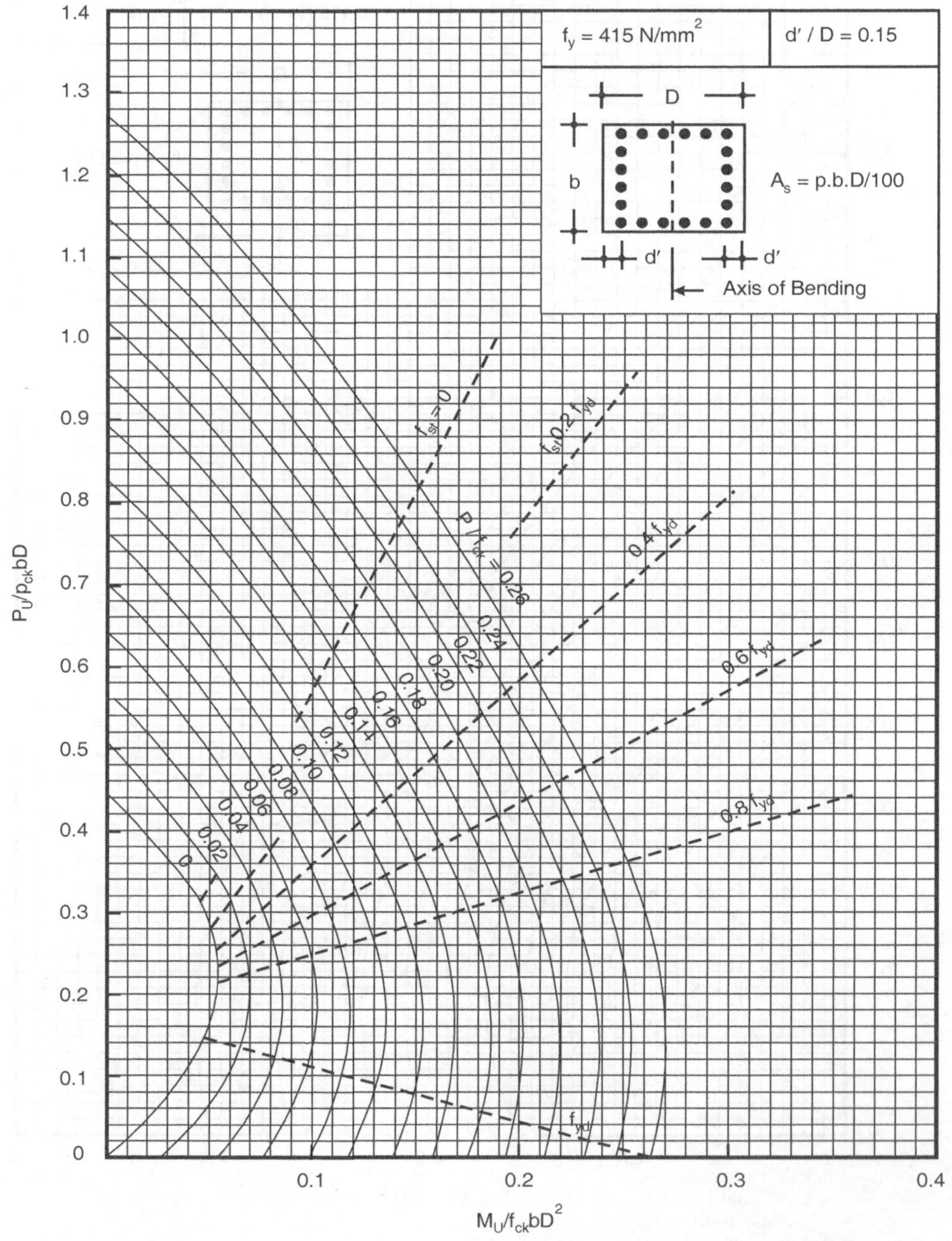

Chart 21.20. Compression with Bending – Rectangular Section – Reinforcement Distributed Equally on Four Sides

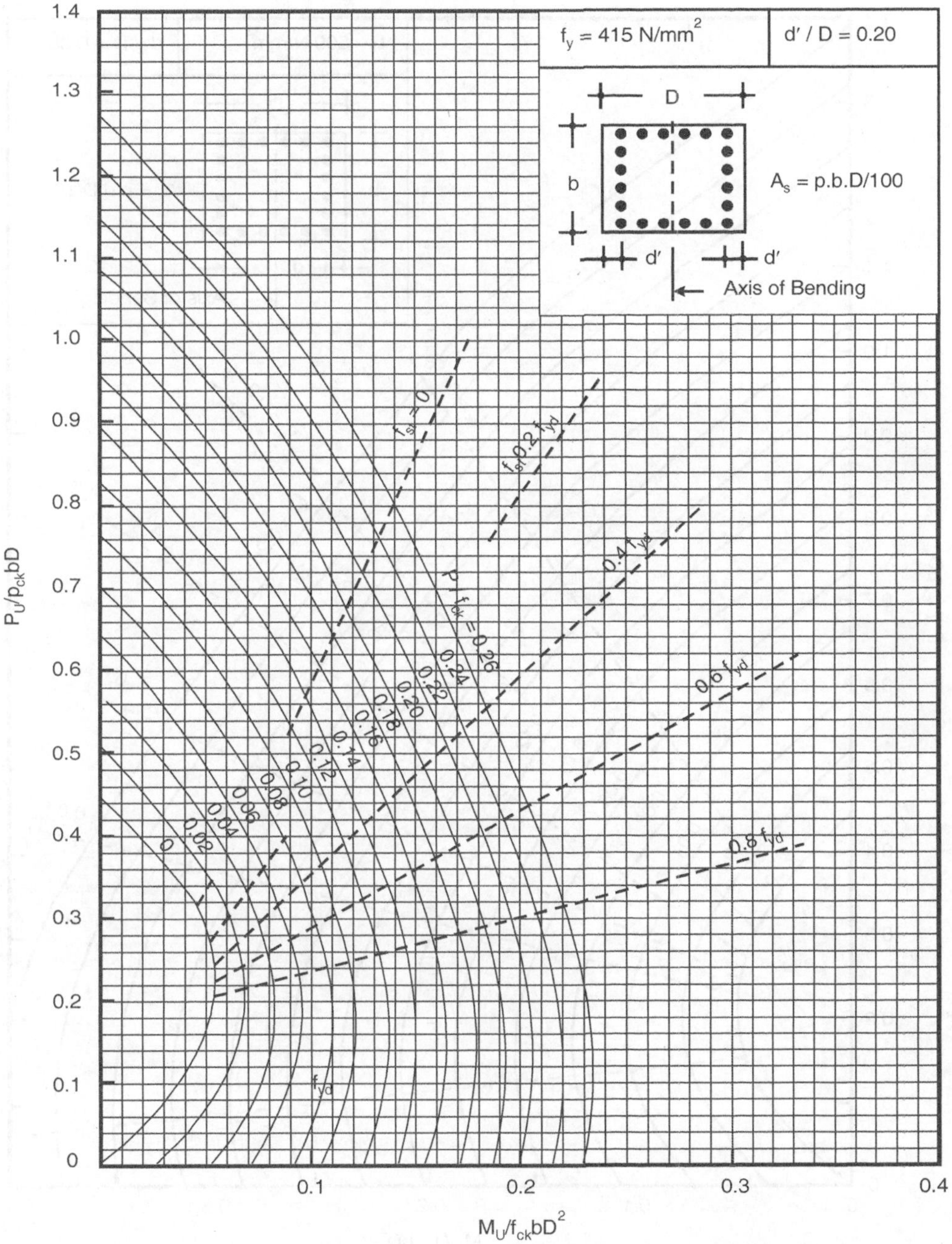

Chart 21.21. Compression with Bending – Rectangular Section – Reinforcement Distributed Equally on Four Sides

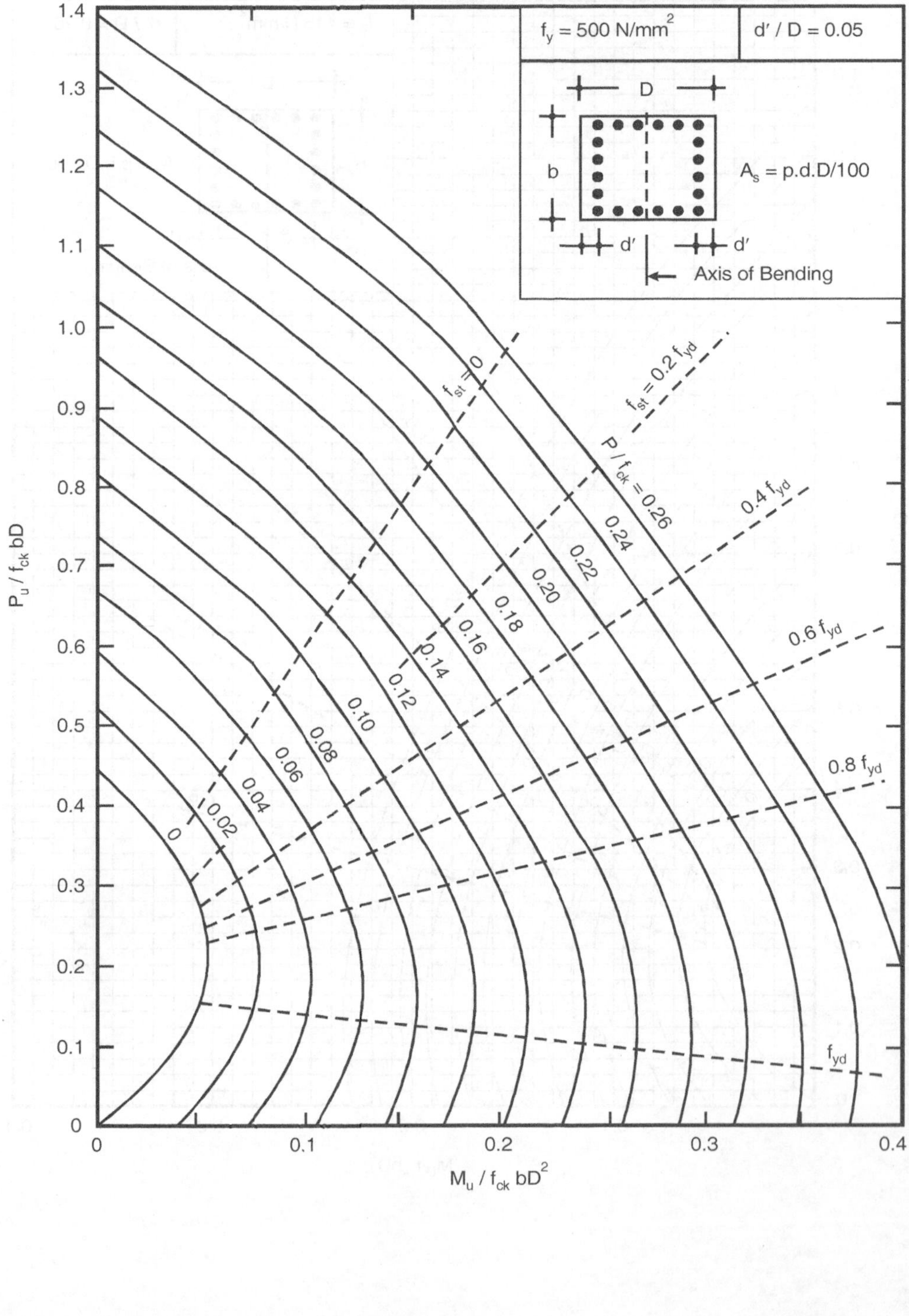

Chart 21.22. Compression with Bending – Rectangular Section – Reinforcement Distributed Equally on Four Sides

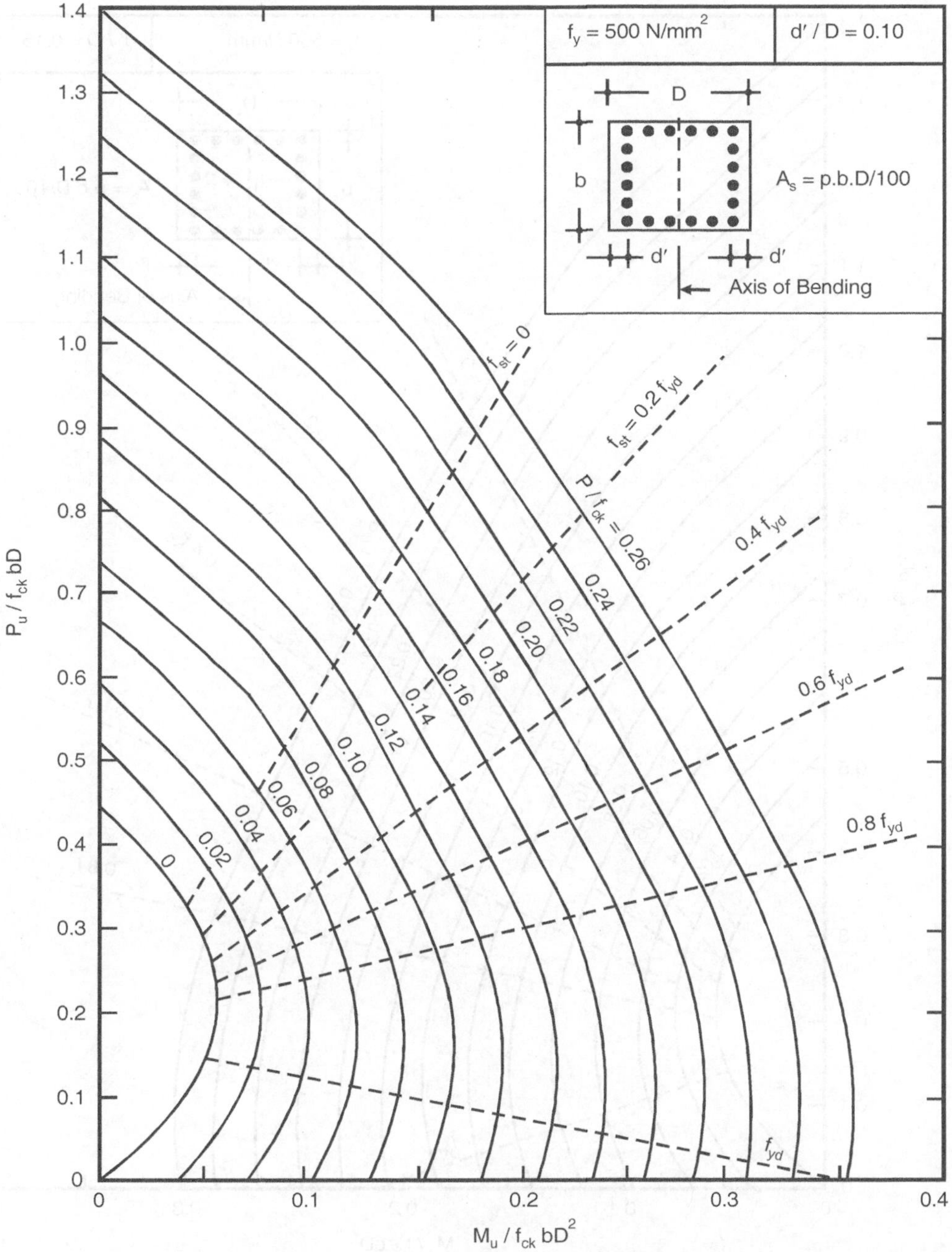

Chart 21.23. Compression with Bending – Rectangular Section – Reinforcement Distributed Equally on Four Sides

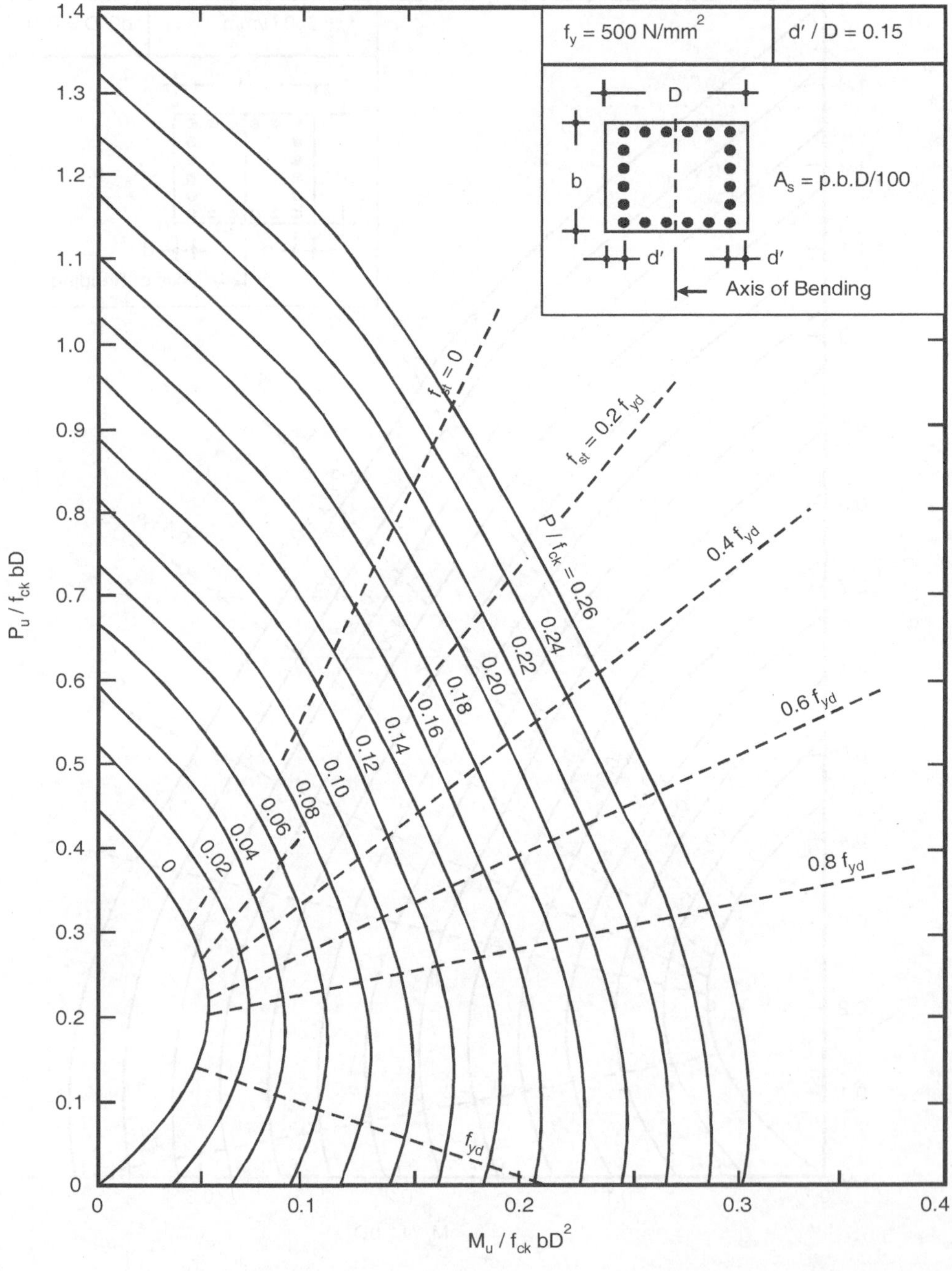

Chart 21.24. Compression with Bending – Rectangular Section – Reinforcement Distributed Equally on Four Sides

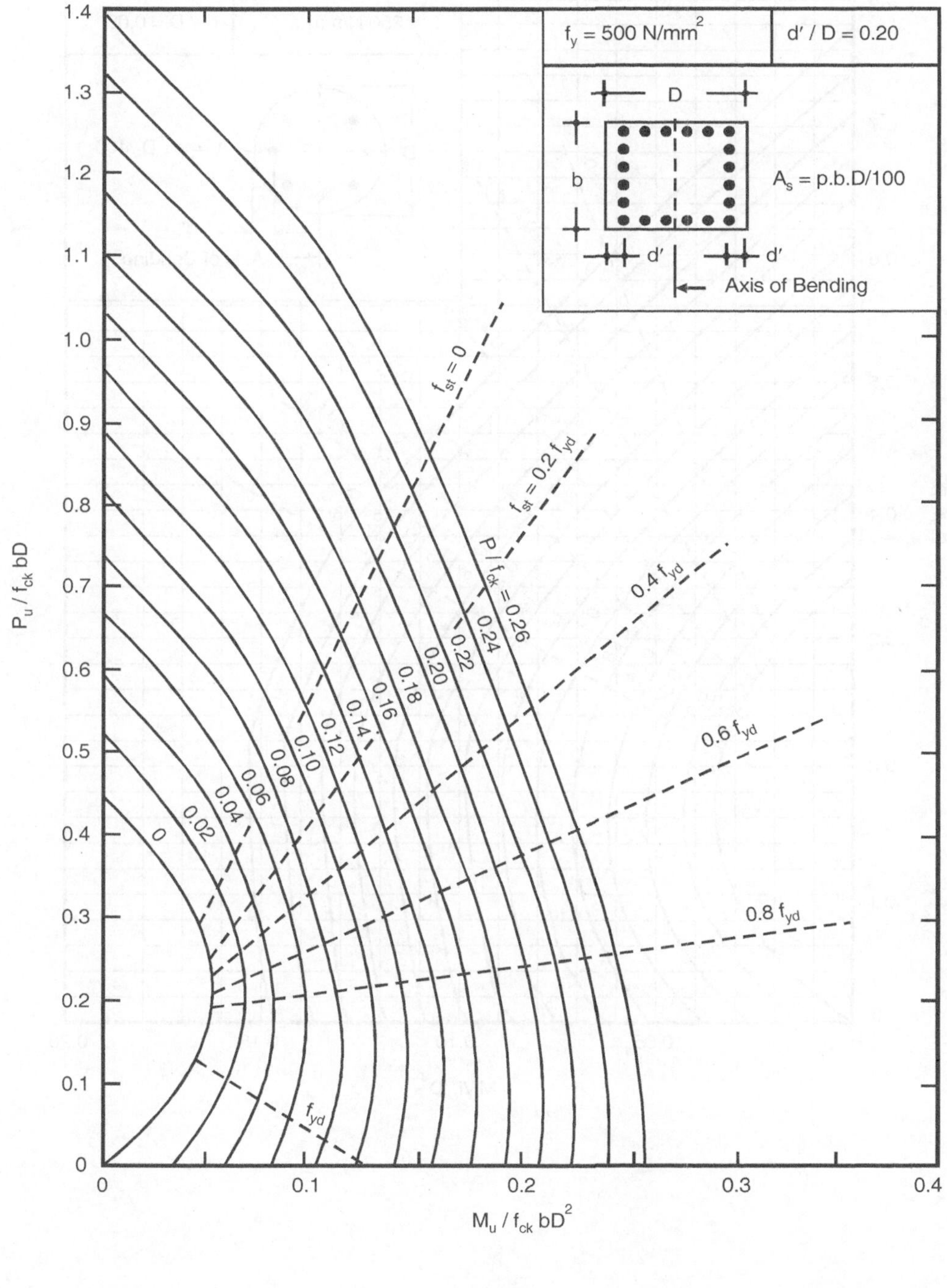

Chart 21.25. Compression with Bending – Circular Section

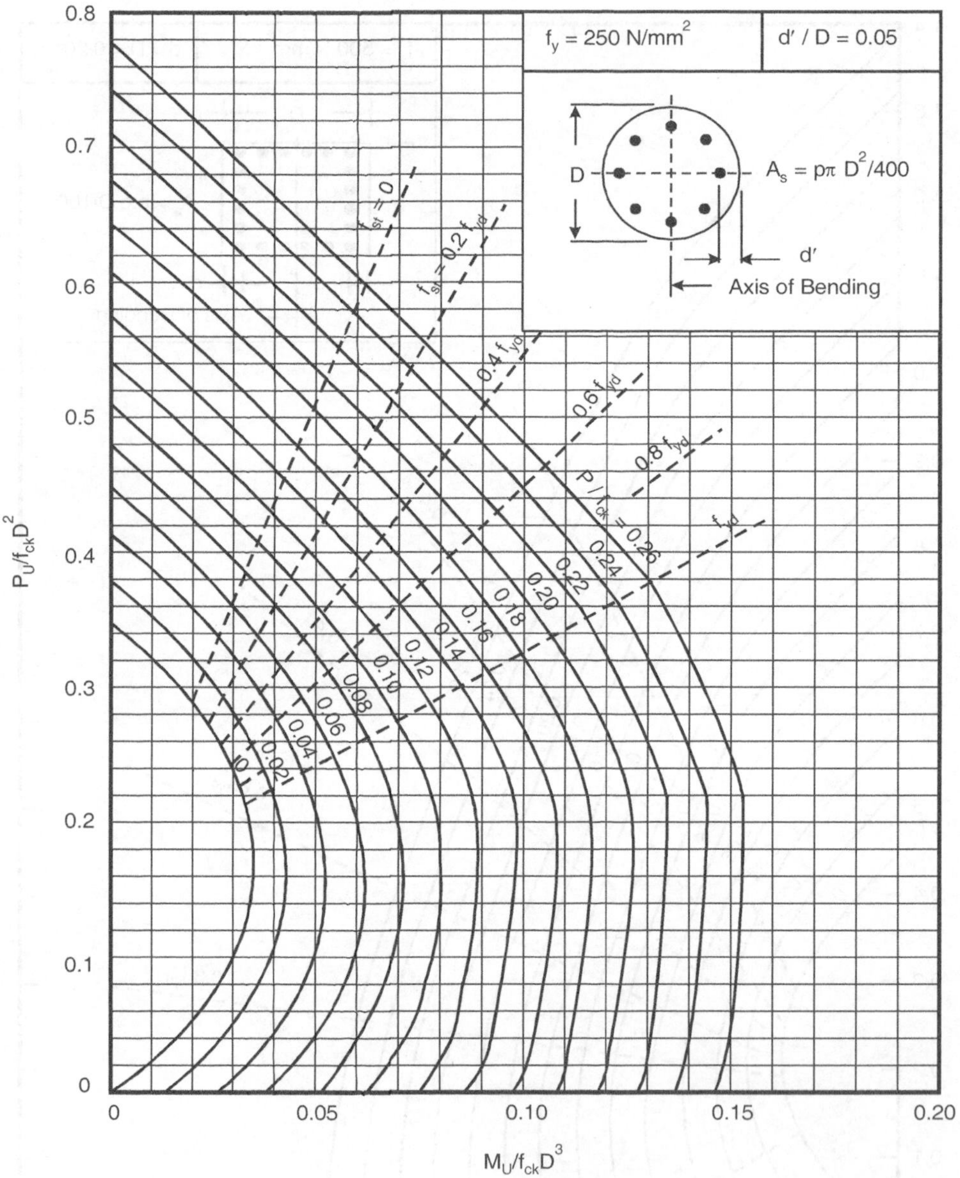

Chart 21.26. Compression with Bending – Circular Section

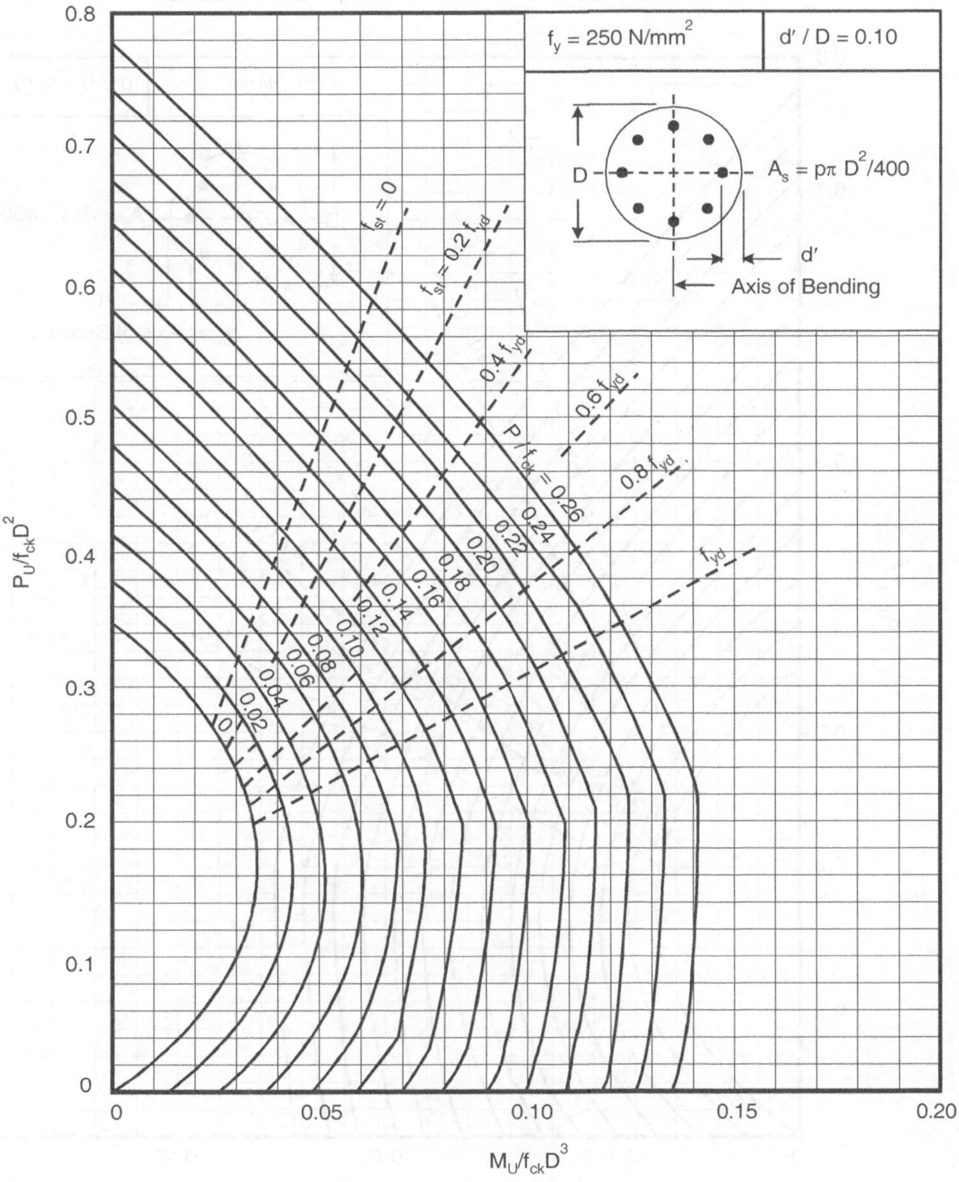

Chart 21.27. Compression with Bending – Circular Section

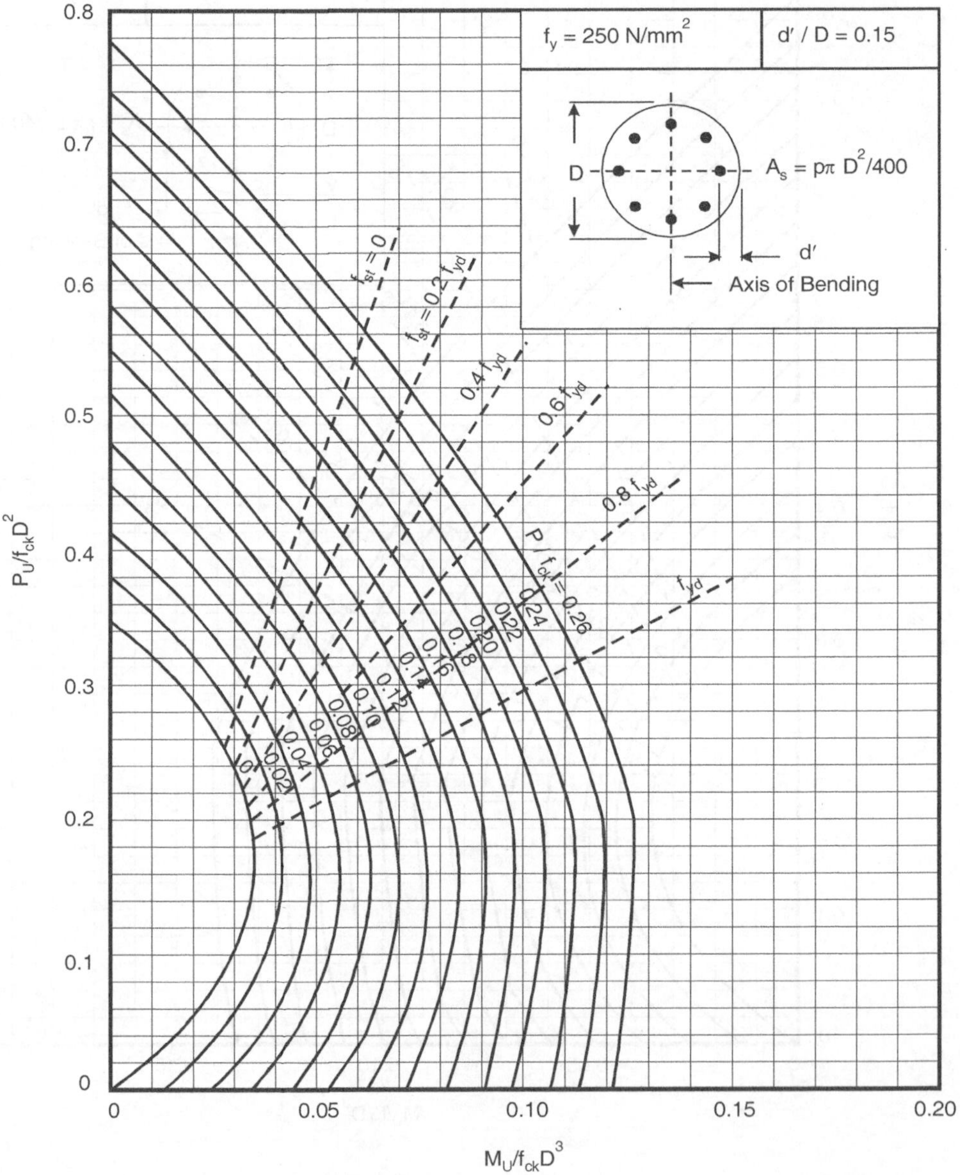

Chart 21.28. Compression with Bending – Circular Section

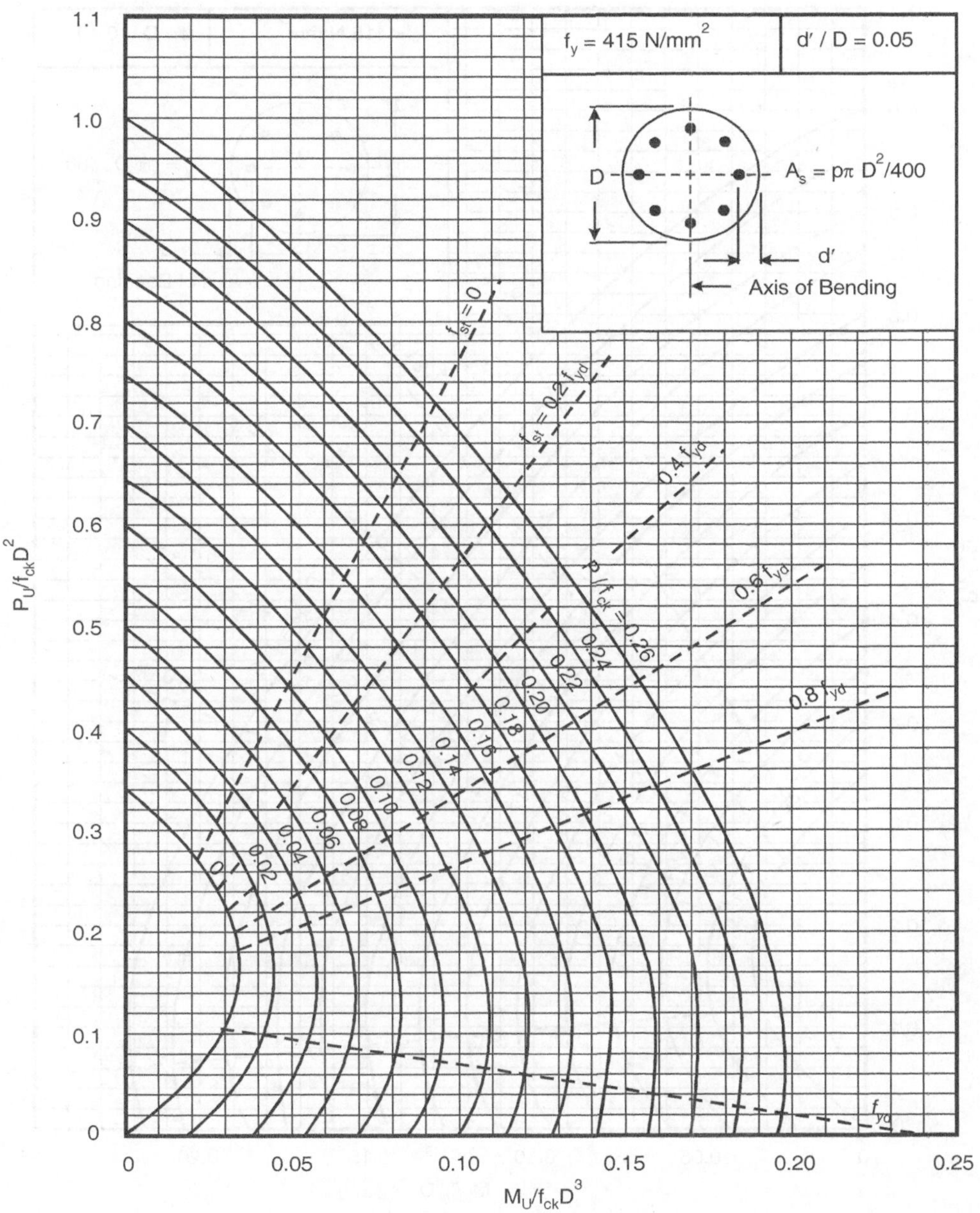

Chart 21.29. Compression with Bending – Circular Section

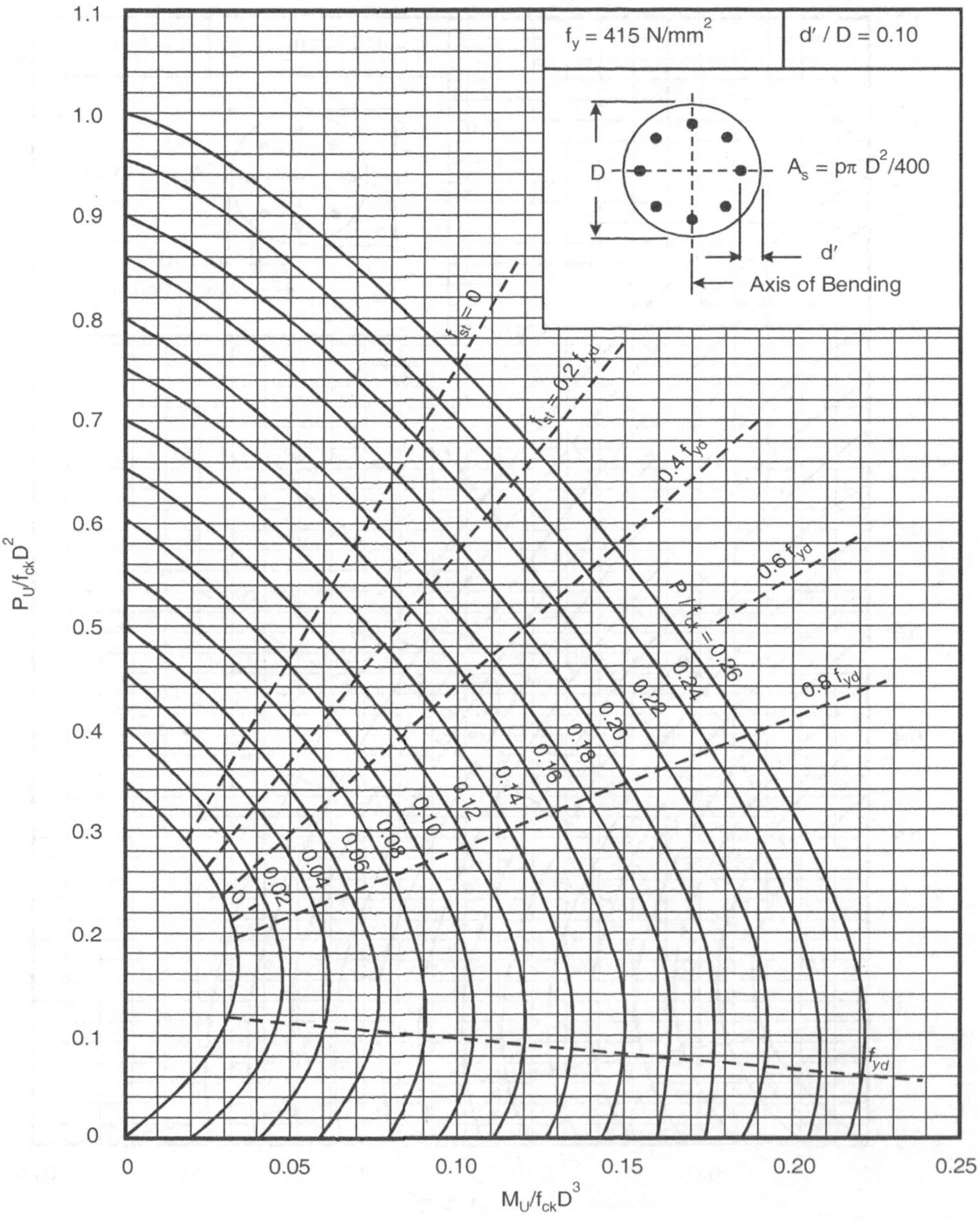

Chart 21.30. Compression with Bending – Circular Section

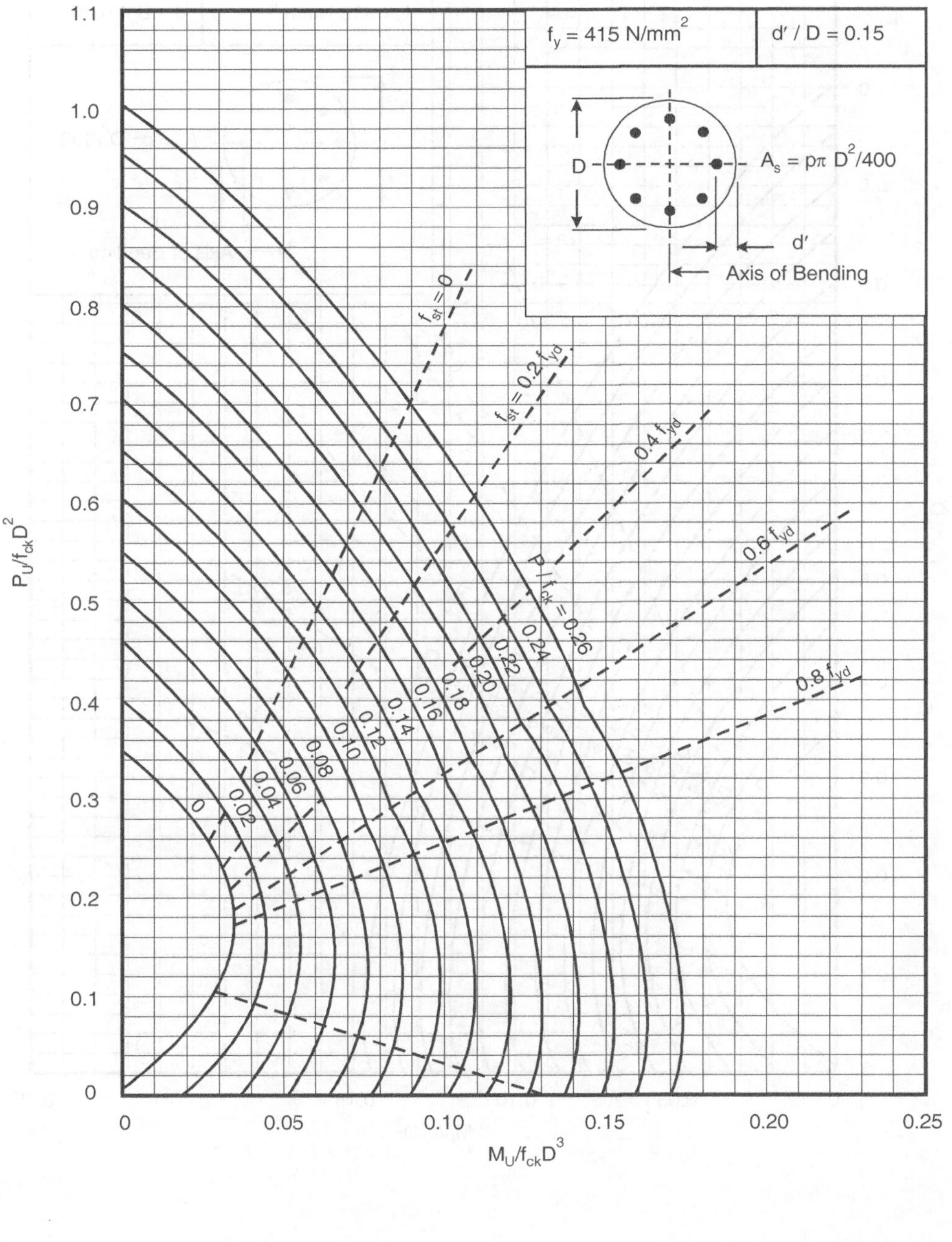

Chart 21.31. Compression with Bending – Circular Section

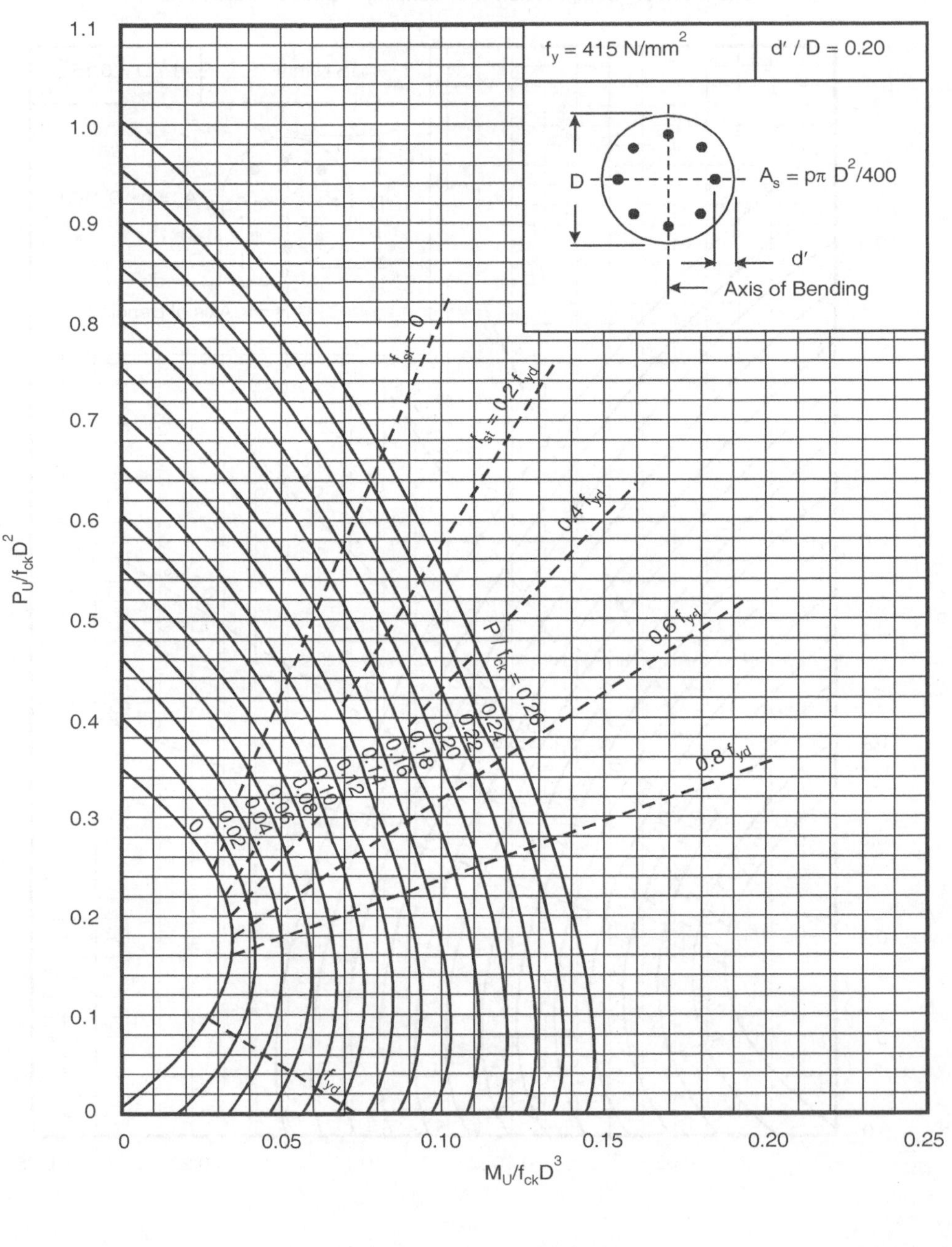

Chart 21.32. Compression with Bending – Circular Section

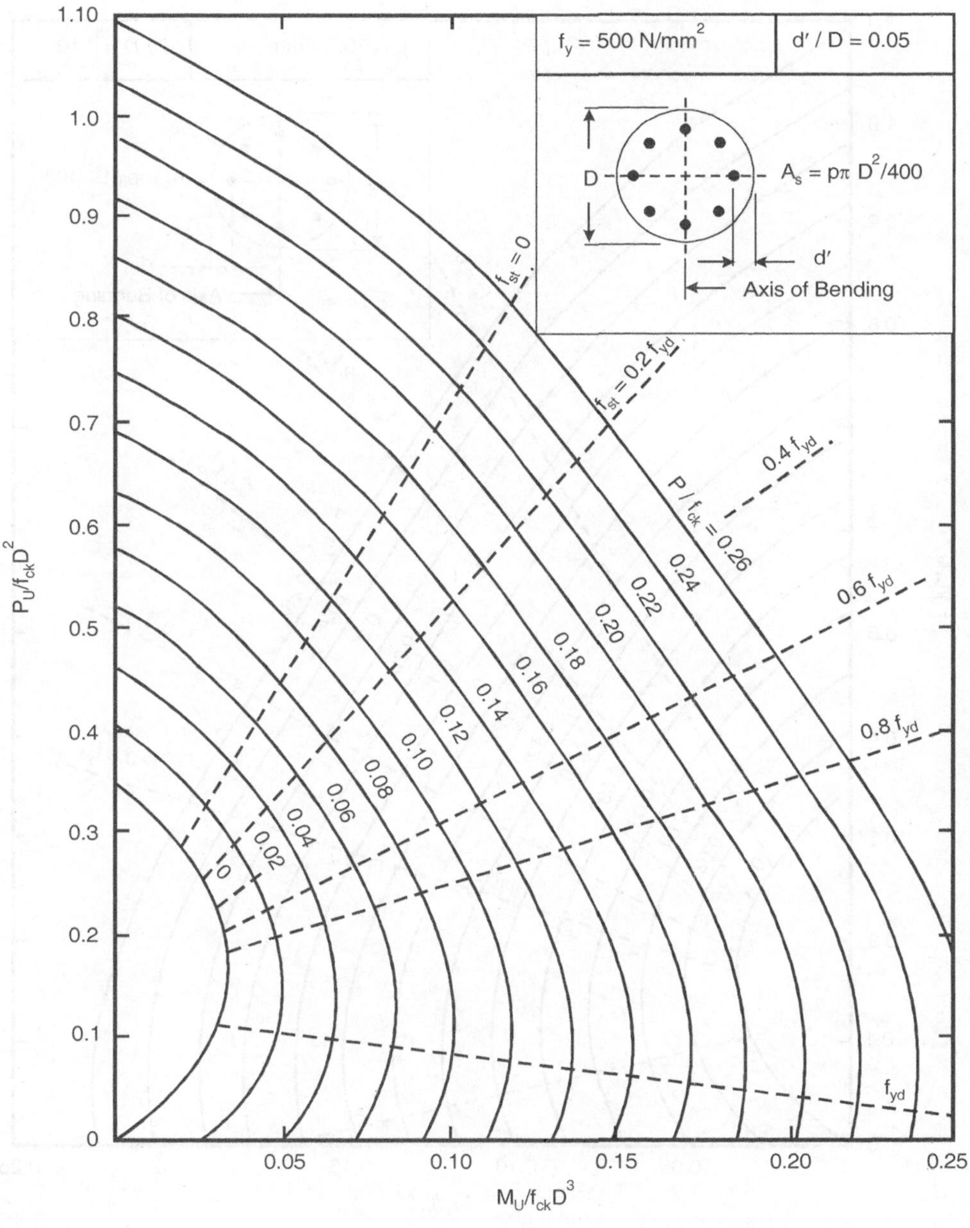

Chart 21.33. Compression with Bending – Circular Section

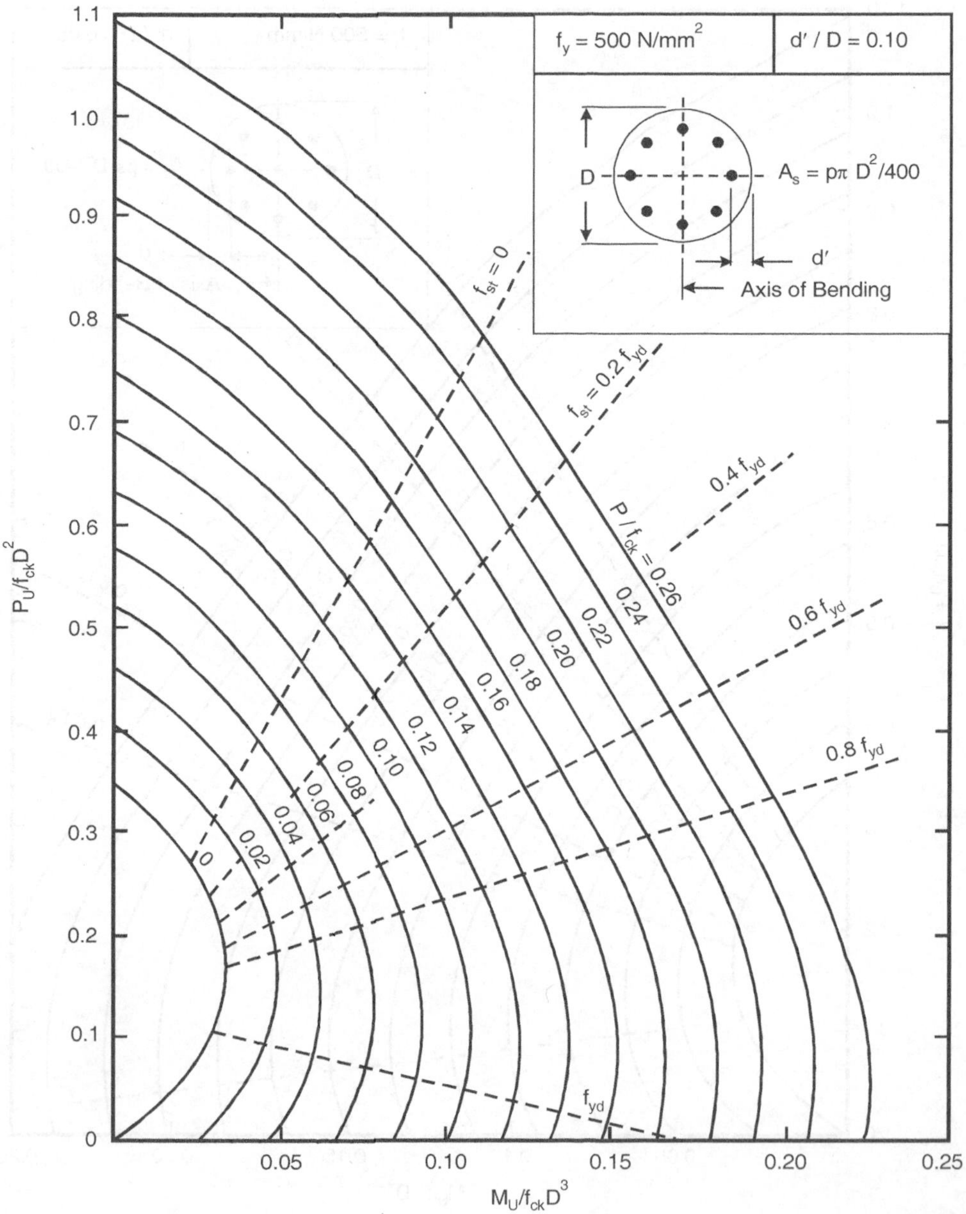

Chart 21.34. Compression with Bending – Circular Section

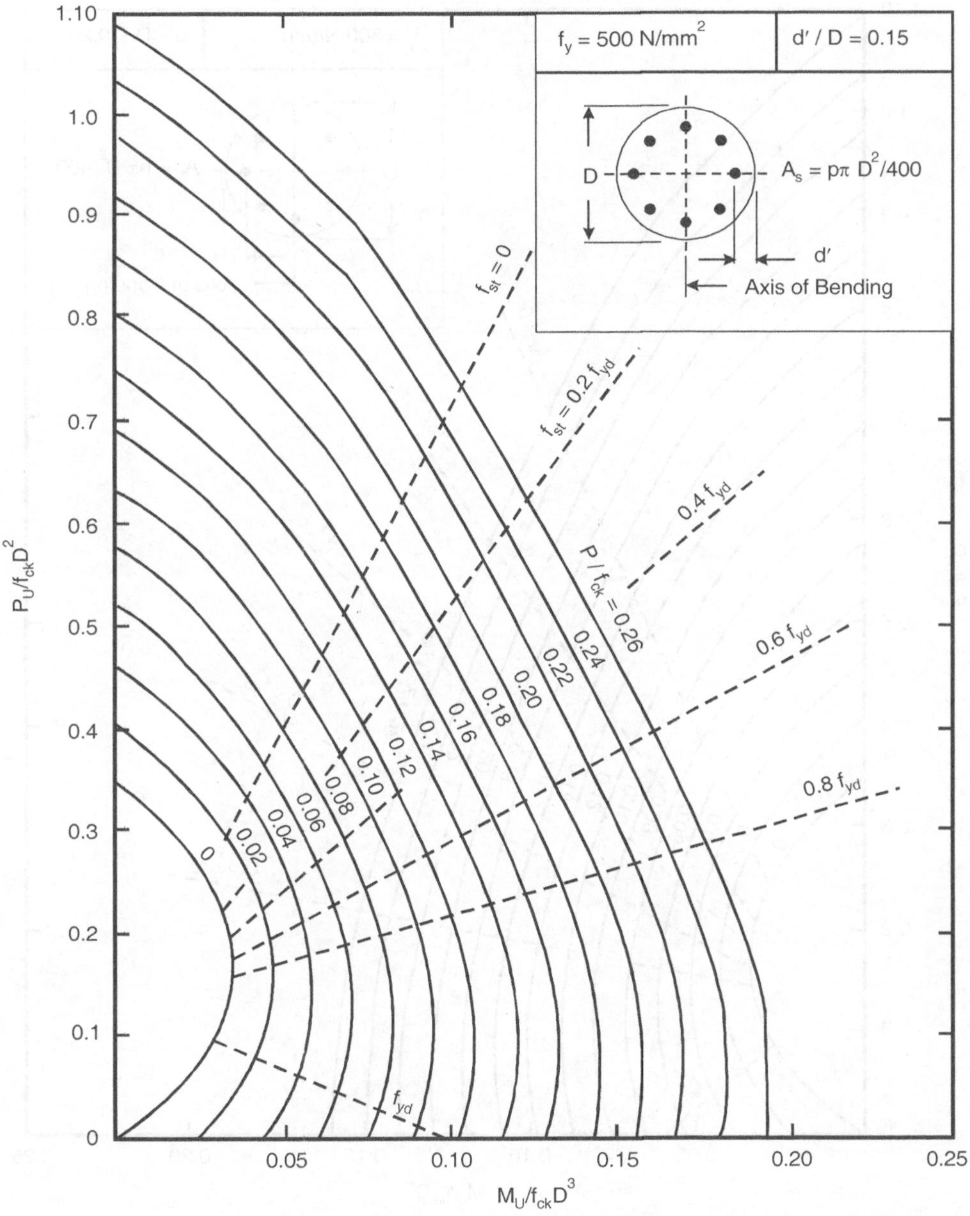

Chart 21.35. Compression with Bending – Circular Section

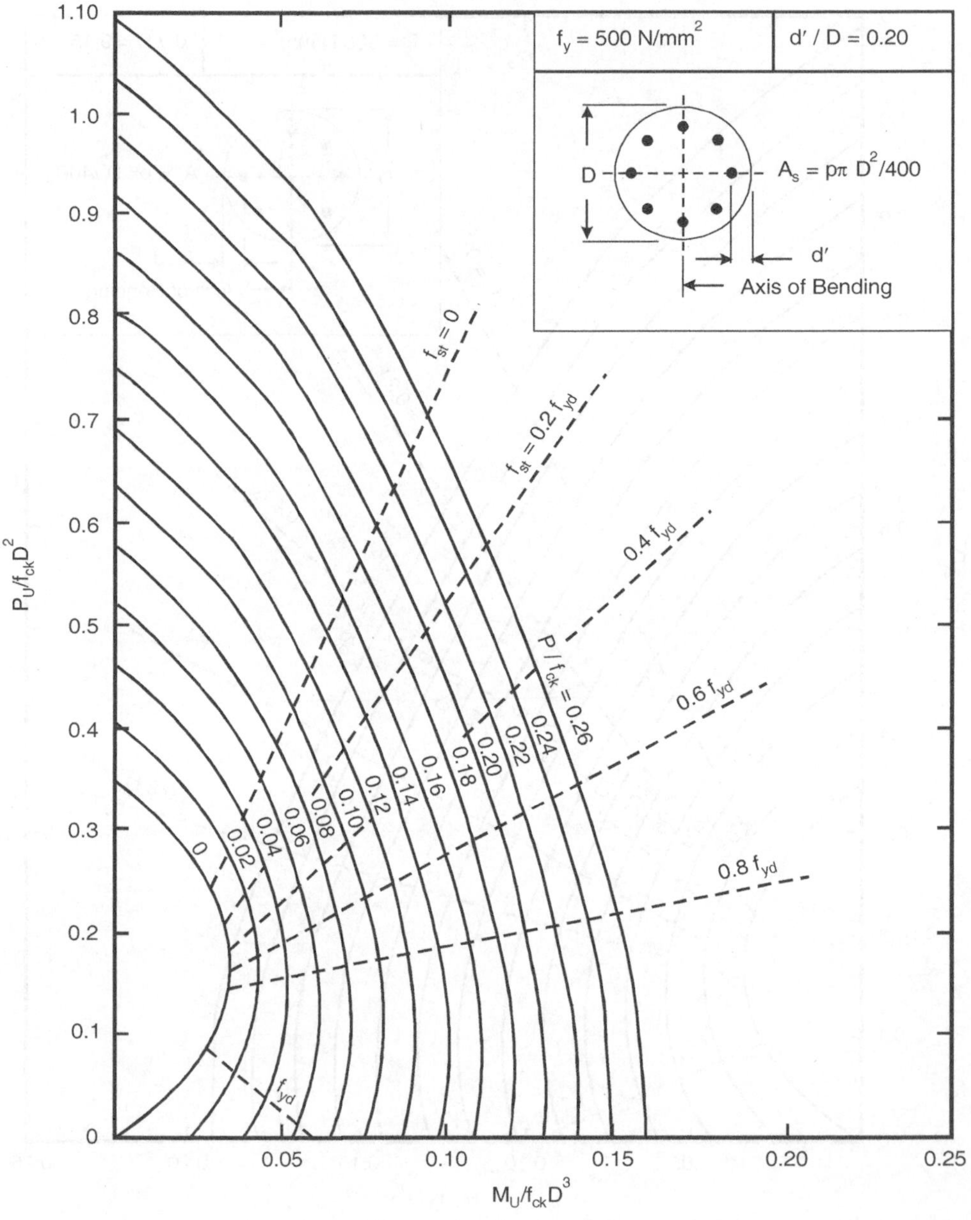

Chart 21.36. Values of P_{uz} for Compression Members

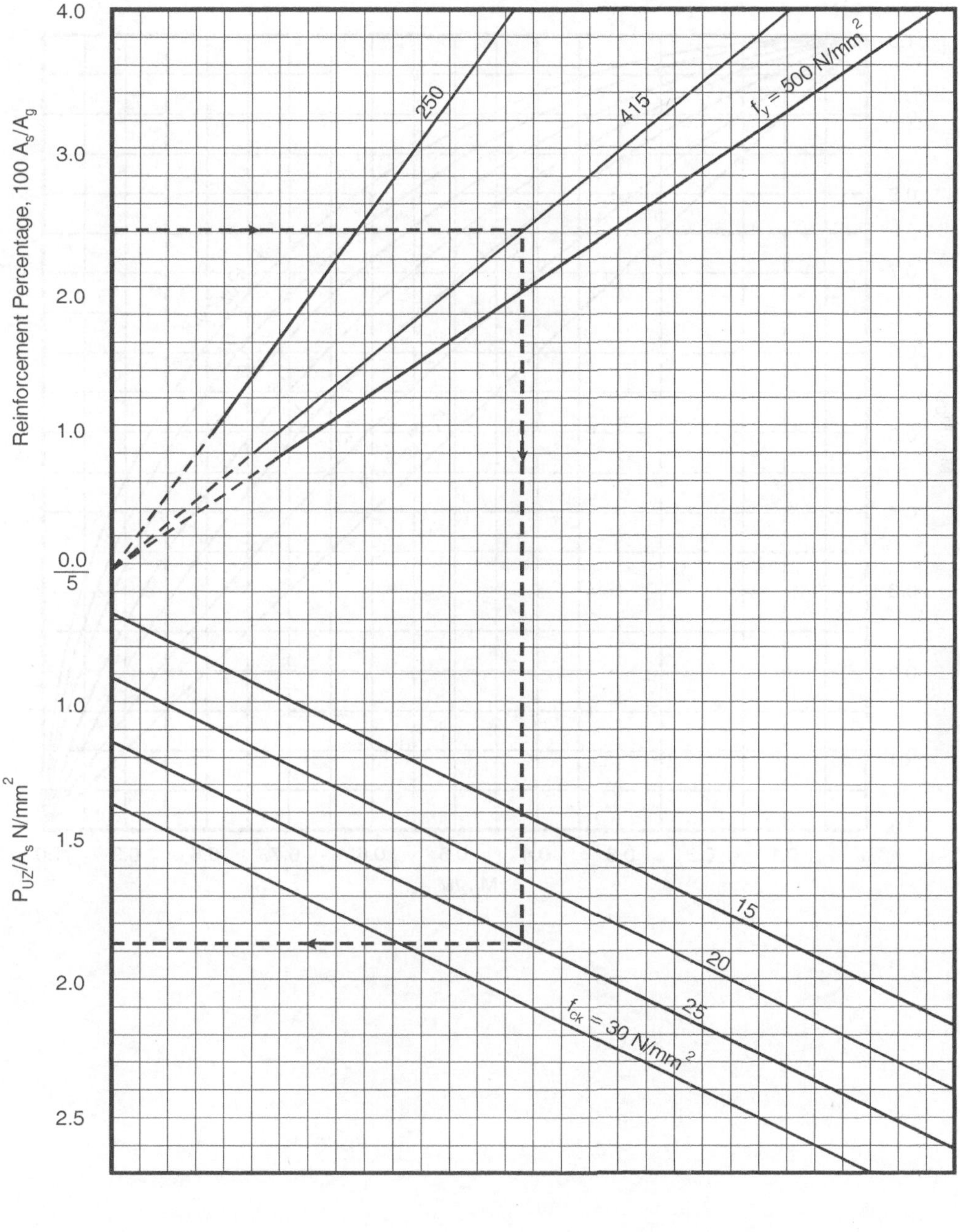

Chart 21.37. Biaxial Bending in Compression Members

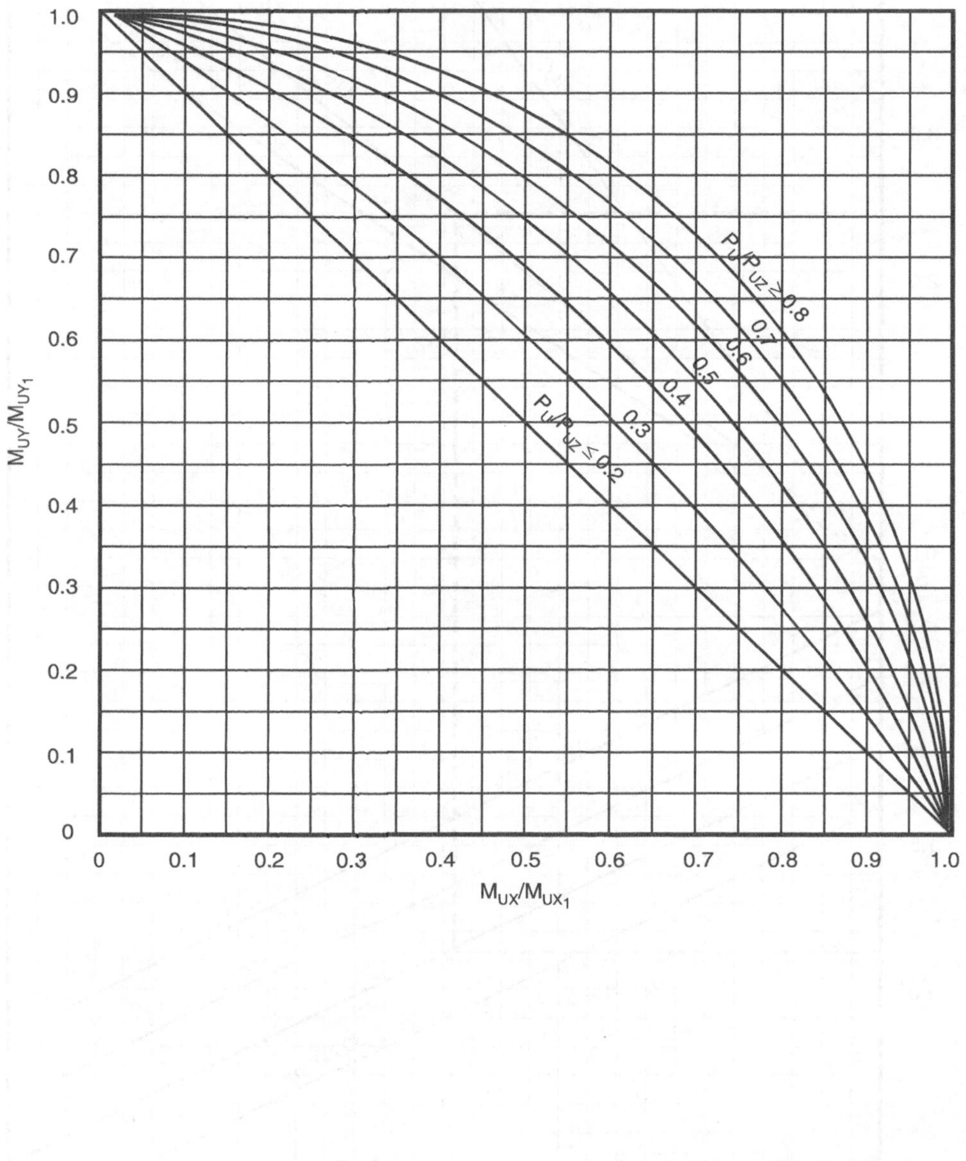

EXERCISE 21

1. Determine the cross-section and the reinforcement for an axially loaded column with the following data.

   Factored load = 4000 KN

   Concrete = M 20

   Characteristic strength of reinforcement = 415 N/mm^2

   Unsupported length of the column = 3 m

2. Determine the reinforcement to be provided in a square of column subjected to uniaxial bending with the following data.

   Size of the column = 400 × 400 mm

   f_y = 415 N/mm^2, Concrete = M 20

   Factored load = 2000 KN, Factored moment = 100 KN.m

   The reinforcement is only on two sides.

3. Determine the reinforcement to be provided in a circular column with the following data.

   Diameter of the column = 300 mm, Grade of concrete = M 20, f_y = 415 /mm^2, Factored load = 800 KN, Factored moment = 40 KN.m

   Lateral reinforcement should be (*i*) Hoop reinforcement (*ii*) Helical reinforcement

4. Determine the reinforcement to be provided in a short column subjected to biaxial bending with the following data.

   Size of the colum = 300 × 300 mm

   Concrete = M 20, f_y of steel = 415 N/mm^2

   P_u = 600 KN, M_{ux} = 60 KN.m. M_{uy} = 80 KN.m

5. Design the R.C.C. column for an axial factored load of 5000 KN. Use M 20 concrete, and HYSD Fe 415 steel. Effective length of column = 4 m.

6. Determine the reinforcement to be provided in a square column 300 mm × 300 mm subjected to uniaxial bending with the following information.

   Concrete = M20, steel f_y = 415 N/mm^2, Factored load = 1000 KN, Factored moment = 120 KN.m. The reinforcement should be on four faces equally.

CHAPTER

22

Limit State of Serviceability

22.1 DEFLECTION

Excessive deflections and deformations can impair the appearance and efficiency of a structure and cause discomfort or alarm to the occupants. Excessive deflections can cause cracking and possible separation of plaster finishes, crushing of partition walls or cracking of glazing units. The maximum deflections which are permitted by Code under normal working loads are given, usually in terms of span or height. Experience has indicated that deflections are likely to be satisfactory if certain limiting span to effective depth ratios are not exceeded. The harmful effects of large deflections can be reduced by provision of an initial camber. The precamber of horizontal members should not generally exceed span/250. The acceptable deflection for partitions and finishes depends on their ability to absorb or resist strain. The code specifies that additional deflection after the application of finishes must not exceed either Span/350 or 20 mm, unless provision is made for the movement.

The lateral deflections can be readily restricted to height/500, which ensures virtually complete freedom from difficulties with glazing panels and opening lights, etc.

The limit states of deflection are concerned with the probable maximum loading under normal working conditions, throughout the life of the structure. Since the partial factors of safety for loads are unity, therefore, characteristic loads in this case shall be the normal working loads. The control of deflection for flexural members such as beams and slabs has already been described in chapters of Working Stress Theory.

For more exact calculations of deflection, reference can be made to the procedure given in Appendix *C* of IS : 456 – 2000. The total deflection shall be the sum of the short-term deflection and the long-term deflection.

(*A*) *Short-Term Deflection*

The short-term deflection may be calculated by the usual methods for elastic deflections using the short-term modulus of elasticity of concrete E_c and an effective moment of inertia I_{eff} given by the following equation.

$$I_{eff} = \frac{I_r}{1.2 - M_r / M \; Z/d(1 - x/d) b_w / b}, \text{ but} \qquad \ldots(22.1)$$

$$I_r \leq I_{eff} \leq I_{gr}$$

where I_r = moment of inertia of the cracked section,

M_r = cracking moment, equal to $\dfrac{f_{cr} \cdot I_{gr}}{y_t}$

f_{cr} is the modulus of rupture of concrete,

I_{gr} is the moment of inertia of the gross-section about the centrodial axis, neglecting the reinforcement and y_t is the distance from centroidal axis of gross-section, neglecting the reinforcement, to the extreme fibre in tension,

M = Maximum moment under service loads,

Z = lever arm

x = depth of neutral axis

d = effective depth

b_w = breadth of web
b = breadth of compression face

For continuous beams, deflection shall be calculated using the values of I_r, I_{gr} and M_r modified by the following equation

$$X_e = k_1\left[\frac{X_1+X_2}{2}\right]+(1-k_1)X_0 \qquad ...(22.2)$$

where

X_e = modified value of X
X_1, X_2 = values of X at the supports
X_0 = coefficient given in Table 22.1 and
X = value of I_r, I_{gr} or M_r as appropriate

Table 22.1. Values of Coefficient, k_1

k_2	0.5 or less	0.6	0.7	0.8	0.9	1.0	1.1	.1.2	1.3	1.4
k_1	0	0.03	0.08	0.16	0.30	0.50	0.73	0.91	0.97	1.0

Note : k_2 is given by

$$k_2 = \frac{M_1+M_2}{MF_1+MF_2}$$ where M_1, M_2 = support moments, and MF_1, MF_2 = fixed end moments.

(*B*) *Deflection Due to Shrinkage*

The deflection due to shrinkage a_{cs} may be computed from the following equation.

$$a_{cs} = k_3\,\psi_{cs}\,l^2 \qquad ...(22.3)$$

where k_3 = a constant depending upon the support conditions,
= 0.5 for cantilevers,
= 0.125 for simply supported members,
= 0.086 for members continuous at one end, and
= 0.063 for fully continuous members

ψ_{cs} is the shrinkage curvature equal to $k_4\dfrac{\varepsilon_{cs}}{D}$

where

ε_{cs} is the ultimate shrinkage strain of concrete (0.0003).

$$k_4 = 0.72\frac{p_t-p_c}{p_t} \le 1.0 \text{ for } 0.25 \le p_t - p_c < 1.0$$

$$= 0.65\frac{p_t-p_c}{p_t} \le 1.0 \text{ for } p_t - p_c \ge 1.0$$

where

$$p_t = \frac{100A_{st}}{bD} \text{ and } p_c = \frac{100\,A_{sc}}{bd}$$

and D is the total depth of the section and l is the length of span.

(*C*) *Deflection Due to Creep*

The creep deflection due to permanent loads, a_{cc} (perm) may be obtained from the following equation.

$$a_{cc}\,(\text{perm})\, a = a_{i,cc}\,(\text{perm}) - a_i\,(\text{perm}) \quad ...(22.4)$$

$a_{i,cc}$ (perm) = initial plus creep deflection due to permanent loads obtained using an elastic analysis with an effective modulus of elasticity.

$$E_{ce} = \frac{E_c}{1+\theta}, \ \theta \text{ being the creep coefficient}$$

and a_i (perm) = short-term deflection due to permanent load using E_c.

Elastic theory assumes linear behaviour between stress and strain even at high loads. The concrete in reinforced concrete member, however, cracks at relatively low tensile stresses so that the load deflection curve tends to be bilinear. The short-term deflection is, therefore, determined according to whether or not the member is cracked or uncracked, from the short-term moduli associated with the design of the materials.

22.2 CRACKING

The limit states of local damage or cracking are concerned with damage of the structure which would occur due to the application of normal working load, *i.e.*, loads which are not expected to exceed the characteristic values. The partial factors of safety for load are taken, in general, equal to unity and for material strengths are also unity. Local damage can occur due to either excessive tension or compression in either the concrete or the steel. Tension in concrete can result in the limit state of crack formation. In most structures, it is sufficient to ensure that crack widths do not become excessive but in liquid retaining structures it is often necessary to ensure that the limit state of crack formation is not reached.

For flexural members, the limit state of cracking is controlled with the compliance of spacing requirement of reinforcement.

For compression members, cracks due to bending in a compression member subjected to design axial load greater than $0.2\,f_{ck}A_c$ where f_{ck} is the characteristic compressive strength of concrete and A_c is the area of gross-section of the member, need not be checked. A member subjected to less load than $0.2\,f_{ck}A_c$ may be considered as a flexural member for purpose of crack control.

EXERCISE 22

1. What do you understand by 'Short term Deflection'?
2. Explain deflection due to shrinkage.
3. Discuss briefly cracking of concrete.
4. Discuss briefly Limit State Serviceability conditions.

Development Length and Anchorage

CHAPTER 23

23.1 GENERAL

The development length L_d is given by

$$L_d = \frac{\sigma_s \phi}{4\tau_{bd}} \quad ...(23.1)$$

where

ϕ = diameter of the bar

σ_s = stress in bar

τ_{bd} = design bond stress in limit state method for plain bars in tension are given below:

Grade of concrete	M15	M20	M25	M30	M35	M40
Design bond stress τ_{bd}, N/mm^2	1.0	1.2	1.4	1.5	1.7	1.9

For deformed bars conforming to IS : 1786 – 1979 or IS : 1139 – 1966, the values shall be increased by 60%

Table 23.1.

Development Length for Deformed Bars

f_y = 415 N/mm^2

Tabulated values are in centimetres

BAR	TENSION BARS				COMPRESSION BARS			
DIAMETER	GRADE OF CONCRETE				GRADE OF CONCRETE			
mm	M15	M20	M25	M30	M15	M20	M25	M30
6	33.8	28.2	24.2	22.6	27.1	22.6	19.3	18.1
8	45.1	37.6	32.2	30.1	36.1	30.1	25.8	24.1
10	56.4	47.0	40.3	37.6	45.1	37.6	32.2	30.1
12	67.7	56.4	48.4	45.1	54.2	45.1	38.7	36.1
16	90.3	75.2	64.5	60.2	72.2	60.2	51.6	48.1
18	101.5	84.6	72.5	67.7	81.2	67.7	58.0	54.2
20	112.8	94.0	80.6	75.2	90.3	75.2	64.5	60.2
22	124.1	103.4	88.7	82.7	99.3	82.7	70.9	66.2
25	141.0	117.5	100.7	94.0	112.8	94.0	80.6	75.2
28	158.0	131.6	112.8	105.3	126.4	105.3	90.3	84.2
32	180.5	150.4	128.9	120.3	144.4	120.3	103.2	96.3
36	203.1	169.3	145.0	135.4	162.5	135.4	116.1	108.3

Note: The development lengths given above are for a stress of 0.87 f_y in the bars.

Table 23.2.

Development Length for Fully Stressed Deformed Bars

$f_y = 500\ N/mm^2$

Tabulated values are in centimetres

BAR	TENSION BARS				COMPRESSION BARS			
DIAMETER	GRADE OF CONCRETE				GRADE OF CONCRETE			
mm	M15	M20	M25	M30	M15	M20	M25	M30
6	40.8	34.0	29.1	27.2	32.6	27.2	23.3	21.8
8	54.4	45.3	38.8	36.3	43.5	36.3	31.1	29.0
10	68.0	56.6	48.5	45.3	54.4	45.3	38.8	36.3
12	81.6	68.0	58.3	54.4	65.3	54.4	46.6	43.5
16	108.8	90.6	77.7	72.5	87.0	72.5	62.1	58.0
18	122.3	102.0	87.4	81.6	97.9	81.6	69.9	65.3
20	135.9	113.3	97.1	90.6	108.8	90.6	77.7	72.5
22	149.5	124.6	106.8	99.7	119.6	99.7	85.4	79.8
25	169.9	141.6	121.4	113.3	135.9	113.3	97.1	90.6
28	190.3	158.6	135.9	126.9	152.3	126.9	108.8	101.5
32	217.5	181.3	155.4	145.0	174.0	145.0	124.3	116.0
36	244.7	203.9	174.8	163.1	195.8	163.1	139.8	130.5

Note: The development lengths given above are for a stress of $0.87\, f_y$ in the bars.

For bars in compression, the values of bond stress for bars in tension shall be increased by 25%

The value of the development length corresponding to a stress of $0.87\, f_y$ in the reinforcement, is required for determining the maximum permissible bar diameter for positive moment reinforcement and for determining the length of laps. Values of this development length for different grades of steel and concrete are given in Tables 23.1 to 23.2.

Anchorage value of hooks and bends – In the case of bars in tension, a standard hook (Fig. 23.1) has an anchorage value equivalent to a straight length of 16 ϕ and a 90° bend has an anchorage value of 8 ϕ. The anchorage values of standard hooks and bends for different bar diameters are given in Table 23.3.

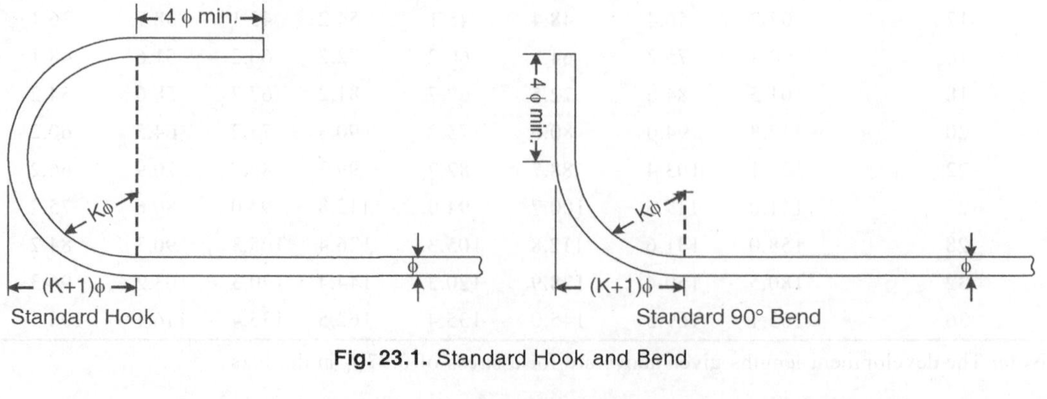

Fig. 23.1. Standard Hook and Bend

Table 23.3. Anchorage Value of Hooks and Bends

Tabulated values are in centimeters

BAR DIAMETER mm	6	8	10	12	16	18	20	22	25	28	32	36
ANCHORAGE VALUE OF HOOK	9.6	12.8	16.0	19.2	25.6	28.8	32.0	35.2	40.0	44.8	51.2	57.6
ANCHORAGE VALUE OF 90° BEND	4.8	6.4	8.0	9.6	12.8	14.4	16.0	17.6	20.0	22.4	25.6	28.8

Type of steel	*Min. value of k*
Mild steel	2
Cold worked steel	4

Note 1 : Table is applicable to all grades of reinforcement bars.

Note 2 : Hooks and bends shall conform to the detail given

EXERCISE 23

1. (a) What do you understand by development length.

(b) Derive an expression for development length in the Limit State method of design.

(c) Is the development length different for bars in tension and compression?

2. Sketch the Standard Hooks for various bend angles.

PART - IV

WATER STORAGE TANKS

Water storage tanks are very important public structures. These tanks may be used for storage of water for domestic use, swimming pools, sewage sedimentation etc. They may be underground, over ground or elevated depending upon the requirement. Their design is entirely different from ordinary structures. The design of such structures is based upon the uncracked section theory to avoid leakage. Richer materials are used to avoid seepage and leakage. Hydrostatic pressures, soil pressures under dry, wet and saturated conditions, wind forces and earthquake forces are main forces which are to be resisted by these structures. Water tightness of joints is very essential for such structures. Working stress method is prevalent for designing such structures. Little is available for design based upon limit state theory particularly for tank proper.

CHAPTER

24

Water Tanks

24.1 INTRODUCTION

Storage of water in the form of tanks for drinking and washing purpose, swimming pools for exercise and enjoyment, and sewage sedimentation tanks are getting increasing importance in the present day life. Water retaining RCC structures commonly know as "Tanks" are classified as follows:-

(*i*) Tanks resting on ground

(*ii*) Tanks constructed underground

(*iii*) Elevated tanks

All the above three types of tanks may be of circular, rectangular, hexagonal or any other shape in plan. Elevated tanks may be Intz type also. These tanks may be with or without a roof covering.

The side walls of tanks are subjected to water pressure from within in case of tanks resting on ground while they may also be subjected to earth pressure (saturated or unsaturated) from outside in the case of underground tanks. While designing underground tanks both conditions of tank full of water and empty have to be considered. In case of tank full, water pressure on side walls has to be considered assuming there is no soil outside of the wall. However, in case when tank is empty, soil pressure is to be considered from outside. If the ground water table is much below the bottom level of the base slab of tank, the base shall be subjected to weight of water from above and soil reaction from below. Generally there is very little difference between the two. The base slab is provided with nominal thickness and minimum steel. But in case the ground water table is above and base slab, it is to be designed for the net uplift exerted by the ground water. The covering roof slab, if any, in both these types are designed in a conventional manner depending upon the shape and the end conditions and the loading to which it may be subjected. The loading may be consisting of self weight or in addition, turfed soil weight.

Elevated tanks are supported on staging which may consist of solid or perforated masonry walls, R.C.C columns braced together or a thin hollow shaft. The actual tank portion is designed for water pressure, live load and self weight of different parts. The staging is to resist wind forces, and earthquake forces in addition to the forces transferred from tank proper. The foundation slab in such cases is generally provided as raft or on piles depending upon the soil conditions. Leakage or seepage is a common problem in water retaining structures. To minimize, it impervious concrete of minimum grade M 25 must be used. The design for water retaining components is based upon no crack theory.

The following three factors must be considered while designing a R.C.C. tank:

(*i*) Strength

(*ii*) Water tightness

(*iii*) Overall stability

24.2 DESIGN REQUIREMENTS OF MATERIALS

24.2.1 Stresses in steel

(*a*) For resistance to cracking. When steel and concrete are assumed to act together for checking the tensile stress in concrete for avoidance of cracks, the tensile stress in the steel will be limited by the requirement that the permissible tensile stress in the concrete is not exceeded so the tensile stress in steel shall be equal to the product of modular ratio of steel and concrete and the corresponding allowable tensile stress in concrete.

(*b*) For strength calculations:- For strength calculations, the permissible stresses in steel reinforcement shall be as given in Table 24.1.

24.2.2 Stresses in concrete

(*a*) For resistance to cracking: For calculations relating to the resistance of members cracking, the permissible stresses in tension (direct and due to bending) and shear shall conforms to the values specified in Table 24.2. The permissible tensile stresses due to bending apply to the face of member in contact with the liquid. In members less than 225 m thick and in contact with the liquid on one side, these permissible stresses in bending apply also to the face remote from the liquid.

Table 24.1. Permissible Stresses in Steel Reinforcement for Strength Calculations

S. No.	*Type of Stress in Steel Reinforcement*	*Permissible Stress in N/mm²*	
		Plain Round Mild Steel Bars	*High Strength Deformed Bars*
(1)	(2)	(3)	(4)
(*i*)	Tensile stress in members under direct tension bending and shear	115	130
(*ii*)	Compressive stress in columns subjected to direct load	125	140

Table 24.2. Permissible Concrete Stresses in Calculations Relating to Resistance to Cracking

S. No.	*Grade of Concrete*	*Permissible Concrete Stresses N/mm²*	
		Direct Tension	*Tension Due to Bending*
(1)	(2)	(3)	(4)
(*i*)	M 25	1.3	1.8
(*ii*)	M 30	1.5	2.0
(*iii*)	M 35	1.6	2.2
(*iv*)	M40	1.8	2.4
(*v*)	M 45	2.0	2.6
(*vi*)	M 50	2.1	2.8

Note:

1. The maximum values of shear stress in concrete shall be as given in IS 456.

Table: 24.3. Permissible Shear Stress in Concrete
(Clause 4.5.2.2, and Table 2)

Sl No.	$100\frac{A}{bd}$	Permissible Shear Stress in Concrete τ_c N/mm² Grade of concrete			
		M 25	M 30	M 35	M 40 and Above
(1)	(2)	(3)	(4)	(5)	(6)
(i)	0.15	0.19	0.20	0.20	0.20
(ii)	0.25	0.23	0.23	0.23	0.23
(iii)	0.50	0.31	0.31	0.31	0.32
(iv)	0.75	0.36	0.37	0.37	0.38
(v)	1.00	0.40	0.41	0.42	0.42
(vi)	1.25	0.44	0.45	0.45	0.46
(vii)	1.50	0.46	0.48	0.49	0.49
(viii)	1.75	0.49	0.50	0.52	0.52
(ix)	2.00	0.51	0.53	0.54	0.55
(x)	2.25	0.53	0.55	0.56	0.57
(xi)	2.50	0.55	0.57	0.58	0.60
(xii)	2.75	0.56	0.58	0.60	0.62
(xiii)	3.00 and above	0.57	0.60	0.62	0.63

NOTE: A_s is that area of longitudinal tension reinforcement which continues at least one effective depth beyond the section being considered except at supports where the full area of tension reinforcement may be used provided the detailing conforms to 26.2.2 and 26.2.3 of IS 456.

(*b*) for strength calculations :- In strength calculations, the permissible concrete stresses shall be in accordance with Table 24.2 and Table 24.3

24.2.3 Stresses due to Drying Shrinkage or temperature change

These stresses may be ignored provided that the permissible stresses in steel and concrete are not otherwise exceeded. Further adequate precautions are taken to avoid cracking of concrete during construction period and until the reservoir is put too use. Shrinkage stresses may however be required to be calculated in special cases, when shrinkage coefficient of 30×10^{-6} may be assumed. The coefficient of concrete may be taken as 11×10^{-6} per degree centigrade. When shrinkage stresses or temperature stresses are allowed, the permissible tensile stresses in concrete (direct and bending) may be increased by 33 1/3 %.

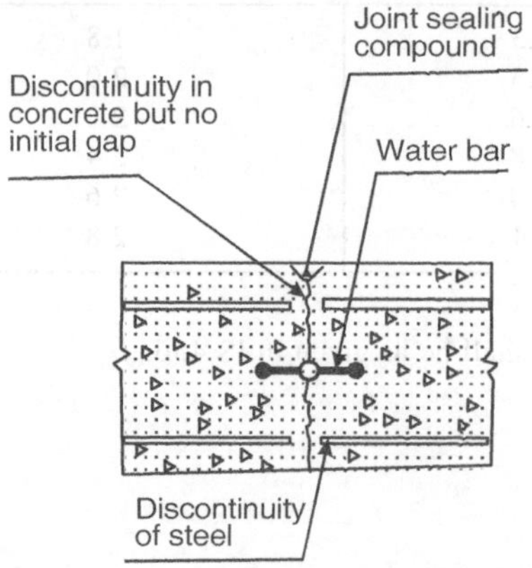

1A Complete Contraction Joint

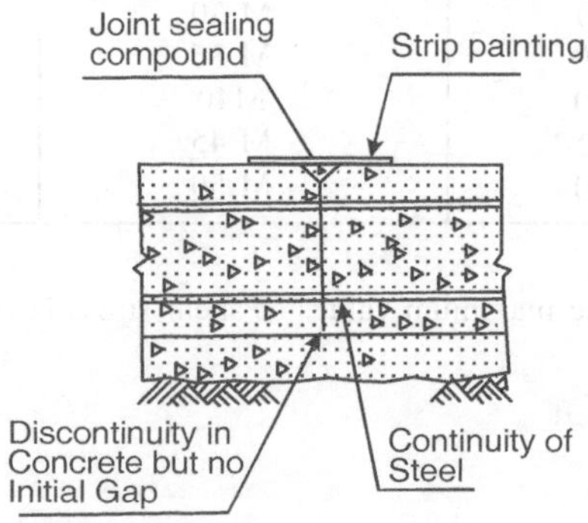

1B Partial Contraction Joint

Fig. 24.1 Typical Contraction Joints.

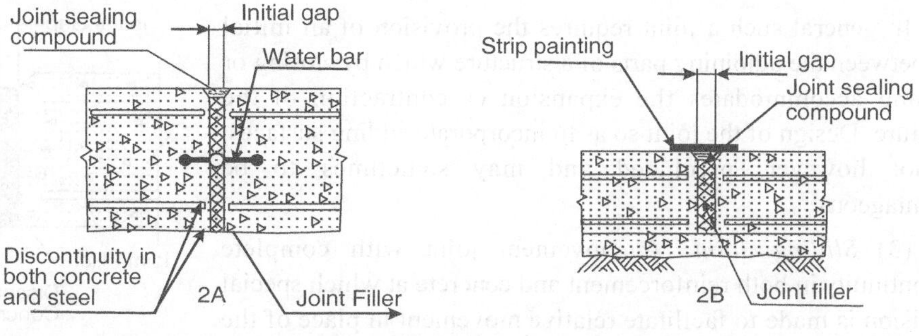

Fig. 24.2 A Typical Expansion Joint.

24.2.4 Concrete

In water retaining structures a dense, impermeable concrete is thus required. Concrete Mix weaker than M 25 is not generally used in tank proper. The minimum quantity off cement in the concrete mix shall not be less than 3000 N/m^3. Concrete mix weaker than M20 is not generally used in staging and foundations.

24.2.5 Joints

(*A*) **Types:** Joints are invariably provided in water tanks such as improvement joints (contraction, expansion or sliding joints) where the length of structure is very large and construction joints to achieve continuity of concrete when whole of the structure cannot be cast at one time. The details of the above joints are as follows:

(*a*) **Movement joints:** There are three categories of movement joints:

(1) *Contraction joint:* A movement joint with a deliberate discontinuity but no initial gap between the concrete on either side of the joint, the joint being intended to accommodates contraction of the concrete. (Fig. 24.3)

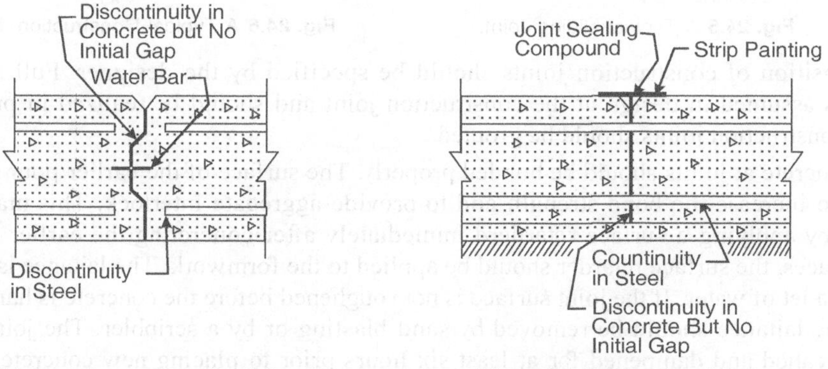

(*a*) Complete contraction joint (*b*) Partial Contraction Joint.

Fig. 24.3 Typical Contraction Joints.

A distinction should be made between a complete contraction joint (Fig. 24.3*a*) in which both concrete and reinforcing steel are interrupted and a partial contraction joint (Fig. 24.3*b*) in which only the concrete is interrupted, the reinforcing steel running through.

(2) *Expansion joint:* A movement joint with complete discontinuity in both reinforcement and concrete and intended to accommodate either expansion or contraction of the structures (Fig 24.4).

In general such a joint requires the provision of an initial gap between the adjoining parts of a structure which by closing or opening accommodates the expansion or contraction of the structure. Design of the joint so as to incorporate sliding surfaces, is not however, precluded and may sometimes to be advantageous.

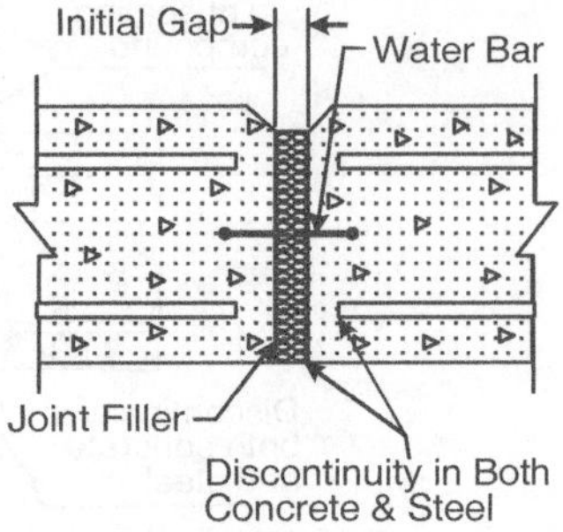

Fig. 24.4 A Typical Expansion Joint.

(3) *Sliding Joint:* A movement joint with complete discontinuity in both reinforcement and concrete at which special provision is made to facilitate relative movement in place of the joint. A typical application is between wall and floor in some cylindrical tank design (Fig. 24.5).

(*b*) **Construction Joint:** A joint in the concrete introduced for convince in construction at which special measures are taken too achieve subsequent continuity without provision for further relative movement is called a construction joint (Fig. 24.6).

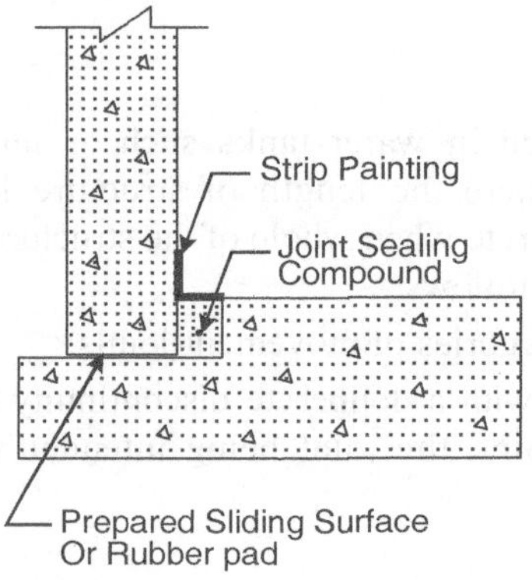

Fig. 24.5 A Typical sliding Joint.

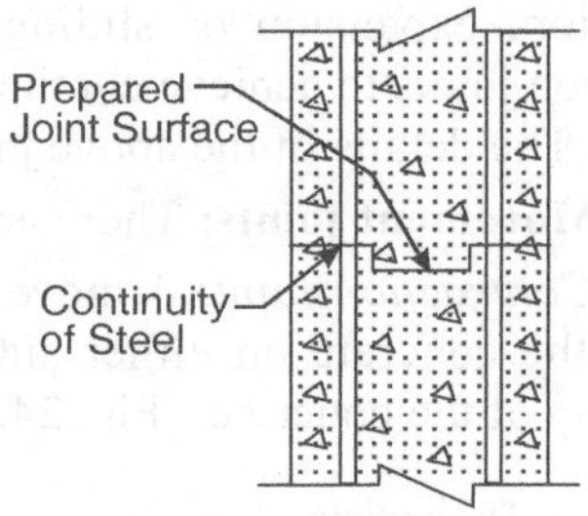

Fig. 24.6 A Typical Construction Joint.

The position of construction joints should be specified by the designer. Full structural continuity is assumed in design at the construction joint and should be realized in practice. If necessary, construction joints should be grouted.

The concrete at joints should be bonded properly. The surface of the earlier pour should be roughened to increase the bond strength and to provide aggregate interlock. This may be best carried out by applying a surface retarderd immediately after concreting the earlier pour. For Vertical surfaces, the surface retarder should be applied to the formwork. The laitance is removed by applying a jet of water. If the joint surface is not roughened before the concrete is hardened; in that case, the laitance should be removed by sand blasting or by a scribbler. The joint surface should be cleaned and dampened for at least six hours prior to placing new concrete. It is not desirable to apply layer of mortar over the old surface.

Design of joints: Design of movement joint should aim at the following desiable properties for its efficient functioning:

(*a*) The joint should accommodate repeated movement of the structure without loss of watertightness.

(*b*) The design should provide for exclusion of grit and debris which would prevent the closing of the joint.

(*c*) The material used in the construction of movement joints should have the following properties:

(1) It should not suffer permanent distortion or extrusion and should not be displaced by fluid pressures.

(2) It should not slump unduly in hot weather or become brittle in cold weather.

(3) It should be insoluble and durable and should not be affected by exposure to light or by the evaporation of solvent or plasticizers.

(4) In special cases, the materials should be non-toxic, taintless or resitant to chemical and biological action as may be specified.

(c) Spacing of joints: Unless alternative effective means are taken to avoid cracks by following for additional stresses that may be induced by temperature or shrinkage changes or by unequal settlement, movement joints should be provided at the following spacing:

(*a*) In reinforced concrete floors, movement joints should be spaced at not more than 7.5m where sliding joints occur at base of the wall in which case correspondence is not so important

(*b*) For floors with only nominal percentage of reinforcement (smaller than the minimum specified), the concrete floor should be cast in panels with sides not more than 4.5m

(*c*) In concrete walls, the vertical movement joints should normally be placed at a maximum spacing of 7.5m in reinforced walls and 6m in unreinforced walls. The maximum length desirable between vertical movement joints will depend upon the tensile strength of the walls, and may be increased by suitable reinforcement. Thus, when a sliding layer is placed at the foundation of a wall, the length of wall that can be kept free of cracks depends upon the capacity of wall section to resist the friction induced at the plane of sliding. Approximately, the wall has to stand the effect of a force at the plane of sliding equal to weight of half the length of wall multiplied by the coefficient of friction.

(*d*) Amongst the movement joints in floors and walls as mentioned above, expansion joints should normally be provided at a spacing of not more than 30m between successive expansion joints or between the end of the structure and the next expansion joint, all other joints being of the contraction type.

(*e*) When, however, the temperature changes to be accommodated are abnormal or occur more frequently than usual as in the case of storage of warm liquids or in unisulated roof slabs, a smaller spacing than 30 m should be adopted (that is a greater propotion of the movement joints should be of the expansion type). When the range of temperature is small, for e.g., in certain covered structures none of the movement joints provided in small structures upto 45m length need be of the expansion type. Where sliding joints are provided between the walls and either the floor or roof, the provision of movement joints in each element can be considered independentally.

24.2.6 Floors

(*i*) Floors of Tanks resting on Ground : If the tank is resting directly over ground, floor may be constructed of concrete with nominal percentage of reinforcement (smaller than the minimum specified in 24.2.9 provided that it is certain that the ground will carry the load without appreciable subsidence in any part and that the concrete floor is cast in panels with sides not more than 4.5m with contraction or expansion joints in between. Where walls or floors are founded on the ground, a layer of lean concrete not less than 75mm thick shall be placed on the ground: In normal circumstances this flat layer of concrete may be weaker than M15 as specified in IS 456. Where, however injurious soils or aggressive ground water are expected, the concrete should not be weaker than M20 as specified in IS 456, and if necessary sulphate resisting or other special cement should be used. Following a layer of lean concrete, the floor shall be cast in a single layer. A separating layer of polyethylene sheet of mass 1Kg/m^2 should be provided in between the floor slab and the layer of lean concrete.

(ii) **Floor of tanks Resting on Supports**

(*a*) If the tanks are supported on walls or other similar supports the floor slab shall be designed as floors in buildings for bending moments due to water load and self weight.

(*b*) When the floor is rigidly connected to the walls (as is generally the case) the bending moments at the junction between the walls and floors shall be taken into account in the design of floor together with any direct forces transferred to the floor shall be taken into the account in the design of floor together with any direct forces transferred to the floor from the wall or from the floor to the wall due to suspension of the floor from the wall. If the walls are non-monolithic with the floor slabs, such as in cases where movement joints have been provided between the floor slabs and walls, the floor shall be designed only for the vertical loads on the floor.

(*c*) In continuous T-beams and L-beams sections at supports shall not exceed the permissible stresses for controlling cracks in concrete. The width of the slab shall be determined in usual manner for calculations of the resitance to cracking T-beam, L-beam sections at supports.

(*d*) The floor slab may be suitably tied to the walls by rods properly embedded in both the slab and the walls. In such cases no separate beam (curved or straight) is necessary under the walls provided the wall of the tank itself designed to act as beam over the supports under it.

(*e*) Sometimes it may be economical to provide the floors of circular tanks, in the shape of dome. In such cases the dome shall be designed for the vertical loads of the liquid over it and the ratio of its ratio of its rise to its diameter shall be supported as its bottom on the ring beam which shall be designed of resultant circumferential tension in addition to vertical loads.

24.2.7 Walls

(i) **Provision of Joints**

Suitable movement or construction joints may be provided as given in 24.2.5.

(ii) **Pressure on Walls**

Pressure due to water or soil (dry or saturated) should be considered in the design of walls. For computation of pressure the density of water is taken as 9810N/m^3 and that of dry soil 16000N/m^3 (Normally). The pressure of saturated soil although depends upon degree of saturation, but the pressure of fully saturated soil may be taken as the sum of water pressure plus the submerged weight of soil, i.e.

$$P = w.h + (w_s - w)\frac{1-\sin\theta}{1+\sin\theta}h$$

Where
p = pressure due to saturated soil
w = density of water
w_s = dry density of soil
ϕ = angle of repose of soil
h = height of saturated soil

(iii) **Design Criteria**

(*a*) On liquid retaining faces, the tensile stresses due to combination of direct horizontal tension and bending action should satisfy the following condition:

$$\frac{\sigma_{t'}}{\sigma_t} + \frac{\sigma_{ct'}}{\sigma_{ct}} \le 1$$

Where $\sigma_{t'}$ = calculated direct tensile stress in concrete
σ_t = permissible direct tensile stress in concrete
$\sigma_{ct'}$=calculated tensile stress due to bending in concrete
σ_{ct}=permissible tensile stress due to bending in concrete

(*b*) Side walls of tanks may be designed as vertical cantilevers or as two way slabs in case of rectangular tanks. The walls may be assumed to be hinged or free at top and hinged or fixed at the bottom.

24.2.8 Roofs

(*i*) Provision of Movement joints. To avoid the possibility of cracking, it is important to ensure that movement joints in the roof correspond with those in the walls, if roof and walls are monolithic. If, however, provision is made by means of a sliding joints for movement between the roof and the wall correspondence of joints is not so important.

(*ii*) Loading. Field covers of liquid retaining structures should be designed for gravity loads, such as the weight of roof slab, earth cover if any, live loads and mechanical equipment. They should also be designed for upward load if the liquid retaining structure is subjected to internal gas pressure.

A superficial load sufficient to ensure safety with the unequal intensity of loading which occurs during the placing of the earth cover should be allowed for in designing roofs. The engineer should specify a loading under these temporary conditions which should not be exceeded. In designing the roof, allowance should be made for the temporary condition of some spans loaded and other spans unloaded even through in the final state the load may be small and evenly distributed.

(*iii*) Water tightness. In case of tanks intended for the storage of water domestic purpose, the roofs must be made water-tight. This may be achieved by limiting the stress as for the rest of the tank, or by the use of the covering of the waterproof membrane or by providing slopes to ensure adequate drainage.

(*iv*) Protection against corrosion. Proctection measures shall be provided to the underside of the roof to prevent it from corrosion due to condensation.

24.2.9 Minimum Reinforcement

The minimum reinforcement in walls, floors and roofs in each of two direction at right angles. Within each surface zone shall not be less than 0.35 percent of the surface zone, cross section as shown in Fig. 24.1 and Fig. 24.2 for high strength deformed bars and not less than 0.64 percent for mild steel reinforcement. It can be further reduced to 0.24 percent for the deformed bars and 0.40 percent for plain round bars for tanks having any dimension not more than 15 m. In wall slab less than 200 mm in thickness, the calculated amount of reinforcement may all be placed in one face. For ground slab less than 300mm thick (Fig 24.2) the calculated reinforcement should be placed in one face as near as possible to the upper surface consistent with the nominal cover. Bar spacing should generally not exceed 300mm or the thickness of the section, whichever is less.

24.2.10 Minimum Cover to Reinforcement

(*a*) For liquid faces of parts of members either in contact with the liquid (such as inner faces of roof slab) the minimum cover to all reinforcement should be 25mm or the diameter of the main bar whichever is greater. In the presence of the sea water and soils and water of corrosive characters the cover should be increased by 12 mm but this additional cover shall not be taken into account for design calculations.

(*b*) For faces away from liquid and for parts of the structure neither in contact with the liquid on any face, nor enclosing the space above the liquid, the cover shall be as for ordinary concrete member.

24.3 TANKS RESTING ON GROUND

24.3.1 Circular tank with Flexible Joint at the Base

The walls of such tanks are designed as vertical cylinders subjected to hydrostatic pressure. The intensity of water pressure at any depth 'h' is equal to 'w h' units per unit area. The corresponding tension/unit height at this level is given by

$$T = \frac{whD}{2}$$

Where D = Internal diameter of the tank

The area of reinforcement per unit height is calculated to resist the above tension as

$$A_{st} = \frac{whD}{2\sigma st}$$

Where σ_{st} = permissible stress in steel (refer 24.2.1)

This hoop reinforcement may be provided preferably near both faces with a minimum clear cover of 25 mm. However, the amount of steel provided should not be less than specified in article 24.2.9.

Thickness of the wall should be computed to resist the hoop tension in the wall such that

Tensile stress in wall concrete = $\dfrac{T}{b.t+(m-1)A_{st}}$ should be less than permissible direct tensile stress in concrete as given in 24.2.

Where t = thickness of wall

m = modular ratio

The bottom slab shall be designed for the difference of pressure due to weight of water acting downward and the upward pressure due to soil. This difference is generally small and nominal thickness (minimum 100 mm for practical consideration) with less than minimum steel as given in 24.2.9 shall be provided in the form of top mesh and bottom mesh preferably.

Example 24.1. Design a circular tank for a capacity of 400 kiloliters with flexible base i.e. walls and base slab are not monolithic with each other. Use Fe 415 steel bars for reinforcement and M25 concrete.

Solution. Let the depth of tank be 3m with free board as 300mm. The diameter of the tank 'D' can be computed as

$$\frac{\pi}{4}D^2 \times (3-0.3) = \frac{400\times10^6}{10^6}$$

$$D = 13.73\text{m}$$

Provide diameter, $D = 13.75$m

(i) Bottom 1 m height of walls

Depth of water = 2.7m

$$\left.\begin{array}{l}T_{max} \text{ at the base for} \\ 1\text{m height of walls}\end{array}\right\} = \frac{whD}{2} = \frac{9810\times2.7\times13.75}{2} = 182098 \text{ N}$$

$$A_{st} = \frac{182098}{130} = 1400 \text{ mm}^2$$

Provide 13 bars of 12 mm ϕ

Use mix M 25. σ_t = 1.3 N/mm²

t = thickness of the walls

$$\text{Tensile stress} = \frac{182098}{1000\,t + (11-1)\,12 \times 113} = 1.3$$

$$t = 126.51 \text{ mm}$$

Provide 140 mm thickness of walls uniform throughout its height

(ii) Middle 1 m height of walls

$$\text{Depth of water} = 1.7\text{m}$$

$$T = \frac{9810 \times 1.7 \times 13.75}{2} = 114654 \text{ N}$$

$$A_{st} = \frac{114654}{130} = 881\, mm^2$$

Provide 8 bars of 12 mm ϕ

(iii) Top 1m height of walls

$$\text{Depth of water} = 0.7 \text{ m}$$

$$T = \frac{9810 \times 0.7 \times 13.75}{2} = 47210 \text{ N}$$

$$A_{st} = \frac{47210}{130} = 363\, mm^2$$

Provide 4 bars of 10 mm ϕ in 0.7 m height and in rest 0.3 m free board height proportional steel may be provided i.e. 4 bars of 8 mm ϕ.

Hoop steel shall be provided equally on both faces.

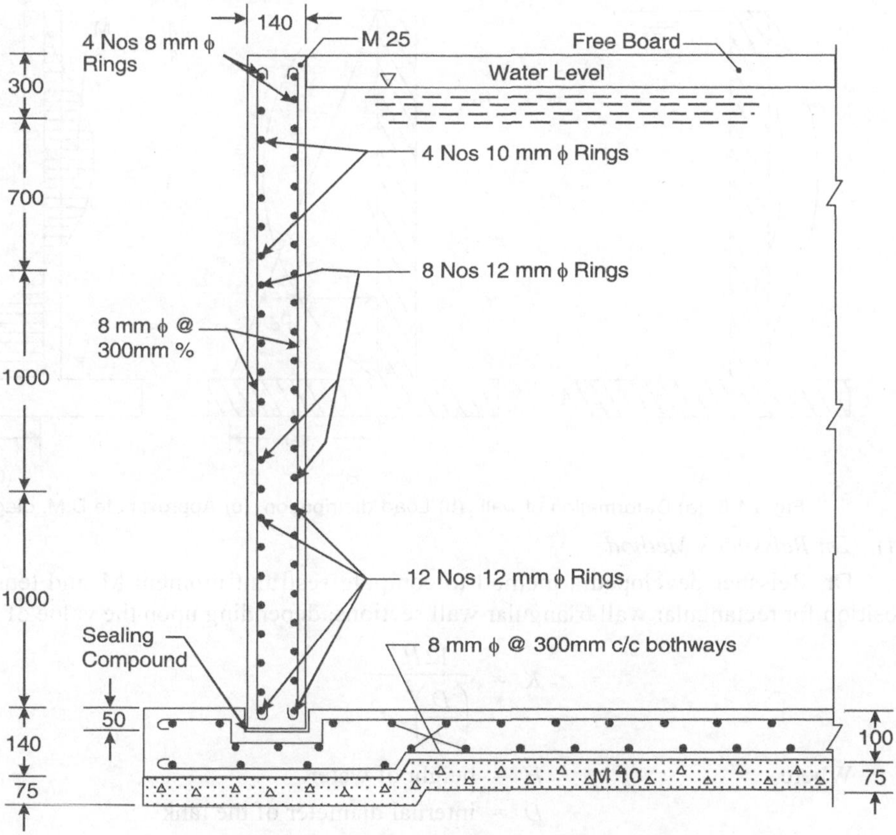

Fig. 24.7. Details of Reinforcement (Dia = 13.75 m)

Vertical Reinforcement (Temperature and Distribution steel)

$$A_{st} = \frac{0.24}{100} \times 140 \times 1000 = 336\ mm^2 \text{ on both faces.}$$

$$= 245 \text{ mm}^2 \text{ on each face}$$

Provide 8 mm ϕ 300 mm c/c on each face.

Base Slab

Provide 100 mm thick slab in M 25,

Minimum distribution and temperature steel

$$= \frac{0.24}{100} \times 100 \times 1000 = 240 \text{ mm}^2 \text{ on both faces}$$

$$= 100 \text{ mm}^2 \text{ on each face}$$

Provide 8 mm ϕ @ 300 mm c/c both ways on top and bottom.

Fig. 24.7 shows the details of reinforcement.

24.3.2 Circular Tanks with Restraint at Base and Roof

Due to restraint at the base and roof, the walls are subjected to horizontal shear, hoop tension and bending moment. In the case the wall will resist the water pressure partly by hoop action and partly by cantilever action and at higher levels; there will be predominant hoop action as shown in Fig. 24.8.

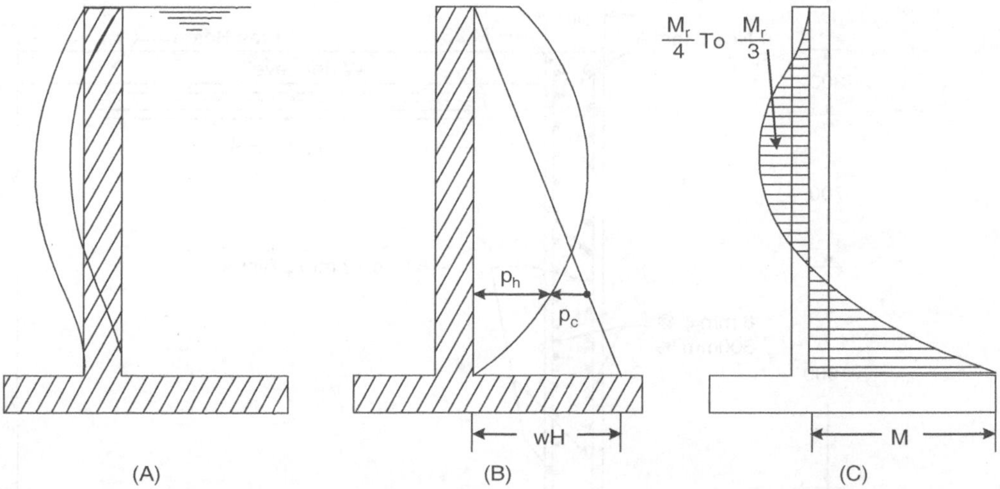

Fig. 24.8. (a) Deformation of wall (b) Load distribution (c) Approximate B.M. diagram.

(*A*) *Dr. Reissner's Method*

Dr. Reissner developed a method to compute restraint moment M_r and tension T and its position for rectangular wall triangular wall sections, depending upon the value of a parameter.

$$K = \frac{12H^4}{\left(\frac{D}{2}\right)^2 . t^2}$$

Where H = height of water

D = internal diameter of the tank

t = thickness of wall

The values of *Mr* and *T* are presented in Tables 24.4 and 24.5 respectively.

Table 24.4. Values of Restraint Moment M_r $(p = wH)$

K	*Rectangular wall section*	*Triangular wall section*
0	0.167 pH^2	0.167 pH^2
10	0.110 pH^2	0.140 pH^2
100	0.0582 pH^2	0.0707 pH^2
1000	0.024 pH^2	0.009 pH^2
10000	0.0085 pH^2	0.99 pH^2
∞	0	0

Table 24.5. Ring Tension in Tank Wall $(p = wH)$

K	*Rectangular wall section*		*Triangular wall section*	
	Max. Tension	*Height from base*	*Max. Tension*	*Height from base*
0	0	—	0	—
10	$0.18\ p\left(\frac{D}{2}\right)$	1.0 H	$0.09\ p\left(\frac{D}{2}\right)$	0.65 H
100	$0.27\ p\left(\frac{D}{2}\right)$	1.0 H	$0.31\ p\left(\frac{D}{2}\right)$	0.58 H
1000	$0.47\ p\left(\frac{D}{2}\right)$	0.47 H	$0.52\ p\left(\frac{D}{2}\right)$	0.44 H
10000	$0.67\ p\left(\frac{D}{2}\right)$	0.31 H	$0.70\ p\left(\frac{D}{2}\right)$	0.30 H
∞	$1.0\ p\left(\frac{D}{2}\right)$	0	$1.0\ p\left(\frac{D}{2}\right)$	0

(B) Carpenter's Method

Carpenter simplified Dr. Reissner's method. The maximum cantilever moment and the position and magnitude of maximum ring tension are given by the following relations.

(*i*) Position of maximum ring tension = $K.H.$

From base where K is a coefficient

Depending upon H/D and H/t

(*ii*) Maximum hoop tension at $K.H.$

Above base

$$= w\,(H - KH)\frac{D}{2} = \frac{wHD}{2}\,(1 - \text{K})$$

(*iii*) Maximum cantilever bending moment

$$= \text{F w H}^3$$

where F is a coefficient depending upon values of H/D and H/t.

The coefficients K and F can be obtained from Table 24.6

Table 24.6. Coefficients F and K

Factors		*F*				*K*			
H/t :		*10*	*20*	*30*	*40*	*10*	*20*	*30*	*40*
	0.2	0.046	0.028	0.022	0.015	—	0.50	0.45	0.40
Values	0.3	0.032	0.019	0.014	0.010	0.55	0.43	0.38	0.33
of	0.4	0.024	0.014	0.010	0.007	0.50	0.39	0.35	0.30
H/D	0.5	0.020	0.012	0.006	0.006	0.45	0.37	0.32	0.27
	1.0	0.012	0.006	0.005	0.003	0.37	0.28	0.24	0.21
	2.0	0.006	0.003	0.002	0.002	0.30	0.22	0.19	0.16
	4.0	0.004	0.002	0.002	0.001	0.27	0.20	0.17	0.14

(C) I.S. Code Method

The bending moment, shear and hoop tension at base for a cylindrical tank wall can be obtained from Tables 24.7 to 24.10 for different loadings and end conditions.

Table 24.7. Tension in Circular Ring Wall, Fixed Base, Free Top and Subject to Triangular Load

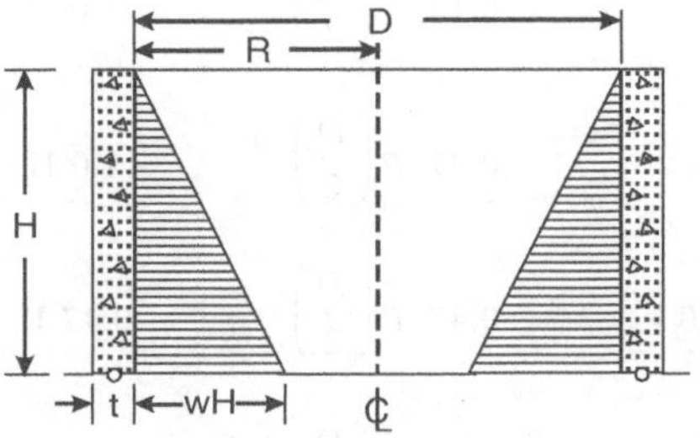

$\frac{H^2}{Dt}$	COEFFICIENTS AT POINT									
	0.0*H*	0.1*H*	0.2*H*	0.3*H*	0.4*H*	0.5*H*	0.6*H*	0.7*H*	0.8*H*	0.9*H*
(1)	(2)	(3)	(4)	(5)	(6)	(7)	(8)	(9)	(10)	(11)
0.4	+0.149	+0.134	+0.120	+0.101	+0.082	+0.066	+0.049	+0.029	+0.014	+0.004
0.8	+0.263	+0.239	+0.215	+0.109	+0.160	+0.130	+0.096	+0.063	+0.034	+0.010
1.2	+0.283	+0.271	+0.254	+0.234	+0.209	+0.180	+0.142	+0.099	+0.054	+0.016
1.6	+0.265	+0.268	+0.268	+0.266	+0.250	+0.226	+0.185	+0.134	+0.075	+0.023
2.0	+0.234	+0.251	+0.273	+0.285	+0.285	+0.274	+0.232	+0.172	+0.104	+0.031
3.0	+0.134	+0.203	+0.267	+0.322	+0.357	+0.362	+0.330	+0.262	+0.157	+0.052
4.0	+0.067	+0.164	+0.256	+0.339	+0.403	+0.429	+0.409	+0.334	+0.210	+0.073
5.0	+0.025	+0.137	+0.245	+0.346	+0.428	+0.477	+0.469	+0.398	+0.259	+0.092
6.0	+0.018	+0.119	+0.234	+0.344	+0.441	+0.504	+0.514	+0.447	+0.301	+0.112
8.0	–0.001	+0.104	+0.218	+0.335	+0.443	+0.534	+0.575	+0.530	+0.381	+0.151
10.0	–0.001	+0.098	+0.208	+0.323	+0.437	+0.542	+0.608	+0.589	+0.440	+0.179
12.0	–0.005	+0.097	+0.202	+0.312	+0.429	+0.543	+0.628	+0.633	+0.494	+0.211
14.0	–0.002	+0.098	+0.200	+0.306	+0.420	+0.539	+0.639	+0.666	+0.541	+0.241
16.0	0.000	+0.099	+0.199	+0.304	+0.412	+0.531	+0.641	+0.687	+0.582	+0.265

Note 1 — *w* = Density of the liquid. **Note 2** — Positive sign indicates tension.

Table 24.8. Moments of Cylindrical Wall, Fixed Base, Free Top and Subject to Triangular Load

Moment = Coefficient × wH^3 kg·m/m

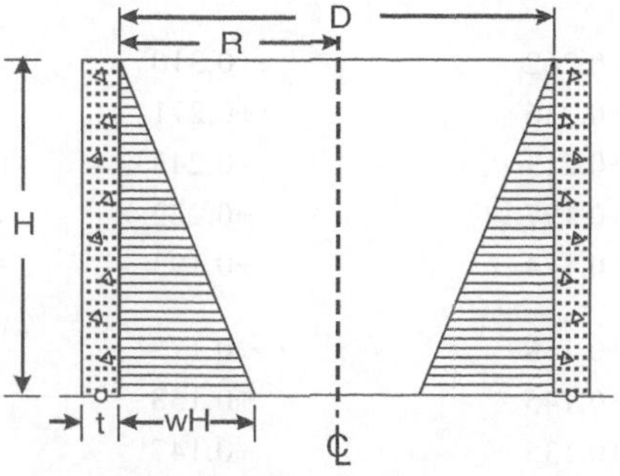

$\frac{H^2}{Dt}$	COEFFICIENTS AT POINT								
	0.1*H*	0.2*H*	0.3*H*	0.4*H*	0.5*H*	0.6*H*	0.7*H*	0.8*H*	0.9*H*	1.0*H*
(1)	(2)	(3)	(4)	(5)	(6)	(7)	(8)	(9)	(10)	(11)
0.4	+0.0005	+0.0014	+0.0021	+0.0007	–0.0042	–0.0150	–0.0302	–0.0529	–0.0816	–0.1205
0.8	+0.0011	+0.0037	+0.0063	+0.0080	+0.0070	+0.0023	–0.0068	–0.0024	–0.0465	–0.0795
1.2	+0.0012	+0.0042	+0.0077	+0.0103	+0.0112	+0.0090	+0.0022	–0.0108	–0.0311	–0.0602
1.6	+0.0011	+0.0041	+0.0075	+0.0107	+0.0121	+0.0111	+0.0058	–0.0051	–0.0232	–0.0505
2.0	+0.0010	+0.0035	+0.0068	+0.0099	+0.0120	+0.0115	+0.0075	–0.0021	–0.0185	–0.0436
3.0	+0.0006	+0.0024	+0.0047	+0.0071	+0.0090	+0.0097	+0.0077	+0.0012	–0.0119	–0.0333
4.0	+0.0003	+0.0015	+0.0028	+0.0047	+0.0066	+0.0077	+0.0069	+0.0023	–0.0080	–0.0268
5.0	+0.0002	+0.0008	+0.0016	+0.0029	+0.0046	+0.0059	+0.0059	+0.0028	–0.0058	–0.0222
6.0	+0.0001	+0.0003	+0.0008	+0.0019	+0.0032	+0.0046	+0.0051	+0.0029	–0.0041	–0.0187
8.0	0.0000	+0.0001	+0.0002	+0.0008	+0.0016	+0.0028	+0.0038	+0.0029	–0.0022	–0.0146
10.0	0.0000	0.0000	+0.0001	+0.0004	+0.0007	+0.0019	+0.0029	+0.0028	–0.0012	–0.0122
12.0	0.0000	–0.0001	+0.0001	+0.0002	+0.0003	+0.0013	+0.0023	+0.0026	–0.0005	–0.0104
14.0	0.0000	0.0000	0.0000	0.0000	+0.0001	+0.0008	+0.0019	+0.0023	–0.0001	–0.0090
16.0	0.0000	0.0000	–0.0001	–0.0002	–0.0001	+0.0004	+0.0013	+0.0019	+0.0001	–0.0079

Note 1 — *w* = Density of the liquid. **Note 2** — Positive sign indicates tension on the outside.

Table 24.9. Shear at Base of Cylindrical Wall

$$v = \text{coefficient} \times \begin{cases} wH^2 \text{ kg (triangular)} \\ pH \text{ kg (rectangular)} \\ M/H \text{ kg (moment at base)} \end{cases}$$

$\frac{H^2}{Dt}$	*Triangular Load Fixed Base*	*Rectangular Load Fixed Base*	*Triangular or Rectangular Load Hinged Base*	*Moment at Edge*
0.4	+0.436	+0.755	+0.245	– 1.58
0.8	+0.374	+0.552	+0.234	– 1.75
1.2	+0.339	+0.460	+0.220	– 2.00

1.6	+ 0.317	+0.407	+0.204	– 2.28
2.0	+0.299	+0370	+0.189	– 2.57
3.0	+0.262	+0.310	+0.158	– 3.18
4.0	+0.236	+0.271	+0.137	– 3.68
5.0	+0.213	+0.243	+0.121	– 4.10
6.0	+0.197	+0.222	+0.110	– 4.49
8.0	+0.174	+0.193	+0.096	– 5.18
10.0	+0.158	+0.172	+0.087	– 5.81
12.0	+0.145	+0.158	+0.079	– 6.38
14.0	+0.135	+0.147	+0.073	– 6.88
16.0	+0.127	+0.137	+0.068	– 7.36

Note 1 — w = Density of the liquid. **Note 2** — Positive sign indicates shear acting inward.

Table 24.10. Tension in Circular Ring Wall, Hinged Base, Free Top and Subject to Triangular Load

T = Coefficient × wHR kg/m

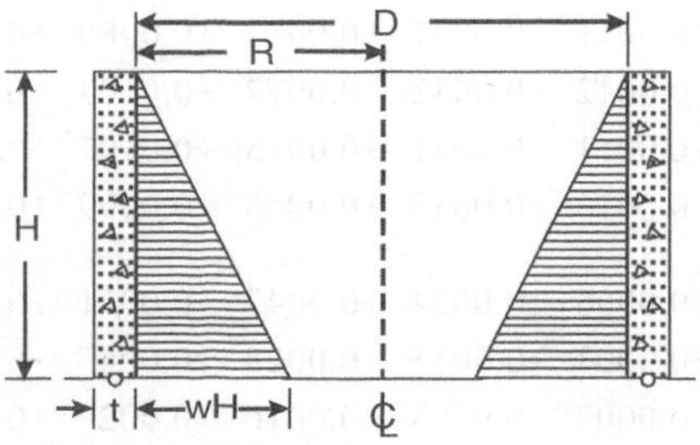

$\frac{H^2}{Dt}$	COEFFICIENTS AT POINT									
	0.0*H*	0.1*H*	0.2*H*	0.3*H*	0.4*H*	0.5*H*	0.6*H*	0.7*H*	0.8*H*	0.9*H*
(1)	(2)	(3)	(4)	(5)	(6)	(7)	(8)	(9)	(10)	(11)
0.4	+0.474	+0.440	+0.395	+0.352	+0.308	+0.264	+0.215	+0.165	+0.111	+0.057
0.8	+0.423	+0.402	+0.381	+0.358	+0.330	+0.297	+0.249	+0.202	+0.145	+0.076
1.2	+0.350	+0.355	+0.361	+0.362	+0.358	+0.342	+0.309	+0.256	+0.186	+0.098
1.6	+0.271	+0.303	+0.341	+0.369	+0.385	+0.385	+0.362	+0.314	+0.233	+0.124
2.0	+0.205	+0.260	+0.321	+0.373	+0.411	+0.434	+0.419	+0.369	+0.280	+0.151
3.0	+0.074	+0.179	+0.281	+0.375	+0.449	+0.506	+0.519	+0.479	+0.375	+0.210
4.0	+0.017	+0.137	+0.253	+0.367	+0.469	+0.545	+0.579	+0.553	+0.447	+0.256
5.0	–0.008	+0.114	+0.235	+0.356	+0.469	+0.562	+0.617	+0.606	+0.503	+0.294
6.0	–0.011	+0.103	+0.223	+0.343	+0.463	+0.566	+0.639	+0.643	+0.547	+0.327
8.0	–0.015	+0.096	+0.208	+0.324	+0.443	+0.564	+0.661	+0.697	+0.621	+0.386
10.0	–0.008	+0.095	+0.200	+0.311	+0.428	+0.552	+0.666	+0.730	+0.678	+0.433
12.0	–0.002	+0.097	+0.197	+0.302	+0.417	+0.541	+0.664	+0.750	+0.720	+0.477
14.0	0.000	+0.098	+0.197	+0.299	+0.408	+0.531	+0.659	+0.761	+0.752	+0.513
16.0	+0.002	+0.100	+0.198	+0.299	+0.403	+0.521	+0.650	+0.764	+0.776	+0.536

Note 1 — w = Density of the liquid. **Note 2** — Positive sign indicates tension.

Hoop tension per metre height = coefficient × $\frac{wHD}{2}$

B.M. per metre = coefficient wH^3

S.F. at the base of the wall = coefficient × wH^2

(*a*) Wall with fixed base and free top subjected to triangular load (Fig. 24.9).

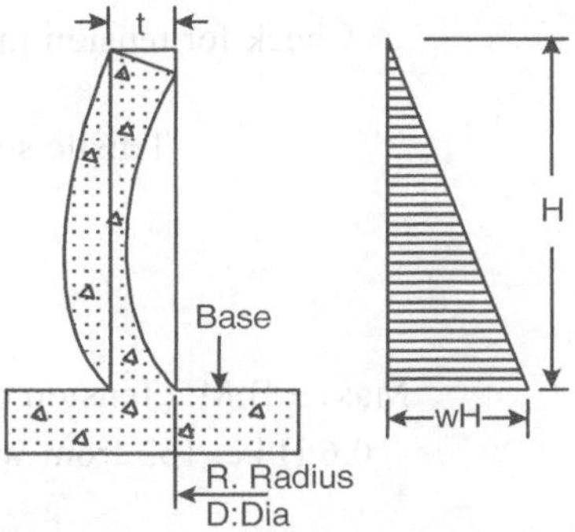

Fig. 24.9

Walls built continuous with their footing are sometime designed as though the base was fixed and the top free, although strictly specking the base is seldom fixed, but it is helpful to start with this assumption and then to go on to the design procedures for other more corrects conditions. Ring tension at various heights of walls of cylindrical tanks fixed at base, free at top, and subjected to triangular load are given in Table 24.9.

(*b*) Wall with hinged Base and free Top, Subjected to Triangular load. (Fig. 24.10)

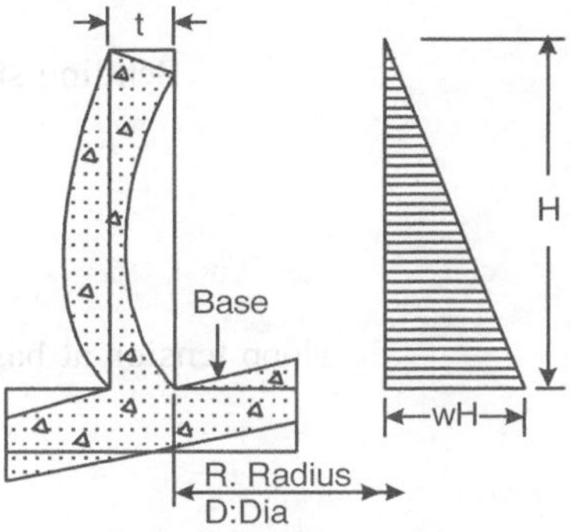

Fig. 24.10

The values given in (A) are based on the assumption that the base joint is continuous and the footing is prevented from even the smallest rotation of kind shown exaggerated in Fig. 24.10. The rotation required to reduce the fixed base moment from some definite value to, say, zero is much smaller than rotations that may occur when normal settlement takes place in subgrade. It may be difficult to predict the behavior of the subgrade and in effect upon the restraint at the base, but it is more reasonable to assume that the base is hinged than fixed, and the hinged-based assumption gives a safer design. Ring tensions and moments at different heights of wall of cylindrical tank are given in Table 24.10. Coefficients for shear at base of cylindrical wall are given in Table 24.9.

Example 24.2. Design a circular tank 13.75 m diameter and 3.0 m height of wall. Free board = 0.3 m. The tank rests on a firm ground. The walls are fixed at base and free at top. Use M25 and Fe 415 rebars.

Solution. Assume thickness of wall = 140 mm

$$\frac{H^2}{Dt} = \frac{2.7 \times 2.7}{13.75 \times 0.14} = 3.79$$

Max. hoop tension at 0.5 H i.e. 135 m from base

$$= 0.415 \times 9810 \times 2.7 \times \frac{13.75}{2} \text{ (Refer Table 24.7)}$$

$$= 75570 \text{ N}$$

$$A_{st} = \frac{75570}{130} = 581 \text{ mm}^2 \text{ on both faces.}$$

$$= 290 \text{ mm}^2 \text{ on each face}$$

Provide 8 mm ϕ 170 mm c/c, A_{st} = 588 mm² on two faces in 1 height from 0.85 m to 1.85 m from base. As the hop tension decreases towards top and bottom of point of maximum hoop tension, the spacing of bars can be increased to 300 mm c/c in the rest.

Check for tension in concrete

$$\text{Tensile stress,} \quad \sigma_t' = \frac{75570}{140\times1000+(11-1)502}$$

$$= 0.52\ \text{N/mm}^2$$

$$< 1.3\ \text{N/mm}^2 \qquad \text{(Table 24.2)}$$

Safe

Max. + B.M. (Tension outside) per metre height at 0.6 H i.e. 162 from water level

$$= 0.00812 \times 9810 \times 2.7^3 = 1568\ \text{Nm.}$$

Max. – B.M. (Tension inside) per metre height at the base

$$= 0.02816 \times 9810 \times 2.7^3 = 5439\ \text{Nm.}$$

$$\text{Bending stress,} \quad \sigma_t' = \frac{M}{Z} = \frac{5439\times1000\times6}{1000\times140\times140}$$

$$= 1.665\ \text{N/mm}^2$$

$$< 1.8\ \text{N/mm}^2$$

Safe (Table 24.2)

$$\text{Hoop tension at base} = 0.0686 \times 9810 \times 2.7 \times \frac{13.75}{2}$$

$$= 12490\ \text{N}$$

$$\text{Tensile stress at base,} \quad \sigma_t' = \frac{12490}{140\times1000+(11-1)\times\frac{50.5\times1000}{300}} = 0.088\ \text{N/mm}^2$$

$$\text{Interaction equation} \quad \frac{\sigma'_t}{\sigma_t} = \frac{\sigma'_{ct}}{\sigma_{ct}} \le 1$$

$$\text{L.H.S.} = \frac{0.088}{1.3} + \frac{1.665}{1.8}$$

$$= 0.0678 + 0.925 = 0.9928 < 1,\ \text{Safe.}$$

$$A_{st}\ \text{for Max.} - \text{B.M.} = \frac{5439\times1000}{150\times0.90\times(140-40)} = 402.9\ \text{mm}^2$$

$$A_{st}\ \text{for Max.} + \text{B.M.} = \frac{1668\times1000}{150\times0.90\times(140-40)} = 123.5\ \text{mm}^2$$

Minimum distribution and temperature steel

$$= (0.24)\times\frac{1}{100}\times140\times1000$$

$$= 326\ \text{mm}^2\ \text{on both faces}$$

$$= 163\ \text{mm}^2\ \text{on each face}$$

Provide 8 mm ϕ @ 150 mm c/c on inside face vertically

Check for shear

$$\text{Max. S.F. at the base} = 0.241 \times 9810 \times 2.7^2 \qquad \text{(Table 24.9)}$$

$$= 17235\ \text{N}$$

$$\text{Nominal Shear stress} = \frac{17235}{1000 \times 100} = 0.172 \text{ N/mm}^2$$

Permissible minimum shear stress for

$$\text{M25} = 0.23 \text{ N/mm}^2 \qquad \text{(Table 24.3)}$$

Safe

Base Slab

Provide 150 mm thick base slab in M25 concrete.

$$\text{Minimum distribution steel} = \frac{0.24 \times 150 \times 1000}{100}$$

$= 360$ mm^2 in one direction on two faces

$= 180$ on each face

Provide 8 mm @ 230 c/c both ways on top and bottom.

Fig. 24.11 shows the reinforcement detail of the tank. It is always preferable to provide a haunch of 150 mm side at the junction of wall with base to avoid leakage of water.

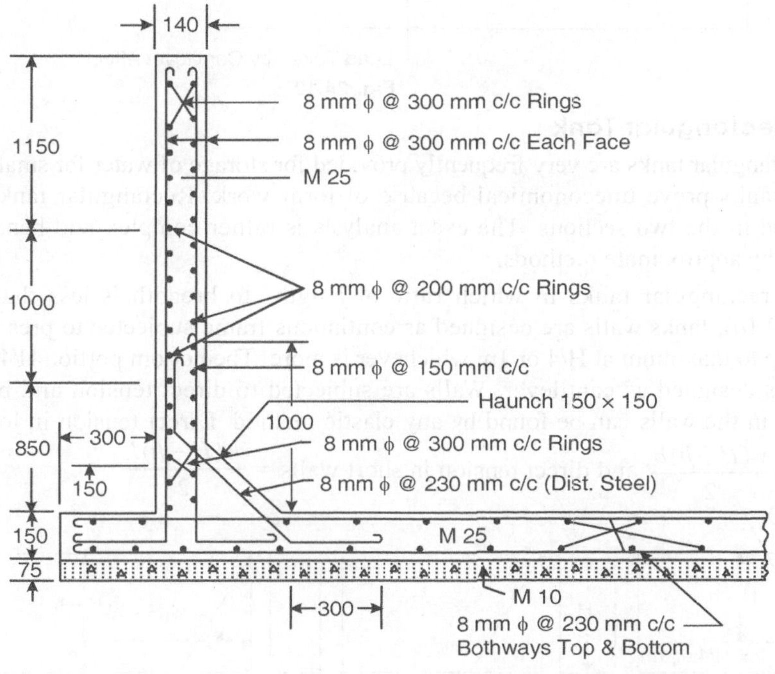

Fig. 24.11 Details of Reinforcement of Circular. Tank with 13.75 m in Diameter

(d) Approximate method of Design of Circular Tanks with Fixed Base

In this method bottom one-fourth height of water in the tank or one meter, whichever is more, shall be taken as cantilever wall. The pressure distribution on this cantilever portion shall be triangular as shown in Fig. 24.12. This assumption is more accurate H^2/Dt is more than 12. When H^2/Dt is between 6 and 12, the cantilever portion may be assumed as one third the height of water or one metre whichever is more. For the rest of the portion, the tank wall may be designed only for hoop tension.

For H^2/Dt less than 6, whole wall may be taken as cantilever fixed at base.

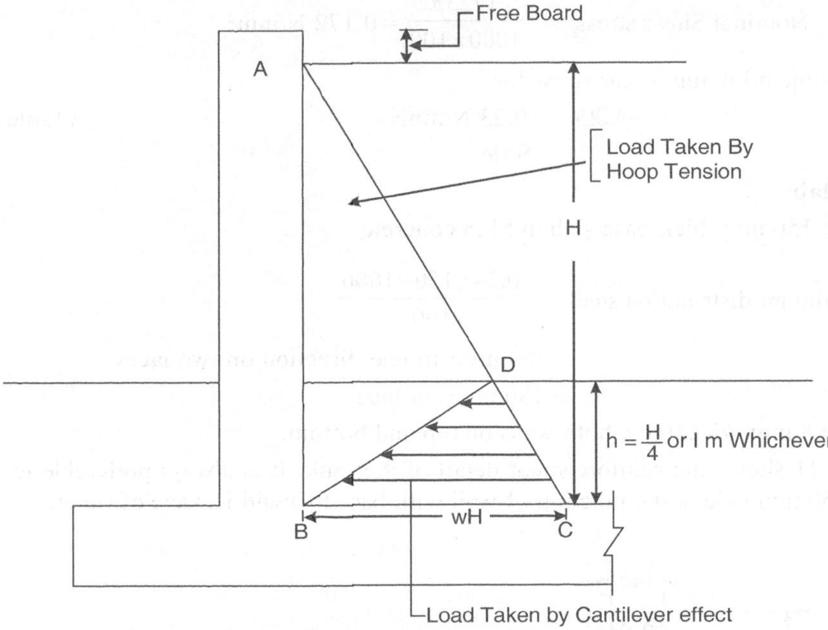

Fig. 24.12

24.3.3 Rectangular Tank

Rectangular tanks are very frequently provided for storage of water for small capacities as the circular tanks prove uneconomical because of form work. Rectangular tanks have moments distributed in the two sections. The exact analysis is rather complex and hence such tanks are designed by approximate methods.

For rectangular tanks in which ratio of length to breadth is less than 2 as shown in Fig. 24.13 (*a*), tanks walls are designed as continuous frame subjected to pressure varying from zero at top to maximum at H/4 or 1m whichever is more. The bottom portion H/4 or 1m whichever is more is designed as cantilever. Walls are subjected to direct tension and bending. Bending moments in the walls can be found by any elastic method. Direct tension in long walls shall be equal to $\dfrac{w(H-h)B}{2}$ and direct tension in short walls = $\dfrac{w(H-h)L}{2}$

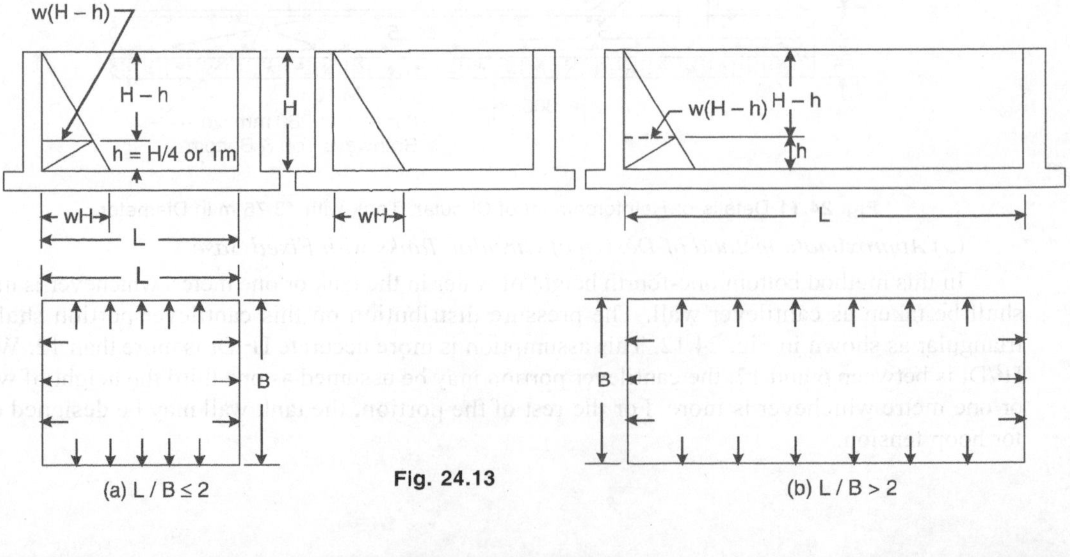

Fig. 24.13

For rectangular tanks in which the ratio of length to breadth is more than 2, as shown in Fig. 24.13(*b*), the long walls are designed as cantilever and the short walls as slabs supported on long walls. Bottom portion of short walls H/4 or 1m, whichever is more is designed as cantilever.

$$\text{Long Wall, Maximum B.M. at base} = wH^3/6$$

$$\text{Short Wall, maximum B.M. at support} = \frac{w(H-h)B^2}{12}$$

$$\text{B.M. at centre} = \frac{w(H-h)B^2}{16}$$

$$\text{Cantilever maximum B.M. for bottom portion} = \frac{wHh^2}{6} \text{ or } \frac{wH \times 1}{6} \text{ (whichever is greater)}$$

In addition to bending moment, short walls and long walls are subjected to direct tension.

$$\text{Long wall, direct tension} = \frac{w(H-h)B}{2}$$

Short walls, Direct tension = $w(H–h)$ on the assumption that end one metre width of long wall contributes to direct tension on the short walls.

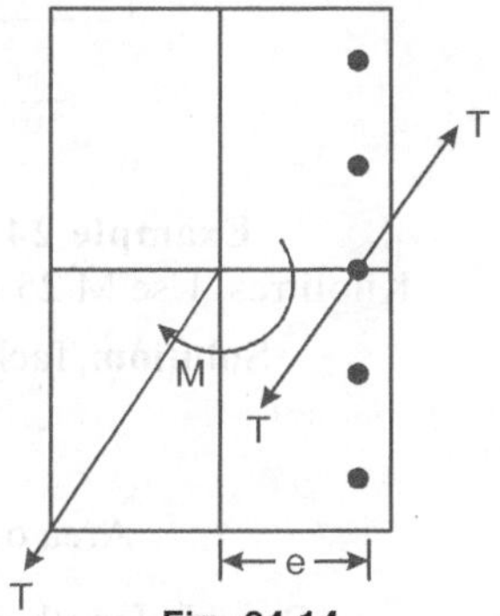

Fig. 24.14

(*a*) Design of Tank wall subjected to B.M and Pull Fig. 24.14 shown a wall subjected to an axial pull 'T and a bending moment 'M' (including in right hand side of the section). Let there be two equal and Opposite forces 'T' imagined to be acting on reinforcement. The moments and various forces as shown are equivalent to a resultant moment = (M – Te).

$$\text{Steel for the resultant moment, } A_{st} = \frac{(M-Te)}{\text{safe stress in steel} \times \text{lever arm}}$$

$$\text{Steel for direct pull, } A_{st2} = \frac{T}{\text{safe stress in steel}}$$

$$\text{Total steel} = A_{st1} + A_{st2}$$

(*b*) Design of Base Slab. The bending moments in a base slab are computed at the end. At ends the bending moment on the long cantilever wall shall be $\frac{wH^3}{6}$ hogging per unit width. The base slab shall also be subjected to a pull equal to $\frac{wH^3}{2}$ (Fig. 24.15). In general the wall hogging moment shall be computed for hydro- static pressure.

This moment shall get distributed in wall and base as per their relative stiffnesses if the walls and roof slab are taken hinged or considered monolithic.

End portion of base slab shall be designed accordingly. In the rest of slab, nominal thickness and steel be provided as these due to self weight of base slab and water on it shall get balanced from equal and opposite soil reaction. The width of base slab which acts as end portion i.e. effective to take care of bending moment shall be such that the end slab and wall are in static equilibrium under the forces acting on wall and slab portion. This is approximately equal to 0.5 H.

Where here H is the height of wall. It may be noted that end base slab must have the same thickness as that of wall at base.

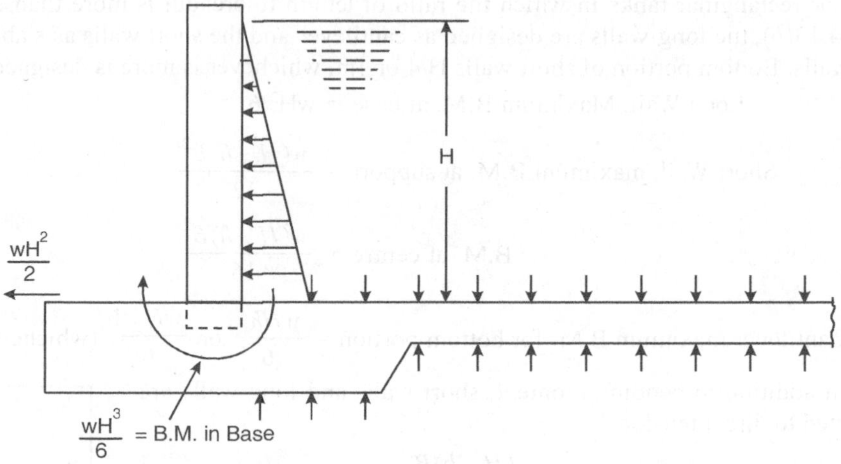

Fig. 24.15

Example 24.3. Design a rectangular tank resting on the ground for a capacity 100 Kilolitres. Use M 25 concrete and Fe 415 rebars.

Solution. Tacking a height of 4 m for tank with a free board of 0.30 m.

$$\text{Effective height} = 4 - 0.3 = 3.7 \text{ m}$$

$$\text{Area of the tank required} = \frac{100}{3.7} = 27 \text{ m}^2$$

Provide length of 6 m and breadth of 5 m

Area provided = 6 × 5 = 30 m² > 27 m²

$$L/B = 6/5 = 1.2 < 2$$

The walls of tanks are to be designed as continuous slab

$$H = \frac{100}{30} = 3.33 \text{ m}$$

$$\text{Adopt depth of water, } H = 3.4 \text{ m}$$

$$\text{Total height of wall} = 3.4 + 0.3 = 3.7 \text{ m}$$

$$\frac{H}{4} = \frac{3.4}{4} = 0.85$$

Bottom 1 m shall be designed as cantilever from top (water surface)

$$\text{Pressure at } (3.4 - 1.0) \quad m = p = 9810 \times 2.4 = 23544 \text{ n/m}^2$$

$$\text{Pressure at} \quad 3.4^m = 9810 \times 3.4 = 33394 \text{ N/m}^2$$

$$H = 1^m$$

$$\text{B.M. at base} = \frac{33394 \times 1}{2} \times \frac{1}{3}$$

$$= 5566 \text{ N.m.}$$

$$A_{st} = \frac{5566 \times 100}{130 \times 0.9 \times 420} = 113 \text{ mm}^2$$

Provide Minimum reinforcement in bottom 1^m height of wall and on each face both ways.

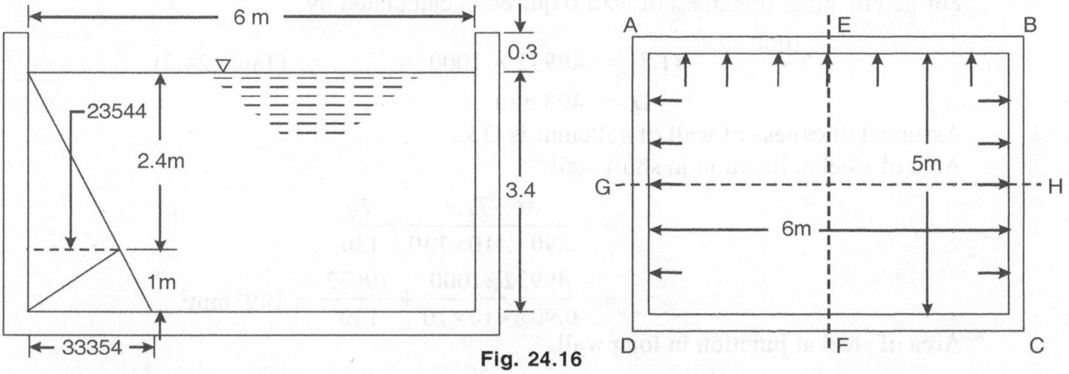

Fig. 24.16

As the top 3 m height of walls constitute a frame, moments in walls shall be computed by moment distribution for one- quarter of the tank only (Fig. 24.16).

Joint	A	
Member	AB	AD
Distribution factor	0.455	0.545
Fixed end moment	$-\frac{36}{12}p$	$+\frac{25}{12}p$
Balancing Moment	$+\frac{5}{12}p$	$+\frac{6}{12}p$
Final	$+\frac{31}{12}p$	$+\frac{31}{12}p$

Considering 1^m wall between 2.4^m to 1.4^m height from water surface

$$\text{Moment at support} = \frac{31}{12}p = \frac{31}{12}\times 23544$$
$$= 60822 \text{ N.m.}$$

$$\text{B.M. at centre of long span} = \frac{23544\times 6^2}{8} - 60822$$
$$= 45126 \text{ N.m.}$$

$$\text{B.M. at centre of short span} = \frac{23544\times 5^2}{8} - 60822$$
$$= 12753 \text{ N.m.}$$

T_S = Pull in short wall $= 23544 \times 5/2 = 58860$ N

T_L = Pull in long wall $= 23544 \times 6/2 = 70632$ N

Assume the thickness of wall = 450mm

And effective cover = 40mm

$$e = \frac{450}{2} - 40 = 185 \text{ mm} = 0.185 \text{ m}$$

$$\text{Net B.M. in short wall} = M - T_S e$$
$$= 60822 - 58860 \times 0.185$$
$$= 49932 \text{ N.m.}$$

$$\text{Net B.M. in long wall} = M - T_L.e$$
$$= 60822 - 70632 \times 0.185$$
$$= 47755 \text{ N.m.}$$

For no cracking, thickness of wall required is calculated by

$$\frac{1000 \times D^2}{6} \times 1.8 = 49932 \times 1000 \qquad \text{(Table 24.2)}$$

$$D = 408 \text{ mm}$$

Assumed thickness of wall of 450cmm is O.k.

Area of steel at junction in short wall

$$= \frac{M - T_S.e}{0.90 \times 410 \times 130} + \frac{T_S}{130}$$

$$= \frac{49932 \times 1000}{0.90 \times 410 \times 10} + \frac{70632}{130} = 1493\text{mm}^2$$

Area of steel at junction in long wall

$$= \frac{m - T_L e}{0.90 \times 410 \times 130} + \frac{T_L}{130}$$

$$= \frac{47755 \times 1000}{0.90 \times 410 \times 130} + \frac{70632}{130} = 1546\text{mm}^2$$

Provide 16 mm ϕ @ 130mm c/c (A_{st} = 1546 mm² in a depth from 1.4 m to 2.4 m horizontally at junction in long and short walls upto 0.3 × 6 = 1.8 m and 0.3 × 5 = 1.5 m respectively and thereafter @ 260 mm c/c. In top 1.4 m depth, provide horizontal steel in proportion to the depth of water i.e. 12 mm θ @ 130 mm c/c (A_{st} = 870 mm² in a similar manner as shown in Fig. 24.15.

Steel at centre of span away from water face

In short wall $= \frac{(12753 - 58860 \times 0.185)1000}{0.90 \times 410 \times 130} + \frac{58860}{130}$

$= 491 \text{ mm}^2$

In long wall $= \frac{(45126 - 70632 \times 0.185)1000}{90 \times 410 \times 130} + \frac{70632}{130}$

$= 1311\text{mm}^2$

Provide horizontally 12 mm ϕ @75mm c/c (=1506mm²) in long wall and 12mm ϕ @ 150. c/c (823mm²) in short wall.

Minimum distribution steel $= \frac{0.24}{100} \times 450 \times 1000 = 1080\text{mm}^2$ on both faces

$= 540 \text{ mm}^2$ on each face

Distribution steel as 12mm ϕ @ 200mm

Cantilever Moment in bottom/m of wall

Cantilever moment $= 3.4 \times 9810 \times \frac{1}{2} \times 1 \times \frac{1}{3} = 5559$ N.m.

Area of steel required $= \frac{5559 \times 1000}{0.90 \times 410 \times 130} = 115 \text{ mm}^2$

$< 900\text{mm}^2$

Provide 12 mm ϕ @ 250 mm c/c each face as minimum distribution steel.

Base Slab

Provide base slab 150mm thick as shown in Fig. 24.15 with minimum distribution steel as

$$= (0.35) \times \frac{150 \times 1000}{100}$$

$= 525 \text{ mm}^2$ on two faces in a direction

$= 262 \text{ mm}^2$ each face in a direction

∴ Provide 10 mm ϕ @ 300 mm c/c both ways at top and bottom of the slab.

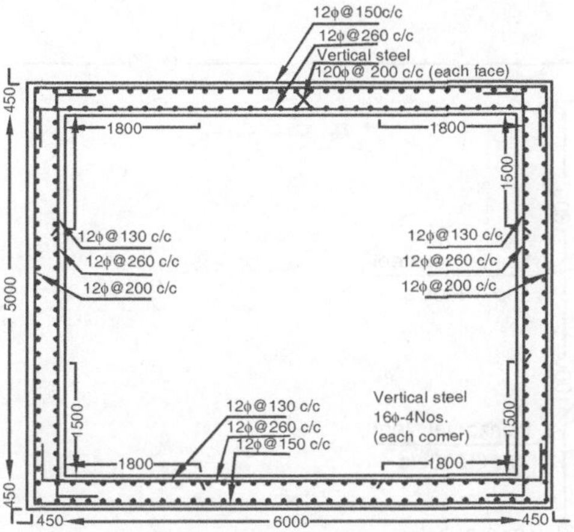

Plan Section of Top 1.6 m

Fig. 24.17 (a)

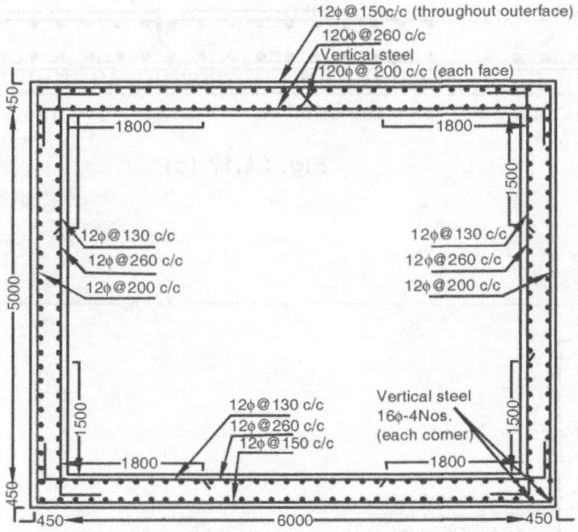

Plan Section between 1.6m to 2.6 m Depth from Top.

Fig. 24.17 (b)

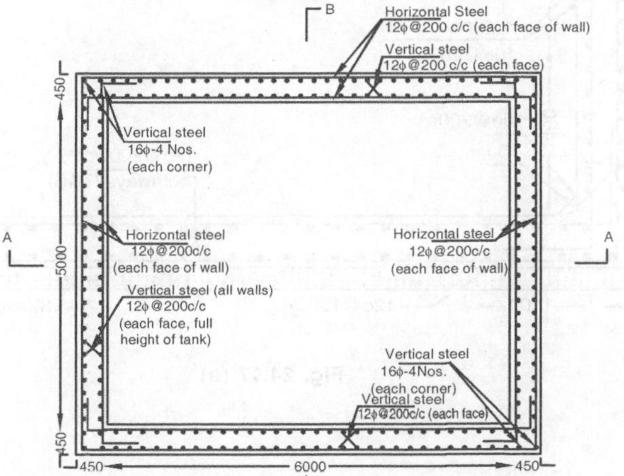

Plan Section of Bottom 1.4m

Fig. 24.17 (c)

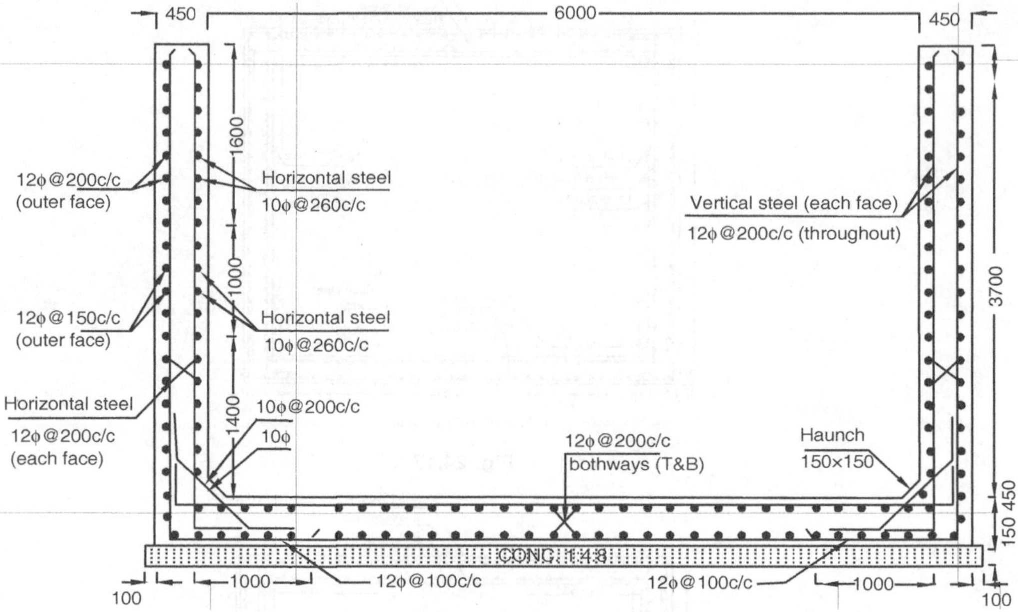

Fig. 24.17 (d)

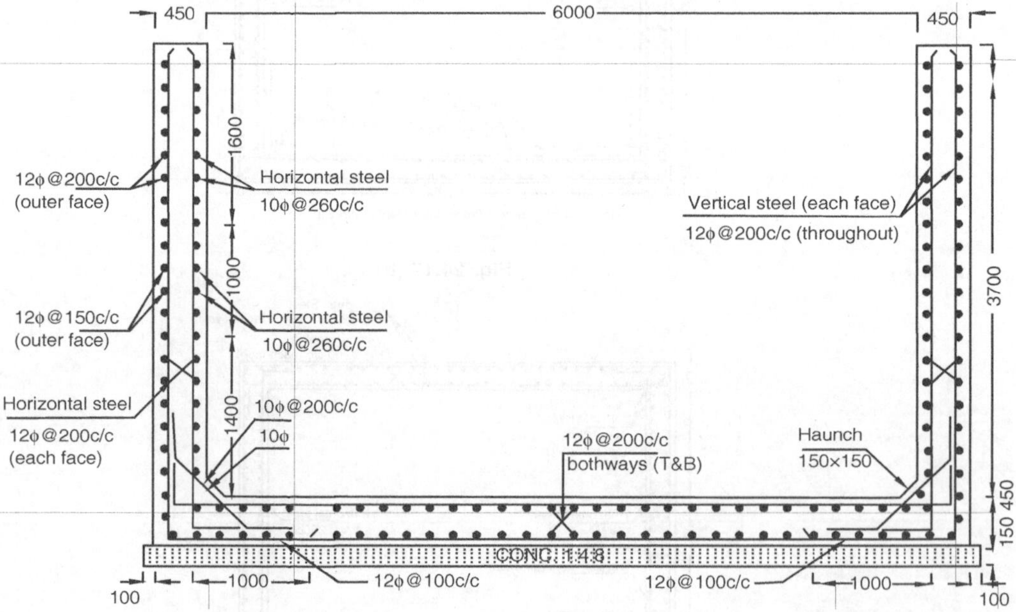

Fig. 24.17 (e)

Table 24.11. Moment Coefficients for Individual Wall Panel, Top and Bottom Hinged, Vertical Edges Fixed

a = height of the wall

b = width of the wall

w = density of the liquid

Horizontal moment = $M_y wa^3$

Vertical moment = $M_x wa^3$

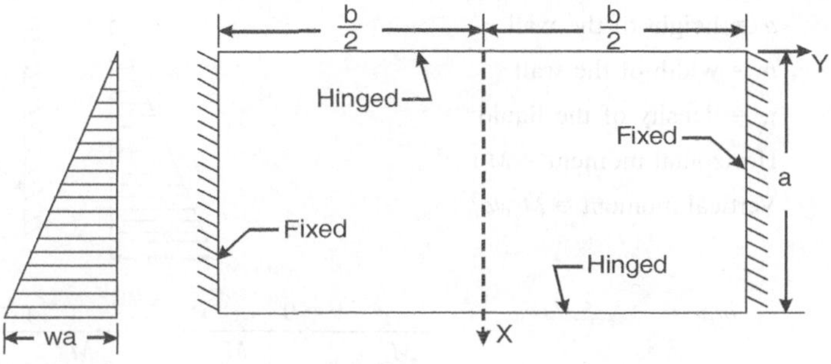

b/a	x/a	$y = 0$		$y = b/4$		$y = b/2$	
		M_x	M_y	M_x	M_y	M_x	M_y
(1)	(2)	(3)	(4)	(5)	(6)	(7)	(8)
3.00	1/4	+0.035	+0.010	+0.026	+0.011	–0.008	–0.039
	1/2	+0.057	+0.016	+0.044	+0.017	–0.013	–0.063
	3/4	+0.051	+0.013	+0.041	+0.014	–0.011	–0.055
2.50	1/4	+0.031	+0.011	+0.021	+0.010	–0.008	–0.038
	1/2	+0.052	+0.017	+0.036	+0.017	–0.012	–0.062
	3/4	+0.047	+0.015	+0.036	+0.014	–0.011	–0.055
2.00	1/4	+0.025	+0.013	+0.015	+0.009	–0.007	–0.037
	1/2	+0.042	+0.020	+0.028	+0.015	–0.012	–0.059
	3/4	+0.041	+0.016	+0.029	+0.013	–0.011	–0.053
1.75	1/4	+0.020	+0.013	+0.012	+0.008	–0.007	–0.035
	1/2	+0.036	+0.020	+0.023	+0.013	–0.011	–0.057
	3/4	+0.036	+0.017	+0.025	+0.012	–0.010	–0.051
1.50	1/4	+0.015	+0.013	+0.008	+0.007	–0.006	–0.032
	1/2	+0.028	+0.021	+0.016	+0.011	–0.010	–0.052
	3/4	+0.030	+0.017	+0.020	+0.011	–0.010	–0.048
1.25	1/4	+0.009	+0.012	+0.005	+0.005	–0.036	–0.028
	1/2	+0.019	+0.019	+0.011	+0.009	–0.009	–0.045
	3/4	+0.023	+0.017	+0.014	+0.009	–0.009	–0.043
1.00	1/4	+0.005	+0.009	+0.002	+0.003	–0.004	–0.020
	1/2	+0.011	+0.016	+0.006	+0.006	–0.007	–0.035
	3/4	+0.016	+0.014	+0.009	+0.007	–0.007	–0.035
0.75	1/4	+0.001	+0.006	+0.000	+0.002	–0.002	–0.012
	1/2	+0.005	+0.011	+0.002	+0.003	–0.004	–0.022
	3/4	+0.009	+0.011	+0.005	+0.005	–0.005	–0.025
0.50	1/4	+0.000	+0.000	+0.000	+0.001	–0.001	–0.005
	1/2	+0.001	+0.005	+0.001	+0.001	–0.002	–0.010
	3/4	+0.004	+0.007	+0.002	+0.002	–0.003	–0.014

Table 24.12. Moment Coefficient for Individual Wall Panel, Top Free, Bottom Hinged, Vertical Edges Fixed

a = height of the wall
b = width of the wall
w = density of the liquid
Horizontal moment = $M_y wa^3$
Vertical moment = $M_x wa^3$

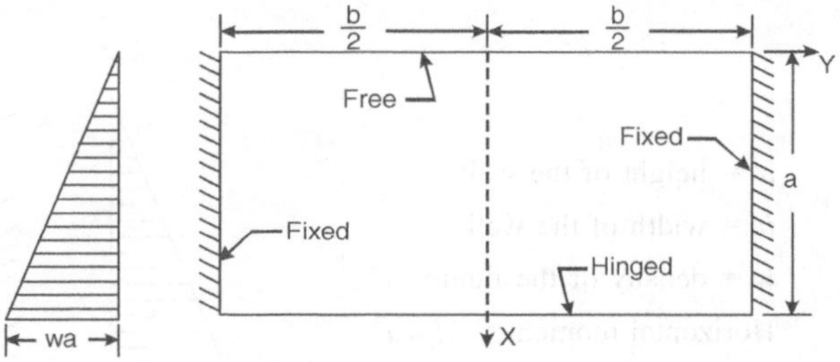

b/a	x/a	$y = 0$		$y = b/4$		$y = b/2$	
		M_x	M_y	M_x	M_y	M_x	M_y
(1)	(2)	(3)	(4)	(5)	(6)	(7)	(8)
3.00	0	0	+0.070	0	+0.027	0	–0.196
	1/4	+0.028	+0.061	+0.015	+0.028	–0.034	–0.170
	1/2	+0.049	+0.049	+0.032	+0.026	–0.027	–0.137
	3/4	+0.046	+0.030	+0.034	+0.018	–0.017	–0.087
2.50	0	0	+0.061	0	+0.019	0	–0.138
	1/4	+0.014	+0.053	+0.010	+0.022	–0.026	–0.132
	1/2	+0.042	+0.044	+0.025	+0.022	–0.023	–0.115
	3/4	+0.041	+0.027	+0.030	+0.016	–0.016	–0.078
2.00	0	0	+0.045	0	+0.011	0	–0.091
	1/4	+0.016	+0.042	+0.006	+0.014	–0.019	–0.04
	1/2	+0.033	+0.036	+0.020	+0.016	–0.018	–0.089
	3/4	+0.035	+0.034	+0.025	+0.014	–0.013	–0.065
1.75	0	0	+0.036	0	+0.008	0	–0.071
	1/4	+0.013	+0.035	+0.005	+0.011	–0.015	–0.076
	1/2	+0.028	+0.032	+0.017	+0.014	–0.015	–0.076
	3/4	+0.031	+0.022	+0.021	+0.012	–0.012	–0.059
1.50	0	0	+0.027	0	+0.005	0	–0.052
	1/4	+0.009	+0.028	+0.003	+0.008	–0.012	–0.059
	1/2	+0.022	+0.027	+0.012	+0.011	–0.013	–0.063
	3/4	+0.027	+0.020	+0.017	+0.011	–0.010	–0.052
1.25	0	0	+0.017	0	+0.003	0	–0.034
	1/4	+0.005	+0.020	+0.002	+0.005	–0.008	–0.042
	1/2	+0.017	+0.023	+0.009	+0.009	–0.010	–0.049
	3/4	+0.021	+0.017	+0.013	+0.009	–0.009	–0.044
1.00	0	0	+0.010	0	+0.002	0	–0.019
	1/4	+0.002	+0.013	+0.000	+0.003	–0.005	–0.025
	1/2	+0.010	+0.017	+0.005	+0.006	–0.007	–0.036
	3/4	+0.015	+0.015	+0.009	+0.007	–0.007	–0.036
0.75	0	0	+0.005	0	+0.001	0	–0.008
	1/4	+0.001	+0.008	+0.000	+0.002	–0.003	–0.013
	1/2	+0.005	+0.011	+0.002	+0.004	–0.004	–0.022
	3/4	+0.010	+0.012	+0.006	+0.004	–0.005	–0.026
0.50	0	0	+0.002	0	+0.000	0	–0.003
	1/4	+0.000	+0.004	+0.000	+0.001	–0.001	–0.005
	1/2	+0.002	+0.006	+0.001	+0.002	–0.002	–0.010
	3/4	+0.007	+0.008	+0.002	+0.002	–0.003	–0.014

Table 24.13. Moment Coefficient for Individual Wall Panel, Top free, Bottom and Vertical Edges Fixed

a = height of the wall

b = width of the wall

w = density of the liquid

Horizontal moment = $M_y wa^3$

Vertical moment = $M_x wa^3$

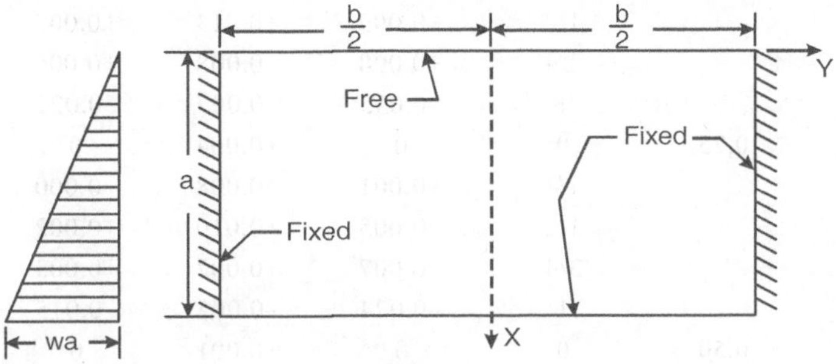

b/a	x/a	$y = 0$		$y = b/4$		$y = b/2$	
		M_x	M_y	M_x	M_y	M_x	M_y
(1)	(2)	(3)	(4)	(5)	(6)	(7)	(8)
3.00	0	0	+0.025	0	+0.014	0	–0.082
	1/4	+0.010	+0.019	+0.007	+0.013	–0.014	–0.071
	1/2	+0.005	+0.010	+0.008	+0.010	–0.011	–0.055
	3/4	–0.033	–0.004	–0.018	–0.000	–0.006	–0.028
	1	–0.126	–0.025	–0.092	–0.018	0	0
2.50	0	0	+0.027	0	+0.013	0	–0.074
	1/4	+0.012	+0.022	+0.007	+0.013	–0.013	–0.066
	1/2	+0.011	+0.014	+0.008	+0.010	–0.011	–0.053
	3/4	–0.021	–0.001	–0.010	–0.001	–0.005	–0.027
	1	–0.108	–0.022	–0.077	–0.015	0	0
2.00	0	0	+0.027	0	+0.009	0	–0.060
	1/4	+0.013	+0.023	+0.006	+0.010	–0.012	–0.059
	1/2	+0.015	+0.016	+0.010	+0.010	–0.100	–0.049
	3/4	–0.008	+0.003	–0.002	+0.003	–0.005	–0.027
	1	–0.086	–0.017	–0.059	–0.012	0	0
1.75	0	0	+0.025	0	+0.007	0	–0.050
	1/4	+0.012	+0.022	+0.005	+0.008	–0.010	–0.052
	1/2	+0.016	+0.016	+0.010	+0.009	–0.009	–0.046
	3/4	–0.002	–0.005	+0.001	–0.004	–0.005	–0.027
	1	–0.074	–0.015	–0.050	–0.010	0	0
1.50	0	0	+0.021	0	+0.005	0	–0.040
	1/4	+0.008	+0.020	+0.004	+0.007	–0.009	–0.044
	1/2	+0.016	+0.016	+0.010	+0.008	–0.008	–0.042
	3/4	–0.003	–0.006	+0.003	–0.004	–0.005	–0.026
	1	–0.060	–0.012	–0.041	–0.008	0	0
1.25	0	0	+0.015	0	+0.003	0	–0.029
	1/4	+0.005	+0.015	+0.002	+0.005	–0.007	–0.034
	1/2	+0.014	+0.015	+0.008	+0.007	–0.007	–0.037
	3/4	+0.006	+0.007	+0.005	+0.005	–0.005	–0.024
	1	–0.047	–0.009	–0.031	–0.006	0	0

(Continued)

1.0	0	0	+0.009	0	+0.002	0	−0.018
	1/4	+0.002	+0.011	+0.000	+0.003	−0.005	−0.023
	1/2	+0.009	+0.013	+0.005	+0.005	−0.006	−0.029
	3/4	+0.008	−0.008	+0.005	+0.004	−0.004	−0.020
	1	−0.035	−0.007	−0.022	−0.005	0	0
0.75	0	0	+0.004	0	+0.001	0	−0.007
	1/4	+0.001	+0.008	+0.000	+0.002	−0.002	−0.011
	1/2	+0.005	+0.010	+0.002	+0.003	−0.003	−0.017
	3/4	+0.007	+0.007	+0.003	+0.003	−0.003	−0.013
	1	−0.024	−0.005	−0.015	−0.003	0	0
0.50	0	0	+0.001	0	+0.000	0	−0.002
	1/4	+0.000	+0.005	+0.000	+0.001	−0.001	−0.004
	1/2	+0.002	+0.006	+0.001	+0.001	−0.002	−0.009
	3/4	+0.004	+0.006	+0.001	+0.001	−0.001	−0.007
	1	−0.015	−0.003	−0.008	−0.002	0	0

Table 24.14. Moment Coefficients for Tanks with Wall free at Top and Hinged at Bottom

a = height of the wall
b = width of the wall
w = density of the liquid
Horizontal moment = $M_y wa^3$
Vertical moment = $M_x wa^3$

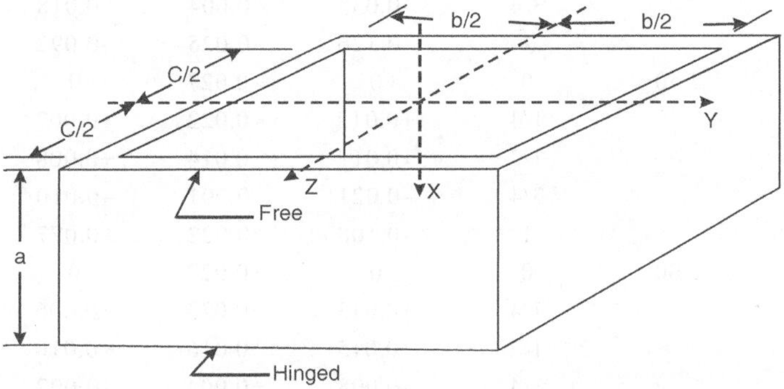

c/a	x/a	$b/a = 3.0$									
		$y = 0$		$y = b/4$		$y = b/2$		$z = c/4$		$z = 0$	
		M_x	M_y	M_x	M_y	M_x	M_y	M_x	M_z	M_x	M_z
(1)	(2)	(3)	(4)	(5)	(6)	(7)	(8)	(9)	(10)	(11)	(12)
3.00	0	0	+0.070	0	+0.027	0	−0.196	0	+0.027	0	+0.070
	1/4	+0.028	+0.061	+0.015	+0.028	−0.034	−0.170	+0.015	+0.028	+0.028	+0.061
	1/2	+0.049	+0.049	+0.032	+0.026	−0.027	−0.137	+0.032	+0.026	+0.049	+0.049
	3/4	+0.046	+0.030	+0.034	+0.018	−0.017	−0.087	+0.034	+0.018	+0.046	+0.030
2.50	0	0	+0.073	0	+0.033	0	−0.169	0	+0.013	0	+0.057
	1/4	+0.028	+0.063	+0.016	+0.033	−0.030	−0.151	+0.009	+0.014	+0.022	+0.050
	1/2	+0.049	+0.050	+0.033	+0.029	−0.025	−0.126	+0.023	+0.017	+0.041	+0.043
	3/4	+0.046	+0.030	+0.037	+0.020	−0.017	−0.084	+0.029	+0.014	+0.040	+0.027
2.00	0	0	+0.075	0	+0.039	0	−0.146	0	−0.005	0	+0.031
	1/4	+0.029	+0.065	+0.017	+0.036	−0.027	−0.133	+0.002	−0.002	+0.013	+0.032
	1/2	+0.050	+0.051	+0.035	+0.032	−0.023	−0.113	+0.016	+0.005	+0.030	+0.029
	3/4	+0.046	+0.031	+0.037	+0.021	−0.016	−0.078	+0.022	+0.008	+0.034	+0.020

(*Continued*)

c/a	x/a	b/a = 3.0									
		y = 0		y = b/4		y = b/2		z = c/4		z = 0	
		M_x	M_y	M_x	M_y	M_x	M_y	M_x	M_z	M_x	M_z
(1)	(2)	(3)	(4)	(5)	(6)	(7)	(8)	(9)	(10)	(11)	(12)
1.75	0	0	+0.076	0	+0.041	0	–0.137	0	–0.018	0	+0.014
	1/4	+0.029	+0.065	+0.018	+0.038	–0.025	–0.125	–0.003	–0.012	+0.007	+0.018
	1/2	+0.050	+0.052	+0.036	+0.033	–0.021	–0.106	+0.011	–0.003	+0.023	+0.020
	3/4	+0.046	+0.031	+0.037	+0.021	–0.015	–0.074	+0.018	+0.004	+0.027	+0.015
1.5	0	0	+0.077	0	+0.043	0	–0.129	0	–0.033	0	–0.006
1.5	1/4	+0.035	+0.010	+0.027	+0.011	–0.007	+0.035	+0.007	+0.006	+0.014	+0.013
	1/2	+0.057	+0.15	+0.045	+0.017	–0.011	–0.057	+0.015	+0.010	+0.027	+0.020
	3/4	+0.051	+0.013	+0.042	+0.014	–0.010	–0.051	+0.019	+0.011	+0.029	+0.017
1.25	0	0	+0.078	0	+0.045	0	–0.122	0	–0.052	0	–0.031
1.25	1/4	+0.035	+0.010	+0.027	+0.011	–0.006	–0.032	+0.003	+0.003	+0.008	+0.011
	1/2	+0.057	+0.015	+0.046	+0.017	–0.011	–0.053	+0.008	+0.006	+0.017	+0.017
	3/4	+0.051	+0.013	+0.042	+0.014	–0.010	–0.048	+0.013	+0.008	+0.021	+0.016
1.00	0	0	+0.079	0	+0.047	0	–0.118	0	–0.074	0	–0.060
1.00	1/4	+0.035	+0.010	+0.027	+0.011	–0.006	–0.029	–0.001	+0.000	+0.002	+0.008
	1/2	+0.057	+0.015	+0.046	+0.017	–0.010	–0.048	+0.002	+0.002	+0.007	+0.014
	3/4	+0.051	+0.013	+0.043	+0.014	–0.009	–0.044	+0.007	+0.004	+0.013	+0.013
0.75	0	0	+0.079	0	+0.047	0	–0.120	0	–0.098	0	–0.092
0.75	1/4	+0.035	+0.010	+0.028	+0.011	–0.005	–0.025	–0.003	–0.005	–0.002	+0.001
	1/2	+0.057	+0.015	+0.046	+0.017	–0.008	–0.042	–0.003	–0.005	–0.001	+0.007
	3/4	+0.052	+0.013	+0.043	+0.014	–0.008	–0.039	+0.002	–0.002	+0.006	+0.007
0.50	0	0	+0.078	0	+0.047	0	–0.30	0	–0.126	0	–0.123
0.50	1/4	+0.036	+0.010	+0.028	+0.011	–0.004	–0.021	–0.004	–0.011	–0.005	–0.008
	1/2	+0.057	+0.015	+0.047	+0.017	–0.007	–0.035	–0.007	–0.016	–0.006	–0.010
	3/4	+0.052	+0.013	+0.043	+0.014	–0.007	–0.033	–0.004	–0.010	–0.001	–0.004

c/a	x/a	b/a = 2.5									
		y = 0		y = b/4		y = b/2		z = c/4		z = 0	
		M_x	M_y	M_x	M_y	M_x	M_y	M_x	M_z	M_x	M_z
(1)	(2)	(3)	(4)	(5)	(6)	(7)	(8)	(9)	(10)	(11)	(12)
2.50	0	0	+0.061	0	+0.019	0	–0.138	0	+0.019	0	+0.061
	1/4	+0.024	+0.053	+0.010	+0.022	–0.026	–0.132	+0.010	+0.022	+0.024	+0.053
	1/2	+0.042	+0.044	+0.025	+0.022	–0.023	–0.115	+0.025	+0.022	+0.042	+0.004
	3/4	+0.041	+0.027	+0.030	+0.016	–0.016	–0.078	+0.030	+0.016	+0.041	+0.027
2.00	0	0	+0.065	0	+0.026	0	–0.118	0	+0.003	0	+0.038
	1/4	+0.025	+0.055	+0.012	+0.027	–0.023	–0.113	+0.005	+0.006	+0.015	+0.037
	1/2	+0.043	+0.046	+0.028	+0.025	–0.020	–0.102	+0.018	+0.011	+0.032	+0.033
	3/4	+0.042	+0.028	+0.031	+0.018	–0.014	–0.070	+0.023	+0.011	+0.034	+0.022
1.75	0	0	+0.067	0	+0.030	0	–0.108	0	–0.006	0	+0.025
	1/4	+0.025	+0.057	+0.013	+0.030	–0.021	–0.104	+0.001	–0.002	+0.010	+0.026

(*Continued*)

	1/2	+0.044	+0.047	+0.029	+0.027	–0.019	–0.096	+0.013	+0.004	+0.025	+0.025
	3/4	+0.043	+0.028	+0.033	+0.019	–0.013	–0.066	+0.019	+0.008	+0.029	+0.019
1.50	0	0	+0.068	0	+0.033	0	–0.100	0	–0.018	0	+0.008
	1/4	+0.026	+0.058	+0.014	+0.032	–0.019	–0.097	–0.033	–0.012	+0.004	+0.013
	1/2	+0.045	+0.047	+0.030	+0.029	–0.018	–0.089	+0.008	–0.002	+0.017	+0.017
	3/4	+0.043	+0.029	+0.034	+0.019	–0.013	–0.063	+0.015	+0.004	+0.024	+0.015
1.25	0	0	+0.069	0	+0.035	0	–0.092	0	–0.030	0	–0.010
	1/4	+0.026	+0.059	+0.015	+0.034	–0.018	–0.089	–0.006	–0.024	–0.002	–0.003
	1/2	+0.045	+0.048	+0.031	+0.031	–0.016	–0.082	+0.003	–0.012	+0.008	+0.007
	3/4	+0.044	+0.029	+0.034	+0.020	–0.012	–0.059	+0.011	–0.002	+0.018	+0.008
1.00	0	0	+0.070	0	+0.037	0	–0.082	0	–0.045	0	–0.032
	1/4	+0.026	+0.060	+0.015	+0.036	–0.017	–0.083	–0.010	–0.036	–0.008	–0.021
	1/2	+0.046	+0.048	+0.031	+0.032	–0.015	–0.077	–0.003	–0.021	–0.001	–0.008
	3/4	+0.044	+0.029	+0.033	+0.021	–0.011	–0.056	+0.006	–0.008	+0.011	–0.000
0.75	0	0	+0.070	0	+0.038	0	–0.082	0	–0.062	0	–0.055
	1/4	+0.025	+0.060	+0.015	+0.037	–0.016	–0.078	–0.014	–0.053	–0.014	–0.042
	1/2	+0.045	+0.047	+0.030	+0.032	–0.014	–0.071	–0.008	–0.035	–0.009	–0.025
	3/4	+0.043	+0.029	+0.033	+0.020	–0.011	–0.054	+0.002	–0.016	+0.005	–0.011
0.50	0	0	+0.069	0	+0.039	0	–0.080	0	–0.081	0	–0.080
	1/4	+0.025	+0.059	+0.014	+0.038	–0.015	–0.075	–0.019	–0.072	–0.019	–0.068
	1/2	+0.044	+0.046	+0.028	+0.032	–0.014	–0.068	–0.014	–0.056	–0.017	–0.048
	3/4	+0.042	+0.028	+0.032	+0.019	–0.010	–0.052	–0.003	–0.030	–0.002	–0.026

		$b/a = 2.0$									
c/a	x/a	$y = 0$		$y = b/4$		$y = b/2$		$z = c/4$		$z = 0$	
		M_x	M_y	M_x	M_y	M_x	M_y	M_x	M_z	M_x	M_z
(1)	(2)	(3)	(4)	(5)	(6)	(7)	(8)	(9)	(10)	(11)	(12)
2.00	0	0	+0.045	0	+0.011	0	–0.091	0	+0.011	0	+0.045
	1/4	+0.016	+0.042	+0.006	+0.014	–0.019	–0.094	+0.006	+0.014	+0.016	+0.042
	1/2	+0.033	+0.036	+0.020	+0.016	–0.018	–0.089	+0.020	+0.016	+0.033	+0.036
	3/4	+0.036	+0.024	+0.025	+0.014	–0.013	–0.065	+0.025	+0.014	+0.036	+0.024
1.75	0	0	+0.048	0	+0.015	0	–0.081	0	–0.001	0	+0.032
	1/4	+0.017	+0.044	+0.007	+0.017	–0.017	–0.085	+0.003	+0.006	+0.012	+0.032
	1/2	+0.034	+0.038	+0.021	+0.019	–0.017	–0.083	+0.015	+0.011	+0.027	+0.029
	3/4	+0.036	+0.024	+0.025	+0.015	–0.012	–0.061	+0.020	+0.012	+0.031	+0.021
1.50	0	0	+0.050	0	+0.019	0	–0.072	0	–0.010	0	+0.018
	1/4	+0.018	+0.046	+0.008	+0.021	–0.015	–0.077	+0.000	–0.002	+0.007	+0.020
	1/2	+0.035	+0.039	+0.022	+0.021	–0.015	–0.076	+0.009	+0.004	+0.020	+0.022
	3/4	+0.036	+0.025	+0.026	+0.016	–0.012	–0.058	+0.016	+0.008	+0.025	+0.017
1.25	0	0	+0.052	0	+0.023	0	–0.064	0	–0.021	0	–0.000
	1/4	+0.009	+0.048	+0.009	+0.024	–0.014	–0.068	–0.002	–0.013	+0.001	+0.005
	1/2	+0.023	+0.041	+0.023	+0.023	–0.014	–0.069	+0.005	–0.004	+0.011	+0.012
	3/4	+0.026	+0.025	+0.026	+0.017	–0.011	–0.054	+0.011	+0.002	+0.016	+0.011
1.00	0	0	+0.054	0	+0.027	0	–0.058	0	–0.037	0	–0.023
	1/4	+0.019	+0.050	+0.010	+0.027	–0.012	–0.062	–0.005	–0.025	–0.005	–0.013

(*Continued*)

	1/2	+0.037	+0.042	+0.024	+0.025	–0.013	–0.064	–0.000	–0.015	+0.001	+0.000
	3/4	+0.037	+0.026	+0.027	+0.018	–0.010	–0.051	+0.006	–0.006	+0.008	+0.004
0.75	0	0	+0.055	0	+0.030	0	–0.058	0	–0.049	0	–0.044
	1/4	+0.018	+0.051	+0.011	+0.029	–0.012	–0.062	–0.009	–0.040	–0.010	–0.031
	1/2	+0.038	+0.043	+0.025	+0.026	–0.012	–0.062	–0.005	–0.029	–0.007	–0.015
	3/4	+0.037	+0.026	+0.027	+0.018	–0.010	–0.049	+0.002	–0.015	+0.001	–0.004
0.50	0	0	+0.054	0	+0.030	0	–0.065	0	–0.064	0	–0.061
	1/4	+0.018	+0.052	+0.011	+0.029	–0.014	–0.068	–0.012	–0.056	–0.014	–0.051
	1/2	+0.038	+0.044	+0.025	+0.025	–0.013	–0.064	–0.010	–0.045	–0.012	–0.034
	3/4	+0.037	+0.026	+0.026	+0.017	–0.010	–0.050	–0.003	–0.026	–0.004	–0.018

		b/a = 1.50									
c/a	*x/a*	*y* = 0		*y* = *b*/4		*y* = *b*/2		*z* = *c*/4		*z* = 0	
		M_x	M_y	M_x	M_y	M_x	M_y	M_x	M_z	M_x	M_z
(1)	(2)	(3)	(4)	(5)	(6)	(7)	(8)	(9)	(10)	(11)	(12)
1.50	0	0	+0.027	0	+0.005	0	–0.052	0	+0.005	0	+0.027
	1/4	+0.009	+0.028	+0.003	+0.008	–0.012	–0.059	+0.003	+0.008	+0.009	+0.028
	1/2	+0.022	+0.027	+0.012	+0.011	–0.013	–0.063	+0.012	+0.011	+0.022	+0.027
	3/4	+0.027	+0.020	+0.017	+0.011	–0.010	–0.052	+0.017	+0.011	+0.027	+0.020
1.25	0	0	+0.031	0	+0.008	0	–0.045	0	–0.005	0	+0.011
	1/4	+0.010	+0.031	+0.005	+0.012	–0.010	–0.050	+0.001	–0.001	+0.004	+0.015
	1/2	+0.024	+0.030	+0.014	+0.014	–0.011	–0.056	+0.007	+0.006	+0.014	+0.020
	3/4	+0.027	+0.021	+0.018	+0.012	–0.010	–0.048	+0.013	+0.006	+0.018	+0.016
1.00	0	0	+0.035	0	+0.013	0	–0.038	0	–0.016	0	–0.006
	1/4	+0.011	+0.034	+0.006	+0.016	–0.008	–0.042	–0.002	–0.010	–0.001	+0.001
	1/2	+0.025	+0.032	+0.015	+0.017	–0.010	–0.049	+0.002	–0.003	+0.006	+0.010
	3/4	+0.028	+0.022	+0.019	+0.014	–0.009	–0.045	+0.008	+0.002	+0.009	+0.010
0.75	0	0	+0.038	0	+0.016	0	–0.034	0	–0.024	0	–0.019
	1/4	+0.011	+0.036	+0.007	+0.018	–0.008	–0.038	–0.005	–0.020	–0.004	–0.013
	1/2	+0.025	+0.033	+0.016	+0.019	–0.008	–0.042	–0.002	–0.014	–0.001	–0.004
	3/4	+0.028	+0.022	+0.019	+0.015	–0.008	–0.041	+0.003	–0.007	+0.002	+0.001
0.50	0	0	+0.040	0	+0.017	0	–0.036	0	–0.030	0	–0.028
	1/4	+0.010	+0.037	+0.007	+0.019	–0.008	–0.040	–0.008	–0.031	–0.007	–0.027
	1/2	+0.024	+0.034	+0.017	+0.020	–0.009	–0.044	–0.006	–0.027	–0.006	–0.020
	3/4	+0.028	+0.022	+0.018	+0.016	–0.008	–0.040	–0.002	–0.018	–0.004	–0.010

		b/a = 1.0									
c/a	*x/a*	*y* = 0		*y* = *b*/4		*y* = *b*/2		*z* = *c*/4		*z* = 0	
		M_x	M_y	M_x	M_y	M_x	M_y	M_x	M_z	M_x	M_z
(1)	(2)	(3)	(4)	(5)	(6)	(7)	(8)	(9)	(10)	(11)	(12)
1.00	0	0	+0.010	0	+0.002	0	–0.019	0	+0.002	0	+0.010
	1/4	+0.002	+0.013	+0.000	+0.003	–0.005	–0.025	+0.000	+0.003	+0.002	+0.013

(*Continued*)

	1/2	+0.010	+0.017	+0.005	+0.006	−0.007	−0.036	+0.005	+0.006	+0.010	+0.017
	3/4	+0.015	+0.015	+0.009	+0.007	−0.007	−0.036	+0.009	+0.007	+0.015	+0.015
0.75	0	0	+0.016	0	+0.007	0	−0.013	0	−0.004	0	+0.003
	1/4	+0.003	+0.017	+0.001	+0.008	−0.004	−0.020	−0.001	−0.005	−0.001	+0.003
	1/2	+0.011	+0.020	+0.006	+0.009	−0.007	−0.033	+0.002	−0.001	+0.005	+0.007
	3/4	+0.016	+0.014	+0.009	+0.009	−0.006	−0.032	+0.004	+0.002	+0.009	+0.008
0.50	0	0	+0.020	0	+0.011	0	−0.011	0	−0.007	0	−0.005
	1/4	+0.003	+0.018	+0.001	+0.010	−0.004	−0.018	−0.002	−0.012	−0.003	−0.007
	1/2	+0.012	+0.016	+0.008	+0.010	−0.006	−0.032	+0.001	−0.009	+0.002	−0.005
	3/4	+0.017	+0.013	+0.010	+0.009	−0.006	−0.031	+0.002	−0.005	+0.006	+0.001

Table 24.15. Moment Coefficients for Tanks with Walls Hinged at Top and Bottom

a = height of the wall

b = width of the wall

w = density of the liquid

Horizontal moment = $M_y wa^3$

Vertical moment = $M_x wa^3$

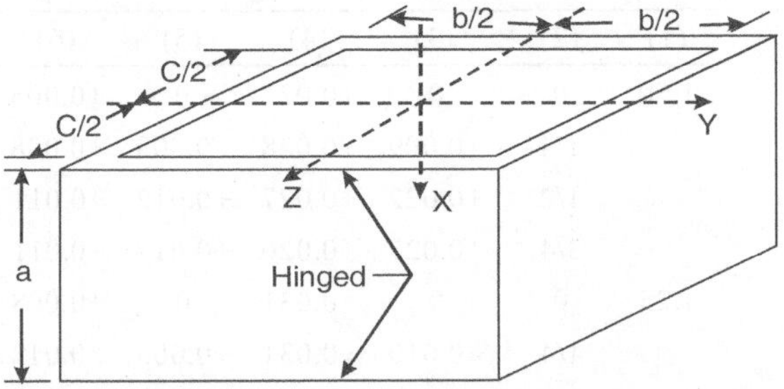

c/a	x/a	b/a = 3.0									
		y = 0		y = b/4		y = b/2		z = c/4		z = 0	
		M_x	M_y	M_x	M_y	M_x	M_y	M_x	M_z	M_x	M_z
(1)	(2)	(3)	(4)	(5)	(6)	(7)	(8)	(9)	(10)	(11)	(12)
3.00	1/4	+0.035	+0.010	+0.026	+0.011	−0.008	−0.039	+0.026	+0.011	+0.035	+0.010
	1/2	+0.057	+0.016	+0.044	+0.017	−0.013	−0.063	+0.044	+0.017	+0.057	+0.016
	3/4	+0.051	+0.013	+0.041	+0.014	−0.011	−0.055	−0.041	+0.014	+0.051	+0.013
2.50	1/4	+0.035	+0.010	+0.026	+0.011	−0.008	−0.039	+0.021	+0.010	+0.031	+0.011
	1/2	+0.057	+0.016	+0.044	+0.017	−0.012	−0.062	+0.036	+0.017	+0.052	+0.017
	3/4	+0.051	+0.013	+0.041	−0.014	−0.011	−0.055	+0.036	+0.014	+0.047	+0.014
2.00	1/4	+0.035	+0.010	+0.026	+0.011	−0.008	−0.038	+0.015	+0.010	+0.025	+0.013
	1/2	+0.057	+0.016	+0.045	+0.017	−0.012	−0.062	+0.028	+0.015	+0.043	+0.020
	3/4	+0.051	+0.013	+0.042	+0.014	−0.011	−0.054	+0.029	+0.013	+0.041	+0.016
1.75	1/4	+0.035	+0.010	+0.027	+0.011	−0.007	−0.037	+0.011	+0.008	+0.020	+0.013
	1/2	+0.057	+0.015	+0.045	+0.017	−0.012	−0.060	+0.021	+0.013	+0.036	+0.020
	3/4	+0.051	+0.013	+0.042	+0.014	−0.011	−0.053	+0.024	+0.012	+0.036	+0.016
1.50	1/4	+0.035	+0.010	+0.027	+0.011	−0.007	+0.035	+0.007	+0.006	+0.014	+0.013
	1/2	+0.057	+0.015	+0.045	+0.017	−0.011	−0.057	+0.015	+0.010	+0.027	+0.020
	3/4	+0.051	+0.013	+0.042	+0.014	−0.010	−0.051	+0.019	+0.011	+0.029	+0.017

(Continued)

1.25	1/4	+0.035	+0.010	+0.027	+0.011	–0.006	–0.032	+0.003	+0.003	+0.008	+0.011
	1/2	+0.057	+0.015	+0.046	+0.017	–0.011	–0.053	+0.008	+0.006	+0.017	+0.017
	3/4	+0.051	+0.013	+0.042	+0.014	–0.010	–0.048	+0.013	+0.008	+0.021	+0.016
1.00	1/4	+0.035	+0.010	+0.027	+0.011	–0.006	–0.029	–0.001	+0.000	+0.002	+0.008
	1/2	+0.057	+0.015	+0.046	+0.017	–0.010	–0.048	+0.002	+0.002	+0.007	+0.014
	3/4	+0.051	+0.013	+0.043	+0.014	–0.009	–0.044	+0.007	+0.004	+0.013	+0.013
0.75	1/4	+0.035	+0.010	+0.028	+0.011	–0.005	–0.025	–0.003	–0.005	–0.002	+0.001
	1/2	+0.057	+0.015	+0.046	+0.017	–0.008	–0.042	–0.003	–0.005	–0.001	+0.007
	3/4	+0.052	+0.013	+0.043	+0.014	–0.008	–0.039	+0.002	–0.002	+0.006	+0.007
0.50	1/4	+0.036	+0.010	+0.028	+0.011	–0.004	–0.021	–0.004	–0.011	–0.005	–0.008
	1/2	+0.057	+0.015	+0.047	+0.017	–0.007	–0.035	–0.007	–0.016	–0.006	–0.010
	3/4	+0.052	+0.013	+0.043	+0.014	–0.007	–0.033	–0.004	–0.010	–0.001	–0.004

		$b/a = 2.5$									
c/a	x/a	$y = 0$		$y = b/4$		$y = b/2$		$z = c/4$		$z = 0$	
		M_x	M_y	M_x	M_y	M_x	M_y	M_x	M_z	M_x	M_z
(1)	(2)	(3)	(4)	(5)	(6)	(7)	(8)	(9)	(10)	(11)	(12)
2.50	1/4	+0.031	+0.011	+0.021	+0.010	–0.008	–0.038	+0.021	+0.010	+0.031	+0.011
	1/2	+0.052	+0.017	+0.036	+0.017	–0.012	–0.062	+0.036	+0.017	+0.052	+0.017
	3/4	+0.047	+0.015	+0.036	+0.014	–0.011	–0.055	+0.036	+0.014	+0.047	+0.015
2.00	1/4	+0.031	+0.011	+0.021	+0.010	–0.008	–0.038	+0.015	+0.009	+0.025	+0.012
	1/2	+0.052	+0.017	+0.036	+0.017	–0.012	–0.061	+0.028	+0.015	+0.042	+0.020
	3/4	+0.047	+0.015	+0.036	+0.014	–0.011	–0.054	+0.029	+0.013	+0.041	+0.016
1.75	1/4	+0.032	+0.011	+0.021	+0.010	–0.007	–0.037	+0.011	+0.008	+0.020	+0.012
	1/2	+0.052	+0.018	+0.037	+0.017	–0.012	–0.059	+0.022	+0.013	+0.035	+0.021
	3/4	+0.047	+0.015	+0.036	+0.014	–0.011	–0.053	+0.024	+0.012	+0.035	+0.017
1.50	1/4	+0.032	+0.011	+0.022	+0.010	–0.007	–0.035	+0.007	+0.006	+0.014	+0.013
	1/2	+0.052	+0.018	+0.037	+0.017	–0.011	–0.057	+0.015	+0.010	+0.027	+0.021
	3/4	+0.047	+0.015	+0.036	+0.014	–0.010	–0.051	+0.019	+0.010	+0.029	+0.017
1.25	1/4	+0.032	+0.011	+0.023	+0.010	–0.006	–0.032	+0.003	+0.004	+0.007	+0.012
	1/2	+0.053	+0.018	+0.038	+0.017	–0.011	–0.053	+0.008	+0.007	+0.018	+0.019
	3/4	+0.048	+0.015	+0.038	+0.014	–0.010	–0.048	+0.014	+0.008	+0.022	+0.016
1.00	1/4	+0.032	+0.011	+0.023	+0.011	–0.006	–0.028	–0.001	+0.000	+0.002	+0.008
	1/2	+0.053	+0.018	+0.038	+0.017	–0.010	–0.048	+0.002	+0.002	+0.007	+0.014
	3/4	+0.048	+0.015	+0.038	+0.015	–0.009	–0.044	+0.007	+0.004	+0.013	+0.013
0.75	1/4	+0.033	+0.011	+0.024	+0.011	–0.005	–0.024	–0.003	–0.005	–0.002	+0.002
	1/2	+0.054	+0.018	+0.039	+0.017	–0.008	–0.041	–0.003	–0.005	–0.000	+0.005
	3/4	+0.049	+0.015	+0.038	+0.015	–0.008	–0.039	+0.000	–0.002	+0.006	+0.006
0.50	1/4	+0.033	+0.012	+0.024	+0.011	–0.004	–0.021	–0.004	–0.011	–0.005	–0.008
	1/2	+0.054	+0.018	+0.040	+0.017	–0.007	–0.035	–0.007	–0.016	–0.006	–0.010
	3/4	+0.049	+0.015	+0.039	+0.015	–0.007	–0.034	–0.004	–0.010	–0.001	–0.004

(Continued)

		b/a = 2.0									
c/a	x/a	y = 0		y = b/4		y = b/2		z = c/4		z = 0	
		M_x	M_y	M_x	M_y	M_x	M_y	M_x	M_z	M_x	M_z
(1)	(2)	(3)	(4)	(5)	(6)	(7)	(8)	(9)	(10)	(11)	(12)
2.00	1/4	+0.025	+0.013	+0.015	+0.009	−0.007	−0.037	+0.015	+0.009	+0.025	+0.013
	1/2	+0.042	+0.020	+0.028	+0.015	−0.012	−0.059	+0.028	+0.015	+0.042	+0.020
	3/4	+0.040	+0.016	+0.029	+0.013	−0.011	−0.053	+0.029	+0.013	+0.040	+0.016
1.75	1/4	+0.025	+0.013	+0.015	+0.009	−0.007	−0.036	+0.011	+0.008	+0.020	+0.013
	1/2	+0.042	+0.020	+0.028	+0.015	−0.012	−0.058	+0.022	+0.013	+0.035	+0.021
	3/4	+0.040	+0.016	+0.029	+0.013	−0.010	−0.052	+0.024	+0.012	+0.035	+0.017
1.50	1/4	+0.025	+0.013	+0.016	+0.009	−0.007	−0.034	+0.007	+0.006	+0.014	+0.013
	1/2	+0.043	+0.020	+0.028	+0.015	−0.011	−0.056	+0.015	+0.011	+0.027	+0.021
	3/4	+0.041	+0.016	+0.029	+0.013	−0.010	−0.010	−0.019	+0.010	+0.029	+0.017
1.25	1/4	+0.026	+0.013	+0.016	+0.010	−0.006	−0.032	+0.003	+0.003	+0.007	+0.011
	1/2	+0.043	+0.020	+0.029	+0.015	−0.010	−0.052	+0.008	+0.007	+0.018	+0.019
	3/4	+0.041	+0.016	+0.030	+0.013	−0.010	−0.048	+0.013	+0.008	+0.021	+0.016
1.00	1/4	+0.026	+0.013	+0.017	+0.010	−0.006	−0.028	−0.001	+0.000	+0.002	+0.008
	1/2	+0.044	+0.020	+0.030	+0.016	−0.009	−0.048	+0.002	+0.002	+0.007	+0.014
	3/4	+0.041	+0.016	+0.031	+0.014	−0.009	−0.044	+0.007	+0.004	+0.013	+0.013
0.75	1/4	+0.027	+0.013	+0.018	+0.010	−0.005	−0.024	−0.003	−0.004	−0.001	+0.002
	1/2	+0.045	+0.020	+0.031	+0.016	−0.008	−0.040	−0.002	−0.004	+0.000	+0.005
	3/4	+0.042	+0.016	+0.032	+0.014	−0.008	−0.041	+0.002	−0.002	+0.005	+0.008
0.50	1/4	+0.027	+0.013	+0.019	+0.010	−0.004	−0.021	−0.004	−0.010	−0.004	−0.007
	1/2	+0.046	+0.020	+0.033	+0.017	−0.007	−0.034	−0.006	−0.015	−0.006	−0.009
	3/4	+0.042	+0.016	+0.032	+0.015	−0.007	−0.037	−0.003	−0.010	−0.002	−0.003

		b/a = 1.5									
c/a	x/a	y = 0		y = b/4		y = b/2		z = c/4		z = 0	
		M_x	M_y	M_x	M_y	M_x	M_y	M_x	M_z	M_x	M_z
(1)	(2)	(3)	(4)	(5)	(6)	(7)	(8)	(9)	(10)	(11)	(12)
1.50	1/4	+0.015	+0.013	+0.008	+0.007	−0.006	−0.032	+0.008	+0.007	+0.015	+0.013
	1/2	+0.028	+0.021	+0.016	+0.011	−0.010	−0.052	+0.016	+0.011	+0.028	+0.021
	3/4	+0.030	+0.017	+0.020	+0.011	−0.010	−0.048	+0.020	+0.011	+0.030	+0.017
1.25	1/4	+0.016	+0.013	+0.009	+0.008	−0.006	−0.029	+0.004	+0.004	+0.009	+0.012
	1/2	+0.029	+0.021	+0.017	+0.012	−0.010	−0.049	+0.009	+0.008	+0.018	+0.019
	3/4	+0.030	+0.017	+0.020	+0.012	−0.009	−0.045	+0.014	+0.009	+0.023	+0.016
1.00	1/4	+0.016	+0.013	+0.010	+0.009	−0.005	−0.025	+0.000	+0.001	+0.003	+0.008
	1/2	+0.030	+0.021	+0.019	+0.012	−0.009	−0.043	+0.003	+0.003	+0.008	+0.014
	3/4	+0.031	+0.017	+0.021	+0.013	−0.008	−0.041	+0.008	+0.005	+0.014	+0.014
0.75	1/4	+0.018	+0.014	+0.011	+0.010	−0.004	−0.021	−0.002	−0.003	−0.001	+0.002
	1/2	+0.032	+0.022	+0.021	+0.014	−0.007	−0.036	−0.002	−0.004	+0.001	+0.005
	3/4	+0.032	+0.018	+0.022	+0.014	−0.007	−0.036	+0.002	−0.000	+0.000	+0.008
0.50	1/4	+0.020	+0.016	+0.013	+0.012	−0.003	−0.017	−0.003	−0.009	−0.004	−0.006
	1/2	+0.035	+0.024	+0.023	+0.018	−0.006	−0.031	−0.006	−0.014	−0.005	−0.007
	3/4	+0.034	+0.020	+0.024	+0.016	−0.007	−0.033	−0.003	−0.008	−0.001	−0.001

(Continued)

c/a	x/a	b/a = 1.0, y = 0, M_x	y = 0, M_y	y = b/4, M_x	y = b/4, M_y	y = b/2, M_x	y = b/2, M_y	z = c/4, M_x	z = c/4, M_z	z = 0, M_x	z = 0, M_z
(1)	(2)	(3)	(4)	(5)	(6)	(7)	(8)	(9)	(10)	(11)	(12)
1.00	1/4	+0.005	+0.009	+0.002	+0.003	–0.004	–0.020	+0.002	+0.003	+0.005	+0.009
	1/2	+0.011	+0.016	+0.006	+0.006	–0.007	–0.035	+0.006	+0.006	+0.011	+0.016
	3/4	+0.016	+0.015	+0.009	+0.007	–0.007	–0.035	+0.009	+0.007	+0.016	+0.015
0.75	1/4	+0.006	+0.010	+0.003	+0.004	–0.003	–0.016	+0.000	+0.000	+0.001	+0.005
	1/2	+0.013	+0.017	+0.008	+0.008	–0.006	–0.029	+0.001	+0.001	+0.005	+0.009
	3/4	+0.017	+0.016	+0.010	+0.008	–0.006	–0.031	+0.004	+0.003	+0.008	+0.010
0.50	1/4	+0.007	+0.011	+0.005	+0.006	–0.002	–0.010	–0.002	–0.005	–0.003	–0.002
	1/2	+0.015	+0.018	+0.010	+0.010	–0.004	–0.021	–0.003	–0.007	–0.003	–0.002
	3/4	+0.018	+0.016	+0.012	+0.010	–0.005	–0.026	–0.001	–0.004	–0.000	+0.001

Table 24.16. Shear at Edges of Wall Panel Hinged at Top and Bottom

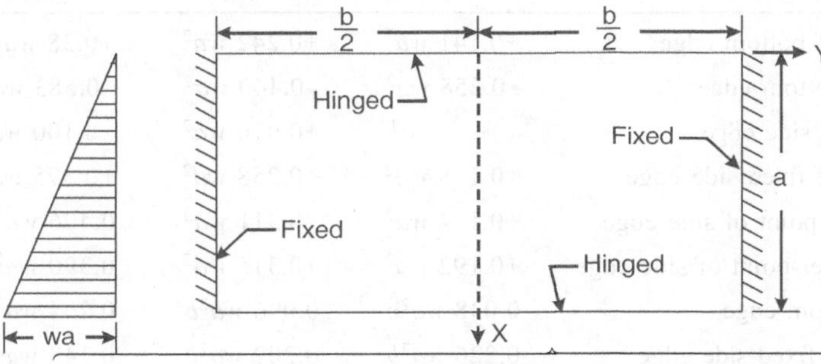

	b/a = $\frac{1}{2}$	1	2	5	10	Infinity
(1)	(2)	(3)	(4)	(5)	(6)	(7)
Mid-point of bottom edge	+0.1407 wa^2	+0.2419 wa^2	+0.3290 wa^2	—	—	+0.3333 wa^2
Corner at bottom edge	–0.2575 wa^2	–0.4397 wa^2	–0.5833 wa^2	—	—	–0.6000 wa^2
Mid-point of fixed side edge	+0.1280 wa^2	+0.2582 wa^2	+0.3604 wa^2	—	—	+0.3912 wa^2
Lower third-point of side edge	+0.1736 wa^2	+0.3113 wa^2	+0.4023 wa^2	—	—	+0.4116 wa^2
Lower quarter-point of side edge	+0.1919 wa^2	+0.3153 wa^2	+0.3904 wa^2	—	—	+0.3980 wa^2
Total at top edge	0.0000 wa^2b	0.0052 wa^2b	0.0538 wa^2b	0.1203 wa^2b	0.1435 wa^2b	0.1667 wa^2b
Total of bottom edge	0.0480 wa^2b	0.0960 wa^2b	0.1818 wa^2b	0.2715 wa^2b	0.3023 wa^2b	0.3333 wa^2b
Total at one fixed side edge	0.2260 wa^2b	0.1994 wa^2b	0.1322 wa^2b	0.0541 wa^2b	0.0271 wa^2b	0.275 wa^3*
Total at all four edges	0.5000 wa^2b	0.5000 wa^2b	0.5000 wa^2b	0.5000 wa^2b	0.5000 wa^2b	0.5000 wa^2b

Note 1 — Negative sign indicates that reaction acts in direction of load.

Note 2 — w = Density of the liquid.

*Estimated.

Table 24.17. Shear at Edges of Wall Panel Free at Top and Hinged at Bottom.

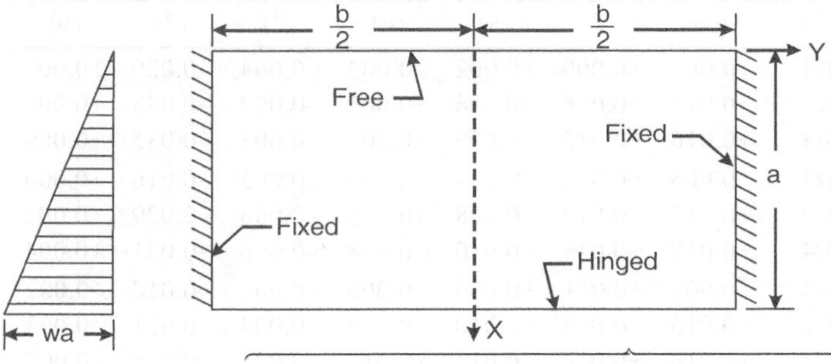

	b/a			
	$\frac{1}{2}$	1	2	3
(1)	(2)	(3)	(4)	(5)
Mid-point of bottom edge	+0.141 wa^2	+0.242 wa^2	+0.38 wa^2 †	+0.45 wa^2 †
Corner of bottom edge	–0.258 wa^2	–0.440 wa^2	–0.583 wa^2	–0.590 wa^2
Top of fixed side edge	wa^2	+0.010 wa^2	+0.100 wa^2	+0.165 wa^2
Mid-point of fixed side edge	+0.128 wa^2	+0.258 wa^2	+0.375 wa^2	+0.406 wa^2
Lower third-point of side edge	+0.174 wa^2	+0.311 wa^2	0.406 wa^2	+0.416 wa^2
Lower quarter-point of side edge	+0.192 wa^2	+0.315 wa^2	0.390 wa^2	+0.398 wa^2
Total at bottom edge	0.048 wa^2b	0.096 wa^2b	0.204 wa^2b	0.286 wa^2b
Total at one fixed side edge	0.226 wa^2b	0.202 wa^2b	0.148 wa^2b	0.107 wa^2b
Total at all four edges	0.500 wa^2b	0.500 wa^2b	0.500 wa^2b	0.500 wa^2b

Note 1 — w = Density of the liquid.
Note 2 — Data are derived by modifying values computed for walls hinged at top and bottom.
*Negative sign indicates that reaction acts in direction of load.
† This value could not be estimated accurately beyond two decimal places.

24.3.4 IS Code Method for Design of Rectangular Tanks

The design of rectangular tank walls basically consists of finding the moments and shears at salient location of individual panels considered fixed along vertical edges but having different edge conditions at top and bottom. The moment coefficients for these panels under hydrostatic pressure are presented in Tables 24.11 to 24.13. These wall panels have been assumed to act as thin plates under the following end conditions:

1. Top hinged, bottom hinged.
2. Top free, bottom hinged.
3. Top free, bottom fixed.

In square tanks, moment coefficients may be taken direct from the above mentioned tables. However, in a rectangular tank, an adjustment has to be made in a manner similar to the modification of fixed end moments in a frame analysts by the method of moment distribution. The moment coefficients computed in the manner described are tabulated in Tables 24.14 and 24.15 for top and bottom edge conditions as shown for single cell tanks with a large number of ratios of *b*/*a* and *c*/*a*, *b* being the larger and c the smaller of horizontal tank dimensions.

The values of shear force along the edge of a tank wall would be required for computation of diagonal tension and bond stresses. Along vertical edges, the shear in one wall will cause axial tension in the adjacent wall and should be combined with the bending moment for the purpose of determining die tensile reinforcement. Shear coefficients for a wall panel (of width *h,* height *a* and subjected to hydrostatic pressure due to liquid of density (w) considered fixed at the two vertical edges and assumed hinged at top and bottom edges arc given in Fig. 24.18 and Table 24.16. Shear coefficients for the same wall panel fixed at the two vertical edges and assumed hinged at the bottom but free at top edge are given in Fig. 24.19 and Table 24.17.

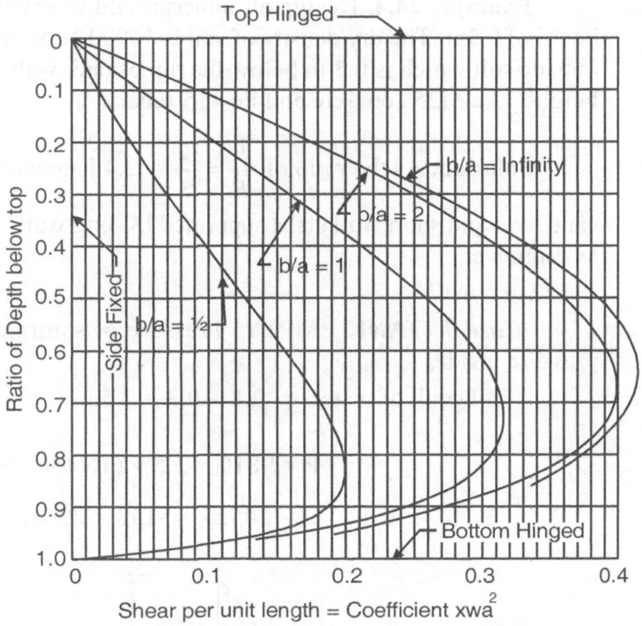

Fig. 24.18 Coefficients for Wall Panel, Fixed at Vertical Edges, Hinged at Top and Bottom Edges

24.4 UNDERGROUND TANKS

When a tank is built underground, the wall should be investigated for both internal water pressure and external earth pressure. The external pressure may be due to dry earth or due to a combination of earth and ground water (i.e. saturated soil). The design principles for such tanks are same as that for tanks resting on ground. In case of tank built below ground with earth covering the roof slab, there will be trapezoidal lateral earth pressure on the wall.

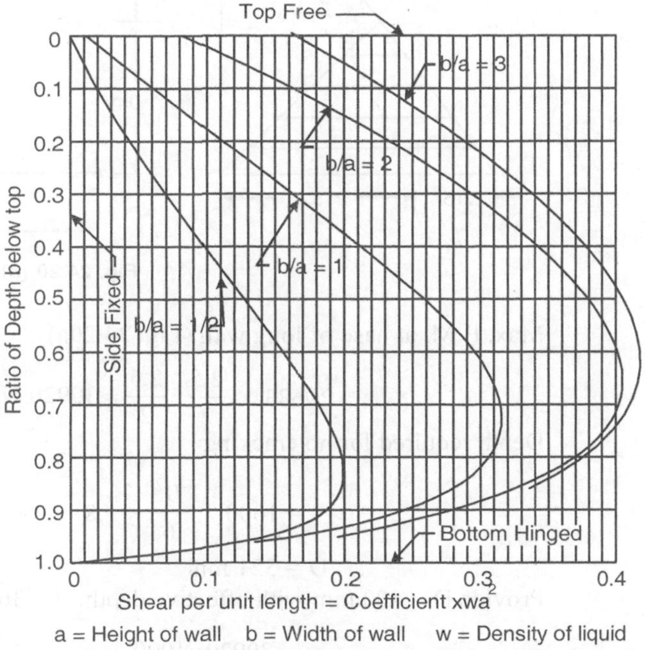

Fig. 24.19. Coefficients for Wall Panel, Fixed at Vertical Edges, Hinged at Bottom Edge and Free at Top Edge

Whenever there is a possibility of ground Water Table to rise above the base slab, not only walls are to be designed for satured soil up to the extent of water above the base slab, but also the base slab is to be designed for the net uplift pressure of water (less weight of slab for tank empty case). In addition, check has to be applied for stability of the tank as a whole against uplift.

Example 24.4. Design an underground reservoir 12m × 5m × 2.8m deep including a free board of 0.3m. The dry density of soil is 16000 N/m^3 and the angle of repose of dry soil is 30. The outside soil which is 0.3 m below the top of tank wall may be taken as fully saturated upto its full height. Use M25 concrete and Fe 415 rebars.

Solution. As the ratio of $\frac{L}{B} = \frac{12}{5} = 2.4$ is greater than 2, the long walls will be designed as cantilever and short wall slabs supported by long walls. Bottom 1 m of short wall shall be designed as cantilever.

Long wall

Case 1 (TANK EMPTY) Pressure of saturated soil acting from outside and no water pressure inside.

$$\text{Depth of water} = 2.8 - 0.3 = 2.5\text{m}$$

$$P = 9810 \times 2.5 + (16000 - 9810)\, 2.5 \times \frac{1 - \sin 30°}{1 + \sin 30°}$$

$$= 24525 + 5158 = 29683 \text{ N/m}$$

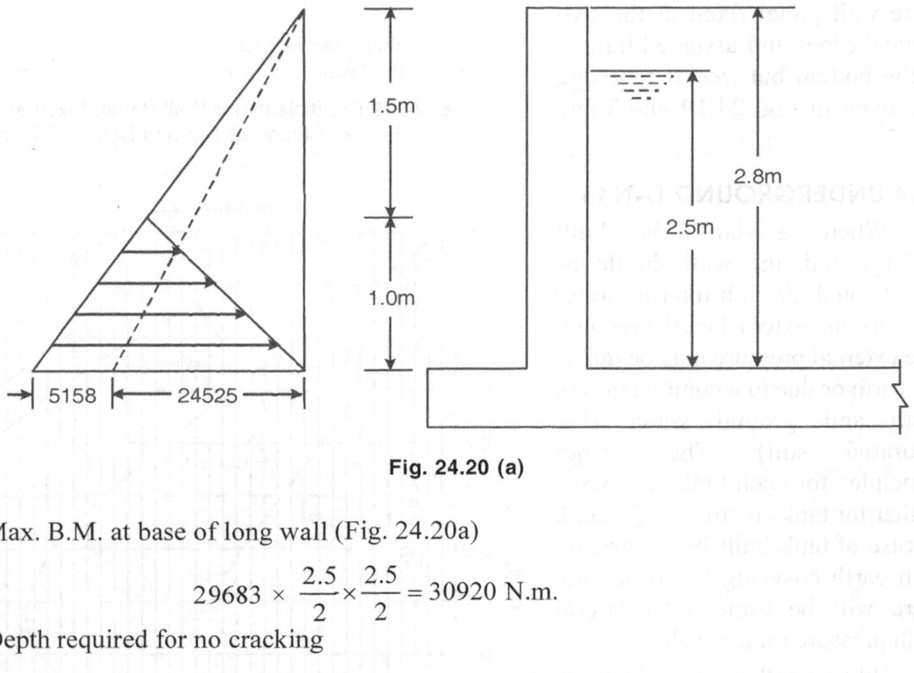

Fig. 24.20 (a)

Max. B.M. at base of long wall (Fig. 24.20a)

$$29683 \times \frac{2.5}{2} \times \frac{2.5}{2} = 30920 \text{ N.m.}$$

Depth required for no cracking

$$= 1000 \times \frac{D^2}{6} \times 1.8 = 30920 \times 1000$$

$$D = 321 \text{ mm}$$

Provide D = 400 mm with effective depth, $d = 360$ mm

$$A_{st} \text{ required} = \frac{30920 \times 1000}{0.90 \times 130 \times 360} = 734 \text{ mm}^2$$

Provide 10 mm ϕ @ 100 mm c/c vertically on outside face up to half the height of wall and thereafter curtail alternate bars.

Direct Compression is caused in long walls because of earth pressure acting on short walls, which is generally of little effect.

Case-II (TANK FULL)

Pressure of water acting from inside and no earth pressure from outside.

$$\text{Max. B.M. due to water pressure} = 9810 \times \frac{2.5^3}{6}$$

$$= 25547 \text{ N.m.}$$

$$< 30920 \text{ N.m.}$$

$\therefore$ thickness of wall provided i.e. O.K.

$$A_{st} = \frac{25547 \times 1000}{0.90 \times 130 \times 300} = 728 \text{ mm}^2$$

Provide 10 mm ϕ @100 mm c/c vertically on the inside face upto half the height of wall and thereafter curtail alternate bars.

$$\text{Area of distributed steel} = \frac{0.24}{100} \times 340 \times 1000 = 816 \text{ mm}^2$$

$$\text{Area of steel/face} = \frac{816}{2} = 408 \text{mm}^2$$

Provide 10mm ϕ @ 950 mm c/c vertically in long walls

Horizontal Steel 10 mm ϕ @ 150mm

Direct Tension in Long Walls.

Since the short walls are to get their support from the long walls, direct tension shall be caused in long walls.

Direct tension at 1m above base = 9810 × (2.5–1.)5.0/2 = 36787 N

Area of steel required in one metre height

$$\frac{36787}{130} = 282 \text{ mm}^2$$

$$< 1190 \text{ mm}^2$$

$\therefore$ No additional horizontal steel is required is 10 mm ϕ @ 150 mm provided minimum Steel

Short Wall

Case-1 (TANK EMPTY) Pressure of saturated soil acting from outside and no water pressure from inside.

Bottom one metre height

$$\text{B.M. (Fig. 24.20}(b)) = (24525+5158)\ \frac{1}{3}$$

$$= 4947 \text{ N.m.}$$

For no cracking,

$$1000 \times \frac{D^2}{6} \times 1.8 = 4947 \times 1000$$

$$D = 128.4 \text{ mm} < 340 \text{ mm}$$

$$A_{st} = \frac{4947 \times 1000}{0.90 \times 130 \times 300} = 141 \text{ mm}^2$$

$$< 595 \text{ mm}^2$$

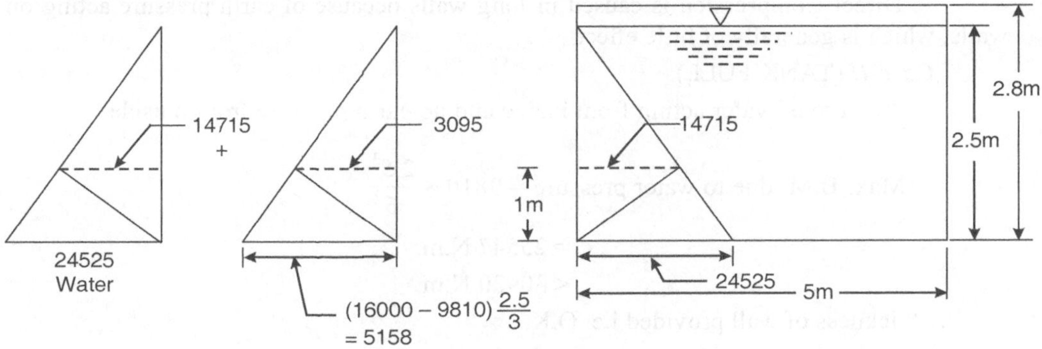

Fig. 24.20 (b)

∴ Distribution steel is O.K.

Wall above 1 m from base

$$\text{Total pressure} = 14715 + 3095 - 17810 \text{ N/m}^2$$

$$\text{B.M. at support} = \frac{pl^2}{12} = \frac{17810 \times (5+0.34)^2}{16} = 42322 \text{ N.m.}$$

$$\text{B.M. at centre of span} = \frac{pl^2}{16} = \frac{17810 \times (5.34)^2}{12} = 31742 \text{ N.m.}$$

Thickness of wall at junction of short wall with long wall for no cracking

$$1000 \times \frac{D^2}{6} \times 1.8 = 42322 \times 100$$

$$D = 375.60\text{mm}$$

Provide haunch inside 100 × 100mm as shown in Fig. 24.20

$$A_{st} \text{ at support} = \frac{42322 \times 1000}{0.90 \times 130 \times 440 - 40}$$

$$= 904 \text{ mm}^2$$

Provide 10 mm ϕ @ 85 mm c/c in one metre height, on inner face horizontally, which may be increased uniformally to 200 mm c/c near top of wall.

Case-II (TANK FULL)

Pressure of water acting from inside and no earth pressure from outside.

Bottom one meter height of wall

$$\text{BM} = \frac{14175}{2} \times 1^2 + 9810 \times \frac{1}{2} \times \frac{1}{3} = 8992.5 \text{ N.m.}$$

Minimum reinforcement is O.K.

$$\text{Wall pressure at 1.5m depth} = 9810 \times 1.5$$

$$= 14715 \text{ N/m}^2$$

$$\text{B.M. at support} = \frac{wl^2}{12} = \frac{14715 \times 5.34^2}{12}$$

$$= 34967 \text{ N.m.}$$

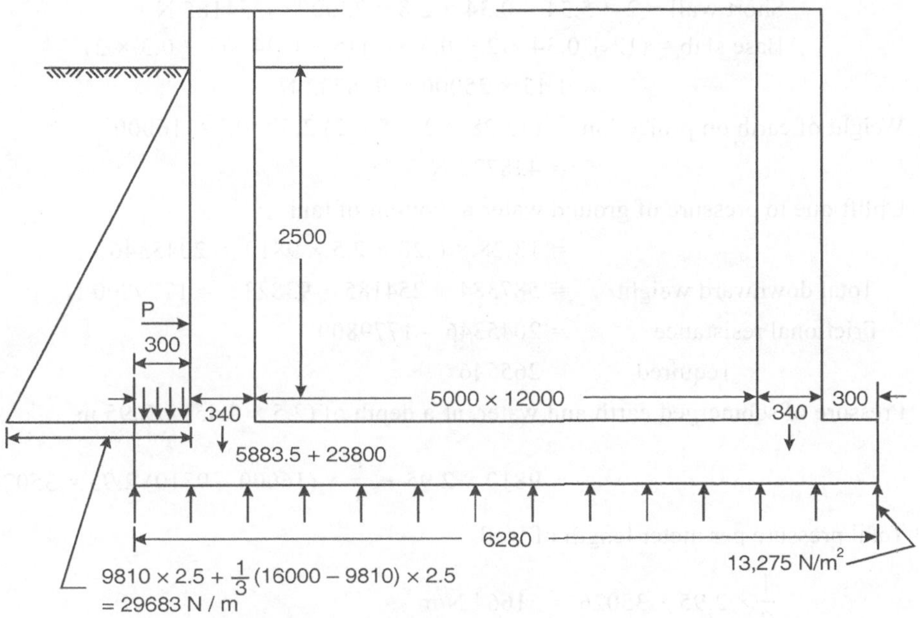

Fig. 24.21

Direct tension from end 1m of long wall

$$= 14715 \times 1 = 14715 \text{ N/m height}$$

Net B.M. at support = (M-T.e)

$$= 34967 - 14715 \left(\frac{0.34}{2} - 0.04\right)$$

$$= 33054 \text{ N.m.}$$

$$A_{st} = \frac{33054 \times 1000}{0.90 \times 130 \times (440 - 40)} + \frac{14715}{130}$$

$$= 820 \text{ mm}^2$$

Provide 10mm ϕ @ 90mm c/c on inside face horizontally which may be increased uniformly to 180 mm c/c near top of wall

Net B.M. at centre of span = $M - T.e$

$$= 26225 - 14715 \times 0.13 = 24312 \text{ N.m.}$$

$$A_{st} = \frac{24312 \times 1000}{0.90 \times 130 \times 300} + \frac{14715}{130} = 805 \text{mm}^2$$

Provide 10mm ϕ @ 100mm c/c on outer face horizontally which may be uniformally increased to 200mm c/c near top of wall.

BASE SLAB (450 mm thick, M20)

Check against Uplift

Projection of 0.3 n beyond the face of wall is provided to add the stability uplift.

Weight of vertical walls

$$\text{Long walls} = 2 \times 12.3 \times 0.34 \times 2.8 \times 25000 = 587384 \text{ N}$$

Short wall = 2 × 5.34 × 0.34 × 2.8 × 25000 = 254185 N

Base slab = (12 + 0.34 × 2 + 0.3 × 2) (5 + 0.34 × 2 + 0.3 × 2)

× 0.45 × 25000 = 938232 N

Weight of earth on projections = (13.28 × 2 + 5 × 2) 2.5 × 0.3 × 16000

= 438720 N

Uplift due to pressure of ground water at bottom of tank

= 13.28 × 6.28 × 2.5 × 9810 = 2045346

Total downward weight = 587384 + 254185 + 938232 = 1779800 N

Frictional resistance = 2045346 – 1779800

required = 265546

Pressure of submerged earth and water, at a depth of (2.5 × 0.45) = 2.95 m

$$= 9810 \times 2.95 + \frac{1}{3} \times (16000 - 9810)\,2.95 = 35026 \text{ N/m}^2$$

Total pressure per meter length of wall

$$\frac{1}{2} \times 2.95 \times 35026 = 51664 \text{ N/m}$$

The angle of repose of soil will be low as it is saturated. Assuming coefficient of friction as 0.15 frictional resistance per metre length of wall

= 0.15 × 51664

= 7750 N

Total frictional resistance from 4 side walls

= 7750(5.68 + 12.68)2 = 284480 N

> 265546 N

Safe.

Consider one metre length of slab, upward pressure of water per square metre

= 2.5 × 9810 = 24525 N/m^2

Self weight of slab = 1 × 1 × 0.45 × 25000 – 11250 N/m^2

Net upward pressure = 24525 – 11250 = 13275 N/m^2

Weight of wall per metre run = 0.34 × 2.8 × 25000 = 23800 N/m

Weight of earth on projection = 16000 × 2.5 × 0.3 = 12000 N/m

Net unbalanced force = 13275 × 6.28 - 2 (23800 + 12000) = 11767 N

$$\text{Reaction on each wall} = \frac{11767}{2} = 5883.5 \text{ N}$$

B.M. at edge of cantilever portion

$$= \frac{13275 \times 0.3^2}{2} + 29683 \times \frac{2.5}{2}\left(\frac{2.5}{3} + 0.225\right) - \frac{12000 \times 0.3}{2}$$

= 597 + 39268 – 1800

= 38065

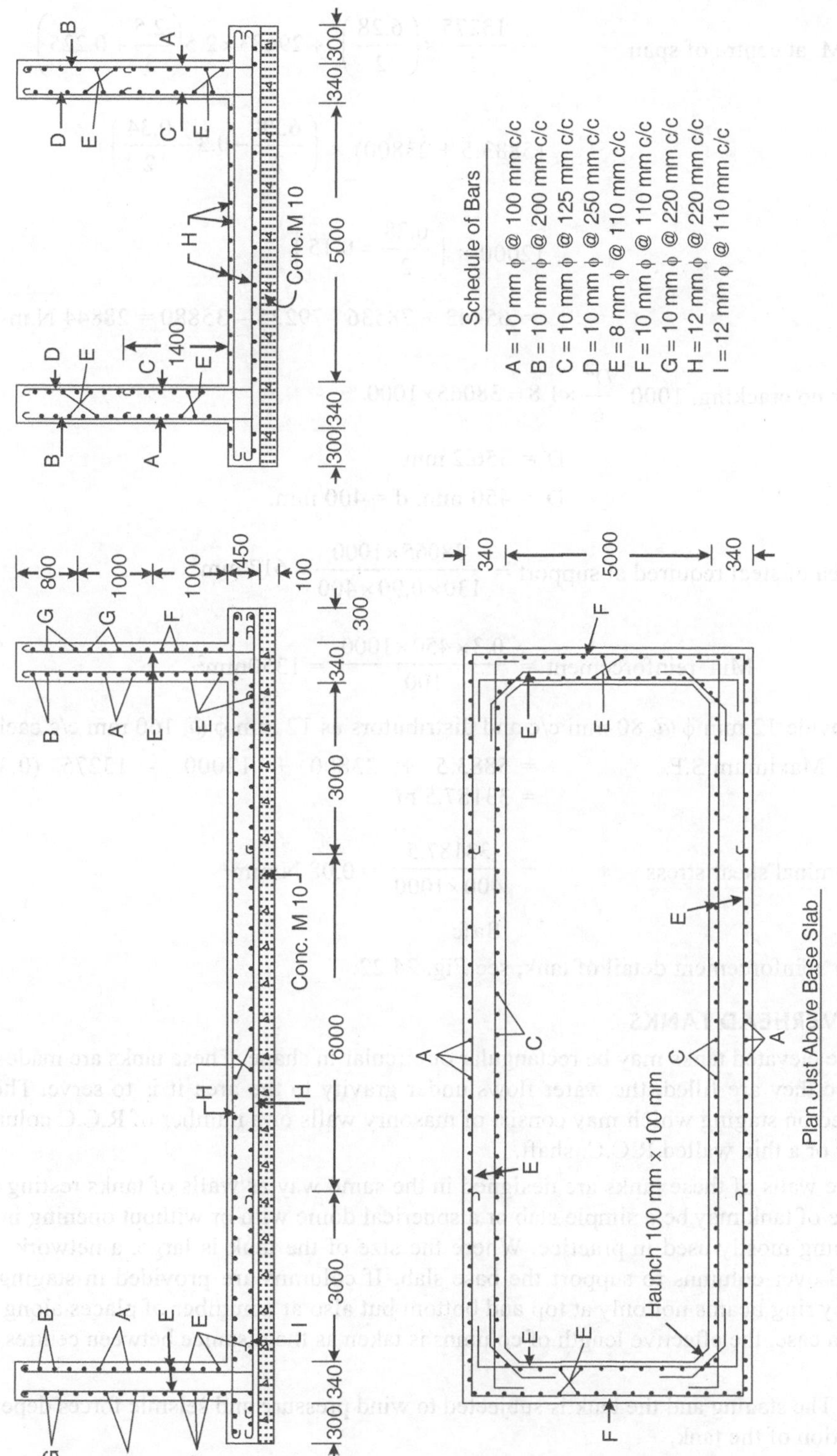

Fig. 24.22. Under ground rectangular tank.

B.M. at centre of span $= \frac{13275}{2} \times \left(\frac{6.28}{2}\right)^2 + 29683 \times 2.5\left(\frac{2.5}{3} + 0.225\right)$

$- (5883.5 + 23800) \times \left(\frac{6.28}{2} - 0.3 - \frac{0.34}{2}\right)$

$-12000 \times \left(\frac{6.28}{2} = 0.15\right)$

$= 65443 + 78536 - 79225 - 35880 = 28844$ N.m

For no cracking, $1000\ \frac{D^2}{6} \times 1.8 = 38065 \times 1000$

D = 356.2 mm

D = 450 mm, d = 400 mm.

Area of steel required at support $= \frac{38065 \times 1000}{130 \times 0.90 \times 400} = 617$ mm^2

Min. reinforcement $= \frac{0.3 \times 450 \times 1000}{100} = 1350$mm^2

Provide 12 mm ϕ @ 80 mm c/c and distributors as 12 mm ϕ @ 160 mm c/c each face.

Maximum S.F. = 5883.5 + 23800 + 12000 – 13275 (0.3 + 0.34)
= 33187.5 m

Nominal shear stress $= \frac{33187.5}{400 \times 1000} = 0.08$ N/mm^2

Safe

For reinforcement detail of tank, see Fig. 24.22.

24.5 OVERHEAD TANKS

The elevated tanks may be rectangular or circular in shape. These tanks arc made elevated so that once they are filled, the water flows under gravity to the area it is to serve. The tanks are supported on staging which may consist of masonry walls or a number of R.C.C columns braced together or a thin walled R.C.C. shaft.

The walls of these tanks are designed in the same way as walls of tanks resting on ground. The base of tank may be a simple slab or a spherical dome with or without opening in the centre, latter being mostly used in practice. Where the size of the tank is large, a network of beams is provided over columns to support the base slab. If columns are provided in staging, these are braced by ring beams not only at top and bottom but also at a number of places along the height. In such a case, the effective length of columns is taken as the distance between centres of adjacent bracings.

The staging and the tank is subjected to wind pressure and seismic forces depending upon the location of the tank.

24.5.1 Analysis of Staging for Wind Forces

Let *'P'* be the total wind force acting on the rank and the staging above ground level acting at a height *'h'* above the ground level. This force produces tension on windward side columns and compression on the leeward side columns. This axial force. *F,* is proportional to the distance of the column from the *C.G.* of the column group.

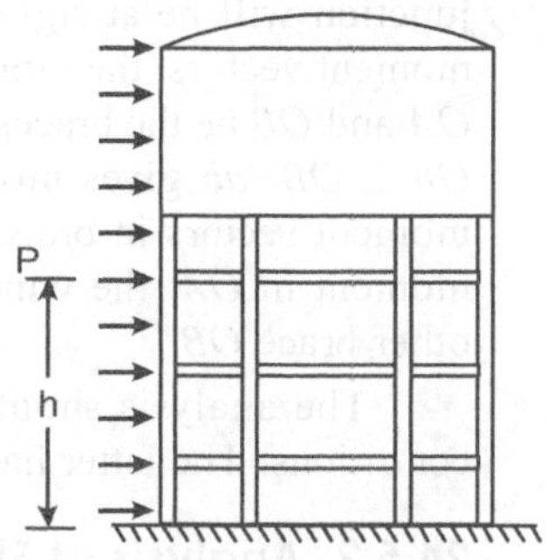

Let r_1, r_2 ...be the distance of the column from the *CG* of the column group, measured parallel to the direction of wind.

The forte, *F,* in any individual column is computed by

$$F = \frac{(Ph)r}{\Sigma r^2}$$

where $\Sigma r^2 = r_1^2 + r_2^2 + r_3^2 + ...$

and r = distance of individual column from *CG* of column group (Fig. 24.23).

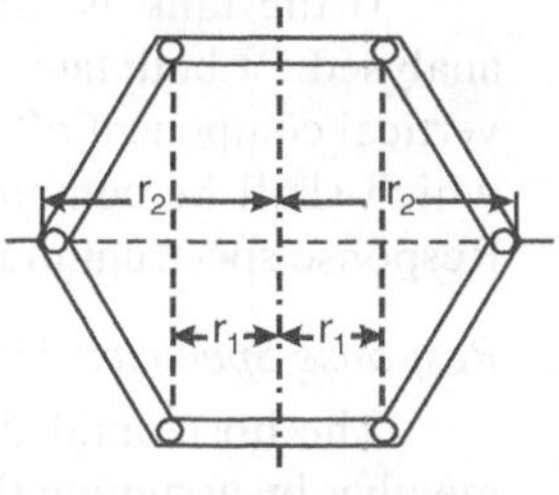

Fig. 24.23

It may be noted that while computing the value of *P,* the surface of tank proper and staging shall be considered. Full wind shall be assumed to act on the faces of columns and braces on windward side.

Whereas 50% of the wind shall be assumed to act on the faces of all such columns and braces which are exactly behind these facing the direct wind.

The columns are also subjected to bending moment due to the horizontal wind force acting on the column. It is assumed that all the columns share this horizontal shear force equally and that the point of contraflexure lies at mid height of the column between centres of adjacent braces.

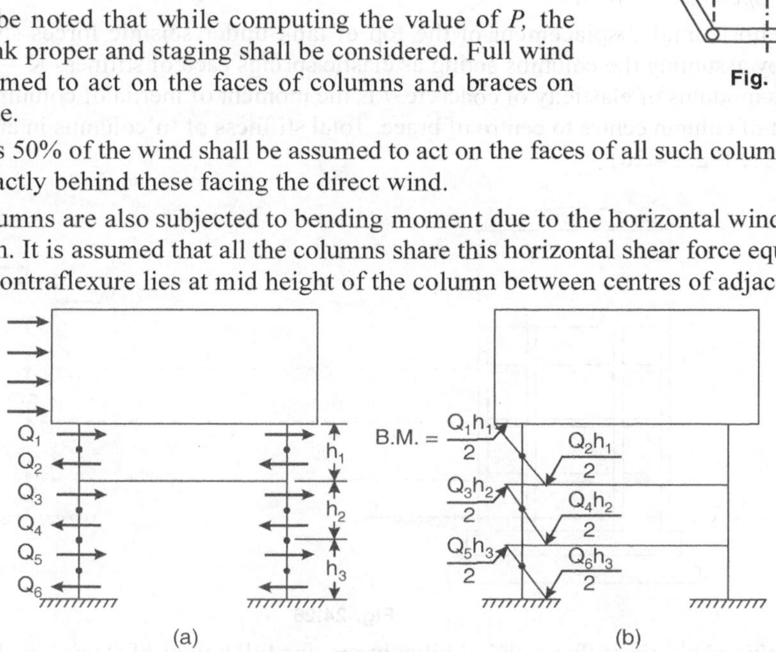

Fig. 24.24

The bending moment in an individual column shall be equal to horizontal shear in that multiplied by half the column height between braces, as shown in Fig. 24.24. The columns shall be designed for axial force and bending moment.

The bracing is designed for bending moment and shear. Same reinforcement is generally provided at top and bottom face of the brace section as the wind may blow from any side. Circular tanks are sometimes provided with inclined columns, in such cases, the vertical component of each column is taken by columns as axial force and horizontal component by braces. The moments in the braces meeting at a column (Fig. 24.25) can be computed as follows.

The axes of moments in column above and below a particular brace level will be acting at right angle to the direction of wind. The axes of moments acting in the two braces meeting at the junction will be at right angles to their axes. By triangle law of moment vectors, the moments in the braces can be determined. Let OA and OB be the braces meeting at column O. Draw $Oa \perp OA$ and $Ob \perp OB$. ab gives moment in the column and Oa and Ob are moment vectors in braces OA and OB respectively. For maximum moment in OA, the wind must be blowing all right angles to the other brace OB.

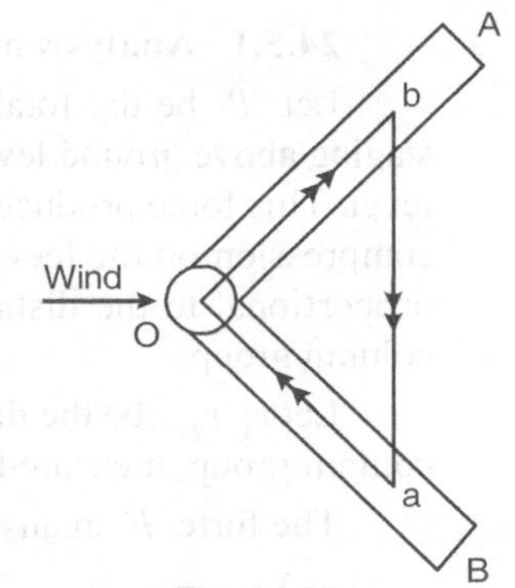

Fig. 24.25

The analysis should be carried out for tank full and empty conditions. The latter has been found to be the worst case.

24.5.2 Analysis of Staging For Seismic Forces

If the tank is located in seismic prone areas, it should be analysed for both tank full and empty conditions. The horizontal or vertical component of earth-quake forces on tank shall be dependent upon the total weight of tank which shall be computed by adding weight, of tank proper and one-third weight of staging. Response spectrum method is used for computing the seismic force:

Response Spectrum Method

The horizontal displacement of the top of tank under seismic forces shall be cantilever member by assuming the columns acting as elastic springs each of stiffness K = 12E I/h^3 where E is young's modulus of elasticity of concrete. I is the moment of inertia of column section and h is the height of column centre to centre of brace. Total stiffness of 'n' columns in any height 'h' = 12 $El\ n/h^3$, (See Fig. 24.26).

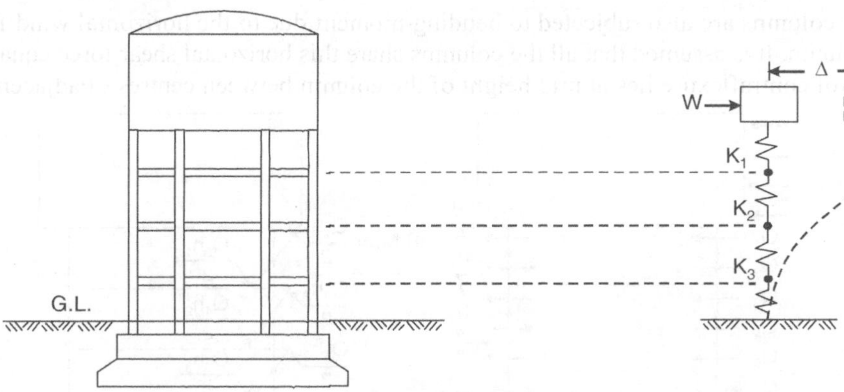

Fig. 24.26

Combined elastic stiffness 'K' of all columns for full height of staging is determined from

$$1/K = 1/K_1 + 1/K_2 + 1/K_3 + 1/K_4 + \ldots$$

Natural time period 'T' in seconds of the tank to vibrate under seismic forces is obtained from

$$T = 2\pi\sqrt{\frac{W}{g.K}}$$

The horizontal and vertical components of seismic forces can be computed by the following formulae:

F_h = Horizontal force acting on the C.G. of tank

$= \propto_h . w$

F_v = Vertical forces acting on the C.G. of tank

$= \propto_v.w$

where $W = W_f$ for tank full

$= W_e$ for tank empty

$\propto_h$ = Horizontal seismic coefficient

$$= \frac{Z.I.}{2R}\frac{S_a}{g}$$

Provided that for any structure with time period ≥ 0.1 sec, the values of $\propto_h$ for a structure will not be less than $\frac{Z}{2}$ whatever be the value of $\frac{I}{R}$, where

Z = Zone factor (Table 24.18 and appendix F and G)

I = Importance factor (Table 24.19)

R = Response spectrum factor (Table 24.20)

$\frac{S_a}{g}$ = Average response acceleration coefficient for rock and soil sites as given in Fig. 24.25

$$T = \text{Time period} = \frac{0.09\,h}{\sqrt{d}}$$

where h = height of structure in metres

d = base dimension in metres in the direction under consideration

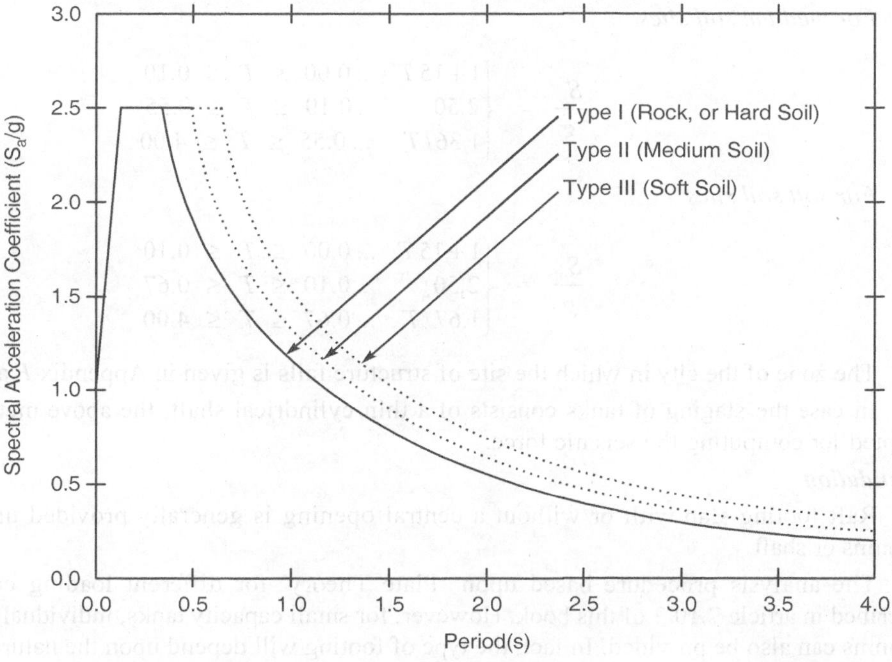

Fig. 24.27. Response spectra for Rock and Soil sites for 5 per cent damping

Table 24.19. Zone Factor 'Z'

Seismic Zone	II	III	IV	V
Seismic Intensity	Low	Moderate	Severe	Very severe
Z	0.1	0.16	0.24	0.36

Table 24.20. Importance Factor '*I*'

S.No.	*Structure*	*I*
1.	Important Services and community Buildings (schools, monumental structures, emergency buildings like telephone exchange, television and radio stations, fire stations, cinemas, assembly halls, subway station and power stations)	1.5
2.	All other buildings	1.0

Table 24.21. Response Spectrum Factor '*R*'

S.No.	*Building Frame System*	*R*
1.	Ordinary R.C.C. moment resisting frames	3.0
2.	Special R.C.C. moment resisting frames	5.0

For rocky or hard soil sites

$$\frac{S_a}{g} = \begin{cases} 1+15T \ldots 0.00 \le T \le 0.1 \\ 2.5 \quad \ldots 0.10 \le T \le 0.55 \\ 1.00/T \ldots 0.40 \le T \le 4.00 \end{cases}$$

For medium soil sites

$$\frac{S_a}{g} = \begin{cases} 1+15T \quad \ldots 0.00 \le T \le 0.10 \\ 2.50 \quad \ldots 0.10 \le T \le 0.55 \\ 1.36/T \quad \ldots 0.55 \le T \le 4.00 \end{cases}$$

For soft soil sites

$$\frac{S_a}{g} = \begin{cases} 1+15T \quad \ldots 0.00 \le T \le 0.10 \\ 2.50 \quad \ldots 0.10 \le T \le 0.67 \\ 1.67/T \quad \ldots 0.67 \le T \le 4.00 \end{cases}$$

The zone of the city in which the site of structure falls is given in Appendix *F* and *G*.

In case the staging of tanks consists of a thin cylindrical shaft, the above method may be adopted for computing the seismic force.

Foundation

Raft footing slab with or without a central opening is generally provided under R.C.C. columns or shaft.

The analysis procedure based upon 'Plate Theory' for different loading conditions is described in article 7.10.2 of this book. However, for small capacity tanks, individual footings for columns can also be provided. In fact, the type of footing will depend upon the nature of soil and type of staging. In case of low lying areas of low safe bearing capacity with high ground water table, pile footings are provided. In any case of foundation slab, lean mix of 1:4:8, 150 mm thick may be provided as levelling course.

24.6 INTZE TYPE TANKS

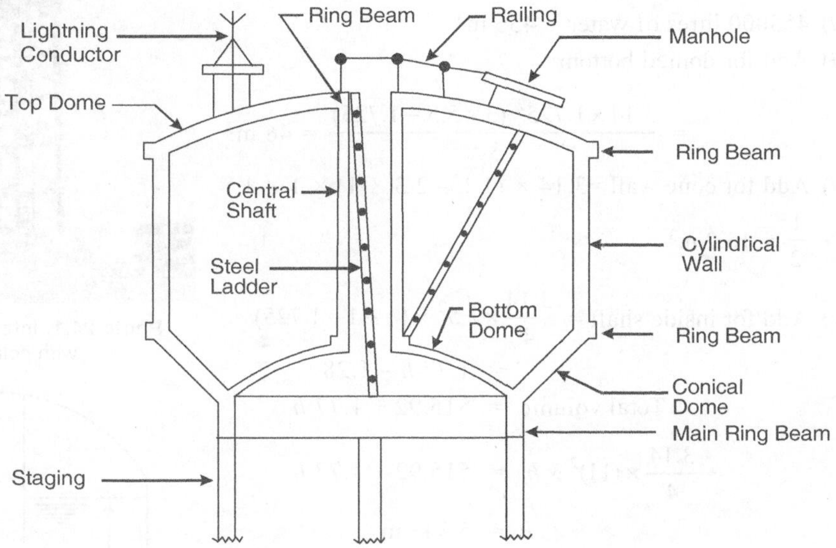

Fig. 24.28. Intze Tank

Most frequently Intze type tanks (Fig. 24.28) are used for capacities ranging from 227.5 Kilolitres (50,000 gallons) to about 1825 Kilolitres (4 lac gallons). For such capacities, simple slab base in tank proper prove highly uneconomical.

24.7 DESIGN CONSIDERATIONS

Design requirements for tank proper shall be the same as for tanks resting on ground but design requirements for staging portion and foundation slab shall be the same as given in rest of a book depending upon which method of design (Working Stress or Limit State) is used. If staging consists of R C.C. columns or a thin cylindrical shaft, minimum percentage vertical steel shall be 0.8% of cross-sectional area of concrete required to transmit direct load using mild steel bars and 0.64% of cross-sectional area of concrete using high strength ribbed tor steel bars. Horizontal rings may be 6 mm dia. bars, @ 200 mm *c/c*. As an alternative, 0.3% of gross cross sectional area of concrete may be provided in each direction *i.e.* horizontally and vertically using mild steel and 0.24% using H.S.Ribbed tor steel. The minimum thickness of cylindrical shaft should be 100 mm.

Example 24.5. Design a Intz type elevated water tank for a capacity of 455 *KL* (1 lac gallons).The diameter at the centre line of the staging shaft may be adopted as 8.50 m. The internal diameter of the lank = 11.0 *m*. The height upto the Full Supply Level (FSL) is 30.48 m (100′-0") from 'ground level'. The other dimensions are as follows:

(*i*) radius of bottom dome = 5.5 m

(*ii*) radius of top dome = 7.5 m

(*iii*) rise of bottom dome = 1.725 m

(*iv*) rise of top dome = 2.17 m

(*v*) external diameter of central shaft inside tank = 15 m

The safe bearing capacity of soil may be adopted as 87.4 KN/m^2. The tank site is in Delhi- Use M30 concrete.

Solution. Let h be the depth of water (Fig. 24.29) for cylindrical portion of the tank. Further the density of water is taken as 10000 N/m^3.

(*i*) 455000 litres of water = 455 m^3

(*ii*) Add for domed bottom

$$= \frac{3.14 \times 1.725^2\ (3 \times 5.5 - 1.725)}{3} = 46 \text{ m}^3$$

(*iii*) Add for cone wall $= 3.14 \times (1.1 - 2/3 \times 1) \times 1 \times 1 \times \frac{1}{2} = 16.2 \text{ m}^3$

(*iv*) Add for inside shaft $= \frac{3.14}{4} \times 1.5^2 \times (h + 1 - 1.725)$

$$= 1.77\,h - 1.28$$

$$\text{Total volume} = 515.92 + 1.77\,h$$

$$\frac{3.14}{4} \times (11)^2 \times h = 515.92 + 1.77\,h$$

$$h = 5.343 \text{ m}$$

Provide $h = 5.4$ m and free board = 0.6 m.

Photo 24.1. Intze Type OHSR with columns

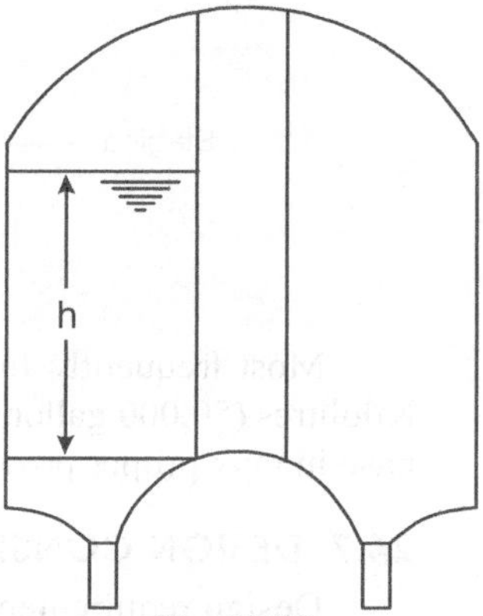

Fig. 24.29

Design of Top Dome (M–30 mix)

Thickness = 75 mm

(*a*) Dead load = $0.075 \times 25000 = 1875$ N/m^2

(*b*) Live load = $(75 - 345 \times 2.17^2/11^2) \times 10 = 616$ N/m^2

Total = 2491 N/m^2

Say = 2500 N/m^2

$\sin\theta = 5.275/7.5 = 0.703,\ \theta = 44.695°$

$\cos\theta = 0.7109$

At Crown

N = Tangential force, T = Circumferential force

$$N = T = \frac{1}{2} wR = \frac{1}{2} \times 2500 \times 7.5 = 9375 \text{ N/m}$$

Circumferential stress = Tangential stress

At Springing

$$= \frac{9375}{75 \times 1000} = 0.125 \text{ N/mm}^2$$

$$N = \frac{wR}{1 + \cos\theta} = \frac{2500 \times 7.5}{1 + 0.7109} = 10960 \text{ N/m.}$$

$$\text{Tangential stress} = \frac{10960}{75 \times 1000} = 0.146 \text{ N/mm}^2$$

$$\text{Circumferential stress} = \frac{w.R}{A}\left(\frac{1-\cos\theta-\cos^2\theta}{1+\cos\theta}\right)$$

$$= \frac{2500\times 7.5}{75\times 1000}\left(\frac{1-0.7109-0.7109^2}{1+0.7109}\right)$$

$$= 0.0316 \text{ N/mm}^2$$

Stresses in top dome are very small

Check for Shear

$$\text{Surface area of dome} = 2 \times 3.14\, Rh$$

$$= 2 \times 3.14 \times 7.5 \times 2.17 = 102.26 \text{ m}^2$$

$$\text{Weight of top dome} = 102.26 \times 2500 = 255650 \text{ N.}$$

$$\text{Load/m of periphery, } W = \frac{255650}{3.14\times 11} = 7400 \text{ N}$$

$$\text{Shear stress} = \frac{7400}{75\times 1000} = 0.099 \text{ N/mm}^2 \text{ (Safe)}$$

$$\text{Horizontal thrust} = W\cot\theta$$

$$= 7400 \times 0.7109/0.703 = 7485 \text{ N.}$$

$$\text{Area of steel required} = 0.24 \times 75 \times 1000/100 = 180 \text{ mm}^2$$

Provide 8 mm ϕ @ 250 mm c/c both ways as shown in fig. 24.30(a).

Design of Top Rib Beam (M–30 mix).

Size 300 mm × 200 mm

Horizontal thrust from top dome = 7485 N

Hoop Tension = $7485 \times 11 \times \frac{1}{2} = 41170$ N

Area of steel required = 41170/140 = 294 mm^2

Provide 10 mm ϕ 4 Nos.

as shown in Fig. 24.30(a)

Provide 8 mm ϕ stirrups @ 300 mm c/c

Equivalent concrete area = 300 × 200 + (13–1) × 4 × 50 = 62400 mm^2

Tensile stress = 41170/62400 = 0.66 N/mm^2 <1.5 N/mm^2 (Safe)

Design of Outer Wall (M-30 mix). (Fig. 24.30a)

Thickness at base = 150 mm tapered to 100 mm at top.

It is assumed that the wall is hinged at base and free at top (for worst conditions of hoop stress).

$$h = 5.4 - 1 = 4.4 \text{ m}$$

$$p = w.h = 10000 \times 4.4 = 44000 \text{ N, } D_{base} = 11 \text{ m.}$$

$$\frac{H^2}{Dt} = \frac{4.4^2}{11\times 0.15} = 11.733,\ w.h.R = 44000 \times 5.5 = 242000 \text{ N/m}$$

Depth from F.S.L (m)	Coeff. K	Hoop tension T = k.w.h.R (N)	Area of steel cm^2		Plain Steel			Tor Steel		
			Plain T/115 (mm^2)	Tor T/150 (mm^2)	Dia	Spacing (Fe250)	Nos.	Dia	Spacing (Fe415)	Nos.
0.44	0.097	23474	204	180	8	142	3	8	300	2
0.88	0.198	47916	417	368	8	110	4	8	250	2
1.32	0.304	73568	640	565	10	110	4	10	200	2
1.76	0.420	101640	884	781	10	80	5	10	250	2
2.20	0.545	131900	1147	1015	12	80	5	10	200	2
2.64	0.665	160950	1400	1238	12	73	6	12	80	3
3.08	0.749	181020	1575	1393	16	110	4	12	160	3
3.52	0.718	173750	1511	1336	12	73	6	12	170	3
3.96	0.475	115000	1000	884	12	88	5	12	220	2
4.40	—	—	—	—	12	88	5	12	220	2

Moments in Wall

$H^2/D.t = 11.733$

Max. B.M. $= 0.0122 \times 10000 \times 4.4' = 10352$ N.m.

Area of steel required for bending

$$= \frac{10392000}{0.87 \times 125 \times 130}$$

$= 765\ mm^2$ on water face,

Provide 10mm ϕ @ 100 mm c/c on water face, vertically and curtail alternative bar after half height & 10 RC 200c/c on outer face.

Minimum distribution & temperature steel

$= 0.35 \times 150 \times 1000/100$

$= 525\ mm^2$ half on each face.

Provide 8 mm ϕ @ 192 mm c/c each face horizontally and outer face).

$$\text{Average force acting at } 0.7\text{ H} = \frac{160950 + 181020 + 173750}{3}$$

$= 171907$ N

Thickness at that section $= 100 + 0.7 \times 50 = 135$ mm.

Equivalent concrete area $= 135 \times 1000 \times (9 - 1) \times 1575$

$= 147600\ mm^2$

Max. tensile stress in concrete $= 171907/147600$

$= 1.165\ N/mm^2$

$< 1.5\ N/mm^2$ (Safe)

Shear at base $= 0.078 \times 10000 \times 4.4^2 = 15100$ N

$$\text{Shear stress at base} = \frac{15100}{1000 \times 120} = 0.126 N/mm^2$$

Design of cone Top (M-30 mix). Fig. 24.30 (a)

Size 250 mm × 750 mm (area 187500 mm^2)

1. Load from roof		= 7400 N/m
2. Load from top rib beam	= 0.3 × 0.2 × 25000	= 1500N/m
3. Weight from vertical wall	= 4.7 × 0.125 × 25000	= 14700N/m
4. Self weight of cone wall	= 0.25 K × 0.75 × 25000	= 4700N/m
5. Weight of water column	= 4.4 × 0.1 × 10000	= 4400 N/m
	Total	= 32700 N/m
	Say w	= 33000 N/m

Horizontal thrust = $w \tan \theta$ = 33000 × 1 = 33000 N.

$$\text{Hoop tension} = 33000 \times (11 - 0.1) \times \frac{1}{2} = 179850 \text{ N.}$$

Since the cone top does not come in direct contact with water and the thickness is also more than 225 mm, the stress in bars may be taken as 125 N/m^2 and in *H45D* steel as 140 N/mm^2

Area of plain steel required = 179850/125 = 1440 mm^2

Provide 20 mm φ 4 Nos. and 12 mm φ 2 Nos.

or 179850/140 = 1284 mm^2, Provide 16 mm φ 6Nos. and 10 mm φ 2Nos.

8 mm φ 2 legged stirrups @ 300 mm c/c

Check for tensile stress

Equivalent concrete area = 250×250 + 9 – 1 ×1357 = 187500 + 10904

= 182656 mm^2

Tensile stress = 179850/198404 = 0.906 N/mm^2, 1.5 N/mm^2 (Safe)

Design of cone Wall (M – 30 mix). Fig. 24.30(a)

Thickness = 300 mm

Slant length = 1.42 m

Diameter at mid height = (11 + 805)/2 = 9.75 m

Loading

(i) Weight of conical wall = 3.14 × 9.75 × 1.42 × 0.3 × 25000

= 326050 N.

(ii) Weight of water column directly resting on conical wall

(a) 3.14 × 10.0 × 4.4 × 10000 × 1.0 = 1382500

(b) $3.14 \times 9.67 \times \frac{1}{2} \times 1 \times 1 \times 10000 = 151500$

(iii) Load from cone top = 33000 × 3.14 × 11 = 1139800

Total = 2999850

Say = 3000000 N.

Max. meridional thrust in cone wall

$$= \frac{3000000}{3.14 \times 9}$$

= 150750 N/m

$$\text{Comp. stress} = \frac{150750}{300 \times 1000} = 0.5025 \text{ N/mm}^2\text{, O.K.}$$

Let at any height 'x' above ring beam, diameter of the conical wall = D_x

Height of water = $(5.4 - x)$ m

$D_x = 9 + 2x$

Weight of cone wall/m² surface

$= 0.3 \times 25000 = 7500$ N/m

W_n = Load acting normal to the surface

$= 7500 \sin\theta + (5.4 - x) \times 10000$

$= 59302 - 10000x$

Hoop tension $= W_n \sec\phi\, D_x/2$

$$= \frac{1}{2} \times (59302 - 10000x) \times (9 - 2x) \times 1.42$$

$= 378940 + 20310 \times - 14200\, x^2$

Max. hoop tension at, $x = 0 = 378940$ N

Max. hoop tension at, $x = 1\text{m} = 385050$ N

$$\text{Area of steel required} = \frac{385050}{\frac{1}{2}(125 + 115)}$$

$= 3210$ mm² (Plain).

or $$\frac{385050}{\frac{130 + 140}{2}} = 2832 \text{ mm}^2 \ (\textit{Tor})$$

Steel in Full slant length $= 2832 \times 1.42 = 4032$ mm² (HYSD bars)

Provide 16 mm ϕ 20 Nos. (10 nos. on inner face and 10 Nos. on outer)

Temp. and distribution steel = $0.35 \times 300 \times 1000 = 1050$ mm² on each face.

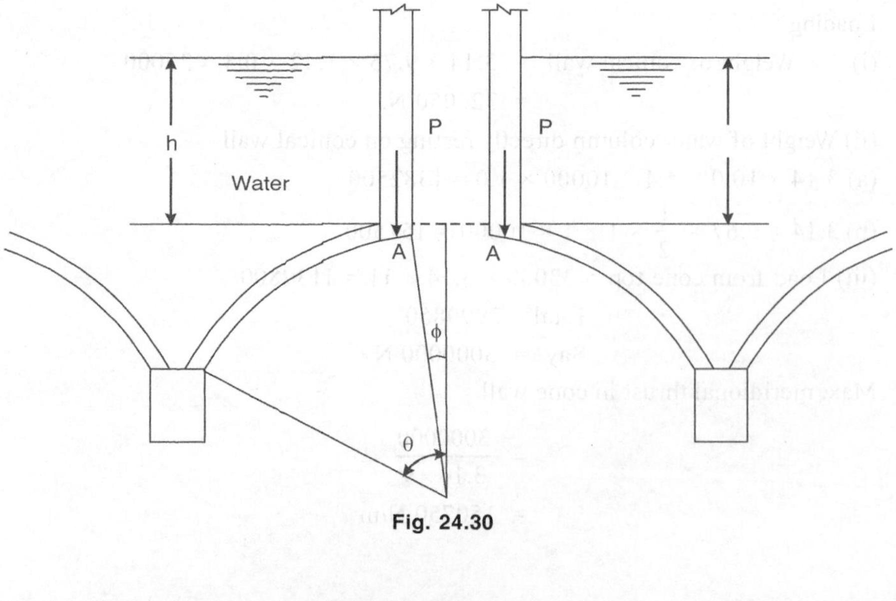

Fig. 24.30

Check for Bending :

$$W = 326050 + 1382500 + 151500 = 1860050 \text{ N}$$

$$\text{Max. B.M.} = \frac{1860050 \times 1}{8 \times 3.14 \times 9.5} = 7800 \text{ N.m.}$$

$$\text{Area of steel required} = \frac{7800 \times 1000}{125 \times 0.87 \times 250} = 290 \text{ mm}^2 \text{ (Plain)}$$

or
$$\frac{7800 \times 1000}{130 \times 0.9 \times 250} = 276 \text{ mm}^2 \text{ (HYSD bars)}$$

Provide 10 mm ϕ @ 250 mm *c/c* on each face or 10 mm ϕ @ 300 mm *c/c* on each face.

Check for Tensile Stress

$$\text{Equivalent area of concrete} = 300 \times 1000 + (9 - 1) \times 201 \times 20$$
$$= 332160 \text{ mm}^2$$

$$\text{Tensile stress in cone wall} = \frac{385050}{33460} = 1.16 \text{ N/mm}^2 < 1.5 \text{ N/mm}^2 \text{ (Safe)}$$

Design of Central Shaft (*M–30* mix) 120 mm thick

$$\text{Height of Central shaft} = 5.4 + 0.6 + 2.17 - 1.725 = 6.445 \text{ m.}$$

If
$$h = \text{height of water around shaft}$$
$$= 5.4 - 1.725 = 3.675 \text{ m}$$
$$D_{outer} = 1.5 \text{ m}, D_{inner} = 1.26 \text{ m}, t = 120 \text{ mm.}$$

$$\text{Max. hoop compression} = 10000 \times 3.675 \times 1.5 \times \frac{1}{2}$$
$$= 27562.5 \text{ N}$$

$$\text{Comp. stress} = \frac{27562.5}{1000 \times 120} = 0.23 \text{ N/mm}^2$$

$$\text{Minimum distribution steel} = 0.24 \times 120 \times 1000/100$$
$$= 288 \text{ mm}^2$$

i.e. 144 mm² on each face.

Provide 8 mm ϕ @ 200 mm c/c both ways

Also provide 8 mm ϕ upto 1.5 m high @ 200 mm c/c from dome to central shaft. Also provide 150 × 150 circular beam around lop opening with 10 mm ϕ 4 Nos and 8 mm ϕ stirrups @ 250 mm *c/c*.

Design of Ring Beam at Junction of Central Shaft and Bottom Dome

(*a*) *Design of Ring Beam at A* : (Fig. 24.28)

$$\sin \phi_0 = 0.75/5.5 = 0.1363$$
$$\phi_0 = 7.84°, \cos \phi_0 = 0.9906$$
$$\sin \theta = -3.95/5.5 = 0.718$$
$$\theta = 46°, \cos \theta = 0.695$$
$$\text{Self weight of shaft} = P = 0.12 \times 25000 \times 6.445 = 19350 \text{ N.}$$

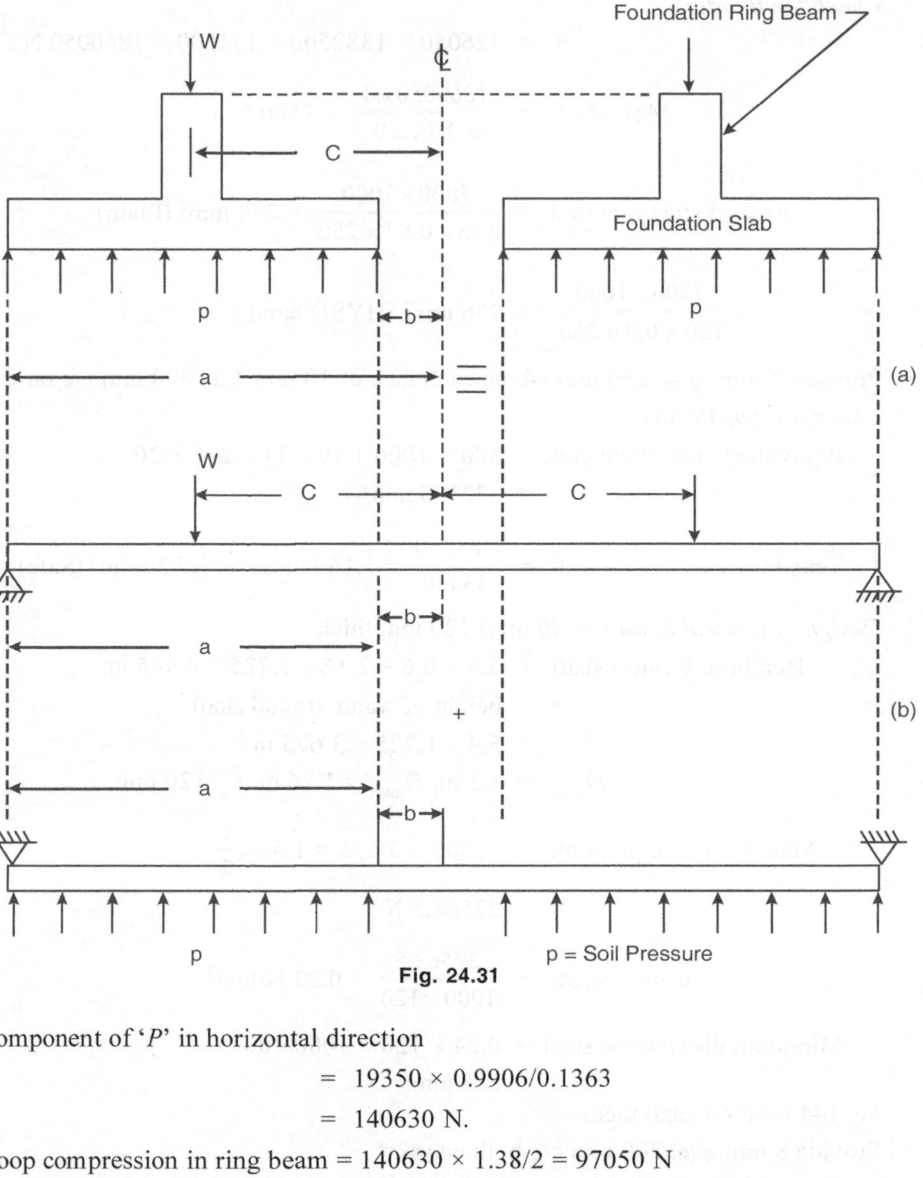

Fig. 24.31

Component of 'P' in horizontal direction

$$= 19350 \times 0.9906/0.1363$$

$$= 140630 \text{ N.}$$

Hoop compression in ring beam = 140630 × 1.38/2 = 97050 N

Hoop compressive stress = 97050 / (150 × 150) = 4.313 N/mm²

$$< 8 \text{ N/mm}^2 \text{ Safe.}$$

B.M. in base of shaft will be torsion in the ring beam.

$$\text{Torsional moment} = 0.0015 \times 10000 \times 3.675^3 = 745 \text{ N.m.}$$

$$\text{Shear stress} = \frac{M_t\,(3 + 2\,b/D)}{b^2 D} = \frac{745 \times 1000\,(3 + 2 \times 150/150)}{150 \times 150 \times 150}$$

$$= 1.103$$

$$< 2.2 \text{ N/mm}^2$$

Area of steel required for torsion $= \dfrac{M_t \cdot s}{0.8 X_1 Y_1 \cdot \sigma_{st}}$

$$= \frac{745 \times 1000 \times 100}{0.8 \times 120 \times 120 \times 130}$$

$$= 50 \text{ mm}^2$$

Minimum area of steel required $= \dfrac{0.35}{100} \times 150 \times 150$

$$= 78.75 \text{ mm}^2$$

Provide 8 mm ϕ with stirrups 8 mm ϕ @ 125 mm c/c as shown in Fig. 24.30 (a).

Design of Bottom Dome (M-30 mix) 175 mm thick

Death weight of dome $= \dfrac{175}{1000} \times 25000 = 4380 \text{ N/m}^2$

$$h = 3.675 \text{ m}, \quad a = 5.5 \text{ m}$$

(A) At Springing

(*i*) Stresses due to water :

$$N_\phi = -wa/6 \left[3h + a \cdot \frac{1 - \cos\theta}{1 + \cos\theta} (1 + 2\cos\theta) \right]$$

$$= -\frac{10000 \times 5.5}{6} \left[3 \times 3.675 + 5.5 \times \frac{1 - 0.695}{1 + 0.695} (1 + 2 \times 0.695) \right]$$

$$= -123000 \text{ N/m}$$

$$N_\theta = +wa/6 \left[3h + a \cdot \frac{1 - \cos\theta}{1 + \cos\theta} (5 + 4\cos\theta) \right]$$

$$= +\frac{10000 \times 5.5}{6} \left[3 \times 3.675 + 5.5 \times \frac{1 - 0.695}{1 + 0.695} (5 + 4 \times 0.695) \right]$$

$$= +172000 \text{ N/m}$$

(*ii*) Stresses due to shaft load P and self weight of dome p

$$N\phi = -p.a \left[\frac{\cos\phi_0 - \cos\theta}{\sin^2\theta} \right] - \frac{P \cdot \sin\phi_0}{\sin^2\theta}$$

$$= -4380 \times 5.5 \left[\frac{0.9906 - 0.695}{0.718^2} \right] - \frac{19350 \times 0.01363}{0.718^2}$$

$$= -13810 - 5110 = -18920 \text{ N/m.}$$

$$N\theta = p.a \left[\frac{\cos\phi_0 - \cos\theta}{\sin^2\theta} - \cos\theta \right] + \frac{P \sin\phi_0}{\sin^2\theta}$$

$$= 4380 \times 5.5 \left[\frac{0.9906 - 0.695}{(0.718)^2} - 0.695 \right] + \frac{19350 \times 0.1363}{(0.718)^2}$$

$$= -3062.9 + 5110 = 2047.1 \text{ N/m}$$

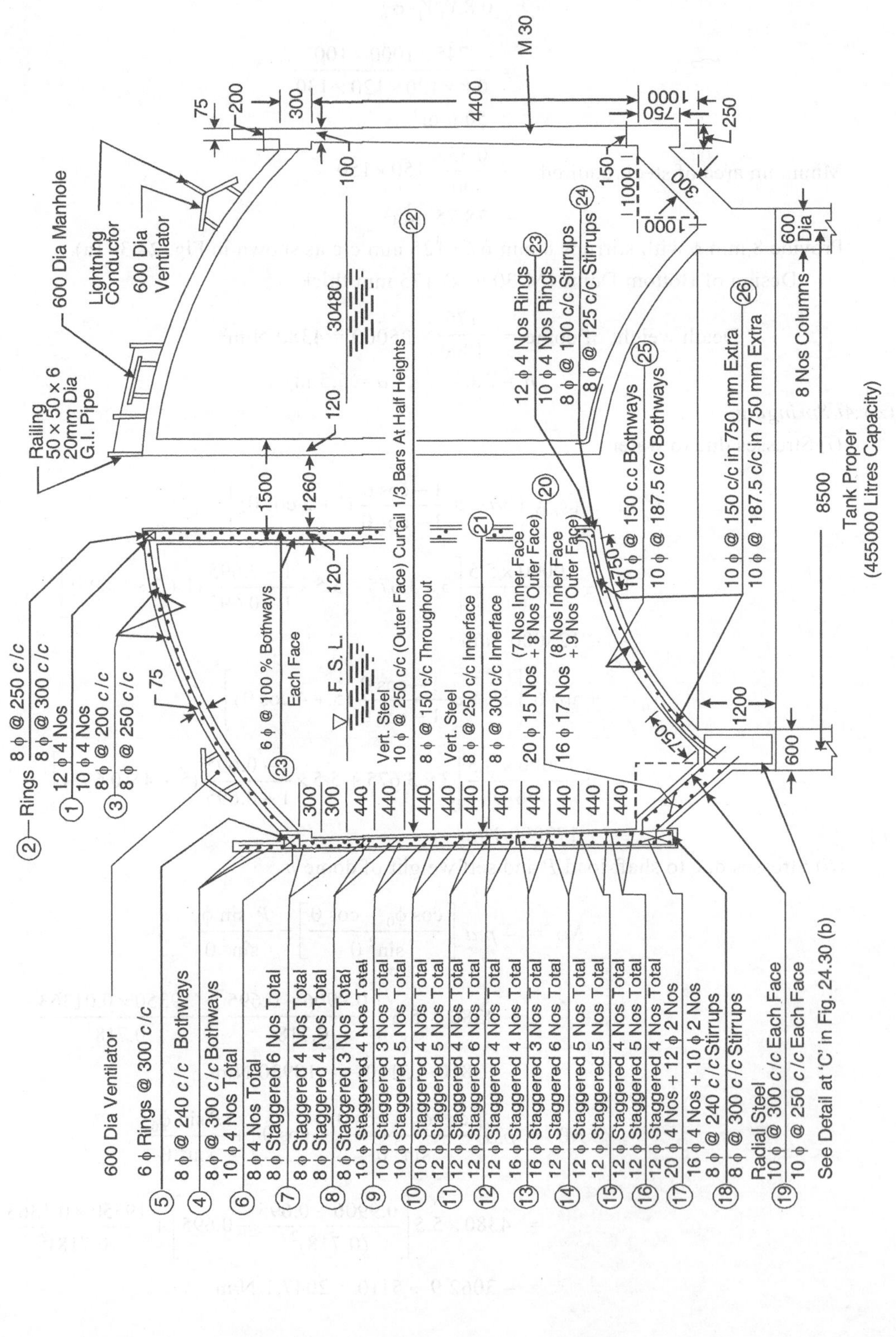
600 Dia Manhole
Lightning Conductor
600 Dia Ventilator
Railing 50 × 50 × 6 20mm Dia G.I. Pipe
75
200
300
4400
M 30
100
1000
750
250
150
1000
1000
300
30480
120
1500
1260
120
(2) Rings 8 ϕ @ 250 c/c
8 ϕ @ 300 c/c
(1) 12 ϕ 4 Nos
10 ϕ 4 Nos
8 ϕ @ 200 c/c
(3) 8 ϕ @ 250 c/c
75
(23) 6 ϕ @ 100 % Bothways Each Face
F. S. L.
(22) Vert. Steel 10 ϕ @ 250 c/c (Outer Face) Curtail 1/3 Bars At Half Heights
8 ϕ @ 150 c/c Throughout
Vert. Steel 8 ϕ @ 250 c/c Innerface
(21) 8 ϕ @ 300 c/c Innerface
(20) 20 ϕ 15 Nos (7 Nos Inner Face + 8 Nos Outer Face)
16 ϕ 17 Nos (8 Nos Inner Face + 9 Nos Outer Face)
(23) 12 ϕ 4 Nos Rings
10 ϕ 4 Nos Rings
8 ϕ @ 100 c/c Stirrups
(24) 8 ϕ @ 125 c/c Stirrups
750
(25) 10 ϕ @ 150 c.c Bothways
10 ϕ @ 187.5 c/c Bothways
(26) 10 ϕ @ 150 c/c in 750 mm Extra
10 ϕ @ 187.5 c/c in 750 mm Extra
8500
Tank Proper (455000 Litres Capacity)
600 Dia
8 Nos Columns
1200
600
750
300
300
440
440
440
440
440
440
440
440
440
440
440
600 Dia Ventilator
(5) 6 ϕ Rings @ 300 c/c
8 ϕ @ 240 c/c Bothways
(4) 8 ϕ @ 300 c/c Bothways
10 ϕ 4 Nos Total
(6) 8 ϕ 4 Nos Total
8 ϕ Staggered 6 Nos Total
(7) 8 ϕ Staggered 4 Nos Total
8 ϕ Staggered 4 Nos Total
(8) 8 ϕ Staggered 3 Nos Total
10 ϕ Staggered 4 Nos Total
(9) 10 ϕ Staggered 3 Nos Total
10 ϕ Staggered 5 Nos Total
(10) 10 ϕ Staggered 4 Nos Total
12 ϕ Staggered 5 Nos Total
(11) 12 ϕ Staggered 4 Nos Total
12 ϕ Staggered 6 Nos Total
(12) 12 ϕ Staggered 4 Nos Total
16 ϕ Staggered 4 Nos Total
(13) 16 ϕ Staggered 3 Nos Total
12 ϕ Staggered 6 Nos Total
(14) 12 ϕ Staggered 5 Nos Total
12 ϕ Staggered 5 Nos Total
(15) 12 ϕ Staggered 4 Nos Total
12 ϕ Staggered 5 Nos Total
(16) 12 ϕ Staggered 4 Nos Total
20 ϕ 4 Nos + 12 ϕ 2 Nos
(17) 16 ϕ 4 Nos + 10 ϕ 2 Nos
8 ϕ @ 240 c/c Stirrups
(18) 8 ϕ @ 300 c/c Stirrups
Radial Steel
10 ϕ @ 300 c/c Each Face
(19) 10 ϕ @ 250 c/c Each Face
See Detail at 'C' in Fig. 24.30 (b)

Fig. 24.32 (a)

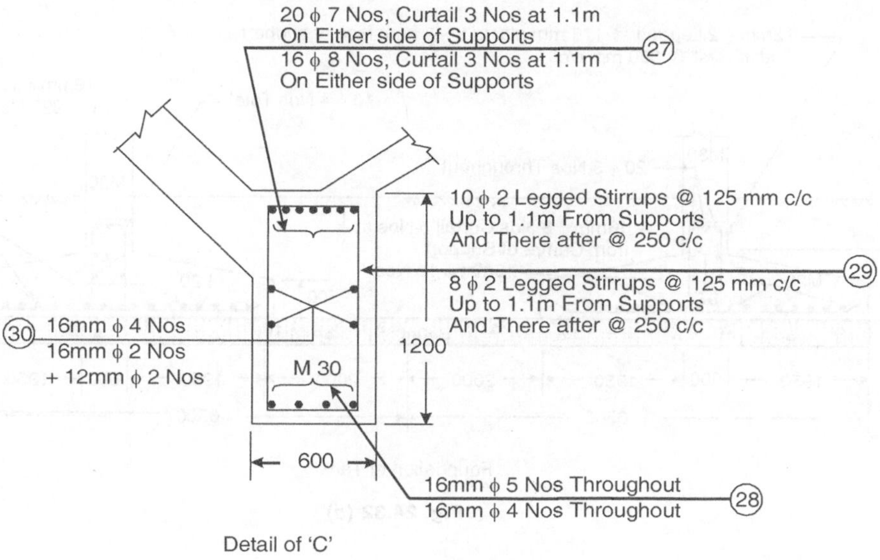

Fig. 24.32 (b)

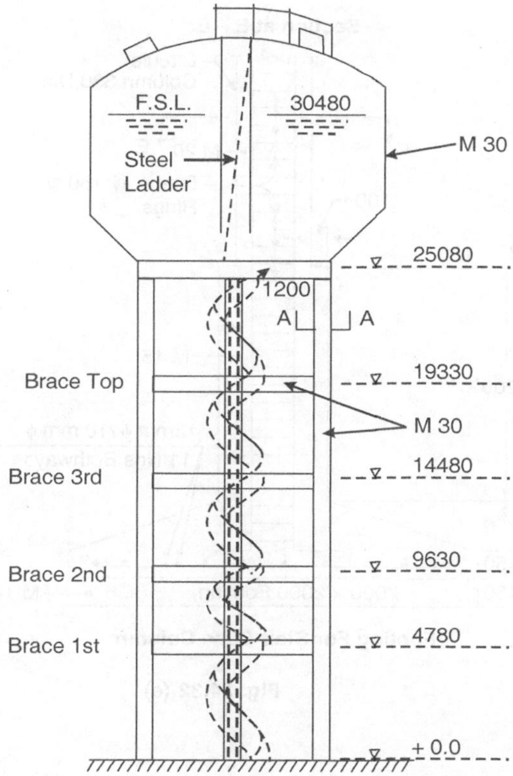

Fig. 24.32 (c)

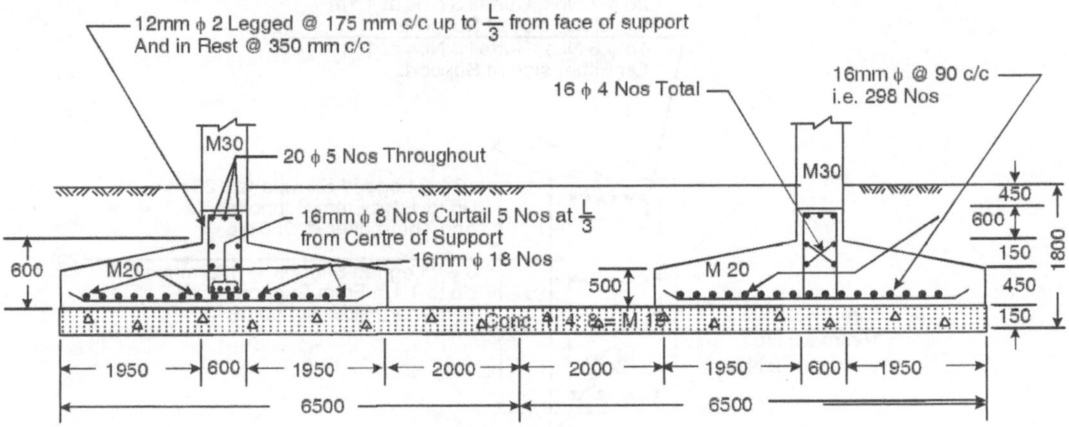

Foundation of Tank

Fig. 24.32 (d)

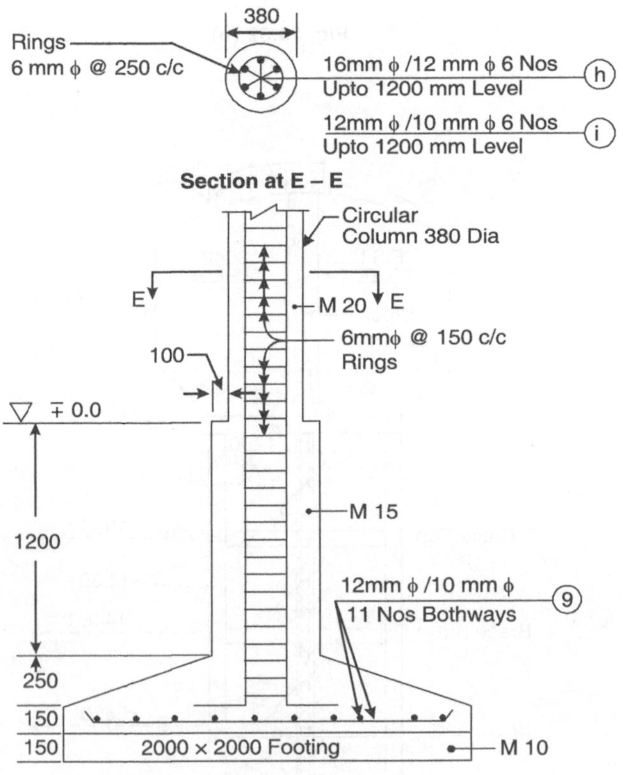

Footing For Stair Case Column

Fig. 24.32 (e)

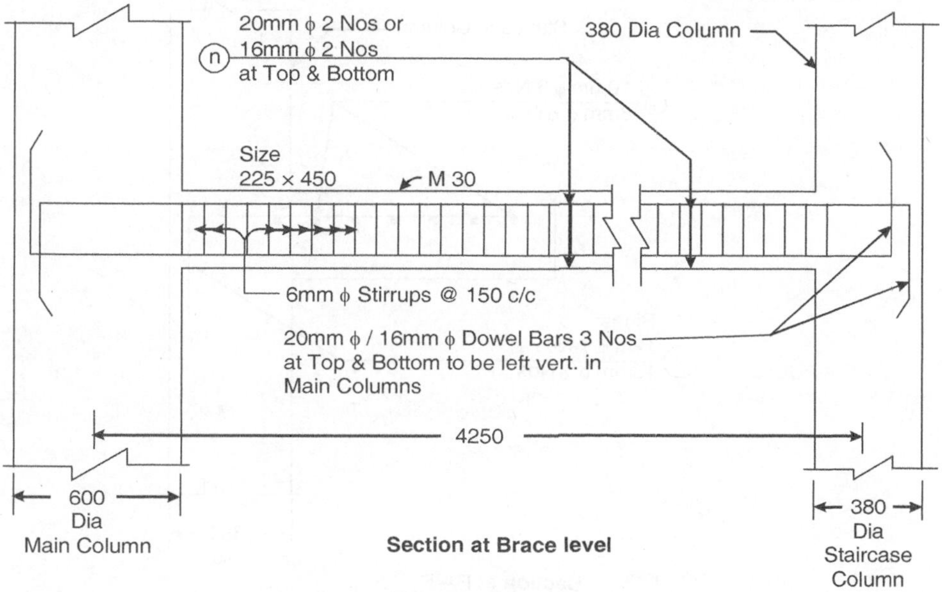

Section at Brace level

Fig. 24.32 (f)

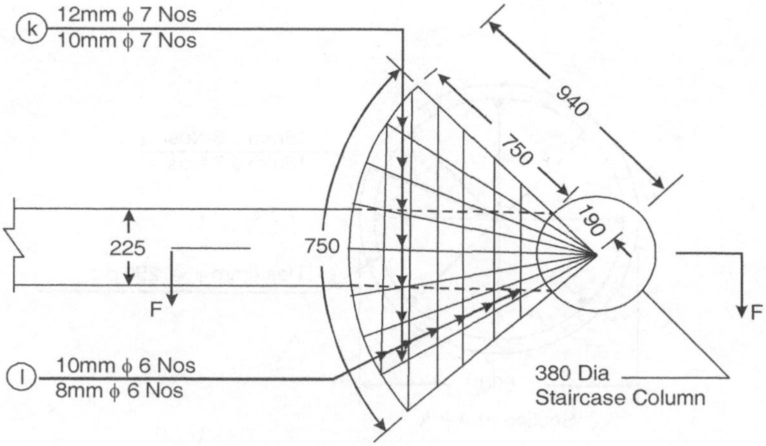

Detail of Landing

Fig. 24.32 (g)

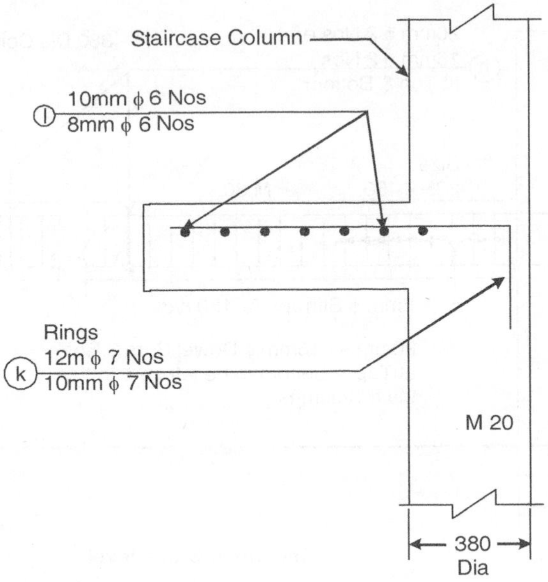

Section at F – F

Fig. 24.32 (h)

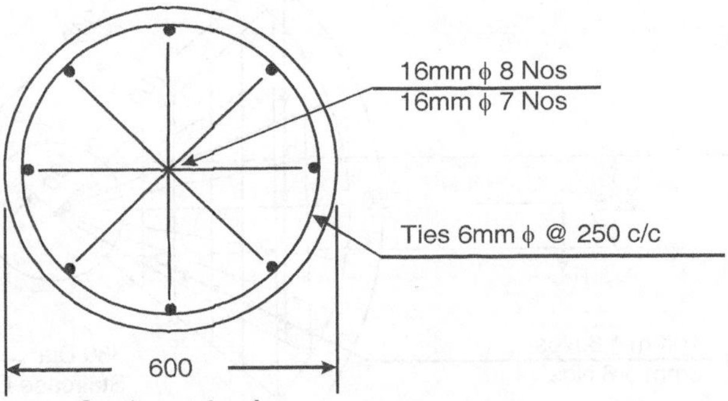

Section at A – A

Cross Section of Main Column (M 20)

Fig. 24.32 (i)

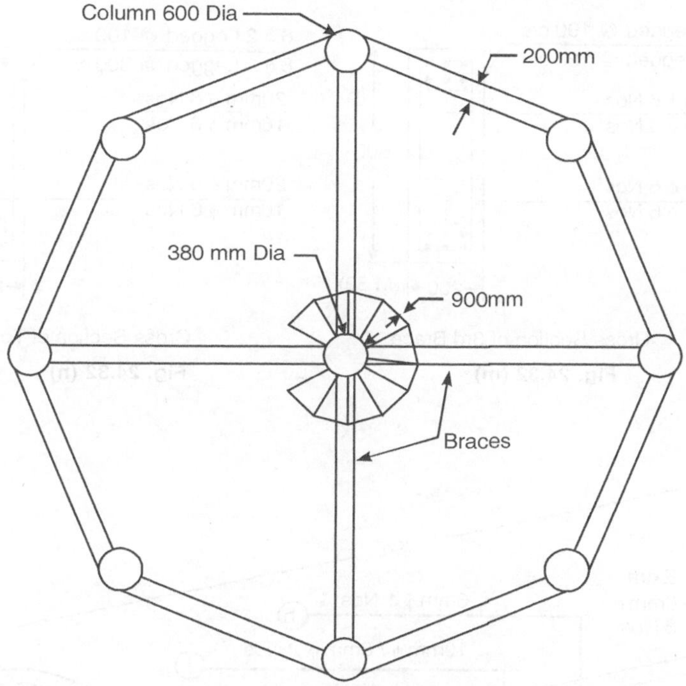

Plan Showing Central Staircase

Fig. 24.32 (j)

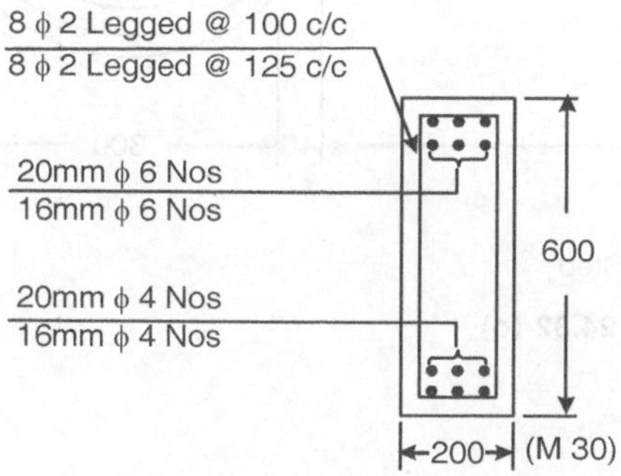

Cross Section of 1st Brace

Fig. 24.32 (k)

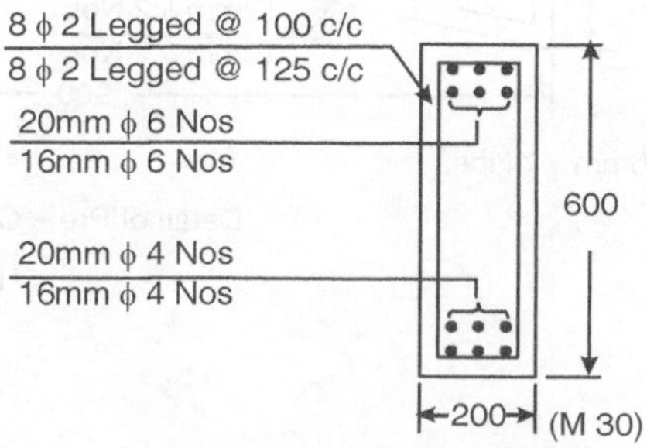

Cross Section of 2nd Brace

Fig. 24.32 (l)

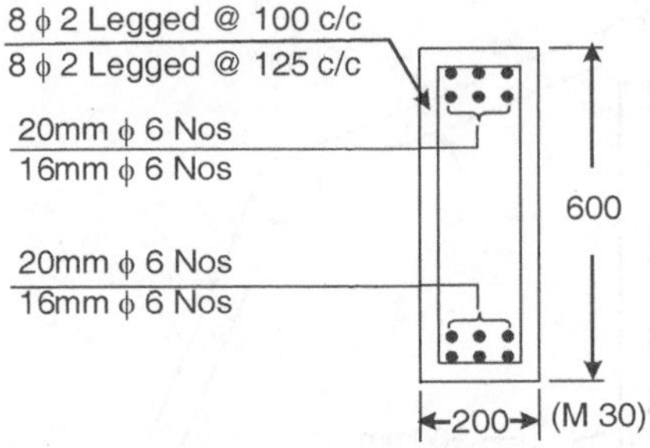

Cross Section of 3rd Brace

Fig. 24.32 (m)

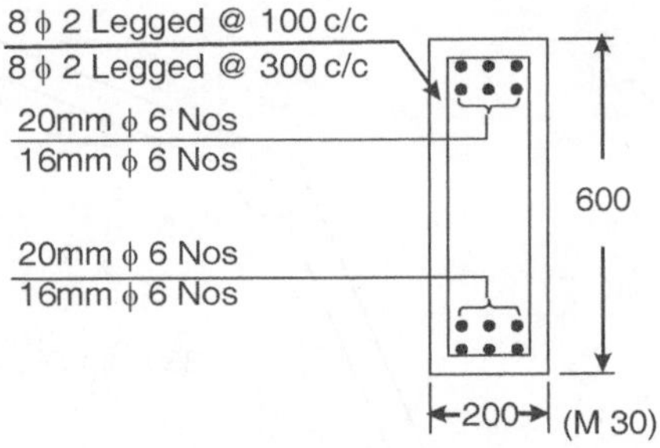

Cross Section of Top Brace

Fig. 24.32 (n)

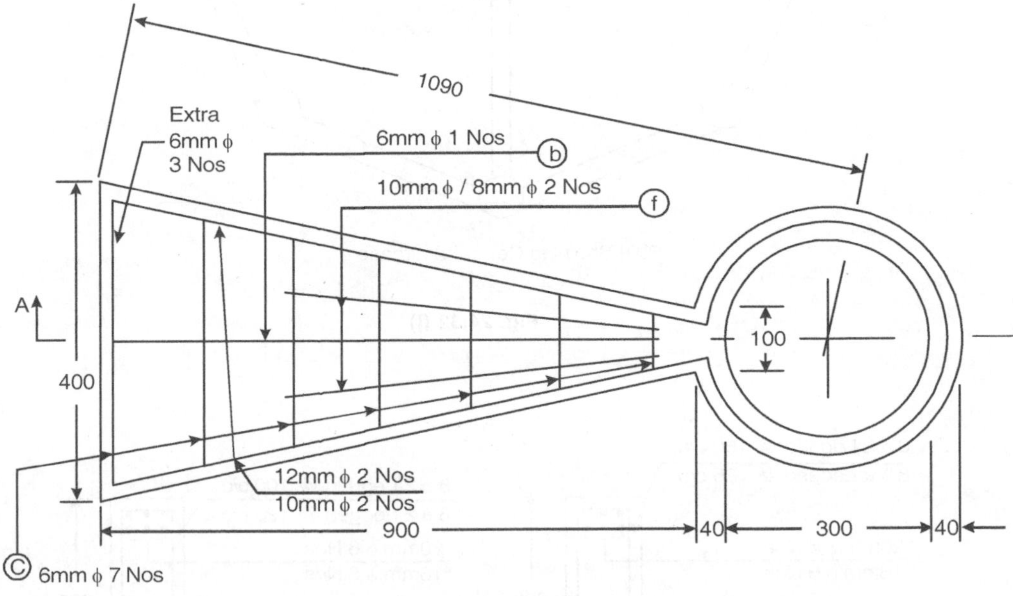

Detail of Pre – Cast Step

Fig. 24.32 (o)

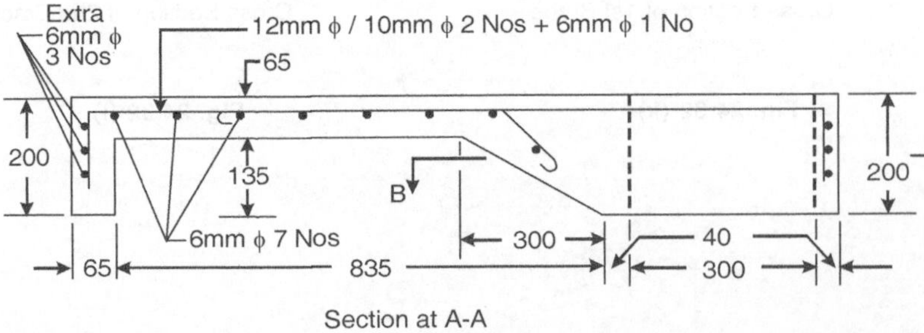

Section at A-A

Fig. 24.32 (p)

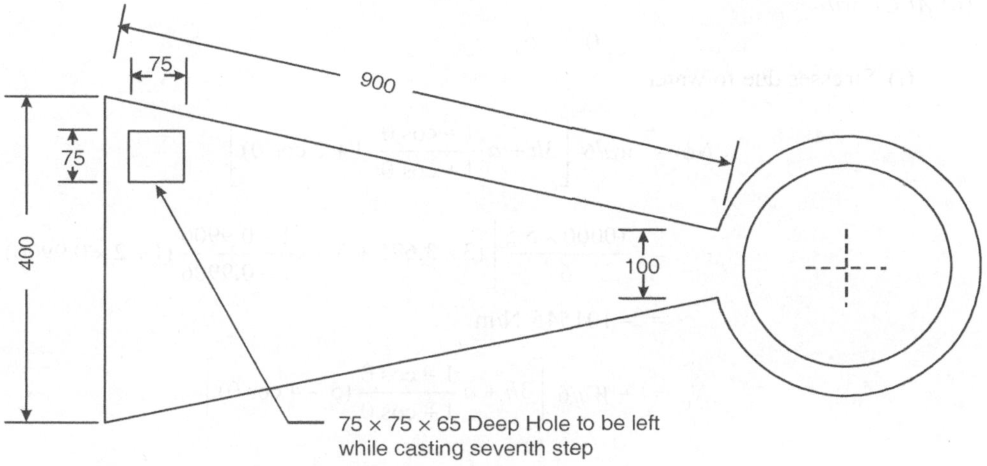

Fig. 24.32 (q)

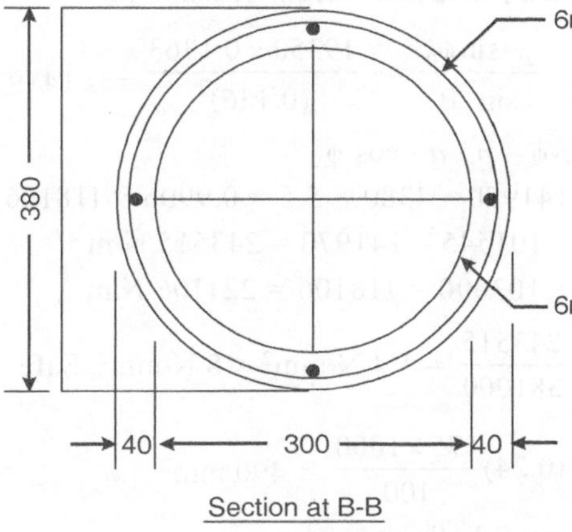

Fig. 24.32 (r)

$$\text{Total } N_\phi \text{ at springing} = -123000 - 18920 = -141920 \text{ N/m}$$

$$\text{Total } N_\theta \text{ at springing} = 172000 + 2047.1 = 174047.1 \text{ N/m}$$

$$\text{Assuming minimum distribution steel} = \frac{0.24 \times 175 \times 1000}{100}$$

$$= 420 \text{ mm}^2$$

$$\text{Stress at springing} = \frac{174047.1}{175 \times 1000 + (9-1)\,420}$$

$$= \frac{174047.1}{178360}$$

$$= 0.972 \text{ N/mm}^2 < 8 \text{ N/mm}^2 \text{ (comp.) Safe.}$$

(*b*) *At Crown*

$$\theta = \phi_0$$

(*i*) Stresses due to water

$$N\phi = wa/6\left[3h + a\cdot\frac{1-\cos\theta}{1+\cos\theta}(1+2\cos\theta)\right]$$

$$= \frac{10000\times 5.5}{6}\left[(3\times 3.675 + 5.5\times\frac{1-0.9906}{1+0.9906}(1+2\times 0.9906)\right]$$

$$= -\,101545 \text{ N/m}$$

$$N_\theta = +\,Wa/6\left[3h + a\frac{1-\cos\theta}{1+\cos\theta}(5+4\cos\theta)\right]$$

$$= +\frac{10000\times 5.5}{6}\left[3\times 3.675 + 5.5\times\frac{1-0.9906}{1+0.9906}(5+4\times 0.9906)\right]$$

$$= +\,103000 \text{ N/m}$$

(*ii*) Stresses due to shaft load *P* and self weight of dome *p*.

$$N\phi = -\frac{p\cdot\sin\phi_0}{\sin^2\theta} = -\frac{19350\times 0.1363}{(0.136)^2} = -\,141970 \text{ N/m}$$

$$N_\theta = N\phi - p\cdot a\cdot\cos\phi_0$$

$$= 141970 - 4380\times 5.5\times 0.9906 = 118106 \text{ N/m}$$

$$\text{Total } N\phi \text{ at crown} = -\,101545 - 141970 = 243515 \text{ N/m}$$

$$\text{Total } N_\theta \text{ at crown} = +\,103000 = 118106 = 221106 \text{ N/m}$$

$$\text{Max. comp. stress} = \frac{243515}{381000} = 1.4 \text{ N/mm}^2 < 8 \text{ N/mm}^2\text{, Safe}$$

$$\text{Min steel} = (0.24)\frac{175\times 1000}{100} = 490 \text{ mm}^2$$

Provide 10 mm ϕ 1505 mm *c/c* both ways.

Also provide 10 mm ϕ @ 150 mm *c/c* from ring beam into dome at springing at top face as shown in Fig. 24.32 (a).

Design of Bottom Ring Beam (M–30)

Size 600 mm × 1200 mm

Tensile force from bottom dome at centre line of dome

$$= \frac{141920\times 0.695\times 7.9\times 3.14}{3.14\times 8.5}$$

$$= 91670 \text{ N/m}$$

Circumference of ring beam = 3.14 × 8.5 = 26.7 m.

Self weight of beam = 0.6 × 1.20 × 25000 = 18000 N/m

$$\text{Comp. force from cone wall at centre line of beam} = \frac{3000000\times 0.1}{3.14\times 10.1\times 7.9} = 120880 \text{ N/m}$$

$$\text{Net comp. force} = 120880 - 91670 = 29210 \text{ N/m}$$

$$\text{Hoop comp.} = 29210 \times 0.5 \times 1 = 248285.0 \text{ N}$$

$$\text{Hoop stress} = \frac{248285}{600 \times 1000} = 0.413 \text{ N/mm}^2$$

which is negligible.

Vertical load on ring beam from dome

$$= N\phi \sin \phi$$

$$= 141920 \times 0.718 = 101900 \text{ N/m.}$$

Weight of cone wall on ring beam

$$= 150750/1.42$$

$$= 106160 \text{ N/m}$$

$$\text{Total U.D.L.} = 101900 + 106160 = 226060 \text{ N/m say } 230{,}000 \text{ N/m}$$

$$\text{Number of supports} = 8, \quad \theta = \frac{360}{8} = 45°$$

$$\text{Max. –ve B.M. at support} = 0.066 \times 230000 \times 4.25^2 \times 3.14/4$$

$$= 215350 \text{ N.m.}$$

$$\text{Max. –ve B.M. at mid span} = 0.03 \times 230000 \times 4.25^2 \times 3.14/4$$

$$= 97900 \text{ N.m.}$$

$$\text{Max. twisting moment} = 0.005 \times 230000 \times 4.25^2 \times 3.14/4$$

$$= 16310 \text{ N.m.}$$

$$d = \sqrt{\frac{215350 \times 1000}{1.3 \times 600}} = 526 \text{ mm.}$$

Provide overall depth of 1200 mm because of shear consideration.

$$M_t = T\left(\frac{1 + d/b}{1.7}\right) = \frac{16310\left(1 + \frac{1200}{600}\right)}{1.7}$$

$$= 25500 \text{ N.m.}$$

$$\text{Total –ve B.M.} = 215350 + 25500 = 240850 \text{ N.m.}$$

$$\text{Total +ve B.M.} = 97900 + 25500 = 123400 \text{ N.m.}$$

Area of steel required at support

$$\text{HYSD steel} = \frac{240850 \times 1000}{130 \times 0.87 \times 1150} = 1857 \text{ mm}^2$$

Provide 20 mm ϕ, 6 Nos., Curtail 3 Nos. at 1.1 m on either side of column.

Area of steel required at mid span

$$\text{HYSD steel} = \frac{123400 \times 1000}{130 \times 0.90 \times 1150} = 917 \text{ mm}^2$$

Provide 16 mm ϕ 5 Nos.

$$\text{Face reinforcement} = \frac{0.1}{100} \times 600 \times 1200 = 720 \text{ mm}^2$$

Provide 16 mm ϕ 4 Nos.

$$\text{Equivalent shear from torsion} = 1.6\ T/b = 1.6\ \frac{16310 \times 1000}{600}$$

$$= 43500 \text{ N.}$$

$$\text{Shear at edge of beam} = w(1/2\ R - 0.3) = 230000\ (1/2 \times 4.25 \times 3.14/4 - 0.3)$$

$$= 314865 \text{ N.}$$

$$\text{Total shear force} = 314865 + 43500 = 358365 \text{ N}$$

$$\tau_v = \text{Nominal shear stress} = \frac{358365}{600 \times 1150} = 0.519 \text{ N/mm}^2$$

$$\frac{100\ A_s}{bd} = \frac{100 \times 4 \times 200}{600 \times 1150} = 0.116$$

τ_c = Permissible shear stress= 0.220 N/mm^2 > τ_v

Shear strength of concrete = 0.220 × 600 × 1150 = 151800 N

Design shear = 358365 – 151800 = 206565 N

Provide 10 mm ϕ 2 legged stirrups @

$$= \frac{140 \times 78.5 \times 2 \times 1150}{206565}$$

= 122.3 mm *c/c* say 125 mm *c/c* upto 1.1 m from supports

Check for minimum shear reinforcement

$$A_{sw} = \frac{T \cdot a_v}{b_1\ d_1} \sigma_{sv} + \frac{V \cdot S_v}{25\ d_1 \cdot \sigma_{sw}}$$

$$= \frac{16310 \times 125 \times 1000}{400 \times 1140 \times 140} + \frac{358365 \times 125}{2.5 \times 1140 \times 140}$$

$$= 144 \text{ mm}^2,$$

Provide 10 mm ϕ 2 legged @ 125 mm *c/c* is O.K.

For Tor steel

$$8 \text{ mm } \phi \text{ 2 legged @ } \frac{230 \times 2 \times 50.0 \times 1150}{206565} = 128 \text{ mm,}$$

i.e. @ 125 mm *c/c* in 1.1 m and in rest @ 250 mm *c/c*.

Design of Staging (M–20)

F.S.L. = 30.48 m (100′ – 0″)

Concrete for columns = *M* 20

Diameter of columns = 600 mm

Nos. of braces = 4

Size of brace = 200 mm × 600 mm

Distance between *c/c* of brace *i.e.* one storey

$$\text{height} = \frac{30.48 - 5.4 - 1.2 \times 0.3 + 0.2}{5} = 4.85 \text{ m}$$

Design for Axial Load

(*i*) Load on column from ring beam

$$= \frac{230000 \times 3.14 \times 8.5}{8} = 767340 \text{ N.}$$

(*ii*) Self weight of column $= \dfrac{3.14 \times 0.6^2}{4} \times 26.34 \times 25000 = 186200$ N.

(*iii*) Weight of braces, 4 (3.34 – 0.6) × 0.2 × 25000 = 32900 N.

(*iv*) Misc. loads = 30000 N

(*v*) Total load = 1016440 N.

$$\text{Area of concrete actually provided} = \frac{3.14 \times 600^2}{4}$$

$$= 2826000 \text{ mm}^2$$

Min. A_{sc}. (plain steel) = (0.8/1000) × 204000 = 1632 mm²

Provide 16 mm ϕ = 8 Nos.

Min. A_{sc}. (tor steel) = 0.8 × 1632 = 1308.16 mm²

Provide 16 mm ϕ = 7 nos

Ties

Provide 6 mm ϕ @ 250 mm *c/c* throughout

Check for Wind Loads

Projected area of tank portion subjected to wind

$= 11.3 \times (2.17 \times 2/3 + 7.2) = 97.7 \text{ m}^2$

Wind force on tank = 1500 × 0.7 × 97.7 = 102600 N.

Centre of action from top of foundation = 8.646/2 + 23.88 = 28 m.

Wind force acting on staging = 5 × 0.6 (23.98 – 4 × 0.6) × 1500 × 0.7 + 4 × 0.6 × 9.1 × 1500 × 0.8

= 68050 + 26210 = 94260 N

acting at (24/2) = 12 m.

Overturning moment due to wind about base

= 102600 × 28.2 + 94260 × 12

= 28,93,330 + 11,31,120 = 40,25,000 N.m.

$\Sigma r^2 = 2 \times 4.25^2 + 4 \times 3^2 = 72.125 \text{ m}^2$

Max. vertical load on remotest column

$$= \frac{4025000 \times 4.25}{72.125} = 237200 \text{ N.}$$

Total axial downward load on column

$= 10{,}20{,}000 + 2{,}37{,}200 = 12{,}57{,}200$ N.

Total stress $= 1026000 + 94260 = 196860$ N.

Wind shear per column $= 196860/8 = 24610$ N.

Wind moment $= 24610 \times 4.85/2 = 59680$ N.m.

$$A_{eq} = \frac{3.14 \times 600^2}{64} + (1.5 \times 13 - 1)\, 8 \times 201 = 282600 + 29748$$

$$= 312348 \text{ mm}^2$$

$$I_{eq} = \frac{3.14 \times 600^4}{64} + 4 \times 201\,(250 \times 0.707)^2 + 2 \times 201 \times 250^2$$

$$= 6.3585 \times 10^9 + 9.29489 \times 10^8$$

$$= 7.28798 \times 10^9 \text{ mm}^4$$

(*a*) *Tank full*

$$\text{Comp. force} = \frac{1257200}{312348} = 4.025 \text{ N/mm}^2$$

$$\text{Bending stress} = \frac{59680 \times 1000 \times 300}{7.28798 \times 10^9} = 2.4567 \text{ N/mm}^2$$

Max. comp. stress $= 4.0250 + 2.4567 = 6.4817$ N/mm^2

Min. stress $= 4.0250 - 2.4567 = 1.5683$ N/mm^2

Interaction, Eq. $\dfrac{\sigma_{cc,cal}}{\sigma_{cc}} + \dfrac{\sigma_{cbc,cal}}{\sigma_{cbc}} \le 1.0$

Increase the allowable stresses by 50% since wind is considered.

$$\therefore \quad \text{L.H.S.} = \frac{4.0250}{1.5 \times 5} + \frac{2.4567}{1.5 \times 7}$$

$$= 0.537 + 0.234 = 0.771 < 1.0 \text{ Safe.}$$

(*b*) *Tank empty :*

Direct load excluding weight of water per column

$= 1020000 - 4550000/8 - 237200 = 214000$ N.

$$\text{Direct stress} = \frac{214000}{312348} = 0.6853 \text{ N/mm}^2$$

Bending stress $= 2.4567$ N/mm^2

Min. stress $= 0.6853 - 2.4567$

$= -1.7714$ N/mm^2 $< 1.5 \times 1.8$ N.mm^2 < 2.7 N/mm^2, O.K.

(*i.e.*, Tensile)

Seismic Analysis

1. Weight of tank when full of water less live load on top dome.

$= 230000 \times 3.14 \times 8.5 - 61.6 \times 102.26$

$= 6{,}079{,}000$ N

2. Weight of staging $= 186200 \times 8 + 32900 \times 8 \times 4 = 2542,400$ N.

W_f = for tank full $= 6079000 + 1/3 \times 2542400 = 6926600$ N

W_e = for tank empty $= 6,079,000 + 1/3 \times 2,542,400 - 4,550,000$

$= 2,376,600$ N

E for M 20 concrete $= 2.55 \times 10^4$ N/mm^2

Column stiffnesses.

$$K = 8 \times 12\, EI/h^3 = 8 \times 12 \times 2.55 \times 10^4 \times 7.28798 \times 10^9/(4850)$$

$$= 1.5638 \times 10^5 \text{ N/mm}$$

Combined stiffness of spring $1/K = 1/K_1 + 1/K_2 + 1/K_3 + 1/K_4 + 1/K_5$

$$= \frac{5}{1.5638 \times 10^5} = 3.197 \times 10^{-5}$$

Natural time period (T)

(*a*) Tank full $T_f = 2 \times 3.14 \sqrt{W/(g.k)}$

$$= 2 \times 3.14 \sqrt{\frac{6926600}{9810} \times 3.197 \times 10^{-5}}$$

$= 0.944$ seconds.

(*b*) Tank empty $T_e = 2 \times 3.14 \sqrt{2376600 \times 3.197 \times 10^{-5}/9810}$

$= 0.552$ sec.

To compute coefficient α_h

$$\alpha_h = \frac{ZI}{2R} \cdot \frac{S_a}{g}$$

$Z = 0.24$ for Delhi in seismic zone IV (Ref. Appendices F and G)

$I = 1.5$ from Table 24.19

$R = 5.0$ from Table 24.20 for special R.C.C. moment resisting frames adopting ductile behaviour requirements for reinforcements

For $\frac{S_a}{g}$ for Medium soils

$T_f = 0.944$ sec $\quad \frac{S_a}{g} = \frac{1.36}{0.944} = 1.44$

$T_e = 0.552$ sec $\quad \frac{S_a}{g} = \frac{1.36}{0.552} = 2.46$

(*a*) *Tank full of water*

$$\therefore \quad \alpha_{hf} = \frac{0.24 \times 1.5}{2 \times 5} \times 1.44 = 0.05184$$

$$\alpha_{vf} = \frac{0.05184}{2} = 0.02592$$

(*b*) *Tank* (*Empty*)

$$\alpha_{he} = \frac{0.24 \times 1.5}{2 \times 5} \times 2.46 = 0.08856$$

$$\alpha_{ve} = 0.4428$$

(*a*) *Tank full*

$$\text{Horizontal force} = 0.05184 \times 6926600 = 359075 \text{ N}$$

$$\text{Vertical force} = \frac{1}{2} \times 359075 = 179537 \text{ N}$$

(*b*) *Tank empty*

$$\text{Horizontal force} = 0.08856 \times 2376600 = 210472 \text{ N}$$

$$\text{Vertical force} = \frac{1}{2} \times 210472 = 105236 \text{ N}$$

(*a*) *Tank full*

Overturning moment due to seismic lateral force

$$= 359075 \times 28.2 = 10125915 \text{ N.m.}$$

(*b*) *Tank Empty*

Overturning moment due to seismic lateral force

$$= 210472 \times \frac{2}{3} \times 28.2 = 3956874 \text{ N.m.}$$

Note: In case of empty, the C.G. of weight of tank proper & one third that of staging is assumed to be at $\frac{2}{3}$ of 28.2 *i.e.* height of C.G. when tank is full of water. However for more accuracy, actual C.G. of weight of tank proper & one third weight staging may be computed.

To compute vertical force on remotest column

(*a*) *Tank full*

$$\text{Vertical force} = \frac{10125915}{72.125} \times 4.25 = 596674 \text{ N}$$

(*b*) *Tank empty*

$$\text{Vertical force} = \frac{3956874 \times 4.25}{72.125} = 233160 \text{ N}$$

(*a*) *Tank*

$$\text{Total axial force} = 1020000 + 596674 + 179557 = 1796231 \text{ N}$$

$$\text{Shear force per column} = \frac{359075}{8} = 44884 \text{ N}$$

$$\text{Seismic Moment in Column} = \frac{44884 \times 4.85}{2} = 108845 \text{ N.m}$$

$$\text{Bending stress} = \frac{108845}{7.28798} \times \frac{300 \times 1000}{10^9} = 4.48 \text{ N/mm}^2$$

$$\text{Direct stress} = \frac{1796231}{312348} = 5.75 \text{ N/mm}^2$$

Interaction Equation

$$\frac{5.74}{1.33 \times 5} + \frac{4.48}{1.33 \times 7} = 0.865 + 0.48$$

$$= 1.345 > (1 \times 1.15)$$

The factor 1.15 may be taken as age factor for concrete for 6 months *i.e.* the tank will take at least six month for completion.

Use M30 concrete in staging

$$\frac{5.75}{1.33 \times 8.0} + \frac{4.48}{1.33 \times 10} = 0.54 + 0.337$$

$$= 0.877 < 1.15$$

Therefore O.K.

(*b*) *Tank empty* :

$$\text{Seismic shear/column} = 210472/8 = 26309 \text{ N}$$

$$\text{B.M. per column} = 26309 \times 4.85/2 = 63800 \text{ N.m}$$

$$\text{Total axial force} = 1020000 - 4550000/8 - 233160$$

$$= 112854 \text{ N.}$$

$$\text{Direct stress} = 112854/312348 = 0.361 \text{ N/mm}^2$$

$$\text{Bending stress} = \frac{63800 \times 1000 \times 300}{1.33 \times 8.0 7.28798 \times 10^9} = 2.62 \text{ N/mm}^2$$

$$\text{Resultant stress (max.)} = 2.981 \text{ N/mm}^2$$

$$< \frac{2}{3} \times 1015 \times 3.0 = 2.3 \text{ N/mm}^2 \quad \text{Safe.}$$

Use M30

By limit state method of design (Ref. Chart 21.19)

$$P_u = 1.5 \times 1796231 = 2694346$$

$$M_u = 1.5 \times 108845 = 163268 \text{ N.m.}$$

$$\frac{F_u}{f_{ck} \cdot D^2} = \frac{2694346}{30 \times 600^2} = 0.25$$

$$\frac{M_u}{f_{ck} \cdot D^3} = \frac{163268 \times 1000}{30 \times 600 \times 600 \times 600} = 0.0250$$

$$\frac{p}{f_{ck}} = 0.01\, p = 30 \times 0.01 = 0.30$$

$$A_s = \frac{0.30 \times 3.14 \times 600^2}{100 \times 1.33 \times 4} = 637 \text{ mm}^2$$

16 mm ϕ 7 Nos. O.K.

DESIGN OF BRACES

Size 200 × 600 mm. (4 Nos.)

Wind force on tank portion = 102600 N.

Wind force on staging portion = 94260 N.

$$\text{Max. design force on lowest brace} = 102600 + \frac{94260 \times 4}{5}$$

$$= 178008 \text{ N}$$

This will decrease by $\frac{94260}{5}$ = 18850 N for every successive brace.

Max. shear force due to seismic force = 359075 N.

>> 178008 N

∴ Design share force for all braces = 359075 N

$$\text{Shear per column } `S\text{'} = \frac{359075}{8} = 44884.4 \text{ N}$$

say 45000 N

$$\text{Max. B.M. per column } `M_c\text{'} = \frac{S.h}{2} = 45000 \times \frac{4.85}{2}$$

$$= 109125 \text{ N.m}$$

$$\text{Moment in any brace beam } `M_b\text{'} = 2M_c \operatorname{cosec} \theta$$

$$= 2 \times 109125 \times 1.414$$

$$= 308606 \text{ N.m.}$$

$$A_s = \frac{M_b \times 10^3}{1.5 \times 230 \times 570 \times 0.9} = 1744 \text{ mm}^2$$

Provide 20 mm ϕ 6 Nos. top & bottom.

$$\text{Maximum shear force} = \frac{2M_b}{L} = \frac{2 \times 308606}{4.45} = 138700 \text{ N}$$

$$\text{Shear stress} = \frac{138700}{200 \times 570} = 1.22 \text{ N/mm}^2$$

$$p = \frac{6 \times 314 \times 100}{200 \times 570} = 1.65$$

Permissible shear stress = 0.5 N/m²

Design shear force = (1.22 – 0.5) × 200 × 570

= 82080 N

$$\text{Providing 8 mm } \phi \text{ 2 legged @ } \frac{0.87 \times 230 \times 2 \times 50 \times 570}{82080}$$

= 139 mm c/c

say 125 mm c/c + rest @ 200 mm c/c

Design of Braces Holding the Central Stair Case Column

Size of brace = 225 × 450 (M20 concrete)

No. of braces = 5

Projected length of column brace to brace = 4.74 m

Wind force = 0.7 × 1500 × 4.74 × 0.45 = 2300 N.

Force on stage = 1/4 × 1.8 × 4.74 × 1500 = 3200 N.

Total = 5500 N.

Shear Force = 5500/2 = 2750 N.

B.M. in brace = 2750 × 2 × 4.25 = 23400 N.m.

$$d = \sqrt{\frac{23400 \times 1000}{0.9 \times 225}} = 340 \text{ mm.}$$

$$D = 450 \text{ mm.} \qquad \therefore\ d = 400 \text{ mm.}$$

$$A_{st} = \frac{23400 \times 1000}{230 \times 0.9 \times 400} = 288 \text{ mm}^2$$

Provide 16 mm ϕ 2 Nos. at top and 2 Nos. at bottom

Or 20 mm ϕ 2 Nos. at top and 2 Nos. at bottom.

Provide 6 mm ϕ 2 legged stirrup @ 150 mm c/c throughout.

Design of Stair Case Column (M-20 mix) Fig. 24.32 (*c*)

Size = 380 mm diameter

Inner core = 300 mm Precast rings 40 mm thick around.

Length of column = 24.18 m.

Weight of one step = 400 N.

Dead load of steps = 24.18 × 400/0.2 = 48500 N.

L.L. on steps = 120 × 0.25 × 0.9 × 3000 = 81000 N.

Load from braces = 4 × 0.45 × 0.25 × 25000 × 4.25 = 48000 N.

Load from self weight = 24.18 × 3.14 × 0.38^2 × 25000/4 = 68500 N.

Total load = 246000 N.

Area of concrete required = 246000/4 = 61500 mm^2

$$\text{Area of concrete provided} = \frac{3.14 \times 380^2}{4} = 113300 \text{ mm}^2$$

Min. area of steel = 0.008 × 0.8 × 61500 = 400 mm^2 Tor

Provide 12 mm ϕ 6 Nos. or 16 nos. upto 12 m. and then 12 mm ϕ 6 Nos. 10 mm ϕ 6 Nos.

Ties/Rings 6 mm ϕ @ 250 mm *c/c*.

Load from col. = 246000 N.

Self weight @ 15% = 40000 N.

Total = 286000 N = 286 KN.

$$\text{Area of footing required} = \frac{286}{87.4} = 3.27 \text{ m}^2$$

Provide 2 m × 2 m square footing.

$$\text{Net upward pressure} = 246000/4 = 61500 \text{ N/m}^2$$

$$\text{Max. B.M.} = 61500 \times 2\ (2 - 0.30)^2/8 = 44500 \text{ N.m.}$$

$$d = \sqrt{\frac{44500 \times 1000}{0.90 \times 380}} = 360 \text{ mm}$$

$$D = 400 \text{ mm} \qquad d = 325 \text{ mm.}$$

$$A_{st} = \frac{44500 \times 1000}{0.9 \times 230 \times 325} = 660 \text{ mm}^2$$

Provide 10 mm *Q* 11 Nos. both ways and 12 mm ϕ 11 Nos. both ways.

Design of Precast Steps M-20

Precast steps will consist of cantilever slab 400 mm wide at outer edge and 100 mm at support edge.

Length of step = 900 mm

Thickness = 65 mm at outer edge upto the plain 300 mm from inner edge, which increases to 175 mm at support.

(*a*) Live load on slab 0.9 × 0.25 × 3000 = 800 N

(*b*) Self weight (say) = 400 N

(*c*) Miscellaneous = 100 N

Total u.d.I. = 1300 N

Bending Moment (considering it as a point load acting at outer edge of cantilever step) :

$$\text{B.M. at 0.3 m from edge} = 1300 \times 0.6 = 780 \text{ N.m.}$$

$$d = \sqrt{\frac{780 \times 1000}{1.2 \times 200}} = 57 \text{ mm}$$

∴ Provide 175 mm at the built in edge tapered to 65 mm within 300 mm and then 65 mm throughout to free end is O.K.

$$A_{st} = \frac{780 \times 1000}{140 \times 0.87 \times 57} = 112.4 \text{ mm}^2$$

Provide 2 Nos. 12 mm ϕ at edge + 1 No. 6 mm ϕ at centre or 2 Nos. 10 mm ϕ at edge + 1 No. 6 mm ϕ at centre

Also provide 7 Nos. 6 mm ϕ as distributors. See drawing for further details of reinforcement.

Design of Landing Slab

Self weight = 0.1 × 0.75 × 0.425 × 25000 = 798 N

L.L. = 0.75 × 0.425 × 3000 = 956 N

Total u.d.1. = 1754 N

Assumed to be adding at 0.7 × 750 × 630 mm from built in end. Max. B.M. at built in end = 1754 × 0.525 = 920.85 mm

$$d = \sqrt{\frac{920.85 \times 1000}{1.2 \times 100}} = 87.6 \text{ mm}$$

$$D = 100 \text{ mm} \qquad d = 85 \text{ mm}$$

$$A_{st} = \frac{920.85 \times 1000}{140 \times 0.87 \times 85} = 88.94 \text{ mm}^2$$

Provide 12 mm ϕ 7 Nos. or 10 mm ϕ 7 Nos. (HYSD) Distributors 10 mm ϕ 6 Nos. or 8 mm ϕ 6 Nos. (HYSD)

Design of Foundation

Bearing capacity of soil $= 87.4$ KN/m^2

D.L. + L.L. at base of columns $= 1{,}020{,}000 \times 8$

$= 8{,}160{,}000$ N

Assuming self weight of foundation @ 10%

$= 816{,}000$ N

Total weight on soil $= 8{,}976{,}000$ N

$= 8976$ KN

Area of foundation required $= \dfrac{8976}{87.4} = 102.7$ m^2

Width of footing required $= \dfrac{102.7}{\pi \times 8.5} = 3.846$ m

Provide circular raft footing with 13.0 m external diameter and 4.0 m internal diameter.

Area of footing provided, $\dfrac{\pi}{4}(13.0^2 - 4.0^2) = 120.16$ m^2

Section Modulus $= \dfrac{\pi}{4}\left[\dfrac{13.0^4 - 4.0^2}{13.0}\right] = 213.75$ m^3

Pressure on soil due to D.L. + L.L. $= \dfrac{8976}{120.16} = 74.7$ KN/m^2

Pressure distribution on soil

(*A*) *Under Earthquake Effects.*

(*i*) *Tank Full*

$$\text{Direct load} = 8160 + 179.532 = 8339.532 \text{ KN}$$

$$\text{Overturning moment} = 10125.915 \text{ KN.m}$$

$$\text{Pressure} = \frac{8339.532}{120.16} \pm \frac{10125.915}{213.75}$$

$$= 69.4 \pm 47.37 = 116.77 \text{ KN/m}^2 \text{ or } 22.03 \text{ KN/m}^2$$

$$< 87.4 \times 1.5 \textit{ i.e. } 131.1 \text{ KN/m}^2$$

Safe.

(*ii*) *Tank Empty*

$$\text{Direct load} = 8160 - 179.532 - 4550 = 3430.5 \text{ KN}$$

$$\text{Overturning moment} = 3956.874 \text{ KN.m}$$

$$\text{Pressure} = \frac{343.5}{120.16} \pm \frac{3956.874}{213.75}$$

$$= 28.55 \pm 18.51$$

$$= 47.06 \text{ KN/m}^2$$

$$< 87.4 \times 1.5 \text{ KN/m}^2$$

Safe.

(B) Under Wind Effect

(i) Tank Full

$$\text{Direct load} = 8476 - \frac{102.26 \times 616}{1000} = 8913 \text{ KN}$$

$$\text{Overturning moment} = 4025.0 \text{ KN.m.}$$

$$\text{Pressure} = \frac{8913}{120.16} \pm \frac{4025}{213.75}$$

$$= 74.176 \pm 18.83 = 93.00 \text{ KN/m}^2$$

Safe.

(ii) Tank Empty

$$\text{Direct load} = 8913 - 4550 = 4463 \text{ KN}$$

$$\text{Overturning moment} = 4025.0 \text{ KN.m.}$$

$$\text{Pressure} = \frac{4463}{120.16} \pm \frac{4025}{213.75}$$

$$= 37.14 \pm 18.83 = 55.9 \text{ or } 18.31 \text{ KN/m}^2$$

Safe.

Design Pressure of soil on foundation slab can be worked out as follows :

(*i*) Pressure on soil due to D.L. + L.L. = 74.7 KN/m^2

(*ii*) Pressure on soil under earth quake effects

(Tank full) = 116.77 KN/m^2 (max.)

= 28.08 (min)

(*iii*) Pressure on soil under wind effects

(Tank full) = 93.00 KN/m^2 (max.)

= 55.396 (min).

The max. pressure on soil is 116.77 KN/m^2 under earth quake effects which is 56.31% higher as compared to 74.7 KN/m^2 *i.e.* pressure on soil under D.L. + L.L. The increase in permissible stresses in materials (concrete and steel) when the earth quake effects are considered is $33\frac{1}{3}\%$. For design purposes, increase the pressure due to D.L. + L.L. to such an extent that the max. pressure in wind/earth quake is within $33\frac{1}{3}\%$. This is done to make the analysis simple. By doing this, the foundation slab shall be designed under a uniform design pressure. However, exact analysis can be made under varying loads for better accuracy. Practically it has been observed that the difference in the analysis by following the two procedures is negligible.

Let X = the increase in pressure due to D.L. + L.L.

$$1.33\,(74.7 + X) = 116.77$$

$$X = 13.1 \text{ KN/m}^2$$

$\therefore$ Design pressure, $p = 74.7 + 13.1 = 87.8 \text{ KN/m}^2$

$= 100 \text{ KN/m}^2$ (say)

so that balance pressure is less than 25%.

Design of Foundation slab

For analysing the foundation slab for radial and circumferential moments and shear forces, the two cases V (Fig. 7.24) and VI (Fig. 7.25) shall be superimposed as shown in (Fig. 24.31).

Fig. 24.31 (*a*) Foundation slab under Ring load from above and soil pressure from below with central hole.

Fig. 24.31 (*b*) Foundation slab simply supported with W distributed total load along circumference of a concentric circle.

Fig. 24.31 (*c*) Foundation slab simply supported at outer edge with a central hole and loaded uniformly with soil pressure.

Where a = outer radius of slab

b = radius of central hole

c = radius of foundation ring beam centre line *i.e.* load circle.

W = Total load on foundation ring beam

p = soil pressure/unit area

p = 100 KN/m², a = 6.5 m, b = 2.0 m, C = 4.25 m.

Section at inner edge of the slab ($r = b = 2.0$ m)

Case-I. Forces of upward soil pressure on slab.

M_r = radial moment = 0

M_θ = circumferential moment

$$-\frac{pr^2}{16}+\frac{p\cdot b^2}{4}\left\{\log_e\frac{r}{a}+\frac{3}{4}\left(-\frac{1}{3}+\frac{a^2}{b^2}+\frac{a^2}{r^2}\right)-\left(\frac{a^2+r^2}{r^2}\right)\left(\frac{b^2}{a^2-b^2}\right)\log_e\frac{a}{b}\right\}$$

$$=\frac{-100\times 2^2}{16}+\frac{100\times 2^2}{4}\left\{\log_e\frac{2}{6.5}+\frac{3}{4}\left(-\frac{1}{3}+\frac{6.5^2}{2.0^2}+\frac{6.5^2}{2.0^2}\right)\right.$$

$$\left.-\left(\frac{6.5^2+2.0^2}{2^2}\right)\left(\frac{2^2}{6.5^2-2.0^2}\right)\log_e\frac{6.5}{2.0}\right\}$$

= 1274 KN.m

$$\text{S.F.}=\frac{1}{2}p\cdot r-\frac{1}{2}p\frac{b^2}{r}=\frac{1}{2}\times 100\times 2-½\times\frac{100\times 2^2}{2}=0$$

Case-II Forces of downward ring beam load

W = 120.16 × 100 = 12016 KN

M_r = 0

$$M_\theta=\frac{W}{4\pi}\times\frac{a^2}{a^2-b^2}\left(1+\frac{b^2}{r^2}\right)\left[\log_e\frac{a}{c}+\frac{1}{2}-\frac{C^2}{2a^2}\right]$$

$$=\frac{12016}{4\pi}\times\frac{6.5^2}{6.5^2-2^2}\left(1+\frac{2^2}{2^2}\right)\left[\log_e\frac{6.5}{4.25}+\frac{1}{2}-\frac{4.25^2}{2\times 6.5^2}\right]$$

= 1420 KN.m.

S.F. = 0

Net $M_\theta = 1420 - 1274 = 146$ KN m.

S.F. = 0

Factored moment $M_{u\theta} = 1.5 \times 146 = 219$ KN.m.

Adopt mix M20 and D = 500 mm at inner edge $\therefore$ $d = 425$ mm.

$$\frac{M_{u\pi}}{bd^2} = \frac{219000 \times 1000}{1000 \times 425^2} = 1.20$$

From Table 18.5 of Chapter 18.

$$p = 0.4\%$$

$$A_{st} = \frac{0.4 \times 1000 \times 425}{100} = 1700 \text{ mm}^2$$

Provide 16 mm ϕ @ 110 mm c/c circumferentially

Section at inner edge of ring beam

$r = 4.25 - 0.3 = 3.95$ m (Assuming width of Foundation ring beam as 600 mm)

Case-I: Forces of upward soil pressure on slab.

$$M_\theta = -\frac{pr^2}{16} + \frac{pb^2}{4}\left[\log_e \frac{r}{a} + \frac{3}{4}\left(-\frac{1}{3} + \frac{a^2}{b^2} + \frac{a^2}{r^2}\right) - \frac{a^2 + r^2}{r^2} \cdot \frac{b^2}{a^2 - b^2} \log_e \frac{a}{b}\right]$$

$$= -\frac{100 \times 3.95^2}{16} + \frac{100 \times 2^2}{4}\left[\log_e \frac{3.95}{6.5} + \frac{3}{4}\left(-\frac{1}{3} + \frac{6.5^2}{2^2} + \frac{6.5^2}{3.95^2}\right) - \frac{6.5^2 + 3.95^2}{3.95^2} \cdot \frac{2^2}{6.5^2 - 2^2} \log_e \frac{6.5}{2}\right]$$

$$= 777.3 \text{ KN.m.}$$

$$M_r = -\frac{3}{16pr^2} + \frac{pb^2}{4}\left[\log_e \frac{r}{a} + \frac{3}{4}(1 + \frac{a^2}{b^2} - \frac{a^2}{r^2}) + \frac{a^2 - r^2}{r^2} \cdot \frac{b^2}{a^2 - b^2} \log_e \frac{a}{b}\right]$$

$$= -\frac{3}{16} \times 1000 \times 3.95^2 + \frac{100 \times 2^2}{4}\left[\log_e \frac{3.95}{6.5} + \frac{3}{4}\left(1 + \frac{6.5^2}{2^2} - \frac{6.5^2}{3.95^2}\right) + \frac{6.5^2 - 3.95^2}{3.95^2} \times \frac{2^2}{6.5^2 - 2^2} \log_e \frac{6.5}{2.0}\right]$$

$$= 322 \text{ KN.m.}$$

$$\text{S.F.} = \frac{1}{2}pr - \frac{1}{2}p\frac{b^2}{r} = \frac{1}{2} \times 100 \times 3.95 - \frac{1}{2} \times \frac{100 \times 2^2}{3.95}$$

$$= 146.86 \text{ KN.}$$

Case-II: Forces in Slab due to downward ring beam load.

$$M_\theta = \frac{W a^2}{4\pi a^2 - b^2}\left[\left(1+\frac{b^2}{r^2}\right)\left(\log_e\frac{a}{c}+\frac{1}{2}-\frac{c^2}{2a^2}\right)\right]$$

$$= \frac{12016}{4\pi}\times\frac{6.5^2}{6.5^2-2^2}\left[\left(1+\frac{2^2}{3.95^2}\right)\left(\frac{6.5}{4.25}+\frac{1}{2}-\frac{4.25^2}{2\times 6.5^2}\right)\right]$$

$$= 943.2 \text{ KN.m.}$$

$$M_r = \frac{W}{4\pi}\times\frac{a^2}{a^2-b^2}\left[\left(1-\frac{b^2}{r^2}\right)\left(\log_e\frac{a}{c}+\frac{1}{2}-\frac{C^2}{2a^2}\right)\right]$$

$$= \frac{12016}{4\pi}\times\frac{6.5^2}{6.5^2-2^2}\left[\left(1-\frac{2^2}{3.95^2}\right)\log_e\frac{6.5}{4.25}+\frac{1}{2}-\frac{4.25^2}{2\times 6.5^2}\right]$$

$$= 558.4 \text{ KN.m.}$$

S.F. = 0

Net $M_r = 558.4 - 322 = 236.4$ KN.m.

Net $M_\theta = 943.2 - 777.3 = 165.9$ KN.m

Net S.F. = 146.86 KN

Take $D = 600$ mm, $d = 535$ mm.

Radial reinforcement Fig. 24.32 (*d*)

$$\frac{M_{ur}}{bd^2} = \frac{236.4\times 10^6\times 1.5}{1000\times 535\times 535} = 1.23$$

From Table 18.5 of chapter 18, $p = 0.4\%$

$$A_{st} = \frac{0.4}{100}\times 1000\times 535 = 2140 \text{ mm}^2$$

Provide 16 mm ϕ @ 90 mm *c/c* radially.

Circumferential reinforcement Fig. 24.32 (*d*)

$$\frac{M_{u\theta}}{bd^2} = \frac{169.9\times 10^6\times 1.5}{1000\times 525\times 525} = 0.93$$

From Table 18.5 of chapter 18, $p = 0.30\%$

$$A_{st} = \frac{0.30}{100}\times 1000\times 535 = 1605 \text{ mm}^2$$

Provide 16 mm ϕ @ 125 mm *c/c* circumferentially.

Check for shear

$$\text{Nominal shear stress} = \frac{146.86\times 1000\times 1.5}{1000\times 535} = 0.41 \text{ N/mm}^2$$

$$< 0.42 \text{ N/mm}^2\text{, minimum for M20}$$

Section at outer edge of slab ($r = 6.5$ m)

Case-I Forces of upward soil pressure on slab.

$$M_r = 0$$

$$M_\theta = -\frac{p \cdot r^2}{16} + \frac{p \cdot b^2}{4}\left[\log_e \frac{r}{a} + \frac{3}{4}\left(-\frac{1}{3} + \frac{a^2}{b^2} + \frac{a^2}{r^2}\right) - \frac{a^2 + r^2}{r^2} \cdot \frac{b^2}{a^2 - b^2}\log_e \frac{a}{b}\right]$$

$$= -\frac{100 \times 6.5^2}{16} + \frac{100 \times 2^2}{4}\left[\log_e \frac{6.5}{6.5} + \frac{3}{4}\left(-\frac{1}{3} + \frac{6.5^2}{2^2} + \frac{6.5^2}{6.5^2}\right) - \frac{6.5^2 + 6.5^2}{6.5^2} \times \frac{2^2}{6.5^2 - 2^2}\log_e \frac{6.5}{2.0}\right]$$

$$= 553.73 \text{ KN.m.}$$

$$\text{S.F.} = 0$$

Case-II Forces of downward ring beam load

$$M_r = 0$$

$$M_\theta = \frac{W}{4\pi}\left[\log_e \frac{C}{r} + \frac{1}{2} + \frac{a^2}{a^2 - b^2}\left(\frac{r^2 + b^2}{r^2}\right)\left(\log_e \frac{a}{c} + \frac{1}{2} - \frac{C^2}{2a^2}\right) - \frac{C^2}{2r^2}\right]$$

$$= \frac{12016}{4\pi}\left[\log_e \frac{4.25}{6.5} + \frac{1}{2} + \frac{6.5^2}{6.5^2 - 2^2}\left(\log_e \frac{6.5}{4.25} + \frac{1}{2} - \frac{4.25^2}{2 \times 6.5^2}\right) - \frac{4.25^2}{2 \times 6.5^2}\right]$$

$$= 687.95 \text{ KN.m.}$$

$$\text{S.F.} = 0$$

Net $\quad M_\theta = 687.95 - 553.73 = 134.22$ KN.m.

$\quad M_{u\theta} = 1.5 \times 134.22 = 201.33$ KN.m.

Take $\quad D = 450$ mm

$\quad d = 375$ mm

$$\frac{M_{u\theta}}{bd^2} = \frac{201.33 \times 10^6}{1000 \times 375^2} = 1.43$$

$$p = 0.47\%$$

$$A_{st} = \frac{0.47 \times 1000 \times 375}{100} = 1763 \text{ mm}^2$$

Provide 16 mm ϕ @ 110 mm *c/c*.

Foundation Ring Beam 600 × 1200 (M15)

$$\text{Design Soil Pressure} = \frac{12016}{\pi \times 8.5} = 450.2 \text{ KN/m}$$

$$\text{Maximum negative B.M. at support} = 0.066 \times 450.2 \times 4.25^2 \times \frac{\pi}{4}$$
$$= 421.52 \text{ KN.m.}$$

$$\text{Max. positive B.M. at mid span} = 0.03 \times 450.2 \times 4.25^2 \times \frac{\pi}{4}$$
$$= 191.53 \text{ KN.m.}$$

$$T_u = \text{Max. torsional moment at } 9\frac{1}{2}^{o} = 0.005 \times 450.2 \times 4.25^2 \times \frac{\pi}{4}$$
$$= 31.92 \text{ KN.m.}$$

$$M_t = T_u\left(\frac{1 + D/b}{1.7}\right)$$

$$= 31.92\left(\frac{1 + \dfrac{1.2}{0.6}}{1.7}\right) = 56.33 \text{ KN.m.}$$

$$\text{Total negative B.M. at support} = 421.52 + 56.33$$
$$= 477.85 \text{ KN.m.}$$

$$\text{Max. S.F. at face of support} = \frac{150.2}{2} \times \left(4.25 \times \frac{\pi}{4} - 0.6\right)$$
$$= 616.3 \text{ KN}$$

$$\text{Max. S.F. at } 9\frac{1}{2}^{o} \text{ from centre line support} = 450.2\left(\frac{4.25 \times \pi}{4 \times 2} - 4.25 \times \frac{\pi}{180} \times 9.5\right)$$
$$= 434 \text{ KN}$$

$$\text{Equivalent shear due to torsion} = 1.6\frac{T}{b}$$
$$= 1.6 \times \frac{31.92}{0.6}$$
$$= 85.12 \text{ KN.}$$

Factored Forces

Negative B.M. at support = 1.5 × 477.85 = 716.77 KN.m.

Positive B.M. at mid span = 1.5 × 191.53 = 287.3 KN.m.

Max. S.F. at face of support = 1.5 (616.3)
= 742.64 KN.

$$\text{Max. S.F. at point of Max. torsion } \left(9\frac{1}{2}^{o}\right)$$
$$= (434 + 85.12) \times 1.5$$
$$= 778.88 \text{ KN.}$$

At Support

$$\frac{M_u}{bd^2} = \frac{716.77 \times 10^6}{600 \times (1150)^2} = 0.90$$

From Table 18.5

$$p_t = 0.3\%$$

$$A_{st} = \frac{0.3 \times 600 \times 1150}{100} = 2070 \text{ mm}^2$$

Provide 20 mm ϕ 8 Nos., curtail 5 Nos. at L/3 from centre line of supports (steel at bottom face of beam)

At mid Span

$$\frac{M_u}{bd^2} = \frac{287.3 \times 10^6}{600 \times (1150)^2} = 0.362$$

From Table 18.5

$$p_t = 0.85\%$$

$$A_{st} = \frac{0.85 \times 600 \times 1150}{415} = 1414 \text{ mm}^2$$

Provide 20 mm ϕ 5 Nos. throughout (steel at top face of beam).

Check for Shear

(*i*) At face of support.

$$\text{Nominal shear stress} = \frac{742.64 \times 1000}{600 \times 1150} = 1.1 \text{ N/mm}^2$$

$$\frac{A_s \times 100}{bd} = \frac{8 \times 314 \times 100}{600 \times 1150} = 0.36$$

Permissible shear stress (Ref. Table 19.1)

$$= 0.36 \text{ N/mm}^2$$

$$\therefore \quad \text{Design shear} = (1.1 - 0.36) \times 600 \times 1150$$

$$= 524400 \text{ N.}$$

Provide 12 mm ϕ 2 legged stirrups @ spacing $= \dfrac{0.87 \times 415 \times 2 \times 113 \times 1150}{524400} = 178.9$ mm

i.e. provide 175 mm *c/c* upto *L*/3 from face of support and in rest @ 350 mm *c/c*

Min. shear requirement

$$\text{Spacing} = \frac{A_{sv} \cdot f_y \times 0.87}{0.4 \times b}$$

$$= \frac{2 \times 113 \times 415 \times 0.87}{0.4 \times 600}$$

$$= 340 \text{ mm}$$

O.K.

Reinforcement details of Foundation Ring Beam are shown in Fig. 24.32 (*d*).

(*ii*) Check for shear at point of maximum torsion.

$$\text{Nominal shear stress} = \frac{778.88 \times 1000}{600 \times 1150} = 1.12 \text{ N/mm}^2$$

Shear stirrups provided are O.K.

$$\text{Face reinforcement} = \frac{0.1 \times 600 \times 1150}{100} = 690 \text{ mm}^2$$

Provide 16 ϕ 4 Nos.

Reinforcement details of Tank Proper, staging column and braces and foundation are presented in Fig. 24.32 (*a*) to 24.32 (*r*).

Example 24.6. Design a Intze type with staying shalf 1.0 Lac gallons capacity OHSR with staging height 24.39 meters at chandigarh (Seigmic Zone IV)

(A) Computation for radii of top & bottom domes

For Bottom Dome

(*i*) *Radius of Bottom dome*

$$\text{Rise of Bottom dome} = 1.725 \text{ m}$$

$$\text{Dia} = 8.0 \text{ m}$$

$$1.725\,(2R - 1.725) = 4 \times 4$$

$$R = 5.5 \text{ m}$$

(ii) *Radius for Top Dome*

$$\text{Rise of top dome} = 2.4 \text{ m}$$

$$2.40\,(2R - 2.4) = (5.5 \times 5.5)$$

$$R = 7.5 \text{ m}$$

Let h be the depth of the water for cylindrical portion of tank.

$$\text{Density of water} = 10 \text{ KN/m}^3$$

(1) Capacity $= 455 \text{ m}^3$

(2) Add for Domed bottom $= 3.14 \times 1.725^2\,(3 \times 5.5 - 1.725)/\,3$

$= 46$ cum

Add for cone wall $= 3.14(1.1 - 2/3 \times 1)1 \times 1 \times 1/2$

$= 16.2$ cum

Add for inside shaft $= 3.14/4 \times 1.5^2(h + 1 - 1.725)$

$= 1.77h - 1.28$

Total Volume $= 515.92 + 1.77h$

$$314/4 \times 11.1^2 \times h = 515.92 + 1.77h = 5.343$$

$$h = 5.4\text{m and free board} = 0.6\text{m}$$

Design of top dome M25

$$\text{Dead load} = 0.075 \times 25000$$

$$= 1875 \text{ N/ sq. m}$$

$$\text{Live load} = (75 - 345 \times 2.17^2/\,11^2) \times 10$$

$$= 616 \text{N / m}^2$$

Total working u.d.l. $= 2491 \text{N/m}^2$ say 2500N/m^2

$$\sin\theta = 5.275/\,7.5$$

$= 0.703\ \Theta = 44.695°$

$\text{Cos}\ \theta = 0.7109$

At Crown

N = Tangential Force

T = Circumferential Force

$N = T = (1/2)\ WR$

$= (1/2) \times 2500 \times 7.5$

$= 9375$ N/ m

Tangential Force = Circumferential Force

At Springing $= 9375/75 \times 1000 = 0.125\ \text{N/mm}^2$

$N = WR/\ (1+ \text{Cos}\ \theta)$

$= 2500 \times 7.5/\ (1 + 0.7109)$

$= 10960$ N/ m

Tangential stress $= 10960/\ 1000 \times 75$

$= 0.146$ N/ sq. m

Circumferential stress $= (WR/\ A)\ [(1 - \text{Cos}\ \theta - \text{Cos}\ \theta^2)/\ (1+ \text{Cos}\ \theta)]$

$= 2500 \times 7.5/\ 1000 \times 75\ [(1 - 0.7109 - 0.7109^2)/\ (1+ 0.7109)]$

$= 0.0316\ \text{N/ mm}^2$

Stresses are very small.

Check for shear

Surface area of Dome $= 2 \times 3.14 \times Rh$

$= 2 \times 3.14 \times 7.5 \times 2.17$

$= 102.26\ \text{m}^2$

Weight of top dome $= 102.26 \times 2500$

$= 255650$ N

Load/ m of periphery 'W' $= 255650/\ 3.14 \times 11$

$= 7400$ N/ m

Shear stress $= 7400/\ (75 \times 1000)$

$= 0.0999$ N/ sq. mm

Therefore Safe.

Horizontal thrust $= W\ \text{Cot}\ \theta$

$= 7400 \times 0.7109/\ 0.703$

$= 7485$ N/m

Area of steel required $= 0.24 \times 75 \times 1000/100 = 180\text{mm}^2$

Provide 8mm dia tor @ 250mm c/c bothways

Design Of Top Rib Beam150 × 300 Mm

Hoop tension = Horizontal thrust × D/ 2

$= 7485 \times 11 \times 1/\ 2$

$= 41170$ N

Provide 8 mm dia tor 4 Nos.

Stps: 8 mm dia @ 300 mm c/c

Equivalent area of concrete = $150 \times 300 + (13 - 1) \times 4 \times 50$

$= 47400$ sq. mm

Tensile stress $= 41170/47400$

$= 0.868$ N/ sq. mm

< 1.5 N/ sq. m

Therefore Safe.

Design of outer cylinderical wall

Thickness 150 mm at base 100 mm at top. It is assumed that the wall is hinged at base & free at top (for worst conditions of hoop stress)

$$H = 5.4–1 = 4.4\text{m}$$

$$P = WH$$

$$= 10000 \times 4.4 = 44000$$

$$H^2/Dt = 4.4^2/(11 \times 0.15)$$

$$= 11.733$$

$$W.H.R. = 44000 \times 5.5$$

$$= 242000 \text{ N}$$

Depth from *F.S.L.*	Coeff. K	Hoop Tension $T = K.W\,hR$	Area of Steel $As = T/130\text{mm}^2$	Dia Spacing each face	Nos.
0.44	0.097	23474	180	8mm dia tor @ 220 mm c/c	4
0.88	0.198	47916	368	8mm dia tor @ 220mm c/c	4
1.32	0.304	73568	566	10mm dia tor @ 220 mm c/c	3
1.76	0.420	101640	781	10mm dia tor @ 220 mm c/c	4
2.20	0.545	131900	1014	12mm dia tor @ 220 mm c/c	4
2.64	0.665	160950	1238	12mm dia tor @ 180 mm c/c	5
3.08	0.749	181020	1392	12mm dia tor @ 150 mm c/c	6
3.52	0.718	173750	1366	12mm dia tor @ 150 mm c/c	6
3.96	0.475	115000	1366	12mm dia tor @ 150mm c/c	4
4.40				12mm dia tor @ 150mm c/c	4

Moment inwall

$$H^2/Dt = 11.733$$

Triangle load fixed base

$$\text{Max. B.M.} = 0.011 \times 10000 \times 4.4^3 = 9370 \text{ Nm}$$

Thickness of wall 0.9m height from top = 100 + 45 = 145mm

$$\text{Area of steel required} = 9370/0.87 \times 115 \times 130$$

$$= 720\text{sq. mm}$$

Provide 10mm dia @ 100mm c/c on water face (inner face) & curtail alternate at half height

Min distribution & temperature steel

$$= 0.24 \times 150 \times 1000/100 = 360\text{mm}^2$$

Half on each face

Provide 8 mm tor @ 150mm c/c on outer face

Vertically and curtail every third bar half height or 8mm dia @ 150mm c/c on outer face vertically

Average Hoop stress = 160950 +181020+173750/ 3

= 171907 N

Thickness of that section = $100 + 0.7 \times 50 = 135$mm

Equivalent concrete area = $135 \times 1000 + (9 - 1) \times 12 \times 1575$

= 147600 mm^2

Max. Tensile stress = 181882/ 147600

= 1.165 N/ sq. mm

< 1.5 N/ sq. mm

∴ Safe.

Max. shear at base = $0.145 \times 10000 \times 4.4^2$

= 28072 KN

Nominal shear stress = $28072/ 1000 \times 120$

= 0.23 N/ sq. mm

∴ Safe.

Design of cone top beam

Size 250×600

1. Load from roof slab = 7400 N/m
2. Load from top beam = $0.3 \times 0.2 \times 25000$ = 1500 N/m
3. Weight of vertical wall = $4.7 \times 0.125 \times 25000 = 14700$ N/m
4. Self weight of cone top = $0.25 \times 0.6 \times 25000 = 3750$N/m
5 Weight of water column = $4.4 \times 0.1 \times 10000 = 4400$ N/m

Total = 31750N/M say 32000N/m

Horizontal thrust = 32000 N

Hoop tension = $32000\ (11 - 0.1) / 2$

= 174400 N

Or 174400/130 = 1341mm^2, Provide 16mm dia 3 No inner face and 16mm dia 3 Nos On Outer face & 1 Nos on each face

Provide 8 mm dia tor 2 legged @ 200 mm c/c

Check for tensile stress

Equivalent concrete area = $250 \times 600 + 8 \times 201 \times (9-1)$

= 150000 +12864 = 162864mm^2

Tensile stress = 174400/162864 mm^2

= 1.07N/mm^2

< 1.5 N/mm^2

Safe.

Design of cone wall

Thickness = 300 mm

Slant length = 1.42m

Dia at mid height = $(11.1 + 8.5)/2 = 9.75$m

Loading

(I) Weight of conical wall = $3.14 \times 9.75 \times 1.42 \times 0.3 \times 25000$
= 326050 N

(ii) Weight of water column directly resting on conical wall

(a) $3.14 \times 10 \times 4.4 \times 10000 \times 1.0 = 1382500$ N

(b) $3.14 \times 9.67 \times 1/2 \times 1 \times 1 \times 10000 = 151500$ N

(iii) Load from cone wall = $32000 \times 3.14 \times 11.1 = 1115328$ N

Total = 2975378

Say 3000000 N

Max, meridional thrust in cone wall = 3000000/(3.14 × 9)
= 150750 N/m

Comp. Stress = 150750/(300 × 1000)
= 0.5025 N/m²

Weight of cone wall = 0.3×25000 = 7500 N/m

Safe

Let any height '*X*' above ring beam, dia on conical wall = D_X

Height of water = $(5.4 - x)$ m

$D_X = 9 + 2x$

W_n = Load acting normal to the surface
= 7500 Sine +(5.4 – *x*) × 10000
= 59302 – 10000*x*

Hoop tension = *Wn*. Sece θ .$D_x/2$
= 1/2 (59302 – 10000*x*) (9 + 2*x*) × 1.42
= $378940 + 20310x - 14200x^2$

Max. Hoop tension at × = 0 = 378940 N

Max. Hoop tension at × = 1.0^N = 385050 N

Area of steel required = 385050/1/2(150 +190) = 2265mm² (Tor)

A_{st} in slant length = 3210 × 1.42 = 3217mm²

Provide 16 m dia tor 17 No (8 Nos on inner face and 9 Nos on outer face

Temp and distribution steel = 0.002 × 300 × 1000 = 600mm² i.e 300mm² on each face

Check for bending

W = 326050 + 1382500 + 151500
= 1860050 N

Max. B.M. = 1860050 × 1/(8 × 3.14 × 9.5)
= 7800 N.m say 8000Nm

A_{st} = (7800 × 1000)/(140 × 0.9 × 250)
= 247 mm²

10 mm dia @ 300mm c/c on each face

Check for tensile stress

Eq. Area of concrete = 300 × 1000 + (9–1) × 314 × 15
= 337680mm²

tensile stress in cone = 385050/337680

$= 1.14 \text{ N/m}^2$

safe $< 1.5 \text{ N/mm}^2$

Design of central shaft 120 mm THK

Height of central shaft $= 5.4 + 0.6 + 2.17 - 1.725$

$= 6.445\text{m}$

If h = height of water around shaft

$= 5.4 - 1.725 = 3.675\text{m}$

$D_{outer} = 1.5 \text{ m}$ $D_{inner} = 1.26\text{m}$

$T = 120\text{mm}$

Max. hoop compression $= 10000 \times 3.675 \times 1.5 \times 1/2$

$= 27562.5 \text{ N}$

Compression stress $= 27562.5/(1000 \times 120)$

$= 0.23 \text{ N/mm}^2$

Min distribution steel $= 0.00295 \times 120 \times 1000$

$= 354\text{mm}^2$

i.e., 177mm^2 on each face

Provide 8 m dia @250 mm c/c from dome to central shaft. Also provide 150 × 150 circular beam around top opening with 8mm dia 4Nos c/c and 8mm dia @ 300mm c/c

Design of ring beam at junction of central shaft and bottom dome

$\sin\theta = 0.75/5.5 = 0.1363$ $\phi_o = 7.84^0$ $\cos\theta = 0.9906$

$\sin\theta = 3.95/5.5 = 0.718$ $\phi_o = 46^0$ $\cos\theta = 0.695$

Self weight of shaft = p = 0.12 × 25000 × 6.445 = 19350 N

Component of P in horizontal direction

$= 19350 \times 0.9906/0.1363$

$= 140630 \text{ N}$

Hoop compression in ring beam $= 140630 \times 1.38/2 = 97050 \text{ N}$

Hoop compressive stress $= 97050/(150 \times 150) = 4.313 \text{ N/mm}^2$

$< 8 \text{ N/mm}^2$ safe

BM in base of shaft will be torsion in the ring beam

Torsional moment = 0.0015 × 100000 × 3.6753 = 745 Nm

Shear stress $= M_t(3+2b/D)/B2D = 745 \times 1000(3 + 2 \times 150/150)/150 \times 150 \times 150$

$= 1.103$

$< 2.2\text{N/mm}^2$

Area of steel required for torsion = Mt.s/0.8 × 1y1.0st

$= 745 \times 1000 \times 100/0.8 \times 120 \times 120 \times 115$

$= 56.23\text{mm}^2$

Min area of steel required $= 0.3/100 \times 150 \times 150$

$= 67.5\text{mm}^2$

Provide 4 Nos 10mm dia with strrups or 8mm @ 150mm c/c

Design of bottom dome 150mm thick

Dead weight of dome $= 150/1000 \times 25000 = 3750 \text{ N/m}^2$

$H = 3.675$, $a = 5.5\text{m}$

(1) Stress due to water

$N_\varphi = -(W.a/6)\,[3h + a(1-\cos\theta)/(1+\cos\theta)(1+2\cos\theta)]$

$= -(10000 \times 5.0/6)\ [3 \times 3.675 + 5.0 \times 1–0.695 / 1 + 0.695$
$(1+2 \times 0.695)\}$

$= -123000\text{N/m}$

$N_\theta = (+\text{ Wa}/6)\ [3h + a\ (1 - \cos\theta / 1 + \cos\theta)\ (5 + 4\cos\theta)]$

$= (10000 \times 5.5/6)\ [3 \times 3.675 + 5.0 \times 1–0.695$
$1+0.695(5+4 \times 0.695)\} = +172000\ \text{N/m}$

$N_\theta = -\text{p.a}\ [(1 - \cos\theta)/\sin^2\theta] - \text{P.}\sin\phi_{o/}/\sin^2\theta$

$= -3750 \times 5.5\ [(0.9906 - 0.695)/\ 0.718^2]–(19319350$
$\times 0.01363)/\ 0.7187^2$

$= -13810–5110 = -18920\ \text{N/ m}$

$N_\theta = -\text{p.a.}\ [((\cos\phi_o/ - \cos\theta)/\sin^2\theta) - \cos\theta] + \text{P}\sin\phi_o/\sin^2\phi$

$= -3750 \times 5.555\ [(0.9906 - 0.695)/ - (0.695)]$
$+19350 \times 0.1363/0.718^2$

$= -3062.9 + -5110 = -2047.1\ \text{N/ m}$

Total $N\varphi$ at spring $= -123000 - 18920$

$= -141920$ N/ m comp

Total N_θ at springing $= 172000 + 2047.1$

$= 174047.14$ N/ m comp.

Assuming minimum distribution of steel

$= 0.24 \times 150 \times 1000/100 = 360\text{mm}^2$

Stress at springing $= 174047.1/179000$

$= 0.972\ \text{N/mm}^2 < 8\text{N/mm}^2$ (comp) safe

(B) At Crown $\theta = \phi_o$

(1) Stress due to Water

$N_\varphi = (-\text{W.a}/6)\ [3\text{h} + \text{a}\ (1 - \cos\theta)/ 1 + \cos\theta)(1+2\cos\theta)]$

$= (10000 \times 5.5/6)\ \{3 \times 3.675 + 5.5\ 1– 0.9906/1 + 0.9906$
$(1 + 2 \times 0.9906)\}$

$= -\ 101545\ \text{N/m}\ 5 + 4.9906)/1 + 0.99906)\ (1+2 \times 0.9906)$

$= -101545$ N/ m

$N_\theta = (+\text{ Wa}/6)\ [3\text{h}+\text{a}\ (1 - \cos\theta/ 1 + \cos\theta)\ (5 + 4\cos\theta)]$

$= +1000 \times 5.5/6\{3 \times 3.675 + 5.5\ (1–0.990/1 + 0.9906 \times$
$(5 + 4 \times 0.9906)\}$

$= +103000$ N/m

(2) Stress due to shaft load P and self weight of dome p.

$N_\varphi = -\text{P.}\sin\phi_o/\sin^2\theta$

$= -(19350 \times 0.1363/0.1363^2)$

$= -141970$ N/m

$N_\theta = N_\varphi - \text{p.a}\cos\phi_o$

$= 141970 - 3750 \times 5.5 \times 0.9906$

$= 118106$ N/m

$\therefore$ Total N_φ at crown $= -101545 - 141970$

$= 243515$ N/ m comp

Total N_θ at crown $= 103000118106$

$= 221106$ N/ m comp

∴ Min. steel = $0.24 \times 150 \times 1000/100 = 360mm^2$

Provide 10 mm dia tor @ 200 mm c/c

10mm dia @ 200mm c/c from ring beam into dome at springing at top face.

Design of stagging shaft **M 20**

As per IS:11682–1985 (2nd reprint 2001)

Thickness of staging wall = $150 + (D - 6000/120)$

= $150 + 8500 - 6000/120) = 170mm$

Weight of Top dome = $7.400 \times \Pi \times 11 = 256$ KN

Weight of outer cylindrical wall

= $\Pi \times 11.15 \times 5 \times 0.15 \times 25 = 657$ KN

Weight of internal shaft = $\Pi \times 1.35 \times (6+2.4 - 1.725) \times 0.12 \times 25 = 85$ KN

Weight of bottom dome = $2 \times \Pi \times 5.5 \times 1.725 \times 0.15 \times 25 = 223$ KN

Weight of cone wall = $\Pi \times 9.5 \times 1.414 \times 0.3 \times 25 = 316$ KN

Weight of top beam = $0.5 \times 0.6\ \Pi \times 8.5 \times 25 = 200$ KN

Weight of water = $455.000 \times 10 = 4550$ KN

Weight of staging shaft = $\Pi \times 8.5 \times 25.0 \times 0.17 \times 25 = 2836$ KN

Total load at base = 9123 KN say 9200 KN

Permissible axial comp stress = 5 N/mm^2

Bearing area required = $9400 \times 1000/5$

= 1840000 mm^2

Area provided = $\Pi \times 8500 \times 170$

= 4537300 mm^2

Vertical steel in staging shaft

Min steel = $0.25 \times 170 \times 1000/100 = 425$ mm in a direction

= 212.5mm^2 each face

10 dia @ 360mm c/c

As per IS:11682–1985 (Reprint Sept 2001)

Max spacing can be 300 c/c

Hor steel

Min steel staging shaft = $0.2 \times 170 \times 1000/100 = 340$ mm^2

= 170 mm^2 each

4 cm^2 (min as per IS code)

Provide 10 dia tor @ 300mm c/c

Check for wind forces

Wind velocity = 47m/sec

Basic wind pressure = $0.6 \times 47^2 = 1325$ N/m^2 = 1.325 KN/m^2

Design wind pressure with gust = $1.325 \times 1.13 = 1.5$ KN/m^2

Max Horizontal force as tank proper

= $1.5 \times \Pi \times 11.3 \times 0.7 \times 6.6 = 245$ KN

Acting at 28 above base

Max Horizontal force on staging

= $\Pi \times 8.67 \times 25 \times 1.5 \times 0.7 = 715$ KN

Acting at 12.5m above base

OM at base = $246 \times 28 + 715 \times 12.5$

$= 6888 + 8938$

$= 15826$ KNm say 16000 KN

Total weight $= 9200$ KN

Direct stress $= 9200 \times 1000/4537300 = 2.02$ KN/mm^2

$I = \Pi/ 64(D^4 - d^4) = \Pi/64(8.670^4 - 8.330^4) \times 10^{12}$

$= \Pi/64\{5650–4815\}\times10^{12} = 40.97 \times 10^{12}$ mm^4

$Z = \Pi/D/2 = 40.97 \times 10^{12} \times 2/8670 = 9.45 \times 10^9$ mm^3

Tank Full

Bending stress $= 16000 \times 10^6/9.45 \times 10^9 = 1.70$ N/mm^2

Net stress $= 2.02–1.70 = 0.32$N/mm^{2}comp

(a) Tank Empty

Direct stress $= (9200 - 4550) \times 1000/4537300 = 1.02$N/m^2

Bending stress $= 1.76$N/mm^2

Net tension $= 1.70–1.02$

$= 0.68$ N/m^2<1.7N/mm^2

Safe

Check due to earth quake forces

(A)Tank full

$Wf = (9200 - 2836) + 1/3 \times 2836$

$= 6364 + 945 = 7310$KN

$K = 12\, EI / L^3 = 12\times2\times10^5\times40.97\times10^{12}/253^3\times10^9 = 1.68\times10^6$

$T = 2\pi \sqrt{W_f}$ g.K

$T = 2\pi \sqrt{7310} \times 10^6/ 9810 \times 1.68 \times 10^6$

$= 4.18$ sec.

$Sa/g = 1.67/4.18 = 0.42$

$\alpha_{hf} = Z.I/2R.s_a/ g$

$= 0.24 \times 1.5/2 \times 3 \times 0.42$

$= 0.03$

Tank full

Hor force $= \propto_{hw} = 0.03 \times 7310 = 220$ KN

Acting at 28m from base

OM at base due to seismic force

$= 220 \times 28 = 6160$ Knm

Bending stress due to earthquake

$= 6160 \times 10^6/9.45 \times 10^9 = 0.65$N/mm^2

Direct stress $= 9200 \times 10^3/4537300 = 2.02N/mm^2 > 0.65$N/mm^2

Safe

Tank Empty

$We = (6364 - 4550) + 1/3 \times 2836$

$= 1814 + 945 = 2760$ KN

$Te = 2\pi \sqrt{W_e/ g.K}$

$= 2\pi \sqrt{2760} \times 10^6/ 9810 \times 1.68 \times 10^6$

$= 2.5$ sec

$W = 1814 + 2836 = 4650$ KN

Sa/g = 1.67/2.5 = 0.70

Direct stress = 4850000/4537300 = 1.02 N/mm^2

α_{hf} = Z.I/2 R.s_a/ g

= 0.24 × 1.5/2 × 3 × 0.7

= 0.042

Hor force due to seismic effects

= 0.042 × 2760 = 116 KN

OM at base = 116 × 14 = 1624 Kn/m

Bending stress = 1624 × 10^6/9.45 × 10^9 = 0.17 N/mm^2

Direct stress > bending stress

Hence OK

Design of foundation

Total load at base = 9200 KN

Self weight of footing 10% of SI load

= 920 KN

Total load at base = 10120 KN

SBC of soil = 10.8 t/m^2 = 108 KN/m^2

Bearing area required = 10120/108 = 93.7m^2

Bearing area provided = Π/4(14^2 – 3.5^2) = 144.2m^2

Pressure distribution on soil

(A) Under wind effects

(*a*) *Tank Full*

Direct load = 10120

Direct pressure = 10120/144.2 = 70 KN/m^2

OM = 16000 KNm

Section Modulus = {Π/64(14^4–3.5^4} × 2/13.65 = 268m^3

Bending Pressure = 16000/268.0mm^3 = 60.0 KN/m^2

Total Max Pressure any where

70 + 60 = 130<150 KN/m^2

OK

Seismic Forces are much

Smaller than wind forces

Hence safe

Design pressure = 100 KN/m^2

While 30 KN/m^2 can be ignored for increase in permissible in materials in 33 1/3%

OK

Factored design pressure = 100 × 1.5 = 150KN/m^2

Design of foundation slab

a = 7.00 m b = 1.75 m c = 4.25 m

W = 150 × 144.2 = 21630 KN r = 4.5m

$$M_r = (-3/16)\,pr^2 + pb^2/4\,[\log_e (r/a) + (3/4)\,(1 + (a^2/b^2) - (a^2/r^2)) + (a^2 - r^2/r^2) + (b^2/a^2 - b^2)\,\log_e (a/b)]$$

$= (-3/16) \times 150 \times 4.5^2 + 150 \times 1.75^2/4[\log_e (4.5/7.125) + (3/4)(1 + (7.0^2/3.5^2) - (7.0^2/4.5^2)) + (7.0^2 - 4.5^2/ 4.5^2) (1.75^2/7.125^2 - 3.\ 5^2) \log_e (7.0/1.25)]$

$= -569 + 114.84\{-0.41+3/4(1 + 16.0 - 2.42) + (1.42 \times 0.066 \times 1.386\}$

$= 651.0$ KNm

$M_r = (W/4\pi) \{\log_e c/r - 1/2 + (a^2/a^2 - b^2) [(r^2 - b^2/r^2) (\log_e (a/c) + (1/2) - (c^2/2r^2))]$

$= (21346/4\pi) \times (\log 4.25/4.5 - 1/2 + 1.06 \times 0.85 \times (1.647 + 0.5 - 0.207) - 0.5\}$

$= 1190$ KNm

Net Mr $= 1190–651.0 = 539$ KNm

M 20

$d = \sqrt{539 \times 10^6/1000 \times 2.76} = 442$

$D = 600$mm $\quad d = 540$ mm

$K = 442 \times 10^6/1000 \times 540^2 = 1.52$

$A_{st} = 0.49 \times 540 \times 1000/100 = 2646$ mm^2

(20) dia 118mm c/c radially (114 Nos 20 dia 5300 mm + 114 Nos 20 dia 3600 mm)

$M_\theta = (-1/16)\ pr^2 + pb^2/4\ [\log_e(r/a) + (3/4)\ (-1/3(a^2/b^2)-(a^2/r^2)) (a^2- r^2/r^2)) - (b^2/a^2-b^2) \log_e (a/b)]$

$= (-1/16) \times 150 \times 4.5^2 +150 \times 1.75^2/4 \times [\log_e (4.5/7.0) + (3/4)(-1/3 + (7.0^2/4.5^2)) \times 72/1.752 - (1.75^2+4.5^2/4.5^2)) \times (1.75^2/7.125^2-1.75^2 \times \log_e (7.0/1.75)]$

$= -189.84 + 114.84\{-0441 + 3/4(-0.333 + 16.0 + 2.42)- 3.42 \times 0.066 \times 1.386\}$

$= -189.84 +114.84\{-0.441 + 13.56-0.313\}$

$= -189.84 +1470 = 1280$ KNm

Due to conc Ring load

$M_\theta = (W/4\pi) \{\log_e c/r + 1/2 + (a^2/a^2 - b^2) [(r^2+ b^2/r^2) (\log_e (a/c) + (1/2)-(c^2/2a^2))]$

$= (21630/4p) (\log 4.25/4.5 + 1/2 + 7.0^2/7.0^2 1.75^2) (4.5^2 + 1.75^2/4.5^2 (\log_e 7.0/4.25 + 1/2 - 4.25^2/20 \times 7.115^2)$

$= 1722\{-0.057 + 0.5 + 1.066 \times 1.15 \times 0.816)-0.445$

$= 1720$ KNm

Net $M_\theta = 1720 - 1280 = 440$ KNM

$K = 440 \times 106/1000 \times 540^2 = 1.5$

$A_{st} = 0.48 \times 540 \times 1000/100 = 2892$

20dia 120mm c/c average as M_θ goes to 2/3 at free edges.

Design of staircase column **m-20**

Size 380 mm dia

Inner core = 300 mm

Length of column = 25.0 m

Weight of one step = 400N

Dead load of steps = 25 × 400/0.2 = 50000 N
L.L. on steps = 50000 N
Load from braces = 4 × 0.40 × 0.20 × 25000 × 4.25
= 34000 N
Load from self weight = 25 × 3.14 × 0.38² × 25000/4
= 70850 N
Total load = 154850 N = 155 KN
Area of conc. Required = 155000/5
= 31000 mm²
Area of conc. Provided = 314 × 380²/4
= 113300 mm²
Min steel = 0.008 × 31000
= 248mm²

Provide 16mm dia tor 6nos through out

Ties/Rings 8mm dia @ 200mm c/c

Load from col = 155 KN
Self weight @ 10 % = 15.5 KN
Total = 170.5 KN
Area of footing required = 170.5/108 = 1.6 m²

Provide 1.85m × 1.85m square footing

Net upward pressure = 170.5/1.8 × 1.8 = 52.6KN/m²
Max BM = 52.6 × 1.8 × (1.8–0.3/8)2 = 27.0m
d = √ 27000 × 1000/0.87 × 380 = 358 mm
D = 380mm d = 300 mm
A_{st} = 27000 × 1000/0.9 × 230 × 300 = 434 m²

Provide 12mm 9nos c/c bothways

Design of precast steps **m-20**

Precast steps will consist of cantilever slab 400 mm wide at outer edge &100mm at support edge.

Length of step = 900mm
Thickness = 65 mm at outer edge upto plane 300mm from inner edge, which increases to 175 mm at support.

(*a*) Live load on slab = 0.9 × 0.25 × 3000
= 800 N
(*b*) Self weight = 400 KN
(*c*) Miscellaneous = 100 N
Total *u d* l = 1300 N

B.M at 0.3 m

From edge = 1300 × 0.6 = 780 Nm
D = 175 mm at built in edge & 65 mm thin 300 mm and then 65mm throughout the free edge
A_{st} = (780 × 1000)/(1.2 × 200)
= 57 mm²

Provide 175 mm at the built in edge tapered to 65 mm within 300 mm and then 65 mm throughout to free end is OK

$$A_{st} = 780 \times 1000/140 \times 0.87 \times 57 = 112.4 \text{ mm}^2$$

Provide 2 Nos 12 mm dia at edge + 1 No. 6 mm dia at center

Or 2 nos 10 mm dia at edge + 1 No 6 mm dia at center

Also provide 7 Nos 6 mm dia as distributors. See drawing for further details reinforcement.

Design of landing slab = **100 mm**

$$\text{Self weight} = 0.1 \times 0.75 \times 0.425 \times 25000 = 798 \text{ N}$$

$$LL = 0.75 \times 0.425 \times 3000 = 956 \text{ N}$$

$$\text{Total } udl = 1754 \text{ N}$$

Assumed to be adding at 0.7 × 750 × 630 mm for built in end. Max BM at built in end = 1754 × 0.525 = 920.85mm

$$d = \sqrt{} 920.85 \times 1000/1.2 \times 100 = 87.6 \text{ mm}$$

$$D = 100 \text{ mm} \quad d = 85 \text{ mm}$$

$$A_{st} = 920.85 \times 1000/140 \times 0.87 \times 85 = 88.94 \text{ mm}^2$$

Provide 12mm dia 7 nos or 10 mm dia 7 Nos. Distributors 10mm dia 6 Nos. or 8 mm dia 6 Nos

Cantilever slab

$$\text{Dead load} = 200 \text{ Kg}$$

$$\text{Live load} = 300 \text{ Kg}$$

$$\text{Total load} = 500 \text{ Kg}$$

$$\text{Max } BM = 5 \times 0.9^2/2 = 2.025 \text{ KNm}$$

$$Mu\ bd^2 = 1.5 \times 2.025 \times 10^6/1000 \times 80^2 = 0.48$$

$$K = 0.16$$

$$A_{st} = 0.16 \times 1000 \times 80/100 = 126 \text{ mm}^2$$

Provide 8mm dia bars @ 150mm c/c

$$\text{Distt steel} = 8 \text{ mm dia @ 200mm c/c}$$

EXERCISE 24

1. Design circular tank for a capacity of 500 kilolitres with flexible joint at the base. Use (*i*) mid steel reinforcement (ii) HYSD bars.
2. Design a circular tank 16 m diameter and 3 m height of wall. Keep a free board of 0.3 m. The tank is resting on a firm ground. Assume the walls of the tank fixed at base and free at top. Use M 25 concrete and Fe 415 grade reinforcement.
3. Design a rectangular tank resting on ground for a capacity of 80 kilolitres. Use M 25 concrete and Fe 415 grade reinforcement.
4. Design an underground rectangular tank 10 m × 6 m × 3 m deep. The tank is covered at top. Take the density of soil as 16000 N/m^3 and angle of repose as 30°. Use M 25 concrete and Fe 415 grade reinforcement.
5. Design an Intze type over head service reservoir (OHSR) for a capacity of 910 Kilolitres (2 lac gallons). Take inside diameter = 14.5 m for tank proper and 11.5 m *c/c* of columns 12 numbers. FSL = 30.48 m (100 '–0"). Bearing capacity of soil = 85 KN/m^2. Basic wind pressure = 1500 N/m^2. The tank is situated in seismic zone IV. Use M 25 concrete and Fe 415 grade reinforcement.

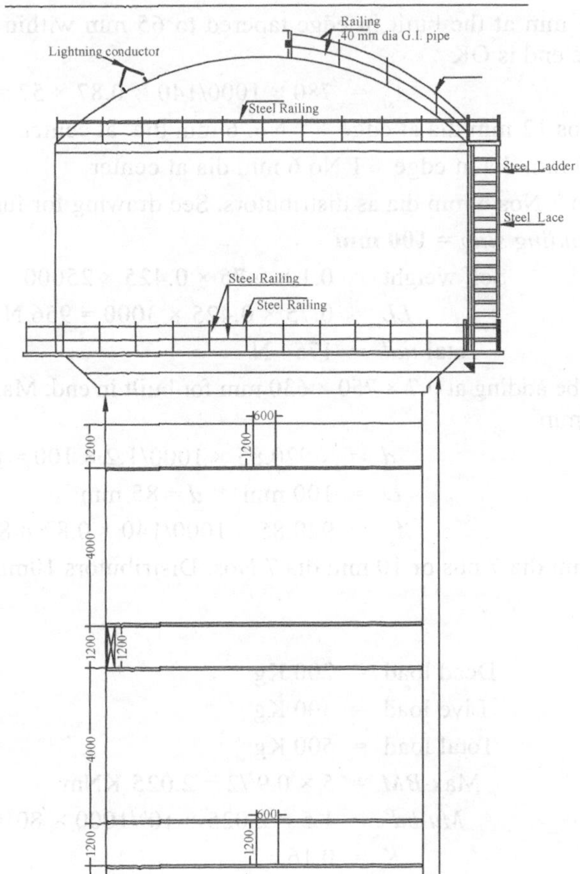

Elevation showing window opening in shaft

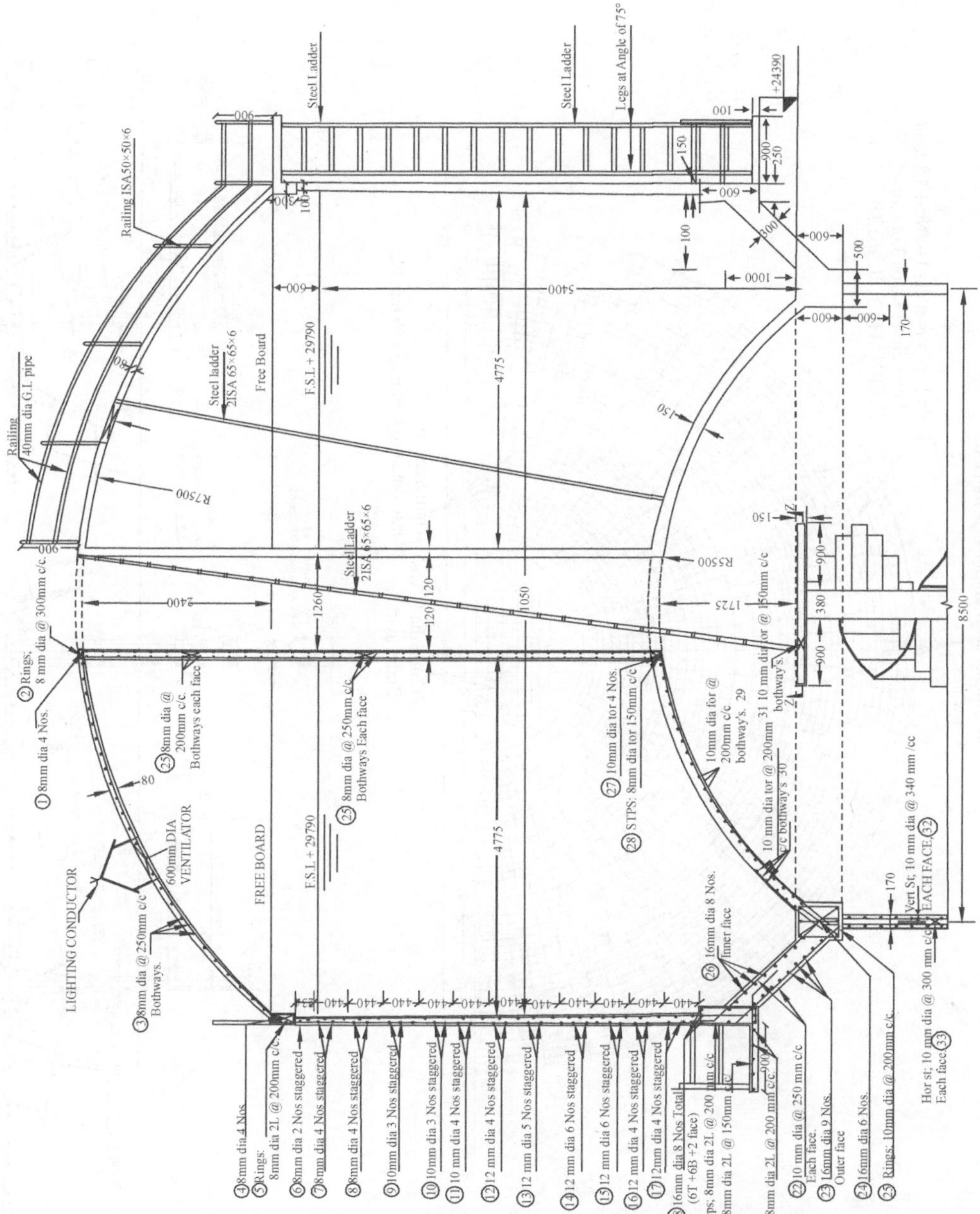

DETAIL OF TANK PROPER

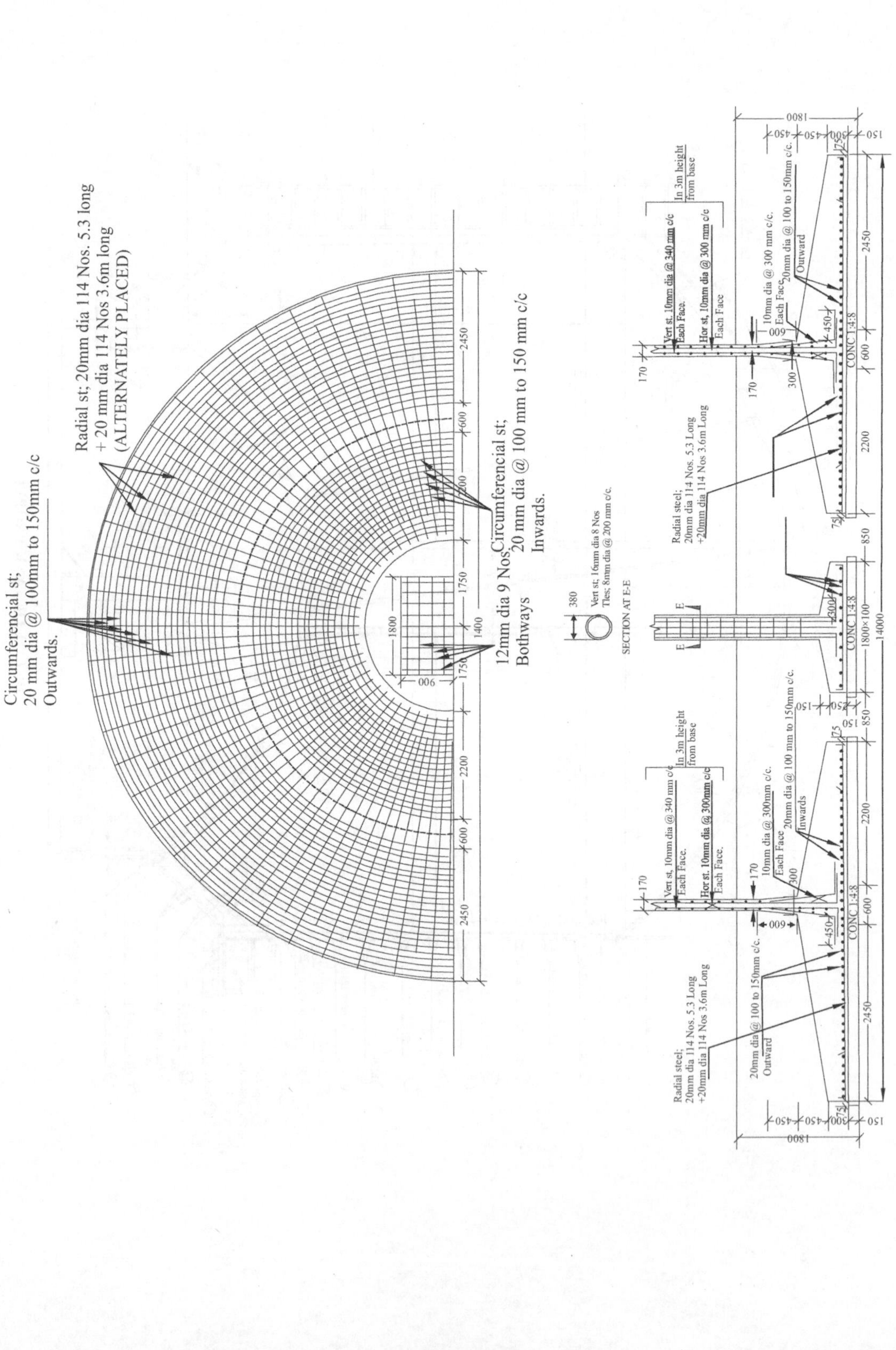
Circumferencial st;
20 mm dia @ 100mm to 150mm c/c
Outwards.
Radial st; 20mm dia 114 Nos. 5.3 long
+ 20 mm dia 114 Nos 3.6m long
(ALTERNATELY PLACED)
Circumferencial st;
20 mm dia @ 100 mm to 150 mm c/c
Inwards.
12mm dia 9 Nos
Bothways
2450
600
2200
1750
1400
1800
900
380
Vert st; 16mm dia 8 Nos
Tles; 8mm dia @ 200 mm c/c.
SECTION AT E-E
Radial steel;
20mm dia 114 Nos. 5.3 Long
+20mm dia 114 Nos 3.6m Long
Vert st, 10mm dia @ 340 mm c/c
Each Face.
Hor st, 10mm dia @ 300 mm c/c
Each Face
In 3m height
from base
10mm dia @ 300 mm c/c.
Each Face
20mm dia @ 100 to 150mm c/c.
Outward
20mm dia @ 100 mm to 150mm c/c.
Inwards
CONC 1:4:8
1800×100
14000
850
170
300
450
75
150
1800

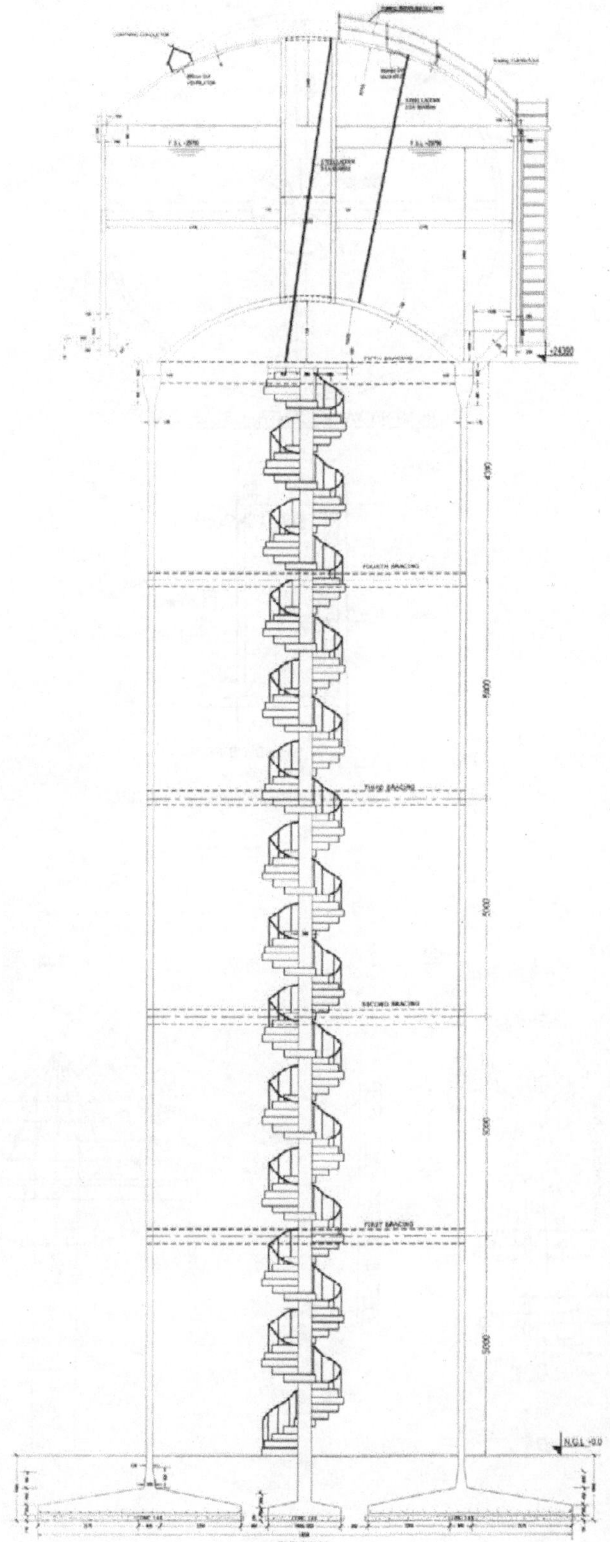
+24380
FOURTH BRACING
SECOND BRACING
FIRST BRACING
5000
N.G.L.-0.0
ELEVATION

PLAN SHOWING CENTRAL STAIRCASE

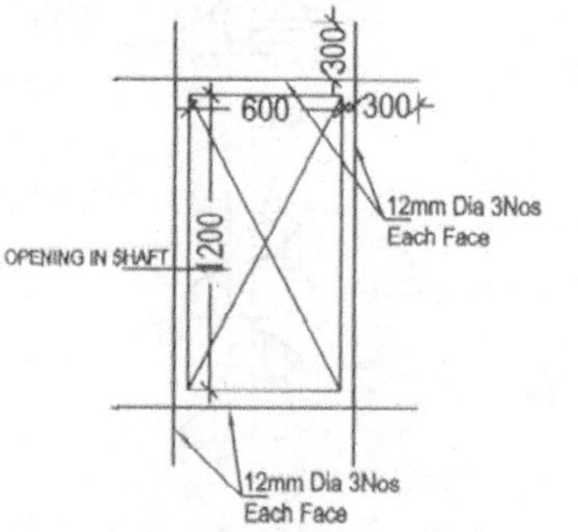

TYPICAL DETAIL OF REINF AROUND WINDOW OPENING IN SHAFT

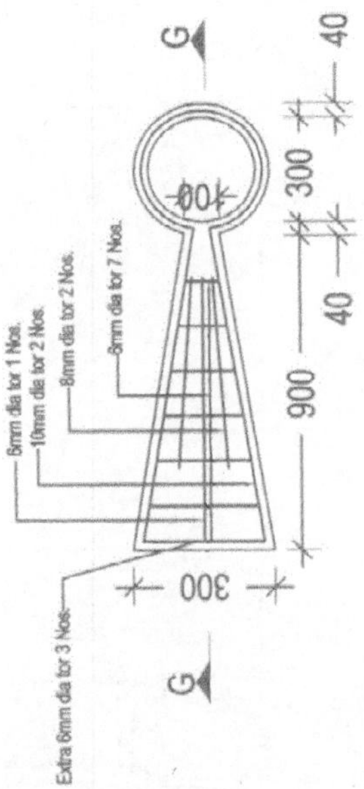

DETAIL OF PRE-CAST STEP

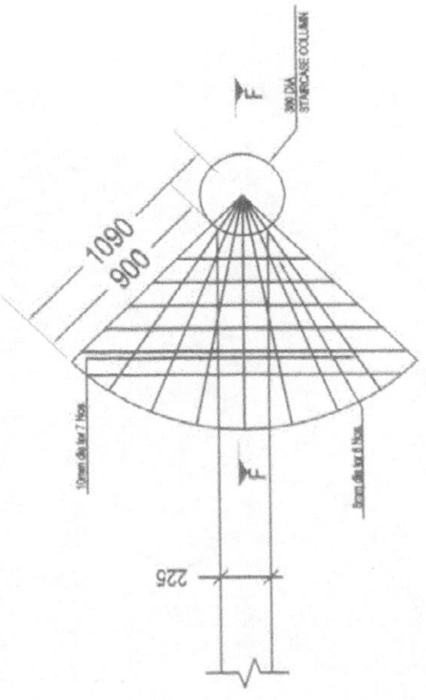

DETAIL OF LANDING

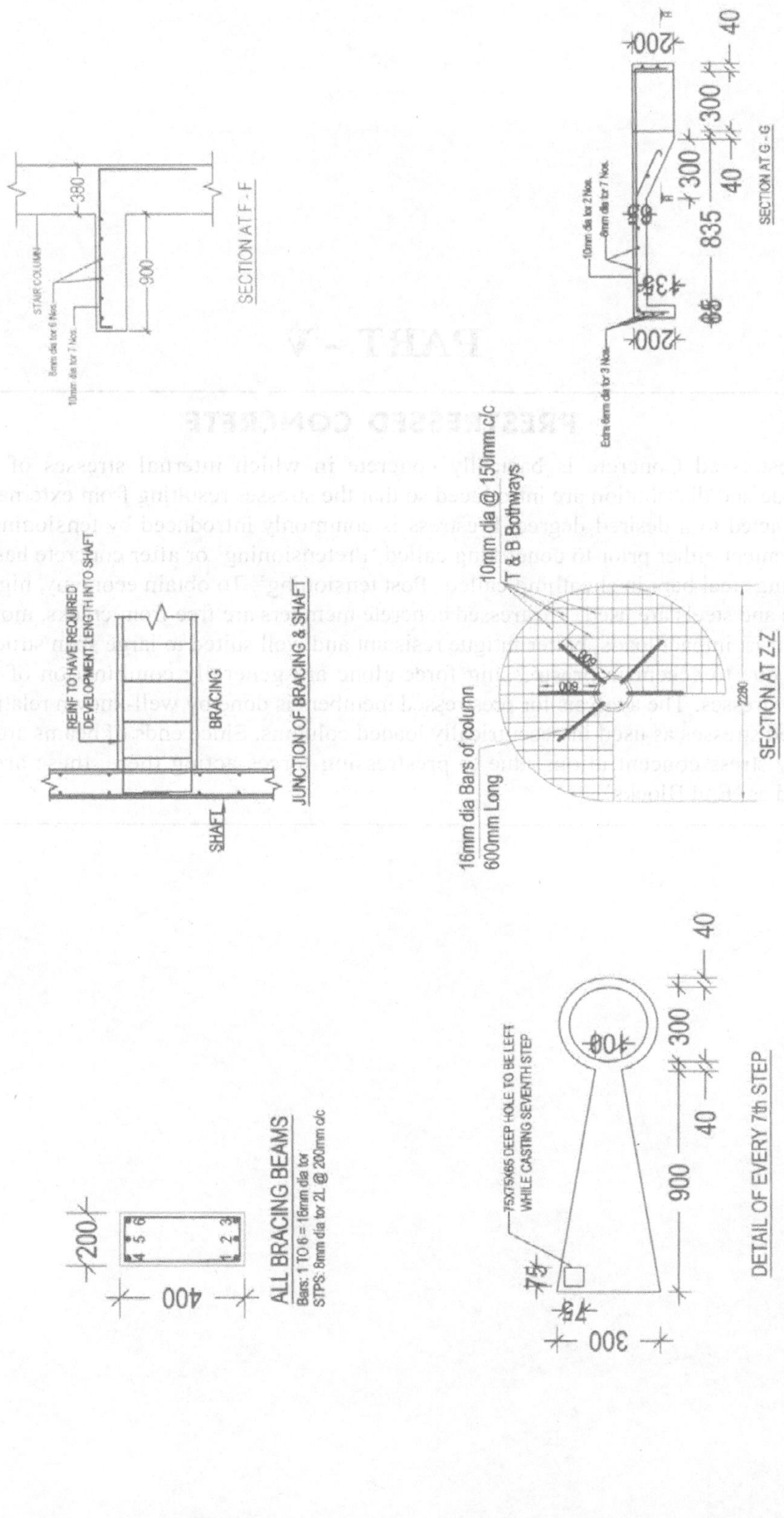
STAIR COLUMN
8mm dia tor 6 Nos
10mm dia tor 7 Nos
380
900
SECTION AT F - F
10mm dia tor 2 Nos
6mm dia tor 7 Nos
Extra 6mm dia tor 3 Nos
200
135
835
300
40
SECTION AT G - G
REINF TO HAVE REQUIRED DEVELOPMENT LENGTH INTO SHAFT
BRACING
SHAFT
JUNCTION OF BRACING & SHAFT
10mm dia @ 150mm c/c T & B Bothways
16mm dia Bars of column 600mm Long
2290
SECTION AT Z-Z
200
400
ALL BRACING BEAMS
Bars: 1 TO 6 = 16mm dia tor
STPS: 8mm dia tor 2L @ 200mm c/c
75X75X65 DEEP HOLE TO BE LEFT WHILE CASTING SEVENTH STEP
75
300
900
100
DETAIL OF EVERY 7th STEP

PART – V

PRESTRESSED CONCRETE

Prestressed Concrete is basically concrete in which internal stresses of a suitable magnitude and distribution are introduced so that the stresses resulting from external loads are counter acted to a desired degree. Prestress is commonly introduced by tensioning the steel reinforcement either prior to concreting called 'Pretensioning' or after concrete has hardened by keeping steel bars in sheathing called 'Post tensioning'. To obtain economy, high strength concrete and steels are used. Prestressed concrete members are free from cracks, more durable, able to resist impact loads, better fatigue resistant and well suited to large span structures. The stresses due to eccentric prestressing force alone are generally combination of direct and bending stresses. The analysis for prestressed members is done by well-known relationship for combined stresses as used in eccentrically loaded columns. Since ends of beams are subjected to heavy stress concentrations, due to prestressing forces acting there, these are specially designed as 'End Blocks'.

This Page has been intentionally left blank

CHAPTER 25

Prestressed Concrete

25.1 INTRODUCTION

Prestressing a structure consists of deliberately creating in it a type of stress opposite to that which the design loads will cause. The technique is of particular importance to concrete which has an inherent weakness of having low tensile strength. If compression is created in a concrete section before or simultaneously with the application of loads, tensile stress could be eliminated or greatly reduced. The process of initially stressing the steel reinforcement in concrete members, is called 'Prestressing'. Prestressed concrete is basically concrete in which favourable compressive internal stresses of suitable magnitude and distribution are initially introduced to counteract partially or wholly the undesirable stresses produced due to external loads. The application of permanent compressive stress increases the apparent tensile strength of concrete because the subsequent application of loads producing tensile stresses will first nullify the compressive prestress.

The process of prestressing is very old as even in olden times shrink fitting of metal strips on wooden wheels and wooden barrels made by force-fitting of metal bands or ropes, were good examples indicating the existence of art of prestressing. In 1886, P.H. Jackson an Engineer of San Francisco obtained patents for tightening steel tie rods in artificial stones and concrete arches to serve as floor slabs. Around 1888, C. E. W. Doehring of Germany secured a patent for concrete reinforced with metal that had tensile stress introduced before the slab was loaded. Freyssinet in 1928 demonstrated the utility of using high strength concrete and steel to account for the various losses of prestress due to creep and shinkage. He later invented a double acting jack for stressing high tensile steel wires. Prestressed concrete started getting used in long span bridges, industrial shell roofs, marine structures, nuclear pressure vessels, water retaining structures, transmission poles, railway sleepers, long span auditorium and gymnasiums.

Research on prestressed concrete showed good amount of loss of prestress due to various causes which necessitated the use of high strength steel and concrete so that loss of prestress is a very small fraction of initial stress induced. High strength concrete is less liable to shrinkage cracks and smaller creep strain.

25.2 TERMINOLOGY

1. *Tendon* : A stretched element used in a concrete member of structure to impart prestress to the concrete. Generally, high tensile steel wires, bars, cables or strands are used as tendons.

2. *Anchorage* : A device generally used to enable the tendon to impart and maintain prestress in concrete. Commonly used anchorages are clips, cones, wedges and anchor plates.

3. *Pretensioning* : A method of prestressing concrete in which the tendons are tensioned before the concrete is placed. In this method, the prestress is introduced to concrete by bond between steel and concrete.

4. *Post-Tensioning* : A method of prestressing concrete by tensioning the tendons against hardened concrete. In this method, the prestress is imparted to concrete by bearing.

5. *Bonded Prestressed Concrete* : Concrete in which prestress is introduced to concrete through a bond between the cables and surrounding concrete. Pre-tensioned members belong to this group.

6. *Non Bonded Prestressed Concrete* : Concrete in which prestress is introduced to concrete through anchorage devices and tendons or cables are not bonded to concrete. Tendons are placed in ducts formed in the concrete members.

7. *Full Prestressing* : Prestressed concrete in which tensile stresses in concrete at working loads are entirely counterbalanced by having sufficiently high prestress.

8. *Partial Prestressing* : Prestressed concrete in which tensile stresses in concrete at working loads are partially counterbalanced.

9. *Axial Prestressing* : Members in which whole of the cross section of concrete has a uniform compressive presstress. In this type of prestressing, tendon run along centre line of section.

10. *Eccentric Prestressing* : Members in which tendons do not run along centre line of section.

11. *Concordant Prestressing* : Prestressing of members in which tendons or cables follow a concordant profile.

12. *Cracking Load* : The load on the structural member corresponding to the first appearance of crack.

13. *Proof Stress* : The tensile stress in steel which produces a residual strain of 0.2 per cent of the original gauge length on unloading.

14. *Creep Coefficient* : The ratio of total creep strain to elastic strain in concrete.

15. *Degrees of Prestressing* : Measure of magnitude of prestressing force related to the resultant stress occurring in the member at working load.

16. *Characteristic Load* : Load which has 95 per cent probability of not being exceeded during life of the structure.

17. *Creep in Concrete* : Increase with time in the strain of concrete subjected to sustained stress.

18. *Transfer* : The act of transferring the stress in prestressing tendons from the jacks or pretensioning bed to the concrete member.

25.3 ADVANTAGES OF PRESTRESSED CONCRETE OVER REINFORCED CONCRETE CONSTRUCTION

1. The use of high strength concrete and steel in prestressed concrete members results in lighter and slender members than could be possible by using reinforced concrete with Mild steel as reinforcement. In prestressed concrete, steel is no longer required for reinforcement, but it is provided there as a means of prestressing. Prestressed concrete needs about one-third the quantity of concrete and one-fourth of steel as compared to R.C.C.

2. In fully prestressed members, free from tensile stresses under working loads, whole of the section is effective whereas in R.C.C., only 30 to 40 per cent of the concrete actively resists bending, thus, material is wasted.

3. In prestressed members, dead loads may be counterbalanced by eccentric prestressing whereas in R.C.C. dead loads are of great importance. There is saving in use of materials in prestressed concrete.

4. Prestressed concrete members possess better resistance to shearing forces due to effect of compressive stresses presence or eccentric cable profiles.

5. Use of high strength concrete and freedom from cracks, in prestressed members, contribute to improved durability under aggressive environmental conditions.

6. Long span structures are possible in prestressed concrete so that saving in weight becomes significant whereas in R.C.C. very heavy sections are required.

7. Factory products (namely, Railway sleepers, electric poles) are possible in prestressed concrete members whereas R. C. C. members have to be cast in situ.

8. Prestressed concrete members are tested before use but R.C.C. members can not be tested as such.

9. Prestressed Concrete structures deflect appreciably before ultimate failure, thus giving ample warning before collapse. No such indication is therein even an under-reinforced R.C.C. beam.

10. Fatigue strength of prestressed concrete members is better chiefly due to the small stress variations in prestressing steel, therefore, it is recommended for dynamically loaded structures.

11. Multiple units are possible by centralising the prestressing work which reduces labour cost and gives better control of the product.

25.4 DISADVANTAGES OF PRESTRESSED CONCRETE CONSTRUCTION

1. The availability of builders, experienced with work of prestressing, is scanty.
2. Initial equipment cost for prestressing is very high.
3. Cost of high strength materials is high.
4. Availability of Engineers, experienced with prestressing work, is scanty.
5. Prestressed sections are brittle because high tension steel is used. All high strength steels have higher carbon content.
6. Slender sections are difficult to transport.
7. Prestressed conccrete sections are less fire resistant.
8. No openings can be made once members are prestressed.

25.5 CLASSIFICATION AND TYPES

Prestressed concrete structures can be classified in a number of ways depending upon the features of design and construction.

25.5.1 Externally or Internally Prestressed

Sometimes it is possible to prestress a concrete member by adjusting its external reactions by use of jacks. Creep and shrinkage in concrete may completely offset the artificial strains unless they can be readjusted. Such structures are called 'externally prestressed'. More frequently used method of pretensioning steel tendons before concreting the member, is called 'Internal prestressing'.

25.5.2 Linear or Circular Prestressing

Prestressing circular structures such as round tanks, silos and pipes where tendons are wound around in circles, is called 'Circular Prestressing'. Prestressing linear members such as beams, sleepers, electric poles, slabs, etc., is called 'Linear Prestressing'. The prestressing cables or tendons in linear prestressed structures are not necessarily straight, they can be either bent or curved but they do not have shape in the form of circular rings.

25.5.3 Pretensioning and Post Tensioning

Pretensioning is used to describe any method of pretressing in which the tendons or cables are tensioned before the concrete is placed. In this method tendons are temporarily anchored and tensioned and the prestress is transferred to the concrete after it has hardened. This method is used for factory products like prestressed railway sleepers and electric poles.

Post tensioning is a method of prestressing in which the tendon is tensioned after concrete has hardened. In this method, tendons are placed in sheathing at suitable place in the member before casting and later after hardening of concrete, tendons are tensioned and anchored by use of anchorage devices.

The distinctive features of Pretensioning and Post tensioning systems are presented in Table 25.1.

Table 25.1. Distinctive Features of Pretensioning and Post-tensioning Systems

Sr. No.	*Pretensioning System*	*Post-tensioning System*
1.	It is cheaper because there is no expenditure on rubber core, sheathing, spacers, etc.	It is costly because of use of rubber core, sheathing, spacers, etc.
2.	It has more certainty about durability.	There is dependence on two anchorages.
3.	The two abutments which hold the wires in tensioned position should be very strong and can not be reused till concrete in the member under construction has hardened enough to take the stresses. Method is suitable for factory production.	Method may be employed for precast or cast-in-situ members.
4.	Size of member is limited because heavier and longer members are more difficult to transport.	Size of members may be any since members can be prestressed in field.
5.	Loss of prestress is more. Wires when released shorten and later on due to shrinkage and creep of concrete further loss of prestress takes place.	Loss of prestress is comparatively less. In post tensioning system, the elastic strain and part of shrinkage and creep takes place, while wires are still being stretched. Thus this loss can be made up by stretching the wires a little further.

25.6 STAGES OF LOADING

For cast-in-situ structure, prestresed concrete has to be designed for atleast two stages; the initial stage during prestressing and the final stage under external loadings. For Pre-cast members, a third stage, that of handling and transportation is also very important.

25.6.1 Initial Stages

A structure is said to be in initial stage of prestressing when it is under prestress but is not subjected to any superimposed external loads. Some periods in the initial stage are important for certain designs and are summarised as follows:

(A) Before Prestressing

Before a concrete member is prestressed, it is quite weak in carrying loads, hence a few things must be ensured e.g., prevention of yielding of supports, provision for shrinkage of concrete, careful curing, avoidance of sudden change in temperature etc.

(B) During Prestressing

Often, the maximum stress to which the tendons will be subjected throughout their service period, occurs during prestressing. The concrete is not fully aged at this stage, crushing of concrete at anchorage is possible if its quality is not proper. Breaking of a wire due to defects in its manufacture or overstressing is not of much importance because there are many more in position. Broken wire may be replaced if total wires are few in number. The order of prestressing various tendons must be properly studied before hand.

(C) At Transfer of Prestress

The transfer of prestress is sudden in pretensioning and gradual in post tensioning systems, but in both cases there are no external loads acting except its self weight. This imposes a worst condition on concrete and may be controlling condition for design. At cambering effect of prestressing, self-weight of member is considered for economy.

(*D*) *Decentering and Retensioning*

If a structure is cast and prestress in situ, it usually becomes self-supporting during or after prestressing. the frame-work can be removed after prestressing and no new load condition is imposed on the structure. Some structures are retensioned, *i.e.*, prestressed in stages. It is important to see the stresses at these stages.

25.6.2 Intermediate Stage

This stage occurs during transportation and erection and is there only in precast members. It should be ensure that such members do not change their support conditions during transportation or erection. For example a simple beam designed for end supported conditions shall break or crack if lifted at mid span or a part of it overhangs from truck during transportation.

25.6.3 Final Stage

This is the stage when actual imposed working loads act on the structure. For prestressed concrete structures, it is often necessary to investigate effects of sustained loads, working loads (such as live loads, lateral loads, i.e., wind and earthquake forces), cracking load and ultimate loads.

(*A*) *Sustained Load*

The Camber or deflection of a prestressed section under actual sustained load (*i.e.*, dead load) is often controlling factor in design. Flexural creep increase due to sustained load.

(*B*) *Working Load*

Prestresses concrete members like R.C.C. members, are designed on the basis of working loads encountered during their service peiod.

(*C*) *Cracking Load*

Cracking load in certain structures such as water tanks, pipes, etc., becomes the controlling load for design. Cracks are more objectionable, because a cracked member may have excessive deflections, corrosion of steel tendons may result in decrease in strength.

(*D*) *Ultimate Load*

It is necessary to know the ultimate strength of a member so that actual margin of safety for overloads can be ascertained. Ultimate load is the maximum load a member can carry before collapse.

25.7 CONCRETE FOR PRESTRESSING

Stronger concrete is generally required for prestressed than for reinforced concrete work. Higher strength is necessary in prestressed concrete in order to minimize the cost, to use commercial anchorages, to have high resistance in tension, shear, bond and bearing. High strength concrete is less liable to shrinkage cracks. Higher modulus of Elasticity and smaller creep strain results in smaller loss of prestress in steel. Generally, 28 to 34 MPa strengths work out for economical designs. Concrete strength of 28 to 40 MPa can be obtained easily without excessive labour and cement. To attain a strength in excess of 34 MPa, it is necessary to use w/c ratio of not more than 0.45 by weight, to facilitate placing a slump of 50 to 100 mm. To obtain 75 mm slump with w/c ratio 0.45 would require about 10 bags of cement per cu. metre of concrete. Since excessive cement tends to increase shrinkage, a lower cement factor is desirable. Good vibration or use of proper admixtures increases workability. Air entrainment of 3 to 5% improves workability and reduces bleeding.

25.8 STEELS FOR PRESTRESSING

High strength steel is mostly used as prestressing material. High strength steels are often obtained by alloying, which permits the manufacture of such steels under normal operation. Carbon is an economical element for alloying since it is cheap and easy to handle. Other alloys include manganese and silicon. The most common method for increasing the tensile strength for prestressing is by cold-drawing, the process of cold-drawing tends to realign the crystals, and the strength is increased by each drawing so that the smaller the diameter of wires, the higher their ultimate unit strength.

High strength steel for prestressing takes one of the three forms : Wires (5-12 mm dia.) strands (2-4 mm dia.) or bars. For post tensioning, wires are widely employed ; they are grouped in parallel into cables. Strands are fabricated in the factory by twisting wires together.

High tensile steel usually contain 0.7-0.8 per cent carbon, 0.6 per cent manganese and about 0.1 per cent silica. Bars are first hot rolled and subsequently heat treated. Cold drawn wires are tempered to improve its ductility properties. Tempering or stress relieving of wires at 150-400° C results in increasing of tensile strength. The wires used individually or in wire cables are generally 5-7 mm in diameter having ultimate tensile strength of about 1500 N/mm². The strands used vary in nominal diameter from 10 to 44 mm. The ultimate tensile strength of high tensile steel varies inversely to diameter of wire. As per IS Code 1343, characteristic strengths for some diameters of wires, bars and strands are given in Table 25.2.

Table 25.2. Characteristic Strength of High Tensile Steel

(a) H.T. Wires

Nominal Diameter mm	*Characteristic Strength N/mm²*		
	Plain wires cold worked stress relieved	*Indented wires*	*Plain wires as drawn*
3	1900	1900	–
4	1750	1750	1750
5	1600	1600	1600
7	1500	–	–
8	1400	–	–

(b) H.T. bars

Nominal Diameter mm	*Characteristic Strength N/mm²*
10 – 40	*1000*

(c) Strands

Designation		*Maximum Breaking load kN*
2.0 mm	2 ply	13.00
2.9 mm	2 ply	26.00
6.3 mm	7 ply	45.36
7.9 mm	7 ply	70.30
9.5 mm	7 ply	95.25
11.1 mm	7 ply	127.00
12.5 mm	7 ply	167.85
15.2 mm	7 ply	231.35

The typical stress-strain curve of H.T. wires vs. Mild steel bars is shown in Fig. 25.1.

The 0.2% proof stress for H.T. wires and alloy bars used for prestressed concrete should not be less than 80% of the minimum ultimate tensile strength. The steel used in prestressing must behave more plastically at stresses near ultimate failure so that there is sufficient warning before collapse. To avoid brittle fracture, I.S. Code prescribes a minimum of 2% for elongation at rupture when tested over a gauge length of 200 mm.

Fibre glass has very high tensile strength (an ultimate tensile strength of 6900 MPa is quite common) and can replace steels but for its use extensive research is under-way.

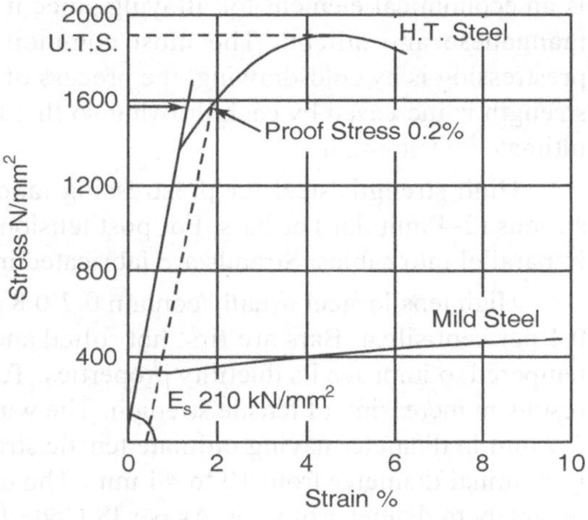

Fig. 25.1. Stress-Strain Curve

25.9 PRESTRESSING SYSTEMS

The various methods of prestressing concrete are as follows :

1. Development of compressive force between the structural member and its abutments by use of jacks.

2. By elongating steel wires or bars, either before concrete is hardened, *i.e.*, Pretensioning or after the concrete has attained its strength, *i.e.*, Post-tensioning, by use of Jacks.

3. Electrical Prestressing dispenses with the use of jacks altogether. The steel is lengthened by heating with electricity which is a post tensioning method.

4. Chemical Prestressing : Chemical reactions taking place in expensive cements can stress the embedded steel which in turn compresses the concrete.

Most widely used methods of Pretensioning and Post-tensioning are described subsequently.

25.9.1 Pretensioning System

Pretensioning means tensioning the tendons or cables before placing the concrete. Hoyer's system is very popular for tensioning. It is also known as Longline method. In this method, the wires are stretched between two points situated at the ends of the yard or stressing bed. It is used to have two vertical posts at each end of the line set at a given distance apart and supporting a horizontal cross head to which the wires are anchored. Anchoring of wires can be done by abutment method or Strut method or mould method as shown in Fig. 25.2. The wires are anchored at one end to a fixed cross head and at the other to a fixed or a movable cross head.

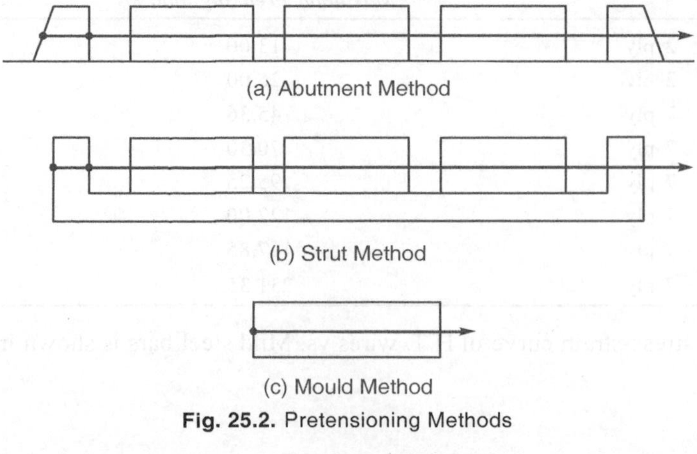

(a) Abutment Method

(b) Strut Method

(c) Mould Method

Fig. 25.2. Pretensioning Methods

The moulds are assembled in line, sides may or may not be in position depending upon the depth of the member to be cast. The wires are threaded through the fixed cross head and then through the end plates of all the moulds which have been drilled or slotted to support the wire in the correct position in the cross section. When the wires are to be tensioned singly or in pairs, they are first anchored at one end and then pulled at the other. When the correct extensions have been given to the wires, the widges are driven home in the gripping devices. The concrete is poured and vibrated until full consolidation has taken place. It is then allowed to harden. The side moulds are stripped as soon as possible to allow for free movement and shrinkage of concrete. When the concrete attains full strength, the tension in the wires is released and then wires cut. As soon as the wires are cut, they may be burnt off flush.

The wires or strands are generally anchored to abutments by steel wedges. Generally strands upto 18 mm diameter and high tensile wires of upto 7 mm diameter anchor themselves satisfactorily by surface bond and interlocking of the surrounding matrix. Strands have better bond characteristics than single wires. Supplementary anchoring devices such as Weinberg Clip and Dorland Clip (Fig. 25.3) are also used.

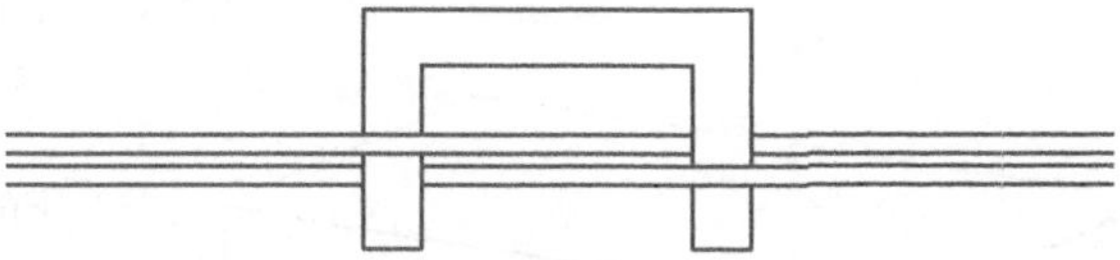

(a) Weinberg Clip

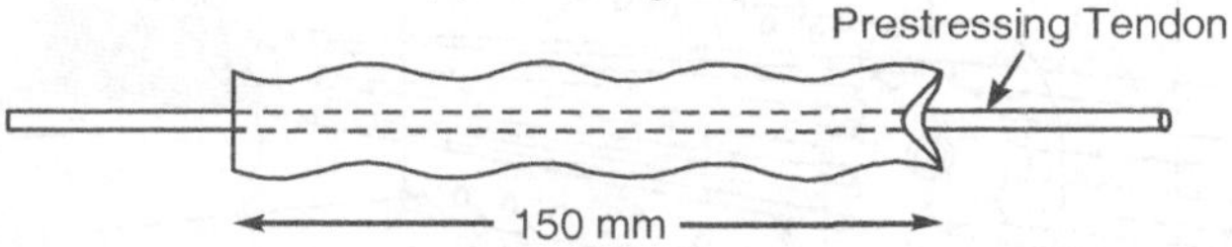

(b) Dorland Clip

Fig. 25.3. Clips

25.9.2 Post-Tensioning System

In post-tensioning, the concrete hardens and ducts for passing tendons are housed in the concrete. When concrete attains sufficient strengths, high tension wires are tensioned by means of jacks bearing on end face of the member and the wires are anchored by wedges or nuts. The space between ducts and tendons is generally grouted. Most widely accepted principles of anchoring the tendons are listed below.

1. Wedge action generating a friction grip on wires.
2. Direct Bearing from heads formed at the wires ends.
3. Winding the wires around the concrete.

(*A*) *Freyssinet System*

This is one of the most widely used post-tensioning methods of prestressing. Steel wires 5 mm to 7 mm, 8, 12 or 16 in number are grouped into cables with a helical spring inside. The spring keeps proper spacing between the wires and the sheathing (Fig. 25.4) prevents a bond between concrete and steel. Such cables are arranged in the mould before concreting. The anchorage consists of a precast concrete cylinder (Fig. 25.5) having corrugations outside, conical hole in the centre, and a heavy hoop reinforcement inside. The cylinders are placed in position before concreting. The plugs (Fig. 25.6) are pushed in position after tensioning of cables to act as wedges. The cables are thus held in position.

Through the hole in the plug (Fig. 25.7), cement mortar is pumped to grout the space between the wires after prestressing. This is an additional help in preventing the wires from slipping.

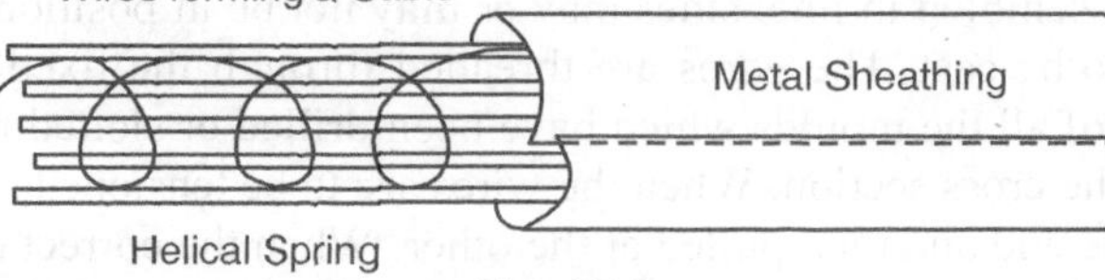

Fig. 25.4.

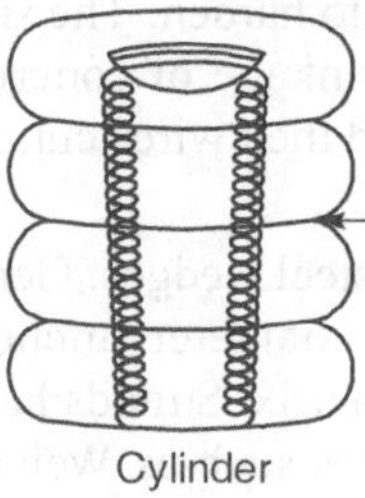

Fig. 25.5.

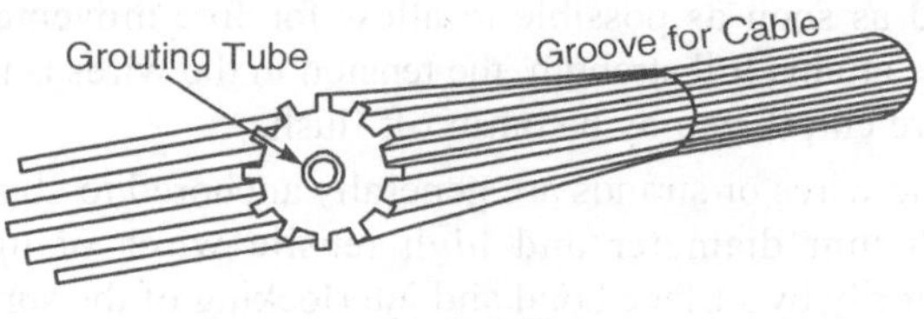

Fig. 25.6

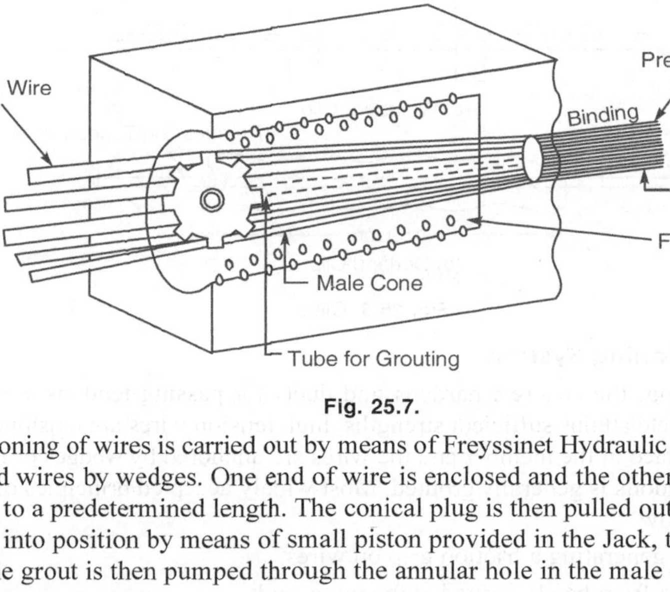

Fig. 25.7.

The tensioning of wires is carried out by means of Freyssinet Hydraulic Jack carrying suitable grooves to hold wires by wedges. One end of wire is enclosed and the other end is pulled till the wires elongate to a predetermined length. The conical plug is then pulled out in this process which is then pushed into position by means of small piston provided in the Jack, thus locking the wires in position. The grout is then pumped through the annular hole in the male cone.

(*B*) *Magnel Blaton System*

The principles on which this method is based are :

(*i*) The wires must be placed in a definite order and not at random.

(*ii*) Between all the wires in a cable, spaces must be left to allow easy injection of cement grout in order to protect wires from corrosion.

(*iii*) Only two wires should be stretched at a time so as to obtain a practically uniform stress in wires.

(*iv*) The anchorage must be strong enough to enable one particular wire to break when stretching force is applied, without damage to locking device.

Professor Magnel and Contractors Blaton in Belgium developed this system. In this system 5 mm and 7 mm diameter wires are used. A cable in this case can consist of any even number of wires upto 64. These wires are arranged four in a group in a horizontal line. Vertical and horizontal spacer (Fig. 25.8) are provided @ nearly 1.5 m c/c to keep the wires in correct position. The cable is then placed in a sheet metal sheath or holes in the concrete member to permit the cables to be passed through the member after the concrete has hardened. The ends of the wires are then attached to steel plates which are in turn placed against a cast steel distributing plate. Each locking plate called a 'Sandwitch plate' as shown in Fig. 25.9 is provided with wedge shaped grooves in each of which two wires are secured with a steel wedge. A stretching force is applied by a jack stretching two wires at a time. Holes are left in the prestressed member to grout the space between the wires.

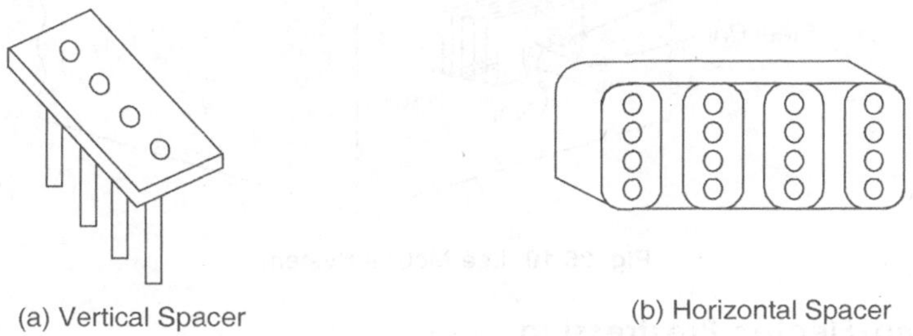

Fig. 25.8.

This system differs from the Freyssinet system inasmuch as the unit of stretching here is only two wires at a time against 8 to 16 in the other. It is claimed that the stresses in the wires in this system is more uniform, but it is doubtful as the elongation of all wires in Freyssinet system is also the same and could be measured easily. All wire grooves in the wedges and the Jacks are of standard size and chances of stretching even in Freyssinet system are remote. Another essential difference is that here regular holes are cast at suitable places by introducing rubber core in the mould. The local bearing pressure between concrete and edges of sandwitch plate is distributed by means of a bearing plate of size dependent upon the number of wires in the cable.

(*C*) *Lee McCall System*

In this system instead of wires, high tensile steel bars varying from 12 mm to 30 mm dia. as tendon are used. Desired length of tendon be obtained by joining two bars with screwed coupling. The diameter of duct is generally kept 12 mm larger than the diameter of bar. There are two ways of anchoring the bars.

(*i*) *Wedge Anchorage* : Plain and unthreaded bar is anchored after tensioning by a sleeve wedge driven home into a machined hole in the end plate.

(*ii*) *Screwed Anchorage* : In this type of anchorage, the bar is threaded on both ends. After tensioning the bar, it is anchored by screwing a nut and washer tightly against the end plate (Fig. 25.10). This system has an advantage over the other method that the stressing can be done in stages as the nut could be tightened at any stage. The member can also be stressed or unstressed when desired. At the untensioned end, only small length of threaded part is required just to take

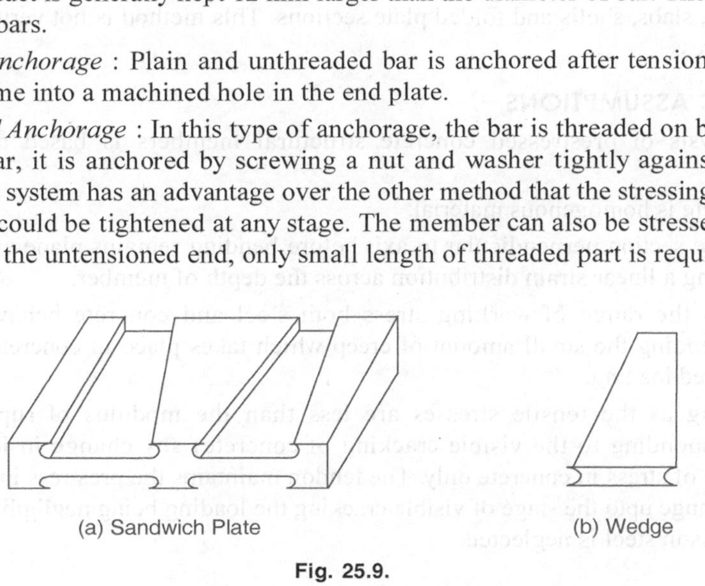

Fig. 25.9.

washer and nut but for the other end where jack is applied, comparatively larger threaded portion is desired.

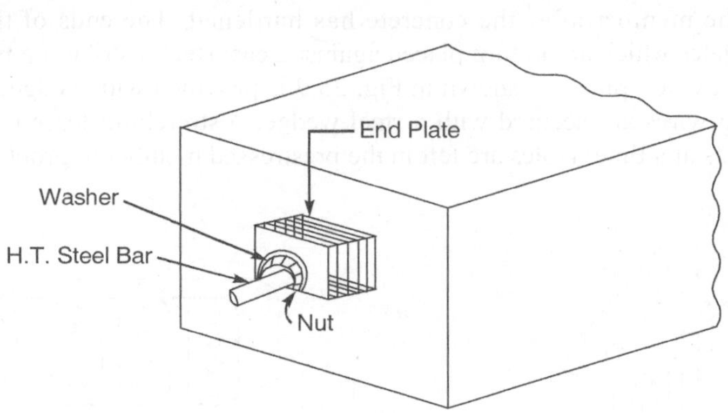

Fig. 25.10. Lee McCall System

25.9.3 Thermo-Electric Prestressing

In this pretensioning method, prestressing by elongation is achieved by heating the H.T. wires by passing electricity. The bar is heated electrically to 300-400° C in 3 to 5 minutes and 0.4 to 0.5 per cent elongation is obtained. As the bars cool, bars tend to shorten which is prevented by the fixed anchors at two ends. By this method, stress upto 500-600 N/mm^2 can be introduced in the bars. As the temperature falls below 90°C, concreting is done.

25.9.4 Chemical Prestressing

With the development of expanding cements by Lossier in 1944, Chemical prestressing became feasible. The expanding cement consists of 7.5% Portland Cement, 15% High Alumina Cement and 10% Gypsum. The linear expansion of cement is about 3-4%. When the concrete expands it is restricted by high strength steel wires and thus compressive stress is induced in concrete and tensile in steel. This system is also called 'Self prestressing'. The tensile stress upto 850 N/mm^2 can be developed in steel by this method. This method is suitable for prestressing in pipe, thin walls, slabs, shells and folded plate sections. This method is not very suitable for high degree of prestress.

25.10 BASIC ASSUMPTIONS

The anlaysis of prestressed concrete structural members is based on the following assumptions.

1. Concrete is homogenous material.
2. A plane section perpendicular to axis before bending remains plane even after bending implying a linear strain distribution across the depth of member.
3. Within the range of working stress both steel and concrete behave elastically not withstanding the small amount of creep which takes place in concrete and steel under sustained loading.
4. As long as the tensile stresses are less than the modulus of rupture of concrete (corresponding to the visible cracking of concrete) any change in loading results in change of stress in concrete only. The tendon maintains the prestress in concrete without any change upto the stage of visible cracking the loading being negligibly small, a change in stress in steel is neglected.

25.11 LOSS OF PRESTRESS

Prestressing force is measured either through the extension of the prestressed wire or by a pressure gauge reading of hydraulic Jack at the time of jacking. The force recorded at the time of jacking at the jack position is not necessarily the amount of force available at all sections of member and at various working stages of the member. Further the effective prestress in concrete does not remain constant but reduces gradually with time from the stage of transfer. This reduction in prestress is called 'Loss of Prestress'.

The various types of losses of prestress may be due to ;

1. Elastic shortening of concrete.
2. Creep and shrinkage in concrete.
3. Creep in Steel.
4. Anchorage slip.
5. Steam Curing.
6. Bending of Member.
7. Friction.

25.11.1 Elastic Shortening of Concrete

(A) Pretensioned Members

Elastic shortening depends upon the modulus of Elasticity of Concrete, steel and initial prestressing force. In pretensioned construction, the prestressing force is measured by jacking against bulk heads. As the prestress is transferred to the concrete, the member shortens and alongwith it prestressed steel shortens and as such there is a loss of prestress in steel. Considering only the shortening of concrete due to the direct load effect of concrete, the loss of prestress is computed as

$$f_s = m \frac{F_1}{A_c + m \cdot A_s} \qquad \text{...(25.1)}$$

where,

f_s = Loss of prestress,

F_1 = Initial prestressing force,

m = Modular ratio = $\dfrac{E_s}{E_c} = \dfrac{\text{modulus of elasticity of steel}}{\text{modulus of elasticity of concrete}}$

A_c = Net area of concrete section,

A_s = Area of prestressing steel.

(B) Post-tensioned Members

In post-tensioned construction, wires are stretched with reaction against the member itself, so the elastic deformation takes place simultaneously and the jacking or the initial stresses are independent of the elastic shortening. If the tensioning of the wires is done in stages, the wires that were anchored will undergo elastic shortening while the jacking is done for the next stage of prestressing. The wires tightened in the very beginning will be subjected to maximum loss of stress and those tightened in the last will not suffer any loss due to elastic shortening of concrete. If the prestressing is done at equal intervals, an average prestressing force could be taken for computing the elastic shortening. If a single tendon is tensioned by jacking against the member, no loss of prestress is considered as the force in the cable is measured after the elastic shortening of concrete has taken place.

25.11.2 Creep and Shrinkage in Concrete

(A) Creep in Concrete

The additional deformation in excess of elastic deformation under sustained long-term loading is called 'Creep'. The amount of creep is often one to two times the elastic shortening. It is not possible to overtension the tendons excessively in order to allow for such loss, because that would mean very high initial stresses in the steel which increase its creep or approach its yield point. The creep of concrete causes loss of stress in tensioned steel.

Creep Coefficient of concrete for loss of prestress

$$\phi = \frac{\text{Creep Strain}}{\text{Elastic Strain}} = \frac{E_c}{E_t} \qquad ...(25.2)$$

$$\text{Creep strain in concrete} = \phi \cdot \varepsilon_c = \phi \cdot \frac{f_c}{E_c} \qquad ...(25.3)$$

$$= \text{Strain in steel.}$$

$$\text{Hence loss of stress in steel} = \phi \cdot \frac{f_c}{E_c} \cdot E_s$$

$$= \phi \cdot f_c \cdot m \qquad ...(25.4)$$

where

m = modular ratio $= \dfrac{E_s}{E_c}$

f_c = stress in concrete at the level of steel.

Creep coefficient value vary depending on the humidity, concrete quality, duration of loading and age of concrete when load is applied. It varies from 1.5 for watery condition to 4.0 for dry condition.

(B) Shrinkage in Concrete

Bulk of shrinkage takes place in the early days of hardening of concrete. The shrinkage depends upon the type of aggregates, cement, method of curing, interval of time, etc. The main reason of shrinkage of concrete is progressive loss of water from concrete. Shrinkage is higher at the surface than at interior portion causing a gradient of strain and leads to cracking of surface. Shrinkage can be reduced by minimising water-cement ratio, proportion of cement and by using well graded larger size of coarse aggregate.

The loss of stress due to shrinkage is higher in pretensioned members than in the post-tensioned members. Due to shrinkage in the pretensioning system, the wires before release are subjected to some compression and the concrete around them to some tension as the wires are anchored and could not shorten in length. The concrete thus tends to slip in early stages and develop fine cracks. The cracks close when the wires are released causing a loss of prestress. A further loss of stress is caused as the shrinkage goes on decreasing the length of the member for a long time, although the effect is small. For ordinary prestressed concrete, an average value of shrinkage strain of 0.0003 appears to be quite reasonable.

Loss of stress due to shrinkage

$$= E_s \times \text{shrinkage strain} \qquad ...(25.5)$$

where E_s = Young's modulus of steel.

IS : 1343 recommends shrinkage strain for

(*i*) Pretensioned member = 0.0003

(*ii*) Post-tensioned member = $\dfrac{0.0002}{\text{Log}_{10}\,(t+2)}$

where

t = age of concrete at transfer in days.

25.11.3 Creep in Steel

The prestressing force applied by steel is proportional to recoverable strain only. This means some part of elongation of steel becomes unrecoverable. Thus, there is a loss of prestress due to creep. The creep of steel could however be reduced considerably by overstressing it by about 10 per cent above the working stress before use. Creep varies from 1 to 5 per cent and an average of 3% should be a fair guess. The Indian Standard Code recommends a value varying from 0 to 90 N/mm^2 for stress in wires varying from 0.5 f_{pu} to 0.8 f_{pu}.

25.11.4 Anchorage Slip

In the post-tensioned system, an allowance should be made for slipping of steel at the time of transferring the tendon force from the jacks to the member. In most of the systems, there is likely a small amount of adjustment of anchorage cones resulting in small amount of relaxation of prestress. An average of about 2.5 mm is estimated as anchorage slip in wedge type of grips; for heavy strands the slip may be 5 mm.

The magnitude of the loss of prestress due to slip of anchorage is calculated as

$$\text{Loss of stress due to anchorage slip} = \frac{P}{A}$$

$$= \frac{E_s \cdot \Delta}{L} \qquad \text{...(25.6)}$$

where,

Δ = slip of anchorage.

L = length of Cable.

A = cross sectional area of tendon.

E_s = modulus of elasticity of steel.

P = prestressing force in cable.

25.11.5 Steam Curing

The steam curing is often used with pretensioning system. The curing period has to be accelerated so as to release the pretensioning bed and the equipment for further fabrication. Raising the temperature of concrete with steam before hardening heats the prestressed steel and thus relaxes some of the prestressing force. On cooling, the concrete contracts and offsets the strain recovery in the steel. Creep characteristics of steel changes at 100° C and consequently there will be a considerable loss of prestress.

25.11.6 Bending of Member

Loss of prestress due to bending depends upon the bending character of the construction. In an unbonded construction when the member bends due to prestressing force, there is bound to be some change in the length of the tendon. Loss of prestress due to bending is illustrated in Fig. 25.11.

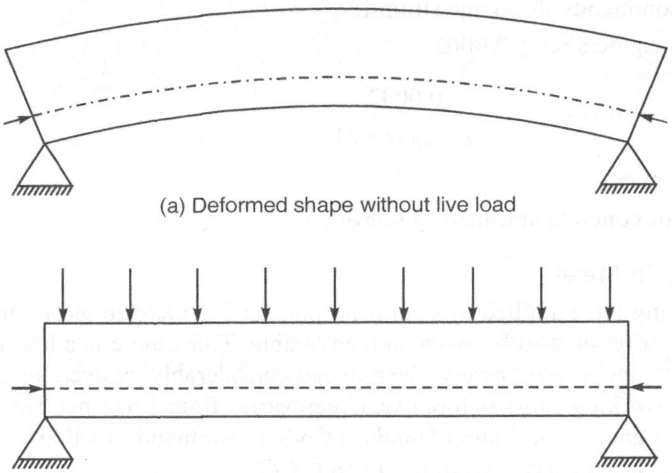

(a) Deformed shape without live load

(b) Shape with live load

Fig. 25.11. Bending of Member

In the bonded construction, though there will be change in the length of the tendon as the member is subjected to live load, it would cause a local effect and no effect on prestressing force. As the live load is removed, the loss due to bending will be recovered.

25.11.7 Friction

There will be considerable movement of sliding of tendon relative to the surrounding duct during tensioning operation. Because the tendon is in direct contact with the duct or with spacers provided, the friction will cause a reduction in prestressing force as the distance from the jack increases. The loss due to friction may be classified as :

(*i*) *Loss Caused by Length Effect* : The ducts for straight cables are mostly not perfectly straight and certainly there exist undulations. There will be loss due to friction beteeen cable and undulation. This loss due to length is also called wobbling effect or wave effect.

The prestressing at distance x from the tensioning end

$$P_x = P_o \cdot e^{-kx} \qquad \text{...(25.7)}$$

where P_o = initial stress in cable

k = friction coefficient for wave effect.

(*ii*) *Loss Due to Curvature Effect* : In case of curved cables, magnitude of the prestressing force, P_x at a distance x from the tensioning end is given as under :

$$P_x = P_o \cdot e^{-(\mu\alpha + kx)} \qquad \text{...(25.8)}$$

where

P_o = prestressing force at jacking end.

μ = coefficient of friction between cable and duct.

α = cumulative angle in radians through which the tangent to the cable profile has rotated between any two points under consideration.

k = friction coefficient for wave effect.

e = 2.718

IS code recommends the following values of μ and K.

(A) Coefficient of friction μ

0.55	for steel moving on smooth concrete.
0.35	for steel moving on steel fixed to duct.
0.25	for steel moving on steel fixed to concrete.
0..18-0.30	for multilayer rope cables in rigid rectangular steel sheaths.
0.15-0.25	for multilayer wire rope cables with spacer plates.

(B) Coefficient of friction for wave effect *K*

0.15 per 100 m	for normal conditions
0.15 per 100 m	for thin-walled ducts where heavy vibrations are encountered.
0	where the clearance between the duct and cable is quite large to eliminate wave effect.

The loss due to length effect and curvature effect can be combined to get the loss of stress at a point x distance from tensioning end due to friction.

Prestress at x distance after loss

$$P_x = P_o\, e^{-(kx + \mu x/R)}$$

$$= P_o (1 - k.\,x - \mu x / R) \quad \text{(expanding and ignoring smaller terms)} \qquad ...(25.9)$$

$$\text{Loss due to friction} = P_o - P_x$$

$$= P_o \{1 - e^{-(kx + \mu x/R)}\}$$

$$= P_o \{1 - (1 - k.\,x - \mu x/R)\}$$

$$= P_o (k\,.\,x + \mu x/R) \qquad ...(25.10$$

I.S. Code recommendation for loss due to friction in cable which is straight or moderately curved

$$P_x = P_o\, e^{-(\mu \cdot \alpha + k_f \cdot x)} \qquad ...(25.11)$$

Expanding $e^{-(\mu \cdot \alpha + k_f \cdot x)}$ and neglecting higher order terms

$$P_x = P_o (1 - \mu \cdot \alpha - k_f \cdot x) \qquad ...(25.12)$$

Loss due to friction

$$= P_o - P_x$$

$$= P_o - P_o (1 - \mu \cdot \alpha - k_f \cdot x)$$

$$= P_o (\mu{\cdot}\alpha + k_f x) \qquad ...(25.13)$$

where,

k_t = friction coefficient for wave effect.

μ = coefficient of friction between cable and duct.

α = cumulative angle in radians through which the tangent to the cable profile has turned between two points being considered.

x = distance from tensioning end.

25.12 TOTAL LOSSES ALLOWED FOR IN DESIGN

It is extremely difficult to calculate the total loss of prestress in the design of prestressed concrete members because loss depends on large number of factors, namely, properties of concrete and steel, level of prestress applied initially, method of prestressing and method of curing. Typical values of losses in pre-tensioning and post-tensioning are given in Table 25.3.

Table 25.3.

Type of Loss	*Pretensioning system %*	*Post-tensioning system %*
Elastic shortening and bending of concrete	3	1
Shrinkage of concrete	7	6
Creep of concrete	6	5
Creep in steel	2	3
Other losses	small	small
Total loss	18	15

25.13 ANALYSIS OF PRESTRESS

The stress due to prestressing force are direct stresses if tendon is placed concentric in the member and combined direct and bending stresses if the tendon is placed eccentrically.

Let P = prestressing force (positive for compression).

e = eccentricity of prestressing force

M = P.e = Moment due to prestressing.

A = area of cross section of concrete member.

I = second moment of area of section about its centroid.

Z_t and Z_b = section modulus for top and bottom fibres.

f_t and f_b = prestress in concrete developed at top and bottom (positive when compressive and negative when tensile)

y_t and y_b = distance of the top and bottom fibres from the centroid of the section

r = radius of gyration.

25.13.1 Concentric Tendon

A member having prestressing tendon concentric to the section, shall have direct stresses due to prestressing as shown in Fig. 25.12.

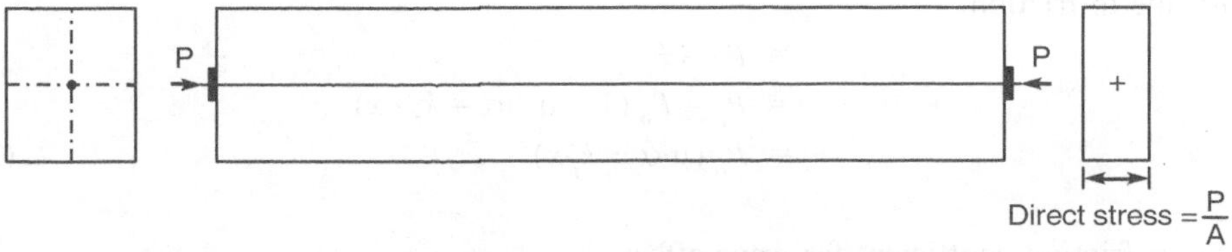

Fig. 25.12. Concentric Prestressing

25.13.2 Eccentric Tendon

Fig. 25.13 shows a concrete member having rectangular section subjected to eccentric prestressing force of magnitude P located at eccentricity e. The stresses developed at top and bottom fibres are given by the relations.

$$f_b = \frac{P}{A} + \frac{P.e}{Z_b} = \frac{P}{A}\left(1 + \frac{e \cdot y_b}{r^2}\right) \quad ...(25.14)$$

$$f_t = \frac{P}{A} - \frac{P \cdot e}{Z_t} = \frac{P}{A}\left(1 - \frac{e \cdot y_t}{r^2}\right) \quad ...(25.15)$$

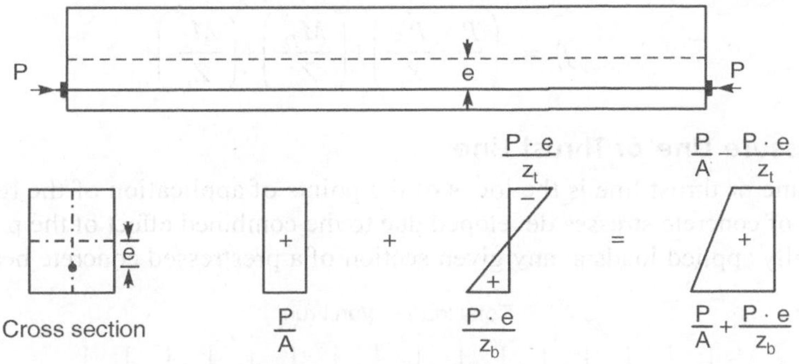

Fig. 25.13. Eccentric Prestressing

25.13.3 Resultant Stresses

Fig. 25.14 shows a concrete member having rectangular section subjected to eccentric prestressing force, dead loads (g) and (q) live loads. The resultant stresses in concrete at any section is obtained by superposing the effect of prestress and flexural stresses developed due to other loads. If M_D and M_L are the dead load and live load moments at any section, the resultant stresses at the top and bottom fibres of concrete at any section are determined from the relations

$$f_b = \left(\frac{P}{A}+\frac{P.e}{Z_b}\right)-\left(\frac{M_D}{Z_b}\right)-\left(\frac{M_L}{Z_b}\right) \quad ...(25.16)$$

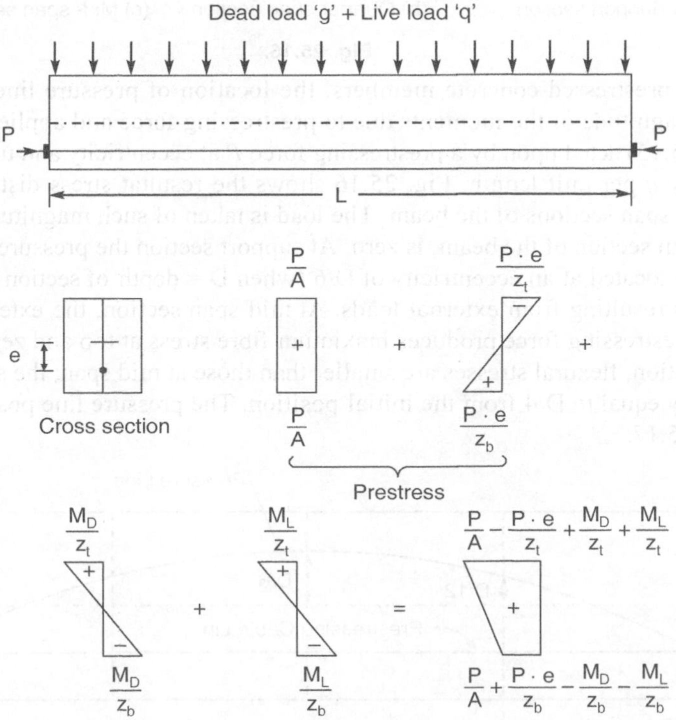

Fig. 25.14. Resultant Stress

$$f_t = \left(\frac{P}{A} - \frac{P.e}{Z_t}\right) + \left(\frac{M_D}{Z_t}\right) + \left(\frac{M_L}{Z_t}\right) \qquad ...(25.17)$$

25.13.4 Pressure Line or Thrust Line

Pressure line or thrust line is the locus of the points of application of the resultant force of the distribution of concrete stresses developed due to the combined effect of the prestressing force and the externally applied loads at any given section of a prestressed concrete beam.

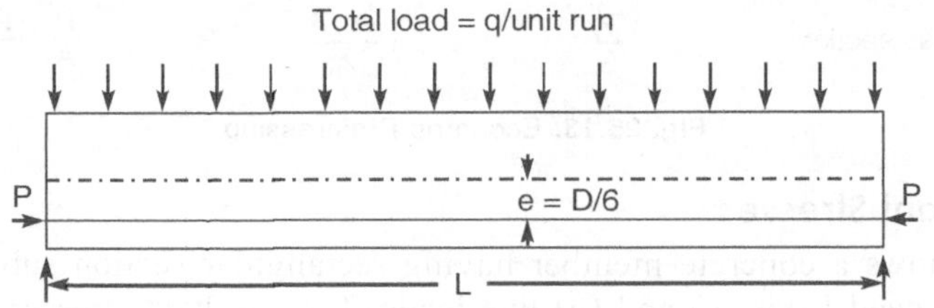

Fig. 25.15.

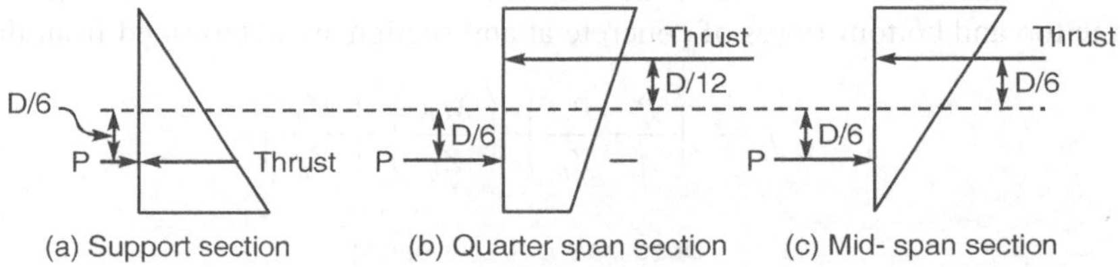

Fig. 25.16.

In case of prestressed concrete members, the location of pressure line depends upon the direction and magnitude of the moments due to prestressing force and applied loads. Consider a member (Fig. 25.15) acted upon by a prestressing force P at eccentricity and uniformly distributed load of intensity q per unit length. Fig. 25.16 shows the resultat stress distribution at support, quarter and mid span sections of the beam. The load is taken of such magnitude that bottom fibre stress at mid span section of the beam, is zero. At support section the pressure line coincides with centroid of steel located at an eccentricity of D/6 (when D = depth of section), since there are no flexural stresses resulting from external loads. At mid span section, the external load combined with effect of prestressing force produces maximum fibre stress at top and zero at the bottom. At quarter span section, flexural stresses are smaller than those at mid span, the shift in pressure line is approximately equal to D/4 from the initial position. The pressure line position in the beam is shown in Fig. 25.17.

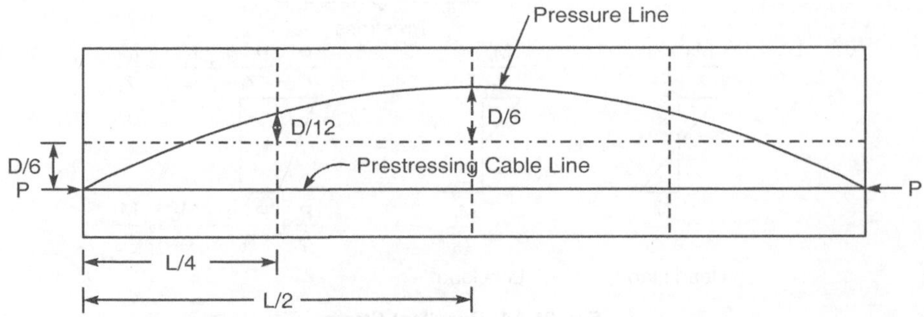

Fig. 25.17. Location of Pressure Line in Prestress members

A change in the external loads in the elastic range of prestressed concrete member results in a shift of the pressure line whereas in reinforced concrete member any such change in loading will result in change in tensile and compressive forces in the section. In prestressed member, forces remain constant, lever arm changes whereas in R.C.C. members forces change and lever arm remains constant.

25.13.5 Concept of Load Balancing

In the concept of load balancing, cable profile in a prestressed concrete member is so selected that transverse component of the tendon prestressing force balances the given external loads. The bending stresses in concrete member shall get neutralized if the cable profile corresponds to the shape of the bending moment resulting from the external loads.

The concept enables to transform a flexural member into a member under direct stress. Under this concept, the tendon is replaced by the forces acting on the concrete which is considered as free body (Fig. 25.18).

Thus, the transverse load on the beam is balanced and becomes $(w - w_b)$ in the downward direction and the beam is subjected to axial force P producing uniform compressive stress P/A on the concrete.

$$f = \frac{P}{A} \pm \frac{(w - w_b)\, L^2 y}{8I} \qquad \text{...(25.18)}$$

where $$w_b = \frac{8\, P.h}{L^2}$$

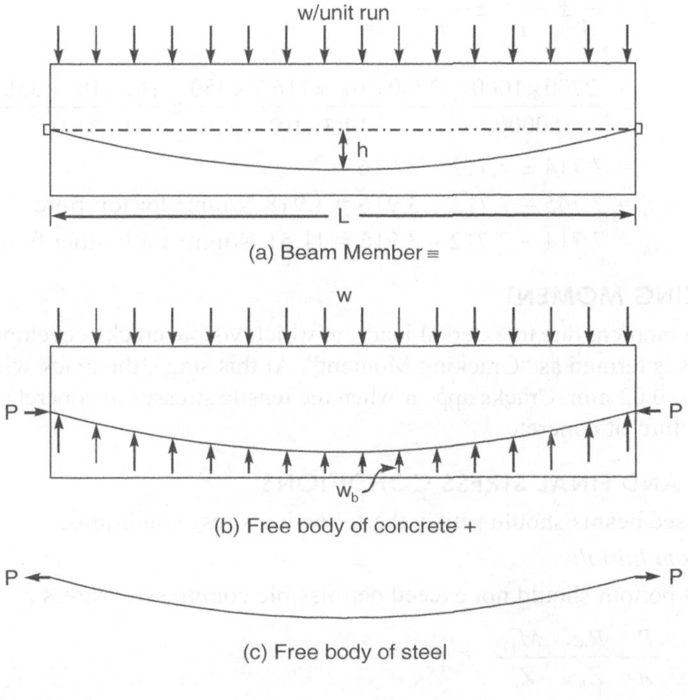

Fig. 25.18. Free Body Diagram for Prestressed Beam under Lateral Loading '*W*'

Example 25.1. A prestressed beam 500 × 700 mm has a simple span of 8m and is loaded with a *u.d.l* of 20 kN/m including its own weight (Fig. 25.19). The prestressing tendon is located at lower third point and produced an effective prestress of 2700 kN. Compute the fibre stresses in the concrete at the mid span.

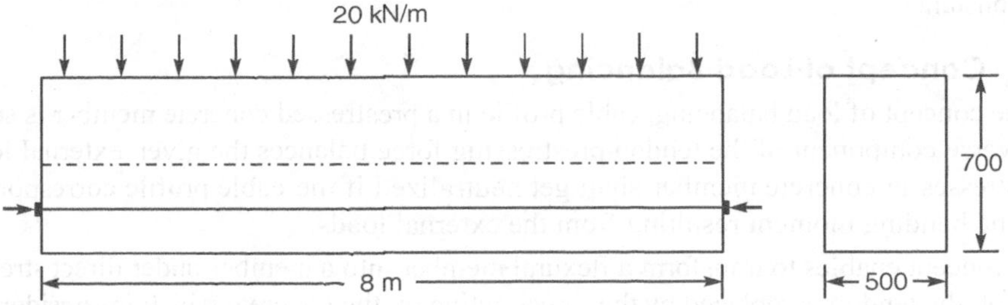

Fig. 25.19.

Solution.

$P = 2700$ kN $\qquad A = 500 \times 700 = 350000$ mm^2

$$e = 350 - \frac{700}{3} = 116.7 \text{ mm} \qquad I = \frac{bd^3}{12} = \frac{500 \times 700^3}{12} = 14.3 \times 10^9 \text{ mm}^4$$

$y = 350$ mm.

$$M = \frac{wl^2}{8} = \frac{20 \times 8^2}{8} = 160 \text{ kN.m.}$$

$$f = \frac{P}{A} \pm \frac{P.e.y}{I} \pm \frac{M.y}{I}$$

$$= \frac{2700 \times 1000}{350000} \pm \frac{2700 \times 10^3 \times 116.7 \times 350}{14.3 \times 10^9} \pm \frac{160 \times 10^6 \times 350}{14.3 \times 10^9}$$

$$= 7.714 \pm 7.712 \pm 3.916$$

$f_t = 7.745 - 7.712 + 3.916 = 3.918$ N/mm^2 for top fibre

$f_b = 7.714 + 7.712 - 3.916 = 11.51$ N/mm^2 for bottom fibre.

25.14 CRACKING MOMENT

The bending moment due to external loads, at which visible cracks develop in the prestressed concrete members is termed as "Cracking Moment". At this stage, the crack widths are generally of the order 0.01 to 0.02 mm. Cracks appear when the tensile stresses in concrete exceeds the value of modulus of rupture of concrete.

25.15 INITIAL AND FINAL STRESS CONDITIONS

The prestressed beams should satisfy the following stress conditions.

(*A*) *Strees at Bottom Initially*

The stress at bottom should not exceed permissible compressive stress f_c

$$\frac{P}{A} + \frac{P.e}{Z_b} - \frac{M_D}{Z_b} \leq f_c \qquad \text{...(25.19 a)}$$

(B) Stress at Bottom Finally

After looses, the effective prestressing forces may be taken as $\eta.P$. The value of reduction factor n may be chosen suitable between 0.8 to 0.85. The final stress at bottom should be less than the minimum allowable stress f_t.

$$-\frac{\eta P}{A} - \frac{\eta.P.e}{Z_b} + \frac{M_D}{Z_b} + \frac{M_L}{Z_b} \leq f_t \qquad ...(25.19\ b)$$

(C) Stress at Top Initially

The initial stress at top fibre of concrete should not be less than the minimum permissible value of stress f_t.

$$\frac{P}{A} + \frac{P.e}{Z_t} - \frac{M_D}{Z_t} \leq f_t \qquad ...(25.19\ c)$$

(D) Stress at Top Finally

The stress in the top fibre at final stage should be less than the maximum permissible stress.

$$\frac{\eta P}{A} - \frac{\eta.P.e}{Z_t} + \frac{M_D}{Z_t} + \frac{M_z}{Z_t} \leq f_c \qquad ...(25.19\ d)$$

25.16 KERN DISTANCE

Under initial conditions when dead load and prestressing force is acting on the member, the stress at top should not be less than zero (*i.e.*, no tension allowed) and at bottom compressive stress should not exceed f_c value. To satisfy these requirements, the resultant compressive resistance offered by the section (*C*) as shown in Fig. 25.20, should not be below a certain level.

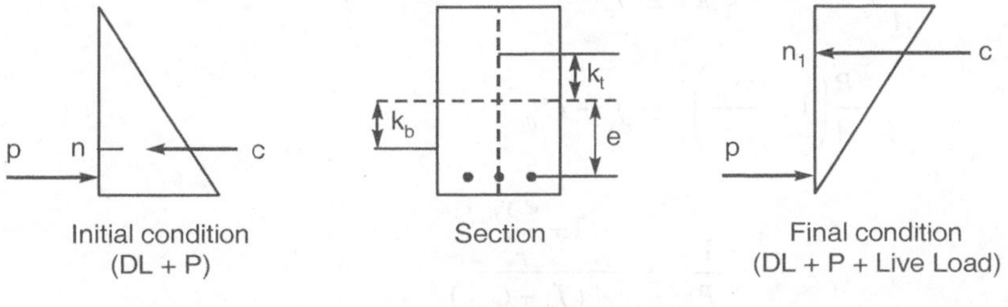

Fig. 25.20. Kern Distances

The point (*n*) below which *C* should not act so that stresses are within permissible limits, is called 'Lower Kern Point'.

For initial condition at top fibre

$$\frac{C}{A} - \frac{CK_b}{A.r^2} \cdot y_t = 0 \qquad ...(25.20)$$

$$K_b = \frac{r^2}{y_t} \qquad ...(25.21)$$

where, K_b = Bottom Kern Point.

Under the applied bending moment, the point of application of *C* reaches at a higher level to provide a larger lever arm so that the Couple provided by *C* and *P* balance the external moment. To satisfy the final condition, the point of application of *C* can not go above certain level. The limiting point (n_1) above which *C* can not go, is called 'Upper Kern Point'.

For Final condition at bottom fibre.

$$\frac{C}{A} - \frac{C.K_t.y_b}{A.r^2} = 0$$

$$K_t = \frac{r^2}{y_b} \qquad \text{...(25.22)}$$

K_t is called Upper Kern distance.

Thus, $$M_D + M_L = P(e + K_t) \qquad \text{...(25.23)}$$

$$M_D = P(e - K_b) \qquad \text{...(25.24)}$$

$$M_L = P(K_t + K_b) \qquad \text{...(25.25)}$$

The solution of Equations 25.19-*b* and 25.19-*d* which represent final stresses at bottom and top of section, provide the value of minimum prestressing force P and eccentricity e. Knowing these two values, the design of the section can be carried out.

Prof. Magnel of Belgium proposed an easy graphical method for determining the value of P and e.

25.17 GRAPHICAL METHOD OF DESIGN

The section of the beam at mid span must satisfy the following conditions.

(*i*) *Bottom Fibre*

(*a*) Immediately the prestress is established the compressive stress in the concrete under dead load, must not exceed f_c.

$$\frac{P}{A} + \frac{P.e.y_b}{I} - C_{db} \le f_c \qquad \text{...(25.26-}a\text{)}$$

$$\frac{P}{A}\left(1 + \frac{e.y_b}{r^2}\right) \le f_c + C_{db} \qquad \text{...(25.26-}b\text{)}$$

$$\frac{1}{P} \ge \frac{1 + \dfrac{e.y_b}{r^2}}{A(f_c + C_{db})} \qquad \text{...(25.26-}c\text{)}$$

where C_{db} = Stress due to dead load at bottom fibre

For $e > \left(-\dfrac{r^2}{y_b}\right)$, $\dfrac{1}{P} = \dfrac{1}{A(f_c + C_{db})}$

For $e = 0$, $\dfrac{1}{P} = 0$

For $e < \left(-\dfrac{r^2}{y_b}\right)$ Eq. 25.26 (*c*) is always satisfied.

The above inequalities when plotted are satisfied in shaded zone of Fig. 25.21 (*a*).

(*b*) After a time, *i.e.*, finally under the effects of dead load and imposed loads, the tensile stress in the concrete must not exceed f_t.

$$-\frac{\eta P}{A}-\frac{\eta P.e.y_b}{I}+C_{1b}+C_{db} \le f_t \qquad ...(25.27\text{-}a)$$

$$\frac{\eta P}{A}\cdot\left|1+\frac{ey_b}{r^2}\right| \ge C_{1b}+C_{db}-f_t \qquad ...(25.27\text{-}b)$$

$$\frac{1}{P} \le \frac{\eta(1+e.y_b/r^2)}{A(C_{1b}+C_{db}-f_t)} \qquad ...(25.27\text{-}c)$$

For $e \le \left(-\frac{r^2}{y_b}\right)$, Equation (25.27 *c*) is never satisfied

For $e = -\frac{r^2}{y_b}$, $\frac{1}{P} \le 0$

For $e = 0$, $\frac{1}{P} \le \frac{\eta}{(C_{1b}+C_{db}-f_t)}$

where, C_{1b} = Stress due to imposed live load at bottom fibre.

The above inequalities when plotted, are satisfied is shaded zone of Fig. 25.21 (*b*).

(*ii*) *TopFibre*

(*a*) Immediately the prestress is established, the tensile stress in concrete under dead loads must not exceed the permissible tensile stress f_t in the concrete. This condition does not apply when the cable is within 'Core' of the section (where Core defines that zone of the beam, corresponding to middle third of a rectangular beam within which the cable must be, if tensile stress in the concrete due to prestressing alone are to be avoided).

$$\frac{-P}{A}+\frac{P.e}{I}\cdot y_t - C_{dt} < f_t \qquad (25.28\text{-}a)$$

$$\frac{P}{A}\left(\frac{e.y_t}{r^2}-1\right) < f_t + C_{dt} \qquad ...(25.28\text{-}b)$$

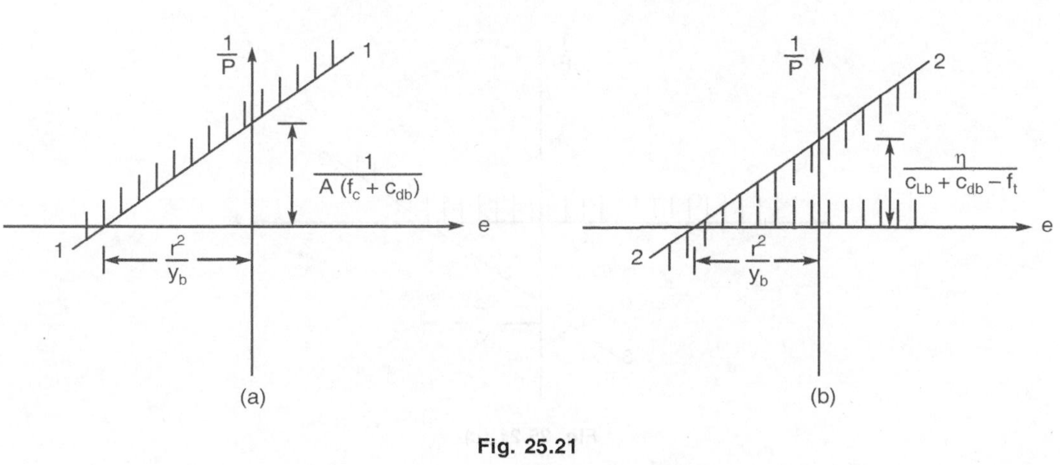

Fig. 25.21

$$\frac{1}{P} \geq \frac{\left(\frac{ey_t}{r^2}-1\right)}{A\,(f_t + C_{dt})} \qquad ...(25.28)$$

where C_{dt} = Stress due to dead loads to top fibre

For $e \leq \frac{r^2}{y_t}$, Eqn. 25.28-*c* is always satisfied.

For $e > \frac{r^2}{y_t}$, Eqn. 25.28-*c* may or may not be satisfied.

For $e = 0$, $\frac{1}{P} \geq -\frac{1}{A(C_{dt} + f_t)}$

For $e = \frac{r^2}{y_t}$, $\frac{1}{P} = 0$

The above inequalities when plotted, are satisfied in shaded zone of Fig. 25.21 (*c*).

(*b*) After a time, *i.e.*, finally under dead and imposed loads combined, the compressive stress in concrete must not exceed the permissible compressive stress f_c in the concrete, if the cable is below the core.

$$\frac{\eta.P}{A} - \frac{\eta\, P.e.y_t}{I} + C_{1t} + C_{dt} \leq f_c \qquad ...(25.29\text{-}a)$$

$$\frac{\eta.P}{A}\left\{1 - \frac{e.y_t}{r^2}\right\} \leq (f_c - C_{1t} - C_{dt}) \qquad ...(25.29\text{-}b)$$

$$\frac{1}{P} \geq \frac{\eta\left\{1 - \frac{e.y_t}{r^2}\right\}}{A\,(f_c - C_{dt} - C_{1t})} \qquad ...(25.29\text{-}c)$$

where C_{1t} = Stress due to imposed loads at top fibre

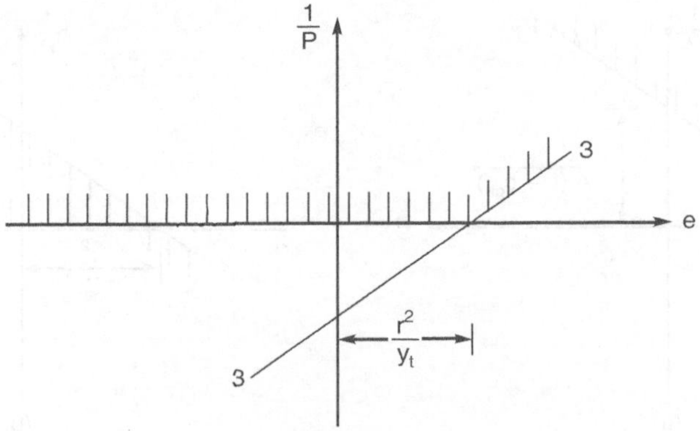

Fig. 25.21 (*c*)

For $e \geq \dfrac{r^2}{y_t}$, Eqn. 25.29-*c* is always satisfied.

For $e = 0$ $\quad \dfrac{1}{P} \geq \dfrac{\eta}{A\,(f_c - C_{dt} - C_{1t})}$

For $e = \dfrac{r^2}{y_t}$ $\quad \dfrac{1}{P} = 0$

The above inequalities when plotted, are satisfied in shaded zone of Fig. 25.21 (*d*).

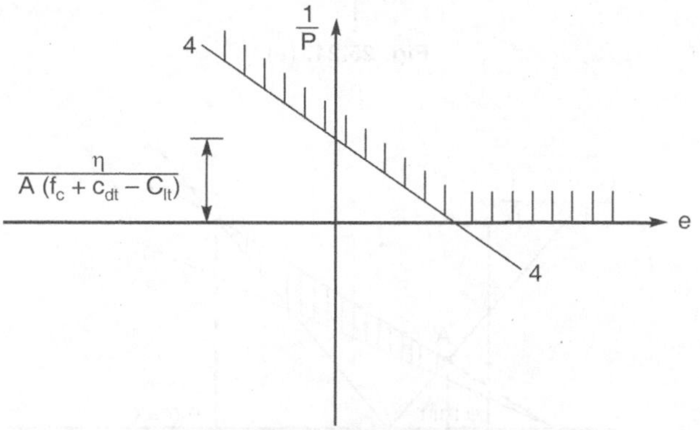

Fig. 25.21 (*d*)

Now for $\quad f_c < (C_{lt} + C_{dt})$ ...(25.30-*a*)

$$C_{dt} + C_{lt} - f_c \leq \frac{\eta P}{A}\left\{\frac{e.y_t}{r^2} - 1\right\} \qquad \text{...(25.30-}b\text{)}$$

$$\frac{1}{P} \leq \frac{\eta\left\{\dfrac{e.y_t}{r^2} - 1\right\}}{A\,(C_{dt} + C_{lt} - f_c)} \qquad \text{...(25.30-}c\text{)}$$

For $e \leq \dfrac{r^2}{y_t}$, Eqn. (25.30-*c*) is never satisfied.

For $e > \dfrac{r^2}{y_t}$, Eqn. (25.30-*c*) may or may not be satisfied.

For $e = 0$, $\quad \dfrac{1}{P} \leq \dfrac{-\eta}{(C_{dt} + C_{lt} - f_c)A}$

$e = \dfrac{r^2}{y_t}$, $\quad \dfrac{1}{P} = 0$

The above inequalities when plotted, are satisfied in shaded zone of Fig. 25.21 (*e*).

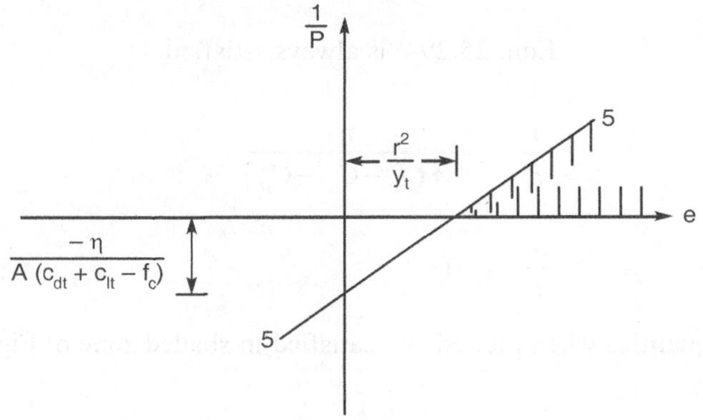

Fig. 25.21. (*e*)

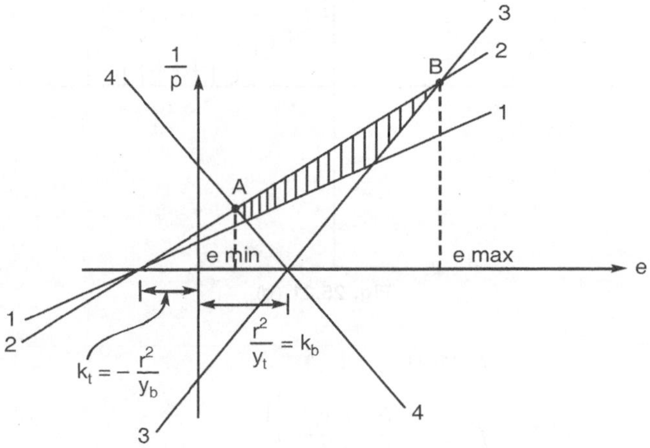

Fig. 25.22.

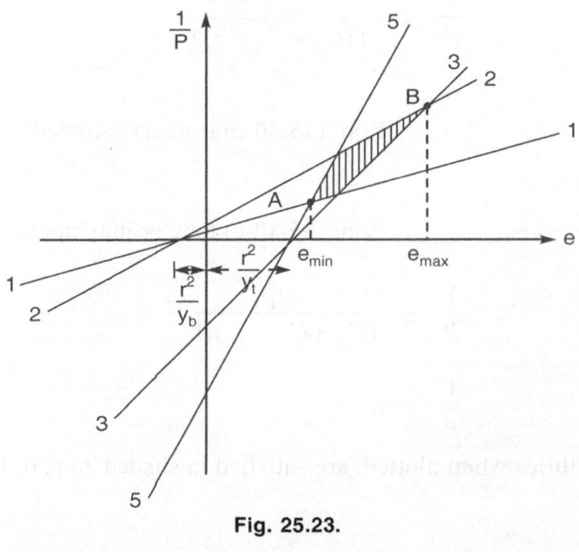

Fig. 25.23.

For $f_c \geq (C_{dt} + C_{lt})$ (Combining Fig. 25.21-a, b, c, d) See Fig. 25.22

For $f_c \leq (C_{dt} + C_{lt})$ (Combining Fig. 25.21-a, b, c, e) See Fig. 25.23

From the above graphs shown in Figs. 25.22 and 25.23, the suitable value of $\frac{1}{P}$ and e can be chosen.

If $e < e_A$, Revise the section.

If $e > e_B$, Adopt $e = e_B$

If $e < e_B$ and $e > e_A$, Actual value of 'e' to be adopted.

25.18 DESIGN OF SECTION

(A) Rectangular Section

The thickness of slab will be 1/30 to 1/40 of span. The width of beam should not be less than 1/30th of effective length between restraints or 1/4 of the depth of beam. The depth of beam will vary from 1/5 to 1/20 of span.

The stress at initial and final stages at top and bottom are rewritten as :

Bottom Initial
$$\frac{P}{A} + \frac{Pe}{Z_b} - \frac{M_D}{Z_b} \leq f_c \quad \text{...(25.31)}$$

Bottom Final
$$-\eta\frac{P}{A} - \frac{\eta Pe}{Z_b} + \frac{M_D}{Z_b} + \frac{M_L}{Z_b} \leq f_t \quad \text{...(25.32)}$$

Top Initial
$$-\frac{P}{A} + \frac{Pe}{Z_t} - \frac{M_D}{Z_t} \leq f_t \quad \text{...(25.33)}$$

Top Final
$$\frac{\eta P}{A} - \frac{\eta P.e}{Z_t} + \frac{M_D}{Z_t} + \frac{M_L}{Z_t} \leq f_c \quad \text{...(25.34)}$$

Now we add Eqns 25.33 and 25.34

$$-\frac{P}{A}(1-\eta) + \frac{Pe}{Z_t}(1-\eta) + \frac{M_L}{Z_t} = f_t + f_c \quad \text{...(25.35)}$$

But for rectangular section $Z_t = Z_b = Z$

Then,

$$-\frac{P}{A}(1-\eta) + \frac{Pe}{Z}(1-\eta) + \frac{M_L}{Z} = f_t + f_c$$

$$\frac{M_L}{Z} = f_c + f_t + \frac{P}{A}\left\{(1-\eta)\left(1 - \frac{e.A}{Z}\right)\right\} \quad \text{...(25.36)}$$

Now take
$$\frac{P}{A} = f_c \text{ approximately}$$

$$\frac{M_L}{Z} = f_c + f_t + \frac{f_c}{2}(1-\eta)\left(1 - \frac{e.A}{Z}\right)$$

or, $$Z = \frac{M_L}{f_c + f_t + \frac{f_c}{2}(1-\eta)\left(1 - \frac{A_e}{Z}\right)} \quad ...(25.37)$$

Taking $$\eta = 0.85 \text{ and } \frac{Ae}{Z} = 2,\ Z = \frac{1}{Y} = \frac{M_L}{0.925\, f_c + f_t} \quad ...(25.38)$$

Now adding Eqns. (25.31) and (25.32)

$$\frac{P}{A}(1-\eta) + \frac{Pe}{Z_b}(1-\eta) + \frac{M_L}{Z_b} = f_c - f_t \quad ...(25.39)$$

$$Z_t = Z_b = Z \quad \text{(For rectangular section).}$$

Hence, $$\frac{P}{A}(1-\eta) + \frac{P \cdot e}{Z}(1-\eta) + \frac{M_L}{Z} = f_c + f_t \quad ...(25.40)$$

Taking

$$\frac{P}{A} = \frac{f_c}{2} \quad \text{approximately,}$$

$$\frac{f_c}{2}(1-\eta) + \frac{f_c . A}{2Z} \cdot e(1-\eta) + \frac{M_L}{Z} = f_c + f_t$$

$$Z = \frac{I}{y} = \frac{M_L}{(f_c + f_t) - \frac{f_c}{2}(1-\eta)\left(1 + \frac{A.e}{Z}\right)} \quad ...(25.41)$$

Expression represented by (25.38) gives bigger section than that given by expression (25.37).

Now if we take $\frac{Ae}{Z} = 2$, then expression (25.38)

becomes $$Z = \frac{I}{y} = \frac{M_L}{\frac{f_c}{2}(3\eta - 1) + f_t} \quad \text{approximately} \quad ...(25.42)$$

Taking $$\eta = 0.85,\ Z = \frac{M_L}{0.775 f_c + f_t} \quad ...(25.43)$$

Trial section for the design may be chosen considering the expression represented by (25.43).

(B) I Section

I section may be of two types.

(*a*) Symmetrical *I*-section.

(*b*) Unsymmetrical *I*-section.

Though in actual practice the flanges are splayed, to avoid sharp corners and to make the process of concrete pouring easy, the properties of section found out by taking equivalent for flanges will not differ more than 1% from actual calculations.

To determine trial depth, two cases arise.

(*i*) When $M_D \geq 20\ \%$ of $(M_D + M_L)$

Then $h = k\sqrt{M_T}$

Where h : approximate trial depth

k : constant varying from 3 to 4

M_T = Total moment $(M_D + M_L)$ in (KN.m)

(*ii*) When $M_D < 20\%$ of M_T ($= M_D + M_L$)

Then

$$h = k.\sqrt{M_L}$$

where, M_L : Applied or live load bending moment.

(*a*) *Symmetrical I-Section*: For symmetrical I section, first find the depth of rectangular section and then increase the depth by 10%.

The thickness of flange may be taken as 10 to 20 per cent of overall depth and thickness of web is taken as 20 to 30 per cent of width of top flange or it may be taken as per the requirement for accommodating the cables (Fig. 25.24)

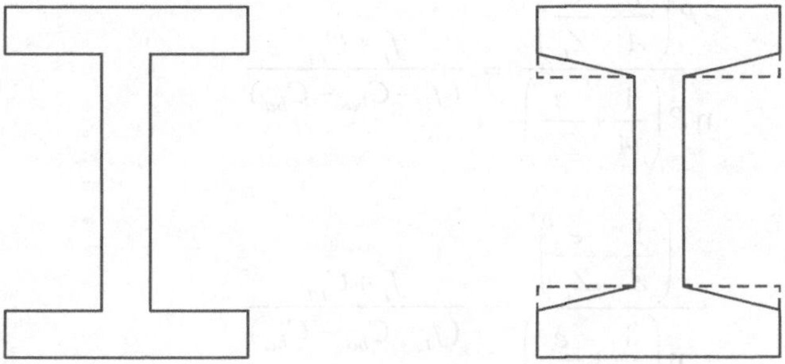

Fig. 25.24.

(*ii*) *Unsymmetrical I-Section*: Adding Eqn. (25.31 and Eqn. (25.32) we get,

$$Z_b = \frac{I}{y_b} = \frac{M_L}{(f_c + f_t) - \frac{P}{A}(1-\eta)\left(1 + \frac{A.e}{Z_b}\right)} \quad ...(25.44)$$

Adding (25.33) and (25.34), we get

$$Z_b = \frac{I}{y_t} = \frac{M_L}{(f_c + f_t) + \frac{P}{A}(1-\eta)\left(1 - \frac{A.e}{Z_t}\right)} \quad ...(25.45)$$

Putting $\frac{P}{A} = \frac{f_c}{2}$ in 25.44 and 25.45 we get,

$$Z_b = \frac{I}{y_b} = \frac{M_L}{\frac{f_c}{2}(3\eta - 1) + f_t} \quad ...(25.46)$$

$$Z_t = \frac{I}{y_t} = \frac{M_L}{\frac{f_c}{2}(1+\eta) + f_t} \quad ...(25.47)$$

Taking $\eta = 0.85$

From Eqn. 25.46 $$Z_b = \frac{I}{y_b} = \frac{M_L}{0.775 f_c + f_t} \quad \text{...(25.48)}$$

From Eqn. 25.47 $$Z_t = \frac{I}{y_t} = \frac{M_L}{0.925 f_c + f_t} \quad \text{...(25.49)}$$

The Eqns. 25.46 and 25.47 show that the bottom flange should be made heavier than top flange. Top flange is kept wider than bottom flange keeping in view the prevention of buckling.

The thickness of bottom flange is determined as per the requirement of accommodating tendons. Though Eqns. 25.46 and 25.47 show that bottom flange is heavier but in practice upper flange is kept heavier so that C.G. of section shifts upward and maximum amount of eccentricity is available; thus keeping the prestressing force to be minimum.

For minimum prestressing force, Eqns 25.32 and 25.33 should be satisfied.

Solving Eqns. 25.32 and 25.33,

$$\frac{P\left(\frac{1}{A} - \frac{e}{Z_t}\right)}{\eta P\left(\frac{1}{A} + \frac{e}{Z_b}\right)} = \frac{f_t + C_{td}}{(f_t - C_{bd} - C_{ba})}$$

or

$$\frac{\left(\frac{1}{A} - \frac{e}{Z_t}\right)}{\eta\left(\frac{1}{A} + \frac{e}{Z_b}\right)} = \frac{f_t + C_{td}}{(f_t - C_{bd} - C_{ba})} \quad \text{...(25.50)}$$

From this Eqn. 25.50 value of *e* can be found out. If available, eccentricity is equal to or less than this value, then the actually available *e* is put in Eqn. 25.32 or 25.33 and *P* can be obtained.

But if *e* available is more than the needed, the section may be revised.

25.19 GENERAL DESIGN REQUIREMENTS

1. *Areas and Moments of Inertia of Sections*

In pretensioned members where the prestressing tendons are single wires distributed on the cross section or strands of wires of relatively small cross sectional area, the allowance for tendons need not to be made in calculation of area, centroid and moment of intertia of the section. Where allowance is made, it should be equal to $(m - 1) \times$ area of steel tendons, where *m* is the modular ratio.

In post-tensioned member deduction for prestressing tendons, cable ducts and such other opening in the concrete, is made while calculating area, centroid and moment of intertia of the cross section.

2. *Modulus of Elasticity*

(*a*) *Modulus of elasticity of concrete* : The modulus of elasticity of concrete E_c shall be given by the following formula unless otherwise determined by tests.

$$E_c = 5700 \cdot \sqrt{f_{ck}} \ \text{(N/mm}^2\text{)}$$

f_{ck} = characteristic strength of concrete N/mm^2

(*b*) *Modulus of Elasticity of Steel* : Modulus of elasticity of steel, to be used in design, is determined by tests on the samples of steel.

The following values may be adopted where it is not possible to determine the value by performing the tests.

Type of steel	*Modulus of Elasticity* E_s *(kN/mm²)*
Plain drawn wires (cold drawn or cold worked)	210
High tensile steel bars (rolled or heat treated)	200
Strands	195

3. *Permissible Stresses*

(*i*) Steel

At the time of tensioning, the maximum tensile stress immediately behind the anchorages shall not exceed :

(*a*) 80% of ultimate tensile strength of the wire or bar steel without a definite yield point, and

(*b*) The yield stress for steel with a guaranteed yield point.

After accounting for losses due to creep and shrinkage of the concrete, friction effects and relaxation of steel, the maximum tensile stress (f_t) at the point of maximum bending shall not exceed.

(*a*) 60% of the ultimate strength of wire or bar for steel without a definite yield point.

(*b*) 80% of the yield stress for steel with guaranteed yield point.

(*ii*) *Concrete*

The stresses in concrete due to prestressing operation and working loads should not exceed the values given below. Unless otherwise specified, the cube strength refers to that of concrete obtained at 28 days when ordinary portland cement is used and at 7 days when rapid hardening cement is used, on cube of size 15 cm.

(*a*) *Compressive Stress Due to Initial Prestress*

The maximum permissible compressive stress due to bending and direct force at the time of transfer of prestress F_r shall be determined from a graph in which a straight line joins a point given by 0.54 F_r for a concrete with f_{ck} of 30 N/mm² to a second point giving a permissible stress of 0.37 F_r for concrete with f_{ck} of 60 N/mm² [Fig. 25.25 (*B*)]. The above F_r values apply to post tensioned work. For pretensioned work the variation represented by Fig. 25.25 (*A*) will apply.

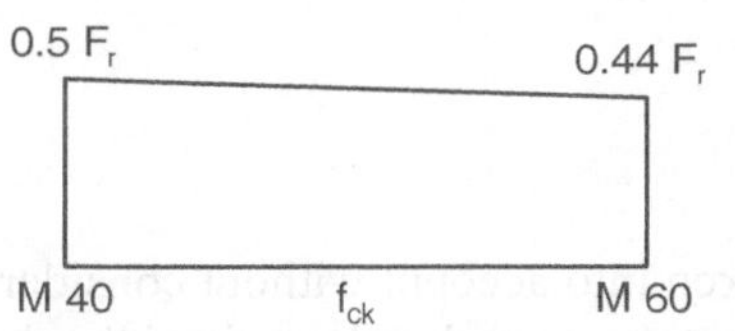

Fig. 25.25 (A). Pretensioned work

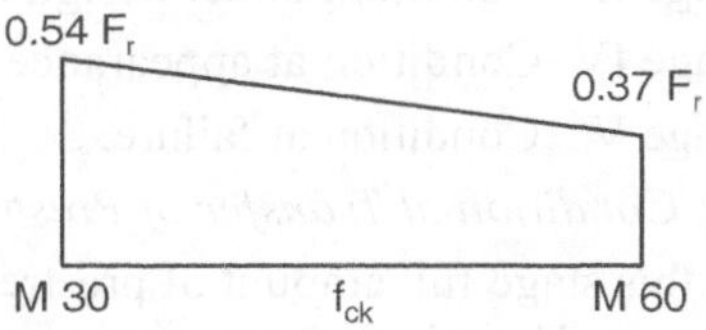

Fig. 25.25 (B). Post Tensioned Work

For the purpose of Fig. 25.25, F_r shall be the cube strength of concrete at transfer which in no case shall be less than half the corresponding f_c.

(*b*) *Compressive Strength Due to Final Prestress*

The maximum permissible compressive stress f_c, prestress and service loads after deduction of full losses in the specified prestress shall be determined by straight line relation as in Fig. 25.26 but different stress limits shall apply to the concrete of structure depending on whether it falls in a part of section where the compressive stresses are not likely to increase in service (Zone I) or in a part of section where the compressive stresses are likely to increase in service (Zone II).

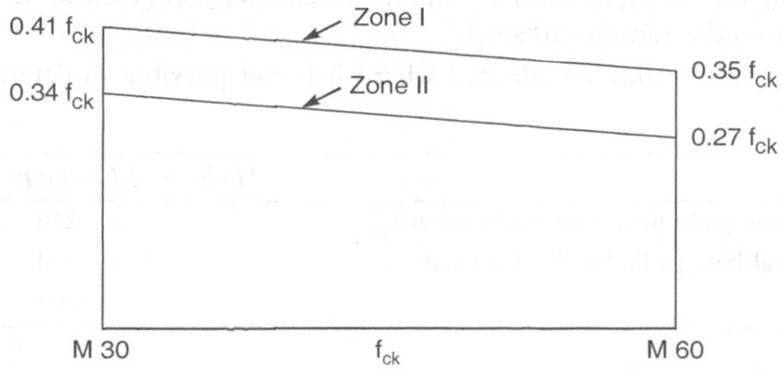

Fig. 25.26. Computation of Max. Permissible Compressive Stress in Flexure due to Final Prestress

(*c*) *Flexural Stresses*

(*i*) In structures subjected to dynamic loading under the severest combination of service loads, all direct force and bending stresses in any section shall remain compressive. In extreme fibre, closest to the line of action of prestressing steel the maximum compressive stress shall not be less than 5% of the maximum compressive stress permitted on that fibre unless otherwise approved by the appropriate authority. In the latter case the minimum compressive stress may be reduced upto a limit of zero.

(*ii*) On the other structures under the same circumstances a tensile stress (f_t) may be permitted not exceeding one tenth of the maximum permissible compressive stress or 140 N/mm^2 for pretensioned and 100 N/mm^2 for post-tensioned whichever is less.

(*iii*) In case of members assembled out of precast blocks no tensile stress shall be permitted at any stage of loading upto cracking.

25.20 STAGES OF LOADING TO BE CONSIDERED IN DESIGN

To determine all possible extremes of stresses, the various loading conditions shall be investigated in which they occur.

The different stages of loading, to be investigated are given below.

Stage I Condition at transfer of prestress.

Stage II Condition during handling and erection.

Stage III Condition under design load.

Stage IV Condition at appearance of first crack.

Stage V Condition at failure.

Stage I : *Condition at Transfer of Prestress*

At this stage full amount of prestressing force is taken into account without considering any loss of stress. The relief of stresses caused by dead load may be considered provided the dead load effect is not interrupted during hoisting and handling.

If members are assembled out of precast blocks then no tension is allowed at this stage. However, in other cases tensile stress shall not exceed 0.1 times maximum permissible stress or 1.4 N/mm^2 and 1 N/mm^2 for pretensioned or post-tensioned work whichever is less.

Stage II : *Condition during Handling and Erection*

The residual prestress after different losses at the time of the particular operation, shall be considered with the handling and erection stresses. Thus computed compressive stress should be less than 0.5 times the cube strength of concrete at that stage. The tensile stress should be less than 0.5 times the permissible tensile stress.

The point from where the member can be lifted or slung for erection should be specified.

Stage III : *Condition Under Design Load*

The worst combination of dead load, live load, wind load, seismic load and temperature should be considered to find stresses at this stage. The stresses should be within permissible stresses.

Stage IV : *Condition at First Visibility of Cracks*

This stage needs the calculation of loading producing a tensile stress anywhere in member equal to the modulus of rupture of concrete. The load thus computed shall not be less than 1.1 times dead plus 1.25 times live load placed in such a way to produce maximum stresses in concerned member.

The modulus of rupture of concrete shall be taken as $0.54\sqrt{F_c}$ where, F_c in N/mm^2.

The cracking load for a prestressed member assembled out of precast blocks is computed on the assumption that concrete can not take up any tensile stress.

Stage V : *Condition at Failure*

The ultimate loads assumed for the purpose of determining the condition at failure shall not be less than the following.

(*a*) For statically determinate structure subjected to dynamic loading such as bridge .... $1.5DL + 3\ L\ L$.

(*b*) For statically determinate structure and subjected to dynamic load such as building .... $1.5DL + 2.5\ L\ L$ or $2\ (DL + LL)$ whichever is greater.

25.21 REINFORCEMENT IN PRESTRESSED CONCRETE

(*A*) *Transverse Reinforcement*

Shear and torsion govern the amount and spacing of transverse reinforcement. When the web is thin and cables are located in the web, the transverse reinforcement shall be provided.

In case of member subjected to dynamic load at least 0.3 per cent of sectional area in plan shall be provided in the web. This reinforcement may be reduced to 0.2%. If the depth of web is not more than 4 times its thickness. These values can be reduced to 0.2% and 0.15% respectively if high strength tensile reinforcement is used.

The transverse reinforcement shall not be less than 0.1% of the sectional area of web in the plan if depth of web is more than 4 times its thickness.

Reinforcement shall be provided in the form of links in helix to resist heavy compression or shock loading.

(*B*) *Longitudinal Reinforcement*

A minimum longitudinal reinforcement equal to 0.2% of the total concrete area shall be provided in all cases except in the case of pretensioned units of small sections.

This reinforcement may be reduced to 0.15% if high strength deformed bars are used.

Longitudinal distribution reinforcement not less than 0.5% of the area of the web shall be provided on each face, when depth of web exceeds 50 cm. The spacing of such reinforcement shall not exceed 20 cm.

(*C*) *Continuity*

In case of continuous prestressed concrete member support moments and moments due to prestressing shall be taken into consideration.

(D) Butted Assembly

In case of a butted assembly provision shall be made for the transfer of all shear stresses. Whenever shear stresses exceed the permissible limits, provision for the same including keying of all abutting faces shall be made.

(E) Transmission Zone in Pre-Tensioned Member

The length should be determined from tests carr ied out under most unfavourable conditions. If the concrete has strength at transfer not less than 35 N/mm^2, the transmission length for plain and indented wire, crimped wire and strands may be taken as 100ϕ, 650ϕ, 35ϕ, respectively. Where ϕ is the diameter of the tendon. The development of stress in tendon may be assumed to vary parabolically along the length of member.

(F) Concrete Cover and Spacing of Prestressing Steel

As per IS : 1343 in pretensioned work, the concrete cover measured from outside of prestressing tendon should not be less than 20 mm.

In post-tensioned member, the minimum clear cover when cables and large sized bars are used, should be equal to the least of the following :

(*i*) 30 mm

(*ii*) Size of cable or bar.

The minimum requirement shall be increased by 10 mm when member is exposed to weather.

The portland cement concrete generally protects embedded cable or tendons against corrosion from various environmental agencies. However carbonation of hydrated cement progressed from surface to interior of the concrete gradually thus reducing effective protection against rusting. Many codes have provided minimum concrete cover.

The minimum clear spacing between single wire in pretensioned system should not be less than the following :

(*i*) 3 × Diameter of single wire used.

(*ii*) 1 ½ times maximum size of aggregate.

The minimum clear spacing between cables or large bars shall not be less than :

(*i*) 40 mm

(*ii*) Maximum dimension of the cable or bar.

(*iii*) 6.5 mm in excess of the largest size of aggregate used.

The grouping of cable is permissible but not more than two cables should touch each other in vertical plane and minimum spacing in vertical plane between such grouped cable should be 50 mm.

25.22 TYPES OF FLEXURAL FAILURE

Percentage of reinforcement, degree of bond between tendons and concrete, ultimate strength of the tendons and compressive strength of concrete are the principal controlling factors affecting type of flexural failure of prestressed concrete members subjected to bending loads.

The behaviour of prestressed concrete in post cracking stage becomes same of reinforced concrete member.

The various types of failure of prestressed concrete member are the following:

(1) *Failure of Steel in Tension*

The sudden failure of prestressed member without any warning is due to fracture of steel in tension zone when the percentage of steel provided in the concrete member is so low that it is not in a position to take up the additional tensile stress transferred on to it by the cracked concrete.

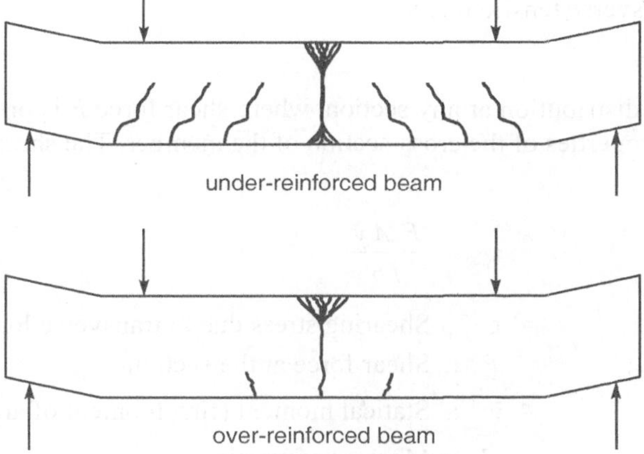

Fig. 25.27. Fluxural Failure of Prestressed Beams

This type of failure can be prevented by providing a certain minimum percentage of steel in the cross section.

Indian standard code recommends a minimum of 0.2 per cent of cross sectional area in longitudinal direction in case of pretensioned unit of small sections. This can be reduced to 0.15 per cent when high yield strength deformed reinforcement is used.

(2) *Failure of Under-reinforced Sections*

The failure of under-reinforced sections takes place due to excessive elongation of steel followed by crushing of concrete.

When percentage of steel, in the member, is so provided that increase in load caused excessive elongation of steel thus raising neutral axis upwards to the compression face at critical section. This causes gradual reduction of compression zone and crushing of cencrete takes place when the area of concrete reduces to such an extent that it is unable to take resultant internal compressive force.

This type of failure is desirable as there is considerable warning before impending failure because of large deflection and wide cracks at failure (Fig. 25.27)

(3) *Failure of Over-reinforced Sections*

Failure due to sudden crushing of concrete takes place when percentage of steel in prestressed beam is higher than a certain percentage making the section over reinforced.

With the gradual increase of load the tensile stress in steel increases but remains well within the tensile strength and in many cases it may well be within the proof stress of the tendons. At the same time top compressive fibres undergo excessive plastic deformation under heavy stresses. With further increase in loads, failure takes place by crushing of concrete.

In structural concrete member such failure is undesirable since there is sudden failure without any warning in the form of excessive deflection and cracking of concrete.

(4) *Other Modes of Failure*

If members are not adequately designed for shear, the member may fail in shear before full flexural strength is attained. Web shear cracks may develop if the principal stresses are excessive and thin web is used.

In pretensioned member the failure of bond between steel and surrounding concrete may take place due to insufficient transmission lengths at the ends of members.

In post-tensioned member anchorage failure may occur if the end block is not properly designed to resist transverse tensile force.

25.23 SHEAR

The shear stress distribution at any section, where shear force F is present, is a function of the shear force, and properties of the cross-section of the member. The shear stress at any point is given by

$$\tau_v = \frac{F A y}{I b} \qquad \text{...(25.51)}$$

where

τ_v : Shearing stress due to transverse load.

F : Shear force at the section.

$A \bar{y}$: Statical moment (first moment of area).

I : Moment of inertia.

b : Breadth of section at the given point.

This shear stress causes principal tensile stresses on the diagonal plane. Diagonal tension cracks first appear in the region of high shear stresses.

In prestressed concrete member, a direct stress in the axial direction also act in conjunction with shear stress. There will also be compressive stress in direction perpendicular to the axis of member if vertical prestressing is adopted.

When curved cable is used to apply axial prestressing making an angle θ with horizontal then the component $P \sin \theta$ will act opposite to shear force thus reducing the net shear force at the section. But when there is only dead load, there might be reversal of direction of shear force if $P \sin \theta$ is greater than shear force due to dead load.

Net shear force to be resisted by concrete at the section (Fig. 25.28)

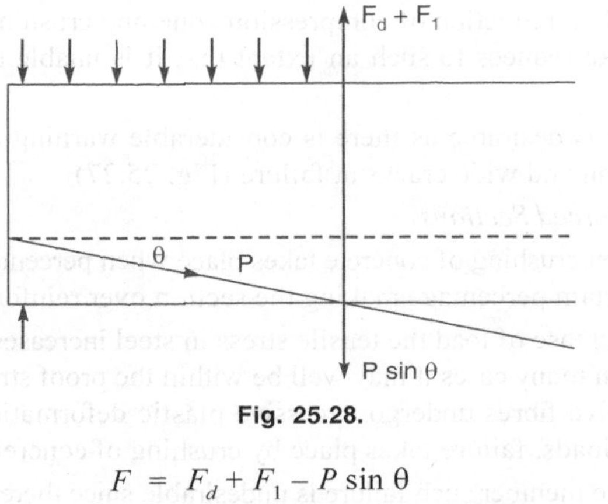

Fig. 25.28.

$$F = F_d + F_1 - P \sin \theta \qquad \text{...(25.52)}$$

where, F_d = shear force due to dead load.

F_1 = shear force due to live load.

A typical element subjected to two-dimensional stress system is shown in Fig. 25.29.

The principal stress developed is given by

$$f_{\substack{\max \\ \min}} = \left[\left\{\frac{f_x + f_y}{2}\right\} \pm 1/2\sqrt{(f_x - f_y)^2 + 4\tau_v^2}\right] \qquad \text{...(25.53)}$$

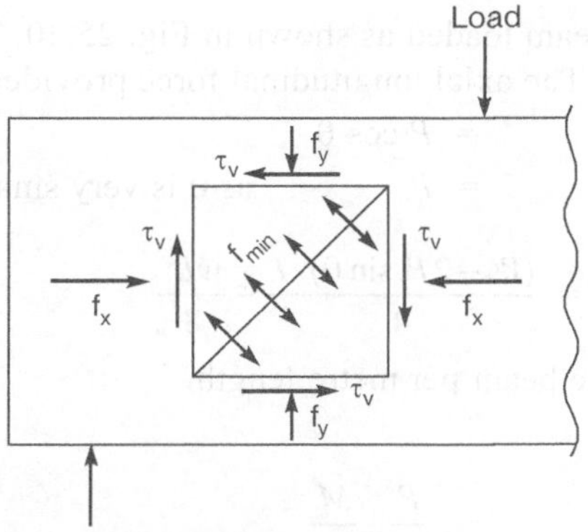

Fig. 25.29. Stress System

f_x and f_y are direct compressive stress caused by horizontal and vertical prestressing. These reduce the principal tensile stress to large extent and in some cases even eliminate tensile stress.

Both major and minor principal stress become compressive and the risk of diagonal tension cracks in concrete is eliminated.

In general the following factors tend to reduce the shearing stress in the prestressed concrete beam.

(1) The vertical component of the prestressing force transmitted through curved cables.

(2) The presence of horizontal or axial prestressing.

(3) The presence of vertical or transverse prestressing.

Vertical prestressing is not generally adopted since the length of cable being too short the loss of prestress due to anchorage slip is excessive.

Post-tensioning is generally uneconomical for vertical prestressing. A viable alternative is to use pretensioned vertical wires closely spaced to achieve the direct prestress.

25.24 CONCEPT OF LOAD BALANCING

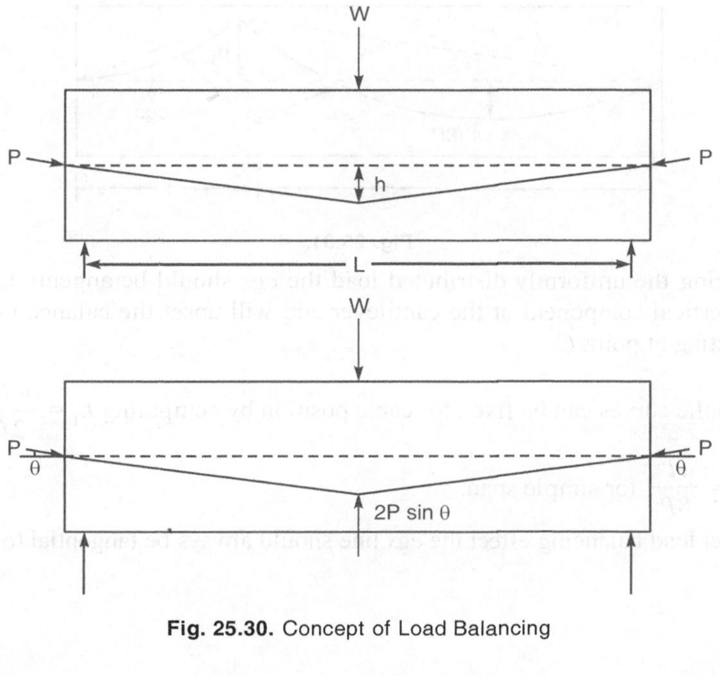

Fig. 25.30. Concept of Load Balancing

Consider a prestressed beam loaded as shown in Fig. 25.30. The net downward load at the centre will be $(W - 2P\sin\theta)$. The axial longitudinal force provided by the tendon

$$= P\cos\theta$$

$$= P \qquad \text{as } \theta \text{ is very small angle.}$$

$$\text{Net Bending Moment } M = \frac{(W - 2P.\sin\theta)\cdot L}{4} + \frac{wL^2}{8} \qquad \text{...(25.54)}$$

where w is the dead load of the beam per metre length.

Extreme fibre stress

$$= \frac{P}{A} \pm \frac{M}{Z} \qquad \text{...(25.55)}$$

For complete balancing, the load

$$W = 2P\sin\theta \text{ (Since } \theta \text{ is very small)} \qquad \text{...(25.56)}$$

$$\sin\theta \simeq \tan\theta, \text{ Hence}$$

$$W = 2P\tan\theta$$

$$= 2P.\frac{h}{L/2} = \frac{4Ph}{L} \qquad \text{...(25.57)}$$

or,
$$P = \frac{WL}{4h} \qquad \text{...(25.58)}$$

When prestressing force P is such that it completely balances the load, then stress at any section $= \dfrac{P}{A}$.

Now we consider the cantilever portion of any overhanging beam as shown in the Fig. 25.31

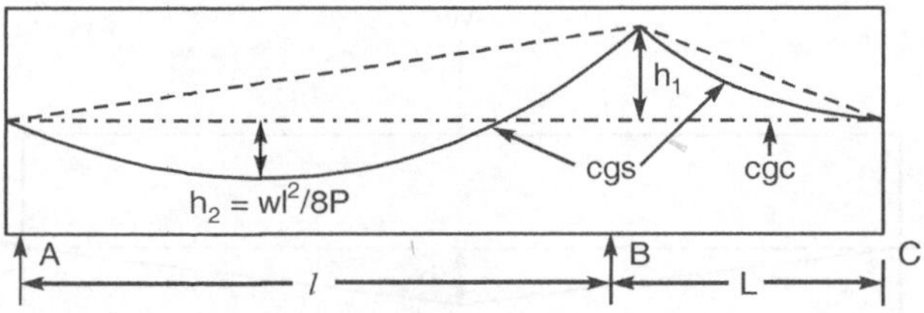

Fig. 25.31.

For balancing the uniformly distributed load the *cgs* should be angential to *cgc* at end *C* otherwise the vertical component at the cantilever end will upset the balance unless there is an external load acting at point *C*.

Two parabolic curves can be fixed for cable position by computing $h_1 = \dfrac{wL^2}{2P}$ for cantilever portion and $h_2 = \dfrac{wl^2}{8P}$ for simple span.

To get exact load balancing effect the *cgs* line should always be tangential to *cgc* for the ends of beam.

25.25 END BLOCKS

In post-tensioned members the wires or cables are stressed and then anchored by mechanical means bearing against the end of beams. This results in large concentrated forces over comparatively short area on the bearing surface in the anchorage zone or end block and the stress distribution is three-dimensional and very complex. These highly discontinuous forces develop transverse and shear stresses.

St. Venant's principle suggests that the stress distribution will be linear and can be computed from simple bending theory at a distance equal to depth of beam.

The zone between the end of beam and the section where only longitudinal stress exists is called end block or anchorage zone (Fig. 25.32).

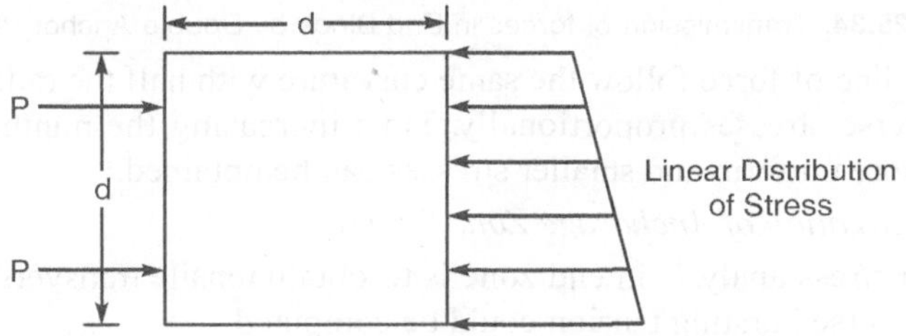

Fig. 25.32

Over a large length in end zone the transverse stresses developed are tensile in nature. Since concrete is weak under tension. Hence it becomes necessary to know the stress distribution properly in end zone to provide reinforcement properly to sustain the transverse tensile stresses.

Besides these, there are other transverse tensile stresses causing spalling of concrete.

(A) Stress Distribution in End Block

The line of forces shown in the Fig. 25.33 can be considered as indivisual fibres acting as curved struts inserted betwen end force 2 P and main body of beam to get the physical concept of state of stresses in the transverse direction. (Normal to planes parallel to top and bottom surfaces of the beam).

The curvature of struts is convex towards the central line in the zone *A* causing compressive stresses. In zone *B* the tensile stresses are developed as the curvature is reversed in this zone.

The struts are straight in zone *C* causing no transverse stresses, and only longitudinal stresses develop in this zone.

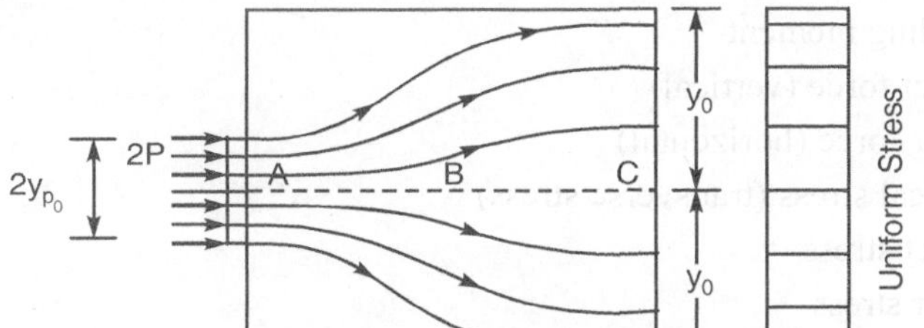

Fig. 25.33. Transmission of Forces in End Block by Single Anchor Plate

Let us consider the same total force through two zones whose positions are symmetrical in upper and lower halves of beam as shown in Fig. 25.34.

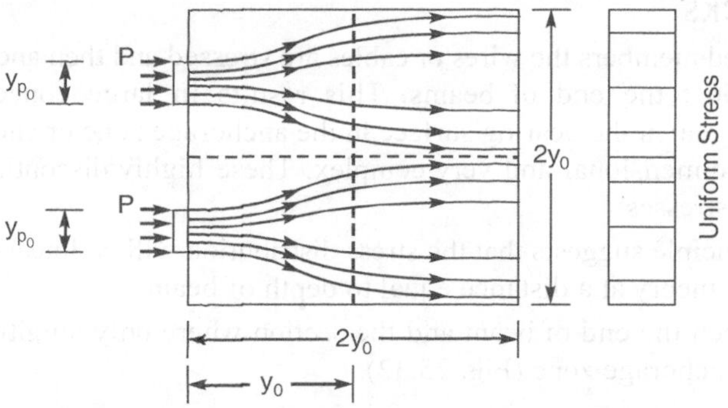

Fig. 25.34. Transmission of forces in End Block by Double Anchor Plates

In this case the line of force follow the same curvature with half the radius of curvature thus reducing the transverse stresses proportionally. Thus increasing the number of application of prestressing force more uniform and smaller stresses can be obtained.

(*B*) *Methods of Investigation of Anchorage Zone Stresses*

The purpose of stress analysis in end zone is to obtain tensile transverse stress distribution from which the transverse bursting tension could be computed.

(*i*) *Magnel Method*

The basis of this method is that the end block, of length equal to depth of beam at the end of beam, is considered as deep vertical beam subjected to concentrated loads on one side due to anchorages and normal and tangential distributed loads from linear direct stress distribution and parabolic shear stress distribution from the other side.

The forces acting on the end block and stresses acting on any point on the horizontal axis are shown in the Fig. 25.35. The stress distribution across the section can be approximated by

$$f_v = k_1\left(\frac{M}{bh^2}\right) + k_2(H/bh) \quad \text{...(25.59)}$$

$$\tau = k_3(V/bh) \quad \text{...(25.60)}$$

$$f_h = \frac{P}{bh}\left(1+12\frac{e'^2}{h'^2}\right) \quad \text{...(25.61)}$$

where,

M = Bending moment

H = Direct force (vertical)

V = Shear force (horizontal)

f_v = Vertical stress (transverse stress)

f_h = Direct stress

τ = Shear stress

The directions as shown in the Fig. 25.35 can be taken as positive.

M is positive when producing compression at nearest end of beam.

V is positive when acting away from the end of the beam.

H is positive when acting vertically downwards.

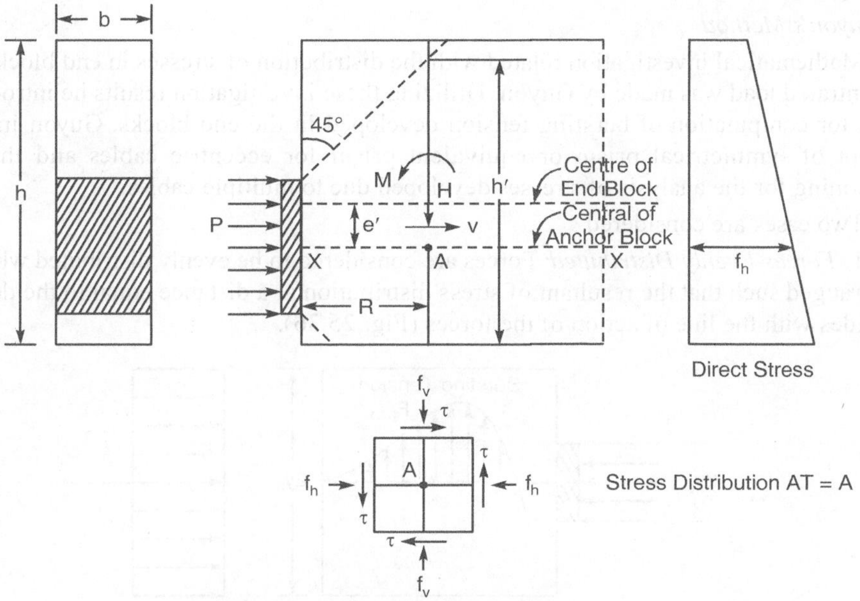

Fig. 25.35. Forces Acting on End Block

The concentrated load is considered to be dispersed at 45° to compute the direct stress f_h. The depth of section intercepted between the dispersion lines is considered to be effective and used in calculations.

The principal stresses acting at the point are computed by using the following equations:

$$f_{\substack{max \\ min}} = \frac{(f_v + f_h)}{2} \pm \frac{1}{2}\sqrt{(f_h - f_v)^2 + 4\,\tau^2} \quad ...(25.62)$$

$$\tan 2\theta = \left(\frac{2\tau}{f_v - f_h}\right)$$

The coefficients k_1, k_2 and k_3 are given in Table 25.4 for varying distance from the end face of the beam. Knowing the distribution of principal tensile stress on the required axis, the bursting tension is calculated and reinforcement is provided to take care of this tension.

Table 25.4. Coefficients for Stresses in End Blocks

Distance from far end (x/h)	K_1	K_2	K_3
0.0	20.00	-2.00	0.00
0.1	9.72	0	1.46
0.2	2.56	1.28	2.05
0.3	– 1.96	1.96	2.06
0.4	– 4.32	2.16	1.73
0.5	– 5.00	2.00	1.25
0.6	– 4.48	1.60	0.77
0.7	– 3.24	1.08	0.38
0.8	– 1.76	0.56	0.13
0.9	– 0.52	0.16	0.02
1.0	0	0	0

(*ii*) *Guyon's Method*

Mathematical investigation related with the distribution of stresses in end block subjected to concentrated load was made by Guyon. Utilizing these investigation results he introduced design tables for computation of bursting tension developed in the end blocks. Guyon introduced the concept of symmetrical prism or equivalent prism for eccentric cables and the method of partitioning for the analysis of stresses developed due to multiple cables.

Two cases are considered.

1. *Forces Evenly Distributed*: Forces are considered to be evenly distributed when the forces are arranged such that the resultant of stress distribution at a distance equal to the depth of block coincides with the line of action of the forces (Fig. 25.36).

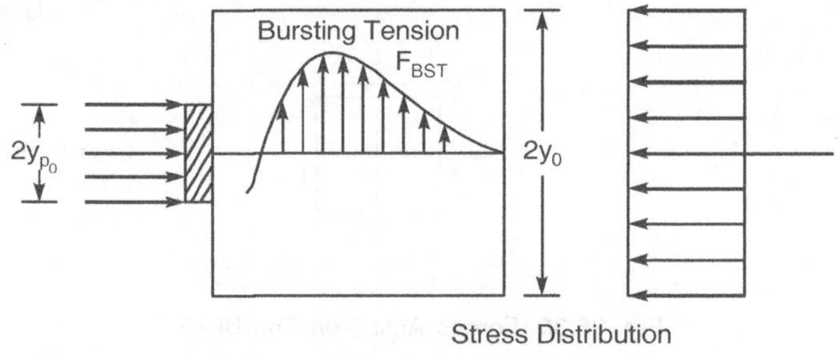

Fig. 25.36. Evenly Distributed Force System

Symmetric prism method is used for eccentric forces and multiple cables. Symmetrical prism can be defined as the prism of concrete of side equal to two times the distance of prestressing cable from the nearest free edge (Fig. 25.37).

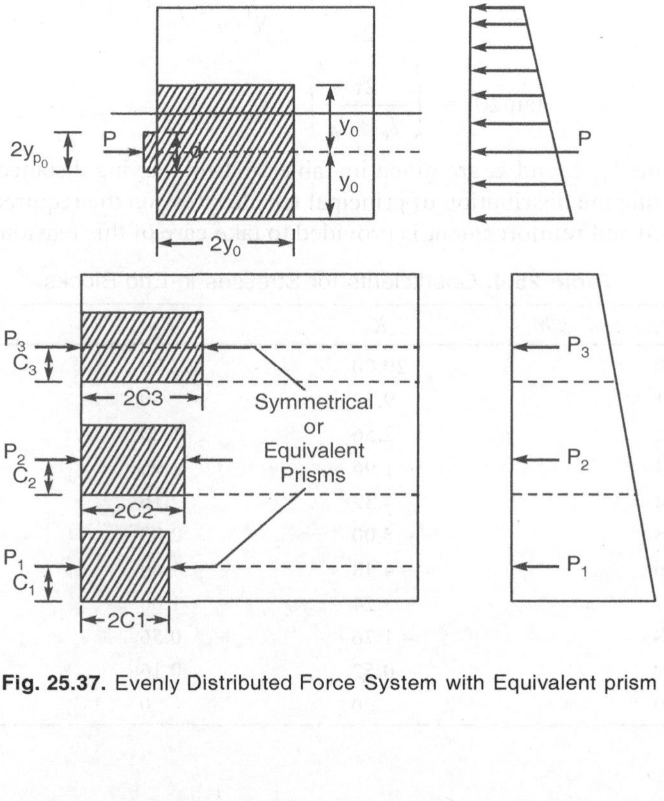

Fig. 25.37. Evenly Distributed Force System with Equivalent prism

According to Guyon

$$\text{Bursting tension } F_{bst} = 0.3\ P\left\{\left(1-\frac{Yp_o}{Y_o}\right)^{0.58}\right\} \quad ...(25.64)$$

where P = Anchorage force

Y_{P_o}/Y_o = Distribution ratio.

Coefficients given in Table 25.5 under the category of distributed axial forces can be used to find the position of zero stress, maximum transverse stress and its magnitude for the forces that are evenly distributed.

2. *Forces Not Evenly Distributed.* When the end forces cannot be arranged evenly, Guyon recommends that the transverse tensile stresses should be investigated along successive resultants.

(*i*) Resultant of all forces

(*ii*) Resultant of smaller group of forces.

(*iii*) Line of action of indivisual forces.

The line of action of the resultant force is taken as the axis of an equivalent prism of length and depth equal to twice the distance of axis from the free edge or the adjoining equivalent prism. Use the coefficients given in Table 25.5 under the category of concentrated eccentric force and eccentric shear force. Since the coefficients are applicable for forces at interval of 1/8th of the prism depth, the end forces have to be replaced by a statically equivalent system of normal and shear forces acting at these regular intervals (Fig. 25.38).

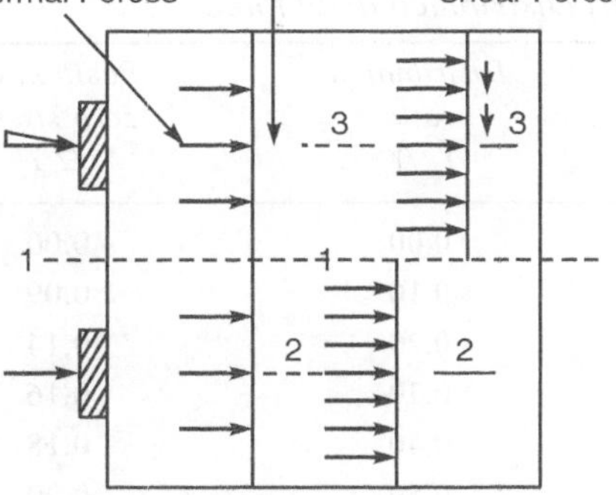

Fig. 25.38. Normal and Shear Forces Distribution

(*iii*) *Zielinski and Rowe's Method*

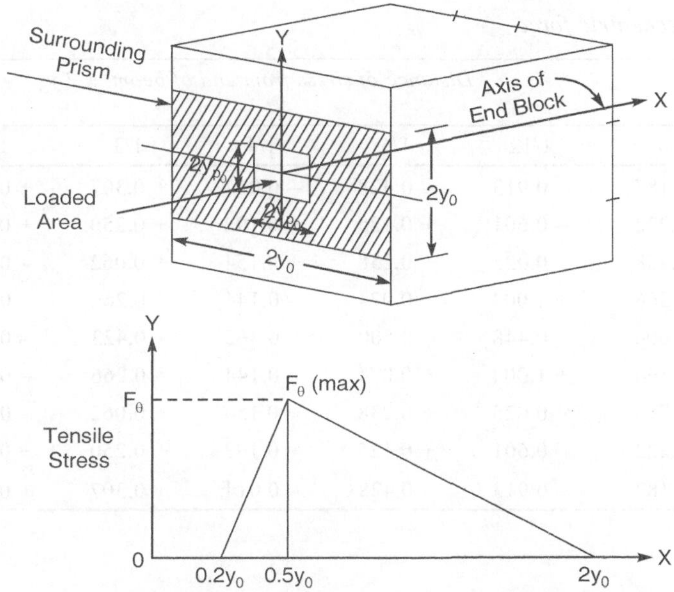

Fig. 25.39. Tensile Stress Distribution in end Block

Zielinkski and Rowe developed empirical relations to compute maximum transverse tensile stress and bursting tension when an end block is subjected to a concentrated load at the end face (Fig. 25.39).

The distribution of transverse stress is affected by the ratio Y_{po}/Y_o. But the position of zero stress and maximum stress is not affected by this ratio. The transverse tensile stress is zero at 0.2 Y_o and it is maximm at a distance of 0.5 Y_o. After this stress starts decreasing and became zero at 2 Y_o distance, as shown in Fig. 25.39.

The notations used are as follows:

$2\,Y_o$ = Side of surroundings prism.

$2\,Y_{po}$ = Side of loaded or punching area.

$$Y_{po}/Y_o = \frac{\text{Side of loaded area}}{\text{Side of surrounding prism}}$$

Table 25.5. Vertical Stresses Along Axis at Ends of Prestressed Beams (Guyon) (Ratios of local stress to average stress across whole section)' + = compressive; – = tension

(*a*) *Distributed axial force*

Distribution ratio Y_{po}/Y_o	*Position of zero stress* $x/2\,Y_o$	*Position of maximum stress* $x/2\,Y_o$	*Ratio of maximum tensile stress to average stress*
0.00	0.00	0.17	0.50
0.10	0.09	0.24	0.43
0.20	0.14	0.30	0.36
0.30	0.16	0.36	0.33
0.40	0.18	0.39	0.27
0.50	0.20	0.43	0.23
0.60	0.22	0.44	0.18
0.70	0.23	0.45	0.13
0.80	0.24	0.46	0.09

(*b*) *Concentrated eccentric force*

Eccentricity of force	*Distance of stress from end of beam,* $x/2\,y_o$						
$e/2y_o$	0	1/12	1/6	1/4	1/3	1/2	3/4
+ 1/2	– 2.187	– 0.913	– 0.428	– 0.014	+ 0.307	+ 0.399	+ 0.192
+ 3/8	– 1.222	– 0.601	+ 0.125	+ 0.192	+ 0.250	+ 0.242	+ 0.122
+ 1/4	–0.758	– 0.025	+ 0.238	+ 0.154	+ 0.062	– 0.024	– 0.016
+ 1/8	– 0.566	+ 1.004	+ 0.074	– 0.144	– 0.266	– 0.262	– 0.128
0	0.000	– 0.448	– 0.500	– 0.462	– 0.423	– 0.314	– 0.161
– 1/8	– 0.566	+ 1.004	+ 0.074	– 0.144	– 0.266	– 0.262	– 0.128
– 1/4	– 0.758	– 0.025	+ 0.238	+ 0.154	+ 0.062	– 0.024	– 0.016
– 3/8	– 1.222	– 0.601	+ 0.125	+ 0.192	+ 0.250	+ 0.242	+ 0.122
– 1/2	– 2.187	– 0.913	– 0.428	– 0.014	+ 0.307	+ 0.399	+ 0.192

(c) Eccentric shear force

Eccentricity of force $e/2y_o$	Distance of stress from end of beam, $x/2\ y_o$						
	0	1/12	1/6	1/4	1/3	1/2	3/4
+ 1/2	–	0	0	0	0	0	0
+ 3/8	+ 5.66	+ 2.96	+ 0.87	+ 0.19	– 0.05	– 0.14	– 0.07
+ 1/4	+ 4.00	+ 3.10	+ 1.52	+ 0.44	– 0.22	– 0.20	– 0.10
+ 1/8	+ 5.66	+ 2.96	+ 0.87	+ 0.19	– 0.05	– 0.14	– 0.07
0	–	0	0	0	0	0	0
– 1/8	– 5.66	– 2.96	– 0.87	– 0.19	+ 0.05	+ 0.14	+ 0.07
– 1/4	– 4.00	– 3.10	– 1.52	– 0.44	+ 0.22	+ 0.20	+ 0.10
– 3/8	– 5.66	– 2.96	– 0.87	– 0.18	+ 0.05	+ 0.14	+ 0.07
– 1/2	–	0	0	0	0	0	0

f_v = Transverse tensile stress.

f_c = Uniform compressive stress in the surrounding prism = $\dfrac{\text{Prestress}}{(2Y_o)^2}$

T = Applied tendon jacking force (applied compressive force on end block)

P_{bst} = Bursting tension.

The equations recommended by Zielinski and Rowe's are

$$f_{v(max)} = f_c\left(0.98-0.825\left(\frac{Y_{po}}{Y_o}\right)\right) \quad ...(25.65)$$

This equation is valid for the ratio $\dfrac{Y_{p_o}}{Y_o}$ varying from 0.3 to 0.7 only.

$$P_{bst} = T\,(0.48 - 0.4\,Y_{po}/Y_o) \quad ...(25.66)$$

If some tension is allowed in the concrete the corrected value of bursting tension can be obtained as

$$P_{bst}\ \text{(corrected)} = P_{bst}\left[1-\left(\frac{f}{f_{v\,max}}\right)^2\right] \quad ...(25.67)$$

where f_1 : permissible tension in concrete.

The reinforcement is provided to resist the bursting tension in the zone varying from 0.2 Y_o to 2 Y_o. The reinforcement is provided in both directions *i.e.*, both in the direction of depth as well as in the direction of width.

(iv) Indian code Provision

Indian code provision is also based on the work of Zielinski and Rowe.

The bursting tension is obtained by the following expression.

$$P_{bst} = T\left\{0.32-0.3\frac{(Y_{po})}{(Y_o)}\right\} \quad ...(25.68)$$

where T = Tendon jacking force and $\frac{Y_{p_o}}{Y_o}$ = Distribution ratio.

The reinforcement is designed to sustain this bursting tension and is distributed in the zone of 0.2 Y_o to 2.0 Y_o from the loaded face of the end block.

The reinforcement is designed considering its design strength of 0.87 f_y. The stress is limited to a strain of 0.1% when concrete cover provided is less than 50 mm.

The end block is divided into a series of symmetrically loaded prisms, if groups of anchorages are encountered. Each symmetrically loaded prism is analysed for bursting tension separately.

(*v*) *Approximate Method*

The end zone of the beam, subjected to large concentrated forces due to end anchorage, has very complex variation of stresses and analysis required should essentially be three-dimensional. Some authors have attempted to evolve a two-dimensional approach for the problem.

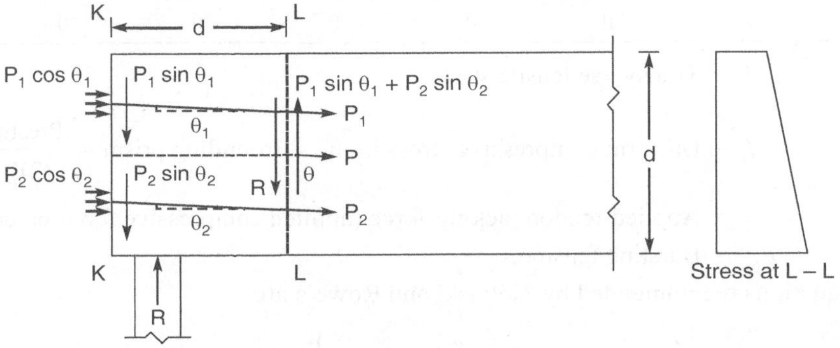

Fig. 25.40.

Approximate method treats the anchorage forces at the end of beam as a column load acting on the footing. This method is most commonly used for the analysis of the end block of the beam.

Fig. 25.40 shows two cables inclined at angles θ_1 and θ_2 with anchoring forces P_1 and P_2 respectively. The resultant of these two forces is P acting on the end face of the end block (*LL*) at an eccentricity e and at an angle θ with horizontal. R is the support reaction acting at the end block.

Since the horizontal component of resultant force *i.e.*, $P \cos \theta$ acts at eccentricity e, the resulting stress diagram will be trapezoidal as shown in Fig. 25.40. The bending stress due to dead load and live load can be neglected as the moment caused by these loads is very small in the end zones.

The external forces on the end block are listed as follows:

1. Horizontal Components of anchorage forces at face *KK*
   $= P_1 \cos \theta_1 + P_2 \cos \theta_2$
   $= \Sigma P_1 \cos \theta_1$
2. Horizontal component of resultant force acting on face *LL* at an eccentricity '*e*'
   $= P \cos \theta$
3. Reaction at support = R
   This is usually considered to be acting at a distance of 30 cm from the free edge.
4. Vertical forces at face *KK* = $P_1 \sin \theta_1 + P_2 \sin \theta_2$
   acting downwards
   $= \Sigma P_1 . \sin \theta_1$
5. Vertical forces at face *LL* = $P \sin \theta$ acting upwards.

Stresses at Any Point in the End Block

For calculating stresses at any point in the end block of the beam by approximate method, the forces acting on one side of the horizontal and vertical planes passing through the point under consideration are considered.

Bending moments are also considered to contribute bending stresses in the anlaysis. Since the depth of beam is large, the Deep Beam Theory is used for finding the bending stresses.

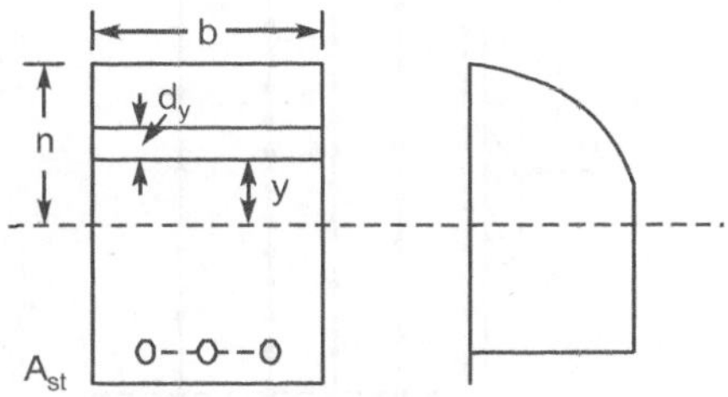

Fig. 25.41.

The shear stresses can be calculated by the following equation (Fig. 25.41).

$$\tau_v = \frac{V}{Ib}\int_y^n b.ydy$$

$$= \frac{V}{I}\left[\frac{n^2}{2}-\frac{y^2}{2}\right]$$

$$= \frac{V}{2I}(n^2 - y^2) \qquad ...(25.69)$$

Where I = Moment of inertia of section

$= 1/3\ bn^3 + m\ A_{st}\ (d - n)^2$

Principal stresses at any point in the end beam can be obtained by determining normal and shear stresses acting vertically and horizontally on the planes passing through that point. The principal stresses can be calculated by the use of Mohr's circle or by using the formula given below.

$$\text{Principal stresses} = 1/2\left[(f_x + f_y) \pm \sqrt{(f_x - f_y)^2 + 4\tau^2}\right] \qquad ...(25.70)$$

τ = Shear Stress

f_x, f_y = Normal stresses in x and y direction.

Reinforcements (generally vertical) are provided in the direction of principal tensile stress. It is found that principal stresses are usually maximum at central axis and at two third of depth from free end (Fig. 25.42).

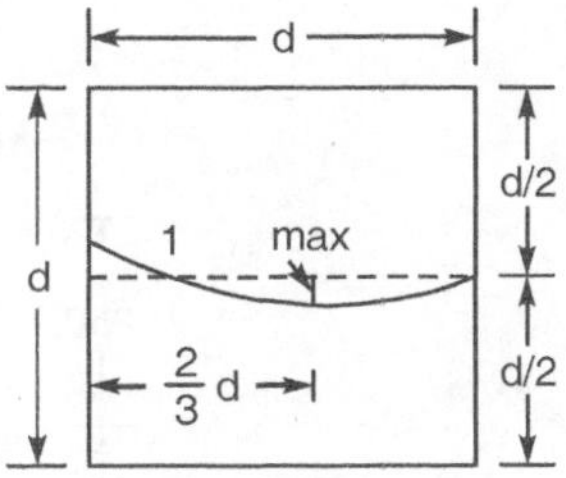

Fig. 25.42. Principal Stress Distribution

25.26 END ZONE REINFORCEMENT

The main reinforcement provided for taking care of bursting tension is determined by the transverse tensile stress distribution on the critical axis. The critical axis generally coincides with the line of action of the largest individual force.

The arrangement of reinforcement in end block for plate and embedded type of anchorage is shown in the Fig. 25.43.

Mats, helics, loops are provided in perpendicular direction. The helical reinforcement is considered more effective than mat reinforcement.

Suitable reinforcement in the form of hair pin bars should be provided to prevent failure of corner zone from spalling tension developed at the corners.

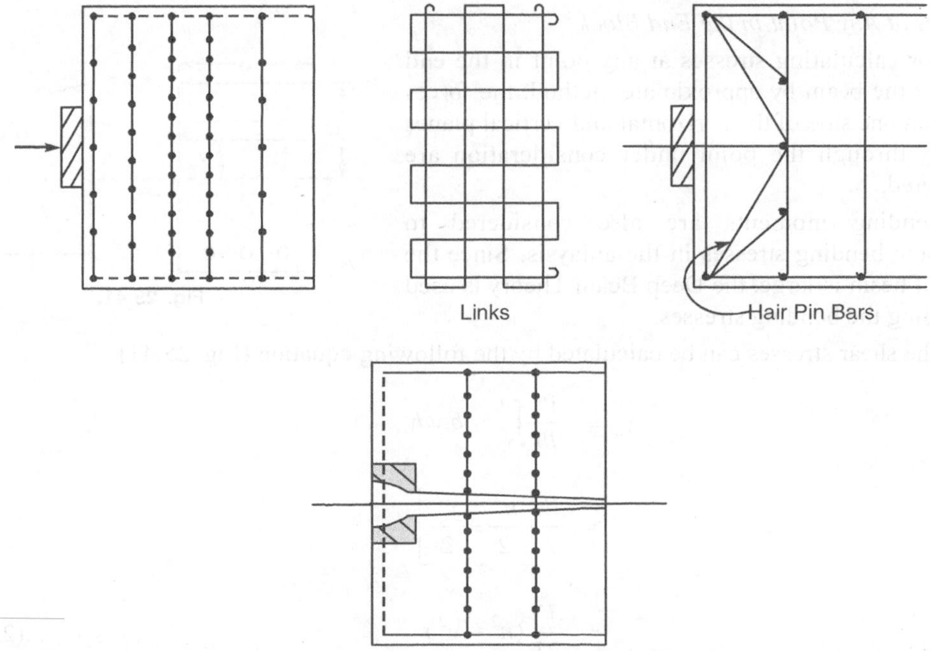

Fig. 25.43. Reinforcement Arrangement in End Blocks

It is required to provide a little extra reinforcement if the situation is doubtful.

Example 25.2. Design a suitable end block for a beam 600 mm × 900 mm and subjected to two cable anchorages. Each anchorage is exerting 1000 kN. The cables are located 250 mm on either side of centre of the beam. Cables have slope 1 in 20 at the end. Reaction at the left support is 160 kN. The centre line of support is considered to be at 30 cm from the free end of the beam. Permissible Bearing Stress of concrete may be taken as 0.8 × 40 = 32 N/mm² as concrete used is M 40.

Solution.

Size of Bearing Plate (Fig. 25.44)

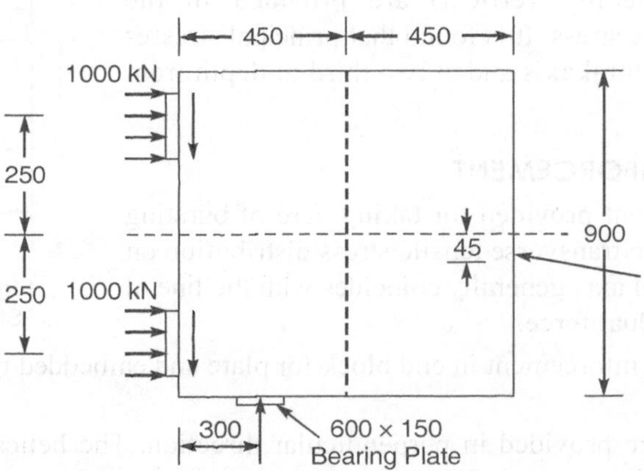

Fig. 25.44.

$$\text{Load at support} = 16\ t = 160 \text{ KN}$$

$$\text{Bearing plate area required} = \frac{160{,}000}{32} = 5000 \text{ mm}^2$$

Adopt 600 mm × 150 mm, (Width of beam = 600 mm)

For Anchorage: Load per anchor = 1000 KN

$$\text{Bearing Area Required} = \frac{1000000}{32}$$

$$= 31250 \text{ mm}^2$$

Allow 50 mm diameter for each duct tube.

$$\text{Area of each duct tube} = \pi/4 \times 50^2$$

$$= 1960 \text{ mm}^2$$

$$\text{Total area required} = 31250 + 1960$$

$$= 33210 \text{ mm}^2$$

Adopt 200 mm × 200 mm providing

$$\text{Area} = 40000 \text{ mm}^2 > 33210 \text{ mm}^2$$

Forces Acting on the End Block : (Fig. 25.45)

$$\text{Length of end block} = 900 \text{ mm}$$

$$\text{Eccentricity of resultant forces in the cable at 900 mm from the end} = \frac{900}{20} = 45 \text{ mm}$$

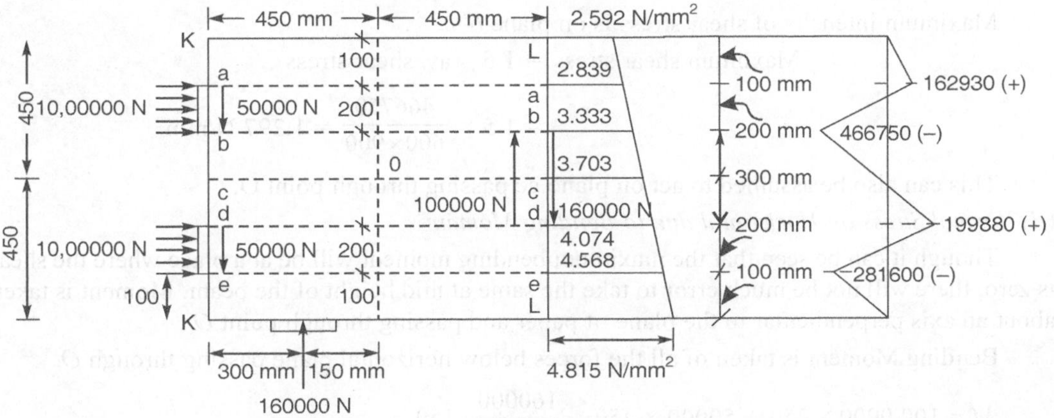

Fig. 25.45.

Stress Across Secton *LL*

$$= \frac{P}{A}\left\{1 \pm \frac{6e}{d}\right\} = \frac{200{,}0000}{600 \times 900}\left\{1 \pm \frac{6 \times 45}{900}\right\}$$

$$= 3.7 \times (1 \pm 0.3)$$

$$\text{Top fibre stress} = 2.592 \text{ N/mm}^2$$

$$\text{Bottom fibre stress} = 4.815 \text{ N/mm}^2$$

$$\text{Shear at each cable anchorage due to slope of cable} = \frac{100{,}0000}{20} = 50{,}000 \text{ N}$$

$$\text{Shear for two cables} = 2 \times 50000$$
$$= 100000 \text{ N}$$

HORIZONTAL PLANE

Consider horizontal plane through 0, *i.e.*, mid depth of beam in plane $C - C$.

(*i*) *Shear above CC due to Horizontal Forces*

We have to determine the shear force on horizontal section *i.e.*, section parallel to face *KL*. Net shear force acting on horizontal planes is required.

$$\text{Shear on plane } a\text{–}a = \left\{\frac{2.592 + 2.839}{2}\right\} \times 100 \times 600$$
$$= +162930 \text{ N}$$

$$\text{Shear on plane } b\text{–}b = \left\{\frac{2.592 + 3.333}{2} \times 300 \times 600\right\} - (100{,}0000)$$
$$= -466750 \text{ N}$$

$$\text{Shear on plane } d\text{–}d = \left\{\frac{2.592 + 4.074}{2} \times 600 \times 600\right\} - (100{,}0000)$$
$$= +199880 \text{ N}$$

$$\text{Shear force on plane } e\text{–}e = \left\{\frac{2.592 + 4.568}{2} \times 800 \times 600\right\} - 2 \times 100{,}0000$$
$$= -281600 \text{ N}$$

Maximum intensity of shear stress is on plane *b–b*.

$$\text{Maximum shear stress} = 1.5 \times \text{av. shear stress}$$
$$= 1.5 \times \frac{466750}{600 \times 900} = 1.297 \text{ N/mm}^2$$

This can also be assumed to act on plane cc passing through point O.

(*ii*) *Normal Stress on Horizontal due to Bending Moment*

Though it can be seen that the maximum bending moment will be at a place where the shear is zero, there will not be much error to take the same at mid height of the beam. Moment is taken about an axis perpendicular to the plane of paper and passing through point *O*.

Bending Moment is taken of all the forces below horizontal plane passing through *O*.

$$M = 100{,}0000 \times 250 + 50000 \times 450 - \frac{160000}{2} \times 150$$
$$-\frac{160000}{2} \times 450 + 50000 \times 450 - \left\{600 \times \left(\frac{3.703 + 4.815}{2} \times 450\right) \times \frac{2 \times 4.815 + 3.703}{4.815 + 3.703} \times \frac{450}{3}\right\}$$
$$= -22993250 \text{ N.mm}$$

(– Shows that B.M. is clockwise)

Treating the footing as deep beam, the maximum tensile stress across the horizontal plane (Fig. 25.46).

$$f_t = \frac{4M}{bd^2} = \frac{4 \times 22993250}{600 \times 900 \times 900}$$
$$= 0.19 \text{ N/mm}^2$$

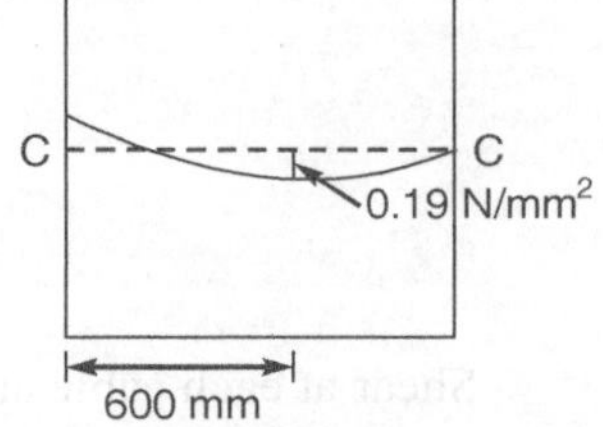

Fig. 25.46.

This value of stress occurs at a distance of 2/3 d = 600 mm from the free edge, on the plane CC passing through O.

(*iii*) *Normal Stress due to Direct Stresses Causing Compression Across the Horizontal Plane Through O.*

$$\text{Compression} = 50000 - \frac{10,0000}{2} + \frac{160000}{2}$$

$$= 80000 \text{ N downward.}$$

(Portion of end block above the horizontal plane through 0 is considered)

$$\text{Compressive Stress} = \frac{80000}{600 \times 900} = 0.148 \text{ N/mm}^2$$

VERTICAL PLANE

(*iv*) *Normal Stress on the Vertical Plane Passing Through O due to Bending Moment.*

Considering the forces to the left of the vertical plane passing through O.

Bending Moment = 10,0000 × 450 – 160000 × 150

= 210,0000 N.mm

Applying the Deep Beam Theory,

Maximum tensile stress due to bending

$$f_t = \frac{4M}{bd^2} = \frac{4 \times 210,0000}{600 \times 900^2} = 0.173 \text{ N/mm}^2$$

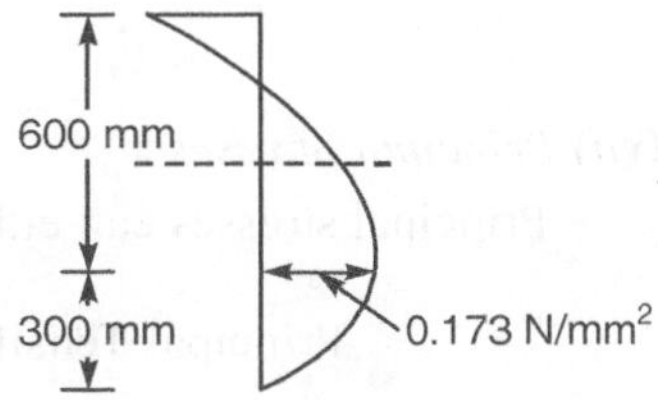

Fig. 25.47.

This maximum value acts at lower one-third. At middle section the stress is very small and may be neglected (Fig. 25.47).

(*v*) *Shear due to Vertical Forces on a Vertical Plane Passing Through O*

Right half is considered

Shear force due to support reaction = 160000 N upwards

Shear force due to inclination of cable = 50000 + 50000 = 10,0000 N

Net shear force on vertical plane to right half of the end block

= 160000 – 100000

= 60000 N.

Maximum shear stress = 1.5 × av. shear stress

$$= 1.5 \times \frac{60000}{600 \times 900} = 0.166 \text{ N/mm}^2$$

The vertical and horizontal shear stress are complimentary. Hence net shear stress in the element at O (Fig. 25.48), $\tau = 0.166 + 1.297$

$= 1.463$ N/mm²

(*vi*) *Normal Stress Causing Compression on the Vertical Plane Through 'O'*

Slope of cable = 1 in 20

$\theta = \tan^{-1}(1/20)$

$= 2.86° = 2° \ 51' \ 44.6''$

$$\text{Compressive Stress} = \frac{P_1 \cos\theta + P_2 \cos\theta}{600 \times 900}$$

$$= \frac{100,0000 \times 2 \times 0.998}{600 \times 900}$$

$$= 3.699 \text{ N/mm}^2$$

The normal and shear stresses on the element are shown in Fig. 25.48.

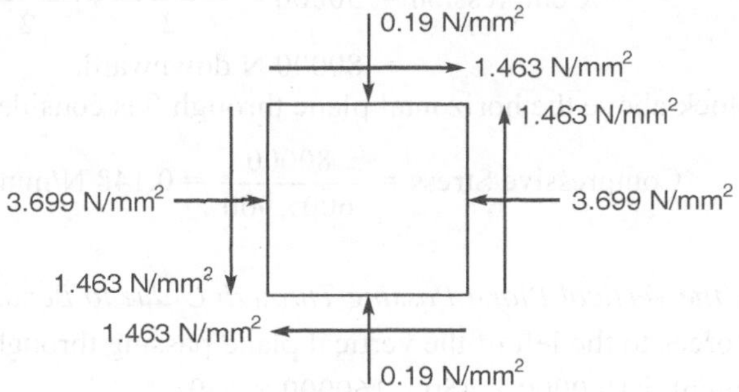

Fig. 25.48.

(*vii*) *Principal Stresses*

Principal stresses can either be found by Mohr's Circle or by equation as given below.

$$\text{Principal Tensile Stress} = 1/2 \{(3.699 + 0.190) - \sqrt{(3.699 - 0.19)^2 + 4 \times 1.463^2}\}$$

$$= -0.34 \text{ N/mm}^2 \text{ (Tensile)}$$

$$\tan 2\theta = \frac{2\tau}{0.190 - 3.699} = \frac{2 \times 1.463}{0.190 - 3.699} = 0.834$$

$$2\theta = 39.82$$

$$\theta = 19.91° \text{ to the vertical.}$$

The principal tensile stress act at 19.91° to the vertical. Stress is small, but it is better to provide vertical reinforcement specially beyond the support say 16 mm bars at 80 mm interval.

25.27 EXAMPLES

Example 25.3. A rectangular beam 15 meter long and cross section 32 cm × 55 cm is post-tensioned by straight tendons tensioned to a stress of 10500 Kg/cm² at the jacking end. Calculate the wobble friction loss at the mid span and at the end remote from jack.

Wobble Coefficient K_f = 0.30 per 100 meter.

Solution.

Loss due to wave effect

$$= P_o - P_x$$

$$= P_o \cdot K_f . x$$

where P_o = prestress at jacking end.

P_x = prestress at a distance x from jacking end.

K_f = Wobble Coefficient.

(*a*) At mid span, $x = 15/2 = 7.5$ m

Loss of stress $= P_o\, K_f x$

$$= 10500 \times \frac{0.30}{100} \times 7.5$$

$= 236.25 \text{ kg/cm}^2$

Percentage loss of prestress $= \dfrac{236.25}{10500} \times 100 = 2.25\%$

(*b*) At the end remote from the jacking end,

$x = L = 15 \text{ m}$

Loss of stress $= P_o K_f x$

$= 10500 \times \dfrac{0.3}{100} \times 15$

$= 472.5 \text{ kg/cm}^2$

Percentage loss of prestress $= \dfrac{472.5}{10500} \times 100 = 4.5\%$

Example 25.4. A prestress concrete beam is continuous over two spans shown in the Fig. 25.49. Its curved tendon is to be tensioned from both ends. Compute the loss due to friction from one end to centre of the beam (from *A* to *E*).

Deviation of tangents to cable profile, α between *A* and *B*, *B* and *C*, *C* and *D*, and *D* and *E* are 0, 0.167, 0 and 0.100 radians respectively.

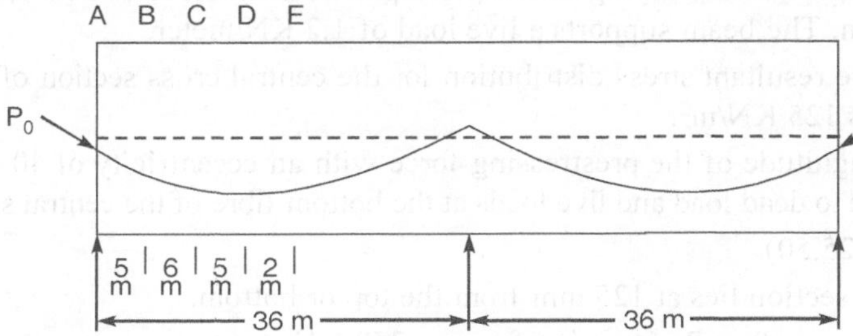

Fig. 25.49.

Coefficient of friction between cable and duct $\mu = 0.4$

Average wobble or length effect is represented by $K_f = 0.25$ per 100 meter

Solution. Let the prestress at the jacking end be P_o.

Loss due to friction $= P_o (\mu\, \alpha + K_f \,.\, x)$

where α = cumulative angle in radian through which the tangent to cable profile has turned between two points being considered.

Loss between $AB = P_o \left(0.4 \times 0 + \dfrac{0.25}{100} \times 5\right)$

$= 0.0125\ P_o$

Therefore Prestress at $B = (1 - 0.0125)\ P_o$

$= 0.9875\ P_o$

Loss of prestress between $B\,C = (0.9875\ P_o)\left(0.4 \times 0.167 + \dfrac{0.25}{100} \times 6\right)$

$= 0.08078\ P_o$

$$\text{Therefore Prestress at } C = (0.9875 - 0.08078)\,P_o = 0.9067\,P_o$$

$$\text{Loss between } CD = (0.9067\,P_o)\left(0.4 \times 0 + \frac{0.25}{100} \times 5\right) = 0.011334\,P_o$$

$$\text{Prestress at } D = (0.9067 - 0.011334)\,P_o = 0.8954\,P_o$$

$$\text{Loss of Prestress between } DE = (0.8954\,P_o)\left(0.4 \times 0.100 + \frac{0.25}{100} \times 2\right) = 0.04029\,P_o$$

$$\text{Therefore Prestress at } D = (0.8954 - 0.04029)\,P_o = 0.8551\,P_o$$

$$\text{Therefore, Percentage loss from } A \text{ to } E = \frac{(1-0.8551)\,P_o}{P_o} \times 100 = 14.49\%$$

Example 25.5. A rectangular beam 100 mm wide by 250 mm deep, spanning over 8 meter is prestressed by a straight cable carrying an effective prestressing force of 250 KN located at an eccentricity of 40 mm. The beam supports a live load of 1.2 KN/meter.

(*a*) Calculate the resultant stress distribution for the central cross section of the beam. The density of concrete is 125 KN/m^3.

(*b*) Find the magnitude of the prestressing force with an eccentricity of 40 mm which can balance the stress due to dead load and live loads at the bottom fibre of the central section of beam.

Solution. (Fig. 25.50)

(*a*) The C.G. of section lies at 125 mm from the top or bottom.

$$P_o = \text{Prestressing force} = 250 \text{ KN} = 250 \times 10^3 \text{ N.}$$

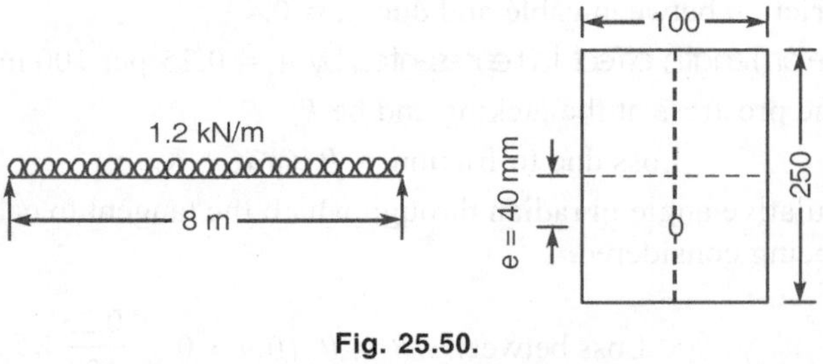

Fig. 25.50.

$$\text{Dead load of the beam} = 0.1 \times 0.250 \times 1 \times 25000 \text{ N/meter} = 625 \text{ N/meter} = 0.625 \text{ KN/meter}$$

$$\text{Live load on the beam} = 1.2 \text{ KN/meter}$$

$$\text{Area of Section, } A = 250 \times 100 = 2.5 \times 10^4 \text{ mm}^2$$

$$\text{Sectional modulus } Z = Z_t = Z_b = \frac{bd^2}{6} = \frac{100 \times 250^2}{6}$$

$= 104 \times 10^4\ mm^3$

$$\text{Dead load moment } (M_D) = \frac{0.625 \times 8^2}{8} \text{ KN.m.}$$

$= 5$ KN.m.

$$\text{Live load moment } (M_A) = \frac{1.2 \times 8^2}{8} \text{ KN.m.}$$

$= 9.6$ KN.m.

$$C_{td} = C_{bd} = \frac{M_D}{Z} = \frac{5 \times 10^6}{104 \times 10^4} = 4.8 \text{ N/mm}^2$$

$$C_{ta} = C_{ba} = \frac{M_A}{Z} = \frac{9.6 \times 10^6}{104 \times 10^4} = 9.2 \text{ N/mm}^2$$

$$\text{Direct stress} = \frac{P_o}{A} = \frac{250 \times 10^3}{2.5 \times 10^4} = 10 \text{ N/mm}^2$$

$$\text{Extreme stresses due to eccentricity of cable} = \frac{P_o e}{Z}$$

$$= \frac{250 \times 10^3 \times 40}{104 \times 10^4} = 9.62 \text{ N/mm}^2$$

$$\text{Stress at top most fibre} = C_{td} + C_{ta} - \frac{P_o.e}{Z} + \frac{P_o}{A}$$

$= 4.8 + 9.2 - 9.62 + 10$

$= 14.4$ N/mm² (Compression)

$$\text{Stress at bottom most fibre} = -C_{bd} - C_{ba} + \frac{P_o.e}{Z} + \frac{P_o}{A}$$

$= -4.8 - 9.2 + 9.62 + 10$

$= 5.6$ N/mm² (Compression)

(*b*) To balance the stresses due to dead and live loads at the bottom fibre means that stress at bottom fibre is zero.

Let the force required $= P_o$

Stress at bottom fibre:

$$-C_{bd} - C_{ba} + \frac{P_o.e}{Z} + \frac{P_o}{A} = 0$$

$$-4.8 - 9.2 + \frac{P_o.40}{104 \times 10^4} + \frac{P_o}{2.5 \times 10^4} = 0$$

$$P_o\left(\frac{40}{104} \times 10^{-4} + \frac{1 \times 10^{-4}}{2.5}\right) = 14$$

$P_o = 178431$ N $= 178.431$ KN.

Example 25.6. A prestressed concrete beam is to be designed to support a superimposed load of 4.4 KN/meter over a span of 5 meter. The section of T beam is shown in the Fig. 25.51.

Permissible stress in concrete (compression) = 15 N/mm²

Permissible stress in tension = 0 N/mm²

Take $\eta = 0.80$

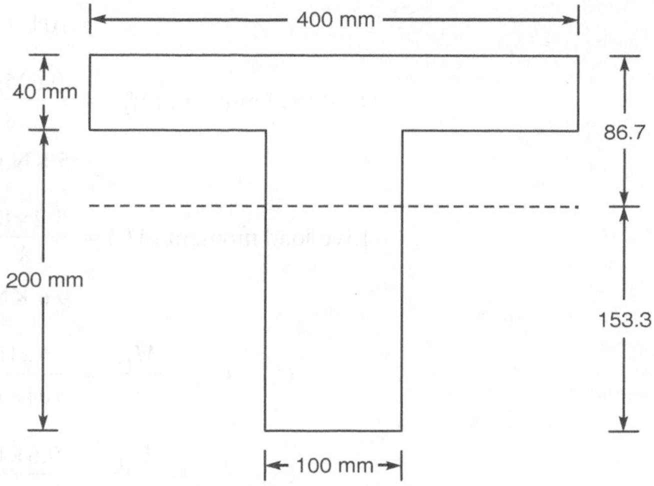

Fig. 25.51.

Solution.

Area of the section

$= 400 \times 40 + 200 \times 100$

$= 16000 + 20000$

$= 36000 \text{ mm}^2$

$$\text{Dead load of beam} = 36000 \times 10^{-6} \times 1 \times 24000$$
$$= 864 \text{ N/meter}$$

$$\text{Depth of centroidal axis from top fibre} = \frac{400 \times 40 \times 20 + 100 \times 200 \times 140}{36000} \text{ mm}$$
$$= 86.7 \text{ mm}$$

Moment of inertia about centroidal axis,

$$I_{xx} = 400 \times 40 \times (66.7)^2 + \frac{100 \times 200^3}{12} + (200 \times 100)(153.3 - 100)^2$$
$$= 1.947 \times 10^8 \text{ mm}^4$$

$$Z_t = \frac{I_{xx}}{Y_t} = \frac{1.947 \times 10^8}{86.7} = 2246110 \text{ mm}^3$$

$$Z_b = \frac{1.947 \times 10^8}{153.3} = 1270305 \text{ mm}^3$$

$$\text{Bending moment due to dead load } M_D = \frac{864 \times 5^2}{8} \text{ N.meter}$$
$$= 2700 \text{ N.meter}$$
$$= 2.7 \times 10^6 \text{ N.mm}$$

Bending moment due to live load

$$M_A = \frac{4400 \times 5^2}{8} \text{ N.meter}$$
$$= 13750 \text{ N.m.}$$
$$= 13.750 \times 10^6 \text{ N.mm}$$

$$C_{dt} = \frac{M_D}{Z_t} = \frac{2.7\times10^6}{2246110} = 1.20 \text{ N/mm}^2$$

$$C_{db} = \frac{M_D}{Z_b} = \frac{2.7\times10^6}{1270305} = 2.13 \text{ N/mm}^2$$

$$C_{at} = \frac{M_A}{Z_t} = \frac{13.75\times10^6}{2246110} = 6.12 \text{ N/mm}^2$$

$$C_{ab} = \frac{M_A}{Z_b} = \frac{13.75\times10^6}{1270305} = 10.82 \text{ N/mm}^2$$

$$\text{Direct stress} = \frac{P_o}{A} = \frac{P_o}{36000} \text{ N/mm}^2$$

Let the eccentricity = e below centroidal axis.

Top Initial stress

$$C_{dt} + \frac{P_o}{A} - \frac{P_o.e}{Z_t} = 0$$

$$1.2 + \frac{P_o}{36000} - \frac{P_o.e}{2246110} = 0$$

$$P_o\left(\frac{1}{36000} - \frac{e}{2246110}\right) = -1.2 \quad ...(1)$$

Top final stress

$$C_{dt} + C_{at} + \frac{\eta P_o}{36000} - \frac{\eta P_o.e}{Z_t} = 15$$

$$1.2 + 6.12 + \frac{0.8P_o}{36000} - \frac{0.8P_o.e}{2246110} = 15$$

$$\frac{0.8P_o}{36000} - \frac{0.8P_o.e}{2246110} = 15 \quad ...(2)$$

Bottom initial stress

$$-C_{db} + \frac{P_o}{A} + \frac{P_o.e}{Z_b} = 15$$

$$-2.13 + \frac{P_o}{36000} - \frac{P_o.e}{1270305} = 15$$

$$\frac{P_o}{36000} + \frac{P_o.e}{1270305} = 17.13 \quad ...(3)$$

Bottom final Stress

$$-C_{db} - C_{ab} + \frac{P_o}{A} + \frac{P_o.e}{Z_b} = 0$$

$$P_o\left(\frac{1}{36000}+\frac{e}{1270305}\right) = 2.13 + 10.82 = 12.95 \qquad ...(4)$$

(1)/(4)

$$\frac{\dfrac{1}{36000}-\dfrac{e}{2246110}}{\dfrac{1}{36000}+\dfrac{e}{1270305}} = -\frac{1.2}{12.95}$$

$$-2.6 \times 10^{-6} - 7.3 \times 10^{-8}\,e = 27.8 \times 10^{-6} - 44.52 \times 10^{-8}\,e$$

$$(44.52 - 7.3) \times 10^{-8}\,e = (27.8 + 2.6) \times 10^{-6}$$

$$e = 81.72 \text{ mm}$$

$$P_o = 139596 \text{ N}$$

Check of Stress

Top final, $1.2 + 6.12 + \dfrac{0.8\times139596}{36000} - \dfrac{0.8\times139596\times81.72}{2246110} = 3.56 \text{ N/mm}^2 < 15 \text{ N/mm}^2$

Bottom initial, $-2.13 + \dfrac{139596}{36000}+\dfrac{139596\times81.72}{1270305} = 10.72 \text{ N/mm}^2 < 15 \text{ N/mm}^2$

Therefore, Safe.

Example 25.7. A prestressed concrete beam 120 mm wide and 300 mm deep is prestressed by a cable which has an eccentricity of 100 mm at centre of the span of beam 6 m long. If the beam supports two point loads of 10 KN each at one third span points, determine the magnitude of the prestressing force in the cable for load balancing for the following cases.

(*a*) Considering live load but neglecting the self weight of the beam.

(*b*) Considering both self-weight of beam and live loads.

Unit weight of concrete = 24000 N/m^3

Solution. (Fig. 25.52)

(*a*) Let the prestressing force be P.

For load balancing, $P \sin\theta = 10 \times 10^3$

or, $\sin\theta \approx \tan\theta = \dfrac{100}{2000} = \dfrac{1}{20}$

Hence, $P.\dfrac{1}{20} = 10 \times 10^3$

or $P = 20 \times 10 \times 10^3 = 2 \times 10^5$ N

$= 200$ KN

(*b*) Self weight of beam $= 0.120 \times 0.300 \times 24000$

$= 864$ N/meter

B.M. at middle third point due to dead load as well as live load

$$= (10 \times 10^3 \times 2) + \left(\frac{864\times6}{2}\times2-864\times2\times1\right)$$

$$= 2 \times 10^4 + 3456$$

$$= 23456 \text{ N.meter}$$

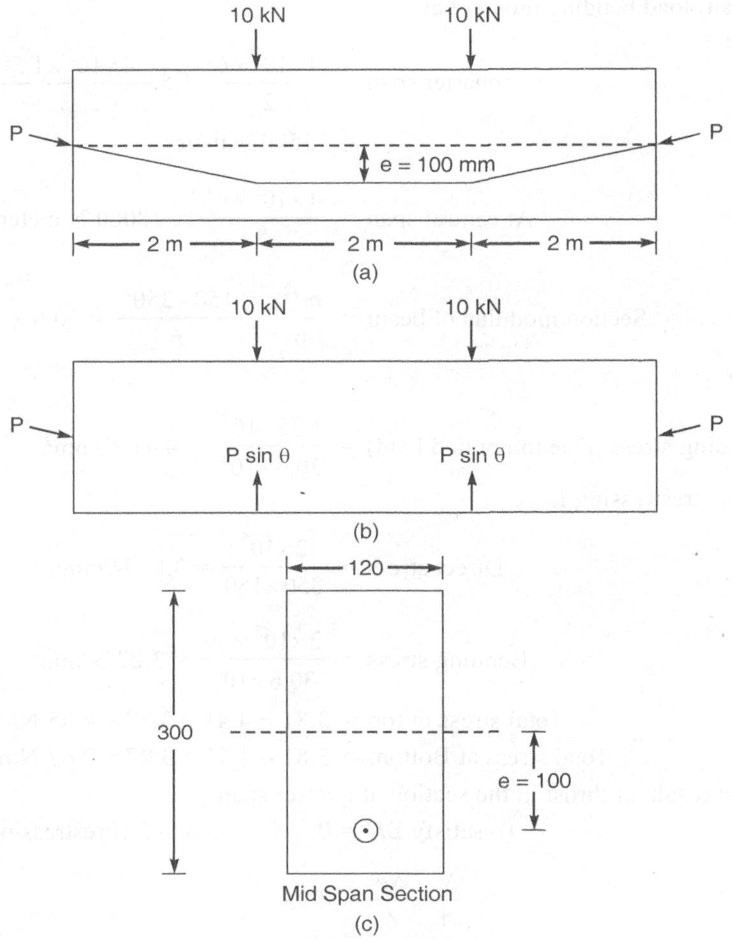

Fig. 25.52.

For load balancing

$$P.e. = 23456 \times 10^3$$

$$P = \frac{23456 \times 10^3}{100} = 234560 \text{ N.} = 234.56 \text{ KN}$$

Example 25.8. A prestressed beam with width of section 150 mm and depth of section as 350 mm is used over a span of 6 meter and supports a uniformly distributed load of 4 KN/meter including self-weight of beam. Beam is prestressed by a cable carrying a force 200 KN and at an eccentricity of 50 mm. Determine the location of thrust line giving position at quarter and central span sections.

Solution.

Prestressing force,	P = 200 KN
Eccentricity	e = 50 mm
(Live + Dead) load on beam	= 4 KN/meter
Span of beam	= 6 meter

(Live + Dead) load bending moment at

$$\text{quarter span} = \frac{4\times10^3\times6}{2}\times1.5-\frac{4\times10^3\times1.5^2}{2}$$

$$= 13500 \text{ N.meter}$$

$$\text{At central span} = \frac{4\times10^3\times6^2}{8} = 18000 \text{ N.meter.}$$

$$\text{Section modulus of beam} = \frac{bd^2}{6} = \frac{150\times350^2}{6} = 30.6\times10^5 \text{ mm}^3$$

At Quarter Span

$$\text{Bending stress (Due to applied load)} = \frac{1.35\times10^7}{30.6\times10^5} = 4.41 \text{ N/mm}^2$$

Stress due to prestressing force

$$\text{Direct stress} = \frac{2\times10^5}{350\times150} = 3.81 \text{ N/mm}^2$$

$$\text{Bending stress} = \frac{2\times10^5\times50}{30.6\times10^5} = 3.27 \text{ N/mm}^2$$

$$\text{Total stress at top} = 3.81 + 4.41 - 3.27 = 4.95 \text{ N/mm}^2 \text{ (Comp)}$$

$$\text{Total stress at Bottom} = 3.81 - 4.41 + 3.27 = 2.67 \text{ N/mm}^2 \text{ (Comp)}$$

Let N be the resultant thrust in the section at quarter span

To satisfy $\Sigma H = 0$, $N = P$ (Prestressing force)

$$\frac{N}{A}+\frac{Ne}{Z} = 4.95$$

$$2\times10^5\left(\frac{1}{150\times350}+\frac{e}{30.6\times10^5}\right) = 4.95$$

e = 17.5 mm above central line of cross section.

Other Way to Get Location

Shift to pressure line from cable line

$$= \frac{M}{P} = \frac{\text{Moment due to } (DL+LL)}{\text{Prestressing force}}$$

$$= \frac{1.35\times10^7}{2\times10^5} \text{ mm} = 67.5 \text{ mm}$$

Therefore $e = 67.5 - 50 = 17.5$ mm.

At Central Span

Shfit of pressure line above the cable line

$$= \frac{M}{P} = \frac{1.8\times10^7}{2\times10^5} = 90 \text{ mm}$$

Therefore $e = 90 - 50 = 40$ mm

Other Way

Bending stress due to $(DL + LL)$

$$= \frac{1.8\times10^7}{30.6\times10^5} = 5.9 \text{ N/mm}^2$$

$$\text{Total stress at top} = 5.9 + 3.81 - 3.27 = 6.44 \text{ N/mm}^2 \text{ Compression}$$

$$\text{Total stress at bottom} = -5.9 + 3.81 + 3.27 = 1.18 \text{ N/mm}^2 \text{ Compression}$$

Let the resultant thrust $N (= P)$ act at e above central axis.

$$\frac{N}{A} + \frac{N.e}{Z} = 6.44$$

$$2 \times 10^5 \left(\frac{1}{350\times150} + \frac{e}{30.6\times10^5}\right) = 6.44$$

$$e = 40 \text{ mm}$$

Example 25.9. A concrete beam of rectangular section is shown in Fig. 25.53. The span of beam is 10 meter. The beam is subjected to uniformly distributed load of 4 KN/meter. Find the effective force necessary in the cable for zero shear stress in the beam. The cable is provided in parabolic shape with eccentricity of 100 mm at centre and zero at ends.

Solution. (Fig. 25.53)

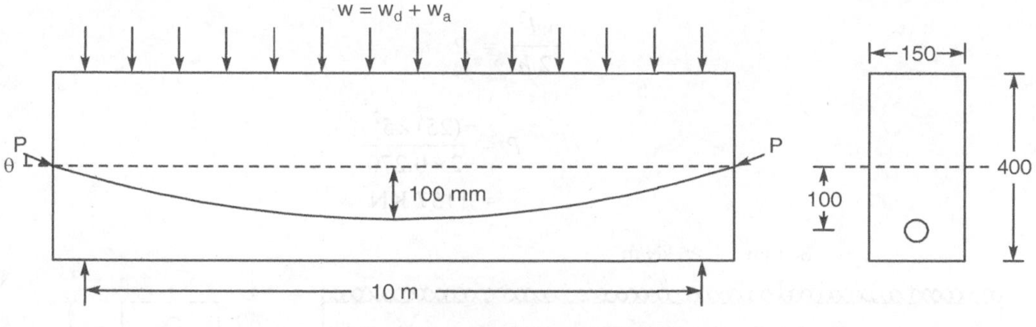

Fig. 25.53.

$$\text{Area of the section} = 150 \times 400 = 6 \times 10^4 \text{ mm}^2$$

$$\text{Dead load } w_d = 0.15 \times 0.40 \times 1 \times 24000 = 1.44 \text{ KN/meter.}$$

$$\text{Live load } w_a = 4.0 \text{ KN/meter.}$$

$$\text{Total loading on beam} = 1.44 + 4.00 = 5.44 \text{ KN/meter.}$$

To obtain zero shear stress, the uniformly distributed loading must be completely balanced by upward force transmitted by tensioned tendon.

Let the tension required in cable be P.

$$P = \frac{w_b . l^2}{8h}$$

where $w_b \rightarrow$ Upward transmitted force by tendon tensioned to force P.

For Load Balancing

$$w_b = w_d + w_a$$

$$= 5.44 \text{ KN/meter}$$

$$P = \frac{5.44 \times 10^3 \times 10^2}{8 \times 0.100} = 680,\ 000 \text{ N}$$

$$= 680 \text{ KN.}$$

Now, the stress across the section $= \dfrac{P \cos \theta}{A}$ (Since θ is very small hence $\cos \theta \simeq 1$)

$$= \frac{p}{A} = \frac{680 \times 10^3}{150 \times 400}$$

$$= 11.3 \text{ N/mm}^2$$

Example 25.10. A double cantilever beam is having a uniformly distributed load of 25 KN/meter (including self weight). Find out the required prestressing force to balance the loading. The concrete cover needed is equal to 80 mm.

Solution.

Dead load of beam = 0.7 × 0.3 × 24000 = 5040 N/m

To balance the uniformly distributed load the *cgs* should be tangential to *cgc* at the ends.

To obtain least amount of prestressing force, the maximum available eccentricity should be used.

Hence available, $h = 350 - 80 = 270$ mm

$$\frac{wl^2}{2\,h} = P$$

or,

$$P = \frac{(25) \times 5^2}{2 \times 0.270}$$

$$= 1157 \text{ KN}$$

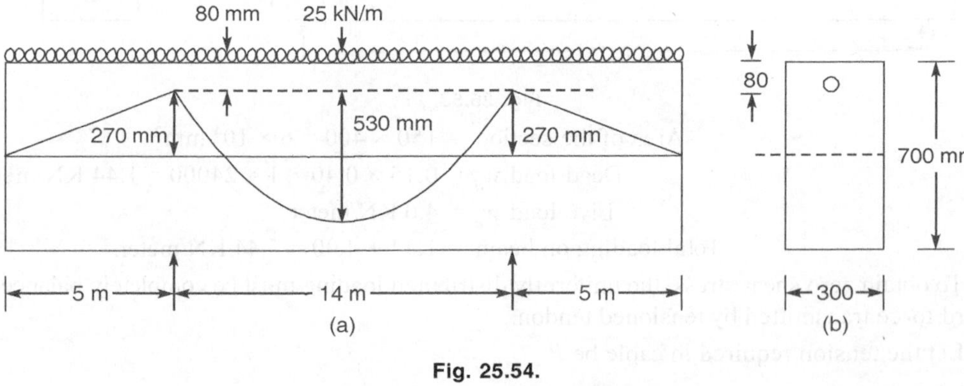

Fig. 25.54.

For this value of prestressing force the sag of parabola at mid span can be obtained.

$$h = \frac{wl^2}{8\,P}$$

$$= \frac{25 \times 14^2}{8 \times 1157} = 0.529 \text{ meter} = 530 \text{ mm.}$$

Depth of section required = 80 + 530 + 80 = 690 mm < 700 mm (provided).

Example 25.11. A prestressing force of 250 KN is transmitted through a distribution plate 120 mm wide and 120 mm deep, the centre of which is located at 100 mm from the bottom of an end block having a section 120 mm wide and 300 mm deep. Evaluate the position and maximum tensile stress on horizontal section passing through the centre of distribution plate using the following methods:

1. Magnel's method.
2. Guyon's method.
3. Zielinski & Rowe's method.

Also find out the area of steel necessary to resist the largest tensile force resulting from any of these methods.

Yield stress of steel = 260 N/mm^2

Solution.

Magnel's Method (Fig. 25.55)

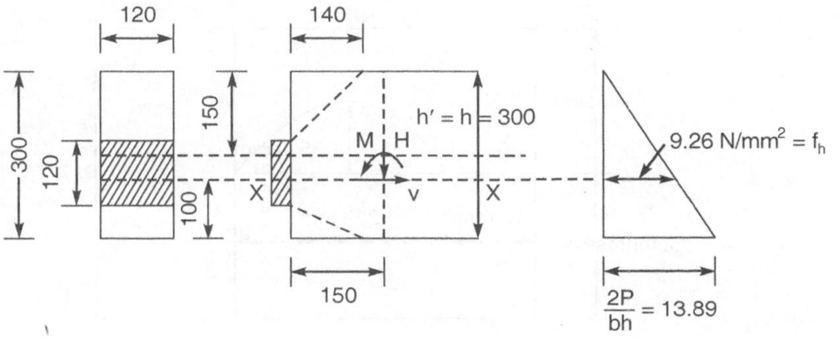

Fig. 25.55.

Direct Stress Distribution

$$\text{Direct stress} = \frac{250\times10^3}{120\times300}\left(1\pm\frac{6\times50}{300}\right)$$

$$= \frac{250\times10^3}{120\times300}(1\pm1)$$

At top, stress = 0

$$\text{At bottom, stress} = \frac{2\times250\times10^3}{120\times300} = 13.89 \text{ N/mm}^2$$

Stresses on X–X

Bending moment, assuming half prestress acting on upper half of distribution plate

$$M = (1/2 \times 9.26 \times 200 \times 120) \times (1/3 \times 200) - \left(\frac{250}{2}\times10^3\times30\right)$$

$$= 3658 \times 10^3 \text{ N.mm}$$

Shear force (horizontal)

$$V = 1/2 \times 9.26 \times 200 \times 120 - \frac{250\times10^3}{2} = -13880 \text{ N}$$

Direct force (vertical)

$$H = 0$$

At $x = 0.5h = 0.5 \times 300 = 150$ mm from end face

$f_h = +9.26$ N/mm²

Vertical tensile stress on horizontal plane is given by

$$f_v = K_1\left(\frac{M}{bh^2}\right) + K_2\,(H/bh)$$

$$= (-5)\left(\frac{3658 \times 10^3}{120 \times 300^2}\right) + 0$$

$$= -1.69 \text{ N/mm}^2$$

Sheer stress, $\tau_v = K_3\left(\frac{V}{bh}\right) = 1.25\left(-\frac{13880}{120 \times 300}\right) = -0.482 \text{ N/mm}^2$

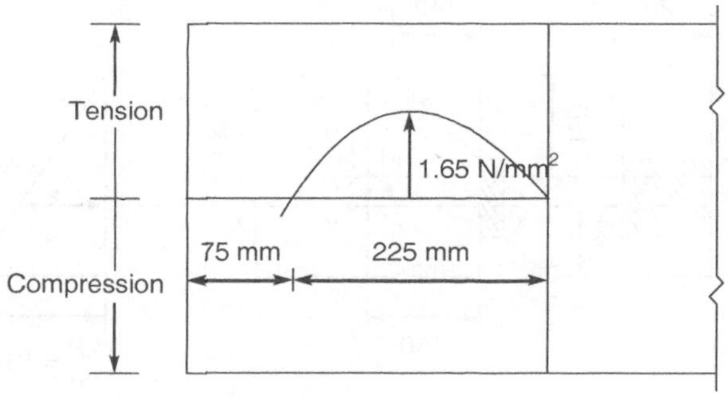

Fig. 25.56.

Principal Tensile Stress

$$f_{\substack{\max\\ \min}} = \frac{f_v + f_h}{2} \pm \tfrac{1}{2}\sqrt{(f_h - f_v)^2 + 4\tau^2}$$

$$f_{\max} = \left(\frac{9.26 - 1.69}{2}\right) + \tfrac{1}{2}\sqrt{(9.26 + 1.69)^2 + 4(-0.482)^2}$$

$$= 3.785 + 5.435 = 9.22 \text{ N/mm}^2 \text{ (Comp)}$$

$$f_{\min} = \frac{9.26 - 1.69}{2} - \tfrac{1}{2}\sqrt{(9.26 + 1.69)^2 + 4(-0.482)^2}$$

$$= 3.785 - 5.435 = -1.65 \text{ N/mm}^2$$

Total splitting tension, $P_{bst} = 0.3P\left[1 - \left(\frac{Y_{p_0}}{Y_0}\right)^{0.58}\right]$

$$= 0.3 \times 250 \times 10^3\,[1 - (0.6)^{0.58}]$$

$$= 19231.5 \text{ N}$$

however using the approximate formula for area of stress diagram.

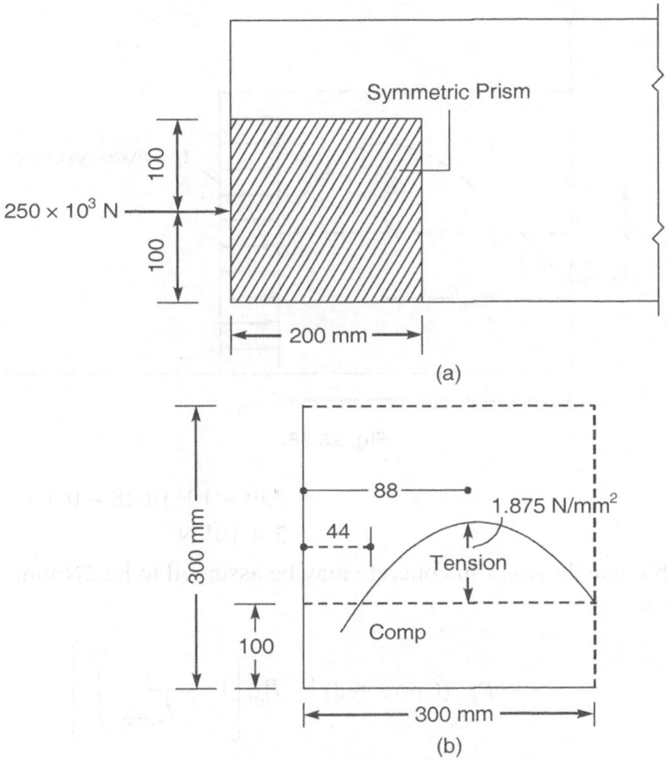

Fig. 25.57.

$$\text{Bursting tension} = \left(\frac{2}{3} \times 256 \times 1.875\right) \times 120 = 38400 \text{ N}$$

Zielinski and Rowe's Method

$$\text{Applied jacking force } T = 250 \text{ KN}$$

$$\text{Side of loaded area } 2Y_{p_o} = 120 \text{ mm}$$

$$\text{Side of surrounding prism } 2Y_0 = 120 \text{ mm}$$

$$\text{Distribution Ratio} = \frac{Y_{p_o}}{Y_0} = 1.0$$

$$\text{Average compressive stress } f_c = \frac{250 \times 10^3}{120 \times 120} = 17.36 \text{ N/mm}^2$$

$$\text{Maximum transverse tensile stress, Eqn. 25.65, } f_{v,\max} = f_c\left[0.98 - 0.825\frac{(Y_{p_o})}{Y_0}\right]$$

$$= 17.36\,[0.98 - 0.825 \times 1.0]$$

$$= 2.69 \text{ N/mm}^2$$

Acting at a distance = 0.5 × 60 = 30 mm from the end face

Bursting tension without allowing any tension, Eqn. 25.66,

$$P_{bst} = T\left[0.48 - 0.4\frac{(Y_{p_o})}{Y_o}\right]$$

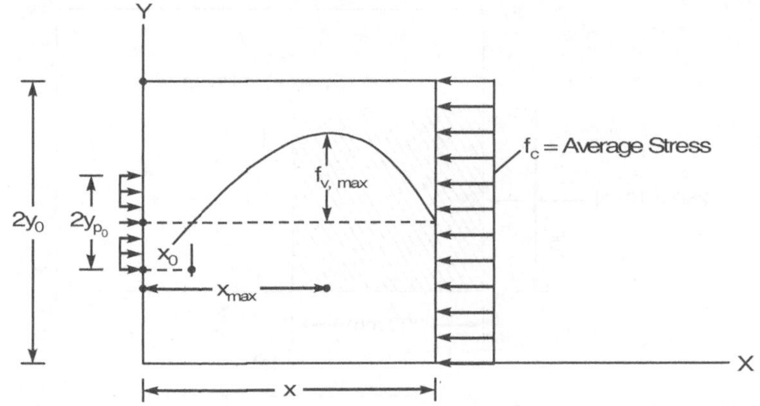

Fig. 25.58.

$$= 250 \times 10^3\ [0.48 - 0.4 \times 1.0]$$
$$= 2 \times 10^4 \text{ N}$$

The permissible tensile stress in concrete may be assumed to be 2N/mm²

From Eqn. 25.67

$$P_{bst}\ (\text{Corrected}) = P_{bst}\left[1 - \left(\frac{f}{f_{v,max}}\right)^2\right]$$

$$= 2 \times 10^4 \left[1 - \left(\frac{2}{2.69}\right)^2\right]$$

$$= 8944 \text{ N}$$

From all the above three methods maximum bursting tension obtained is 38400 N. Hence

Area of steel $A_{st} = \dfrac{38400}{260} = 147.7 \text{ mm}^2$

Comparative Analysis of Magnel, Guyon and Zielinski and Rowe's Method

A comparative analysis of salient features of some of the important investigation based on two dimentional approach in the Table form is given below.

Authors	$\frac{x}{Y_o}$	$\frac{y_{po}}{Y_o}$	$\frac{x_o}{Y_o}$	$\frac{x_{max}}{Y_o}$	$\frac{f_{v,max}}{f_c}$
Magnel	1.0	0.3 – 0.7	0.5	1.0	0.43–0.42
Guyon	1.0	0.1–0.9	0.2–0.45	0.5–0.9	0.42–0.06
Rowe	1.0	0.1–0.4	–	–	–

The above difference is attributed to the simplifying assumption made by different persons investigating for solution of two dimensional elasticity problem.

In view of widely varying results and due to the heterogenous nature of concrete, it is advisable to adopt the worst distribution of transverse tensile stress for design of end block reinforcement.

25.28 CABLE PROFILES

The layout of simple beams is controlled by the maximum moment at mid section and end sections so that, after these two sections are designed, other sections can often be determined by inspection. It sometimes happens, however, that intermediate points along the beam may also be critical, and in many instances it would be desirable to determine the permissible and desirable profile for the cables. The do this, a limiting zone for the location of centre of gravity of steel is first obtained, then the cables are arranged so that their centroid will lie within the zone.

The method described here is intended for simple beams, but it also serves as an introduction to the solution of more complicated layouts, such as cantilever and continuous spans, where cable location cannot be easily determined by inspection. The method is a graphical one; giving the limiting zone within which the centre of gravity of steel must pass in order that no tensile stresses will be produced. Compressive stresses in concrete are not checked by this method. It is assumed that the layout of the concrete sections and the area of prestressing steel have already been determined. Only the profile of the centre of gravity of steel is to be located.

Referring to Fig. 25.59, having determined the layout of concrete sections, we proceed to compute their kern points, thus yielding two kern lines, one top and one bottom Fig. 25.59(*c*). For variables sections, these kern lines would be curved.

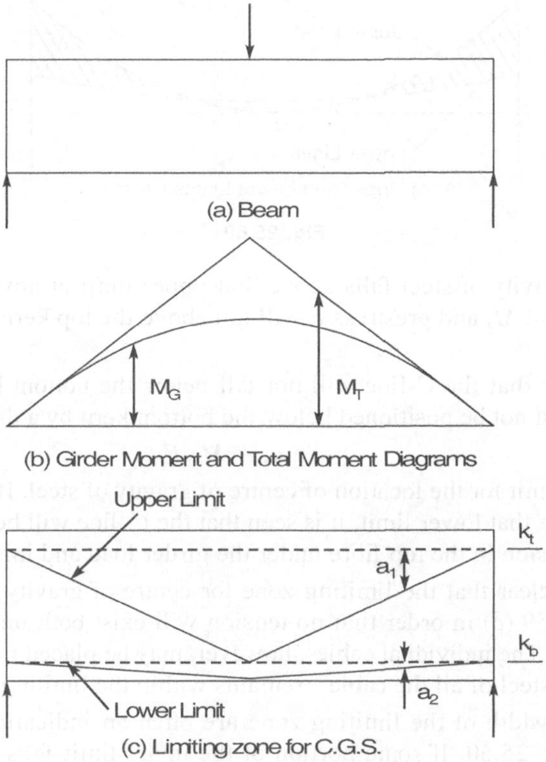

Fig. 25.59.

For a beam loaded as shown in (*a*), the minimum and maximum moment diagrams for the girder load and for the total working load respectively are marked as M_G and M_T in (*b*). In order that, under the working load, the centre of pressure *i.e.*, the *C*-line, will not fall above the top kern line, it is evident that the centre of gravity of steel must be located below the top kern at least a distance, $a_1 = M_T/F$.

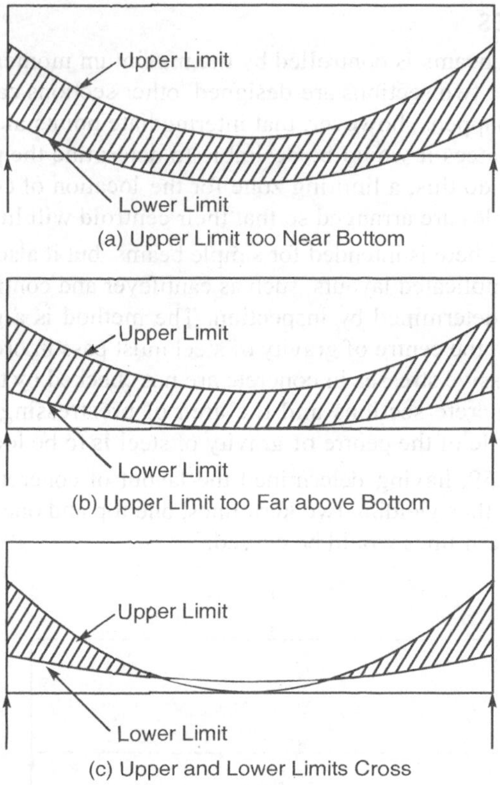

Fig. 25.60

If the centre of gravity of steel falls above that upper limit at any point, then the *C*–line corresponding to moment M_T and prestress *F* will fall above the top kern, resulting in tension in the bottom fibre.

Similarly, in order that the C-line will not fall below the bottom kern line, the centre of gravity of steel line must not be positioned below the bottom kern by a distance greater than

$$a_2 = M_G/F_O$$

which gives the lower limit for the location of centre of gravity of steel. If the centre of gravity of steel is positioned above that lower limit, it is seen that the *C*–line will be above the bottom kern and there will be no tension in the top fibre under the girder load and initial prestress F_O.

Thus, it becomes clear that the limiting zone for centre of gravity of steel is given by the shaded area in Fig. 25.59 (*c*) in order that no tension will exist both under the girder load and under the working load. The individual cables, however, may be placed in any position so long as the centre of gravity of steel of all the cables, remains within the limiting zone.

The position and width of the limiting zone are often an indication of the adequacy and economy of design, Fig. 25.60. If some portion of the upper limit falls outside or too near the bottom fibre, in (*a*), either the prestress *F* or the depth of beam at that portion should be increased. On the other hand, if it falls too far above the bottom fibre, in (*b*), either the prestress or the beam depth can be reduced. If the lower limit crosses the upper limit, in (*c*), it means that no zone is available for the location of cente of gravity of steel, and either the prestress *F* or the beam depth must be increased or the girder moment must be increased to depress the lower limit if that can be done. On the other hand, the case shown in Fig. 25.59 (*c*) may be very satisfactory when we are allowing tensile stress in concrete.

25.29 LIMIT STATE METHOD OF DESIGN

25.29.1 Introduction

The structural design of prestressed concrete members may be based on limit state concepts. In this method of design, the structure shall be designed to withstand safely all loads liable to act on it throughout its life; it shall also satisfy the serviceability requirements, such as limitations on deflection and cracking. The acceptable limit for the safety and serviceability requirements before failure occurs is called a 'Limit State'. The aim of design is to achieve acceptable probabilities that the structure will not become unfit for the use for which it is intended, that is, that it will not reach a limit state. All relevant limit states shall be considered in design to ensure an adequate degree of safety and serviceability. In general, the structure shall be designed on the basis of the most critical limit state and shall be checked for other limit states. For ensuring the specified objective, the design should be based on characteristic values for material strengths and applied loads, which take into account the variation in the material strengths and in the loads to be supported. The characteristic values should be based on statistical data if available; where such data are not available, they should be based on experiences. The 'design values' are derived from the characteristic values through the use of partial safety factors, one for material strengths and the other for loads. In the absence of special considerations, these factors should have the values as explained in Chapter 17.

25.29.2 Limit State of Collapse

(*A*) *Flexure*

(*i*) *Assumptions*: Design for the limit state of collapse in flexure shall be based on the assumptions given below:

(*a*) Plane sections normal to the axis remain plane after bending.

(*b*) The maximum strain in concrete at the outermost compression fibre is taken as 0.0035 in bending.

(*c*) The relationship between the compressive stress distribution in concrete and the strain in concrete may be assumed to be rectangle, trapezoid, parabola or any other shape which results in prediction of strength in substantial agreement with the results of tests. For design purposes, the compressive strength of concrete in the structure shall be assumed to be 0.67 times the characteristic strength. The partial safety factor r_m = 1.5 shall be applied in addition to this.

(*d*) The tensile strength of the concrete is ignored.

(*e*) The stress in bonded prestressing tendons, whether initially tensioned or untensioned, and in additional reinforcement are derived from the representative stress-strain curve for the type of steel used given by the manufacture or typical curves given in Fig. 25.61 for prestressing tendons. For design purposes, the partial safety factor r_m equal to 1.15 shall be applied.

(*f*) If tendons are unbonded in post-tensioned members, the stress in the tendons may be obtained from a rigorous analysis or from tests.

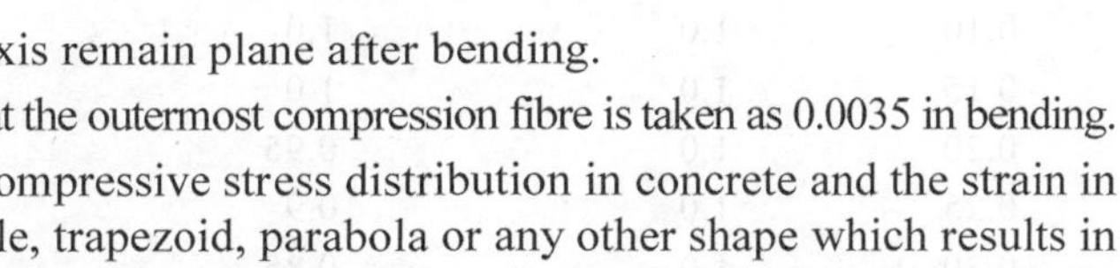

Fig. 25.61. Representative Stress-Strain Curve for Prestressing Steels

(ii) Moment of Resistance for Rectangular and T-Sections: The moment of resistance of rectangular and *T* sections in which neutral axis lies within the flange may be obtained as follows:

$$M = f_{pu} \cdot A_p \,(d - 0.42\, x_u) \qquad ...(25.71)$$

where M = moment of resistance of section

f_{pu} = ultimate tensile stres in the tendons

A_p = area of pretensioning tendons

d = effective depth and

x_u = neutral axis depth

Table 25.6. Conditions at Ultimate Limit State for Rectangular Beams with Pretensioned Tendons having Effective Bond

$\dfrac{A_p \cdot f_p}{b.d.f_{ck}}$	Stress in Tendon as proportion of the design Strength $\dfrac{f_m}{0.87 f_y}$		Ratio of the Depth of Neutral Axis to that of the centroid of the tendon in the tension zone x_u/d	
	Pre-Tensioning	Post-tensioning with effective bond	Pre-tensioning	Post-tensioning with effective bond
(1)	(2)	(3)	(4)	(5)
0.025	1.0	1.0	0.054	0.054
0.05	1.0	1.0	0.109	0.109
0.10	1.0	1.0	0.217	0.217
0.15	1.0	1.0	0.326	0.316
0.20	1.0	0.95	0.435	0.414
0.25	1.0	0.9	0.542	0.488
0.30	1.0	0.85	0.655	0.558
0.40	0.9	0.75	0.783	0.653

Table 25.7. Conditions at Ultimate Limit State for Post-Tensioned Rectangular Beams having Unbonded Tendons

$\dfrac{A_p \cdot f_p}{b.d.f_{ck}}$	Stress in Tendons as a Proportion of the Eff. Prestress f_m/f_p for values of l/d. $\dfrac{\text{Effective span}}{\text{Effective Depth}}$			Ratio of Depth of Neutral Axis to that of the Centroid of the tendons in the tension zone x_u/d for values of l/d $\dfrac{\text{Effective span}}{\text{Effective Depth}}$		
	30	20	10	30	20	10
(1)	(2)	(3)	(4)	(5)	(6)	(7)
0.025	1.23	1.34	1.45	0.10	0.10	0.10
0.05	1.21	1.32	1.45	0.16	0.16	0.18
0.10	1.18	1.26	1.45	0.30	0.32	0.36
0.15	1.14	1.20	1.36	0.44	0.46	0.52
0.20	1.11	1.16	1.27	0.56	0.58	0.64

For pretensioned members and for post tensioned members with effective bond between the concrete and tendons, values of f_{pu} and x_u are given in Table 25.6. The effective prestress after all losses should not be less than 0.45 f_p, where f_p is the characteristic stregth of prestressing steel.

For post-tensioned members with unbonded tendons the values of f_{pu} and x_u are given in Table 25.7.

For flanged sections in which the neutral axis lies outside the flange, the moment of resistance shall be worked out as per assumptions made in article 25.29.2.*a*(*i*).

(B) Compression

Prestressed concrete compression members in framed structures, where the mean stress in the concrete section imposed by tendons is less than 2.5 N/mm², may be analysed as reinforced concrete compression members in accordance with general theory of reinforced concrete.

(C) Tension

Tensile strength of tension members shall be based on the design strength (0.87 times the characteristic strength of prestressing tendons) and the strength developed by any additional reinforcement. The additional reinforcement may usually be assumed to be acting at its design stress (0.87 times the characteristic strength of reinforcement); in special cases it may be necessary to check the stress in the reinforcement using strain compatability.

(D) Shear

The ultimate shear resistance of the concrete alone, V_c, should be considered at both sections uncracked and cracked in flexure, the less value taken and, if necessary, shear reinforcement provided.

(*i*) *Sections Uncracked in Flexure*: The ultimate shear resistance of a section uncracked in flexure, $V_c = V_{co}$ is given by

$$V_{co} = 0.67\, b.D \sqrt{f_t^2 + 0.8 f_{cp} \cdot f_t} \quad \text{...(25.72)}$$

where, b = breadth of the beam which for *T*, *I* and *L* beams should be replaced by breadth of rib b_w,

D = Overall depth of member;

f_t = maximum principal tensile stress given by $0.24\sqrt{f_{ck}}$

taken as positive; and f_{cp} = compressive stress at centroidal axis, taken as positive in flanged members where the centroidal axis occurs in flange, the principal tensile stress should be limited to $0.24\sqrt{f_{ck}}$ at the intersection of the flanged web; in this calculation, 0.8 of the stress due to prestress at this intersection may be used, in calculating V_{co}.

For a section uncracked in flexure and with inclined tendons or vertical prestress, the component of prestressing force normal to the longitudinal axis of the member may be added to V_{co}.

(*ii*) *Sections Cracked in Flexure*: The ultimate shear resistance of section cracked in flexure $V_c = V_{cr}$, is given by

$$V_{cr} = \left(1 - 0.55\frac{f_{pe}}{f_p}\right) \cdot \tau_c \cdot b \cdot d + M_o \cdot \frac{V}{M} \quad \text{...(25.73)}$$

where, f_{pc} = effective prestress after all losses have occured, which shall not be put greater than $0.6 f_p$;

f_p = characteristic strength of prestressing steel;

τ_c = ultimate shear stress capacity of concrete obtained from Table 25.8.

b = breadth of member, which for flanged sections shall be taken as breadth of the web b_w;

d = distance from the extreme compression fibre to the centroid of the tendons at the section considered

M_o = moment necessary to produce zero stress in the concrete at the depth given by $M_o = 0.8\, f_{pt}.\, I/r$

where, f_{pt} is the stress due to prestress only at depth d and distance y from the centroid of the concrete section which has the second moment of area I and

V and M = shear force and bending moment respectively, at the section considered due to ultimate loads

$$V_{cr} \nless 0.1\, b.d\sqrt{f_{ck}}$$

Table 25.8. Design Shear Strength of Concrete τ_c (N/mm²)

$100\dfrac{A_p}{bd}$	Concrete Grade M30	M35	M40 and Above
(1)	(2)	(3)	(4)
0.25	0.37	0.37	0.38
0.50	0.50	0.50	0.51
0.75	0.59	0.59	0.60
1.00	0.66	0.67	0.68
1.25	0.71	0.73	0.74
1.50	0.76	0.78	0.79
1.75	0.80	0.82	0.84
2.00	0.84	0.86	0.88
2.25	0.88	0.90	0.92
2.50	0.91	0.93	0.95
2.75	0.94	0.96	0.98
3.00	0.96	0.99	1.01

Note: A_p is the area of prestressing tendon.

The value V_{cr} calculated at a particular section may be assumed to be constant for a distance equal to $d/2$, measured in the direction of increasing moment, from that particular section. For a section cracked in flexure and with inclined tendons, the component of prestressing force normal to the longitudinal axis of the member should be ignored.

(*iii*) *Shear Reinforcement*: When V, the shear force due to ultimate loads, is less than V_c, the shear force which can be carried by the concrete, minimum shear reinforcement should be provided in the form of stirrups such that :

$$\frac{A_{sv}}{bS_{sv}} = \frac{0.4}{0.87\, f_y} \qquad ...(25.74)$$

where,

A_{sv} = total cross-sectional area of stirrup legs effective in shear:

b = breadth of the member which for T, I and L beams should be taken as the breadth of the rib, b_w

S_v = stirrup spacing along the length of the member; and

f_y = characteristic strength of the stirrup reinforcement which shall not be taken greater than 415 N/mm².

However, shear reinforcement need not be provided in the following cases:

(*a*) where V is less than 0.5 V_c; and

(*b*) in members of minor importance.

When the shear force V, due to the ultimate loads, exceeds V_c, the shear force which can be carried by concrete, shear reinforcement shall be provided such that

$$\frac{A_{sv}}{S_v} = \frac{V - V_c}{0.87 f_y d_t} \qquad ...(25.75)$$

In rectangular beams, at both corners in the tensile zone, a stirrup should pass around a longitudinal bar, a tendon or a group of tendons having a diameter not less than the diameter of the stirrup. The depth d_t is then taken as the depth from the extreme compression fibre either to the longitudinal bars or to the centroid of the tendons whichever is greater.

The spacing of stirrups along a member should not exceed 0.75 d_t nor 4 times the web thickness for flanged members. When V exceeds 1.8 V_c, the maximum spacing should be reduced to 0.5 d_t. The lateral spacing of the individual legs of the stirrups provided at a cross-section should not exceed 0.75 d_t.

Maximum Shear Force – In no circumstances should the shear force V, due to ultimate loads, exceed the appropriate values given in Table 25.9 multiplied by bd.

(*E*) *Torsion*

In general, where the torsional resistance or stiffness of members has not been taken into account in the analysis of the structure, no specific calculations for torsion will be necessary; adequate control in torsional cracking being provided by the required nominal shear reinforcement. Where the torsional resistance or stiffness of members is taken into account in the analysis, the members shall be designed for torsion.

Application of Design Rules for Torsion – The design rules laid down below apply to :

(*a*) beams of solid rectangular cross-section ($D > b$),

(*b*) hollow rectangular beams with $D > b$ and with a wall thickness $t \geq b/4$

(*c*) T-beams and I-beams

In all these cases the average intensity of prestress in the concrete shall be less than 0.3 f_{ck}.

25.29.3 Longitudinal Reinforcement

The longitudinal reinforcement shall be designed to resist an equivalent ultimate bending moment M_{e1} given by ;

$$M_{e1} = M + M_t \qquad ...(25.76)$$

where,

M = applied ultimate bending moment at the cross-section acting in combination with T;

$M_t = T\sqrt{1 + 2D/b}$, the sign of M_t being the same as that of M;

D = overall depth of the beam; and

b = breadth of the member which for T and I beams shall be taken as the breadth of the web, b_w.

Where the numerical value of M is less than that of M_t, the members shall also be designed to withstand a moment M_{e2} given by

$$M_{e2} = M_t - M \qquad ...(25.77)$$

the moment M_{e2} being taken as acting in the opposite sense to the moment M.

Where the numerical value of M is less than or equal to that of M_t, the beam shall be designed to withstand an equivalent transverse bending moment M_{e3} (not acting simultaneously with M_{e1}), given by

$$M_{e3} = M_t\left(1+\frac{x_1}{2e}\right)^2\left(\frac{1+2b/D}{1+2D/b}\right) \qquad ...(25.78)$$

and acting about an axis at right angles to the axis of M, where x_1 is the smaller dimension of a closed hoop used as torsional shear reinforcement and e is defined in Eqn. 25.79.

Table 25.9. Maximum Shear Stress

Concrete Grade	M 30	M 35	M40	M45	M50	M55 and over
Maximum Shear Stress, N/mm^2	3.5	3.7	4.0	4.3	4.6	4.8

25.29.4 Transverse Reinforcement

(A) Torsional Moment and Shear Carried by Concrete

The reduced torsional moment carried by the concrete, T_{c1} is given by

$$T_{c1} = T_c\left(\frac{e}{e+e_c}\right) \qquad ...(25.79)$$

where,

$$T_c = 1.5b^2D\left(1-\frac{b}{T}\right)\lambda_p\sqrt{f_{ck}}$$

$$e = T/V$$

$$e_c = \frac{T_c}{V_c}$$

$$\lambda_p = \sqrt{1+12f_{cp}/f_{ck}}$$

In the above expression,

T_c = torsional moment carried by concrete,

b = breadth of member, which for T and I beams shall be taken as the breadth of the web, b_w

D = overall depth of beam,

f_{ck} = characteristic compressive strength of concrete.

T = torsional moment applied to a cross-section under ultimate load conditions,

V = shearing force at a cross-section calculated for the specified ultimate loads,

V_c = theoretcial shear strength at cross-section, assuming the most unfavourable conditions for inclined cracking, that is, smaller of V_{co} and V_{cr} and

f_{cp} = average intensity of effective prestress in concrete.

(B) The Shear Force Carried by the Concrete V_{ci} is Given by:

$$V_{ci} = V_c\left(\frac{e}{e+e_c}\right) \quad ...(25.80)$$

where,

V_c = smaller of V_{co} and V_{cr} obtained as in Eqns. 25.72 ad 25.73.

(C) Design of Transverse Reinforcement

The area of cross section, A_{sv} of the closed stirrup enclosing the corner longitudinal bars shall be taken as the larger of the following two values :

$$A_{sv} = \frac{M_t S_v}{1.5 b_1 d_1 f_y} \quad ...(25.81)$$

and $$A_{sv} = A_v + 2A_t$$

where

$$A_v = \frac{(V - V_{cl})\, S_v}{0.87 f_y d_1} \quad \text{and} \quad A_t = \frac{(T - T_{cl})\, S_y}{0.87\, b_1 d_1 f_y}$$

In the above expressions,

M_t is as defined earlier

S_v = spacing of the stirrup reinforcement,

b_1 = centre to centre distance between corner bars in the direction of the width,

d_1 = centre to centre distance between corner bars in the direction of the depth,

f_y = characteristics strength of shear reinforcement,

V, V_{cl} and T and T_{cl} are defined earlier.

(D) Minimum Reinforcement

The value of A_{sv} shall not be taken lesser than given by :

$$\frac{A_{sv}}{b\, S_v} = \frac{0.4}{0.87 f_y} \quad ...(25.82)$$

There shall be at least one longitudinal bar not less than 12 mm in diameter in each corner of the stirrups.

(E) Distribution of Torsion Reinforcement

When a member is designed for torsion, torsion reinforcement shall be provided as below :

(*a*) All transverse reinforcement provided for torsion shall be in the form of closed stirrups perpendicular to the axis of the members.

(*b*) The spacing, S_v, of the stirrups shall not exceed $(x_1 + y_1)/4$ or 200 mm whichever is smaller, where x_1 and y_1 are respectively, short and long dimensions of the stirrup.

(*c*) Each end of the bar forming the stirrup shall be anchored in accordance with as in R.C.C. construction.

(*d*) Torsional reinforcement shall be continued to a distance not less than $(D + b_w)$ beyond the point at which it is no longer theoretically required, where D is the overall depth and b_w is the effective width of the web of a flanged member.

25.29.5 Limit State of Serviceability : Deflection

25.29.5.1 Type 1 and Type 2 Members

(*A*) *Short-Term Deflection*

The instantaneous deflection due to design loads may be calculated using elastic analysis based on the uncracked section and the modulus of elasticity of concrete.

(*B*) *Long-Term Deflection*

The total long-term deflection due to the prestressing force, dead load and any sustained imposed load may be calculated using elastic analysis, taking into account the effects of cracking and of creep and shrinkage. Due allowance shall be made for the loss of prestress after the period considered. The deflections should comply with the following limits:

Limit State of Serviceability

Deflection: The deflection of a structure or part thereof shall not adversely effect the appearance or efficiency of the structure or finishes or partitions. The deflection shall generally be limited to the following :

(*a*) The final deflection due to all loads including the effects of temperature, creep and shrinkage and measured from the cast level of the supports of floors, roofs and all other horizontal members, should not normally exceed span/250.

(*b*) The deflection including the effects of temperature, creep and shrinkage occuring after erection of partitions and the application of finishes should not normally exceed span/350 or 20 mm whichever is less.

(*c*) If finishes are to be applied to prestressed concrete members, the total upward deflection should not exceed span/300, unless uniformity of camber between adjacent units can be ensured.

25.29.5.2 Type 3 Members

Where the permanent load is less than or equal to 25 per cent of the design imposed load, the deflection may be calculated as above in (*A*). When the permanent load is more than 25 per cent of the design imposed load, the vertical deflection limits for beams and slabs may generally be assumed to be satisfied provided that the span to effective depth ratios are not greater than the values obtained as below :

(*a*) Basic values of span to effective depth ratios for spans upto 10 m

Cantilever	7
Simply supported	20
Continuous	26

(*b*) For spans above 10m, the values in (*a*) may be multiplied by 10/span in metres, except for cantilever in which case deflection calculations should be made.

25.29.6 Limit State of Serviceability : Cracking

In members made up of precast units, no tension shall be allowed at any stage at mortar or concrete joints. For member which is free of joints, the tensile stress shall not exceed the values specified below for the 3 types of members.

(*a*) TYPE 1– No tensile stress.

(*b*) TYPE 2– The tensile stress shall not exceed 3 N/mm^2. However, where part of the service loads is temporary in nature, this value may be exceeded by 1.5 N/mm^2, provided under the permanent component of the service load, the stress remains compressive.

(*c*) TYPE 3– For type 3 members in which cracking is permitted, it may be assumed that the concrete section is uncracked and that hypothetical tensile stresses exist at the maximum size of cracks. The hypothetical tensile stresses for use in these calculation for members with either pre-tensioned or post-tensioned tendons are given in Table 25.10 modified by coefficients given in Fig. 25.62.

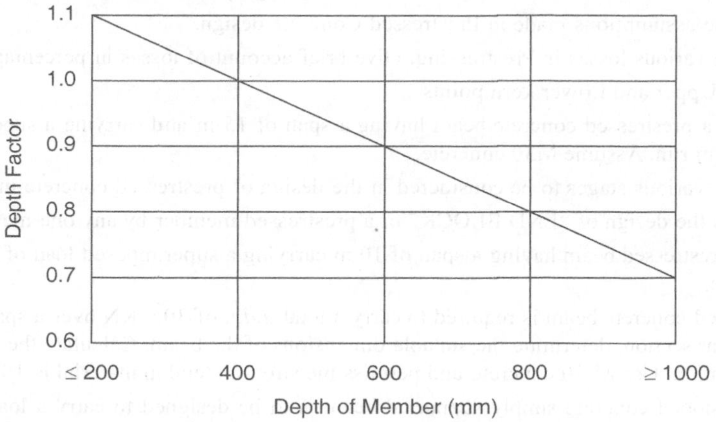

Fig. 25.62. Depth Factor for Tensile Stress for Type-3 Members

Table 25.10. Hypothetical Flexural Tensile Stress for Type-3 Members

Type of tendons	*Limiting crack width mm*	*Stress of Concrete for grade M30*	*M35*	*M40*	*M45*	*M50 and above*
(1)	(2)	(3)	(4)	(5)	(6)	(7)
Pre-tensioned tendons	0.1	–	–	4.1	4.4	4.8
	0.2	–	–	5.0	5.4	5.8
Grouted post-tensioned tendons	0.1	3.2	3.6	4.1	4.4	4.8
	0.2	3.8	4.4	5.0	5.4	5.8
Pre-tensioned tendons distributed in the tensile zone and positioned close to the tension faces of concrete	0.1	–	–	5.3	5.8	6.3
	0.2	–	–	6.3	6.8	7.3

Note: The values in Table 25.10 shall be multiplied by the factors obtained from the figure depending on the depth of the member.

Note: When additional reinforcement is distributed within the tension zone and positioned close to the tension face of concrete, the hypothetical tensile stresses may be increased by an amount which is proportional to the cross-sectional area of the additional reinforcement expressed as a percentage of the cross-sectional area of the concrete. For 1 per cent of additional reinforcement, the stress may be increased by 4N/mm^2 for members with pre-tensioned and grouted post-tensioned tendons and by 3N/mm^2 for other members. For other percentages of additional reinforcement the stresses may be increased in proportion excepting that the total hypothetical tensile stress shall not exceed 0.25 times the characteristic compressive strength of concrete.

EXERCISE 25

1. (*a*) Define Prestressing and list its advantages and disadvantages over other types of construction.

   (*b*) Describe Freyssinet System of prestressing with Sketches.
2. Give a comparison of Pre-tensioning and Post-tensioning Systems.
3. (*a*) State the assumptions made in Prestressed Concrete design.

   (*b*) Explain various losses in Prestressing. Give brief account of losses in percentage form.
4. (*a*) Define Upper and Lower kern points.

   (*b*) Design a prestressed concrete beam having a span of 15 m and carrying a superimposed load of 40 KN/m run. Assume M30 concrete.
5. (*a*) List the various stages to be considered in the design of prestressed concrete members.

   (*b*) Explain the design of "END BLOCK" of a prestressed member by any one method.
6. Design a prestressed beam having a span of 10 m carrying a superimposed load of 20 KN/m. Use M 30 concrete.
7. A prestressed concrete beam is required to carry a total *u.d.l.* of 30.0 KN over a span of 12 m. Using a rectangular section, determine the suitable dimensions of the beam. Calculate the size of the tendon and its location. Use M 30 concrete and permissible stress in tendon material is 1000 N/mm^2.
8. A post-tensioned concrete simply supported beam is to be designed to carry a load of 10 KN/m in addition to its self weight, over a span of 8 m. Use M 30 concrete and permissible stress in cable material as 1200 N/mm^2. Determine the magnitude of prestressing and also location of the cable.
9. A rectangular beam 12 m long and cross section 300 × 600 mm is to be post-tensioned by straight tendons tensioned to a stress of 1200 N/mm^2 at the jacking end. Calculate the wobble friction loss at the mid span and at the end remote from jack.
10. A prestressed concrete beam is to be designed to support a load of 5 KN/m over a span of 6 m. The section of the *T*-beam is shown in Fig. 25.63. The permissible stress in compression in concrete is 25 N/mm^2 and tension is not to be permitted. Take $\eta = 0.80$.

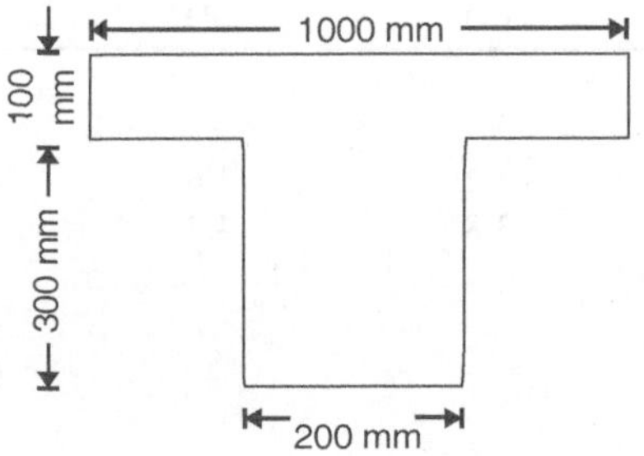

Fig. 25.63.

Masonry Structures

26.1 INTRODUCTION

Masonry is an assemblage of units namely bricks made up of clay or dressed stones, properly bonded together with mortar such as mud mortar, lime surkhi mortar or cement mortar. Columns, piers, buttresses, retaining walls are composed of brick or stone masonry units. Buildings upto four storey height and short span arch bridges are very common in India and other countries. Brick units are more popular but in hilly terrains, stone units are also made use of. Masonry units can be any one of the following types :

(*a*) Common brunt clay building bricks
(*b*) Brunt clay fly ash building bricks
(*c*) Pulverized fuel ash lime bricks
(*d*) Stones in regular sized units
(*e*) Sand lime bricks
(*f*) Concrete blocks (solid or hollow)
(*g*) Lime based blocks
(*h*) Brunt clay hollow bricks
(*i*) Gypsum partition blocks
(*j*) Autoclaved cellular concrete blocks
(*k*) Concrete stone masonry bricks.

IS : 1905 'code of practice for structural use' and SP20 : 1991 'Handbook of Masonry design' or NBC of India are reference codes.

26.2 MORTAR

Mix proportions and compressive strengths of some of the commonly used mortars are presented in Table 26.1.

Table 26.1 Mix Proportions and Strength of Mortars for Masonry

Sl. No.	*Grade of Mortar*	*Mix Proportions (by Loose Volume)*					*Minimum Compressive Strength at 28 Days in N/mm²*
		Cement	*Lime*	*Lime Pozzolana Mixture*	*Pozzolana*	*Sand*	
(1)	(2)	(3)	(4)	(5)	(6)	(7)	(8)
1	H1	1	$\frac{1}{4}$ C or B	0	0	3	10
2(a)	H2	1	$\frac{1}{4}$ C or B	0	0	4	7.5
2(b)		1	$\frac{1}{4}$ C or B	0	0	$4\frac{1}{2}$	6.0
3(a)	M1	1		0	0	5	5.0
3(b)		1	1 C or B	0	0	6	3.0
3(c)		0	0	1 (LP-40)	0	$1\frac{1}{2}$	3.0
4(a)	M2	1	0	0	0	6	3.0

(*Contd.*)

Sl. No.	*Grade of Mortar*	*Mix Proportions (by Loose Volume)*					*Minimum Compressive Strength at 28 Days*
		Cement	*Lime*	*Lime Pozzolana Mixture*	*Pozzolana*	*Sand*	
(1)	(2)	(3)	(4)	(5)	(6)]	(7)	(8)
4(b)		1	2 B	0	0	9	2.0
4(c)		0	1 A	0	0	2	2.0
4(d)		0	1 B	0	1	1	2.0
4(e)		0	1 C or B	0	2	0	2.0
4(f)		0	0	1 (LP-40)	0	$1\frac{3}{4}$	2.0
5(a)	M3	1	0	0	0	7	1.5
5(b)		1	3 B	0	0	12	1.5
5(c)		0	1 A	0	0	3	1.5
5(d)		0	1 B	0	2	1	1.5
5(e)		0	1 C or B	0	3	0	1.5
5(f)		0	0	1 (LP-40)	0	2	1.5
6(a)	L1	1	0	0	0	8	0.7
6(b)		0	1 B	0	1	2	0.7
6(c)		0	1 C or B	0	2	1	0.7
6(d)		0	0	1 (LP-40)	0	$2\frac{1}{4}$	0.7
6(e)		0	0	1 (LP-20)	0	$1\frac{1}{2}$	0.7
7(a)	L2	0	1 B	0	0	3	0.5
7(b)		0	1 C or B	0	1	2	0.5
7(c)		0	0	1 (LP-7)	0	$1\frac{1}{2}$	0.5

Notes: 1. Sand for making mortar should be well graded. In case sand is not well graded, its proportion shall be reduced in order to achieve the minimum specific strength.

2. For mixes in Sl. No. 1 and 2, use of lime is not essential from consideration of strength as it does not result in increase in strength. However, its use is highly recommended since it improves workability.

3. For mixes in Sl. No. 3(a), 4(a), 5(a) and 6(a), either lime C or B to the extent of $\frac{1}{4}$ part of cement (by volume) or some plasticizer should be added for improving workability.

4. For mixes in Sl. No. 4(b) and 5(b), lime and sand should first be ground in mortar mill and then cement added to coarse stuff.

5. It is essential that mixes in Sl. No. 4(c), 4(d), 4(e), 5(d), 5(e), 6(b), 6(c), 7(a) and 7(b) are prepared by grinding in a mortar mill.

6. Mix in Sl. No. 2(b) has been classified to be of same grade as that of Sl. No. 2(a), mixes in Sl. No. 3(b) and 3(c) same as that in Sl. No. 3(a), mixes in Sl. No. 4(b) to 4(f) same as that in Sl. No. 4(a), even though their compressive strength is less. This is from consideration of strength of masonry using different mix proportions.

7. A, B and C denote eminently hydraulic lime, semi-hydraulic lime and fat lime respectively, as specified in appropriate standards listed in Part 5 'Building Materials'.

26.3 DESIGN CONSIDERATIONS

Masonry structures gain stability from support offered by cross walls, floors, roof and other elements. Load bearing walls are structurally more efficient when load is uniformly distributed and the structure is so planned that eccentricity of loading on members is as small as possible. Lateral supports for a masonry element, such as, load bearing wall or column are intended to limit slenderness of that element to prevent or reduce the possibility of buckling of member due to vertical loads and to resist the horizontal components of the forces so as ensure stability against sliding and overturning. R.C.C. floor/roof slab irrespective of direction of span, must bear on wall or cross wall, to the extent of at least 90 mm. A wall or column subject to vertical and lateral loads may be considered to be provided with adequate lateral support from consideration of stability, if the construction providing the support is capable of resisting simple static reactions at the point of lateral support to all the lateral loads plus 2.5 per cent of the total vertical load. In case of load bearing buildings upto four storeys, stability requirements may be deemed to have been met with if,

(*a*) height to width ratio of building does not exceed 2.

(*b*) cross walls acting as stiffening walls are continuous from outer wall to outer wall or outer wall to load bearing wall, and of thickness and spacings as given in Table 26.2.

Table 26.2 Thickness and Spacing of Stiffening Walls

Sl.No.	*Thickness of Load Bearing Wall to be Stiffened*	*Height[(1)] of Storey Not to Exceed*	*Stiffening Wall[(1)]*		
			Thickness not less than		*Maximum Spacing*
			1 to 3 storeys	*4 to 6 storeys*	
	mm	*m*	*mm*	*mm*	*m*
(*1*)	(*2*)	(*3*)	(*4*)	(*5*)	(*6*)
(*i*)	100	3.2	100	—	4.5
(*ii*)	200	3.2	100	200	6.0
(*iii*)	300	3.4	100	200	8.0
(*iv*)	Above 300	5.0	100	200	8.0

[(1)] Storey height and maximum spacings as given are centre-to-centre dimensions.

In case of halls exceeding 8.0 m in length safety and adequacy of lateral supports shall always be checked by structural analysis. Maximum overhanging length of bearing wall which could effectively function as flange should be 12 times the thickness of wall or *H*/6, whichever is less in case of *T*/*I* shaped walls and 6 times the thickness or *H*/16, whichever is less in case of *L*/*U* shaped walls, where *H* is total height of the wall above level being considered as shown in Fig. 26.1.

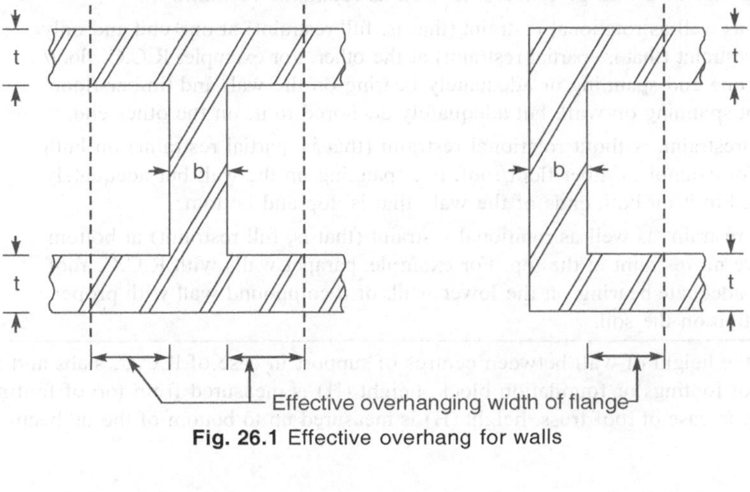

Fig. 26.1 Effective overhang for walls

In case of external walls of basement and plinth, stability requirements may be deemed to have been met with if

(*a*) bricks used in basement and plinth have a minimum crushing strength of 5 N/mm^2 and mortar used in masonry is of grade M1 or better.

(*b*) Clear height of ceiling in basement does not exceed 2.6 m.

(*c*) Adequate cross walls

(*d*) Minimum thickness of basement walls is in accordance with requirements as given in Table 26.3.

Table 26.3 Minimum Thickness of Basement Walls

Sl.No.	*Height of the Ground Above Basement Floor Level*		*Minimum Nominal Thickness of Basement*
	Wall loading (permanent load) less than 50 kN/m	*Wall loading (permanent load) more than 50 kN/m*	
(1)	*(2)*	*(3)*	*(4)*
(i)	Up to 1.4 m	Up to 1.75 m	300 mm
(ii)	Up to 2 m	Up to 2.5 m	400 mm

A free standing wall such as compound wall or parapet wall which is acted upon by wind force tending to overturn the wall, is safe and stable if the resistance provided the weight of the wall (or the resistance provided by cement mortar used) is at least 1.5 times the tendency.

The effective height of a brick masonry wall shall be taken as given in Table 26.4 and shown in Figs. 26.2 and 26.3 for walls and columns standing in different situations respectively.

Table 26.4 Effective Height of Walls

Sl.No. *(1)*	*Condition of Support* *(2)*	*Effective Height* *(3)*
(i)	Lateral as well as rotational restraint (that is, full restraint) at top and bottom. For example, when the floor/roof spans on the walls so that reaction to load of floor/roof is provided by the walls, or when an R.C.C. floor/roof has bearing on the wall (minimum 9 cm), irrespective of the direction of the span foundation footings of a wall give lateral as well as rotational restraint.	0.75 *H*
(ii)	Lateral as well as rotational restraint (that is, full restraint) at one end and only lateral restraint (that is, partial restraint) at the other. For example, R.C.C. floor/roof at one end spanning or adequately bearing on the wall and timber floor/roof not spanning on wall, but adequately anchored to it, on the other end.	0.85 *H*
(iii)	Lateral restraint, without rotational restraint (that is, partial restraint) on both ends. For example, timber floor/roof, not spanning on the wall but adequately anchored to it on both ends of the wall, that is, top and bottom.	1.00 *H*
(iv)	Lateral restraint as well as rotational restraint (that is, full restraint) at bottom but have no restraint at the top. For example, parapet walls with R.C.C. roof having adequate bearing on the lower wall, or a compound wall with proper foundation on the soil.	1.50 *H*

Notes: 1. H is the height of wall between centres of support in case of R.C.C. slabs and timber floors. In case of footings or foundation block, height (H) is measured from top of footing or foundation block. In case of roof truss, height (H) is measured up to bottom of the tie beam. In case of beam and slab-construction, height should be measured from centre of bottom slab to centre of top beam. All these cases are illustrated by means of examples shown in Fig. 26.2.

2. For working out effective height, it is assumed that concrete D.P.C., when properly bonded with masonry, does not cause discontinuity in the wall.
3. Where memberane type damp-proof course or termite shield causes a discontinuity in bond, the effective height of wall may be taken to be greater of the two values calculated as follows :
   (*a*) consider H from top of footing ignoring D.P.C. and take effective height as 0.75 H.
   (*b*) consider H from top of D.P.C. and take effective height as 0.85 H.
4. When assessing effective height of walls, floors not adequately anchored to walls shall not be considered as providing lateral support to such walls.
5. When thickness of a wall bonded to a pier is at least two-thirds of the thickness of the pier measured in the same direction, the wall and pier may be deemed to act as one structural element.

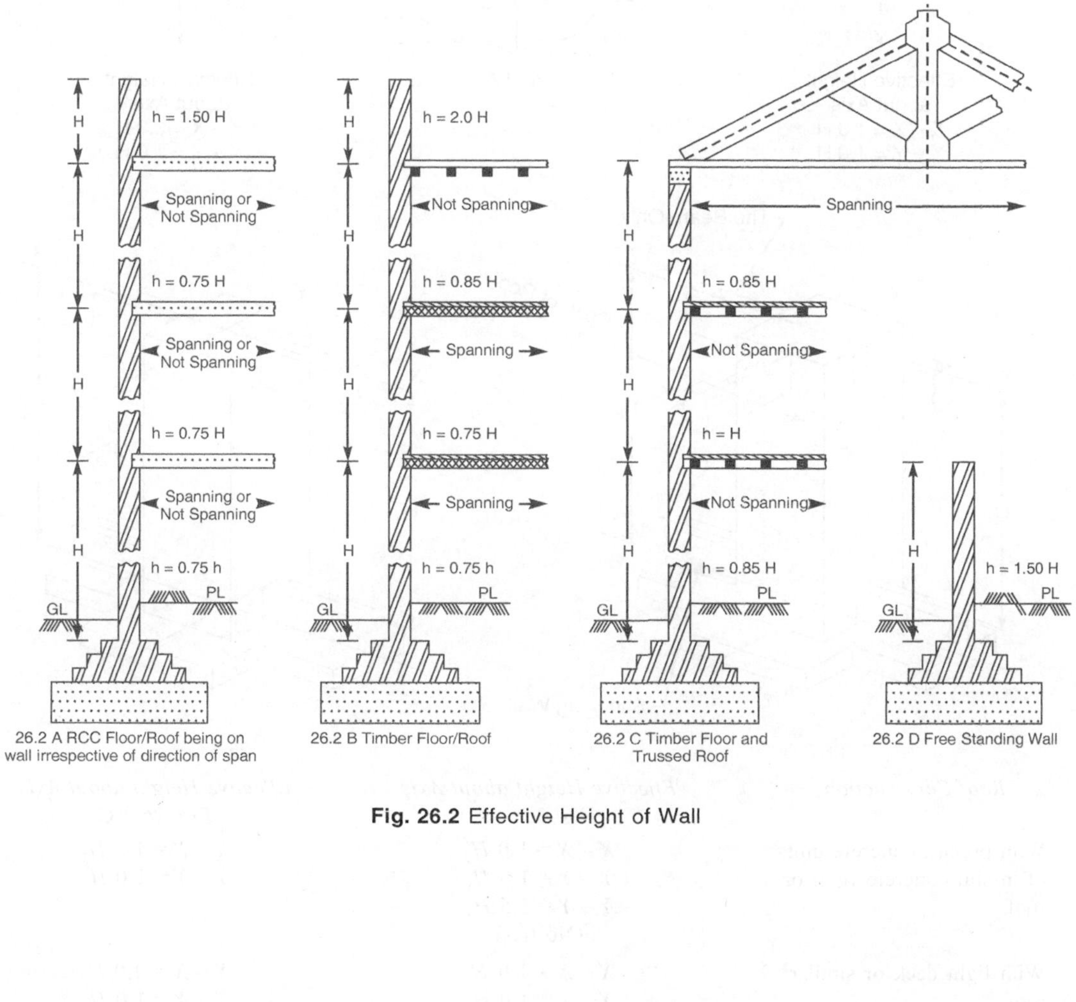

Fig. 26.2 Effective Height of Wall

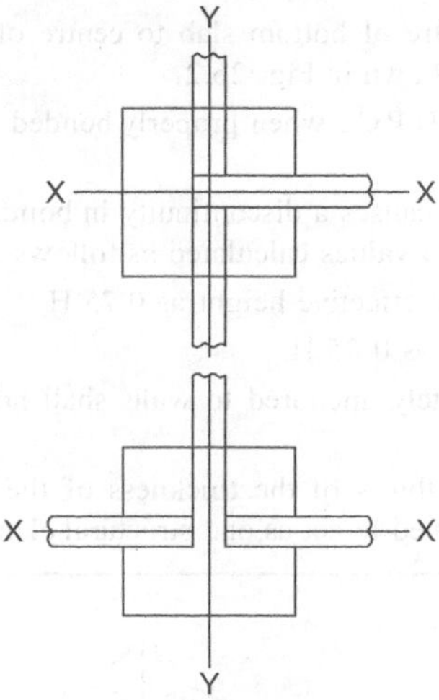

Effective Height
about Axis
X – X = 1.0 H
Y – Y = 1.0 H

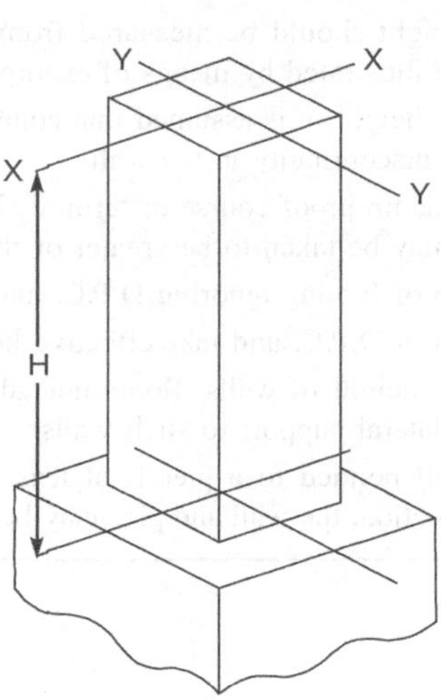

26.3 A

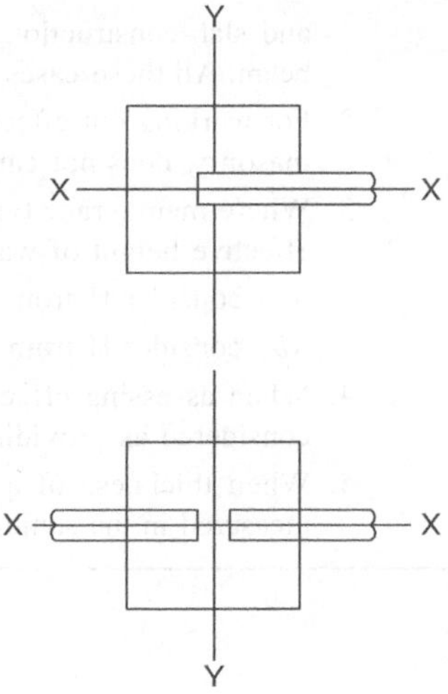

Effective Height
about Axis
X – X = 2.0 H
Y – Y = 1.0 H

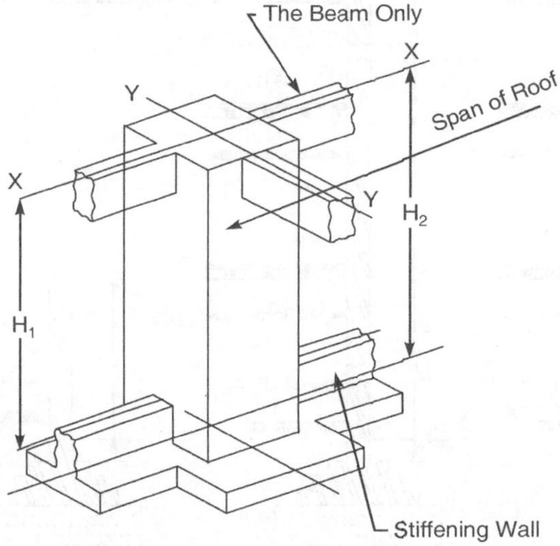

26.3 B

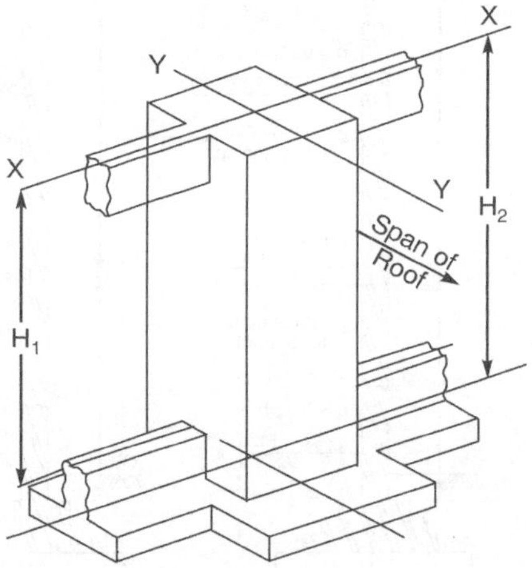

26.3 C

Roof Construction	*Effective Height about Axis Fig. 26.3 B*	*Effective Height about Axis Fig. 26.3 C*
With precast concrete units of in-situ concrete floor or roof	$X - X = 1.0\ H_2$ $Y - Y = 1.0\ H_1$ $Y - Y = 1.5\ H_1$ (No ties)	$X - X = 1.5\ H_2$ $Y - Y = 1.0\ H_1$
With light deck or similar roof	$X - X = 1.0\ H_2$ $Y - Y = 1.0\ H_1$ $Y - Y = 2.0\ H_1$ (No ties)	$X - X = 2.0\ H_2$ $Y - Y = 1.0\ H_1$

Fig. 26.3 Example of Effective Height of Columns

The effective length of a wall to be considered for its stability should be taken as given in Table 26.5 and Fig. 26.4.

Table 26.5. Effective Length of Walls

Sl.No. (*1*)	*Conditions of Support* (*2*)	*Effective Height* (*3*)
(*i*)	Where a wall is continuous and is supported by cross wall and there is no opening within a distance of H/8 from the face of cross wall (see Fig. 26.4) or Where a wall is continuous and is supported by piers/buttresses	0.8 *L*
(*ii*)	Where a wall is supported by a cross wall at one end and continuous with cross wall at other end or Where a wall is supported by a pier/buttress at one end and continuous with pier/buttress at other end	0.9 *L*
(*iii*)	Where a wall is supported at each end by cross wall or Where a wall is supported at each end by a pier/buttress	1.0 *L*
(*iv*)	Where a wall is free at one end and continuous with a pier/buttress at the other end or Where a wall is free at one end and continuous with a pier/buttress at the other end	1.5 *L*
(*v*)	Where a wall is free at one end and supported at the other end by a cross wall or Where a wall is free at one end and supported at the other end by a pier/ buttress	2.0 *L*
	where L = Length of wall from or between centres of cross wall, piers or buttress and H = Actual height of wall between centres of adequate lateral support.	

Note: In case there is an opening taller than 0.5 H in a wall, ends of the wall at the opening shall be considered as free. Cross walls shall.

Slenderness ratio which is the ratio of effective height or effective length to effective thickness of a masonry unit, is very important for stability of the unit. For solid walls, the effective thickness is the actual thickness of wall or column but for solid walls adequately bonded into piers, buttresses, effective thickness for determining the slenderness ratio based on effective height should be taken as actual thickness of wall multiplied by the stiffening coefficient as given in Table 26.6.

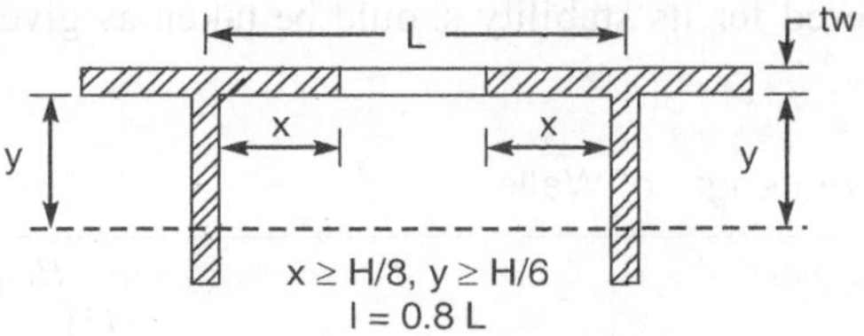

Wall is continuous at both ends and is supported by cross walls of thickness tw/2 or 100 mm whichever is more, length of cross wall is not less than H/6, opening in wall is not less than H/8 from cross wall

26.4 Case 1

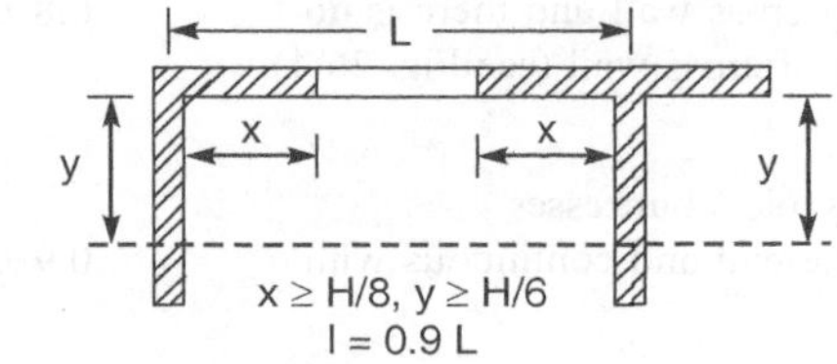

Same as case 1 except that one end of the wall is discontinuous

26.4 B Case 2

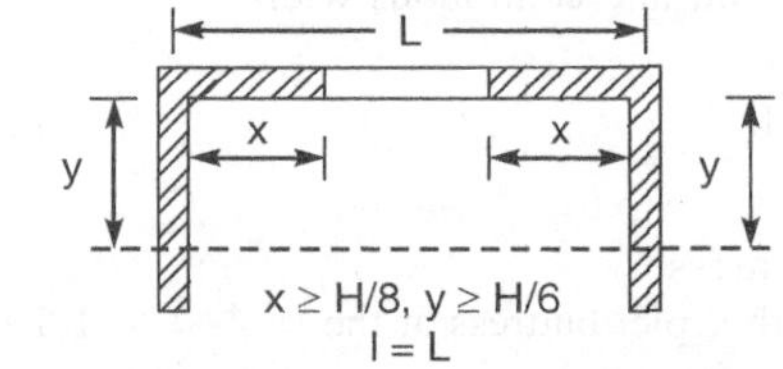

Same as case 1 except that wall is discontinuous on both ends

26.4 C Case 3

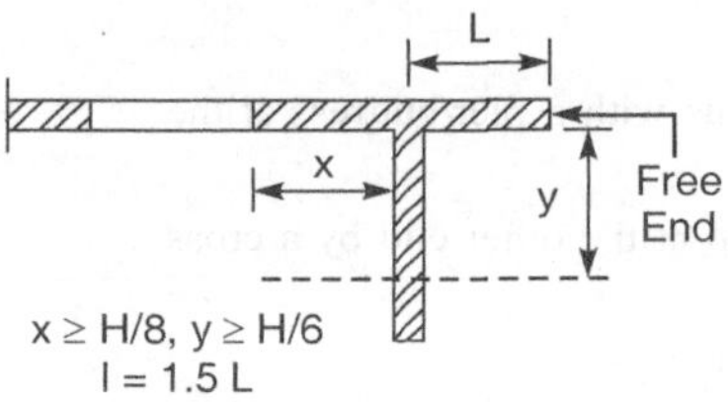

One end of the wall is free, other is supported by a cross wall and is continuous. There being no opening within H/8 from cross wall

26.4 D Case 4

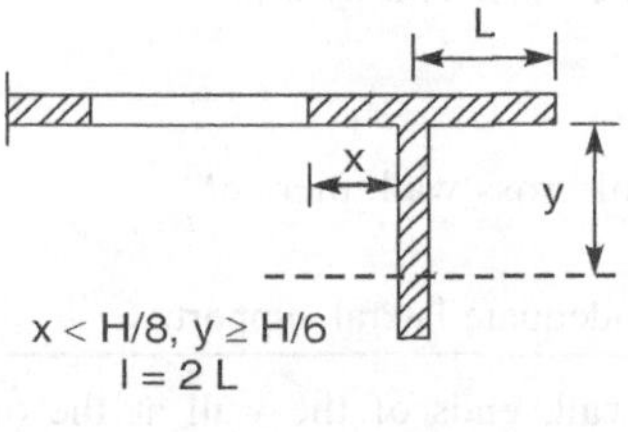

Same as case 4 but opening is within H/8 from cross wall and thus that end is taken as discontinuous

26.4 E Case 5

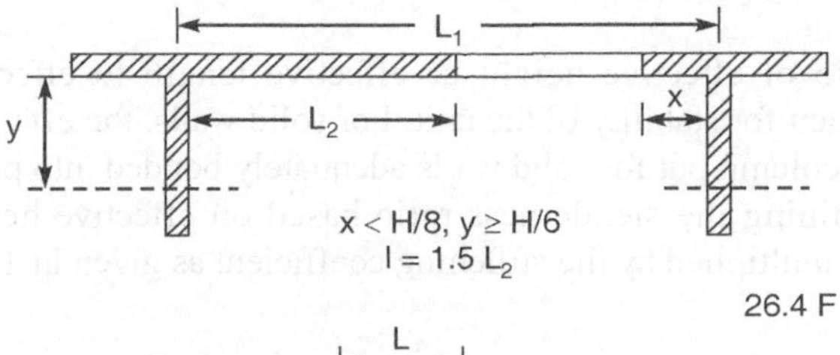

This illustration is with an opening which is within H/8 from cross wall

26.4 F Case 6

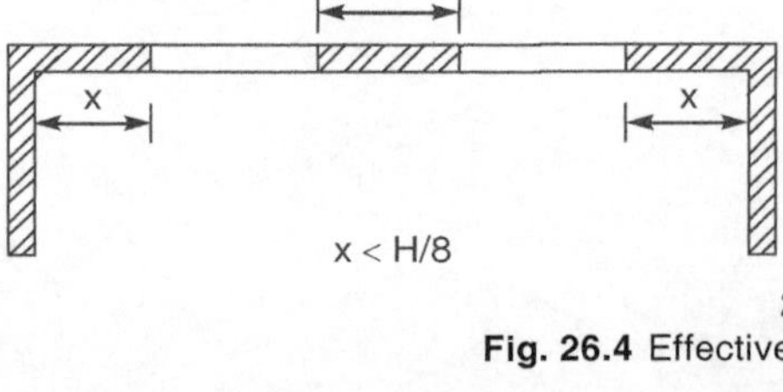

Wall length is between two opening which are closer than H/8 from cross walls

26.4 G Case 7

Fig. 26.4 Effective Length of Wall

Table 26.6 Stiffening Coefficient for Walls Stiffened by Piers, Buttresses or Cross Walls

Sl.No.	*Ratio* $\frac{S_p}{w_p}$	*Stiffening Coefficient*		
		$\frac{t_p}{t_w} = 1$	$\frac{t_p}{t_w} = 2$	$\frac{t_p}{t_w} = 3$ *or more*
(1)	(2)	(3)	(4)	(5)
(*i*)	6	1.0	1.4	2.0
(*ii*)	8	1.0	1.3	1.7
(*iii*)	10	1.0	1.2	1.4
(*iv*)	15	1.0	1.1	1.2
(*v*)	20 or more	1.0	1.0	1.0

where S_p = Centre-to-centre spacing of the pier or cross wall,

t_p = Thickness of pier,

t_w = Actual thickness of the wall proper,

and w_p = Width of the pier in the direction of the wall or the actual thickness of the cross wall.

Note: Linear interpolation between the values given in this table is permissible but not extrapolation outside the limits given.

For cavity walls with, effective thickness is taken as two-thirds of the sum of actual thickness of the two leaves.

Maximum slenderness ratio for load bearing walls should be taken as given in Table 26.7. Slenderness ratio for columns should not exceed 12.

Table 26.7 Maximum Slenderness Ratio for a Load Bearing Wall

Number of Storeys	*Maximum Slenderness Ratio*	
	Using Portland Cement or Portland Pozzolana Cement in Mortar	*Using Lime Mortor*
(1)	(2)	(3)
Not exceeding 2	27	20
Exceeding 2	27	13

Permissible compressive stresses in masonry depends upon the basic ultimate compressive strength of the masonry unit and the type of mortar used and are given in Table 26.8 and further multiplying this value by the factors known as stress reduction factor (k_s), area reduction factor (k_a) and shape modification factor (k_p).

Table 26.8 Basic Compressive Stresses for Masonry (After 28 days)

Sl. No.	*Mortar Type (Ref Table 26.1)*	*Basic Compressive Stresses in N/mm² Corresponding to Masonry Units of which Height to Width Ratio does not Exceed 0.75 and Crushing Strength, in N/mm², is not Less than*											
		3.5	*5.0*	*7.5*	*10*	*12.5*	*15*	*17.5*	*20*	*25*	*30*	*35*	*40*
(1)	*(2)*	*(3)*	*(4)*	*(5)*	*(6)*	*(7)*	*(8)*	*(9)*	*(10)*	*(11)*	*(12)*	*(13)*	*(14)*
(i)	H1	0.35	0.50	0.75	1.00	1.16	1.31	1.45	1.59	1.91	2.21	2.5	3.05
(ii)	H2	0.35	0.50	0.74	0.96	1.09	1.19	1.30	1.41	1.62	1.85	2.1	2.5
(iii)	M1	0.35	0.50	0.74	0.96	1.06	1.13	1.20	1.27	1.47	1.69	1.9	2.2
(iv)	M2	0.35	0.44	0.59	0.81	0.94	1.03	1.10	1.17	1.34	1.51	1.65	1.9
(v)	M3	0.25	0.41	0.56	0.75	0.87	0.95	1.02	1.10	1.25	1.41	1.55	1.78
(vi)	L1	0.25	0.36	0.53	0.67	0.76	0.83	0.90	0.97	1.11	1.26	1.4	1.06
(vii)	L2	0.25	0.31	0.42	0.53	0.58	0.61	0.65	0.69	0.73	0.78	0.85	0.95

Notes: 1. The table is valid for slenderness ratio up to 6 and loading with zero eccentricity.
2. The values given for basic compressive stress are applicable only when the masonry is properly cured.
3. Linear interpolation is permissible for units having crushing strengths between those given in the table.
4. The permissible stress for random rubble masonry may be taken as 75 per cent of the corresponding stress for coarsed walling of similar materials.
5. The strength of ashlar masonry (natural stone masonry of massive type with thin joints) is closely related to intrinsic strength of the stone and allowable working stress in excess of those given in the table may be allowed for such masonry at the discretion of the designer.
6. For calculation of basic compressive stress of stabilized mud block having thickness 100 mm or more, reference to specialist literature may be made.

Table 26.9 Stress Reduction Factor for Slenderness Ratio and Eccentricity

Slenderness Ratio	*Eccentricity of Loading Divided by the Thickness of the Member*					
	0	*1/24*	*1/12*	*1/6*	*1/4*	*1/3*
(1)	*(2)*	*(3)*	*(4)*	*(5)*	*(6)*	*(7)*
6	1.00	1.00	1.00	1.00	1.00	1.00
8	0.95	0.95	0.94	0.93	0.92	0.91
10	0.89	0.88	0.87	0.85	0.83	0.81
12	0.84	0.83	0.81	0.78	0.75	0.72
14	0.78	0.76	0.74	0.70	0.66	0.66
16	0.73	0.71	0.68	0.63	0.58	0.53
18	0.67	0.64	0.61	0.55	0.49	0.43
20	0.62	0.59	0.55	0.48	0.41	0.34
22	0.56	0.52	0.48	0.40	0.32	0.24
24	0.51	0.47	0.42	0.33	0.24	—
26	0.45	0.40	0.35	0.25	—	—
27	0.43	0.38	0.33	0.22	—	—

Notes: 1. Linear interpolation between values is permitted.

2. Where in special cases the eccentricity of loading lies between 1/3 and 1/2 of the thickness of the member, the stress reduction factor should vary linearly between unity and 0.20 for slenderness ratio of 6 and 20 respectively.

3. Slenderness ratio of a member for sections within 1/8 of the height of the member above or below a lateral support may be taken to be 6.

Table 26.10 Shape Modification Factor for Masonry Units

Height to Width Ratio of Units (as Laid)	*Shape Modification Factor (k_p) for Units Having Crushing Strength in N/mm² is*			
	5.0	*7.5*	*10.0*	*15.0*
(1)	*(2)*	*(3)*	*(4)*	*(5)*
Up to 0.75	1.0	1.0	1.0	1.0
1.0	1.2	1.1	1.1	1.0
1.5	1.5	1.3	1.2	1.1
2.0 to 4.0	1.8	1.5	1.3	1.2

Note: Linear interpolation between values is permissible.

Increase in permissible compressive stresses are allowed in case of members subjected to eccentric vertical loads and lateral loads as given below :

(*a*) When resultant eccentricity ratio exceeds 1/24 but does not exceed 1/6, 25 per cent increase in permissible compressive stress is allowed in design.

(*b*) When the resultant eccentricity exceeds 1/6, 25 per cent increase in permissible stress is allowed but the area of the section under tension shall be disregarded for computing the load carrying capacity of the member.

As a general rule, design of masonry should be based on the assumption that masonry is not capable of taking any tension. However, in case of lateral loads normal to the plane of wall, which causes flexural tensile stress as, for example, panel, curtain and free standing walls, flexural stresses may be permitted as per the mortar used.

For Grade M1 or better mortar,

(*a*) 0.07 N/mm² for bending in vertical direction where tension developed is normal to bed joints.

(*b*) 0.14 N/mm² for bending in longitudinal direction for tension developed is parallel to bed joints provided crushing strength of masonry units is not less than 10 N/mm².

For Grade M2, 0.05 N/mm² in case (*a*) and 0.10 N/mm² in case (*b*) as above.

Permissible shear stress in walls built in mortar lesser than Grade M1 resisting horizontal forces in the plane of the wall, should be taken as $f_s = 0.1 + f_d/6$ subject to maximum of 0.5 N/mm² where f_d = compressive stress due to dead loads in N/mm².

Free standing walls in seismic zone II may be proportioned with the help of Table 26.11 provided the mortar used is of grade not lesser than M1. If there is horizontal damp proof course near the base of the wall, the approximate height to thickness ratio given in Table 26.11 be reduced by 25 per cent.

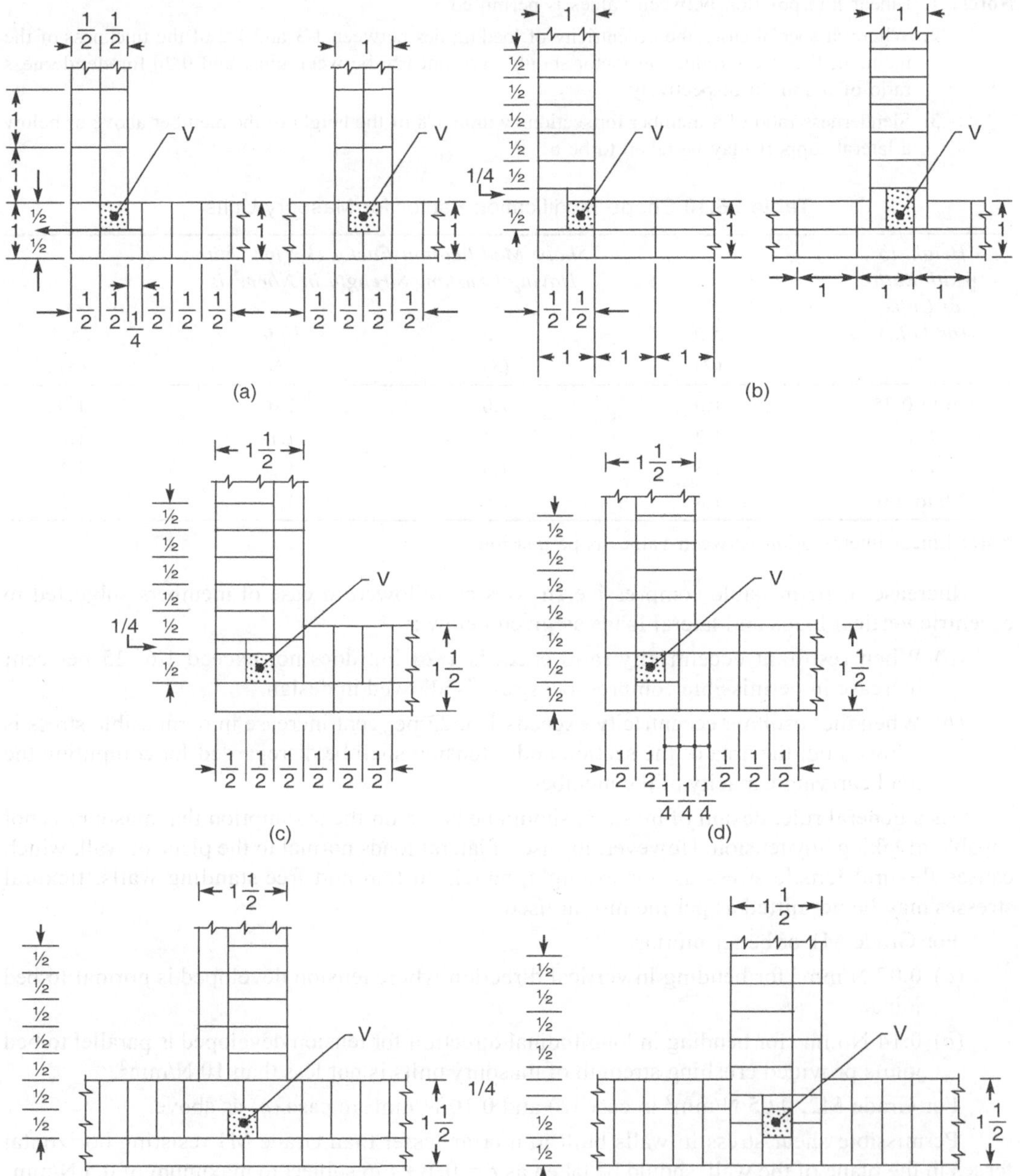

1 — One-brick length, ½ — Half brick length, V — Vertical steel bar with mortar/concrete filling in pocket

(a) and (b) — Alternate courses in one brick wall

(c) and (d) — Alternate courses at corner junction of 1½-brick wall

(e) and (f) — Alternate courses at T-junction of 1½-brick wall

Fig. 26.9 Typical Details of Providing Vertical Steel Bars in Brick Masonry

Table 26.11 Height to Thickness Ratio of Free Standing Walls Related to Wind Speed

Design Wind Pressure N/mm^2 (*1*)	*Height to Thickness Ratio* (*2*)
Up to 285	10
575	7
860	5
1150	4

Notes: 1. For intermediate values, linear interpolation is permissible.

2. Height is to be reckoned from 150 mm below ground level or top of footing/foundation block, whichever is higher, and up to the top edge of the wall.

3. The thickness should be measured including the thickness of the plaster.

For making the building more stronger to resist earthquake shocks, the junctions of walls (weakest points) be reinforced by corner reinforcement as given in Fig. 26.9.

26.4 EXAMPLES

Example 26.1. Design the brick masonry 230 mm thick wall which is subjected to compressive load from floor/roof slabs and its self weight of 100 kN/m run. Minimum eccentricity of 1/24 of thickness of wall may be considered in design. Effective height of wall (actual length) = 3 m. Design its footing also if the Net SBC of soil is 100 kN/m^2.

Solution. Comp. Load at base of wall = 100 kN/m., Using CM 1 : 6 in first class clay bricks with crushing strength of 10 N/mm^2.

Reference Table 1, CM 1 : 6 is M2 grade of mortar, Basic Comprssive Stress of M2 mortar = 0.81 N/mm^2.

Taking effective height of wall as actual height of wall = 3 m = 3000 mm

$$\text{Slenderness ratio} = \frac{3000}{230} = 13$$

Refer to Table 26.9,

For slenderness ratio 12 $\left(\frac{1}{24}\text{ Eccentricity}\right)$, reduction factor = 0.83.

For slenderness ratio 14 $\left(\frac{1}{24}\text{ Eccentricity}\right)$, reduction factor = 0.76.

$\therefore$ For slenderness ratio 13,

$$\text{Reduction factor} = \frac{0.83 + 0.76}{2} = 0.795$$

$\therefore$ Permissible axial compression for CM 1 : 6 and slenderness ratio 13

$$= 0.81 \times 0.795 = 0.644\ \text{N/mm}^2$$

Axial compressive stress at base section of wall

$$= \frac{100 \times 1000}{1000 \times 230} = 0.434\ \text{N/mm}^2$$

$$< 0.644\ \text{N/m}^2 \quad \therefore \text{Safe}$$

Load on soil assuming 15% as self weight of footing

$$= 100 \times 1.15 = 115\ \text{kN/m}$$

$$\text{Net SBC of soil} = 100\ \text{kN/m}^2$$

$\therefore$ Width of footing required $= \dfrac{115}{100} = 1.15$ m

Provide footing width = 1.2 m as shown in Fig. 26.10.

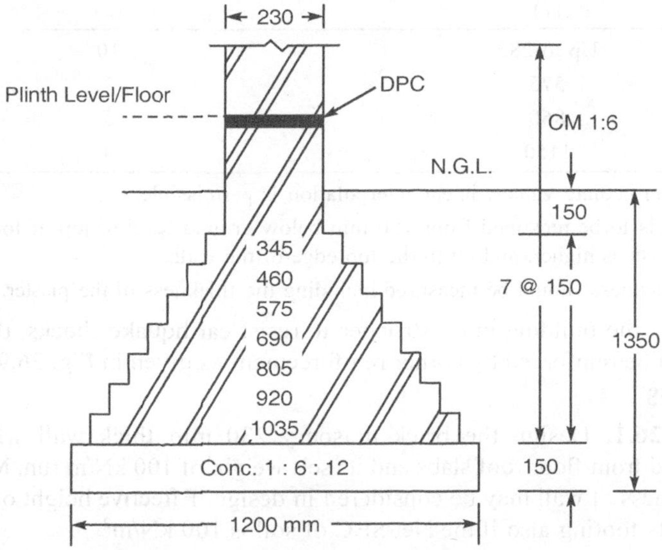

Fig. 26.10 Wall section and Footing details

Example 26.2. Design a brick masonry pier/column which carries a superimposed axial compressive load of 100 kN at base of column. Take effective height of column as 3.0 m. Design the footing if Net SBC of soil is = 100 kN/m^2.

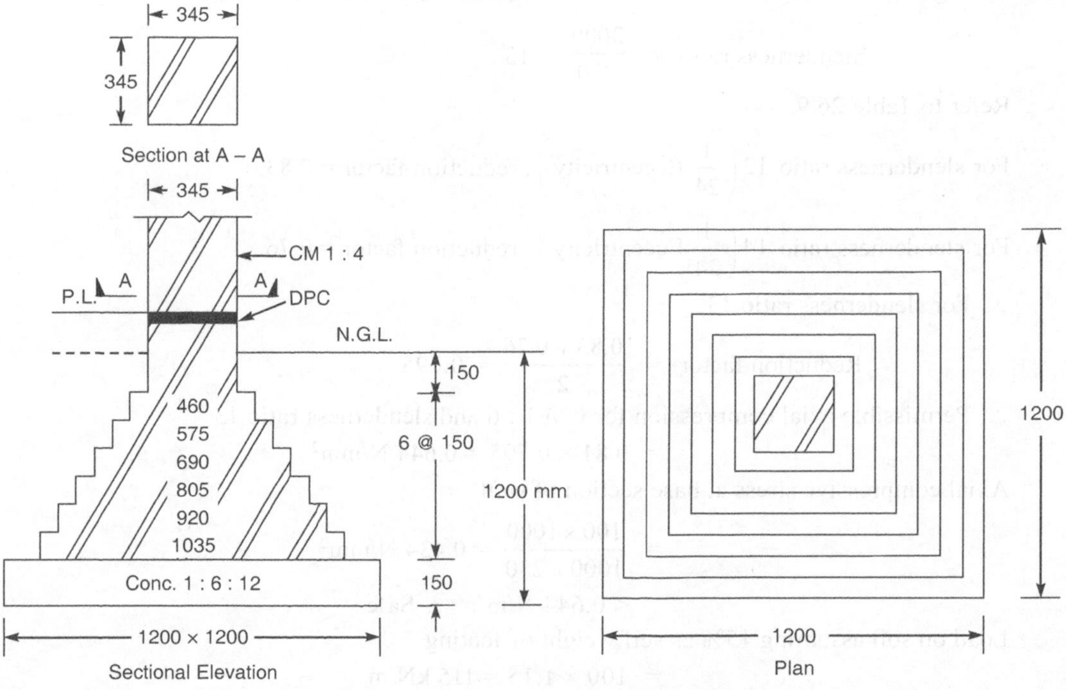

Fig. 26.11 Square column with square footing

Solution. Total SI load at base of column = 100 kN

Using H2 grade of mortar (CM 1 : 4)

Permissible axial compressive load for a short column (slenderness ratio ≤ 12) from Table 26.8 = 0.96 N/mm².

$$\text{Area of section required} = \frac{100 \times 1000}{0.96} = 104166 \text{ mm}^2$$

Provide 345 × 345 mm square section of pier (Area provided = 119025 mm²).

$$\therefore \quad \text{Slenderness ratio} = \frac{3 \times 1000}{345} = 8.695$$

∴ It is a short column.

Load on soil assuming 15% as self weight of footing = 100 × 1.15 = 115 kN.

$$\text{Net SBC of soil} = 100 \text{ kN/m}^2$$

$$\text{Bearing area required} = \frac{115}{100} = 1.15 \text{ m}^2$$

Provide square footing 1.20 m × 1.20 m (Area provided = 1.2 × 1.2 = 1.44 sq. m > 1.15 m²) as shown in Fig. 26.11.

Example 26.3. Design a circular brick masonry chimney 60 m in height subjected to a wind velocity of 50 m/sec and a gust factor of 1.20 (Fig. 26.12). Gross SBC of soil = 120 kN/m².

Solution. Wind velocity = 50 m/sec

Wind pressure = 0.6×50^2

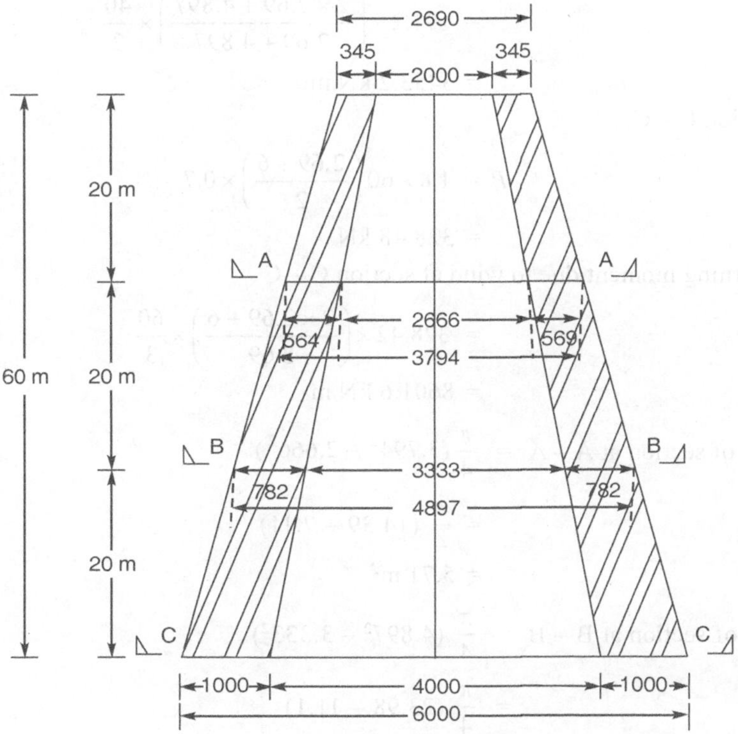

Fig. 26.12 Sectional elevation of chimney

$$= 1500 \text{ N/m}^2$$

Wind pressure including gust factor

$$= 1500 \times 1.2$$
$$= 1800 \text{ N/m}^2$$
$$= 1.8 \text{ kN/m}^2$$

Horizontal wind force 'P'.

(*i*) At section A – A,

$$P = 1.8 \times 20 \times \left(\frac{2.69 + 3.794}{2}\right) \times 0.7$$
$$= 81.7 \text{ kN}$$

Overturning moment due to wind pressure at section A – A

$$= 81.7 \times \left(\frac{2 \times 2.69 + 3.794}{2.69 + 3.794}\right) \times \frac{20}{3}$$
$$= 770.7 \text{ kN.m}$$

(*ii*) At section B – B,

$$P = 1.8 \times 40 \times \left(\frac{2.69 + 4.897}{2}\right) \times 0.7$$
$$= 191.2 \text{ kN}$$

Overturning moment due to wind at section B – B

$$= 191.2 \times \left(\frac{2 \times 2.69 + 4.897}{2.69 + 4.897}\right) \times \frac{40}{3}$$
$$= 3453.2 \text{ kN.m}$$

(*iii*) At section C – C,

$$P = 1.8 \times 60 \left(\frac{2.69 + 6}{2}\right) \times 0.7$$
$$= 328.48 \text{ kN}$$

Overturning moment due to wind at section C – C

$$= 328.42 \times \left(\frac{2 \times 2.69 + 6}{8.69}\right) \times \frac{60}{3}$$
$$= 8601.6 \text{ kN.m}$$

$$\text{Area of section at A – A} = \frac{\pi}{4}(3.794^2 - 2.666^2)$$
$$= \frac{\pi}{4}(14.39 - 7.11)$$
$$= 5.71 \text{ m}^2$$

$$\text{Area of section at B – B} = \frac{\pi}{4}(4.897^2 - 3.333^2)$$
$$= \frac{\pi}{4}(23.98 - 11.1)$$
$$= 10.11 \text{ m}^2$$

$$\text{Area of section at C – C} = \frac{\pi}{4}(6^2 - 4^2)$$
$$= 15.7 \text{ m}^2$$

$$\text{Section modulus at A – A} = \frac{\pi}{64}(3.794^4 - 2.666^4) \times \frac{2}{3.794}$$
$$= 4.05 \text{ m}^3$$

$$\text{Section modulus at B – B} = \frac{\pi}{64}(4.897^4 - 3.33^4) \times \frac{2}{4.897}$$
$$= 9.06 \text{ m}^3$$

$$\text{Section modulus at C – C} = \frac{\pi}{64}(6^4 - 4^4) \times \frac{2}{6}$$
$$= 17.0 \text{ m}^3$$

Weight of chimney up to section A – A

$$= \pi \times \text{Average Dia} \times \text{Average thickness} \times \text{height} \times \text{density}$$
$$= \pi \times \left(\frac{2.345 + 3.23}{2}\right) \times \left(\frac{0.564 + 0.345}{2}\right) \times 20 \times 20$$
$$= 1591.25 \text{ kN}$$

Weight of Chimney up to section B – B

$$= \pi \times \left(\frac{2.345 + 4.115}{2}\right) \times \left(\frac{0.345 + 0.782}{2}\right) \times 40 \times 20$$
$$= 4572.1 \text{ kN}$$

Weight of chimney up to base section C – C

$$= \pi \times \left(\frac{2.345 + 5}{2}\right) \times \left(\frac{0.345 + 1}{2}\right) \times 60 \times 20$$
$$= 9306 \text{ kN}$$

Maximum pressure at section A – A

$$= \text{Direct pressure} \pm \text{bending pressure}$$
$$= \frac{1591.25}{5.71} \pm \frac{770.7}{4.05}$$
$$= 278.7 \pm 190.3 = 88.4 \text{ kN/m}^2 \text{ comp.}$$

Maximum pressure at section B – B

$$= \frac{4572.1}{10.11} \pm \frac{3453.2}{9.06}$$
$$= 452.23 \pm 381.14$$
$$= 71.09 \text{ kN/m}^2 \text{ comp.}$$

Maximum pressure at base section C – C

$$= \frac{93.06}{15.7} \pm \frac{3453.2}{17.0}$$
$$= 592.74 \pm 203.1$$
$$= 795.1 \text{ kN/m}^2 \text{ comp.}$$

Assuming the depth of foundation for chimney as 2.5 m and diameter of base 12 m and average density of concrete footing and refilled soil as 21 kN/m^3.

$$\therefore \quad \text{Weight of foundation} = \frac{\pi}{4} \times 12^2 \times 2.5 \times 21$$

$$= 5935 \text{ kN}$$

$$\text{Total load on soil} = 9306 + 5935 = 15241 \text{ kN}$$

$$\therefore \quad \text{Area of base} = \frac{\pi}{4} \times 12^2 = 113 \text{ m}^2$$

$$\text{Section modulus} = \frac{\pi}{32} \times 12^3 = 169.56 \text{ m}^3$$

$$\therefore \quad \text{Pressure at base} = \frac{15241}{113} \pm \frac{3453.2}{169.56}$$

$$= 134.9 \pm 20.36$$

$$= 155.26 \text{ kN/m}^2 \text{ less than}$$

$$\text{Permissible gross SBC of soil} = 120 \times 1.5 = 180 \text{ kN/m}^2$$

$\therefore$ Safe.

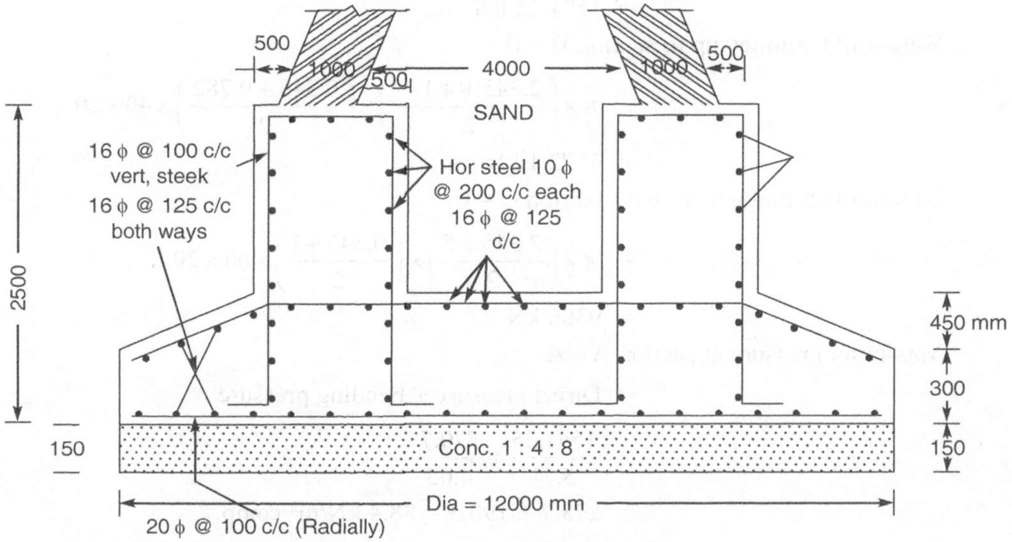

Fig. 26.13 Foundation raft slab details

Direct pressure due to super structure

$$= \frac{9306}{113} = 82.35 \text{ kN/m}^2$$

$$\text{Factored pressure} = 82.35 \times 1.5 = 123.6 \text{ kN/m}^2$$

Design of Footing Slab,

$$\text{Maximum B.M.} = 123.6 \times \left(\frac{1 + 1.72}{2}\right) \times 2.5 \times 0.65 \times 2.5$$

$$= 682.9 \text{ kN.m}$$

M20 concrete

$$d = \sqrt{\frac{682.9 \times 10^6}{1000 \times 2.76}} = 498 \text{ mm}$$

Provide $D = 750$ mm; $d = 690$ mm,

$$k = \frac{682.9 \times 10^6}{1000 \times 690^2} = 1.43$$

For HYSD bars of 500 grade steel,

$$A_{st} = 0.45 \times 690 \times \frac{1000}{100} = 3105 \text{ mm}^2$$

Provide 20 mm ϕ @ 100 mm c/c radially (Fig. 26.13).

$$\text{Distt. st} = 0.2 \times 750 \times \frac{1000}{100} = 1500 \text{ mm}^2$$

Provide 16 ϕ @ 125 mm c/c

Maximum shear force at section $d/2$ from face of pedestal

$$= 123.6 \times \left(2.5 - \frac{0.69}{2}\right)$$

$$= 123.6 \times 2.155$$

$$= 266 \text{ kN}$$

$$\text{Nominal shear stress} = \frac{266 \times 1000}{675 \times 1000} = 0.4 \text{ N/mm}^2$$

$\therefore$ Safe.

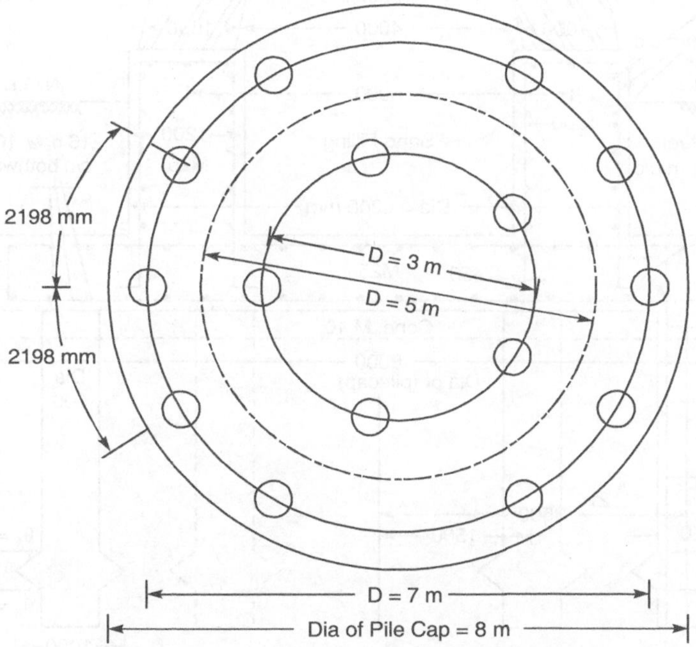

Fig. 26.14 Pile cap with arrangement of piles

In case SBC of soil is not adequate (poor soil with water) or the bearing area is not available in required size, the piles 500 mm dia single under-reamed 12 m long as shown in Fig. 26.14 are provided.

$$\text{Working capacity of single pile} = \left[42 + \left(\frac{12 - 3.5}{0.3}\right) \times 2.4\right] \times 0.75 \times 1.0$$

$$= 82.5 \text{ tonnes} = 825 \text{ kN}$$

Self weight of foundation if depth of embedment is kept 1.5 m

$$= \frac{\pi}{4} \times 8^2 \times 1.5 \times 21$$

$$= 1583 \text{ kN}$$

$$\text{Total load on soil} = 9306 + 1583 = 10889 \text{ kN say } 11000 \text{ kN}$$

$$\text{No. of piles required} = \frac{11000}{825 \times 0.9} = 14.81 \text{ say } 15$$

Arrange 15 nos. piles in 2 rows with $D = 7$ m and $D = 3$ m. Provide 10 nos. piles with $D = 7$ m and 5 nos. with $D = 3$ m. Since, the spacing of inner row of piles is less than twice the diameter of under-reamed bulb, the working capacity of piles shall be reduced by 10%.

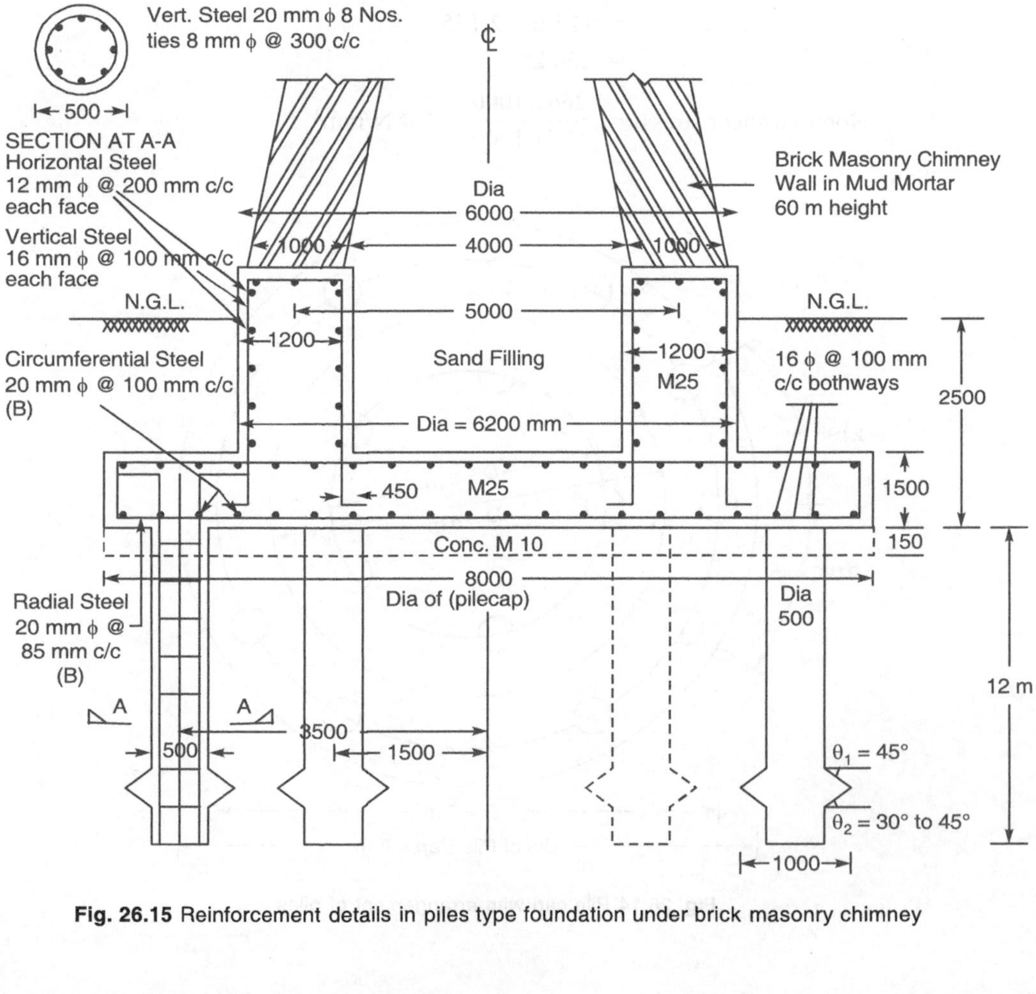

Fig. 26.15 Reinforcement details in piles type foundation under brick masonry chimney

∴ Reduced working capacity of pile = 825 × 0.9 = 732.9 kN.

$$\Sigma r^2 = 2 \times 3.5^2 + 4 \times 2.83^2 + 4 \times 1.08^2 + 1 \times 1.5^2 + 2 \times 0.48^2 + 2 \times 1.21^2$$

$$= 108.84 \text{ m}^2$$

$$\text{Maximum load on a pile} = \frac{11000}{15} \pm \frac{3553.2 \times 3.5}{108.84} = 733 + 114.25$$

$$= 847 \text{ kN} < 732.5 \times 1.5 = 1048.75 \text{ kN}$$

∴ Safe.

$D = 1000$ mm; $d = 960$ mm.

$$\text{Maximum radial moment} = 733 \times \frac{0.4^2}{2} = 58.64 \text{ kN.m}$$

$$\text{Maximum shear stress} = \frac{733 \times 1000}{2198 \times 1440} = 0.23 \text{ N/mm}^2$$

Permissible value of shear stress for 0.25% minimum reinforcement as per Table 2.2

$$A_{st} = \frac{58.64 \times 10^6}{230 \times 0.9 \times 940} = 301 \text{ mm}^2$$

Minimum reinforcement from shear considerations

$$= 0.25 \times \frac{1440 \times 1000}{100} = 3600 \text{ mm}^2$$

Provide 20 mm dia bars @ 85 mm c/c radially at bottom.

$$\text{Provide circumferential steel} = \frac{0.2}{100} \times 1500 \times \frac{1000}{100} = 3000 \text{ mm}^2$$

Provide 20 mm ϕ @ 100 mm c/c at bottom.

For top face reinforcement, half the minimum reinforcement be provided both ways, *i.e.*, 1800 mm^2.

Provide 16 mm ϕ @ 100 mm c/c both ways.

Since, the load on pile due wind blowing is $\frac{114.25}{733} \times 100 = 15.6\%$ of that due to self weight and can be neglected as increase in permissible stresses in this case is $33\frac{1}{3}\%$.

Reinforced details are shown in Fig. 26.15.

Example 26.4. Design of boundary wall (Fig. 26.16) in open terrain for a height of 3.60 m free standing for Central Jail, Rohtak (Haryana, India). Wind speed be taken as 47 m/sec. Gust factor for terrain = 1.15, SBC of soil = 100 kN/m^2 at 1.5 m depth.

Solution. As per Table 26.11, the thickness of boundary wall should be actual height of wall divided by 4, *i.e.*, 915 mm.

Considering per metre length of wall for safety and stability conditions.

Self weight of wall at base section A – A (N.G.L.)

$$= 3.6 \times 0.9 \times 20$$

$$= 64.8 \text{ kN/m}$$

$$\text{Basic wind velocity of Rohtak} = 47 \text{ m/sec}$$

$$\text{Basic wind pressure} = 0.6 \times 47^2$$

$$= 1325 \text{ N/m}^2$$

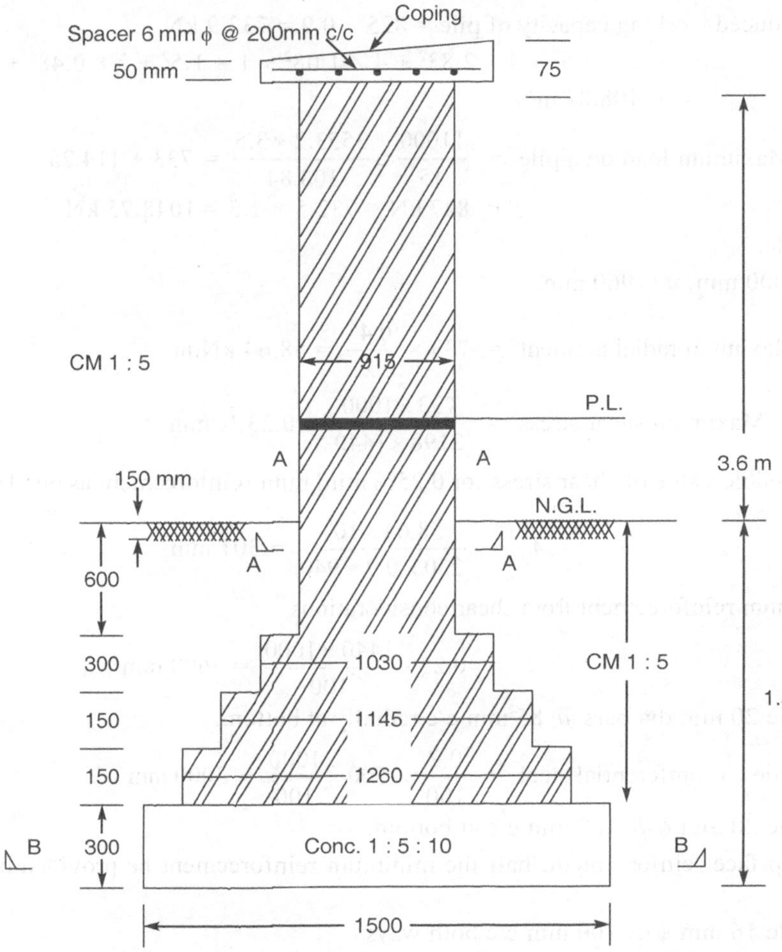

Fig. 26.16 Boundary wall section

$$= 1.325 \text{ kN/m}^2$$

Design wind pressure including gust factor

$$= 1.15 \times 1.325$$

$$= 1.524 \text{ kN/m}^2$$

Taking max. over turning moment at A – A

$$= \frac{1.524 \times (3.6 + 0.15)^2}{2}$$

$$= 10.72 \text{ kN.m}$$

Max. shear force at A – A $= 1.524 \times 3.6$

$$= 5.486 \text{ kN}$$

Maximum pressure at section A – A of wall

$$= \frac{W}{A} \pm \frac{M}{Z}$$

$$= \frac{64.8 \times 1000}{915 \times 1000} \pm \frac{10.72 \times 10^6 \times 6}{1000 \times 915^2}$$

$$= 0.07 \pm 0.0768$$

$$= 0.14 \text{ N/mm}^2 \text{ (comp)} - 0.0068 \text{ N/mm}^2 \text{ tension}$$

As per CM 1 : 5 provided, tensile stresses permitted in wall is 0.14 N/mm^2 which is greater than actual tensile stress developed i.e., 0.0068 N/mm^2.

The S.B.C. of soil is given as 100 kN/m^2 at a depth of 1.5 m below N.G.L., hence the wall section has to be increased in steps to 1260 with P.C.C. 300 mm thick of 1500 mm width as shown in Fig. 26.16.

Consider base section B – B at resting place of 1.5 m below N.G.L. Provide footing as shown in Fig. 26.16 with minimum three off sets and P.C.C. base as 1500 mm.

$\therefore$ Self weight of footing $= 1.5 \times 1.5 \times 18$

(Assuming density of brick masonry and earth refill as 18 kN/m^2) = 40.5 kN/m.

$$\therefore \text{ Maximum pressure on soil } = \frac{(64.8 + 40.5) \times 1000}{1500 \times 1000} \pm \frac{10.72 \times 10^6 \times 6}{1000 \times 1500^2}$$

$$= 0.07 \pm 0.028 = 0.098 \text{ N/mm}^2$$

$$= 98 \text{ kN/m}^2 < 100 \text{ kN/m}^2$$

$\therefore$ Safe.

The thickness of the wall can be reduced provided the safety and stability of wall under design loads is safe and tensile stresses if developed, are within safe limits of those permitted values because normally boundary walls are considered as non-structural members and expansion joints are invariably provided at internals of 30 m. Depending upon the importance of the boundary wall, R.C.C. columns are also provided at intervals of 3 metres to 4 metres. In case the terrain is undulating or low lying, boundary wall is placed on R.C.C. beams resting on R.C.C. pillars. The space below R.C.C. beams till the natural surface is either not filled or filled with brick masonry wall with spread footing. In situations of marshy land, R.C.C. continuous reinforced slab is provided as base or pile foundation is recommended.

Concrete Bridges

CHAPTER 27

27.1 DEFINITION

A bridge is a structure providing safe passage over an an obstruction (namely railway track, road, river, nallah, etc) without closing the way or operations beneath.

27.2 COMPONENTS OF BRIDGE

The main components of concrete bridge structure are:- (Fig 27.1)

1. Decking or superstructure consisting of deck slab and beam girders longitudinal and cross beams.
2. Bearing for decking
3. Abutments and piers
4. Wing walls or return walls
5. Approach slab
6. Hand rails, parapets and guard stones.

27.3 SCOUR DEPTH

Scour of a river bed occurs when there is a high flood discharge and the velocity of water exceeds the limiting velocity that can be resisted by the particles of the bed material.

Mean scour depth can be computed by the Eq. 1 for natural streams in all soils

$$dsm = 1.34\ (D_b^2/K_{sf})^{0.33}$$

H.F.L. meters

D_b = Discharge in m^3 per meter width of river i.e effective linear water way

K_{sf} = Silt factor (See Table 27.1)

Table 27.1 Slit factors for different Bed Materials

Type of Bed Material	*Size of Particle (mm)*	*Silt factor K_{sf}*
Fine Silt	Upto 0.125	0.6
Medium Silt	0.250	0.85
Standard Silt	0.320	1.00
Medium Sand	0.50	1.25
Coarse Sand	0.75	1.50
Heavy sand	1.30	2.00
Boulders	upto 250 mm	6.00

27.4 LINEAR WATER WAY

When the water course to be crossed a river with well defined natural streams the linear waterway is full width of channel or stream.

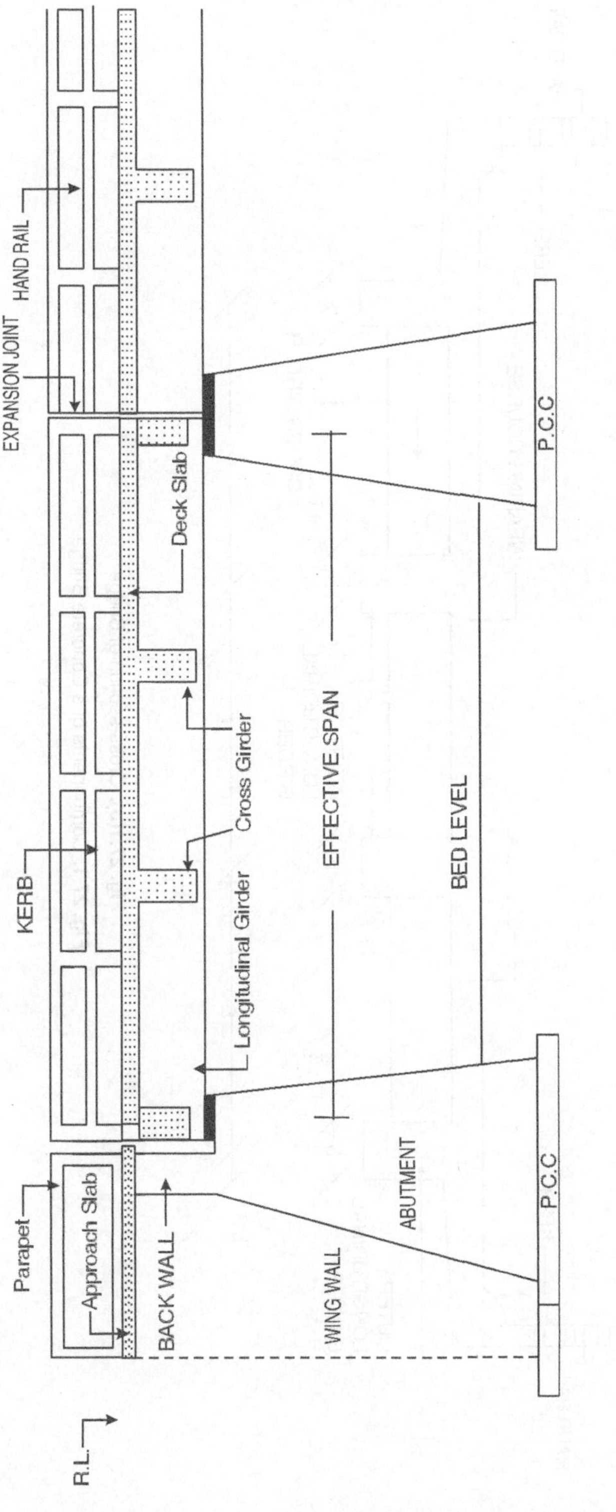

Fig. 27.1 (a). Components of Bridge

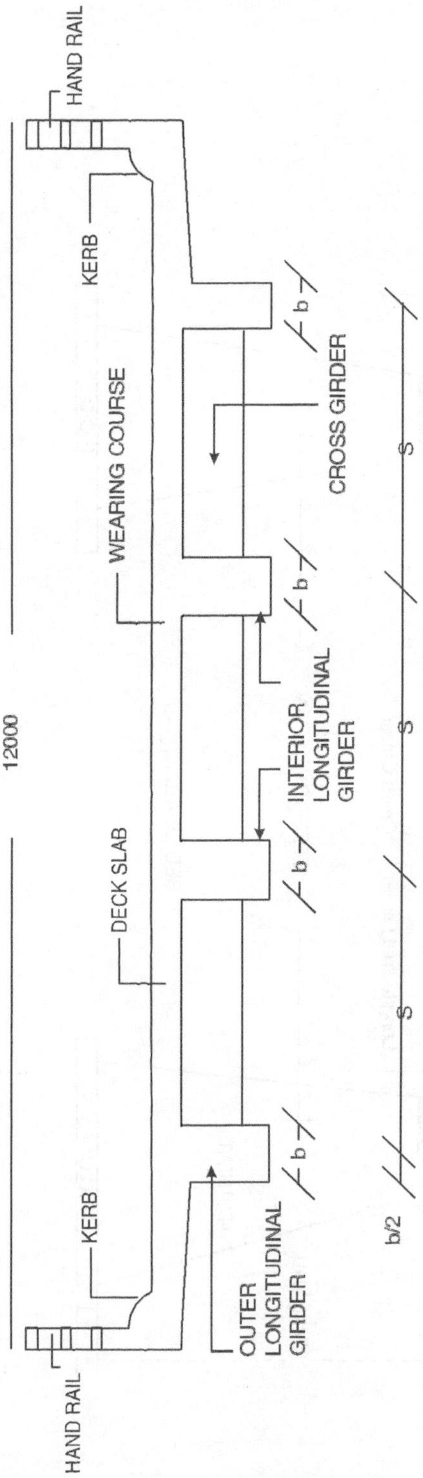

Fig. 27.1(b). Cross-section of Bridge

Fig. 27.1. Components of a Concrete Bridge

For large alluvial effective streams with undefined banks, the effective linear water way can determined by Lacey's formula

$$W = C\sqrt{Q}$$

where W = effective linear waterway (m)

Q = Designed maximum discharge (m^3/sec)

C = a constant taken as 4.8 for regime channels & 4.5 to 6.3 according to local conditions.

27.5 LOADS TO BE CONSIDERED

When designing road bridges and culverts, the following loads, forces and stresses should be considered

(*a*) Dead Load (DL)
(*b*) Live Load (LL)
(*c*) Snow Load
(*d*) Impact (I)
(*e*) Wind Load (WL)
(*f*) Longitudinal forces caused by the tractive efforts or by braking of vehicles
(*g*) Centrifugal forces due to curvature
(*h*) Horizontal forces due to water current
(*i*) Buo yancy
(*j*) Earth pressure including live load surcharge
(*k*) Temperature effects
(*l*) Deformation effects
(*m*) Secondary effects
(*n*) Erection stresses
(*o*) Wave pressure
(*p*) Grade effect
(*q*) Forces and effects due to earthquake

All bridge members should be designed to sustain safely and combination of the above forces that can coexist.

Typical combinations of loads and forces to be considered in design and allowable stresses for certain combinations of loads are presented in Table 27.2.

(A) Dead loads:- The dead load carried by a bridge consists of self weight of member components and permanent load on it namely wearing course & parapets or hand rails etc.

(B) Live loads:- (IRC standard Live Loads) There are four types of standard loadings for which road bridges are designed.

(a) IRC class AA. This loading is adopted for national state highway bridge, within certain municipal limits, existing and complex industrial areas

For single lane bridges with carriageway 3.8m,

TABLE 27.2. LOAD COMBINATIONS AND PERMISSIBLE STRESSES

	G	Q	G_S	Q_{im}	F_{im}	V_C	W	F_{we}	(F_a or F_b) &/or F_r			F_{ef}	G_b	F_{ep}	F_{te}	F_d	F_S	F_{er}	F_{eq}	F_{wp}	G_e	%	
	Dead Load (G)	Live Load (Q)	Snow Load (G_s)	Vehicle Impact (Q_{im})	Impact Floating Bodies (F_{im})	Vehicle Collisioni Load (V_c)	Wind (W)	Water Current (F_{we})	Tractive (F_a)	Braking (F_b)	Bearing Friction (F_f)	Centrifugal Force (F_{ef})	Buoyancy (G_b)	Earth Pressure (F_{ep})	Temperature (F_{te})	Deformation Effects (F_d)	Secondary Effects (F_s)	Erection Effects (F_{er})	Seismic (F_{eq})	Wave Pressure (F_{wp})	Grade Effect (G_e)	Permissible Stresses (%)	Remarks
I	1	1	*	1				1	1	1	1	1	1	1							1	100	Service Condition
IIA	1	1	*	1				1	1	1	1	1	1	1	1	1	1				1	115	Service Condition
IIB	1	0.5		1				1	0.5	0.5	1	0.5	1	1	1	1	1				1	115	Service Condition
IIIA	1	1	*	1			1	1	1	1	1	1	1	1	1	1	1			1	1	133	Service Condition
IIIB	1	0.5		1			1	1	0.5	0.5	1	0.5	1	1	1	1	1			1	1	133	Service Condition
IV	1	1	*	1			1	1	1	1	1	1	1	1		1	1			1	1	133	Service Condition
V	1					1																150	Service Condition
VI	1	0.5		1				1	0.5	0.5	1	0.5	1	1	1	1	1		1	1	1	150	Service Condition
VII	1	1	*	1	1		1	1	1	1	1	1	1	1	1	1	1				1	133	Service Condition
VIII	1						1	1			1		1	1				1			1	133	Construction Condition
IX	1							1			1		1	1				1	0.5		1	150	Construction Condition

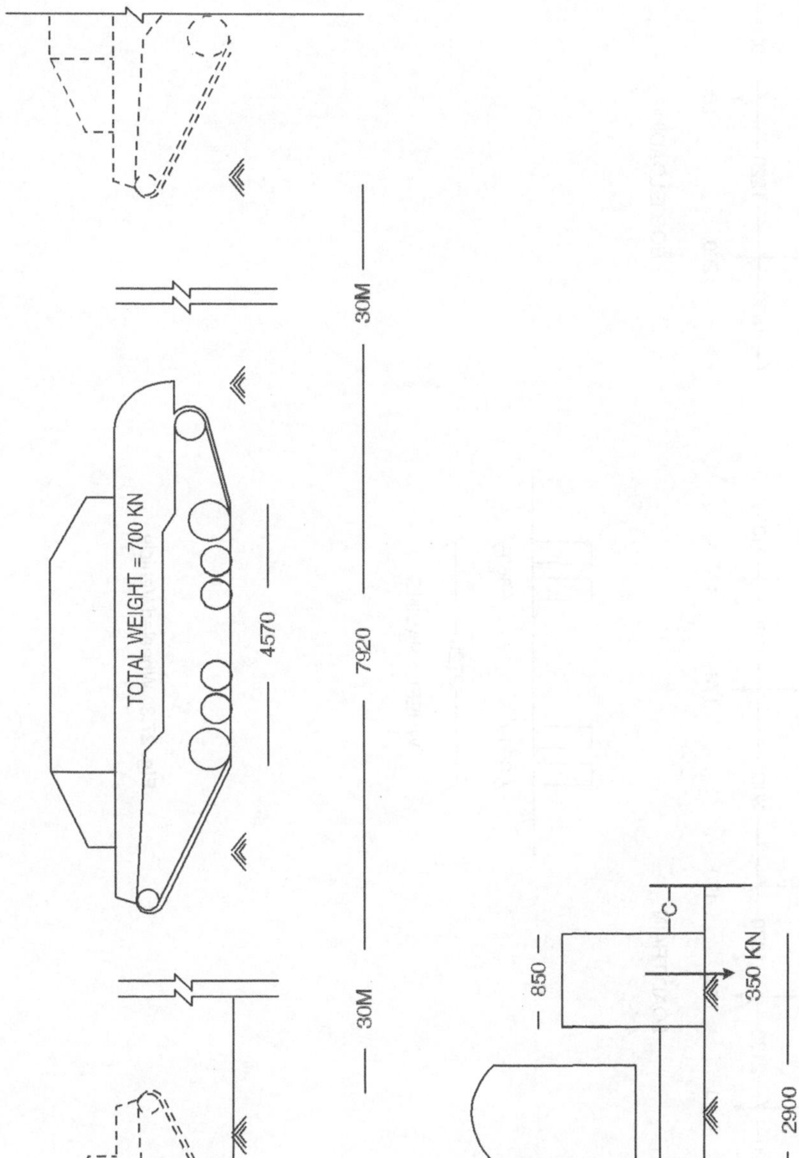

Fig. 27.2. Tracked Vehicle

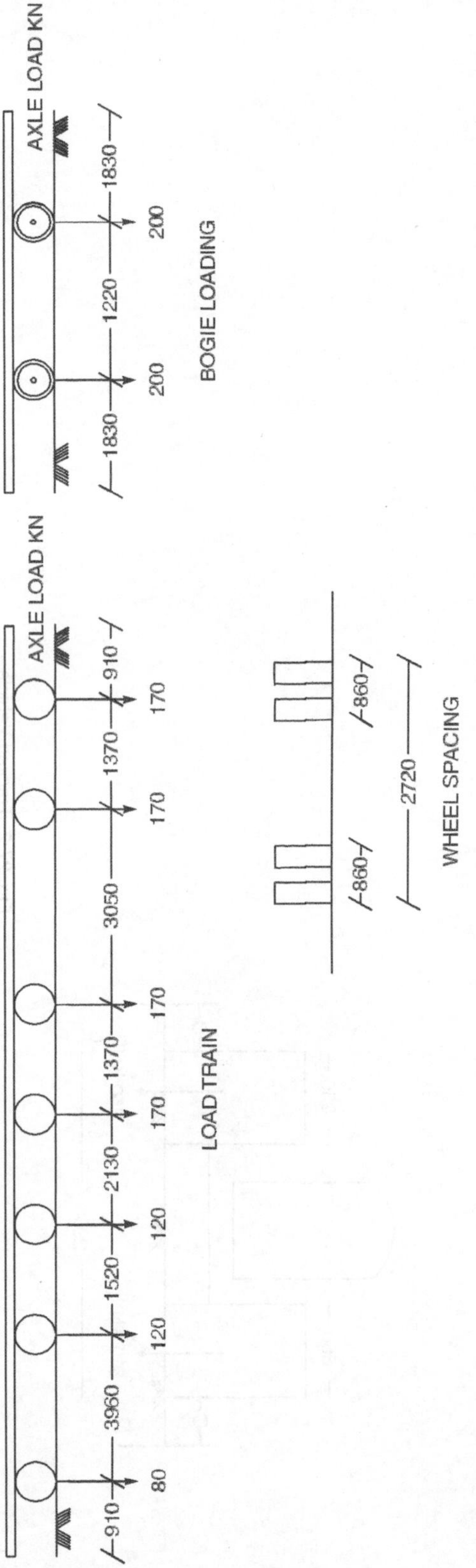

Fig. 27.3. Wheeled Vehicle

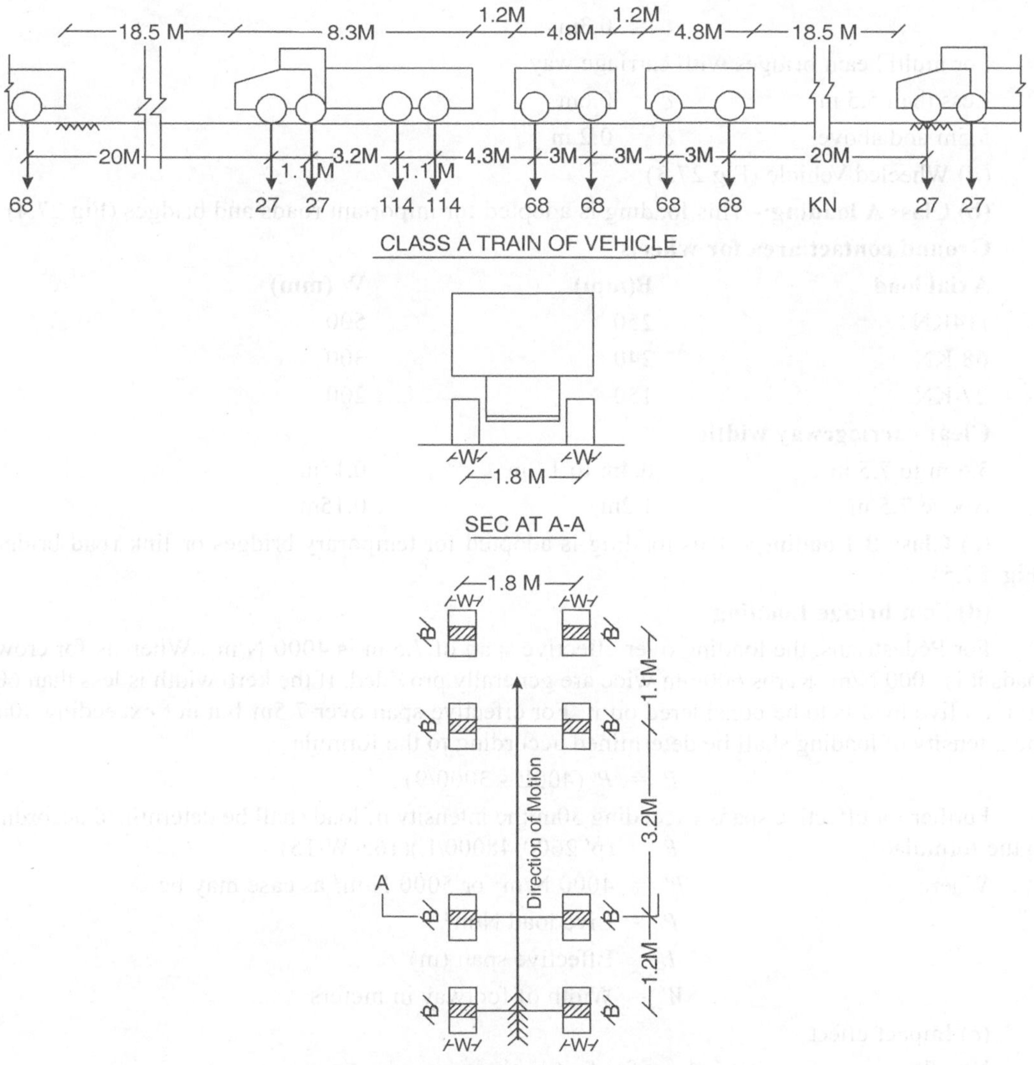

Fig. 27.4(a). Class "A" Train of Vehicle

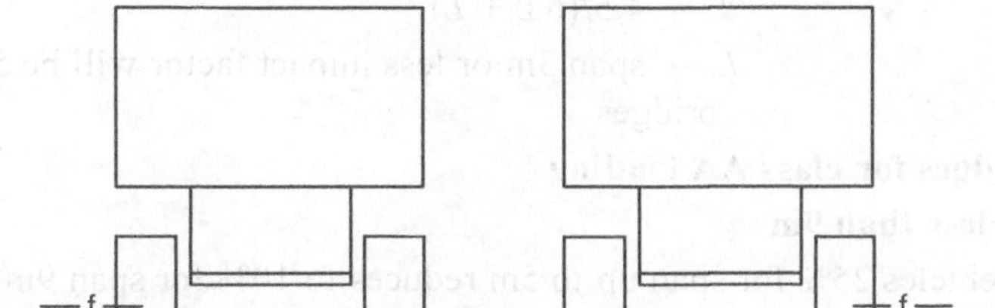

Fig. 27.4(b). Clearances

$c = 0.3\text{m}$

For multi beam bridges with carriage way

Less than 5.5 m $c = 0.6\text{m}$

5.5m and above $c = 0.2\text{ m}$

(*ii*) Wheeled Vehicle (Fig 27.3)

(b) Class A loading:- This loading is adopted for important roads and bridges (Fig 27.4)

Ground contact area for wheels

Axial load	**B(mm)**	**W (mm)**
114KN	250	500
68 KN	240	300
27 KN	150	200
Clear carriageway width		
5.5 m to 7.5 m	0.4m to 1.2m	0.15m
Above 7.5 m	1.2m	0.15m

(c) Class B Loading:- This loading is adopted for temporary bridges or link road bridges (Fig. 27.5).

(d) Foot bridge Loading

For Pedestrians, the loading over effective span of 7.5 m is 4000 N/m^2. Whereas for crowd loads it is 5000 N/m^2 Kerbs 600mm wide are generally provided. If the kerb width is less than 600 mm, no live load is to be considered on it. For effective span over 7.5m but not exceeding 30m, the intensity of loading shall be determined according to the formula

$$P = P' (400\text{L} - 3000/9)$$

Further for effective spans exceeding 30m the intensity of load shall be determined according to the formulae $P = (p' 2600\text{-}48000/\text{L})(165\text{-W}/15)$

Where $P' = 4000\text{ N/m}^2$ or 5000 N/m^2 as case may be

P = Live load N/m^2

L = Effective span (m)

W = Width of footway in meters

(*c*) Impact effect

No allowance of impact is kept for foot bridges

For road bridges (RCC bridges)

$$I = 4.5/(6.0 + L)$$

Where L = span 3m or less impact factor will be 50% for concrete bridges

For road bridges for class AA loading

(*a*) For span less than 9m

(*i*) Tracted vehicles 25% for span up to 5m reduces to 10% for span 9m

(*ii*) Wheeled vehicles 25%

(*b*) For span larger than 9m

(*i*) Tracked vehicles 10% up to 40m, span beyond 40m see Fig 27.6

(*ii*) Wheeled vehicles 25% up to 12m & beyond 12m see Fig 27.6

(D) Centrifugal Force

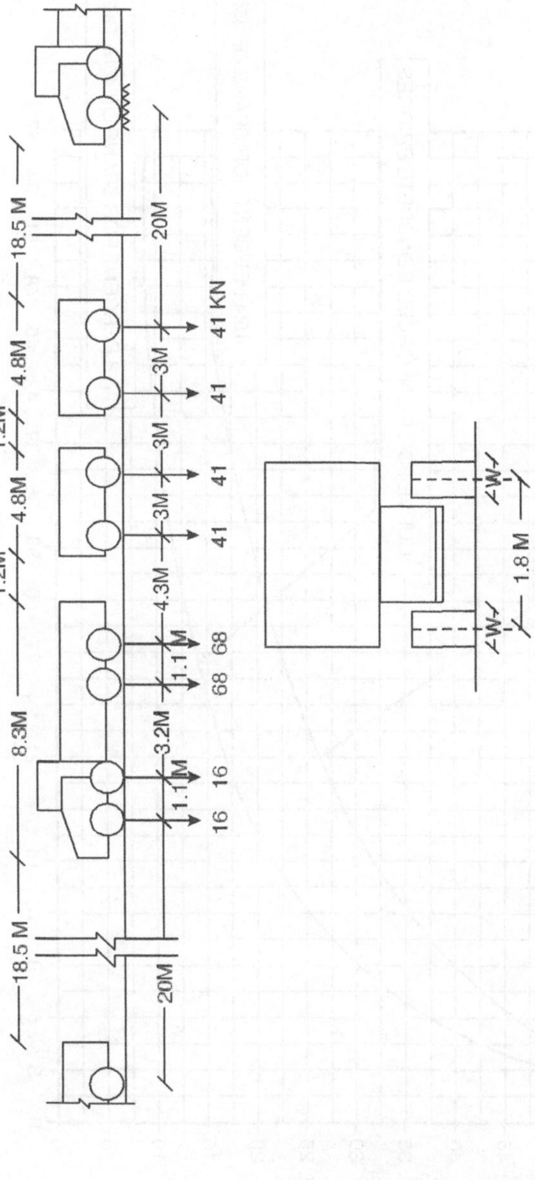

Fig. 27.5. Class B Loading

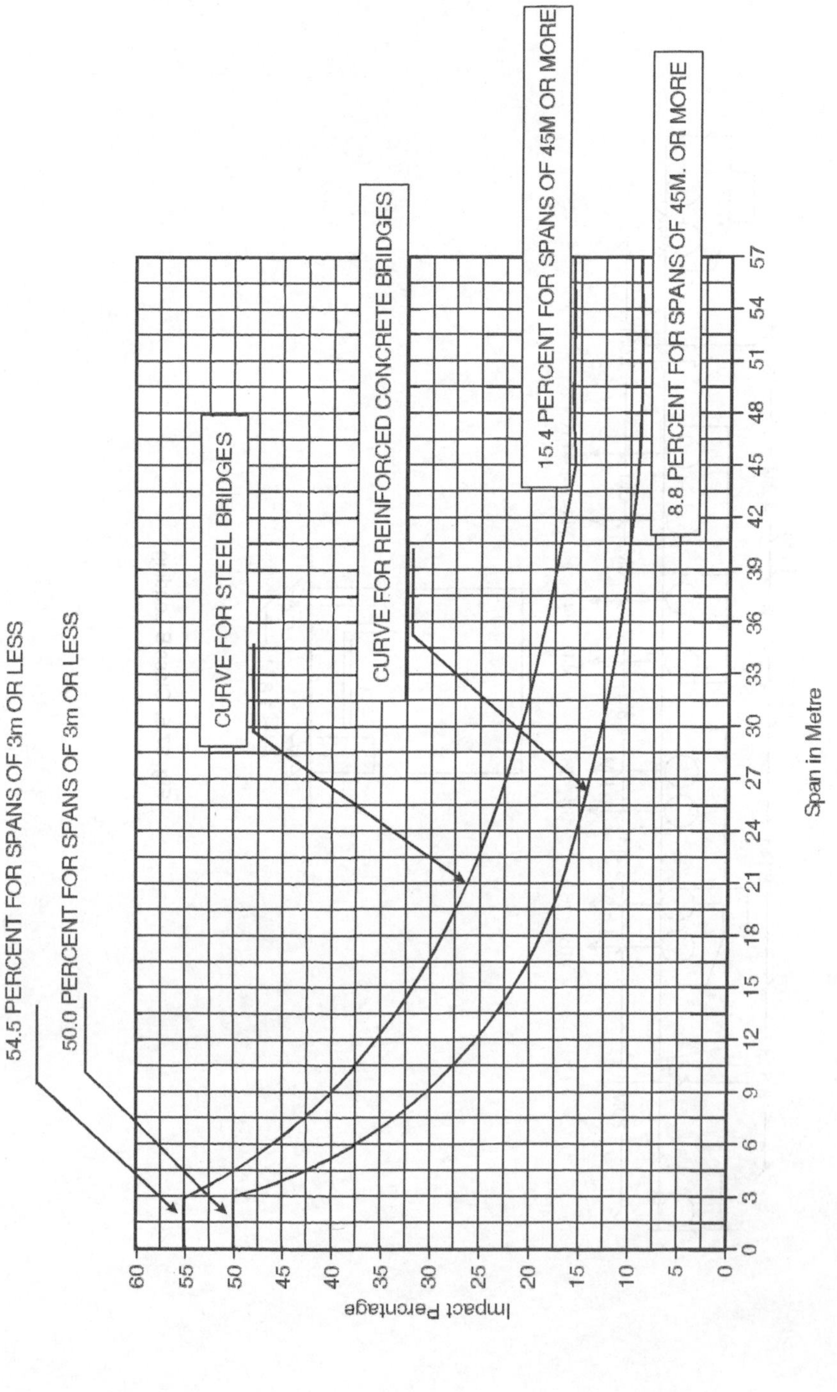

Fig. 27.6. Impact percentage for highway bridges for Class A and Class B loading (Clause 211.2)

For road bridges on curves, the centrifugal force shall be determined from the formulae

$$C = W.V^2/12.95R$$

Where C = Centrifugal Force (KN)

W = Live Load (KN)

V = Speed of vehicle KM/Hour

R = Radius of curvature (meters)

Centrifugal force is acting at a height of 1.2m above road surface.

(E) Wind Forces

The wind forces shall be assumed to be acting horizontally with intensity as given in Table 27.3 and acting at 1.5m above road or deck level.

Table 27.3 Wind Velocities and Wind Pressures

H m	V km/hr	P kN/m²	H m	V km/hr	P kN/m²
0	80	0.40	30	147	1.41
2	91	0.52	40	155	1.57
4	100	0.63	50	162	1.71
6	107	0.73	60	168	1.83
8	113	0.82	70	173	1.93
10	118	0.91	80	177	2.02
15	128	1.07	90	180	2.10
20	136	1.19	100	183	2.17
25	142	1.30	110	186	2.24

Note: H = average height in meteres of the exposed surface above the mean retarding surface

V = horizontal velocity of wind in kilometres per hour at height *H*

P = horizontal wind pressure kN/m² at height *H*.

$$\text{Wind pressure} = 0.6\ V^2\ \text{Kg} \quad \text{where V in meter/sec}$$

Where Kg = Gust factor = 1.15

The lateral wind pressure against any moving load shall be considered acting at 1.5m above load way and 3000 N/linear meter ordinary highway bridges 2 for 4500 N per linear meter for bridgs carriying.

A wind pressure of 2400 N/sqm of unloaded structure shall be used if it produces large stresses than those produced by combined wind forces.

(F) Horizontal forces due to water current

On piers parallel to the direction of the water current, the intensity of pressure is given by Eq 27.1.

$$P = \text{K.W}\ (V^2/2g)$$

Where P = Intensity of water pressure KN/m²

W = Unit weight of water KN/m³

V = Velocity of current m/sec

G = acceleration due to gravity (m/sec²)

K = Constant depending on shape of pier: 1.5 for square ended piers, 0.66 for circular peirs or for piers with semicircular cut waters. 0.56 to 0.9 for triangular cutwaters & 1.25 for type of piers

(G) Buoyancy effect

In case of sumbersible bridges, full buoyancy effect is to be considered for super structure alonen, abutments & piers

For high level bridges, buoyancy effect due to submerged part of the sub structure & foundation should be considered. For piers submerged in water & foundation in sand, full buoyancy should be considered. For other soils, some proportion be taken into account

(H) Earth Pressures

According to Coulomb's theory the horizontal thrust P is given by the Eq 27.2 P = 0. wh^2{Cosecθ × Sin (θ – ϕ) / $\sqrt{\text{Sin}(\theta + z)}$. Sin (θ – S)/Sin (θ – δ)}

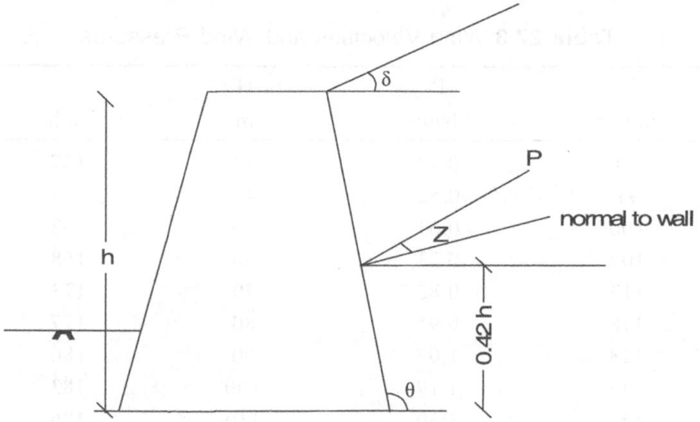

Fig. 27.7.

Where P = Total active pressure acting at height of 0.42h inclined at an angle *z* to the normal to the wall (Fig 27.7) on earth side.

W = Unit weight of earth fill (KN/m^3)

h = height of wall (Meters)

Θ = Angle subtended by earth side wall with horizontal on earth side.

Φ = angle of internal friction of the earth fill

δ = Angle of surcharge

For θ = 90° and *Z* = δ

The earth pressure *P* (Same as in Rankine's theory) is given Eq 27.3

P = 0.5 w.h.cosq {Cos*d* – cos^2*q*/cos *d* + cos^2*s* – cos^2*q*} ...(27.3)

When Surcharge angle δ = 0, Eq 27.3 takes the shape of eq 27.4

$$P = 0.5 \text{ wh}^2 (1 - \text{Sin}\theta/1 + \text{Sin}\theta)$$

As per clause 714.4 of IRC Bridge code, all abutments are designed for a live load surcharge equivalents to 1.2m height of earth fill

Wing /return walls should be designed for full height of retained & live load surcharge of 600mm height of earth fill

(I) Temperature effects:-

Temperature variations effect should be accounted for in design of bridges. The coefficient of thermal expansion per degree centrigrade shall be taken as 11.7 × 10^{-6} for steel and for reinforced concrete structures and 10.8 × 10^{-6} for plain concrete structures.

(J) Deformation stresses

These stresses are produced due to eccentricity of connections. All bridges should be designed & construted in such a manner that the stresses are minimum

(K) Erection Stresses:-

These stresses are induced in members during errection of bridge & should be taken into account in the design

(L) Seismic Force:-

IS: 1893-2002 & IRC 6-2000 have divided the country in for seismic zones II to V (Zone V being reverest) Bridges in seismic zones II & III need not be designed for seismic forces if (a) Span is less than 15m

Total bride length is less than 60m

Vertical seismic coefficient should be considered in case of bridges built in zones IV & V and shall be taken half of the horizontal seismic coefficient. Both horizontal and vertical seismic forces should be considered simultaneously.

Horizontal seismic force should be computed by the following Eq 27.5

$$F_{eq} = Ah \times (\text{Dead load} + \text{Appropriate live load}) \quad ...(27.5)$$

F_{eq} = seismic force developed

A_h = Horizontal seismic coefficient

= Z.I./2.R (Sa/g) ...(27.6)

where z = zone factor as per table 27.4

I = Importance factor

For Important bridges 1.5

Other bridges 1.0

T = Fundamental time period of bridge (Seconds) for horizontal vibrations

Sa/g = to be taken from Fig 27.8 depending upon actual time period. In the absence of data Sa/g may be taken as 2.5

R = Response reduction (2.5)

Table 27.4 Zone factor Z

Zone No.	Zone factor
V	0.36
IV	0.24
III	0.16
II	0.10

For rocky and Hard Strata

Sa = 2.5 0.0</0.55

g = 1.36/T 0.55</T</0.40

For soft soils

Sa = 2.5 0.0</T</0.67

g = 1.67/T 0.67</T</4.00

These horizontal forces due to seismic effect should be assumed acting at center of gravity of all the loads under considerations. The seismic force due to live load should not be considered when acting in the direction traffic but must be considered in direction perpendicular to the traffic.

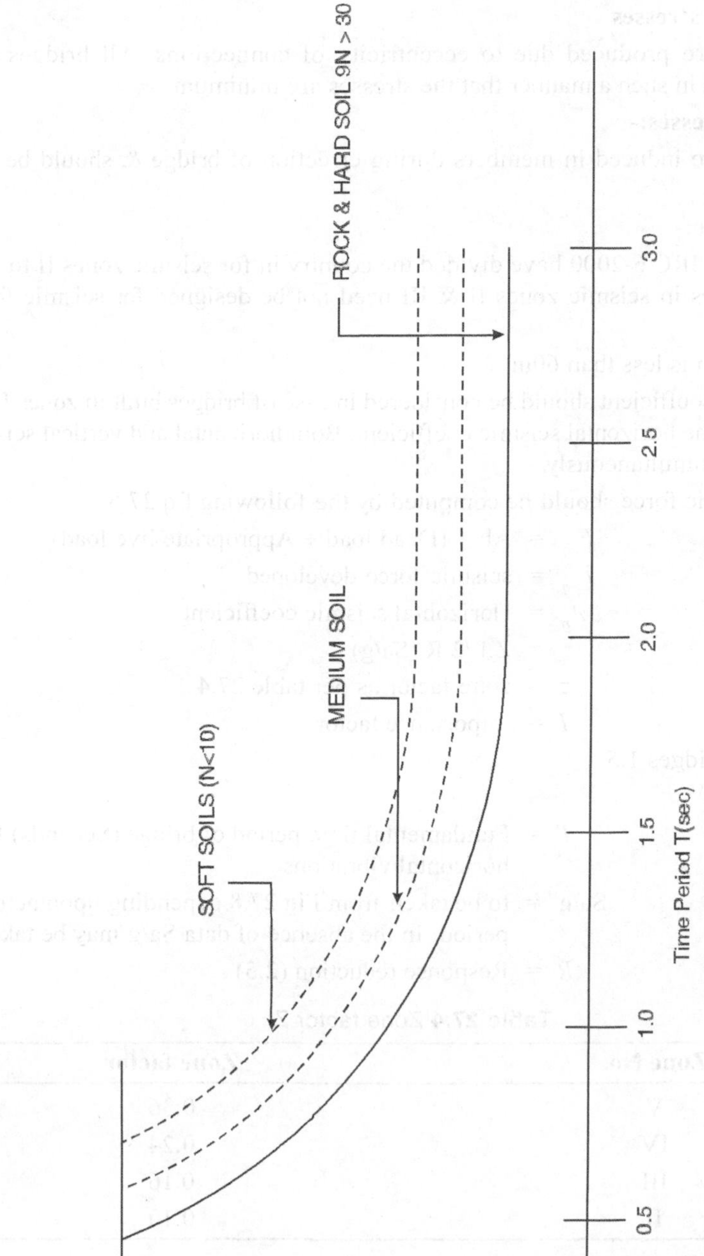

Fig. 27.8. Response Spectra

Use of unreinforced masonry & concrete arches should be avoided in zone V. It may be noted that parts of the structure (incase of abutments & piers) embeded in soild should not considered to produced any seismic forces.

(M) Snow load

Snow load of 900 Kg/m^3 where applicable should be considered in bridge deck if snow accumulation is equal 25cm.while designing the structure for wheeled vehicle and 50cm for designing the structure for tracted vehicle.

In case of snow accumulation more than 50cm, actual snow load be considered without live load.

27.6 DESIGN OF DECK SLAB BRIDGES

The deck slab bridge should be designed as one way slab to carry dead loads and the prescribed live load with impact. This slab should be designed for worst live load either on elane of IRC 70R/class AA tracked or wheeled or two lanes of class A load. The distribution steel is either 0.12% of gross cross sectional area or that required for 0.3 live load moment plus 0.2 dead load moment, which ever is larger. The Ministry of surface transport (Roads wing) Govt of India (MOST) has publised set of standard plans for 3.0 m to 10.0m spanning bridges with reinforcement detils. The overall width of bridge is generally kept as 12.0 meters and made in concrete M25 (M30 for 10.0 meter span) and rebars of grade. Fe 415 with minimum cover to reinforcement as 50mm salient details of general arrangement are shown in Fig 27.9. Wearing course over the slab bridge is generally of asphalted concrete 56mm thick or cement concrete 75mm thick. Top of deck slab is sloped in the transverse direction 2.5% (Camber) for strom water drainage. The size & reinforcement for various spans of bridge are given in Table 27.5.

Table 27.5. Typical size reinforcement for RC Slab Bridges

Effective Span (m)	Slab Thickness		Bottom Force reinforcement		Top force reinforcement	
	D1	D2	Along X	Along Y	Along X	Along Y
3.0	450	300	16 dia 150	8 dia 150	10 dia 250	10 dia 250
4.0	500	350	16 dia 140	10 dia 150	10 dia 250	10 dia 250
5.0	550	400	20 dia 150	10 dia 125	10 dia 225	10 dia 225
6.0	600	450	20 dia 125	12 dia 150	10 dia 220	10 dia 220
7.0	650	500	20 dia 110	12 dia 150	10 dia 200	10 dia 200
8.0	700	600	20 dia 100	12 dia 140	10 dia 140	10 dia 140
9.0	820	670	25 dia 130	12 dia 130	10 dia 130	10 dia 130
10.0	900	750	25 dia 125	12 dia 115	10 dia 120	10 dia 125

A wheel load is practically considered as a concentrated load on slab. This load gets dispersed with its effects span wise (dispersed length of span) and widthwise (effective width) as shown in Fig 27.10.

The effective width of dispersion can be computed by using the following formulae

$$b_{ef} = \gamma . x(1 - x/L) + b_1$$

Where bef = effective width of dispersion

L = effective span of simply supported slab (clear span in case of continuous slab)

X = distance of the center of gravity of the where from the nearest support

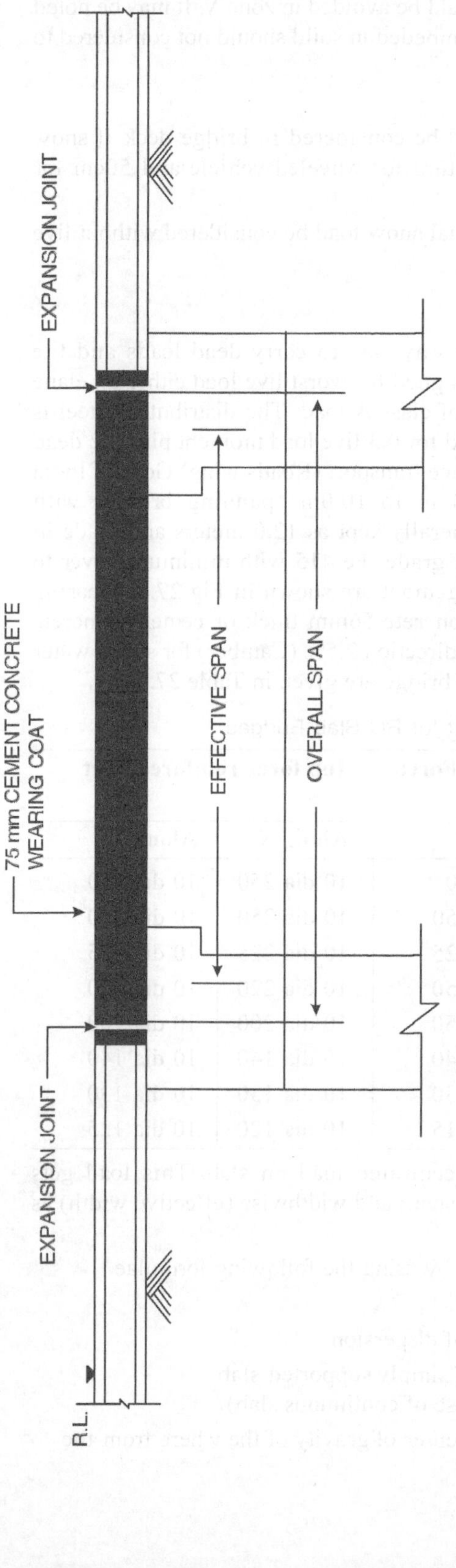

Fig. 27.9(a). General Arrangements of Deck Slab Bridge: ELEVATION

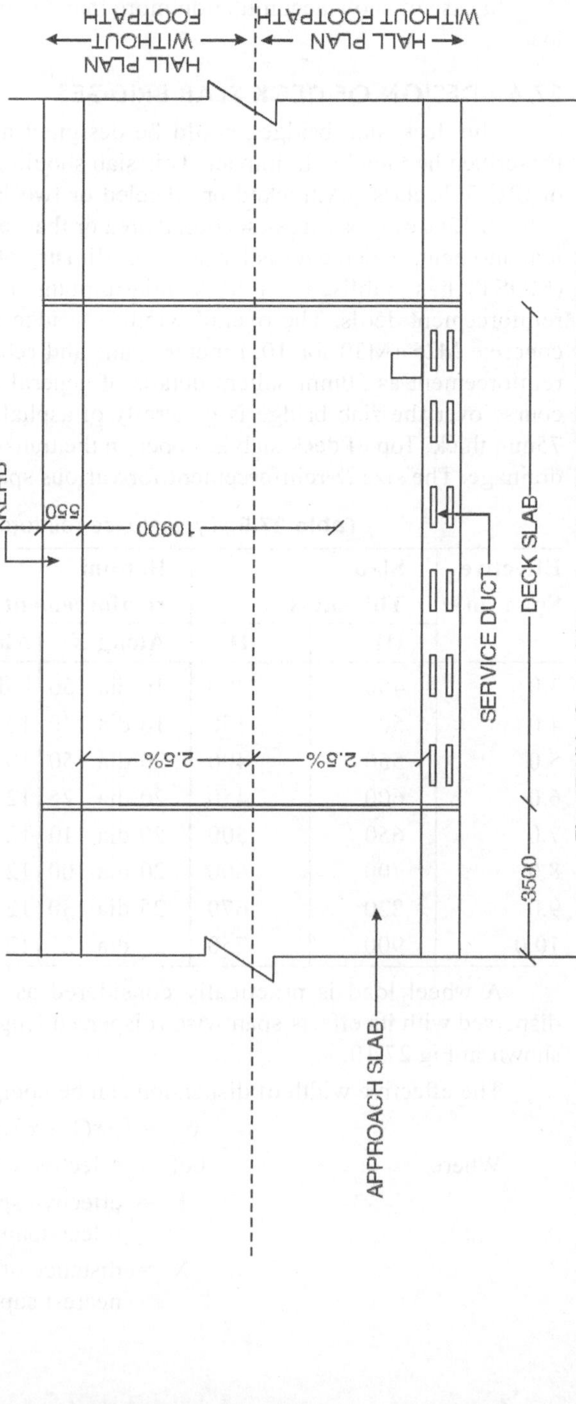

Fig. 27.9(b). General Arrangements of Deck Slab Bridge: PLAN

A = a constant having values depending upon B/1 values (see Table 27.5)

B_1 = width of dispersion area of the wheel load on slab through the wearing coat

Table 27.6 Values of γ for slab (IRC : 21)

B/1	γ Simply supported	Continuous	B/1	γ Simply supported	Continuous
0.1	0.40	0.40	1.1	2.60	2.28
0.2	0.80	0.80	1.2	2.64	2.36
0.3	1.16	1.16	1.3	2.72	2.40
0.4	1.48	1.44	1.4	2.80	2.44
0.5	1.72	1.68	1.5	2.84	2.48
0.6	1.96	1.84	1.6	2.88	2.52
0.7	2.12	1.96	1.7	2.92	2.56
0.8	2.24	2.08	1.8	2.96	2.60
0.9	2.36	2.16	1.9	3.00	2.60
1.0	2.48	2.24	2.9 & above	3.00	2.60

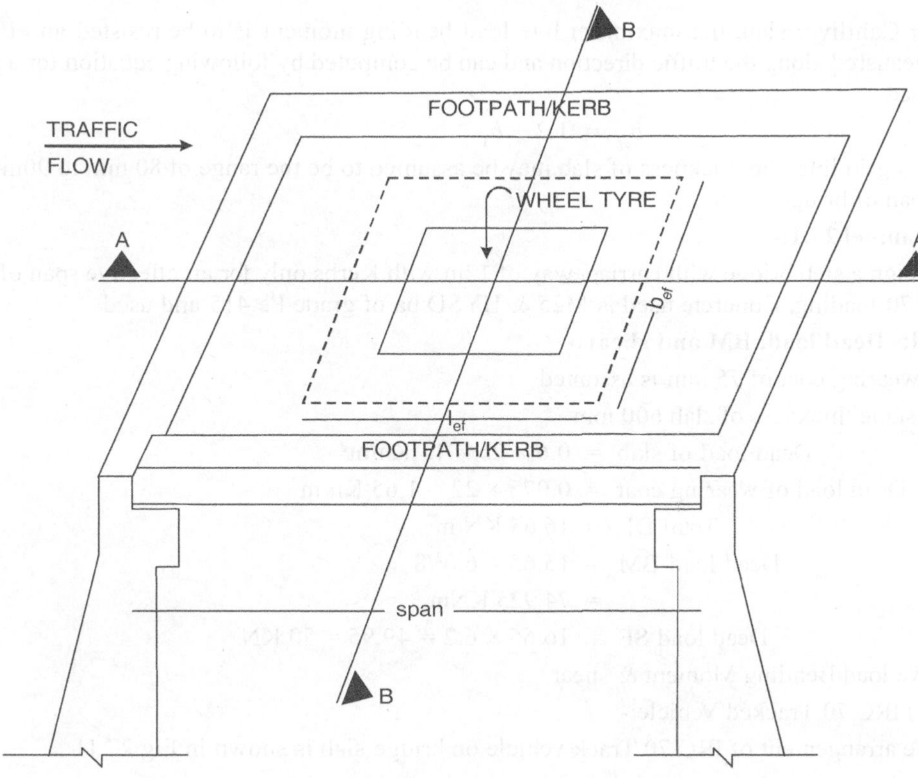

Fig. 27.10.(a). Dispersion of Tyre Pressure

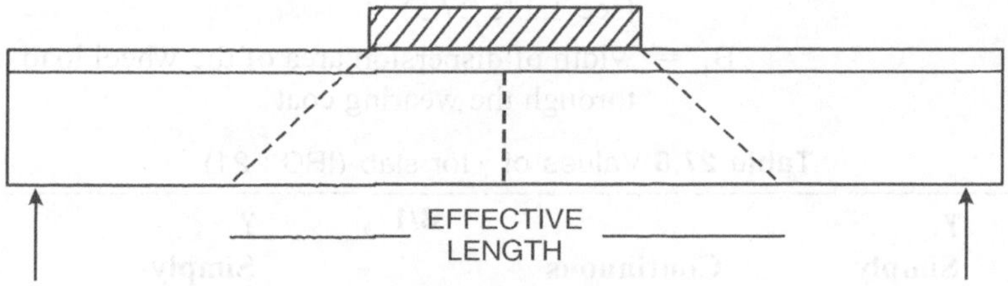

Fig. 27.10(b). Section at AA

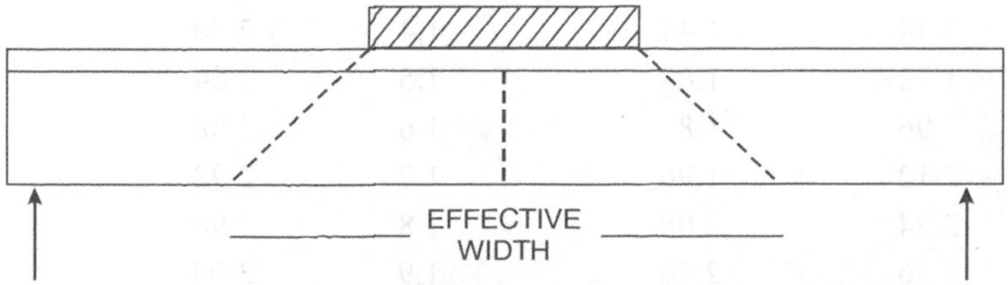

Fig. 27.10(c). Section at BB

For Cantliver slab, the maximum live load bending moment is to be resisted an effective width meausred along the traffic direction and can be computed by following equation for a single wheel load

$$b_{ef} = 1.2 + b_1$$

As a grid line, the thickness of slab may be assumed to be the range of 80 mm to 90mm per meter span of bridge.

Exampel 27.1:-

Design a slab bridge with carriageway of 12m with Kerbs only for an effective span of 6.0m for IRC 70 loading. Concrete used is M25 & HYSD ba of grade Fe 415 and used

Sol:- Dead load, BM and shear:-

A wearing coat of 75 mm is assumed

Assume thickness of slab 600 mm

$$\text{Dead load of slab} = 0.6 \times 25 = 15 \text{ Kn/m}^2$$

$$\text{Dead load of wearing coat} = 0.075 \times 22 = 1.65 \text{ Kn/m}^2$$

$$\text{Total DL} = 16.65 \text{ KN/m}^2$$

$$\text{Dead load BM} = 15.65 \times 6.0^2/8$$

$$= 74.925 \text{ KNm}$$

$$\text{Dead load SF} = 16.65 \times 6.2 = 49.95 = 50 \text{ KN}$$

Live load Bending Moment & shear

(A) IRC 70 Tracked Vehicle:-

The arrangement of IRC 70 Track vehicle on bridge slab is shown in Fig 27.11.

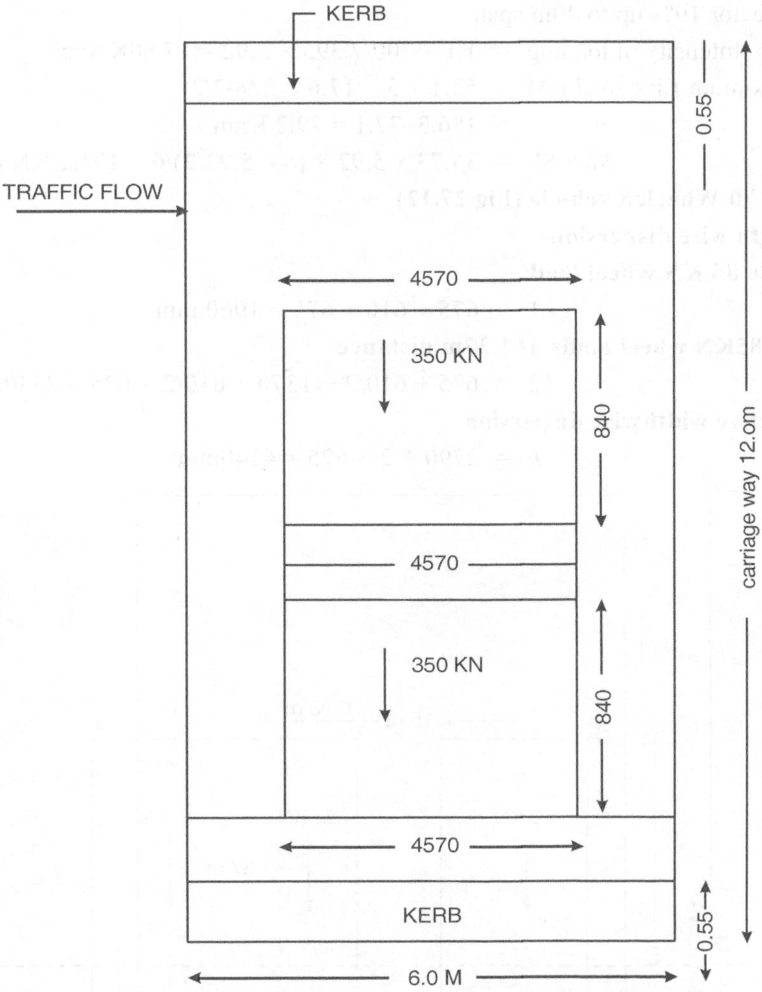

Fig 27.11. Arrangement of IRC 70 Tank Vehicle on deck slab

Dispersed Tracked load length

$= 4.57 + 0.675 \times 2 = 5.92\text{M}$

The effective width of dispersion for a single track (wheel)

Width of deek slab $= 12\text{m}$

$L = 6\text{m}$

$K = \text{B}/1 = 12/6 = 2$

From table $\gamma = 3.00$

Center of gravity, distance x = of the track (wheel) from the support = 6.2 = 3m

$b_1 = 0.84 + 2 \times 0.075 = 0.99$

$bef = \text{g.n}\,(1\text{–x}/1) + b_1$

$= 3.0 \times 3 \times (1 - 3/6) + 0.99$

$= 4.5 + 0.99 = 5.49\text{m}$

Effective width of dispersion for two wheels

$= 2.06 + 2.174 + 3.165 = 7.395\text{m}$

Impact factor 10% up to 40m span

$$\text{Intensity of loading} = 1.1 \times 700/7.395 \times 5.92 = 17.60\text{KN/m}^2$$

$$\text{Maximum Live load BM} = 52.1 \times 3 - 17.6 \times 2.962/2$$

$$= 156.3 - 77.1 = 79.2 \text{ Knm}$$

$$\text{Max SF} = 33.73 \times 5.92 \times (6 - 5.92/2)/6 = 101.2 \text{ KN/m}$$

(B) IRC 70 Wheeled vehicle (Fig 27.12)

a) Length wise dispersion

***i)* Single 85 KN wheel load**

$$L1 = 675 + 610 + 675 = 1960 \text{ mm}$$

***ii)* Two 85KN wheel loads at 1.37m distance**

$$L2 = 675 + 610/2 + 1370 + 610/2 + 625 = 3330\text{mm}$$

b) Effective widthwise dispersion

$$b = 2790 + 2 \times 675 = 4140\text{mm}$$

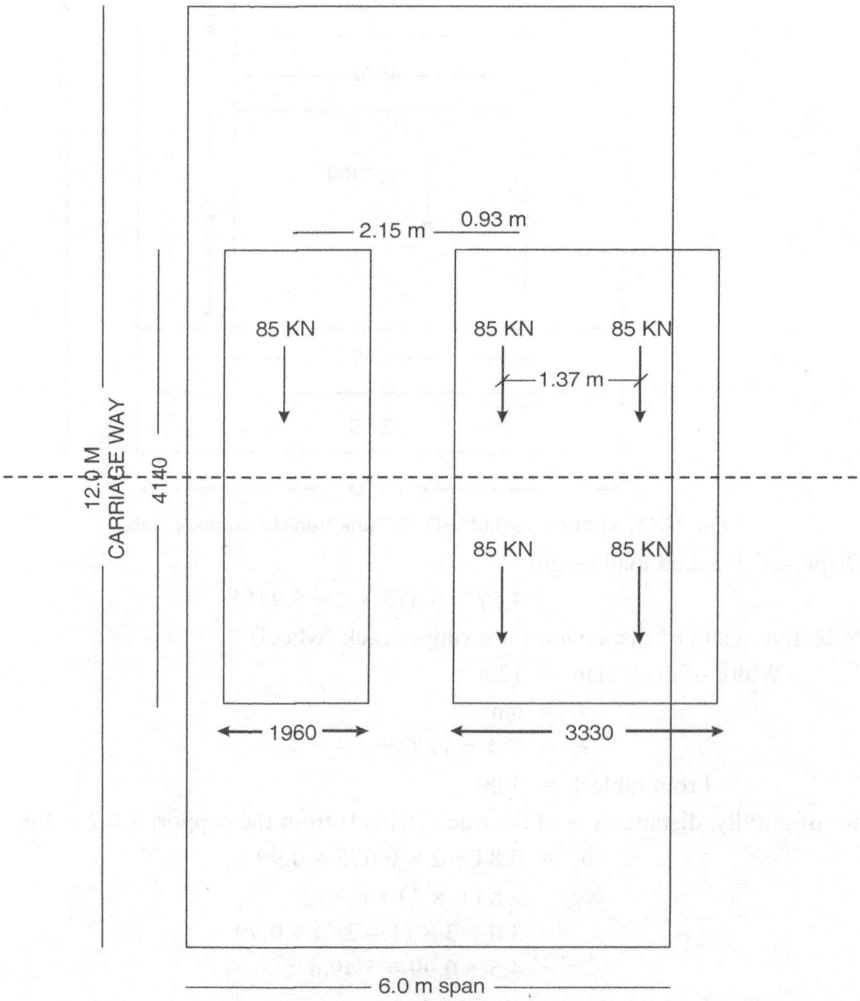

Fig 27.12. IRC 70 Wheeled Vehicle

Single axel load of 170 bm equivalent

$$Udl = 170 \times 2/3.33 \times 4.14 = 24.66 \text{ KN/m}^2$$

$$RA = 21.95 \times (6 - 0.9175) \times 1.835 + 24.55 \times 3.165 \times 1.5825/6$$

$$= 204.72 + 123.51/6 = 54.7 \text{ KN}$$

$$\text{Total udl load} = 21.95 \times 1.835 + 24.66 \times 3.165$$

$$= 40.28 + 78.05 = 118.33$$

$$RB = 118.33 - 54.7 = 63.63 \text{ Knm}$$

$$\text{Max BM at midspan} = (63.63 \times 3 - 24.66 \times 3.0 \times 1.5) \times 1$$

$$= 190.89 - 110.9) = 79.92 \text{ Knm}$$

$$\text{Impact} = A/B + L = 4.5/6 + 6 = 0.375$$

c) Two lanes of class A loadings (Fig 27.13)

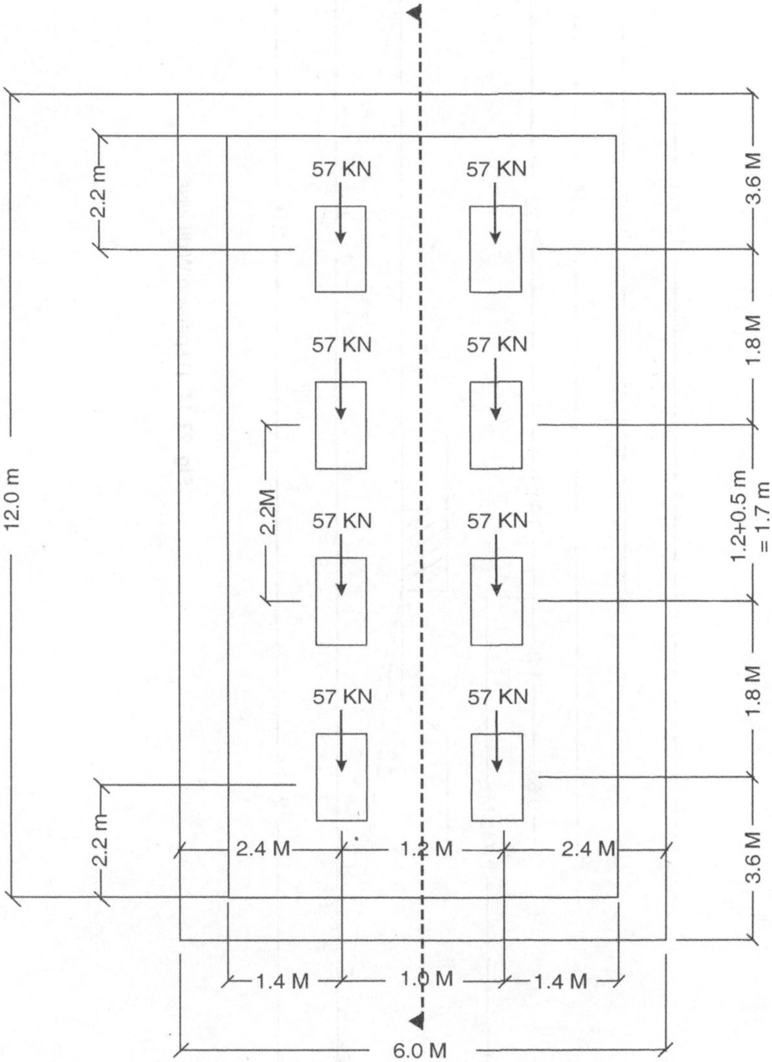

Fig. 27.13. Two Lanes of Class 'A' Loadings

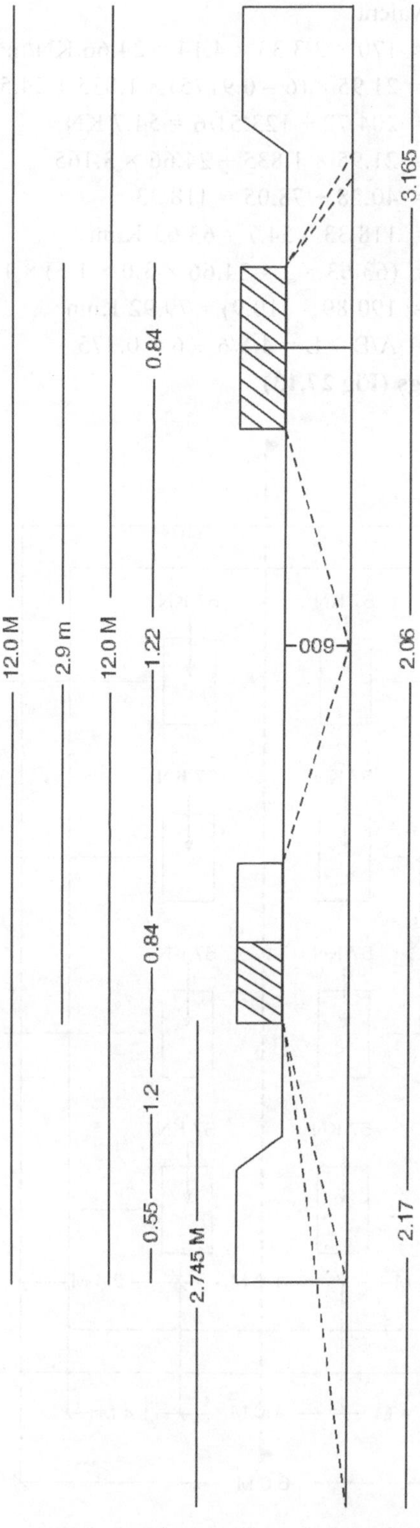

Fig. 27.14. Dispersion Widthwise

(a) Dispersion along length of one wheel

L = 1.2 + 0.25 + 0.675 × 2 = 2.8m

(b) Effective width (Fig 27.14)

Bef = g.x (1–x/l) + b1

= 2.88 × 2. 4 (1 – 2.4/6) + 0.25

= 4.4m

Total Dispersion widthwise

= 2.2 + 1.2 + 2.2 = 5.6 m

Eqivalent *udl* = 57 × 8/3.8 × 5.6 = 21.5KN/m^2

Reaction at A = 21.5 × 2. 8 × 3.0/6 = 60.2KN

Max BM at mid span = 60.2 × 3 – 21.5 × 2.8^2/2

= 180.6 – 84.28 = 96.32 KNm

Impact = 0.375

Max BM at mid span including impact

= 96.32 × 1.375 = 132.44 KNm

Max SF = (4 × 57 + 57 × 4 × 4.8/6) × (1.375/5.6)

= 536.8/5.6 KN = 95.9 KN

Max BM due to DL + LL including impact

= 74.925 + 132.44 = 207.365 KNm

Effective depth required

= 207.365 × 10^6/1.1 × 10^3 = 434.2mm

D = 600mm to 450mm at kerbs

Average d = 600 + 450/2 – 40 = 485m

Area of steel required

Ast = 207.365 × 10^6/200 × 0.9 × 485 = 2375mm^2

Provide 20mm dia tor @ 125mm c/c

Thus providing an area of steel reinforcement

= 314 × 1000/125 = 2512mm^2

P = 2572 × 100/100 × 485 = 0.578%

Distt steel = 0.12 (600 + 450/2) × 1000/100 = 630mm^2

Provide 12mm dia tor 175mm c/c and at top 10mm dia tor @ 220mm c/c bothways (As provided = 78.5 × 1000/220 = 365 mm^2 as against 1/2 × 630 = 315mm^2)

Check for

Max SF DL + LL including impact

= 50 + 114.9 = 164.9 KN/m

Nominal shear stress = 164.9 × 1000/100 × 485 = 0.34 N/mm^2

Permissible shear stress = 0.32 N/m^2

The difference is too small & can be neglected

Design of kerb

The Kerb may be designed for a live load of 5 KN/m^2

Height of Kerb = 450 + 450 = 900mm

Width of Kerb = 550mm

Live load per meter of kerb = 5 × 0.9 = 4.5 Kn/m

Dead load of kerb = 0.9 × 0.55 × 25 = 12.375 KN/m

Weight of railings = 0.5 KN/m

Total DL + LL = 17.375 KN/m say 18

Max BM $18/6^2/8$ = 81 Knm

Concrete M35

$$D = 81 \times 106/\ 1000 \times 1. = 271.4 \text{ mm}$$

$$D \text{ provided} = 900 \text{ mm } d = 850\text{mm}$$

$$\text{Ast} = 81 \times 106/200 \times 0.9 \times 860 = 523\text{mm}^2$$

Provide 12mm dia tor 5 Nos at top & bottom

Provide shear stirrups 10mm dia tor 2L @ 280mm c/c

The reinforcement arrangement in deck slab is shown in Fig 27/15.

27.7 DESIGN OF ABUTMENT (Fig 27.15)

An abutment will carry the superstructure deck of the bridge and will also retaining the earth from approach abutment. Abutment may be plain or reinforced concrete. The abutment may be in plaon or reinforced concrete. The abutment may be either solid type, couterfore type, hollow type. An abutment is to resist the following forces or loads.

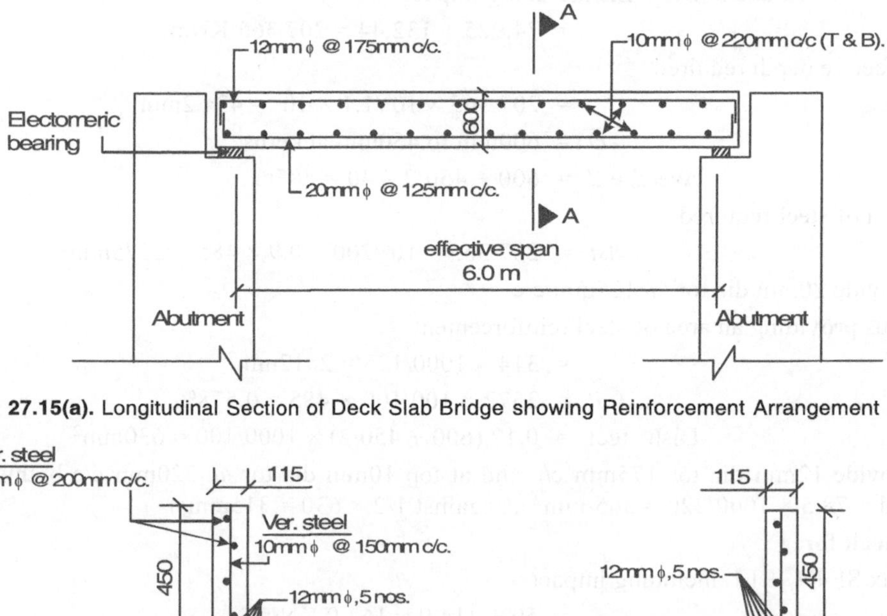

27.15(a). Longitudinal Section of Deck Slab Bridge showing Reinforcement Arrangement

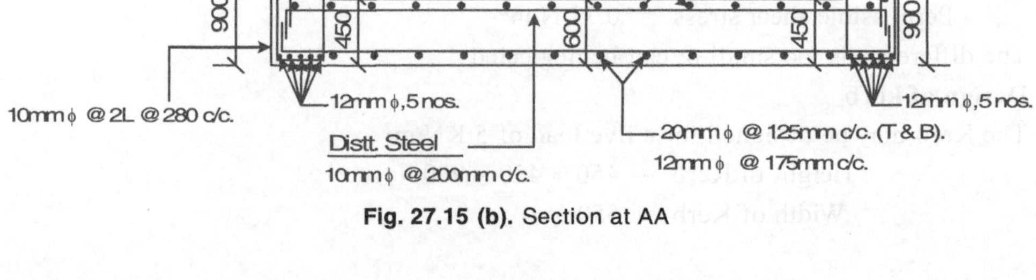

Fig. 27.15 (b). Section at AA

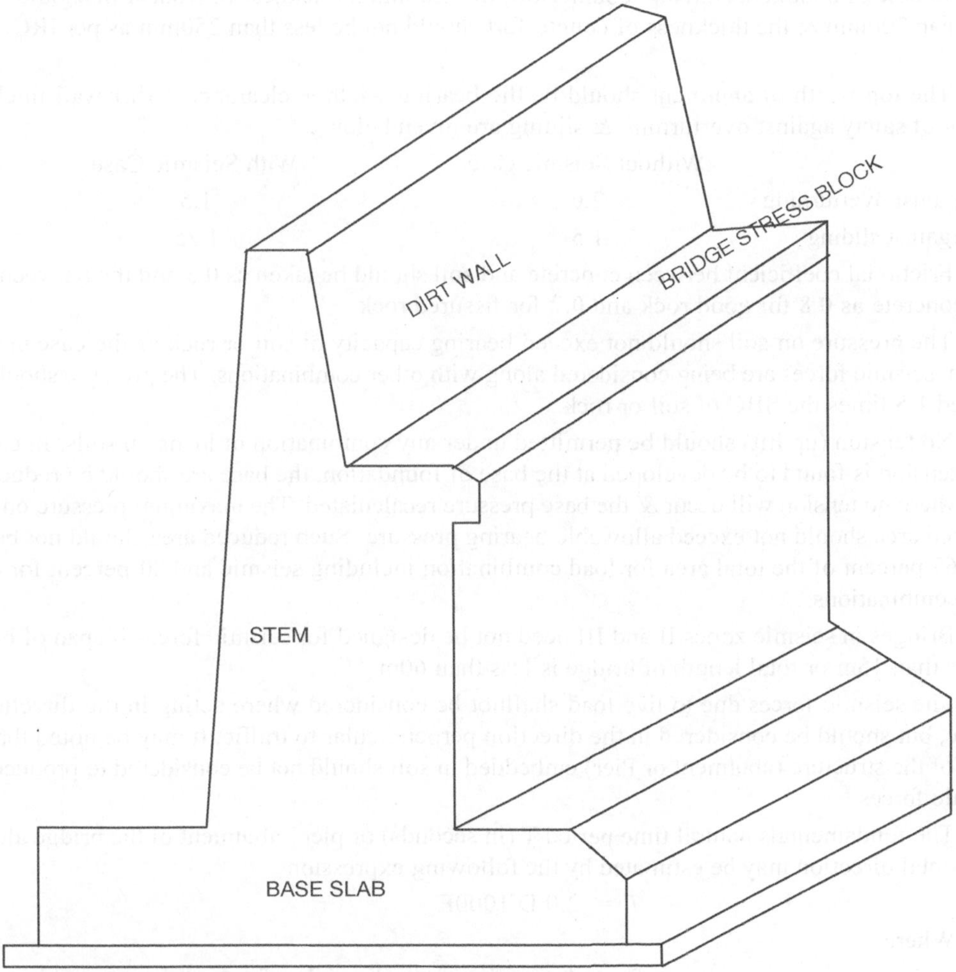

Fig. 27.16. Abutment

1. Self weight of abutment
2. Earth pressure acting at height of 0.4 height of earthfill.
3. Dead of bridge deck
4. Live load with impact with surcharge load equivalent to 1.2m height of earthfill
5. Live load with brakin effects
6. Scour from high floods
7. Water current forces
8. Buoyancy
9. Wind Forces.
10. Seismic forces:- While calculating seismic forces horizontally and vertically, Following Formulae as per IRC 6 : 2000 is to be used

$$\alpha_h = \frac{Z \cdot I}{2 \cdot R} \cdot \frac{S_a}{g}$$

In case of abutments having counterfort, the minimum thickness of front wall should not be less than 200mm & the thickness of counterfort should not be less than 250mm as per IRC : 78 – 2000

The top width of abutment should be the bearing width + clearance + dirt wall thickness Factor of safety against overturning & sliding are given below

	Without Seismic case	With Seismic Case
(*i*) Against overturning	2.0	1.5
(*ii*) Against sliding	1.5	1.25

Frictional coefficient between concrete and soil should be taken as 0.5 and the between rock and concrete as 0.8 for good rock and 0.7 for fissured rock

The pressure on soil should not exceed bearing capacity of soil or rock as the case may be. When seismic forces are being considered along with other combinations, The pressure should not exceed 1.5 times the SBC of soil or rock.

No tension (up lift) should be permitted under any combination of loads on soils. In case of rock tension is found to be developed at the base of foundation, the base are should be reduced to size where no tension will occur & the base pressure recalculated. The maximum pressure on such reduced area should not exceed allowable bearing pressure. Such reduced area should not be less than 67 percent of the total area for load combination including seismic and 80 percent for other load combinations.

Bridges in seismic zones II and III need not be designed for seismic forces if span of bridge is less than 15m or total length of bridge is less than 60m

The seismic forces due to live load shallnot be considered where acting in the direction of traffic, but should be considered in the direction perpendicular to traffic. It may be noted that the parts of the structure (abutment or Pier) embedded in soil should not be considered to produce any seismic forces.

The fundamentals natural time period T (in seconds) or pier / abutment of the bridge along a horizontal direction may be estimated by the following expression

$$T = 2.0 \text{ D/1000F}$$

Where

D = Approximate dead weight of supersturcutre and live load in KN

F = Horizontal force in KN required to applied at the center of man of super structure for one mm horizontal deflection at top of the pier / abutment along the direction of horizontal force.

Maximum scour depth below height F/Load level (HFL) for the design of piers and abutments having individual foundations shall computed from the following equation.

$$\text{Mean scour depth} = d_{sm} = 1.34(Db^2 / K_{sf})^{1/3}$$

Where Db = the design discharge for foundation per meter width of effective waterway

Kef = Silt factor (Table 27.1)

Ksf can be more depending upon the size boulders & gravel

Maximum scour depth = 2.0dsm for piers

= 1.27 d_{sm} for abutments on lowest level

= 2.0 d_{sm} with scour all around

Maximum scour depth for design should be taken as 0.9 × maximum scour depth.

Example 27.2. Design an abutment in concrete for a 6m span solid slab bridge with the following data

Carriageway = 12.0m

Overall length = 6.5m

Loading : IRC 70 Tracked or wheeled or two lanes of class A

Back fill:- Density of soil = 18 Kn/m^3

Angle of repose = 30°

Approach slab: 300 mm thk

Seismic zone *V*

Max Load from structure (Deck slab)

(*a*) Dead Load = 50KN/m

(*b*) Live load = 115 KN/m

SBC of soil strata = 450 Kn/m^2

Coeff of friction between concrete & soil = 0.5

Bed level = 91.9m

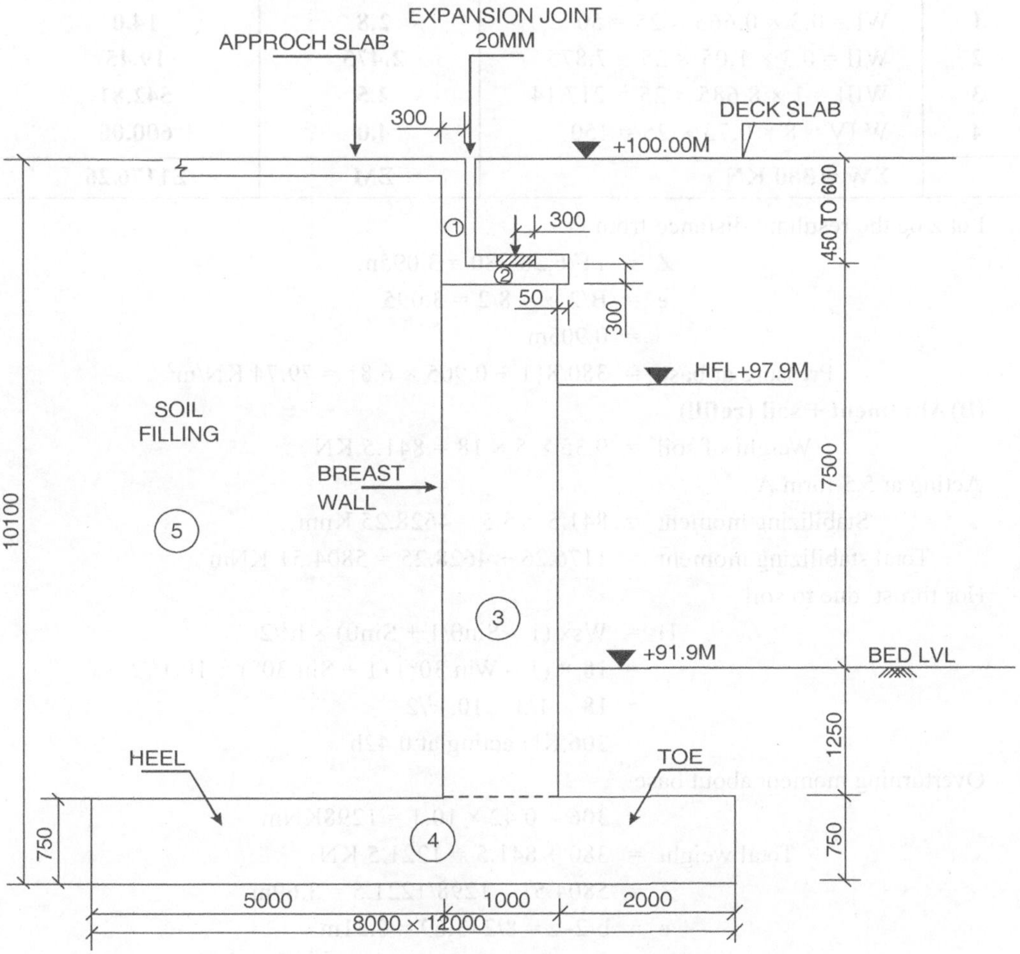

Fig. 27.17. Assumed Section Abutment

The section of abutment chosen is shown in Fig. 27.17.

Deck level = 100.00m

H.F.L = 97.9m

Max Discharge = 90 cumes

Discharge/m = $Qb = 90/5.4 = 16.66$ cumes/m

Taking silt factor = 1.0

Normal scour = $1.345\,(Qb/K)^{0.33} = 1.34\,(16.66/1)^{0.33} = 3.43$m

Max scour = $2 \times 3.43 = 6.86$m

Scour below bed level = $6.86 - (97.9 - 91.9) = 0.86$m

For checking stability of the abutment one has to consider various cases as given below per meter river of wall

(A) Abutment constructed alone

SNO	WEIGTH/LOAD (KN)	LIVER ARM (m)	MOMENT ABOUT (Knm)
1	WI = 0.3 × 0.665 × 25 = 50	**2.8**	**14.0**
2	WII = 0.3 × 1.05 × 25 = 7.875	**2.475**	**19.45**
3	WIII = 1 × 8.685 × 25 = 217.14	**2.5**	**542.81**
4	W IV = 8 × 0.75 × 25 = 150	**4.0**	**600.00**
	ΣW = 380 KN	**ΣM**	**Σ1176.26**

Let z be the resultant distance from O

Z = 1176.26/380 = 3.095m

e = B/2–z = 8/2 = 3.095

= 0.905m

Pressure at base = 380/8{1 + 0.905 × 6.8} = 79.74 KN/m^2

(B) Abutment + soil (refill)

Weight of soil = 9.35 × 5 × 18 = 841.5 KN

Acting at 5.5 form A

Stabilizing moment = 841.5 × 5.5 = 4628.25 Knm

Total stabilizing moment = 1176.26 + 4628.25 = 5804.51 KNm

Hor thrust due to soil

H = Wsx (1 – Sinθ/1 + Sinθ) × $h^2/2$

= 18 × (1 – Win 30°)/(1 + Sin 30°) × $10.1^2/2$

= 18 × 1/3 × $10.1^2/2$

= 306 KN acting at 0.42h

Overturning moment about base

= 306 × 0.42 × 10.1 = 1298KNm

Total weight = 380 + 841.5 = 1221.5 KN

Z = 5804.51 – 1298/1221.5 = 3.69m

e = b/2–z = 8/2–3.69 = 0.31m

Pressure at base = 1221.5/8 {1 + –6 × 0.31/8}

= 188.2 KN/m^2

FOS against overturning = 5804.51/1298 = 4.47m > 2.0

FOS against sliding = 0.5 × 1221.5/306 = 1.996 > 1. 5

(C) Abutment + soil (refill) + LL on Approach slab

In this case of live load on approach slab, one has take height of soil 1.2m extra than the actual height of soil

Overturning moment = 306 × 0.42 (10.1 + 1.2) = 1452.Knm

Max Live load = 700 KN/m = 58.33

Σ W = 1221.5 + 58.33 = 1279.83 say 1280 KN

z = 5804.51 – 1452.3/1280

= 3.40m

e = B/2-z

= 8/2–3.4 = 0.6m

Pressure at base = W/B{1 + –6xe/B}

= 1280/8 {1 + 6 × 0.6/8}

= 232 Kn/m^2

FOS against overturning = Σ W × 0.5/H

= 1280 × 0.5/306

= 2.09 > 1.50

(D) Abutment + soil + DL of deck + No. Live load

DL of deck slab = 50KN/m

Σ W = 1221.5 + 50 KN/m

Stabilizing moment due to DL of deck slab

= 50 × 2.3 = 115 Knm

Total stabilizing moment = 5804.5 + 115 = 5919.51 KNm

Overturning moment = 1298 KNm

Z = 5919.51 – 1298/1271.5 = 3.634m

e = B/2-*z*

= 8/2 – 3. 634 = 0.366m

= 1271.5/8{1 + 6 × 0.366/8} = 202.56 KN/m^2

(E) Abutment + soil + DL of deck slab + live load with impact

Live load with impact = 700 × 1.1/ 2 × 12 = 385/12 KN = 32.1Knm

Live load with impact due to IRC 70 wheeled vehicle

OMe = {170 + 170 × 4.63/6 + 170 × 1.58/6 + 170 × 0.21/6} 1/12 × 1.375

= 1.375 × 1/12 (170 + 131.2 + 44.76 + 5.95)

= 40.32 Knm

Stabilizing moment due to live load with impact

= 40.32 × 2.3 = 92.74Knm

Σ *W* = 1271.5 + 40.32 = 1311.82 kN

Total stabilizing moment = 5919.51 + 92.74 = 6012.25 KNm

Z = 6012.25 – 1298/1311.82 = 3.594m

Pressure at base = 11311.82/8.0 {1 + 6 × 0.406/8} = 213.9 KN/m^2

FOS against overturning = 6012.25/1298 = 4.63 > 2.0

FOS agianst sliding = 0.5 × 1311.82 / 306

= 2.14 > 1.50

(F) Abutment + soil + DL of deck slab + live load & braking (Longitudinal tractive forces)

Hor force due to braking @ 20% of live load = 40.32 × 20/100 = 8.064 Kn/m

Overturning moment due to horizontal force of braking

= 8.064 × 9.435 = 76.08 KN.m/m

z = 6012.5 – 1298 – 76.08/1311.82 = 3.536m

e = 8.0/2 – 3.536 = 0.464m

Pressure at base = 1311.82/8 {1 + 6 × 0.464/8}

= 221.04 Kn/m^2

Fos against overturning = 6012.5/1298 + 76.08 = 4.37 > 2.0

Fos against sliding = 1311.82 × 0.5/(306 + 8.064) = 2..09> 1.5

(G) Force due to temperature variation & shrinkage

Assuming moderate climate variation and shrinkage

Assuming moderate climate variation in temperature is taken as +17°C

Coeff of thermal expansion = 11.7×10^{-6} per°C

Strain due to temperature variation

= $17 \times 11.7 \times 10^{-6} = 1.99 \times 10^{-4}$

Strain due to concrete shrinkage

= 2.0×10^{-4}

Total strain due to temperatuer & shrinkage

= (1.99 + 2.0) × 10–4 = 3.99×10^{-4}

Hor deformation of deck due to temperature and shrinkage affecting one abutment

= $3.99 \times 10^{-4} \times 6600/2$

= 1.315mm

Strain in bearing = Deformation/Electrometric thickness

= 1.315/65 – 5 × 3

= 0.0236

Assuming shear modulus G = 1N/mm^2 and number of bearing 12 Nos of size 309 × 488

Hor force due to strain in longitudinal direction at bearing level

= 1.10 × Strain × 6 × area of plans in bearing × No. of bearing / 1000 × width

= 1.1 × 0.063 × 1 × 308 × 488 × 12/1000 × 12 = 4.38 KN/m

Overturning moment = 4.348 × 9.435 = 41KNm

(H) Extending case (F) further

z = 6012.5 – 1298 – 16.0841 / 1311.82 = 3.5m

e = 8.0/2 – 3.5 = 0.5m

Pressure at base = 131182/8 { 1 + 6 × 0.5/8} = 225.46 Kn/m^2

FOS against overturning

= 6012.5/1298 + 76.8 + 41 = 6012.5 / 1415.08

= 4.25 > 2.0

FOS against sliding = 1311.82 × 0.5/306 + 8.06 + 4.348

= 1311.82 × 0.5/318.41

= 2.06 > 1.5

(I) Abutment + soil + DL of deck slab + seismic forces in the direction of traffic

(Live load in this case is not to be considered)

To complete the weight of abutment that is not embedded in soil below bed level.

SNO	WEIGTH/LOAD (KN)	LIVER ARM (m)	MOMENT ABOUT (Knm)
1	W I 5.0 KN	**2.8**	**14.0**
2	W II = 0.3 × 1.05 × 25 = 7.875 KN	**2.475**	**19.45**
3	W III = 1 × 7.135 × 25 = 178.375 KN	**2.5**	**445.9375**
4	W IV = 8.1 × 5 × 18 = 729 KN	**5.5**	**4009.5**
	DL of deck = 50KN	**2.3m**	**115.0**
		ΣM	**1176.26**

Total weight load above bed level

= 5.0 + 7.875 + 178.375 + 729 + 50 = 970.25 KN

To compute sesmic forces as per IRC : 6 : 2000

Hor seismic coefficient for sesmic zone V

Ah = Z.I/2R. Sa/g

= 0.36 × 1.5 × 2.5/2 × 2. 5 = 0.27

Hor force due to earthquake

= 0.27 × 970.25 = 261.9675 KN

Vertical forces due to earthquake

= Half of hor force

= 261.9675/2

= 130.98 KN

Hor thrust due to soil = 306 KN

OM due to soil = 306 × 0.5 × 10.1

= 1545.3 Knm

To compute CG of all loads weights above bed level, taking moments about bed level

SNO	WEIGTH/LOAD (KN)	LIVER ARM (m)	MOMENT ABOUT (Knm)
1	W I = 5.0 KN	7.8	39.0
2	W II = 0.3 × 1.05 × 25 = 7.875 KN	7.285	57.37
3	W III = 1 × 7.135 × 25 = 178.375 KN	3.567	636.26
4	Soil Part = 729	4.05	2952.45
5	DL of deck = 50 KN	7.5	375.0

Let y be height above bed level of CG of W1, W2, W3, Soil part & DL

= 5.0 + 7.875 + 178.375 + 729 + 50)

xy = 39.0 + 57.37 + 636.26 + 2952.45 + 375

= 970.25 y = 4060.08

y = 4.18m

OM due to earthquake about O = 261.9675 × (4.1872)

= 1618.96 KNm

In fact one has to consider earthquake forces & overturning moments (worst) in case (D)

ΣW = 1271.5 + vertical component of earthquake

= 1271.5 + 130.98

= 1402.48 KN or 1140.52 KN

Hor forces due to earthquakes = 261.9675 KN

Hr force due to soil = 306 KN

Total hor force due to soil and earthquake

= 306 + 261.9675 = 567.9675 KN

Total overturning moment about O due to soil & earthquake

= 1545.3 + 1618.96 = 3164.26 Knm

z = 5919.5 – 3164.26/1402.48 = 1.9645m

e = 8.0/2–1.9645 = 2.0355

Pressure at base = 1402.48/8{1 + 6 × 2.0355/8}

= 175.31 × {1 + 1.526)

= 442.858 or –92.21 KN/m^2

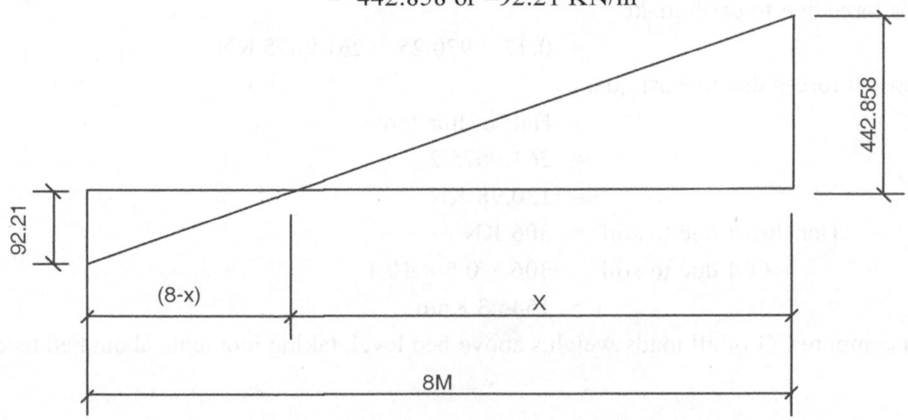

Fig. 27.18

Pressure distribution is shown in Fig 27.18

$$\text{SBC of soil} = 450 \times 1.5 = 675 \text{ KN/m}^2$$

$$442.858/x = 92.21/8 - x$$

$$= 3542.86 - 442.828x = 92.21x$$

$$= 535.07x = 3542.86$$

$$x = 6.62\text{m}$$

Therefore 82.75% of base in under compression & balance 17.25% is under up lift As per IRC : 6 : 2000, compression and 80% or more is acceptable

Section modulus under compa area

$$= 1 \times 6.622/6 = 7.3\text{m}^3$$

$$\text{Pressure at base} = 1402.48/6.62 \{1 + 6 \times 2.0355/6.62\}$$

$$= 211.85 \{1 + 1.849)$$

$$= 602.5 \text{ KN/m}^2 < 675\text{KN/m}^2$$

Hence safe

(J) Abutment + soil + DL of deck slab + semic force in direction perpendicular to

Traffic + LL From case E

Total load W with earthquake = 1311.82 + 130.98 = 1442.80 KN

OM due to soil direction traffic = 1545.3 KNm

OM due to earthquake in direction perpendicular to traffic

$$= 1618.96 \times 12 = 2428.44 \text{ Knm}$$

$$\text{Pressure at base} = W/A + M_T + M_P/Z_P$$

$$= 1442.8 \times 12/12 \times 8.0 + 545.3 \times 6/12 \times 82 + 2428.44 \times 6/1 \times 122$$

$$= 180.35 + 144.87 + 103.27$$

$$= 428.49 \text{ KN/m}^2 < 450 \times 1.5 = 675 \text{ kN/m}^2$$

Design of stem or Breast wall

$$H = 10.1\text{M}$$

$$\text{Max BM} = 18 \times (1 - \sin 30/1 + \sin 30) \times 1012/2 \times 0.42 \times 10.1$$

$$= 1298.2 \text{ KNm}$$

M30 concrete

$$d = 1298.2 \times 10^6/100 \times 1.65 = 887 \text{ mm}$$

$$D = 1000 \text{ mm} = 950\text{mm}$$

$$\text{ast} = 1298.2 \times 10^6/ 200 \times 0.9 \times 950 = 7590\text{mm}^2$$

Provide 28 dia tor @ 80mm c/c

Upto 2.5m height @ 16mm c/c for next 2.5m & 320mm c/c in top balance portion on soil face

$$\text{Distt steel} = 0.12 \times 1000 \times 100/100 = 1200\text{mm}^2$$

Provide vertical 16mm dia tor @ 150mm c/c on river bed side

Provide 12mm dia tor 150mm c/c horizontally each face

Design of Toe

$$\text{Max design pressure} = 225.46^2/2 = 450.92 \text{ Knm}$$

M30 concrete

$$D = 450.92 \times 106/1000 \times 1.65 = 522\text{mm}$$

$$D = 750\text{mm} \quad d = 700\text{mm}$$

$$Ast = 450.92 \times 10^6/200 \times 0.9 \times 700 = 3578\text{mm}^2$$

Provide 25 mm dia tor @ 125mm c/c

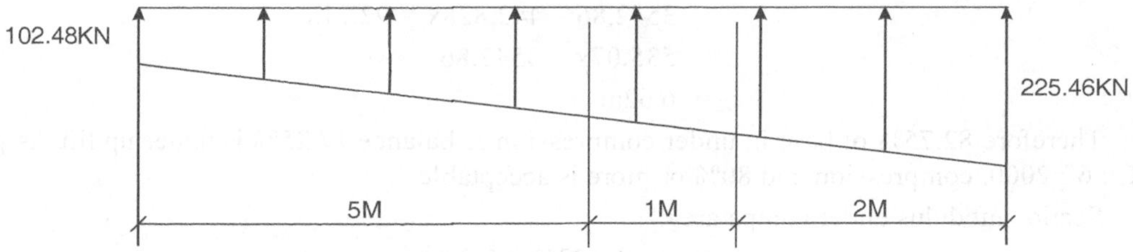

Fig. 27.19. Pressure Distribution

Design of Heel

$$\text{Max weight on it} = 18 \times 9.35 = 168.3 \text{ Kn/m}$$

$$\text{Self wt of slab} = 0.75 \times 25 = 18.70$$

Upward pressure on heel

$$\text{Design downward pressure} = 187 - (102.48 + 173.1/2)$$

$$= 187 - 137.79 = 49.21 \text{ say } 50 \text{ Kn/m}^2$$

$$\text{Max BM} = 50 \times 5^2/2 = 625 \text{ KNm}$$

$$D = 625 \times 1000/1.65 = 615 \text{ mm}$$

$$D = 750\text{mm} \quad d = 700 \text{ mm}$$

$$Ast = 625 \times 10^6/200 \times 0.9 \times 700 = 4960\text{mm}^2$$

Provide 25 dia tor @ 100 mm c/c

$$\text{Distt steel} = 0.12 \times 750 \times 1000/100 = 900\text{mm}^2$$

Design of stress Block or AButment cap 300mm thk

Provide 10% steel equally distributed at top & bottom & bothways

$$\text{Reinforcement} = 1.0 \times 300 \times 1000/100 = 3000\text{mm}^2$$

$$= 750\text{mm}^2 \text{ each way T \& B}$$

Provide 12 dia tor @ 100mm c/c T & B bothways

In addition provide 8mm dia tor @ 100 c/c bothways at 20mm from top face

12 dia tor @ 100 c/c each face bothways

The reinforcement arrangement in abutment section is shown in Fig. 27.20.

27.8 DESIGN OF PIER

Piers are provided to break large span into smaller spans for economy purpose. These piers may be solid, cellular, tressle or hammerhead. The loads and forces to be considered for design of piers are written below.

1. Dead load of superstructure deck and pier itself.
2. Live load
3. Impact
4. Bouyancy
5. Wind on moving loads
6. Water current
7. Longitudinal, tractive and braking forces
8. Seismic effects

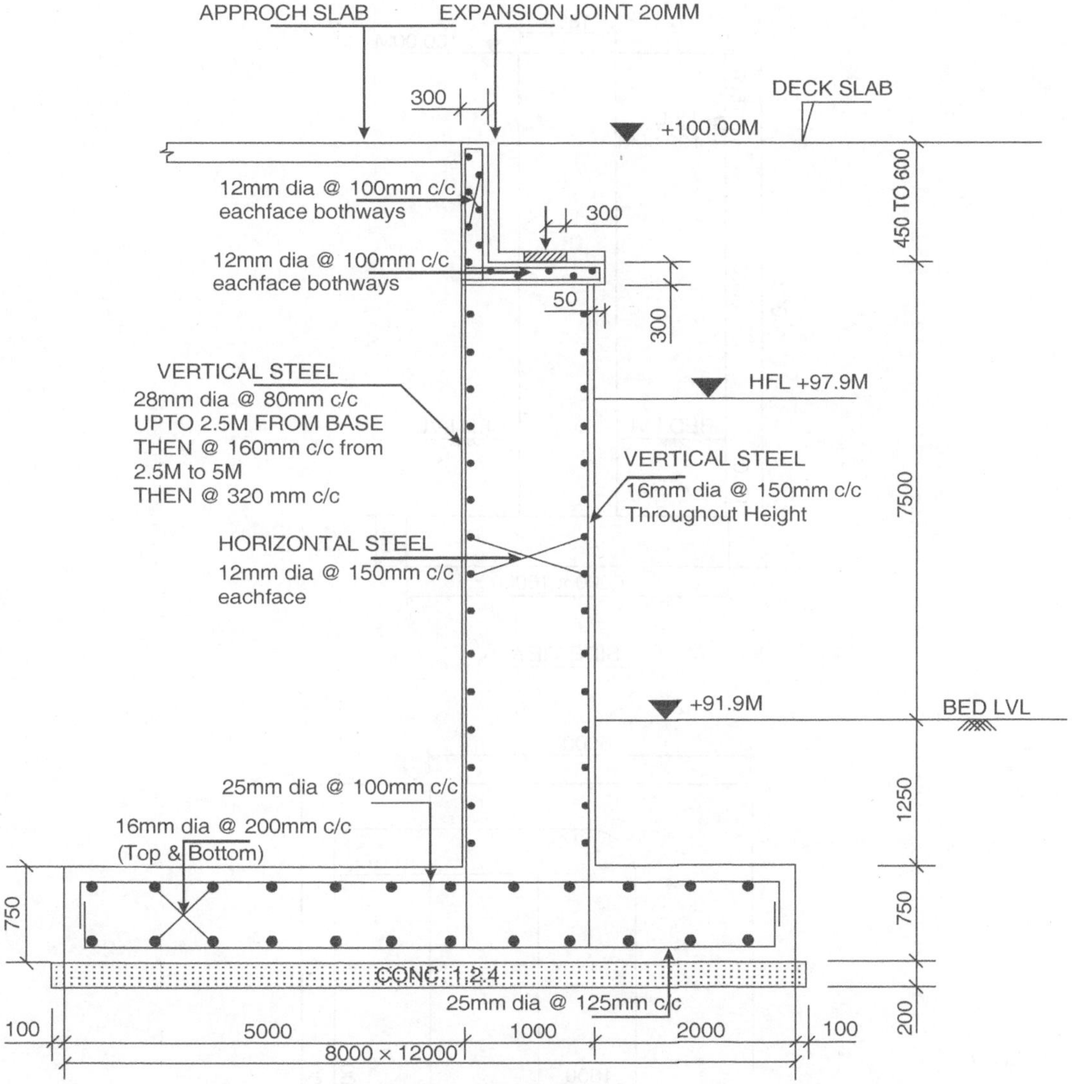

Fig. 27.20.

Design pier for a T-beam bridge 18.0 m span in concrete M30 as shown in Fig 27.21 with following data

(*a*) Dead load from each span = 1324KN

(*b*) Reaction due to Live Load on one span = 850KN

(*c*) Max discharge of river rom two spans 18.0m wide cumecs

(*e*) Live load : IrC 70 Tracked or wheeled or two lanes of class A

(*f*) Carriage way = 12m

(*g*) SBC of soil = 400 KN/m^2

(*h*) Velocity of water = 3.0m/sec

(A)Pressure on soil due to dead weight and self weight of

$$\text{pier} = \{1.8 \times 12 + \Pi / 4 \times 1/82\} \times 9 \times 25$$
$$= (21.6 + 2.54) \times 9 \times 25 = 5432.26\text{KN}$$

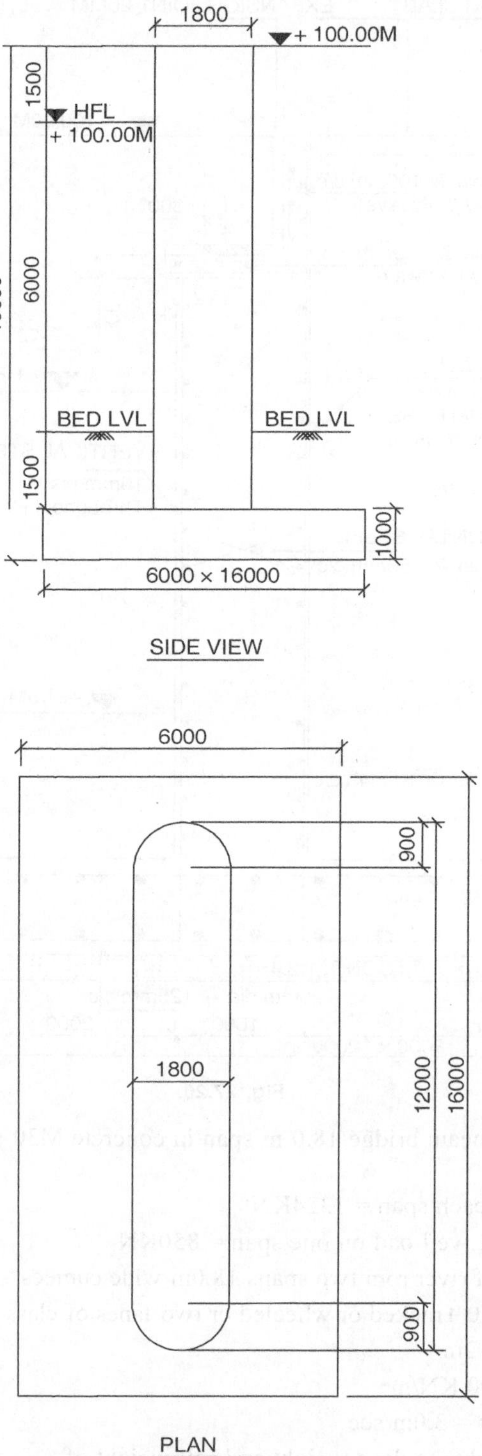
1800
+ 100.00M
1500
HFL
+ 100.00M
10000
6000
BED LVL
BED LVL
1500
1000
6000 × 16000
SIDE VIEW
6000
900
1800
12000
16000
900
PLAN

Fig. 27.21.

DL from bridge = 2 × 1324 = 2648 KN

Self weight of base slab = 16 × 6 × 1 × 25 = 2400 KN

Total direct load at base slab = 5432.26 + 2648 + 2400 = 10480.26 say 105500KN

Direct Pressure at base = 16500/6 × 16 = 109.375 KN/m^2

(B) Effect of buoyancy

Submerged volume of pier = 7.5 × 1.8 5 12 + 16 × 6 × 1

= 162 + 96 = 258cum

Reduction in weight of pier & its base due to buoyancy

= 258 × 10 = 2580KN

Upward Pressure due to buoyancy on base

= 2580/6 × 16 = 26.82 KN/m^2

(c) Pressure due to eccentrically of live load

Moment due to eccentrically = 850 × 0.45 = 382.5 Knm

Section modulus of Pier base

= 16 × 6^2/6 = 96m^3

Pressure at base due to eccentrically of live load

= 382.5/96 = 3.98 KN/m^2

(d) Pressure due to longitudinal forces

(*i*) Due to tractive effort or braking

Longitudinal force = 0.2 × 850 = 170 KN

Moment at base = 170 × 10 = 1700 Knm

Pressure at base = 1700/96 = 17.7 Kn/m^2

(*ii*) Due to resistance in bearing movement due to temperature. For worst case live may be on left span, the coeff of friction on left bearing & right bearing may be taken as 0.25 and 0.225

Total resistance by left side bearing

= 0.25 (1324 + 850) = 543.5KN

Total resistance by right side bearing

= 0.225 (1374) = 287.9 KN

Unbalanced force at bearing

= 543.5 – 297.9

= 245.6 KN

Moment at base = 245.6 × 10

= 2456 Knm

Pressure at base = 2456/96 = 25.58 Kn/m^2

(e) Pressure due to wind forces

Assuming the deck slab + railings to be 1.2m thick

Wind pressure for 10m height from

Table = 0.91 Kn/m^2

Max Force due to wind

= 1.2 × 18 × 0.91 = 19.656 KN

Moment at base = 19.656 × 10 = 196.56 KN

Pressure at base = 196.56/96

= 2.05 Kn/m^2

(f) Pressure due to water Current

Intensity of pressure = $0.5\ KV^2$

= $0.5 \times 0.66 \times 3.0^2 = 2.97\ KN/m^2$

Force due to water current = $13.8 \times 6 \times 2.97$

= 245.92 KN

This force acts at a height of $2/3 \times 6 = 4.0^m$

Above base

Moment at base = $245.92 \times 4.0 = 993.68$ KNm

Pressure at base = $+993.68 \times 6 = 3.88\ KN/m^2$

If the current direction varies by 20 degrees

Pressure parallel to pier = $2.97 \times \cos 20$

= $2.79\ KN/m^2$

Pressure perpendicular to pier = $2.97 \times \text{Sin } 20$

= $10.02\ KN/m^2$

Pressure at base due to Component parallel to pier

= $+ 3.88 \times 2.79 = 3.64 KN/m^2$

Force perpendicular to pier = $1.02 \times 13.8 \times 6 = 84.46$ KN

Moment at base = $84.46 \times 4.0 = 337.84$ KN/m

Pressure at base due to Component pezrpendicualr to pier

$$= \frac{33.764 \times 6}{16 \times 6^2} = 3.52 KN/m^2$$

The summary presumes at base is presented in Table 27.7

Table 27.7. Summary of Pressures on Pier

Sr.No.	Loads		Pressure No.Flood	KN/m^2 During Flood
1. Dead load and Self weight			+ 109.375	+ 109.375
2. Buoyancy			–	–26.82
3. Eccetric live load			+3.98	+3.98
4. Longitudinal Forces				
(a) Tractive Effort			+ 17.70	+ 17.70
(b) Bearing Resistance			+ 25.58	+ 25.58
5. Wind Forces			+ 2.05	+ 2.05
6. Water Current				+ 85.48
Net		Maximum	158.685	244.165
		Maximum	66.95	48.255

27.9 DESIGN OF T-BEAM BRIDGE

Example 27.4. Design a R.C.C T-Beam 40° Skew bridge with following data Fig (27.22)

(*a*) Carriage way 10900mm

(*b*) With Kerbs 550mm each side

(*c*) IRC 70 Tracked & wheeled vehicle

(*d*) Class A two Lanes

(*e*) Span 18m Skew 40° i.e. span of Girder beams 24 meters

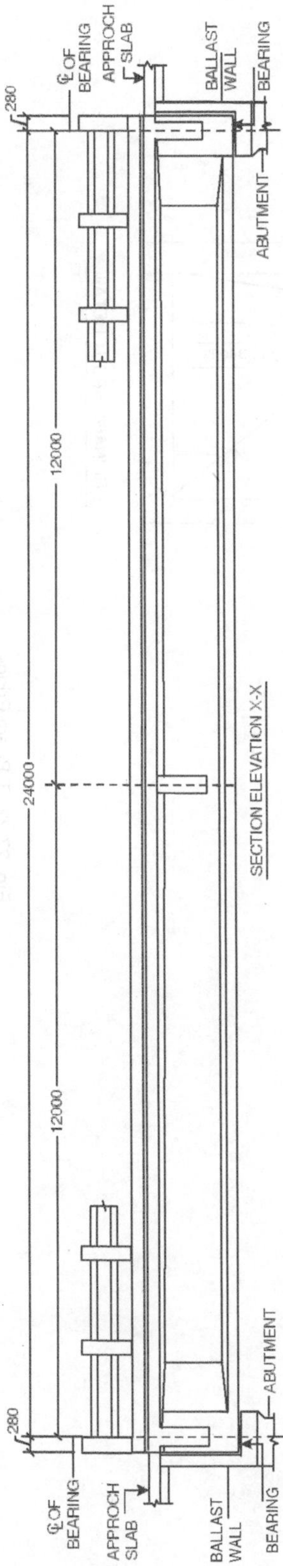

SECTION ELEVATION X-X

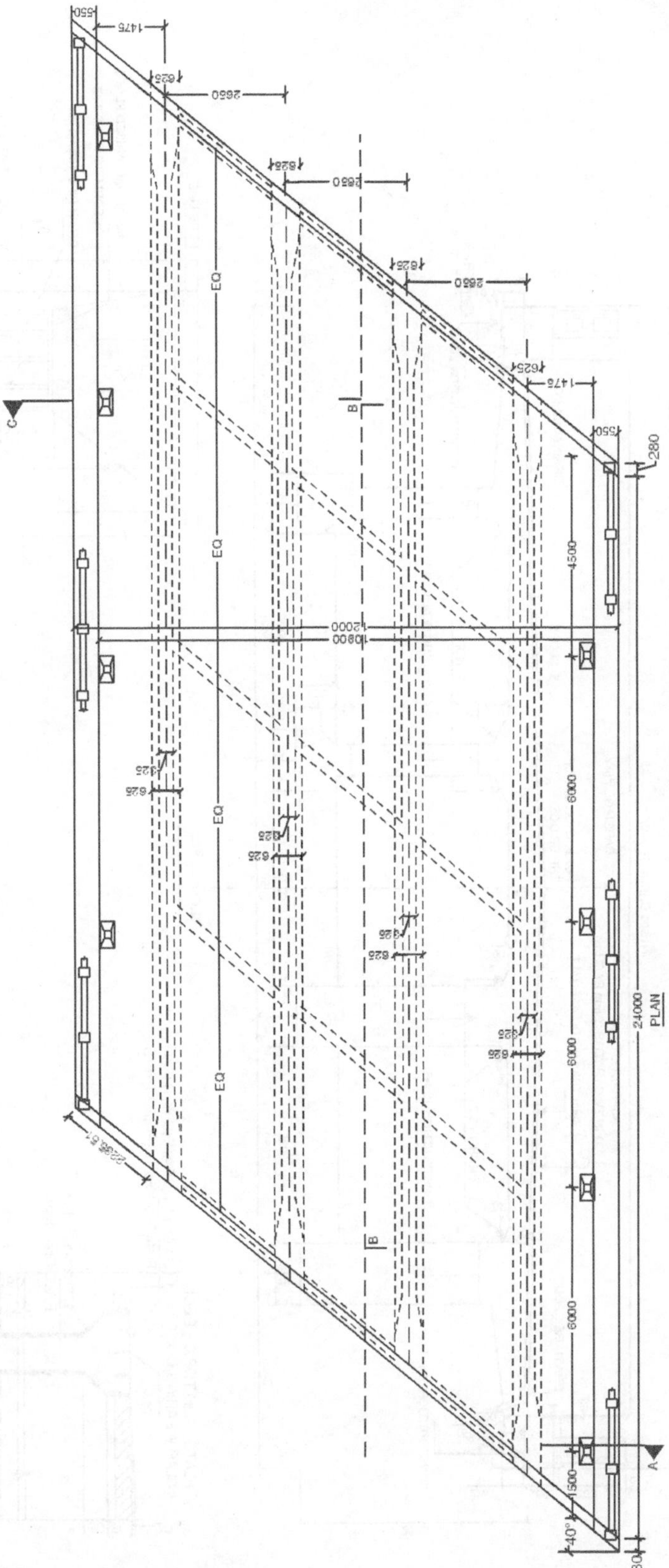

PLAN

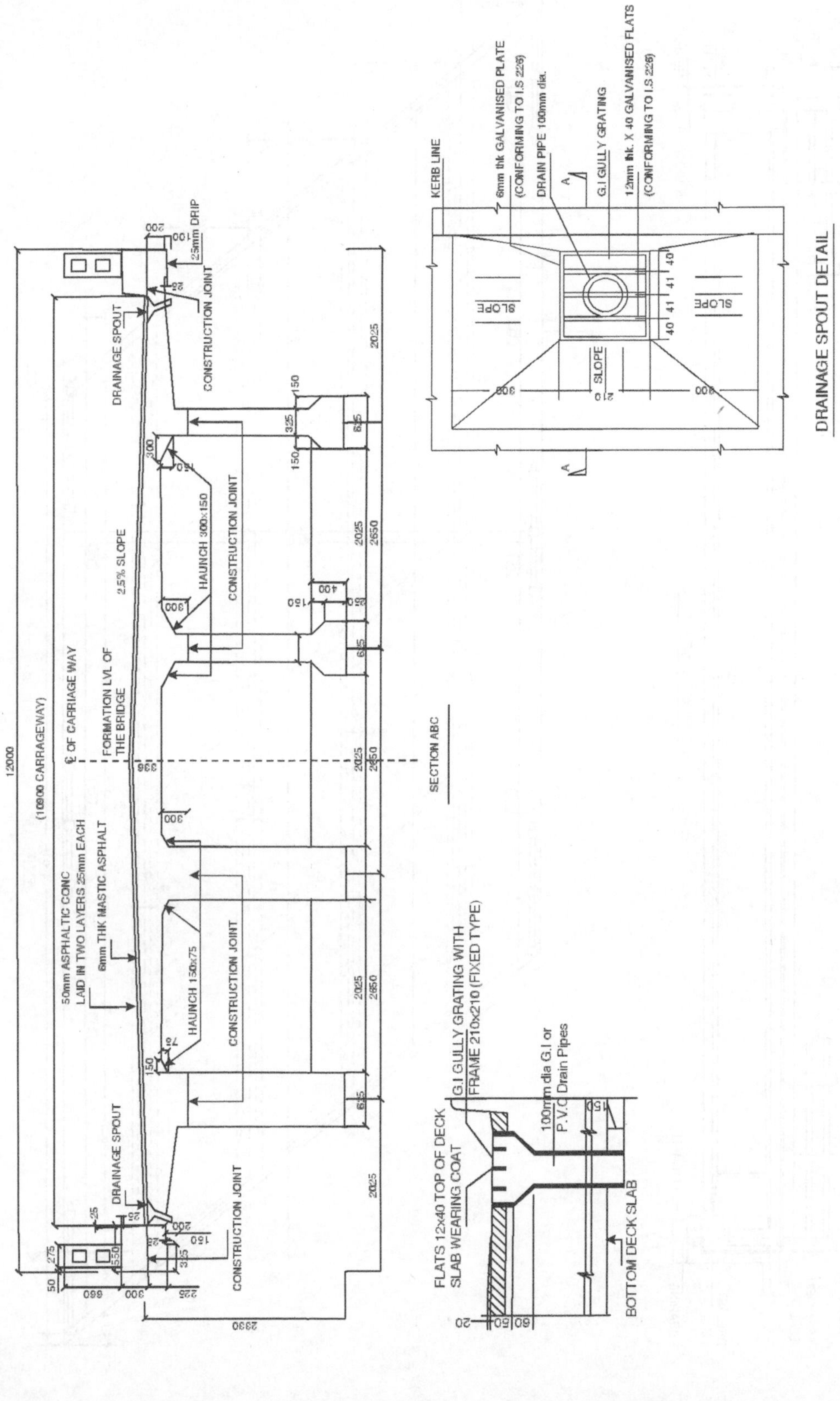

Fig. 27.22. T-Beam Bridge

DESIGN OF CANTILEVER SPAN

(*I*) Moment due to dead load

The total maximum moment due to dead load per meter width of cantilever slab is computed as follows

SNO	DESCRIPTION	LOAD KN	LEVER ARM (m)	MOMENT ABOUT a (tm)
1	Hand rails (approx)	3.00		
2	**Kerb 0.5 × 0.53 × 25 =**	**6.625**	**1.76**	**11.66**
3	Wearing course			
	1.183 × 0.075 × 22 = 1.95 say 2	2	**1.1**	**2.20**
4	Slab			
	0.2 × 1.495 × 25 =	7.5	**1.062**	**8.00**
	0.15 × 1.495/2 × 25	2.8	**0.850**	**2.40**
	SW = 21.925 KN say 22 KN		**SM = 29.774**	**Say 30KN**

Weight of 2 Nos horizontal rial beam = 0.175 × 0.2 × 25 × 2 = 1.75 KN/m

Weight of pillar = 0.275 × 0.25 × 25 × 0.85/ 1.87 = 0.78 KN/m

Total Load = 2.53 SAY 3.00Kn/M

(*ii*) Moment due to LL

Due to the specified minimum clearance, IRC 70 Tracked vehicle will not operate on cantilever slab IRC 70 wheeled loading can travel on this part Effective width of dispersion be

$$b_e - 1.2x + b_w$$

$$b_w = 0.110 + 2 \times 0.075 = 0.56\text{m}$$

Hence b_e = 1.2 × 0.7825 + 0.56 = 1.50

Live load perm including impact = 170/2 × 1.25(0.7825 + 0.3125)/1.50

Max moment due to LL including impact = 70.83 × 1.095 = 77.6KN

Total moment due to DL + LL = (30 + 77.6) = 107.6

M35 concrete

D = 350mm tapered to 200mm d = 350 – 30–20/2 = 310mm

$d = 1076 \times 10^6/1000 \times 1.35 = 282\text{mm}$

$\text{Ast} = 107.6 \times 10^6/200 \times 0.9 \times 310 = 1928\text{mm}^2$

Provide 20mm dia tor @ 220mm c/c + 12mm dia tor @ 220c/c

Ast provided (314 + 113/2) × 1000/100 = 1940mm^2

Provide bottom reinforcement 12mm dia @220mm c/c

Distt st = $0.12 \times (200 + 350/2) \times 1000/100 = 330\text{mm}^2$

Provide 100mm dia @ 200mm c/c (*i*) under main reinforcement at top and similarly 10mm dia @ 200mm c/c

(d) Max SF = 170/2 × 1.25/1.50 = 70.83 KN

Nominal shear stress = $70830/1000 \times 310 = 0.23\ \text{N/mm}^2$

Safe

DESIGN OF DECK SLAB

Slab thickness (an) = 237 + 300/2 = 268.5 say 270mm

Span in transverse direction = 2.65m

Effective span in transverse direction = 2.65 – 0.625 = 2.02m

Span in longitudinal direction = 11.75 – 0.3 = 11.45m

Since ratio of effective span in longitudinal direction to that in transverse direction is more than 2, slab is assumed to be supported on adjacent longitudinal griders

(*i*) Maximum BM due to DL

Weight of deck slab = $0.27 \times 25 = 6.75$ KNm

Weight of waring coat = $0.075 \times 22 = 1.65$

Total weight = 8.40 KN/m say 9.00 KN/m

Moment in transverse span = $9 \times 2.02^2/10 = 3.7$KNm

(*ii*) Live load BM due to IRC 70 tracked vehicle

Impact factor fraction = 25%

Width of spread of load along longitudinal direction

= 4.59 + 0.240 + 0.240 = 5.05 say 5m

Width of spread of lad along transverse dircetion

= 840 + 240 + 240 = 1320m = 1.32m

Total load on the spanper track perm including impact

= $700/2 \times 5.0 \times 1.32 = 66.3$ KN/m²

Reaction = $33.3 \times 1.32/2 = 43.76$KN

Max BM due to this IRC70 tracked vehicle with deck slab simply supported

Total = 66.3 KN/m

= $43.76 \times 1.325 - 66.3 \times 0.66 \times 0.33$

= 42.6 – 14.4 = 28.2 KNm

(*iii*) Live load BM due to IRC 70 wheeled vehicle when deck slab in simply supported

Impact factor fraction = 25%

Width of sperad of wheeled load in longitudinal direction

= 610 + 240 + 240 = 1090mm

Width of spread in transverse direction = 410 + 240 + 240

= 890mm = 0.9^{m}

Max wheeled load perm including impact/m = $170/2 \times 1.25 = 106.25$ KN

Max moment = $106.25/2 \times 1.1325 - 106.25.0.9 \times 0.45 \times 0.45^2/2$

= 2.265 = 60.2 – 12.05 = 48.20 KNm

Design LL moment due to DL + LL including impact when deck slab in assumed simply supported = 28.2 + 48.20 = 76.40 KNm

Assuming 50% of moment is taken as-ve moment at beam ends & 50% on mid span of deck slab consuming it continuous over many spans

D = $38.2 \times 106/1000 \times 1.35 = 168$m provide is $d = 240$

D = 270mm average (300 tapered to 237mm)

Ast = $38.2 \times 10^6/20 \times 0.9 \times 234 = 907$mm²

Provide 10mm dia @ 200mm c/c

The reinforcement arrangement in shown in Fig 27.23.

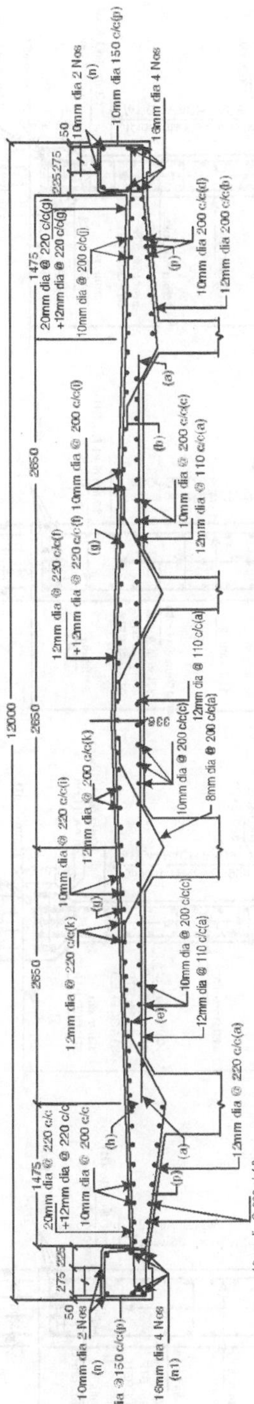

CROSS SECTION A B C (SHOWING REINF)

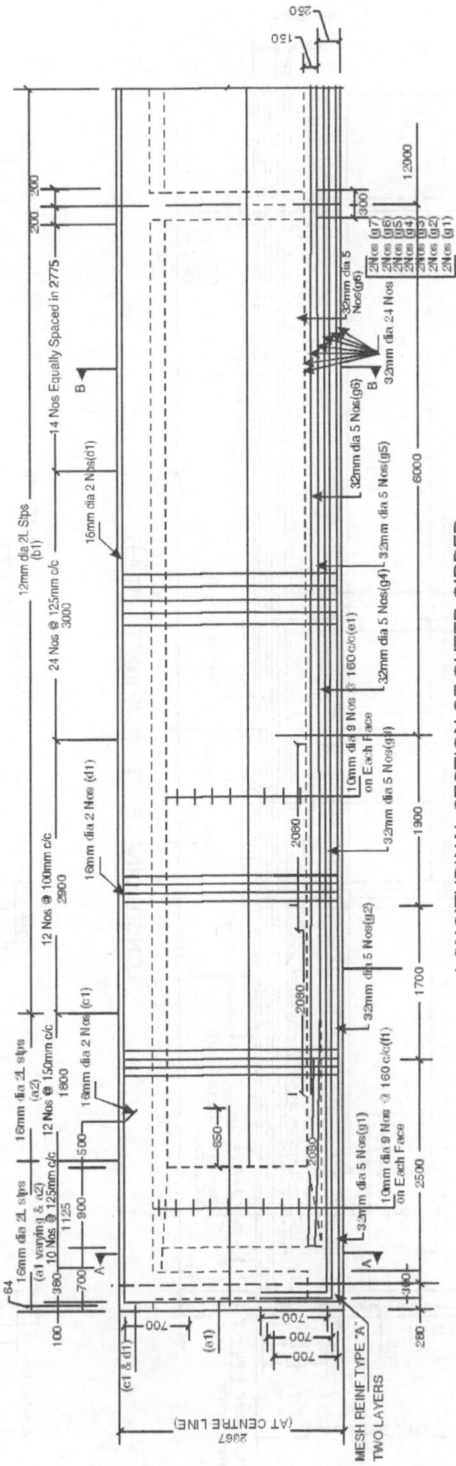

LONGITUDINAL SECTION OF OUTER GIRDER

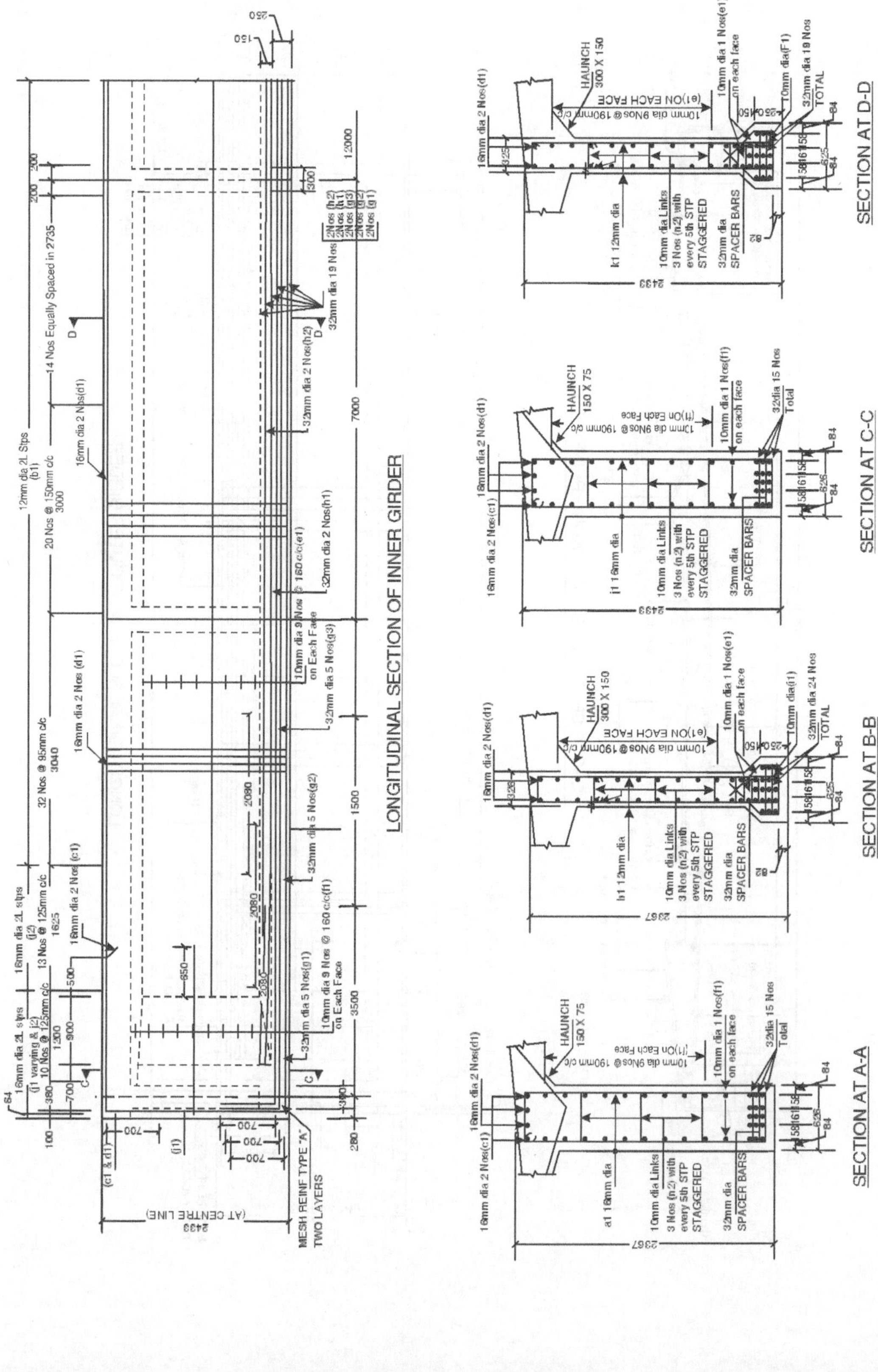
LONGITUDINAL SECTION OF INNER GIRDER
SECTION AT A-A
SECTION AT B-B
SECTION AT C-C
SECTION AT D-D

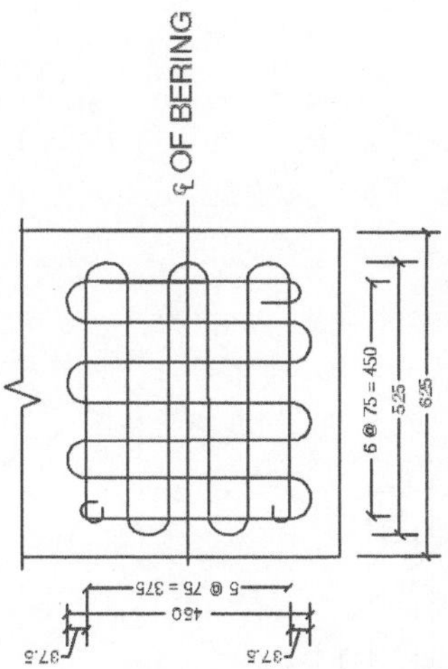

8 mm dia MESH Reinf Type-A

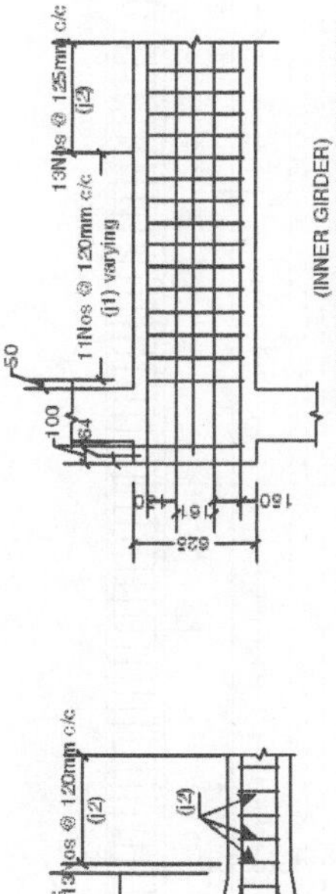

(INNER GIRDER)
PLAN SHOWING BULB REINF AT THE END
SIMILAR SCHEME OF DETAILING WILL BE
OLLOWED FOR OUTER GIRDER

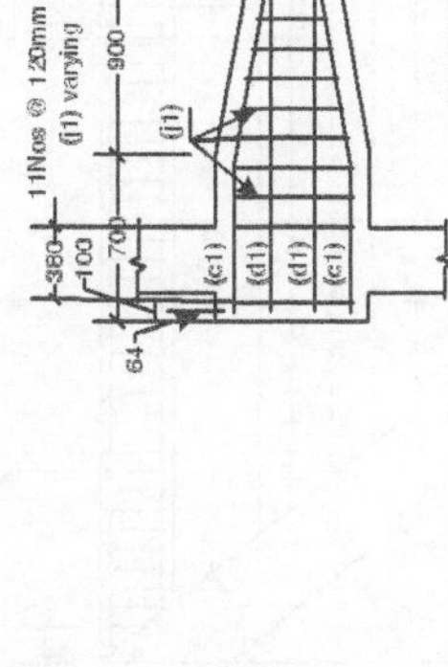

(INNER GIRDER)
PLAN SHOWING WEB REINF (TOP) AT THE END

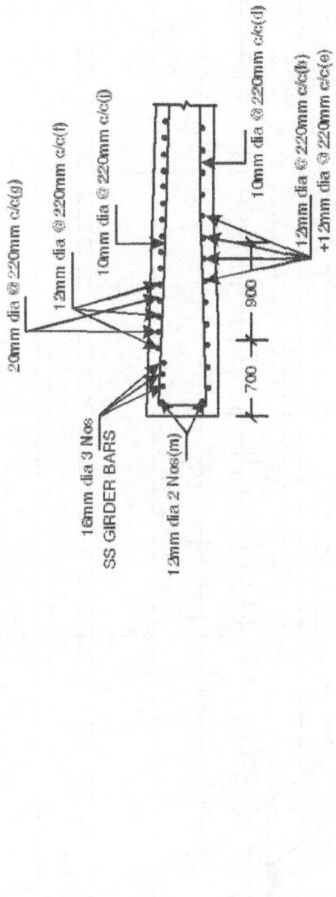

SECTION P-Q

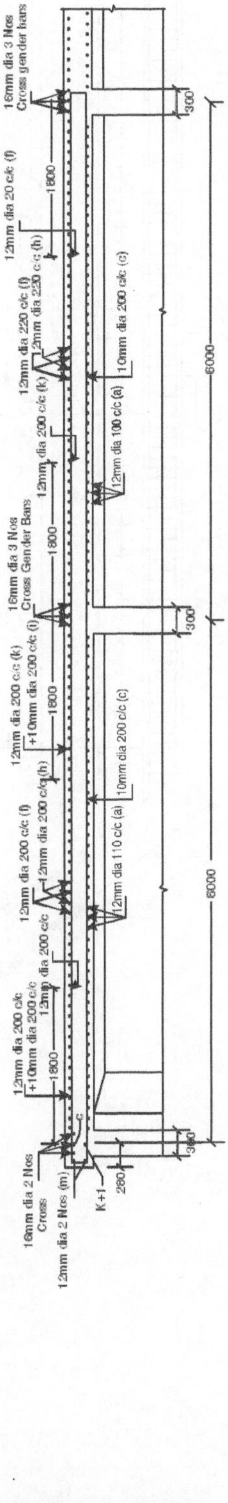

SECTION XYZ

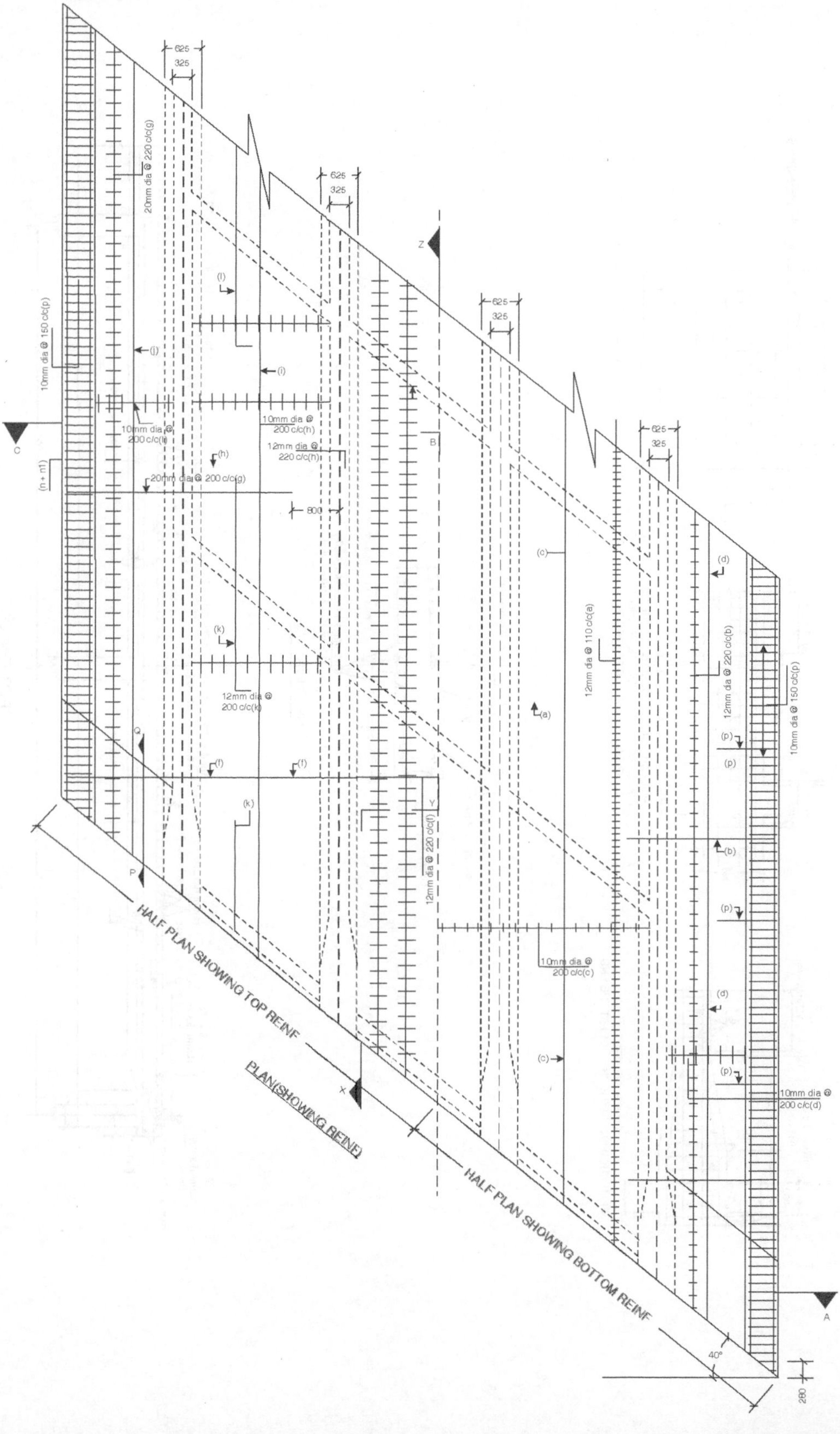
625
325
20mm dia @ 220 c/c(g)
10mm dia @ 150 c/c(p)
(n + n1)
C
(j)
(i)
(l)
10mm dia @ 200 c/c(h)
12mm dia @ 220 c/c(h)
10mm dia @ 200 c/c(l)
(h)
20mm dia @ 200 c/c(g)
800
B
Z
(c)
12mm dia @ 110 c/c(a)
(d)
12mm dia @ 220 c/c(b)
10mm dia @ 150 c/c(p)
(k)
12mm dia @ 200 c/c(k)
(a)
(p)
Q
(f)
(k)
Y
12mm dia @ 220 c/c(f)
P
(b)
10mm dia @ 200 c/c(c)
HALF PLAN SHOWING TOP REINF
PLAN(SHOWING REINF)
X
10mm dia @ 200 c/c(d)
HALF PLAN SHOWING BOTTOM REINF
A
40°
280

Fig. 27.23.

INTERMEDIATE LONGITUDINAL GIRDERS

Effective span = 23.5 + 0.5 = 24m

Slab thickness average = 336 + 300/2 = 318mm

Width of rib to be same as in 24m spacing bridge of standard plant MOST

(SD/211) = 625MM

Spacing of main longitudinal beams = 2.65m c/c

Overall detph of beam = 2460mm

(*i*) B.M due to dead load

SNO	ITEM	DETAILS	WEIGHT KN
1	Wearing courcse	**0.075 × 2.65 × 22**	**4.4**
2	Deck slab	0.318 × 2.65 × 25	**21.07**
3	T-rib	(2.466 – 0.3 – 0.3) × 0.325 × 25 + 0.625 × 0.25 × 25 + 0.15 × 0.15/2 × 2 × 25	**19.65**
4	Fillets	2 × 0.5 × 0.3 × 0.15 × 25	**1.125**
5	Cross beams (Providing cross grider at quarter & mid span & Ends)	5 × (2.46 – 0.318 – 0.4) × 0.3 × 25/24	**2.72**
			48.96 say 50

Max BM = $50 \times 24^2/8 = 3600$ KNm

(*ii*) BM due to live load 2 braces of class A loading

Maximum live load BM may occur under class A two lane loading

Impact factor fraction = 0.15

Impact factor = 1.15

RA = 2 × W × 2.05/2.65 + 2 × W × (2.65 – 2.4) 2.65 = 1.548 W + 0.19 W

Arrangement of wheels in longitudinal direction for maximum bending moment is given below

Reactin on intermediate grider due to 27 KN

Axel load including impact = 1.7 × 13.5 × 1.15 = 26.4 KN

Reaction of 68 KN axel load including impact = 1.7 × 34 × 1.15 = 66.5 KN

CG of load system = {66.5 × (3 + 6 + 9) + 115 × (13.3 + 14.5) + 26.4 (17.7 + 18.8}/548.8

= 1197 + 3197 + 964/548.8 = 5558

598.8 = 9.76m

For maximum moment, resultant of laods & one of heavier load should be equidistance from of span

Reactin at left hand end support

= 66.5 (0.55 + 3.55 + 6.55 + 9.55) + 115 (13.85 + 15.05) + 26.4 (18.25 + 19.35)/24

= 1343.3 + 3323.5 + 972.64/24 = 5660/24 = 236 KN

Max BM at load 115 KN load

$= 236 \times (12 - 1.3) - 26.4 \times (4.4 + 5.5)\ 115 \times 1.2)$

$= 236 \times 10.7 - 26.4 \times 9.9 - 115 \times 1.2$

$= 2325.2 - 261.36 - 138$

$= 2125.84$ say 2130

***(iii)* Live BM due to tracked vehicle**

Transversely load on one grider

$= 1.2 \times 60.74 = 73$ KN

Max BM $= 73 \times 4.57 \times 12/24 \times 12 - 73 \times 2.285 \times 1.425$

$= 2002 - 237.7 = 1746.3$ KNm

***(iv)* Live BM due to wheeled IRC70**

C.G of load $= 170(1.37 + 4.42 + 5.79) + 120\ (7.92 + 9.44) + 80 \times 13.4/1000$

$= 1968.6 + 2083.2 + 1072/1000$

$= 5123.8/1000 = 5.123$m

Reaction at LHS end $= 1000 \times 12.3/24 = 512.5$ KN

Max BM under 170 KN LOAD

$= 487.5 \times 12 - 170 \times 4.27 - 170 \times 3.4 - 170 \times 0.3$

$= 5850 - 810.9 - 578 - 51$

$= 4410.1$ KNm

Actual load BM on grider $= 4410.1 \times 0.5 \times 1.15$

$= 2206$ KNm

Total BM due to DL + LL + Impact

$= 3600 + 2206 = 5806$ KNm

M35

MOR as T section $= 12 \times 30$

$d = 2433 - 30 - 10 - 5 \times 32$

$= 2233$

Ast $= 5806 \times 10^6/200 \times 0.9 \times 2233 = 14444\text{mm}^2$

Provide 32mm dia 19 Nos (As provided $= 19 \times 800 = 15200\text{mm}^2$)

COMPUTAION OF MAXIMUM S.F

(*a*) Max SF due to DL $= 50 \times 24/2 = 600$ KN

(*b*) Due to live loads

Due to tracked vehicle $= 700/2 = 350$ KN

Max SF $= 350 \times (24 - 2.285)/24 = 308.5$ KN

Max SF including impact $= 308.5 \times 1.15$

$= 355$ KN

***(ii)* Max SF due to IRC 70 vehicle loads**

Max SF/reaction end $= \{85 \times (23.9 + 22.53 + 19.48 + 18.11) + 60 \times (15.98 + 14.46) + 40 \times 10.5\}/24$

$= (7142 + 1826.4 + 420)/24$

$= 9388.4/24 = 391.2$ KN

Max SF including impact $= 391.2 \times 1.15 = 450$ KN

***(iii)* Max SF due to two class A loads**

Max Sf/reaction at end = {115(23.9 + 22.7) + 66.5 (18.4 + 15.4 + 12.4 + 9.4}/24
= (5359 + 33697.4)/24
= 378 KN say 380 KN

Max SF including impact = 380 KN

Max SF for desing = 450 KN

Nominal shear stress = $450 \times 1000/\ 325 \times 2167 = 0.64$ N/mm^2

For M35

Permissible shear stress for p = $15 \times 800 \times 100/325 \times 2167 = 1.75 = 0.5$ N/m^2

Design shear = $0.14 \times 325 \times 2107$N

Provide 10mm dia tor @ $200 \times 2 \times 78.50 \times 2167/0.14 \times 325 \times 2107 = 790$ mm c/c

Min shear reinf

Asv/b.sv<0.4/fs

$S_V = 2 \times 78.5 \times 415/325 \times 0.4 = 501$

Provide 10 dia tor 2L @ 190mm c/c is shown in the drawing (Fig. 27.23)

DESIGN OF END LONGITUDINAL GIRDERS (2 NOS)

***(i)* Moments due to dead load**

Dead load of cantilever portion = 22 KN/m

Dead load of deck slab between end & intermediate girders

Wearing course = $2.65/2 \times 0.075 \times 22 = 2.2$ KN/m

Slab $(0.237 + 0.3/2) \times 25$ = 7.8 KN/m

Self weight of griders = $2.0 \times 0.325 \times 1.0 \times 25 = 16.25$ KN/m
= $0.325 \times 1.73 \times 25 + 0.4 \times 0.625 \times 25$
= $14.0 + 6.0 = 20$ Kn/m

Weight of haunch = $0.025 \times 0.25/2 \times 25 = 0.15$

Total dead load = 52.15 say 53 KN/m

Max BM due to dead load = $53 \times 24^2/8 = 3816$ KNm

***(ii)* BM due to live load**

Max live load from cantilver slab = 70.83 KN say 71 KN

Outer to outer width of IRC70 wheeled vehicle = 2790mm

Tyre width = 800mm

Center to center distance = 2790 – 860 = 1930mm

Live load on end grider from bridge deck side = $170/2 \times 1.615/2.65 = 51.8$ KN

Live load including impact = $51.8 \times 1.15 = 59.6$ say 60 KN

Tyre length = 610mm

After dispersions, Tyre length will become = 610 + 200 + 200
= 1010 say 1m

(a) Total live load on end grider due to axal load of 170 KN = 71 + 60 = 131

To compute

Max BM at mid span

(b) Total live load on end grider due to axel load of 80 KN = $80/170 \times 131 = 61.65$ say 62 KN

Design Calculations for 18m Spanning Skew Bridge (40° Skew) Bye Pass Patiala

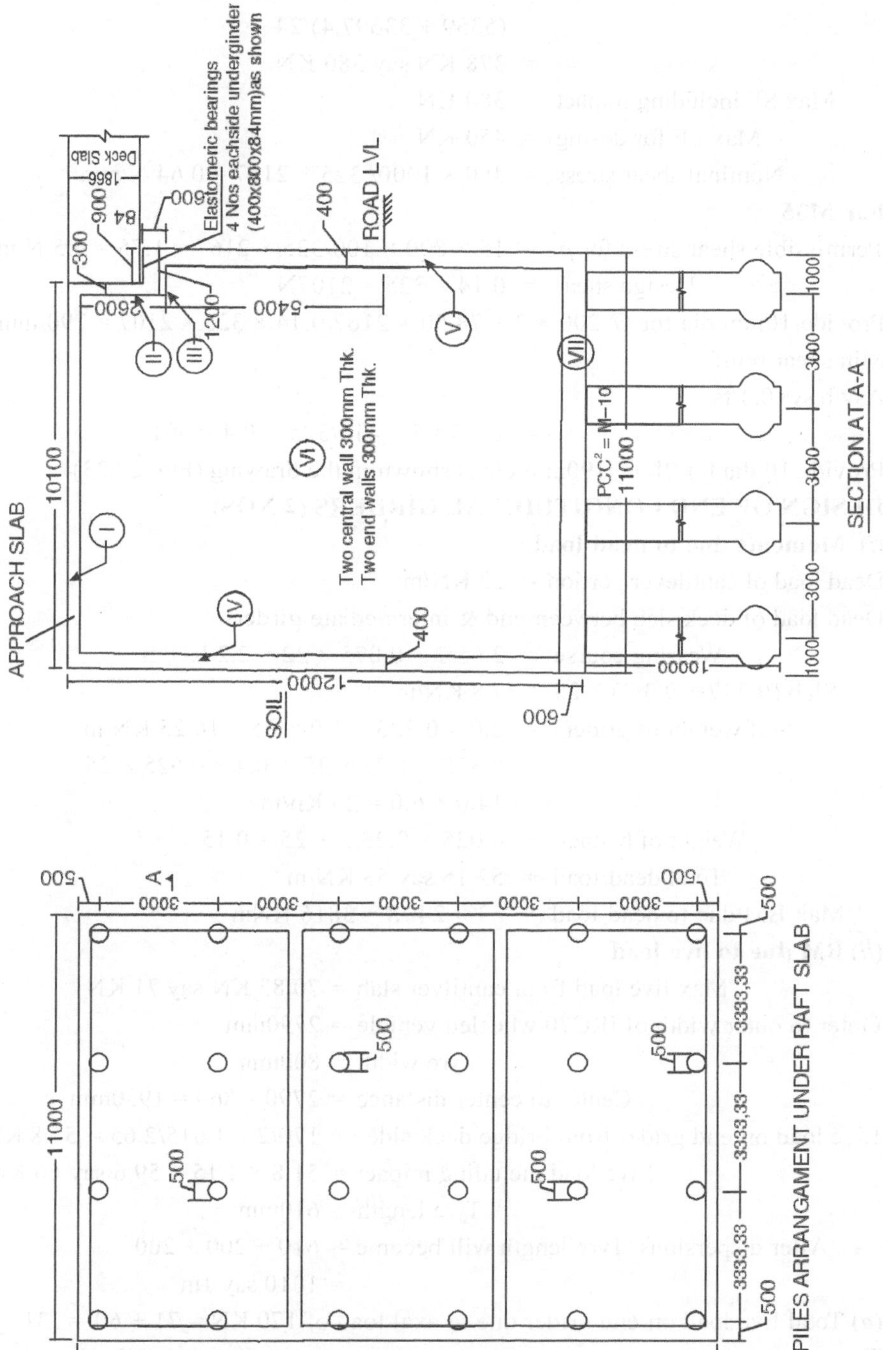

SECTION AT A-A

PILES ARRANGAMENT UNDER RAFT SLAB

Max BM under 170 KN load for live load case

$= 4410.1 \times 1.15 \times 131/170$

$= 3908$ KNm

Max DL + LL moment $= 3600 + 3908$

$= 7508$ Knm

$d = 2367 - 30 - 5 \times 32 - 10 = 2167$ mm

$Ast = 7508 \times 10^6/200 \times 0.9 \times 2167 = 19248mm^2$

Provide 32mm dia tor 24 Nos (Ast provided = $24 \times 800 = 19200m^2$) see Fig. 27.23

The arrangement of reinforcement in cross girders beams is shown in Fig 27.24. The Bar bending schedule for this bridge deck slab & beams is presented in Fig. 27.25.

DL from superstructure = 2554 KN

Live load including impact = 850 KN

Self weight of Abutment	W(KN)	Liverrarm (m)	Moment about O (Knm)
Element I = 0.3 × 9.4 × 16 × 25	= 1128	5.9	6655
Element II = 2.466 × 0.3 × 16 × 25	= 296	1.05	311
Element III = 1.2 × 0.6 × 16 × 25	= 288	0.6	173
Element IV = 12 × 0.4 × 16 × 25	= 1920	10.8	20736
Element V = 9 × 0.4 × 16 × 25	= 1440	0.2	288
Element VI = {4 × 0.3 × 10.2 × 9 × 25}			
{4 × 9.4 × 2.16 × 0.3 × 25}	= 3364	5.5	18502
Element VII = 11 × 0.6 × 16 × 25	= 2640	5.5	14520
ΣW	= 11076		

Longitudinal Forces

(*i*) Force due to braking due to 70R wheeled Vehicle $= 0.2 \times 1000 = 200$ KN

This force acts at 1.2 m above road level

Force on one abutment wall = 100 KN

Force on one abutment wall including impact $= 100 \times 1.675 = 187.5$ KN

Acting at 13.2m above point 0

$= 187.5 \ \Sigma \ 13.2 = 2475$ KNm

(*ii*) Force due to temperature variation & shrikage

Assuming moderate climate, variation in temperature is taken as +- 17°c as per clause 218.5 of Bridge code

Coeff of thermal expansion = 11.7 × 10–6/° C

Strain due to temp variation $= 17 \times 11.7 \times 10^{-6}$

$= 1.99 \times 10^{-4}$

From 220.3 strain due to concrete shrinkage $= 2.\ 0 \times 10^{-4}$

Total strain due to temperature & Shrinkage

$= 1.99 \times 10^{-4} + 2.0 \times 10^{-4} = 3.99 \times 10^{-4}$

Horizontal deformation of deck due to temp and shrinkage on abutment

= 3.99 × 10 – 4 × 24000 = 9.60mm

Strain in bearin = Deformation/Electrometric = 9.6/84 – 6 × 5 = 9.6/54 = 0.178

Assuming G = 1 N/mm²

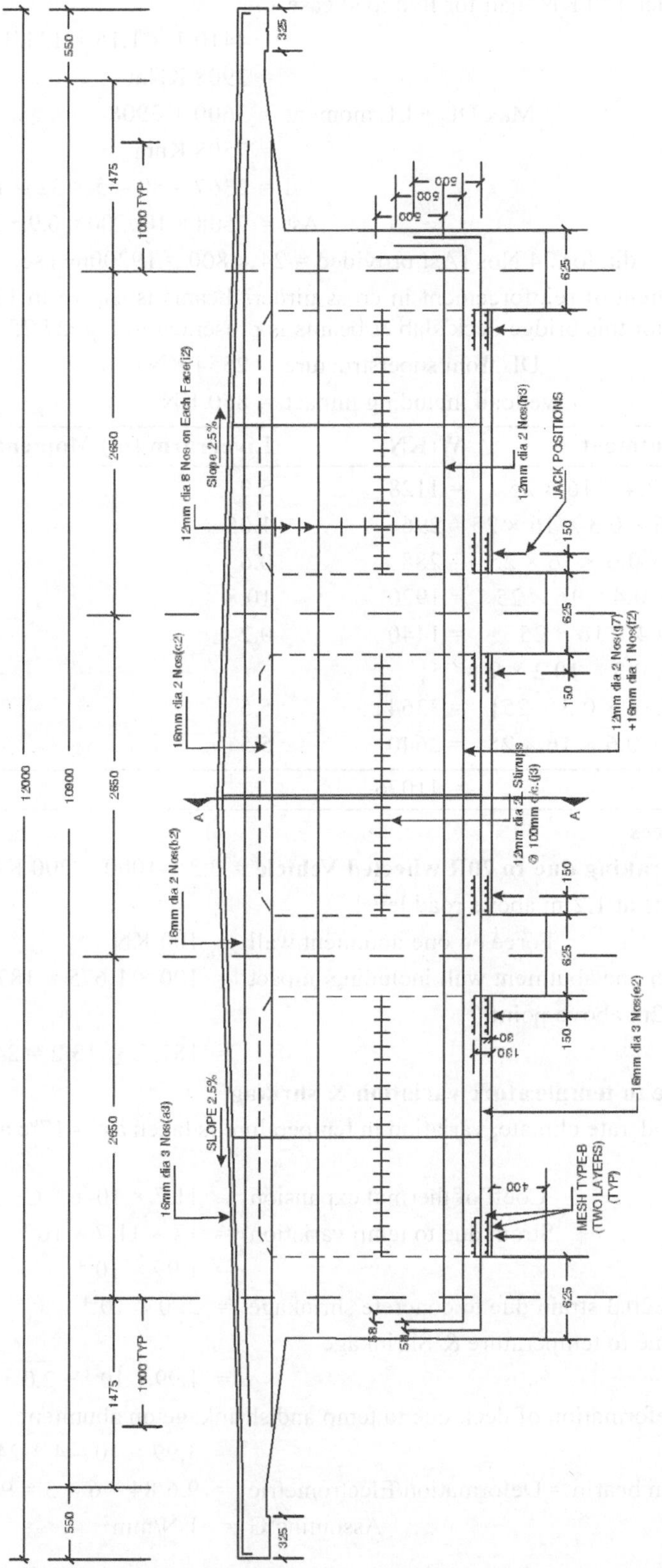
12mm dia 8 Nos on Each Face(i2)
Slope 2.5%
16mm dia 2 Nos(c2)
16mm dia 2 Nos(b2)
SLOPE 2.5%
16mm dia 3 Nos(a3)
12mm dia 2 Nos(h3)
JACK POSITIONS
12mm dia 2 Nos(g7)
+16mm dia 1 Nos(f2)
12mm dia 2L Stirrups
@ 100mm c/c (j3)
16mm dia 3 Nos(e2)
MESH TYPE-B
(TWO LAYERS)
(TYP)
A
A
12000
10900
550
1475
1000 TYP
2650
2650
2650
325
625
150
600
100
130
80
58

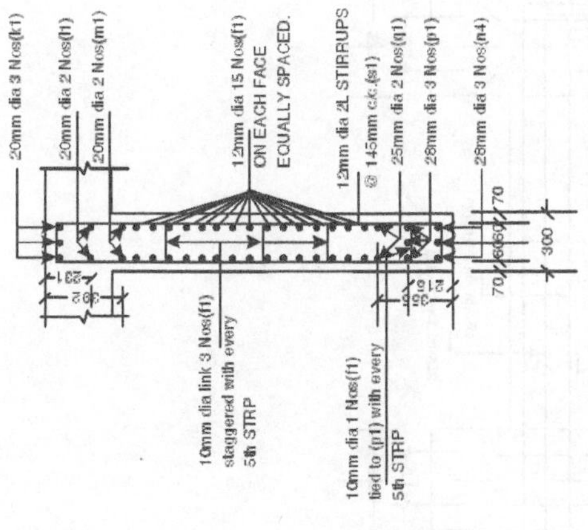

SECTION AT B-B

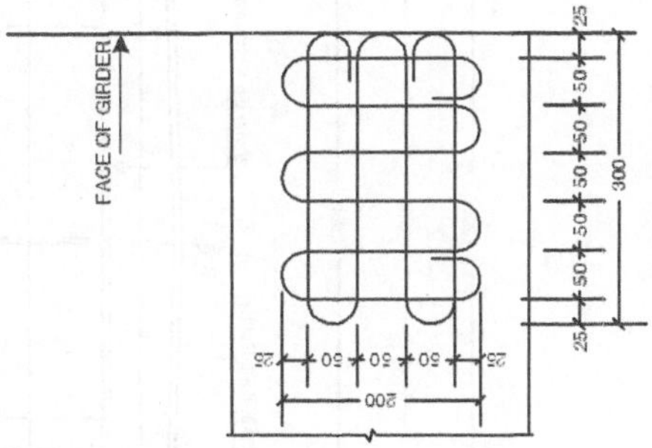

8mm dia MESH REINFTS TYPE-B

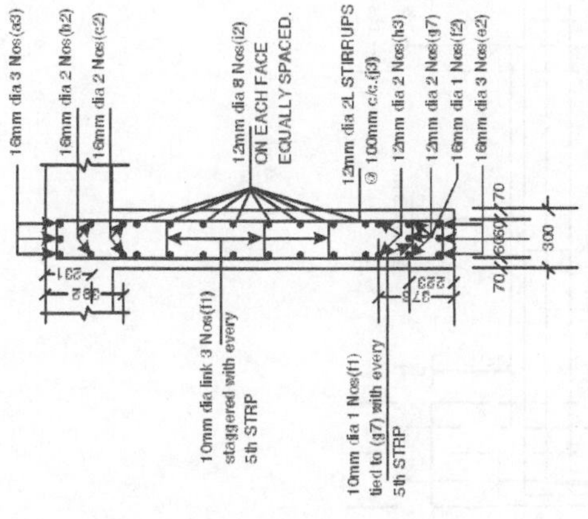

SECTION AT A-A

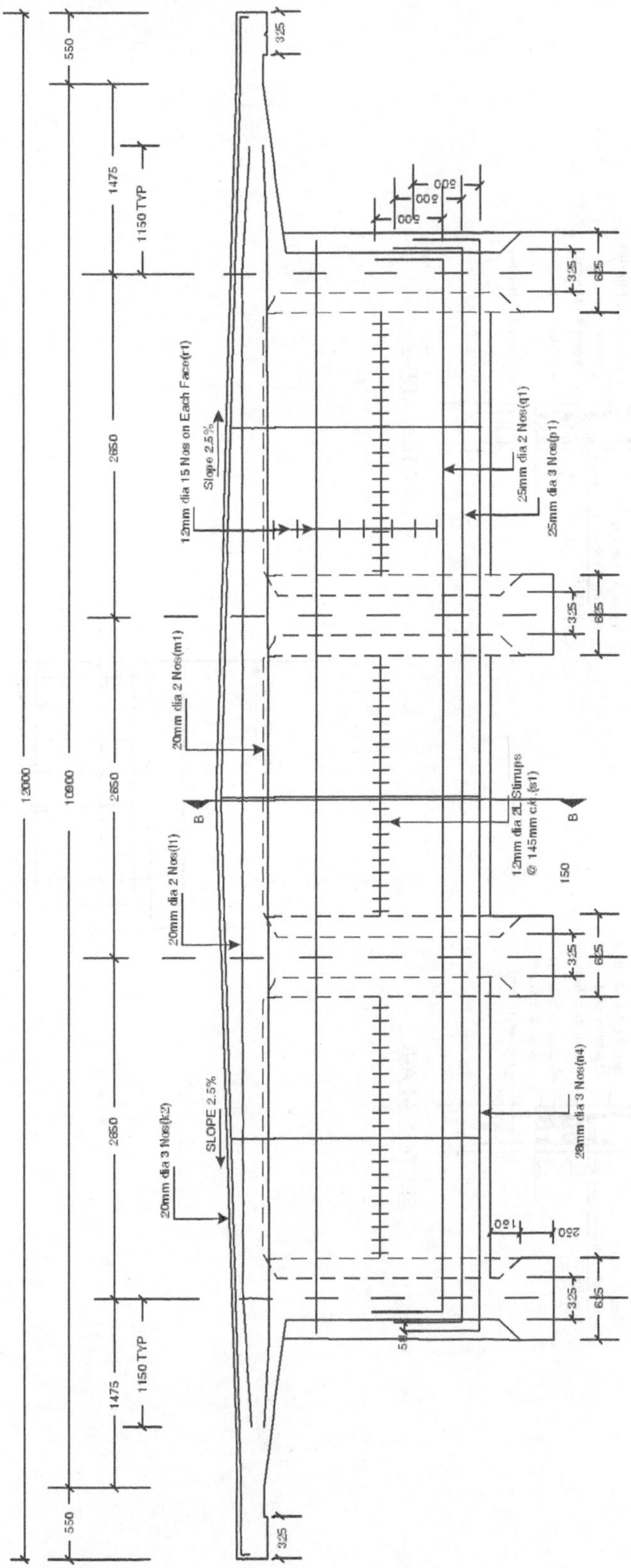
12000
10900
550
1475
1150 TYP
2650
325
625
12mm dia 15 Nos on Each Face(r1)
Slope 2.5%
25mm dia 2 Nos(q1)
25mm dia 3 Nos(p1)
20mm dia 2 Nos(m1)
20mm dia 2 Nos(l1)
B
12mm dia 2L Stirrups
@ 145mm c/c (s1)
150
20mm dia 3 Nos(k2)
SLOPE 2.5%
28mm dia 3 Nos(n4)
600
250
150

Fig. 27.24.

S.No.	TYPE OF BAR	SHAPE OF BAR (NOT TO SCALE) DIMENSIONS ARE IN mm	DIA OF BARS IN mm	LENGTH OF EACH BARS IN mm	NO OF BARS	TOTAL LENGTH IN mm	WEIGHT IN Kg
		SLAB					
1	a	6575	12	8575	214	1835.05	1651.55
2	b	150, 530, 1245, R=120, 1692, 500, 140, 137, R=120	12	2976	190	565.44	508.90
3	c	150, L-Varying from 109 to 212, 150, L, L, 24460	10	24960	39	973.44	584.06
4	d	L-Varying from 68 to 270, 140, R=100, 140, L, 901, 21360, 901, L, 650, Varying from 0 to 70, 650, R=100	10	24960	16	399.36	239.62
5	e	150 Varying from 249 to 333, R=120, Varying from 1543 to 1692, 500, 140, Varying from 137 to 198, Varying from 249 to 333, R=120	12	2987	32	95.58	86.02
6	f	5952, 75, R=120, 75, 149	12	12054	108	1301.83	1171.65
7	g	3826, 96, 120	20	3946	212	836.55	2091.38
8	h	3976, 3976, 99, R=120	12	7952	106	842.91	758.62
9	i	24460, L, L=Varying from 109 to 212, L, 150, 150	10	24960	39	973.44	584.06
10	j	24460, L, L=Varying from 68 to 270, L, 75, 75	10	24888	16	398.21	238.93
11	k	3330, L, 150, L=Varying from 109 to 212	12	3588	72	259.34	232.50
12	l	6200	12	6200	36	223.20	200.88
13	m	11900	12	11900	4	47.60	42.84
14	n	24460	10	24460	4	97.84	58.70
15	n1	24460	16	24460	8	195.68	313.90
16	p	405, 380, 380, 453, 500, Varying from 37 to 54	10	2208	328	724.38	434.52
17	q	393, 150, 860, 950, 150, 433, R=50	8	2110	236	497.96	199.18
		LONGITUDINAL GIRDER					
1	a1	L1-Varying from 2237 to 2241, L1, L2, L2-Varying from 2229 to 2233, Varying from 193 to 493	16	5476	44	240.94	385.51
2	a2	2233, 2237, 193	16	5176	48	248.45	397.52
3	b1	2233, 2237, 193	12	5096	268	1365.73	1229.16
4	c1	708, R=96, Elevation, 636, 500, 500	16	2693	16	43.09	68.94
5	d1	24432, 708, R=96, 708	16	25720	8	205.76	329.22
6	e1	21360	10	21360	80	1708.80	1025.28
7	f1	160, 913, 650, 636	10	2199	160	351.84	211.10
8	g1	716, R-192, 24176, 716	32	25608	20	512.16	3226.61
9	g2	716, R-192, 24176, 716	32	25480	20	509.60	3210.48

B) M.S. BARS (MESH REINFORCEMENT IN LONGITUDINAL GIRDERS)

MESH IN GIRDERS	1	u	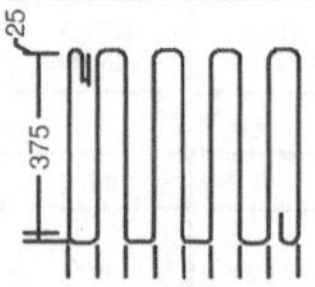	8	3550	16	56.80	22.72
	2	v	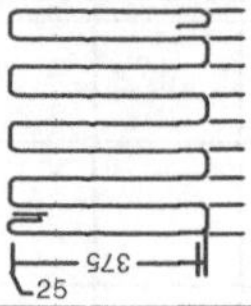	8	3506	16	56.10	22.44

B) M.S. BARS 9MESH REINFORCEMENT IN CROSS GIRDERS)

19	u1		8	1453	24	34.87	13.95
20	v1	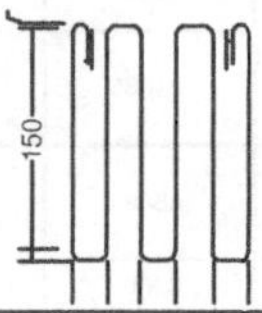	8	1510	24	36.24	14.50

S.No.	TYPE OF BAR	SHAPE OF BAR (NOT TO SCALE) DIMENSIONS ARE IN mm	DIA OF BARS IN mm	LENGTH OF EACH BARS IN mm	NO OF BARS	TOTAL LENGTH IN mm	WEIGHT IN Kg
10	g3		32	25352	20	507.04	3194.35
11	g4	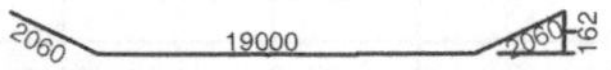	32	23160	6	138.96	875.45
12	g5	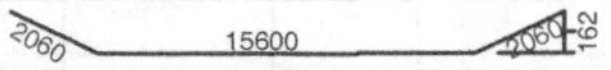	32	19760	4	70.04	497.95
13	g6		32	16160	4	64.64	407.23
14	h1	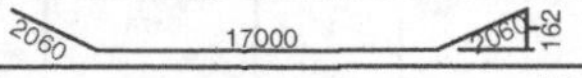	32	21160	4	84.64	533.23
15	h2	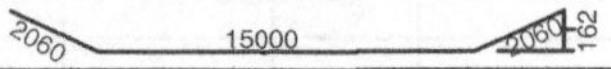	32	18160	4	72.64	457.24
16	i1	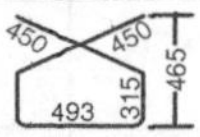	10	1733	408	707.06	424.24
17	j1	L1-Varying from 2303 to 2307 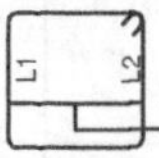 L2-Varying from 2295 to 2299 Varying from 193 to 493	16	5608	48	269.18	430.69
18	j2	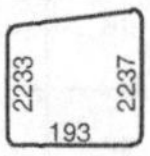	16	5308	52	276.02	441.63
19	k1	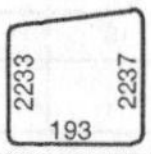	12	5228	264	1380.19	1242.17
20	n2	 L-Varying from 213 to 513	10	583	72	41.98	25.19
21	n3	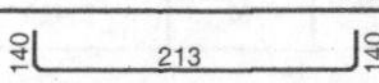	10	433	384	166.27	99.76

S.No.	TYPE OF BAR	SHAPE OF BAR (NOT TO SCALE) DIMENSIONS ARE IN mm	DIA OF BARS IN mm	LENGTH OF EACH BARS IN mm	NO OF BARS	TOTAL LENGTH IN mm	WEIGHT IN Kg
		CROSS GIRDERS					
1	a3	95, 5952, 5952, 152	16	12094	6	72.56	116.10
2	b2	1000, 7950, 1000, 41	16	9950	4	39.80	63.68
3	c2	412, 9132, 412, 64	16	9956	4	39.82	63.71
4	e2	508, R=64, 8475, 508	16	9395	6	56.37	90.16
5	f2	508, R=64, 8383, 508	16	9303	2	18.61	29.78
6	g7	508, R=48, 8379, 508	12	9319	4	37.28	33.55
7	h3	508, R=48, 8287, 508	12	9227	4	36.91	33.22
8	i2	233, R=48, 8443, 233	12	8837	32	282.78	254.50
9	j3	Varying from 1852 to 1942, 176	12	4386	126	552.64	497.38
10	k2	95, 5952, 5952, 156	20	12144	3	36.43	91.08
11	l1	1150, 7950, 1150, 50	20	10250	2	20.50	51.25
12	m1	1175, 7908, 1175, 110	20	10250	2	20.52	51.30
13	n4	514, R=168, 8131, 514	28	8334	3	26.80	128.64
14	p1	512, R=100, 8012, 512	25	8886	3	26.66	101.31
15	q1	512, R=100, 7896, 512	25	8770	2	17.54	66.65
16	r1	529, R=48, 8151, 529	12	9137	30	274.11	246.70
17	s1	Varying from 1848 to 1942, 176	12	4382	51	223.48	201.13
18	t1	140, 196, 140	10	416	168	69.89	41.93

Fig. 27.25. Schedule of Bars in Deck Slab and Girder Beams

Horizontal force due to strain in longitudinal direction at bearing level

= 1.1 × strain × G × (area of plate in bearing) × no. Of bearings /1000

= 1.1 × 0.178 × 1.0 × 400 × 800 × 4/1000 = 250 Kn

Vertical reactin due to bracking = 200 × (1. 2 + 2.4)//24 = 30 KN

Earth pressure

Active earth pressure = 0.5 × wh^2.Ka × 16

= 0.5 × 18 × 13.2^2 × 1/3 × 16 = 8364 KN

Acting at heigh above base = 0.42 × 13.2 = 5.54 m

Over turning moment about base = 8364 × 5.54 = 46337 KNm

LIVE LOAD SURCHARGE + APPROACH SLAB

Horizontal froce due to LL surcharge = 700 × 1.25 × 0.2 = 187.5 KN

This force acts at 13.2m above base

For Piles

$$\Sigma r^2 = 2 \times 6 \times 5^2 + 2 \times 6 \times 1.65^2$$
$$= 300 + 33 = 333\text{m}^2$$

Working capacity of pile 500mm dia 10m long and single under reamed

$$= \{42 + 10.35/0.3 \times 2.4\} \times 0.75 \times 0.9$$
$$= 63.45t = 634.5 \text{ KN}$$

Check for stability under DL + Imposed LL

SNO	DETAILS	FORCE KN		ARM M	MOMENT ABOUT KNM	
		V	H		MV	MH
1.	DL from super structure	2554		0.45	1150	
2.	Hor force due to lump & shrinkage		250	9.6		2400
3.	Active earth pressure		8364	5.54		46337
4.	Hor force due to LL surcharge & approach slab		187.5	13.2		2475
5.	Vert load due to LL surcharge &	850*		5.9	5015*	
6.	Self wt of abutment	11076				
	Σ Forces (span unloaded conditions)	14480	8801.5		72350	51212
7.	LL from super str. including impact	1187.5		9.6	11400	
8.			0.45		13.5	
	Vert force due to braking	30				
9			187.5	9.6		1800
	Hor load due to braking					
		14847.5			72570	53025

FOS against overturning = 72350/51212 = 1.41 < 1.5

Piles are provided

Span Unloaded Condition

Overturning moments = 2400 + 46337 + 2475 = 51212 Knm

C.G of self weight of abutment from O = 66185/11076 = 5.975m

As the stabilizing moments are more than overturning moments, there is no effect of OM on pile

Load on piles = 14480KN

Max load on apile = 14480/24 = 603.3 KN < 634.5 KN

Span Loaded Conditions

Load on pile = 14847.5/24 = 618 KN

Say 620 KN

Piles resistance + Frictional resistance of weight of abutment

Max lateral load = 8801.5 KN < (45 × 20) + (14817.5 × 0.6)

= 900 + 8890.5 = 9790.5 KN

safe

Check for Earthquake Forces

As per IS : 1893 – 2002 (Part I)

Patiala is in seismic zone III

Ah = Z.I/2R.Sa/g

= 0.16 × 1.5 × 2.5/ 2 × 2.5 = 0.12

A) Seismic forces in the direction of traffic

Self weight of abutment I to VI = 11076 – 2640

(*i.e.*, that part of abutment which is not embedded in ground)

= 9436 KN

Hor force due to earthquake

= 9436 × 0.12

= 1132.3 KN

Earthquake acts at 0.5 × 11.4 = 5.7m above base slab

Lever arm = 5.7 + 0.6 = 6.3m

OM due to earthquake = 1132.3 × 6.3

= 7134 Knm

Unloaded Condition

ΣW = 2554 + 850 + 9436 = 1284kn

ΣM_V = 1150 + 5015 + 66185 – 14520 = 57830 KN

ΣM_H = 2400 + 46337 × 0.5/0.42 + 2475 + 7134 = 67172 KNm

Max load on pile = 12840/24 + (67172 – 57830/33) × 5

= 535 + 140 = 675 < 634.5 × 1.5 = 950 KN

B) Seismic Forces in the direction perpendicular to the traffic

Hor force due to earthquake = 1132.3 KN

OM due to earthquake = 7134 KNm

Unloaded condition/Loaded condition

ΣM_V = 57830 KN

ΣM_H = 2400 + 2475 + 7134 = 12009 KNm

$M_V >> M_H$

Load on piles = 12840/24 KN = 535 KNm

Design of base slab of abutment

Max load on pile = 620 KN

Max BM = 620 × 5.23/6 = 540.44 KNm

Portion of raft slab to resist the BM = 3300 × 600mm

M35 concrete

d = 540.4 × 10^6/3300 × 1.575 = 322.4mm

D = 600 mm *d* = 540 mm

Ast = 540.4 × 10^6/200 × 0.9 × 540 × 3.33 = 1685mm^2

Provide 16 dia tor @ 100m c/c

Distt st = 0.12 × 600 × 1000/100 = 720mm^2 = 360mm each face

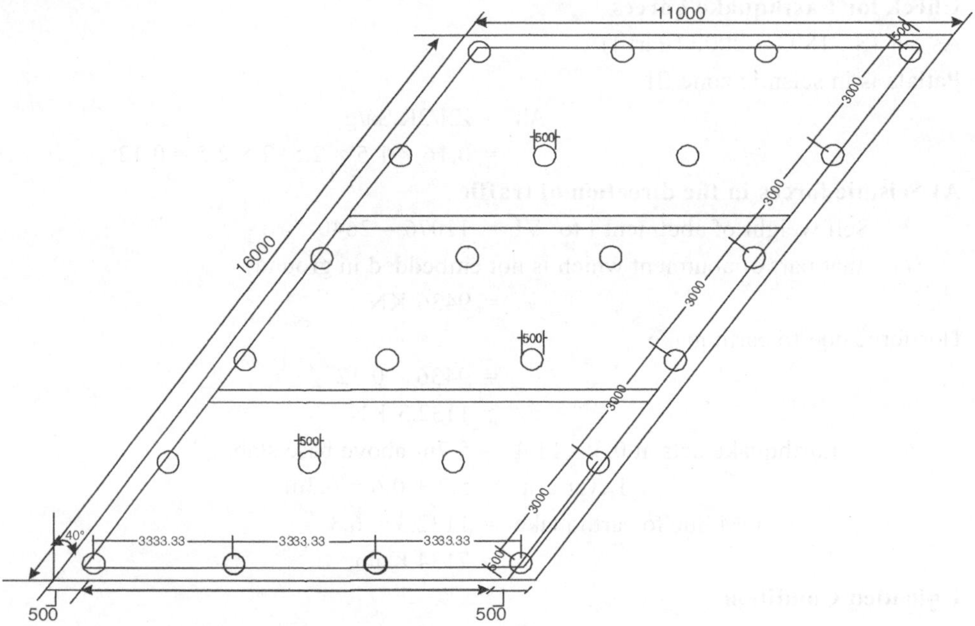

PILES ARRANGAMENT UNDER RAFT SLAB

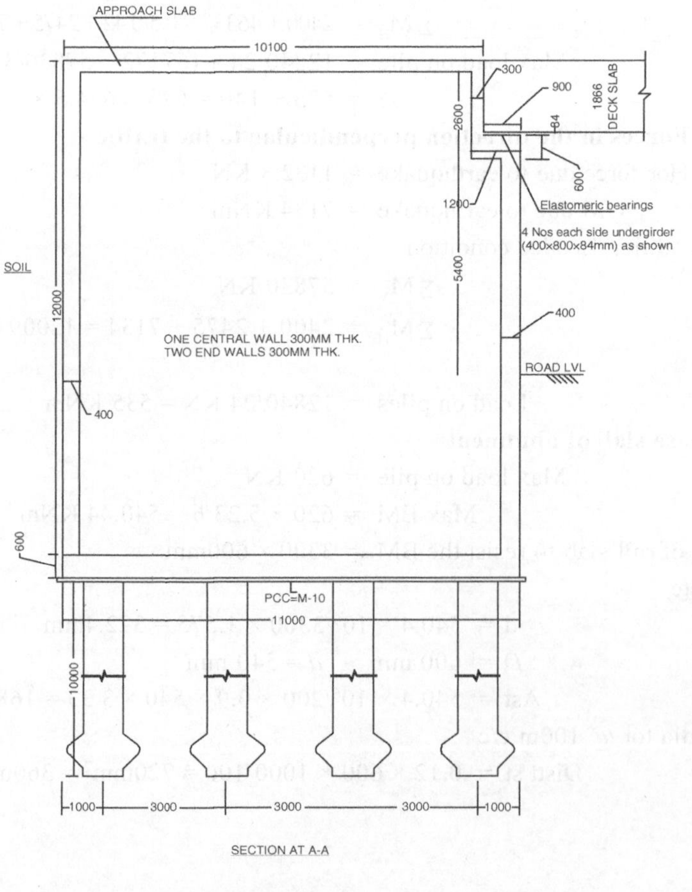

SECTION AT A-A

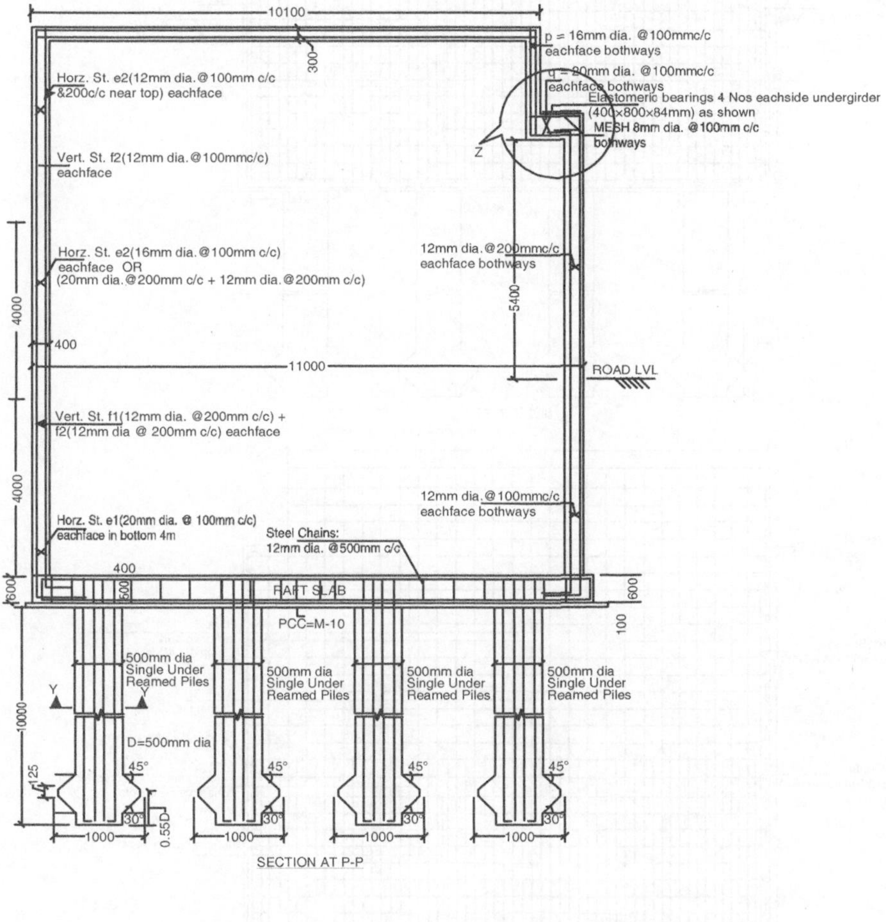

SECTION AT P-P

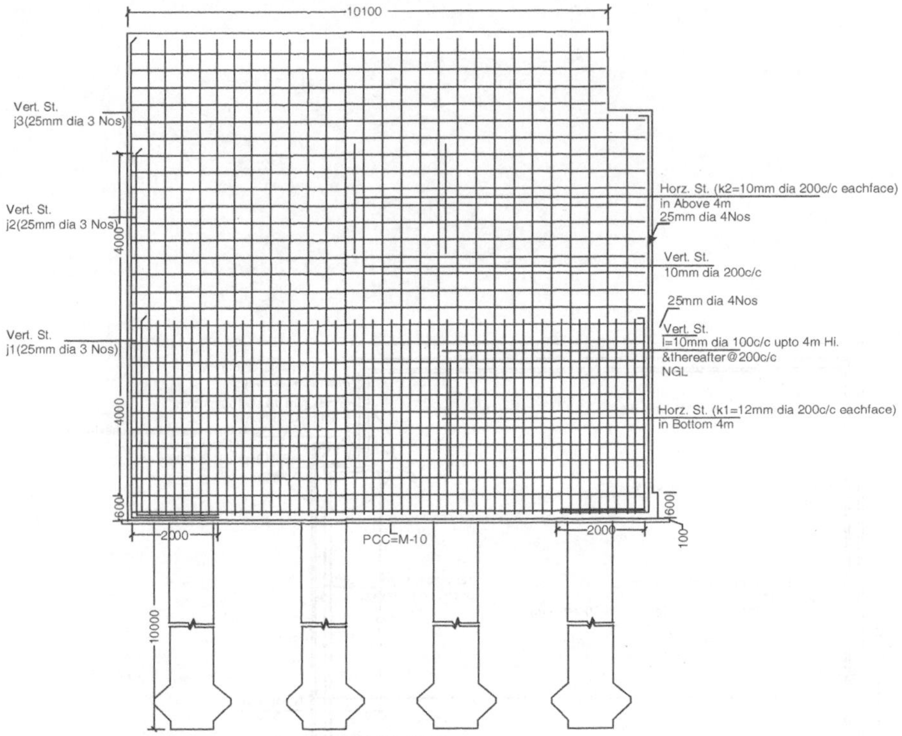

SECTION AT R-R

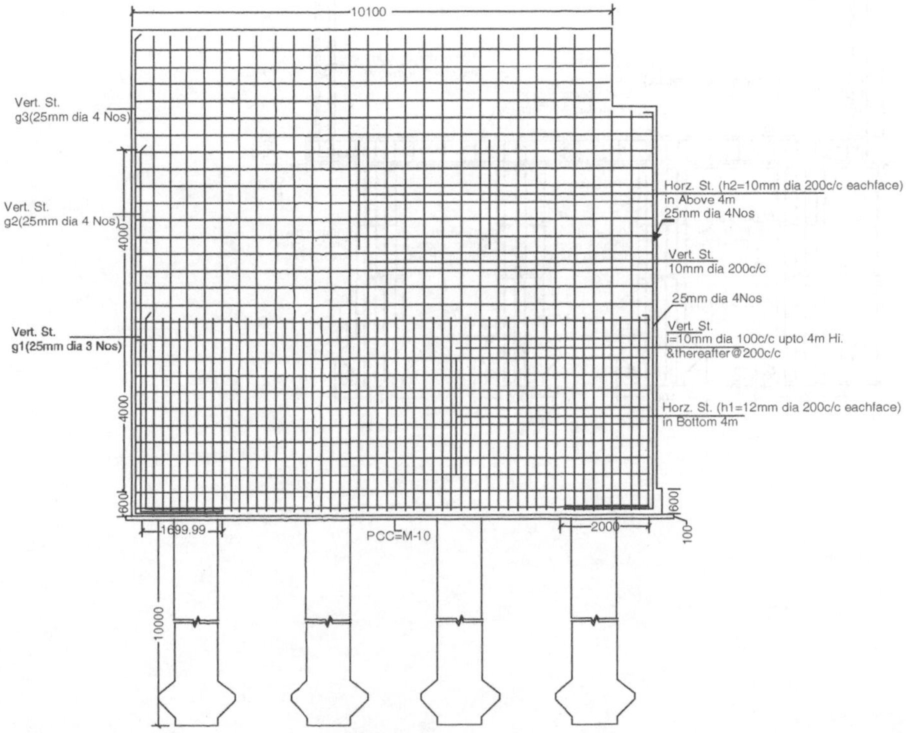

SECTION AT Q-Q
(Reinf of Central Counterfort Wall)

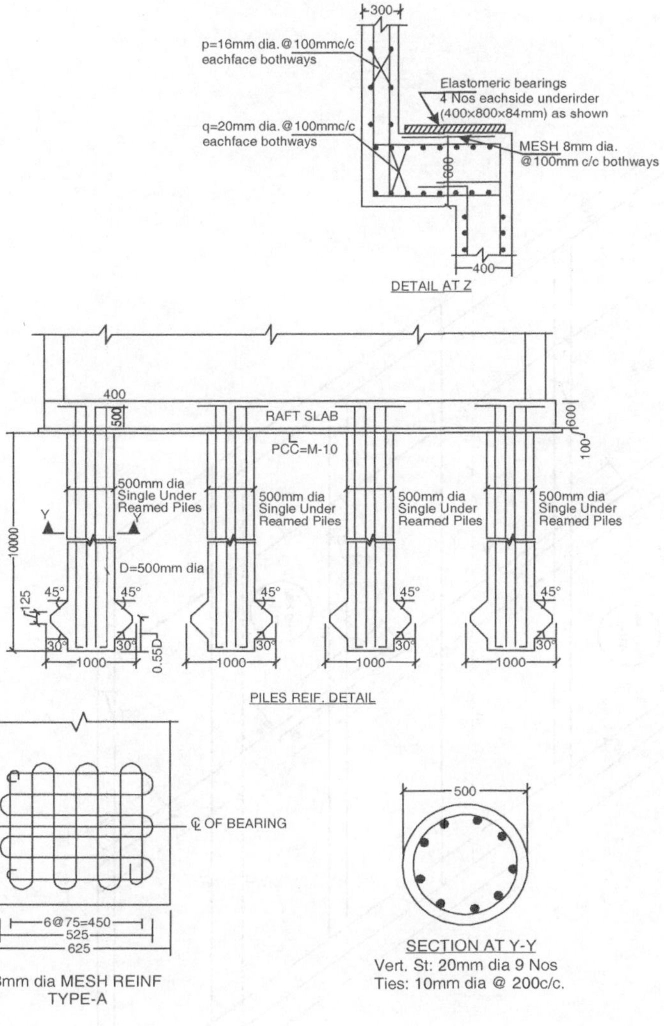

300
p=16mm dia. @100mmc/c eachface bothways
Elastomeric bearings 4 Nos eachside underirder (400×800×84mm) as shown
q=20mm dia. @100mmc/c eachface bothways
MESH 8mm dia. @100mm c/c bothways
600
400
DETAIL AT Z
400
500
RAFT SLAB
600
PCC=M-10
100
500mm dia Single Under Reamed Piles
500mm dia Single Under Reamed Piles
500mm dia Single Under Reamed Piles
500mm dia Single Under Reamed Piles
Y
Y
10000
D=500mm dia
45°
45°
45°
45°
45°
125
30°
30°
30°
30°
30°
0.55D
1000
1000
1000
1000
PILES REIF. DETAIL
37.5
450
5@75=375
37.5
℄ OF BEARING
6@75=450
525
625
8mm dia MESH REINF TYPE-A
500
SECTION AT Y-Y
Vert. St: 20mm dia 9 Nos
Ties: 10mm dia @ 200c/c.

Fig. 27.26(a).

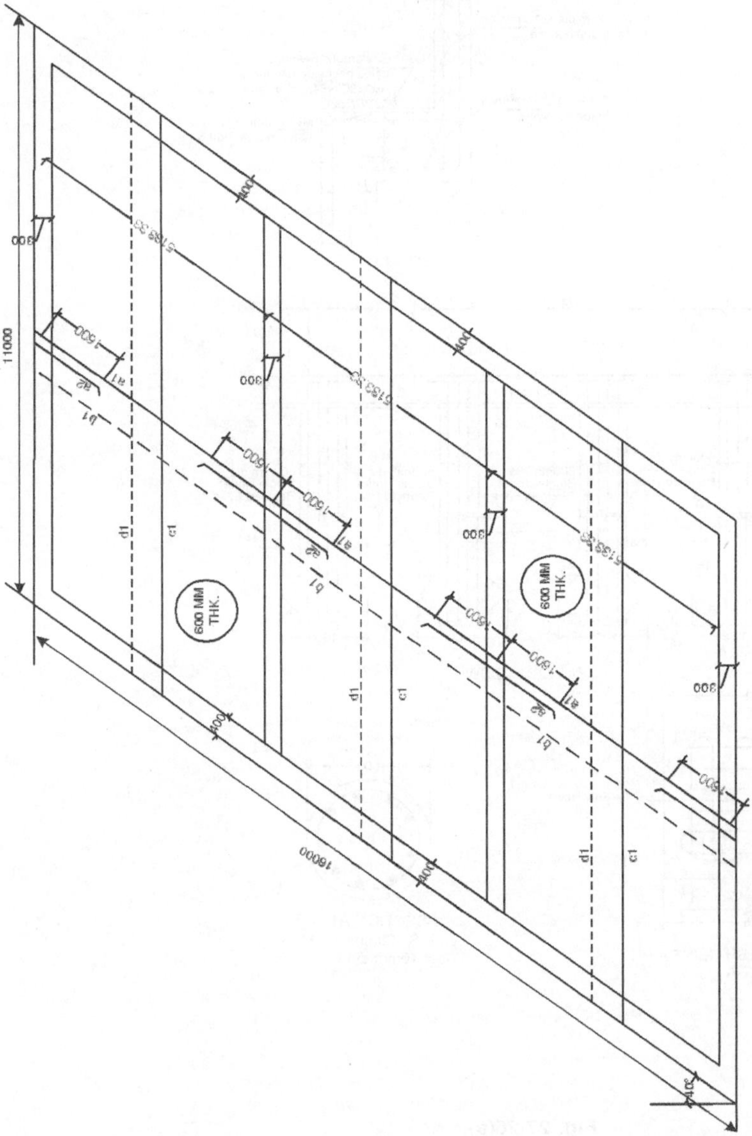

RAFT SLAB DETAIL

THICKNESS OF RAFT
SLAB = 600mm

a1 = 16mm dia @ 200c/c	B
a2 = 16mm dia @ 200c/c	B
b1 = 16mm dia @ 100c/c	T
c1 = 12mm dia @ 200c/c	B
d1 = 12mm dia @ 200c/c	T

NOTE:
Steel Reinf chair; 12mm dia @500 c/c.

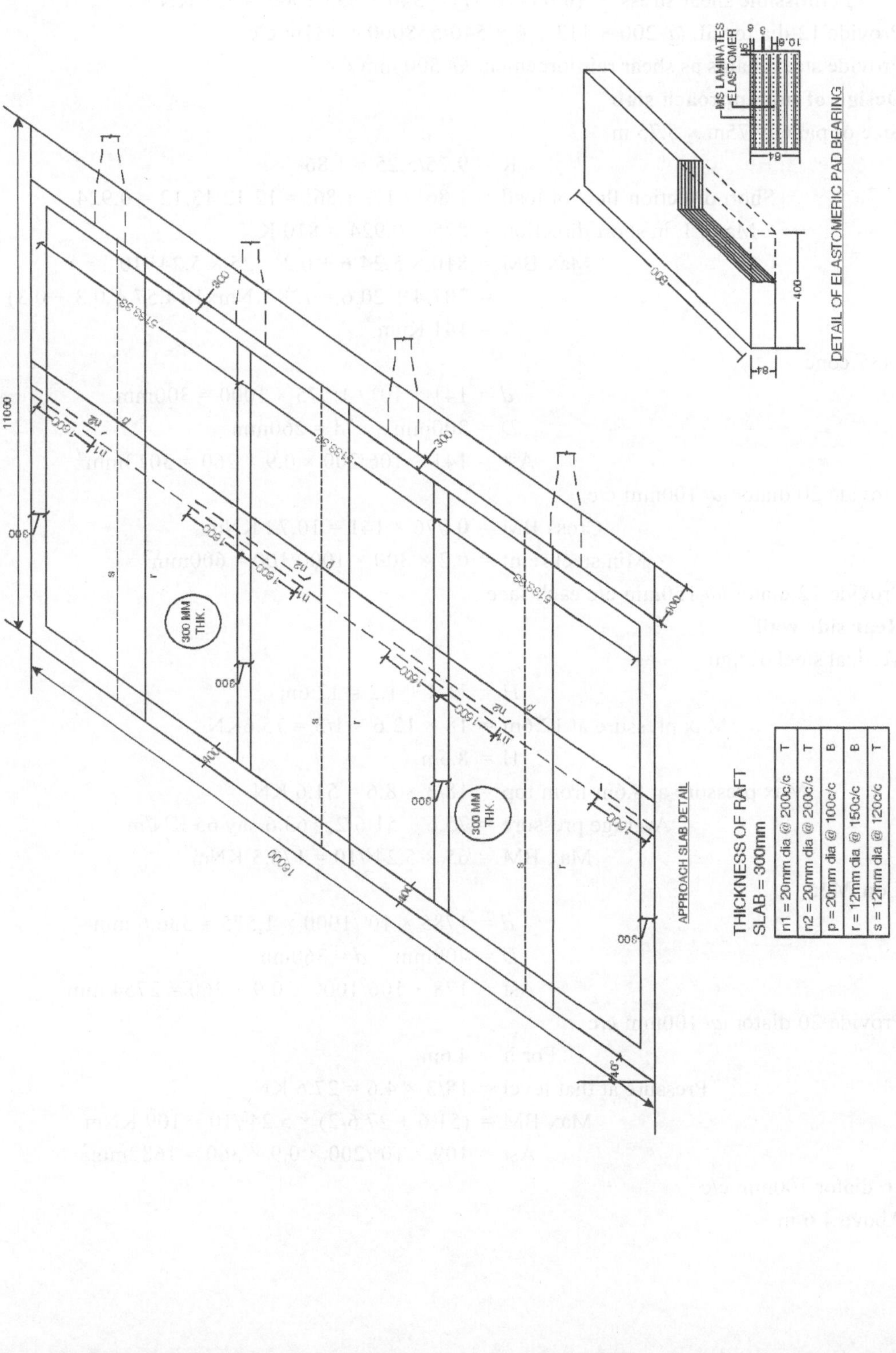

DETAIL OF ELASTOMERIC PAD BEARING
MS LAMINATES
ELASTOMER
APPROACH SLAB DETAIL
300 MM THK.
THICKNESS OF RAFT SLAB = 300mm
n1 = 20mm dia @ 200c/c T
n2 = 20mm dia @ 200c/c T
p = 20mm dia @ 100c/c B
r = 12mm dia @ 150c/c B
s = 12mm dia @ 120c/c T

Fig. 27.26(b)

Provide 12 dia tor @ 200mm c/c

Max shear stress = (620 + 620) × 1000/3333 × 540 = 0.68 N/m^2

P = 2010mm^2 × 100/1000 × 540 = 0.37

Permissible shear stress = (0.68 – 0.37) × 540 × 333/1000 = 558 KN

Provide 12 dia tor 6L @ 200 × 113 × 6 × 540/558000 = 131m c/c

Provide steel chains as shear reinforcement @ 500 mm c/c

Design of top approach slab

Size of panel 9.75m × 5.25 m

R = 9.75/5.25 = 1.86

Short direction flow of load = 1.86^4 / 1 + 1.86^4 = 12.12/13.12 = 0.924

Max LL in short direction = 875 × 0.924 = 810 KN

Max BM = 810 × 5.24/6 + 0.3 × 25 × 5.24^2/10

= 707.4 + 20.6 = 728 KNmx1/(4.57 + 0.3 + 0.3)

= 141 Knm

M35 conc

d = 141 × 10^6 / 1.575 × 1000 = 300mm

D = 300mm d = 260mm

Ast = 141 × 106/200 × 0.9 × 260 = 3013mm^2

Provide 20 diator @ 100mm c/c

Cross BM = 0.076 × 141 = 10.72 KNm

Min steel reinf = 0.2 × 300 × 1000/100 = 600mm^2

Provide 12 diator @ 150mm c/c each face

Rear side wall

Vertical steel design

H = 11.4 + 1.2 = 12.6m

Max pressure at 12.6m = 18 × 12.6 × 1/3 = 75.6KN

H = 8.6m

Max pressure at 8.6m from top = 18/3 × 8.6 = 51.6 KN

Average pressure = 75.6 + 51.6/2 = 63.6 say 65 KN/m

Max BM = 65 × 5.24^2/10 = 178.5 KNm

Con M35

d = 1785 × 10^6/1000 × 1.575 = 336.6 mm

D = 400mm d = 360mm

Ast = 178 × 106/1000 × 0.9 × 360 = 2754 mm^2

Provide 20 diator @ 100mm c/c

For h = 4.6m

Pressure at that level = 18/3 × 4.6 = 27.6 KN

Max BM = (51.6 + 27.6/2) × 5.24^2/10 = 109 KNm

Ast = 109 × 10^6/200 × 0.9 × 360 = 1682mm^2

16 diator 100mm c/c

Above 4.6 m

12 diator @ 100 c/c

Min shear reinf

Asv/b.sv>0.4/fy

Providing 10 dia tor 2 legged @

78.5 × 2/300 × sv>0.4/415

$$S_V = 522\text{mm}$$

Provide 10 diator 2L@200 mm c/c each face

Front wall

$$\text{Min steel reinf} = 0.2 \times 400 \times 1000/100 = 800\text{mm}^2$$

Provide 12 diator @ 100mm c/c upto 4m from bottom & thereafter 12 diator @ 200mm c/c

Design of Central vertical walls (2 Nos) as counterforts

$$\text{Max BM at base} = 18/3 \times 12.6^3/6 \times 5.33 = 10662\text{Knm}$$

$$D = 11000\text{mm} \quad d = 10500\text{mm}$$

$$\text{Max SF} = 18/3 \times 1/2 \times 12.6^2 \times 5.33 = 2538 \text{ KNm}$$

$$\text{Ast} = 10662 \times 106/ 200 \times 0.9 \times 10500 = 5641 \text{ mm}^2$$

Provide 25 mm diator 12 Nos

$$\text{Max shear stress} = 2538 \times 1000/300 \times 10500 = 0.8\text{N/mm}^2$$

$$P = 11 \times 615 \times 100/10500 \times 300 = 0.21$$

$$\text{Permissible shear stress} = 0.25 \text{ N/mm}^2$$

$$\text{Design shear force} = (0.8 - 0.25) \times 300 \times 10500/1000 = 1732.5\text{KN}$$

Provide 12 diator 2 Legged @ 113 × 2 × 200 × 10500/1732500 = 273mm c/c

Say 200mm c/c

Minimum shear reinforcement

12 diator 2L@ 113 × 2/300 × sv>/0.4/415

$$S_V = 781\text{m}$$

Provided 200 mm c/c is OK

Front wall

$$\text{Min steel reinf} = 0.2 \times 400 \times 1000/100 = 800\text{mm}^2$$

Privide 12 diator @ 100mm c/c upto 4m & thereafter @ 200mm c/c

End counterfort vertical walls

Assuming that 60% of pressure on central counterfort wall is to be resisted by end counterfort wall

$$\text{Ast} = 5641 \times 0.6 = 3384.6\text{mm}^2$$

Provide 25 diator 9 Nos (Asf = 491 × 9 = 4419m^2)

Hor & vertical reinforcement same as in central walls.

CHAPTER 28

Design Examples on Limit State Method

28.1 INTRODUCTION

Bureau of Indian Standards in its special publication SP : 16 has presented Charts and Tables to facilitate the designing of concrete members namely slabs, beams and columns based on limit state method prescribed in IS : 456. Design problems for beams reinforced in tension only, beams reinforced in tension and compression, slabs one way and two way, columns subjected to uniaxial eccentricity, footings and stair cases, have been presented in this chapter using the said Charts or Tables. Special care is taken to use minimum grade of concrete M20 as per IS : 456-2000. However, concretes of grade M20 to M60 are very commonly used in construction of multistorey framed structures. Similarly, steel reinforcement of yield strength Fe 415 and Fe 500 are commonly being used.

28.2 DESIGN EXAMPLES

Example 28.1. Design a R.C. rectangular beam for a simply supported span of 5 m and carrying a S.I. load of 20 kN/m inclusive of self weight of the beam. Use M20 concrete and HYSD bars of grade Fe 415.

Solution. Total factored *u.d.l* on beam = 1.5 × 20 = 30.0 kN/m

$$\text{Max. factored BM} = \frac{wl^2}{8}$$

$$= \frac{30 \times 5^2}{8} = 93.75 \text{ kN.m}$$

Assuming the width of beam to be 300 mm.

For balanced section,

$$d = \sqrt{\frac{M}{k \cdot b}} = \sqrt{\frac{93.75 \times 10^6}{2.76 \times 300}} = 336.5 \text{ mm}$$

Providing overall depth D = 525 mm; d = 525 – 35 = 490 mm.

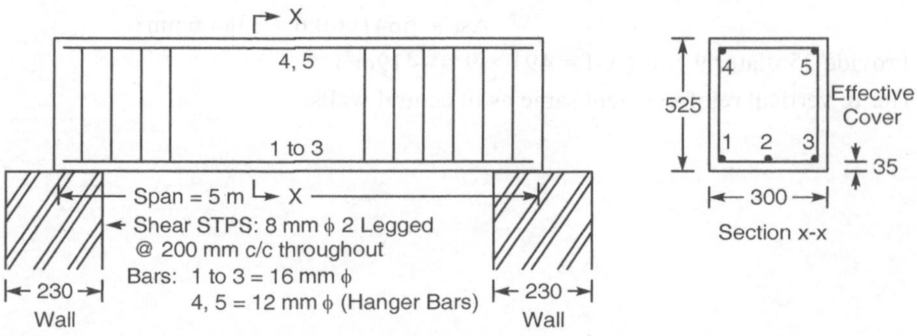

Fig. 28.1

$$\therefore \qquad k = \frac{93.75 \times 10^6}{300 \times 490^2} = 1.30$$

From Table 18.5A

$$A_{st} = 0.392 \times \frac{300 \times 490}{100} = 576 \text{ mm}^2$$

Provide 16 mm ϕ 3 nos. (thus providing A_s = 603 sq. mm) as shown in Fig. 28.1.

Check for shear at critical section, *i.e.*, at a distance 0.49 m from support.

$$\text{Max. factored S.F.} = \frac{wl}{2} - 0.49 \times 20000$$

$$= \frac{20000 \times 5}{2} - 9800 = 40200 \text{ N}$$

$$\text{Nominal shear stress} = \frac{40200}{300 \times 490} = 0.273 \text{ N/mm}^2$$

$$p = \frac{603 \times 100}{300 \times 490} = 0.41$$

$$\text{Permissible shear stress} = 0.43 \text{ N/mm}^2$$

∴ Safe.

Minimum shear reinforcement 8 mm ϕ 2 logged at

$$s_v = \frac{A_{sv} \cdot f_y}{0.4b} = \frac{2 \times 50 \times 415}{0.4 \times 300} = 345 \text{ mm c/c}$$

∴ Provide 8 mmϕ 2 legged @ 200 mm c/c throughout.

Example 28.2. Design a cantilever beam over a span of 3.0 m carrying a S.I. load of 20 kN/m. Use M20 concrete and HYSD (Fe 415) bars. Size of beam may be taken as 300 × 700 mm (Fig. 28.2).

Solution. Self weight of beam = 0.3 × 0.7 × 25 = 5.25 kN.m²

S.I. load = 20.00 kN.m²

Total working *u.d.l* = 25.25 kN.m

Factored *u.d.l* = 25.25 × 1.5 = 37.875 kN.m

Max. factored B.M. at built-in-end

$$= \frac{37.875 \times 3^2}{2} = 170.4375 \text{ kN.m}$$

For limiting resistance, $d = 700 - 40 = 660$ mm

$$M = kbd^2 \qquad k = 2.76 \text{ (Table 18.3)}$$

$$= \frac{2.76 \times 300 \times 660^2}{10^6} = 360.67 \text{ kN.m} > 170.4375 \text{ kN.m}$$

$$\therefore \qquad \text{Actual } k = \frac{170.4375 \times 10^6}{300 \times 660^2} = 1.30, \; P_t = 392 \text{ (Table 18.5A)}$$

$$A_{st} = 0.392 \times 300 \times \frac{660}{100} = 776.16 \text{ mm}^2$$

Provide 16 mm ϕ 4 nos. on top face (thus providing A_{st} = 804 mm²) as shown in Fig. 28.2.

$$\text{Max. factored S.F.} = 37.875 \times 3 = 111.625 \text{ kN}$$

$$\text{Nominal shear stress} = \frac{111.625 \times 1000}{300 \times 660} = 0.564 \text{ N/mm}^2$$

$$p = \frac{804 \times 100}{300 \times 660} = 0.4$$

Permissible shear stress = 0.43 N/mm²

Design factored S.F. = $(0.564 - 0.43) \times 300 \times 660 = 26532$ N

Providing 8 mm ϕ 2 legged @

$$\frac{0.87 \times f_y \cdot A_{sv} \cdot d}{V_s} = \frac{0.87 \times 415 \times 2 \times 50 \times 660}{26532} = 898 \text{ mm c/c}$$

Maximum spacing = $0.75 \times d = 0.75 \times 660$
= 495 mm

For minimum shear reinforcement, spacing of 8 mmϕ 2 legged @

$$\frac{0.87 \cdot A_{sv} \cdot f_y}{0.4b} = \frac{0.87 \times 2 \times 50 \times 415}{0.4 \times 300} = 300 \text{ mm}$$

$\therefore$ @ 200 mm c/c throughout.

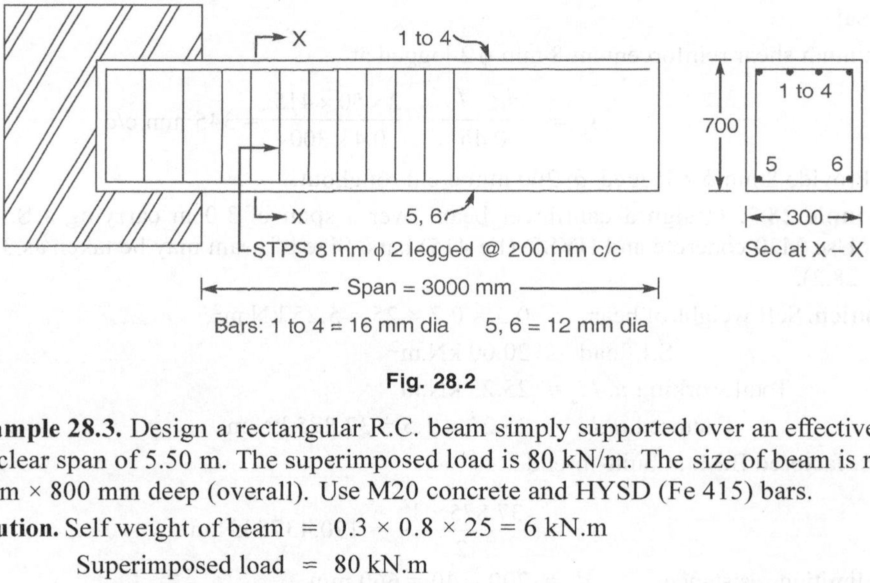

Fig. 28.2

Example 28.3. Design a rectangular R.C. beam simply supported over an effective span of 6 m and clear span of 5.50 m. The superimposed load is 80 kN/m. The size of beam is restricted to 300 mm × 800 mm deep (overall). Use M20 concrete and HYSD (Fe 415) bars.

Solution. Self weight of beam = $0.3 \times 0.8 \times 25 = 6$ kN.m

Superimposed load = 80 kN.m

Total working *u.d.l* = 86 kN.m

Factored *u.d.l* = $86 \times 1.5 = 129$ kN.m

$$\text{Max. factored B.M.} = \frac{129 \times 6^2}{8} = 580.5 \text{ kN.m}$$

For limiting resistance of beam, assuming effective depth of beam to be 750 mm.

$$M_{u,\,\text{lim}} = M_1 = kbd^2 \qquad k = 2.76 \text{ (Table 18.3)}$$

$$= \frac{2.76 \times 300 \times 750^2}{10^6} = 465.75 \text{ kN.m} < 580.5$$

It is doubly reinforced beam, *i.e.*, a beam reinforced with compression reinforcement also.

$$k = \frac{580.5 \times 10^6}{300 \times 750^2} = 3.44$$

For $$\frac{d'}{d} = \frac{50}{750} = 0.06$$

(From Table 18.8A) $$A_{st} = 1.15 \times 300 \times \frac{750}{100} = 2587.5 \text{ mm}^2$$

$$A_{sc} = 0.207 \times 300 \times \frac{750}{100} = 466 \text{ mm}^2$$

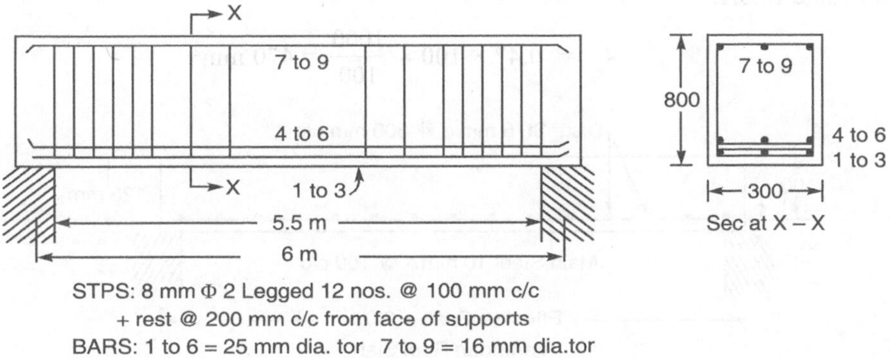

Fig. 28.3

Provide 25Φ 6 nos. on tension face and 16 mmΦ 3 nos. on compression face.

$$\text{Max. factored S.F.} = 129 \times \frac{5.50}{2} = 354.75 \text{ kN}$$

$$\text{Nominal shear stress} = \frac{354750}{300 \times 750} = 1.576 \text{ N/mm}^2$$

$$p = \frac{491 \times 6 \times 100}{300 \times 750} = 1.31$$

$$\text{Permissible shear stress} = 0.68 \text{ N/mm}^2$$

$$\therefore \quad \text{Design factored S.F.} = (1.576 - 0.68) \times 300 \times 750$$

$$= 201600 \text{ N}$$

Providing 8Φ 2L @ $$\frac{0.87 \times 415 \times 2 \times 50 \times 750}{201600} = 134 \text{ mm c/c}$$

Say provide 12 nos. @ 100 mm c/c and balance at 200 mm c/c.

Minimum shear reinforcement spacing

$$= \frac{0.87.A_{sv} \cdot f_y}{0.4b} = \frac{2 \times 50 \times 415 \times 0.087}{0.4 \times 300} = 300 \text{ mm c/c}$$

Example 28.4. Design a verandah slab to cover an effective span of 3.0 m for concrete grade M20 and HYSD (Fe 415) bars. It carries a superimposed load of 5.0 kN/m including finishes.

Solution. Assuming thickness of slab as 125 mm.

$$\text{Self weight of slab} = 0.125 \times 25 = 3.75 \text{ kN/m}^2$$

$$\text{Superimposed load} = 5.00 \text{ kN/m}^2$$

$$\text{Total working } u.d.l = 8.75 \text{ kN/m}^2$$

$$\text{Factored } u.d.l = 8.75 \times 1.5 = 13.125$$

$$\text{Maximum factored B.M.} = \frac{13.125 \times 3^2}{8} = 14.76 \text{ kN.m}$$

$$d = 125 - 25 = 100 \text{ mm}$$

$$k = \frac{14.76 \times 10^6}{1000 \times 100^2} = 1.476$$

From Table 18.5A,

$$A_{st} = 0.47 \times 100 \times \frac{1000}{100} = 470 \text{ mm}^2$$

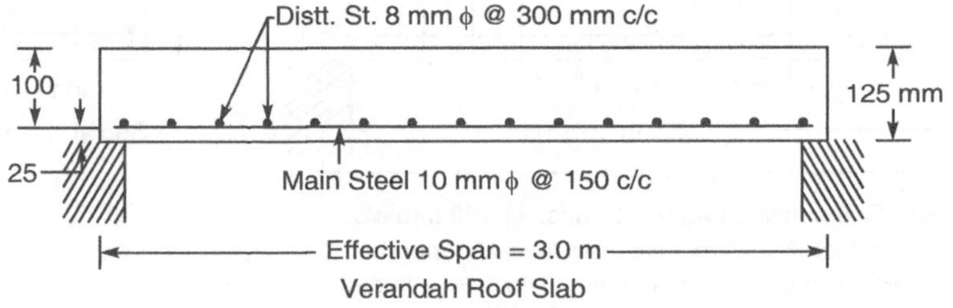

Fig. 28.4

$$\text{Provide 10 mm}\phi \ \frac{78.5 \times 1000}{470} = 167 \text{ mm say 150 mm c/c } (A_{st} = 523 \text{ mm}^2).$$

$$\text{Distt. steel} = 0.12 \times 125 \times \frac{1000}{100} = 150 \text{ mm}^2$$

$$\text{Provide 8 mm}\phi \ @ \ \frac{50 \times 1000}{150} = 333 \text{ mm c/c say 300 mm c/c.}$$

$$\text{Maximum shear force} = \frac{13.125 \times 3}{2} = 19.6875 \text{ kN/m}$$

$$\text{Nominal shear stress} = \frac{19.6875 \times 1000}{1000 \times 100} = 0.196 \text{ N/mm}^2$$

$$p = \frac{523 \times 100}{1000 \times 100} = 0.523$$

$$\text{Permissible stress} = 0.48 \times 1.3 = 0.624 \text{ N/mm}^2$$

∴ Safe.

The reinforcement arrangement is shown in Fig. 28.4.

Example 28.5. Design a two way slab for a room having effective spans in two directions as 4 m and 5 m. Take live load = 3 kN/m² and finishes as 1.5 kN/m². Use concrete M20 and HYSD bars (Fe 415).

Solution. Assuming thickness of slab as 125 mm and effective depth as 100 mm.

$$\text{Self weight of slab} = 0.125 \times 25 = 3.125 \text{ kN/m}^2$$

$$\text{Live load} = 3.000 \text{ kN/m}^2$$

$$\text{Finishes} = 1.500 \text{ kN/m}^2$$

$$\text{Total working } u.d.l = 7.625 \text{ kN/m}^2$$

$$\text{Factored } u.d.l = 7.625 \times 1.5 = 11.44 \text{ kN/m}^2$$

$$r = \frac{l_y}{l_x} = \frac{5.0}{4.0} = 1.25$$

$$\text{Short span B.M.} = 0.089 \times 11.44 \times 4^2 = 16.29 \text{ kN.m}$$

$$\text{Long span B.M.} = 0.057 \times 11.44 \times 4^2 = 10.43 \text{ kN.m}$$

Computation of areas of steel,

For short span, $k_1 = \dfrac{16.29 \times 10^6}{1000 \times 100^2} = 1.629$

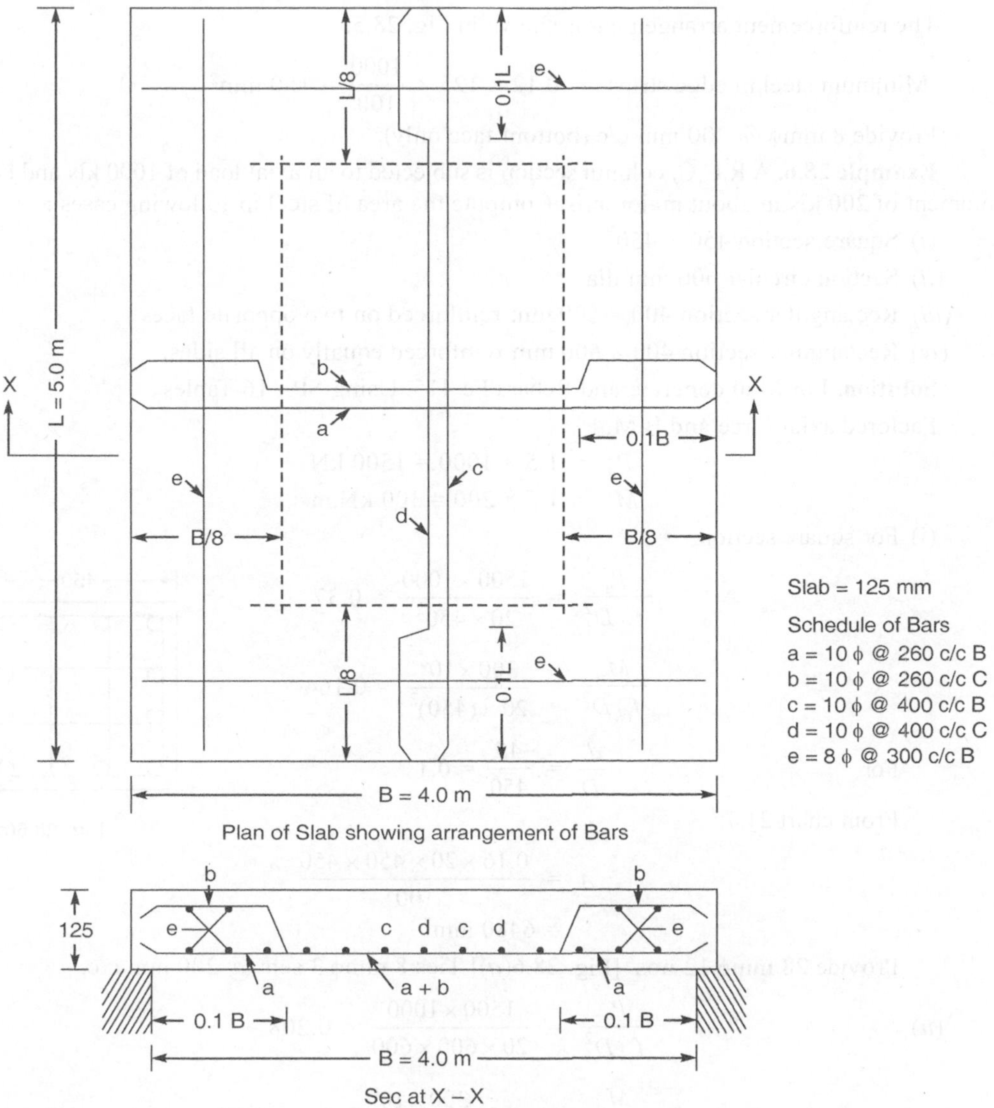

Fig. 28.5

From Table 18.5A $A_{st} = 0.503 \times 100 \times \frac{1000}{100} = 503 \text{ mm}^2$

Provide 10 mmϕ @ 130 mm c/c (thus providing $A_{st} = 603 \text{ mm}^2$).

For long span, $d = 100 - 10 = 90$ mm

$$k = \frac{10.43 \times 10^6}{1000 \times 90^2} = 1.29$$

From Table, 18.5A

$$A_{st} = 0.389 \times 90 \times \frac{1000}{100} = 350 \text{ mm}^2$$

Provide 10 mmϕ @ $\frac{78.5 \times 1000}{350} = 224$ mm c/c say @ 200 mm c/c.

The reinforcement arrangement is shown in Fig. 28.5.

$$\text{Minimum steel in edge strips} = 0.12 \times 125 \times \frac{1000}{100} = 150 \text{ mm}^2$$

Provide 8 mmϕ @ 300 mm c/c (bottom face only).

Example 28.6. A R.C.C. column section is subjected to an axial load of 1000 kN and bending moment of 200 kN.m about major axis. Compute the area of steel in following cases :

(*i*) Square section 450 × 450

(*ii*) Section circular 600 mm dia

(*iii*) Rectangular section 400 × 600 mm reinforced on two opposite faces

(*iv*) Rectangular section 400 × 600 mm reinforced equally on all sides.

Solution. For M20 concrete and Rebars Fe 415. Using SP : 16 Tables.

Factored axial force and B.M,

$$P_u = 1.5 \times 1000 = 1500 \text{ kN}$$

$$M_u = 1.5 \times 200 = 300 \text{ kN.m}$$

(i) For square section,

$$\frac{P_u}{f_{ck} D^2} = \frac{1500 \times 1000}{20 \times 450^2} = 0.37$$

$$\frac{M_u}{f_{ck} D^3} = \frac{300 \times 10^6}{20 \times (450)^3} = 0.164$$

For $\frac{d'}{D} = \frac{45}{450} = 0.1$

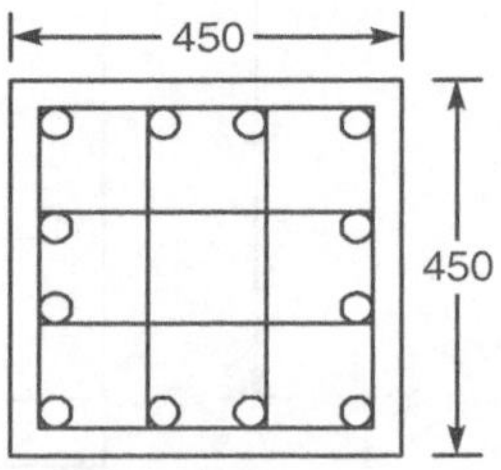

Fig. 28.6(a)

From chart 21.7,

$$A_s = \frac{0.16 \times 20 \times 450 \times 450}{100}$$

$$= 6480 \text{ mm}^2$$

Provide 28 mmϕ 12 nos. [Fig. 28.6(*a*)] Ties 8 mmϕ 3 sets @ 200 mm c/c.

(*ii*) $$\frac{P_u}{f_{ck} D^2} = \frac{1500 \times 1000}{20 \times 600 \times 600} = 0.208$$

$$\frac{M_u}{f_{ck} D^3} = \frac{300 \times 10^6}{20 \times 600 \times 600 \times 600} = 0.07$$

From curves,

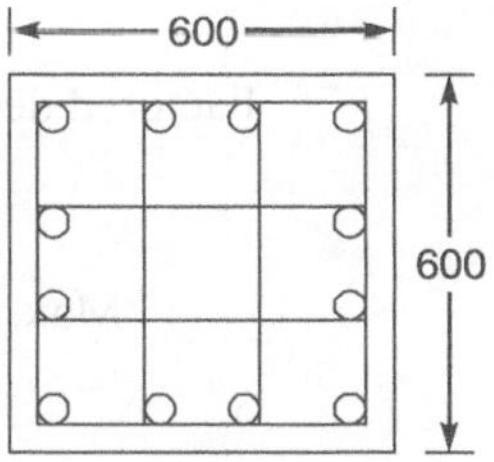

Fig. 28.6 (b)

$$A_s = \frac{0.02 \times 20}{100} \times \frac{\pi}{4} \times 600^2$$
$$= 1130.40 \text{ mm}^2$$

Min. steel in a column section

$$= 0.8 \times 0.008 \times \frac{\pi}{4} \times 600^2$$
$$= 1808 \text{ mm}^2$$

Provide 16 mm ϕ 12 nos. [Fig. 28.6(*b*)]; Ties : 8 mm ϕ 3 sets @ 200 c/c.

(*iii*) Rectangular section 400 × 600, with reinforcement on two opposite faces.

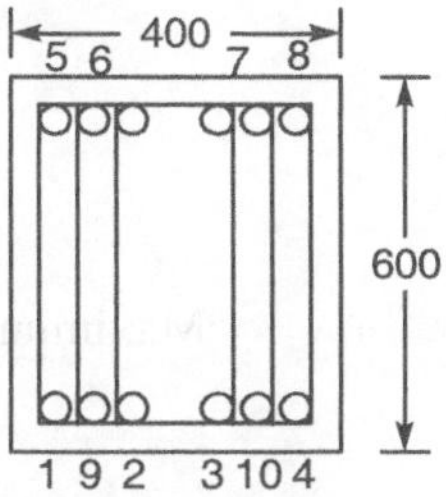

Fig. 28.6 (c)
1 to = 20 ϕ
9 to 12 = 16 ϕ

$$\frac{P_u}{f_{ck}bD} = \frac{1500 \times 1000}{20 \times 400 \times 600} = 0.3125$$

$$\frac{M_u}{f_{ck}bD^2} = \frac{300 \times 10^6}{20 \times 400 \times 600^2} = 0.104$$

$$\frac{d'}{D} = \frac{45}{600} = 0.075$$

From curves,

$$A_s = \frac{0.06 \times 20 \times 400 \times 600}{100} = 2880 \text{ mm}^2$$

Provide 20ϕ 8 nos. + 16ϕ 4 nos., (20ϕ 4 nos. + 16ϕ 2 nos.) each face; Ties : 8 mmϕ 3 sets @ 200 mm c/c [see Fig. 28.6(*c*)].

(*iv*) Rectangular section 400 × 600 with reinforcement equally on four sides,

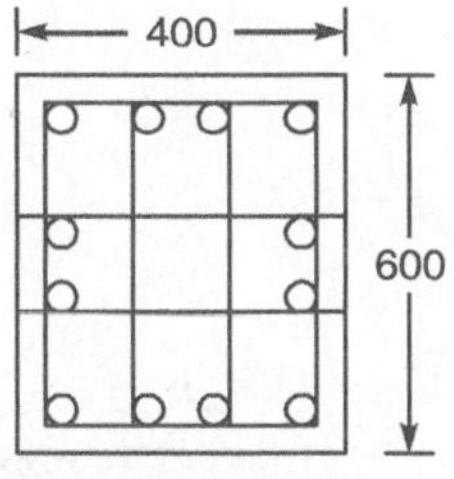

Fig. 28.6 (d)

$$\frac{P_u}{f_{ck}bD} = \frac{1500 \times 1000}{20 \times 400 \times 600} = 0.3125$$

$$\frac{M_u}{f_{ck}bD^2} = \frac{300 \times 10^6}{20 \times 400 \times 600^2} = 0.104$$

$$\frac{d'}{D} = \frac{45}{600} = 0.075$$

$$A_s = \frac{0.07 \times 20 \times 400 \times 600}{100} = 3360 \text{ mm}^2$$

Provide 20ϕ 12 nos., Ties : 8 mm ϕ 3 sets @ 200 mm c/c [see Fig. 28.6(*d*)].

Example 28.7. Design a square footing for square column 600 × 600 mm carrying an axial load of 900 kN. Conc. M20 and HYSD (Fe 415) bars. SBC of soil = 110 kN/m^2 at depth of 1.2 m from N.S.L.

Solution. Axial load on column = 900 kN

Self weight of footing 10% of S.I. load = 90 kN

Total axial load = 990 kN

SBC of soil = 110 kN/m^2

$$\text{Bearing area required} = \frac{990}{110} = 9 \text{ m}^2$$

Provide 3 m × 3 m base

$$\text{Factored design pressure} = \frac{900 \times 1.5}{3 \times 3} = 150 \text{ kN/m}^2$$

$$\text{Max. factored BM} = \frac{150 \times 1.2^2}{2} = 108 \text{ kN.m}$$

$$d = \sqrt{\frac{108 \times 10^6}{2.76 \times 1000}} = 198 \text{ mm}$$

$$D = 600 \text{ mm from shear considerations}$$

$$d = 600 - 60 = 540 \text{ mm}$$

$$K = \frac{108 \times 10^6}{1000 \times 540^2} = 0.37$$

$$\text{Maximum factored S.F.} = 150 \times \left(1.2 - \frac{0.54}{2}\right) = 139.5 \text{ kN}$$

$$\text{Nominal shear stress} = \frac{139.5 \times 1000}{1000 \times 540} = 0.258 \text{ N/mm}^2$$

$$A_{st} = 0.12 \times 540 \times \frac{1000}{100} = 648 \text{ mm}^2$$

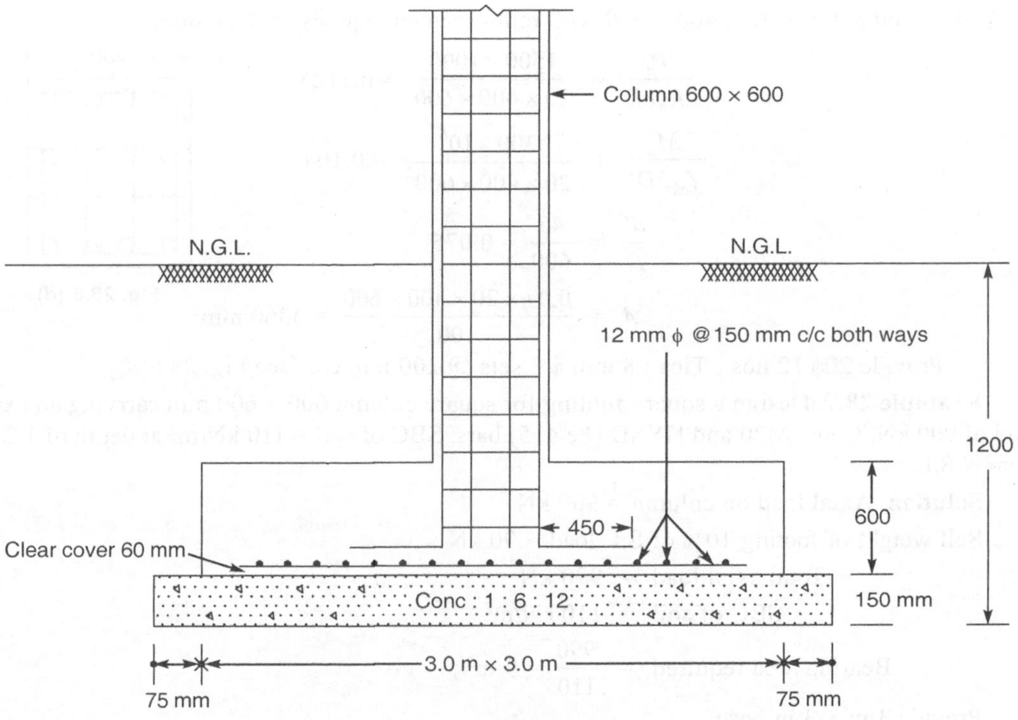

Fig. 28.7

Provide 12 mm ϕ @ 150 mm c/c bothways $\left(A_{st} \text{ provided} = \dfrac{113 \times 1000}{150} = 753 \text{ mm}^2\right)$

$$p = \frac{753 \times 100}{1000 \times 540} = 0.139$$

Permissible shear stress = 0.28 N/mm². The arrangement of reinforcement is shown in Fig. 28.7.

Example 28.8. Design a single flight stair (Fig. 28.8) to climb a roof slab 3 m high. The rise and tread of steps may be taken as 200 mm and 250 mm respectively. Width of stair is 0.9 m. Take live load = 3000 N/m². Finishes = 750 N/m². Use M20 concrete and HYSD (Fe 415) bars.

$$\text{No. of steps} = \frac{3 \times 1000}{200} = 15$$

Solution. Effective span = 15 × 250 = 3750 mm = 3.75 m

Assume thickness of slab = 150 mm

Self weight of slab/horizontal metre

$$= 1 \times 0.15 \times 25000 \left(\frac{\sqrt{25^2 + 20^2}}{25}\right) = 4802.4 \text{ N/m}^2$$

Weight of steps/horizontal metre

$$= \frac{1}{2} \times 0.25 \times 0.2 \times \frac{1 \times 25000}{0.25} = 2500 \text{ N/m}^2$$

$$\text{Live load} = 3000 \text{ N/m}^2$$

$$\text{Finishes} = 750 \text{ N/m}^2$$

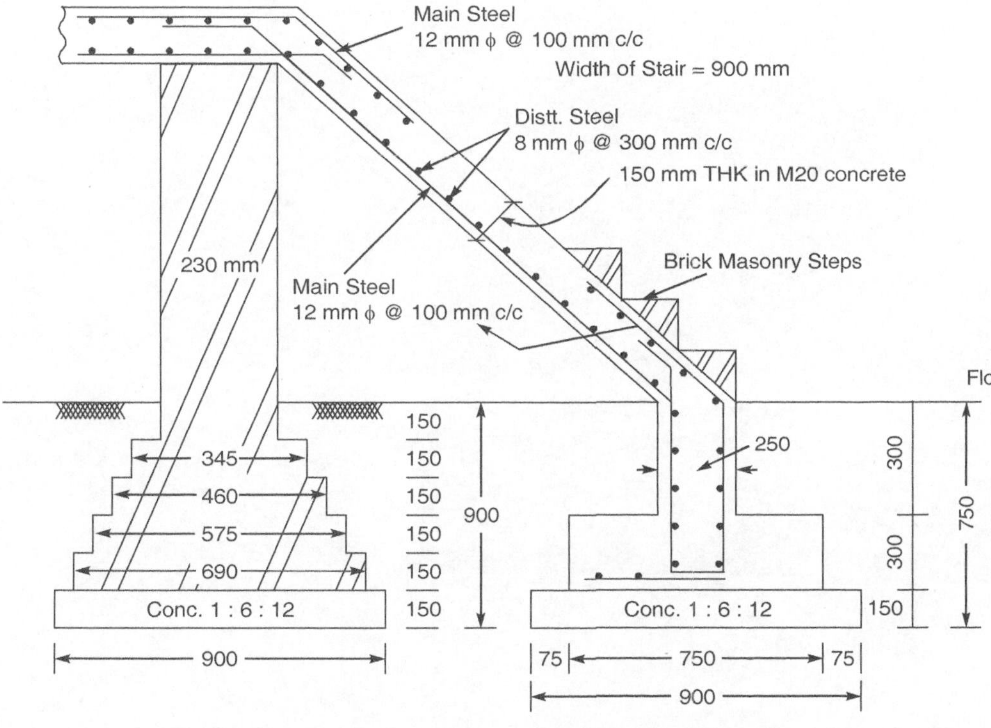

Fig. 28.8. Reinforcement arrangement in single flight stair

$$\text{Total working } u.d.l = 11052.4 \text{ N/m}^2$$

$$\text{Factored } u.d.l = 1.5 \times 11052.4 = 16580 \text{ N/m}^2$$

Load per metre on waist slab of 0.9 m width

$$= 16580 \times 0.9 = 14920.74 \text{ N/m say } 15000 \text{ N/m}$$

$$\text{Maximum factored BM} = \frac{15000 \times 3.75^2}{8} = 26367 \text{ N.m}$$

$D = 150$ mm; $d = 120$; using SP : 16 charts,

$$k = \frac{26367 \times 10^6}{900 \times 120^2} = 2.03$$

$$A_{st} = \frac{0.65 \times 900 \times 120}{100} = 972 \text{ mm}^2$$

Provide 12 mm ϕ @ 100 mm c/c (A_{st} provided = 1131 mm^2)

$$\text{Distribution Steel} = 0.12 \times 150 \times \frac{900}{100} = 162 \text{ mm}^2$$

Provide 8 mm ϕ @ 300 mm c/c

Reinforcement arrangement is shown in Fig. 28.8.

CHAPTER 29

Yield Line Theory

29.1 INTRODUCTION

The yield line theory is largely based upon the yield lines (or crack lines) that develop in any reinforced concrete slab (Rectangular, circular, square or any other geometrical shape in plan) before its final collapse. This stage reaches under loads approaching collapse loads or ultimate loads that the slab can carry.

The collapse loads, moments and shears can be calculated from the crack line pattern, developed in slab, under idealised support conditions and subjected to only uniformly distributed loads.

The crack lines or yield lines propagate with the increase in deflection till the slab is divided or broken into a number of segments.

This method is considered to be safe and conservative.

Yield line is a line in the plane of slab across which reinforcing bars have yielded and about which excessive deformations or plastic solutions under constant limit moment or ultimate moment continues to occur leading to failure.

29.2 BASIC ASSUMPTIONS

1. The reinforcing bars are fully yielded across the yield lines at failure.
2. The yield lines divide the slab into various segments who in turn behave elastically.
3. The entire deformations take place only in the yield lines and the individual segments of the slab are plane segments in the collapse condition.
4. The bending and twisting moments have the maximum values and are uniformly distributed along the yield lines.
5. The yield lines are straight lines at the intersection of individual inclined segments.

29.3 YIELD LINE DEVELOPMENT

Depending upon the boundary conditions of the slabs, the yield line patterns can be located for:

(A) One way simply supported slab

The yield line pattern in a simply supported one way slab is at the mid of span and is on the lower side because of the positive maximum moment which is sagging in nature. The yield line is refered as **positive yield line.** and is perpendicular to the direction of reinforcement, as shown in Fig. 29.1(a).

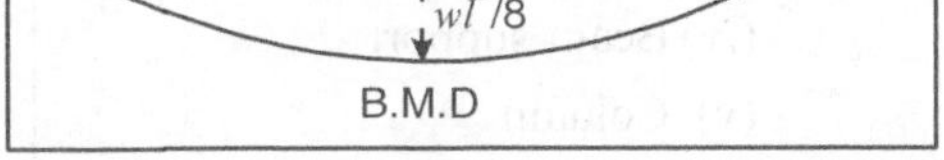

Fig. 29.1 (a)

(B) One way continuous slab (*i.e.*, fixed end conditions)

Due to continuity, both positive and negative moments on bottom face at centre and top face near supports give respectively the **positive** and **negative** yield lines, as shown in Fig 29.1(b).

Both the yield lines are perpendicular to reinforcing bars.

(C) Two way slabs.

These slabs can be square or rectangular for which the yield lines are not perpendicular to the reinforcement. Depending upon the end conditions, either positive or both positive and negative yield lines occur. The initiation, growth and development of yield lines is shown in Fig. 29.1(c) and Fig. 29.1 (d).

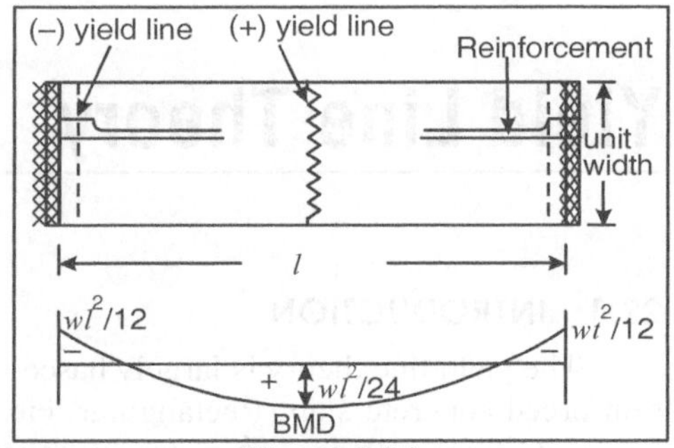

Fig. 29.1 (b)

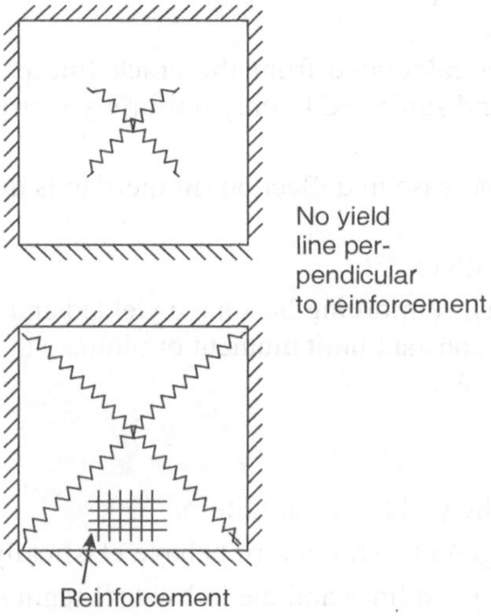

Fig. 29.1 (c). Square Slab simply supported

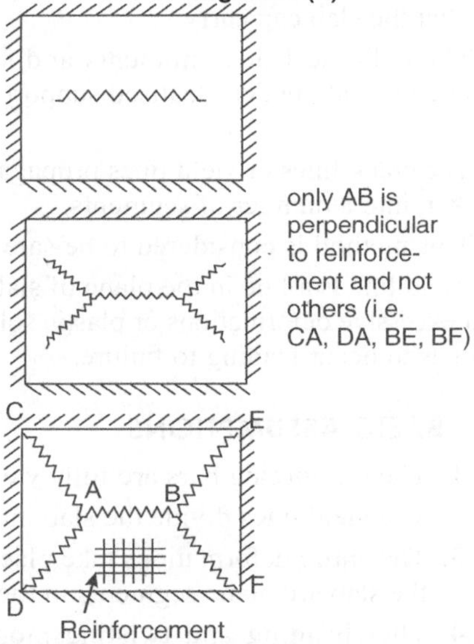

Fig. 29.1 (d). Rectangular Two-way slab simply supported

29.4 SIGN CONVENTIONS

The various sign conventions normally used for yield line patterns and supports/edges are shown in Fig. 29.2.

(*a*) Yield lines
- (*i*) positive
- (*ii*) negative

(*b*) Supports or edges
- (*i*) Free or unsupported edge
- (*ii*) Simply supported edge
- (*iii*) Fixed or continuous edge
- (*iv*) Beam support
- (*v*) Column

(*c*) Point load

(*d*) Axis of rotation

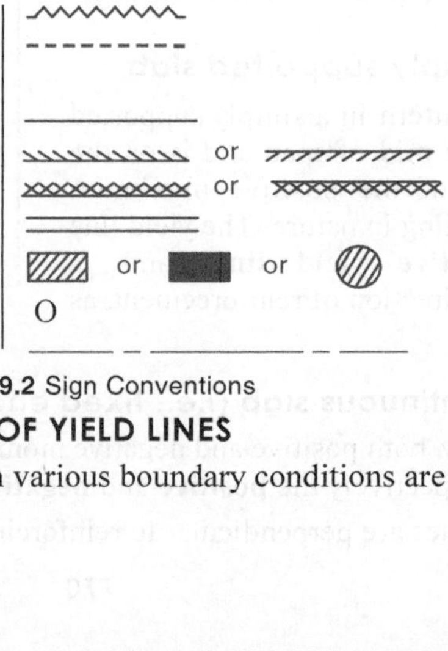

Fig. 29.2 Sign Conventions

29.5 CHARACTERISTIC FEATURES OF YIELD LINES

The yield line patterns for slabs with various boundary conditions are selected with the help of following characteristics:-

1. The yield lines are usually straight lines and normally act as plastic hinges of a collapse mechanism.
2. The yield lines normally terminate i.e end-up at boundary of the slab or at the intersection of other yield lines.
3. The yield lines act as axes of rotation for movement of the adjoining segments.
4. Each segment of slab will rotate in a rigid body motion.
5. The axis of rotation generally lie along the lines joining the supports or pass through the column supports.
6. For a fixed or continuous edge of slab, an yield line will form along the support.
7. The yield lines or yield lines produced pass through the axes of rotation of adjacent slab segments, for the mechanism to develop.

29.6 YIELD LINE PATTERNS

The various possible yield line patterns for slabs of various shapes and different support conditions are given in Fig. 29.3 (a) for square slabs, Fig. 29.3 (b) for rectangular slabs and Fig. 29.3 (c) for other shapes or non-rectangular slabs.

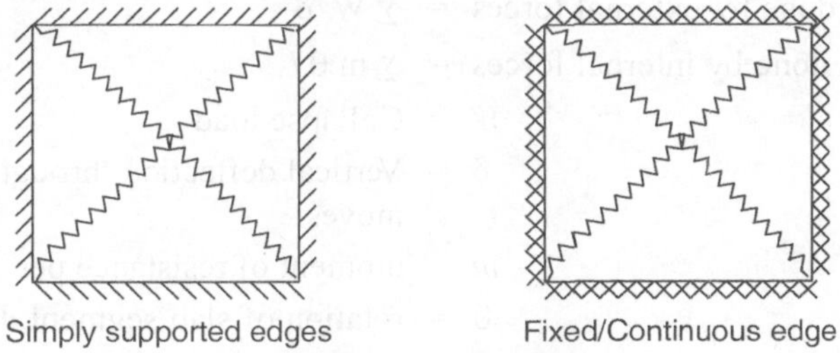

Fig. 29.3 (a). Square Slabs

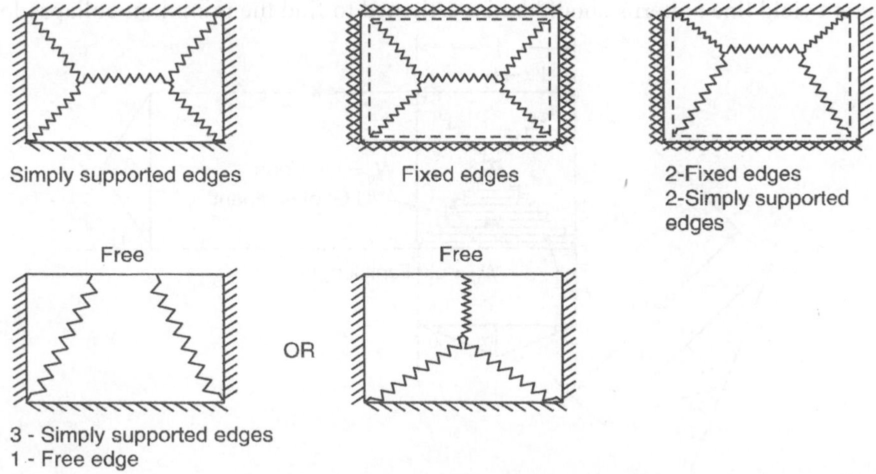

Fig. 29.3 (b). Rectangular Slabs

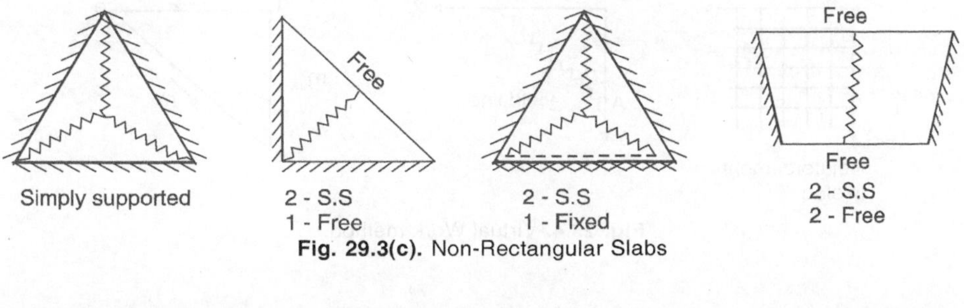

Fig. 29.3(c). Non-Rectangular Slabs

29.7 METHODS OF ANALYSIS

There are two methods of analysis of slabs.

1. Virtual Work Method
2. Equilibrium Method

Both these methods should give exactly same results because these are based upon same assumptions or characteristics of a yield line theory. To achieve a collapse mechanism, a yield line pattern is assumed in either method. The above methods help us in determining the exact location of the yield lines. Yield lines normally form along lines of maximum moment where neither shear nor torsion is present.

1. Virtual Work Method

In this method, a work equation is used to determine the **location of yield lines and the collapse loads.** The equation is formed by equating the total work done by external forces and internal forces. The yield lines divide the slab into various segments leading to its collapse. External work is done by collapse loads during simultaneous rotations of rigid slab segments and the internal work is done by the bending moments on all yield lines (Fig. 29.4.)

Work done by external forces = $\Sigma W. \delta$

Work done by internal forces = $\Sigma m.\theta.l$

Where W = Collapse load

δ = Vertical deflection through which collapse load moves

m = moment of resistance per unit length

θ = rotation of slab segment due to δ

l = length of yield line

Different yield line patterns should be investigated to find the minimum collapse load.

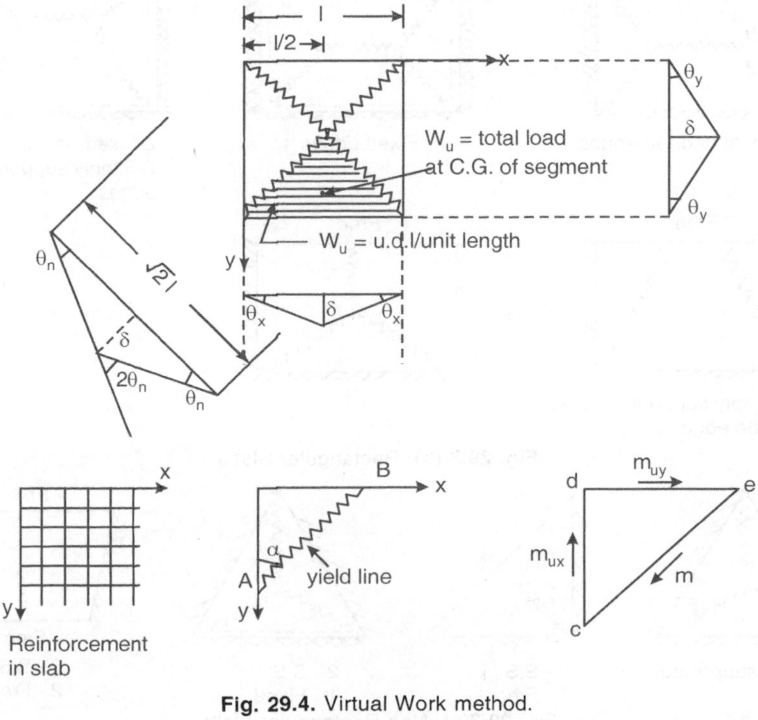

Fig. 29.4. Virtual Work method.

2. Equilibrium Method

Equilibrium equation is used to determine the location of yield lines and the collapse loads. It is obtained by equating external loads to internal forces while keeping the deflection same for both.

In the following articles, only virtual work method is used to analyse one-way and two-way slabs.

29.8 ONE-WAY SLABS

Consider a continuous one-way slab as shown in Fig. 29.5. It has both positive and negative steel reifnorcement at bottom face and at top face. Consider one-span length '*l*' and 1.0 m wide strip of slab

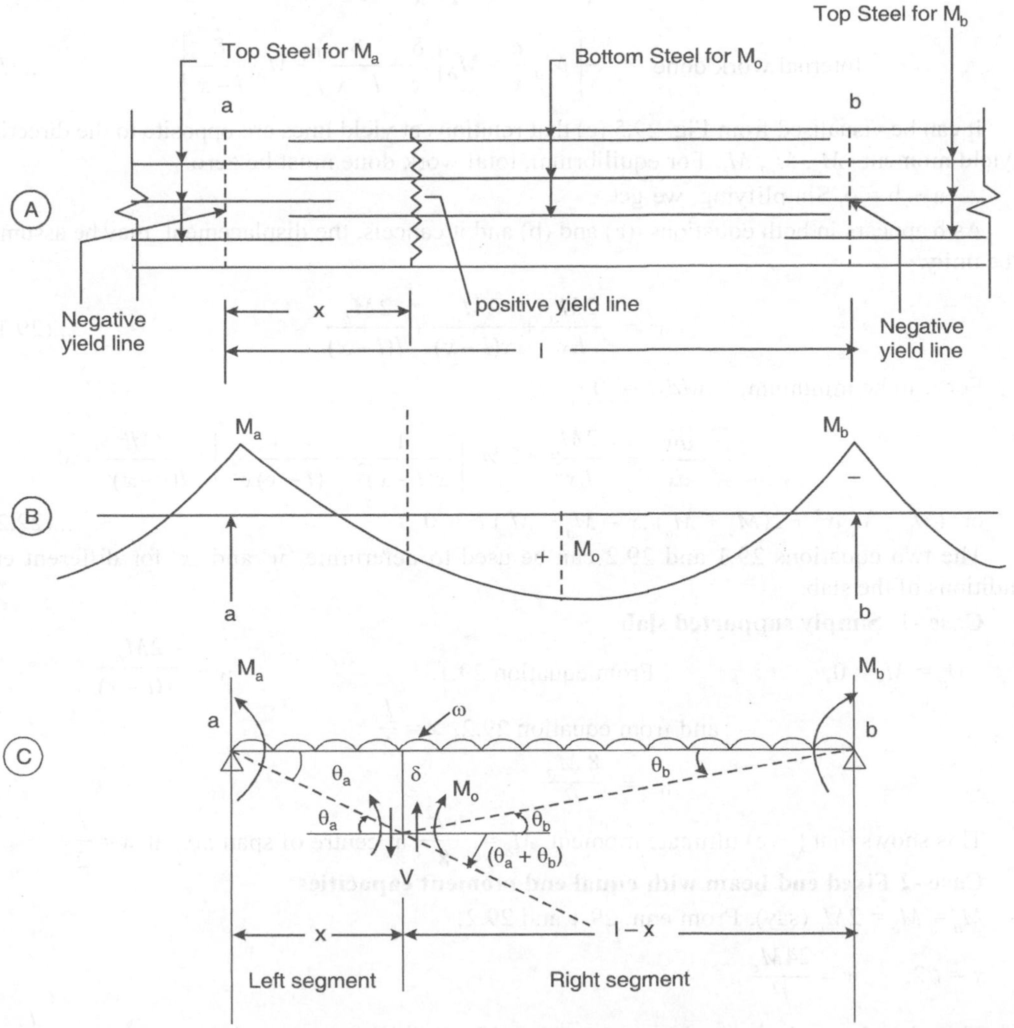

Fig. 29.5. One-way continuous slab

Per unit width of slab, the mid-span positive bottom reinforcement has ultimate moment of resistance of M_o, the negative reinforcement at supports 'a' and 'b' has ultimate moment of resistance, M_a and M_b respectively. At collapse load of 'w' per unit length of span; three yield lines form along the length of slab at right angles to reinforcement one each yield line along left and right supports and one at distance '*x*' from left support.

Consider the rigid body rotations of the slab segments for span 'ab' and are shown in Fig. 29.5 (c).

$$\text{External work done} = \left(\frac{1}{2} w.x\right)\delta + \left[\frac{1}{2} w(l - x)\right]\delta \qquad ....(a)$$

$$\text{Internel work done} = -\left[M_a.\theta_a.1 + M_o(\theta_a + \theta_b)1 + M_b\theta_b.1\right]$$

From Fig. 29.5, $\theta_a = \dfrac{\delta}{x}$ and $\theta_b = \dfrac{\delta}{l - x}$

$$\therefore \quad \text{Internal work done} = -\left[M_a \frac{\delta}{x} + M_o\left(\frac{\delta}{x} + \frac{\delta}{l - x}\right) + M_b.\frac{\delta}{l - x}\right] \qquad ...(b)$$

It can be visualised from Fig. 29.5 (c) that rotations at yield lines are opposite to the direction of yield moments M_a, M_o, M_b. For equilibrium, total work done must be zero.

∴ a + b = o, Simplifying, we get

As δ appears in both equations (a) and (b) and it cancels, the displacement may be assumed to be unity.

$$w = \frac{2M_a}{l.x} + \frac{2M_o}{x(l - x)} + \frac{2M_b}{l(l - x)} \qquad ...(29.1)$$

For w to be minimum, $dw/dx = 0$

$$\therefore \quad \frac{dw}{dx} = \frac{2M_a}{l.x^2} + 2M_o\left[\frac{1}{x(l - x)^2} - \frac{1}{(l - x)x^2}\right] + \frac{2Mb}{l(l - x)^2} = 0$$

or $(M_b - M_a)x^2 + 2(M_a + M_o)l.x - (M_a + M_o)\, l^2 = 0$ ...(29.2)

The two equations 29.1 and 29.2 can be used to determine '*w*' and '*x*' for different end conditions of the slab.

Case -1 Simply supported slab

$M_a = M_b = 0$, From equation 29.1, $w = \dfrac{2M_o}{x(l - x)}$

and from equation 29.2, $x = \dfrac{l}{2}$

$$\therefore \quad w = \frac{8\,M_o}{l^2}$$

This shows that (+ve) ultimate moment, $M_o = \dfrac{wl^2}{8}$ at centre of span *i.e.*, at $x = \dfrac{l}{2}$

Case -2 Fixed end beam with equal end moment capacities

$M_a = M_b = 2M_o$ (say), From eqn. 29.1 and 29.2,

$x = l/2$, $\quad w = \dfrac{24M_o}{l^2}$

This also shows that +ve ultimate moment, $M_o = wl^2/24$ at centre of span, *i.e.*, at $x = \dfrac{l}{2}$

Example 29.1. The factored moment capacities for intermediate span '*ab*' of a one-way continuous slab are : $M_a = -100$ kN.m/m width, $M_b = -120$ kN.m/m and $M_o = +80$ kN.m/m. If the span of slab is 6m, determine (*i*) the location of plastic hinges (*ii*) collapse load. Also draw the B.M.D. and S.F.D.

Sol.

$$M_a = -100 \text{ kN.m/m}$$
$$M_b = -120 \text{ kN.m/m}$$
$$M_o = +80 \text{ kN.m/m}$$

(*i*) Location of plastic hinge '*x*' from left support from equation 29.2

$$(120 - 100)x^2 + 2(100 + 80)\,6x - (100 + 80) \times 6 \times 6 = 0$$

or $\quad 20x^2 + 2160x - 6480 = 0$

or $\quad x^2 + 108x - 324 = 0$

Solving, we get $\quad x = 2.921\text{m}$

(*ii*) Collapse load '*w*' from equation 29.1 by substituting $x = 2.921$ m.

$$w = \frac{2\times100}{6\times2.921} + \frac{2\times80}{2.921(6-2.921)} + \frac{2\times120}{6(6-2.921)}$$

$= 42.193$ KN/m/m width of slab

The BMD and SFD are shown in Fig. 29.6.

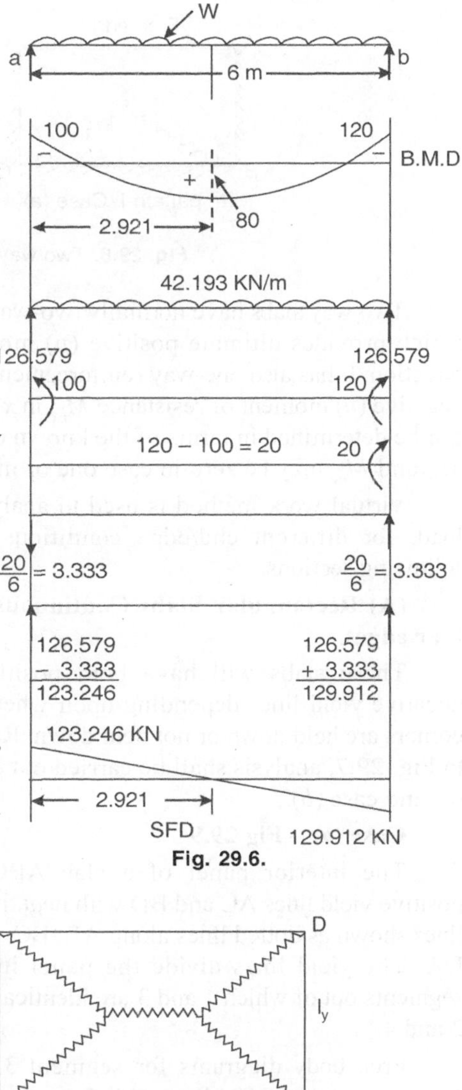

Fig. 29.6.

29.9 TWO-WAY SLABS

The yield line patterns for different end supports and with or without reinforcement are shown in Fig. 29.7 for two way slabs supported on all edges and in Fig. 29.8 for two way slab supported on three edges.

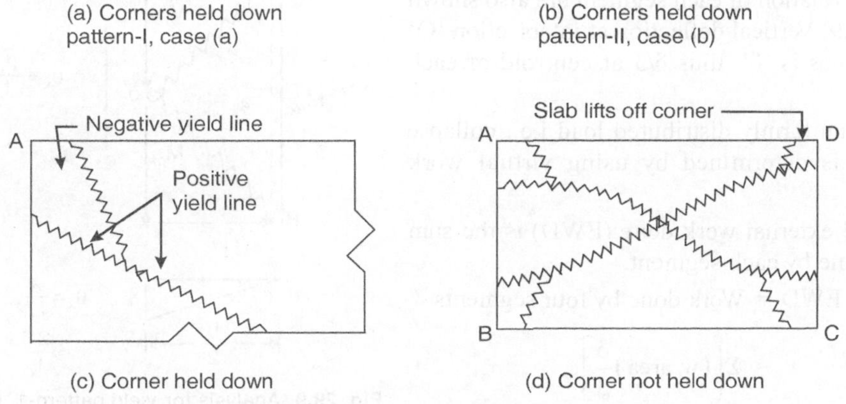

Fig. 29.7. Two-way slabs supported on all edges.

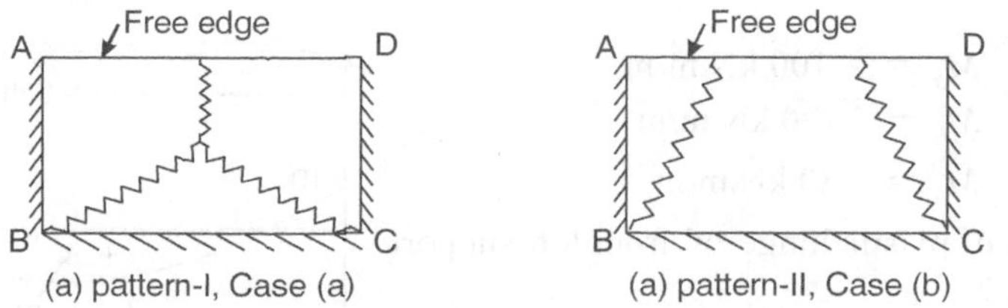

(a) pattern-I, Case (a) (a) pattern-II, Case (b)

Fig. 29.8. Two way slabs supported on three edges

Two-way slabs have normally two-way reinforcement near the bottom face in the central zone which provides ultimate positive (p) moment of resistance M_{px} in x-direction and M_{py} in y–direction. It has also one-way reinforcement near top face across the edges which provides ultimate negative (n) moment of resistance M_{nx} in x–direction and M_{ny} in y-direction. The collapse load 'w' can be determined in terms of the known quantities L_x, L_y, M_{px}, M_{py}, M_{nx} and M_{ny}. The values of M_{nx} and M_{ny} may be zero in case one or more edges are free or simply supported.

Virtual work method is used to analyse different yield line patterns, to determine collapse load, for different end/edge conditions in the following sections.

(A) Rectangular Slabs Continuous on all four edges.

These slabs will have both positive and negative yield lines depending upon whether the corners are held down or not held down. Referring to Fig. 29.7, analysis shall be carried out for case (a) and case (b).

Case (a) – Fig 29.9

The interior panel of a slab ABCD has positive yield lines AC and BD with negative yield lines shown as dotted lines along AB, BC, CD and DA. The yield lines divide the panel into four segments out of which 1 and 3 are identical as also 2 and 4.

Free body diagrams for segment 3 (BCO) and segment 4 (ABO) along with the moments and respective rotation of each segment are also shown in Fig. 29.9. Vertical deflection at intersection 'O' of yield lines is 'δ', thus δ/3 at centroid of each segment.

w = uniformly distributed load i.e., collapse load and is determined by using virtual work method.

Total external work done (EWD) is the sum of workdone by each segment.

∴ EWD = Work done by four segments

$$= \Sigma\left[(w.\text{ area})\frac{\delta}{3}\right]$$

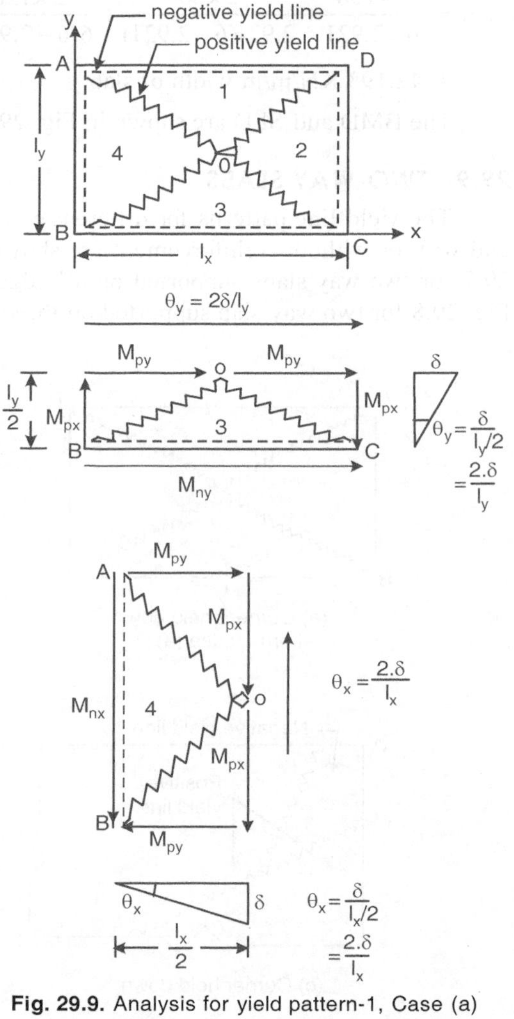

Fig. 29.9. Analysis for yield pattern-1, Case (a)

$= w.\ \text{area of four segments}.\ \dfrac{\delta}{3}$

$$= w\,(l_x.l_y)\frac{\delta}{3} \qquad ...(i)$$

$IWD =$ Total internal work done by yield moments

$$= 2M_{py}.l_x.\theta_y + 2M_{px}.l_y.\theta_x + 2M_{ny}.l_x.\theta_y + 2M_{nx}.l_y.\theta_x \qquad ...(ii)$$

Equating (*i*) and (*ii*) and solving for 'ω' after substituting $\theta_x = 2.\delta/l_x$ and $\theta_y = 2.\delta/l_y$,

$$w = 12\left[\frac{M_{px}+M_{nx}}{l_x^2}+\frac{M_{ny}+M_{py}}{l_y^2}\right] \qquad ...(29.3)$$

This equation 29.3 can be used to determine the collapse load for square slabs with (*a*) simply supported edges and (*b*) slab fixed at all edges, as under :

(*a*) **Simply supported square slab**

$lx = ly = 1$ (say) $\qquad M_{nx} = M_{ny} = 0$,

$M_{px} = M_{py} = M$(say)

Eq. 29.3 gives $w = 24\ M/l^2$ or $M = wl^2/24$

(*b*) **Square slab fixed at all edges**

Let the support moments be equal to mid-span moments.

So $M_{px} = M_{py} = M_{nx} = M_{ny} = M$ (say)

∴ Equ. 29.3 gives $w = 48\ \text{m}/l^2$ or

$M = wl^2 / 48$

Case (b) Fig. 29.10

The interior panel ABCD of a slab has positive yield lines AE, BE, EF, CF, and DF with negative yield lines along edges AB, BC, CD and DA shown as dotted lines. In this case, the location of E or F from vertical edges AB and CD is also to be determined in addition to collapse load.

The yield lines divide the panel into four segments, out of which 1 and 3 are identical as also 2 and 4.

Free body diagrams for segment 3 (BCFE) and segment 4(ABE) along with the moments and respective rotations of each segment are also shown in Fig. 29.10.

Vertical deflection of E and F, the points of intersection of yield lines, is 'δ' thus for (*i*) 2 and 4 is δ/3 and (*ii*) for 1 and 3, centroid can be found either from trapezoidal area or by dividing the segment into triangles BEE′, CF′F and rectangle EE′ F′ F.

Total External Work Done (EWD) is the sum of work done by each segment.

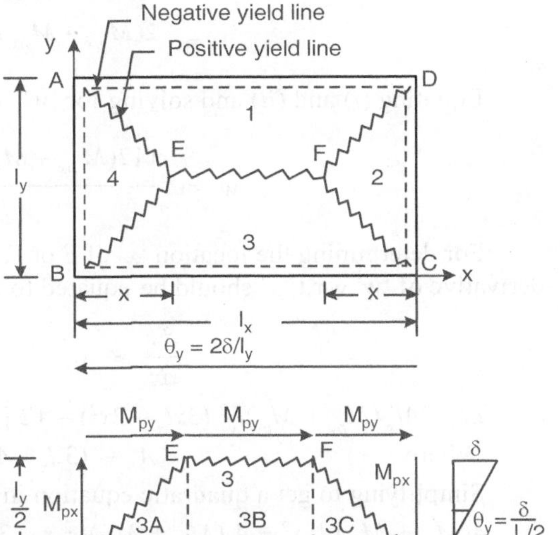

Fig. 29.10. Pattern–II, Case (b)

$\therefore$ EWD = Work done by four segments

$= W_1 + W_2 + W_3 + W_4$

$= 2\,W_3 + 2W_4$

$= 2(W_{3A} + W_{3B} + W_{3C}) + 2W_4$

$$= 2\left[w\left(\frac{1}{2}.x.\frac{l_y}{2}\right)\frac{\delta}{3} + w(l_x - 2x)\frac{l_y}{2}.\frac{\delta}{2} + w.\frac{1}{2}\left(x.\frac{l_y}{2}\right)\frac{\delta}{3}\right] + 2\left[w\left(\frac{1}{2}x.l_y\right)\frac{\delta}{3}\right]$$

$$= w\frac{\delta}{6}\left(3l_x.l_y - 2x.l_y\right) \qquad ...(i)$$

IWD = Total Internal Work Done by yield moments for each segment

$$= 2(M_{ny}.l_x.\theta_y) + 2(M_{py}.l_x.\theta_y) + 2(M_{nx}.l_y.\theta_y) + 2(M_{px}.l_y.\theta_x)$$

$$= 2(M_{px} + M_{nx})l_y.\frac{\delta}{x} + 2(M_{py} + M_{ny})l_x.\frac{2\delta}{l_y} \qquad ...(ii)$$

Equating (*i*) and (*ii*) and solving for '*w*'

$$w = \frac{\left[12(M_{px} + M_{ux})ly^2 + 24x.lx(M_{py} + M_{uy})\right]}{ly^2(3x.lx - 2.x^2)} \qquad ...(29.4)$$

For determining the location '*x*' of E or F, the collapse load should be minimum and for that derivative of '*w*' w.r.t '*x*' should be equated to zero.

$$\therefore \qquad \frac{dw}{dx} = 0$$

i.e., $24l_x\,(M_{py} + M_{ny})l_y^{\,2}\,(3xl_x - 2x^2) - 12\,[M_{px} + M_{nx})\,b^2{}_y + 2xl_x\,(M_{py} + M_{ny})]\,A = 0$

Where $\qquad A = (3.l_x - 4.x)l_y^{\,2}$

Simplifying to get a quadratic equation in '*x*' as

$4(M_{py} + M_{ny})\,l_x\,x^2 + 4\,(M_{px} + M_{nx})l_y^{\,2}\,x - 3\,(M_{px} + M_{nx})\,l_x\,l_y^{\,2} = 0 \qquad ... (29.5)$

It is important to note that for

a) $x = l_x/2$ *i.e.,* for yield line pattern of case (*a*)

Eq. 29.5 gives

$$\frac{M_{px} + M_{nx}}{M_{py} + M_{ny}} = \frac{l_x^{\,2}}{l_y^{\,2}} \qquad ...(29.5(a))$$

If this eq. 29.5 (*a*) is satisfied means point of interesection 'E' of yield lines shifts to mid-span 'o' of l_x depicting yield line pattern of case (*a*).

b) $x < l_x/2$ *i.e.,* for yield line pattern of case (*b*), eq. 29.5 gives

$$\frac{M_{px} + M_{nx}}{M_{py} + M_{ny}} < \frac{l_x^{\,2}}{l_y^{\,2}} \qquad ...(29.5\ (b))$$

Following steps can be followed to know the yield line pattern in which a two way slab continuous on all four edges is likely to collapse.

i) Calculate left hand side (I.H.S) of eq. 29.5 (*a*)

ii) Calculate right hand side (RHS) of eq. 29.5 (*a*)

iii) If LHS = RHS, then yield line pattern of case (*a*) develops *iv*) If LHS < RHS, then yield line pattern of case (*b*) develops.

v) If $l_x = l_y$, then slab is square and R.H.S. =1, so for squae slab, LHS shall be equal to 1.0 for yield line pattern of Case (*a*)

vi) If the edges are simply supported, then $M_{nx} = 0$ and $M_{ny} = 0$ can be substituted in eq. 29.5 (*a*) to know the pattern for square or rectangular two-way slab.

Example 29.2 : A rectangular slab 5m × 6m is simply supported all around and is reinforced with 10 mm dia. @ 200 mm c/c in short span and 10mm dia. @ 225 mm c/c in long direction. Using yield line theory, determine the superimposed uniformly distributed load the slab can carry safely. Use concrete of grade M20 and steel of grade Fe415. Overall depth of slab is 150 mm. Fig. 29.11.

Sol. Given data (*i*) $l_x = 6$m, $l_y = 5$m.

(*ii*) All edges are simply supported i.e., no negative moments or $M_{nx} = M_{ny} = 0$

(*iii*) Consider 1m width of slab *i.e.*, b = 1000 mm

(*iv*) $D = 150$ mm, Take cover = 30mm. $d_2 = 150 - 30 = 120$mm for steel of long direction and $d_1 = 120 + 10 = 130$ mm for steel of short direction.

v) For width = 1000 mm, Area of steel (A_t) in

a) short span = 10 φ 200 c/c

$$A_t = \left(\frac{\pi}{4}\times 10\times 10\right)\times\frac{1000}{200} = 392.5 \text{ mm}^2$$

b) Long span = 10 φ @ 225 c/c

$$A_t = \left(\frac{\pi}{4}\times 10\times 10\right)\times\frac{1000}{225}$$

$$= 348.9 \text{ mm}^2$$

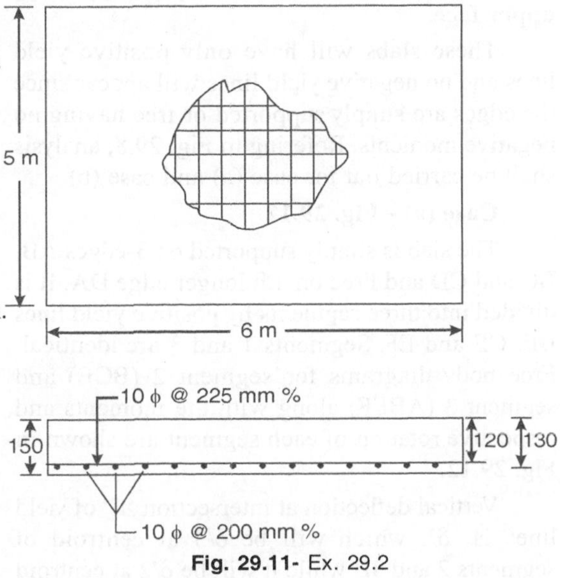

Fig. 29.11- Ex. 29.2

Calculation of M_{py} and M_{px}.

Assume the section to be balanced section.

$$M_u = 0.87\,\sigma_y\,A_t\left[d - \frac{\sigma_y . A_t}{\sigma_{ck} . b}\right]$$

∴ M_{py} in short direction

$$= 0.87 \times 415 \times 392.5\left[130 - \frac{415\times 392.5}{20\times 1000}\right]$$

$$= 17268472.\text{N.mm}$$

$$= 17.27 \text{ KN.m}$$

M_{px} in long direction

$$= 0.87 \times 415 \times 348.9\left[120 - \frac{415\times 348.8}{20\times 1000}\right]$$

$$= 14205675 \text{ N.mm}$$

$$= 14.21 \text{ KN.m}$$

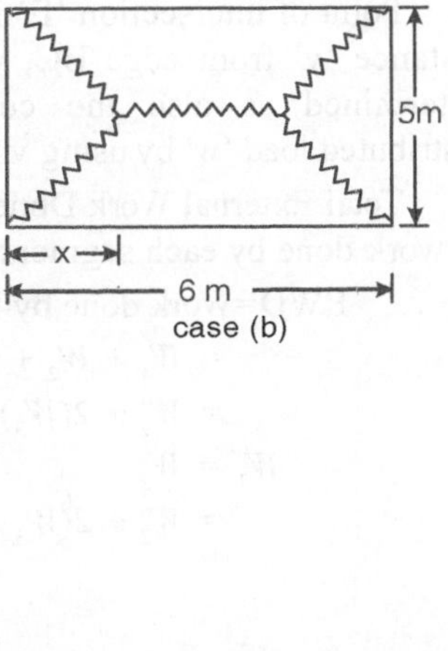

case (b)

Eq. 29.5 (a)
$$\frac{M_{px} + M_{nx}}{M_{py} + M_{ny}} = \frac{l_x^2}{l_y^2}$$

$$\text{L.H.S.} = \frac{14.21 + 0}{17.27 + 0} = 0.823$$

$$\text{R.H.S.} = \frac{6\times 6}{5\times 5} = 1.44$$

LHS < RHS, so yield pattern of case (*b*) prevails. So both '*x*' and '*w*' are to be calculated

For '*x*' eq. 29.5

$$\left[4(M_{py}+M_{ny})l_x\right]x^2+\left[4(M_{px}+M_{nx})l_y^{\,2}\right]x-\left[3(M_{px}+M_{nx})l_x\,l_y^{\,2}\right]=0$$

$$\left[4(17.27+0)6\right]x^2+\left[4(14.21+0)5^2\right]x-\left[3(14.21+0)6\times5^2\right]=0$$

$$414.48x^2 + 1421x - 6394.5 = 0$$

Solving, $\quad x = \mathbf{2.5715m}$

For '*w*' eq. 29.4

$$w = \frac{[12(M_{px}+M_{nx})l_y^{\,2}+24.x.l_x(M_{py}+M_{ny})}{l_y^{\,2}(3xl_x-2x^2)}$$

$$= \frac{12(14.21+0)(5\times5)+24\times2.5715\times6(17.27+0)}{(5\times5)(3\times2.5715\times6-2\times2.5715\times2.5715)}$$

$$= \mathbf{12.8946 \approx 12.90\ KN/m^2}$$

(B) Rectangular slabs simply supported at 3-edges and free at the 4th long edge at upper face.

These slabs will have only positive yield lines and no negative yield lines will appear since the edges are simply supported or free having no negative moments. Refering to Fig. 29.8, analysis shall be carried out for case (a) and case (b)

Case (a) - Fig. 29.12

The slab is simply supported on 3-edges AB, BC and CD and Free on 4th longer edge DA. It is divided into three segments by positive yield lines BE, CE and EF. Segments 1 and 3 are identical. Free body diagrams for segment 2 (BCE) and segment 3 (ABEF) along with the moments and respective rotation of each segment are shown in Fig. 29.12.

Vertical deflection at intersection 'E' of yield lines is 'δ', which will be δ/3 at centroid of segments 2 and 3B while it will be δ/2 at centroid of segment 3A.

Point of intersection 'E' is assumed to be at distance 'y' from edge BC, which needs to be determined as also the collapse uniformly distributed load '*w*' by using virtual work method.

Total External Work Done (EWD) is the sum of work done by each segment.

$\therefore$ EWD=Work done by the three segments.

$$= W_1 + W_2 + W_3$$

$$= W_2 + 2(W_3)$$

$\because \quad W_1 = W_3$

$$= W_2 + 2(W_{3A} + W_{3B})$$

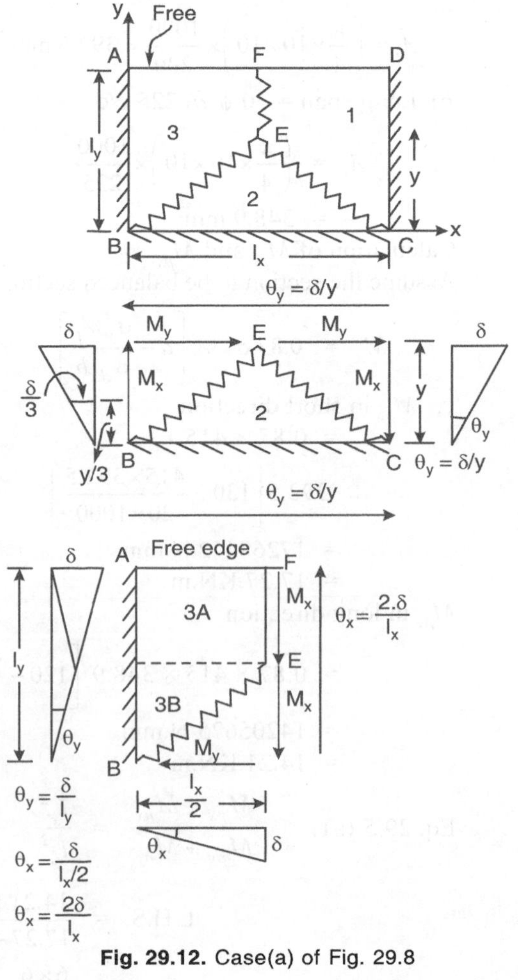

Fig. 29.12. Case(a) of Fig. 29.8

$$= \left[w\left(\frac{1}{2}l_x.y\right)\frac{\delta}{3}\right] + 2\left[w.\frac{l_x}{2}(l_y - y).\frac{\delta}{2} + w\left(\frac{1}{2}.y.\frac{l_x}{2}\right)\frac{\delta}{3}\right]$$

$$= w.l_x.\frac{\delta}{6}(3l_y - y) \qquad ...(i)$$

IWD = Total internal work done by yield moments

$$= 2M_x.l_y.\theta_x + M_y.l_x.\theta_y$$

$$= 2M_x.l_y.\frac{2\delta}{l_x} + M_y.l_x.\frac{\delta}{y} \qquad ...(ii)$$

Equating (*i*) and (*ii*) and solving for *w*,

$$w = \frac{24M_x.l_y.y + 6M_y.l_x^2}{l_x^2(3.y.l_y - y^2)} \qquad ...(29.6)$$

For getting the vlaue of '*y*', derivative of '*w*' w.r.t '*y*' should be equal to zero for '*w*' to be minimum.

$$\therefore \qquad dw/dy = 0$$

$$\text{or} \quad (24M_x.l_y)\left[l_x^2(3y.l_y - y^2)\right] - (24M_x.y.l_y + 6M_y.l_x^2)\left[l_x^2(3l_y - 2y)\right] = 0$$

Simplifying to get a guadratic in '*y*'

$$(4M_x.l_y)y^2 + (2M_y.l_x^2)y - (3M_y.l_x^2.l_y) = 0 \qquad ...(iii)$$

$$\text{or} \quad y = \frac{-2\dfrac{M_y}{M_x}l_x^2 + \sqrt{\left(\dfrac{M_y}{M_x}\right)^2 4.l_x^4 + 48\left(\dfrac{M_y}{M_x}\right)l_x^2 l_y^2}}{8.l_y} \qquad ...(iv)$$

y will always be less than l_y *i.e.* (*iv*) < l_y

$$\text{Solving we get} \quad \frac{M_y}{M_x} < 4\left(\frac{l_y}{l_x}\right)^2 \qquad ...(29.7)$$

Case (b) - Fig. 29.13

The slab is simply supported on 3-edges AB, BC and CD and free on 4th longer edge DA. It is divided into three segments 1, 2, 3 by positive yield lines BF and CE. Segments 1 and 3 are identical. Free body diagrams for segment 2 (BCEF) and segment 3(ABF) along with moments and respective rotation of each segment are shown in Fig. 29.13. For making the calculation for position of centroid, segment 2 is subdivided into three parts 2A, 2B and 2C. Vertical deflection at intersections E and F, where yield lines meet the free edge, is 'δ', which will be $\frac{\delta}{3}$ at centroid of segments 1, 3, 2A and 2C while it will be δ/2 at centroid of segment 2B. Points E and F are assumed to be at distance '*x*' from edge, CD and AB respectively, which needs to be determined as also the collapse uniformly distributed load '*w*' by using virtual work method.

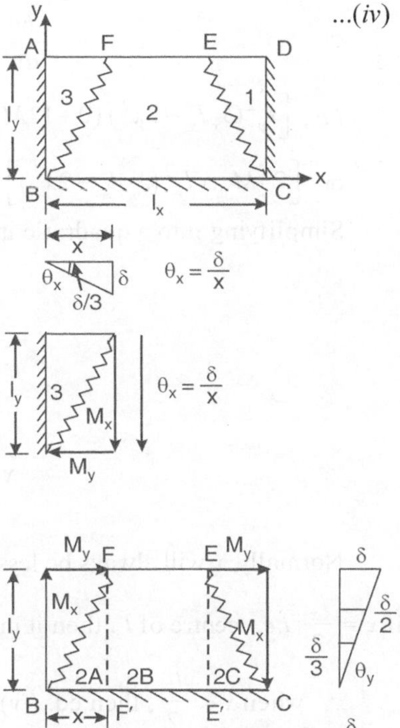

Fig. 29.13. Case(b) of Fig. 29.8

Total External Work Done (EWD) is the sum of work done by each segment.

$$\therefore \quad \text{EWD} = W_1 + W_{2A} + W_{2B} + W_{2C} + W_3$$

$$= 2W_3 + 2.W_{2A} + W_{2B}$$

$$= 2\left[\left(\frac{1}{2}.l_y.x\right)w.\frac{\delta}{3}\right] + 2\left[\left(\frac{1}{2}l_y.x\right)w.\frac{\delta}{3}\right] + \left[(l_x - 2x).l_y.w\frac{\delta}{2}\right]$$

$$= w.l_y.\frac{\delta}{6}(3l_x - 2x) \qquad ...(i)$$

IWD = Total internal Work Done by yield moments for each segment.

$$= W_1 + W_{2A} + W_{2B} + W_{2C} + W_3$$

$$= 2\, W_3 + 2.W_{2A} + W_{2B}$$

$$= 2.M_x.l_y.\theta_x + 2.M_y x.\ \theta_y + 0$$

$$= 2M_x.l_y.\frac{\delta}{x} + 2M_y.x.\frac{\delta}{l_y} \qquad ...(ii)$$

Equating (*i*) and (*ii*) and solving for '*w*', we get

$$w = \frac{12M_x.l_y^{\,2} + 12M_y.x^2}{l_y^{\,2}(3xl_x - 2x^2)} = \frac{A}{B} \qquad ...(29.8)$$

For determining the location '*x*' of E or F, the collapse load should be minimum and for that derivative of '*w*' w.r.t. '*x*' should be equal to zero.

$$\therefore \quad \frac{dw}{dx} = 0 = \text{B}.dA - A.\ dB$$

i.e., $\left[l_y^{\,2}(3_x l_x - 2x^2)(0 + 12M_y.2x)\right] - \left[12M_x.l_y^{\,2} + 12M_y.x^2\right]\left[l_y^{\,2}(3.l_x - 4x\right] = 0$

or $\left[24M_y.x.l_y^{\,2}(3x\,l_x - 2x^2)\right] - \left[12M_x l_y^{\,2} + 12M_y.x^2\right]\left[l_y^{\,2}(3l_x - 4x)\right] = 0$

Simplifying into a quadratic in '*x*'

$$\left[3.l_x\left(\frac{M_y}{M_x}\right)\right]x^2 + (4.l_y^{\,2})x - 3l_x.l_y^{\,2} = 0 \qquad ...(iii)$$

$$\therefore \quad x = \frac{-4.l_y^{\,2} + \sqrt{16l_y^{\,4} + 36\left(\frac{M_y}{M_x}\right)l_x^{\,2}.l_y^{\,2}}}{6.l_x(M_y / M_x)} \qquad ...(iv)$$

Normally, *x* will always be less than $\frac{l_x}{2}$ *i.e.,* pattern - II or Case (b) under consideration. But if $x = \frac{l_x}{2}$ *i.e.,* centre of l_x, then it may represent pattern -I or Case (a) discussed earlier.

$\therefore$ when $x < \frac{l_x}{2}$, then eq. (*iv*) $< \frac{l_x}{2}$, solving

$$16.l_y^{\,4} + 36(M_y / M_x)l_x^{\,2}.l_y^{\,2} < \left[3(M_y / M_x)l_x^{\,2} + 4.l_y^{\,2}\right]^2$$

or $$12\left(\frac{M_y}{M_x}\right)l_x^2 l_y^2 < 9\left(\frac{M_y}{M_x}\right)^2 .lx^4$$

or $$\frac{M_y}{M_x} > \frac{12.l_y^2}{9.l_x^2}$$

or $$\frac{M_y}{M_x} > \frac{4}{3}\left(\frac{l_y}{l_x}\right)^2 \qquad ....(29.9)$$

Following steps can be followed to know the yield line pattern in which a two-way slab, simply supported on 3-edges and free at 4th edge, is likely to collapse.

(*i*) Calculate M_y/M_x and $\left(\frac{l_y}{l_x}\right)^2$

(*ii*) If M_y/M_x is less than $4\left(\frac{l_y}{l_x}\right)^2$, then the yield line pattern will be case (a) and *y* will be calculated from eq. (*iv*) and collapse load '*w*' from eqn. 29.6

(*iii*) If *My*/*Mx* is greater than $\frac{4}{3}\left(\frac{l_y}{l_x}\right)^2$, then the yield line pattern will be case (b), and *x* will be calculated from eq. (*iv*) and collapse load '*w*' from eqn. 29.8

Example 29.3. A square slab 4m × 4m is simply supported on 3-sides and is free on 4th side. Determine the collapse load if the moment capacities are equal in both directions.

Sol. Given data $L_x = L_y = 4$m, $M_x = M_y = M$ (say), first of all, find the yield line pattern whether it collapses with yield lines of case (*a*) or case (*b*)

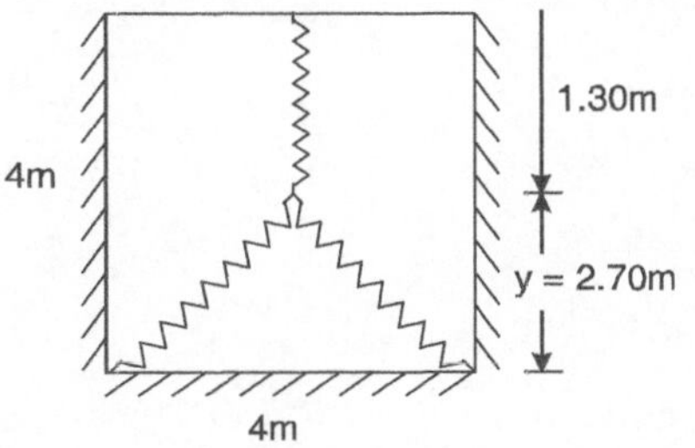

Fig. 29.14.

(*i*) M_y/M_x = M/M = 1 and $\left(\frac{l_y}{l_x}\right)^2 = (4/4)^2 = 1$

(*ii*) For Case (a) $M_y/M_x < 4\left(\frac{l_y}{l_x}\right)^2$

i.e., 1 < (4 × 1) – – – True

∴ Collapse mechanisum of Case (a) governs (Fig. 29.14).

(*iii*) Substitute the above values in eq. (*iv*) of Case (*a*), we get

$$y = \frac{-2.\frac{M_y}{M_x}.l_x^2 + \sqrt{4\left(\frac{M_y}{M_x}\right)^2 l_x^4 + 48\left(\frac{M_y}{M_x}\right)l_x^2 l_y^2}}{8.ly}$$

$$= \frac{-2\times1\times4\times4+\sqrt{4\times1\times4^4+48\times1\times4^2\times4^2}}{8\times4}$$

= 2.70 m

(*iv*) Use eq (29.6) for determining the collapse load '*w*'

$$w = \frac{24.M_x.l_y y + 6M_y.l_x^2}{l_x^2(3.y.l_y - y^2)}$$

$$= \frac{24\times M\times4\times2.7+6\times M\times4\times4}{4\times4(3\times2.7\times4-4\times4)} = \mathbf{0.88375\ M}$$

Example 29.4. A rectangular slab 6m long × 3m wide is simply supported on 3-sides and free on 4th longer side. Calculate the collapse load if the moment capacities are same in both the directions.

Sol. Given data:

$$L_x = 6\text{m}$$
$$L_y = 3\text{m}$$
$$M_x = M_y = M$$

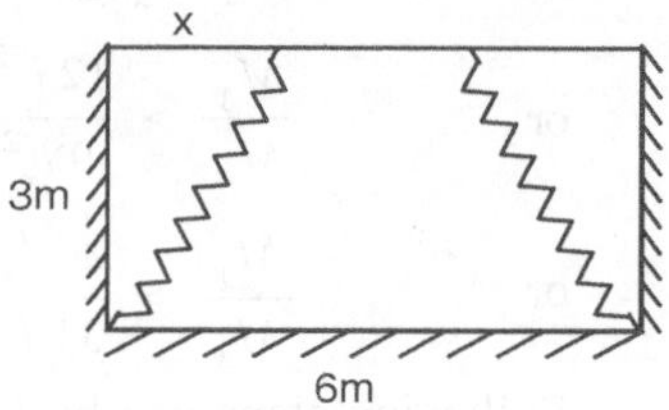

Fig. 29.15.

First of all, find the yield line pattern whether it collapses as case (a), or case (b)

(*i*) $M_y/M_x = 1$ and $(l_y/l_x)^2 = (3/6)^2 = 1/4$

(*ii*) $M_y/M_x \not< 4(l_y/l_x)^2$ – it is not case (a)

(*iii*) $M_y / M_x > \frac{4}{3}(l_y / l_x)^2$

$$1 > \left(\frac{4}{3} \times \frac{1}{4}\right)$$

$> \frac{1}{3}$ - So it is case (b) shown in Fig. 29.15.

(*iv*) Substitute the above values in eq. (*iv*) of case (*b*)

$$x = \frac{-4.l_y^{\,2} \pm \sqrt{16.l_y^{\,4} + 36\left(\frac{M_y}{M_x}\right) l_x^{\,2}.l_y^{\,2}}}{6.l_x\left(\frac{M_y}{M_x}\right)}$$

$$= \frac{-(4 \times 3 \times 3) \pm \sqrt{16 \times 81 + 36 \times 1 \times 36 \times 9}}{6 \times 6 \times 1}$$

$$= \frac{-36 \pm 113.84}{36}$$

$$= \frac{-149.84}{36} \quad \text{or} \quad \frac{77.84}{36}$$

$$= -4.16\text{m} \quad \text{or} \quad 2.16\text{ m}$$

$\therefore$ Adopt $\boldsymbol{x = 2.16}$**m** $\quad \because$ it cannot be negative

(*v*) Use eq. (29.8) for determining the collapse load '*w*'.

$$w = \frac{12M_x.l_y^{\,2} + 12.M_y x^2}{l_y^{\,2}(3.x.l_x - 2x^2)}$$

$$= \frac{12 \times M \times 9 + 12 \times M \times 2.16^2}{9(3 \times 2.16 \times 6 - 2 \times 2.16^2)}$$

$$= \mathbf{0.616\ M}$$

29.10 CIRCULAR SLAB SIMPLY SUPPORTED ALL AROUND

In a circular slab simply supported all around, only positive radial yield lines, starting from the centre 'o', will appear.

If it is continuous all around, then negative yield lines all along the circumference may also appear.

δ = Vertical deflection at centre 'o'
= 1 (assumed)

$M_x = M_y = M$(say)

Consider point 'P' at distance 'r' from centre. P will take position P' after unit deflection at 'o'. Normal at P' will pass through C as shown in Fig. 29.16.

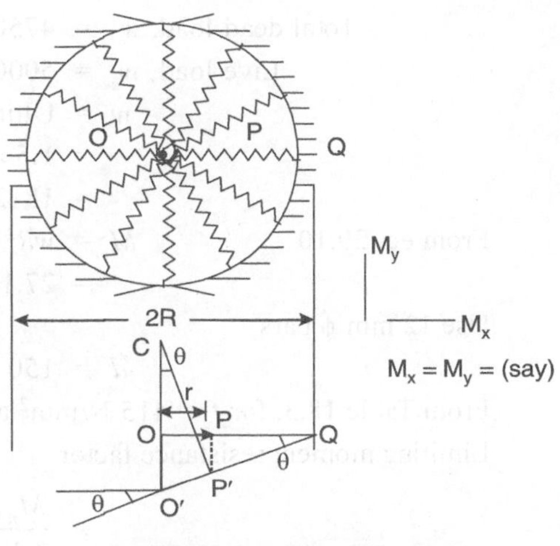

Fig. 29.16. Case(b) or Fig. 29.8

$$\theta = \frac{OO'}{OQ} = \frac{1}{R}$$

$$CP' = \frac{r}{\theta} = R.r$$

Change of slope at P/unit of arc, in the tangential direction $= \dfrac{1}{CP'} = \dfrac{1}{R.r}$

$\therefore$ Total change of slope at P in whole circle

$$= \left(\frac{1}{R.r}\right) \times \text{Circumference at } P.$$

$$= \frac{1}{R.r}.2\pi r$$

$$= \frac{2\pi}{R}$$

$\therefore$ Internal Work done $= \Sigma M.l.\theta$

$$= M.R.\frac{2\pi}{R} = 2\pi.M. \qquad ...(i)$$

External Work $= w.\pi R^2.\delta/3 = w.\pi R^2.1/3$ ...(*ii*)

$\because$ at Collapse, a flat cone will be formed and the mean deflection will be $\delta/3$ or 1/3 because $\delta = 1$ (assumed)

External Work = Internal work

$w.\pi R^2/3 = 2\pi M$

So $M = WR^2/6$ or $w = \dfrac{6M}{R^2}$ (29.10)

Example 24.5. Design a circular slab 6.0 in diameter simply supported all around. Take live load as 5000N/m² and load factors as 1.5 and 2.2 for dead and live loads respectively. Use M20 concrete and steel of grade Fe 415.

Sol. Thickness of slab $= \dfrac{\text{Span}}{40} = \dfrac{6000}{40} = 150$ mm

Self Weight/m² $= 25000 \times \dfrac{150}{1000} \times 1 \times 1$

$= 3750$ N/m²

Finishes $= 1000$ N/m²

$\therefore$ Total dead load, w_d = 4750 N/m^2

Live load, w_e = 5000 N/m^2

w = Ultimate load

= 1.5 × 4750 + 2.2 × 5000

= 18125 N/m^2

From eq. 29.10 $M = wR^2/6 = 18125 \times 3 \times 3/6 = 27187.5$ N.m/m

= 27.188 KN.m/m ...(*a*)

Use 12 mm ϕ bars

$d = 150 - 20 - 12 = 118$ mm

From Table 18.3, for $f_y = 415$ N/mm^2 and $f_{ck} = 20$ N/mm^2,
Limiting moment resistance factor

$$= \frac{M_u, \lim}{bd^2} = 2.76 \text{ N/ mm}^2$$

$= 2.76 \times 10^3$ KN/m^2

$\therefore$ Limiting moment of slab, M_u, lim

$$= (2.76 \times 10^3) \times 1 \times \left(\frac{118}{1000}\right)^2$$

= 38.43 KN.m/m ...(*b*)

Actual moment M = 27.188 is less than M_u, lim = 38.43 KN.m/m, hence the section is to be designed as under-reinforced,

using eq. 18.18, $$M_u = 0.87 f_y \left(\frac{p_t}{100}\right)\left(1 - 1.005 \frac{f_y}{f_{ck}} \cdot \frac{p_t}{100}\right) bd^2$$

$$\frac{27.188 \times 1000 \times 1000}{1000 \times 118 \times 118} = 0.87 \times 415 \left(\frac{p_t}{100}\right)\left(1 - 1.005 \times \frac{415}{20} \times \frac{p_t}{100}\right)$$

$0.54081 = p_t(1 - 0.2085375\, p_t)$

or $0.2085375\, p_t^2 - p_t + 0.54081 = 0$

Solving, $$p_t = \frac{1 \pm \sqrt{1 - 4 \times 0.2085375 \times 0.54081}}{2 \times 0.2085375} = \frac{1 \pm \sqrt{0.5488834}}{0.417075}$$

$$= \frac{1 \pm 0.7408666}{0.417075}$$

$\therefore$ $$p_t = \frac{0.2591334}{0.417075} \text{ or } \frac{1.7408666}{0.417075}$$

= 0.6213 or 4.174

Adopt = 0.6213% and neglecting 4.174%

$\therefore$ Area of steel $$= \left(\frac{0.6213}{100}\right)(1000 \times 118) = 733.15 \text{ mm}^2$$

Provide 12mm ϕ @ 150 mm centre to centre bothways (Area provided = 754 mm^2)

29.11 TRIANGULAR SLAB SIMPLY SUPPORTED ON TWO SIDES

PQR is a triangular slab simply supported on sides PQ and QR and free on 3rd side PR as shown in Fig. 29.17. The yield line meets free edge at S. ϕ is unknown. vertical deflection at S is δ, assumed to be unity. Projection of QS on sides QR is Q $a = x$ and on PQ is $Sa = y$.

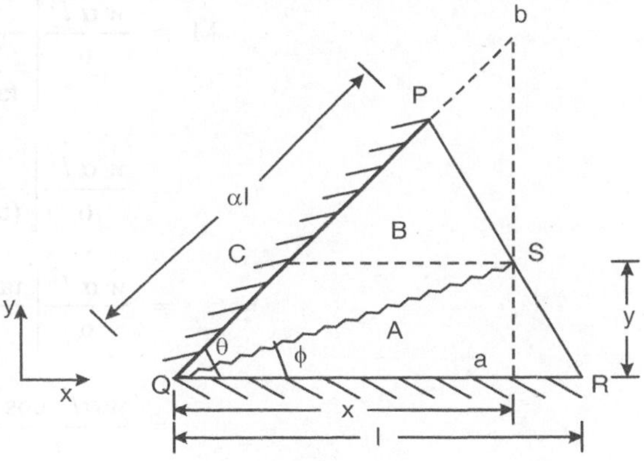

Fig. 29.17

Coordinates of S be (xy) with respect to Q.

$$\begin{aligned}
\text{EWD} &= \text{External Work Done by Segments A and B} \\
&= W_A + W_B \\
&= \text{Work done by PQR} \\
&= w\left[\frac{1}{2} l.\alpha l.\sin\theta\right]\frac{\delta}{3} \\
&= \frac{\omega.\alpha.l^2.\text{Sin}\,\theta}{6} \quad \text{where } \delta = 1 \text{ (assumed) and deflection at centroid of segments A and B will be } 1/3. \qquad ...(i)
\end{aligned}$$

$$\begin{aligned}
\text{IWD} &= \text{Internal Work Done by yield moments for each segment.} \\
&= \Sigma\left[(M.\theta)_A + (M.\theta)_B\right] \\
&= \Sigma\left[(M_x.\theta_x + M_Y.\theta_y)_A + [M_x.\theta_x + M_y.\theta_y)_B\right] \\
&= \Sigma(M_x.\theta_x + M_y.\theta_y)_A + \Sigma\left(M_x\theta_x + M_y.\theta_Y\right)_B \\
&= M.x.\frac{\delta}{Sa} + M.x.\frac{\delta}{Sb} + M.y.\frac{\delta}{Sc} \\
&= M.x.\frac{1}{x.\tan\phi} + M.x.\frac{1}{x\tan\theta - x\tan\phi} + M.y.\frac{1}{x - y\cot\theta} \\
&= \frac{M}{\tan\phi} + \frac{M}{\tan\theta - \tan\phi} + \frac{M}{\cot\phi - \cot\theta} \\
&= M\left[\frac{1}{\tan\phi} + \frac{1 + \tan\phi\tan\theta}{\tan\theta - \tan\phi}\right] \qquad ...(ii)
\end{aligned}$$

$$\text{EWD} = \text{IWD}$$

$$\therefore \quad \frac{\omega\alpha l^2.\sin\theta}{6} = \text{M}\left[\frac{1}{\tan\phi} + \frac{1 + \tan\phi.\tan\theta}{\tan\theta - \tan\phi}\right]$$

$$\therefore \quad M = \frac{w\,\alpha\,l^2}{6}\left[\frac{\text{Sin}\,\theta}{\dfrac{1}{\tan\phi}+\dfrac{1+\tan\phi.\tan\theta}{\tan\theta-\tan\phi}}\right]$$

$$= \frac{w\,\alpha\,l^2}{6}\left[\frac{\tan\phi.\sin\theta\,(\tan\theta-\tan\phi)}{(\tan\theta-\tan\phi)+\tan\phi\,(1+\tan\phi.\tan\theta)}\right]$$

$$= \frac{w\,\alpha\,l^2}{6}\left[\frac{\tan\phi\,\sin\theta(\tan\theta-\tan\phi)}{\tan\theta\,(1+\tan^2\phi)}\right]$$

$$= \frac{w.\alpha l^2.\cos\theta}{6}\left[\frac{\tan\theta.\tan\phi-\tan^2\phi}{1+\tan^2\phi}\right]$$

$$= \frac{w.\alpha l^2.\cos\theta}{6}\left[\frac{\tan\theta.\tan\phi-\tan^2\phi}{\dfrac{1}{\text{Cos}^2\,\phi}}\right]$$

$$= \frac{w.\alpha l^2.\cos\theta}{6}\left[\tan\theta.\sin\phi.\cos\phi-\sin^2\phi\right]$$

$$= \frac{w.\,\alpha\,l^2}{6}\left[\frac{\sin\theta.\sin 2\phi}{2}-\cos\theta.\sin^2\phi\right] \qquad ...(29.11)$$

To locate the position of 'S' *i.e.*, angle ϕ, derivative of M with respect to ϕ should be equal to zero.

$$\therefore \quad \frac{dM}{d\phi} = \frac{w.\alpha\,l^2}{6}[\sin\theta.\cos 2\,\phi-2\cos\theta\sin\phi\cos\phi]=0$$

$$\therefore \quad \sin\theta.\cos 2\,\phi = 2\cos\theta.\sin\phi\cos\phi$$

$$\text{or} \quad \frac{\sin\theta}{\cos\theta} = \frac{2\sin\phi\cos\phi}{\cos 2\,\phi}=\frac{\sin 2\,\phi}{\cos 2\,\phi}=\tan 2\,\phi$$

$$\therefore \quad \tan\theta = \tan 2\,\phi$$

$$\text{or} \quad \theta = 2\,\phi$$

$$\text{or} \quad \phi = \frac{\theta}{2} \qquad ...(29.12)$$

This shows that yield line bisects the angle θ, the angle between supported sides of slab, at S on free edge. Substitute value of $\phi = \dfrac{\theta}{2}$ in eq. 29.11

$$\therefore \quad M = \frac{w.\alpha\,l^2}{6}\left[\frac{\sin\theta\sin\theta}{2}-\cos\theta.\sin^2\left(\frac{\theta}{2}\right)\right]$$

$$= \frac{w.\alpha\, l^2}{6}\left[\frac{(2\sin\frac{\theta}{2}.\cos\frac{\theta}{2})(2.\sin\frac{\theta}{2}.\cos\frac{\theta}{2})}{2} - \cos\theta.\sin^2\left(\frac{\theta}{2}\right)\right]$$

$$= \frac{w.\alpha\, l^2}{6}\left[2.\sin^2\left(\frac{\theta}{2}\right)\cos^2\left(\frac{\theta}{2}\right) - \cos\theta.\sin^2\left(\frac{\theta}{2}\right)\right]$$

$$= \frac{w.\alpha\, l^2}{6}\sin^2\left(\frac{\theta}{2}\right)\left[2.\cos^2\left(\frac{\theta}{2}\right) - \cos\theta\right]$$

$$= \frac{w.\alpha\, l^2}{6}\sin^2\left(\frac{\theta}{2}\right)\left[2.\cos^2\left(\frac{\theta}{2}\right) - 2\cos^2\left(\frac{\theta}{2}\right) + 1\right]$$

$$= \frac{w.\,\alpha l^2.\sin^2\left(\frac{\theta}{2}\right)}{6} \quad ...(29.13)$$

Example 24.6. A triangular slab PQR is simply supported along PQ = 6m, QR = 4m and is free along the edge PR. The horizontal and vertical reinforcements at the bottom of the slab provide ultimate moment capacities as 60 KN.m/m each. Determine the yield line pattern and the uniformly distributed collapse load.

Sol. Given data. (Fig. 29.18)

$$\propto l = 4\text{m},\ l = 6\text{m},\ M = 60\text{K.m/m}$$

$$\theta = 40°$$

$$\therefore \quad \propto = \frac{4}{6} = \frac{2}{3}$$

From eq. 29.12, ϕ, the angle of yield line is Q S

$$\phi = \frac{\theta}{2} = \frac{40}{2} = \mathbf{20°}$$

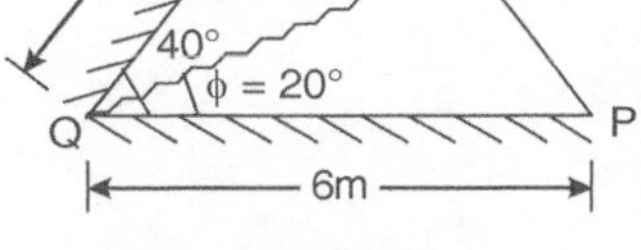

Fig. 29.18.

Eq. 29.13, $$M = \frac{w.\propto l^2}{6}\sin^2\left(\frac{\theta}{2}\right)$$

$$\therefore \quad w = \frac{6M}{\propto l^2.\sin^2\left(\frac{\theta}{2}\right)} = \frac{6\times 60}{\left(\frac{2}{3}\times 6\times 6\right)\left(\sin^2 20°\right)} = \frac{15}{\sin^2 20°} = \frac{15}{(0.3420)^2}$$

$$= \mathbf{128.24\ KN/m^2} \quad \textbf{Ans.}$$

EXERCISE 29

1. A simply supported two way slab 6.0 m × 6.0m carries uniformly distributed load of 5 KN/ m² at collapse. Find the moment of resistance if it is same in both the directions.
2. A continuous two way slab 4m × 6m has yield moments of 15.00 KN.m/m for positive yield lines and 18.00 KN.m/m for negative yield lines, equal in both the directions. Find the ultimate load at collapse.

3. Design a rectangular slab 6m × 4m simply supported on all edges, subjected to live load of 5 KN/m^2. Load factors are 1.5 and 2.2 for dead load and live load respectively. The ultimate moment in longer direction is half of that in short direction. Use M20 concrete and reinforcing bars of grade Fe 415.
4. A rectangular slab 6 m × 4 m is simply supported on 3-sides and is free on fourth long side. Calculate the collapse load if the moment capacity in x-direction is half of that in y-direction.
5. Design a circular slab 6m diameter, fixed all around the edges. Take live load as 5 KN/m^2 and load factors of 1.5 and 2.2 for dead and live loads respectively. Use M25 concrete and Re-bars of grade Fe 500.
6. A triangular slab, isotropically reinforced in two directions at bottom has ultimate moment capacities as 50 KN.m/m each. The slab PQR is simply supported on PQ = 6m, QR = 6m, included angle 50° is free along edge PR. Determine the yield line pattern and uniformly distributed collapse load.

References

1. A.C.I. Committee, *Concrete Masonry Structures – Design and Construction,* Journal of the American Concrete Institute, June 1970.
2. A.M. Neville, *Properties of Concrete* Pitman Publication, 1972.
3. B.C. Punmia, *Reinforced Concrete Structures,* Vol. 1, 2, Standard Publishers, Delhi, 1980.
4. B.P. Hughas, *Limit State Theory for Reinforced Concrete,* Pitman Publication, 1971.
5. Charles E. Reynolds and James C. Steedman, *Reinforced Concrete Designer's Handbook,* 8th Edition, A view Point Publication, 1976.
6. Design Aids for Reinforced Concrete to IS : 456-2000, SP-16.
7. Fredericks Merrit, *Building Construction Handbook,* 3rd Edition, McGraw Hill Book Co., New York, 1975.
8. IS : 456-2000 Code of Practice for Plain and Reinforced Concrete.
9. IS : 1983-2002 (Part I) Indian Standard Criteria for Earthquake Resistant Design of Structures.
10. IS : 1904-1966 Code of Practice for Structural Safety of Buildings : Foundations.
11. IS : 3370 (Parts I, II-1965, IV-1967) Code of Practice for Concrete Structures for the Storage of Liquids.
12. K.L. Rao, *Calculation, Design and Testing of Reinforced Concrete,* Charotar Book Stall, Anand (W.R.) India, 1976.
13. National Building Code of India – 1970, Group 2.
14. O.P. Jain and Jaikrishna, *Plain and Reinforced Concrete Structures,* Vol. I, II Nem Chand Bors., Roorkee.
15. S.K. Mallik and A. P. Gupta, *Reinforced Concrete,* Oxford I.B.H. Publication, Calcutta, 1980.
16. S.K. Mallik and K.S. Rangasani, *Reinforced Concrete Design,* Khanna Publishers, Delhi, 1976.
17. V.N. Vazirani and M. M. Ratwani, *Concrete Structures,* Vol. IV Khanna Publishers, Delhi.

APPENDIX-A

Imposed Floor Loads for Different Occupancies

Ref. IS : 875 (Part 2) – 1987

S.No.	*Occupancy Classification*	*Uniformly Distributed Load (UDL)* **kN/m²**	*Concentrated Load* **kN**
1	2	3	4
(i)	RESIDENTIAL BUILDINGS		
(a)	Dwelling houses:		
	(1) All rooms and kitchen	2.0	1.8
	(2) Toilet and bath rooms	2.0	–
	(3) Corridors, passages, staircases including fire escapes and stores rooms	3.0	4.5
	(4) Balconies	3.0	1.5 per metre run concentrated at the outer edge
(b)	Dwelling units planned and executed in accordance with IS : 8883–1978* only:		
	(1) Habitable rooms, kitchens, toilet and bathrooms	1.5	1.4
	(2) Corridors, passages and staircases including fire escapes	1.5	1.4
	(3) Balconies	3.0	1.5 per metre run concentrated at the outer edge
(c)	Hotels, hostels, boarding houses, lodging houses, dormitories, residential clubs:		
	(1) Living rooms, bed rooms and dormitories	2.0	1.8
	(2) Kitchens and laundries	3.0	4.5
	(3) Billiards room and public lounges	3.0	2.7
	(4) Store rooms	5.0	4.5
	(5) Dining rooms, cafeterias and restaurants	4.0	2.7
	(6) Office rooms	2.5	2.7
	(7) Rooms for indoor games	3.0	1.8
	(8) Baths and toilets	2.0	–

	(9) Corridors, passages, staircases including fire escapes, lobbies – as per the floor serviced (excluding stores and the like) but not less than	3.0	4.5
	(10) Balconies	Same as rooms to to which they give access but with a minimum of 4.0	1.5 per metre run concentrated at the outer edge
(d)	Boiler rooms and plant rooms – to be calculated but not less than	5.0	6.7
(e)	Garages :		
	(1) Garage floors (including parking area and repair workshops) for passenger cars and vehicles not exceeding 2.5 tones gross weight including access ways and ramps – to be calculated but not less than	2.5	9.0
	(2) Garage floors for vehicles not exceeding 4.0 tonnes gross weight, including access ways and ramps) – to be calculated but not less than	5.0	9.0
(ii)	EDUCATIONAL BUILDINGS		
	(a) Class rooms and lecture rooms (not used for assembly purposes)	3.00	2.7
	(b) Dining rooms, cafeterias and restaurants	3.0 †	2.7
	(c) Offices, lounges and staff rooms	2.5	2.7
	(d) Dormitories	2.0	2.7
	(e) Projection rooms	5.0	–
	(f) Kitchens	3.0	4.5
	(g) Toilets and bathrooms	2.0	–
	(h) Store rooms	5.0	4.5
	(j) Libraries and archives:		
	(1) Stack room/stack area	6.0kN/m^2 for a minimum height of 2.2 m + 2.0 kN/m^2 per meter height beyond 2.2 m	4.5
	(2) Reading rooms (without separate storage)	4.0	4.5
	(3) Reading rooms (without separate storage)	3.0	4.5
	(k) Boiler rooms and plant rooms – to be calculated but not less than	4.0	4.5
	(m) Corridors, passages, lobbies, staircases including fire escapes – as per the lab floors serviced (without accounting for storage and projection rooms) but not less than	4.0	4.5

(n)	Balconies	Some as rooms to which they give access but with a minimum of 4.0	1.5 per metre run concentrated at the outer edge
(iii) INSTITUTIONAL BUILDINGS			
(a)	Bed rooms, wards, dressing rooms, dormitories and launges	2.0	1.8
(b)	Kitchens, laundries and laboratories	3.0	4.5
(c)	Dining rooms, cafeterias and restaurants	3.0 †	2.7
(d)	Toilets and bathrooms	2.0	–
(e)	X-ray rooms, operating rooms, general storage areas – to be calculated but not less than	3.0	4.5
(f)	Office rooms and OPD rooms	2.5	2.7
(g)	Corridors, passages, lobbies and staircases including fire escapes – as per the floor serviced but not less than	4.0	4.5
(h)	Boiler rooms and plant rooms – to be calculated but not less than	5.0	4.5
(j)	Balconies	Same as the rooms to which they give access but with a minimum of 4.0	1.5 per metre run concentrated at the outer edge
(iv) ASSEMBLY BUILDINGS			
(a)	Assembly areas :		
	(1) with fixed seats †	4.0	–
	(2) without fixed seats	5.0	3.6
(b)	Restaurants (subject to assembly), museums and art galleries and gymnasia	4.0	4.5
(c)	Projection rooms	5.0	–
(d)	Stages	5.0	4.5
(e)	Office rooms, kitchens and laundries	3.0	4.5
(f)	Dressing rooms	2.0	1.8
(g)	Lounges and Billiards rooms	2.0	2.7
(h)	Toilets and bathrooms	2.0	–
(j)	Corridors, passages, staircases including fire escapes	4.0	4.5
(k)	Balconies	Same as rooms to which they give access but with a minimum of 4.0	1.5 per metre run concentrated at the outer edge

	(m) Boiler rooms and plant rooms including weight of machinery	7.5	4.5
	(n) Corridors, passages subject to loads greater than from crowds, such as wheeled vehicles, trolleys and the like. Corridors, staircases and passages in grandstands	5.0	4.5
(v) BUSINESS AND OFFICE BUILDINGS			
	(a) Rooms for general use with separate storage	2.5	2.7
	(b) Rooms without separate storage	4.0	4.5
	(c) Banking halls	3.0	2.7
	(d) Business computing machine rooms (with fixed computers or similar equipment)	3.5	4.5
	(e) Records/files store rooms and storage space	5.0	4.5
	(f) Vaults and strong room to be calculated but not less than	5.0	4.5
	(g) Cafeterias and dining rooms	3.0 †	2.7
	(h) Kitchens	3.0	2.7
	(j) Corridors, passages, lobbies and staircases including fire escapes – as per the floor serviced (excluding stores) but not less than	4.0	4.5
	(k) Bath and toilet rooms	2.0	–
	(m) Balconies	Same as rooms to which they give access but with a minimum of 4.0	1.5 per metre fun concentrated at the outer edge
	(n) Stationary stores	4.0 for each metre of Storage height	9.0
	(p) Boiler rooms and plant rooms – to be calculated but not less than	5.0	6.7
	(q) Libraries	*see* SI. No. (ii)	
(vi) MERCANTILE BUILDINGS			
	(a) Retail shops	4.0	3.6
	(b) Wholesale shops – to be calculated but not less than	6.0	4.5
	(c) Office rooms	2.5	2.7
	(d) Dining rooms, restaurants and cafeterias	3.0 †	2.7
	(e) Toilets	2.0	–
	(f) Kitchens and laundries	3.0	4.5
	(g) Boiler rooms and plant rooms – to be calculated but not less than	5.0	6.7

(h) Corridors, passages, staircases including fire escapes and lobbies	4.0	4.5
(j) Corridors, passages staircases subject to loads greater than from crowds, such as wheeled vehicles, trolleys and the like	5.0	4.5
(k) Balconies	Same as rooms to which they give access but with a minimum of 4.0	1.5 per metre run concentrated at the outer edge
(vii) INDUSTRIAL BUILDINGS		
(a) Work areas without machinery/equipment	2.5	4.5
(b) Work areas with machinery/ equipment §		
(1) Light duty — To be calculated but not less than	5.0	4.5
(2) Medium duty — To be calculated but not less than	7.0	4.5
(3) Heavy duty — To be calculated but not less than	10.0	4.5
(c) Boiler rooms and plant rooms – to be calculated but not less than	5.0	6.7
(d) Cafeterias and dining rooms	3.0	2.7
(e) Corridors, passages and staircases including fire escapes	4.0	4.5
(f) Corridors, passages, staircases subject to machine loads, wheeled vehicles – to be calculated but not less than	5.0	4.5
(g) Kitchens	3.0	4.5
(h) Toilets and bathrooms	2.0	–
(viii) STORAGE BUILDINGS | |		
(a) Storage rooms (other than cold storage) warehouses – to be calculated based on the bulk density of materials stored but not less than	2.4 kN/m^2 per each meter of storage height with a minimum of 7.5 kN/m^2	7.0
(b) Cold storage – to be calculated but not less than	5.0 kN/m^2 per each metre of storage height with a minimum of 15 kN/m^2	9.0
(c) Corridors, passages and staircases including fire escapes – as per the floor serviced but not less than	4.0	4.5
(d) Corridors, passages subject to loads greater than from crowds, such as wheeled vehicles, trolleys and the lime	5.0	4.5
(e) Boiler rooms and plant rooms	7.5	4.5

* Guide for requirements of low income housing.

† Where unrestricted assembly of persons is anticipated, the value of UDL should be increased to 4.0 kN/m^2.

‡ With fixed seats implies that the removal of the seating and the use of the space for other purposes is improbable. The maximum likely load in this case is, therefore, closely controlled.

§ The loading in industrial buildings (workshops and factories) varies considerably and so there loadings under the terms 'light', 'medium' and 'heavy' are introduced in order to allow for more economical designs but the terms have no special meaning in themselves other than the imposed load for which the relevant floor is designed. It is, however, important particularly in the case of heavy weight loads, to assess the actual loads to ensure that they are not in excess of 10 kN/m^2, in case where they are in excess, the design shall be based on the actual loadings.

|| For various mechanical handling equipment which are used to transport goods, as in warehouses, workshops, store rooms, etc., the actual load coming from the use of such equipment shall be ascertained and design should cater to such loads.

APPENDIX-B

SI Units Equivalents

SI.No	*Quantity*	*Metric Units*	*SI Units*
(1)	Length	1 m	1000 mm
		1 cm	10 mm
(2)	Area	1 m^2	10^6 mm^2
		1 cm^2	100 mm^2
(3)	Volume	1 m^3	10^9 mm^3
(4)	Mass / Weight	1 kg	9.81 N say 10 N
		1 t	10 KN
(5)	Linear Density	1 kg/m	10 N/m
(6)	Density	1 kg/m^3	10 N/m^3
(7)	Stress and Pressure	1 kg/cm^2	0.1 N/mm^2
		1 kg/m^2	10 N/m^2
		1 t/m^2	10 KN/m^2
(8)	Bending Moment	1 kg.m	10 N.m
(9)	Section Modulus	1 cm^3	10^3 mm^3
(10)	Moment of Inertia	1 cm^4	10^4 mm^4

Newton (*N*) : Newton is that force which, when applied to a body having a mass of 1 kilogram, gives it an acceleration of one metre per sec^2

1 KN = 1000 N = 100 kg = 225. 0 lbsf

APPENDIX-C

Properties of Round Bars

Reinforcement : Cross-sectional Area of Specific Bars, Weight and Perimeter

Designation	*Diameter mm*	*Weight per meter Newton*	*Cross-sectional areas mm²* 1	2	3	4	5	6	7	8	9	*Perimeter* mm
ISRO 5	5	1.5	20	39	59	79	98	118	137	157	177	15.7
ISRO 6	6	2.2	28	56	85	113	141	170	198	226	254	18.0
ISRO 8	8	4.0	50	100	151	201	251	302	352	402	452	25.1
ISRO 10	10	6.2	78.5	157	236	314	393	471	550	628	707	31.4
ISRO 12	12	8.9	113	226	339	452	565	678	791	904	1018	37.7
ISRO 16	16	15.8	201	402	603	804	1005	1206	1407	1608	1810	50.3
ISRO 20	20	24.7	314	628	942	1257	1571	1885	2199	2513	2827	62.8
ISRO 25	25	38.5	491	981	1473	1963	2454	2845	3436	3927	4418	78.5
ISRO 28	28	48.0	616	1232	1847	2463	3079	3695	4310	4926	5542	88.0
ISRO 32	32	63.1	804	1608	2413	3217	4021	4825	5630	6434	7238	100.5
ISRO 40	40	98.7	1257	2513	3770	5026	6283	7540	8796	100053	11310	12.5

Reinforcement : Per Metre Width of Slab $(mm)^2$

Bar Size	*Bar Spacing in millimeters*									
mm	75	100	125	150	175	200	225	250	275	300
6	377	283	226	188	162	141	126	113	103	94
8	670	503	402	335	287	251	223	201	183	168
10	1047	785	628	524	449	393	349	314	286	262
12	1508	1131	905	754	646	565	503	452	411	377
16	2681	2011	1608	1340	1149	1005	894	804	731	670
20	4189	3142	2513	2094	1795	1571	1396	1257	1142	1047
25	6545	4909	3927	3272	2805	2454	2182	1963	1785	1636

APPENDIX-D

Unit Weights of Building Materials and Stored Materials

		Material	*Nominal size or thickness*	*kN*	*Per*
1.		Acoustical Material			
		Eelgrass	10	5.70×1^{-2} to 7.65×10^{-2}	m^2
		Glass Fibre	10	3.8×10^{-2}	"
		Hair	10	19.10×10^{-3}	"
		Mineral Wool	10	13.45×10^{-2}	"
		Slag Wool	–	2.65	m^3
		Cork	–	2.35	"
2.		Aggregate, Course			
	(a)	Broken Stone Blast			
		Dry well – shaken	–	15.70 to 18.35	m^3
		Prefectly wet	–	18.85 to 21.95	"
		Shingles 3 to 38 mm	–	14.35	"
	(b)	Broken Bricks			
		Fine	–	14.20	"
		Course	–	9.90	"
		Foam Slag	–	6.85	"
		Cinder	–	7.85	"
3.		Aggregate, Fine			
		Sand :			
		Dry, clean	–	15.10 to 15.70	"
		River	–	18.05	"
		Wet	–	17.25 to 19.60	"
		Brick dust (Surkhi)	–	9.90	"
4.		Aggregate, Organic	–		
		Saw dust, loose	–	1.55	"
5.		Bitumen	–	0.102	"
6.		Bricks			
		Common brunt clay bricks	–	15.70 to 18.85	"
		Engineering Bricks	–	21.20	"
		Refractory bricks	–	17.25 to 19.60	"
		Bricks chips & Broken bricks	–	9.90 to 14.2	"
7.		Cement			
		Ordinary & Aluminous	–	14.10	"
		Rapid hardening	–	12.55	"

8.	Cement Concrete, Plain			
	Aerated	–	7.45	m^3
	No fines, with heavy aggregates	–	15.70 to 18.80	”
	No fines, with light weight	–	8.65 to 12.55	”
	With brunt clay aggregate	–	17.25 to 21.20	”
	With clinker aggregate	–	12.25 to 17.25	”
	With sand and grvel or crushed natural stone aggregate	–	22.00 to 23.50	”
	With saw dust	–	6.30 to 16.50	”
	With formed slag aggregate	–	9.40 to 18.05	”
9.	Cement Concrete, Prestressed	–	23.50	”\
10.	Cement Concrete, Reinforced			
	With sand and gravel or crushed natural stone aggregate	–		
	with 1% Steel	–	22.75 to 24.2	”
	with 2% Steel	–	23.25 to 24.8	”
	with 5% Steel	–	24.80 to 26.50	”
11.	Cement Mortar	–	20.40	”
12.	Cement Plaster	–	20.40	”
13.	Cork	–	2.35	”
14.	Glass			
	Sheet	2.0	0.049	”
		2.50	0.062	”
		3.0	0.074	”
		5.0	0.123	”
		5.5	0.134	”
		6.5	0.167	”
		4.0	0.098	”
15.	Lime	–	18.80	m^3
16.	Masonry, Stone			
	Dry rubble	–	20.40	”
	Marble dressed	–	26.50	”
	Sand stone	–	22.10	”
17.	Excavated Materials			
	Earth, dry loose	–	12.60	”
	Compact	–	15.20	”
	Earth (Moist) loose	–	14.10 to 15.70	”
	Compact	–	17.30 to 18.10	”
18.	Liquids			
	Waterfresh	–	9.81	”
	Salt	–	10.00	”
	Vegetable Oil	–	9.2	”
19.	Steel	–	77.0	”
20.	Timber	–		
	Fir	–	4.56	”

	Deodar	–	5.35	”
	Chir	–	5.64	”
	Sisso	–	7.70	”
	Sal	–	8.48	”
	Teak	–	6.28	”
21.	Tiles			
	Mangalore pattern	–	0.02 to 0.03	”
22.	Ceilings			
	Plaster on tile	1.3 cm	0.25	m^2
	Plaster on wood lath,		0.39	”
	Suspended metal lath	2.5 cm	0.74	”
	and cement plaster			
	or Gypsum plaster		0.49	”
23.	Agricultural and Food Product			
	Butter		8.45	m^3
	Drinks in bottles	–	7.35	m^3
	Eggs packed	–	2.95	”
	Flour in sacks up to			
	1 m height	–	2.20 to 5.90	”
	Fruits	–	3.50	”
	Rice	–	6.55	”
	Salt	–	7.05 to 9.40	”
	Seeds	–	3.90 to 7.85	”
	Sugar	–	7.85	”
	Manures	–	11.75 to 17.65	”

Index